TUNNELS AND UNDERGROUND CITIES: ENGINEERING AND
INNOVATION MEET ARCHAEOLOGY, ARCHITECTURE AND ART

PROCEEDINGS OF THE WTC2019 ITA-AITES WORLD TUNNEL CONGRESS, NAPLES, ITALY, 3-9 MAY, 2019

Tunnels and Underground Cities: Engineering and Innovation meet Archaeology, Architecture and Art

Volume 6: Innovation in underground engineering, materials and equipment - Part 2

Editors

Daniele Peila
Politecnico di Torino, Italy

Giulia Viggiani
University of Cambridge, UK
Università di Roma "Tor Vergata", Italy

Tarcisio Celestino
University of Sao Paulo, Brasil

CRC Press
Taylor & Francis Group
Boca Raton London New York

CRC Press is an imprint of the
Taylor & Francis Group, an **informa** business
A BALKEMA BOOK

Cover illustration:

View of Naples gulf

First published in paperback 2024

First published 2020
by CRC Press/Balkema
4 Park Square, Milton Park, Abingdon, Oxon, OX14 4RN

and by CRC Press/Balkema
2385 NW Executive Center Drive, Suite 320, Boca Raton FL 33431

CRC Press/Balkema is an imprint of the Taylor & Francis Group, an informa business

Publisher's Note
The publisher has gone to great lengths to ensure the quality of this reprint but points out that some imperfections in the original copies may be apparent.

ISBN: 978-0-367-46871-2 (hbk)
ISBN: 978-1-03-283939-4 (pbk)
ISBN: 978-1-003-03163-5 (ebk)

DOI: 10.1201/9781003031635

Typeset by Integra Software Services Pvt. Ltd., Pondicherry, India

Visit the Taylor & Francis Web site at
http://www.taylorandfrancis.com

and the CRC Press Web site at
http://www.crcpress.com

Tunnels and Underground Cities: Engineering and Innovation meet Archaeology,
Architecture and Art, Volume 6: Innovation in underground engineering,
materials and equipment - Part 2 – Peila, Viggiani & Celestino (Eds)
© 2020 Taylor & Francis Group, London, ISBN 978-0-367-46871-2

Table of contents

Tunnels and Underground Cities: Engineering and Innovation meet Archaeology,
Architecture and Art, Volume 6: Innovation in underground engineering,
materials and equipment - Part 2 – Peila, Viggiani & Celestino (Eds)
© 2020 Taylor & Francis Group, London, ISBN 978-0-367-46871-2

Preface

The World Tunnel Congress 2019 and the 45th General Assembly of the International Tunnelling and Underground Space Association (ITA), will be held in Naples, Italy next May.

The Italian Tunnelling Society is honored and proud to host this outstanding event of the international tunnelling community.

Hopefully hundreds of experts, engineers, architects, geologists, consultants, contractors, designers, clients, suppliers, manufacturers will come and meet together in Naples to share knowledge, experience and business, enjoying the atmosphere of culture, technology and good living of this historic city, full of marvelous natural, artistic and historical treasures together with new innovative and high standard underground infrastructures.

The city of Naples was the inspirational venue of this conference, starting from the title Tunnels and Underground cities: engineering and innovation meet Archaeology, Architecture and Art.

Naples is a cradle of underground works with an extended network of Greek and Roman tunnels and underground cavities dated to the fourth century BC, but also a vibrant and innovative city boasting a modern and efficient underground transit system, whose stations represent one of the most interesting Italian experiments on the permanent insertion of contemporary artwork in the urban context.

All this has inspired and deeply enriched the scientific contributions received from authors coming from over 50 different countries.

We have entrusted the WTC2019 proceedings to an editorial board of 3 professors skilled in the field of tunneling, engineering, geotechnics and geomechanics of soil and rocks, well known at international level. They have relied on a Scientific Committee made up of 11 Topic Coordinators and more than 100 national and international experts: they have reviewed more than 1.000 abstracts and 750 papers, to end up with the publication of about 670 papers, inserted in this WTC2019 proceedings.

According to the Scientific Board statement we believe these proceedings can be a valuable text in the development of the art and science of engineering and construction of underground works even with reference to the subject matters "Archaeology, Architecture and Art" proposed by the innovative title of the congress, which have "contaminated" and enriched many proceedings' papers.

Andrea Pigorini   Renato Casale
SIG President   *Chairman of the Organizing Committee WTC2019*

Tunnels and Underground Cities: Engineering and Innovation meet Archaeology,
Architecture and Art, Volume 6: Innovation in underground engineering,
materials and equipment - Part 2 – Peila, Viggiani & Celestino (Eds)
© 2020 Taylor & Francis Group, London, ISBN 978-0-367-46871-2

Acknowledgements

REVIEWERS

The Editors wish to express their gratitude to the eleven Topic Coordinators: Lorenzo Brino, Giovanna Cassani, Alessandra De Cesaris, Pietro Jarre, Donato Ludovici, Vittorio Manassero, Matthias Neuenschwander, Moreno Pescara, Enrico Maria Pizzarotti, Tatiana Rotonda, Alessandra Sciotti and all the Scientific Committee members for their effort and valuable time.

SPONSORS

The WTC2019 Organizing Committee and the Editors wish to express their gratitude to the congress sponsors for their help and support.

Tunnels and Underground Cities: Engineering and Innovation meet Archaeology,
Architecture and Art, Volume 6: Innovation in underground engineering,
materials and equipment - Part 2 – Peila, Viggiani & Celestino (Eds)
© 2020 Taylor & Francis Group, London, ISBN 978-0-367-46871-2

WTC 2019 Congress Organization

HONORARY ADVISORY PANEL

Pietro Lunardi, President WTC2001 Milan
Sebastiano Pelizza, ITA Past President 1996-1998
Bruno Pigorini, President WTC1986 Florence

INTERNATIONAL STEERING COMMITTEE

Giuseppe Lunardi, Italy (Coordinator)
Tarcisio Celestino, Brazil (ITA President)
Soren Eskesen, Denmark (ITA Past President)
Alexandre Gomes, Chile (ITA Vice President)
Ruth Haug, Norway (ITA Vice President)
Eric Leca, France (ITA Vice President)
Jenny Yan, China (ITA Vice President)
Felix Amberg, Switzerland
Lars Barbendererder, Germany
Arnold Dix, Australia
Randall Essex, USA
Pekka Nieminen, Finland
Dr Ooi Teik Aun, Malaysia
Chung-Sik Yoo, Korea
Davorin Kolic, Croatia
Olivier Vion, France
Miguel Fernandez-Bollo, Spain (AETOS)
Yann Leblais, France (AFTES)
Johan Mignon, Belgium (ABTUS)
Xavier Roulet, Switzerland (STS)
Joao Bilé Serra, Portugal (CPT)
Martin Bosshard, Switzerland
Luzi R. Gruber, Switzerland

EXECUTIVE COMMITTEE

Renato Casale (Organizing Committee President)
Andrea Pigorini, (SIG President)
Olivier Vion (ITA Executive Director)
Francesco Bellone
Anna Bortolussi
Massimiliano Bringiotti
Ignazio Carbone
Antonello De Risi
Anna Forciniti
Giuseppe M. Gaspari

Giuseppe Lunardi
Daniele Martinelli
Giuseppe Molisso
Daniele Peila
Enrico Maria Pizzarotti
Marco Ranieri

ORGANIZING COMMITTEE

Enrico Luigi Arini
Joseph Attias
Margherita Bellone
Claude Berenguier
Filippo Bonasso
Massimo Concilia
Matteo d'Aloja
Enrico Dal Negro
Gianluca Dati
Giovanni Giacomin
Aniello A. Giamundo
Mario Giovanni Lampiano
Pompeo Levanto
Mario Lodigiani
Maurizio Marchionni
Davide Mardegan
Paolo Mazzalai
Gian Luca Menchini
Alessandro Micheli
Cesare Salvadori
Stelvio Santarelli
Andrea Sciotti
Alberto Selleri
Patrizio Torta
Daniele Vanni

SCIENTIFIC COMMITTEE

Daniele Peila, Italy (Chair)
Giulia Viggiani, Italy (Chair)
Tarcisio Celestino, Brazil (Chair)
Lorenzo Brino, Italy
Giovanna Cassani, Italy
Alessandra De Cesaris, Italy
Pietro Jarre, Italy
Donato Ludovici, Italy
Vittorio Manassero, Italy
Matthias Neuenschwander, Switzerland
Moreno Pescara, Italy
Enrico Maria Pizzarotti, Italy
Tatiana Rotonda, Italy
Alessandra Sciotti, Italy
Han Admiraal, The Netherlands
Luisa Alfieri, Italy
Georgios Anagnostou, Switzerland

Andre Assis, Brazil
Stefano Aversa, Italy
Jonathan Baber, USA
Monica Barbero, Italy
Carlo Bardani, Italy
Mikhail Belenkiy, Russia
Paolo Berry, Italy
Adam Bezuijen, Belgium
Nhu Bilgin, Turkey
Emilio Bilotta, Italy
Nikolai Bobylev, United Kingdom
Romano Borchiellini, Italy
Martin Bosshard, Switzerland
Francesca Bozzano, Italy
Wout Broere, The Netherlands
Domenico Calcaterra, Italy
Carlo Callari, Italy

Luigi Callisto, Italy
Elena Chiriotti, France
Massimo Coli, Italy
Franco Cucchi, Italy
Paolo Cucino, Italy
Stefano De Caro, Italy
Bart De Pauw, Belgium
Michel Deffayet, France
Nicola Della Valle, Spain
Riccardo Dell'Osso, Italy
Claudio Di Prisco, Italy
Arnold Dix, Australia
Amanda Elioff, USA
Carolina Ercolani, Italy
Adriano Fava, Italy
Sebastiano Foti, Italy
Piergiuseppe Froldi, Italy
Brian Fulcher, USA
Stefano Fuoco, Italy
Robert Galler, Austria
Piergiorgio Grasso, Italy
Alessandro Graziani, Italy
Lamberto Griffini, Italy
Eivind Grov, Norway
Zhu Hehua, China
Georgios Kalamaras, Italy
Jurij Karlovsek, Australia
Donald Lamont, United Kingdom
Albino Lembo Fazio, Italy
Roland Leucker, Germany
Stefano Lo Russo, Italy
Sindre Log, USA
Robert Mair, United Kingdom
Alessandro Mandolini, Italy
Francesco Marchese, Italy
Paul Marinos, Greece
Daniele Martinelli, Italy
Antonello Martino, Italy
Alberto Meda, Italy

Davide Merlini, Switzerland
Alessandro Micheli, Italy
Salvatore Miliziano, Italy
Mike Mooney, USA
Alberto Morino, Italy
Martin Muncke, Austria
Nasri Munfah, USA
Bjørn Nilsen, Norway
Fabio Oliva, Italy
Anna Osello, Italy
Alessandro Pagliaroli, Italy
Mario Patrucco, Italy
Francesco Peduto, Italy
Giorgio Piaggio, Chile
Giovanni Plizzari, Italy
Sebastiano Rampello, Italy
Jan Rohed, Norway
Jamal Rostami, USA
Henry Russell, USA
Giampiero Russo, Italy
Gabriele Scarascia Mugnozza, Italy
Claudio Scavia, Italy
Ken Schotte, Belgium
Gerard Seingre, Switzerland
Alberto Selleri, Italy
Anna Siemińska Lewandowska, Poland
Achille Sorlini, Italy
Ray Sterling, USA
Markus Thewes, Germany
Jean-François Thimus, Belgium
Paolo Tommasi, Italy
Daniele Vanni, Italy
Francesco Venza, Italy
Luca Verrucci, Italy
Mario Virano, Italy
Harald Wagner, Thailand
Bai Yun, China
Jian Zhao, Australia
Raffaele Zurlo, Italy

Innovation in underground engineering, materials

and equipment - Part 2

Tunnels and Underground Cities: Engineering and Innovation meet Archaeology, Architecture and Art, Volume 6: Innovation in underground engineering, materials and equipment - Part 2 – Peila, Viggiani & Celestino (Eds)
© 2020 Taylor & Francis Group, London, ISBN 978-0-367-46871-2

The design approach of cut & cover excavation in hyperbaric condition applied for Napoli/Cancello high speed railway

G. Lunardi, G. Cassani, A. Bellocchio & C. Nardone
Rocksoil S.p.A., Milan, Italy

M. Cafaro & G. Ghivarello
Salini Impregilo, Milan, Italy

R. Sorge & F. Carriero
Astaldi, Rome, Italy

ABSTRACT: In the frame of the high speed railway line Napoli-Bari, it is foreseen the realization of a cut and cover tunnel which partially develops below the ground water level. The top down construction will be performed using the compressed air in order to achieve a dry base during the soil excavation and the cast of the internal structures. All the benefits of this technology are discussed, including the aspects related to the environment. The practical application of the technique has revealed numerous detailed problems from a structural point of view: together with the additional load on structures, the limitation of the air losses has been a relevant topic. The measures implemented for the maintenance of the prescribed pressure are presented in the paper describing the solution developed to minimize the dissipation of compressed air through production joints. The air losses are estimated after a brief survey of the formulations available in literature.

1 INTRODUCTION

The project in question is part of the transport network upgrading along the transversal axis Naples - Benevento - Foggia - Bari. The works are aimed at giving adequate response to the changing mobility needs of travelers and goods and constitute a fundamental element for the development of southern Italy, for a better economic and social integration in the country and in Europe.

In this sense, the construction of the high capacity railway line Naples-Bari, together with the activation of the Rome-Naples high-speed railway system, will favor the integration of the railway infrastructure of the South-East with the Connecting Directories to northern Italy. This upgrading will have a paramount impact for the socio-economic development of the South, reconnecting two areas, Campania and Puglia.

The enhancement of the railway axis connecting the Tyrrhenian and the Adriatic coastlines will also allow the creation of a "tripole" (Rome, Naples and Bari) which will constitute one of the largest metropolitan systems in Europe. On the international front, as part of the new structure of the trans-European corridors (TEN-T) defined by the European Commission on 19 October 2011, the development of the Naples-Bari route, which specifically falls within the Corridor 5 Helsinki - Valletta, has been identified as a priority.

The rehabilitation and development of the Naples-Bari itinerary involves the doubling of the single-track railway sections and the change of the current alignment, in order to ensure the connections speeding up and the increase of the railway transport offer (Figure 1).

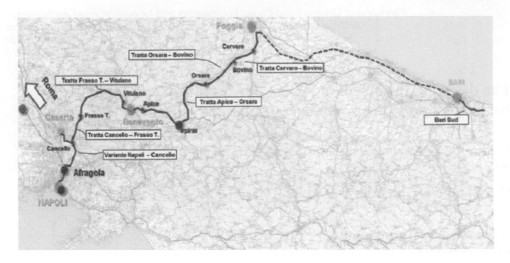

Figure 1.   Transport network upgrading along the transversal axis Naples - Benevento - Foggia - Bari.

The entire work was divided into functional sections for realization purposes. The project in question is part of the first section (Naples-Cancello) particularly strategic in the overall arrangement of metropolitan, regional and long-distance connections.

In fact, the variant in the stretch between Naples and Cancello, allows bringing the tracks of the line to the new station of Naples Afragola, which in the future will become the station for passenger interchange between regional and AV services increasing overall accessibility to the railway transport in the Naples hub.

The line, in the section of interest develops for about 15.5 km through the municipalities of Casoria, Casalnuovo, Afragola, Caivano and Acerra (Figure 2). The chainage starts, in the south, from km 0+000.00 (coinciding with km 241+727 of the historical line) and ends, in the north, at km 15+585,066 (coinciding with km 229+530 of the historical line).

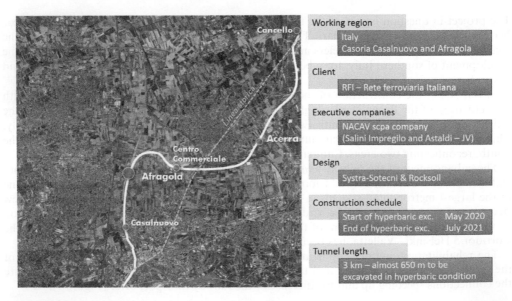

Figure 2.   Section Overview.

Although, in general, the preferred solution is that of railway embankment, in some particular points the construction of viaducts or tunnels is envisaged in order to resolve specific interferences and to better integrate with the environment.

The tunnel cross section presents a geometrical variability as summarized below:

1. "Parapioggia" (Figure 3)
   In the first 180 m the tunnel has a rectangular section, in this stretch the tunnel presents a single tube section.
2. Top down tunnel with single-tube section (Figure 4)
   For about 300 m the tunnel continues with a top down section characterized by diaphragm walls horizontally restrained by roof and foundation r.c. slabs. In this stretch the section is a single tube section.
3. Top down tunnel with double-tube section (Figure 5)
   In the following stretch, the Circumvesuviana line joins the new Cassino line, therefore the tunnel presents a double-tube section and is constructed following the top down method
4. Top down tunnel with double-tube section and intermediate slab (Figure 6)
   In this stretch, due to the considerable excavation height an intermediate slab is foreseen to restrain horizontally the structure
5. Casalnuovo Station
   In correspondence of Casalnuovo station, the tunnel maintains the same structural concept. It is different from the adjacent stretches only in terms of width that is greater due to architectural purposes.
6. Top down tunnel with double-tube section
   In the last portion, the surface level decrease allowing to foresee only two horizontal slabs that correspond to the double-tube section

The tunnel interferes with the phreatic level in the first stretch, from pk 0+550 to pk 1+600. In order to perform all the excavation and construction activities in dry condition the tender design foresaw the execution of jet grouting plug. The soil treatment reduces the soil perme-

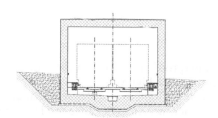

Figure 3. "Parapioggia".

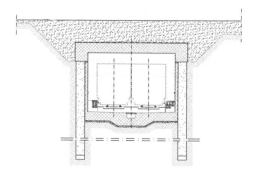

Figure 4. Top down tunnel with single-tube section.

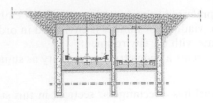

Figure 5. Top down tunnel with double-tube section.

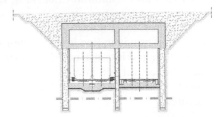

Figure 6. Top down tunnel with double-tube section and intermediate slab.

Table 1. Comparison of the alternative solutions.

	COMPRESSED AIR	JET GROUTING
Aquifer pollution	✓	✗
Duration of works	✓	✗
Flexibility	✓	✗
Production plant cost	✗	✓
Soil improvement cost	✓	✗
Soil to landfilling	✓	✗

ability, allowing creating a waterproof layer. Due to the water height to be counter-balanced the thickness of this treatment reached high values.

With regard to the bid issued by Italferr in 2016, it was requested to the Competitors to study an original technical solution which could solve the interference with the groundwater, par-ticularly with the possibility to get an easy compartmentalization between the following work-ing phases. For this purpose, NACAV scpa company, a JV between Salini Impregilo and Astaldi, decided to choose an original technical solution, already successfully used abroad, consisting in making use of pressurized air dig. Such choice allowed NACAV scpa company to get the job. Table 1 summarize the comparison of the alternative solutions.

That is the use of compressed air, whose application has a double advantage:

– a higher flexibility of the system in order to adapt to ground water level fluctuation;
– no impact on the environment due to the exclusion of any contamination of the ground-water, which serves as a reservoir for the water supply of the city.

2 COMPRESSED AIR EXCAVATION

2.1 Famous historical applications

The use of compressed air for excavation under water table has been used for long time especially for the mechanized excavation (TBM). In the field of artificial tunnels this technique has been developing in recent years, particularly in the countries of northern Europe, as, in certain contexts, it represents a valid alternative to traditional solutions, bringing improvements and advantages.

Historically, this type of technology finds its most widespread applications in the construction of underwater foundations and in particular through the use of pneumatic caissons. For example, the most important geotechnical applications include the foundations of the Eiffel Tower. In Italy, on the other hand, a well-known example is represented by the bridge of the Hach Industry in Rome for which the compressed air foundation was used for the first time in Italy, a caisson technique whose origin should be searched in France in 1841. Another mention must be made regarding the metro line 4 in Paris for which compressed air was applied for the Seine under-passing.

Among the underground tunnel in hyperbaric condition projects developed abroad are mentioned below (Schwarz & Hehenberger, 2004):

- Chlodwig-Platz, Cologne, Lot South - North-South suburban railway, Cologne Oil-free screw compressor;
- Stans/Terfens Tunnel - Inntal, construction lot H4-3 oil-free screw compressor;
- Vomp/Terfens tunnel - Inntal, construction lot H5 oil-free screw compressor;
- Allmend Tunnel Lucern - Central railway - double track - construction on lower level oil-free screw compressors;
- BF Olympiapark North, underground railway line 3 North - Ventilation;
- Fritzens/Braumkirchen - Inntal tunnels, construction lots H7-1, H7-2, H-3 oil-free screw compressors;
- AUDI Tunnel in Ingolstadt (Germany).

2.2 *General operating principles*

The excavation with compressed air of artificial tunnels built with the top-down method, as in the case of the GA01, generally requires a similar methodology dictated by technological needs, even if, depending on the site and the specific characteristics of the work, small variations are found in the examined cases.

In general, a r.c. wall in c.a. is constructed at the entrance of the tunnel and two watertight doors are installed for people and vehicles access. These accesses lead to a watertight seal chamber where the air is gradually pressurized to reach the design pressure value.

Through two more doors, the workers and the vehicles can access to the working construction site inside the tunnel, which is in a hyperbaric environment at the established pressure.

Through other two doors, both the vehicles and the workers can enter and exit (separately) in the compensation chamber, which must therefore be subjected to time periods of compression (entry) or decompression (exit).

The pressure value to be applied in the tunnel is equal to the corresponding value of hydrostatic pressure acting on the excavation bottom due to the water head at the time of the excavation under the top slab.

As the excavation proceeds, the pressure to be applied in the tunnel must be adjusted to the value of hydrostatic pressure to be balanced at the different advancement steps.

The internal pressure is thence an additional load acting on the structure in the temporary phase that should be considered in the structural analyses. It is to be remarked that in general the air pressure acts in the opposite direction of the most demanding loads: for both the retaining walls and the top slab the internal pressure partially counteract the earth and water pressure acting inward the tunnel. In other words, this methodology does not require a strengthening of the structure and furthermore, assuming the air pressure is maintained during the whole excavation process, it reduces the internal actions associated to the temporary phase.

While the compressed air system is active the global equilibrium toward the possible uplift should be checked: the internal pressure, determined to counteract the water pressure at the bottom of the excavation, can lead to a structure uplift in case of low backfilling. This condition should be taken into account performing a specific check, which should consider the internal pressure of the air, the weight of the structure, the backfilling load, the anchoring strength of the diaphragm walls in the ground.

The tunnel construction phases will thence be:

- Excavation to the diaphragm wall execution level;
- Realization of the diaphragm walls;
- Cast of the r.c. roof slab;
- Backfilling above the roof slab;
- Excavation below the roof slab, between the diaphragm walls, to the foundation level with the application of the prescribed air pressure;
- Cast of the internal structures and finishes installation;
- Deactivation of the internal air pressure.

3 THE APPLICATION TO THE CASE OF STUDY

3.1 Technical and methodological aspects

In the case of study, the maximum excavation level varies according to the T.O.R. of the alignment. Moreover, the excavation section changes due to the section type variability foreseen by the design and to the fact that the tunnel locally widens for the STI exits, for the wastewater lifting plants and the relative inspection chambers and for the niches.

This configuration would require the adjustment of the air pressure of compressed air at each advance. For logistic purpose and to avoid excessive air losses at the tunnel face, the tunnel has been divided into compartments (n.14, Figure 7) limited on the sides and on the top by the tunnel structures and at the front by specific diaphragm walls.

The diaphragm walls that constitute the transversal partitions are executed together with the longitudinal ones and are placed at a variable distance each other, determined considering the expected losses.

At each of the 14 compartments will correspond a uniform design value of the air pressure (maximum value of the water head in the stretch individuated by the compartment) and fixed volume. To define correctly the design pressure to be maintained in each compartment, a specific piezometric monitoring has been performed during the development of the design, see Table 2. Since the records showed a certain variability due to weathering and to seasonal fluctuation, the maximum water level recorded has been considered. Furthermore, an addition possible raising of 1 m is precautionary assumed.

Considering the maximum hydraulic head in each compartment, he design pressure required for lowering the water table below the excavation surface ranges between 0.2 and 1 bar. The range already takes into account the variability of the excavation level in correspondence of a local structures such as the wastewater lifting plant.

The fact that the maximum pressure required by the system is equal to 1 bar confirms the effective advantage of the solution. The safety of workers in hyperbaric conditions represents an important aspect to be considered in the design phase since can have strong impact on the indirect costs of the project. Despite the regulatory environment is still developing on this topic and is not homogeneous comparing the references from different countries, the safety measures to be implemented always depend on the working pressure. In addition, all the standards available on the subject consider 1 bar as a limit for the first level of safety equipment since the human beings can easily adapt to such a pressure.

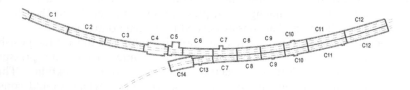

Figure 7. General plan with compartments.

Table 2. Piezometric readings from the geotechnical campaign.

	PK	Head mslm	H max mslm	H min mslm	08/03	22/03	23/03	16/04	14/05	01/06
					m from head					
E1PZ	0+735,35	17.44	14.39	14.04			3.4	3.20	3.05	3.13
E3PZ	1+150.98	22.89	15.15	15.04		7.85		7.81	7.74	7.72
E4PZ	1+276.58	26.35	15.05	14.77	11.58			11.49	11.3	11.24
E5PZ	1+705.99	34.60	15.51	15.20		19.4		19.26	19.09	19.02

3.2 Construction phases

Once the slab is cast and backfilled according to the design final ground level, the compressed air plant shall be switched on and, when the design pressure value is achieved the tunnel excavation could start.

The excavation will be performed without any partialization of the face and proceeding by compartments: the excavation of a compartment will begin only after completion of the final structures of the previous one. The excavation muck will be transported out of the hyperbaric work area by means of a conveyor belt devised to ease the working process.

The choice of this system requires a certain space for its installation and, consequently, leads to postpone the application of the compressed air to the second compartment.

After the excavation of the first compartment and the cast of the internal structures the compressed air system is switched on. In order to guarantee the pressurization of the next compartment, some holes are provided in the partition wall before its demolition that is followed by the excavation. The compartment will be excavated up to its end, proceeding simultaneously in the two tubes if the tunnel present a double tube section.

At the end of the excavation process the waterproofing system will be installed and fixed to the structures in order to limit the air losses in the next phase. Once the cast of the internal structure is complete it's possible to proceed with the same construction process for the next compartment.

3.3 Construction details

In order to minimize the air losses during the excavation with compressed air, the construction joints have been studied in detail to improve the tightness both to water and to air. The calculations for the sizing of the plants depend on the equivalent overall permeability to air of the surfaces exposed to the air pressure. For the case of study, considering the average permeability of the soil to water measured by means of specific tests in situ, the following assumptions were made:

– During the excavation the structures and the soil present an equivalent overall permeability to air equal to 10^{-6} m/s;
– After the cast of the internal structures the equivalent overall permeability is assumed equal to at 10^{-8} m/s.

These values have been estimated considering all the precautions actuated to reduce the air losses and should be confirmed before the start of the excavation by means of a field test.

In order to optimize the design and to reduce the cost related to the compressed air plant the tightness of the structure should be studied in detail focusing on the weak points of the system with respect of the air leaks that are:

– The connection between the roof slab and the diaphragms;
– The longitudinal joints between diaphragms: this context shows the greatest number of discontinuities (every 2.5 m);
– Joints (casting joints and expansion joints) present on the slabs (linings, top and foundation slabs);
– Interface between the waterproofing sheets and the internal structures on the base slab and on the linings at the end of each compartment.

The precautions to be taken to improve the hydraulic seal of the structural joints and of the construction joints in the different parts of the structure must therefore be studied in order to obstruct also the air passage.

For the joint between the roof slab and the diaphragm wall, a waterstop is installed between the latter and the linings in order to stop a possible ascent of the ground water (Figure 8). Moreover, the waterproofing layer installed on the roof slab is mechanically fixed to the diaphragms below the contact section with the roof slab abutment providing protection against the rainwater inlet. This mechanical fixing also guarantees an excellent seal against compressed air, limiting considerably the air losses.

For the joint between the partition walls and the roof slab a waterstop is installed at the top of the capping beam as shown in Figure 9.

To limit the air losses that could occur at the joint between two adjacent diaphragm walls, in addition to the particular dovetail shape, a waterstop is inserted. This detail is foreseen for all the diaphragm wall, both the one of the tunnel and those of the partitions.

In correspondence of the construction or expansion joints, the detail studied to guarantee water tightness perform very well also as a barrier against air. In fact, while for the construction joints the continuity of the membrane is foreseen, in correspondence of the expansion joints a waterstop which allows to compensate the differential shrinkage of the structures is inserted guaranteeing continuity of the waterproofing system (Figure10).

At the end of each compartment, the waterproofing membrane is mechanically fixed to the linings and to the foundation slab to allow a perfect seal of the already executed portion: this allow considering for that part a lower permeability to air as cited before.

The TNT layer placed between the PVC and the lean concrete to protect the waterproofing membrane constitute an escape route for the air. In correspondence of the structural joints, the waterproofing package is fixed to the r.c. structure by anchored waterstops, but it is also necessary to secure it outward (on foundation slab and diaphragm walls).

Since the operations (excavation, laying of the waterproofing and casting of the internal structures) will be carried out compartment by compartment, it is envisaged to create this kind of joint at the end of each compartment (Figure 11).

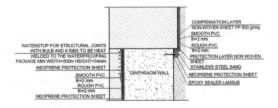

Figure 8.   Joint between the roof slab and the diaphragm wall.

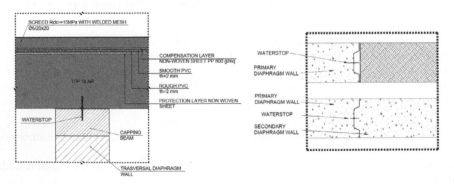

Figure 9.   joint between the partition walls and the roof slab (left) joint between adjacent diaphragm walls (right).

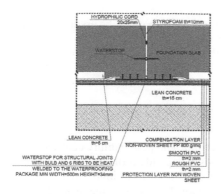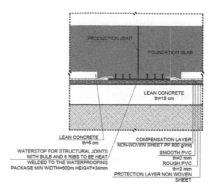

Figure 10. Longitudinal joint - foundation slab.

In detail, the waterproofing in correspondence of the lining is fixed to the diaphragm walls and the one in correspondence of the foundation slab is fixed to a specially made joist.

4 AIR LOSSES EVALUATION

The plant for the compressed air should be dimensioned considering all the factors that determine the required flow. In particular, once the volume of work is put under pressure, the main contributions are:

- Losses through structural joints
- Losses due to the opening and closing of access doors for vehicles and personnel
- The forced aspiration required to maintain healthy conditions in the workplace.

While the last contribution is a requirement dictated by the workers' safety manuals, the others must be estimated for a correct sizing of the plant (ITA-AITES WG5, 2015).

Of the two, the one linked to the opening and closing of the access doors has a minimal influence since it is foreseen a compression/decompression chamber for muck, means of transport and workers.

The contribution that have to be evaluated is therefore the one due to losses through ground and structural joints. For this purpose, the formulation of Schenck and Wagner, developed to evaluate air leaks during tunnel excavation, is applied (Semprich & Scheid, 2001).

$$Q = 2 \cdot k_a \cdot \frac{(P_1 - P_2)}{\gamma_w \cdot L} \cdot A \cdot \frac{P_1}{P_2}$$

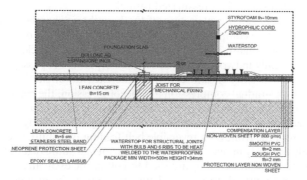

Figure 11. Mechanical fixing at the end of each compartment - Foundation slab.

- k_a is the permeability coefficient of the ground in the air that can be assumed equal to 70kw
- kw is the permeability coefficient of soil to water
- P_1 is the absolute pressure inside the tunnel that corresponds to the pressure necessary to lower the pitch below the excavation bottom plus the atmospheric pressure (Ptun + 1) atm
- P_2 is atmospheric pressure equal to 1 atm
- γ_w is the specific weight of water
- L is he length of the air path that corresponds to the tunnel overburden
- A is the area through which losses occur

It should be noted that it is not possible to determine the permeability coefficient of the soil to air by means of laboratory tests, since it is strongly dependent on the in-situ boundary conditions. For this reason, the aforementioned relation is used to estimate this coefficient starting from the water permeability of the medium. Obviously, the assessments made here shall be confirmed in the execution pháse through specific in situ tests performed before the start of the excavation. On the basis of these results, it will be possible to adjust the design pressures in order to better control the lowering of the water table (Bull, 2003).

In order to guarantee a correct application of the internal pressure, the system also provides a series of instruments that allow a constant monitoring of losses: in every instant the incoming and outgoing air volume and the actual air pressure will be measured.

The Schenck and Wagner formulation is valid for laminar flow, which can be foreseen in material where $k_w < 1 \cdot 10^{-3}$ m/s (Jardine & McCallum, 2001). Given that in our case study, the permeability of the soil that varies between $1 \cdot 10^{-4}$ m/s and $1 \cdot 10^{-6}$ m/s the basic requirement is met.

The above formulation refers to the tunnel excavation with the traditional method, therefore there is no structure that prevents the escape of the air but the ground itself. For an artificial tunnel the volume of soil to be excavated is enclosed in a box consisting of diaphragm walls (laterally and frontally) and top slab. At the bottom of the excavation there is no structural element, but considering that the soil is saturated its permeability to air is very low.

In light of these considerations, the permeability value assumed for the calculation of the losses is equal to $1 \cdot 10^{-6}$ m/s before the final linings are casted and equal to $1 \cdot 10^{-8}$ m/s when the internal structures are finished.

Precautionary considering the maximum permeability until the installation of the mechanical fixing at the end of the compartment, the required air flow rate estimated considering one compartment not lined and a cumulative loss in the previous ones ranges between 45 and 95 m³/min.

For the sizing of the plants it is considered appropriate to consider, in addition to the strictly necessary equipment, an additional compressor able to guarantee 50% of the estimated flow.

5 TECHNOLOGICAL ASPECTS

5.1 *Hyperbaric chamber*

Salini Impregilo and Astaldi designed a concrete structure for the access of personnel and equipment for the excavation of the pressurized tunnel.

For the case of study the transversal section of the hyperbaric chamber, 40 m long and 13 m wide, has been divided in three compartments: two for the accumulation of the excavated material (compartment A and B, see Figure 12) and one for the passage of vehicles and personnel (Compartment C, divided vertically in two volumes in order to separate the passage of vehicles and personnel). Considering that to entry in and exit from the excavation site it is necessary to follow specific procedures for the gradual compression and decompression, the construction site has been studied with the aim of minimizing the number of passages. In particular, in order to manage the muck transport it is envisaged the use of an extensible conveyor belt, hanged which will discharge the material coming from the front on a reversible belt feeding two shuttle belts. The shuttle belts will store the muck alternatively in one of the two compartments addressed to this specific function (Comp. A and B). When compressed they are used to collect the excavated material from the belt conveyor, and when depressurized, to load and transport away the ground, loading by a wheel loader. They are built with concrete walls and closed by self-sealing steel doors. One door, on the

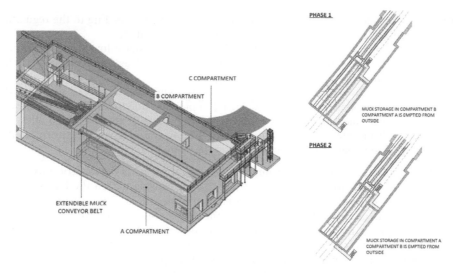

Figure 12. Hyperbaric Chamber.

atmospheric pressure side is large enough to accommodate the wheel loader; the other one, located on the pressurized side, presents the minimum dimensions to let in the conveyor belt and is controlled by hydraulic jacks.

The chamber dedicated to the equipment transfer is also equipped on both ends with self-sealing steel doors fitted with a remote-control device, to operate the door from the control cabin.

The personnel transfer chamber is a classical hyperbaric chamber steel made, including a main and an access chamber. The purpose of the access chamber is to make sure that at any time it is possible to shelter the crew retreating from the front, or to have a rescue team compressed for help the compressed crew.

A control cabin will be supplied and will be designed and equipped to manage all the information and controls for the excavation of the tunnel and the operations of the hyperbaric chambers, as well as the compressed air supplies. The hydraulic power pack to operate the jacks of the doors of the locks will be installed in the control cabin.

The hyperbaric equipment includes: all the 6 doors with relevant operating jacks and seals, the man-lock and relevant control panel, the hydraulic power pack, several steel plates, 600 x 600 mm size with relevant fixing frames, for concrete wall penetrators.

All pressurized equipment will be designed and constructed to meet the CE 12110 standard for air locks and compressed air shields. In particular, all parts which could not be hydraulically tested will be calculated with a safety coefficient 2 relative to the expected service pressure (in this case is 1 bar and safety coefficient 2).

The two ground locks and the equipment lock will be equipped with a closed circuit color video system which will be displayed by a screen installed into the control cabin.

A pressure recorder, installed in the control cabin, will be able to control: the pressure of the tunnel, the pressure of the man lock, the pressure in the ground locks and the pressure in the equipment lock. There are 3 pressure sensors installed in each of the different chambers and the recording interval is from 1 sec to 12 h.

5.2 Compressors and pipelines for compressed air

Compressors have been provided, placed outside, to supply and maintain the air at the desired pressure in the tunnel according to the design prescriptions. The compressed air system will have, in addition to the compressors:

- filtration system suitable for ensuring breathable air certified according to the regulations;
- cooling system to feed air in the tunnel at the correct temperature;
- extensible steel pipeline to carry the air to the front and a spare line placed in a position faraway from the operating area with the aim of having at least one operational pipeline even in case of damage to the other;
- control system, located in the tunnel near to the working front, able to maintain the air pressure established, compensating the decrease in pressure due to any leaks to the outside;
- system with a controlled air flow, allocated at the beginning of the tunnel, to allow the air to flow outwards in order to ensure an renewal of the air, according to the norm, that takes into account the number of people operating in the tunnel.

Inside the tunnel, in a closed environment, the operation with the electric motors of the equipment causes the air to heat up. To contain this air heating within the limits of comfort for people an air cooling and recirculation system will be provided. This system will consist of a cooling coil and an electric fan placed towards the tunnel entrance in order to recirculate the air from the working front. The cooled air that passes through the electric fan will be addressed directly to the working front by a flexible tube. This system therefore ensures that the personnel work in an environment with the correct temperature and humidity values.

In order to ensure the continuous operation of the "vital" systems inside the tunnel in case of shut-down of the power supply line, emergency generators located outside are designed to satisfy all the functions considered particularly sensitive for the operating personnel like breathing air, lighting, opening of doors, fire system, etc.

Every opening, even the small ones, that remain cause air leaks from the already excavated tunnel. To counter the consequent decrease of the desired pressure a sensor is installed near the working front which will automatically manage the operation of the compressors. This sensor can be adjusted so as to ensure the correct value of the air pressure for each working field.

6 CONSTRUCTION MACHINERY AND PLANTS

6.1 Conveyors

Inside the tunnel, in hyperbaric conditions, the fundamental criterion is to maintain conditions of wellbeing for the workers that are reflected in the air parameter control systems. A solution in this direction is the adoption of only electric motors in the tunnel for all the operating machines. Therefore the transport of excavated material along the tunnel is implemented with the use of a system of conveyor belts equipped with electric motors that are stretched to follow the excavation front. This system will be equipped with all the safety measures suitable for the particular environmental conditions in the tunnel as non-flammable components, temperature sensors, etc.

6.2 Excavators

The excavation of the materials and the consequent loading on the belt conveyor system is provided by Shaeff ITC machines. Indeed these machines combine the excavation function and the loading through a conveyor able to feed the conveyor belt system with continuity. Infact the conveyor belt system is designed with mobile elements able to follow the excavator and then receive anytime the excavated material.

6.3 Electric construction vehicles

All the operating machines and the equipment used inside the tunnel will be equipped with electric motors. This solution allows zero emissions into the tunnel environment and thus ensures a perfectly breathable air quality. Depending on the availability from the market and the required functions, some machines will be powered through the 380-volt power line while others will be equipped with rechargeable batteries; for this case an external installation will be provided for recharging the spare batteries.

6.4 *Molds*

The project foresees that, after having affixed the waterproof sheath under the base slab and the walls, a concrete base slab and a finishing concrete layer on the walls will be carried out. In particular, for the walls works a series of formwork sets are provided. The formworks will be equipped with anchors at the base and upper section and with wheels suitable for permitting translation. They will also be equipped with wall-mounted vibrators and concrete filling mouths positioned at various levels to facilitate the regular laying and a perfect surface finishing.

7   CONCLUSIONS

The choice adopted by Salini Impregilo and Astaldi for the excavation of the artificial tunnel Casalnuovo has proved innovative and represents an avantgarde in the national landscape.

The chosen solution allows controlling the water table, avoiding any risk connected to the transport of fine materials in case of a drowning and those induced by possible heaving due to jet grouting and above all the ground water pollution.

Following a careful design procedure, the safety of workers is guaranteed in any construction phase and assured by a redundancy of safety measures.

According to this technology Rocksoil has developed the structural design taking into account the additional actions acting on the bearing elements. In addition to it, all the construction joints have been studied in detail in order to limit as much as possible the air losses during construction.

Finally, the whole team, including both constructors and designers have worked side by side in order to define the construction process that could guarantee the success of the excavation and assure the workers safety.

REFERENCES

Semprich, S. & Scheid, Y. 2001. Unsaturated flow in a laboratory test for tunnelling under compressed air. *Proceedings of 15th Int. Conf. Soil Mech. And Geot. Eng.*, Vol. 2, 1413–1417. Instanbul: Balkema.

Schwarz, J. & Hehenberger, F. 2004. Compressed air for top down. *Proceedings Tunnels & Tunnelling International*, 36 (2).

Bull, J.W. 2003. *Numerical analysis and modelling in geomechanics*. London and New York: Spon Press of the Taylor & Francis Group.

Jardine, F.M. & McCallum, R.I. 1992. Engineering and Health in Compressed Air Work. *Proceedings of the International Conference*. Oxford.

ITA-AITES WG5, 2015. *Guidelines for good working practice in high pressure compressed air.*

Tunnels and Underground Cities: Engineering and Innovation meet Archaeology,
Architecture and Art, Volume 6: Innovation in underground engineering,
materials and equipment - Part 2 – Peila, Viggiani & Celestino (Eds)
© 2020 Taylor & Francis Group, London, ISBN 978-0-367-46871-2

Innovative TBM transport logistics in the constructive lot H33 – Brenner Base Tunnel

A. Lussu
Brenner Base Tunnel, BBT-SE, Innsbruck, Austria

C. Kaiser & S. Grüllich
Arge Strabag-Salini/Impregilo, Innsbruck, Austria

A. Fontana
ÖBA-ARGE Pini & Partners, Innsbruck, Austria

ABSTRACT: The construction of the Tulfes Pfons lot, as part of the Brenner Base Tunnel (BBT), includes a 15 km long tunnel section excavated with an open gripper TBM. Due to the very difficult logistical conditions (steep access tunnel with a downward gradient of 11 %, a 90° turn to the main tubes with a small cross-section of 8 m diameter), an alternative TBM logistical supply solution was developed in collaboration with the contractor. Multi Service Vehicle (MSV) is a laser-controlled custom vehicle on rubber tires that can operate without rails. The MSV is able to handle both the steep access (11 %) and the 90° turn. In addition to the safety-related advantages, it was proven that the MSV system can keep pace with conventional TBM supply systems. After 12 km of excavated tunnel with a top performance of 61 m/day and monthly performances of up to 800 m/month in hard rock conditions with up to 1200 m overburden, it was clear that the use of the MSV has been a complete success. MSVs are becoming a cost efficient and technically approved alternative to conventional TBM supply equipment on rail.

1 INTRODUCTION

The Tulfes Pfons construction lot is the most northern construction lot and is considered one of the most complex lots of the Brenner Base Tunnel. It connects the Brenner base tunnel to the existing Lower Inn Valley tunnel which was built from 1989–1994. This construction lot with a total length of 42 km was awarded in June 2014 to the STRABAG/Salini/Impregilo joint venture, under the technical leadership of STRABAG. The construction works started in July 2014. The excavation works for the seven conventional tunnel drives using drill and blast method started in September 2014 with an official groundbreaking ceremony. In August 2018, 26 km from a total of 27 km conventionally driven tunnels have been excavated and 12 km from a total of 15 km with an open gripper TBM. The peak of tunnel excavation occurred between 2016 and 2017, when all eight drives were being excavated simultaneously:

2 PROJECT DESCRIPTION

The Tulfes-Pfons construction lot includes the following sections (see Figure 1):

- 15 km of exploratory tunnel A= 50 m²,
- 9.7 km of rescue tunnel (A= 35 m²) for the Lower Inn Valley tunnel with 28 cross-passages while rail operations were ongoing,

- 3.8 km and 1.9 km of connection tunnels (A= 115 m²) to link the Brenner Base Tunnel with the existing Lower Inn Valley tunnel,
- 3.2 km of the Innsbruck emergency stop (A=78 m²),
- 3.8 km of main tunnels for the Brenner Base Tunnel (A= 70 m²) and
- several access tunnels, cross-chambers and cross-passages .

The 15-km stretch of exploratory tunnel to be driven southwards is a challenge in itself. The exploratory tunnel is being used during the construction phase of the Brenner Base Tunnel to investigate the geological conditions in order to use this information for the tenders process of the main tunnels to be excavated above. In the operating phase, it will be used as a service and drainage tunnel.

In order to guarantee the best possible level of prospection, an open gripper TBM with a cutter head diameter of 8 m was stipulated in the tender. The local geological conditions can be constantly recorded and documented behind the shield all along the tunnel length. The visual documentation by the geologist is supported by probe drilling, lateral prospective drilling and ongoing seismic tests. These measures are helping to control the ongoing tunneling works as well as the prospection for the subsequent excavation of the parallel main tubes.

The excavation of the exploratory tunnel with a maximum overburden of 1,200 m using a gripper TBM presents a special challenge for the joint venture with regards to construction

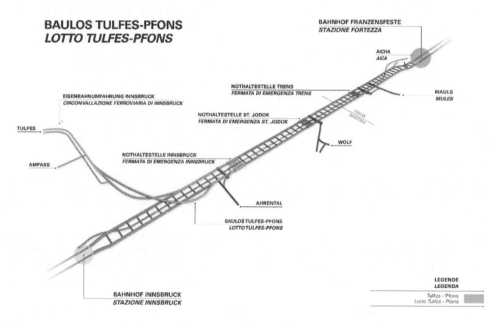

Figure 1. Tunnels to be excavated in construction lot H33 [BBT].

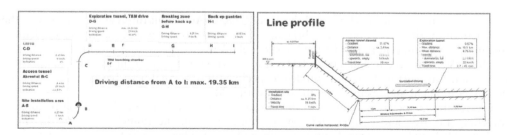

Figure 2. Plan view and longitudinal profile of the access tunnel and exploratory tunnel [ROWA].

engineering, safety engineering and logistical aspects. This article will deal primarily with logistics challenges and its safety related requirements.

3 PARTICULARITIES OF THE TBM LOGISTICS AT LOT TULFES PFONS

The challenges of all BBT construction lots is that the supply for the main tunnels has to be managed through long access tunnels running perpendicular to the main tunnel itself. For example the Ahrental access tunnel was driven from west to east from the northern Wipptal valley down to the level of the tunnel.

The access tunnel is 2.4 km long and has a maximum slope of 11.5 % (Figure 2) and cross section of approximately 92 m². At its end, the access tunnel terminates with a 90° turn into the exploratory tunnel which runs from north to south. About 1.2 km south of this junction, the assembly/launching chamber for the TBM was built.

4 CONTRACTOR'S SOLUTION - MULTI SERVICE VEHICLE

Detailed cycle analyses and studies that were carried out by the contractor showed that an optimal excavation cycle required 95 tons of consumables to be transported with a single vehicle from the portal to the TBM. This corresponds with a double stroke of 3.4 m of tunnel length. When the MSVs for this project were procured (Spring 2015), 55 tons was the maximum load capacity for MSVs with acceptable track accuracy for the curve demands of the tunnels. Therefore the length of the MSVs was limited due to trajectory deviation.

Therefore decisive developments jointly with ROWA, the MSV manufacturer, were necessary to improve the track accuracy and increase the load capacity of the vehicles accordingly. The 90° turn at the end of the access tunnel and the centered entrance to the TBM back-up gantries were the key challenges. The development of a "virtual rail" was the only way to improve the track accuracy and keep the trajectory deviation of the multiple axles to a minimum, while maintaining the 30 m curve radius. It was necessary to solve the following issues:

- Development of a software that would collect and evaluate the vehicle data from different steering sensors and distance sensors (on the front and sides of the vehicle). The steering of the MSV's uses a complex, newly developed algorithm which considers the length of the platform cars, tow bar angles between the platform cars, steering angles of the axles and the axle spacing.
- The operational reliability of the sensors which were required to scan the surrounding of the MSVs. Some of the sensors available on the market proved not to be suitable for the extreme environmental conditions of a tunnel excavation (dust, humidity, surface vs. underground temperature differences, driving in a smoke-filled tunnel in case of fire, etc.)
- Suitability of the hardware components and/or the mechanical and structural parts (for example, the steering system, the shock absorbers, the braking system, axles, motors, chassis, wheels, electrical and hydraulic systems, steering and regulation systems, main onboard computer etc.).

It took six months for BBT SE and the JV of STRABAG/Salini/Impregilio, jointly with the pertinent authorities (fire and rescue brigades & the inspector for worker safety) to implement this innovative solution to handle the logistical challenges. A team that included members of the Joint Venture, the manufacturer ROWA and in-house experts from the JV headquarters in Vienna, Stuttgart and Milan compiled a specifications manual with basic requirement for. design, equipment and operation of the self-steering MSVs.

The result was newly developed MSVs consisting of 5 mechanically linked platform cars, travelling rail-free and carrying 95 tons of material from the tunnel portal to the TBM backup. All 14 axles are steered automatically and independently so as to follow the new "virtual rail". The first axle defines the track of the "rail" and the following axles follow it; any additional deviation is recognized by the various sensors and automatically compensated by the control software.

A further challenge for the working group was the slope of the lateral access tunnel. A fully loaded MSV was required to travel a 2.4 km stretch with a 11.5 % slope in traffic along with other vehicles in safe manner. Therefor a triple-redundant breaking system was developed and installed (see also chapter 6.5).

5 SAFETY

During numerous discussions the emergency services specified two requirements with regards to the use of the MSV's

- Requirement for the fire brigade to use the MSV's for rescue and firefighting services with all necessary equipment
- Evacuation of the workers from the exploratory tunnel in case of smoke (driving with maximum safety distance precautions due to poor visibility)

The requirements for the fire brigade train were taken from the tender documents and adapted in a special fire brigade MSV (see Reichel, 2019).

If visual conditions are insufficient for safe driving, the driver will steer the rescue vehicle via a monitor showing the stretch in front of the vehicle. Radar sensors show the distance to the tunnel walls and the driver must therefore steer the vehicle in a way to always remain in the largest possible distance to the tunnel walls. The radar sensor also scans the area in front of the vehicle for obstacles.

6 TECHNICAL DESCRITPION OF THE MSV

The technical data and the various configuration of the MSV's being used in the Construction Lot Tulfes Pfons are listed as follows.

Figure 3.   Layout of the ROWA Multi Service Vehicle [ROWA].

6.1 *Configuration of the MSV's supplying the TBM*

The TBM is supplied by three MSV long trains, each of them carries consumables for two 1,7m strokes, i.e. a total length of 3,4m. The consumables include:

- Steel Ribs (UNP140/160/180, TH29), wire mesh (AQ60), elements of conveyor belt frames, anchor bolts, rails, pipes – pre loaded on pallets
- Invert segments: 2 elements, 5 m in width, 1.7 m in length, 1.4 m in height, weight: 14 t per segment
- Covers for the invert segments: 2 elements, 3.18 m in width, 1.7 m in length, 0.3 m in height, weight approx. 4 t
- 1 container with back fill grout of the invert segments 1.5 m^3, empty weight ~1.85 ton, full 4.15 t

Table 1. Technical data of the ROWA Train [ROWA].

Installed diesel power:	approx. 460	kW
Maximum speed on a 0.67% upward gradient; full train:	25	km/h
Maximum speed on a -11% downward gradient; full train:	15	km/h
Maximum speed on a 11% upward gradient; empty train:	15	km/h
Maximum speed on a 11% upward gradient; full train (exception):	5	km/h
Weight of the empty train:	approx. 35	t
Loaded weight on the way towards the TBM:	max. 95	t
Weight of the full train (towards the TBM):	max. 132	t
Load on the return trip:	max. 17	t
Weight of the train on the return trip:	max. 52	t
Maximum axle load	approx. 90	kN
Maximum wheel load	approx. 45	kN
Minimum curve radius	30	m
Length:	52.4	m
No. of steered and braked axles	14	
No. of driven/powered axles	7	

- 1 shotcrete tank: volume 8 m³, weight 25 t per full bucket, 5 t empty

Figure 4. MSV configurations [ROWA].

1 short train for special transports (for example to transport drums of conveyor belt and medium voltage cable)

- Vehicle load capacity: about 30 t
- Tare weight: about 18 t
- Length: 22.9 m

Figure 5. Special transports [ROWA].

6.2 *Special vehicles (Self rescue and Rescue/firefighters)*

1 short train for self-rescue

- Autonomous rescue container for 25 people for 8 hrs stay on a platform car
- Always parked about 50–100 m behind the TBM

Figure 6. Special vehicle (Self Rescue MSV) [ROWA].

1 short train for/firefighter/rescue MSV)

- Each of the two cars holds an isolated rescue container with an integrated driver's cabin and self-contained air supply (regeneration technology)

- Six people plus the driver can fit in the rescue container of the motor vehicle. Nineteen people plus the driver can fit in the rescue container of the trailer vehicle. The air supply is meant to last 10 hours for max. 20 people.
- The vehicle is always ready for use at the Ahrental portal

Figure 7.   Special vehicle (Fire fighter/rescue MSV) [ROWA].

6.3   Driving modes for the MSV's

6.3.1   Normal driving mode
This is the driving mode used during normal operations. In this mode, the driver simply regulates the vehicle's speed or uses the brake. The actual steering of the vehicle is done in all sections of the tunnel (site installation area, Ahrental access tunnel, assembly/launching chamber, and exploratory tunnel) by the intelligent steering system.

The driver has to choose the listed tunnel sections by using joystick in the operators cabin. For all tunnel sections a predefined track is specified.

6.3.2   Manual mode
The manual mode is an "emergency -mode" which is only needed if there are no guide plates installed at the tunnel walls. Since the steering cannot be adjusted automatically by the lateral sensors, the driver must steer the vehicle manually. In order to maintain sufficient track accuracy at low speed, a trim function has been built in.

6.3.3   Manoeuvre mode
The manoeuvre mode is required when a MSV must be driven out of a special situation, for example, if it comes too close to the tunnel wall or if it is stopped due to an obstacle detected by the safety controls. In this case, the driver must select the manoeuvre mode on the touch screen. Having done this, the driver can select and control individual axles or groups of axles to manoeuvre the train out of the situation. In this mode, the vehicle's top speed is only 1 km/h.

6.4   Train control and steering system

The following sensors are required to scan the surrounding of the MSV's:

- Laser scanners (in the front of both driver's cabins, with a range of about 26 m, tested on autonomous driven wheel loaders and other underground machinery)
- Radar sensors (in the front of both driver's cabins, with a range of about 40 m tested on road vehicles with ACC, functional even in poor visibility conditions and able to detect objects even at greater distances)
- Laser sensors (on the sides of the vehicle, in order to constantly monitor the distance to the tunnel wall), each platform car has 3 laser sensors with a range of about 3 m,
- Ultrasonic sensors (on the sides of the vehicle, in order to constantly monitor the distance to tunnel wall, when laser sensor functionality is limited by smoke or dust, range: about 5 m, tested on other underground machinery to detect objects at short distances)

On-site tests were carried out before production began to assess how the surface of the tunnel walls would be detected by the sensors. Reflecting strips and guide plates were installed on the tunnel walls to test how the identification of the track could be improved, which was acknowledged as a problematic issue. Reflecting guide plates placed on the tunnel wall at intervals of 3 m to 6 m provided the best results.

The tunnel wall or the reflecting plates on the wall are identified by the radar sensors and laser scanners installed below the driver's cabin. Based on that information the steering software creates a "virtual rail", which can run in the middle of the tunnel or at a certain distance to the tunnel wall. An electronic map of position and distance information leads the first axle, which is virtually steered as if along an actual rail. The virtual rail is also used to define the track position of all the following axles. Due to a complex, newly developed algorithm (input value for example the tow bar angle, steering angle and speed) all axles follow exactly the track of the previous axle. The tolerances and inaccuracies resulting from the sensor system, which are the reason for the track deviations, are equalized by the continuous measurement of the ultrasonic and laser sensors on the sides of the platform cars. Since the lengths of the platform cars and the distances between the following axles are known, the newly developed algorithm ensures that the distance of the following axle to the tunnel wall is the same as distance of the previous axle.

The steering sensors (front and sides) are redundant, so there are multiple sensors per platform car. Using the virtual rail it is possible to indicate a pre-defined track in the various tunnel sections to be followed by the self-steering MSV's in automatic mode. However, the driver of the vehicle always maintains control over the speed, steering and if necessary emergency braking.

6.5 Braking system/Braking levels

The MSV's has independent braking systems (sustained-action braking system, service/foot-brake and parking break), which engage and release in different ways

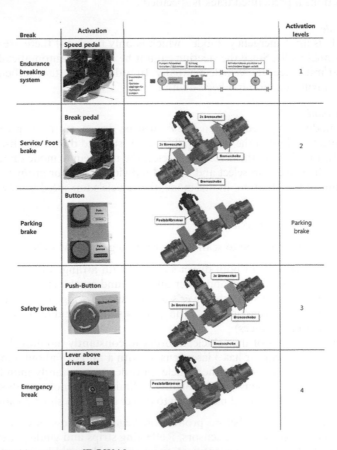

Figure 8. MSV Braking systems [ROWA].

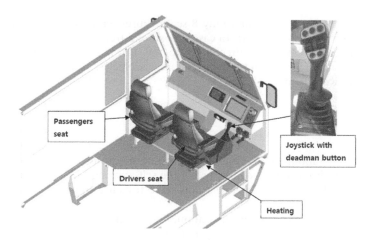

Figure 9.   Layout of the driver's cabin [ROWA].

7   COMPARISON OF THE MSVS WITH A NORMAL BOUND TBM SUPPLY SYSTEM

After the successful completion of a 12 km TBM drive, it was concluded that MSVs could be more advantageous than conventional rail bound logistics systems for this construction lot. In the special conditions of the Brenner Base Tunnel, with its steep access tunnel and long tunnel drives, the following advantages of the MSVs as compared to rail-bound supply trains were determined:

- Continuous supply transport from the site installation area via the access tunnel and the exploratory tunnel to the TBM without re-loading in the tunnel.
- Avoiding the use of over 10 standard trucks which would be required in case of handing over the consumables to a rail bound train. This means less traffic in the access tunnel which is already congested with trucks to supply the drill and blast tunnel drives and inner lining works.
- There is no need to install rails behind the TBM, making access to the TBM much more flexible. The TBM can be reached with normal vehicles as well. For safety reasons, the

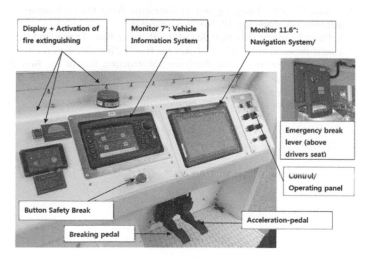

Figure 10.   Layout of panel in Driver's cabin [ROWA].

number of vehicles is restricted and only 8-seater buses with specially trained drivers are used to transport workers and staff. In case of a smoke-free accident on the TBM, a rescue vehicle or ambulance can drive directly up to the TBM. Rescue times are significantly reduced and rescue services avoid having to change vehicles, especially since they have all their equipment with them.

8 SUMMARY

Till 31.08.2018 12 km (about 80 %) of the excavation of the exploratory tunnel in construction lot Tulfes Pfons have been completed, the use of the self-steering MSV's can already be determined as a success. The three MSV long trains have travelled approx. 115,000 km without major problems.

The availability of the MSV's is currently about 99 %, which means that standstill times due to breakdowns were significantly less compared to a rail-bound supply system. Furthermore all the standstill times required for to maintain and repair the rail track and switches are eliminated while using MSV's.

On-site tests have shown that the MSV's perform well as a rescue vehicle.

What's new in these MSVs:

- The first worldwide use of self-steering MSVs (almost autonomous driving in a tunnel project
- The first use of a virtual rail created by a tight sensor network with laser, ultrasonic and radar measurements to create an electronic map with positioning and distance information
- The development of a complex control software in order the minimize track deviations of the various axles which are steered independently but in total synchronic
- An extension of the MSV length (up to ten platform cars) and the consequent increase is load capacity at any time due to the currently installed software and algorithms.

The self-steering MSV's currently used in the Brenner Base Tunnel are the first step in developing driver-less tunnel supply vehicles, which will certainly be the future of supply logistics in tunnel construction sites. It takes courage from all project parties involved to share the risk to use new technologies.

REFERENCES

Bergmeister, K.; Reinhold, C. 2017. Learning and optimization from the exploratory tunnel – Brenner Base Tunnel. *Geomechanics & Tunneling* 10, No. 5, S. 467–476

Bergmeister, K. 2019. Tunnel construction methods of the Brenner Base Tunnel and their logistical challenges – A multi-criteria decision approach. *ITA-AITES World Tunnel Congress 2019*, Naples, Italy.

Reichel, E. 2019. Novel safety concept at the Austrian construction lots of the Brenner Base Tunnel. *ITA-AITES World Tunnel Congress 2019*, Naples, Italy.

Rowa 2018. Rowa Tunnelling Logistics AG, Wangen, Swiss

Schwarz, C. 2019. Seismic prediction of the rock quality at the Brenner Base Tunnel. *ITA-AITES World Tunnel Congress 2019*, Naples, Italy.

Insam, R.; Carrera, E.; Crapp, R.: Das Entwässerungssystem des Brenner Basistunnels. *Swiss Tunnel Congress*, Luzern, 2018.

Tunnels and Underground Cities: Engineering and Innovation meet Archaeology,
Architecture and Art, Volume 6: Innovation in underground engineering,
materials and equipment - Part 2 – Peila, Viggiani & Celestino (Eds)
© 2020 Taylor & Francis Group, London, ISBN 978-0-367-46871-2

Experiences from tunnel boring when hard-to-very hard rock with focus in performance predictions and cutter life assessments

F.J. Macias

JMConsulting – Rock Engineering, Oslo, Norway

ABSTRACT: The use of hard rock tunnel boring machines (TBMs) has become widely and generally used with success but in too many cases, due to inappropriate assessments, with undesirable consequences. A process of great complexity is involved during tunnel boring. When hard-to-very hard rock (i.e. low-to-extremely low boreability), the complexity is accentuated becoming, in many cases, critical for the achievement of the final schedule and reasonable tunnelling cost. Performance predictions and costs estimates have a major influence on the planning and risk management of TBM excavation projects. A proper understanding of tunnel boring and wear processes in hard rock enhances an appropriate applicability of the models for performance prediction and cutter life assessments. The paper compiles experiences and outcomes from research and consulting collaborations on several hard rock TBM projects and during the revise and extend of the current version of the NTNU prediction model for TBM performance and cutter life.

1 INTRODUCTION

The Tunnel Boring Machine (TBM) approach has become widely used and is currently an important method employed by the tunnelling industry for civil and mining infrastructures. Nowadays, the TBM method, due to the technology development, is applicable in an increasingly wider range of rock mass conditions: excavations can now be carried out in almost all rock conditions using this method, given certain economic constraints.

A process of great complexity is involved when tunnel boring due to the interaction between the rock mass and the machine. The prediction of performances (e.g. Penetration rate) and cutter life are not straightforward issues and involve major risk assessments. When hard-to-very hard rock (i.e. low-to-extremely low boreability) the complexity is accentuated becoming, in many cases, critical for the achievement of the final schedule and reasonable tunnelling cost

Understanding of tunnel boring and wear processes thus enhances performance prediction and cutter life assessments in hard rock tunnel boring project. Performance predictions and costs estimates are often decisive in the selection of excavation methods and have a major influence on the planning and risk management of TBM excavation projects.

The paper compiles experiences and outcomes from research and consulting collaborations on several hard rock TBM projects and during the revise and extend of the current version of the NTNU prediction model for TBM performance and cutter life.

2 HARD ROCK TUNNEL BORING

2.1 *Rock boreability in hard rock*

'Rock mass boreability' is a comprehensive parameter of rocks under excavation and expresses the result of the interaction between a given rock mass and a TBM. Boreability can be defined as the resistance (in terms of ease or difficulty) encountered by a TBM as it penetrates a rock mass composed of intact rock containing planes of weakness.

Penetration rate and cutter wear are influenced by intact rock and the rock mass properties. Intact rock properties are typically defined in terms of strength, abrasivity, porosity, schistosity and rock petrography.

Rock mass fracturing (e.g. planes of weakness or discontinuities, orientation...) is found to be the geological factor that exerts the greatest influence on net penetration rate, and which consequently also has a major impact on tunnelling costs in hard rock (Bruland, 2000). A high rock mass fracturing parameter means greater rock mass boreability during hard rock TBM excavation. In addition, the penetration rate may also be influenced by in situ rock stress, groundwater and other factors.

Hard rock refers to a minimum UCS values > 50 MPa (High rock strength) and typically UCS values > 100 MPa (Very high strength) according to the classification given by ISRM (1978).

2.2 Performance prediction and cutter life models for hard-to-very hard rock

The net penetration rate (m/h) is determined from both rock mass properties and machine parameters and it is the main factor used for predictions (i.e. advance rate, cutter consumption and the costs) of hard rock TBM tunnelling projects.

Advance rate determines the total boring advance achieved over a period (e.g. days, weeks, months...). Advance rate is given from the net penetration rate (m/h) and considering the machine utilization. The machine utilization is net boring time of the total available time expressed in percentage. Much of the available time is used for other activities than boring (e.g. re-gripping, cutter change and inspection, repair and service of the TBM and back-up, rock support, transport system installation and delays, tunnel service installation and delays, surveying and others). The final goal of performance prediction for hard rock TBMs is the estimation of time and cost.

Estimations should in addition consider assembly and disassembly of the TBM and the back-up, permanent rock support and lining, excavation of niches or branching, boring through and stabilizing zones of poor rock quality, probe drilling and pre-injection works, major machine breakdowns, dismantling of installations, additional time for unexpected rock mass conditions.

Several prediction models for estimates of performance and cutter wear in hard rock tunnel boring have been developed in recent decades. Models used to estimate penetration rates include those of CSM model (Rostami and Ozdemir, 1993; Rostami, 1997, Yagiz, 2014), the Gehring model (Gehring, 1995; Wilfing, 2016), the NTNU model (Bruland, 2000; Macias, 2016), the QTBM (Barton, 2000), RME (Bieniawski et al., 2006), Gong and Zhao (2009), Hassanpour et al. (2011) and Farrokh et al. (2012).

These models adopt very different approaches and their input parameters, especially in terms of rock mass properties, exhibit substantial variation. This makes balanced comparisons problematic. However, under ordinary conditions, the results of the models may exhibit satisfactory agreement. A further issue for consideration is the scope of applicability of the different models.

3 EXPERIENCES WHEN HARD-TO-VERY-HARD ROCK

3.1 Introduction

Tunnelling technologies continue to improve, and the TBM sector is no exception. Continuous updating and a better understanding of the tunnel boring process should be applied to the development and revision of prediction models.

Research and consulting collaborations on several hard rock TBM projects during the revise and extend the NTNU prediction model for TBM performance and cutter life (Macias,

2016) result in new experiences and outcomes giving a better understanding of the tunnel boring process. Latest and current consulting collaborations carry out by the undersigned are being used for validation and further development.

3.2 Rock boreability and TBM predictions

There is no single parameter that can fully represent the properties of jointed rock masses. Different parameters have different emphases and can only provide a satisfactory description of a rock mass in an integrated form (Singh and Goel, 2011).

To evaluate the influence of rock mass when hard-to-very hard rock can be decisive on the tunnel boring and it is not always easy or straightforward. A comprehensive understanding of the rock mass boreability requires different approaches to its evaluation (e.g. use of both, chip analysis and tunnel face inspection, to support engineering geological back-mapping).

Excavation costs predictions for hard rock TBM projects incorporate a geological risk, which becomes of major importance, from a cost point of view, when hard-to-very hard rock and degrees of fracturing are low.

The establishment of dimensional-related definitions for the wide variety of intact rock and rock mass discontinuities has proved to be problematic. Figure 1 presents a proposed length-based classification of the main types of discontinuities from an engineering geological perspective according to the NTNU back-mapping methodology.

The influence of geology on the prediction performance of the NTNU model increases as the degree of fracturing in the rock mass decreases. Based on mapping analysis and experience, the influence of the average spacing has been extended (up to 480 cm) within the latest version to date of the NTNU model (Macias, 2016) including a fracture class (s_f) "Class 1", which is intermediate between the formers fissure classes 0 and 0-I. Table 1 shows the fracture class terminology as defined by the average spacing between fractures (Macias, 2016).

Low values of degree of fracturing (e.g. St 0, 0-I and St I-) have a dramatic influence on excavation cost estimates. However, if the rock mass is highly fractured, then variations in degree of fracturing do not significantly influence excavation costs.

Figure 2 illustrates relative excavation costs as determined by the degree of fracturing for a 7 m diameter TBM with standard machine specifications and rock properties (medium drillability and abrasivity) by using the NTNU model. The reference excavation cost value applied is for fracture class St I.

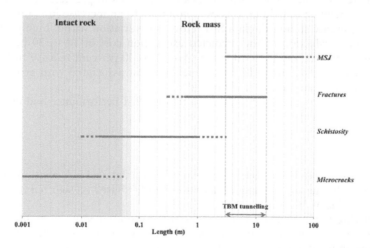

Figure 1. The main types of intact rock and rock mass discontinuities encountered during back-mapping in hard rock TBM tunnelling. Types are categorised according to length based on the NTNU methodology. The shaded area indicates where data have been derived from laboratory tests carried out on intact rock (Macias, 2016).

Table 1. Fracture class terminology as defined by the average spacing between fractures (Macias, 2016).

Fissure class (St) (Bruland, 2000)	Fracture Class (Sf) (Macias, 2016)	Average spacing between fractures a_f (cm)	Range class (cm)	Degree of fracturing
0	0	∞	480 – ∞	Non-fractured
0	1	320	240 – 480	Extremely low
0 - I	2	160	120 – 240	Very low
I -	3	80	60 – 120	Low
I	4	40	30 – 60	Medium
II	5	20	15 – 30	High
III	6	10	7.5 – 15	Very high
IV	7	5	4 – 7.5	Extremely high

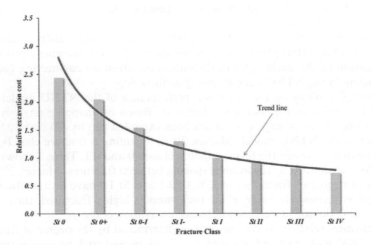

Figure 2. The relationship between relative excavation costs and degree of rock fracturing.

At lower degrees of fracturing, results indicate increments in predicted excavation costs of 20 to 70 per cent between individual classes, amounting to an increase in approx. 150 per cent relative to the reference standard, class St I. However, variation within highly fractured rock masses results in a reduction of the predicted excavation cost of 10 to 30 per cent between individual classes (amounting to a reduction of approx. 40 per cent relative to the reference). Important to bear in mind that additional time and cost associated with ground stability in highly fractured and faulted zones is not included.

Special care should thus be taken when predicting TBM performance and excavation costs in rock masses exhibiting low levels of fracturing.

3.3 Cutter life

The NTNU cutter life model has been reviewed based on detailed information about geology, rock mass, drillability testing and instantaneous cutter life based on selected tunnel sections at several projects (Macias, 2016).

Basic cutter ring life was back-calculated from the instantaneous cutter life parameter for every tunnel section, taking correction factors for TBM diameter, cutterhead rpm, number of cutters, abrasive minerals content and cutter thrust into consideration.

A variation factor of 15% of the cumulative distribution of CLI and quartz content is applied. In order to account for uncertainties arising from ring steel quality, set bearing capacity and other properties, a variation factor of 10% is used.

The updated version has been modified in the light of minimum cutter ring life values and data obtained from previous versions of the model (Figure 3).

The increment of the predicted basic cutter ring life in the updated version compared with the previous (Bruland, 2000) has been up to around 20%. This might be due to the effects of changes of cutter tip design and/or improvements in cutter technology over the last two decades.

Figure 4 shows plots of basic cutter ring life (h) for 432 and 483 mm cutter diameters obtained from the updated (2016) version of the NTNU model.

Calculations for 508 mm (20 inch) cutter diameters have not been included due to insufficient data, although it can be noted that results indicate a longer basic cutter ring life, as expected, for highly abrasive rocks (exhibiting extremely and very low CLI values).

3.4 Other geological factors influencing cutter wear

In fractured rock masses, or in situations where extremely good rock chipping occurs, the cutters will be exposed to large instantaneous loads. Entacher et al. (2013) measured momentary

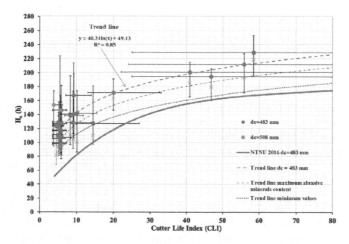

Figure 3. A plot showing basic cutter ring life obtained using the 2016 version of the NTNU model for 483 mm and 508 mm diameter cutters. Results from the uncertainty analysis are also plotted.

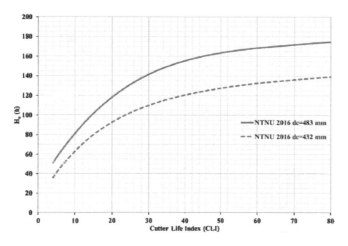

Figure 4. Plots showing basic cutter ring life (h) for 432 and 483 mm cutter diameters obtained from the updated (2016) version of the NTNU model.

loads in individual cutter discs up to 3.5 times higher than a nominal cutter load and up to 10 times higher than the average load in fractured rock.

Under such conditions, a cutter ring exhibits a tendency to chip along its edge. Extensive ring chipping and high cutter thrust may result in bevel edge wear, loosened rings and blocked bearings.

Additional loads will result in higher abrasion on protruding cutters when the difference in diameter between adjacent cutters is too large as a result of deficient wear height control. Heavy vibration of the cutterhead results in high lateral forces on the cutters, which in turn causes additional abrasion.

Fractured rock masses promote greater penetration rates and thus effectively prolong cutter life (in m/c and sm³/c). However, due to the aforementioned effect of fractures, higher levels of cutter consumption, measured in hours, will be expected. This effect will be more dominant in rocks exhibiting low drillability and high abrasivity.

Figure 5 shows cutter life data (sm³/c) and wear patterns taken from several tunnel sections exhibiting a variety of rock mass conditions. The contact skarn in Figure 5 is a tunnel section transitioning to a fractured rock type, which produces large instantaneous loads on the cutters resulting in a high rate of cutter replacement due to bearing set problems such as blockage.

Higher amount of chipping is related to fractured rock while blocked cutters occur in rock types transition and high cutter thrust.

3.5 *Influence of Cutterhead velocity influence on penetration*

The operational parameters (i.e. applied cutter thrust and cutterhead velocity), in combination with the rock boreability, have a great influence during tunnel boring in hard rock.

Efficient boring should be considered to improve the advance rate along the tunnel by applying an optimal net penetration rate considering cutter consumption, potential damage of cutters, probability of bearing failure and cutterhead damages, as well as energy consumption.

A reduction in cutterhead velocity (rpm) values may improve boring efficiency and reduce excavation costs. Lower cutterhead rpm will promote lower cutter rolling distances and velocities for a given section of tunnel, resulting in significantly higher cutter ring life and a reduction in potential damage to cutters. Moreover, for a given thrust level and fewer revolutions of the main bearing reduce the probability of bearing failure and cutterhead damages, as well as energy consumption.

An 'RPM test' measures cutterhead penetration over a given period at a variety of cutterhead velocities under constant cutterhead thrust. The aim of the test is to evaluate the

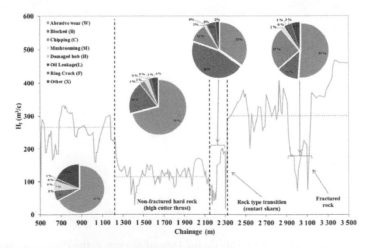

Figure 5.   Cutter life data and wear patterns from several tunnel sections.

influence of cutterhead velocity (rpm) on penetration rate (mm/rev) for maximum net penetration rates (m/h) for a given machine, geology and thrust level (Macias, 2016).

The 'RPM' tests revealed that in general, a lower cutterhead velocity would result in an increase in penetration rate up to a given value beyond which it decreases. The values concerned are dependent on cutterhead design, rock mass properties and level of thrust.

The resulting values obtained from the 'RPM tests' include values of penetration rate (PR, in mm/rev) and net penetration rate (NPR, in m/h) for each cutterhead rpm at constant level of thrust. The 'RPM tests' usually result in a cutterhead velocity value and a maximum NBPR for a given machine, geology and level of thrust. For higher and lower cutterhead rpm values, the NPR is reduced. Lower cutterhead rpm result in an increase in PR up to a given value beyond which it decreases dramatically. This also causes the NPR to be reduced.

An increase in penetration rate (PR), without compromising the net penetration rate (NPR), results in improved tunnelling efficiency. The 'optimal' cutterhead velocity is thus defined as the cutterhead rpm value, which achieves maximum penetration rate while maintaining the 'optimal' level of net penetration rate for the geological conditions and level of thrust in question. Net penetration rate is assumed 'optimal' for a 5% from the maximum value. Figure 6 shows an example of an 'RPM' test in which the main parameters are exemplified.

The variation between the initial and optimal cutterhead rpm and PR values provides an indication of improvements in boring efficiency achieved during the tests for a given machine, geology and level of thrust. The results indicate that the tendency for all normalised 'RPM tests' is similar.

The influence of the operational parameters, cutter thrust and cutterhead velocity (rpm), in hard rock tunnel boring efficiency can be analysed on the basis of field trials or 'on-site' testing (Penetration tests and 'RPM tests'). 'On site' testing, involving penetration and RPM tests, can be used to evaluate the influence of the operational parameters and determine the 'optimal' values on a given geology (rock boreability) and for a given machine.

3.6 *Cutter thrust level influence on cutter consumption*

The influence of applied cutter thrust on cutter life has been analysed by Macias (2016) for tunnel sections within three different projects exhibiting hard rock conditions (high levels of strength, high abrasivity and slightly fractured rock mass properties).

The results of the analysis reveal the influence of gross cutter thrust levels in highly abrasive rock types where $4.5 < CLI < 5.9$. The use of basic cutter ring life (hours) allows the

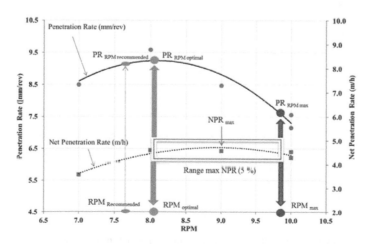

Figure 6. An example of 'RPM test' results. Vertical lines (from right to left) denote the start of the test, the cutterhead rpm reference and the cutterhead rpm optimal (from Macias, 2016).

comparison between the data from different TBM cutterhead diameters, cutterhead rpm levels or amount of abrasive minerals.

Figure 7 indicates that the increment of the applied cutter thrust will reduce considerably the basic cutter ring life while a more limited increment will result from a reduction of the applied cutter thrust. It is important to consider that, a variation of the thrust will result in variations on the net penetration and therefore in cutter life (expressed in m/c or m^3/c).

3.7 *Length Factor*

The length of the tunnel has great influence on the performances due to problems with the tunnelling system (transport system, supply delays, ventilation and/or water). In addition, longer tunnels have greater likelihood to have end rock mass qualities and, normally, a more deficient geology investigation.

It has been emphasized by Barton (2000) that the utilization is time-dependent with, in reality, a deceleration gradient. The machine utilization, and thus advance rate, is not a constant in tunnel length and therefore time.

Tunnel length exerts an important influence on the time taken to carry out tunnelling activities (Barton. 2000). During the initial "learning curve" period of a tunnelling project, certain operations take a relatively long time as the crew builds up its skills levels, and the tunnelling quality assurance system evolves. For long headings (>8 km), miscellaneous factors put increasing demands on available tunnelling time (Bruland, 2000).

The longer the tunnel, the higher the probability of problems arising linked to factors such as muck transport and ventilation, electricity and water supply systems, as well as other supply delays. Waiting times for transport will increase substantially if the capacity of the transport system is inadequate.

Problems linked to ventilation, electricity and water supplies will continue to increase with increasing tunnel length. Moreover, the transport system will be dependent on penetration rate and the amount of material that requires transportation. The higher the penetration rate, the greater the likelihood of downtime and other problems linked to supply delays and other issues with the transport system.

The NTNU model database is based on tunnels up to 10 kilometres in length. Time consumption data used in previous versions of the model were averaged over total tunnel length. The latest version of the NTNU model (Macias, 2016), includes additional time consumption (h/km) related to tunnel length with influence in machine utilization and therefore advance

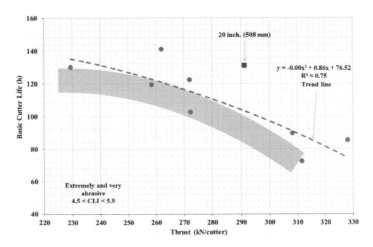

Figure 7. The influence of cutter thrust levels on cutter life (19-in. cutters) for highly abrasive rocks (4.5<CLI<5.9). The 20-in. cutter diameter value is deliberately excluded from the data used to construct the trend line (from Macias, 2016).

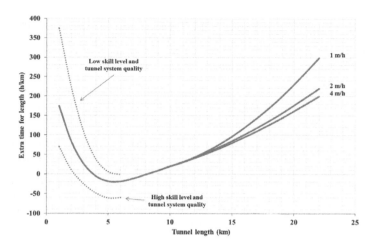

Figure 8.　Additional time (h/km) plotted against tunnel length (Macias, 2016).

rate factor (Figure 8). Boundary limits have been included for low and high skills levels and tunnel system quality. The values for every kilometre correspond to the extra time during the last km.

The term 'skills levels' refers not only to crew members, but also to equipment manufacturers and others.

4　CONCLUSIONS

Tunnelling technologies continue to improve, and the TBM sector is no exception. Currently, the TBM method, due to the technology development is applicable in an increasingly wider range of rock mass conditions, given certain economic constraints.

A process of great complexity is involved when tunnel boring due to the interaction between the rock mass and the machine. The prediction of performances (e.g. Penetration rate) and cutter life are not straightforward issues and involve major risk assessments. When hard-to-very hard rock (i.e. low-to-extremely low boreability) the complexity is accentuated becoming, in many cases, critical for the achievement of the final schedule and reasonable tunnelling cost.

Continuous updating, likely resulting in a better understanding of the tunnel boring process, should be applied to the development and revision of prediction models.

The paper has concisely compiled experiences and outcomes from research and consulting collaborations on several hard rock TBM projects and during the revise and extend of the NTNU prediction model for TBM performance and cutter life (Macias, 2016): Rock boreability when low degree of fracturing, update cutter life assessments, other geological factors influencing cutter life, influence of applied cutter thrust on cutter life and tunnel length influence on advance rate, excavation time and cost.

Latest and current consulting collaborations carry out by the undersigned are being used for validation and further development.

REFERENCES

Barton, N. 2000. TBM tunnelling in jointed and faulted rock. *A. A. Balkema*, Rotterdam (2000). ISBN 90 5809 341 7.
Bieniawski, Z.T., Celada, B., Galera, J.M. & Alvarez, M. 2006. Rock mass excavability indicator: new way to selecting the optimum tunnel construction method. *Tunnelling and Underground Space Technology* Vol. 21, no.3–4 (2006), pp 237.

Bruland, A. 2000. Hard Rock Tunnel Boring: Vol 1–7, Background and Discussion. PhD thesis. *Norwegian University of Science and Technology (NTNU)*, Trondheim, Norway, 2000.

Entacher, M., Winter, G. & Galler, R. 2013. Cutter force measurement on tunnel boring machines – Implementation at Koralm tunnel. *Tunnelling and Underground Space Technology* Vol. 38 (2013), pp 487–496.

Farrokh, E., Rostami, J. & Laughton, C. 2012. Study of various models for estimation of penetration rate of hard rock TBMs. *Tunnelling and Underground Space Technology* Vol. 30 (2012), pp 110–123.

Gehring, K. 1995. Leistungs- und Verschleißprognosen im maschinellen Tunnelbau. *Felsbau*, Vol. 13, no. 6 (1995), pp 439–448 (in German).

Gong, Q.M. & Zhao, J. 2009. Development of a rock mass characteristics model for TBM penetration rate prediction. *International Journal of Rock Mechanics & Mining Sciences* Vol. 46 (2009), pp 8–18.

Hassanpour, J., Rostami, J. and Zhao, J. 2011. A new hard rock TBM performance prediction model for project planning. *Tunnelling and Underground Space Technology* Vol. 26 (2011), pp. 595–603.

ISRM International Society for Rock Mechanics. 1978. Suggested methods for the quantitative description of discontinuities in rock masses. – Commission on Standardization of Laboratory and Field Tests, Document No. 4, *International Journal of Rock Mechanics & Mining Science*, Vol. 15 (1978), pp 319–368.

Macias, F.J. 2016. Hard Rock Tunnel Boring: Performance predictions and cutter life assessments. PhD thesis, *Norwegian University of Science and Technology (NTNU)*, Trondheim, Norway (2016).

Macias, F.J., Andersson, T. & Eide, L.N.R. 2017. Project control by using the NTNU model methodology: The new Ulriken Tunnel. *Proceedings of the World Tunnel Congress 2017-Surface Challenges-Underground solutions*. Bergen, Norway.

Rostami, J. & Ozdemir, L. 1993. A new model for performance prediction of hard rock TBMs. *Proceedings, rapid Excavation and Tunnelling Conference (RETC 1993)*, Boston, Massachusetts, USA (1993), pp 793–809.

Rostami, J. 1997. Development of a force estimation model for rock fragmentation with disc cutters through theoretical modelling and physical measurement of crushed zone pressure, PhD Thesis. *Colorado School of Mines*, Golden, Colorado, USA (1997).

Singh, B. & Goel, R.K. 2011. Engineering Rock Mass Classification. *Elsevier Inc* (2011). ISBN 978-0-12-385878-8.

Wilfing, L. 2016. The Influence of Geotechnical Parameters on penetration Prediction in TBM Tunneling in Hard Rock. PhD Thesis, *Technical University of Munich*, Munich, Germany (2016).

Yagiz, S. 2014. Modified CSM model for predicting TBM performance in rock mass. *LAP (Lambert Academic Publishing)*, 2014.

Tunnels and Underground Cities: Engineering and Innovation meet Archaeology,
Architecture and Art, Volume 6: Innovation in underground engineering,
materials and equipment - Part 2 – Peila, Viggiani & Celestino (Eds)
© 2020 Taylor & Francis Group, London, ISBN 978-0-367-46871-2

A study of the analysis method of the cause of tunnel lining deformation using "TCI"

K. Maeda & Y. Shigeta
Pacific Consultants Co., Ltd., Tokyo, Japan.

S. Kaise, K. Maegawa, Y. Maeda & T. Ito
NEXCO Research Institute Japan, Tokyo, Japan.

H. Yagi
Central Nippon Expressway Co., Ltd., Nagoya, Japan.

ABSTRACT: Deformation causes of tunnel lining include external causes in which forces are applied from outside the tunnel (hereinafter referred to as "external force") and internal causes due to the material, construction etc. However, there are many unclear aspects in the characteristics of cracks classified according to their causes. Thus in this research, we set thresholds related to the possibility of deformation caused by external force from the cracks development plans of the linings in the road tunnels managed by Nippon Expressway Company (hereinafter referred to as "NEXCO") of Western, Central and Eastern Japan using Tunnel-Lining Crack Index (hereinafter referred to as "TCI"). Moreover, we analyzed the relationship between the set thresholds and the tunnels where external force deformations were observed.

1 INTRODUCTION

The causes of cracks in tunnel lining can be roughly divided into external causes such as the effect of external forces and internal causes due to the material, construction, etc. Generally speaking, cracks are generated by the complex interplay of deformation causes both external and internal. Therefore, the forms of tunnel lining cracks are extremely diverse. However, research/analysis of the characteristics of cracks and estimation of the deformation causes are important for assessing the stability and safety of a tunnel.

This paper conducts analysis of cracks information obtained from the inspections of road tunnels managed by NEXCO using Tunnel-Lining Crack Index (TCI), understands the characteristics of both cracks caused by internal causes and cracks caused by external causes, and examines an objective estimation method of deformation causes.

2 OUTLINE OF TCI

This section outlines the TCI that was used to quantitatively evaluate the features of cracks by investigating the crack deployment diagrams.

Rock mass properties (such as deformation modules and coefficient of permeability) are largely influenced by density, orientation, and width of cracks (joints) in the rock mass, then an index called "crack tensor" that can comprehensively quantify this influence has been studied in the field of Rock Mechanics. The index, TCI has been proposed in the preceding studies for evaluating cracks on lining concrete and the concept of this "crack tensor" has been introduced in the TCI. Width, length, and orientation of cracks on the lining surface are parameters of TCI, which makes possible to decide the "maximum width of crack", "maximum length of crack", "distribution of cracks", and "general orientation of cracks". Fundamental equation and conceptual image are shown respectively in Equation (1) and Figure 1.

$$F_{ij} = \frac{1}{A} \sum_{k=1}^{n} \left(t^{(k)} \right)^{\alpha} \left(l^{(k)} \right)^{\beta} \cos \theta_i^{(k)} \cos \theta_j^{(k)}$$ (1)

A: Area of lining concrete (A=Ls × La)
Ls: Longitudinal length of lining concrete (i.e., span length in general)
La: Transversal length of lining concrete
N: Number of cracks
l(k): Length of crack (k)
t(k): Width of crack (k)
i(k): Angle between the normal vector of crack (k) and x_i axis
j(k): Angle between the normal vector of crack (k) and x_j axis
α: Weighting factor on width of crack
β: Weighting factor on length of crack
F_0: Magnitude of TCI
F_{11}: Longitudinal component of TCI
F_{22}: Transversal component of TCI
$F_{12} = F_{21}$: Shear component of TCI

F_{11} and F_{22} given by Equation (1) are the longitudinal and transversal components of TCI, respectively. Deterioration index (F_0) of lining concrete is expressed by summation of these two components ($F_0 = F_{11} + F_{22}$), which is the invariant of tensor. This deterioration index, F_0 is used as an evaluation value of the TCI, representing the progress of deformation. Furthermore, the features of cracks can be characterized and understood by each component (i.e., F_{11}, F_{22}, F_{12}, and F_{21}).

In this study, the weighting factor (α) for width and that (β) for length were set to be identical and to be 1.0 based on the preceding studies.

Since TCI is a tensor, the major orientation of average crack over the lining span shown in Figure 2 can be given by Equation (2).

$$\theta = \alpha = \tan^{-1} \left(\frac{F_{12}}{r + (F_{11} - (F_0/2))} \right)$$ (2)

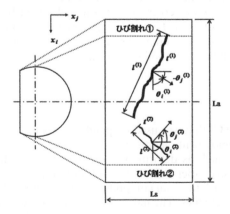

Figure 1.   conceptual diagram of TCI.

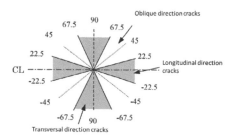

Figure 2.   Definition of the major orientation of crack.

θ = Major orientation of cracks
α = Angle between x axis and a line connecting two points (F_{11} and F_{12}), on which TCI circle crosses x axis
r = Absolute value of (F_{11} - F_{22})

3   OVERVIEW OF EXAMINATION

3.1   *Examination method*

This analysis was conducted with the longitudinal direction component F_{11} and transverse direction component F_{22} of TCI. Figure 3 shows an example of the analysis. Horizontal axis of the graph was set as F_{11}, and vertical axis was set as F_{22}. The intersections of F_{11} and F_{22} calculated at each span of one lining were plotted on the graph. When the plots are distributed below the dotted line on the graph ($F_{11}=F_{22}$ line), it is the longitudinal direction cracks predominant type, and when they are distributed above the dotted line it is the transverse direction cracks predominant type. When the plots are distributed on, or near the dotted line, it is the diagonal cracks predominant type. From this graph, it is inferred, that the following deformation causes can be estimated.

$F_{11}>F_{22}$: Longitudinal direction cracks on the crown due to loosening pressure, longitudinal direction cracks on the side wall due to water pressure or plastic pressure, etc.

$F_{11}=F_{22}$: Diagonal direction cracks due to landslide or side pressure topography, etc.

$F_{11}<F_{22}$: Transverse direction cracks due to sinking or rising of roadbed, etc.

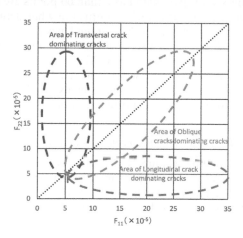

Figure 3.   Analysis of deformation causes.

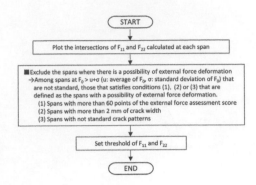

Figure 4.   Conceptual diagram of TCI.

Table 1 .   Averages and Standard Deviations of Subject Tunnels and Each Lining Method.

Lining Method		Lagging support Method	NATM
Span		3,339	6,020
F_{11}	Average u	6.31	3.15
	Standard Deviation σ	4.89	4.26
F_{22}	Average u	4.26	1.92
	Standard Deviation σ	3.83	3.47

In this examination, a distribution diagram of F_{11} and F_{22} in the subject tunnels excluding the spans where there is a possibility of external force deformation, and a threshold that includes this is set. The fact that what are included within the average u ± standard deviation σ among TCI components are deformation from internal causes (standard forms of cracks) shown in the existing study is utilized for this analysis. Thus, among spans at $F_0 > u + σ$ (u: average of F_0, σ: standard deviation of F_0) that are not standard, in other words spans with extensive damages by cracks, those that satisfies conditions (1), (2) or (3) that are shown in Figure 4 are defined as the spans with a possibility of external force deformation.

Here, the external force assessment score is the indicator that translates the deformation situation of lining into points, and spans with more than 60 points are considered to be inspection priority spans. Moreover, standard crack patterns are what defined in an existing study as cracks originating from construction, and having tendency for generating two lines of longitudinal direction cracks with the lagging support method, and having tendency for generating on line of longitudinal direction crack with NATM.

Thus, if an appropriate threshold can be set, it is inferred to be possible to set a threshold that enables estimation of deformation cause not to be external force deformation, in other words to be material degradation, installation-origin, etc. only by using TCI.

3.2 Averages and standard deviations of subject tunnels and each lining method

Table 1 shows the analysis subject tunnels. 50 tunnels lined with the lagging support method and lined with NATM where proximity visual inspection is conducted were selected respectively from the tunnels managed by NEXCO. The table shows their respective F_{11} and F_{22} average and standard deviation of subject span according to the method used.

4 ANALYSIS RESULT

4.1 *Threshold of Lagging Support Method*

Figure 5 shows the distribution of F_{11} and F_{22} of spans of the lagging support method that exclude the aforementioned conditions (1), (2) and (3) that have the possibility of being external force type. At this time, when a u+2σ line is drawn for both F_{11} and F_{22} as their thresholds, it becomes clear that the distribution is mostly below the thresholds.

Figure 6 shows the spans distribution in the area below these thresholds. 3,083 spans are below the threshold at this time, which is 92.3% of the entire 3,339 subjects. External force deformation is generally considered to be about 5% of entire tunnels. Thus these thresholds are considered to be broadly appropriate, and it is inferred that the deformation causes of the spans below the thresholds could be estimated to be material degradation, installation-origin, etc.

4.2 *Threshold of NATM*

The same analysis conducted for the lagging support method was also conducted for NATM. As a result, the number of spans that are distributed below the threshold was 5,460 when both F_{11} and F_{22} thresholds were set at u + 2σ as shown in Figure 7, which is 90.7% of the entire 6,020 subject spans. While these thresholds are slightly on the safety side, they are inferred to be broadly appropriate as was the case for the lagging support method.

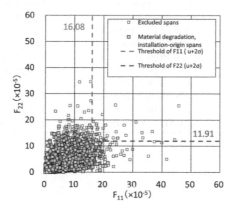

Figure 5.   On setting of threshold.

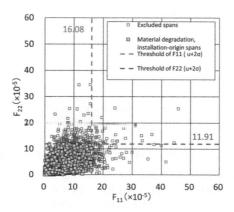

Figure 6.   Thresholds of lagging support method.

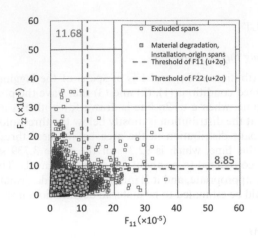

Figure 7.   Thresholds of NATM.

Table 2.   Tunnels with external force deformation (lagging support method).

No.	Tunnel Name	Deformation Cause	Deformation Situation
1	Tunnel A	Loosening Pressure	Longitudinal and Transverse Cracks
2	Tunnel B	Loosening Pressure	Longitudinal and Transverse Cracks
3	Tunnel C	Rising of Roadbed	Transverse and Diagonal Cracks
4	Tunnel D	Rising of Roadbed	Transverse Cracks
5	Tunnel E	Plastic Pressure	Longitudinal Cracks on the Shoulder
6	Tunnel F	Plastic Pressure	Longitudinal Cracks on the Shoulder
7	Tunnel G	Landslide Side Pressure	Longitudinal Cracks on the Shoulder
8	Tunnel H	Landslide Side Pressure	Longitudinal Cracks on the Shoulder
9	Tunnel I	Rising of Roadbed	Transverse Cracks
10	Tunnel J	Rock Mass Slippage	Transverse and Diagonal Cracks

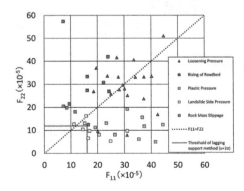

Figure 8.   Relationship between external force deformation and threshold.

4.3   *Comparison of external force deformation and threshold (Lagging Support Method)*

Figure 8 shows the distribution of F_{11} and F_{22} of the lagging support method tunnels where external force deformation occurred in the past listed in Table 2 according to the deformation causes. When the aforementioned F_{11} and F_{22} thresholds u + 2σ of lagging support method are shown in this graph, it results in only one span among the spans where external force deformation occurred is included within the thresholds. Thus, it becomes clear that most of external force deformations are distributed outside the area of the thresholds.

Therefore, it is inferred that it is appropriate for the threshold that allows the deformation causes to be material degradation, installation-origin, etc. to be below $u + 2\sigma$ for both F_{11} and F_{22}.

5 CONCLUSION

This examination revealed that there is a possibility for setting thresholds that enable estimation of deformation causes to be material degradation, installation-origin, etc. by using F_{11} and F_{22}. In future, it is inferred that there is a necessity for understanding the relationship between the external force deformation and progression and set a standard for appropriate health level determination. Moreover, we hope to improve the accuracy by increasing the number of analyzed tunnels and establish a deformation cause estimation method.

NEXCO must manage deteriorating road tunnels and devise repair plans. Meanwhile, the decrease in the number of technicians due to population decrease, birthrate decline and aging is also the cause for concern. In future, we believe that by establishing a deformation cause estimation method for further optimization of tunnel inspection that reduces discrepancy caused by the difference in the skill of technicians, we will be able to contribute to the selection of spans that require careful proximity visual inspection and tapping inspection and to the devising of the repair plan for large-scale restoration works.

Tunnels and Underground Cities: Engineering and Innovation meet Archaeology, Architecture and Art, Volume 6: Innovation in underground engineering, materials and equipment - Part 2 – Peila, Viggiani & Celestino (Eds)
© 2020 Taylor & Francis Group, London, ISBN 978-0-367-46871-2

The present situation of countermeasure of heaving in the expressway of Japan

K. Maegawa, T. Ito & S. Kaise
NEXCO Research Institute Japan, Machida-city, Japan

H. Yagi
Central Nippon Express Company Limited, Nagoya-city, Japan

S. Kunimura & Y. Okui
OYO Corporation, Saitama-city, Japan

ABSTRACT: On the Expressway of Japan, many structures are aging and renewal construction for a longer life is being promoted. The Main contents of the tunnel are to repair and reinforcement of the lining and countermeasure for heaving. The project scale is planned to be 130 km in 15 years from 2015 to 2030. In most cases construction will be carried out while opening one lane of two. As a result, the countermeasures works in a narrow space and the speed of progress is slow. Many problems remain with respect to the countermeasure construction of heaving. Therefore, innovations such as further efficient construction in a narrow space, rapid construction technique, new technology countermeasure of heaving not dependent on additional invert-concrete, construction machinery and materials, etc. are required.

1 INTRODUCTION

Since the opening of the Meishin Expressway in 1963, about 9,000 km of expressways have opened in Japan, out of which the tunnels (about 1,850 in number) have the total length of about 1,760 km.

Approximately 30% of the tunnels have been in use for over 30 years, showing the signs of aging.

NEXCO 3, the company which maintains Japan's expressways with the business license of the Ministry of Land, Infrastructure, Transport and Tourism, has been carrying out renewal constructions to extend the life of the structures.

The renewal constructions for the tunnel structures consists mainly of recovering lining strength reduced by cracks and other reasons, repair of the lining against external forces that occurred after the opening the tunnel, unexpected at the time of construction, countermeasures for heaving on parts that are hinder are expected to hinder passing vehicles due to road surface uplift or reinforcement work.

The tunnel renewal construction project is to cover 130 km in the 15 years of 2015 to 2030.

This report discusses the tunnel renewal constructions, focusing on the current state of countermeasures for heaving.

2 IMPLEMENTATION OF THE COUNTERMEASURES FOR HEAVING IN JAPAN

2.1 *Causes of deformation in cases using invert-concrete*

Figure 1 shows the tunnels with evident heaving, based on the survey of the relationship between heaving and the geological features. The survey was limited to expressway tunnels.

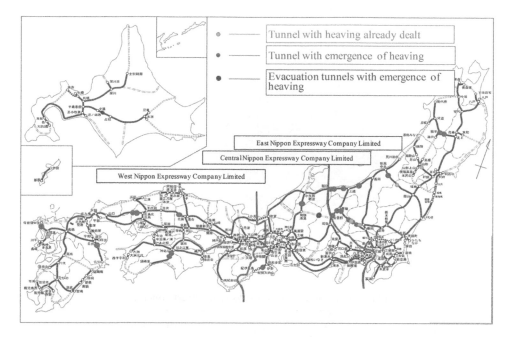

Figure 1. The tunnels under the jurisdiction of NEXCO, with evident heaving (as of September 1, 2016).

Table 1. Types of rocks and geological features likely to cause heaving.

1 Rock types clearly identified in design guidelines and others	Tuff, mudstone, shale, clay slate, serpentinite, weathered crystalline schist, solfataric clay, etc.
2 Rock types which need invert-concrete for sediment, portal, etc., based on current standards	Silt, disintegrated granite soil, talus, debris flow deposits, loam, embankment, weathered soil, top soil, etc.
3 Rock types without clear mentions of lithology but invert-concrete is considered necessary with topography including fault fracture zone	Fault, fault (estimated fault), fault fracture zone, a landslide deposits, landslide colluvium, fracture zone, etc.

Table 1 shows the result of the analysis of the geological features where heaving occurs and the types of rocks which are likely to cause heaving.

Although there are cases in which heaving occurred and progressed rapidly, destroying the invert-concrete of the tunnels, most heaving cases in Japan showed swelling, with expansile clay minerals such as smectite absorbing water and expanding, causing the road surface bulging.

Based on this fact, after 1998, construction of tunnels in the bedrock with rock types which show strength deterioration or expansion in the long term are carried out with invert-concrete installed even for medium hard rocks.

Figure 2 shows the analysis result of the rock types of the tunnels where road surface uplifts are recognized, by geological profile.

As shown in Figure 2, the sections where the road surface uplift occur have rock types which show strength deterioration and expansion in the long term. Many of these sections did not require installation of invert-concrete according to the previous standards 1)

But when the current construction standards are applied, the total length of the sections with the rock types which require invert-concrete accounts for about 90% or more of those sections.

2.2 Choosing invert-concrete for countermeasures

Table 2 shows the cases in which heaving such as uplift of pavement slabs occurred on expressways after they opened, and reinforcement work with invert-concrete were carried out.

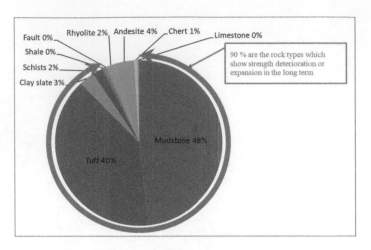

Figure 2. Ratio of rock types where uplift occurred.

Table 2. Cases of countermeasures for expressways using invert-concrete.

No.	Names of the tunnels	Total length	Total length of coutermeasures with invert-concrete	Support method	Other countermeasures
1	A Tunnel	1990 L=683m	1997 L=379.5m	NATM	rock bolt (preliminary reinforcement) reinforcement of the sections beneath the roadbed with rock bolt and micropiles before the invert-concrete was installed
2	B Tunnel	1990 L=2,609.5m	1997 L=505.5m	NATM	rock bolt (preliminary reinforcement) reinforcement of the sections beneath the roadbed with rock bolt and micropiles before the invert-concrete was installed
3	C Tunnel	1992.2 L=684m	1998.11 L=70m	NATM	rock bolt (preliminary reinforcement)
4	D Tunnel	1991 L=1,234m	2002 L=133.65m	NATM	
5	E Tunnel	1993 L=3,998m	2003 L=2,674.7m	NATM	
6	F Tunnel	1991 L=1,234m	2008 L=148m	NATM	
7	G Tunnel	2002 L=2,051m	2013 L=41.5m	NATM	
8	H Tunnel	1991 L=3,191m	2013 L=40.5m	NATM	
9	I Tunnel	1996 L=2,600m	2015 L=126m	NATM	

"A Tunnel" and "B Tunnel" were the first expressway tunnels to be implemented with the countermeasures, where the sections beneath the roadbed were reinforced with micropiles. Although the slowing of the displacement speed was recognized, it took installment of additional invert-concrete to stop the progress (Figure 3). For this reason, additional installation of invert-concrete has become the basic countermeasure for heaving on Japanese expressways.

3 SURVEY AND MEASUREMENT OF DEFORMATION

Survey of heaving by NEXCO clarifies the purposes of survey for each stage of heaving, by grading the conditions into three stages.

In the "latent risk stage," heaving has not yet occurred but the tunnels have geological features or potential. In the "progressive risk stage," heaving has been recognized for the first time or heaving is not serious but follow-up observations are carried out. In the "countermeasures stage," heaving has progressed or deformation is serious, requiring immediate response and countermeasures. Table 3 shows the purposes for survey on each stage and others.

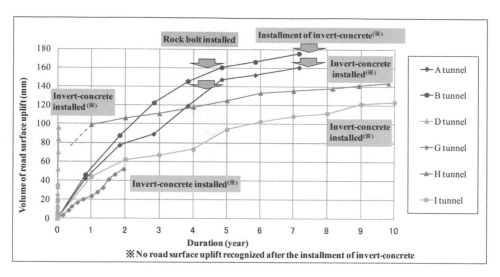

Figure 3. Relationship between the volume of road surface uplift before countermeasures and the years since measurement started.

Table 3. The state of deformation and purposes of survey.

Stages of deformation	State of deformation	Purposes of survey and overview
Latent risk stage	Heaving have not yet occured in tunnels but they have geological features or potential for heaving	To measure the normal level (initial level) before heaving occurs. Survey and measurement of the current road surface elevation and other conditions in sections with geological features with the potential for heaving
Progressive risk stage	Heaving has been recognized for the first time or heaving is not serious but follow-up observations are carried out	To continue observation of the degree of deformation and to grasp its progress. Survey and measurement of displacement of road surface elevation, inner space, etc. and presumption of deformation causes inside sections with heaving found by users and through patrols and tunnel inspections
Counter-measures stage	Heaving is obvious and progress of deformation is serious, requiring immediate response and countermeasures	Survey and measurement of sections with existing deformation to presume the deformation causes, predict future deformation and design countermeasures

In "latent risk stage," the objective of the survey is to measure the initial value before heaving occurs, and survey mainly consists of the leveling of the whole tunnels and collection of data at the time of construction.

In "progressive risk stage" which comes before "countermeasures stage," the survey is aimed at measuring the progressiveness of heaving since its occurrence and setting the schedule for countermeasures. Unlike "latent risk stage," the risk is observed on a continuous basis, with leveling or convergence measurement carried out four times a year, for example.

In "countermeasures stage," the purpose of the survey is to collect data for planning countermeasures with various surveys and bedrock sample tests.

4 PLANNING FOR COUNTERMEASURES

4.1 The structure of countermeasures

Cross section of the additional invert-concrete as countermeasures is decided based on numerical analysis using survey results. An example of the cross section is shown in Figure 4.

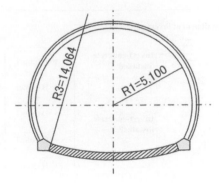 

Figure 4. Cross section example of the additional invert-concrete.

Figure 5. Countermeasure using invert struts.

As shown in Figure 4, the bottom tip of lining is cut slantingly and invert is applied to it so that stress transmission to the lining will be smooth.

Figure 5. Countermeasure using invert struts. In many other cases, invert struts (Figure 5) are installed. In most of these cases, the countermeasures are adopted not necessarily from the design aspect. They are applied, for example, for deformation control when unexpected external force is applied due to excavation around the invert.

4.2 Circumstances around expressways

In Japan, expressways are an indispensable part of neighboring residents' life. Due to social needs and limited traffic capacity of detour roads, it is difficult to close the entire expressways for renovation. Therefore, when carrying out the countermeasures for heaving, it is common to leave open one lane on the two-lane roads.

As a result, the countermeasure works are usually conducted in narrow spaces, slowing the construction speed.

4.3 Designing countermeasures

The numerical analysis for designing countermeasures for heaving is carried out using finite element method (FEM) or finite difference method (FDM). These methods are used to presume the mechanism of deformation by analyzing the current deformation state of batholith and lining, stress condition of lining and more and to reproduce the current state of deformation as well as confirming the effectiveness of the invert reinforcement work against the progress of deformation in future.

Generally, two-dimensional numerical analysis is used to examine tunnel structures and three-dimensional numerical analysis is used to study the effects of countermeasures.

Figure 6 shows the general procedures of numerical analysis.

(1) As the analysis of survey results, deformation mechanism is presumed by grasping the state of heaving and deformation of lining, based on examinations of existing deformation, geological survey, etc.

The state of geological feature distribution around the tunnels is grasped for modeling, based on bowling survey results and face observation results.

(2) For reproduction analysis, the situation at the time of the tunnel building (displacement measurement result) is reproduced. The bedrock's modulus of deformation and state of initial stress are grasped roughly with inverse analysis using measured values.

(3) For the reproduction analysis of deformation, the bedrock properties are set for reproduction of current deformation conditions. Displacement volume (the uplift volume,

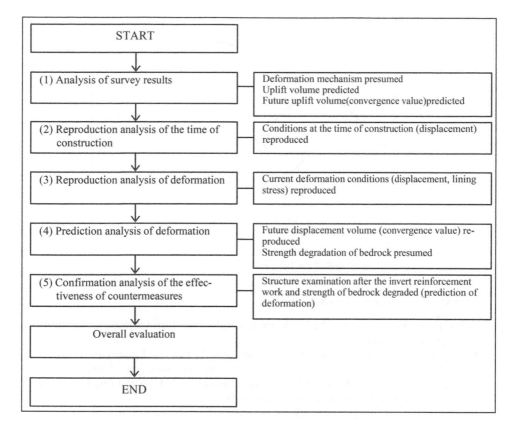

Figure 6. Process of analysis using structure examination with two-dimensional analysis.

displacement volume of inside section and underground), stress value (lining stress) and range of looseness based on deformation survey are reproduced.

The ranges of looseness is calculated on the whole area surrounding the tunnel, the area lower than SL and area lower than pavement slab in order to presume and establish the ranges of looseness which can reproduce the measuring results of deformation survey and tendencies of deformation.

(4) For prediction analysis of deformation, displacement volume is presumed from changes in measured values due to aging, and the material properties value is identified so that it may match the estimated value.

(5) As confirmation analysis of the effectiveness of countermeasures, invert reinforcement work is modeled and the tunnel structure with invert and concrete lining is examined with the calculation using the material properties value of the prediction analysis of deformation.

For two-dimensional analysis for tunnels, over-the-counter software or original software developed by user companies are used. For three-dimensional analysis, the FLAC 3D (ITASCA), analysis software of FDM, is often used. This is presumably because the shear strength reduction method (SSRM), to be mentioned later, can be used for reinforcement design.

4.3.1 *Modeling for bedrock*

4.3.1.1 ELEMENT

Dynamics model for bedrock consists of elastoplastic body, viscoelastic body and viscoelastic plastic body.

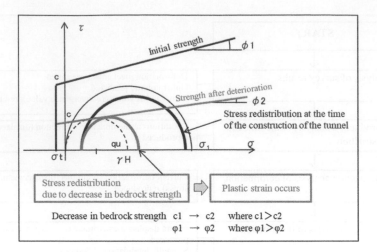

Figure 7.   Conceptual diagram of SSRM for bedrock.

Input material properties value of bedrock are generally based on various survey results and bedrock sample examinations.

The material properties which can reproduce measured values and predicted values of road surface in different stages (at the time of construction, reproduction of present state and future prediction) are set by inverse analysis with modulus of deformation, Poisson's ratio, cohesion and an angle of internal friction as parameters.

For modeling of weakening of bedrock, SSRM is used, as shown in Figure 7.

Generally for modeling, 5D (D : tunnel representative diameter) to the sides is taken for lateral domain, about 4D for lower domain and 4D for upper domain with big earth covering, and to the ground level when the earth covering is small.

4.3.2 *Modeling for existing lining and additional invert-concrete*

For shotcrete, existing lining and additional invert-concrete, modeling is done with elastic body and elastoplastic body (strain softening) etc.

For rock bolt, modeling is generally not done except in cases applied as supplementary construction method of countermeasure works.

There is no set modeling for methods of linking existing lining with additional invert-concrete, due to lack of sufficient knowledge. However, when the axial tension is excellent, rigid connection is applied.

The input material properties value is the shape (thickness of lining, cross-sectional area, geometrical moment of inertia and weight per unit volume) and strength (the elastic modulus, the compressive strength and the tensile strength).

For these, actual measurement are available at times, but they are usually set based on the cross-sectional form and thickness of component and design value of component (concrete strength and the elastic modulus) at the time of tunnel construction.

5 COUNTERMEASURES

5.1 *Construction process*

As mentioned above, countermeasure works are generally carried out with lane restrictions.

Figure 8 shows the cross section of the general construction with lane restrictions and Figure 9 shows the construction process.

For preparatory operations necessary for constructions with lane restrictions, soldier piles are built and protection fences are installed. These operations are not necessary for constructions with entire road closure.

Excavation work and installment of soldier piles are especially time-consuming because they are conducted in narrow spaces with lane restrictions, allowing the use of only few kinds of building machines and requiring extreme caution with passing vehicles nearby and countermeasure works on heaving of bedrock with soft rock or harder.

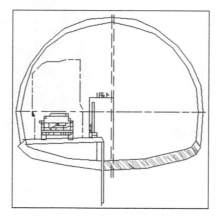

Figure 8.   Cross section of the construction with lane restriction.

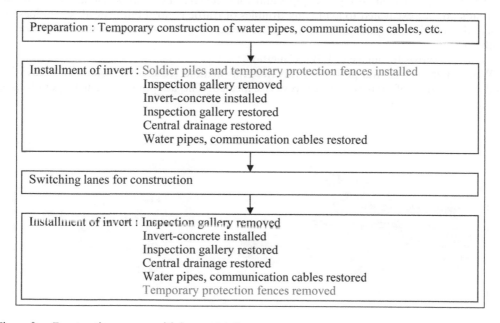

Figure 9.   Construction process with lane restrictions.

5.2 Technical development of construction technique

5.2.1 Cases of actual operation

Countermeasure works with additional invert-concrete have been carried out at "J Tunnel", with the total length of 1,960 meters, from 2016 to 2018,

Efficiency of countermeasures with lane restrictions has improved by changing 0.45m³ tip of the excavator, without changing the entire machinery. Figure 10.

Figure 10. 0.45m³ Excavator attachment(commercially available machine).

5.2.2 Method of invert of hybrid structure using steel pipes and concrete (IHUPC)

For places where it is physically difficult to improve the vehicles' safety and driving environment and to secure sufficient passage width during construction period, a new method of invert of hybrid structure using steel pipes and concrete (IHUPC) was developed. Central part of roads are not excavated to avoid any troubles of traffic control but instead, steel pipes are inserted, and for the side walls, cast-in-place concrete is installed after the excavation of batholith.

Figure 11 shows the image of IHUPC method.

With leaving the central part non-excavated, this method reduces the volume of excavation and the speedup of construction process is expected. The adoption in actual operations is expected in the future.

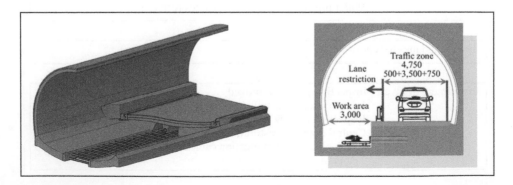

Figure 11. The image of method of IHUPC (patent method in Japan).

6 ISSUES TO BE TACKLED

6.1 *Accelerating the process*

In the 15 years from 2015 to 2030, 130 km of tunnel renewal construction must be carried out.

Given the labor shortage due to the disaster reconstruction of Great East Japan earthquake in 2011 and the preparation for 2020 Tokyo Olympics, the speed for the tunnel renewals needs to be about 50 meters per month in order to accomplish 130 km in 15 years.

Current construction methods with lane restrictions only allows for the progress speed of 10 to 20 meters per month can, so substantial speedup is needed.

For the speedup, followings are being considered. In the places where heaving has actually occurred, adoption of IHUPC method and the capacity improvement of the construction machinery are considered on the premise of implementing countermeasures with additional invert-concrete. For preventive countermeasures, new methods to take place of additional invert-concrete need to be established, including reinforcement of parts under the roadbed.

A committee of experts has been discussing these issues.

6.2 *Removal of lane restrictions during holidays*

Some expressways have substantially more traffic during holidays than on weekdays. Lane restrictions during holidays on these expressways cause traffic jams and its social impact is serious.

This is also true for the countermeasures for heaving in tunnels. The lane restriction of one lane out of two that begins on Monday needs to be re-moved by the end of Friday.

When a pilot construction project was executed in the above-mentioned "J Tunnel", the work for the 10-meter lane including demolition and excavation of the pavement, installment of additional invert-concrete, back-filling and temporary paving were all done on weekdays partly due to the good conditions of roadbed with no spring water from underneath.

The implementation process is being studied for further specdup.

7 CONCLUSIONS

Expressways of Japan are currently renewed to extend the structures' life.

The renewal of tunnel structures mainly consists of repair and reinforcement of the lining and countermeasures for heaving on the parts which hinder or are expected to hinder passing vehicles due to road surface uplift.

The tunnel renewal construction is scheduled to last 15 years, covering 130 km.

For the survey and design of the countermeasures, the conditions of heaving are graded into three stages. The purposes of the survey for each stage are clarified before scheduling for countermeasures and deciding on the survey items needed as data for design.

Numerical analysis are carried out using FEM or FDM for designing the measures. Although some issues still need to be tackled, the procedures and methods of numerical analysis for the design techniques are mostly established.

The circumstances of Japanese expressways usually do not allow for the entire road closure and the constructions are usually done while leaving one lane out of two opened.

Therefore, countermeasure works are implemented in narrow areas and the speed is slow.

To renew 130 km of tunnels in 15 years, construction speed needs to be 50 meters per month. Using the present construction techniques and with lane restrictions, the speed of only 10 to 20 meters per month can be achieved, so substantial speedup is necessary.

For the speedup of countermeasure works, various means are adopted such as the use of efficient building machines fit for work in narrow spaces and dividing roads into three sections to leave the middle section unexcavated which results in reduction of the excavation volume.

However, in order to achieve the desirable speedup, further efforts are needed, including improvement of building machine capability, establishment of new methods to replace the installment of additional invert-concrete and more.

Tunnels and Underground Cities: Engineering and Innovation meet Archaeology,
Architecture and Art, Volume 6: Innovation in underground engineering,
materials and equipment - Part 2 – Peila, Viggiani & Celestino (Eds)
© 2020 Taylor & Francis Group, London, ISBN 978-0-367-46871-2

Back analysis of ground settlements induced by TBM excavation for the north extension of Paris metro, line 12

S. Mahdi, O. Gastebled & S. Khodr
TRACTEBEL ENGIE, Paris, France

ABSTRACT: The north extension of Paris metro line 12 consists in 3.8 km of double track tunnel and 3 stations, constructed in two main stages. The first stage, between 2010 and 2011, included the tunnel boring with an EPB TBM. The monitoring of settlements was particularly dense and provided a wealth of topographic data. A back analysis based on non-local least square method showed that a very good agreement could be found between a calibrated Peck formula and the empirical data. An inverse analysis allowed to deduce the characteristic trough width parameter of each formation. Eventually, a detailed back-analysis based on finite element modelling allowed to derive a law relating confinement to volume loss. This allowed establishing a predictive model based on Peck formulation complemented by two empirically calibrated laws which is shown to match closely the full data set and could be used on projects currently under study.

1 PRESENTATION OF NORTH EXTENSION OF PARIS METRO LINE 12

The extension works of Paris metro line 12 started in 2008 with the construction of the first of three stations, Front-Populaire, which was completed in 2012. Between 2010 and 2011, 3.8 km of 9.15 m diameter tunnels were bored by an EPB shield named Elodie in two drives. This first construction stage allowed early commissioning of an initial one stop extension by the end of 2012. Since 2013, the two other stations, Aimé-Cesaire and Mairie-d'Aubervilliers are under work. The commissioning of the complete extension is planned by the end of 2019.

   In the first construction stage, the owner, RATP, appointed XELIS as design engineer and site supervisor, while TRACTEBEL was appointed as geotechnical reviewer. The works were contracted to a consortium of companies: EIFFAGE and VINCI. Contrary to the first drive, the track alignment of the second drive that extends between the TBM launching shaft located adjacent to the Aimé-Césaire station and a receiving shaft located 900 m beyond the Mairie-d'Aubervilliers station, remained under three main streets in Aubervilliers (Av. Victor Hugo, Bvd Anatole France and Bvd Pasteur), which allowed dense and continuous monitoring of the settlements.

2 BACK-ANALYSIS: OBJECTIVE AND APPROACH

The continuous growth and densification of the urban periphery of Paris requires the development and modernisation of its underground metro network. Existing lines 4, 11, 12 and 14 are currently subject to extension works. Furthermore, four new automatic metro lines, part of the Grand Paris Express project (GPE), are either under construction (south of circle line 15 and suburban line 16) or under design/tendering (east and west of circle line 15, airport line 17 and Saclay line 18).

   The planned tunnels of line 16, line 15 east and line 17 are to be bored in the same geological context as the north extension of line 12, i.e. the Plaine de France basin. This makes the latter a valuable feedback case, on which to improve design choices and risk assessment while boring new tunnels in similar conditions. This paper presents an approach which not only aims at retro-fitting the monitoring settlement data of line 12, but also offers a calibrated predictive tool for the study of the impact on the urban environment of the coming tunnelling works in the vicinity.

3   GEOLOGICAL AND GEOTECHNICAL CONTEXT

The geology encountered by the line 12 extension, Figure 1, consists from grade down, in quaternary deposits (fills and alluvions) and the succession of Eocene strata:

– Priabonian marls (Marnes infragypseuses)
– Bartonian marls and silty sands (Calcaire de Saint-Ouen and Sables de Beauchamp)
– Lutetian calcareous marlstones and limestones (Marnes et Caillasses and Calcaire grossier)

The main geotechnical characteristics derived from the contractual geological baseline report (GBR) are summarized in Table 1:

– Quaternary deposits are relatively weak
– The Ducy formation at the base of Calcaire de Saint-Ouen stratum appears to be altered by gypsum dissolution phenomena
– The silty sand stratum, Sables de Beauchamp, is particularly compact and stiff, with a slightly softer clayey intercalation in the middle of the layer.

In terms of hydrogeology, le TBM bores under the water table, with a water height at the axis varying from 10 m at the launch shaft to 20 m at the exit shaft.

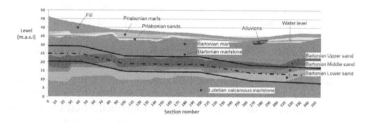

Figure 1.   Geotechnical profile from GBR.

Table 1.   Geotechnical characteristics from GBR.

| Formation | Pressure meter | | | Effective | | Short term | |
	E_m [MPa]	Pl^* [MPa]	α [-]	c' [kPa]	φ' [°]	E_{Young} [MPa]	K_0 [-]
Fills	8.5	0.6	0.50	0	25	10	0.58
Alluvions	4	0.4	0.50	0	10	5	0.83
Priabonian marls/Marnes Infragypseuses	18	1.8	0.50	20	25	25	0.58
Priabonian sands/Sables de Monceau	18	1.8	0.50	20	25	25	0.58
Bartonian marls/Calcaire de Saint-Ouen	12	1.5	0.50	30	25	80	0.58
Bartonian marlstone/Calcaire de Ducy	6.5	0.7	0.50	30	25	20	0.58
Bartonian Upper Sables de Beauchamp	38	4.0	0.33	20	30	190	0.50
Bartonian Middle Sables de Beauchamp	28	3.1	0.50	60	15	140	0.74
Bartonian Lower Sables de Beauchamp	35	4.1	0.33	20	30	160	0.50
Lutetian calcareous marlstone/ Marnes et Caillasses	48	4.5	0.50	50	30	240	0.50

4 BACK-ANALYSIS METHODOLOGY

4.1 Back-analysis fundamentals

The settlement trough induced by a TBM can generally be described by a Gaussian curve, Peck (1969) and Schmidt (1969). O'Reilly and New (1982) characterised the shape of the trough using only two parameters, i.e. the distance of the inflexion point from the axis (i) and the maximum settlement at the axis (s_{max}). The transversal distribution of the settlements is then defined by Equation (1). The first step of the back-analysis consists in determining the empirical parameters s_{max} and i by curve fitting on site monitoring data. The following complementary parameters are also derived:

– the trough width parameter, k_{eq}, $= i/z_0$, whith z_0 is the depth of tunnel axis, Equation (2),
– the volume loss parameter, defined as the ratio between le volume of the settlement trough and the excavated section, see Equation (3)

$$s = s_{max}.exp\left(-\frac{y^2}{2.i^2}\right) \tag{1}$$

$$i = k_{eq}.z_0 \tag{2}$$

$$V_{Loss} = \frac{s_{max} \cdot i \cdot \sqrt{2\pi}}{V_t} \tag{3}$$

with V_t is the excavated section of the tunnel, i.e. 65.76 m^2 for a bored diameter of 9.15 m.

According to Chiriotti (2000), the equivalent trough width parameter k_{eq} can be expressed as a weighted average of the width parameters of each ground layer making the overburden from the tunnel axis up. The second step of the back-analysis consists in taking the geological profile into account to derive a characteristic value of width parameter for each ground layer.

The settlement trough may also be assessed using finite element analysis, 2D or 3D, which gives also access to the deformations of the ground mass. The results should however be treated with caution as those methods tend to overestimate the trough width and to underestimate the maximum settlement and maximum inclination, ITA/AITES (2006). Nonetheless, compared to Peck empirical approach, numerical analysis has the advantage of deriving the relationship between volume loss and confinement pressure applied by the TBM.

The practice in France is to derive the moduli used in the constitutive models of the finite element analysis from pressure meter data, based on empirical correlations which depend on the nature of the ground and on the type of works (tunnelling, retaining walls, foundations…). Similarly, to Peck's parameters, the parameters of these moduli correlations can be fitted on site monitoring data. The third step of the back-analysis will therefore consist in retro-fitting the correlations of the equivalent moduli for finite element analysis and derive simple laws relating volume loss to confinement pressure.

4.2 Step 1 – Fitting of Peck's curves on site monitoring data

Available site monitoring data originate from 300 survey points located on the facades of buildings and 163 survey points located on the road, with an average density of 0.25 points per meter along the tunnel axis. Curve fitting has been carried out through an advancing process in 5m steps, thus allowing to consider the variation in tunnel depth and in encountered geology: a total of 360 fitting sections has been processed. Each section uses a 50 m long and 40 m wide data screening. Fitting of Peck's parameters is achieved using a non-local non-linear least square method, see § 4.2.2.

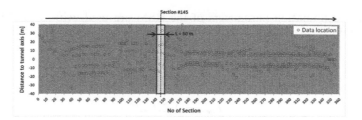

Figure 2.   Advancing fitting process.

4.2.1 *Solving for least square by nonlinear regression*

Peck's curve (equation 1) is fitted on the monitoring data by the least square method using nonlinear regression. The minimum is found through an iterative procedure starting from the initial conditions s_{max_0} = 5 mm and $i_{\_0}$ = 10 m. Iteration is considered converged when the parameter increments are less than 0.1%.

4.2.2 *Non-local fitting by data weighting*

Considering that the settlement trough in a cross-section results from the cumulated effects of tunnel advance on a certain influence length and that the survey points are irregularly located, it was chosen to carry out fitting at a given section using data in the vicinity of the section, weighted according to their axial distance to the considered section. Data weighting is achieved using a normal distribution (Gaussian curve) with a standard deviation of 10 m, which means that survey points further than 25m away from the section become negligible.

Taking into account that the resolution of the monitoring data is in the order of 1mm, it was also chosen to filter out the "noise" away from the tunnel axis using a transversal weighting of the data, equally based on a normal distribution with a standard deviation of 5m. Only the sections with at least 3 measures per meter within 10m of the tunnel axis are considered meaningful and are processed.

This non-local advancing process results in a smooth and continuous distribution of Peck's parameters along the tunnel axis, with an overlap in a range of 8m.

4.3   *Step 2 – Assessment of trough width parameter of each layer*

The 1.8km long geotechnical and tunnel profiles are discretised in 5m steps, similarly to the advancing fitting process. The equivalent trough width parameter can be expressed as a weighted average of the characteristic width parameter of each of the $p=m+n$ overburden layers, Chiriotti (2000):

$$k_{eq} = \frac{(1-\lambda).\sum_{i=1}^{m} z_i.k_i + \lambda.\sum_{j=m+1}^{n} z_j.k_j}{(1-\lambda).\sum_{i=1}^{m} z_i + \lambda.\sum_{j=m+1}^{n} z_j} \tag{4}$$

with λ a weight depending on the distance of the layer from the tunnel axis. Layers closer than one diameter from the tunnel are weighted with $1-\lambda$, while layers further away are weighted with λ. In accordance with Chiriotti (2000), we use λ = 65 %.

The inverse problem searching for characteristic width parameters (one per geotechnical unit) matching best the 360 fitting sections is solved using a probabilistic method presented in Mahdi (2016). To simplify the problem, strata have been grouped in 4 significant geological units.

4.4   *Step 3 – Constitutive models, volume loss and confinement pressure*

2D finite element analysis is used to calibrate the deformation parameters of the Hardening Soil Model (HSM) against the volume loss measured for a known confinement pressure applied at the face. Once calibrated, the finite element model can provide, for a given geotechnical configuration, the evolution law of volume loss against variation of the confinement pressure.

The following deformation parameters of the HSM are used, with α the Menard's rheological factor and μ a correlation factor depending on the type of loading (stand-up time and stress path):

$$E_{50} = \mu \frac{E_m}{\alpha} \tag{5}$$

$$E_{oed} = 1,2.E_{50} \tag{6}$$

$$E_{ur} = 3.E_{50} \tag{7}$$

The fixed ratios chosen for the determination of E_{oed} and E_{ur} are common values for sands and marlstones encountered in the Plaine de France. The stress dependency of soil stiffness has not been considered.

5 BACK-ANALYSIS RESULTS

5.1 Data relevance

Before any data treatment, the relevance of data points is verified by checking if settlement appearance coincides with the arrival of the TBM. Figure 3 shows that the great majority of data points are consistent with expected TBM advance rate. Only 9 survey points displaying heave instead of settlement have been discarded for further back-analysis.

5.2 Considered sections for fitting

Figure 4 shows in blue the sections used for fitting and in red the sections deemed too poor in data to allow meaningful fitting. The data are particularly poor in the first 150m (27 discarded sections), just enough for fitting between 150m and 800m (between section 27 and section 170) and particularly dense between 800m and 1800m. In total 329 sections, i.e. 92 %, are processed.

5.3 Peck's parameters

Solving for least square converged in 315 sections out of 329, i.e. 96%. In sections with a wealth of data, curve fitting is particularly successful as can be seen in Figure 5.

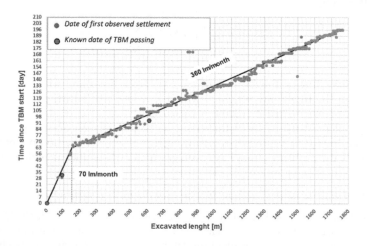

Figure 3. TBM advance rate derived from first settlement detection.

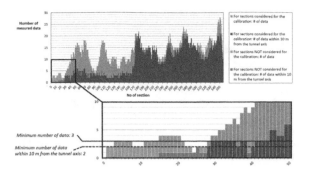

Figure 4.   Number of survey points for each fitting section.

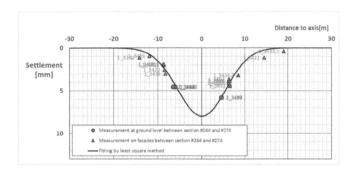

Figure 5.   Example of Peck's curve fitting on data around section #269.

5.3.1 *Equivalent trough width*

The equivalent trough width parameters (k_{eq}) resulting from the fitting process are shown in Figure 6. Over the first half of the drive, the TBM bores with an overburden made of calcaire de Saint Ouen and fills. Fitted k_{eq} values are then quite high (0.6 on the average) and display some local anomalies (values around 1). The results over the first half must be considered with caution as data density is not high enough to achieve good reliability of the fitting. Over the second half of the drive, the TBM bores full face in the Sables de Beauchamp with an overburden composed mainly of Calcaire de Saint Ouen. The k_{eq} values fitted in this denser zone are more stable and within the range [0.3, 0.5], in accordance with reported empirical values for this type of ground.

5.3.2 *Characteristic trough width parameters of individual layers*

Solving for the inverse problem formulated by Equation (4) gives probabilistic values of trough width parameters of individual layers, as presented in Table 2. Solved values fall within the empirical range. Reassessing the apparent k_{eq}, Figure 6, from this data set falls within +/- 0.1 of the fitted k_{eq} value in 76% of the considered sections.

5.3.3 *Volume loss and maximum settlement*

The volume loss and maximum settlement resulting from the fitting process are shown in Figure 7. These values correspond to a well-controlled TBM drive, with volume loss always remaining well below 0.3% and maximum settlement below 10mm, expect on the first 300m (from section 1 to section 60) from the launching shaft and within 50m of the reception shaft.

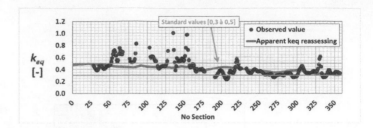

Figure 6. Comparison between observed k_{eq} values fitted along the entire TBM drive and reassessed based of characteristic trough width parameters given in Table 2.

Table 2. Characteristic trough width parameters of ground layers in Plaine de France.

Formation	Solved values
Fills	0.30
Alluvions	0.30
Marnes Infragypseuses	0.50
Calcaire de Saint-Ouen	0.55
Sables de Beauchamp	0.20

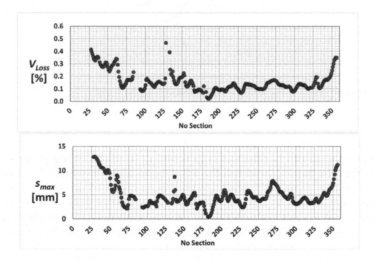

Figure 7. V_{Loss} and s_{max} values as fitted along the entire TBM drive.

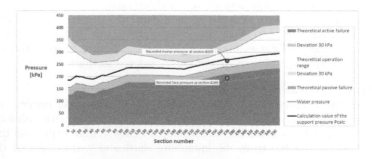

Figure 8. Operation range of the confinement pressure along the TBM drive.

5.4 Back-analysis based on 2D finite element modelling

5.4.1 Confinement pressure upper and lower bounds
Before carrying out finite element analysis, face stability conditions are considered to define theoretical upper and lower bounds to the confinement pressures. The approach developed by Anagnostou & Kovari (1994) and completed by de Broere (2001) is adopted to define the active and passive failure bounds with safety factors of 1.05 on pore pressure and of 1.5 on earth pressure, see Figure 8. A deviation of +/-30 kPa on the control of the confinement pressure in a EPB is further introduced, DAUB (2016).

The full confinement profile applied on site is not available. However, the measured confinement pressure and the grouting pressure at section #269 are reported to be 1.95 bar and 2.66 bar respectively. While the applied confinement pressure is relatively low, just above pore pressure, the applied grouting pressure reaches values high enough to minimise settlements. This is in accordance with observations made by Aristaghes (2001):

- Face support pressure ensures tunnel face stability but has only a negligible effect on settlement control
- When implemented, support pressure along the shield tail plays a sensitive role for settlement control
- Annular gap grouting pressure keeps on the action of the shield tail support pressure when the latter is implemented, or can even heaves the previously unconfined soil along shield tail

5.4.2 Correlation between pressure meter moduli and HSM moduli
Knowing the confinement pressure for section 269, the correlation factor μ is calibrated by finite element analysis against fitted volume loss. The agreement is obtained for $\mu = 4$ both in Calcaire de Saint Ouen and in Sables de Beauchamp formations.

The correlation factor μ is also calibrated in 5 other sections, #6, #32, #60, #150 and #340 which cover the various geotechnical contexts encountered by the TBM. It has been done assuming that the apparent confinement pressure is given by the black line in Figure 8. Again, overall agreement is obtained for $\mu = 4$ both in Calcaire de Saint Ouen and in Sables de Beauchamp formations.

5.4.3 Curve fitting for volume loss as a function of confinement pressure
On the 6 modelled sections, the response of volume loss to confinement ratio has been studied and it was concluded that power laws $Vloss = A.(P_c / \sigma_v)^B$ can relate these parameters, see Figure 9.

Two slightly different families of curves can be distinguished: the a-curve corresponds to boring with the upper half in Calcaire de Saint Ouen, $A = 0.1$ and $B=2.2$, while the b-curve corresponds to boring full face in Sable de Beauchamp, $A = 0.04$ and $B=3.1$.

Once these laws are calibrated, it becomes possible to determine the volume loss associated to the confinement ratio derived from by the confinement pressure given by the black line in Figure 8, and to confront the calculated settlement value with the observed values.

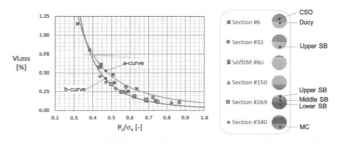

Figure 9.   Evolution of V_{Loss} with confinement ratio.

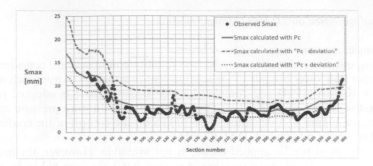

Figure 10. Comparison between observed S_{max} and calculated S_{max}.

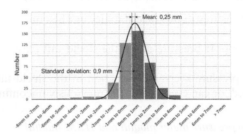

Figure 11. Difference between observed settlement at survey points location and calculated settlement based on proposed model parameters.

Using V_{Loss} laws and k values previously presented, Figure 10 shows a very good agreement between the observed settlement values and calculated ones. The observed maximum settlements along the entire tunnel alignment are well framed by the calculated values obtained considering the theoretical confinement deviation of +/-30kPa.

It is noteworthy that 100m around section #182 where an 8 levels building is located, settlements are very low, which seems to indicate that confinement pressure has been purposely increased to pass this sensitive building.

These calculation process and parameters allow to achieve an estimate very close to the field data. Indeed, the difference between recorded settlement values at the 463 survey points on site and calculated values can be characterized with a normal distribution, nearly centered and with a narrow standard deviation, below 1 millimeter, see Figure 11.

6 MAIN CONCLUSIONS

The extension works of Paris metro line 12 includes a 3.8 km 9.15 m diameter tunnel that was bored by an EPB shield between 2010 and 2011. The back-analysis of settlement monitoring and the resulting proposed predictive model presented in this paper aim to improve design choices and risk assessment while boring new tunnels in similar geological context, i.e. the Plaine de France basin.

The proposed model is particularly suited to carry out vulnerability analysis of planned tunnels of line 16, line 15 east and line 17 of the future Grand Paris metro, that are to be bored in the same geological context.

The back-analysis of monitored settlements has been performed by the least square method using nonlinear regression and non-local fitting by data weighting using 463 survey points data. It shows settlement troughs in near greenfield condition are well described by Peck 's formula.

The volume loss depends on the confinement ratio, see Figure 9, and commonly obtained values are around 0.1% to 0.2%. Observed settlement trough wide parameter values are between 0.3 and 0.4, see Figure 7, and k values presented in Table 2 combined with Equation (4), Chiriotti (2000), give a good estimate of the apparent trough width parameter at surface k_{eq}.

Once observed Peck's parameters has been defined, 2D FEM analysis has been carried out to define modelling parameters that allow the best fitting over these data.

While FEM gives an erroneous estimate of the settlement trough width, it gives however a good estimate of the volume loss, from 2D FEM and HSM soil model, considering a correlation factor for the pressure meter μ =4 both in Calcaire de Saint Ouen and in Sables de Beauchamp formations, see §4.4.

The confinement pressure used in the FEM model is set to the grouting pressure, which can be chosen between 0.5 and 1 bar above the face stability pressure. While this calibration is carried out on a single fully instrumental section, the generalization of the deduced empirical law for volume loss yields an excellent match of maximum settlement deduced from monitoring data over the full length of the drive, within the usual deviation on EPB face pressure.

REFERENCES

Anagnostou G. & Kovari K. 1994. The Face Stability of Slurry-shield-driven Tunnels. *Tunnelling and Underground Space Technology*, Vol. 9.

AristaghesP & Autuori P. 2001. Calcul des tunnels au tunnelier. *Revue Française de Géotechnique*. No 97

BroereW. 2001. Tunnel face stability and new CPT Applications. PhD Thesis, Technical University of Delft.

Chiriotti E. Marchionni V. Grasso P. 2000. Porto Light Metro System, Lines C, S and J. Interpretation of the Results of the Building Condition Survey and Preliminary Assessment of Risk. Methodology for Assessing the Tunnelling Induced Risks on Buildings along the Tunnel Alignment. *Normetro – Transmetro, Internal technical report.*

DAUB. 2016. Recommendations for Face Support Pressure Calculations for Shield Tunnelling in Soft Ground.

ITA/AITES Report. 2006. Settlements induced by tunneling in Soft Ground. *Tunnelling and Underground Space Technology*. No 22.

Mahdi S. Houmymid F.Z. Chiriotti E. 2016. Use of numerical modelling and GIS to analyse and share the risks related to urban tunnelling - Greater Paris - Red Line – South section, ITA-AITES WTC 2016.

O'Reilly M.P. and New B.M. (1982). Settlements above tunnels in the United Kingdom - Their magnitudes and predictions, *Tunnelling* 1982.

Peck R.B. 1969. Deep excavations and tunnelling in soft ground, 7th International Conference on Soil Mechanics and foundation Engineering, Mexico City, State-of-the-art Volume.

Schmidt B. 1969. Settlements and Ground Movements associated with Tunnelling in soils, PhD Thesis University of Illinois, Urbana.

Tunnels and Underground Cities: Engineering and Innovation meet Archaeology,
Architecture and Art, Volume 6: Innovation in underground engineering,
materials and equipment - Part 2 – Peila, Viggiani & Celestino (Eds)
© 2020 Taylor & Francis Group, London, ISBN 978-0-367-46871-2

Addressing the radial joints behaviour of Steel Fibre Reinforced Concrete (SFRC) segments under concentrated loads

M. Mangione, M. Ceccarelli & A. Pillai
Arup, London, United Kingdom

C. Agostini
S.G.S. Studio Geotecnico Strutturale, Rome, Italy

ABSTRACT: Precast concrete tunnel segments are subjected to high concentrated compressive stresses generated by the transfer of hoop loads across the radial joints. This effect is emphasized by the joint opening associated with diametrical deformations resulting by tolerances during lining installation and deformations imparted by external actions. These concentrated stresses produce tensile bursting forces at radial joints that may lead to critical joint damage, especially in SFRC segments. The bursting stress intensity and distribution depends strictly on the contact area at the joint location. While the bursting stresses can be evaluated by means of analytical formulas, the estimation of the joint rotation and contact area is more complex due to the non-linear behaviour of the concrete material. This paper provides an analytical method for the evaluation of radial joint behaviour and it is tested against advanced numerical modelling using geometrical properties of the segmental lining of the Airport Link Project, one of the largest designed and built SFRC segments.

1 INTRODUCTION

Mechanized tunnelling is widely adopted in different ground conditions due to its efficiency and enhanced safety. In TBM driven tunnels, the lining consists of a ring composed by precast concrete segments assembled within the TBM shield.

The tunneling industry in several countries is moving towards a preference in use of SFRC segments, where conventional reinforcement is eliminated in favour of steel fibres added to the concrete mix (fib 2017). The use of SFRC improves manufacturing efficiency and optimises production cost. From a structural perspective, steel fibres enhance the post-crack properties of concrete, leading to a more ductile material behaviour. Such ductility is due to the ability of the fibres to transfer tensile stresses across a cracked section, providing improved toughness of the material, which leads to less damage due to impact, improved quality of the finished segments, structural capacity in tension and limitation to crack development.

Tunnel segments are subjected to many loading conditions from the precast plant (e.g. demoulding, stacking and handling) to the tunnel installation (e.g. TBM thrust, tail injection). In long term conditions, precast segments are loaded by compressive hoop forces, arising from surrounding soil and groundwater, which generate high concentrated compressive stresses in correspondence of radial joints. The stress concentration is emphasized by the joint opening and the consequent reduction of contact area between two adjoining segments (Figure 1a). The opening of radial joints is associated with diametrical deformations resulting by assembly tolerances of the ring within the TBM and deformations imparted by external actions.

The concentrated compressive hoop force at radial joints redistributes over the thickness of the segment generating orthogonal tensile bursting stresses (Figure 1b) in accordance with the theory firstly developed by Leonhardt (1964). Bursting stresses may lead to concrete cracks

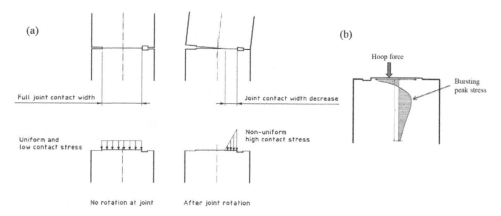

Figure 1. Effect of contact area on compressive stress distribution (a) Bursting stress distribution at the radial joint (b).

which may affect serviceability and ultimate limit state requirements. In traditionally reinforced segments, reinforcement at the joints is included to meet the demand for bursting stresses. However, deeper understanding of the behaviour of the radial joints is a most significant benefit for the structural design of SFRC segments, to ensure that the tensile forces arising by the concentrated joint stresses are duly assessed and the need for localised reinforcement is eliminated or reduced.

Bursting stresses are assessed by means of analytical approaches, e.g. the Leonhardt's theory or equations proposed by standard codes, such as Eurocode 2 or ACI-318. Whereas the intensity and distribution of the bursting stress can be deducted from these references, they do depend on the definition of the zone of contact between segments at the radial joints. There is no clear guidance in international codes and manuals on how to estimate the contact area accounting for tolerance and future deformation on the rings and this may lead to practical limits on evaluating the ultimate capacity of the SFRC segments at the radial joints.

The estimation of the contact area is function of the ovalization and the joint behaviour, which is influenced by the non-linearity in the concrete material. At the ultimate limit state, for example, the crushing of concrete enhances the joint closure, so the compressive hoop force is distributed over a larger surface of the joint. The assessment of the joint contact area is also dependent on the joint type (convex or flat joint). This paper will focus on flat-flat joints.

This paper aims to provide an alternative approach to represent the radial joint behaviour, compare it to the current methodology available to the tunnelling community, and present results from numerical modelling with an advanced material constitutive law to justify the presented approach. In particular, Chapter 2 briefly gives an overview of the analytical methods from literature commonly used for the evaluation of the contact area and highlights some simplifications of these approaches relative to the evaluation of the contact area. Chapter 3 describes the alternative method, developed and proposed by the Authors to overcome such limitations. Finally, the results in terms of joint rotation-contact area derived by both literature approaches, Authors approach and numerical modelling on a case study are presented and discussed in Chapter 4.

2   RADIAL JOINT BEHAVIOUR: EXISTING DESIGN APPROACHES

For the design of the ring, the radial joint local behaviour for an ovalised geometry must be properly addressed especially in case of SFRC only lining.

The ovalization of the ring can be caused by poor build, tolerances during installation within the TBM shield and ground loading. This initial deformation is typically expressed as diametrical distortion, defined in project specifications or guidelines such as the BTS Specification for Tunnelling (BTS 2010). Additional distortions may be imposed by Clients as a long-term performance requirement.

Establishing the contact area at radial joints is complex due to the effect of the diametrical deformation of the ring and the associated rigid body rotation of segments which results in an opening of the joints. As the hoop load is transferred, the concentrated contact stress produces strains over the segment length which lead to localised segment shortening which results in an increase in contact area (closing effect) until a state of equilibrium is achieved. The extent of joint closure is a function of lining geometry, concrete stiffness and magnitude of applied hoop load.

Analytical approaches have been proposed by different Authors, e.g. Janssen (1983), Blom (2002), Tvede-Jensen et al. (2017) to describe the behaviour of flat-flat joints.

The radial joint behaviour can be idealized in three stages, which are function of the magnitude of the hoop force applied. Along with the increase of joint rotation, the radial joint at first exhibits a linear-elastic behaviour (Stage 1 - linear) with a constant rotational stiffness: at this stage, the radial joint is fully closed. Then a non-linear elastic behaviour (Stage 2 – geometrical non-linearity) occurs when the joint starts opening. A third stage is reached when the concrete elastic strain is overcome (Stage 3 – concrete non-linearity).

The three analytical approaches, mentioned above, have been derived considering that the joint deformed zone presents a constant strain and it extends for a length equal to the joint thickness. Janssen (1983)'s method is based on the elastic theory and it does not account for the non-linear behaviour of the concrete, i.e. Stage 3.

This method has been further expanded by Blom (2002), who adds to Janssen's approach the third stage considering a bi-linear stress-strain relationship for the concrete (i.e. elastic perfectly plastic behaviour). The rotational stiffness is non-linear and the maximum stress is equal to characteristic compressive strength of the concrete, until the ultimate strain is reached.

Recently, Tvede-Jensen et al. (2017) refined Blom's method, modifying the third stage. They considered the parabola-rectangle stress-strain relationship and, more importantly, they accounted for the increased of compressive strength and strain limit due to partial loaded area and confined concrete effects, respectively. To include those effects, Eurocode 2 equations have been adopted.

Details about the mathematical formulations of the three approaches can be founded in Tvede-Jensen et al. (2017).

The aforementioned methods are widespread in the current design practice; however, they present some limitations. In particular, the spreading of compressive stress from the joint face to the full thickness of the segment is not accounted, assuming that the strains caused by the hoop load are only localised in close vicinity of the joint loaded area. The shortening of the segment, due to the compressive force is therefore not considered, providing a conservative estimate of joint closure and associated bursting stresses. The approach proposed by the Authors and described in the following chapter aims to address this aspect.

3 PROPOSED APPROACH: THE JOINT CONTACT METHOD

The proposed method, hereinafter the "Joint Contact Method", has been developed by the Authors and it has been applied successfully on several projects worldwide. The Joint Contact Method was developed to assess the contact area and stress distribution at radial joints to perform an optimised design of joints, in term of bearing and bursting requirements.

The contact area is the result of the rotation at segment joints and deformation of the concrete. As mentioned above, already after the ring installation, joint opening and rotation are expected because of assembly tolerances of the ring within the TBM and deformations imparted by external actions.

The initial joint opening ("birdsmouthing") of the unloaded ring can be calculated assuming an elliptical deformation of the ring and infinitely stiff segments so that all deformation occurs by rotation of the joints as shown in Figure 2. This initial opening of the joints is coun-teracted, to some extent, by the closing effect associated with the transfer across the joint of hoop forces arising from the external loads. A distribution of compressive stresses from the hoop loads shall be defined so that the resulting strains imparting a shortening/deformation of the concrete within the contact area can be calculated.

As per the methods presented before, the Joint Contact Method considers three possible scenarios for the radial joint behaviour for an applied hoop load. When joint closure occurs, it is assumed that any component of the hoop load above the value that leads to closure will have a constant stress distribution along the joint thickness (Case 1 in Figure 3). However, the closure of the joint may not occur if the concrete segments are significantly stiff, the imposed initial deformations are high, or the maximum hoop force is relatively low in magnitude. In this instance, the joint will close partially, and the contact area will be lower than the available joint thickness (Case 2 and 3 in Figure 3). If compressive stresses exceed the design bearing capacity of the concrete section, the contact stress is expected to distribute over a larger area (Case 3 in Figure 3). The bearing strength is calculated with the partial loaded areas rule fol-lowing Eurocode 2.

The eccentric hoop force at the radial joint produces strains and deformation, leading to shortening and section rotation, which can be derived using the theory of elasticity. A critical consideration for the approach is that the stress distribution within the segment is not constant over the length of the segment. To calculate segment deformation, the segment is ideally divided in two different regions, D and B (Figure 4) according to Leonhardt's strut and tie theory.

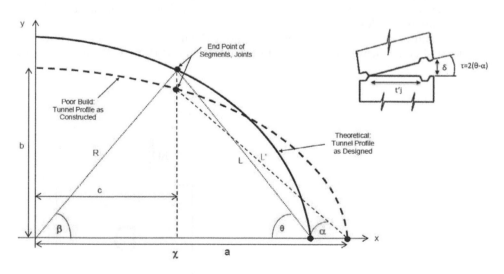

Figure 2.   Schematic radial joint opening due to assumed elliptical deformation.

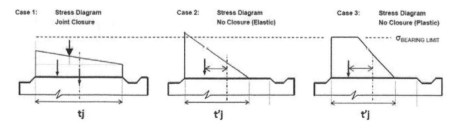

Figure 3.   Possible configurations of contact area depending on joint closure.

In the case of segmental lining, the so-called D-region (where D stands for discontinuity or disturbance) refers to the first part of the segment, where the beam theory is not valid due to the high compressive force transferred at the radial joint. The D-region is assumed to extend over a length equal to the segment thickness and the method assumes that the variation of hoop force and bending moment imparted by external load is negligible within this region. Therefore, the stress at the boundary necessary for the equilibrium of the D-region can be easily calculated (Figure 4) and the stresses distribution within this region can be derived through Leonhardt's strut and tie approach. In the D-region, stress and strain distribution are non-linear, however the mean circumferential strains can be conservatively computed as:

$$\varepsilon_{A,mean} = 0.5\left(\varepsilon_A + \varepsilon_{A'}\right) \tag{1}$$

$$\varepsilon_{B,mean} = 0.5\left(\varepsilon_B + \varepsilon_{B'}\right) \tag{2}$$

i.e. a linear variation between the edges AB and A'B' (Figure 4) is assumed. This approach is considered conservative as it produces a lower joint closure and therefore a higher concentration in stresses.

In the B-region (where B stands for Bernoulli or beam), the Bernoulli hypothesis of planar section is considered valid and the internal state of stress can be derived from the sectional forces. The present method assumes that in the B region the eccentricity of the load is reduced progressively up to the distance L from the joint, where an inflection point (i.e. zero bending moment) occurs, resulting in a constant stress distribution. Such length can be reduced to allow for a more robust design.

The mean values of strain can be derived in both D and B-regions from stresses at the boundaries and consequently the movement of points A and B (Figure 4) can be computed.

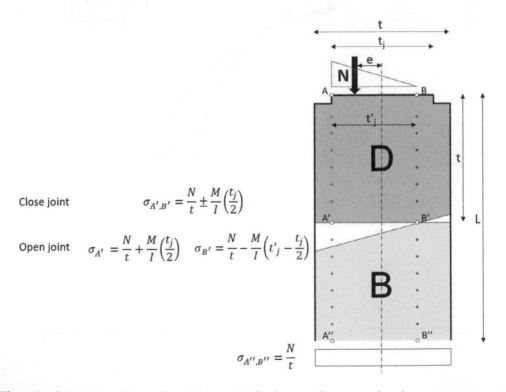

Close joint $\qquad \sigma_{A',B'} = \dfrac{N}{t} \pm \dfrac{M}{I}\left(\dfrac{t_j}{2}\right)$

Open joint $\quad \sigma_{A'} = \dfrac{N}{t} + \dfrac{M}{I}\left(\dfrac{t_j}{2}\right) \quad \sigma_{B'} = \dfrac{N}{t} - \dfrac{M}{I}\left(t'_j - \dfrac{t_j}{2}\right)$

$\sigma_{A'',B''} = \dfrac{N}{t}$

Figure 4.   Assumption of circumferential stress distribution over the segment length.

The rotation of the joint imparted by external loads is then given by the differential movement between the two points over the contact length. The so obtained rotation is used to recalculate the joint closure and the contact area.

As a result, the Joint Contact Method aims to provide a rational approach, accounting for the mechanical response of the whole segment and not only a small portion near to the joint as per Janssen (1983), Blom (2002) or Tvede-Jensen et al. (2017). Moreover, plastic concrete behaviour at high stresses is adopted along with the bearing strength gain due do the partially loaded area. These key features allow for a more realistic assessment of radial joint behaviour, extremely beneficial in case of SFRC-only lining.

However, the structural behaviour of the ring is complex and geometrical as well as material non linearity is difficult to incorporate in an analytical design approach, and only advance numerical modelling can give insight on the complex radial joint response. The Joint Contact Method has been verified with both 2D and 3D finite element during detailed design of segmental lining. In this paper, the Joint Contact Method and its assumptions were tested against advanced numerical modelling using the geometrical properties of the lining of the Airport Link Project.

4 CASE STUDIES

In order to better understand the development of the contact area with the increase of tunnel ovalisation and to validate the method proposed above, a parametric study for different load conditions has been carried out with DYNA-LS.

The non-linearity has been accounted by using advanced non-linear numerical analysis where all segments were modelled as solid elements. The software DYNA-LS has been used to run such model while an elastic perfectly plastic constitutive law for the Steel Fibre Reinforced Concrete was adopted, including the failure surface developed by Ottosen (1977), which offers a more realistic approximation to the different triaxial stress states that lead to failure in concrete.

The advanced numerical tool has been used in a full soil structure interaction model (Figure 5) to recreate the Airport Link Project ring. The ring is loaded with ground stresses and is deformed diametrically to match the design ovalisation of the ring.

This approach using DYNA-LS can give a better and more realistic estimation of the joint contact area as the whole ring is modelled accounting for the real thickness of the segments and the ground stiffness. The loading conditions are applied as external loads and the constitutive model can provide a more realistic contact area including softening effects of the concrete at high compressive stress regime (localised crushing) as well as softening due to the non-linear stress-strain relationship of the concrete in tension.

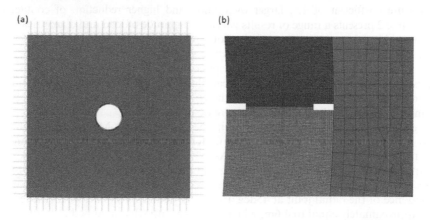

Figure 5.   Ground and ring model in DYNA-LS (a). Radial joint detail (b).

4.1 FE models features

The Airport Link geometry has been considered with an internal diameter of 11.35m and thickness of segmental lining equal to 400mm. A recess of 80mm at both sides of the segment has been considered for gasket positioning and caulking groove.

The SFRC has been modelled using an elasto-plastic constitutive model where the following properties have been applied:

- Characteristic compressive strength f_{ck} = 50.0 MPa;
- Characteristic residual tensile strength $f_{R3,k}$ = 2.235 MPa.

Two different overburden levels have been analysed, considering two lateral pressure coefficients:

Table 1. Loading conditions considered for the DYNA-LS analysis.

	Vertical Stress	Horizontal stress (k_m = 0.8)	Horizontal stress (k_m = 1.5)
Mid-level (34m of depth)	681 kPa	545 kPa	1022 kPa
Deep-level (68m of depth)	1360 kPa	1088 kPa	2040 kPa

The ground load is applied as boundary condition: the loads are applied at the time step 0 and are kept constant during the analysis. At the same time, the proper initial value of the internal stress in the whole domain is imposed.

As the method proposed in the paper is focused on the structural response on the lining, for sake of simplicity the surrounding ground has been assumed as linear elastic material with an elastic modulus of 100MPa, and a Poisson's ratio of 0.2. The interface between lining and soil is set as a full slip condition, whereas the joint-joint interface is characterized by a frictional coefficient equal to 0.6.

It is important to highlight that, due to model computing limitations, at the time step 0 the tunnel is undeformed, and it gets deformed with the gradual application of the external loads. Therefore, construction tolerances and joint opening related to poor building are forced in the loaded ring, which is an accepted approximation for the purpose of defining the contact area.

4.2 Results

The results obtained with the advanced numerical analyses are satisfactory. As expected, the higher is the deviatoric stress the higher is the ovalisation of the tunnel. In fact, in the cases of lateral pressure coefficient of 1.5, larger ovalisation and higher reduction of contact area are observed. Table 2 presents a range of results varying magnitude of load and diametrical distortion.

As mentioned above, the concrete behaviour plays a significant role in the overall joint response. In fact, in the case of lateral pressure coefficient equal to 1.5, the deeper tunnel exhibits a wider area of plastic stress in correspondence of the radial joint, whereas, the lower hoop stress in the mid tunnel stay in the linear elastic branch. This justifies the difference of contact areas between the two models: i.e., the concrete plastic behaviour enhances the joint closure and consequently the contact area increase (Table 2).

A first check on the assumptions adopted in the Joint Contact Method has been made. As shown in Figure 6, the disturbed area of circumferential stresses due to the hoop load transferred at the joint presents a limited extent. As expected, the so-called D-region extends for a length within the segment thickness. Moreover, an important assumption of the Joint Contact Method was the location of the inflection point, where the bending moment is zero, occurs in correspondence of the radial joint at 45deg above the tunnel springline, so L (the length of the chord) is approximately equal to 4.6m, which is in line with what assumed in the Joint Contact Method.

Table 2. Hoop force, ovalisation and contact length reached in the last analysis step.

	Deep level $k_m = 1.5$	Deep level $k_m = 0.8$	Mid level $k_m = 1.5$	Mid level $k_m = 0.8$
Hoop force (MN/m)	10.39	7.69	5.36	4.00
Diametrical displacement (mm)	104.4	48.8	47.7	23.8
ovalisation (%)	(0.92%)	(0.43%)	(0.42%)	(0.21%)
Contact length (mm)	169	200	140	212

The Joint Contact Method has been tested with the numerical modelling evidence and the methods from literature, presented in Chapter 2. The joint rotation arising in DYNA-LS are plotted against the contact area (Figure 7) and are compared to the analytical solutions by Janssen (1983) Tvede-Jensen et al. (2017) and the method proposed by the Authors. The plot shows a similar behaviour of the three approaches. In particular, the Joint Contact Method aligns better with the numerical data at low angles, confirming that the current approaches are conservative for joint closure. In joint opening, the response of the concrete within the ring is stiffer than what the Joint Contact Method produces. Further analysis will be required to check the Joint Contact Method with different tunnels size, loading conditions and concrete properties.

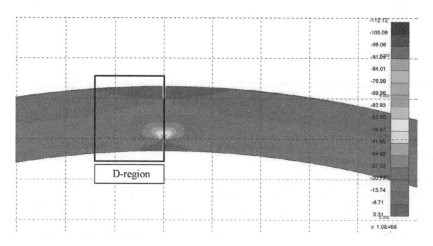

Figure 6. Circumferential stress in correspondence of the joint at the crown.

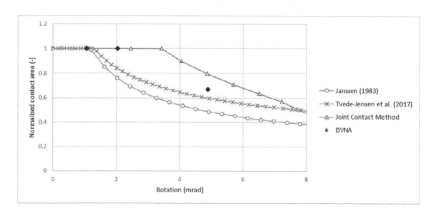

Figure 7. Comparison of DYNA-LS results with analytical formula for hoop force of 5200 kNm/m.

5 CONCLUSION

The behavior of radial joints is the key point of structural design of segmental lining, especially with steel fibre reinforced concrete (SFRC). A reliable assessment of the behaviour of the radial joints can ensure that the tensile bursting forces arising by the concentrated stresses are properly assessed, leading to a partial or complete removal of traditional reinforcement.

Assessing the contact area at radial joints is complex, there is limited guideline and its value is a critical parameter for the design. The Authors have developed a close form approach, tested in a number of projects over the last decade. However, the Joint Contact Method here presented has been tested with numerical analysis. In fact, numerical modelling can address properly the joint behaviour, in term of contact area and bursting stress for different loading conditions. These preliminary results are quite satisfactory: the analytical Joint Contact Method seems to represent well the radial joint non-elastic behaviour by providing a good match with the results obtained by the numerical analysis. A good estimation of the joint contact area which takes into account the non-elastic behaviour of the concrete allows to perform a reliable design avoiding over-conservatism in the estimation of the bursting stresses and consequently in the reinforcement design. This aspect is of even more important in case of SFRC-only ring design, where the tensile resistance is only provided by the post-crack properties of the fibres.

To justify full applicability of the method, the Authors plan to run further analyses with different tunnel geometries and load case scenarios to cover a wider spectrum of design situations, validated by non-linear numerical analyses.

ACKNOWLEDGEMENT

The Authors would like to acknowledge the Arup tunnelling team for the support and valuable input and the Arup Advance Technology and Research team for the contribution to the numerical modelling.

REFERENCES

ACI 318M-14. Building Code Requirements for Structural Concrete. American Concrete Institute.
BTS 2010. Specification for Tunnelling, Third Edition. Thomas Telford.
Blom, C.B.M., 2002a. Design philosophy of concrete linings for tunnels in soft soils. PhD thesis Technische Universiteit Delft.
EN 1992- 1-1 2004. Eurocode 2: Design of concrete structures – Part 1-1: General rules and rules for buildings.
Fib CEB-FIP 2010. Model Code for Concrete Structures 2010.
Fib CEB-FIP 2017. Bulletin 83 Precast tunnel segment in fibre-reinforced concrete.
Janssen, P., 1983. Tragverhalten von Tunnelausbauten mit Gelenktübbings. PhD thesis Technische Universität Braunschweig.
Leonhardt, F. 1964. Prestressed concrete, design and construction. Wilhelm Ernst & Sohn, Inc., Berlin and Munich.
Ottosen, N.S., (1977) "A Failure Criterion for Concrete," Journal of the Engineering Mechanics. Division, Volume 103, Number 4, July/ August, pages 527–535.
Tvede-Jensen, B., Faurschou, M. and Kasper, T., 2017. A modelling approach for joint rotations of segmental concrete tunnel linings. Tunnelling and Underground Space Technology, 67, pp.61–67.

Tunnels and Underground Cities: Engineering and Innovation meet Archaeology,
Architecture and Art, Volume 6: Innovation in underground engineering,
materials and equipment - Part 2 – Peila, Viggiani & Celestino (Eds)
© 2020 Taylor & Francis Group, London, ISBN 978-0-367-46871-2

Fiber glass and "green" special composite materials as structural reinforcement and systems; use and applications from Milan Metro, Brenner Tunnel up to high speed train Milan – Genoa

G. Manuele, M. Bringiotti & G. Laganà
Maplad Srl – Catania, Italy

D. Nicastro
GeoTunnel Srl – Genoa, Italy

ABSTRACT: Glass Fibre-Reinforced Polymer (GFRP) is nowadays a common practice in tunnelling; they are used in several applications as provisional structures, in a safe and cost effective way. Classical example is the so called "soft eye" where a TBM has to pass. GFRP bars have a much higher tensile strength than steel rebars, they are easily machined and can be broken down into small pieces by the cutter head and can then be transported by the conveyor system together with the spoil. We'll describe other different applications of this kind of material as structural elements in the conventional tunnelling, manual constructed reinforcement cages in big launching shafts, ending in special products in the railway industry, where the anti-galvanic corrosion properties are unique. Maplad is working also on special basalt fiber polymer elements, which show increased mechanical and physical properties, most probably the product of the next future.

1 GLASS FIBER REINFORCED POLYMER PROFILES

1.1 *The material*

Fiber reinforced plastic (FRP), also known as fiber reinforced polymer, is a composite material made up of a polymer matrix blended with certain reinforcing materials, such as fibers. The fibers are generally basalt, carbon, glass or aramid; in certain cases, asbestos, wood or paper may be used as the fibers. Composite material is greater than the sum of its parts. The matrix, which is the core material devoid of fiber reinforcement, is hard but comparatively weaker, and must be toughened through the addition of powerful reinforcing fibers or filaments. It is the fiber which is critical in differentiating the parent polymer from the FRP.

The matrix is composed of unsaturated resin as the polymer (normally used are polyester or vinylester and sometime, for particular applications, different resins are used, but hardly for profiles intended for the construction market).

Catalysts and additives are added in the batch with the resin.

1.2 *Pultrusion process*

In the traditional pultrusion process the fibres are pulled from a creel through a resin bath and then on through a heated die. The die completes the impregnation of the fiber, controls the resin content and cures the material into its final shape as it passes through the die. This cured profile is then automatically cut to length. Fabrics may also be introduced into the die to provide fiber direction other than at 0°.

To make the rebars, it is used a different process: the fibers are pulled from creel through a resin bath, cross a ring used to define the diameter of the rebar and they are wound by a

Table 1. Comparison between GFRP and other traditional materials.

Property	GFRP	Steel	Aluminum	PVC	Unit
Density	1.8	7.8	2.8	1.4	g/cc
Tensile strength	300–600	370–500	200–400	40–60	MPa
Flexural resistance	400–450	330–500	200–400	70–100	MPa
Elastic modulus	25–30	210	70	2.8–3.3	GPa
Bending modulus	15–20	210	70	2.8–*3.3	GPa
Tensile elongation	1.5–2.0	13–35	5–35	10–80	%
Impact resistance	200	400	200	85–95	MPa/m^2
Thermal conductivity	0,25–0,35	100–230	100–230	0,15–0,25	W/m°C
Dielectric capacity	5–15	-	-	40–50	KV/mm
Volumetric resistance	10^{10}–10^{14}	0,2–0,8	0,028	>10^{16}	ωcm

transversal thread that reduces the section so as to create the improved adhesion of the profile. Catalysis does not take place inside the molds but through ovens.

Pultrusion is a continuous process, generating a profile of constant cross-section.

1.3 *Comparison with other materials used in construction and advantages of GFRP*

The following table, Table 1., shows the average values of the main mechanical and physical characteristics of the profiles in GFRP in comparison with other materials mostly used in constructions.

We can deduced that the GFR profiles are light materials with elevated mechanical features able to replace the steel profiles, but at the same time electrically insulating, with low thermal conductivity resistant to chemical agent; similar characteristics to plastic materials.

The GFRP profiles are also characterized by UV resistance, discoloration resistance and, being a light material, easy to install and assemble.

2 GFRP IN THE TUNNELING'S INDUSTRY

2.1 *Slope and face stabilization – Adeco Technique*

The ADECO-RS (Analysis of Controlled Deformations in Rocks and Soils) is a design philosophy that places at the center of the design of an underground work the deformations that occur in the middle in which the excavation proceeds, analyzing them in depth and identifying their more effective systems to control them.

It was born about 30 years ago from a theoretical/experimental research during which modern constructive technologies were developed - including the reinforcement of the core-face with fiberglass reinforcement - and they were severely tested in the field.

The method focuses attention on the study of the Deformative Response considering:

- the medium through which construction takes place,
- the action taken in order to accomplish the excavation and
- the reaction (or Deformation Response) produced following the above-mentioned action.

The medium is the terrain which, in depth, is subject to triaxial stress states.

The action is produced by the advancement of the excavation front at a determined speed V and causes a stress perturbation in the surrounding soil both in the transverse and longitudinal direction altering the pre-existing tension states.

The speed rate V depends as well to the excavation system used (mechanized or conventional): high speed rates reduce the propagation of the perturbation, influencing the Deformation Response which conditioned by the choice of the excavation system.

The ADECO-RS works on the base of three different components of the Deformative Response:

Figure 1. TOTO S.p.A., job site La Spezia. SIG at visit, drilling and manual long bar junction procedure.

- the extrusion, its primary component, which largely gets within the core and manifests itself, in correspondence with the surface delimited by the excavation face, in a longitudinal direction to the axis of the tunnel;
- preconvergence, identified as a secondary component of the Deformative Response;
- convergence, identified as the third component of the Deformative Response.

According to the ADECO-RS the convergence, therefore, is only the last stage of a very complex deformation phenomenon that originates upstream of the excavation face in the form of extrusion and preconvergence of the advance core and then evolve downstream of the same in the form of convergence of the cavity.

The novelty of the ADECO-RS is to always advance to full section and stabilize the excavation by first intervening on the ground upstream of the front using the core as a "control tool" upstream and the immediate closure of the pre-covering with the inverted arch as a downstream control instrument.

The ADECO-RS, having understood the true genesis and evolution of the Deformative Response, concentrates all efforts on the control of the extrusion which, being the "initial stage" and the source of the deformation process, if properly maintained in the elastic field, evolves towards preconvergence and convergence phenomena also in the elastic field, thus allowing to minimize the thrusts on short and long-term coatings.

The same method identifies three categories of fundamental tensile-deformational behavior:

- category A or stable core-face behavior;
- category B or short-term stable core-front behavior;
- category C or unstable core-face behavior.

It is therefore evident that in order to stabilize a tunnel during excavation in the short and long term (Figure 1), type B and type C behaviors must be reported to category A, intervening on the stiffness of the advance core by means of conservative pre-assembly of the cable and, subsequently, regulating, downstream of the excavation face, the extruding manner of the core-face, by closing and stiffening the first phase covering, close to the front, with the inverted arch.

The analysis and the control of the Deformative Response play a fundamental role as indispensable steps to design and correctly realize a work in the underground:

- the analysis must be performed using suitable analytical or numerical calculation tools based on the forecasts made and the designer must also make the necessary operational choices, in terms of systems, phases, digging cadences, consolidation and stabilization tools;
- control takes place at the "construction moment", when, by proceeding with the excavation, the design choices are made and verified by the measurement of the Deformative Response of the means to the actions implemented.

It follows that to correctly design and build a work in the underground is essential:

- in the planning phase:
  - preliminary study of the tenso-deformative behavior (Deformative Response) of the ground, in the absence of stabilization works;

- define the type of pre-containment or containment actions necessary to regulate and control the Deformative Response of the excavation vehicle;
- choose the type of stabilization work;
- to compose, according to the expected behavior of the ground, the typical sections defining, in addition to stabilization interventions more appropriate to the context in which it is expected to operate, phases, cadences and times of implementation of the same;
- sizing and checking, the selected interventions to achieve the desired behavior of the excavation and the necessary safety coefficient of the work, also providing the tensile-deformative behavior of the same thus stabilized;

- under construction:
  - verify, during construction, that the tunnel's behavior during the excavation is the same as that foreseen by analytical way during the design phase. Then proceed with the development of the project by balancing the weight of the interventions between the core-front and the perimeter.

2.2 *Diaphragms walls and piles – Soft-eye Technique*

For several decades, TBM's have been used for the construction of tunnels. Depending on the local situation, the TBM may be placed at the start or at the end of its drive; for example maybe in a precut in the open terrain or maybe by lowering it into an excavation shaft down to the tunnel level. This latter technique is used mostly in congested city areas. A few years ago, starting and receiving a TBM in an excavation shaft required extensive measures such as breaking through the walls of the shaft, which are secured out of steel reinforced concrete. This preparation work needed time and has been expensive. In recent years however the use of Soft-Eyes in these areas are becoming more and more popular. A Soft-Eye may for example be a diaphragm wall or bore piles reinforced with Glass Fiber Reinforced Polymer bars (GFRP) instead of reinforcement out of steel. Also an anchored tunnel face with GFRP anchors will not obstruct the TBM head driving through. The use of GFRP products in tunnelling is getting more and more common in Southeast Asia and is widely applied in Europe and Japan nowadays.

Soft-Eyes consist usually of bore piles or diaphragm walls, which are locally reinforced with GFRP bars. The sections below and above the tunnel are reinforced conventionally. Depending on the designer and contractors preferences, full rectangular sections are built out of GFRP bars and the fiber reinforcement follows more closely the tunnel section resulting in a circular arrangement of the GFRP links or may be a circular sections.

Both possibilities have their advantages. While a rectangular arrangement saves time during the design and assembly of the cages, following more closely the tunnel section thus reducing the material costs for the GFRP bars. Often applied as a compromise, where the vertical bars cover a rectangular section, while the shear links follow the circular layout. Experience shows that this approach decreases the material costs for the GFRP material by less than 5% still maintaining the detailed design and managing the assembly of the cage to be efficient. Building the corresponding reinforcement cages out of GFRP bars on site requires the same working procedures as for an equal steel cage.

The necessary bars are tailor made and delivered to site where the assembly takes place. The bars are fixed together with binding wire, cable binders or similar products. U-bolts are used for clamping bars together when high loads have to be transferred over a connection.

This is a connection between the vertical GFRP bars and the corresponding steel bars, which have to carry the dead load of the reinforcement cage during the lifting process and lowering of the cage into the trench. Welding as is commonly done with steel reinforcement but not possible with GFRP bars.

2.3 *Railways and subways*

The polymeric nature of the materials used for the production of glass fiber reinforcement polymer profiles as well as the insulating characteristics, the chemical resistance, atmospheric agents and the mass pigmentability allow installation and use with practically zero maintenance, thus producing high technical/economic advantages compared to the traditional use of aluminum, steel, wood or PVC.

The elements lend themselves easily to normal assembly and coating operations by means of connections by bolts, screws and rivets or simply by gluing and painting.

The traditional solutions in steel or wood type materials, although apparently cheaper, require assembly operations with heavy vehicles due to their high weight as well as to painting and/or surface treatments that make them partially resistant to the aggressions of the typical environments in which they are used.

3 APPLICATIONS

3.1 *Isarco, Rebars*

The Brenner Base Tunnel is the central element of the new railway line that connects Munich to Verona and will represent the longest underground railway link in the world with its 64 km. The construction lot called "Sottoattraversamento Isarco" is the extreme southern part of the Base Tunnel before access to the Fortezza station (BZ).

The construction of the works is technically very complex: the tunnels of the main tubes and interconnections will pass below the Isarco River, the A22 motorway, the SS12 state road and the Verona - Brennero historic railway line.

Before starting the tunnel construction work, a series of preliminary surface activities must be carried out, including the displacement of the SS12 national road, the construction of two bridges, over the Isarco river and the Rio Bianco, and the construction of the loading area/ unload on the A22 which will be necessary for the transport and supply of construction materials. As part of the implementation phases of the intervention, the definitive deviation of the historic Verona-Brennero railway line for a stretch of about 1 km is also required. The construction of 4 deep wells of about 30 ml includes the temporary reinforcement in GRFP, the assembly of which is made with straight bars and curvilinear elements (Figure 2).

3.2 *M4, Reinforcement for soft-eye*

MM Line 4: Usage of GFRP Rebar Cages for Tunnel Boring Machine "Soft-eye" openings Metro Line 4 will be serving the densely populated areas in city centre of Milan. In order to minimize disruption caused by construction activities, it has been designed to be compatible with other modes of transport and maintain sufficient groundwater level. Metro Line 4 will have twin tunnels with single tracks in each direction. Extensive use of tunnel boring machines (TBM) will be required. Metro Line 4 will have a total of 21 stations, including interchange stations on Lines 1, 2 and 3. The 21 stations, including the terminal, are San Cristoforo FS, Segneri, Gelsomini, Frattini, Tolstoi, Washington-Bolivar, Foppa, Parco Solari, S. Ambrogio, De Amicis, Vetra, S. Sofia, Sforza-Policlinico, San Babila, Tricolore, Dateo, Susa, Argonne, Forlanini FS, Q.re Forlanini and Linate Airport.

The stations are built in open construction pits: An open central shaft and blind-hole side tunnel technique will be implemented to facilitate passage of the TBM and minimize excavation.

Figure 2.   Isarco shaft consolidated with GRFP i-BARS in the bypass tunnel area.

TBMs can not cut through steel-reinforced concrete drilled shaft walls as the steel bars get caught in the shovels of their shield. In addition, the steel bars can not be cut into pieces small enough to allow their transport by the TBM's conveyor belt system. As a result, the conventional construction method with steel-reinforced drilled shaft walls needs the manual removal of the steel reinforcement in the path of the TBM. Not only is this time-consuming and expensive in itself, it also required the stoppage and retraction of the TBM in front of each shaft wall. Finally, to ensure that neither the soil nor potential groundwater outside the shaft wall would collapse into the opened hole, complex and expensive soil stabilization measures are required outside the wall. All these time-consuming and costly measures are not required when the areas of the launch shaft head walls to be penetrated by the TBM are reinforced with glass fibre-reinforced polymer (GFRP). Even though these bars have a much higher tensile strength than steel rebars, they are easily machined and can be broken down into small bar segments by the cutter head of the TBM. These segments can then be transported by the machine's conveyor system together with the excavated soil. The TBM does not have to be stopped, and soil stabilization measures are not required, as the soil is always stabilized by the TBM. The resulting savings in the overall construction time and cost are substantial.

Construction of the first two shafts for the project, at Argonne and Frattini Stations, was opened for bids in January 2015. In both cases, GFRP reinforcement was specified in the bid documents. In early July 2015, MAPLAD was awarded the contract to deliver the soft-eyes GFRP rebar cages (Figure 3).

3.3 *Cociv, i-PIPE profile for face stabilization*

The new high-speed railway line called Terzo Valico develops for a total of 53 km, 36 km of which in the tunnel, and covers 14 municipalities in the provinces of Genoa and Alessandria and the regions of Liguria and Piemonte.

In detail, the line, starting from the railway junction of Genoa (Bivio Fegino), develops almost entirely in tunnels (Galleria di Valico and Galleria Serravalle) up to the Piana di Novi, with the exception of a short section in the open air at Libarna. The Valico Tunnel, about 27 km long, has four intermediate adit tunnels, both for construction and safety reasons.

By the most advanced safety standards, the sections in the tunnel will be largely made of two single track tunnels side by side and joined together by transversal connections so that each can serve as a safety tunnel for the other. From the exit of the Serravalle tunnel the line develops mainly outdoors until you enter the Pozzolo Gallery, at the exit of which the line develops outdoors until it joins the existing line Pozzolo Formigaro - Tortona (to Milan); in the uncovered section between Novi Ligure and Pozzolo Formigaro, the construction of the artificial tunnel link from and to Turin on the current Genoa-Turin line is planned.

Figure 3.   Metro Milano breakthrough in "GFRP mode".

Figure 4. Left: Consorzio Tunnel Giovi (Pizzarotti S.p.A. & Collini S.p.A.), GFRP at the front face. Right: Oberosler S.p.A.. radial and face long bars for front excavation and subsequent cavern enlargement.

For the sections to be made in traditional excavation, the consolidation of the fronts is done by stabilizing the core-face with fiberglass reinforcement and the design and construction phases are managed through the ADECO-RS method.

MAPLAD is presently working in practically all the contracts (Figure 4).

4 RESEARCH APPLIED ON NEW MATERIALS – IBAR® BS BY MPLD

4.1 Basalt fibers

Basalt fiber is a material made from extremely fine fibers of basalt, which is composed of the minerals plagioclase, pyroxene and olivine.

It is similar to carbon fiber and fiberglass, having better physic mechanical properties than fiberglass, but being significantly cheaper than carbon fiber.

Basalt fibers are 100% natural and inert. Tested and proven to be non-carcinogenic and non-toxic and easy to handle. In contrary, fiberglass is made from a mixture of many materials, some of which are not environmentally friendly.

Since basalt is the product of volcanic activity, the fiberization process is more environmentally safe than that of glass fiber. Basalt continuous filament is a green product. Abundant in nature so can never deplete the supply of basalt rock.

The "greenhouse" gases that might otherwise be released during fibre processing were vented millions of years ago during the magma eruption so won't affect the current pollution scenario.

Further, basalt is 100 percent inert, that is, it has no toxic reaction with air or water and is non-combustible and explosion proof.

4.1.1 Advantages

Superior Thermal Protection: Maplad's Basalt has a thermal range of -260 °C to +982 °C (1800 °F) and melting point of 1450 °C. Fibers are ideal for fire protection and insulative applications.

Durable: Tough and long-lasting, fibers deliver acid, alkali, moisture and solvent resistance surpassing most mineral and synthetic fibers. They are immune to nuclear radiation, UV light, biologic and fungal contamination.

They're stronger and more stable than alternative mineral and glass fibers, with tenacity that exceeds steel fibers many times over.

Additionally, basalt fibers are naturally resistant to ultraviolet (UV) and high-energy electromagnetic radiation, maintain their properties in cold temperatures, and provides better acid resistance.

In Table 2 and 3 some technical data are reported.

4.1.2 Mechanical and physical characteristics of basalt fiber reinforced polymer rebar

Basalt rebar has a lower Young's modulus compared with steel, but is 15–30% higher than fiberglass rebar. It is strong in tension and has very little stretch.

Table 2. Comparison between basalt fiber and glass or carbon fiber.

Capability	Basalt fiber	E-Glass fiber	S-Glass fiber	Carbon fiber	Unit
Tensile strength	3000–4840	3100–3800	4020–4650	3500–6000	MPa
Elastic modulus	79.3–93.1	72.5–75.5	83–86	230–600	GPa
Elongation at break	3.1	4.7	5.3	1.5–2.0	%
Diameter of filament	6–21	6–21	6–21	5–15	μm
Tex	60–4200	40–4200	40–4200	60–2400	
Temperature of Application	(-260)-(+500)	(-50)-(+380)	(-50)-(+300)	(-50)-(+700)	°C

Table 3. Breaking strength of basalt fiber for different diameter.

Capability						Unit
Filament diameter	5	6	8	9	11	μm
Breaking strength to weight ratio of elementary fibers	215	210	208	214	212	kg/mm^2

Table 4. Comparison between basalt rebar and fiber glass rebar.

Properties	Glass Rebar	Basalt Rebar	Unit
Elastic modulus	>30000	>50000	N/mm^2
Elongation at break	>2	>2.5	%
Fiber content	>60	>70	%
Shear strength	>16	>20	Ksi

If rebar is subjected to beyond the spec limits then it will break rather than stretch. The rebar placement design needs to allow for this.

The structural engineering needs to consider "tensile modulus". In a properly structurally engineered design, the rebar will not be subjected to anything like the force needed to break it.

The thermal expansion coefficient is very close to that of concrete (whereas steel is very different). This helps a lot to avoid concrete cracking.

Table 4 is a short properties comparison sum up.

4.2 Green pultrusion

A strongly growing demand for reinforced concrete reinforcements in GFRP has been reconditined on the market in the last decade,

Especially for underground works realized by mechanized excavation, the traditional reinforcement is a limitation because it cannot be easily demolished.

In this specific field of application, experimentation on the applicable materials aims above all at the development of innovative and sustainable pultrusion processes.

Hence the "Green Pultrusion" project fielded by the Universities of Catania and Palermo and by some Italian companies headed by Maplad, a leading company in the sector of pultruded elements production.

The phases of the project include the design and construction of a prototype pultrusion plant dedicated to the production of environmentally friendly products, using raw materials of natural origin, totally reusable with a recycling process with low environmental impact.

The steps of the project include the study of the characteristics of natural fibers in relation to their use in the developed pultrusion process, the optimization of a recyclable and

Figure 5. Left: basalt rebars. Right: basalt fibers clothed in a reel.

environmentally friendly resin system and the creation of a recycle loop with a low environmental impact.

The project also includes advanced process control and automation.

Industrial research will therefore focus on the characterization of the fibers, on the stretching and alignment processes of the same as well as on the development of the resin formulation and its recycling process.

In this phase the study of the mechanical characteristics of the fibers is essential because the resistance properties of the natural fibers do not easily allow direct use in the normal pultrusion processes.

The same pultrusion process will be developed specifically to manage both the wettability and premixing phase of the fibers as well as the extraction, ironing and cooling of the pultruded material to guarantee a correct and valid compatibilization of the composite.

The entire process is designed in every single phase according to the chemical-physical properties of the selected natural fibers and to the characteristics of the required polymer fiber/matrix composite.

The project will realize eco-friendly composites with low environmental impact and easily recyclable.

This last aspect is an absolute novelty that, in perspective, can lead to significant advantages in terms of containment of disposal costs as well as representing a new opportunity for the development of recycled products.

In the usual pultrusion processes the reinforcement is impregnated with a resin and pulled through a heated mold in which the resin polymerizes. Almost all pultruded products containing glass fibers and thermosetting resins are non-renewable materials.

The only replacement of glass fibers with natural fibers can not fully realize the scope in order to obtain a truly "green" composite because of the thermoset ones have the greatest environmental impact.

This is especially true with compared to the energy parameter (MJ/ton) required associated to production (Cumulative Energy Demand).

LCA (Life Cycle Analysis) studies of Sachsenlinen show that, for the production of a panel of 1.2 m^2, glass fibers get an energy consumption of 76.08 MJ and the thermosetting resin 84.42 MJ. Therefore, although the replacement of glass with vegetable fibers involves a reduction to 12.58 MJ for the contribution of the reinforcement, the contribution share of the matrix maintaining the traditional resin would always be at a value of 81.04 MJ using fossil-based resins. The cause of the full influence of the matrix is to be found in energy-consuming operations connected to the extraction of fossil products.

The project faces the challenge of introducing an eco-innovative business concept in the pultrusion industry, from the use of sustainable raw materials to the use of a new ecological recycling strategy.

It aims to reduce energy consumption through the use of low impact materials and through the implementation of advanced control systems on the plant.

Greenpultrusion also provides for the installation, along the process line, of a series of self-correcting sensors such as to make the process itself innovative not only in terms of biocompatible and eco-sustainable products, but also in terms of intelligent management.

In Figure 5 can be seen basalt rebars under test and basalt fiber product clothed in a reel.

5 CONCLUSION

The use of fiberglass profiles in the construction market, and especially in large underground infrastructures, is now a consolidated fact.

This has happened thanks to the enormous industrial developments following a phase of applied research, carried out in concert between the Academic World and specialized companies.

So we got an entailed technological development and an increase in production also through the now irrefutable awareness that the use of these materials derives a productive and performance benefit.

Moreover, the decrease in the times related to some constructive processes of underground works and the elimination of problems due to the use of reinforcing steel, are some factors that have allowed the development of these technologies.

The presence of products with mechanical characteristics comparable to traditional steel profiles would allow a greater diffusion of the composites, without obviously significantly increasing the cost of construction of the works with the advantage of considerably reducing maintenance costs.

The profiles derived from the use of basalt fibers would allow to increase the mechanical characteristics compared to those made with glass fibers, without however reaching the costs of those made with carbon fibers.

Furthermore, basalt fibers are made from raw materials that are readily available and are not being depleted.

Another factor of considerable importance is the fact that the basalt fibers are a green material, both for intrinsic properties and because the process of extraction of the raw material and processing do not produce pollution.

The latter, however, must be used with a matrix having the same characteristics of low environmental impact. The use of a bio resin, that is coming from naturally occurring materials, allows the creation of a profile that can be defined as Green.

REFERENCES

Afeltra, R. 2017. *Development of a new technology for production of SKEletons in composite materials for realization of pre-cast tunnel segments* Composite Solutions 11, 16
Carrino, L. & Caprino, G. 1995. *Tecnologia della pultrusione – Aspetti tecnici e considerazioni di mercato*, Editrice Promaplast, Varese, Italy
Cicala, G., Kumar, S., Blanco, I., Manuele, G. & Recca, A. 2018. *Novel pultrusion process for bended rebars for civil engineering application*, Catania, Italy
Kovári, K. 1994. *On the Existence of NATM, Erroneous Concepts behind NATM*, Tunnel, No. 1
Lunardi, P. 2001. *The ADECO-RS approach in the design and construction of the underground works of Rome to Naples High Speed Railway Line: a comparison between final design specifications, construction design and "as built"*. AITES-ITA World Tunnel Congress su "Progress in tunnelling after 2000", Milano, 10–13 giugno, Vol. 3, 329–340
Martel, J., Roujon M. & Michel D. 1999. *TGV Méditerranèe – Tunnel Tartaiguille: méthode pleine section*. Proceedings of the International Conference on "Underground works: ambitions and realities" Parigi, 25–28 October
Rabcewicz, L. 1964. *The New Austrian Tunnelling Method*, Part 1, Water Power, November 1964, 453–457, Parte 2, Water Power, dicembre 1964, 511–515
Rabcewicz, L. 1965. *The New Austrian Tunnelling Method*, Part 3, Water Power, January 1965, 19–24.
Starr, Trevor F. 2000. *Pultrusion for Engineers*, CRC Press

*Tunnels and Underground Cities: Engineering and Innovation meet Archaeology,
Architecture and Art, Volume 6: Innovation in underground engineering,
materials and equipment - Part 2 – Peila, Viggiani & Celestino (Eds)
© 2020 Taylor & Francis Group, London, ISBN 978-0-367-46871-2*

Sequential excavation method – Single shell lining application for the Brenner Base Tunnel

T. Marcher
Skava Consulting ZT GmbH, Innsbruck, Austria

T. Cordes & K. Bergmeister
Brenner Base Tunnel BBT SE, Innsbruck, Austria

ABSTRACT: The Brenner Base Tunnel is a railway tunnel between Austria and Italy through the Alps with a total length of 64 km. The construction lot Tulfes-Pfons includes 38 km of tunnel excavation work. This lot consists of several structures such as the 9 km long Tulfes emergency tunnel. Such a service (non-public) tunnel does not necessarily require a tunnel lining system with two shells (primary shotcrete lining and secondary cast in place lining). Under certain boundary conditions a single shell lining approach can be applied. The required conditions and limitations for a single lining approach are reflected and a proposal for structural verification is provided. For verification approach a novel constitutive model for the shotcrete design is used.

1 OVERVIEW

The Brenner Base Tunnel (BBT) is a flat railway tunnel between Austria and Italy. It runs from Innsbruck to Fortezza. Including the Innsbruck railway bypass the entire tunnel system through the Alps is 64 km long. It is the longest underground rail link in the world.

This tunnel consists of a system with two single-track main tunnel tubes 70 meters apart that are connected by crosscuts every 333 metres (see Figure 1). A service and drainage gallery lies about 10 -12 meters deeper and between the main tunnel tubes.

It is constructed ahead of the main tunnels and will be used as an exploratory tunnel for them. Four connection tunnels in the north and south link to the existing lines and also belong to the tunnel system, with a total length of approx. 230 km.

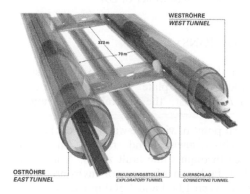

Figure 1. Brenner Base Tunnel System with two Main Tunnels and the Exploratory Tunnel, see Eckbauer et al., (2014).

Three emergency stops, each about 20 km apart, are planned in Ahrental, St. Jodok and Trens. The emergency stops serve to rescue passengers from trains with technical difficulties. In addition, all emergency stops are accessible through driveable approach tunnels. A detailed project description is provided e.g. in Eckbauer et al. (2014).

The two-track Inn Valley Tunnel has been already completed in 1994 and will be integrated into the overall tunnel system. This will reduce the travel time from Germany greatly. The 8 km long section of the Inn Valley Tunnel between the Tulfes portal and the link with main Brenner tubes is a part of the BBT system and will, for reasons of safety, be retrofitted with a rescue tunnel.

Excavation works of the construction lot "Tulfes-Pfons" will last until Spring of 2019. The construction lot includes 38 km of tunnel excavation work. It consists of several structures such as the Tulfes emergency tunnel, the Connection Ahrental access tunnel – Innsbruck emergency stop, the Innsbruck emergency stop with central tunnel and ventilation structures, the Main tunnel tubes, various Connecting tunnels and Ahrental-Pfons exploratory tunnel (see Figure 2).

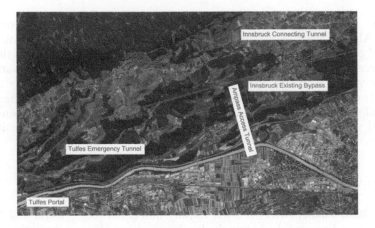

Figure 2.   Construction lot Tulfes-Pfons of the Brenner Base Tunnel project.

The paper focuses on the experiences with the emergency tunnel. The emergency tunnel is being driven parallel to the existing Innsbruck railway bypass; it will be 9 km long and the ex-cavation cross-section is 35 m². The drill-and-blast excavation work (SEM) on this tunnel starts from three points at the same time: from Tulfes westwards (already completed), from the Ampass access tunnel eastwards and again westwards. The excavation works for this emergency tunnel were completed in Summer of 2017.

2   GEOLOGICAL CONDITIONS

This tunnel system within the Central Eastern Alps is crossing the collision zone of the European and the Adriatic (African) plate. The main lithological units are the *Innsbrucker* quartz phyllite, the *Bündner* schist, the central gneisse and the *Brixener* granite (see Figure 3).

From the hydrogeological point of view the water ingress is very limited in the homogenous sections of the phyllites, schists, gneisses and granites. The amount of water is expected to increase only with advance through some fault zones. A lowering of the water table at the surface is prohibited in these fault zones because of environmental aspects.

The overburden varies between 1,000 and 1,500 m over most of the tunnel. The maximum of 1,800 m will be reached in the central gneisses.

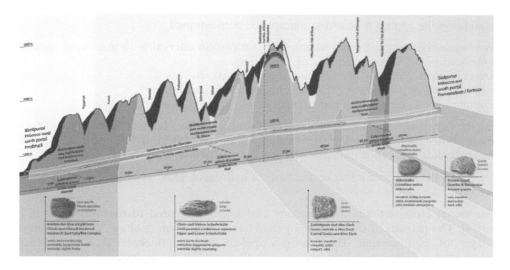

Figure 3. Longitudinal profile of Geology with main Lithological units, see Eckbauer et al. (2014).

3 OPTIMISATIONS

Such an emergency tunnel does not necessarily require a tunnel lining system with two shells (double shell lining system), but under certain boundary conditions the support can be achieved by a single shell lining approach. This is especially valid for non-public tunnels where lower levels of water tightness are acceptable. Single shell lining systems offer the most efficient lining design as they take both the temporary and long-term loads. Additionally, the construction is very fast compared to a double shell (or even composite lining systems) where multiple stages of construction are required.

The tunnel cross-sectional area remains unchanged compared to the tender design (see Figure 4). The permanent shotcrete lining shall have a minimum thickness of d = 25 cm with two layers of reinforcement. The effectively applied shotcrete thickness shall be determined based on back-calculations taking into account all available on-site information. Additionally the following requirements have been defined:

- increased shotcrete quality
- design for service life of 200 years (same as for main tunnels)
- increased concrete cover of 55mm
- alkali-free accelerator to be used

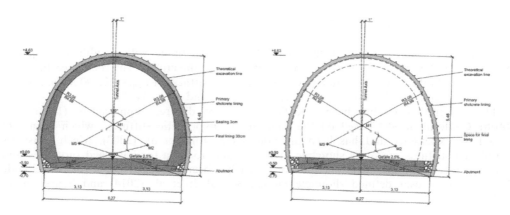

Figure 4. Regular cross section of the emergency tunnel (left: tender design/right: execution engineering).

In addition the applied monitoring concept has been adopted:

- five measuring bolts per cross section with installation interval 5–10 m in weak rock/fault zones and 10–30 m in competent rock formations
- timely implementation of the first convergence measurements
- measurement of the shotcrete thickness by means of tunnel scan
- continuous tests for stiffness and strength development of shotcrete

For the permanent shotcrete lining the verification analyses have to be carried out for the ultimate limit state (ULS) and the serviceability limit state (SLS) as well as for the durability of the shotcrete. This verification assessment is done in two steps:

1. Design of the shotcrete lining on the basis of hydrogeological, geological and geotechnical forecasts.

The tunnel cross-sectional area remains unchanged compared to the tender design (see Figure 4). The permanent shotcrete lining shall have a minimum thickness of d = 25 cm with two layers of reinforcement. The effectively applied shotcrete thickness shall be determined based on back-calculations taking into account all available on-site information. Additionally the following requirements have been defined:

- increased shotcrete quality
- design for service life of 200 years (same as for main tunnels)
- increased concrete cover of 55mm
- alkali-free accelerator to be used

In addition the applied monitoring concept has been adopted:

- five measuring bolts per cross section with installation interval 5–10 m in weak rock/fault zones and 10–30 m in competent rock formations
- timely implementation of the first convergence measurements
- measurement of the shotcrete thickness by means of tunnel scan
- continuous tests for stiffness and strength development of shotcrete

For the permanent shotcrete lining the verification analyses have to be carried out for the ultimate limit state (ULS) and the serviceability limit state (SLS) as well as for the durability of the shotcrete. This verification assessment is done in two steps:

2. Design of the shotcrete lining on the basis of hydrogeological, geological and geotechnical forecasts.
3. Verification on the basis of the measured (monitored) deformations, the actual shotcrete thicknesses and the determined shotcrete properties (modulus of elasticity, strengths).

4   TUNNEL BOUNDARY CONDITIONS

The reference calculation (see Figure 5) considers a service tunnel with a width and height of W/H = 14.4/10.9 m and an overburden of 650 m above the crown. The rock mass has been modelled as Mohr-Coulomb material with a rock mass stiffness of E_{rm} = 12,000 MPa, Poisson ratio v = 0.2 and a rock mass strength of c_{rm} = 2.0 MPa and φ_{rm} = 34.3°.

The shotcrete itself has been modelled using two different approaches: (1) using a Mohr Coulomb model based on the concept of "hypothetic stiffness" which is in fact based on experience and has been introduced e.g. John et al. (2003), which considers a reduced stiffness of young shotcrete of 5 GPa and 15 GPa for hardened shotcrete, and (2) the shotcrete model mentioned above which has been implemented in the model code PLAXIS as a user defined shotcrete model, Schädlich et al. (2014) and Saurer et al. (2014).

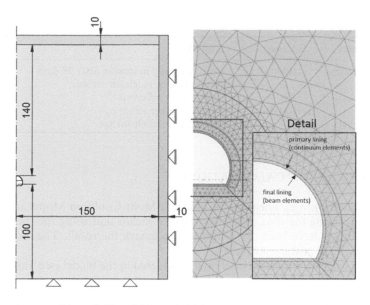

Figure 5. Boundary conditions (left) and FE mesh (right).

5 CONSTITUTIVE MODEL OF SHOTCRETE

The theoretical background to the novel constitutive model has already been presented in Schädlich et al. (2014) and validated against a practical engineering task, which is an executed high-speed railway tunnel project in Germany, built according to the SEM design philosophy, see Saurer et al. (2014). Fundamental equations are provided in Schädlich et al. (2014). The model has been implemented in the finite element software PLAXIS 2D 2012, see Brinkgreve et al. (2012). Note that compression is defined as a negative number.

6 MODEL PARAMETERS

The model parameters for the shotcrete model are presented in Table 1.

Table 1. Parameters of the shotcrete model and values considered in the example calculation.

Name	Unit	Value	Remarks
E_{28}	[GPa]	30	Young's modulus after 28 days
v	[-]	0,2	Poisson's ratio
$f_{c,28}$	[MPa]	33	Uniaxial compressive strength after 28 days
$f_{t,28}$	[MPa]	3,3	Uniaxial tensile strength after 28 days
ψ	[°]	0	Angle of dilatancy
E_1/E_{28}	[-]	0,6	Ratio of Young's modulus after 1 day and 28 days
$f_{c,1}/f_{c,28}$	[-]	-2*	Ratio of f_c after 1day and 28days
f_{c0n}	[-]	0,15	Normalized initial yield stress (compr.)
f_{cfn}	[-]	1	Normalized failure strength (compr.)
f_{cun}	[-]	1	Normalized residual strength
ε_{cp}^{P}	[-]	-0,03, -0,002, -0,001	Plastic peak strain in uniaxial compression at shotcrete ages of 1 hour, 8 hours and 24 hours
$G_{c,28}$	[kN/m]	6,1	fracture energy in compression after 28 days
f_{tun}	[-]	1	Normalized residual tensile strength

(Continued)

Table 1. (*Continued*)

Name	Unit	Value	Remarks
$G_{t,28}$	[kN/m]	0,03	fracture energy in tension after 28 days
φ^{cr}	[-]	2,0	Ratio of creep vs. elastic strains
t_{50}^{cr}	[days]	3,0	Time at 50 % of creep
$\varepsilon_{\infty}^{shr}$	[-]	0,0005	Final shrinkage strain
t_{50}^{shr}	[days]	70	Time at 50 % of shrinkage

7 SHOTCRETE LINING CALCULATIONS

Initially, calculations using a linear elastic model, a Mohr-Coulomb Model and the shotcrete model without applying the time dependency, which means that using the common approach with reduced stiffness have been performed to benchmark the result. The results are marked as 'calculations with hypostatic stiffness'.

The following calculation phases have been considered in the model (see Table 2):

Table 2. Calculation phases.

Phase	Name/Description
1	in-situ stress state
2a	Stress release of the Top Heading
2b	Excavation of the Top Heading, installation of the Shotcrete in the Top Heading
3a	Stress release of the bench
3b	Excavation of the Bench, installation of the Shotcrete in the Bench

The shotcrete strength has been considered according to the J2 curve described above. In order to consider the effect of time in the model, two additional analyses have been performed marked with 'calculations with time-dependent stiffness'.

Initially, a benchmark model using linear elastic material, Mohr-Coulomb material and the UDM shotcrete model with the hypothetic stiffness approach (i.e. with E = 5/15 GPa) has

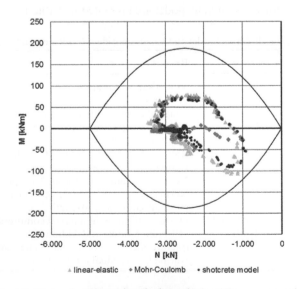

Figure 6.   Comparison of effects of actions using the hypothetic stiffness approach.

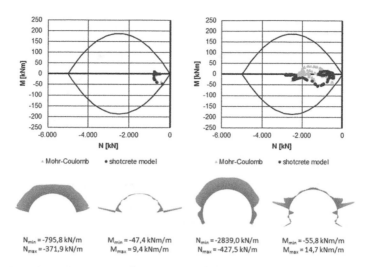

Figure 7. Effects of actions using the UDM for shotcrete using the hypothetic stiffness approach (lower part) and comparison with MC results as an interaction plot (upper part) for both excavation of crown and crown + bench.

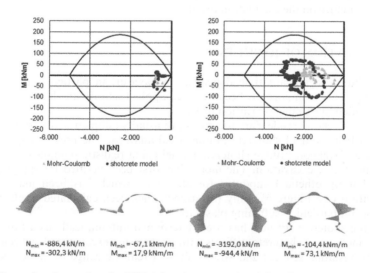

Figure 8. Effects of actions using the UDM for shotcrete using full stiffness approach with creep and shrinkage (lower part) and comparison with MC results as an interaction plot (upper part) for both excavation of crown and crown + bench.

been calculated to compare the resulting effects of actions in the elastic mode. (i.e. ignoring implicit time dependency).

As can be observed in Figure 7 the effects of actions remain within the elastic limits and results from all three models are comparable. Hence the model allows comparing the results in terms of time effects shown hereafter as a next step.

As a next step, the time dependency of the shotcrete strength and stiffness have been included in the model, while still considering E = 15 GPa as maximum stiffness instead of consideration of implicit creep and shrinkage. Here, the boundary conditions as shown in the first chapters have been considered.

As a last step, the effects of shrinkage and creep have been modelled using the equations described in Schädlich et al. (2014), considering realistic parameters for both, material model

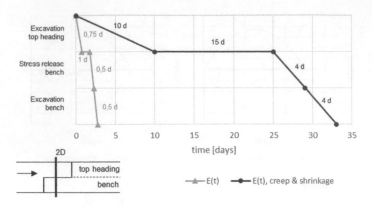

Figure 9.   Construction time considered in the calculation phases.

parameters and construction time. For this model, the parameters as stated in Table 1 have been considered.

However, note that for fully time dependent calculations, boundary conditions in terms of construction time and the time dependent behaviour of shotcrete are important factors that have a major influence on the calculation results.

8   INTERPRETATION OF RESULTS

Shotcrete exhibits a significant time dependent behaviour, in particular during the initial hours of curing. This is important, because once applied as primary lining, the shotcrete is immediately loaded due to the excavation process.

In practical engineering crude simplifications are usually adopted with respect to modelling the mechanical behaviour of shotcrete in numerical analysis. The present paper presents the application of a novel constitutive shotcrete model using realistic boundary conditions for a shotcrete tunnel lining excavation. The model has first been verified by applying a classical application of a hypothetic E-modulus for shotcrete, which has been manually changed during calculation (see Figure 6). Only the strength has been calculated implicitly based on elastoplastic strain hardening/softening plasticity.

Thereafter the shotcrete stiffness has been determined automatically as a function of time. An automatic increase of shotcrete stiffness with progressing excavation has been applied (see Figure 9). The resulting stresses and strains show a realistic range of values (see Figures 7 and 8).

The benefit of such calculation is that there is no longer need for manual adaptation of strength and stiffness with time. Of course, such an advanced constitutive model requires both a higher number of parameters and more detailed knowledge about construction time.

9   CONCLUSION

The present paper illustrates the required boundary conditions and limitations for the single shotcrete lining approach. A proposal for geomechanical and structural verification is provided. For the verification approach both, the use of a novel constitutive model with elastoplastic strain hardening/softening plasticity and time dependent strength and stiffness, creep and shrinkage are considered.

The paper illustrates a successful implementation of such a "Single Shell Lining Concept" for the Emergency Tunnel of the Brenner Base Tunnel Project.

REFERENCES

Bergmeister, K. 2015. Life cycle design and innovative construction technology. *Swiss Tunnel Congress.*

Brinkgreve R. B. J., Engin E., and Swolfs W. M. 2012. Finite element code for soil and rock analyses. User's Manual. Plaxis bv, The Netherlands.

Eckbauer, W., Insam, R. and Zierl, D. 2014. Planning optimisation for the Brenner Base Tunnel considering both maintenance and sustainability. *Geomechanics and Tunnelling*, No. 5.

Eurocode 2: EN 1992- 1-1. 2004. Design of concrete structures – Part 1-1: General rules and rules for buildings [Authority: The European Union Per Regulation 305/2011, Directive 98/34/EC, Directive 2004/18/EC].

Insam, R. and Rehbock-Sander, M. 2016. The Brenner Base Tunnel. *RETC conference proceedings, San Francisco.*

John M., Mattle B., Zoidl T. 2003. Berücksichtigung des Materialverhaltens des jungen Spritzbetons bei Standsicherheitsuntersuchungen für Verkehrstunnel, in: DGGT (Hsg.): *Tunnelbau*, S 149–149188.

Saurer E., Marcher T., Schädlich B., and Schweiger H.F. 2014. Validation of a novel constitutive model for shotcrete using data from an executed tunnel, *Geomechanics and Tunnelling* 7, No. 4, 353–361.

Schädlich B., and Schweiger H.F. 2014. A new constitutive model for shotcrete, *Numerical methods in Geotechnical Engineering*, Hicks, Brinkgreve, Rohe (eds.) 103–108.

Tunnels and Underground Cities: Engineering and Innovation meet Archaeology, Architecture and Art, Volume 6: Innovation in underground engineering, materials and equipment - Part 2 – Peila, Viggiani & Celestino (Eds)
© 2020 Taylor & Francis Group, London, ISBN 978-0-367-46871-2

Water tightness by final lining with programmed waterproofing construction joints in conventional tunneling

A. Marchioni & S. Miliziano
Università La Sapienza di Roma, Rome, Italy

ABSTRACT: In conventional tunneling, as an alternative to the use of a PVC membrane installed between primary and secondary supports, water tightness could be guaranteed by adopting waterproofing construction joints in the final lining and if necessary in the invert arch. Using this technology, tightness could be more effectively ensured. The absence of the PVC sheet, involving the long-term durability of primary supports using alkali-free sprayed concrete and steel ribs with a circular section, permits to consider, for the long-term stability, both primary and final lining as a composite single resistant structure. The study illustrates how it is possible ensure effective and durable water tightness by eliminating the traditional membrane. It describes the operative facilities and quantifies, from the results of interaction numerical analyses for three different geotechnical contexts, the percentage of concrete saved.

1 INTRODUCTION

In conventional tunneling the purpose of primary support is to guarantee the stability of the cavity in short-term conditions. In soft soils and weak rocks, the most common primary lining combines metallic ribs and sprayed concrete. The final lining consists of reinforced or plan cast-in-situ concrete and is designed to withstand all soil-structure interaction loads without taking into account the presence of primary supports as a consequence of concrete degradation (spritz-beton is usually accelerated by silicates) and steel corrosion (ribs are not adequately protected by the concrete cover).

To ensure water tightness, a PVC waterproofing sheet is usually installed between the two supports. The waterproofing sheet is installed on the primary lining surface which is uneven and it makes difficult the application made by human hands. The waterproofing capacity of the membrane is undermined by the surface irregularities and the damage can occur during the installation of final lining due to the poorly welded components. The experience shows that this waterproofing system is not effective and infiltrations of water occur during the tunnel life even when using complex double PVC strata.

More effective water-tightness could be achieved by adopting the waterproofing construction joints (WPCJ) technology, currently employed in other civil construction fields where experience has proven that the use of WPCJ is very effective, durable, easy to install and repair.

In conventional excavated tunnels, the final lining could be designed by foreseeing a numbers of programmed joints where deformation could occur and where an appropriate water-stop could be positioned. The use of this technology is particularly simple and can be customized in underground tunnel projects where the invert arch and the final lining are built in cyclic repeated steps of short portions of cast-in-place concrete.

Another important advantage that can be achieved by eliminating the PVC membrane is the possibility of considering the two supports as a unique resistant structure, thus saving material. In fact, by doing away with the PVC sheet, shear stress and normal stress transfer between the two linings occurs, ensuring a complete interaction between the two supports.

However, in order to consider a double shell lining as a compound structure, the primary lining must be as durable as the secondary lining.

The main goal of sprayed concrete lining is to rapidly achieve the initial strength to guarantee the stability of the cavity in short-term conditions. Traditionally, shotcrete's initial mechanical properties are achieved by the injection of high-alkaline additives that negatively affect the development of long-term strength and reduce durability. Using alkali-free accelerators, the improved concrete quality in long-term conditions is testified by laboratory test results and by practical applications at job-sites worldwide (Leikauf and Oppliger, 1998).

The long term durability of primary lining is also reduced by the corrosion of the steel ribs that are not adequately protected. Iron protection is achieved by guaranteeing an adequate thickness of the concrete cover that depends on environmental conditions and, in any case, it cannot be less than 5 centimeters for the concrete surfaces exposed to the ground. The open profiles (IPE, HE, IPN), widely used in tunneling, create 'shadows', volumes that are hard to fill. The use of circular close profiles removes any related problem guaranteeing the full contact between steel ribs and shotcrete, both inside and outside. In fact, the shotcrete can easily fill the gap between excavation profile and the tubular ribs.

In this study, WPCJ technology is described and possible arrangements for tunneling applications are illustrated. Technical, operational and economic advantages are discussed.

The results of interaction numerical analyses, carried out for three different geotechnical contexts, are shown; the structural calculations show that the thickness of the final lining can be considerably reduced using WPCJ technology and by considering primary and secondary lining as a single composite resistant structure for all the life design of the tunnel.

2 WATERPROOFING CONSTRUCTION JOINTS

A well designed and well casted-in-place concrete can be considered impermeable. However, the cracks that can occur during the curing process and during the construction's life (due to temperature variations, settlements, vibrations, earthquake, etc.) can easily impair water tightness; thus, a PVC membrane is usually installed to guarantee waterproofing. The WPCJ is based on the control of these undesired cracks by introducing working joints placed between two concrete castings (where two different concrete castings meet) and also inside a single concrete casting. To ensure the fissures' occurrence only along the joints, they must be properly designed and installed at the right distances; otherwise, random cracks may occur. The absence of random cracks guarantees the water cannot seep appreciably through the concrete nor where the cracks are forced to occur thanks to the presence of waterstops. Waterstops can be made of rubber or metal bars. With this technology, the concrete, in addition to have static functions, can render the internal support impermeable.

The WPCJ is currently employed for many types of new-build structures and typical applications include standard residential constructions, swimming pools, railway stations, water

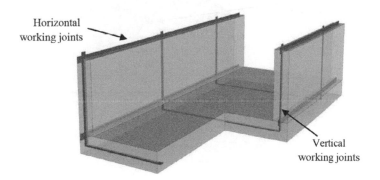

Figure 1.   Scheme of programmed waterproofing construction joints.

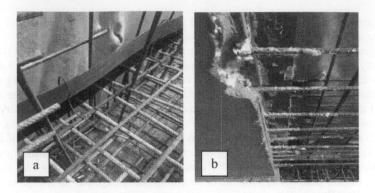

Figure 2.   a) Horizontal working joints; b) Vertical working joints.

treatment constructions and cut-and-cover tunnels. Figure 1 shows an example of the application of WPCJ where the weakened zones between the different slabs are properly sealed by working joints. Figure 2a and 2b illustrate details regarding the horizontal and vertical working joints, respectively.

The concrete used for this application must be impermeable. Long exposure to water leads to surface deterioration that waterproofing aims to solve. Concrete mix-design limits water seepage through the concrete and prevents starting of damages. At least 30 centimeters of thickness are essential to guarantee sealing; obviously, the slab's thickness may be over 30 cm in order to fulfill static requirements.

2.1   Applicability in full face conventional tunneling – secondary lining and invert arch

Conventional tunneling is built in a cyclic execution process of repeated steps of excavation followed by the application of relevant primary support and final lining, all of which depend on existing ground conditions and ground behavior (ITA Work Group – Conventional Tunneling, 2009). The WPCJ can be easily customized in full face conventional tunneling without interfering with the typical construction steps and the installation of waterstops can be part of the cyclic execution process. Figure 3 provides an example of the working joints' scheme observing the construction steps (the joints are placed among cast-in-place concrete). Transversal waterproofing construction joints can be easily installed between two segments of invert and between segments of final lining while longitudinal waterproofing construction

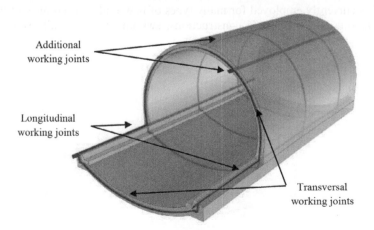

Figure 3.   Construction steps with programmed waterproofing construction joints.

joints are installed between the invert and the final lining. For the typical tunnel geometries/ dimensions, each segment of the final lining could be too big and random cracks could occur; in this case, additional working joints, both transversal and longitudinal, can be easily installed inside a single lining segment, to control the fissure formation and guarantee water-proofing (Figure 3).

From a construction point of view, there are significant benefits such as time saving for in stalling the working joints (connected to concrete reinforcement) compared to the time spent to install the waterproofing membrane. Furthermore, for later construction work in the case of concrete damage or water infiltration, an injection system can be used to seal cracks, voids and joints instead of expensive repairs necessary to fix the conventional PVC waterproofing membrane.

It is worth highlighting that for the design of the programmed joints a new calculation method should be developed. In fact, usually, for ordinary applications, the slabs equipped with waterproofing construction joints are disconnected from other structures using a plastic sheet, while in the tunneling application we propose of avoiding any disconnection between the two supports.

3 DURABILITY OF PRIMARY LINING

To take into account the presence of a primary lining, the steel ribs and the shotcrete have to preserve their initial mechanical characteristics for the duration of the tunnel's expected design life. Recently, it is emerging a tendency to replace the heavy open profiles of rolled steel arches with a lighter tubular steel cross section and reinforced shotcrete (Zenti et al. 2012). The behavior of the tubular steel cross section has been investigated by numerical ana-lyses, laboratory tests and field tests by Zenti et al. (2012). The benefits connected to this new primary lining technology, compared with the typical open profiles, are both static and oper-ational. The tubular ribs are easier to handle during transport and installation. From a static point of view, better performances are guaranteed by the opportunity of completely filling the profile with spritz-beton. In addition, thanks to the geometry, the sprayed concrete can easily fill the gap between the still ribs and the excavation profile, adequately protecting the iron from corrosion and ensuring durability (Figure 4).

Shotcrete is used worldwide in underground tunnel projects in many ground conditions as a temporary lining. Traditionally, the requirements of shotcrete are spraying performance (low dust and low rebound) and quick development of initial strength. All these conditions are achieved by the use of high-alkaline content accelerators that, in turn, reduce the development of long-term strength and durability; thus, the primary lining cannot be considered as a per-manent support.

Currently alkali-free additives can be employed, that can effectively control the rebound and, at the same time, provide high early and long-term strength and enhanced durability. According to several studies, the strength development of a sprayed concrete added with dif-ferent alkali-free accelerators is consistent with the standard requirements of the Austrian Guideline for sprayed concrete (Bernhard et al. 1998).

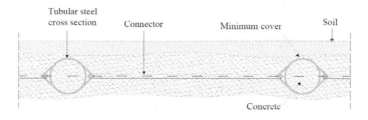

Figure 4. Primary lining section (adapted from Zenti et al. 2012).

For some time, different types of fibers have been used to reinforce shotcrete in tunneling applications. The fibers determine a considerable improvement in the post cracking behavior of concrete. Several investigations have demonstrated the relevant effect of fiber addition in improving ductility and punching resistance of tunnel segments. Fibers arrange themselves in three-dimensional directions inside the concrete matrix and they are able to absorb the tensile stress induced by shrinkage (Coppola et al. 2017) also guaranteeing the durability of shotcrete.

The adoption of tubular ribs, associated with the use of reinforced fiber alkali-free sprayed concrete, can improve the durability of the primary supports throughout the tunnel's design life.

4 MATERIAL SAVING USING WPCJ AND DURABLE PRIMARY SUPPORTS

As already stated, the WPCJ solution implies that the primary lining and secondary lining can be taken to be a unique resistant compound structure by avoiding the use of a water-proofing membrane that disconnects the two supports. These features reduce the required structural capacity of the final lining and consequently its thickness. In order to evaluate the benefits that can be obtained employing WPCJ associated with durable primary support, a structural project is developed referring to three different geotechnical and geometrical contexts and the results obtained by the two different approaches are compared. To evaluate the state of stress on the linings and to develop structural calculations, the soil-structure interaction has been studied performing numerical analyses by the finite difference code FLAC 7 (Itasca, 2011).

For the mechanical behavior of the soil the very simple linearly elastic, perfectly plastic model with the Mohr-Coulomb yield criterion and zero dilatancy (not associated flow rule) has been adopted. The use of this soil model is typical in tunnel engineering and it offers the additional advantage of being characterized by only a few parameters with a clear meaning and relatively easy to determine (Kovári, 1977).

4.1 Soil-structure interaction numerical analyses

To assess and quantify the material saving, three different geotechnical and geometrical situations are studied. The bending moment, shear force and axial force distribution on the linings have been obtained using the stress relaxation method (Panet, 1995). To evaluate the benefits of the WPCJ changing the in situ stress three different contexts have been selected (see Table 1). For each case an anisotropic initial state of situ stress was simulated supposing a coefficient of earth pressure at rest $K_0 = 0.7$. A poisson's ratio $v' = 0.3$ and a costant stifness $E's = 5$ GPa have been supposed for the ground. The cover above the tunnel and the vertical state of stress (σ'_v) are referred to the center of the tunnel.

The relaxation factors have been calculated through axial-symmetric modelling of a cylindrical tunnel built in a homogenous and isotropic soil.

The case studied is a full-face excavation conventional tunnel with a cross section of 13.20 m equivalent diameter. The following typical construction process has been simulated in the analyses:

– face excavation of 1 m and installation of the primary lining;
– realization of the invert between 8 and 16 m behind the face;
– realization of the final lining between 32 and 40 m behind the face.

For the axial-symmetric analyses the structural supports have been simulated as continuum elements of volume of 25 cm of thickness, t, with elastic behavior and constant stiffness. The equivalent stiffness, E_{EQ}, for the primary lining (PL), invert arch (IA) and final lining (FL) have been calculated and reported in Table 2.

Table 1. Geotechnical and geometrical contexts.

Contexts	Cover	σ'_v	c'	φ'
	(m)	(kPa)	(kPa)	(°)
1	50	1000	150	30
2	100	2000	300	30
3	200	4000	600	30

Table 2. Equivalent stiffness.

Supports	t	E_{EQ}
	(cm)	(GPa)
PL	25	4
PL+IA	25	61
PL+IA+FL	25	111

Table 3. Structural supports thicknesses.

Contexts	PL	IA	FL
	(cm)	(cm)	(cm)
1	35	40	40
2	35	70	60
3	35	100	100

Once the relaxation factors for the three cases studied have been calculated, the cross section (2D) numerical analyses were carried out to obtain the state of stress in the supports and to make structural calculations to determine the optimized lining thickness. Taking into account both layers, due to the high distance from the face at the moment of building, the secondary lining results essentially unloaded and no static requirements are necessary. In any case, as already stated, to guarantee the correct waterproofing function a thickness of 30 cm should be adopted. In a conventional design, in long-term conditions, the final lining has a static function and its thickness is calculated without considering the presence of primary supports. The primary lining initially introduced into the model is eliminated with to simulate the long-term conditions.

In both the design approaches (conventional and WPCJ) in the cross section analyses, all the structural elements were simulated as beam elements, attached to the continuum elements adopted for the ground, with linear elastic behavior with Young's modulus $E_{lin} = 31$ GPa and Poisson ratio $v_{lin} = 0.2$. The thicknesses of structural supports are obtaining by an iterative process of optimization and reported in Table 3. The soil parameters are the same used for the axial-symmetric analyses.

The presence of primary lining in long-term conditions involves a thickness reduction of final lining and invert arch depending on ground conditions, geometry and in situ stress.

According to the European Standard (Eurocode 2-1, 2004), the verifications of ultimate limit state have been carried out to compare the thickness required to withstand the load in long-term conditions in both design approaches (there is no difference in short term conditions). A concrete strength class C25/30 has been supposed.

Without considering the presence of primary supports, for the contexts 1, 2 and 3, 40 cm, 60 cm and 100 centimeters of thickness for the final lining are required to withstand the load in long-term conditions, respectively. Therefore, 10 cm, 30 cm and 70 cm of material saving is obtained in the final lining employing the WPCJ and ensuring the durability of primary supports (Figures 5b, 6b, 7b). The presence of primary lining also reduces the thickness necessary for the invert (10–30 cm of material savings: Figures 5a, 6a, 7a).

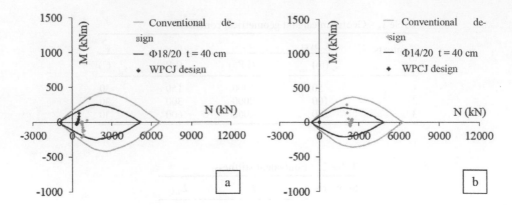

Figure 5.   a) Invert arch_ Context 1. b) Final lining_ Context 1.

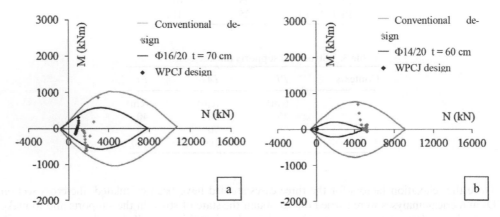

Figure 6.   a) Invert arch_ Context 2. b) Final lining_ Context 2.

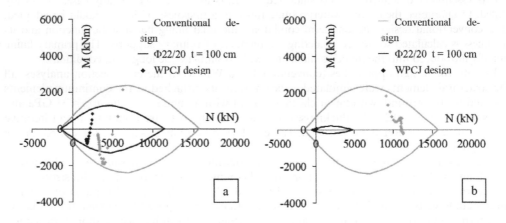

Figure 7.   a) Invert arch_ Context 3. b) Final lining_ Context 3.

Φ XX/YY t = ZZ

XX = Diameter of concrete reinforcement (mm)

YY = Distance between each concrete reinforcement (cm)

ZZ = Concrete thickness of supports (cm)

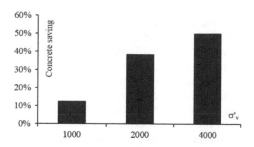

Figure 8.   Percentage of concrete saving changing the vertical stress in situ.

The large number of variables (ground mechanical parameters, geometry section, initial stress in situ and construction's step) does not permit a full parametric study to be performed. However, thanks to the three contexts studied, it has been possible understand the trend of the benefits changing the stress in situ. The results of the analyses carried out show that the advantages associated with the employment of WPCJ in term of percentage of concrete saving grow with increasing the stress in situ (Figure 8). In context n. 3 (highest cover) the material saving obtained using the WPCJ was 50% and also the required concrete reinforcement was less than the amount necessary using conventional tunnel's design.

5   DISCUSSION AND CONCLUDING REMARKS

In conventional tunneling, a PVC membrane installed between primary and secondary supports is used to ensure water tightness. Alternatively, water tightness could be guaranteed, at least in principle, introducing waterproofing construction joints (WPCJ). This technology uses the same concrete of the final lining to obtain impermeability: the final lining is equipped by a number of joints where deformation can occur and where appropriate waterstops can be placed. The use of this technology is particularly simple and easily adaptable in underground conventional tunnel projects where the invert and final lining are built in short cast-in-situ reinforced concrete segments.

This type of technology avoids the onerous installation of PVC sheet and offers operational and economic advantages due to the rapidity of waterstop installation and the ease of repair during the tunnel design life.

The WPCJ technology, by removing any disconnection between the layers, permits to consider the primary lining and secondary lining as a unique resistant compound structure. However, as a long-term support, for the primary lining should be used durable shotcrete and the metallic ribs should be protected by an adequate cover of good quality concrete. All these features can be achieved using tubular steel ribs and alkali-free shotcrete reinforced with metallic fibers.

The results of the analyses carried out for three different cases showed that using the WPCJ technology there is considerable saving of material growing as the depth of tunnel increases. For the deepest tunnel studied there was a 50% reduction in the concrete used and a sizeable reduction of the reinforcement was also obtained. The reduction of the thickness supports implies appreciable further time and cost saving due to the reduction in the excavated soil volume.

The saving of materials depends on ground properties, in-situ stress, geometry cross sections, construction sequences, etc., and has to be analyzed case by case. However, the many operational advantages offered by the use of this technology during the construction phase, ensure cost and time savings regardless of the variables involved.

Perhaps comparable advantages would be obtained when ground exhibits a time dependent behavior, such as viscosity, or due to consolidation but further insights are necessary.

The calculation method for the design of the distances among the joints should be adapted to this specific civil construction. For tunnel applications, to exploit to the full the advantages obtained from this sealing system, the final lining cannot be disconnected from the primary lining. In other civil work applications, a disconnection stratum, usually a sheet of plastic material, is installed between interacting structures (i.e. between bulkhead and internal impervious lining). The conventional calculation method needs to be adapted: a new model that can take into account the strong interaction between the two supports, needs to be developed to properly design the position of the waterproofing construction joints.

Need to try to use this promising technology in tunnelling!

BIBLIOGRAPHY

Bernhard Leikauf & Max Oppliger. 1998. Alkali-free accelerators for Sprayed Concrete. *Chimia 52 (1998)*: 222–224

Eurocode 2. 2011. Design of concrete structures - Part 1-1: General rules and rules for buildings.

Itasca. 2011. FLAC 2D fast lagrangian analysis of continua v. 7.0". *User's Manual*, Itasca Consulting Group, Minneapolis, Minnesota, USA.

ITA Working Group Conventional Tunneling. 2009. *General report on conventional tunneling Method.*

Coppola, L., Buoso, A., Coffetti, D., Kara, P., Lorenzi, S. & D'Alessandro, F. 2017 - The effect of sodium silicate on the behavior of shotcretes for tunnel lining. *Journal of Scientific Research & Reports* 14(2): 1–8.

Kovári, K., 1977. The elasto-plastic analysis in the design practice of underground openings. *Finite elements in geomechanics*, 377–412.

Panet M. 1995. Calcul des tunnels par la méthode de convergence-confinement. Press de l'Ecole Nationale des Ponts et Chaussées, Parigi.

Zenti, C.L., Lunardi, G., Rossi, B. & Gallovich A. 2012. "A new approach in the design of first lining steel rib", *Tunnelling and Undergroudn Space for a Global Society: World Tunnel Congress 2012, Bangkok 21–23 May.*

*Tunnels and Underground Cities: Engineering and Innovation meet Archaeology,
Architecture and Art, Volume 6: Innovation in underground engineering,
materials and equipment - Part 2 – Peila, Viggiani & Celestino (Eds)
© 2020 Taylor & Francis Group, London, ISBN 978-0-367-46871-2*

Soil-building interaction in mechanized tunneling: A comparison of modeling approaches

A. Marwan
Institute for Structural Mechanics, Ruhr-Universität Bochum, Germany
Department of Civil Engineering, Faculty of Engineering, Minia University, Minia, Egypt

A. Alsahly
Institute for Structural Mechanics, Ruhr-Universität Bochum, Germany

M. Obel & P. Mark
Institute of concrete structures, Ruhr-Universität Bochum, Germany

G. Meschke
Institute for Structural Mechanics, Ruhr-Universität Bochum, Germany

ABSTRACT: During the planning phase of tunneling projects, it is crucial to ensure that existing buildings are protected from damage when tunneling in urban environments. Tunneling inevitably causes ground movements whose impact on above-ground structures must be assessed. Various approaches that differ in precision and complexity may be employed to predict settlements, depending on the magnitude of expected settlements and the vulnerability of the structures with regard to tunneling induced damage. This contribution presents a three-step damage assessment concept adjustable to the level of detail necessary. First, ground movements are predicted using analytical or numerical approaches. Second, the above-ground structures are idealized by means of surrogate beam-, slab- or 3D-models. Lastly, structural damage is assessed according to strain information or tilt. This method enables the evaluation of the potential damage to above ground structures associated with various tunnel alignments during the planning phase. BIM concepts offer opportunities to streamline and simplify this process by using geometrical BIM sub-models as a basis for performing structural calculations. This paper aims to provide recommendations for a sufficient level of detail (LOD) of surface structures for the assessment of the induced damage in computational simulations of mechanized tunneling process.

1 INTRODUCTION

The determination of tunnel alignment during planning, a priori accounts for existing surface structures, in particular during tunneling in urban areas. The assessment of building soil interaction is of great interest especially for historical/important buildings. If damage is predicted to occur, counter measures could be applied to control ground deformations (i.e. ground improvement or changing tunnel alignment). The simplest approach of damage assessment uses the analytical equations for settlement prediction without the consideration of soil-building interactions [Peck (1969)] and the relative displacement with respect to the structure length determines the expected damage according to the structural system [Burland et al. (2001)]. Generally, such approach leads to a more conservative damage assessment. A more improved damage assessment is proposed by Franzius (2003) which represents the building as an elastic beam and accounts for the soil-building interactions in the analytical settlement prediction.

Building response to tunnel-induced settlements is indeed a fully coupled 3D problem. Numerical models of the tunneling process and its interaction with the buildings are more addressed in recent publications [Burd et al. (2000), Giardina (2013), Fargnoli et al. (2015), Bilotta et al. (2017) and Obel et al. (2018)]. In numerical simulations, the buildings are integrated as substitute models or via detailed representation of the main structural components.

The uses of Finite Element (FE) simulations form an essential part in the design of modern tunneling projects. However, these models are often time consuming to construct and require data from many different sources with different formats. Therefore, a Building Information Management (BIM) based methodology is presented from which it is possible to automatically generate FE models. The models presented here are based on an integrated product model developed within the context of the German Research Foundation funded Collaborative Research Center 837. This product model has been extended to allow for the automatic generation of an FE model of tunneling simulation with the required level of detail of surface structures depending on the level of expected settlements and the structural response with regard to tunneling induced damage. The model is able to automatically incorporate the results of FE simulations into a coherent visualization scheme in which other data is shown [Alsahly et al. (2018)].

In this context, this paper deals with the settlement predictions, structural modeling of buildings and damage assessment. A process-oriented simulation, based on BIM concepts, is used to properly evaluate the mutual interaction between surface structures and tunneling process. The staged analysis procedure is presented to provide a design approach for risk of damage assessment during tunneling in urban areas as well as a strategy for the optimal use of numerical simulations for the damage assessment during tunneling. The analysis performed presents a step forward for detailed evaluation of tunneling-building interaction.

2 CONCEPT OF DAMAGE EVELUATION

The assessment of building damage requires the definition of three basic steps; surface settlement prediction, representation of building and the method of damage assessment. Each of these steps can be performed at different stages that represent varying accuracy, see Figure 1. This can be summarized according to Obel et al. (2018) as:

a. Methods of settlement prediction:
    a.1 Analytical ground settlement prediction (green field) [Peck (1969)].
    a.2 Numerical ground settlement prediction (green field) [Franzius (2003)].
    a.3 Numerical ground settlement prediction with building substitute stiffness [Bilotta et al. (2017)].
    a.4 Numerical ground settlement prediction with detailed building representation [Fargnoli et al. (2015)] .

b. Basics for structural idealization:
    b.1 Outer geometry of building (3D city model) [Mark et al. (2012) and Schindler et al. (2016)].
    b.2 Design and building data.

c. Methods of damage assessment:
    c.1 Slope of relative displacement [Maidl et al. (2011)]
    c.2 Beam model with damage related to the critical tensile strains [Burland (1974)]
    c.3 2D FE models of the building façade [Giardina (2013)]
    c.4 3D FE models of the detailed building [Fargnoli et al. (2015)]

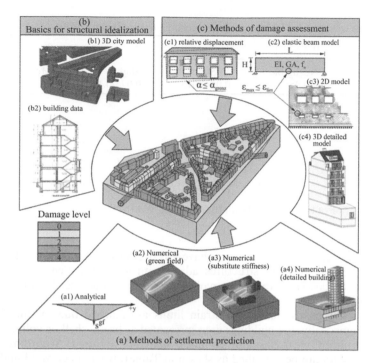

Figure 1.   Evaluation methods for structures in the vicinity of tunnel alignments.

For the aforementioned steps, damage assessment concepts with rising accuracy can be introduced, which can be applied during the early design stage of tunnel alignments in urban environments with numerous buildings. According to the status of the building and the expected damage level, three possible approaches can be performed as:

I. Analytical settlement prediction (a1) with damage assessment using the limits of relative displacement or beam model with the critical tensile strains (c1/c2).
II. Numerical settlement prediction for the green field (a2) followed by damage assessment using separate 2D simulation of the building façade or 3D detailed simulation (c3/c4).
III. Numerical settlement prediction including building with substitute stiffness or with detailed discretization (a3/a4) followed by damage assessment as before (c3/c4). For the detailed building model, damage can be evaluated by the maximum tensile principal strains or a particular damage model.

It can be noted that simple approaches (i.e. I and II) are sufficient for the assessment of noncritical scenarios. While other situations, e.g. masonry structures, important/historical building or shallow tunnels, require more elaborate and reliable alternatives. The challenge is to balance time and accuracy of damage assessment to maximize efficiency e.g. keeping the costs for compensation measures as low as possible.

3   BUILDING MODELS FOR DAMAGE ASSESSMENT

The response of a building is mainly governed by its structural system and the properties of construction material. For low-rise structures, buildings with different materials can be encountered during tunneling in urban areas (e.g. masonry, wooden, concrete or steel). To realistically capture the overall building behavior, the main load-carrying components have to be identified with its interconnection with the foundation system. Generally, the interaction between a structure and the ground is mainly influenced by the flexural stiffness of the

Table 1. Damage category and limiting strains, according to [Boscardin (1989)].

Damage level	0	1	2	3	4
Damage category	Negligible	V. Slight	Slight	Moderate	Sever
Strain limits [%]	0-0.05	0.05-0.075	0.075-0.15	0.15-0.3	>0.30

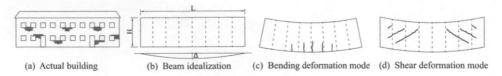

    (a) Actual building   (b) Beam idealization   (c) Bending deformation mode (d) Shear deformation mode

Figure 2. Idealization of building as a beam [Burland and Worth (1974)].

building [Mair (2013)]. Masonry structures have less overall stiffness compared with concrete/ steel structures. Thus, masonry structures adapt with surface settlements by introducing minor repairable cracks. However, more evolution of crack pattern leads to loss of structural integrity and failure.

Damage occurs when the maximum strain limits are reached due to tunneling induced deformation. Numerically, the buildings are represented as beam, shell or volume elements, in which either simplified linear elastic or nonlinear materials are used. According to the strain limits, damage classes can be identified as shown in Table 1, where in class 2 buildings start to show slight damage. Class 3 affects the serviceability limits that requires repair and class 4 affects the structural integrity and requires major repair.

Burland and Worth (1974) related the damage in a building to the critical tensile strains. The concept of critical tensile strain is applied to a simple structure (i.e. uniform, weightless, elastic beam). Using the assumption of a circular ground deformation, the beam undergoes two possible extreme modes (i.e. bending mode and shear mode), from which, the maximum tensile strain is determined, see Figure 2.

Liu (1997) and Bloodworth (2002) utilized 2D and 3D models to study tunneling interaction with masonry building. The latter is modeled via 2D shell elements which represent the main load carrying components (i.e. vertical brick walls) with no consideration of the flooring system. In addition, simple material behavior has been adopted in their simulation with linear elastic, no tension relation. Similar concepts are adopted in literature. In 3D simulations, Netzel & Kaalberg (2000) modeled masonry buildings via shell elements to represent the external façades and the internal walls while roofing system is neglected. In [Franzius (2003)], buildings were modeled via elastic beams in 2D analyses or via shell elements in 3D.

3.1 Building representation by substitute stiffness

A straightforward representation of existing buildings can be achieved by building substitute stiffness Schindler et al. (2014). The substitute stiffness accounts for building rigidity while numerical discretization is performed with simple geometries (i.e. 2D shells or 3D solids). The building mass and rigidity are determined from its specific structural elements (e.g. foundation, roofing system and walls). In addition, Schindler et al. (2014) included other characteristic feature (i.e. external dimensions, year of construction and location) and provided characteristic curves for masonry buildings for rapid evaluation of its mass. Different approaches with various assumptions are presented in literature. In [Obel et al. (2018)], two approaches are adopted for the upper and lower limits calculations of the substitute stiffness of the building. The lower limit is determined from the moment of inertia of the facade. The upper limit of the substitute stiffness includes, in addition to the facade, the stiffness of the roofing system (i.e. slabs and beams). Generally, the respective material of each structural element is accounted by the corresponding elasticity modulus.

4 NUMERICAL SIMULATION OF TUNNELLING PROCESS

The process oriented finite element model has been developed within the framework of the object-oriented finite element code Kratos [Dadvand et al. (2010)]; an open-source framework for the development of numerical methods for multiphysics simulations. It is denoted as "ekate" (Enhanced Kratos for Advanced Tunneling Engineering). In this model, all relevant components involved in mechanized tunneling has been considered i.e. shield machine, hydraulic jacks, lining, soil medium, ground water, annular gap grouting and face support pressure – and their relatively complex interactions. A brief illustration of the model is presented whereas a detailed description can be found in [Nagel et al. (2010) & Meschke et al. (2011)].

Figure 3 shows the main components involved in mechanized tunneling (left) and their representation in the finite element model (right). Shield skin tapering and its weight are considered. The conical shield skin interacts with the excavated soil by means of the surface to surface contact condition. The shield is supported as well on the lining by means of hydraulic jacks. The jacks are modeled by truss elements. In order to realistically simulate shield advancement, particularly in a curved tunnel routes, an automatic steering algorithm is employed in order to control jacks movements [Alsahly et al. (2016)]. Then the jacks are retrieved to assemble a new ring. The sequential excavations are repeated by pushing the jacks onto the newly installed rings, removing the excavated soil elements. After segments erection and shield advancement, the resulting annular gap is usually filled with pressurized grouting mortar.

Tunneling in urban areas requires a particular interest for the mutual interaction between the ground and surface structures, where surface deformations are mainly dependent on the building stiffness as well as its position relevant to tunnel path. Therefore, buildings are integrated in the numerical tunneling model, in which buildings are designated from the 3D city model and simulated with volume elements with substitute stiffness or with detailed discretization of the main structural components as shown in Figure 4. However, the use of detailed discretization of surface structures, in particular for large models with several surface structures, requires high computation costs. For this reason, the accuracy of building discretization is usually reduced in which buildings are represented by its substitute stiffness in order to get the global response. Building, that requires particular interest, can be further discretized with a higher level of detail. A node to volume LAGRANG tying algorithm is used to impose a deformation constraint in which the buildings bases and surface settlements are tied regardless of the FE discretization.

Such a sophisticated model requires considerable effort to generate. To manage such aspect, the model has been incorporated into the so-called Tunneling Information Model (TIM), which is a Building Information Modeling (BIM) based product, for a more automatized process for the generation and execution of FE simulations [Alsahly et al. (2018)].

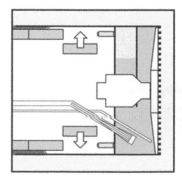 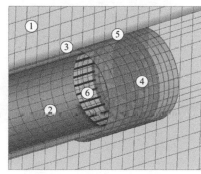

| 1 Surrounding soil |
| 2 Lining |
| 3 Tail gap grouting |
| 4 TBM |
| 5 TBM skin friction |
| 6 Hydraulic jacks |

Figure 3.   Simulation of mechanized tunneling process; (left) components involved in mechanized tunneling and (right) finite element simulation model 'ekate'.

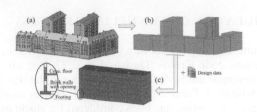

Figure 4. Integration of buildings in numerical simulation; (a) 3D city model, (b) volume geometries with substitute stiffness and (c) detailed discretization of the main structural components.

4.1 *Coupling numerical simulation with BIM*

Building Information Modeling (BIM) methods address the problems generated by decentralized data management. BIM methods use standardized exchange formats such as the industry Foundation Classes (IFC) to ensure that a coherent data exchange exists between all models and information sources within a project [Building Smart (2015)]. BIM models organize data on geometrical and spatial levels and, by modifying IFC's, are able to easily augment a main model with project specific elements. Such an element typically consists of a visual component that is linked to the main model geometry and an information component that is linked to the element geometry. Information is always accessed through a geometrical model and is intuitively organized. Additionally, BIM concepts are able to address the entire lifecycle of a building model, from planning to operation stages, which is critical for highly process-oriented projects, such as Tunneling.

Although BIM methods have been originally applied to Buildings, they have also been applied to tunneling projects [Hegemann et al. (2012)]. In Schindler et al. (2014) an academic BIM model tailored to fit the needs of a tunneling project has been implemented using data

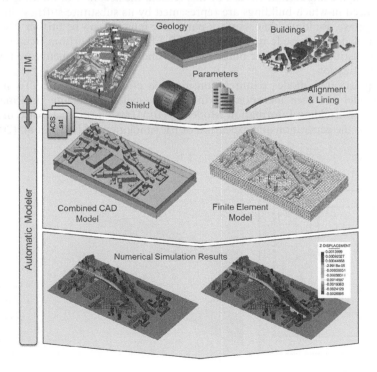

Figure 5. Schematic showing the various components of the Tunnel Information Model (TIM) and coupling with numerical simulation.

taken from the Wehrhahn-line project in Düsseldorf, Germany. This model includes tunneling related geometrical models (tunnel, tunnel boring machine, boreholes, ground and city models), property and city data, and measurements (machine data and settlement). Machine data and settlements can be shown by both tabular and geometric representations. Not only does the TIM provide a data management platform, but it also allows the user to visually interact with and analyze the data through animations or by sequentially time-stepping through processes.

5 CASE STUDY

The proposed concepts for the investigation of the soil-building interaction and the prediction of expected damage using TIM concepts are demonstrated with a numerical simulation of a reference project. The latter is a twin-tube tunnel that passes under an urban area. The outer diameter of the tunnel is $D_A = 10.97$ m and it is driven by two hydroshield machine with an overburden between ca. 4 and 20 m. Both tunnel tubes have 50 cm thick reinforced concrete segmental tunnel lining. In order to reduce the numerical effort in this case study, two selected section of the project will be modeled under the construction of one tunnel only. The topology data of the subsoil, the geotechnical properties of the soil layer, and existing infrastructure by means of substitute models for buildings (as a simplified approximation), have been directly included, through the BIM, into the presented numerical simulation.

The first section, shown in Figure 6, passes under a steel frame warehouse. The presented case study at this section reveals the merits of the BIM-FEM coupling and shows that it is feasible to conveniently perform an automatic numerical simulation for a tunneling project with minimum user intervention. This section is equipped with a sensor field to monitor the settlements during the construction phase of the tunnel. A comparison between the measurements and the prediction of the numerical simulation for the settlements perpendicular to tunnel axis is as well presented in Figure 6.

The second section is the investigation of the tunnel passes under 11 residential buildings as illustrated in Figure 7. The structural damage of these buildings is investigated using the multi

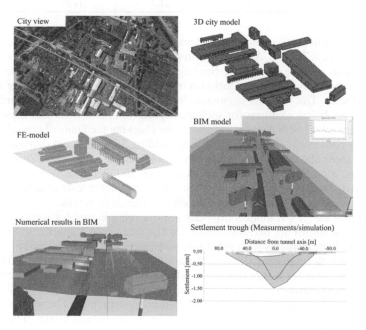

Figure 6.   Schematic showing the BIM-FEM technology applied to a reference project including building models.

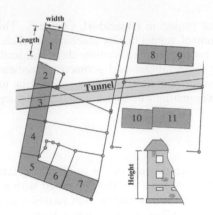

Figure 7.   Plan view on the investigated structures and the tunnel alignment.

Table 2.   Dimensions and substitute stiffness for the structures.

| Building No. | Length [m] | width [m] | Height [m] | building stiffness [GNm2] | |
				lower limit	upper limit
1	17.1	10	13.8	312	4810
2	10	10.15	14.5	334	6118
3	20.1	10.15	14.5	269	7817
4	20.1	10.15	14.5	269	7817
5	10	16	14.7	377	8081
6	10	10	14.7	774	4715
7	11	18.4	16.15	462	13016
8	10.5	15.95	20.1	1484	18513
9	10.5	15.95	20.1	1484	18513
10	10.7	18.2	21.1	1831	20725
11	10.7	21.7	21.1	1831	20725

stage procedure according to section 2. The relevant dimensions and the limit values of the substitute stiffness are given is Table 2. A lower bound solution assumes that only the walls perpendicular to the tunnel axis produce an effective bending stiffness, and an upper bound can be defined if all the main structural components are included in a shear-stiff manner (i.e. adding the floors and foundation stiffness). The simplest approach for damage assessment is primarily utilized. It incorporates the analytical solutions for settlement prediction, and then uses these results as an input for the beam model with damage assessment by checking the limits of critical tensile strains. This resembles a conservative solution, in which, building 3 experience a damage class 3 as it is directly located above the tunnel axis, while other buildings are in uncritical situation. Moderate structural damage is predicted with the beam idealization and therefore further detailed investigations are necessary. Only building 3 will be investigated with higher level of details. Table 3 shows the structural idealization of building 3 with the different levels of details.

At this level, the use of FE models provides a more reliable tool for settlements prognosis. The numerical simulations are performed first for the green field scenario, and then, buildings are included as volume elements with substitute stiffness. Later, only the building 3 that is predicted to experience damage is modeled with the highest level of detail. The settlement trough is then used as an input for a separate 2D model of the masonry façade of building 3 which leads to an improved damage of class 1. The third combination of the prognosis is taken into account by the direct incorporation of the soil-building interaction. The main structural components of building 3 is discretized and integrated in the numerical model. The resulted maximum principal tensile strains are less than 0.03% and consequently the expected damage is expected to be negligible (damage class 0).

Table 3. Model complexity relevant for damage assessment.

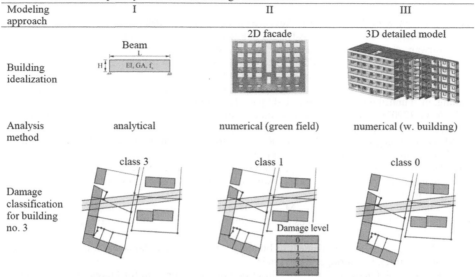

Modeling approach	I	II	III
Building idealization	Beam	2D facade	3D detailed model
Analysis method	analytical	numerical (green field)	numerical (w. building)
Damage classification for building no. 3	class 3	class 1	class 0

6 CONCLUSIONS

Within this contribution, a new approach with different level of details for surface structures idealization and damage assessment are presented. This approach provides a flexible damage assessment concept that can be adopted during planning phase. In the first step, all building are included, with simplified damage assessment. Here, all buildings can be characterized according to the expected damage level employing the proposed a conservative approach. In the next step, an improved numerical prediction of deformation and strains are adopted for a more reliable damage prediction. This reduces the computational costs by adopting the sufficient level of detail according to the predicted damage level and the status of the building.

The proposed approaches are integrated into an automatized process for the generation and execution of FE simulations. Furthermore, it has been incorporated into the so-called Tunneling Information Model (TIM), which is a Building Information Modeling (BIM) based product model that has been developed in the context of the German Research Foundation funded Collaborative Research Center 837. The applicability and the effectiveness of the proposed framework to automatically generate a fully 3D finite element model for a selected level of detail have been demonstrated by means of a reference project.

The presented case study reveals the merits of the proposed approach and shows that it is feasible to conveniently perform an automatic numerical simulation for a tunneling project with minimum user intervention. It should be emphasized, that this proposed approach serves as a predictive tool during the design phase. The investigation of various driven paths, i.e. tunnel alignments, during the design phase can be performed with minimum user effort. Alternative methods of rising accuracy can be introduced and consecutively applied, which enables to evaluate tunnel alignments in urban environments with numerous buildings.

ACKNOWLEDGEMENTS

Financial support was provided by the German Research Foundation (DFG) in the framework of project C1 and D3 of the Collaborative Research Center SFB 837 Interaction modeling in mechanized tunneling. This support is gratefully acknowledged.

REFERENCES

Alsahly, A., J. Stascheit, & G. Meschke 2016. Advanced finite element modeling of excavation and advancement processes in mechanized tunneling. Advances in Engineering Software, Vol. 100, pp 198 –214.

Alsahly, A., Marwan, A., Gall, V. E., Scheffer, M., König, M., Meschke, G. 2018. BIM-to-FEM: Incorporating Numerical Simulations into BIM Concepts with Application to the Wehrhahn-Line Metro in Düsseldorf, in 2018 World Tunnel Congress (ITA-AITES WTC 2018)

Bilotta, E., Paolillo, A., Russo, G., Aversa, S. 2017. Displacements induced by tunneling under a historical building, in: Tunnelling and Underground Space Technology, Vol. 61, pp. 221–232.

Bloodworth, A. G. (2002). Three-Dimensional Analysis of Tunnelling Effects on Structures to Develop Design Methods. Ph. D. thesis, University of Oxford.

Boscardin, M. D., Cording, E. J. 1989. Building Response to Excavation-Induced Settlement, in: Journal of Geotechnical Engineering, Vol. 115, Iss. 1, pp. 1–21.

Burd, H. K., Houslby, G. T., Augarde, C. E., Liu, G. 2000. Modelling tunnelling-induced settlement of masonry buildings, in: Proceedings of the Institution of Civil Engineers-Geotechnical Engineering, Vol.143, Iss. 1, pp.17–30.

Burland, J. B., Standing, J. R., Jardine, F. M. 2001. Building response to tunneling, Case studies from construction of the Jubilee Line Extension, London. Volume 1: Projects and Methods. London, Telford.

Burland, J. B., Wroth, C. P. 1974. Settlement of buildings and associated damage, in: Conference on the Settlement of Structures, Cambridge, pp. 611–654.

Dadvand, P., Rossi, R., & Oñate, E. 2010. An Object-oriented Environment for Developing Finite Element Codes for Multi-disciplinary Applications. Archives of Computational Methods in Engineering.

Fargnoli, V., Gragnano, C. G., Boldini, D., Amorosi, A 2015. 3D numerical modelling of soil–structure interaction during EPB tunneling, in: Géotechnique, Vol. 65 (2015), Iss. 1, pp. 23–37.

Franzius, J. N. 2003. Behaviour of buildings due to tunnel induced subsidence, Ph.D. thesis, Imperial College, London.

Giardina, G. 2013. Modelling of settlement induced building damage. Ph.D. thesis, TU Delft.

Hegemann, F., Lehner, K., König, M. 2012. IFC-based Product Modeling for Tunnel Boring Machines, in eWork and eBusiness in Architecture, Engineering and Construction, Edited by C. Gudnason and R. Scherer. Reykjavík, Iceland.

Liu, G. 1997. Numerical Modelling of Damage to Masonry Buildings Due to Tunnelling. Ph. D. thesis, University of Oxford.

Maidl, B., Herrenknecht, M., Maidl, U., Wehrmeyer, G. 2011. Mechanised Shield Tunnelling, Berlin, Ernst & Sohn, ISBN 9783433600757.

Mair, R. J. 2013. Tunnelling and deep excavations: Ground movements and their effects, in proceedings of the 15th European Conference on Soil Mechanics and Geotechnical Engineering, pp. 39–70.

Mark, P., Niemeier, W., Schindler, S., Blome, A., Heek, P., Krivenko, A., Ziem, E. 2012. Radarinterferometrie zum Setzungsmonitoring beim Tunnelbau, in: Bautechnik 89, Heft 11, S. 764–776.

Meschke, G., Nagel, F., Stascheit, J. 2011. Computational Simulation of Mechanized Tunneling as Part of an Integrated Decision Support Platform, in Journal of Geomechanics 11, 519–528.

Nagel, F., Stascheit, J. & Meschke, G., 2010. Process-oriented numerical simulation of shield-supported tunnelling in soft soils. Geomechanics and Tunnelling 3, No. 3, pp. 268–282.

Netzel, H. & Kaalberg, F. 2000. Numerical damage risk assessment studies on adjacent buildings due to TBM-tunnelling in Amsterdam, in proceedings of the GeoEng2000.

Ninić, J., Stascheit, J. & Meschke, G., 2014. Beam–solid contact formulation for finite element analysis of pile–soil interaction with arbitrary discretization. Numerical and Analytical methods in Geomechanics, 38, pp 1453–1476.

Obel, M., Marwan, A., Alsahly, A., Freitag, S., Mark, P., & Meschke, G. 2018. Schadensbewertungskonzepte für innerstädtische Bauwerke bei maschinellen Tunnelvortrieben. Bauingenieur. In press.

Peck, R. B. 1969. Deep excavations and tunneling in soft ground, in: 7th International Conference on Soil Mechanics and Foundation Engineering, Mexico, pp. 225–290.

Schindler, S., Hegemann, F., Alsahly, A., Barciaga, T. Galli, M., Lehner, K., Koch, C. 2014. An interaction Platform for mechanized Tunneling: Application on the Wehrhan Line in Düsseldorf (Germany), in Geomechanics and Tunneling, 7(1): 72–86.

Schindler, S., Hegemann, F., Koch, C., König, M., Mark, P. 2016. Radar interferometry based settlement monitoring in tunnelling: Visualisation and accuracy analyses, in: Visualization in Engineering, Vol. 4, Iss. 7, pp. 1–16.

Tunnels and Underground Cities: Engineering and Innovation meet Archaeology,
Architecture and Art, Volume 6: Innovation in underground engineering,
materials and equipment - Part 2 – Peila, Viggiani & Celestino (Eds)
© 2020 Taylor & Francis Group, London, ISBN 978-0-367-46871-2

Soil improvement in tunnel face using foam reagents in EPB TBM

S.V. Mazein
Russian Tunneling Association, Moscow, Russia

A.N. Pankratenko, A.G. Polyankin & E.A. Sharshova
National university of science and technology MISiS, Moscow, Russia

ABSTRACT: The purpose of the study was to select the methods of material selection (foam reagents) for the soil conditioning during the tunneling with EPB shields in various Moscow geological conditions. The report contains: the results of laboratory studies on the additives selection for each type of soil and recommendations for optimal use of them; the results of full-scale testing checking the effect of additives-conditioners on the technological properties of the soil under various hydro-geological conditions at three construction sites using EPB TBM Ø 10.82 m. In the course of the full-scale experiment, the laboratory selection correctness of foam agents for conditioning the soil was verified.

1 INTRODUCTION

Currently in Moscow, designing and construction of tunnels for various with the use of TBM complexes, using the change in soil properties in the tunnel bottom (EPB-technology). Soil conditioning during the penetration is carried out using foam reagents, additives for polymers and bentonite.

The addition of such reagents enlarges the range of geological conditions, where clay casting can be successfully applied. This thesis confirms one of the first publications on the use of foam in the chamber of the EPB shield as the main additive (Herrenknecht & Maidl, 1995), as well as information on the use of chemical foam to improve the drilling of EPB shields in granular soils (Quebaud & all, 1998).

When building tunnels in complex geological and urban planning conditions in Moscow, it is necessary to correctly select the consumption of additives-conditioners in order to avoid emergency situations. In EPB technology consumption is predetermined by: water concentration of reagent C (%), expansion (magnification) foam FER, foam/soil volume ratio FER and soil volume (FIR,%) (Sebastiani & all, 2018).

Effectiveness of additives-conditioners are largely assessed the maximum decrease in cutter-head torque (Roby & Willis, 2014).

The aim of the study is selection of materials (foam reagents) for conditioning soil with EPB shields tunneling in different mining and geological conditions of Moscow city.

2 LABORATORY TECHNIQUE

The task of the selection of additives-air conditioners for each characteristic soil type includes a study of the influence of additives on the technological properties of soil with getting:

– results of laboratory studies on the selection of additives for each type of soil and recommendations for the optimal consumption;

– results of full-scale tests of the additives-conditioners impact on the technological properties of soil in different hydrogeological conditions.

The need for soil conditioning is dictated by, first of all, increase the plasticity of the soil developed to reduce resistance when cutterhead running and foam-soil maintaining uniformity while shield stopping.

2.1 Foam generation

For laboratory preparation and investigation of the multiplicity and stability of the foam, a whipping method is chosen to ensure the stability of the foam in a closed volume.By the method of whipping, the foam is formed by repeated impacts on the surface of the solution with a plate with holes attached to the rod. Whipping of the solution can be carried out both manually and with the help of simple mechanical devices.

Usually 30 strokes are made within 30 seconds, provided that the height of the plate is constant from the bottom of the graduated cylinder. Instead of a perforated plate, a wire mesh can be used to ensure greater uniformity of bubble sizes.

In the laboratory, the initial volume of obtained foam V_f is measured from the volume of the aqueous solution of the foaming agent V, then the container with the foam is closed with a lid, then the volume of the precipitate V_p is measured several times at intervals of 1 minute, then at least-to 50% loss in the foam volume.

The index of multiplicity FER is calculated:

$$FER = V_f / (V - V_f) \tag{1}$$

Then a graph of the time variation in the volume of the foam V_f and the volume of the precipitate V_p is plotted (Wu & all., 2018), the half-life of the foam is determined by the volume of the outflow of the liquid (sediment) T_{50}.

The durability of the foam mixture in full-scale conditions can be estimated by the achieved minimum EPB pressure value in the face during the stoppage of the shield.

2.2 Foam-soil mobility

The minimum valid value of the criterion of foam-soil mobility usually taken indicators slump cone SC = 10 cm (Raynaud & Guerry, 2018) and blurry cone BC = 20 cm. The maximum valid value for the criterion of mobility indicators accepted SC = 15 cm (Zhang & all., 2017) and BC = 25 cm. Indicators of SC (GOST 10181-2000) and BC (GOST 310.4-81) are correlated to any foam-soil composition (Figure 1).

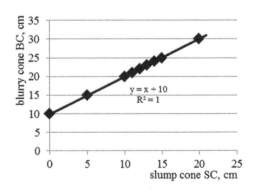

Figure 1.   Dependence of the blurry cone BC and slump cone SC.

For laboratory experiments on mobility take test on blurry cone for vibration table (1-st method) because of the great foam-soil mobility and laboratory materials savings.

Laboratory equipment for mixing foam is a mixer (blender) with a speed control.

The mixer is connected via a wattmeter, which allows you to measure the power consumption during the preparation of the mixture. During the mixing, which for each mixture is 3 minutes, every minute the readings of the wattmeter P are taken.

To analyze the mobility of the foam (2nd method), the average value in watts P is taken and the mixing power dP (W) is found:

$$dP = P - P_o,$$ (2)

where P_o - idling power.

For laboratory experiments, samples of the 2 most common soil types, medium-sized sand, and soft plastic loam were taken.

The mobility of the foam in full-scale conditions can be estimated by reducing the torques of the rotor of the shield and the screw conveyor.

3 THE RESULTS OF LABORATORY RESEARCH

3.1 Foam-soil mobility with additives

Tests for two methods for determining the mobility of foams (loam and sand) show good convergence of results, methods for measuring the mixing power dP (2 method) and blurry cone BC with shaking (1 method) have a high mutual correlation with respect to loams and sands (Figure 2).

Between the indicators, a power law dependence is tracked, with the best informative value of the first method being obtained for more rigid sand mixtures (steeper slope of the curve (2)).

For a more accurate determination of mobility, the first method (BC) is recommended for use in low-mobility soils (low wet loam and medium-sized sands (2)), and the second method (determination of mixing power dP) for flowing soils (water-saturated loams and silty sands (1)).

3.2 Foam-soil mobility with water

Moisture of loams additionally affects the mobility in the presence of conditioning additives (Figure 3). Optimal conditioning parameters (minimum $dP = 10$ W and BC $= 20$ cm) for loam are achieved with its humidity of about 30%.

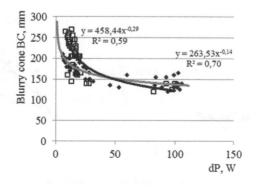

Figure 2.   Dependence of the blurry cone BC and slump cone SC.

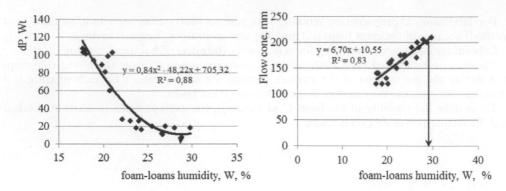

Figure 3. Effect of humidity of a mixture of loam with foam (an average content of additives of 2.3 g/l) on the mobility indices of the mixture.

3.3 *Featured additives consumption*

Based on the results of laboratory experiments aimed at studying the loam mobility, we recommend the following additive consumption Q:

- for ACP 143: $Q = 3\ldots4$ g/l;
- for CLB F5: $Q = 3\ldots4$ g/l;
- for SLF 41: $Q = 3\ldots6$ g/l.

Moisture in the sands affects their mobility when adding a foaming agent and a polymer to bind water.

The optimal conditioning parameters ($dP = 10$ W and BC = 20 cm) for sand are reached at its humidity from 8% to 12% (Figure 4). Sand with a polymer (0.8 g/l) at a humidity of $W = 15\%$ obtained the same mobility as sand without polymer at a moisture content of $W = 8\%$.

Based on the results of laboratory experiments aimed at studying the mobility of sand by two methods, we recommend the following additive consumption Q:

- for ABR 5: $Q = 2\ldots2.5$ g/l;
- for CLB F5: $Q = 1\ldots3.5$ g/l;
- for SLF 41: $Q = 1.5\ldots3$ g/l.

The results of laboratory experiments on the selection of additives to the main two types of soil make it possible to give estimated values of additives consumption, which are checked by tunneling with the help of EPB TBM.

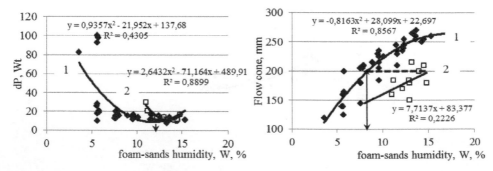

Figure 4. Influence of foam-sand humidity on indices of its mobility: 1 - averaged content of additives (1.5 g/l); 2 - the same + averaged polymer content (1 g/l) for water binding.

4 OBJECT OF RESEARCH AND RESULTS

A feature of full-scale tests is that it is possible to slightly change the consumption of additives in order not to disturb the technological process of development and transportation of soil during drilling.

Adjustment of the additive consumption to the volume of conditioned ground is made only by the TBM operator (he changes the consumption of foaming agent, water and compressed air).

We present the results of full-scale verification tests of the effect of additives and conditioners on the technological properties of the soil in the various hydro-geological conditions of three construction distances using Ø 10.82 m EPB shield.

The main enclosing soils, for which the properties have been investigated by conditioning with additives, are the following layers (from top to bottom):

- fine clayey sand saturated with water,
- sandy, soft-plastic clay-loam;
- saturated with water sand of medium size;
- semi-hard and hard heavy clay with mica.

All these layers are simultaneously developed by a large diameter TBM rotor, forming a conglomerate mixture. Characteristics of conglomerate can be contingently attributed to the prevailing kind of soil (sand or loam).

Table 1 shows the content of soil types for three distances of the tunnel and the applied additives-conditioners.

In Figure 5 shows the geological profile of the tunnel with a diameter of 10.82 m and a length of about 1200 p.m.

To analyze the process of conditioning the soil, we use data stored in the TBM computer in the form of tabular discrete data (every 10 seconds) of procedural control, and processed data (minimum, average and maximum values).

To determine the ratio of materials, we analyzed the consumption of soil and technological media (foam agent, water, air, bentonite). To determine the uniformity of the foam-soil, we

Table 1. Contents of the soil species and applicable additives (%).

Soil species	clay	loam	sand	additives
Distance 1	10%	20%	70%	SLF 41
Distance 2	25%	30%	45%	ACP 143 + ABR 5
Distance 3	10%	25%	65%	SLF 41 + ABR 5

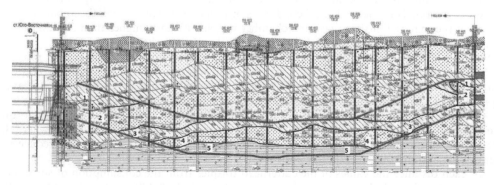

Figure. 5. The longitudinal geological profile along the tunnel axis: 1 – loam; 2 – sand; 3 – loam; 4 – sand; 5 - clay.

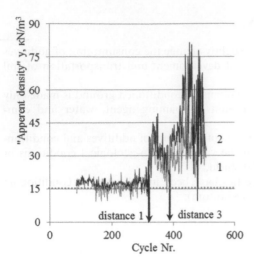

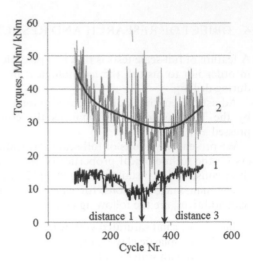

Figure 6. The graph of the "apparent density" of the foam along the drill route: 1 - during the stop, 2 - during the drilling.

Figure 7. Charts: torques of the cutterhead (1, MN·m) and of the screw (2, kN·m) along the tunnel

analyzed the ground pressure along the face. To determine the mobility of the foam, we analyzed the forces of rotation of the rotor and screw, which correlate with the power when stirring.

To analyze the achievement of the foam homogeneity at the tunneling, a graph of the "apparent density" is shown (Figure 6). The methodology we used was described by "Geodata S.p.A" (Guglielmetti and all., 2008). Results are consistent with earlier studies (Mori & all).

Tunnel distance 1 of the tunnel was characterized by a predominant content of sand (up to 70%). Here, the apparent density (γ) of the material in the excavation chamber at different levels corresponded to the norm and was maintained at a constant level. The SLF 41 additive was supplied to the chamber. The stable apparent density of the resulting foam-soil is explained by the fact that the sand is best liable to be loosened with foam reactants. This fact is indicated by the results of laboratory tests (lower consumption of additives and heightened mobility of the foam).

Tunnel distance 2 was characterized by an approximately equal ratio of sandy and clay soils. Here, the apparent density (γ) of the material in the excavation chamber was slightly above the level of the lower limit. Because of the increased content of clay soils, a foam reagent for clay soils ACP 143 and a foam reagent with anti-wear additive ABR 5 were fed in a ratio of approximately 3:1.

Tunnel distance 3 was characterized by a predominant content of sand (up to 65%), which largely corresponds to the geological conditions of tunnel distance 3. Here, the apparent density (γ) of the material in the excavation chamber was significantly higher than the level of the lower limit and fluctuated abruptly from cycle to cycle.

Because of the increased content of sandy soils, the addition of SLF 41 (which replaced ACP 143), and the foam reagent with the anti-wear additive ABR 5 in a ratio of approximately 3:1 were supplied. In the actual conditions with the use of various additives, the average values of the rotor torques and the screw conveyor for the drilling cycle are reduced.

As can be seen from Figure 7, the changes in the cutterhead torques and the screw torque are identical along the penetration path, except for the end of the tunnel distance 1, where the cutterhead torque has decreased due to the wear of the cutting tool. To further reduce the wear of the cutting tool, the additive ABR 5 was applied.

For each distance the values of the cutterhead torques from the consumption of soil conditioning materials and also the optimum consumption of the additive were determined: distance 1 (Figure 8a); distance 2 (Figure 8b); distance 3 (Figure 8c).

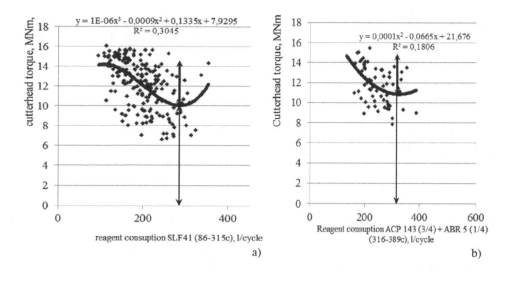

a)

b)

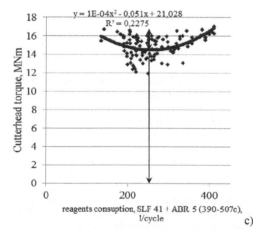

c)

Figure 8.   Cutterhead torques and reagents flow (a – distance 1, b – distance 2, c – distance 3).

On distance 1, the maximum effect of reducing the cutterhead torque (by 4.8 [MN·m]) occurs due to the addition of SLF41 (average flow rate 196 [l/cycle]) at a flow rate of 290 [l/cycle] = 1.7 [l/m³], which is included in the recommended range of 1.5 ... 3 [g/l] for the conditioning of sands. The additive is selected correctly.

On distance 2, ACP 143 and ABR 5 additives with a flow rate of 320 [l/cycle] = 1.9 [l/m³] are much less reduced the cutterhead torque (by 3.9 [MN·m]), which is much below the recommended value of laboratory studies range 3 ... 4 [g/l] ACP 143 for the conditioning of loam. The reagents are not selected optimally.

On distance 3, the maximum effect of reducing the cutterhead torque does not occur (only 2 [MN·m]), due to the additives SLF 41 and ABR 5 with a flow rate of 250 [l/cycle] = 1.5 [l/m³]. It is slightly below the recommended range of 1.5 ... 3 [g/l] SLF 41 for laboratory conditioning of sand. The reagents are not selected optimally.

For distances 2 and 3, the recommendations of the additive manufacturers should be taken into account:

– ACP 143 cannot be mixed with other ground conditioners and chemical additives, this can lead to immediate gel formation;

– ABR 5 is used for tunneling in strong rock to reduce cutting tool wear, dust suppression, vibration reduction and rock resistance during cutterhead rotation. That is, the additive was used outside the geological conditions of the recommended application.

5 CONCLUSION

1. Methods of selection of materials (foam reagents) for soil conditioning during tunneling with EPB shields are selected
2. During the full-scale experiment, the correctness of the laboratory selection of foam agents for conditioning the soil was verified.
3. The effect of additives-conditioners on the cutterhead torque, which characterizes the foam-soil mobility, is estimated.
4. The effect of reducing the rotor torque in dependence on the consumption of additives was investigated.
5. It is indicated that it is necessary to use additives in the geological conditions of the recommended application and in the recommended combinations.

REFERENCES

Herrenknecht, M. & Maidl, U. 1995. Applying Foam for an EPB Shield Drive in Valencia. *Tunnels* (5), (1995), pp. 10–19.
Quebaud, S., Sibai, M. & Henry, J.-P. 1998. Use of Chemical Foam for Improvements in Drilling by Earth pressure Balanced Shields in Granular Soils. *Tunn. Undergr. Sp. Tech.* 13 (2) (1998), pp. 173–180.
Sebastiani D., Miliziano S., Zanetto G. & Ginanneschi R. 2018. Effectiveness of Foam Injection during Mechanized Excavation of Tunnels with TBM-EPB Technology. *Paper proceedings ITA – AITES World Tunnel Congress*, 2018. Dubai, UAE. S. 3813–3826.
Roby J. & Willis D. 2014. Raising EPB Performance in metro-sized machines/*Tunnels for a better Life. Proceedings of the World Tunnel Congress '2014*: Iguassu Falls, Brazil. 09–15 May, 2014. Arsenio Negro, etc. (Editors). – San Paulo: CBT/ABMS.
Wu Y, Mooney MA & Cha M. 2018. An experimental examination of foam stability under pressure for EPB TBM tunneling. *Tunn. Undergr. Sp. Technol.* 2018; 77: 80–93. doi:10.1016/j.tust.2018.02.011.
Raynaud, S. & Guerry C. 2018. How to make ground conditioning safer in case of high water ingress. *Paper proceedings ITA – AITES World Tunnel Congress*, 2018. Dubai, UAE. S. 2979–2986.
Zhang S-C, He S-H, Zhu Z-P & Li C-H. 2017. Research on soil conditioning for earth pressure balance shield tunneling in Lanzhou sandy pebble strata with rich water. *Yantu Lixue/Rock Soil Mech.* 2017; 38: 279–86. doi:10.16285/j.rsm.2017.S2.039.
GOST 10181-2000. Concrete mixtures. Methods of testing (Russian).
GOST 310.4-81 Cements. Methods of bending and compression strength determination (Russian).
Guglielmetti, V., Grasso, P., Mahtab, A. & Xu, S. 2008. Mechanized tunnelling in urban areas. Design methodology and construction control. Taylor&Francis Group.
Mori L, Alavi E, Mooney M. 2017. Apparent density evaluation methods to assess the effectiveness of soil conditioning. *Tunn. Undergr. Sp. Technol.* 2017; 67: 175–86. doi:10.1016/j.tust.2017.05.006.

Tunnels and Underground Cities: Engineering and Innovation meet Archaeology,
Architecture and Art, Volume 6: Innovation in underground engineering,
materials and equipment - Part 2 – Peila, Viggiani & Celestino (Eds)
© 2020 Taylor & Francis Group, London, ISBN 978-0-367-46871-2

Ventilation system for the excavation of 4 tunneling faces with "Plenum" metallic tank

N. Meistro, F. Poma, U. Russo, F. Ruggiero & O. Urbano
COCIV, Consorzio Collegamenti Integrati Veloci, Genova, Italy

ABSTRACT: the construction of new high-speed freight railway line between Milan and Genoa, the "Terzo Valico dei Giovi", foresees the realization of 4 adits to speed up the tunnels excavation guaranteeing accessibility and escaping routes from tunnels both during the construction phase and during the railway operation. The activities performed into the tunnels require that underground air has to be replaced by fresh air in order to allow an healthy workplace. The reduced size of the adit has forced the designers to provide an innovative automatic air distribution system that consists of a metal tank called "Plenum" fed by only 2 air ducts, from which 4 distinct ventilation stations pick up the amount of air required for each excavation face.

1 PROJECT DESCRIPTION

The railway Milan–Genoa, part of the High Speed/High Capacity Italian system (Figure 1), is one of the 30 European priority projects approved by the European Union on April 29th 2014 (No. 24 "Railway axis between Lyon/Genoa – Basel – Duisburg – Rotterdam/Antwerp) as a new European project, so-called "Bridge between two Seas" Genoa – Rotterdam. The new line will improve the connection from the port of Genoa with the hinterland of the Po Valley and northern Europe, with a significant increase in transport capacity, particularly cargo, to meet growing traffic demand.

The "Terzo Valico" project is 53 Km long and is challenging due to the presence of about 36 km of underground works in the complex chain of Appennini located between Piedmont and Liguria. In accordance with the most recent safety standards, the under-ground layout is formed by two single-track tunnels side by side with by-pass every 500 m, safer than one double-track tunnel in the remote event of an accident.

The layout crosses the provinces of Genoa and Alessandria, through the territory of 12 municipalities.

To the South, the new railway will be connected with the Genoa railway junction and the harbor basins of Voltri and the Historic Port by the "Voltri Interconnection" and the "Fegino Interconnection". To the North, in the Novi Ligure plain, the project connects ex-isting Genoa-Turin rail line (for the traffic flows in the direction of Turin and Novara – Sempione) and Tortona – Piacenza –Milan rail line (for the traffic flows in the direction of Milan- Gotthard).

The project crosses Ligure Apennines with Valico tunnel, which is 27 km long, and ex-its outside in the municipality of Arquata Scrivia continuing towards the plain of Novi Ligure under passing, with the 7 km long Serravalle Tunnel, the territory of Serravalle Scrivia (Figure 2). The underground part includes Campasso tunnel, approximately 700 m long and the two "Voltri interconnection" twin tunnels, with a total length of approximate-ly 6 km.

Valico tunnel includes four intermediate adits, both for constructive and safety reasons (Polcevera, Cravasco, Castagnola and Vallemme). After tunnel of Serravalle the main line runs outdoor in cut and cover tunnel, up to the junction to the existing line in Tortona (route to Milan); while a diverging branch line establishes the underground connection to and from Turin on the existing Genoa-Turin line.

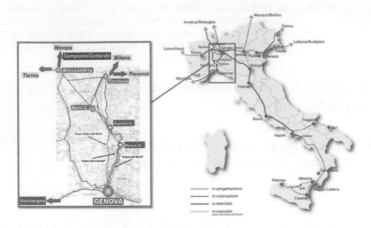

Figure 1. High-speed Italian system.

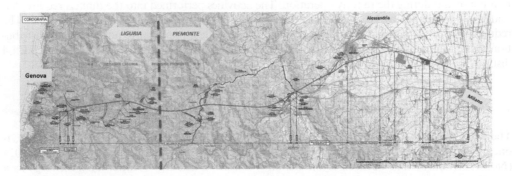

Figure 2. Terzo Valico project.

From a construction point of view, the most significant works of the Terzo Valico are represented by the following tunnels:

- Campasso tunnel 716 m in length (single-tube double tracks)
- Voltri interconnection even tunnels 2021 m in length (single-tube single track)
- Voltri interconnection odd tunnels 3926 m in length (single-tube single track)
- Valico tunnel 27250 m in length (double tube single track)
- Serravalle tunnel 7094 m in length (double tube single track)
- Adits to the Valico tunnel 7200 m in length
- Cut and cover 2684 m in length
- Novi interconnection even tunnels 1206 m in length (single-tube single track)
- Novi interconnection odd tunnels 958 m in lenght (single-tube single track)

The project standards are: maximum speed on the main line of 250 km/h, a maximum gradient 12,5 ‰, track wheelbase 4,0 – 4,5 m, 3 kV DC power supply and a Type 2 ERTMS signalling system.

2 INTRODUCTION

The design of the new high-speed railway line Milan-Genoa, known as "Terzo Valico dei Giovi", involves the construction of four intermediate access windows along the basis tunnel to reduce tunnel excavation times, in order to guarantee access and evacuation routes both during the construction phase and during railway operation.

Since the railway line consists of two parallel single-track tunnels, the access galleys terminates with a connecting chamber, from which the two tunnels branch off, for four excavation faces.

Since the railway line consists of two parallel single-track tunnels, the access tunnel terminates with a grafting chamber, from which the two tunnels branch off, for a totale of 4 excavation faces.

The activities carried out during the excavation, to maintain a healthy place of work, require an appropriate exchange of air to permit a correct dilution of the exhaust gas of the operating means, of the powders produced by the workings, of the natural gases that can be emitted from the rock mass and the maintenance of an acceptable level of temperature.

Since the different operations, carried out underground, can produce many different quantitative of pollutants, it is necessary to install independent ventilation stations, each for each excavation face, also to compensate different demands of flow and pressure for every ventilation station.

The reduced cross-section of the access tunnels does not allow the installation of four ventilation pipes, each for each tunnel, with a diameter sufficient to minimize the pressure drops occured, also given the length of the access windows that result in all cases above 1500 meters, and consequently the fan powers are kept within acceptable values.

In a first phase, our ventilation plant designers, had considered to build a false ceiling that, together with the upper part of the tunnel, composes a ventilation rigid conduit, operating in depression, connected with appropriate metal fittings to the individual ventilation stations, installed in the grafting chamber, which would fed the usual flexible ventilation tube up to the work fronts.

After new evaluations, it was considered that this kind of installation would have greatly reduced the height of the access tunnels, making the vehicles transit difficult along them, in particular for the excavation and consolidation machinery.

This problem was even more evident with regard to the Polcevera access tunnel, because it was excavated by a TBM, with an internal diameter equal to 8.65 and therefore with an extremely reduced internal working height.

After that, another design approach considered the installation of two external ventilation stations, to send up the needed air to the grafting chamber, by means of two flexible ventilation tubes of 2400 mm in diameter. Then each pipe uses a booster station for each front, and metal fittings to divide the flows, possibly integrated by regulation shutters.

However our technicians disagree with this solution because, for previous experiences, the air flows are difficult to control, even by the shutters, as the air prefers the path with less resistance.

Furthermore, to avoid a collapse of incoming duct and a short circuit of air flow in the underground (the return air flow from the work face could be picked up by the booster station), as indicated by the "*Recommandation Suisse SIA196 edition 1998*", the arrival duct and the inlet of the booster station must include a gap, which must be between 0,5 and 1,0 times of duct diameter, and the incoming air flow must be greater of $10 \div 20\%$ of the boosted air flow.

It should be noted that the amount of design air required for the excavation face is 44 m³/s, consequently the required air surplus is between 17 and 34 m³/s, needing a power evaluated in about 150: 300 kW, considering the pressure and air losses along the pipes installed into the access tunnel and without considering the losses produced by the air flows regulation shutters.

Our designers have therefore studied the installation of an innovative system, fundamentally consisting of a metal tank, slightly overpressure, from which every single ventilation unit, physically connected to the tank, could recall the amount of air needed to every front, as if it were in the open air.

The regulation of the ventilation stations, both the external and the four internal ones, carried out by means of inverters, each equipped with its own computerized controller, as well as the installation of suitable flow and pressure gauges, allowed to manage, with the use of a computerized controller, the whole ventilation system, modifying the working parameters of the two external stations in a completely automatic manner, according to the air requests of the four stations placed in the grafting chamber.

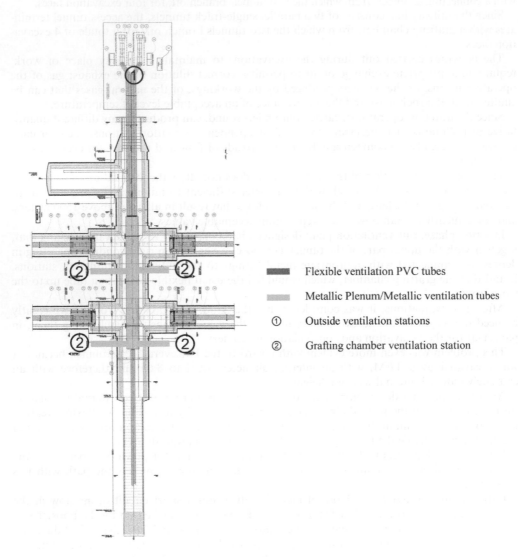

Figure 3. Grafting chamber with ventilation system layout.

Legend:
- ▬ Flexible ventilation PVC tubes
- ▬ Metallic Plenum/Metallic ventilation tubes
- ① Outside ventilation stations
- ② Grafting chamber ventilation station

3 DETAILED DESCRIPTION OF VENTILATION SYSTEM

The tunnels system leaded by the Polcevera access tunnel consists of:

- Access tunnel, 1820 m long; considered 1900 m long for installation calculation;
- Grafting chamber, 180 m long;
- Up track tunnel, to Genova, 2705 m long;
- Down track tunnel, to Genova, 1373 m long. At the end the tunnel is connected with another grafting chamber (420 m long), for a total length of 1793 m;
- Up track tunnel, to Milano, 2727 m long;
- Down tunnel, to Milano, 2727 m long.

As already described in the previous paragraph, each working face of tunnel needs 44 m³/s, which means a requirement of 50 m³/s at ventilation stations, considering the losses along the tunnel, both friction and for air leaks. Therefore, the ducts along the access tunnel delivery each 100 m³/s of fresh air to "Plenum". The design of entire ventilation system in the underground has been calculated keeping into account these values.

Using the guidelines indicated in the "Recommandation Suisse SIA196 edition 1998", we have considered to use a duct, installed in the access tunnel, with characteristics as follow:

- friction coefficient   0,015
- losses surface   5 mm²/m²

Likewise, we have considered a duct, installed in the line tunnels, as follow:

- friction coefficient   0,018
- losses surface   15 mm²/m²

Therefore, we have defined the entire ventilation system, mainly consisting of:

- 2 ventilation stations installed on the outside, each consisting of two fans mounted in series, diameter 1800 mm, power 315 kW; total installed power 1,260 kW; each engine is controlled by inverters, installed in containers;
- 2 flexible ventilation PVC tubes, diameter 2400 mm, 1900 m long each;
- 4 ventilation stations installed into the grafting chamber, consisting of 2 1600 mm diameter fans and 200 kW power each; total installed capacity in underground 1600 kW;
- 4 flexible ventilation tubes, diameter 2200 mm, maximum length 2690 m;
- Metallic Plenum. Dimensions: 34.7 m long, 6 m wide and 3 m high; internal volume of 588 m³; internal working pressure ≤ 500 Pa;

Once the concept of the plant as a whole has been defined, it has been decided to identify suppliers with a know-how capable of developing what is determined in the specifications of the project itself, in collaboration with our designers.

In particular, it was necessary to mount a metal structure, the Plenum, maintaining its geometry taking into account its weight and the operating pressures, as well as any over or under pressure deriving from occasional operating events or deriving from situations of emergency.

The ventilation system computerized control system has to be able to react within a reasonable time to the variations following the changes in flow and pressure due to the different adjustments made to the ventilation units placed at the service of the individual tunnels. Different requests of flow and pressure are due to the different phases of the tunnel realization: hammer excavation, drill and blast excavation, shortcrete, consolidation, casting of the final coating ...

The study of the control system permitted to provide the data necessary for the size of the plenum and of the accessory elements for connection with the ventilation units. The Figure 5, elaborated by the control system supplier, show the pressure distribution, positive or negative, into the plenum and the other ducts, at the maximum design performance of the system, giving all required data for Plenum design.

Once the dimensions of the Plenum were identified, to allow its transport from the production plant to the construction site and to facilitate assembly operations, it has been realized in elements, with a length of 2.3 m, width 6.4 m and height 3, 0 m, subsequently connected on site, by means of bolted flanges and suitably sealed by gaskets.

The whole structure is made up of 19 segments, of which some are special elements: the head, the tail and the elements predisposed for the separation of the connecting pipes with the individual ventilation units. All the walls constituting the tank surfaces have been stiffened by ribs to avoid the vibrations produced by possible fluctuations in the airflow and pressures inside the tank; in addition, the connection pipes with the individual stations are dynamically insulated with joints made of elastic material, also to compensate for any misalignments.

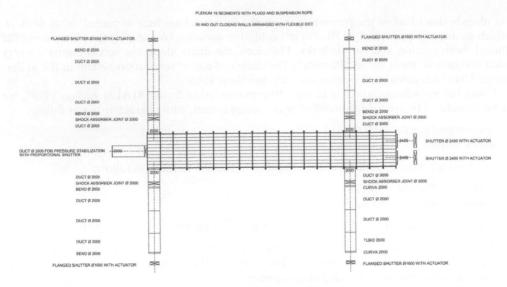

Figure 4.   Metallic structure for air flow distribution (bill of material).

The bolted flanges have been sized so that they also perform a structural function; moreover, four struts also complete the supporting frame, in order to maintain its geometry even during lifting and assembly. Finally, the side paneling is made up of elements 600 mm wide, five for each face, reinforced by diagonal elements.

Each piece of material is suspended at the vault of the grafting chamber by means of 4 wire ropes, using resin blocks; each rope is able to support, during operation, a load equal to the weight of each single piece, equal to 1500 kg. On the side walls of the elements there are provided bearing beams, positioned at a height greater than the barycentric axis to stabilize the lifting and suspension operations, to which the tie rods are connected. The tie rods are composed of two elements in wire rope and a central shock absorber, designed not to transmit any vibrations to the anchor bolts.

From an operational point of view of the Plenum, all the air inlets and outlets are equipped with a closing damper, to allow maintenance operations, even individually/singularly, on the supply pipes from the outside and from those supplying fresh air at the excavation faces, as well as in the event of the shutdown of one or more ventilation stations. On the wall opposite to the air inlet from outside, a 2,000 mm diameter damper is used, which maintains the overpressure inside the plenum, of a proportional type. This output is also used to supply fresh air to the pressurized cabin, obtained in the final part of the grafting chamber, where the electricity supply and distribution booths are located inside the basement and the control stations of the ventilation stations.

The definition of the control software of the entire ventilation system involved/entailed and important commitment for the supplier, even though our plant office request was originally quite simple: the external machines had to be automatically controlled according to the sum of request of air from the four underground stations.

The control software designer immediately highlighted two main difficulties to be solved in order to create a system that:

1. The distance between the outside stations and the plenum, as mentioned at 1900m, results in a delay in starting up, as the demand changes, due to the inertia of the air, which could cause under pressure or over pressure within the Plenum itself; therefore, in addition to installing flow and pressure sensors in strategic areas of the system, it is also necessary that the acceleration and deceleration ramps of the individual ventilation units are controlled by the control system, which continuously reads the flow and pressure parameters. When the

final adjustment of the secondary machines is reached, a closed logic cycle, precisely based on the measurements reported by the various sensors distributed along the ventilation units: the primary ones, placed to the outside, and the secondary ones installed underground.

2. The air flow rates required at the working edges are directly set on the control inverters of the secondary units. To reduce the system adjustment times and to avoid performances fluctuations, the control software, must therefore foresee the losses, under pressure and flow rate, spot for the various pipelines, over time variables as a consequence of the variations in the length of the pipes, which follow advancement, and for the wear conditions/ state of wear, function of the operating system.

Therefore, the software must keep in its memory/must remember the history of the detected parameters, in order to predict the fan rotation speeds. At the time the regime conditions are reached, the losses are calculated and recorded for future additions.

To evaluate and to solve the above points, the software designer, which is also the supplier of ventilation stations and all monitoring system (pressure and air flow), has used its own proprietary software, able to simulate the distribution of pressures inside the ducts and the Plenum, in different conditions. The Figure 5, given as an example, show the pressure distribution, simulating the situation of maximum demand for fresh air at the working faces.

Under normal/standard operating conditions, the overflow valve is adjusted in order to provide a sufficient amount of air needed to the pressurize of the cabin containing the electrical cabins/substations for power supply and distribution in the basement.

Considering that in this case the pressure drops keep/remain constant, the system controls/ checks that the pressure level inside the Plenum, in the proximity of the outlet, remains automatically constant, and that it is approximately at 200pa, a value considered suitable for the functioning of the system.

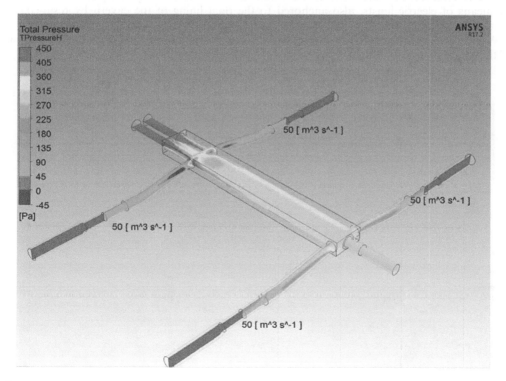

Figure 5. Pressure distribution simulation.

All the difficulties encountered and exposed by the supplier of the adjustment software and the monitoring hardware, have led to the ventilation system being used, in an initial phase, manually controlled on the basis of the flow and pressure parameters found. The automatic system was installed and activated only at a later stage, which was also activated with the installation of additional monitoring sensors at various parts of the Plenum.

Currently the system has been operating for several months automatically, with satisfaction of the personnel/workers operating in the basement.

4 ASSEMBLY SEQUENCE

As previously mentioned, the elements of the ventilation system have been designed in order to be safely assembled and at the same time to minimize the downtimes of tunnel work.

The access tunnel, at the initial stage, foreseen a traditional ventilation system which had a single ventilation station consisting of two series fans of 200 kW and 1,600 mm diameter (this equipment has been subsequently utilized in underground and integrated into the final complex plant system). The fresh air brought up to the cavern by means of a flexible ventilation PVC tube with a diameter of 2,200 mm was utilized for the excavation of both the tunnel access and the cavern.

Along the access tunnel a new flexible ventilation PVC tube with a diameter of 2,400 mm was stretched out up to the cavern; once the connection to the ventilation station succeed was possible to dismantle the existing 2,200 mm flexible PVC and switching it with another 2,400 mm PVC tube. Later on two definitive ventilation stations 315 + 315 kW and diameter 1,800 mm each were installed.

Just after the installation of all the anchor bolts, according to the scheme provided by the Plenum builder, the mounting stage of the plenum within the cavern was carried out during the holiday period (interruption of excavation activities)

The 19 steel forms, which constitute Plenum structure, were brought sequentially to height by means of electric hoists, also anchored to the final lining of the cavern by resin-coated bolts.

Working at height with the hydraulic baskets was possible to proceed with the installation of final tie rods and the bolted connection with the previous section.

Figure 6. Plenum into the chamber.

At the same time, another work team proceeded to assemble the ventilation station on a special metal platform. Also in this case, since there is not enough space to mount the ventilation unit directly in height, a particular steel structure has been designed and realized; the columns of the platform were each made up of two hinged pieces, folded under the platform itself.

In this way the ventilation station was practically mounted on the ground; at this point, with two cranes, the platform complete with a fan was raised up. The hinged legs started opening by gravity effect and once the final height was reached, the hinges were locked by means of bolted plates. Finally, the steel support of the columns were fixed to the ground and four struts were fixed, at the level of the platform, to the side walls of the tunnel.

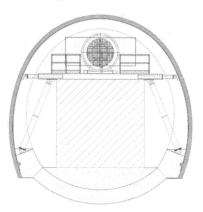

Figure 7. Supporting platform for underground ventilation station.

Subsequently, the shut-off valves and the connection pipes to the ventilation units were installed. Once all the electrical connections had been completed, the flexible PVC tubes coming from the external stations could also be connected and the system activated.

All operations were completed within the holiday period of the construction site.

5 CONSIDERATIONS AND ARRANGEMENTS FOR GAS RISK

Polcevera tunnels, according the NIR n. 28 "*Underground works. Excavation with methane presence*", emitted by the public health services of Regions of Emilia-Romagna and Tuscany regions are classified 1a on the base of the possible discoveries of methane during excavations activities. So there is the possibility to carry out eventual survey analysis, especially in fault zone in compliance with the geological profile.

This classification abovementioned means that the equipment in the tunnel must be in an explosion-proof configuration (Atex) up to a distance of 500 m from the front face during the methane presence investigation phases. Moreover, even if the probability of methane discoveries is low, in this case, the NIR 28 foresee:

- a complete separation of the electrical systems exposed to the gas flow
- to ventilate the underground atmosphere with fresh air, coming from outside, in order to dilute the gas coming from the rock mass, until it is exhausted.

In order to comply with these requirements, all the underground electrical equipment has been placed in a pressurized volume, obtained in the final part of the cavern which is isolated from the surrounding environment through a double metal wall septum. The gap between the two walls, which houses the load-bearing structure, is kept under pressure by means of the tunnel compressed air system, supplied by an external compressor station.

Figure 8.   Ventilation unit with pressurized chamber, under test.

The overpressure level is continuously monitored by specific equipment connected to the tunnel gas monitoring system.

Also the fans installed in correspondence of the cavern can be invested by the return airflow, with potential presence of gas. In order to maintain the correct airflow necessary to clean off the atmosphere in the tunnel affected by the methane discoveries, all the ventilation system, outdoor and indoor stations must remain active.

After checking the feasibility with the Gas Monitoring Manager, the ventilation units were installed inside a metal casing, constituted by a double pressurized chamber, in order to create a double safety barrier. In consideration that, the air that flows inside the fans coming from the outside, results without gas presence and, therefore, can be utilized to dilute the gas concentration of the tunnel atmosphere. See Figure 8 for reference.

The two chambers are continuously monitored by specific equipment, connected to the tunnel gas monitoring system. In case of the gas concentration reaches a dangerous level, all the electrical underground system will be stopped; consequently also the ventilation stations, located in underground area, will be shut off.

6   CONCLUSIONS

The presence of four work faces belonging to the grafting chamber, the reduced size of the access tunnel, the need to manage the ventilation system, individually for each excavation face, the probable release of gas from rocks, they forced plant designers to evaluate different solutions which are normally used.

The various system solutions were analyzed and evaluated mainly with the aim of guaranteeing the correct air supply, to maintain safe e comfortable working conditions. On the other hand, the designers took into account the energy efficiency of each solution, to minimize the running costs.

The difference in power demand between the traditional solution, with boosters, and the one adopted, calculated for the Polcevera site, quantifiable in 350 kW on average, involves a reduction in annual consumption of 3 million kWh, corresponding to an emission reduction of around 1.500 tonnes of carbon dioxide per year.

Tunnels and Underground Cities: Engineering and Innovation meet Archaeology, Architecture and Art, Volume 6: Innovation in underground engineering, materials and equipment - Part 2 – Peila, Viggiani & Celestino (Eds)
© 2020 Taylor & Francis Group, London, ISBN 978-0-367-46871-2

Big data and simulation – A new approach for real-time TBM steering

G. Meschke, B.T. Cao & S. Freitag
Institute for Structural Mechanics, Ruhr University Bochum, Germany

A. Egorov, A. Saadallah & K. Morik
Chair of Artificial Intelligence, Technical University Dortmund, Germany

ABSTRACT: A new concept, which combines machine and monitoring data analysis with numerical simulation of mechanized tunneling processes, is presented in this work. The main idea is to fuse sensors data collected from the TBM and the settlement measurements with finite element simulations capturing the physical behavior of the tunneling process. Both data and simulation based prediction models are developed to assist the TBM steering with respect to the settlement behavior in real-time. The real-time machine and monitoring data analysis offers the possibility of detecting anomalies, e.g. in the soil layers, which are not considered a priori in the simulation models. First results show that most anomalies created by concrete walls in the ground can be detected up to 1 m ahead of the cutting wheel by using the specific torque (a combination of torque and penetration rate data) together with time series preprocessing and drift detection methods.

1 INTRODUCTION

In mechanized tunneling, the prediction of tunneling induced settlements is a prerequisite for the design and the steering of the tunnel boring machine (TBM), targeted at the minimization of the risk of damage of existing surface structures in particular in case of tunneling with low overburden. In order to reduce surface settlements, the machine driver can adjust process parameters such as the face support pressure, the grouting pressure and the advance speed during the tunnel advancement (Maidl et al. 2012). Currently, decisions affecting the steering of TBMs are based upon engineering expert knowledge and monitoring data. However, using monitoring data implies that information (data) related to already passed situations is used to extrapolate on the future behavior of the soil-tunnel interactions. In this work, a new concept of exploiting both numerical simulations of mechanized tunneling processes and measurement data in the context of machine learning is presented with the aim to support the drive of the TBM. The idea is to fuse information from the sensors at the TBM and the settlement measurements with finite element simulations capturing the physical behavior of the shield tunneling processes. Additionally, the adjustment of the process parameters in further advancement steps also depends strongly on the prediction of expected geological conditions in front of the machine, which can be vastly improved by applying real-time data analyses on the available measurement data from the TBM. The benefit of the approach is the possibility to actively identify anomalies, which are not considered a priori in the simulation models.

Computational methods such as the finite element (FE) method are by now well developed and have been successfully applied for numerical simulation in mechanized tunneling (Komiya 2009, Do et al. 2014). These numerical tools are becoming an integrated part of the tunneling design process in order to predict the long-term ground. Recent research on numerical simulation methods in mechanized tunneling can be found in (Meschke 2018, Hofstetter et al. 2017).

As part of the subproject C1 of the collaborative research center SFB 837 "Interaction Modeling in Mechanized Tunneling" established at the Ruhr University Bochum, a concept for the support of the TBM steering is developed, which combines monitoring data and realistic numerical simulations, see (Meschke et al. 2013). At the heart of this novel approach is a holistic, process-oriented numerical model of the shield tunneling process, which enables the prediction of relevant steering parameters in further time steps. For real-time prediction purpose, surrogate models are employed instead of the computationally expensive finite element model to reduce the computation time (Cao et al. 2016, Freitag et al. 2018).

On the other hand, monitoring data from TBM sensors is primarily collected and used for the assessment of the tunneling process through data analyses with computational intelligence approaches, which can provide lessons, knowledge and experiences for future tunnel project applications. A fuzzy-based expert system, which uses fuzzy measured values in the form of linguistic variables in combination with fuzzy rules to propose suitable steering parameters, has been developed in (Oberste-Ufer 2010). In (Nellessen 2005), a neuro-fuzzy system has been trained with real data from a tunnel project and used for the prediction of surface settlement in further excavation steps. The data were largely processed in the form of dimensionless parameters. A similar monitoring-based procedure for settlement prediction by means of Relevance Vector Machine is shown in (Wang et al. 2016). In (Salimi et al. 2016), an Adaptive Neuro-Fuzzy Inference System (ANFIS) and Support Vector Regression (SVR) were used to analyze the TBM performance in hard rock using data from two tunnel projects in Iran. The interactions of ground conditions and machine behavior for hydroshield tunneling in soft soils are investigated based on the data from a metro project in Germany (Düllmann 2014). After post processing the data, a prototype for the predictions of forces and cutting wheel torque has been developed for future projects. In (Maher 2013), machine learning techniques have been applied to automatically identify parameters essential to the penetration rate of an Earth Pressure Balance TBM. Similar to most analyses, the data in (Maher 2013) is also averaged per ring rather considering within smaller motion comparing to the original measurement data which are acquired normally every 10 to 15 seconds.

The rest of the paper is structured as follows. Section 2 presents the numerical model used to simulate the tunneling process and surrogate models employed for real-time prediction purpose. Section 3 summarizes the available measurement data in mechanized tunneling and data processing methodologies. The concept of combining measurement data and numerical simulations is described in Section 4. Section 5 is devoted for an application example using real data from a tunnel project in Germany. In this section, the anomaly condition in the ground is actively identified in real-time based on the drift detection concept. Finally, some conclusions and future works are discussed in Section 6.

2 NUMERICAL SIMULATION IN MECHANIZED TUNNELING

The process-oriented FE simulation model *ekate*, developed at the Institute for Structural Mechanics at Ruhr University Bochum allows to simulate complex interactions between the TBM, the surrounding subsoil and the surface infrastructure for arbitrary curved tunnel alignments, see (Alsahly et al. 2016a). It is based upon the object-oriented finite element framework KRATOS and takes into consideration all relevant components of the mechanized shield tunneling (soil and groundwater conditions, tunnel lining, the TBM with shield and hydraulic jacks, tail void grouting and various types of support) and their (time dependent) mutual interactions. In this model, partially and fully saturated soils are modeled as three-phase materials, accounting for the solid, the pore water and the air as distinct phases according to the theory of mixtures. Two elastoplastic models are available for the consideration of the inelastic response of soft soils: the Drucker-Prager model and the Clay and Sand model. The TBM is represented as a deformable body moving through the soil and interacting with the ground through frictional surface-to-surface contact. This allows for more realistic deformation of the surrounding soil due to the tapered geometry of the shield. The tunnel advance is modeled with a steering algorithm, where the shield is pushed forward by hydraulic jacks thrust. The excavation and the construction processes are modeled by means of the deactivation of soil elements and installation of tunnel lining and grouting elements. Additionally, different types of heading face support can be taken into account. The simulation model is directly linked to a tunnel information model (TIM) (Alsahly et al. 2016b) (Figure 1a). The relevant information (geology, alignment, lining, material and process parameters) is automatically extracted from the TIM model and subsequently a FE simulation of the tunnel drive is executed. The FE simulation model has been validated and tested at the metro project Wehrhahn-Line in Düsseldorf (Alsahly et al. 2016a, Ninic & Meschke 2015).

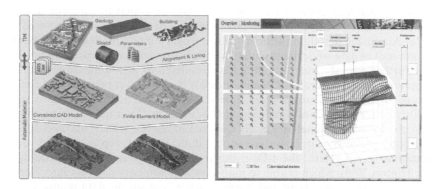

Figure 1.   a) Automated generation of simulation models using the TIM; b) Simulation and monitoring-based assistance system to support the real-time control of tunneling processes.

Mechanized tunneling simulations using the process-oriented FE model described above are time-consuming and therefore best employed in the design stage of a tunneling project. To support the steering phase during the tunnel construction in real-time, surrogate models are required. In (Ninic & Meschke 2015), an Artificial Neural Network has been adopted as a surrogate model with low dimensional outputs for deterministic real time analyses of mechanized tunneling. With the purpose to deliver real-time predictions of expected surface settlements at multiple surface locations, the benefits of Recurrent Neural Network (RNN) and Proper Orthogonal Decomposition (POD) approaches are combined within hybrid RNN-POD surrogate models for deterministic input-output mapping, see (Cao et al. 2016), and interval input-output mapping, see (Freitag et al. 2018). Based on the FE simulation model and the developed surrogate models, a simulation-and-monitoring-based assistance system, named SMART, has been implemented to support the real-time control of tunneling processes (Figure 1b). The SMART application aims to provide a very quick prediction of the settlements in the subsequent excavation stages system behavior corresponding to an arbitrary change of certain process parameters. Currently, the support pressure $^{[n]}SP$ and the grouting pressure $^{[n]}GP$ of each construction step n (tunnel ring), which constitute the main operational parameters to control the settlements caused by the tunnel drive, are considered as input parameters.

3   BIG DATA IN MECHANIZED TUNNELING

In this work the data set of a real-world tunnel project, the Wehrhahn-Linie metro project in Düsseldorf in Germany, is utilized. The tunnel was excavated mainly by a TBM with a cover depth approximately between 12 and 16 meters below the surface. Up to 352 measured values are recorded per measuring interval (15 s) for machine control which leads to 4618764 values per sensor for the whole project timeline. Such a sequence of the values for each sensor is called a *time series*. The available data can be used for different tasks, such as: exploring the sensor data and discovering correlated groups of sensors to remove the redundant information; creating a model to predict the important features in tunneling process (e.g. surface settlements, penetration rates).

As a first step, sensors with missing values, zero variance and nominal data types are excluded. As a result, 331 sensors are remaining compared with the original 352. In this work the more frequent available data is used. The raw measurements contain too much noise and hence the data is averaged in minor time steps (3 minute, 30 minutes, 1 hour etc.). While the excavation and the construction of a whole tunnel ring is in the range of hours, the standstill time for machine maintenance may be up to several days. Finally, all the time series data is normalized.

During the manual inspection of the time series data measured by multiple sensors of the TBM, groups of sensors, which measure almost identical data values, are discovered as shown in Figure 2a. Automatic detection of these groups, which contain redundant information, can reduce the amount of data while maintaining the prediction accuracy depending on the trained model. For further analysis, each time series is considered as high-dimensional example with each time stamp being a feature. A tunneling process is assumed in progress and firstly 10,000 time points of the time series are retrieved. For automatic detection of redundant sensors, the dimensionality of the time series is reduced at the first step. Then the resulting embedding is clustered in the low-dimensional space. As distance measures, which are used by clustering algorithms, converge to zero in high-dimensional data, the first step is required to allow for better clustering results. Here, the t-Distributed Stochastic Neighbor Embedding (t-SNE) (Maaten & Hinton 2008), which is able to map very high-dimensional data into 2 or 3 dimensional space while preserving original distance measure between the data points, is employed. Afterwards the clusters are validated by means of the Spearmann correlation between each point of the cluster on the original data. The points with a relative correlation value less than (1-1/cluster size) are removed in an iterative procedure. From the resulting clusters, one representative is selected and other redundant sensors are removed.

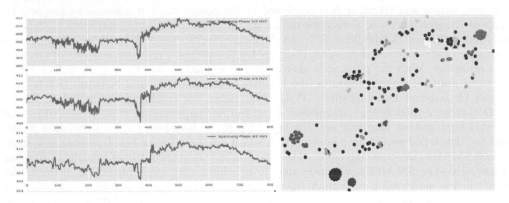

Figure 2. Clustering process for the automatic detection of redundant sensors. a) Various sensor measurements with similar time series response; b) Clustering of the sensors in 2D (blue are outliers, other colors are clusters).

The groups of clusters are discovered without any prior knowledge of its number. Furthermore, the detection of outliers, namely data points with no cluster membership, is of high interest. As in case of TBM, sensors that measure time series data different to other sensors need to be considered on their own. Hence, in this work we consider density-based clustering approaches such as HDBSCAN (Campello et al. 2013). Figure 2b presents the validated result of clustering applied to low-dimensional embedding retrieved with t-SNE. To the end, this reduced data set of 99 time series will be used for further analysis. In case of sensor failure, one could use another redundant one from the same cluster. Furthermore, the clusters found during this exploration step can be used to extend domain expert knowledge and the whole procedure can be applied to various tunneling projects with different sensors.

The 99 time history series can be considered as inputs to predict the future responses such as penetration rates and settlements. In order to verify the prediction quality of the reduced data set, the performance of the penetration rate prediction was considered using both full and reduced data sets. Training a model with raw data to predict the penetration rate for the next 30 minutes interval, SVR with Radial Basis Function kernel showed minor improvement of the Root Mean Squared Error (RMSE) value from 4.95 to 4.44 using the reduced set of features while the predicted value for the penetration rate is from 0 to 57 [mm/rotation]. Depending on the model choice and the number of further steps for the prediction, although the result changed slightly, it stayed surprisingly stable when considering that the selection of the sensors was performed without taking a specific learning task into account and removing ⅔ of the sensors.

4 DATA AND SIMULATION COMBINATION APPROACH

The main objective of the approach is the fusion of data from available measurements and from simulation-based prediction models to support the TBM drive. The approach makes an attempt to improve the simulation model by extracted data from measurement. Vice versa, the simulation results are used to enrich the learning model based on measurement data. The application of machine learning approaches shown in Section 3 can be connected to the numerical model described in Section 2. In the following, several ways are presented to combine machine and simulation data for better models.

Within the hybrid surrogate model RNN-GPOD to predict the complete surface settlement field, the input parameters of the numerical surrogate model are averaged time series values over the measurements which are collected from the TBM during the tunneling process. While measurements data from various sensors installed in the TBM are produced every 10 to 15 seconds, data from monitoring of surface settlements are collected at larger time intervals, e.g. one measurement per day. All process information is employed in machine learning approaches to reveal more hidden information or significant relationships affecting the soil-structure interactions. To this end, a surrogate model based on TBM real measurement data is developed to deliver the settlement prediction at the monitoring points. The predictions from the numerical model and from machine learning are fused to establish a hybrid prediction model for the expected settlements at the monitoring points. In case of limited real surface measurements, more training examples with surface settlements obtaining from numerical simulations can be added to train the data-based model. To predict the complete surface settlement field, the generated predictions are further processed using the GPOD model. The combination concept is visualized in Figure 3.

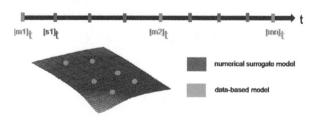

Figure 3. Combination of data-based and numerical surrogate models to predict surface settlement field.

Another possible combination of machine learning (ML) and simulation data is to process the measured sensor data for the training of a model to predict the penetration rate, which can be considered as another input in the numerical surrogate model for a more precise surface settlement prediction.

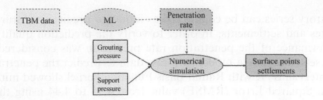

Figure 4. TBM machine data can be used to predict penetration rate and feed this output to the numerical simulation model.

The measurement data from the TBM can also be used to improve the simulation by identifying wrong assumptions in the design phase of the tunneling project. The geology conditions of a tunnel project are often estimated based on a limited number of boreholes. Within the longitudinal profile of the tunnel project, the estimated locations of soil layer change or the locations of unexpected objects are essential for the process parameters adjustment. Therefore, earlier detection and identification of anomaly conditions can help to improve the simulation and adapting the prediction models that may result in a safer, more efficient construction process.

5 ANOMALY DECTECTION USING MEASUREMENT

The concept of using measurement data to detect the anomaly condition in subsurface has been tested on the data set of a real-world tunnel project, the Wehrhahn-Linie metro project in Düsseldorf in Germany. The metro line has overall 6 new stations in which the station facilities were constructed by the cut-and-cover method. The construction of a station is supported by employing concrete diaphragm walls within the area of the station. During the excavation process, the TBM advances through these concrete walls. Hence, Page-Hinkley Test (PHT) (Page 1954) is applied to detect concept drifts when the TBM is approaching those diaphragm walls as a testing case. The positions, where an alarm is triggered, are then compared with the actual location of the walls. Two focused aspects are the correctness of the alarms and the distance from triggered points to the location of the walls.

Objects in the ground or change of soil layers produce changes in the underlying distribution of the measured values. In machine learning, this is called a concept drift and it can be detected using the concept drift detection (CDD) approaches, which will trigger a warning in case the online acquired measurement data exceeds the tolerance setting by history data trend (Gamma, et al. 2014). There is a range of various CDD methods. In this work PHT is applied to detect changes in the data distribution. A concept drift is detected if $m_T > \lambda$ where λ is given by the user and with

$$m_T = \alpha \cdot m_{T-1} + (x_T - \bar{x}_T - \delta), \tag{1}$$

where

$$\bar{x}_T = \frac{1}{T} \sum_{t=1}^{T} x_t, \tag{2}$$

and T is the current data point index. In other words, PHT considers the distance of the current value to the previous mean of the time series. With δ it is possible to allow some slight deviations from the mean. After a detected concept drift, the collected statistics are reset. Düllmann (2014) showed that torque data gives some hints on soil anomalies. Although the change is detectable, there is a lot of noise in the values and some outliers that make finding a threshold value rather difficult. However, Düllmann (2014) suggests using a combination of two sensors as followed

$$Specific\ Torque = \frac{Raw\ torque}{Raw\ penetration\ rate} \tag{3}$$

In this work, the idea of anomaly detection is extended by applying PHT in real-time and without the usage of manually defined threshold after considering all the data. Figure 5 shows a comparison of the two suggested features for anomaly detection while the data is averaged over 3 minutes and smoothed with median filter with the moving TBM. Firstly, PHT uses the raw torque data as shown in Figure 5a as objective parameter to predict the presence of the wall. The detection is not very accurate because of a large number of noisy signals. The number of false alarms is rather high. The analysis is re-performed with specific torque data. The results show that due to less noise and few outliers the model can accurately detect the presence of the concrete wall in real-time in the ground domain as shown in Figure 5b.

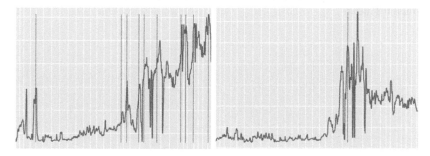

Figure 5. Comparison of data representations averaged over 3 minutes and smoothed with median filter.a) Alarms (red lines) using raw torque data. b) Alarms (red line) using specific torque data.

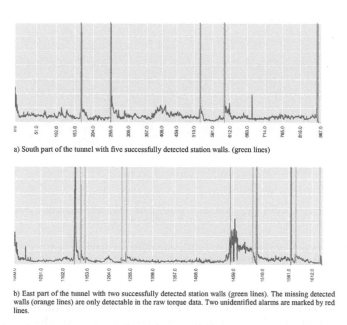

a) South part of the tunnel with five successfully detected station walls. (green lines)

b) East part of the tunnel with two successfully detected station walls (green lines). The missing detected walls (orange lines) are only detectable in the raw torque data. Two unidentified alarms are marked by red lines.

Figure 6. Anomalies detected using an online concept drift detection method. a) South part of the tunnel with five successfully detected station walls. (green lines). b) East part of the tunnel with two successfully detected station walls (green lines). The missing detected walls (orange lines) are only detectable in the raw torque data. Two unidentified alarms are marked by red lines.

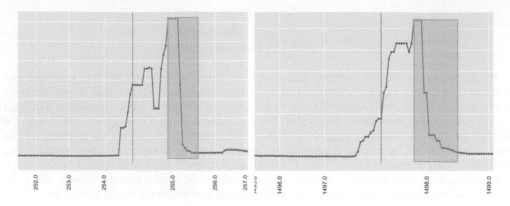

Figure 7. Detection (red line) of the concrete walls in front of the TBM. Each wall (green block) has a width of around 0.8 meter.

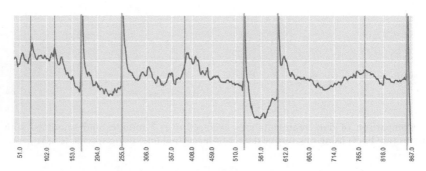

a) Contact force cutting wheel after exponential smoothing and normalization allows for detecting concrete walls (green lines) next to some further unidentified anomalies.

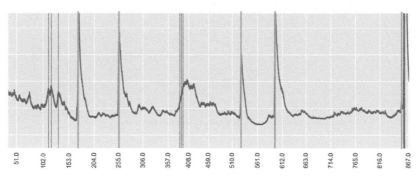

b) Power consumption of the cutting wheel after exponential smoothing and normalization allows for detecting concrete walls (green lines) next to some further unidentified anomalies.

Figure 8. Possible detection of the concrete walls using different machine sensors. a) Contact force cutting wheel; b) Power consumption of the cutting wheel.

The above procedure was then applied throughout the whole tunneling project (south and east part) and the result is presented in Figure 6. The combination of PHT and specific torque detected all concrete walls in the south part and two out of nine in the east part with a distance up to 1 meter before the cutting wheel of the TBM approaches the walls. In summary, only 13 alarms were fired and all of them were around the concrete walls. The number of false alarms was reduced compared to raw torque data where 49 alarms were fired with many of them in between the stations with no known anomalies in the ground. It was possible to reduce the number of alarms on raw torque data to 20 with the usage of exponential smoothing filter to reduce the high noise level. The combination of appropriate preprocessing and concept drift detection algorithm allows for the accurate detection of anomalies in real-time.

As shown in Figure 7 with the more high-frequent data it is possible to detect anomalies in front of the TBM a certain distance up to 1 meter with the assumption that the peak points in data series representing the moment, when the cutting wheel touches the concrete wall. This assumption relies on the fact that the torque is increased significantly and the penetration rate is reduced when the cutting wheel touches the concrete wall, which results in a jump of the specific torque according to Equation 3.

Although the specific torque feature achieved the highest accuracy so far, the discovery of further similar features is of great interest for the tunneling as e.g. some anomalies can be seen only in the raw torque data and hidden in the penetration and specific torque data. Figure 8 shows the possible detection of the concrete walls when using the data from several other machine sensors, e.g. the contact force and the power consumption of the cutting wheel. Some further unidentified anomalies, e.g. the one around ring number 400 in Figure 8, are more visible by evaluating these sensors compared to the specific torque, see Figure 6a.

6   CONCLUSION

In this work, several concepts based on the state-of-the-art machine learning techniques were introduced for mechanized tunneling in order to improve the predictions of the tunneling induced settlements. Clustering the time series sensor data in the low-dimensional embedding allowed for discovery of redundant sensors, which can be removed without a significant loss in prediction quality. Based on this result, more intensive construction of time series features for penetration rate prediction, which will be used as an input for an improved numerical simulation model, can be performed faster.

Furthermore, a method based on measurement data from the TBM to detect geological anomaly conditions ahead of the machine is presented. The method enables to create a real-time warning system. A drift detection algorithm is adopted to process the machine data. The application example on a real-world tunnel project in Germany shows the promising applicability of the proposed approach. Concrete walls existing in the ground of the south part of the tunnel project can be detected correctly from the data behavior of the specific torque parameter. Although the detection using this parameter in the east part of the project is not completely successful, information from other sensors reveal the existence of these concrete walls. This motivates for a further analysis combining all available data sensors in the future. In addition, the approach is planned to be combined with simulation-based approaches for a more accurate detection of various anomalies in the subsoil.

ACKNOWLEDGEMENTS

The authors gratefully acknowledge the financial support of the Mercator Research Center Ruhr (MERCUR) under the project "Fusion of Machine Learning and Numerical Simulation for Real-Time Steering in Mechanized Tunneling" and the finical support of the German Research Foundation (DFG) in the framework of project C1 within the Collaborative Research Center SFB 837 "Interaction Modeling in Mechanized Tunneling". A special thanks to the Capital City of Düsseldorf, Germany (Traffic Department) for their support of the research project by providing essential project data.

REFERENCES

Alsahly, A & Stascheit, J. & Meschke, G. 2016a. Advanced finite element modeling of excavation and advancement processes in mechanized tunneling. *Advances in Engineering Software* 100: 198–214.

Alsahly, A & Gall, V.E. & Marwan, A. & Vonthron, A. & Ninic, J. & König, M. & Meschke, G. 2016b. From Building Information Modeling to real-time simulation in mechanized tunneling: An integrated approach applied to the Wehrhahn-line Düsseldorf. *World Tunnel Congress ITA-AITES WTC 2016*.

Campello, R. J. & Moulavi, D. & Sander, J. 2013. Density-based clustering based on hierarchical density estimates. *In Pacific-Asia conference on knowledge discovery and data mining*. Springer, Berlin, Heidelberg.

Cao, B.T. & Freitag, S. & Meschke, G. 2016. A hybrid RNN-GPOD surrogate model for real-time prediction in mechanised tunnelling. *Advances Modeling in Engineering Sciences* Paper 3: 1–22.

Do, N.A. & Dias, D. & Oreste, P. & Djeran-Maigre, I. 2014. Three-dimensional numerical simulation for mechanized tunneling in soft ground: the influence of the joint pattern. *Acta Geotechnica* 9: 673–694.

Düllmann, J. 2014. Ingenieurgeologische Untersuchungen zur Optimierung von Leistungs-und Verschleissprognosen bei Hydroshildvortrieben im Lockergestein. Dissertation, Ruhr Universtiy Bochum.

Freitag, S. & Cao, B.T. & Ninic, J. & Meschke, G. 2018. Recurrent Neural Network and Proper Orthogonal Decomposition for real-time prediction in mechanised tunnelling with interval data. *Computers & Structures*, in press.

Gamma, J. & Zliobaite, I. & Bifet, A. & Pechenizkiy, M & Bouchachia, A. 2014. A survey on concept drift adaptation. *ACM Computing Surveys (CSUR)* 46 (4):Article 44.

Hofstetter, G. & Bergmeister, K. & Eberhardsteiner, J. & Meschke, G. & Schweiger, H.F. 2017. *Proceedings of the IV International Conference on Computational Methods in Tunnelling and Subsurface Engineering (EURO:TUN 2017)*. Innsbruck, Austria: Studia Universitätverlag Innsbruck.

Komiya, K. 2009. FE modelling of excavation and operation of a shield tunnelling machine. *Geomechanics and Tunneling* 2: 199–208.

Maaten, L.V.D. & Hinton, G. 2008. Visualizing data using t-SNE. *Journal of machine learning research* 9: 2579–2605.

Maher, J. 2013. A Machine learning approach to predicting and maximizing penetration rates in Earth Pressure Balance Tunnel Boring Machines, in *Special Interest Group in Computer Science Education (SIGCSE)*.

Maidl, B. & Herrenknecht, M. & Maidl, U. & Wehrmeyer, G. 2012. *Mechanised Shield Tunnelling*. Ernst & Sohn: Berlin.

Meschke, G & Freitag, S. & Ninic, J. & Cao, B.T. 2013. Simulations- und monitoring-basierte Prozesssteuerung im maschinellen Tunnelbau. In *Ingenieurwissen und Vorschriftenwerk, 17. Dresdner Baustatik-Seminar*.

Meschke, G. 2018. From advance exploration to real time steering of TBMs: A review on pertinent research in the Collaborative Research Center "Interaction Modeling in Mechanized Tunneling". *Underground Space* 3(1): 1–20.

Nellessen, P. 2005. Vortriebssynchrone Prognose der Setzungen bei Flüssigkeitsschildvortrieben auf Basis der Auswertung der Betriebsdaten mit Hilfe eines Neuro-Fuzzy-Systems. Dissertation, Ruhr University Bochum.

Ninic, J. & Meschke, G. 2015. Model update and real-time steering of tunnel boring machines using simulation-based meta models. *Tunnelling and Underground Space Technology* 45: 138–152.

Oberste-Ufer, K. 2010. Steuerung von Tunnelvortriebsmaschinen durch Einsatz eines fuzzy-basierten Expertensystems. Dissertation, Ruhr Universtiy Bochum.

Page, E. S. 1954. Continuous inspection schemes. *Biometrika* 41(1/2): 100–115.

Salimi, A. & Rostami, J. & Moorman, C. & Delisio, A. 2016. Application of non-linear regression analysis and artificial intelligence algorithms for performance prediction of hard rock TBMs. *Tunnelling and Underground Space Technology* 58: 236–246.

Wang, F. & Lu, H. & Gou, B. & Han, X. & Zhan, Q. & Qin, Y. 2016. Modeling of shield-ground interaction using an adaptive relevance vector machine. *Applied Mathematical Modeling* 40: 5171–5182.

*Tunnels and Underground Cities: Engineering and Innovation meet Archaeology,
Architecture and Art, Volume 6: Innovation in underground engineering,
materials and equipment - Part 2 – Peila, Viggiani & Celestino (Eds)
© 2020 Taylor & Francis Group, London, ISBN 978-0-367-46871-2*

Feasibility study of EPB shield automation using deep learning

S. Mokhtari
Research Assistant Professor, Colorado School of Mines, Golden, Colorado

M.A. Mooney
Professor, Colorado School of Mines, Golden, Colorado

ABSTRACT: Data-driven modelling is considered as a promising approach to better understand the operation of EPB shield machines (EPBM) for ultimate use in automation. Recent
advances in deep learning has made it possible to detect the faintest patterns of latent relations
in EPBM operation data. This paper aims to apply deep learning for an automated recommendation of EPBM controllable parameters. Stacked autoencoders are employed for dimensionality reduction which can, automatically, extract the most effective information measures
(features) from EPBM operational data. Extracted features, at any given time are used to predict the controllable parameters for the next time step. EPBM data from the University Link
U230 project in Seattle is employed to validate the proposed framework. The results indicate
that stacked autoencoder can effectively extract features that represent the entire dataset and
the predictive data-driven framework can successfully detect the trends in EPBM controllable
parameters.

1 INTRODUCTION

1.1 *Background*

Earth pressure balance (EPB) shield machines (EPBM) collect detailed operational and performance data during excavation, including alignment information, time history of components' performance or in-service capacities (e.g., thrust jack hydraulic pressures, cutterhead
torque, etc.), and condition sensor data. Recorded EPBM data can be viewed as the controllable parameters, responses or reactions and status information. Controllable parameters such
as cutterhead revolution speed, and individual thrust jacks are controlled by the operator
during excavation. EPBM responses or reactions are the recordings such as cutterhead torque,
screw conveyor torque, advance rate and chamber and component temperatures. Other status
information such as rolling, pitching, deviations, and depth are the parameters that are also
affected by operator's decision and are project dependent. During excavation, the operator
considers the EPBM reactions and status information as well as the previously selected controllable factors to determine new EPBM parameters. Correspondingly, recorded data should
contain implicit information about the logic of EPBM operation. However, complex interactions between numerous components of an EPBM makes it difficult to find the logic and
patterns in EPBM recordings.

  Machine learning based data-driven modeling is a promising approach that can be used to
detect signatures of different behaviors and patterns of input-output relationships, including for
EPBM excavation data. To construct a data-driven model, a set of descriptors (features) is
required to represent the data for the machine learning algorithm. EPBM recorded data can be
used as features for data-driven modeling, however, inclusion of all EPBM recordings is computationally expensive as there are often thousands of sensors. The presence of irrelevant and/or
redundant features can also reduce the prediction performance of models. A common practice

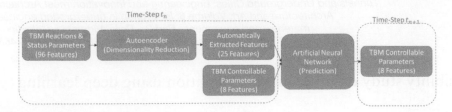

Figure 1. Overall procedure of the proposed method.

in many similar pattern recognition problems is to use intuition or statistical feature evaluation techniques to select a subset of the available features. Such methods are usually subjective and cannot ensure optimal feature set. Furthermore, feature selection techniques discard the suboptimal features and, therefore, ignore some information content of the dataset.

In this paper, we aim to emulate the EPBM operator by predicting future controllable parameters based on the current EPBM reactions and status information. For this purpose, automated feature extraction techniques can be employed to provide a better representation of EPBM reactions and status data. Deep neural network structures such as stacked autoencoders can be employed for automated feature extraction. Autoencoders try to reproduce the inputs using less neurons in middle layers. In this way, deep autoencoders automatically perform dimensionality reduction and provide the most effective features that can represent the dataset, without discarding information.

Automatically extracted features along with the controllable parameters at any time-step can then be used to predict controllable parameters for the next time step. The overall procedure of the study is depicted in Figure 1.

The premise of this study is developed and evaluated using real-world tunneling data collected during the University Link U230 tunnel project carried out in Seattle. A brief description of the EPBM components and project characteristics along with data preparation are presented in Section 2. The background of feature selection and the application of deep autoencoders for automated feature extraction are presented in Section 3. The configuration of deep autoencoders, parameter calibration and implementation of the automated feature extraction framework for EPBM recorded data is also presented in Section 3. Finally, the EPBM parameter prediction results using the extracted features are discussed and concluded in Section 4.

2 PROJECT DATA AND EPBM OPERATION

The data used in this paper was collected during the University Link (U230) tunnel project in the city of Seattle. The project includes 1.2 km (3800 ft) of twin tunnel construction with excavated diameter of 6.44 m. The tunnel was constructed in glacial and non-glacial sediments (Northlink Tunnel Partners, 2009) and the ground cover varies between 13 to 130 feet. Before tunneling, a geological survey was performed and at least three types of deposits were observed: clays, silts, and sand and tills. The geological map of the project is presented in Figure 2. In this figure, tunneling direction is from right to left.

2.1 EPB EPBM Background

A Hitachi Zosen (Hitz) EPBM was used to excavate the northbound tunnel in 2011 and the southbound in 2012. Both excavations began from the Capital Hill station. The Hitz EPBM included forward and rear bodies, connected with passive articulation jacks (see Figure 3). Sixteen thrust jacks in four groups advance the EPBM. The front end of the thrust jacks was pin-connected to the front shield. Each thrust jack could exert a 2500kN force, providing a total thrust capacity of 40MN. Eight variable frequency electric drive (VFD) motors employed in parallel (one lead, and 7 to follow) to rotate the cutterhead, providing a torque

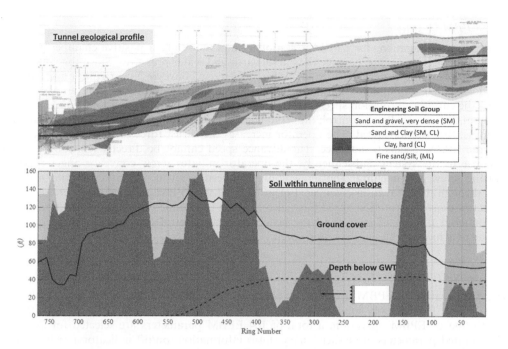

Figure 2. Tunnel geological profile (top, from Irish, 2009) and geology within the tunneling profile (Mooney et al., 2018).

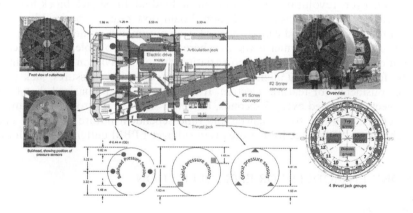

Figure 3. The general arrangement of Hitz EPB EPBM and its instrumentation, modified from (Mori et al., 2016).

capacity of 3.107MN-m at maximum rotation speed of 2.2 RPM. The cutterhead was equipped with knife edge bits and scrapers and had the opening ratio of 45%. The EPBM was instrumented with six chamber pressure sensors mounted to the bulkhead at three different elevations (Figure 3).

The excavate material was extracted from the excavation chamber onto a belt conveyor using a two-piece 800 mm diameter ribbon screw auger. The first screw conveyor was 11.9m and the second one was 16.9m in length, inclined at 18° and 3° from horizontal, respectively, and were driven by 80kN-m capacity torque motors capable of a peak rotation speed of 18.3 rpm.

2.2 EPBM Operation

During ring installation, the EPBM is stationary and maintains a positive chamber pressure and thrust force to counterbalance the lateral earth and water pressures. Since the cutterhead is not rotating, the cutterhead torque is zero. The screw conveyor auger rotation speed and torque are also zero and the screw conveyor gates are closed. When starting an excavation cycle, the cutterhead rotation speed is increased to a desired value (around of 2 rpm) and kept constant through feedback control; the servo-control system will adjust the torque to maintain or adjust the rotation speed. The operator increases the thrust forces via hydraulic pressure, and the EPBM begins to advance. The advance speed cannot be directly controlled; it is a response to the system. Subsequently, the operator initiates screw conveyor auger rotation and the screw gate opening. The screw conveyor speed is set to maintain constant material flow through the excavation chamber. For instance, if the EPBM advance rate increases, then the screw conveyor rotation speed must also increase to maintain constant excavation chamber material flow. Similar to the cutterhead, the servo-control motors deliver the required torque to achieve the desired rotation speed.

2.3 Data Preparation

The investigation of EPBM operation in this analysis includes the excavation data only; the data from EPBM steady-state (no excavation) is not considered. In accordance with the description of EPBM operation, a total of 104 EPBM parameters are selected for analysis. The omitted parameters are either binary status information (on/off indicators) or have constant values. Constant parameters (zero standard deviation) do not change during excavation, therefore, cannot provide further information about fluctuations of other parameters during excavation. Eight controllable parameters including: foam flow rate, cutterhead revolution speed, screw conveyors revolution speeds and four individual thrust jack block hydraulic pressures are selected as target values for prediction. Foam flow rate can be used if the EPBM is using manual foam injection and cannot be used if on automated mode where foam injection ratio (FIR) is constant. The former is the case here. Other controllable factors such as water or additive injection ratios had little or no variation during excavation and are omitted. The remaining 96 features, including alignment information, chamber pressures, jack strokes and so forth, are considered for feature selection. It should also be emphasized that EPBM data for U230 project is collected every 5 seconds, however, selected controllable parameters may not take effect instantly due to the inertia and time-dependent nature of EPBM operations. To address this issue, EPBM recordings are averaged over EPBM cutterhead revolutions.

3 DATA-DRIVEN MODELS FOR EPBM OPERATION

3.1 Automated Feature Extraction

A data-driven EPBM operation model should be able to consider the current state of excavation and predict future controllable parameters. Current state of excavation can be described in terms of EPBM reactions and in-service capacities as well as the project-dependent parameters at the location of excavation, such as depth, pore water pressure and alignment information. In order to consider the current state of excavation in a data-driven model, a set of descriptors (features) should represent the relevant information as input for the machine learning technique. EPBM recorded data can be used for this purpose, however, inclusion of all EPBM recordings is computationally expensive. Also, the presence of irrelevant and/or redundant features can reduce the prediction performance of models. The common practice in many pattern recognition studies is to select a subset of features that provides the most accurate representation of the entire recorded data. However, feature selection is usually intuitive or based on simple statistical feature evaluation techniques that makes the results unreliable and less accurate. Another drawback of feature subset selection is that suboptimal features are discarded from analysis and, consequently, some information is ignored or lost.

The recent advances in deep learning techniques has made it possible to detect the faintest signatures of different patterns in collected datasets. Deep architectures have provided state-of-the-art results in different machine learning problems over the past few years (Baldi 2012). Such methods can be employed to detect complicated input-output relationships, automatically, without human intervention and intuition (unsupervised). Therefore, we employed an automated feature extraction framework using a deep neural network structure, namely deep autoencoders. Autoencoders can be used to perform dimensionality reduction by training a multilayer neural network that has a small central layer to reconstruct the input vectors (Hinton and Salakhitdinov, 2006). A schematic autoencoder structure is depicted in Figure 4. As presented, the central layers of the artificial neural network (ANN) have fewer number of neurons compared to the input layer. Moreover, the output vector of the ANN is the same as the input vector. In such structure the ANN learns to reconstruct the input using fewer nodes. The values of the middle nodes can be considered automatically extracted features. The first half of the ANN that transforms and reduces the dimensions of the input data is called the encoder and the second part that recovers the data from the code is called the decoder. The fundamentals of autoencoders are very similar to those of ANNs and details of their configuration and training can be found in Hinton and Salakhitdinov (2006).

In our application, the autoencoder network can perform automated feature extraction to yield the best representation of EPBM parameters. For this purpose, EPBM reactions (features that represent EPBM response to selected controllable parameters) and general status features (project-dependent) are used as inputs and outputs of an autoencoder network. A total of 96 features were available for feature extraction. The network is designed to have 2 hidden layers in the encoder and decoder. These layers have 75 and 55 hidden nodes, respectively, to ensure gradual transition. The middle layer that will determine the number of extracted features has 25 hidden nodes. 70% of the entire data was assigned for training and the remaining 30% was used for validation to ensure sufficient training and to measure over-fitting. The autoencoder network was trained using backpropagation and gradient descent until no significant improvement was observed in prediction accuracy of the validation dataset.

The automatically extracted features (25 features) can reconstruct the inputs through the decoder network. To check the quality of the trained network for dimensionality reduction, the coefficients of determination (R^2) between the original inputs (all 96 EPBM recordings) and the reconstructed features are calculated. The resulting R^2 values are presented in Figure 5.

The R^2 values that convey the goodness of fit for the original and reconstructed features is relatively high in most cases. Therefore, the automatically extracted features can represent the original information, accurately. Such dimensionality reduction is essential for data-driven modeling because the inclusion of all EPBM data is practically impossible and the presence of redundant and irrelevant data can reduce the predictive performance. Other features that could not be reconstructed accurately were studied further. During data preparation, non-

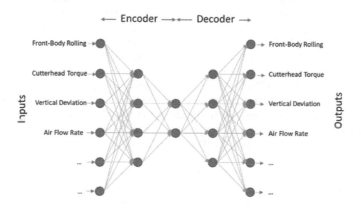

Figure 4. Schematic representation of autoencoder networks for feature extraction.

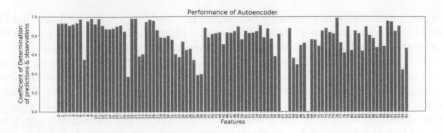

Figure 5. Coefficient of determination between measured and predicted EPBM parameters.

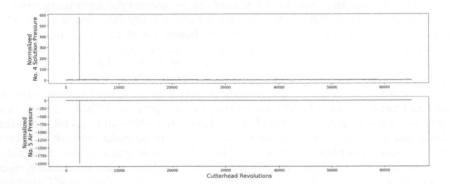

Figure 6. Outlier data in EPBM features.

varying parameters were identified by their zero standard deviation and excluded from analysis. However, standard deviation was not a sufficient indicator of non-varying data. Plots of two features (feature number 61 and 68 in Figure 5) that were not reconstructed accurately (have zero R^2 values) are presented in Figure 6. The presence of outlier data caused some non-varying features to pass through the filter.

3.2 Predicting EPBM Controllable Parameters

We sought to predict future EPBM controllable parameters using the current state of excavation. The current state of excavation is partially defined by the EPBM responses as well as project-dependent parameters. However, current EPBM controllable parameter values are another set of important features that define the current state of excavation. After all, the EPBM operator will consider current controllable parameter values to set future values. Moreover, controllable parameter values cannot change drastically from one time-step to the next; changes must be gradual. Therefore, it can be inferred that there is a time-dependence in the time history of EPBM controllable parameters. Considering the discussion above, we designed a data driven model that accepts the current state of excavation in terms of current values of control parameters, EPBM responses and other project dependent parameters and predicts the controllable parameters for the next time step. A schematic of the proposed data-driven model is presented in Figure 7. It should be emphasized that due to the inertia and complicated interaction of numerous components of EPBMs, selected response parameters may not take effect instantly. Further, EPBM recordings are averaged over cutterhead revolutions, e.g., typically every 30 sec.

As presented in Figure 7, the automatically extracted features (explained in Section 3) along with the controllable parameters of the current time step (current cutterhead revolution) are used as input for an ANN to predict controllable parameters of the next time-step. For this purpose, a feedforward neural network with 2 hidden layers was considered. The first and

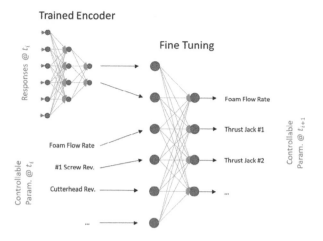

Figure 7. Overall design of the data-driven model for EPBM operation.

second hidden layers have 20 and 13 hidden nodes, respectively. The output layer has 8 nodes to predict the cutterhead revolution speed, including foam flow rate, screw conveyors rotation speed, and all jack block hydraulic pressures for next time step. Similar to the procedure for autoencoder training, 70% of the data was assigned for training and the remaining 30% was used for validating the model. The ANN was trained using gradient descent until no further improvement was observed in prediction accuracy of the validation dataset.

Measured EPBM parameters and the corresponding predicted values for the first 5000 cutterhead revolutions are plotted in Figure 8. In this figure, model predictions are plotted in orange while the measurements are presented in blue. It should be mentioned that all parameters (vertical axis in Figure 8) are normalized using Equation 1 to facilitate the training of the network.

$$X_i' = \frac{X_i - mean(X_i)}{std(X_i)} \tag{1}$$

where, X_i' is the standardized i^{th} feature, X_i is the ith feature and $mean()$ and $std()$ are the average and standard deviation operators, respectively. Standardizing is essential since features have different units and magnitudes and will implicitly affect the weights and importance of the features during training.

From Figure 8, the data-driven model can correctly predict the trends of EPBM controllable parameters. The discrepancies between measured and predicted values are mostly in extreme values. The trends of hydraulic pressure in jack blocks are mostly identified by the model, however, the model predictions seem to overestimate the lower pressures. Due to the presence of outliers in measured features a comprehensive data preparation and outlier detection is recommended for future studies.

4 DISCUSSION AND CONCLUSIONS

A data-driven framework of EPBM operation is proposed in this paper. The excavation data collected by EPBM during operation is used to quantify and express the current status of excavation. For this purpose, 96 features describing the EPBM status (torques, flow rate, etc.) and the excavation condition (depth, water pressure, alignments, and etc.) was used as input to an autoencoder neural network. The autoencoder performs dimensionality reduction on input data and extracted 15 new features that best represent EPBM behavior at each timestep.

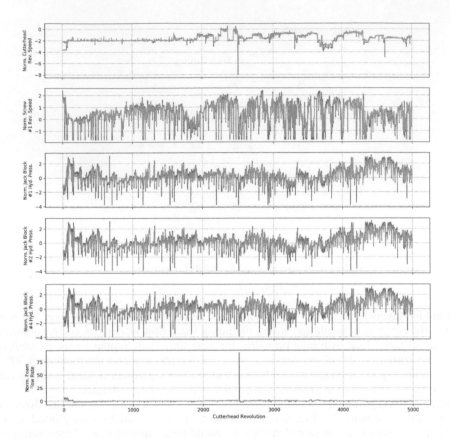

Figure 8.   Measured and predicted EPBM controllable parameters.

The results of the stacked autoencoder shows that such deep learning structures can be successful in reducing dimensions and extracting features that can effectively represent the entire dataset for data-driven modelling. Automatic feature extraction results are not sensitive to multicollinearity in dataset and the redundant features cannot affect the extracted features. The results of autoencoders also revealed that EPBM recoding include significant measurement errors and outlier data. therefore, comprehensive data preparation and outlier detection studies is recommended for future studies. Extracting better feature (higher accuracy in reconstructing original features) may be possible using deeper networks. However, deeper autoencoders might encounter the vanishing gradient problem and therefore need to be trained with a greedy layer-wise strategy. Layer-wise training and deeper networks were avoided in this study to keep the framework simple.

The prediction results indicate that the proposed framework can successfully detect the trends in controllable parameters. The foam flow rate and cutterhead revolution data seems to have significant outlier data. The predicted cutterhead revolution speed show more fluctuations compared the measures data. Based on EPBM operation (Section 2), such behavior was expected because the cutterhead revolution speed is kept relatively constant during excavation. The thrust jack block hydraulic pressures are overestimated for lower pressure values during start up and ramp down. We deem that prediction during these areas of operation is of less importance for optimizing operations.

The results also indicate a time dependent correlation between EPBM parameters. The current model only considers the relationship between adjacent cutterhead revolutions. Deep learning structures such as Long-Short Term Memory (LSTM) networks can model and consider arbitrary time intervals and may provide further insight for automated EPBM operation.

Finally, it is worth noting that this technique was used to predict operator performance and not necessarily optimal performance.

REFERENCES

Baldi, P. 2012. Autoencoders, unsupervised learning, and deep architectures. In Proceedings of *ICML workshop on unsupervised and transfer learning*: 37–49.

Hinton, G.E., & Salakhutdinov, R.R. 2006. Reducing the dimensionality of data with neural networks. *science* 313(5786): 504–507.

Irish, R.J. 2009. *Prebid Engineering Geologic Evaluation of Subsurface Conditions for the University Link Light Rail Tunnels, Capitol Hill Station to Pine Street Stub Tunnel, And Capitol Hill Station Excavation and Support Contract U230, Seattle, Washington.* Denver, CO, R J Irish Consulting Engineering Geologist Inc.

Mooney, M., Yu, H., Mokhtari, S., Zhang, X., Zhou, X., Alavi, E., Smiley, L., & Hodder W. 2018. EPB TBM Performance Prediction on the University Link U230 Project. In Proceedings of *North American Tunneling*.

Mori, L. 2016. *Advancing understanding of the relationship between soil conditioning and earth pressure balance tunnel boring machine chamber and shield annulus behavior* (Doctoral dissertation, Colorado School of Mines. Arthur Lakes Library).

Northlink Tunnel Partners. 2009. *Geotechnical Baseline Report- University Link 230* 6: 66. Seattle, WA.

*Tunnels and Underground Cities: Engineering and Innovation meet Archaeology,
Architecture and Art, Volume 6: Innovation in underground engineering,
materials and equipment - Part 2 – Peila, Viggiani & Celestino (Eds)*
© 2020 Taylor & Francis Group, London, ISBN 978-0-367-46871-2

A new approach for qualifying blasting works in underground

P. Montagneux
UG Director activity – Asian Business Developer for EPC Groupe, France

P. Bouffard Vercelli
Innovation & Software development engineer for EPC Groupe, France

ABSTRACT: EPC Groupe (French explosives manufacturer for 125 years) has developed a new monitoring software for qualifying, and then, improving quality and efficiency of drilling and blasting jobs in tunneling works. Everybody knows that in tunnels built with Drill & Blast method, everything starts with a good quality of drilling and then blasting. If these two first tasks are not well done or controlled, the downstream operations will be impacted with low efficiency and then over costs. One of the great difficulty is to get all informations (advance, over and under breaks, fragmentation, and all KPI's linked to the downstream operations) into one tools in order to have a quick overview of the quality of D&B works, and then, correct as soon as possible, leverages on the drilling pattern, blast pattern and/or timing sequence. With the help of modern tools (using of digital tools for measuring many parameters), and the help of photogrammetry (using simply the camera of a smartphone), it could be easier to get in one box, all information's related to D&B efficiency linked to downstream operation measured with KPI's. This paper presents the tools, and how it is used to qualify the quality of D&B operations in different audits driven in UG mines and tunnels works.

1 INTRODUCTION

There are several methods to build a tunnel. Two big types of excavation methods distinguish themselves. One is called traditional and use mechanical equipment in soft rock, or drill & blast excavation in hard rock, and, another one is called mechanized and use TBM whatever geology encountered.

1.1 *Facts*

1.1.1 *Importance of underground excavation in the world*
AITES recently realized a study market tunnels which specifies that at the world level, the construction of tunnels and subterranean spaces represented 86 billion of euros in 2016, which represents an increase of 23 % with regard to 2013. In terms of length of tunnel bored, the annual average is about 5 200 km for all the types of tunnels

The future trends of tunnel construction are amplified by the sustainable development.

In the developed countries, the rhythm of the building work of tunnels is rather constant with major projects in Europe such as Big Paris, Crossrail2 and HS2 in the United Kingdom, Brenner, Lyon-Turin under the Alps.

This sustainable development process and the necessity of communication between countries are not confined in Europe: we observe a large development of mass transit system in all the cities of China, but also a big development to come in India as well as to Qatar and in Saudi Arabia.

1.1.2 Predominance of TBM excavation

The tunnel boring machine market is largely driven by the increase in spending on road and railway infrastructure. Countries such as China, Japan, Italy, South Korea, Norway, and Germany have vast tunnel networks. The key reason is to connect major cities within the country, sometimes even international, with minimum commute time. This largely optimizes the transportation costs and connects cities to boost commerce. The other method for building tunnels is by using explosives;

The TBM industry use of such fears to convince clients and contractors to use TBM technology for boring underground excavation; *"explosives are highly dangerous, and it is often difficult to control the explosion. They are also a threat for men and machines working in the tunnel. Slight deviation in the charge of explosives can completely jeopardize the project. Thus, tunnel boring machines are the ideal alternatives for tunneling operations"*. In spite of the use of marketing arguments of which most are fallacious, it is necessary to have the intellectual honesty to admit that it has been quite a while since the conventional method *(i.e Drill & Blast method)* did not produce technological revolution, making her more competitive and thus more attractive.

1.2 Reasons of hope for D&B method

Explosives manufacturers are not to stand idly by during all these years. Some research and development were done on the products, equipments but It will have been necessary to wait for the beginning of the 21th century, with the advent of the new digital tools and the considerable increase of the computing powers of computers, to see the light at the end of the tunnel.

1.2.1 Recent evolution for D&B method in the last 20th years

The most dangerous products, as dynamites or similar products, were replaced hopefully, most often by emulsions (packaged or in bulk), which ones are generally 10 times less sensitive than NGL based products.

Bulk emulsion is now for more than 15 years all around the world, the reference were explosives are needed in construction work, bringing safety and high performance; and when country regulator considers matrix emulsion (raw material) as an oxidizer (class 5), it gives very positive solution for products storage on site.

Automatic drilling machine arrived early in the 90's, bringing new equipment which allow to drill blast holes with high accuracy, automatically and then with less workforces. In tunnel excavation using drill & blast method, automatic jumbos are the reference today.

Non-electric detonators replaced advantageously at the end of the 90's, electric initiation system, bringing more safety in tunnel.

Electronic detonators arrived in the market at the beginning of the 21th century, and in spite today of a pricing little bit more expensive compare to conventional initiating system, they could bring very interesting solutions in urban civil works (cities as Monaco, Hong Kong or Singapore encourage the use of such products to guaranty the timing for blasting works, mastering seismic effects on neighborhood structures).

1.2.2 The best is yet to come

Wireless electronic detonator is probably the most interesting thing to come soon. Most part of problems encountered during charging in underground come from wire, when using electronic detonators, as energy and communication is driven through the wires. In difficult geology and or with the use of fiber shotcrete, wires are often injured and could complicate communication with the detonator. By avoiding any wire, the use of electronic detonator will be safer and faster, and will open new windows, for a joint use with equipment, opening the way for a complete automation for explosives loading.

OnlineMWD (Measure While drilling) transfers drilling data in real time to the office to geologist's tablet by sing a WLAN connection. OnlineMWD is a great tool in order to know with high accuracy the rock behind the face. However, these data are never used for optimizing blast pattern and blast sequence in real time. New algorithms are arriving to transform all geological data in blast energy model needed to break the rock. And for sure, as for wireless electronic detonator, it will open windows for developing new generation of automatic charging equipments.

Photogrammetry is revolutionizing the way of realizing tunnel profile and, get in real time all information on excavated volume, over and under break, tunnel advance, granulometry, muck pile shape, etc..., with a simple camera (a smartphone for example) and a minimum light (car's lights for example).

When all these tools will work jointly or on a same equipment, it will be the time to have an automatic drilling and charging equipment. In this case, such technological breakthrough, will bring sufficient time savings on the tunnel excavation cycle, giving D&B method new reasons to be competitive compare to TBM excavation method. This time is coming soon.

2 DRILLING AS THE FIRST LINK IN THE CHAIN, AND BLASTING, THE RELATED LINK

Whichever method is used (conventional or mechanized method), tunnel excavation is a permanent cycle with successive imbricated tasks: surveying, drilling, charging, blasting, ventilating, scaling, mucking, shotcreting, and bolting (sometimes heavy rock reinforcement is needed) and do over.

Figure 1.   tunnel cycle management.

As the first task on this cycle, drilling play an important role. If drilling is not performed well, whatever are the quality and the quantity of explosives that it will be used, the result of the blast will not be optimal; it's that expert called "as planned as drilled".

Which parameters are important to be managed? There are several parameters for drilling that it is important to be sure that they are well implemented: hole mark vs planned, hole length, burden and space at the bottom face, hole inclination and angle, hole missing.

2.1   Triptych of underground blasting

There are 3 fundamentals elements to design a blast in underground: <u>energy</u> distribution in <u>space</u> and in <u>time</u>. These three elements work closely together and cannot be considered in theory separately.

As the 3D face geometry is responsible of the profile excavation (contour, soil and tunnel advance), the energy distribution play an important role in fragmentation, muck pile shape, and certainly, because it is closely link to 3D face geometry, in tunnel advance and contour quality control.

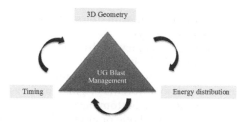

Figure 2.   triptych of UG blasting.

The timing (how the energy is deliver on time), play a role in fragmentation, muck pile spreading, contour quality control, ground vibration and air blast, and for sure in tunnel advance by giving optimal free face to the next series of blasting holes.

Drilling pattern, charging and firing pattern give all theoretical parameters of a blast. If these documents are done generally by blasting engineers, they are implemented on field by experienced operators. It is very important that no frontier exists between the theory and the reality: field controls are very important to be certain that parameters are correctly implemented, and, to link blast results to those parameters as an iterative process.

But for sure, to be efficient, tools and time are needed to process all measurements, control and analysis. It's the price to pay to have efficient blasting, and finally save money.

To be honest, in tunneling excavation, time is essential, and the main goal is to reduce cycle time as short as possible. In this frame of mind, project managers are not predisposed to spend time in measurement and analysis, as there are not considered their project as a laboratory. In this case, efficient and easy tools are the only answer to give to them, and proven added value is the key.

2.2   3D face geometry management

It is important to well mange the 3D face geometry, but it exists different way depending which type of equipment is used.

The figure at the opposite shows how a drilling pattern should be considered on the field.

The important thing is to understand that a drill plan should be designed by considering it in three dimensions (3D).

By avoiding that, a lot of blast dysfunctions may occur.

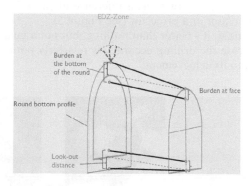

Figure 3.   theoretical drill plan design.

2.2.1   Using automatic drilling machine

When automatic drilling machine is used, drill pattern is memorized into the machine, and jumbo operator should drill more than 80% in automatic mode (bottom holes and some contour holes need to be carefully done in semi-automatic mode to avoid damage on the boom, or to take account of rock face reality).

When automatic jumbo is used, the accuracy of drilling is very high, around a centimeter, and if the blast plan is properly planned, the blast result shouldn't be affected by drilling.

In semi-automatic mode, the jumbo should correct automatically length, inclination and angles when the operator moves the boom from its original mark, assuring to a certain extent, that drilling pattern is well performed at the bottom face.

In manual mode, there is no help from robotic or electronic tools; the experience and feeling of drilling operator is essential, but not a guarantee.

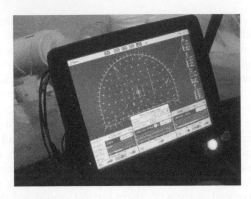

Figure 4.   computerized drilling.

2.2.2 *Using manual drilling machine*

Unfortunately, automatic drilling machine are not always used, because of availability on the country, project budget, or other reasons.

Manual drilling machine is often used is small to medium underground mines, small tunneling projects, or bigger projects in remote and exotic area. In this case it is important to control the implementation of drill holes on face, quality of holes drilled (angles, inclination) and length, and compare it to drill pattern planed.

The main problem is this case is how to mark the drilling pattern on the face to give references for drillers. Sometimes, drillers don't use any mark on face and drill with their own feeling and experience, other times a grid is painting on the face helping drillers to have some references, or hole mark is painting (with as a reference of the axis of the tunnel or not) or projected with the help of a laser or a projector (see Figure 5).

Whatever the system is used, it's better than nothing, but not a guarantee that the drill plan is correctly done. In this case the drilling accuracy is relatively poor, adding errors from the drill plan implementation to its achievement.

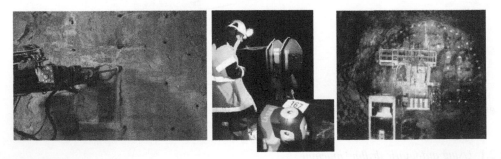

Figure 5.   grid paint of face (left) — laser system projecting hole mark according to drilling pattern (center) — drill plan projected directly on the face with a digital projector (right).

By measuring the accuracy, it's then possible to better manage.

2.2.3 *The importance of measuring drilling accuracy*

All those parameters affect the drilling advance per blast. The drilling advance is a percentage of meters of tunnel effectively blasted, compare to tunnel drilled. Normally, a good productivity should be at least more than 92 to 95%, a very good one reach 100%. Below 90%, something should be analyzed and done, if not over costs will affect the whole excavation budget.

Figure 6.   measurement for efficient management.

There is a famous proverb saying, 'you can't manage what you can't measure". At the time of a precise management on project construction, measuring to better understand, act and master the costs, this proverb was never so true.

Hopefully, automatic drilling machine record all parameters effectively done and it is possible to check the quality of drilling accuracy (but this job is not however done in real time) after job is realized.

By importing drilling data from jumbo, and by overlaying these data on a 3D model obtained from photogrammetry for example, it's possible then to compare with the theoretical drilling pattern, to appreciate the work quality (see Figure 7).

For manual drilling machine, it exists equipment to added in order to track and trace every essential hole parameters (inclination, angle and length, drilling parameters). Without these equipments, it is difficult then and it takes time to measure drilling parameters.

Drilling accuracy is certainly responsible to a half part for a minimum to contour quality, tunnel advance, soil flatness, and in a certain measure rock fragmentation, and ground vibration.

Figure 7.   view of as planned (green point) and as drilled (red point) with a semi-automatic drilling machine.

Figure 8.   additional equipment installed on manual drilling machine (Bever Control AG).

It's the reason why, it is so important to have a continuous view along the excavation project, about this essential parameter.

2.3   *Some blast malfunctioning*

As Drilling and blasting should represents at least 20 to 25% of the total excavation costs (see Figure 9: *typical costs repartition in D&B tunnel excavation*Figure 9), to take care of these direct costs is not a harmless thing. But we can see that consequences on the downstream operations, which costs for some represents between 15% to more than 30% of the total excavation costs, could be enormous.

There is a lot of examples of malfunctioning it could occur in tunneling or underground mining blast:

- Some are related to the management of tunnel profile (under and over break, soil quality and altitude, advance),
- Some are related to fragmentation (poor or excessive blast fragmentation),
- Some are related to muck pile spreading,

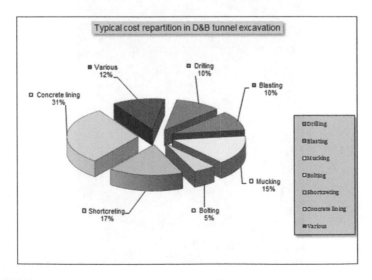

Figure 9.   typical costs repartition in D&B tunnel excavation.

- Others are related to blast nuisances (over ground vibration and air blast, excessive blast fumes),
- In underground mining, problems related to blast ore dilution are also mentioned,

All these problems have a common link: over costs generated directly or, indirectly on downstream operations. We can talk about:

- Augmentation of cycle time due to poor charging productivity (muck pile spreading, over size etc…), over time spending in scaling or reboring tunnel profile, decrease of tunnel round advance due to excessive ground vibrations,
- Augmentation of concrete volume used in the whole project,
- Loss of ore due to bad management drilling and blasting in ore body,
- Poor tunnel advance (increase about 10 to 30% of drilling and blasting costs)

The chart above (see Figure 10) shows how quality of contour combined with efficiency of tunnel round advance, influence jointly overall costs in tunnel construction.

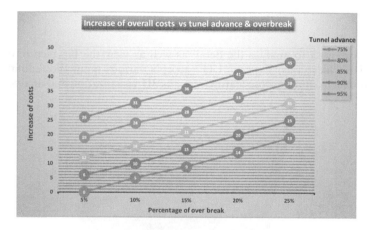

Figure 10.   over costs function of drill advance and over break.

3   EXPERTIR® UG – A NEW MODERN TOOLS TO QUANTIFY AND QUALIFY BLASTING WORKS

3.1   Users's requirements

Today, to design a drilling and blasting pattern, it exists some tools using most of the time theoretical models (JKSIM blast® for example) or a combination of theoretical and experienced model (in house software provided by computerized jumbo manufacturers).

More generally, these software give a theoretical pattern to be done without taking in account the reality of the face (geology, geometry etc…).

User's requirement is exactly the same of what was developed for open pit blast design: from laser scanning or pictures taken from a drone, the blasting software is taking in account the reality of field to make the design efficient and safe.

3.2   Software concept

Expertir UG© had been developed from scratch with at least two goals in mind. First it provides tooling for visualization and measurements of 3D geometric data. It can be a gallery shapes obtained by photogrammetry or laser survey as well as drilling plans or reports with bore holes data.

Then it is also a solution to aggregate those data and more in a single place thus allowing the extraction of relevant Key Performance Indicators (KPI's) like advance, over break volumes, explosives quantities...

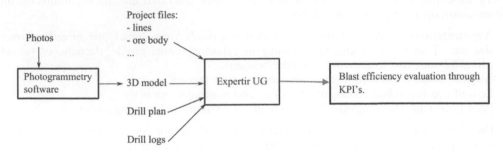

Figure 11.   software overview.

This tool allows both to analyze the detail as well as view the big picture on a complete gallery scale. To achieve this, Expertir UG© accepts various input files coming from other software or equipment. But the core feature remains 3D visualization and analysis.

Figure 12.   3D model computed from pictures get with a classic camera (smartphone).

The main input is the 3D model upon which other geometric data are related. Photogrammetry yields a good accuracy (sub centimeter compared to laser) with the advantage over laser to offer both position and color information. Another software is used to produce a 3D model from the photos taken underground. The quality of the produced model is inherent to the quality of the picture. Unlike the laser photogrammetry doesn't work in complete darkness and descent amount of light is required. Tests have been successful using a vehicle headlight.

The Figure 12 shows a 3D tunnel displayed in Expertir UG®. Displaying the texture allows the user to better understand the model and the geology (water, rock structure...).

By importing drilling data and overlaying with planned data, the 3D model can easily shows potential quality deviation.

3.3   *Advance following*

The measurement of blast advance is automated when models are imported. A mean distance between closed face is computed between each model. A 2D view of the gallery shows a symbolic advance along the gallery line, whatever its path. The 3D model still must be georeferenced before import.

Figure 13.   import drilled data in 3D model.

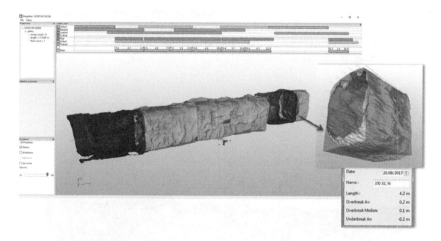

Figure 14.   round advance and volumes measurement.

3.4   *Over/underbreak measurements*

There are different ways of defining the over break and underbreak figures. Since contour holes are not bored exactly parallel to the gallery desired axis (due to drill rig mechanical constraints), the software considers both cases:

- Over break/under break with drilled geometry: allowing to access the efficiency of the blast only
- Over break/under break with project contour geometry: to access global quality of the drill and blast process.

The software allows to view the evolution of these values along the blast as well as provides numerical values for both volumes.

In addition, the imported 3D model can be seen in fake color to show the distance from the planned contour thus allowing a quick highlight of problematic areas.

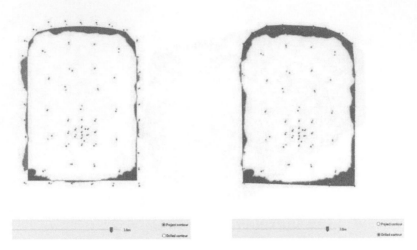

Figure 15. over break and under break measurement as planned vs drilled.

3.5 *Production hole and ore body*

In underground mining blast production (sub level stopping, sub level caving), it is very important to control drilling quality along the ore body shape.

By importing data from multiple source, Expertir UG® lets the user visualize within the software the position of production holes relative to the ore body.

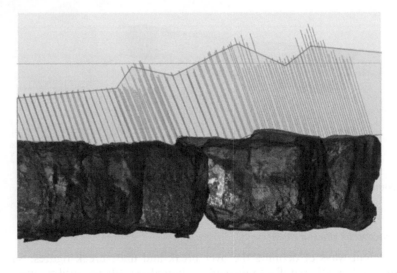

Figure 16. position of drilled hole in ore body in production blasting.

3.6 *KPIs*

In addition to informations extracted from the geometry analysis, users can input other blast related values of interest. Clients having different needs the user interface is flexible to adapt to various input types.

Finally, all theses Key Performance Indicator (KPI) may be exported in an open format for further specific manipulation.

A KPI could be tunnel advance (% of tunnel blasted/tunnel drilled), accuracy of contour blasted (% of over break or underbreak), shape of the muck pile (related to wheel loader efficiency), fragmentation (related to wheel loader efficiency), seismic impact, air pressure impact etc... Depending the main objective chosen by the user, it becomes "easy" to qualify a blast within these KPI's.

KPI's is the only way to prove the positive or negative influence of drilling & blasting parameters changes.

4 CASE STUDY

In Sweden we experienced this tool and some view of the results are given here after:

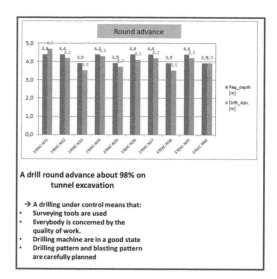
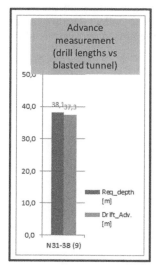

Figure 17.   using EXPERTIR UG® for advance measurement in a project in Sweden.

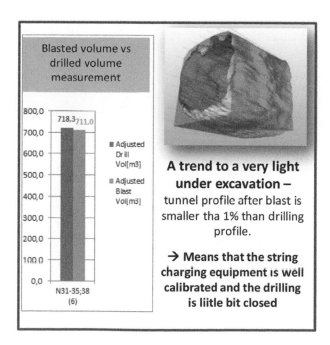

Figure 18.   blasted volume vs drilled volume measurement.

5 CONCLUSION

To manage correctly rock excavation using drill&blast method, it's necessary to input some key measured data during drilling and blasting cycle time, and link those ones to downstream operation to evaluate their negative or positive impact on them.

This is the only way to evaluate any changes on drilling and blasting pattern on operations, and potential money savings could be important.

To achieve that, EPC groupe had developed **EXPERTIR UG®**, a 3D data base object in which one all rounds are collected (with measured or input data), for fine analysis, with one objective, save money on the whole excavation cycle. With the help of new numeric tools, it's now easier to collect data for better management.

Now the famous proverb could be "we can manage as we are measuring".

Tunnels and Underground Cities: Engineering and Innovation meet Archaeology, Architecture and Art, Volume 6: Innovation in underground engineering, materials and equipment - Part 2 – Peila, Viggiani & Celestino (Eds)
© 2020 Taylor & Francis Group, London, ISBN 978-0-367-46871-2

Line C in Rome: Strain measurements in precast TBM lining segments using embedded Smart Wireless Monitoring System

S. Moretti & M. Leonardi
IMG S.r.l., Rome, Italy

E. Romani & I. Mammone
Metro C S.C.p.A., Rome, Italy

ABSTRACT: Line C is the third line of the Rome underground system. The T3 stretch is presently under construction: it runs for a length of about 2.8 km right underneath the historic city center of Rome. Two EPB TBMs with a 6.70 m cut diameter are being used to excavate the tunnels at a depth between 30 and 60 m below the ground level. Strain and temperature measurements have been acquired to collect data in continuous mode, starting from the early curing stages. Wireless and low power consumption data-loggers and vibrating wire strain gauges, both embedded in each segment of the instrumented ring, allow remote monitoring of all the main events in the TBM lining, without any interference with the ordinary underground activities. In order to check the strain behavior of the lining, single instrumented precast segments were tested through special full-scale loading tests.

1 INTRODUCTION

1.1 *Line C – Rome Underground*

Rome's Line C subway, under construction by General Contractor Metro C S.c.p.A. since 2007, is a fully automated underground railway line characterized by an automatic train control system. The line, with its 30 stations, crosses the Capital from the south-east side to the north-west area stretching over a distance of 25,6 km. The construction of Line C will almost double the area covered by the current underground network and it represents one of the most important works ever built in an urban context. The first part to the city center, in operation and under construction, develops for 21 km, 8.5 above ground and 12.5 underground, with a total amount of 24 Stations. The underground part has been excavated by EPB TBMs with a 6.70 m cut diameter, at a depth between 20 and 60 m from ground level.

1.2 *The under construction T3 Stretch*

The T3 Stretch, presently under construction, runs for 2.8 km from San Giovanni Place to Venezia Place in a heavily urbanized area and below extremely important historical monuments (it is an UNESCO World Heritage Site; Figure 1). Twin tunnels are located at a depth between 30 and 60 m below the ground level.

The monitoring system has been designed to provide all the elements necessary to perform an analysis of the interaction between works in progress, the surrounding soil volumes and the anthropic preexistences, in the most complete and quick way.

Instrumentation is organized in monitoring sections distributed along the metro line. The distribution criteria, typology and geometry of monitoring sections is function of the specific contour conditions along the stretch: geotechnical features of the soils, preexisting buildings and monuments, tunnels depth etc. The following Figures 2, 3 and Table 1 show a typical full

Figure 1. An example of interaction between Line C and historical monuments.

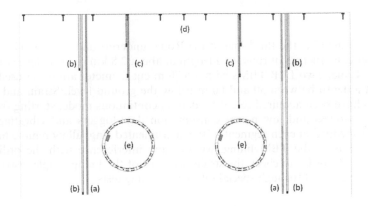

Figure 2. Tunnel line - Typical full monitoring section. Inclinometer (a); Piezometer (b); rod extensometer (c); leveling pin (d); instrumented ring (f).

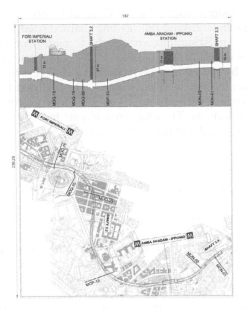

Figure 3. General layout of T3 stretch and distribution of monitoring rings in the tunnel lining (Table 1).

Table 1. T3 Stretch – Monitoring rings of TBM tunnel lining.

Monitoring section	Even Track	Odd Track	Head
Name	km	km	m
MON-01*	0+209,916		23,97
MON-05*		0+350,367	26,32
MOP-13	1+168,100	1+126,680	44,86
MOQ-05	1+855,363	1+804,772	42,00
MOQ-10	2+069,557		33,19
MOQ-15	2+230,238	2+214,201	27,15

* Monitoring rings installed at the date of this paper.

monitoring section of TBM line with a couple of instrumented rings inside the tunnel lining, and its distribution along the T3 Stretch of the Line C.

2 MONITORING OF TBM LINING

The past experiences gained during the construction of the Line C tunnels, with the installation of over 60 segmental lining rings instrumented with vibrating wire strain gauges, and the current integration with the most modern technologies of communication and data transmission, has allowed to refine the methodology of structural monitoring for the TBM tunnel linings of the T3 Stretch. At the same time, this new monitoring system provide important information to the project design, making possible the acquisition of the deformation history of the structures continuously: from the preliminary phases of concrete casting and maturation up to the operation phase of the metro line.

2.1 The Smart Wireless Monitoring System (SWMS)

The Smart Wireless Monitoring System (SWMS) is a structural monitoring system applied to the deformation control of TBM tunnel linings, specially designed and realized to achieve the following goals:

- Eliminate any operational interference with ordinary underground operations, during the installation phases. All the workings have been completed in the precast segment factory,
- Record and transfer monitoring data (ε and T °C) during all the main phases of the structure: from the early construction stages and the curing stage at the manufacturer stock place till the assembly in the tunnel during ring erection, advance of the TBM and tail grouting. The system has been designed also to analyze long-term behavior of the tunnel lining,
- measurement certainty through redundant systems to ensure data access.

The SWMS experimental system, applied for the first time in Metro C to the instrumented rings of the T3 Stretch, consists of (Figure 4):

- Monitoring devices embedded in each segment (radial and longitudinal strain gauges),
- WDG – peripheral wireless data-loggers installed in the inner face of each segment,
- GTW – gateway. Data gathering, storage and export base-station for each instrumented ring.

The peripheral data-loggers acquire physical data by sensors, that are recorded in the integrated memory, and transmitted to the base-station in wireless mode. Data-loggers continue to sample also without radio coverage; indeed they are able to store up and transmit the data packets when the coverage becomes available again.

The gateway can be equipped with different communication solutions, which allow remote visualization and download data from the tunnels. (e.g. radio network, cable, fiber optic etc.).

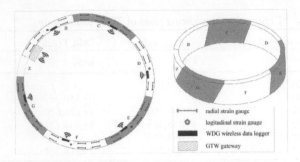

Figure 4.   Instrumented ring with Smart Wireless Monitoring System (SWMS).

2.2 *Instrumentation layout*

The deformation measurement in the segments of the tunnel is performed by vibrating wire strain gauges anchored to the steel reinforcement of the monitored structure. These instruments allow to monitor the strain state in the lining rings and their interaction with the surrounding soil.

Each strain gauge has a built-in thermistor to provide temperature information for thermal corrections and at the same time to follow the thermal behavior after the casting phase of the segmental tunnel lining.

Each lining ring (5.80 m internal diameter and 30 cm thickness), is made up of n. 6 segments beyond the key element (Figure 4). In the instrumented rings, each segment, excluding the key, is instrumented with n. 6 radial vibrating wire strain gauges. Another one is positioned longitudinally, to check the thrust cylinders effect during the installation phases of the ring, for a total of n. 42 strain gauges.

The automatic data acquisition system consist of peripheral wireless data-loggers (WDG), installed in the inner face of each segment and deputy to acquire physical data by sensors (Figure 6). The WDG unit is designed to be customizable for different projects, in terms of number and type of sensor signals.

The data acquisition unit is contained in a plastic box with a high protection degree. The cap of the box, equipped with screws and O-ring closing, can be opened by the inner side of the segment (detail a in Figures 5, 6). The box shape is compatible with geometric gaps of the segment reinforcement. The box is equipped with only one input, connected to a junction box by multicore cable. The junction box is IP68 and contains multiplexer where all the signal cables of the embedded instruments are connected (detail b in Figure 5).

The WDG system is equipped with a waterproof safety box directly connected to the junction box (detail c in Figures 5, 6). This secondary access allows to communicate via cable between data-logger/gateway when the wireless connection does not work (e.g. underwater segment). At the same time, the gateway via this cable is able to power the peripheral data-logger inside the segments (e.g. over battery life). Both the accesses in the inner side of the segment are closed with quick-setting cement, to allow the right working of the suction plate of the erector during the installation phases of instrumented ring (Figure 7).

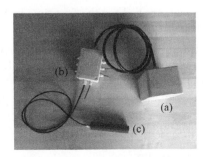

Figure 5.   WDG - Wireless data-logger system. (a) data-logger box, (b) junction box, (c) safety box.

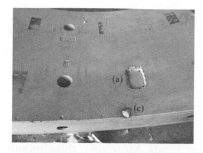

Figure 6. Inner side of instrumented segment. (a) access to data-logger box, (c) access to safety box.

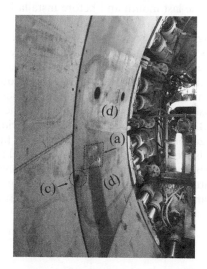

Figure 7. Installation of instrumented ring. (a) datalogger box, (c) safety box, (d) suction plate foot-prints of erector. Dotted areas indicate the cementation of monitoring boxes access.

The receiving unit (GTW - gateway), one for each instrumented ring, will manage all the peripheral wireless data-loggers embedded in the segments (Figure 4). The gateway is a data-logger equipped with a radio in coordinator configuration and an embedded Linux computer with a solid state hard disk, to provide customizable advanced network services and connect-ivity such as FTP server, modem, fiber optic or cable, according to the site needs. The gateway can be directly powered when is in continuous configuration, or equipped with a battery and a timer to temporarily acquire data (e.g. 1 hour a day) by the site.

2.3 *Monitoring data*

The first part of TBM twin tunnel excavation in T3 Stretch, about 400 meters long from 3.3 Shaft to Amba Aradam – Ipponio Station (Figure 3), has been completed in the begin of August 2018 after two months of advancement for each of the two tunnel. In this area the first two experimental instrumented rings have been installed (Table 1), using the new Smart Wire-less Monitoring System. These rings have been pre-cast in the factory in March 2018. Since then the WSMS System is recording in real time, continuously and in wireless mode deform-ations and temperatures of the TBM segmented lining, during the main phases of its work.

A full set of monitoring data is shown in Figure 8, where is collected the main strain behav-ior of the segmented ring MON-01, installed in the even track of the T3 stretch and with 24 m depth from ground level (Tab. 1, Figure 3).

For simplicity, only C, E and G segments data are presented, split up in longitudinal (b), intrados (c) and extrados (d) strain gauges (Figure 8). For each instrument group the average temperature is shown as well (dotted line). The main working phases are readable in the graph (a): TBM advance vs time, dates of concrete casting and assembly of instrumented ring.

Conventionally the strain is either positive (tensile), or negative (compressive). It is assumed that zero reading of strain gauges is the value recorded at the time of ring installation. All monitoring data are already corrected by temperature effects.

The first critical stage is recorded during the concrete casting (indicated by the dot line in Figure 8a), where the mechanical action of the concrete on instruments and its thermal release clearly influence all the measurements. More than three months elapse between the casting phase and the lining ring installation. In the first two months all the instruments register a general trend of compressive strain, around 200 µε, probably related to the curing stage at the manufacturer stock place. In the last month and before installation, a general stabilization of the readings has been observed.

The example in Figure 9 highlight in a weekly window the main effects of the concrete casting on extrados strain gauges of the segments. At the beginning the measurements suffer an important offset caused by mechanical actions on instruments (casting, formwork vibrations), then they tend to stabilize after that concrete thermal effects ran out.

The second critical stage is related to the assembly of the segments in the tunnel during ring erection (indicated by the dash line in Figure 8a), advance of the TBM and tail grouting. In these steps and during the TBM advance, the segmented lining ring start to work recording a permanent strain until stabilization, that is function of the TBM face distance. After stabilization it is possible to analyze also long-term behavior of the tunnel lining.

The example in Figure 10 emphasize the strain behavior of the ring during and after the assembly stage in the tunnel. Data are focused on longitudinal strain gauges oriented towards the thrust direction of the TBM. The data show the cyclical noise of the TBM thrust, registered by strain

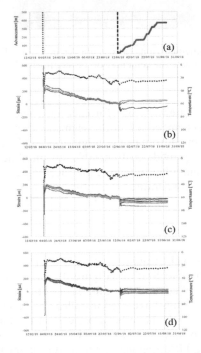

Figure 8. T3 stretch – even track. Instrumented ring MON-01. SG - strain gauges data of segments C, E, G (Figure 4). (a) Tbm advance, dot line: ring casting; dash line: ring installation. (b) longitudinal SG; (c) intrados SG; (d) extrados SG. Strain: µε > 0 traction; µε < 0 compression; Average temperature in dot line.

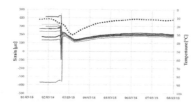

Figure 9. Extrados strain gauges. Zoom on the concrete casting phase and the beginning of curing stage. Average temperature in dot line.

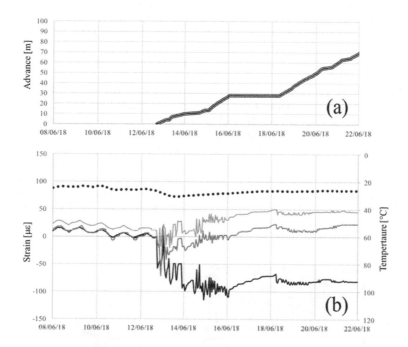

Figure 10. Longitudinal strain gauges. (a) Tbm advance; (b) emphasis during and after the assembly stage of the instrumented ring in the tunnel. Average temperature in dot line.

gauges with a sampling every 30 min. The thrust effects on lining rings tends to decrease with the distance, to become negligible over 30 m the TBM face (5 diameters).

The main stress is recorded after the early stages of the assembly, when the TBM face is less than 12 m to the lining ring (2 diameters). Beyond this distance, and until the completion of the first excavation phase, a general stabilization of the data on all the instruments is observed (Figure 8). With the last TBM breakthrough of Amba Aradam – Ipponio Station occurred on 03 August 2018, starts the long-term behavior analysis of the tunnel lining, currently under observation.

2.4 *Loading test on precast segments – preliminary results*

The behavior of precast tunnel segments, and the relations between load and first crack formation and the evolution of the crack opening, have been observed through special full-scale load tests. The experimental tests were carried out in the Material and Structure Laboratory of the Civil Engineering Department of the University of Rome "Tor Vergata", and affect any reinforcement class of the precast segments.

In order to check the behavior of the strain gauges under external stresses, a sample segment has been tested through a bending test (Figure 11).

The monitoring layout and the steel reinforcement class of the tested segment are the same of the instrumented ring installed in the tunnel lining (Figure 12). The bending test was performed by applying a normal force on the convex surface of the segment by hydraulic jack. The segment was placed on cylindrical supports and the load, applied at mid-span, was transversally distributed adopting a steel beam.

Loads were applied with increasing steps up to a maximum of 750 kN. During the load test all the development of cracking patterns has been studied (number, size, geometry, cracks opening vs load, etc.). The vertical displacements of the segment were measured with three potentiometer wire transducers (WDT), while two LVDTs (MDT) were adopted for measuring the cracks opening at mid-span in the inner side of the segment (Figure 11).

At the time of writing, laboratory results are not available yet, while the preliminary results of embedded strain gauges are plotted in Figure 13 (strain vs load) and Figure 14 (strain vs time).

The higher response is registered in the intrados side with the strain gauge SG36, located in the central part of the segment. The tensile strain increase with the stress until to about 2400 με for 750 kN (maximum load), and recording of 750 με of residual strain after the load is removed (Figure 14). A similar behavior is followed by SG40, but with a lower magnitude, recording the higher value around 900 με and about 400 με of residual strain (Figure 14).

The other instruments do not seem to react to the load increasing steps, particularly for those located on the extrados side of the segment. These instruments reply only after the maximum load, with a residual tensile strain probably related to the final cracking pattern (Figure 14).

Figure 11. Full scale bending test on segments. Laboratory instruments: MDT – Mechanical displacement transducer (n. 2), WDT - Wire displacement transducer (n. 3)

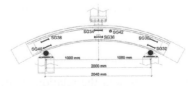

Figure 12. Instrumented test segment. Monitoring layout.

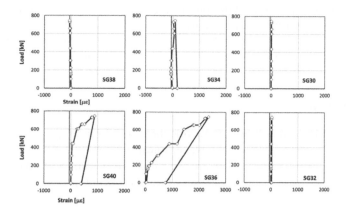

Figure 13.   Bending test. Strain vs load diagrams.

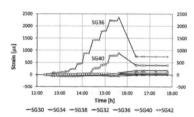

Figure 14.   Bending test. Strain vs time diagram.

3   CONCLUSIONS

The current tunneling works for the T3 stretch of Line C of Rome Underground have provided the chance to test the strain behavior of the segmental pre-cast lining using the experimental embedded Smart Wireless Monitoring System (SWMS).

The SWMS is a structural monitoring system with low energy consumption, specially designed and realized to achieve the following goals:

- Eliminate any operational interference with ordinary underground operations, during the installation phases of the lining. All the workings have been completed in the precast segment factory,
- Record and transfer monitoring data (ε and T °C) in continuous mode during all the main phases of the structure: starting from the curing stage at the manufacturer stock place and followed by the main events until the instrumented ring was set in place in the tunnel, performing also long-term behavior analysis of the tunnel lining,
- measurement certainty through redundant systems to ensure data access.

Referring to the last tip, the SWMS System introduce new solutions designed to guarantee data access in any environmental conditions: a) wireless mode, b) via cable through the safety box (direct access to embedded data-logger), c) via multicore cable behind the data-logger box (direct access to embedded instruments).

Data analysis show that every action could have a substantial influence on the segmental loading that cannot be neglected in the structural design. During the concrete casting of the segments, strain and temperature measurements have been acquired to collect data in continuous mode from the early construction stages and the curing stage at the manufacturer stock place until the assembly in the tunnel during ring erection, advance of the TBM and tail grouting, performing also long-term behavior analysis.

Finally, in order to check the behavior of the monitoring strain gauges under external stresses, a sample segment has been tested through a bending test and the preliminary results have been shared.

ACKNOWLEDGEMENTS

The authors would like to thank Eng. Franco Robotti and Marco Arrigoni of Agisco Group S.r.l. for the fruitful partnership and valuable technical support.

A special thanks to Eng. Valerio Foti and the whole TBM team of Astaldi S.p.A. for their inestimable availability in all the main phases of the experimental project.

REFERENCES

Bilotta E., Russo G., Viggiani C. 2006. Ground movements and strains in the lining of a tunnel in cohesionless soil. In Bakker et al (eds.), *Proceedings of the 5th International Conference of TC28 of the ISSMGE on Geotechnical Aspects of Underground Construction in Soft Ground, The Nederlands, 15–17 June 2005*. London: Taylor & Francis Group.

Cignitti F., Sorge R., Meda A., Nerilli F., Rinaldi Z. 2012. Numerical analysis of precast tunnel segmental lining supported by full-scale experimental tests. In Giulia Viggiani (ed.), *Proceedings of the 7th International Symposium on Geotechnical Aspects of Underground Construction in Soft Ground, Roma, Italy, 17–19 may 2011*. London: Taylor & Francis Group.

Pelizza S., Peila D., Sorge R., Cignitti F. 2012. Back-fill grout with two components mix in EPB tunneling to minimize surface settlements: Rome Metro – Line C case history. In Giulia Viggiani (ed), *Proceedings of the 7th International Symposium on Geotechnical Aspects of Underground Construction in Soft Ground, Roma, Italy, 17–19 may 2011*. London: Taylor & Francis Group.

Sorge R., Moretti S., Tripoli O. 2009. Rome Metro, Line C. *The automation of the monitoring system in an urban project: management and data sharing. Proceedings 13th ASITA National Conference, Bari, Italy, 1–4 December 2009*.

Sorge R., Moretti S., Tripoli O. 2012. Line C in Rome: remote monitoring system. In Giulia Viggiani (ed.), *Proceedings of the 7th International Symposium on Geotechnical Aspects of Underground Construction in Soft Ground, Roma, Italy, 17–19 may 2011*. London: Taylor & Francis Group.

*Tunnels and Underground Cities: Engineering and Innovation meet Archaeology,
Architecture and Art, Volume 6: Innovation in underground engineering,
materials and equipment - Part 2 – Peila, Viggiani & Celestino (Eds)
© 2020 Taylor & Francis Group, London, ISBN 978-0-367-46871-2*

The Rijnlandroute bored tunnel – Continuously improving the mechanized tunneling process

H. Mortier
Dimco, Zwijndrecht, Belgium

M. Brugman
Arthe C&S, Houten, The Netherlands

B. Peerdeman
RHDHV, Rotterdam, The Netherlands

T. Schubert
Vinci Construction Grand Projets, Rueil-Malmaison, France

ABSTRACT: For the Rijnlandroute project, the design and execution of the bored tunnel needed to implement several adaptions to the segmental lining, break-in procedure, gasket design and cross passages concept in order to cope with the encountered challenges. These challenges were due to specific circumstances such as a limitation of the gaskets' temperature rise during a fire scenario, but also promises regarding damage free lining or a safer cross passage construction sequence, put forward during tender stage, led to necessary developments. Finally, difficulties encountered during the first period of the execution generated some alterations. This article will enlighten how all these aspects had their influence on the design and execution of the project.

1 INTRODUCTION

The Rijnlandroute project comprises among others the new link between the existing A4 and A44 highways in the delta region near the city of Leiden in the Netherlands. The necessity for this new link originated due to the large amount of daily commuters seeking a traffic-jam-free passage through the heart of the city. The new link was planned in a rather rural environment just south of the actual city centre. As the inhabitants of the communities of Voorschoten and Leiden were strongly opposed to another highway, the project could only move forward once all stakeholders and technical considerations had been addressed, which meant a solution comprising a bored tunnel and a long open cut was proposed.

Figure 1. Overall view of new N434 link.

The bored tunnel passes underneath a waterway, a railway, sports fields and several monumental, historical buildings. The open cut lies in the open field and crosses a waterway by means of an aqueduct and the A44 by an underpass. Although the inhabitants claimed for a bored tunnel solution for the entire length of the new link, this option was abandoned as the connection with the existing A44 motorway wasn't compatible with a bored tunnel solution up to this interchange. The concept of the open cut was as such that the nearby residents would not hear, see nor smell the new link.

2 SEGMENTAL LINING

2.1 Segment geometry

The internal diameter of the bored tunnel equals 9,79m while the lining has a thickness of 400mm. So called universal rings with an average width of 2006mm and a tapering of 34mm are adopted. The rather strange average width results from the contractual requirement that the distance in between two consecutive cross passages can never exceed 250m. A sensitivity analysis of possible ring joint thicknesses (due to compressed interstitial material) in combination with the 3D alignment of both tubes and the contractor's choice to have the break through always at the heart of the segment led to this value of segment width. Every ring consists of 7 individual segments, each covering approximately 51,4° of the ring.

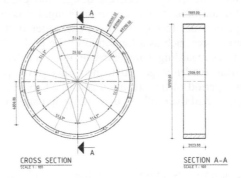

Figure 2.   Ring geometry - general.

The segments are 'cut' under an angle of 8° except from the last segment (A7) and the contact surfaces of the neighbouring segments (A1 and A6) – the so-called counter keys – having a cut angle of 12°. Further on the first and last installed segment of the ring have a trapezium shaped form while all other segments are parallelogram shaped.

2.2 Damage free lining

Comol5 investigated the track records regarding damage to the segmental lining that was observed during the construction phase of the existing bored tunnels in the delta region of the Netherlands. Main causes for damages were inaccurate handling and introduction of unallowable peak stresses when applying the push rams. These peak stresses were very often generated due to an unevenness in the ring joints. This unevenness is frequently related to the small shear keys. As these shear key segments are installed at the end of the ring installation process, it is sometimes hard to push the wedge-shaped segment to its full width into the ring. Furthermore, the small shear key has only one push ram acting on its perimeter. This means that when the adjacent segment of the next ring is being installed, temporarily the shear key is not held in position by any push ram or partly overlapping segment, and thus can be pushed out of its ring. By implementing the large key segment with its increased cut angles, an easy installation of the key segment is assured. The trapezium/parallelogram shaped segments on the other hand ensure that one push ram fixes one segment at all times during the installation process and thus prevents unallowable movements of the key segment.

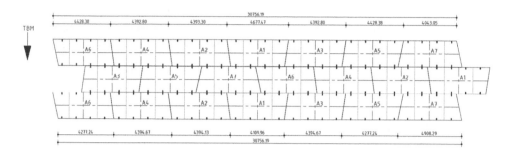

Figure 3. Ring geometry – developed view on intrados.

The damage due to high local peak stresses are countered by an increased splitting reinforcement on the one hand and the use of Datwyler Performance corner sealing gaskets on the other hand. The increased stiffness of the sealing gasket in the corner areas of the segments is a problem that has been recognized since a few years and that led to damage of the segmental lining in several projects in the past. The gaskets are built up by several closed 'chambers' and as such providing a high compressibility. This chamber structure can't be applied at the extremity of every segment side because the vulcanized connection of such sections at the corners are very difficult to produce, especially for the trapezoidal segments used in our project. Therefore, traditionally, the corners of the gasket profiles consist of full sections of solid rubber. As the compressibility of these full sections is much smaller than the regular gasket cross section, the corner pieces behave much stiffer and by this transfer significantly higher forces into the concrete segments for the same gap closure. Datwyler has very recently developed a solution adopting a prismatic full section corner piece. By adopting cutting angles adjustable to the design of the segment geometry, the amount of (full section) rubber is reduced in the corner areas. By doing this, the amount of injected rubber during vulcanization is reduced significantly. Investigation was done to verify the necessity to further decrease the amount of rubber in the corners by cutting the wings of the anchored gasket, but this wasn't the case for the maximum expected thrust forces.

Figure 4. Datwyler 'Performance corner': adapted chamfer (left) and injection compound volumes for 90° and 50° cutting angles (right).

2.3 Definition of the rebar cages

To define the rebar configuration of the standard type segmental lining, six normative cross sections were investigated. These cross sections were chosen on the basis of soil cover and geotechnical layering as can be seen in figure 5.

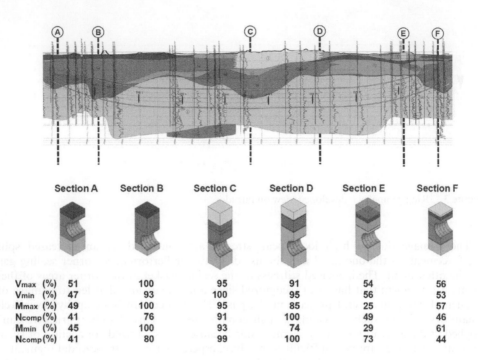

		Section A	Section B	Section C	Section D	Section E	Section F
Vmax	(%)	51	100	95	91	54	56
Vmin	(%)	47	93	100	95	56	53
Mmax	(%)	49	100	95	85	25	57
Ncomp	(%)	41	76	91	100	49	46
Mmin	(%)	45	100	93	74	29	61
Ncomp	(%)	41	80	99	100	73	44

Figure 5.   Normative sections and CP's along tunnel alignment and respective sectional forces.

The finite element software DIANA FEA is used. A 3D continuum model comprising 6 consecutive rings is adopted. The internal lining forces are derived from the middle two rings. The reinforcement quantities can be kept close to the minimum reinforcement ratio on basis of a cover of 70mm.

At both extremities of the bored tunnel, the tunnel drives are installed at close distance. Merely 2,6m or 0,25.D (D = tunnel outer diameter) spacing in between both tubes is available.

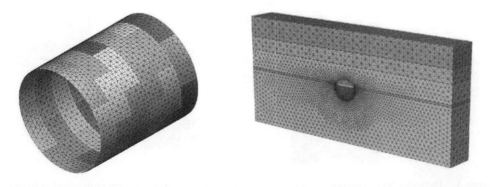

Figure 6.   Calculation model of 6 consecutive rings and soil-structure model.

A staged calculation comprising all construction steps of the two tubes showed that the flexural moments and normal forces in the lining of the firstly installed tube increased by respectively 40 and 10% at the moment the TBM of the second tunnel drive passed by. This increase was caused by the volume loss in the soil mass related to the conicity of the TBM and slightly larger cutting wheel. The same calculations showed that the tail void grouting, when applied in an accurate way, balances this adverse effect. In order to check whether the chosen standard rebar cage could cope with these sectional forces increase, the maximum sectional forces during the construction stages of the first tube were multiplied by 1,10 (normal forces) and 1,40 (bending moments). Still, the standard rebar cage was sufficient. This is not surprising, when looking at the values in the table of figure 5 one can see that the sectional forces at the extremities of the tunnels are significantly smaller than the normative ones used for the standard rebar cage definition.

The lining of the main tunnels at the location of the cross passages however needed a much heavier rebar cage. Two out of the eight cross passages were considered to be normative and structurally calculated. Cross passage 2 was located in clayey soil while all other CP's were located in the Pleistocene sand layer. Cross passage 5 on the other hand was the deepest CP. All construction steps (see also chapter 3) were run through using two different calculation models. In the first approach, a continuous tube of significant length and a cross passage having half of its actual length, due to symmetry aspect, were modelled. The results were only used to define the influence area of the cross passage. The influence area appeared to be approximately 8m out of the center of the CP. It can be expected that the influence area of a continuous tube will be exaggerated compared to a model comprising individual rings. This approach led to the more refined calculation where individual segments, joints, dowels and bicones (see next paragraph) were modelled. To reduce calculation times, the model was cut along additional planes of symmetry. A main tunnel length of 12m from the center of the CP was used. The highest bending moments were encountered in the ring that was cut through at the deepest cross passage located in the sand. The bending moment was more than 20% larger than the normative one found for the standard lining type sections. Therefore, a second rebar cage was designed.

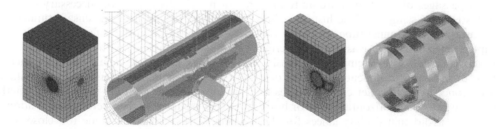

Figure 7.   Calculation models: continuous tube (left) and refined model (right).

2.4   Segmental lining connectors

In the ring joints of the standard lining type four thermoplastic dowels type SOF-FIX Anix 110 per segment are foreseen. At the locations of the cross passages and the junctions of the main tunnels with the shafts, the shear forces in between consecutive rings is much higher than the capacity of these dowels. At the locations of the cross passages one ring is cut when realizing the door opening. The existing hoop forces in the cut lining tend to 'close'. This tendency is partly mitigated by shear connectors, the so called bicones, installed in the ring joint. Per segment, four bicones type SOF shear 375 are installed. Depending on their relative stiffness compared to the strutting installed at the opening, the bicones closest to the opening are capable to transfer approximately 25 to 35% (per side) of the total hoop force to the adjacent rings.

At the junction of the tunnel with the shafts large shear forces and bending moments are introduced in the tunnel lining due to the sharp difference in support stiffness of the piled shaft and the existing soil mass. At these locations installing 'plain' bicones would solve the shear transfer but wouldn't prevent the ring joints to open due to the overall bending of the tunnel tube. Therefore, several consecutive rings are joined by longitudinal fixation. In order to avoid a third type of segment moulds (besides the moulds comprising dowels and those comprising bicones), the longitudinal fixation is implemented using hollow bicones as anchor plates for threaded steel bars type M25. Well-defined tolerances of the threaded lengths of the bars and the bicones enable the use of a limited set of different bar lengths compatible with tapered uniform tunnel rings instead of the theoretical 28 different lengths. The colour code applied on segments and bars facilitates the execution on site.

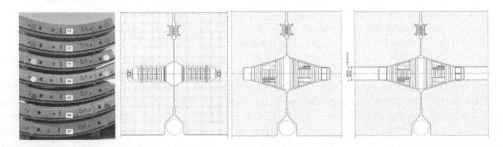

Figure 8.   Connection details a) color code post-tensioning b) dowels c) bicones d) bicones including post-tensioning.

3   CROSS PASSAGES

Contrary to the cross passages in previous bored tunnel projects, the clearance profile of the cross passages of the Rijnlandroute tunnel are not merely defined by the necessary space for pedestrians passing by. Case histories show that people fleeing from an unsafe tunnel tube towards the safe tunnel tube are hit by cars because the traffic in the safe tube wasn't interrupted yet. Therefore it was decided to create enough buffer capacity within the cross passages to temporarily 'store' fleeing people until the safe tube is really safe. The required surface for this storage capacity depends on traffic density in the tunnel and cross passages' interdistance. This concept led to wider cross passages than usually adopted. Comol5 investigated whether oval (large axis in horizontal sense) or horseshoe cross sectional shapes were better to adopt in order to avoid that drilling lances for the freezing works had to be made too close to the invert or crown of the main tunnel. However, purpose made equipment made it feasible to opt for the traditional circular cross section. The governing width, as this width varied in function of the length of the cross passage, was implemented for all eight cross passages in order to reuse the formwork multiple times.

Normally, the installed bicones in the ring joints transfer the hoop forces of the cut ring towards the adjacent rings. Depending on the forces during temporary stages, the amount of

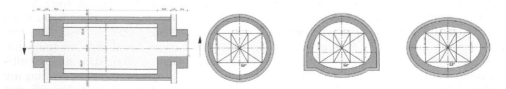

Figure 9.   Cross passage concepts a) storage capacity b) circular c) horseshoe d) oval.

bicones required per segment can mount up significantly and very often the installation of additional steel struts becomes necessary. Moreover, for the final stage of the tunnel structure, a complicated and robust concrete collar has to be made at the outside of the main tunnel lining in the more adverse working environment of the frozen soil mass. As the concrete works comprising the fire resistant doors at the inside of the main tunnels form a robust collar themselves, Comol investigated whether these structures could be realized before the breakthrough of the main tunnel lining. The structurally required dimensions of the collar had to allow the gantries to be pulled back after finalization of the first tunnel drive. In this way, the collars could be made directly after the passage of the TBM + gantries. Further on the collar dimensions should preferably not interfere with the freezing lances as this would complicate the already heavily reinforced collar solution.

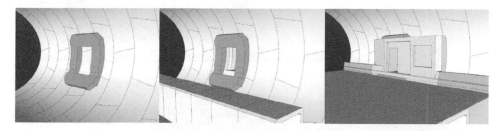

Figure 10. Internal collar execution steps: structural collar (left) – installation cable ducts (middle) – final stage (right).

The concrete collar is connected with the main tunnel lining using a large, but still feasible, amount of bonded fasteners. All construction stages as well as the final stage were calculated using three different models. At first a continuum model was used to have an idea about the bending moments to be expected. The second model was a simple framework model of the curved shaped collar. Analytically defined maximum hoop forces were introduced as load on the upper beam, while the lower beam was supported by springs as shown in figure 11. Finally a third model was elaborated. This was a framework model representing five consecutive rings, comprising the joints and connectors with their relevant characteristics. The upper and lower beam of the collar are connected to the entire rings at both sides of the breakthrough, while the columns are modelled as composite beams together with these same rings. All structural parts were designed using the normative sectional forces resulting out of one of these three different models.

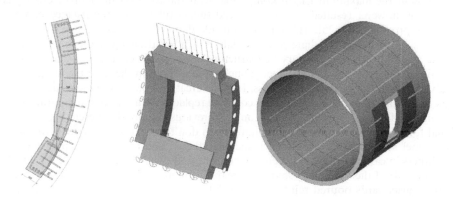

Figure 11. Internal collar: connection with bonded fasteners (left) – simple framework model (middle) – refined framework model (right).

4 BREAK-IN PROCEDURE USING A PERMANENT BELL STRUCTURE

Several concepts do exist to ensure a sound water tightness when breaking into the soil mass at the start of the tunnel drive, at least when tunnelling below the existing ground water level. The principle always comes down to have a sufficient length of tail void grouted before entering into the surrounding soil mass. A temporary dewatered cofferdam nor a starting plug of low strength mortar was adopted because of the absence of an existing water tight soil layer below tunnel invert level and the need for very expensive retaining walls due to the extremely soft upper soil layers.

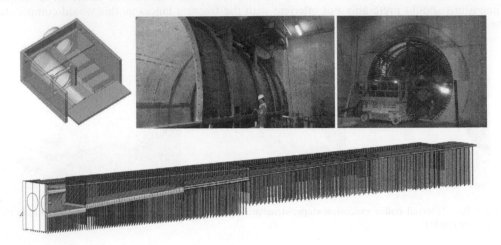

Figure 12. Break-in a) 3D visualization of bell installed for 2nd drive b) steel bell c) concrete bell d) 3D overall view of launching shaft and eastern access.

Therefore a bell solution was adopted. In this concept, a circular steel structure is installed inside the launching shaft. Afterwards the TBM is slid into the bell and starts installing the first (dummy) rings. Before the TBM passes the front retaining wall of the shaft, already 1 or 2 dummy rings are fixed by tail void grouting inside the bell and by this ensuring the required water tightness. Once the TBM has advanced sufficiently far into the soil, the steel bell can be dismantled and be reassembled for the second tunnel drive. The dummy rings are removed as well. It is clear that the steel bell construction needs to be robust to cope with the applied tail void injection pressures. Usually it is made out of a series of manageable heavy components.

Comol5 extended the length of the bored tunnel part, compared to the reference design, in order to obtain the maximum length compatible with the foreseen number of cross passages. This extension however resulted in closely spaced tunnel tubes. An interdistance between both tunnels of approximately 0,25D (D = tunnel outside diameter) was encountered. This together with the presence of an intermediate concrete slab, only a few decimeters above the tunnel crown turned the assembly of the temporary bell structure into a costly challenge. That's why Comol5 chose to fill up the entire space in between launching shaft bottom slab and the intermediate concrete slab by an unreinforced concrete mass. It is like creating a starting plug inside the launching shaft. By doing this, the steel bell structure could be replaced by a reusable temporary formwork and the intermediate slab could be poured on this previously realized concrete mass without any additional formwork. Furthermore the dummy rings don't have to be removed - this means the circular tunnel starts already inside the launching shaft – and the concrete plug increases the water tightness in case cracks would occur in the front wall during the passage of the TBM.

The floor slab of the launching shaft was realized by an unreinforced underwater concrete slab and an afterwards poured reinforced top slab. To define the connection with the lateral diaphragm walls, it was assumed that during temporary stages both slabs acted as a composite structure (big lever arm for upper rebar connection), while in permanent stage the water

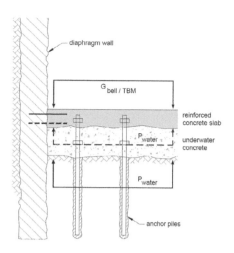

Figure 13.   Calculation model floor slab launching shaft.

pressure acts at the interface in between both slabs while no downward loads from bell nor TBM are present. This enabled the execution to choose up to the last moment to implement a temporary or permanent bell solution.

5   HORIZONTAL STABILITY OF THE SHAFTS

The horizontal stability of the launching and receival shafts are complex issues when situated in weak soil layers combined with high phreatic heads. Not only considerable earth and water pressures, even increased by the temporary embankments, have to be countered, but the forces exerted by the TBM have to be taken into account as well. Although adding the TBM forces in total to the determined earth and water pressures is conservative, this conservatism was judged wise as some aspects like the horizontal excentricity of the TBM or the jacking of the TBM inside the shaft before the actual start of the mining process were not envisaged as well.

The resulting horizontal loads acting on the front wall can be balanced by four resisting forces. These are the soil friction forces on the lateral diaphragm walls of the shaft ($F_{fr;dw}$), the soil friction forces on the temporary lateral sheet pile walls of the adjacent cut-and-cover tunnel sections ($F_{fr;shp}$), the passive earth pressure acting on the compartment screen in between the launching shaft and cut-and-cover tunnel ($F_{pe;cs}$) and the earth pressure acting on the end faces of the lateral diaphragm walls ($F_{pe;dw}$). All available resisting components were introduced in the calculation model with a stiffness compatible with their individual load-deformation behavior.

Firstly the Service Limit State (SLS) loads were introduced into the calculation model. Almost all resistance was found by mobilizing the soil friction on the diaphragm walls. When increasing the horizontal loads by a load factoring of respectively 1,5 – 2 and 3 it can be seen that the soil friction on the sheet piles and finally the passive earth pressures on compartment screen and diaphragm walls are used to greater extent. Even in case of the load factor = 3, the encountered deformations are within acceptable limits.

As the friction on the diaphragm walls is, even in the SLS scenario, fully mobilized, this friction couldn't be used for a second time when designing the diaphragm walls as retaining walls. Further on, first calculation results made clear that the separate diaphragm wall panels would behave like books on a book shelve when subjected to these horizontal loads and this led to unacceptable deformations. Therefore it was decided to create a stiffening top beam on top of these diaphragm wall panels linking them together. This mechanism introduced quite substantial tensile forces at the extremity of each panel situated the farthest from the front shaft wall. To cope with these tensile forces, additional rebar had to be installed in the diaphragm cages.

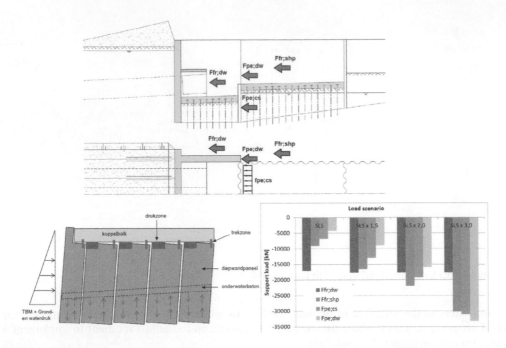

Figure 14. Horizontal stability a) schematic view resisting forces b) structural system of diaphragm wall panels with capping beam c) load distribution.

6 DESIGN OF THE SOFT EYE

Mining through the frontal diaphragm walls is enabled by adopting GFRP rebar, the so-called 'soft eye'. The rebar cages comprise traditional steel rebar above level -6m NAP and below level -18m NAP. The 12m long section in between, where the tunnel is situated, consists of GFRP rebar. As the Young's modulus of GFRP is merely 40 GPa, the stiffness of the GFRP reinforced diaphragm wall equals 5GPa compared to the value of 12 GPa adopted for the cracked steel reinforced diaphragm wall. The tensile strength of the GFRP rebar is strongly depending on the production process and thus supplier specific. Further on, this strength decreases with time due to the alkaline environment of concrete. The lifetime of the soft eye was set equal to a maximum of 3 years. Comol5 chose for the Italian supplier ATP because they offered fully closed stirrups and oval bars (75-40mm), both aspects reducing the total amount of rebar/layers required and thus improving the bentonite-concrete exchange process. The adherence of these bars is assured by 'sanding' the outer surface before hardening of the epoxy-matrix during the production process. As there was some doubt about the resulting adherence of these bars in a diaphragm wall

Figure 15. The soft-eye a) steel/GFRP repartition in rebar cages b) oval GFRP bars with sand coating c) pull-out test set-up.

set-up, where a remaining bentonite film could decrease this adherence significantly, a pull-out test was organized. Therefore, short bars with a well-defined adherence length were installed in the upper part of the diaphragm wall cages. After hardening of the diaphragm walls, these bars were pulled out using hollow jacks. Where similar tests in regular concrete resulted in adherence strengths of 9 MPa, our tests simulating the diaphragm wall process gave values of 4 to 5 MPa. These values were sufficiently higher than the required value of 1 MPa according the design. To reduce the amount of stirrups and create sufficient safety regarding brittle fracture (GFRP shows a linear stress-strain behavior until breakage, so no warning by extreme deformations), the concrete strength of the diaphragm walls was increased from C30/37 to C35/45.

7   BEHAVIOUR RUBBER GASKETS SEGEMENTAL LINING DURING FIRE CASE

The client requested that the rubber gaskets of the segmental lining wouldn't be heated up above 80° Celcius during a two hours lasting fire case scenario according the RWS curve. This criterion was based on outdated values regarding immersed tunnel joints and thus too stringent as the EPDM rubber gaskets can easily sustain a temperature increase up to 130° Celcius. Comol5 tried to comply to this requirement by submitting a numerical analysis of the segmental joint assuming full concrete to concrete contact and calibrated on the measurement data of the sensors during the fire tests on the segmental lining. This approach resulted in a maximum temperature at the joint of 64°C, taking into account the recess in the ring joints where cables will be installed.

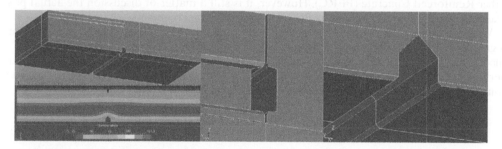

Figure 16.   FEM model: entire model (top left) – detail gasket (middle) – detail cable recess (right) – temperatures after 400 minutes (bottom left).

The owner wanted Comol5 to verify whether a possible ring joint gap opening, in combination with the small circumferential recess along the segment's perimeter would not lead to significant temperature increases due to convection and/or direct radiation on top of the heat conduction through the concrete. Therefore Comol5 performed additional CFD calculations assuming gap widths of respectively 0mm, 3mm, 6mm and 20mm, the latter a rather unrealistic value used to better grasp the parameter's influence. The results of the 0mm model confirmed the earlier obtained result of 64°C, while the results for the gap widths of 3mm and 6mm showed a rather beneficial effect due to the insulating aspect of the air in the gap. Only when modelling an exaggerated gap width of 20mm, the influence of convection was detected and the gasket temperature rose to approximately 180°C.

8   CONCLUSION

The Rijnlandroute project introduces many small improvements to the nowadays common practice within the field of mechanized tunneling. Contrary to the strong persuasion of many structural and geotechnical engineers, it can be seen that the civil construction business is a continuously improving and innovating world.

Tunnels and Underground Cities: Engineering and Innovation meet Archaeology,
Architecture and Art, Volume 6: Innovation in underground engineering,
materials and equipment - Part 2 – Peila, Viggiani & Celestino (Eds)
© 2020 Taylor & Francis Group, London, ISBN 978-0-367-46871-2

An experimental study on the post-cracking behavior of Polypropylene Fiber Reinforced Concrete core samples drilled from precast tunnel segments

A. Mudadu, G. Tiberti, A. Conforti, I. Trabucchi & G.A. Plizzari
University of Brescia, Brescia, Italy

ABSTRACT: In the scientific community and among designers there is a growing interest on macro-synthetic fibers for use in underground structures, especially for precast tunnel segmental linings. Moreover, in the last decade, important research efforts have been devoted to the development of new types of structural macro-synthetic fibers, which are now able to impart significant toughness and ductility to concrete. The enhanced post-cracking strengths of concrete, which are provided by the presence of polypropylene fibers, can be included in analytical and numerical approaches for designing precast tunnel segments in Polypropylene Fiber Reinforced Concrete (PFRC). However, it is still a matter of discussion the actual post-cracking performances exhibited by PFRC in precast tunnel segments.

Within this framework, an experimental study on core samples drilled from precast tunnel segments in PFRC (considering different positions and directions) was carried out in order to shed some new lights on fiber distribution and orientation. The actual PFRC post-cracking performances evaluated by uniaxial tensile tests on drilled core samples from segments, were compared against the results of typical flexural standard tests on notched beams.

1 INTRODUCTION

In the last decade, important research efforts have been devoted to the development of some types of structural macro-synthetic fibers, which are now able to impart significant toughness and ductility to concrete, as well as fiber long-term behavior is guaranteed. Moreover, in the scientific community and among designers there is a growing interest on macro-synthetic fibers for use in underground structures in combination with traditional reinforcement, especially with regard to precast tunnel segmental linings (Conforti et al., 2017; Gilbert and Bernard, 2015); since macro-synthetic fibers do not suffer corrosion problems.

The enhanced concrete post-cracking properties due to polypropylene fibers should be considered in the design process of PFRC precast tunnel segments, either by means of non-linear analytical approaches or finite element (FE) models (ITA Report n.16, 2016; *fib* bulletin n.83, 2017).

Nowadays it is a matter of fact that the post-cracking performance is strongly influenced by the fiber orientation and distribution, as proven in previous research works (Tiberti et al., 2017; Mudadu et al., 2018) at sample scale. Analogously, at structural scale, the load-bearing capacity of the statically indeterminate structure is considerably influenced by fiber efficiency due to the redistributing capacity of a cracked cross-section (Kooiman, 2000; Lofgren, 2005; Bentur and Mindess, 2005). In this regard, *fib* Model Code 2010 (2013) (hereafter MC2010) recognizes that the behavior observed in standard tests can deviate substantially from the behavior of the corresponding FRC at the structural level (Laranjera et al., 2011; Sanal et al., 2013). Thus, MC2010 (2013) suggests the introduction of an orientation factor (K) for taking into account the actual fiber orientation and distribution in the structure with respect to that of small samples generally used for standard tests. Thereby, assuming a structure, the proper knowledge of fiber distribution and orientation relative to the direction of the tensile stresses becomes of paramount importance; nevertheless, these parameters are somewhat intricate to predict since they are affected by several factors: mould geometry, constrictions (wall effect), flow length, flow thickness, concrete workability and fiber length. Different methods, either "destructive" or "non-destructive", can be used in order to count fibers in cross-sections, specimens or structural elements.

The fiber distribution in precast elements could be studied using drilled cores from FRC's elements, with the fiber orientation determined using the above reported methods and the fiber content determined, for instance, by crushing the cores. Using this technique, Kooiman (2000) studied the fiber orientation and distribution in precast tunnel segments reinforced with 60 kg/m^3 of steel fibers. The test results showed that the quality of the composite material varied considerably over the segment thickness since bleeding occurs during compaction. Moreover, the orientation was found to vary significantly over the segment. Due to this orientation but also as a result of the direction of the applied loads, the fiber efficiency in tunnel segments varies. Grünewald (2004), using X-ray images from the drilled cores of self-compacting SFRC precast tunnel lining segments, observed that fibers were distributed homogeneously along the height of the cores. Furthermore, deformation-controlled splitting tensile tests were carried out and slices were sawn from cylinders in order to take X-ray photographs. The orientation of the fibers was determined and related to the performance of the cylinders (splitting tensile test). Experimental results proved that the orientation of the fibers was perpendicular to the flow due to the production process. During the concrete flow through the pipe, its walls oriented the fibers; afterwards concrete drops vertically into the mould and distributes in a circular stream pattern. Moreover, at the casting point, fiber content was higher, which causes the orientation number to increase. More recently, Carmona et al. (2016) studied the actual fiber contents obtained by means of crushed cores drilled from different points of three full-scale tunnel lining segments; higher value of fiber content was reached in the center zone, where the concrete drops from the hopper and the last layer does not flow.

Within this framework, this research study would like to shed some new lights on fiber orientation and distribution on precast tunnel segments reinforced with polypropylene fibers. Core samples were drilled at different location, directly from three precast tunnel segments, whose were tested under flexure and point load tests in a previous research (Conforti et al., 2017). It is worth noting that such cylinders were drilled along two different direction: tangentially and radially to the central axis of the segment. A total amount of 22 samples were tested under Uniaxial Tensile Tests (UTTs) and the post-cracking performances of segments were also compared against the results of typical flexural standard tests on notched beams.

2 EXPERIMENTAL PROGRAM

Core samples were extracted along two different directions (tangential and radial to the central axis) from three precast tunnel segments; eventually, they were submitted to uniaxial tensile test (UTTs) to evaluate the impact of fiber orientation and distribution on the post-cracking strength. The results obtained from UTTs were compared to those obtained from standard tests carried out on small samples cast with segments.

2.1 Materials and specimens geometry

In a previous research full-scale counter-key segments characterized by three different reinforcement configurations were cast and tested under flexural and point load tests (Conforti et al., 2017). The tunnel segments present an internal diameter of 3200 mm, a thickness of 250 mm (external diameter of 3700 mm), leading to a lining aspect ratio equal to 12.8. Three of these segments were selected for the research study herein presented; they were characterized by two different reinforcement solutions: PP fibers only (PFRC) and a combination of PP fibers and conventional reinforcement (hybrid solution, RC+PFRC). The former was adopted in two segments reinforced only by 10 kg/m³ ($V_f = 1.10\%$) of macro-synthetic PP fibers; whereas the latter represents a combination of conventional reinforcement and 10 kg/m³ of macro-synthetic PP fibers. In both reinforcement configurations macro-synthetic embossed PP fibers having a length of 54 mm (L_f) and a diameter (φ_f) of 0.81 mm (leading to an aspect ratio (L_f/φ_f) of 67) were adopted; the tensile strength (f_{uf}) and Young modulus (E) of fibers were about 552 MPa and 6 GPa, respectively.

The segments were cast by means of precast steel moulds and consolidated through the vibration mould system. A local ready mix concrete supplier produced three concrete batches according (for more details concerning the mix-design adopted, see Conforti et al, 2017). The target cylindrical compressive strength at 28 days was of about 48 MPa, corresponding to a C40/50 concrete class according to Eurocode 2 (2004).

Each segment was cast by a singular concrete batch; for each batch, six cubes (150 mm side) for measuring the compressive strength, ten small beams (150x150x550 mm, Figure 1a) for the evaluation of PFRC residual flexural tensile strengths according to EN 14651 (2005) were produced. Both cubes and beams were cured under the same environmental conditions adopted for the full-scale specimens. On the other hand, the UTTs were carried out on drilled cores (hereafter named "Drilled") having a diameter (φ_{cyl}) of 94 mm and a height (h_{cyl}) of 200 mm (Figure 1b). As shown in Figure 1b, the "Drilled" core samples were obtained directly from the precast tunnel segments by drilling along local tangential direction (hereafter named "tangential sample, T") and radially (named "radial sample, R") to the segment axis (Figure 1b).

By taking as reference the loading zone of TBM thrust shoes, three different zones have been distinguished in the segment (Figure 1b): middle zone (MZ); load zone (LZ) and corner zone (CZ). In order to simplify the drilling process of the tangential sample, a slice having a length equal to the height of the core (about 300 mm) was obtained from the segment by means of a peak-shaped diamond circular saw; afterwards cores were properly drilled from the slice.

The combination of the previously mentioned parameters defines a specific series, identified according to the notation T-X-Y-Z-W-J-K-#, where: T represents the type of reinforcement solution, X the $f_{cm,cube}$ value, Y the fiber dosage (V_f), Z the fiber aspect ratio (L_f/Φ_f), W the fiber filament ultimate tensile strength (f_{uf}), W the segment designation, J the location of the core, K the drilling direction and # the number of the cylinder.

All cores extracted were sawed to obtain cylinders with height of about 200 mm. Moreover, cylinders surfaces were prepared carefully through a grind and polish process for obtaining parallel specimens ends, in order to guarantee an adequate bonding with the platens and to avoid, as much as possible, any eccentricity. In order to acquire the stress-crack opening relationship, a triangularly shaped notch was made in the middle section through a lathe to a depth of 5 mm. A detail of the notched specimens is shown in Figure 2b.

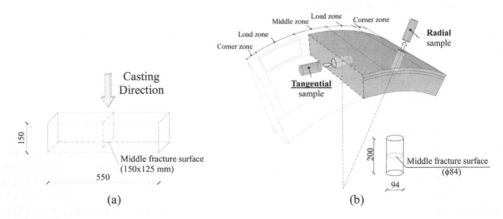

(a)

(b)

Figure 1. Casting directions of beam according to EN 14651 (2005) (a) and drilled cores samples from tunnel segments (b) [dimensions in mm].

Table 1. Main mechanical properties of PFRCs and specimens tested.

Concrete designation	Type and # of specimens tested		Concrete mechanical properties		
	# Beam EN14651	# Cores Drilled	$f_{cm,cube}$ [MPa]	f_{ctm} [MPa]	E_c [MPa]
PFRC	10	16	60.1 (0.03)	3.62*	36700*
RC+PFRC	10	6	60.0 (0.02)	3.61*	36700*

* Value retrieved according to *fib* Model Code 2010 (2013)

The mean compressive cubic strength ($f_{cm,cube}$), the mean tensile strength (f_{ctm}) the Young modulus (E_c) were retrieved according to MC2010 (2013), as listed in Table 1. In addition, the specific number of specimens tested under 3PBTs and UTTs is summarized in Table 1 as well.

2.2 *Three Point Bending Test: test set-up and loading history*

3PBTs were performed on notched beams according to European Standard EN14651 (2005), as suggested by the *fib* Model Code 2010 (2013) for determining post-cracking material properties.

All the specimens were tested with an INSTRON hydraulic servo-controlled (closed-loop) testing machine with MTS control, having a capacity of 500 kN. The tests set-up is shown in Figure 2a. The Crack Mouth Opening Displacement (CMOD) was measured by means of a clip gauge (having a range of measurement of 5 mm and a sensitivity of 2.5 mV/V), positioned astride the notch (with a depth of 25 mm) at midspan. Additional Linear Variable Differential Transformers (LVDTs) were used to measure the Crack Tip Opening Displacement (CTOD) as well as the vertical Load Point Displacement (LPD).

The tests were carried out by using the CMOD as a control parameter (up to 4 mm). Two loading rates were employed, namely 0.05 mm/min (up to CMOD=0.1 mm) and 0.2 mm/min (up to CMOD=4 mm); afterwards, tests continued under LPD control until descending to 1% of the maximum load (P_{max}) or a CTODm of 8 mm.

2.3 *Uniaxial Tensile Test: test set-up and loading history*

In Figure 2b, the experimental set-up and the instrumentation employed are depicted. The test were carried out by using a very stiff hydraulic servo-controlled (closed-loop) testing machine, whose axial stiffness is sufficient to ensure stable tests without any snap-back in the averaged displacement transducer signal. The tests were controlled by means of the mean value of the Crack Opening Displacement (CODm), obtained from three clip-gauges (having a range of measurement of 5 mm and a sensitivity of 2.5 mV/V) placed astride the notch, around the cylindrical specimen at 120 deg. Three additional Linear Variable Differential Transformers (LVDTs, having a range of 20 mm and a sensitivity of 80 mV/V), were placed radially at 120 deg astride the notch to measure the mean displacement (δm) on a base length of 40 mm (Figure 2b and c). The test setup is well-explained in detail in previous research (Mudadu et al., 2018).

Four loading rates were adopted during the test: 3 μm/min (up to COD_m=0.03 mm), 9 μm/min (up to COD_m=0.1 mm), 30 μm/min (up to COD_m=0.5 mm) and 90 μm/min (up to COD_m=3.5 mm). Afterwards, tests continued under stroke control at a rate of 600 μm/min until descending to 1% of the maximum load (P_{max}). The value of all the measuring devices were recorded at a rate of 1 Hz.

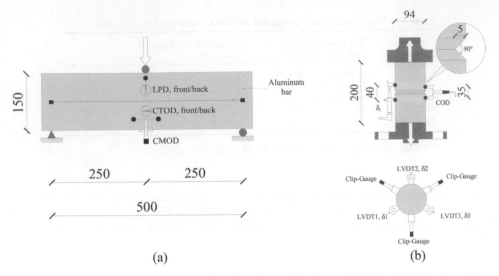

Figure 2. Test set-up and instrumentation layout for 3PBTs (a) and UTTs on drilled core sample (b).

3 POST-CRACKING UNIAXIAL CONSTITUTIVE LAWS

In this section, uniaxial post-cracking tensile behavior obtained from both 3PBTs and UTTs are presented and discussed. The former are indirectly obtained through inverse analysis procedure (Roelfstra and Wittmann, 1986), while the latter are directly obtained from experimental results (Mudadu et al., 2018).

3.1 Post-cracking tensile law from 3PBTs

Table 2 lists the main results of flexural tests on notched beams in terms of limit of proportionality (mean value, f_{Lm}) and post-cracking nominal residual strengths (mean values, f_{R1m}, f_{R2m}, f_{R3m}, f_{R4m} corresponding to CMOD values of 0.5, 1.5, 2.5 and 3.5 mm, respectively). It worth mentioning the significant structural performances provided by PP fibers at both serviceability limit states and ultimate limit states.

As an example, Figure 3 shows the nominal stress vs. CMOD (Crack Mouth Opening Displacement) curves from ten small beams (150x150x550 mm), tested according to EN14651 (2005) for the fracture characterization of PFRCs adopted.

The stress crack width (σ-w) laws of PFRCs were evaluated through the inverse analysis procedure (Roelfstra and Wittmann, 1986); by assuming a multi-linear post-cracking cohesive law (according to the fictitious crack model, FCM (Hillerborg et al., 1976), non-linear numerical analyses were performed for determining the best-fitting law. The numerical analyses were

Table 2. Residual flexural tensile strengths (EN 14651, 2005) of PFRC and RC+PFRC segments.

Segment designation	f_{Lm} [MPa]	f_{R1m} [MPa]	f_{R2m} [MPa]	f_{R3m} [MPa]	f_{R4m} [MPa]
PFRC	4.90 (0.05)	2.97 (0.12)	4.01 (0.16)	4.61 (0.15)	4.83 (0.16)
RC+PFRC	4.47 (0.11)	3.05 (0.26)	4.22 (0.28)	4.66 (0.29)	4.79 (0.28)

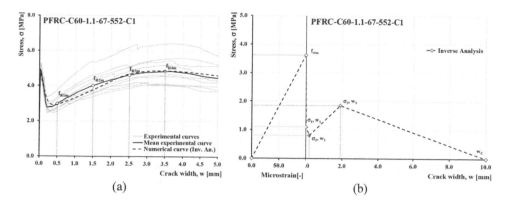

(a)   (b)

Figure 3.   Typical experimental results from 3PBT on standard specimens cast with PFRC tunnel segment with evidence numerical flexural curve obtained from Inverse Analysis (a); post-cracking constitutive law adopted (b).

Table 3.   Main parameters of the σ-w laws obtained from Inverse Analysis based on results from 3PBTs on standard specimens cast with segments.

Segment batch designation	f_{ct} [MPa]	σ_1 [MPa]	w_1 [mm]	σ_2 [MPa]	w_2 [mm]	σ_3 [MPa]	w_3 [mm]	w_c [mm]
PFRC	3.62	1.10	0.023	0.80	0.20	1.85	1.90	10.00
RC+PFRC	3.61	1.50	0.019	0.85	0.15	1.85	1.80	10.00

carried out by using a discrete crack approach; the FE mesh was made by triangular plane stress elements and interface elements for the crack.

The post-cracking constitutive laws were obtained by fitting the numerical curve with the mean flexural experimental one, in order to achieve a maximum error of 5%. The best-fitting numerical curve for specimens C60-1.1-67-585 is plotted with the experimental ones in Figure 3b. The best-fitting post-cracking constitutive laws obtained from the Inverse Analysis is depicted in in Figure 3c.

The main parameters of the post-cracking σ w laws, as shown in Figure 3b, for all the series investigated, are summarized in Table 3. The tensile strength (f_{ct}) was assumed according to Table 1. In fact, in case of drilled cylinders, the tensile strength could be lower as result of the drilling process making these data not reliable. The post-cracking stresses (σ_i) and the corresponding crack openings (w_i) of multi-linear laws adopted are reported as well (Table 3).

3.2   Post-cracking tensile law from UTTs

The experimental results of UTTs carried out on the core samples drilled from the three segments are shown in Figure 4. For each sample, the crack width (*w*) is calculated from the mean displacement δ_m through the base length (l_{meas}) which stands for the base length of the three LVDTs, equal to 40 mm (Figure 2b).

More in detail, Figure 4a, b and c show the results of the UTTs carried out on the samples drilled from the segment named C1 (PFRC), C3 (RC + PFRC) and C4 (PFRC) respectively.

As mentioned above, samples from segments C1 and C3 were extracted along the two orthogonal direction with respect to the axis of the element: tangentially and radially. Moreover, as far as the segment C1 is concerned, the influence of the different drilling position was evaluated as well, since cores were drilled along the load zone as well as the middle zone (Figure 1b).

Within this framework, as sake of simplicity, graphs plot the mean experimental curves retrieved from the samples having both the same drilling direction (tangentially (T) versus radially (R) drilled) and drilling position (load zone (LZ) versus middle zone (MZ)).

It is worth noting that all the curves show an initial load drop generally followed by a strength rise, with exception of tangential cylinders, independently from their position (Figure 4), which exhibit only a slight improvement of resistance beyond the drop, as result of an unfavorable scenario concerning the fiber orientation and distribution.

On the contrary, beyond the load drop, radially drilled core sample are characterized by a noticeable enhancement in term of toughness; the post cracking maximum tensile load is about double with respect to the value at the load drop. Furthermore, it should be noticed that the drilling position strongly affect the specimen performance under UTTs. Basically, cylinders taken out radially from the middle zone exhibit lower post-cracking strength than those belonging to the load zone (Figure 4a).

These tendencies can be also proven by considering as a reference the PFRC segment "C1" in order to compare the experimental results of UTTs for crack openings of 0.5 mm and 2.5 mm, these latter values can be considered significant for the Serviceability Limit State (SLS) and Ultimate Limit State (ULS), respectively.

When considering the "load zone" the mean experimental curve of core samples drilled radially is noticeably higher than that of tangentially drilled core samples over the whole crack width. In fact, when referring to a crack opening of 0.5 mm, the corresponding residual strength $\sigma_{0.5}$ exhibited by series C60-1.1-67-552-C1-LZ-R is equal to 1.31 MPa, while the same parameter for series C60-1.1-67-552-C1-LZ-T was 0.30 MPa. Similarly, when referring to $\sigma_{2.5}$, the corresponding values observed are 1.63 MPa and 0.47 MPa, respectively (Figure 4a).

As mentioned above, the orientation between drilling direction and tensile stress is not the only factor that seems to affect the post-cracking performance. In this regard, it should be

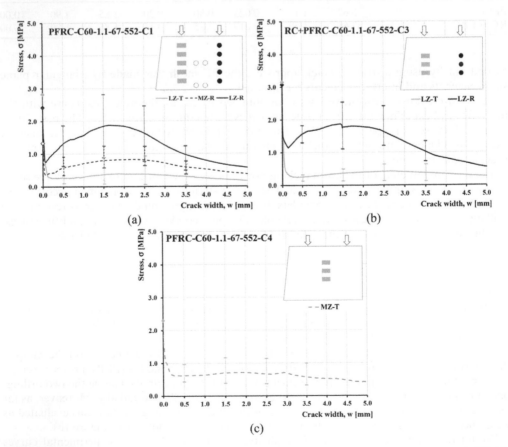

Figure 4. Stress - crack width curves for cores drilled specimens: PFRC-C60-1.1-67-552-C1 (a), RC+PFRC-C60-1.1-67-552-C3 (b), PFRC-C60-1.1-67-552-C4 (c).

noticed the differences between the series C60-1.1-67-552-C1-LZ-R and C60-1.1-67-552-C1-MZ-R. In fact, even though cylinders belonging to these series were both drilled radially, the corresponding post cracking behaviors are still different since they were drilled from different zones (Load zone versus Middle zone). In particular, cylinders from the middle zone have exhibited residual strength of about one half with respect to those taken out from the load zone; for series C60-1.1-67-552-C1-MZ-R $\sigma_{0.5}$ was equal to 0.57 MPa whereas for series C60-1.1-67-552-C1-LZ-R $\sigma_{0.5}$ was equal to 1.31 MPa (Figure 4a).

Furthermore, it is worth noting that in the middle zone both radial and tangential drilled core samples show a similar post-cracking strength. By comparing series C60-1.1-67-552-C1-MZ-R and C60-1.1-67-552-C4-MZ-T (Figure 4a and Figure 4c), it should be noted that the stresses are comparable for the whole range of crack widths, for instance for series C60-1.1-67-552-C4-MZ-T, $\sigma_{0.5}$ and $\sigma_{2.5}$ are equal to 0.62 MPa and 0.66 MPa, respectively whereas they are equal to 0.57 MPa and 0.81 MPa for series C60-1.1-67-552-C1-MZ-R.

Finally, the consistency and repeatability of the above reported results was also proven by considering radially extracted specimens from the load zone of RC+PFRC segment C4 (Figure 4b), whose post-cracking properties are noticeably higher than tangentially drilled core samples.

4   POST-CRACKING TENSILE LAW COMPARISON

The experimental uniaxial post-cracking laws obtained by performing UTTs (see Section 3.2) are herein compared with those determined indirectly by means of the inverse analysis procedure (see Section 3.1). As an example, results of series PFRC-C60-1.1-67-552-C1/C4 are presented. The mean uniaxial tensile laws, as obtained from direct and indirect methods, are compared in Figure 5a and b, respectively for segments C1 and C4. Note that results from UTTs were retrieved by averaging the experimental curves. One should observe that the constitutive laws retrieved indirectly from the 3PBTs, are able to predict only the behavior of cylinders taken out radially in the load zone (Figure 5a).

Particular attention was devoted to the behavior of the samples drilled from the middle zone of the segment, since this latter zone is noticeably stressed during several transitional stages (demoulding, storage, transportation and positioning); hence the representativeness of tests suggested by the current codes have to be proven and will be herein discussed.

Nowadays it is a matter of fact that the post-cracking performance is strongly influenced by the fiber orientation and distribution. In this regard, MC2010 (2013) suggests the introduction of an orientation factor (K), although the method for a proper evaluation of fiber orientation and distribution is not explicitly declared by the code. In this research, as far as the orientation factor (K) is concerned, results from the wide experimental campaign were used to suggest a useful approach. Basically, by considering the uniaxial constitutive law obtained directly from UTTs on drilled core samples and, indirectly, by means of inverse analysis carried out on the 3PBT results, the orientation factor (K) was retrieved for each crack opening as a ratio between the residual strengths obtained from inverse analysis (representative the sample level) and UTTs (representative the structure level).

Due to this ratio, a K-factor higher than the unity indicates that results obtained from the material characterization carried out through standards tests tend to overestimate the actual structural behavior; this case stands for "unfavorable effects" of fiber orientation and distribution at the structural level with respect to the sample level. To the contrary, K-factor negative (less than the unity) stands for "favorable effects".

For precast tunnel segments, when referring to the middle zone, by averaging the K-factor for the whole range of crack openings investigated and shown in Figure 5, a K-factor equal to about 2 was obtained for the middle zone; which means that, in the middle zone, an unfavorable fiber orientation and distribution seems to occur at structural level. Moreover, a similar K-factor for cylinders tangentially extracted from the middle zone (C1, Figure 5a) and radially extracted (C4, Figure 5b) was obtained.

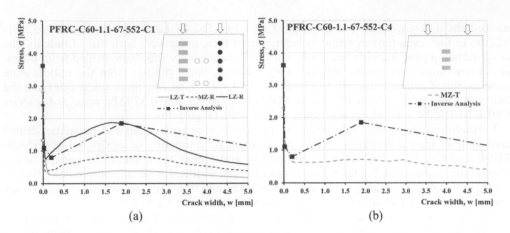

Figure 5. Comparison between uniaxial constitutive law retrieved from UTTs on drilled cores and by means of inverse analysis technique: PFRC-C60-1.1-67-552-C1 (a), PFRC-C60-1.1-67-552-C4 (b).

5 CONCLUSIONS

The present paper concerns the post-cracking behavior of concretes reinforced by polypropylene fibers (PP fibers), by means of bending and uniaxial tensile tests (UTTs). Core samples drilled from precast tunnel segments in PFRC (considering different positions and directions) were tested and the actual PFRC post-cracking performances, as determined by UTTs on drilled core samples, were compared against the results of typical flexural standard tests on notched beams.

Based on results and discussion presented, the following main concluding remarks can be drawn:

1) when considering precast tunnel segments reinforced by PP fibers, cylinders drilled along two different directions (radially and tangentially to the segment axis) in the load zone and tested under UTTs, lead to results noticeably different. Basically, core samples drilled radially exhibit higher post-cracking performance than tangential ones.
2) When considering the middle zone, both radial and tangential cylinders show similar post-cracking behaviors. This tendency could be due to the casting procedure, since concrete is poured in the mold exactly from the middle zone.
3) With exception of cylinders taken out radially from the load zone, uniaxial constitutive laws obtained directly through UTTs on core samples are lower than those retrieved by means of the inverse analysis procedure carried out on the results of 3PBTs on notched beams.
4) As far as the structural behavior is concerned, the influence of fiber distribution and orientation on small-sample performances have to be evaluated through the orientation factor (K). In this regard, when referring to the middle zone, by averaging the K-factor for the whole range of crack openings, a K-factor around 2 was obtained.

ACKNOWLEDGMENTS

The Authors are grateful to Heilderberg Cement Group (Bergamo, Italy) for the facilities made available and to BASF Construction Chemicals Italy for the financial support. The Authors would like to give their appreciation to Engineer Giancarlo Ghiroldi for the assistance in performing the experimental program. A special thank goes to Professor Alberto Meda and Engineer Angelo Caratelli for their kind availability and support during the drilling process of core samples.

REFERENCES

Bentur, A. & Mindess, S. 2007. Fibre Reinforced Cementitious Composites; 2nd edn. *Taylor & Francis*, UK, 2007, ISBN: 978-0-415-25048-1.

Carmona, S., Molins, C., Aguado, A. & Mora, F. 2016. Distribution of fibers in SFRC segments for tunnel linings; *Tunnelling and Underground Space Technology*, 51, pp. 238–249.

Conforti, A., Tiberti, G., Plizzari, G.A., Caratelli, A. & Meda, A. 2017. Precast tunnel segments reinforced by macro-synthetic fibers; Tunnelling and Underground Space Technology, 63, pp. 1–11, DOI: 10.1016/j.tust.2016.12.005.

Di Prisco, M., Plizzari, G. & Vandewalle, L. 2009. Fibre reinforced concrete: new design perspectives; *Material and Structures*, 42, 9, pp. 1261–1281.

EN 14651. 2005. Test method for metallic fibre concrete – measuring the flexural tensile strength (limit of proportionally (LOP), residual); European Committee for Standardization, p. 18.

EN 1992-1-1. 2004. Eurocode 2: design of concrete structures – Part 1: general rules and rules for buildings; European Committee for Standardization, p. 225.

fib bulletin No. 83. 2017. *Precast tunnel segments in fibre-reinforced concrete*; fib WP 1.4.1, p. 168, ISBN: 978-2-88394-123-6.

fib Model Code for Concrete Structures 2010. 2013. Ernst & Sohn, p. 434, ISBN 978-3-433-03061-5.

Gilbert, R. I. & Bernard, E. S. 2015. Time-dependent analysis of macro-synthetic FRC sections with bar re-inforcement; in: ITA/AITES World Tunnel Congress 2015 and 41st General Assembly, May 22–28,2015. Lacroma Valamar Congress Center, Dubrovnik, Croatia.

Grunewald, S. 2004. Performance-based design of self-compacting fibre reinforced concrete, *Ph.D. Thesis*, Technische Universiteit Delft, Duitsland, DUP Science.

Hillerborg, A., Modeer, M. & Petersson, P.E. 1976. Analysis of crack formation and crack growth in concrete by means of fracture mechanics and finite elements; *Cement and Concrete research*, 6, pp. 773–782.

ITA report n. 16. 2016. *Twenty years of FRC tunnel segments practice: lessons learnt and proposed design principles*; p. 71, ISBN 978-2-970-1013-5-2.

Kooiman, A.G. 2000. Modelling Steel Fiber Reinforced Concrete for Structural Design; *Ph.D. thesis*, Delft University of Technology, Optima Grafische Communicatie, Rotterdam, ISBN: 90-73235-60-X.

Laranjera, F., Grunewald, S., Walraven, J., Blom, C., Molins, C. & Aguado, A. 2011. Characterization of the orientation profile of steel fiber reinforced concrete"; *Materials and Structures*, 44, pp. 1093–1111.

Löfgren, I. 2005. Fibre-reinforced Concrete for Industrial Construction: a fracture mechanics approach to material testing and structural analysis; *PhD Thesis*, Department of Civil and Environmental Engineering, Chalmers University of Technology, Göteborg.

Mudadu, A., Tiberti, G., Germano, F., Plizzari, G. & Morbi, A. 2018. The effect of fiber orientation on the Post-cracking behavior of Steel Fiber Reinforced Concrete under bending and Uniaxial Tensile Tests; *Cement and Concrete Composites*, 93, pp. 274–288, DOI: 10.1016/j.cemconcomp.2018.07.012.

Roelfstra, P.E. & Wittmann, F.H. 1986. Numerical method to link strain softening with failure of concrete, in: F.H. Wittmann (Ed.), Fracture Toughness and Fracture Energy, London.

Sanal, I. & Ozyurt Zihnioğlu, N. 2013. To what extent does the fiber orientation affect mechanical performance?; *Construction and Building Materials*, 44, pp. 671–681.

Tiberti, G., Germano, F., Mudadu, A. & Plizzari, G. A. 2017. An overview of the flexural post-cracking behavior of steel fiber reinforced concrete; *Structural Concrete*, 19 (3), pp. 695–718, DOI: 10.1002/suco.201700068.

Tunnels and Underground Cities: Engineering and Innovation meet Archaeology,
Architecture and Art, Volume 6: Innovation in underground engineering,
materials and equipment - Part 2 – Peila, Viggiani & Celestino (Eds)
© 2020 Taylor & Francis Group, London, ISBN 978-0-367-46871-2

Design recommendations for fibreglass reinforced tunnel softeyes

S. Munn, J. Hudler & M. Alexander
Isherwood Geostructural Engineers, Toronto, Ontario, Canada

ABSTRACT: This paper outlines the design challenges and recommendations for use of fibreglass reinforcement in tunnel headwall softeyes as learned through transit construction projects in the Greater Toronto Area (GTA). Tunnel boring machines were used to construct underground transit tunnels requiring the design of shafts with softeyes to facilitate tunneling operations. Variable ground conditions and shaft geometries lead to innovative headwall and tunnel softeye design applications using fibre reinforced polymer (FRP) elements (beams, reinforcing bars, and anchors). This paper outlines the challenges and benefits associated with designing and constructing shafts using various FRP systems including lessons learned throughout design and construction of over 20 structures through the past decade.

1 INTRODUCTION

The City of Toronto has been expanding its metropolitan transit network with a variety of new transit construction projects. A total of 31 new transit stations and 18.6 km (11.5 miles) of twin tunnel underground rail will be added in the Greater Toronto Area (GTA) public transit network with two recent projects: the Toronto York Spadina Subway Extension and the Eglinton Crosstown Light Rail Transit project. (TTC 2017, EllisDon 2016).

Due to existing infrastructure, ground conditions, and spatial constraints of existing surface infrastructure, these projects utilized tunnel boring machines (TBMs) to construct the underground tunnels requiring the design of various types of shafts and headwalls to facilitate underground tunneling operations. To allow for tunneling through the headwalls, these shafts were designed with "softeyes" - a section of the wall without metal reinforcement that allows the TBMs to tunnel through.

2 PURPOSE OF SOFTEYES IN TUNNELING

A softeye is a wall type which can be drilled through and penetrated by a TBM cutterhead. This can be accomplished in a variety of ways including: mass concrete non-reinforced drilled wall, Fibreglass Reinforced Polymer (FRP) secant pile or slurry wall, jet grout block, or ground freezing (Hunt & Finney 2008). In the GTA, we have used FRP secant pile headwalls in the following applications: Launch Shafts, Exit Shafts, Advance Headwalls, Mining Excavation Shafts, and other Access Shafts.

Launch Shaft: This type of shaft accommodates hoisting, assembly, launch of TBMs, and mucking operations (removal of tunneling spoils).

Exit Shaft: Designed to receive, disassemble, and hoist a TBM.

Advance Headwalls: Typically installed at boundaries of future excavations, such as underground subway stations, to allow the TBM to pass through the headwall prior to excavation of the future shaft.

Mining Excavation shafts: Access shafts to accommodate mining operations utilizing a Sequential Excavation Method (SEM) or New Austrian Tunneling Method (NATM). Access

Shafts are used to proactively provide access to the TBM for maintenance of cutter heads at some part through the tunnel drive between the Launch and Exit Shafts.

Each of these provides unique design challenges and requires design capacities, geometries, and other considerations for the uses of FRP reinforced secant pile headwalls

2.1 *FRP in headwall designs*

A fibre reinforced softeye can be accomplished in several ways. In all cases, the steel pile reinforcing typically used for the excavation support system is replaced in the softeye zone with FRP reinforcing. This reinforcing can consist of FRP beams bolted to similar depth steel beams or reinforced cages with fibreglass reinforcing bars through the softeye zone. In Toronto, FRP reinforced softeyes have been designed and used in a wide range of variable ground conditions and adjacent structure geometries, resulting design pressures with K values from 0.25 to 1.0, FRP excavated heights of up to 20 m (65 ft), and unsupported spans of up to 5 m (16 ft).

3 FRP DESIGN CONSIDERATIONS

Fibre reinforced polymer materials are considered suitable for use in softeyes mainly due to their 'cuttability' by the TBM's cutterhead during tunneling operations. FRP exhibits brittle failure which is different from steel's ductility. Brittle failure allows the FRP to break into pieces without obstructing the TBM during operation. Brittle failure is also a main consideration in the design of FRP systems, and one of the reasons why current design codes such as the CAN/CSA-S806-02 apply significantly more conservative factors of safety (FOS = 3 – 4) than other design materials (FOS = 1.5 – 2.0). Figure 1 illustrates the significantly different behavior of FRP and Steel anchor bolts.

The manufacturing process used to produce most fibreglass products, called pultrusion, results in an anisotropic material with directional material properties. Therefore, stresses in the transverse and longitudinal directions must be considered in design. This results in a significantly smaller allowable shear capacity compared to conventional steel beams. FRP also has a significantly lower modulus of elasticity than steel. This leads to specific design considerations when using a composite steel/FRP system like a Steel/FRP headwall. Table 1 outlines a

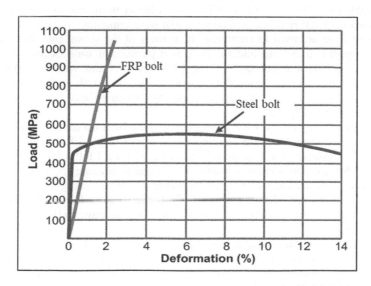

Figure 1.   Comparison of material behavior of FRP and steel bolts (Mohanty et al. 2014).

Table 1. Mechanical property comparison of FRP beams vs. steel beams.

Parameter	Glass FRP shapes		G40.21 Structural steel F_y = 350 MPa
	Lengthwise	Crosswise	
Ultimate tensile strength (MPa)	207	48.3	450
Ultimate compressive strength, (MPa)	20	110	450
Ultimate flexural strength, (MPa)	207	68.9	450
Modulus of elasticity – full section (MPa)	19300		200000
Shear modulus – full section, (MPa)	2930		77000
Ultimate shear strength (MPa)	31		231
Unit weight (kN/m³)	18		77

(Munn & Hudler 2018).

comparison in material property values between a typical FRP pultruded shape (I-beam) versus the typical steel pile secant wall reinforcing used in the GTA. The contrast in the modulus of elasticity of the two materials needs to be accommodated in the design and can lead to different deformation behavior of the two materials resulting in load redistributions in the system (Schürch & Jost 2006). For this reason, the steel/FRP connection is modelled as a hinge since moment transfer is hard to achieve due to strain compatibility issues. This poses a design challenge as the fibreglass beam has low shear capacity. Therefore, the concrete in the secant pile wall acts as additional reinforcing, and the strength should be increased to accommodate shear stresses that exceed the FRP reinforcing capacities. The authors have used concrete strengths ranging from 15 MPa (2200 psi) to 25 MPa (3600 psi) to act as additional shear reinforcing for FRP reinforced secant pile headwalls.

It is important to note the properties of FRP materials depend on the material used to reinforce the polymer (glass, aramid, carbon, etc.) and the manufacturing process used. As illustrated by the variability in the preceding examples, design strengths, design recommendations, and material property values from the specific manufacturers should be consulted when designing FRP systems.

3.1 Fibreglass beams

Secant Pile walls in the GTA are typically reinforced with steel I-Beams in low strength concrete (4 MPa/580 psi). This has led to FRP beams being more commonly used for softeye reinforcing, however FRP reinforced headwalls can have concrete strengths of up to 25 MPa (3600 psi). Excavators are used to trimming the rounds of the secant pile walls to accommodate bracing/support connections and the proposed permanent wall locations, where applicable. Due to the different geometries of the FRP sections and weaker material strength, damage to the FRP materials from the excavator has occurred, such as over-excavation through FRP pile sections and bolted splice connections being sheared off during excavation to the pile face as seen in Figure 2.

To achieve required capacities, FRP beams can also be built up using other FRP shapes to create composite design sections for loading cases where single FRP beams lack capacity. Figure 4 shows an example of a built up pile layout used for a FRP reinforced headwall for a TBM exit shaft. Special consideration needs to be given to bracing connection detailing (I.e. walers and corner bracing, etc.) and excavation activities when using built up FRP connections. In the project example shown in Figures 3- 4, the secant pile wall was not trimmed to the front of the steel pile flanges above the softeye to avoid damage to the built up FRP shape through the softeye.

Since FRP shapes come in a smaller variety of sizes and thicknesses than steel beams, special consideration for splicing beams and preferred pile layout in the headwall needs to be considered. Figure 5 shows examples of FRP/steel softeye splices using either FRP splice plates and bolts or typical steel plates and bolts.

Figure 2. Over-excavation through FRP pile flange (Left) and broken FRP bolts (Right).

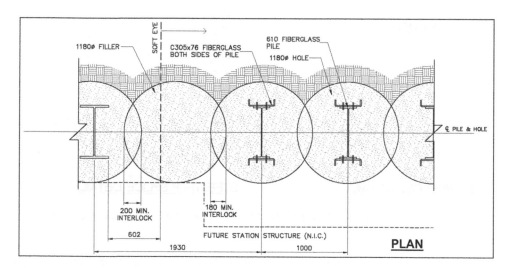

Figure 3. Example pile layout of built up FRP section (units in mm).

3.2 *Fibreglass reinforced cages*

Using reinforced cages with steel/fibreglass for headwall softeyes provides its own unique set of design challenges. Current fibreglass design codes specify conservative allowable stresses for fibreglass design (allowable = 30% ultimate) due to a number of factors (CSA 2009). These include the failure mode of the material, lack of differentiation between temporary and permanent uses, as well as the lack of historical data for use of FRP as a design material compared to other materials. This results in an increased number of bars in the FRP section of the cage to achieve the same required design capacity as the steel section. In the example

Figure 4.   Breakthrough of TBM through FRP beam reinforced softeye.

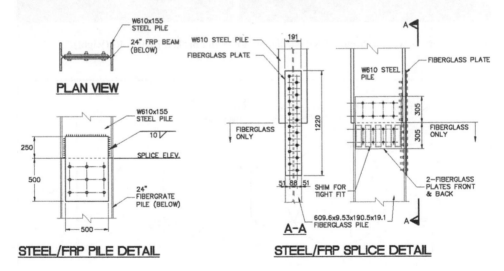

Figure 5.   Steel/FRP splice details – Steel plates and bolts (left) and FRP plates and bolts (right).

illustrated in Figure 6, the initial design had an overall concrete cover of 100 mm (4 inch) but resulted in a small 50 mm (2 inch) cover between the inside of the drill casing and the outside of the cage during installation resulting in the fibreglass cage breaking in several holes during liner extraction. To resolve this issue, additional FRP ties were added to the cage, and an oscillator was used for liner extraction of the remaining piles to limit the torque applied to the cages from continuous rotation in one direction.

Design recommendations for design of FRP cages include:

- Increased cover between drill casing and cage reinforcing – 75 to 100 mm (3 to 4 inch);
- Special attention forces on cages during construction and concrete mix design;
- Use of reverse spiral shear reinforcing on the inside and outside of the cage in lieu of single tie shear reinforcing to limit shear forces applied to the cage during liner extraction; and
- Additional U-bolt reinforcing at splice connections where large forces are transferred;

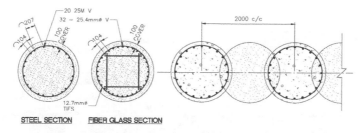

Figure 6. FRP/steel cage reinforcement detail (left) and pile layout example (right).

3.3 Fibreglass anchors

The other strategy used for design of Launch or SEM shaft headwalls is the addition of fibre-glass anchors to decrease bending in the FRP pile section, and decrease the shear at the steel/fibreglass splice connections. These anchors are precluded for use in an Exit shaft due to the nature of the mining activities – the anchors would become damaged as the TBM approached the headwall prior to them becoming redundant resulting in a lack of support for the earth and TBM pressures.

Use of these anchors has resulted in simplified pile layouts due to the decrease in required fibreglass material, as well as a decrease in the required concrete strength to resist the shear at the connections. Figure 7 shows the design detail developed for use of the fibreglass anchors. Due to the limited availability of high tensile capacity fibreglass anchors, three 25 mm (1 inch) FRP bars were used per location to achieve the required design capacities. Due to limited availability of FRP anchor plates and other bolt accessories, the secant pile wall had to be further broken around to allow enough room to separate the FRP bars for three separate anchor connections at the face of the wall. These FRP anchor connections were recessed into the secant pile wall to allow the bars to be trimmed back. The recess was filled with concrete prior to launching the TBM to allow for a continuous mining face. The higher concrete strength of the secant pile wall was used to transfer the forces rather than traditionally welded anchored connections.

3.4 Sequential mining

Use of the FRP anchors has also allowed for larger design heights and different shaft geom-etries since internal bracing of the FRP beams is not required. This has allowed for more open

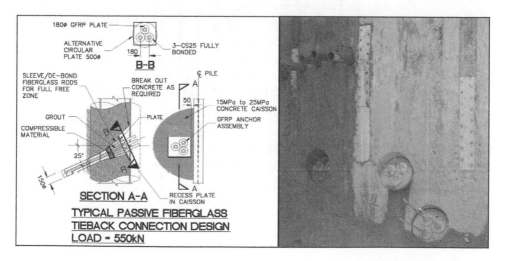

Figure 7. FRP anchor connection detail (left) and example of use at a launch shaft (right).

Figure 8. Mining access shaft headwall with FRP anchors – Sequential mining in progress.

excavations, and sequential mining of SEM headwalls as anchor elevations can be customized to the required geometry. Figure 8 shows a progress photo of the excavation for the SEM access shaft to allow mining for future light rail station using the Sequential Excavation Method. Particular attention needs to be directed at the breakthrough sequence of these head-walls as anchors need to be positioned so the excavation support is stable in all stages. Figure 9 below shows a schematic of the SEM breakthrough. Examples of issues encountered:

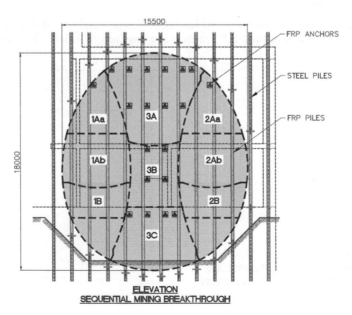

Figure 9. Mining access shaft elevation view – Mining breakthrough stages (Units in mm).

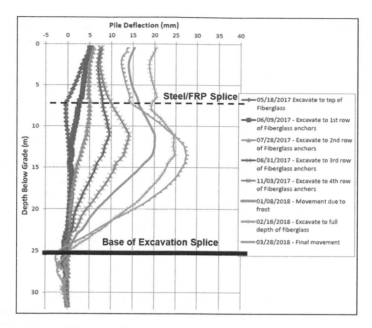

Figure 10.   Steel/FRP pile inclinometer results.

- Cantilever lengths during intermediate stages need to be minimized due to fibreglass strength properties.
- Steel-FRP connection can only transfer shear load
- 3-dimensional geometry needs to be fully understood to avoid conflict between excavation support and the SEM tunnel.

The 18 m fiberglass span of the SEM opening was more than double the span for the headwalls previously designed and constructed, therefore a more comprehensive monitoring program was implemented to confirm the design. The performance of the steel and fiberglass pile was measured using an inclinometer attached to the pile along the full depth (Figure 10). While the movement over the FRP section appears high, it is only 5mm higher than the steel section above which is internally braced. This increased flexibility and resulting movement was expected due to the material properties and use of passive FRP anchors. In addition, the splice was assumed to behave as a pin which only transferred shear to the steel sections above and below the softeye. The overall softeye deflection supports this design assumption by the observed change in deflection pattern at the splice location.

4   GENERAL CONSTRUCTABILITY CHALLENGES

Throughout the design and construction of these FRP reinforced systems, several constructability challenges have been experienced.

4.1   *Hoisting/assembly*

Due to the low modulus of elasticity of the FRP reinforcement, the designer and contractor should work closely to understand the forces which may be encountered during hoisting and placement of the FRP reinforcement to ensure the cage or pile integrity. To minimize these forces, two different systems have been used in the GTA:

- Pre-drilled assembly holes – Vertical splicing over an additional drilled hole followed by vertical hoisting and placement of the reinforcement into the drilled secant pile wall. Also known as "Dummy" holes. (Figure 11).
- Support frames to act as a rigid spine to support the FRP reinforcement during the lifting process. Once the cage is vertical, the spine is disconnected prior to lowering the cage in the drilled hole (Figure 12).

4.2 *TBM forces*

The Shoring wall designer must understand the temporary supernormal loading pressures, which may occur when the TBM is approaching the back of the shoring headwall prior to breakthrough. Earth Pressure Balance Machine's used in the GTA can impose a pressure of up to 200 kPa (30 psi) on the back of the shoring headwall. To ensure adequate FRP Exit Shaft Headwall performance, removable internal bracing can remain in place until TBM makes contact with the back of the headwall and earth pressure forces are negated.

4.3 *Softeye placement accuracy*

Placement of the FRP/steel reinforcing requires coordinating installation of the FRP within tolerances necessary for tunneling by the headwall Shoring Contractor. This can be particularly challenging when trying to mimic the circular tunnel liner shape with the FRP softeye.

4.4 *Monitoring programme*

Based on the design challenges mentioned above, a quality monitoring programme for the temporary FRP walls in softeye applications is prudent. On these projects, a combination of 3D precision survey monitoring, inclinometers, and load cells on the anchors between the tunnels were successfully used.

Figure 11.   Fabrication and Installation of FRP/Steel Beam over Pre-drilled Assembly Hole.

Figure 12.   Steel/FRP Cage Assembly on Spine Support Beam prior to Lifting.

5   CONCLUSIONS

Underground transit infrastructure projects in metropolitan areas provide an excellent opportunity for use of Tunnel Boring Machines (TBMs) to minimize impacts to existing surface and shallow infrastructure and surrounding communities. Use of TBMs requires construction of various Launch, Exit, and Auxiliary Shafts to facilitate tunneling operations requiring the design of shaft headwall softeyes. The use of FRP reinforcement and bracing has proven to be an effective and efficient means of providing structural support of the shaft and suitability for accommodating tunneling operations. The design of FRP systems provides unique construction challenges compared to more common construction materials; therefore, special attention needs to be given to material properties and constructability challenges. Overall, use of FRP has been found to be the most effective solution for softeyes in large diameter tunneling applications in various soil conditions in the Greater Toronto Area.

REFERENCES

Canadian Standards Association (CSA), 2009. CAN/CSA-S806-02. *Design and Construction of Building Components with Fibre-Reinforced Polymers. Rev. 5.* 177p.
Canadian Institute of Steel Construction (CISC), 2010. *Handbook of Steel Construction, Tenth Edition.*
EllisDon, 2016. *Toronto Eglinton Crosstown LRT*, viewed 16 April 2018, http://www.ellisdon.com/pro ject/eglinton-crosstown-lrt/.
Hunt, S. W., Finney, A.J. 2008. Mitigating Tunnel Launch and Reception Challenges. *Society for Mining, Metallurgy & Exploration*, 812–823.
Mohanty, S., Vandergrift, T., Ross, T., Hwi, G., Lee, M. 2014. Fiber Reinforced Polymer Rockbolts for Ground Control in a Strong Jointed Rock Mass. *33rd International Conference on Ground Control in Mining*, 193–198
Munn, S., Hudler, H. 2018. Tunnel Headwall Softeyes: Fiberglass Design Applications and Constructability Challenges. *43rd Annual Conference on Deep Foundations*, 302–311.

Schürch, M., Jost, P. 2006. GFRP Soft-Eye for TBM Breakthrough: Possibilities with a Modern Construction Material. *International Symposium on Underground Excavation and Tunnelling.* 397–404.

Strongwell, 2016. *Strongwell FRP Specifications Section 06600 Fiberglass Reinforced Polymer (FRP) Products and Fabrications,* viewed 17 December 2018, https://www.strongwell.com/wp-content/uploads/2015/08/StrongwellSpecs-FRP-Structural-Shapes-and-Plate.pdf.

TTC, 2017. *Toronto-York Spadina Subway Extension - Overview,* viewed 16 April 2018, https://www.ttc.ca/Spadina/About_the_Project/Overview.jsp.

Tunnels and Underground Cities: Engineering and Innovation meet Archaeology,
Architecture and Art, Volume 6: Innovation in underground engineering,
materials and equipment - Part 2 – Peila, Viggiani & Celestino (Eds)
© 2020 Taylor & Francis Group, London, ISBN 978-0-367-46871-2

Fire resistant high strength concrete concept for inner tunnel linings at the Brenner Base Tunnel

R. Murr & K. Bergmeister
Brenner Base Tunnel BBT-SE, Innsbruck, Austria

ABSTRACT: The geological conditions at the Brenner Base Tunnel, combined with durable tunnel design and service life of 200 years require in special tunnel sections a high strength concrete C40/50 for the inner linings. Additionally to increased compressive strength, the concrete had to combine partially opposing requirements. These requirements include a good workability of fresh concrete, an increased fire protection by microfibers, an unerring achievement of required strength class C40/50, compliance with a maximum component temperature, slow heat development for internal stresses reduction, rapid strength development for a short stripping time and achieving frost resistance concrete close to the portal. In this contribution a new concrete concept is introduced. It focuses on three essential aspects: a) formulation of concrete, b) defining fresh concrete requirements, c) post-treatment to ensure crack prevention. Most important part of the concrete concept was the development of binder concept to combine contradictory requirements by using two certified binders with different, characteristic properties.

1 INTRODUCTION

The Brenner Base Tunnel is the heart of the TEN 1 high speed railway axis Helsinki - La Valetta and ensures a rapid crossing of the main Alpine ridge between Innsbruck (AT) and Franzensfeste (I). In this area, it consists of two separate single-track main tunnel tubes and an exploratory tunnel with a smaller cross-section 12 m below. The exploratory tunnel is intended for detailed exploration of the geology in advance, controlled drainage of the tunnels and a better possibility of servicing the main tunnels. In the Innsbruck bypass area between the Tulfes portal and the Ahrental area, the connection is ensured by two separate, spatially crossing connecting tunnels. The cross section of the connecting tunnel integrates the rescue area, but is separated from the track by a fire protection wall. (see Bergmeister 2011).

2 MOTIVATION

For the inner shell of the connecting main tunnels (Innsbrucker Verbindungstunnel), the use of an inner shell concrete of grade C40/50(90) was necessary due to high rock loads (see Bergmeister, 2019 and Weifner, 2019). In addition, this concrete had to have the properties of increased fire resistance and durability. Since this concrete could not be covered by standard regulations, a holistic concrete concept had to be worked out in which the formulation, paving conditions and necessary after-treatment were regulated and described.

For particularly massive components (e.g. intersection structures, connecting tunnels/connecting ramps or connecting tunnels / evacuation tunnels) an additional alternative concrete composition was developed for grade C40/50(90) with slower strength development and thus reduced hydration heat development, which also showed increased fire resistance. The concretes were examined and tested in the essential parameters (fresh concrete, installation, heat development, strength development).

Table 1. Requirements for the inner shell concrete of the connecting tunnel

Parameter	Request/Default
workability	The expansion dimension (consistency) must be of the consistency class F52 (very soft) (expansion dimension determined according to ÖNORM EN 12350-5), stability and pumpability suitable for transport and installa-tion as inner shell concrete
fresh concrete temperature	Ideal temperature 18 °C, maximum temperature 25 °C at construction site
air content in fresh concrete	3.0 % at construction site (allowable range: 2.5 % to 5.0 %)
compressive strength	Compressive strength on cubs minimum 60 MPa after 90 days
stripping time	In the normal range minimum 2 MPa, in the widening range minimum 9 Mpa
heat development	Temperature rise max. 24 K at the temperature-insulated cube (laboratory test), maximum component temperature target value 50 °C, in hot weather maximum 55 °C
fire prevention (BBG)	Use of a basically tested polypropylene fire protection fibre according to Austrian Society for Construction Technology öbv, 2015, use of air-entraining additive (minimum air-entity content in fresh concrete 2.5 %). The concrete types are given the term BBG in the type designation
durability	is considered verified due to binder contents greater than 350 kg/m³ and a low W/B value of less than 0.50

3 REQUIREMENTS FOR THE CONCRETE FORMULATIONS

The following table summarizes the requirements for the types of concrete used for the inner shell concrete in the highly stressed area of the connecting tunnel.

4 RECIPE FINDING PROCEDURE

In the following, the procedures for the formulation of the two concrete formulations mentioned are schematically explained.

4.1 *Execution recipe C40/50(90) with increased fire resistance (property BBG)*

The execution recipe C40/50(90)/BBG was prepared in 3 steps.

4.1.1 *Step 1:*
Selection and study of heat and strength development of different cement types

Table 2. Cement types

Cement type	Description
CEM II/A-M (S-L) 42.5N	*Portland* composite cement; as classic inner-shell cement, consisting of 80 % to 94% Portland cement clinker and 20 % to 6% blast furnace slag and limestone as additives
CEM III A	Blast furnace cement; consisting of *Portland* cement clinker and 35 % to 64 % blast furnace slag as an additive
CEM I C3A-frei	95 % to 100 % *Portland* cement clinker with low hydration heat develop-ment; due to the absence of a clinker phase (tricalcium aluminate), which contributes substantially to the heat development during the hydrogenation process, this cement type exhibits a lower temperature development with rap-id strength development.
CEM I 52.5N Sp	95 % to 100 % Portland cement clinker with good strength development and high strength performance within the first 28 days, which then largely corre-sponds to the final strength

Table 3. Binder composition of the examined mix designs for the inner shell concrete in the normal range of the connecting tunnel

Mix design	Binder composition[1]
MD 1	CEM II/A-M-(S-L) 42.5 N and AHWZ
MD 2	CEM III/A and CEM I 52.5 N
MD 3	CEM I 42.5 R C3A-frei and AHWZ
MD 4	CEM I 42.5 R C3A-frei and AHWZ and Mikrosilica
MD 5	CEM III/A and CEM I 52.5 N (additional variation)

1) AHWZ is the short form for Chargeable Hydraulically Active Admixture

Table 4. Binder composition of the examined mix design for the inner shell concrete in the widening range of the connecting tunnel (solid structural element)

Mix design	Binder composition
MD 6	CEM III/A

4.1.2 *Step 2:*

With the selected cement and binder combinations, four different mix designs for concretes were developed in the laboratory to meet the requirements for the inner-shell concrete types of the connecting tunnel.

All formulations have been tested with 1.4 kg/m³ basically-tested polypropylene microfibers (BBG fire protection fibers) for increased fire resistance and a maximum total water content of 185 l/m³.

In addition to the workability and fresh concrete properties, the heat development, stripping strength and compressive strength curve with specimen production up to 90 d were tested. In addition, the heat development was determined on a 80x80x80 cm cube on a scale of 1:1.

4.1.3 *Step 3:*

Two selected recipes were used to verify the results on concrete blocks with component-like dimensions on a large technical scale (construction site mixing plant, installation with concrete pump) in order to carry out an evaluation of the installation, processing and temperature development in the component.

On the basis of the determined concrete parameters, the necessary measures for the fresh concrete and after-treatment (curing) in the concrete concept were determined.

4.2 *Formulation C40/50(90)/BBG/EL for solid components (e.g. expansion cross section)*

Based on the findings of the aforementioned formulation, only the favored optimized formulation with the final binder combination was adapted to the special situation of solid construction. The adjustment was made during step 2, where the fresh and hardened concrete properties of the different formulations allowed a single binder combination to be favored by laboratory tests. The type of concrete had to exhibit both increased fire resistance (BBG) and slow strength and temperature development (EL).

5 STEPS IN THE FORMULATION OF THE INNERSHALL CONCRETE OF TYPE C40/50(90)/BBG

5.1 *Step 1: Selection and study of heat and strength development of different cement types*

The choice of binder types concentrated on technically favorable cement types which, in addition to good strength development, exhibit reduced heat generation. The technical

Table 5. Excerpt of binder characteristics (manufacturer's data)

Binder	Manufacturer	Hydration Heat Development partly adiabatic (J/g)		Isotherm (J/g)		Compressive Strength Development (N/mm²)	
		15h	72 h	25 h	72 h	24 h	48 h
CEM II/A-M(S-L) 42.5N	A	215	332	167	267	13	-
CEM II/A-M(S-L) 42.5N	B	245	-	-	-	13	26
CEM I 42.5 R C3A-frei	A	200	332	-	-	15	-
CEM I 42.5 R C3A-frei	B	235	-	-	-	13	28
CEM III A 32.5 N-LH	A	-	-	108	184	6	-
CEM III A 32.5 N-LH	B	190	-	-	-	-	8
CEM I 52.5 N SP	A	259	348	203	311	18	-

Table 6. Selection of binders for further formulation developments

Binder	Manufacturer
CEM II/A-M(S-L) 42.5N WT38	A
CEM I 52.5 N SP	A
CEM I 42.5 R C3A-frei	B
CEM III A 32.5 N-LH	A
AHWZ	A

characteristics - heat development during the hydration process, strength development up to 48 hours - were provided by the manufacturers as data from their own monitoring (see manufacturer data Schretter, Rohrdorfer, 2017).

The strength and heat development characteristics for a C3A-free Portland cement (CEM I 42.5 C3A-free) and a blast furnace cement (CEM III/A) were requested as optimized cement types. In comparison, the characteristic values of the standard cement for an inner-shell concrete production (CEM II/A 42.5N) were also in demand. Furthermore, a hydraulically active, prepared additive type II (AHWZ and a microsilicate for a test) was used for the binder concept.

The following selection was made from the binders available on the basis of their properties and availability:

5.2 *Step 2: Development of the mix design (laboratory tests)*

Laboratory tests were carried out with 5 different concrete formulations on the basis of the specified binders. The binder content had to be sufficiently high and the addition water content correspondingly low in order to achieve a sufficient final strength. The choice of binder preferred a variant with a rapidly hardening component for unerringly achieving the stripping strength of the inner shell concrete in combination with a slowly hardening component, which ensured a steady increase in strength at a later age with moderate strength and temperature development.

Concrete recipe 5 was prepared as a variant of recipe II with the same binder composition but with a slightly increased cement content (+25 kg/m³ CEM III A cement) in order to ensure any required strength reserve at the evaluation age of 90 days. For technical and economic reasons, however, an attempt was made to keep the binder content as low as possible.

Table 7. Concrete mix designs

Mix design	Binder 1[1]	Batch kg/m³	Binder 2[2]	Batch kg/m³	Water Batch kg/m³
MD 1	CEM II/A-M(S-L) 42.5N	300	AHWZ	130	185
MD 2	CEM III/A 32.5 N-LH	320	CEM I 52.5N Sp	80	185
MD 3	CEM I 42.5 R C3A-frei	320	AHWZ	110	185
MD 4	CEM I 42.5 R C3A-frei	280	AHWZ + Microsilica	110 + 20	185
MD 5	CEM III/A 32.5 N-LH	345	CEM I 52.5N Sp	80	185

1) Cement number 1
2) Cement number 2 or Chargeable Hydraulically Active Admixture (AHWZ) or Microsilica

5.2.1 Concrete mix designs

All concretes were produced with the addition of tested fire protection fibers. The dosage was set at 1.40 kg/m³ and corresponds to the minimum dosage of the basic test, including a fixed retention level of 17%.

The aggregate was uniformly designed with the following composition:

Products from one manufacturer were used as a uniform concrete admixture:

The addition of the air-entraining agent is to be recommended from a processing point of view and required for the BBG requirement, since the effect of the fire protection fibers is proven at a minimum porosity in the hardened concrete and the fire protection effect of the fibers is not ensured if the porosity is not reached (see Austrian Society for Construction Technology öbv, 2015).

5.2.2 Fresh concrete test results

The individual recipe optimization in the laboratory was carried out with evaluation of the fresh concrete test results. Compliance with the technical processing parameters for fresh concrete was ensured by the targeted adjustment of the admixture dosage.

5.2.3 Hardened concrete test results

Three main parameters were decisive for the formulation of the hardened concrete parameters. These include temperature development during the hydration process, strength development at a young age and the unerring achievement of the final strength.

The investigation of the temperature development was measured for an early simulation of the installation conditions on test cubes with the dimensions 80x80x80 cm. The dimensions correspond to the maximum shell thickness in the control range of the tunnel. In comparison,

Table 8. Breakdown of the fractions into the total sieve line

Fraction (mm)	Batch (%)
Sand 0/4	40
Gravel 4/8	11
Gravel 8/16	33
Gravel 16/32	16

Table 9. Additive

Function	Name	Batch (M-% of the binder)
Superplasticizer	Premment L 120	about 0.8
Air-entraining agent	Premair LP K 100	about 0.15

Table 10.   Fresh concrete test results

Mix design	Temperature1) Air (°C)	Concrete (°C)	Slump (mm)	Air Content (%)	Density (kg/m³)	Total Water Content (kg/m³)
MD 1	23.6	22.0	540	4.1	2384	186
MD 2	18.9	22.4	520	3.0	2425	185
MD 3	19.3	24.7	540	5.5	2343	183
MD 4	17.3	24.8	540	4.9	2310	189
MD 5	25.9	26.5	520	2.7	2440	187

Table 11.   Temperature conditions and temperature rise of test cubes 80x80x80 cm during the hydration process

Mix design	Start Temperature Air (°C)	Concrete (°C)	Maximum Temperature cube 80x80x80cm (°C)		Maximum Increase in Temperature (K)
MD 1	22	22	48		26
MD 2	22	24	44		20
MD 3	22	24	45		21
MD 4	22	21	42		21
MD 5	22	26	44		18

Table 12.   Temperature conditions and temperature rise of the thermally insulated test cube according to Austrian Society for Construction Technology öbv, 2012

Mix design	Start Temperature Air (°C)	Concrete (°C)	Maximum Temperature cube insolated (°C)		Maximum Increase in Temperature (K)
MD 5	20	25	43		18

measurements were also carried out on temperature-insulated test cubes in accordance with the Austrian Society for Construction Technology öbv, 2012. The results are presented in Tables 11 and 12.

The two other significant parameters for recipe development were the strength; on the one hand, the early strength development during the period of stripping of the inner shell, since from this point on the lining must support itself, and on the other hand, the achievement of the final strength in order to be able to apply the resistance according to the inner shell design. Table 13 shows the strength development during storage in a heating cabinet according to Austrian Society for Construction Technology öbv, 2012, Figure 1 shows the strength development of the test specimens up to the age of 90 days.

5.2.4 *Summary of the mixed design for concrete C40/50(90) with increased fire resistance (property BBG) (laboratory tests)*

The Mix design 2 (MD2) formulation with the binder combination of CEM III/A and CEM I 52.5 N cements has proved to be the optimal solution for agreeing the contradictory requirements - rapid early strength development, low temperature development, unerring strength development with reaching the final strength for strength class C40/50 - see Table 14. The binder combination of Mix design 1 showed a too high temperature development, the binder combinations 3 and 4 showed insufficient strength development to achieve the required final strength reliably, so that the binder content would have had to be increased in total (which in turn would have had an adverse effect on the temperature development).

Table 13.   Stripping time of test specimens (Storage in heat chamber[1])

| Mix design | Stripping time after | | |
	10h (MPa)	11h (MPa)	12h (Mpa
MD 1	-	-	7.1
MD 2	-	-	4.9
MD 3	-	-	5.3
MD 4	-	-	6.1
MD 5	4.1	5.0	6.0

1) The storage of the specimens is temperature-controlled: 0 h to 4 h at 20 °C, 4 h to 8 h at 25 °C, 8 h to 12 h at 30 °.

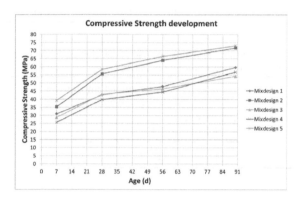

Figure 1.   Compressive Strength development of mix design 1 to 5

Table 14.   Proof of the required target values (Mix design 2)

Parameter	Default Value	Established Value
Minimum strength at stripping time[1]	2 Mpa	4.9 (after 12h)
Maximum heat development[2]	24 K	17 K
Minimum compressive strength after 90 days[3]	60 MPa	71.7 Mpa

1) According to stripping statics and recommendation of the guideline inner shell concrete (see Austrian Society for Construction Technology öbv, 2012)
2) According to guideline inner lining concrete (see Austrian Society for Construction Technology öbv, 2012)
3) According to EN 206-1 for storage according to ONR 23303, see Austrian Standards International 2010 (suitability test, water storage)

Due to the good results in temperature and strength development of Mix design 2 and 5, only Mix design 2 was further pursued and modified for the preparation of the inner shell concrete for solid components C40/50(90)/BBG/EL (Mix design 6, see item 5.2.5). The Mix design 5 with an increased binder content of 25 kg/m³, which was intended as a fallback level, was not required due to the unerring achievement of the strength of the original Mix design 2 formulation.

5.2.5 *Formulation C40/50(90)/BBG/EL for solid components (e.g. expansion cross section)*
For the massive components with increased fire resistance (BBG) and slower strength development (EL), Mix design 2 was further optimized for their requirements.
The binder CEM III/A, which was mainly responsible for controlling the temperature and strength development, was used to 100% and the rapidly strength-developing component

Table 15.   The mix design C40/50(90)/BBG/EL

Rezeptur	Binder 1[1]	Batch kg/m³	Binder 2[2]	Batch kg/m³	Water Batch kg/m³
MD 6	CEM III/A 32.5 N-LH	425	CEM I 52.5N Sp	0	190

1) Cement number 1
2) Cement number 2 or Chargeable Hydraulically Active Admixture (AHWZ) or Microsilica

Table 16.   Fresh concrete test results C40/50(90)/BBG/EL

Mix design	Temperature[1] Air (°C)	Concrete (°C)	Slump (mm)	Air Content (%)	Density (kg/m³)	Total Water Content (kg/m³)
MD 6	17.9	21.4	540	2.6	2390	183

Table 17.   Temperature conditions and temperature rise of the thermally insulated test cube according to Austrian Society for Construction Technology öbv, 2012

Mix design	Start Temperature Air (°C)	Concrete (°C)	Maximum Temperature cube insolated (°C)	(°C	Maximum Increase in Temperature (K)
MD 6	20	24	41		17

Table 18.   Compressive strength at stripping time (storage in heat chamber[1]) and compressive strength value at the age of 28 days, tested on test specimen

Mix design	compressive strength after 20h (MPa)	24h (MPa)	compressive strength after 28d (MPa)
MD 6	7.8	11.8	57.8

1) The storage of the test specimens for stripping resistance is temperature-controlled in accordance with the guidelines of the Austrian Construction Technology Association öbv, December 2012: 0 h to 6 h at 20 °C, 6 h to 12 h at 25 °C, 12 h to 24 h at 30

CEM I 52.5 N was dispensed with in favor of more favorable heat development, see Table 14. This was possible because a longer switch-on time was possible in the expansion cross-sections and thus higher stripping strengths could also be reliably achieved with the blast furnace cement CEM III/A.

The remaining components - aggregates and fire protection fibers - were not changed in type and quantity. The quantity of admixtures added was adapted to the need to achieve the fresh concrete characteristics.

5.3   Step 3: Placement of concrete type C40/50(90)/BBG/EL in the component (abutment test block)

In order to evaluate the workability, stability of the concrete during transport and paving and to re-check the fresh and solid concrete properties of the batches produced in the mixing plant, the concrete formulation Mix design 6, concrete of grade C40/50(90)/BBBG/EL (for solid components), prepared in the laboratory, was additionally tested in a field test. Subsequently, the temperature development in the component was recorded using a data logger, whereby the abutment dimensions amounted to approx. 1.0 m x 1.0 m. In addition, the influence on the temperature development of 2 different after-treatment variants was examined - a) after-treatment only by spraying on a curing agent (corresponds to the standard procedure) and b) after-treatment by spraying on a curing agent and additional covering with fleece and film for reduced heat dissipation and improved after-treatment.

Table 19. Temperature development in concrete components C40/50(90)/BBG/EL, Mix design 6 with post-treatment 1) Curing and 2) Curing + covering with fleece and PE foil

| | Start Temperature | | Maximum Temperature Facing | | | | Maximum of Heat |
Versuch	Air (°C)	Concrete (°C)	Tthe Rock (°C)		Middel (°C)	Airside (K)	Development (h)
1)	17 to 24	18	38		46	37	35
2)	17 to 24	18	-		48	45	37

1) post-treatment only by spraying curing agent (corresponds to the standard procedure)
2) Post-treatment by spraying curing agent and additional covering with fleece and film

The determined fresh and hardened concrete properties of the laboratory tests were confirmed in the placement test. Based on the previous tests, the following target values were defined for the application:

Table 20. Fresh concrete target values at construction site

Parameter	target value
Consistency	59 +- 3cm (minimum 56cm, maximum 62cm)
Air content	3.0% -0.5%, +1.0% (min. 2.5% max. 5.0% - strength!)
Density	according to the mix design +- 50 kg/m³ (with consideration of the air void content in fresh concrete)
Fresh concrete temperature	15 to 22°C (maximum 25°C)

In order to ensure that the limited maximum fresh concrete temperature can be reliably maintained even in the hot summer months, the possibility of cooling was also provided. The cooling was implemented by nitrogen cooling of the cement CEM III/A and water cooling.

In addition, a curing carriage was provided for after-treatment in the tunnel, which ensures optimum curing conditions during the first 36 hours. The curing carriage covers the inner shell concrete with protective fleece and PVC foil. An inflatable rubber bead at the beginning and end of the curing carriage prevents the formation of an air draft on the young concrete surface.

Figures 2 and Figure 3. Curing carriage

Figures 4.   Impression of the dimension and quality of the inner shell

6   CONCLUSION

Due to high rock loads (see Bergmeister, 2019 and Weifner, 2019), the use of inner-shell concrete of type C40/50(90) with special requirements is necessary for tunnel sections of the *Innsbruck* connecting tunnel.

Aim

The task was to develop a concrete with a high strength class C40/50 for inner shells, which exhibits a low heat development in the hydration process but nevertheless permits rapid stripping within 10 to 12 hours. In addition, the concrete should have increased fire resistance and a high durability.

Solution

The unerring achievement of the increased strength required a higher binder dosage with a low total water content compared to standard formulations. The verification age for the strength was increased to 90 days in accordance with the design requirements. By combining a slowly but steadily strength-developing cement with an early high-strength cement, it was possible to develop an optimum binder variant adapted to the requirements. The increased fire resistance was ensured by the addition of basically tested polypropylene microfibers in combination with a microair-entraining agent to control the concrete structure density. Increased durability was ensured by the binder content and the low W/C ratio. The possibility of controlling and limiting the fresh concrete temperature by using nitrogen cooling for the cement allows the maximum component temperature to be limited even in the hot summer months. The provision of a weather-independent after-treatment of the inner shell within the first 36 hours also prevented cracks and increased the surface quality and thus the durability of the concrete.

The inner shells, which have already been produced, are of excellent quality; contain only a small number of superficial blowholes and no cracks along the shell, so that the elaborated concept is confirmed.

REFERENCES

Bergmeister K. 2011. *Brenner Basistunnel - Brenner Base Tunnel - Galleria di Base del Brennero*. Tappeiner Verlag 2011

Bergmeister, K. & Cordes, T. & Murr, R. 2019. Novel semiprobabilistic tunnel lining design approach with improved concrete mixture, *ITA-AITES World Tunnel Congress* 2019, Naples, Italy.

Weifner, T. & Bergmeister, K. 2019. 3D simulations for the Brenner Base Tunnel considering interaction effects, *ITA-AITES World Tunnel Congress* 2019, Naples, Italy.

Austrian Society for Construction Technology öbv, 2015, Guidline „Erhöhter baulicher Brandschutz für unterirdische Verkehrsbauwerke aus Beton"

Herstellerangaben: Zementwerk Schretter & Cie GmbH & CO KG, 2017 Rohrdorfer Gruppe SPZ Rohrdorf, 2017

Austrian Society for Construction Technology öbv, 2012, Guidline „Innenschalenbeton"

Europäische Norm EN 206-1, 2017, „Beton – Teil 1: Festlegung, Eigenschaften, Herstellung und Konformität"

Tunnels and Underground Cities: Engineering and Innovation meet Archaeology, Architecture and Art, Volume 6: Innovation in underground engineering, materials and equipment - Part 2 – Peila, Viggiani & Celestino (Eds)
© 2020 Taylor & Francis Group, London, ISBN 978-0-367-46871-2

Practical experience from the installation of a one-pass lining in Doha's deep tunnel sewer system

A.M. Najder Olliver
Jacobs, Doha, Qatar

T. Lockhart & A. Azizi
Bouygues Travaux Publics, Doha, Qatar

K.G. Singh
Jacobs, Doha, Qatar

ABSTRACT: A one-pass lining solution is being installed for the first time in the Middle East on the Doha South Sewer Infrastructure Project (Contract MTS-01), a deep tunnel sewer system developed by Ashghal (Public Works Authority) in the State of Qatar. With Jacobs as the overseeing Programme Management Consultant, the lining was designed and constructed by Bouygues Travaux Publics and Urbacon Trading & Contracting Joint Venture (BUJV). The one-pass lining comprises a 2.8 mm thick HDPE membrane on the intrados, forming the corrosion protection lining, which is mechanically bonded to a 300 mm thick steel fibre reinforced concrete lining to form a 3.0 m internal diameter tunnel. The HDPE membrane and concrete were cast in a precast segment factory operated directly by BUJV. The welding of approximately 250 km of HDPE membrane joints within the 16.2 km tunnel presented significant challenges in the restricted confines of a TBM backup gantry. The main challenges during the design and development of the one-pass lining are described by Najder Olliver & Lockhart (2017), who conclude that it has demonstrated its potential as a state-of-the-art solution. This follow-up paper presents the practical experience and main challenges encountered in the installation of the lining during construction.

1 INTRODUCTION

All sewer tunnels require both a primary structure tunnel system to withstand the ground and groundwater forces, and an inner lining to protect the durability of the primary system from the effects of corrosive sewer gasses.

The more common solution is a two-pass lining consisting of two parts: a structural precast lining, made up of precast concrete segments, and a thick sacrificial concrete layer with a High-density polyethylene (HDPE) or a Polyvinyl chloride (PVC) membrane cast inside the tunnel using specialised travelling formwork. This more "conservative" approach provides a thick double layer of corrosion protection from aggressive sewer gases, thereby lowering the tunnel's maintenance needs and extending its life cycle.

In the one-pass lining, the HDPE membrane and a thinner sacrificial concrete layer are cast together with the structural part of the segment in a factory. Once installed in the tunnel, the only remaining activity is to weld the joints.

While the one-pass has been applied on projects for at least two decades, it is not used as often as the two-pass because is still considered somewhat "untried" and there is a fear of hidden built-in construction defects. However, it offers significant benefits to schedule, cost, quality and safety during construction if properly executed.

The implementation process and main challenges during the design and development of the one-pass lining on the Doha South Sewer Infrastructure Project (Contract MTS-01), a deep tunnel sewer system developed by Ashghal (Public Works Authority) in the State of Qatar, are described and presented by Najder Olliver & Lockhart (2017), who conclude the following:

- The one-pass lining can improve the quality of the final product whilst reducing construction time and costs, satisfying the required service life, and improving the safety of the tunnel lining installation.
- By simulating the different construction activities through advanced design and testing ahead of construction, combined with additional built-in redundancy and an elevated level of quality control during production and tunnelling, the probability of hidden built-in defects is minimized.

This follow-up paper presents the practical experience and main challenges encountered in the installation of the lining during construction.

2 THE ONE-PASS LINING

The one-pass lining has been designed by BUJV to withstand loads and corrosion for the next 100 years. The internal HDPE membrane from AGRU (Ultra Grip CPL Type 562) and a 120-mm-thick sacrificial layer are provided to protect the 180-mm-thick structural part of the lining from sulphuric acid (H_2SO_4) generated by the sewage environment. General lining information is presented in Table 1.

The high-quality C50/60 concrete contains cement replacement materials such as Ground Granulated Blast furnace Slag (GGBS) and Micro Silica (MS) in addition to Ordinary Portland Cement (OPC) to cope with the high sulphate content in the ground. Due to the high salinity of the groundwater, the segments are made from steel fibre reinforced concrete (SFRC).

The lining is built up from universal tapered rings, with six segments per ring including a large key as shown in Figure 1.

The main purpose of the one-pass lining, as its name explicitly reads, is to remove the need for a secondary, cast-in-situ lining. Therefore, the Corrosion Protection Lining (CPL) is

Table 1. Lining information.

Property	Value
Internal tunnel diameter	3.0 m
External tunnel diameter	3.605 m
Lining thickness	302.5 mm
HDPE membrane thickness	2.8 mm
Concrete thickness	299.7 mm*
Intrados/Median/Extrados radius	1.5/1.651/1.803 m
Ring length	1.2 m
Steel fibre content	40 kg/m^3
Steel fibre type	HE+90/60**
EPDM Gasket	Algaher DV 9 IS
Guiding rod (studded)	Ø40 – 500 mm***
Shear key/dowel	Sof-Fix Opti***
Erector/grout lift socket	Sofrasar Type I***

* Whereof 120 mm is sacrificial for 100 years
** ArcelorMittal
*** Optimas OE Solutions

Figure 1.   Segments stacked per ring.

Figure 2.   HDPE lining installed with the segments (tunnel in construction).

directly placed by the Tunnel Boring Machine (TBM) together with the structural support in the form of integrated segments, as shown in Figure 2.

3   MANUFACTURING OF THE SEGMENTS

The segments were manufactured in an in-house precast factory in Qatar, which was designed, equipped and operated by BUJV, including the concrete batching plant. This provided complete control over the quality of the whole segment production. The initial factory refurbishment setup and training of workforce took four months to complete.

Thirteen sets of six moulds shown in Figure 3 were used to cast an average of 220 segments per day. As shown in Figure 4, a total of 81,947 segments were cast over 18 months for the tunnels, trial rings, mock-up, initial TBM drive and for replacement of rejected segments.

Monitoring of the segments from production up to the installation inside the tunnel was carried out using a barcode system linked to the overall supervision software *Precast Segment Management System* (PSMS) developed by BUJV.

The segment rejection rates were very low (0.12% in the factory and 0.13% on site) with the main reason for rejection being damages occurring as a result of poor handling. No specific concrete or HDPE membrane issues were encountered during the whole production duration.

Figure 3.   Segment moulds in in BUJV's in-house precast factory.

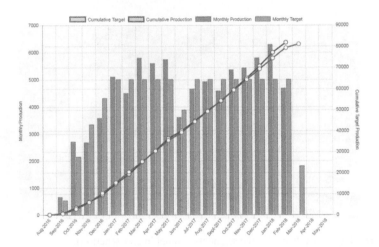

Figure 4.   Monthly production of segments (extract from PSMS).

4   INSTALLATION IN THE TUNNEL

4.1   *Handling and erection*

The TBMs were equipped with a mechanical erector holding the segments with a screw that is inserted into the HDPE socket. The socket is embedded in the segments and its adaptor is welded to the membrane to ensure continuity of the corrosion protection.

The erection process did not induce any significant load on the HDPE membrane since the socket transfers the force directly to the much stiffer concrete.

4.2   *Welding of the joints*

4.2.1   *General*

Before commencing the welding activity, grinding of the surface prior to welding was done. The grinding removes the shallow oxidation and creates a rough surface to ensure an optimal stick of the extruded material. In case of overlapping membrane at the joints, a small gap is created with the width of a grinder disc. Then the welding shoe of 25 mm is set on the extruder to ensure a continuous width to cover the joints. The operator then controls the advance of the welding to keep a continuous seam between the two membranes.

Figure 5.   Mock-up of tunnel lining.

All welds are checked by conducting a spark test following ASTM D6365, where a copper wire is inserted in the joint gap and then covered by extruded material. The integrity of the seam is verified if no spark can be observed when a voltage source is applied along the seam.

The total cumulative length of all joints over the 16 km long tunnel is approximately 250 km. For this reason, a dedicated study for ergonomics and efficiency to optimise the welding activities was carried out by BUJV prior to commencement. As a result, the welding was performed in two phases:

Phase 1: The welding was done during tunnelling by directly following the TBM from a work station on the last backup gantry.

Phase 2: The welding was done when the drive was finished and after the dismantlement of all TBM utilities.

A mock-up consisting of eight rings as seen in Figure 5 was setup in the precast factory to evaluate the working area and to test suitable ergonomic solutions to perform the works.

Operators, specifically trained and certified by the manufacturer of the HDPE membrane, used a "Zero-G" arm to perform the extrusion welding. This focused their attention on the quality of the weld rather than the physical harm induced by the 13-kg weight of the welding tool.

In hydraulic CFD simulations, a roughness factor is used to model the properties of the surface conveying the flow. As the welding is performed manually, the extrusion of the material creates lips on both sides of the welded seam, which resemble a bulge. This could potentially affect the hydraulic performance of the system, however, modelling concluded that lips below 6 mm in size were shown to have negligible influence on the hydraulic performance compared to other alignment parameters (BUJV (2016)). Systematic control of the thickness of the welds was therefore performed on all seams perpendicular to the flow and in the sewage presence area to ensure that the thickness was within limits. Furthermore, the lips became smaller and smaller with continuous improvement of the workmanship.

4.2.2 *Phase 1 welding*
The Phase 1 welding was performed from the last TBM back-up gantry as shown in Figure 6 and Figure 7. The welding activity followed the advance of the machine, which was typically 15 to 20 rings per day. Sixty percent of the welding work was completed during this phase.

4.2.3 *Phase 2 welding*
After a TBM drive is completed, the TBM is removed, utilities are dismantled, and the tunnel is cleaned. Phase 2 of the welding can then start with the following actions:

Figure 6. HDPE welding performed in the TBM backup during Phase 1 using the ergonomic arm "Zero-G".

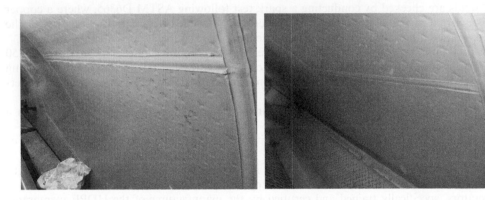

Figure 7. Welded joints (Northern branch, left – Eastern branch, right).

- A mobile work station is setup to process the 40% remaining welds in the lower part of the tunnel which could not be welded during Phase 1 due to presence of the rails and utilities.
- Identified repairs to the lining are performed according to a "snag list".

Thanks to the optimisation of the tools shown in Figure 8 and proper training of the operators, up to 220 linear meters of tunnel, which is equivalent to 1 km of seams, was welded daily.

All circumferential joints (perpendicular to the flow) have been welded, but the longitudinal joints on the bottom 30° which occur in every other ring have been left un-welded to relieve the potential groundwater pressure. In practice, the amount of groundwater infiltration into the tunnels, which is has embedded industry-standard EPDM gaskets, complies with Class 3 of BTS Specification for tunnelling, 3rd edition. This infiltration is also far lower than the expected flowrate of sewage, which is several cubic meters per second in comparison.

As shown in Figure 9, the tunnel is ready for final inspection and handover once Phase 2 is completed. Having the HDPE membrane installed with the segments directly by the TBM has a positive influence on everyone's involvement on the project, since the end product becomes readily visible from the early days of construction.

Figure 8.   HDPE welding during Phase 2 using the ergonomic arm "Zero-G".

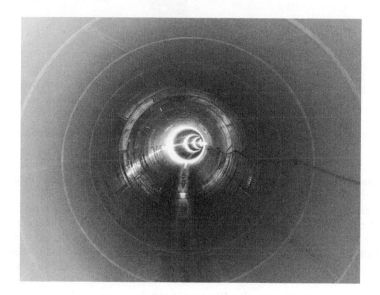

Figure 9.   Tunnel ready for final handover.

4.3   *Damages and Repairs*

During construction, and despite of the special attention paid to the HDPE membrane, defects, damages and snags needing repairs appeared in the tunnels. Fortunately, HDPE is a repair-friendly material, and the same strategy of extrusion welding can be applied to the repairs. Common damages included:

Scratches (Figure 10): These appear in form of shallow lacerations on the membrane and were mainly caused by operational TBM activities and handling. Repairs were carried out by direct extrusion of HDPE on to the membrane, giving the appearance of strips.

Holes in membrane smaller than 15 mm: These were usually caused by the installation of survey equipment, which needed proper anchorage in the concrete, and couldn't be installed

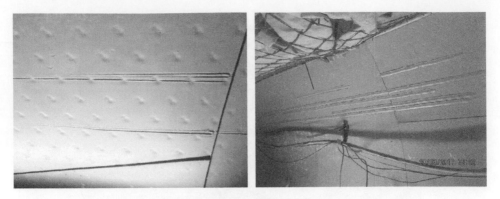

Figure 10. Scratches visible on HDPE membrane (left) and completed repair (right).

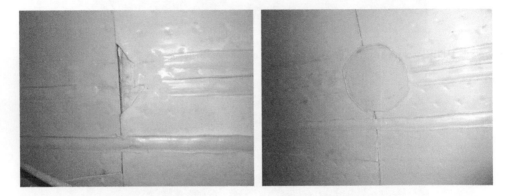

Figure 11. Ripped membrane at a joint (left) and completed repair by installation of a patch (right).

on brackets inside the erector cone. After the hole is plugged using mortar, direct extrusion is used to cover the defective area.

Holes in membrane larger than 15 mm (Figure 11): These were usually caused by TBM activities when the edge was caught by a gantry and ripped off. To repair, a patch of membrane is placed to cover the damage, and extrusion welding then fixes it to the original membrane.

Discoloration of membrane: This was typically caused by rail sleepers as seen in the invert of Figure 8; however, this has no long-term impact on durability, therefore, no repairs were necessary.

To summarise the repairs on the Western branch, where 4,125 rings were installed with approximately 47,000 m² of HDPE membrane, 5,147 repairs have been processed, of which 90% were scratches. There were 323 more significant damages which required the installation of a patch. No bulging of membrane caused by groundwater pressure build-up has been observed anywhere in the tunnel to date, nor has any significant infiltration been detected.

5 TESTING

5.1 *Destructive testing at the precast factory*

Destructive testing was carried out on rejected segments at the precast factory. For each rejected segment, the HDPE membrane was stripped off to check for honeycombing and cracks. All conducted tests showed no cracks or voids under the membrane. The positive

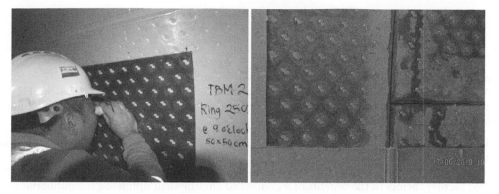

Figure 12. Visual inspection of the concrete intrados after stripping the HDPE membrane.

outcome is a result of the HDPE membrane being carefully placed at the intrados of the segment moulds, accurate control of the concrete quality was maintained and the right level of vibration was applied.

5.2 *Destructive testing inside the tunnels*

At all junctions between the TBM tunnel and mined adits, the segmental lining had to be cut open, and partially removed. This was taken as an opportunity to check the condition of the installed concrete behind the HDPE membrane as well as the annular grout. The membrane was stripped off in the area where the segments had to be demolished later, and a visual inspection with a crack micrometre was conducted as per Figure 12. Coring of segments was also done and sent for analysis.

The testing determined that no cracks were present above the 0.2 mm threshold, and in fact, not even above 0.1 mm. This in turn confirms the prediction of the design where the use of fibre-reinforced concrete and installing the HDPE membrane directly at the bottom of the segment moulds, drastically limits any crack development. This however, also means that the entire adopted sequence of handling, storage, stacking and erection of the segments was successful in preventing any damage to the segments, which is equally important for a damage-free installation.

5.3 *Non-destructive testing*

A few methods of non-destructive testing inside the tunnels have been considered, such as ultrasonic pulse velocity (UPV) testing, but were deemed unsuitable to detect any minor defects behind the HDPE membrane due to the following:

– Effect of the curvature of the segments and presence of joints significantly complicates the signal processing compared to typical test units which are usually flat cast-in-situ slabs, columns or walls.
– Slow acquisition of data is incompatible with the construction programme.
– Spatial resolution is too coarse to detect any defects of 0.2 mm or smaller in size as these methods are normally used to detect larger honeycombing.

As an alternative in the future, a translucent membrane could be used, which would enable a partial view of the concrete to help detect any potential damages. In this way, any remaining doubts with regards to the installed concrete quality, could be eliminated.

6 CONCLUSION

In conclusion, this paper clearly demonstrates that the one-pass lining as a potential state-of-the-art solution presented by Najder Olliver & Lockhart (2017), has evolved into a practical and successful state-of-the-art solution along the 16.2 km long tunnel.

Although numerous challenges linked to the welding and potential damage to the HDPE membrane during the construction period had to be overcome along the way, it can be highlighted that:

– HDPE, like many plastic-based membranes, is repair-friendly using the extrusion welding technique.
– The incorporation of the HDPE membrane directly inside the segment moulds at the precast factory, brings in the benefits of a controlled environment optimised for industrial production of segments, therefore removing the uncertainties associated with cast-in-situ concrete.
– The welds at all joints can be partially performed in hidden time behind the TBM, have no influence on the hydraulic performance of the system and can be designed so that the groundwater pressure can be relieved through the bottom longitudinal joint; additionally, the use of ergonomic tools diminishes the physical strain on the operator, improving the quality and consistency of the welding operation.

Finally, having the HDPE membrane installed with the segments directly by the TBM, has a positive influence on everyone's involvement on the project, since the end product becomes readily visible from the early days of construction.

ACKNOWLEDGEMENTS

This paper is published with the permission of Ashghal. The Authors wish to acknowledge AGRU and their representative in Qatar, QEMS, for their support in developing this one-pass lining solution, providing the technical back-up during design and bringing their expertise into the training of welding operators. Lastly, the authors would like to thank Tom Olliver (Jacobs) for his constructive feedback and support in writing this paper.

REFERENCES

BUJV. 2016. Hydraulic Design Report. Project report ref. MTS01-BUC-GEN-HYD-RPT-00004-04. Public Works Authority. Doha. 279 p.
Najder Olliver, A.M. & Lockhart, T. 2017. Innovative One-pass Lining Solution for Doha's Deep Tunnel Sewer System. *Proceedings of the World Tunnel Congress 2017 – Surface challenges, underground solutions, Bergen, Norway.*

Tunnels and Underground Cities: Engineering and Innovation meet Archaeology,
Architecture and Art, Volume 6: Innovation in underground engineering,
materials and equipment - Part 2 – Peila, Viggiani & Celestino (Eds)
© 2020 Taylor & Francis Group, London, ISBN 978-0-367-46871-2

Assessment of post-blast damage zones in tunneling operations through measurement while drilling records

J. Navarro, P. Segarra, J.A. Sanchidrián, R. Castedo & L.M. Lopez
Universidad Politécnica de Madrid, Madrid, Spain

ABSTRACT: This paper assesses the use of measurement while drilling (MWD) to predict the potential over- and under-excavated zones created in the contour of a tunnel developed by drill and blasting. Scanner profiles of the excavated sections have been compared with the position of contour blastholes to develop an excavated mean distance value (EMD), which may be considered as a macroscopic damage measure. A thorough normalization of the MWD data gathered has been carried to minimize external influences that may hide the actual response of the rig to rock mass. 54 blasts, which comprise around 1700 contour blastholes, have been compared with more than 2200 excavated sections. A non-linear multiple-variable power-form model has been developed to predict the excavated mean distance as function of the normalized penetration rate, hammer pressure, rotation speed, rotation pressure and water flow parameters, and the lookout distance. This combines the rotational, hydraulic and percussive mechanisms of the drill, and the confinement of the rock mass by depth. The model has a determination coefficient of 0.74, with the coefficients of the model strongly significant.

1 INTRODUCTION

In tunneling, overbreak on the contour perimeter is defined as the void created during the excavation in excess of an established perimeter or pay line (Mahtab et al., 1997), and it is usually correlated with the damage extension zone which measures the quality of the blast. Overbreak and underbreak are mainly influenced by the geotechnical condition of the rock mass (rock disturbances and rock strength) and blast design parameters such as the explosive type, the charge concentration, the blast timing, the drill pattern and the drilling deviations (Ibarra et al., 1996; Oggeri and Ova, 2004; Singh et al., 2003; Singh and Xavier, 2005; Hustrulid, 2010; Johnson, 2010). Blasting affects the rock mass structure because of shock wave propagation (vibrations), gas pressure and stress redistribution (Singh et al., 2003; Hu et al., 2014).

The extension of the overbreak has been widely studied by comparing the laser profile of the excavated perimeter with the designed tunnel profile. Kwon et al. (2009) assessed an excavation damage zone in a tunnel construction and determined that in situ stress ratio, Young's modulus and EDZ were the three main parameters in rock mass behavior after blasting. Mandal and Singh (2009) found that the crown is more affected by overbreak, due to the stress conditions in this zone. Kim and Bruland (2009, 2015) estimated a tunnel contour quality index (TCI) based on excavated distances from cross-section scanners, contour roughness and longitudinal overbreak variation in each blasted round. Costamagna et al. (2018) used scanned tunnel profiles to assess the overbreak of the excavated void in relation to the intended theoretical section, and correlated this with the rock mass conditions of each round.

A site investigation for a tunneling project generally includes a description of the rock condition and a rough estimation of the rock mass structural properties. This implies that, during the operation, unexpected anomalies in the rock mass may influence the results of the operation. The monitoring of the performance data of modern drilling machines through the measurement while drilling (MWD) system can gather real time information of the rock mass making it a complementary tool for rock mass characterization. Schunnesson (1996) described the Measure While Drilling (MWD) technique as a drill monitoring system which logs drilling data at predetermined length intervals providing information of the operational parameters involved in the drilling.

Among studies based of use of measure while drilling system to assess geological and geo-mechanical interpretation of the rock mass in rotary percussive drilling, Schunnesson (1997) proposed a methodology to estimate Rock Quality Designation index using percussive drilling. Finfinger (2003) determined a relationship between the drill parameters and the geo-mechanical rock properties in tunnel roofs. Peng et al. (2005) and Tang (2006) developed a method to measure void/fractures in the roof using bolts drilling; they used a normalized feed pressure parameter as a detector of significant discontinuities in the rock. Schunnesson et al. (2012) used Schmidt hammer to calibrate the MWD hardness index provided by Atlas Copco software (Atlas Copco, 2018).

Epiroc AB (Epiroc, 2018), Sandvik (Sandvik, 2018) and Bever Control, in collaboration with AMV (Bever Control, 2018; AMV, 2018) have developed software for their own drilling equipment (Tunnel Manager MWD, iSURE and Bever Control, respectively) as a tool for planning, administration and evaluation of drilling. From the MWD files collected, the blast-holes can be represented in 3D, and hardness and fracturing maps are provided. The theoretical background for this information is however confidential.

Although there are many studies focused on the overbreak control by blasting effect on the one hand and on geological and geo-mechanical interpretation of the rock mass by using MWD on the other, no relation between MWD parameters and overbreak (i.e., under or over excavation with respect to the theoretical tunnel contour line) from blasting exists.

This paper work builds on the methodology developed in Navarro et al., (2018b) to develop drilling index based on MWD parameters to predict potential overbreak zones generated on the perimeter of a tunnel face excavated by blasting. For that, scanner profiles of excavated sections have been compared with the contour blastholes position, to obtain the excavated mean distance (EMD) created by each hole. Since blasting variables (mainly explosive charge and timing) are constant in contour blastholes (as from blast reports) and MWD system can characterize the rock mass condition of the blastholes, the analysis will match high over-excavated zones with highly fractured or softer rock masses before the blast, and low excavated measures with competent rock.

2 DATA OVERVIEW

The study has been developed in the underground extension work of the municipal wastewater treatment plant in Oslo, Norway. The facility is composed of five caverns, a main access gallery of about 850 m length and other sections. The construction was excavated by drill and blasting in competent rock mass, composed by gneiss with small tonalite and quartzite intrusions. 54 blasts have been analyzed from the main gallery making up more than 400 m of excavated rock.

Data from the rig have been used to locate the contour blastholes and to compare them with the excavated profiles. A three-boom jumbo XE3C, manufactured by Atlas Copco, equipped with percussive-rotary top hammer drill, using semi-automatic ABC (Advanced Boom Control) was used to drill the analyzed blasts. Data comprises production face drilling holes of short length, using single rod (5.5 m length and 38 mm diameter) and 46 mm bit. Table 1 lists the parameters monitored by percussive-rotary jumbos; it also gives the units and the acronym used for each of them. They have been described in detail Navarro et al. (2018b).

Table 1. MWD parameters description.

Parameter	Acronym	Unit
Penetration Rate	PR	m/min
Hammer Pressure	HP	bar
Feed Pressure	FP	bar
Damp Pressure	DP	bar
Rotation Speed	RS	rpm
Rotation Pressure	RP	bar
Water Pressure	WP	bar
Water Flow	WF	l/min

3 JUMBO NAVIGATION AND BLASTHOLE POSITIONING

Jumbo navigation is the first operation before starting a new round to follow the correct tunnel layout. The operation is carried out in two steps: jumbo positioning and alignment with the tunnel axis.

The first operation consists of measuring the exact position of the jumbo inside the tunnel to create an absolute coordinate system. For this purpose, the jumbo has a laser scanner on its front side. The position of the jumbo is calculated by trilateration, measuring distances from the laser scanner to target points (with known coordinates) located along the wall side of the tunnel. The absolute coordinates X_{abs}, Z_{abs} obtained are given, in this case under study, in the EUREF 89 Norwegian Transverse Mercator (NTM) projection and the absolute coordinate Y_{abs} is measured as the height above sea level.

The second operation consists on aligning the drill rig with the tunnel axis (i.e. the perpendicular line to the face of a new round). The laser scanner points to the free face in the direction of the tunnel axis and two targets are aligned along one of the booms of the jumbo. For the alignment, the boom rotates until the laser beam passes through both targets. At this stage, the drill rig creates a tunnel reference system $(\vec{x}_t, \vec{y}_t, \vec{z}_t)$ with one axis parallel to the tunnel axis and the other two in the plane of the tunnel free face (Navarro et al., 2018b).

The former operations make the jumbo to be oriented by three angles (γ, θ, ω). To calculate the position of the blastholes, the drill rig rotates the planes formed in the tunnel reference system ($X_t Z_t$, $Y_t Z_t$, $X_t Y_t$), to create a drilling reference system defined by two vertical planes $Y_d Z_d$, $X_d Y_d$ and a horizontal $X_d Z_d$ plane, as correction of the jumbo orientation by angles θ, ω and γ, respectively. Figure 1 shows the rotation carried out by the drill rig.

Blasthole position measured in the drilling reference system is defined by three spherical coordinates (see Figure 2): blasthole length (l_b), azimuth (angle of the horizontal $\vec{x}_d$ axis and the hole projection in the $X_d Y_d$ plane) or lookout direction (L_D) and inclination or lookout angle (L_I). The two later are logged by sensors installed along the boom outside the blasthole and the blasthole length (l_b) corresponds to the drill rod length introduced in the hole. Figure 2 shows the theoretical (X_F, Y_F, Z_F) projections of a blasthole (drilled downwards) in the drilling reference system; they are (Navarro et al., 2018b):

$$X_F = l_b \cdot \sin(L_I) \cos(L_D) \tag{1}$$

$$Y_F = l_b \cdot \sin(L_I) \sin(L_D) \tag{2}$$

$$Z_F = l_b \cdot \cos(L_I) \tag{3}$$

Once the boom is placed in the required position and before starting to drill, the drill rig registers, in the drilling reference system, the collaring position of the blasthole and the azimuth (L_D) and inclination (L_I) angles of the boom (see Figure 2). The end coordinates of the blasthole are theoretically calculated by adding, to their collaring coordinates, the result from equations 1, 2 and 3.

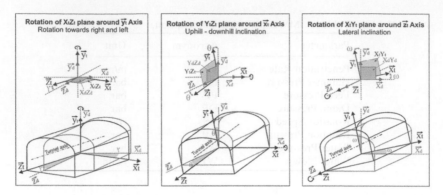

Figure 1. Trasformation from drilling reference system $(\vec{x}_d, \vec{y}_d, \vec{z}_d)$ to tunnel reference system $(\vec{x}_t, \vec{y}_t, \vec{z}_t)$. Left: Rotation towards right and left, or bearing angle, over $\vec{y}_t$; center: uphill- downhill inclination, or elevation angle, rotation over $\vec{x}_t$; right: lateral inclination, roll or bank angle, rotation over $\vec{z}_t$ (Navarro et al., 2018).

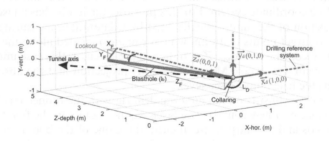

Figure 2. Drilling reference system for blasthole positioning (hole direction in green).

The location of the oriented blastholes in absolute coordinates $(X_{t,NTM}, Y_{t,NTM}, Z_{t,NTM})$ is obtained by adding the NTM coordinates of the jumbo $(X_{NTM}, Y_{NTM}, Z_{NTM})$ to the oriented coordinates of the blastholes (X_t, Y_t, Z_t) in the TRS system.

4 OVERBREAK ASSESSMENT

4.1 *Superposition of excavated profiles and contour blastholes*

During the construction of the access gallery of the underground extension work of Bekkelaget's water reclamation facility, a laser scanner system was used to monitor the excavated void from each blast. Profiles of the excavated void in a direction perpendicular to the tunnel line are collected at steps of 0.2 m from the 3D cloud of points; each profile is dentified by its respective chainage. An interface AutoCAD-Matlab (AutoCad, 2017; Matlab, 2017) has been created to automatically compare the excavated profiles with the contour blastholes for each round. The profile formed by the contour holes (hereinafter named contour profile) is compared with the scanner profiles of the excavated sections to obtain the excavated mean distance between the blasthole and the scanner section at each depth for which MWD data are logged. This distance is considered as an indicator of the resulting damage (i.e. over-excavation). For safety reasons, scanning of the excavated section is done after scaling. This operation obviously modifies somewhat the perimeter excavated and introduce and unknown uncertainty in the analysis.

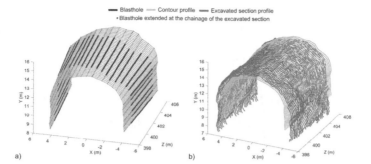

Figure 3. Chainage 399. a) Contour holes profiles and position of the blastholes; b) Overlapping between the contour profiles and the excavated section profiles.

The comparison between excavated and contour profiles must take into account the lookout angle and lookout direction values for each blasthole, which are assumed to increase linearly with depth. The theoretical position of the blasthole at each excavated profile depth is obtained with Eqs. 1 and 2. Variables L_I and L_F are obtained from the MWD files and l_b is the length of the blasthole from the collaring position to the respective excavated profile depth (Z_F, see Figure 2), defined by Eq. 3.

Irregularities on the free face of a new round cause the collars of the blastholes not to be in the same plane, so they have different collar depths. In addition, some excavated profiles (mainly profiles at the beginning of the round) are not influenced by all contour holes of the current blast but for from the previous one. Each blasthole is extended from the foremost collaring hole to the depth of the deepest blasthole of each round to calculate the excavated area for all profiles included in a round. Figure 3a represents the contour profiles (cyan lines), the position of the blastholes (black lines) and their extensions (tiny blue dots).

To carry out the analysis, both contour and excavated profiles must be overlaid. Excavated profiles from AutoCAD files are drawn in a vertical xy plane, where the Y coordinate is referred to the Y_{NTM} absolute coordinate and the hypothetical Z coordinate, i.e. depth of the xy plane, is indicated by the chainage at which it is located. The contour blastholes coordinates in the DRS must be rotated to the TRS. The rotation of the DRS contour blastholes coordinates (X_d, Y_d, Z_d) to the TRS (X_t, Y_t, Z_t) is obtained by introducing the three angles (θ, γ, ω) in a 3D rotation matrix. The translation of the Y_t and Z_t coordinates is carried out by adding the Y_{NTM} and the chainage values of the round studied, respectively. Figure 3b sketches the overlapping of both excavated (red lines) and contour holes (blue lines) profiles for a round.

4.2 Excavated Mean Distance from the blasthole

The Excavated Mean Distance (EMD) corresponds to the area between the midpoint of the spacing on both sides of the hole and the excavated profile, normalized by the distance between the midpoints of the spacing on both sides of the blasthole. Figure 4 defines the calculation of the EMD when two adjacent holes that are on the same side of the excavated profile (EMD 1) and when two consecutive blastholes are located one inside and the other outside of the excavated profile (EMD 2).

The scanner profiles and the EMD values per blasthole are evaluated at 0.2 m intervals. The MWD sample interval is 0.1 m and the collaring chainage of each contour blasthole differs due to irregularities of the free face. Thus, the actual chainage of the MWD logs of each hole varies so that the position of the measurements recorded by the MWD system does not coincide with the depth of the calculated EMD values. A piecewise cubic Hermite interpolating polynomial (Fritsch and Carlson, 1980) is used to interpolate the EMD values at the specific depths of the MWD logs.

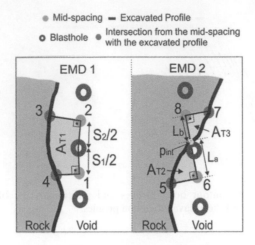

Figure 4.   Calculation of the Excavated Mean Distance. For EMD 1, S_1 and S_2 are the spacing between the current blasthole and the adjacent ones (they are generally around 0.7 m) and A_{T1} is the area excavated by each blasthole defined by points 1, 2, 3 and 4; it is positive when there is over-excavation and negative for under-excavation. For EMD 2, A_{T2} is the area defined by 5, 6 and p_{int}, A_{T3} is the area defined by 7, 8 and p_{int}, p_{int} is the intersection point between the excavated profile and the line joining two adjacent blastholes and L_a and L_b are the distances between p_{int} and the mid-spacing point between the current blasthole and the adjacent one; the total excavated area is obtained by adding both areas with their respective sign.

5   MWD DATA PROCESSING

The response of MWD parameters has been normalized in order to highlight changes in the parameters by the rock effect. Figure 5 shows a flow chart of the filtering and corrections applied to the raw MWD data, with the acronym used after each step.

First correction (i. Filtering of unrealistic values, Figure 5) filters unrealistic high and low performance values of each parameter out. For that the empirical probability distribution function of each MWD parameter is applied with the 95 % confidence interval. The correction is built from the complete data set values (54 blasts, comprising more than 6500 blastholes).

For the second correction (ii. Removing of the ramp-up section, Figure 5), Schunnesson, 2017 and Navarro et al., 2018b defined three operational modes while drilling: collaring, ramp-up, that controls the increase of the drilling pressure to minimize hole deviations, and normal drilling that controls the performance of the parameters to optimize the operation. In this way, only values of the later is retain for the analysis (referred as MWD_{c1}).

Corrections of hole depth and feed pressure (see Figure 5) influences in the data are removed by averaging, for a large amount of data, the response of the parameters.

The correction of the hole length influence (to obtain a signal MWD_{c2}, see Figure 5) is done by:

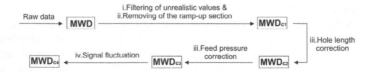

Figure 5.   MWD data processing flow chart.

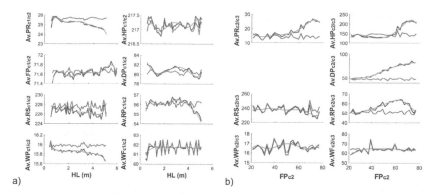

Figure 6. Correction of hole length and feed pressure influence in MWD parameters. a) pass from MWD_{c1} to MWD_{c2}; blue lines: average signals, MWD_{c1} (obtained after filtering unrealistic values and removing ramp-up section); green lines: polynomial regression; red lines: average normalized signals, MWD_{c2}. b) MWD_{c2} conversion to MWD_{c3}; blue lines: average signals, MWD_{c2} (corrected for hole length influence; green lines: polynomial regression; red lines: average normalized signals, MWD_{c3}. Units and acronyms of the parameters in the ordinates are given in Table 1.

$$MWD_{c2}^i = \left[MWD_{c1}^i - MWD_{fit}^i\right] + MWD_{fit}^1; \; with\, i = 1, 2, \ldots, N \tag{4}$$

where i indicates each measurement in a hole log, N being the number of these. MWD_{fit} is a polynomial regression with hole length, of the average value at every 0.1 m hole length for the entire data of each MWD_{c1} parameter (see Figure 5). MWD^1_{fit} is the intercept of the fit, i.e. the value at depth zero.

For the feed pressure influence, Navarro et al. (2018a) pointed to this parameter as the lead parameter that drives the adjustment of the other variables to optimize the drilling. According to this, the feed pressure generates systematic variations in penetration rate, hammer pressure, damp pressure and rotation pressure, that may hide the rock dependence on them. On the contrary, rotation speed, water flow and water pressure are little influenced by the feed pressure, thus being considered independent. The same methodology followed for the hole length influence is now used to correct the feed pressure influence to obtain a signal MWD_{c3} (see Figure 5).

Figure 6 shows the correction of hole depth (graph a) and feed pressure influences (graph b). It represents the average MWD_{c1} (graph a) and MWD_{c2} (graph b) signals (blue lines), the polynomial regression (MWD_{fit}, green lines) and the normalized average signal (red lines). Figure 6b confirm, in line with Navarro et al. (2018a), that penetration rate (PR), hammer pressure (HP), damp pressure (DP) and rotation pressure (RP) parameters have a strong dependence from the feed pressure (FP). For the case of rotation speed (RS), water pressure (WP) and water flow (WF), the influence of feed pressure is considerably less, and these data are not normalized for the subsequent analysis.

Finally, parameters involved in the rotational mechanism of the jumbo (rotation pressure and rotation speed) and the penetration rate have been processed in other to consider fluctuations in the signals. The procedure is carried out for the signal of each hole individually, where the MWD_{c3} values in a hole are divided by the standard deviation of the entire signal of that hole. The resulting signals for these three parameters are MWD_{c4} (PR_{c4}, RS_{c4}, RP_{c4}).

$$MWD_{c4}^i = \frac{MWD_{c3}^i}{std(MWD_{c3})}; \; with\, i = 1, 2, \ldots, N \tag{5}$$

where, i indicates each measurement in a hole log, N being the number of samples per signal.

The processed penetration rate (PR_{c4}), hammer pressure (HP_{c3}) and rotation pressure (RP_{c4}) result in rock dependent parameters. The normalized rotation speed (RS_{c4}) and water flow (WF_{c3}) are independent parameters sensitive to rock variations (Navarro et al., 2018). The response of these normalized parameters can detect variations in the rock and for equal blasting conditions they may explain variations in EMD data. The feed pressure is not considered as it has been used during the normalization.

The lookout distance (i.e. distance from the collaring to the position of the hole at each depth, see Figure 2) is also considered for the analysis as it may reflect the confinement effect by depth. There is a negative influence of the lookout in the EMD which means that the excavated area in relation with the blasthole position decreases with the lookout distance. An increase of confinement with lookout distance is associated with an increase of the difficulty in breaking the rock, resulting then in a decrease of the over-excavated area, hence the EMD, which may be even negative (under-excavation).

A power function of the MWD parameters to predict the EMD is considered:

$$EMD = A_0 + PR_{c4}{}^{A1} \cdot HP_{c3}{}^{A2} \cdot RS_{c4}{}^{A3} \cdot RP_{c4}{}^{A4} \cdot WF_{c3}{}^{A5} \cdot L_{dist}{}^{A6} \qquad (6)$$

where L_{dist} is the lookout distance.

Since EMD has positive and negative values, an additive constant A_0 has been included. The normalized damp pressure and water pressure have been removed since their contributions were found minimal. The non-linear regression model has been programed with Matlab 2016b by a Levenberg-Marquardt non-linear ordinary least squares method (Matlab, 2016). The model coefficients are given in Table 2.

The determination coefficient of the fit R^2 is 0.74. Coefficient values estimated for each parameter (Table 2) are significantly different from zero, i.e. their p-value is <0.0001 for all cases, which means that the analysis is statistically significant. The results of the EMD predicted with the suggested model versus the EMD data are plotted in Figure 7a. The linear regression obtained has a slope of one with a zero-constant term. Figure 7a also shows the upper and

Table 2.   Non-linear regression model coefficients.

		PR_{c4}	HP_{c3}	RS_{c4}	RP_{c4}	WF_{c3}	L_{dist}
	A_0	A_1	A_2	A_3	A_4	A_5	A_6
Coefficient	-2.8432	-0.0341	-0.1188	0.0352	-0.0108	0.0834	-0.0600

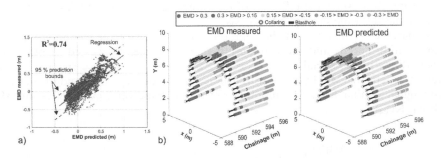

Figure 7.   a) Result of the EMD predicted model: predicted versus measured EMD values; b) Representation of the predicted EMD model (rigth graph) and the EMD measured (left graph).

lower prediction band at a 95% confidence level. An illustration of the application of the model is shown in Figure 7.b; in it, five different over-excavation ranges have been defined. Considering the noise of the MWD data and the additional uncertainty brought by drilling deviations, the scaling before scanning the excavated section, possible variations in the explosive linear density, etc., the quality of the fit is outstanding.

7 CONCLUSION

The overbreak of the remaining rock mass in tunnel blasting has been analyzed trough MWD records, with the purpose of developing a rock index able to identify zones of potentially high geotechnical risk (those for which the over-excavation prediction is high). Fort that, scanner profiles of excavated sections have been compared with the contour blastholes position, to obtain the excavated mean distance (EMD) created by each hole.

Given that blasting factors (mainly explosive charge and timing) are constant in contour blastholes, the overbreak and underbreak are considered mainly influenced by the geotechnical condition of the rock mass. Such rock mass properties are assessed from MWD parameters. A through correction and normalization of MWD data to highlight the dependence of the rock in the parameters.

A non-linear multiple-variable power-form model has been developed to predict the excavated mean distance as function of the normalized penetration rate, hammer pressure, rotation speed, rotation pressure and water flow parameters, and the lookout distance. This combines the rotational, hydraulic and percussive mechanisms of the drill, and the confinement of the rock mass by depth. The model has a determination coefficient of 0.74, with the coefficients of the model strongly significant.

ACKNOWLEDGMENTS

This work has been conducted under the project "TUÑEL" (PCD16264900008) funded by the Centre for Industrial Technological Development (CDTI, Government of Spain). The authors would like to thank OSSA Obras Subterráneas SA, for providing the necessary data and their support in the measuring campaign. The support of MAXAM is also acknowledged.

REFERENCES

AMV, accessed 2018. www.amv.as
Atlas Copco, accessed 2018. www.atlascopco.com
AutoCAD, 2017. Autodesk, Inc.
Beattie, N.C.M., 2009. Monitoring-while-drilling for open-pit mining in a hard rock environment: An investigation of pattern recognition techniques applied to rock identification. *Doctoral Thesis. Dept. of Mining Engineering, Queen's University, Kingston, Canada.*
Costamagna, E., Oggeri, C., Castedo, R., Segarra, P., Navarro, J., 2018. Assessment of contour profile quality in D&B tunneling. *Tunneling and Underground Space Technology* 75: 67–80.
Epiroc, accessed 2018. www.epirock.com
Fritsch, F. N., Carlson, R. E., 1980. Monotone Piecewise Cubic Interpolation. SIAM, *Journal on Numerical Analysis* 17: 238–246.
Hjelme, J.G., 2010. Drill parameter analysis in the Loren tunnel, *M.Sc. thesis in Geosciences, University of Oslo, Department of Geosciences.*
Hu, Y., Lu, W., Chen, M., Yan, P., Yang, J., 2014. Comparison of blast-induced damage between presplit and smooth blasting of high rock slope. *Rock Mechanics and Rock Engineering* 47(1): 1307–1320.
Hustrulid, W., 2010. Some comments regarding development drifting practices with special emphasis on caving applications. *Mining Technology* 119(3): 113–131.
Ibarra, J.A., Maerz, N.H., Franklin, J.A., 1996. Overbreak and underbreak in underground openings part 2: causes and implications. *Geotechnical & Geological Engineering* 14(4): 325–340.

Johnson, J.C., 2010. The Hustrulid bar – a dynamic strength test and its application to the cautious blasting of rock. *Doctoral Thesis, Department of Mining Engineering, University of Utah.*

Kim, Y., Bruland, A., 2009. Effects of drilling and geological parameters on contour quality in a drill and blast tunnel. *Tunneling and Underground Space Technology* 24(5): 584–591.

Kim, Y., Bruland, A., 2015. A study on the establishment of Tunnel Contour Quality Index considering construction cost. *Tunneling and Underground Space Technology* 50: 218–225.

Kwon, S., Lee, C. S., Cho, S. J., Jeon, S. W., Cho, W. J., 2009. An investigation of the excavation damaged zone at the KAERI underground research tunnel. *Tunneling and underground space technology* 24 (1) 1–13.

Mahtab, M.A., Rossler, K., Kalamaras, G.S., Grasso, P., 1997. Assessment of geological overbreak for tunnel design and contractual claims. *International Journal of Rock Mechanics and Mining Sciences* 34 (3): 185.

Mandal, S.K., Singh, M.M., 2009. Evaluating extent and causes of overbreak in tunnels. *Tunneling and Underground Space Technology* 24(1): 22–36.

Matlab, 2017. The MathWorks Inc., Natick, MA.

Navarro, J., Sanchidrián, J.A., Segarra, P., Castedo, R., Paredes, C., López, L.M., 2018a. On the mutual relations of drill monitoring variables and the drill control system in tunneling operations. *Tunneling and Underground Space Technology* 72: 294–304.

Oggeri, C., Ova, G., 2004. Quality on tunneling: ITA-AITES Working Group 16 final report. *Tunneling and Underground Space Technology* 19: 239–272.

Peng, S.S., Tang, D., Sasaoka, T., Luo, Y., Finfinger, G., Wilson, G., 2005. A method for quantitative void/fracture detection and estimation of rock strength for underground mine roof, in: *Proceedings 24th International Conference on Ground Control in Mining, 2–4 August,* Lakeview, WV: 187–195.

Sandvik, accessed 2018. www.sandvik.com.

Schunnesson, H., 1996. RQD predictions based on drill performance parameters. *Tunneling and Underground Space Technology* 11: 345–351.

Schunnesson, H., 1998. Rock characterization using percussive drilling. *International Journal of Rock Mechanics and Mining Sciences* 35: 711–725.

Schunnesson, H., Elsrud, R., Rai, P., 2011. Drill monitoring for ground characterization in tunnelling operations, in: *20th International Symposium on Mine Planning and Equipment Selection, 12–14 October,* Almaty, Kazakhstan.

Schunnesson, H., Poulopoulos, V., Bastis, K., Pettersen, N., Shetty A., 2012. Application of computerized drill jumbos at the Chenani-Nashri tunnelling site in Jammu-Kashmir, India. *Proceedings 21st International Symposium on Mine Planning and Equipment Selection, 28–30 November 2012,* New Delhi, India: 729–751.

Schunnesson, H., 2017. Personal communication at Luleå Techniska Universitet, Sweden.

Singh, P.K., Roy, S.K., Sinha, A., 2003. A new blast damage index for the safety of underground coal mine openings. *Mining Technology* 112(2): 97–104.

Singh, S.P., Xavier, P., 2005. Causes, impact and control of overbreak in underground excavations. *Tunneling and Underground Space Technology* 20: 63–71.

Tang, X., 2006. Development of real time roof geology detection system using drilling parameters during roof bolting operation. *Doctoral Thesis, Department of Mining Engineering,* University of West Virginia.

Tunnels and Underground Cities: Engineering and Innovation meet Archaeology, Architecture and Art, Volume 6: Innovation in underground engineering, materials and equipment - Part 2 – Peila, Viggiani & Celestino (Eds)
© 2020 Taylor & Francis Group, London, ISBN 978-0-367-46871-2

Maintenance of subway tunnel by using maintenance indicator

S. Nemoto, S. Konishi, S. Ito, N. Imaizumi, T. Miura, Y. Enokidani & D. Ogawa
Tokyo Metro Co., Ltd., Tokyo, Japan

K. Fukunaka
SANNO Research Institute, Tokyo, Japan

ABSTRACT: Tokyo Metro Co., Ltd. maintain over 195.1km structures, so our maintenance should be efficient, effective and persuasive for stakeholders. General inspections of tunnels are composed with visual test and hammering test, which determines grades of structural health related to defects. However, it is quite difficult to grade the health equivalently through an entire tunnel because inspections are conducted by human. Moreover, each renovation work is prioritized by experienced employees based on their intuition. To address this issue, we developed an inspection item response model, a mathematical model. In this model, various inspection data is analyzed to calculate maintenance indicators which quantify the health grades of each section of tunnels. This report presents an overview of the maintenance indicator as well as the process, results of those estimations and undertaking of future.

1 INTRODUCTION

Tokyo Metro Co., Ltd. ("Tokyo Metro") currently operates nine subway lines in and around Tokyo, over 195.1 km of track (532.6 km of track if through services are included) and transports roughly 7.24 million passengers per day; Tokyo Metro is an integral part of the massive railway network of the metropolis of Tokyo. However, by 2020, the year of the Tokyo Olympics, 52.6% of Tokyo Metro's tunnels will be at least 50 years old, a fact that heightens the importance of daily maintenance toward continuing to provide services into the future while ensuring safety and security for passengers.

Tokyo Metro conducts normal general inspections every two years and special general inspections every 20 years to evaluate the health grades of cracks, leakage and other deformations in each location.[1] Although the results of these inspections serve as guidelines for assessment, they could also vary depending on the inspection year and environment, and differences between lines. To properly compare lines and changes in conditions over time, quantitative evaluations must be conducted that consider these variations. To address this issue, a newly developed inspection item response model (a mathematical model) was used to statistically process the results of general inspections to calculate an indicator for maintenance ("maintenance indicator") to quantitatively show the health grades of sections within structures. This maintenance indicator serves as a reference when drafting medium- and long-term maintenance plans. This report presents an overview of the calculation of this maintenance indicator, the process of indicator inference as well as the results, and the efforts Tokyo Metro intends to undertake in the future.

2 TOKYO METRO MAINTENANCE

2.1 Subway Tunnel Maintenance

Subway tunnels are maintained through the same cycle—inspection, assessment, planning, and action – as other railway structures. During inspections, health grades are assessed according to the extent of each deformation and the need for action is determined according to the health grades of the deformations. Action (repairs or monitoring, etc.) is then taken after drafting plans that define appropriate periods. Decisions are also made after integrating various inspection results, and major repairs and reinforcement is performed as necessary. There are several types of inspection: general inspections conducted at regular intervals, initial inspections, and unscheduled inspections and individual inspections conducted as necessary. General inspections are conducted to fully understand the condition of structures and determine health grades for them, and can be further classified into normal general inspections (Figure 1) and special general inspections (Figure 2). Normal general inspections are conducted to fully understand the presence of structural deformations, their likelihood of progressing, and other factors. They are generally conducted visually and on foot on a two-year cycle for each line. Special general inspections are conducted to increase the precision of inspections by increasing the precision of health grade assessments. They involve close visual inspections and hammering using scaffolding, and are conducted no less often than once every 20 years. Maintenance Standards for Railway Structures ("Maintenance Standards") require railway operators to determine health grades during general inspections according to the degree to which each deformation has progressed in accordance with the criteria on Table 1. Tokyo Metro determines health grades based on these Maintenance Standards, and stores in a system each deformation's distance in kilometers from the origin of the line, position, inspection year, pictures, health grade assessment and other data.

2.2 *Maintenance Challenges*

As described previously, a method of evaluating health grades for individual deformations has been established, and specific factors are also quantitatively evaluated through salt damage surveys, neutralization projections and the like. However, a specific method for using inspection results to evaluate the health grades of structures in each section or on each line has not been established. In addition, it is often the case that experienced employees use their implicit knowledge to prioritize sections and select sections that require careful maintenance such as long-term measures against deterioration. As the number of these experienced employees is expected to decrease in the future, an indicator must be developed that enables quantitative identification of sections so that all sections that require careful maintenance can be identified without fail, and the basis for those identifications must be clearly specified.

In addition, Tokyo Metro may not always be able to expend the money, personnel and other resources that are currently being devoted to maintenance. Therefore, it is imperative to strive toward the streamlining of maintenance and the appropriate allocation of these limited resources. Table 2 shows the numbers of locations of spalling and peeling per kilometer recorded during normal general inspections for each line. The table illustrates the significant differences in the number of deformations for each line. The construction year, surrounding

Figure 1.   Normal General Inspection

Figure 2.   Special General Inspection

Table 1. Maintenance Standards for Railway Structures

Health grade		Impact on safety of operations, passengers, the public,etc.	Extent of deformation	Actions, etc.
	AA	Significant	Severe	Take action as soon as possible
A	A1	Will eventually become significant. Significant due to effects of abnormal, ecxternal forces.	Deformations, etc. could have negative effects on structual capacity	Take action soon
	A2	Will become significant in the future	Deformations, etc. could have negative effects on structual capacity	Take action when necessary
B		Will become an A-grade deformation if it progresses	Will become an A-grade deformation if it progresses	Monitor or take other action when necessary
C		No impact in its present state	Minor	Do a focused survey as necessary during the following inspection
S		No impact	None	None

Table 2. Locations of B-Rank or Higher Spalling and Peeling per Kilometer in Normal General Inspections

Line	Locations (per km)
A	151.1
B	41.2
C	104.0
D	15.4
E	149.6
F	56.1
G	151.1

environments, differences in construction methods and other factors produce these differences; it is probably not efficient to distribute maintenance resources evenly to each line.

In light of these circumstances, it is considered important to develop indicators that enable quantitative evaluation of sections in the entire system—not only on each line—that require investment.

3 MAINTENANCE INDICATOR

3.1 Overview of Maintenance Indicator

One example of a maintenance indicator is a general indicator that expresses health grades for each section. One way to derive this general indicator is to quantify health grades of deformations determined from the results of general inspections and simply total them for each section.

However, such an indicator would not be appropriate for line-to-line or year-to-year comparisons because it is not possible to eliminate qualitative factors to ensure consistent results, even if there is a clear standard of determining health grades for individual deformations. For example, the same inspector does not necessarily conduct each inspection on a given line, and some variation in judgment still exists even if the inspector is the same. In addition, inspection results can vary depending on the locations and types of deformations as well as the ease of observing deformations owing to differences between factors such as the construction year, construction method and surrounding environments on each line.

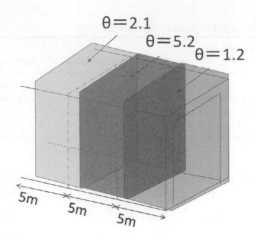

Figure 3. Conceptual Diagram of a Maintenance Indicator

Therefore, it is not appropriate to use the sum of health grades for each section as a maintenance indicator as doing so does not express the true health grade of a tunnel. In other words, qualitative factors in inspections must be considered and accounted for to enable proper comparisons of lines and changes in conditions from year to year.

3.2 *Definition of Maintenance Indicator*

A maintenance indicator can be defined as an indicator that sets forth the observed probability of individual deformations that could potentially exist in tunnel sections divided into set lengths that express their distance in kilometers from the origin of a given line ("distance from origin"), thus expressing the health grade of the structure. Here, the observed probability of individual deformations increases as the health grade of a structure decreases. Figure 3 is an overview of the process.

In addition, the abstract concept of the ease of observing deformations, which is considered a factor that describes the circumstances of observation for each deformation, is defined as a summation by higher-order abstract concepts.

Note that multiple, detailed deformations that were considered for this report were organized and placed into six categories: leakage, cracks, spalling, exposed steel, initial deformations, and other. Inspectors are able to actually observe cracks, leakage and other types of deformations. However, error and the ease of observing deformations are actually abstract concepts that are not observable. It is believed that actually observable deformations are brought about by a combination of error and ease of observing deformations. In addition, differences in this ease of observing deformations can be understood as differences in the ease of reflecting them in the health grades of structures.

3.3 *Calculation Method for the Maintenance Indicator*

Data from the results of assigning health grades in normal general inspections was used to calculate the maintenance indicator. The data was separated into the six health grade categories set out in Maintenance Standards: AA, A1, A2, B, C and S. First, the health grades of the 15 deformations (cracks (left sidewall, upper floor slab, right sidewall), leakage (left sidewall, upper floor slab, right sidewall), spalling/peeling (left sidewall, upper floor slab, right sidewall), steel deterioration (left sidewall, upper floor slab, right sidewall), cold joints, initial deformations, and other) were quantified in binary, assigning 1 point to AA, A1, A2 and B, and 0 to healthy sections with grades of C and S.

Next, each tunnel was divided into an imaginary grid of 5-m sections parallel to the rails, and the number of points were totaled for each deformation in each section. The results were recategorized, assigning 1 to sections with scores of 1 point or higher, and 0 to sections with 0 points. Tables 3 and 4 show the quantification. Next, Formula (1) was used to express the relationship between each deformation observation probability P_{ij} and characteristic values θ_i as a two-parameter logistic model. Here, j expresses the number of variables, and i expresses the number of observations.

$$P_{ij} = \frac{1}{1 + \exp(-a_j(\theta_i - b_j))} \tag{1}$$

Here, a_j is a parameter called "discrimination," and expresses whether the size of the maintenance indicator is more or less likely to impact the observation of deformations. In mathematical terms, discrimination a_j expresses the extent to which the model rises to the upper right, and expresses the clarity of the threshold of deformation observation. Therefore, it is considered to express the extent to which inspectors' technical capacity impacts their judgment of deformations. b_j is a parameter called "difficulty" that is thought to express the ease of deformation observation because it causes functions to shift horizontally. Figure 4 shows the differences between the parameters.

Maintenance indicator θ_i is influenced by the deterioration of structures. Therefore, there are many healthy sections in which structural deterioration is not progressing, but that number stands to decrease as deterioration progresses and more deformations occur to make sections unhealthy. This trend resembles gamma distribution, and as θ_i is a continuous quantity, it is hypothesized to follow gamma distribution.

These models for calculating the total maintenance indicator were named "inspection item response models" (IIRM),[2,3] and Bayesian inference based on Markov chain Monte Carlo (MCMC) methods was used to estimate the parameters. For the prior distribution of the

Table 3. Line Section Scores

Distance from origin	Crack left	Crack upper	Crack right	Leakage left	...
...					...
0k005m	1	0	0	0	...
0k010m	3	2	0	0	...
0k015m	0	0	1	0	...
...	...	...	...	...	...

Table 4. Recategorized Scores

Distance from origin	Crack left	Crack upper	Crack right	Leakage left	...
...					...
0k005m	1	0	0	0	...
0k010m	1	1	0	0	...
0k015m	0	0	1	0	...
...	...	...	...	...	...

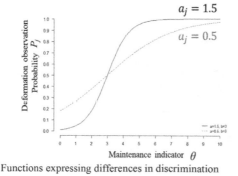

Functions expressing differences in discrimination

Functions expressing differences in difficulty

Figure 4. Inspection Item Response Model

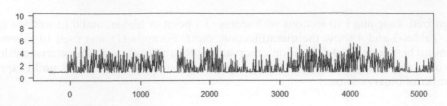

Figure 5. Example of Maintenance Indicator Inference

Bayesian inference, log-normal distribution was used for a_j and gamma distribution for b_j, and a program called Just Another Gibbs Sampler (JAGS) was used to perform the inference. Using IIRM enabled us to calculate characteristic values θ while excluding the distinct characteristics of inspection items as the impact of a_j and b_j, meaning that those characteristic values can serve as general indicators for evaluation under the same standards. Figure 5 shows an example plot of maintenance indicators.

It is worth noting that this method of eliminating variation employs the same concepts applied to TOEFL evaluations.

4 COMPARISON OF MAINTENANCE INDICATORS

The method of maintenance calculated as described above was investigated from two different approaches.

4.1 *Validation*

The first approach was to validate the maintenance indicator. In order to use the maintenance indicator in actual practice, the extent to which it appropriately illustrates potential problems with the health of tunnels must be verified. Toward that end, the visibility of potential problems with the health of target structures to people involved in actual inspections was used as a yardstick for clarifying the corresponding relationship of those potential problems to the maintenance indicator.

As the method of validation, paired comparison tests were administered to 37 seasoned field inspectors. Photographs from a tunnel inspection (Figure 6) was used as the test sample. The specific procedure is as follows:

Step 1: Symbols were extracted and assigned to nine sections along the line used for the paired comparison. Sections with θ values greater than 1 but less than 7 were extracted, and special locations such as station platforms and locations with evidence of repairs were excluded.

A

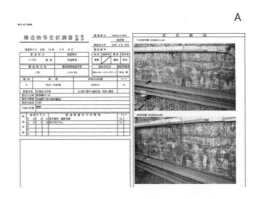

Figure 6. Photographs from an Inspection

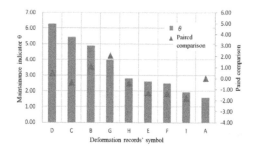

Figure 7. Differences in Risk Assessment

Table 5. Point of mesh per extension

Distance from origin	Spoilling and Peeling	Maintenance Indicator
⋮		
0k005m	1	0
0k010m	1	1
0k015m	0	1
0k020m	1	1
0k025m	0	1
⋮	⋮	⋮

Table 6. Comparison of Maintenance Indicators and Inspection Results on Lines A and B

Line A	Spalling, peeling	
Maintenance indicator	Y	N
At and above threshold	1305	147
Below threshold	343	1620

Inspection rate: 79.2% Precision rate: 89.9%

Line B	Spalling, peeling	
Maintenance indicator	Y	N
At and above threshold	1925	374
Below threshold	136	1379

Inspection rate: 93.4% Precision rate:

Step 2: The test-takers were separated into two groups (A and B), and assigned with five of the extracted sections.

Step 3: The test-takers performed paired comparison of deformation records in the five assigned sections individually, and determined problems with health in terms of civil engineering structures.

Using a common sample between the groups ultimately resulted in a one-dimensional prioritization of all nine locations.

Figure 7 shows the difference between the maintenance indicator and the risk assessment in the paired comparison. The figures expressed as the paired comparison are λ calculated from the results of analysis using the mathematical model in Formula (2). Higher λ indicate determinations of more problematic health by field in It is apparent that the order of θ calculated using IIRM is essentially identical to the determinations of the 37 field inspectors.[5]

$$\log it(P_{ij}) = \lambda_i - \lambda_i \qquad (2)$$

5 COMPARISON WITH INSPECTION RESULTS

The second approach was to perform a comparison with spalling and peeling recorded during inspections. Spalling and peeling can lead to concrete peeling and falling off, and are thus more potentially disruptive to subway operation than any other tunnel deformations.

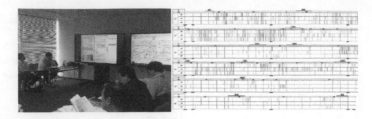

Figure 8. Scenes from a Maintenance Committee Meeting

Therefore, this investigation was conducted in terms of the possibility of using the maintenance indicator to identify these deformations.

As for the methodology, tunnels were divided into 5-m sections, and the presence or absence of spalling and peeling was compared to the maintenance indicator values for each section. As shown on Table 5, 1 was assigned to sections where spalling and peeling were present in normal general inspections and special general inspections, and 0 to sections with none in those inspections. In addition, a threshold was established for the maintenance indicator, and 1 was assigned to sections at and above that threshold, and 0 to sections below that threshold.

For this investigation, two typical subway lines were compared. Table 6 shows the results. The detection rate expresses the rate at which the maintenance indicator can detect spalling and peeling in sections where they are present. The precision rate expresses the rate of precision of the maintenance indicator in sections where the maintenance indicator exceeds the threshold. The detection rate and precision rate exceeded 80% for both lines, which is quite high.

6   EXAMPLES OF APPLICATION IN THE TOKYO METRO NETWORK

After completing normal general inspections in FY 2016, Tokyo Metro established an internal maintenance committee to provide a venue to discuss maintenance policy and other issues for each line. Employees of Tokyo Metro headquarters, site supervision offices and group companies that conduct the actual inspection work attend these maintenance committees to discuss short-term maintenance plans as well as plans for long-term maintenance that incorporate preventive maintenance (Figure 8). Before meeting, the committee uses various approaches to identify tunnel sections that require special attention and thus warrant discussion. The maintenance indicator is one of those approaches. Specifically, the committee establishes a threshold, and identifies sections that exceed that threshold as those that warrant more detailed discussion. They prepare documents that resemble medical charts (Figure 9) to establish the scope of discussion for the meetings. The documents include information on items such as records from construction, numbers and trends of deformations revealed by inspections, and the results of various surveys. In the past, major maintenance plans were mainly based on inspection results largely dependent on the implicit knowledge of veteran engineers, but now Tokyo Metro's maintenance policy for those plans can be determined based on discussions of this maintenance indicator, digitized inspection records, visualization tools and more.

7   CONCLUSION

Tokyo Metro used a newly developed inspection item response model (a mathematical model) to statistically process the results of general inspections to calculate a maintenance indicator to quantitatively show the health grades of sections within structures. The indicator has been successfully validated by comparing the calculation results with veteran engineers and with

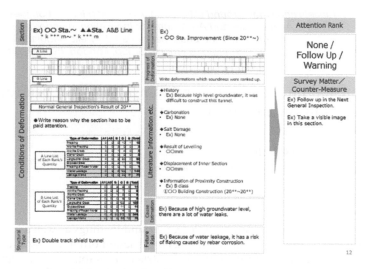

Figure 9. Documents for a Maintenance Committee Meeting

inspection results. The maintenance indicator is also used to identify sections that require special attention and to determine Tokyo Metro's maintenance policy.

We intend to consider applying the maintenance indicator to the voluntary hammering tests that are conducted in addition to legally required inspections. Applying the results of the comparisons of spalling and peeling as discussed in Section 4 should enable the intensive conduction of inspections in sections that require special attention. Tokyo Metro also uses ICT for inspections and recordkeeping in above-ground sections. Therefore, we intend to calculate the same kinds of maintenance indicators based on the results of inspections in these sections as well in an effort to perform maintenance more effectively, efficiently, and explicitly.

It is our hope that this report serves as a reference for efficient tunnel maintenance for other infrastructure operators.

REFERENCES

H. Toyoda, 2002. Introduction to Item Response Theory. Asakura Publishing.

H. TurnerD. Firth, 2012. Bradley-Terry Models in R Journal of Statistical Software, 48, 1–21.

M. Taguchi et al., 2016. Validation of Maintenance Indicators Based on Subway Tunnel Inspection Results 71st JSCE Annual Academic Lectures, VI-682.

Railway Technical Research Institute, 2007. Maintenance Standards for Railway Structures and Commentary (Structures) Tunnels.

Chapter 2. Maintenance Basis: page15 Table2.5.1. Maruzen.

Y. Tashiro et al., 2017. Introducing ICT to Streamline Structural Inspections and Maintenance Plans 24th Joint Symposium on Railway Technology and Policy (J-RAIL 2017), S2-9-2.

Y. Wakui, YoshiyukiS. Wakui, 2003. Illustrated Dictionary of Statistical Analysis Terms Nippon Jitsugyo Publishing

Tunnels and Underground Cities: Engineering and Innovation meet Archaeology,
Architecture and Art, Volume 6: Innovation in underground engineering,
materials and equipment - Part 2 – Peila, Viggiani & Celestino (Eds)
© 2020 Taylor & Francis Group, London, ISBN 978-0-367-46871-2

Robust segmental lining design combining steel fiber reinforced concrete and traditional reinforcement

G.E. Neu, V.E. Gall & G. Meschke

Institute for Structural Mechanics, Ruhr-Universität Bochum, Germany

ABSTRACT: The improper estimation of design ground loadings as well as the exceedance of installation tolerances during segmental lining construction often results in unwanted segment damage. Incurred damages rarely impact the structural stability of the tunnel ring, but can significantly impact the serviceability state of a finished construction. For this reason, existing design codes often provide limits on allowable crack widths. In this contribution, a newly developed Finite Element modeling scheme for hybrid reinforced segmental tunnel linings is proposed with which the non-linear local cracking response can be predicted. By subjecting this model to various loading scenarios (e.g. jack forces, steady-state) and by modeling the exceedance of installation tolerances during the ring-build phase, the response of the segment can be analyzed. Using the proposed method, various traditionally reinforced and fiber-reinforced designs are investigated. Furthermore, hybrid-reinforced designs combining the strengths of both traditional and fiber reinforced concrete are proposed.

1 INTRODUCTION

During the construction of a mechanized tunnel, especially during the installation and loading of segmental linings, it is unavoidable that certain inconsistencies between design assumptions and in-situ conditions will become apparent. Design tunnel loadings may be exceeded due to unexpected ground conditions or segments may be misplaced or mishandled during installation. Both these factors, in turn, may lead to unwanted segment or lining damages. These segment damages rarely impact the structural stability of a tunnel construction, but often lead to excessive cracking which may lead to an exceedance of serviceability limit states such as maximum allowable crack widths. One method with which inconsistencies between design assumptions and construction conditions are accounted for is by means of design tolerances. Properly defined design tolerances, such as those provided by the ÖVBB (2009) or the German Tunneling Committee (2013), can help to ensure that individual segments, or the entire lining ring, do not become unnecessarily damaged during tunnel construction. Segment damages are, however, frequently observed, even if the design tolerances have been explicitly accounted for during design. This indicates that existing design tolerances often cannot account for the complete range of unexpected loadings experienced by tunnel segments. Segment designs which perform well with regard to expected and unexpected loadings should therefore be chosen in tunneling practice.

Unexpected segment damages can be attributed to either improper installation of segments or to the underestimation of expected construction and ground forces during segment design. At the longitudinal joints, for example, segments are subjected to partial area loading stresses which often lead to concrete chipping at the segment corners [Sugimoto (2006)]. Underestimation of the loadings during design, or segment misplacement during installation, can further aggravate the situation and result in the development of intolerable crack formation as a consequence of unexpectedly large splitting stresses. Exceedance of installation tolerances may also lead to unwanted damage in segments during machine advance. Unexpected segment misplacement can lead to the improper bedding of consecutive segments and thus to segment cracking due to unplanned segment loading scenarios. Existing design methods, such as the analytical Strut-and-Tie models used for dimensioning of the highly stressed regions around longitudinal joints or jack pads [German Tunneling Committee (2013), Schlaich & Schäfer (1991)], do not account for these imperfections in segment bedding or contact force. Additionally, these analytical models cannot account for stress redistributions resulting from local concrete cracking. In order to evaluate the performance of a selected segment design with regard to unexpected loadings, more highly-detailed structural models must therefore be used in order to evaluate the non-linear and post-cracking structural response of designed-for structural members.

The goal of this contribution is to improve design methodologies used for segment design by investigating the structural response of various segment designs subjected to unexpected loadings resulting from segment misplacements or the underestimation of ground forces. Specifically, the cracking response of various steel-fiber reinforced (SFRC), traditionally reinforced (RC), and hybrid-reinforced segment designs are investigated. A discrete crack model based on interface elements is used to predict the resulting crack patterns and the corresponding crack width in SFRC and Plain Concrete (PC) [Zhan et al. (2016)] and contact-based constraint condition is used to incorporate discreet reinforcement into the simulation [Gall et al. (2018)].

2 FINITE ELEMENT MODELLING METHODOLOGY

2.1 Modelling of conventional and steel-fiber reinforced concrete

Within the FE simulation, concrete compressive behavior is assumed to be elastic and displacement discontinuities within the SFRC upon cracking are modelled by placing Interface Solid Elements (ISEs) between all linear elastic bulk elements (LE) [Manzoli et al. (2012)]. The material behavior of the ISE's is modified to account for post crack SFRC behavior using a traction-separation law as proposed by [Zhan et al. (2016)]. The traction-separation (t-w) law describes the post- and pre-cracking strength of SFRC as a function of crack opening. It is derived using a multilevel fiber pullout model, and ultimately is derived from analytical fiber pullout tests [Zhan et al. (2014)]. This method can therefore predict not only the crack development, but also the maximum crack width expected in an SFRC structure. The alignment of the steel fibers relative to the casting direction, as well as the approximate size of the casting mold, is taken into account within the above-mentioned model. Secondly, in order to account for the reinforcement bars, the rebar itself is modeled as a simple linear truss and coupled with the concrete matrix using a constraint condition between control points located on the rebar elements and their respective projection points within the elements in which they are embedded [Gall et al. (2018)]. The constraint condition include the bond-slip mechanism as provided in the fib model code 2010 (2013).

2.2 Material Parameters

In this article, the structural response of different segment designs is evaluated which can be either consists of PC, SFRC, RC or combinations of these materials. The general material parameters for the concrete are listed in Figure 1. For design purposes, the actual concrete tensile strength of 3.2 N/mm² is reduced by safety factors in accordance to the Eurocode 2 (2011). The post cracking behavior of SFRC is governed by a traction-separation law, which is integrated in the ISEs. The coefficients for the t-w-relationship of SFRC are calculated depending on the chosen amount, geometry, material properties and orientation of the fibers. All calculations involving fibers use 60kg/m³ of hooked end fibers with a length of 30mm. The fibers have a diameter of 0.35mm and a yield stress of 2200N/mm². It is assumed that in all models and the corresponding casting procedures, the fibers tend slightly to align to the horizontal plane in which 20 % of the fibers lie along the primary casting direction, i.e. the circumferential direction of the fiber, and 40% of the fibers lie in the radial direction. The resulting traction-separation law for this SFRC is given in Figure 1.

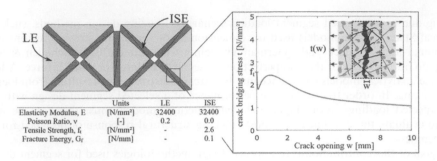

Figure 1. Material parameters (left) and the resulting traction-separation law which governs the post-cracking behavior of the SFRC (right).

The rebars in RC segment designs have a diameter of 10mm and a yield stress of 500 N/mm². The bond-slip parameters used to define the control points of the bond-slip relationship given by fib model code 2010 (2013).

3 INVESTIGATION OF LONGITUDINAL JOINT BEHAVIOR

3.1 *Analytical and numerical modeling of longitudinal joints*

Tunnel lining segments are subjected to many load scenarios that impact the lining locally rather
than structurally. Important examples of such load cases are the local loading of longitudinal joints and the local loadings of segments subjected to jack forces. Both these loadings can be classified as "partial area bearing loads," and are subject to the proofs associated with such loadings within engineering standards [ACI committe 318 (2008), Eurocode 2 (2011)]. These partial area bearing loads act on only a portion of the surface of a segment and lead to two primary sources of failure. The first source of failure results from the exceedance of allowable compressive stresses within the structural member immediately underneath the loaded area. The allowable limits for these applied loads largely govern the size of jack pads and the dimension of the contact area between segments at a longitudinal joint. The second source of failure results from splitting tensile forces that develop as a result of load spreading phenomena. These forces in-turn govern the magnitude and location of splitting tensile reinforcement. Solutions for the evaluation of the magnitude of these splitting stresses Fsd for centric and eccentric loads are provided within the European standards and the guideline from the German Tunneling Committee (2013) by means of simplified strut and tie models, as shown in Figure 2. The strut-and-tie models provided there are based on models proposed by Schlaich & Schäfer (1991) and assume that stresses remain elastic and that the structure can be reduced to a 2D structure.

For the centric loading case, the primary splitting stress Fsd occurs at 0.4*b, where b is the width of the cross section (Figure 2 left). In the eccentric loading case, the total eccentricity of the normal force etot is taken into account, which contains the eccentricity caused by a bending moment e and potential construction tolerances ex (Figure 2 right). This eccentricity also leads to a reduction of the load transferring area. Generally, the eccentric loading result in an increased splitting stress and a shift of these stresses towards the position of the total eccentricity.

On the structural scale, longitudinal joint behavior is comparable to that of a concrete hinge, in that it transfers normal forces, moments, and, through friction, shear forces from segment to segment. As in the case of traditional concrete joints, the moment bearing capacity of longitudinal joints is a function of the transferred normal force [Leonhardt & Reimann (1966)]. In modern lining design, most often either bedded beam or shell models, which

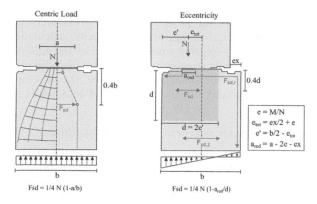

Figure 2. Analytical model for partial area loading for centric (left) and eccentric loads (right).

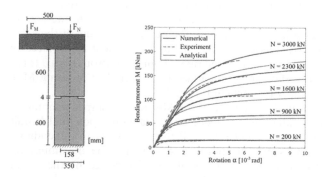

Figure 3. Left: Setup of the joint test [Hordijk & Gijsbers (1996)]; Right: Comparison of the M-α-Relationship at different normal force levels between the experimental, numerical and analytical results.

explicitly account for joint segmentation, are used [German Tunneling Committee (2013)]. In such models, the structural behavior of the longitudinal joint is approximated using non-linear rotational springs, whose behavior is based on concrete-hinge theory [Blom (2002), Tvede-Jensen et al. (2017)]. The applicability of the analytical approach according to Leonhardt & Reimann (1966) for longitudinal joints in tunneling has been validated by experiments performed at the TU Delft [Hordijk & Gijsbers (1996)]. Furthermore, an analytical and experimental investigation, performed within a full-scale test, of the longitudinal joint behavior can be found in [Li et al. (2014)]

Within this contribution, ISE's, as described in Section 2.1, are used to model contact between adjacent segments at the longitudinal joint. In order to validate the proposed modeling scheme, the results of joint rotation tests as carried out by Leonhardt & Reimann (1966) at different normal force levels are used. The layout and geometry of the experiment is shown in Figure 3 (left). Within the experiment, the rotation of the joint between two adjoining segments is measured while the system is subjected to a known external normal force and bending moment.

The results of the test, the predictions of the analytical approach as proposed by Leonhardt & Reimann (1966) and the results measured from the simulations using the proposed interface element are presented in Figure 3 (right). The experimental, analytical, and numerical results are all in good agreement, however, the numerical results fit the data much more accurately than the analytical model. It can be observed that, for smaller normal force levels, the initial rotational stiffness as observed in the experiment is slightly lower than that predicted by the numerical model whereas at higher normal force levels a very good agreement can be obtained. In order to evaluate the applicability of the analytical model (Figure 2), a

2797

comparison between the position of the primary splitting stress predicted by the analytical model and predicted by the numerical results of a linear elastic calculation is carried out. In Figure 4 the position of the primary splitting stress predicted by the analytical model is plotted in the stress state (maximum principle stresses σt) for different bending moments M. A good agreement between the models can be observed. Moreover, the development of the tensile stresses at the upper right edge Fsd,r at a bending moment of 160kNm is visible in Figure 4.

3.2 Influence of imperfections on the structural response of longitudinal joints

Misalignments between adjoining segments are primarily a result of errors during segment installation. Due to the fabrication of segments under controlled conditions, the strict production tolerances in a range of a single millimeter can be fulfilled. Tolerances during the ring installation can, however, not be avoided. A minimum misplacement tolerance of 10mm is typically considered during segment design, with variations based on specific project conditions [Grübl (2006), Kolic (2009)]. Misalignments between adjoining segments can affect the kinematics as well as the load transfer capabilities at the longitudinal joint and therefore influence the structural response of a segment. In order to analyze the kinematics and the load transfer at a misaligned longitudinal joint, the numerical model of the experimental joint test, described in Section 3.1, is used, in which the upper segment is displaced by 20mm to the left (Figure 5 top right) and by 20mm to the right (Figure 5 bottom right). It results in an eccentricity and a reduction of the contact area (see Figure 2). The resulting M-α-relationship for a normal force N of 2300kN is shown in Figure 4.

The general non-linear characteristic of the M-α-relationship is maintained even with present misalignments. The misalignment influences the maximum potential bending moment, which can be transferred at the longitudinal joint. If the eccentricity from bending e and the misalignment ex appear in the same direction (ex = -20mm), the maximum transferable bending moment is reduced around the bending moment ΔM caused by the eccentric transfer of the normal force. Before the segment is subjected to an external bending moment, the eccentric normal force already induces an internal bending moment (see Figure 5, red I) and therefore cannot carry the same amount of external applied bending moment (compare Figure 5, red and green II). If the eccentricity from bending e and the misalignment ex appear in different directions (ex = +20mm), they act against each other (compare Figure 5 green I and II). When the external applied bending moment reaches a value of ΔM, both eccentricities compensate each other (rotation α = 0), and after that, the joint can carry an external bending moment till its general transfer capacity is reached (Figure 5 left). The small deviation in the gradient of the curves can be explained by the reduced contact area in the misaligned cases which results in a higher stress concentration at the joint and therefore an increased resistance against rotations.

Based on the analytical models, misalignments influence the stress state in the segmental lining and therefore the resulting crack patterns. In order to investigate the influence of misalignments on the crack pattern, the rotation test from Section 3.1 is carried out with a PC (properties see Section 2.2) for normal force levels of 1600 and 2300kN and a misalignment of

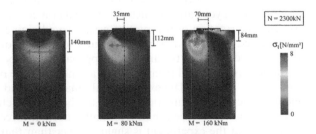

Figure 4. Comparison between the computed maximum principle stress field (linear elastic) and the analytical design model of a specimen subjected to a normal force of 2300kN and different bending moments.

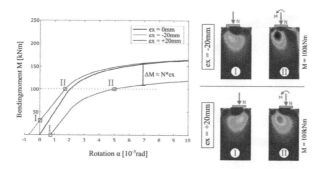

Figure 5. Left: Comparison of the M-α-relationship for different misalignments at a normal force N of 2300kN; Right: Corresponding maximum principle stress fields of two opposed misaligned segments subjected to an external normal force (I) and in combination with a bending moment of 100kNm (II).

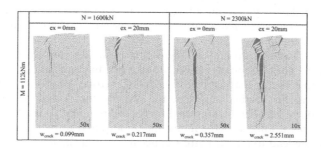

Figure 6. Comparison of resulting crack patterns for a reference (ex=0mm) and misaligned segment (ex=-20mm) subjected to a bending moment of 112kNm and normal forces of 1600kN (left) and 2300kN (right).

ex = -20mm. Figure 6 shows the resulting crack pattern at an external applied bending moment of 112kNm. For both normal force levels, the misalignment greatly affects the resulting crack width whereas the crack pattern remains similar. At a normal force level of 1600kN, the crack width is increased by nearly 46% and exceeds the requirements regarding a maximum allowable crack width according to Eurocode 2 (2011) (w_{max} = 0.2mm). At a normal force level of 2300kN with consideration of a misalignment, the crack can grow through the whole specimen and causes a chipping of the left part. Furthermore, a tendency for chipping for bending dominated problems can be derived. A high normal force in conjunction with an increased M/N ratio can favor a chipping off the segment edges (see Figure 6, N=1600kN / ex=-20mm). Both effects should be considered at the design of the reinforcement layout.

4 CONVENTIONAL, STEEL-FIBER AND HYBRID REINFORCED LINING RESPONSE TO EARTH-PRESSURE LOADINGS

4.1 Holistic Multi-Level Structural Model for a Segmental Tunnel Lining

In order to investigate the influence of steady state loadings on the tunnel lining, especially around the longitudinal joints, a lining ring simulation incorporating an explicit modeling of the longitudinal joints with interface elements is performed. A prescribed displacement Δu is applied on the grouting layer in accordance to the final ground deformations predicted by a fully 3D process oriented tunnel simulation model [Marwan et al. 2017]. The consideration of the structural scale is necessary to incorporate the interdependency between machine advancement and the surrounding soil as well as the time dependent behavior of the grout. The ring has an inner diameter of 9600mm, the thickness of all segments is 500mm with a contact

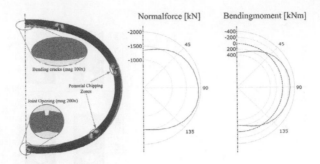

Figure 7. Left: Maximum principal stresses in the segmental lining model and indication of potential failure mechanisms; Right: Distribution of normal force and bending moment around the ring.

length of 250mm at the longitudinal joints and the overburden of the tunnel is 20m. The surrounding soil has an elasticity modulus of 120 N/mm² with a weight of 18kN/m³ and for the lateral earth pressure coefficient Ko a value of 0.5 is assumed. The resulting stress resultants are given in Figure 7.

The highest tensile stresses around a longitudinal joint develop at the joint at 78° (see Figure 7 left), where a normal force of -1934kN and a bending moment of -217kNm are measured. The M/N ratio is bigger than b/6 which results in a rotation of the joint and therefore a reduced contact area. For design purposes, the normal forces and bending moments are increased by a factor of 1.5. The high normal forces are required to increase the stress transmission over the segment joint, and the moments are required to increase the rotation of the joints and to therefore lead to a higher stress concentration at the joint due to a reduced contact area. In this article, the evaluation of the model is limited to the region around the longitudinal joints and therefore no bending cracks are discussed (Figure 7 left).

4.2 Evaluation of proposed segment designs

The structural response of four representative segment designs is investigated and compared in order to arrive a robust segment design. A plain concrete (PC), a conventional reinforced (RC), a steel fiber reinforced (SFRC) and a hybrid reinforced segment (Hybrid) are analyzed. The RC design represents a traditionally reinforced segment with minimum reinforcement as per ZTV-ING (2012) (φ10–10). A concrete cover of 50mm is assumed and a vertical reinforcement bar at the analytical predicted splitting stress position is used for the RC design. The hybrid design combines the minimum reinforcement with a 200mm long SFRC "cap" (Figure 8).

The material properties of all materials can be found in Section 2.2. Rings out of all segment designs are subjected to the loading scenario described in Section 4.1. The resulting crack patterns are shown in Figure 8.

All designs show a similar crack pattern with the tendency to a chipping of the segment corner. In the case of chipping, the SFRC proves to be significantly more effective at reducing crack width. Significantly, the minimum reinforcement provide against splitting stresses in the RC design is not sufficient enough to inhibit the formation of cracks in exceedance of the tolerance [Eurocode 2 (2011)]. The PC and RC designs are both not able to fulfill the serviceability limit of a maximum crack width of 0.2 mm, as they predict crack widths of 0.532 mm and 0.263 mm, respectively. This indicates that SFRC is much better suited to resisting chipping cracks than RC. These findings are in agreement with experimental results as presented by Gong et al. (2017), in which it was observed that longitudinal joints constructed using SFRC segments had a higher overall moment-carrying capacity than traditional RC segments and displayed smaller crack widths at equal M/N ratios. The hybrid design predicts a maximum crack width of 0.094 mm which represents a noticeable increase in comparison to the 0.022 mm crack width as predicted by the full SFRC segment. Nevertheless, the performance of the hybrid design fulfills the serviceability requirements and could certainly be improved by providing a larger SFRC "cap."

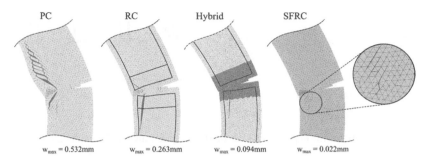

Figure 8. Crack patterns for different reinforcement schemes at the longitudinal joint located at 78°, as measured clockwise from the tunnel crown. All segments are shown at a magnification of 100x.

5 SEGMENT RESPONSE TO HYDRAULIC JACK LOADINGS

5.1 *Experimental set-up*

Beside the local loadings of longitudinal joints, the local loadings of segments subjected to jack forces are one of the most important loading cases for segmental lining design. The magnitude and the arrangement of the jack forces are a major factor to determine the thickness of a segment, since a sufficient load spreading has to be ensured. Hemmy & Falkner (2004) performed experiments on tunnel segments without curvature to analyze the splitting behavior of SFRC. The geometry of the experimental setup is used for the presented calculations (see Figure 9 left).

Two loading cases are investigated: Case I, in which the segment is installed according to plan without any tolerances and is subjected to the ultimate load of F = 9000kN. In Case II, there is an insufficient support on the right side of the segment (Figure 9 right) due to improper segment installation. In this case, only a reduced jack force in accordance to the service load is used (F = 4500kN). Also, it is assumed that due to the bending of the segment, it is in contact with the subsequent segment of the next ring and corresponding boundary conditions are used.

5.2 *Influence of steel-fibers on cracking behavior*

In this section, the structural response of a PC and SFRC segment under the above-mentioned loading scenarios are examined. The material properties are listed in section 2.2. The resulting crack patterns are shown in Figure 10 (loading case I) and Figure 11 (loading case II).

As expected, the crack patterns for load case I are similar to the ones resulting from the centric partial area loading (Figure 10). The characteristic splitting crack develops in the middle of the segment under the loading plates and reaches a crack width of 0.663mm for the PC segment, which exceeds the maximum tolerable crack width of 0.2mm (Figure 10 left). However, no critical state regarding a system failure occur. On the other hand, multiple small cracks with a maximum crack width of 0.0504mm instead of a single crack develops for the SFRC design subjected to load case I, which would fulfill the requirements also in the serviceability state (Figure 10 right).

Insufficient support of a segment during machine advancement resulting from improper segment installation, can lead to a bending of the segment and therefore is characterized by an additional initiation of axial cracks between the load plates (Figure 11). The maximum crack width increases tremendously in comparison to load case I, although only half of the jack forces magnitude is applied. The maximum crack width of the PC segment is 1.26mm, where the crack process through the whole segment and therefore the structural stability cannot be ensured. In contrast, SFRC would reduce the maximum crack width to 0.0898mm and limit the crack growth (Figure 11 right).

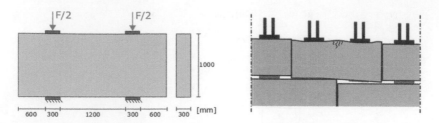

Figure 9. Left: Geometry of the experimental set-up [Hemmy & Falkner (2004)]; Right: Insufficient support of a segment.

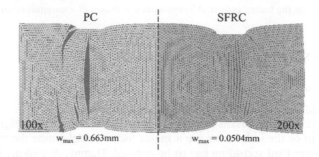

Figure 10. Loading case I: Crack patterns of a PC and a SFRC segment installed according to plan without any tolerances and subjected to the ultimate load of F = 9000kN.

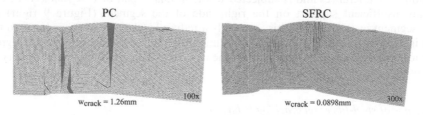

Figure 11. Loading case II: Crack patterns of a PC and a SFRC segment with insufficient support and subjected to the serviceability load of F = 4500kN.

6 CONCLUSION

Within this work, several segmental lining designs are proposed and their structural response with respect to construction related installation problems and unexpected ground loadings is investigated. The presented results demonstrate that segment designs containing SFRC are capable of mitigating the splitting behavior observed at longitudinal joints due to partial area loadings and, in general, show a better crack distribution in connection with a smaller maximum crack width. In order to achieve a favorable post-cracking response, a sufficiently strong SFRC must be used. It is shown that hybrid segment designs can offer a similar performance by combining the advantages of RC and SFRC with a reduction of the material costs. Specifically, the proposed SFRC "cap" design can increase the segments resistance against damages during segment handling. The investigation of misalignments between adjoining segments indicates that the chipping of the segment corners may not necessarily result from imperfections but rather develop due to stress distributions resulting from the applied bending moment to normal force ratio and magnitude. SFRC performs similarly well with regard to pressure jack loadings and, in the case of insufficient bedding during machine advance, dramatically reduces expected crack widths.

ACKNOWLEDGEMENTS

Financial support was provided by the German Research Foundation (DFG) in the framework of project B2 of the Collaborative Research Center SFB 837 Interaction modeling in mechanized tunneling. This support is gratefully acknowledged.

REFERENCES

ACI COMMITTEE 318 (2008). Building Code Requirements for Structural Concrete (ACI 318–08) and Commentary. Technical report, American Concrete Institute, 38800 Country Club Drive, Farmington Hills, MI 48331 U.S.A.

Blom, C. 2002. Design philosophy of concrete linings for tunnels in soft soils, Ph.D. thesis, Delft University

DIN EN 1992-1-1 2011. Eurocode 2 - Bemessung und Konstruktion von Stahlbeton und Spannbetontragwerken – Teil 1–1: Allgemeine Bemessungsregeln und Regeln für den Hochbau.

Gall, V.E., Butt, S., Neu, G.E. & Meschke, G. 2018. An embedded rebar model for computational analysis of reinforced concrete structures with applications to longitudinal joints in precast tunnel lining segments, in: G. Meschke, B. Pichler, J. G. Rots (Eds.), *Computational Modelling of Concrete Structures (EURO-C 2018)*, Taylor & Francis, pp. 705–714.

German Tunnelling Committee (DAUB) 2013, Recommendations for the design, production and installation of segmental rings, Tech. rep., Deutscher Ausschuss für unterirdisches Bauen e. V. (DAUB).

Gong, C., Ding W., Mosalam, K. M., Günay, S. and Soga, K. 2017. Comparison of the structural behavior of reinforced concrete and steel fiber reinforced concrete tunnel segmental joints. *Tunnelling and Underground Space Technology* 68, 38–57.

Grübl, F. 2006. Segmental Rings - Critical Loads and Damage Prevention. *International Symposium on Underground Excavation and Tunnelling: Effect of Groundwater on Tunnel Stability*, Thailand

Hemmy, O. & Falkner, H. 2004. Zum Gebrauchs-und Tragverhalten von Tunnelschalen aus Stahlfaserbeton und stahlfaserverstärktem Beton Deutscher Ausschuss für Stahlbeton (DAfStb)

Hordijk, D. & Gijsbers, F. 1996. Laboratoriumproeven tunnelsegmenten, Tech. Rep. 96-CON-R0708/03, TNO-Bouw, Delft

International Federation for Structural Concrete (fib) 2013, fib Model Code for Concrete Structures 2010, Ernst & Sohn

Kolic, D., Mayerhofer, A. 2009. Segmental Lining Tolerances and Imperfections. *ITA-AITES World Tunnel Congress: Safe Tunnelling for the City and Environment*.

Leonhardt, F. & Reimann, H. 1966. Betongelenke, *Der Bauingenieur* 41, S. 49–56.

Li, X., Yan, Z., Wang, Z. & Zhu, H, Experimental and analytical study on longitudinal joint opening of concrete segmental lining, *Tunnelling and Underground Space Technology*, Volume 46, 2015

Manzoli, O., A. Gamino, E. Rodrigues, & G. Claro 2012. Modeling of interfaces in two-dimensional problems using solid finite elements with high aspect ratio. *Computers and Structures* 94–95, 70–82.

Marwan, A., A. Alsahly, V. Gall, & G. Meschke 2017. Computational modelling for segmental lining installation in mechanized tunneling. In Proceedings of the 4th International Conference on Computational Methods in Tunneling and Subsurface Engineering (EURO:TUN 2017), pp. 153–160.

ÖVBB-Richtlinie: Tübbingsysteme aus Beton. August 2009.

Schlaich, J. & Schäfer, K. 1991: Design and detailing of structural concrete using strut-and-tie models. *The Structural Engineer* 69(6), pp. 113–125.

Sugimoto, M. 2006. Causes of shield segment damages during construction. In International Symposium on Underground Excavation and Tunneling, Bangkok, Thailand, pp. 67–74.

Tvede-Jensen, B. Faurschou, M. & Kasper, T. 2017. A modelling approach for joint rotations of segmental concrete tunnel linings, *Tunnelling and Underground Space Technology* 67, 61–67.

Zhan, Y. & G. Meschke 2014. Analytical model for the pullout behavior of straight and hooked-end steel fibers. *Journal of Engineering Mechanics (ASCE)* 140(12), 04014091(1–13).

Zhan, Y, & G. Meschke 2016. Multilevel computational model for failure analysis of steel-fiber-reinforced concrete structures. *Journal of Engineering Mechanics (ASCE)* 142(11), 04016090(1–14).

Zusätzliche Technische Vertragsbedingungen und Richtlinien für Ingenieurbauten (ZTV-ING): Teil 5 - Tunnelbau 2012

*Tunnels and Underground Cities: Engineering and Innovation meet Archaeology,
Architecture and Art, Volume 6: Innovation in underground engineering,
materials and equipment - Part 2 – Peila, Viggiani & Celestino (Eds)*
© 2020 Taylor & Francis Group, London, ISBN 978-0-367-46871-2

Design of steel fiber reinforced concrete segment with curved radial joints

S.S. Nirmal
Ayesa India Pvt. Ltd., New Delhi, India

ABSTRACT: Mumbai Metro Line 3 is major underground project with 33.5 km of twin tunnels and 27 stations with estimated cost of 2.8 million Euros. The tunnels are being constructed with tunnel boring machines (TBMs). Segments for this project have curved radial joints to minimize crushing of concrete during ovalisation and convergence of segmental lining. Curved radial joints have lower contact widths between segments resulting into higher stresses around area of contact. The existing guidelines do not provide a design approach for curved radial joints of steel fiber reinforced concrete (SFRC) segments. This paper discusses an approach to design radial joints for SFRC segments by combining International Tunnelling Association (ITA) guidelines and Roark's design formulas. The parameters considered for design are included and verified by test results. Likely future improvement in this design approach is optimizing amount of reinforcement required for radial joints by considering tensile capacity of SFRC matrix.

1 INTRODUCTION

Mumbai Metro Line 3 project is divided into seven construction contracts (UGC 01 to UGC 07). The scope of contract UGC 02 is to design and construct four underground stations (CST, Kalbadevi, Girgaon and Grant Road), 5 km long twin TBM bored tunnels and NATM platform tunnels in all stations except CST station.

Figure 1.   Alignment for UGC 02 contract.

The TBMs are driven through the stations followed by widening of these tunnels by NATM construction methodology to construct platform tunnels for this project. During construction of NATM platform tunnels, the originally constructed segmental lining is dismantled. These sacrificial segments are designed with SFRC to save overall quantity of steel required in the project. Since, these segments are designed with curved radial joints, a distinctive design approach is followed. This design approach is based of ITA guidelines on design of SFRC segments. Although, as these guidelines do not explicitly elaborate on the design of segments with radial joints, a combination with Roark's design formulas was used to develop a solution.

SFRC segments were designed for Kalbadevi, Girgaon and Grant Road stations. The geology at each of these stations are different with combination of weak and intact rock types of Basalt and Breccia.

2 GEOLOGY

Geology in Mumbai has volcanic origin dated back to cretaceous period (66 to 67 million years). Majority of rock in basal and breccia with both fresh and weathered strata. Rock is covered by costal alluvial soil deposited by the Arabian sea.

Kalbadevi station has 7m deep alluvial soil layer at top followed by 22m deep weathered basaltic rock and fresh basaltic rock further below. The weathered basaltic rock has intermediate layers of fresh basalt.

Girgaon station has alluvial soil cover of 10m followed by 11m of weathered breccia rock. The fresh breccia rock is present below weathered breccia rock as identified in bore hole investigation.

Grant Road station has 8m of alluvial soil cover at top. Below the alluvial soil, 5m deep weathered breccia is present followed by fresh breccia rock type beneath it.

Water table is approximately 3m to 6m below ground surface level depending on the location. As a conservative approach, ground water table is considered at ground surface level.

3 TYPICAL CONSTRUCTION SEQUENCE

Steel fiber reinforced segments are used as temporary segments in stations and are planned to be dismantled to widen the TBM driven tunnels using NATM method of construction. Further these platform tunnels will be connected to adjacent station box (constructed using top down construction methods) with access tunnels. Planned construction sequence was adopted to develop FE (Finite Element) models and obtain design forces and moments in the SFRC segmental lining.

4 SEGMENTAL LINING CONFIGURATION

Segmental lining for this project is 275mm thick with an internal diameter of 5800mm and 1400mm length. The segmental ring has 5 regular segments with 1 key segment and is constructed with M50 grade of concrete. The lining is designed for a minimum design life of 120 years. The segmental rings are universally tapered rings with 20mm taper on both circumferential joints.

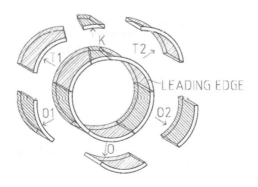

Figure 2.   Configuration of segmental ring lining.

The segments are designed with curved radial joints with radius 2000mm. Curved radial joints has advantages to allow movement in segments joints without damaging the concrete during ovalisation of segmental ring. The curved radial joints used in the project are shown in Figure 3 below.

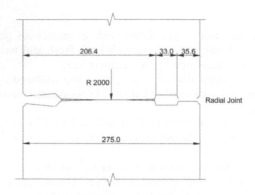

Figure 3.   Details of curved radial joints for segments.

5   STEEL FIBER REINFORCED CONCRETE

Steel fibers increase the post crack tensional capacity of concrete matrix making it as a potential solution to replace conventional reinforcement in segmental lining for tunnels. For the purposes of designing SFRC segments it is assumed to use steel fibers with tensile strength of 1100 MPa with aspect ratio of 80 and dosage of 30 kg/m^3 of concrete. This concrete matrix has characteristic residual flexural strength at 0.5mm and 2.5mm crack of 3.5 N/mm^2 and 2.3 N/mm^2 respectively. These assumptions for concrete mix were made after a comprehensive research done on similar designs used successfully in previous project.

Further, to validate the parameters used in design, tests are performed at construction site with combination of various steel fiber types and their dosages. The concrete mixes satisfying to the assumed design parameters were approved for use in the project. The test results for beam test conducted on SFRC are reproduces in Table 1.

Table 1.   Flexural tensile strength from flexural test of SFRC beam.

Trial Mix	Type (λ/l)	Dosage kg/m^3	f_{r1} MPa	f_{r3} Mpa
TM1	Brand A (80/60)	30	3.97	2.72
TM2	Brand B (80/60)	30	6	3.61
TM3	Brand A (65/35)	35	4.38	2.96

where λ = Slenderness ratio; l = length of fiber; f_{r1} = residual flexural strength at 0.5mm crack; and f_{r3} = residual flexural strength at 2.5mm crack.

6   DESIGN APPROACH

6.1   Design of SFRC segments

FEM analysis was carried out with Plaxis 2D software for each station with their respective geological conditions and construction sequence to obtain the forces and moments in SFRC segmental lining.

Table 2. Design hoop force and moment in SFRC segmental lining.

Station	Hoop force kN	Moment kNm
Kalbadevi	3927	88
Girgaon	2856	200
Grant Road	1911	36

The forces and moments obtained are added with train track loads, additional moments due to ovalisation, TBM ram thrusts, grouting loads and were multiplied with factor of safety to obtain design forces and moments. The design moment and hoop force in lining for each station are shown in Table 2.

Further, load- moment interaction diagrams are developed with design forces and moments for each station with assumed properties of SFRC matrix to study the adequacy of designed segmental lining.

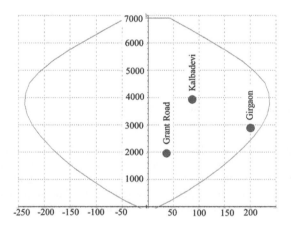

Figure 4. Load moment interaction diagram for SFRC segmental lining.

6.2 *Design of radial joints*

The critical part of design of SFRC segmental lining is design of radial joints. As a result of reduced contact width between segments the bursting stresses generated are higher than those generated for flat joints.

To calculate the bursting stresses the contact width was calculated using Roark's approach (Young, 1989). Roark's approach assumes that the elements in contact are elastic in nature. As per the analysis it is also assumed that the deflection in SRFC segments is within the elastic limit. The contact width is calculated using the approach as below:

$$b = 1.6\sqrt{(N_d K_d C_E)} \tag{1}$$

$$K_d = \frac{D_1 D_2}{D_1 + D_2} \tag{2}$$

$$C_E = \frac{1 - \nu_1^2}{E_1} + \frac{1 - \nu_2^2}{E_2} \tag{3}$$

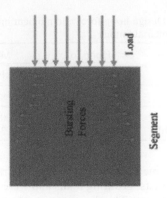

Figure 5.   Typical representation of bursting force.

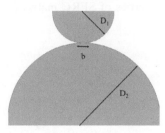

Figure 6.   Roark's approach on contact surface between two curved surfaces.

where b = contact width; N_d = design hoop force; $D_1 = D_2$ = diameter of curved radial joint; $\nu_1 = \nu_2$ = Poisson's ratio of concrete; and $E_1 = E_2$ = modulus of elasticity of concrete.

It was assumed that the SFRC matrix has no tensile capacity against bursting stress and complete stress is taken by reinforcements at radial joints. The requirement for quantity of reinforcement is calculated using design approach mentioned in BS 8110 part 1. The approach is detailed as below-

$$A_{s,\,req} = \frac{F_{bst}L_s}{f_y} \qquad (4)$$

$$\frac{F_{bst}}{N_d} = k \qquad (5)$$

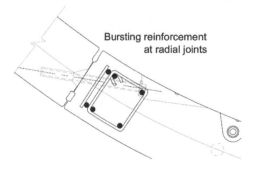

Figure 7.   Reinforcement drawings for curved radial joints.

where $A_{s,req}$ = area of bursting reinforcement required; F_{bst} = bursting force at radial joints; L_s = length of radial joint; f_y = capacity of reinforcement against bursting as per BS 8110 Part 1; and k = ration of bursting force to hoop force as per Table 4.7 of BS 8110 Part 1. The resultant reinforcement drawing for SFRC segments is reproduced in Figure 7.

7 FUTURE PROSPECTS FOR FURTHER DEVELOPMENT

The design of radial joints and calculation for bursting reinforcement ignores the tensile capacity of SFRC matrix. This design can be improved further by assessing the tensile capacity of the SFRC matrix and taking it into account while designing the reinforcement again bursting stresses. The assessment of tensile capacity can be done by theoretical empirical approaches and further validation by testing the SFRC matrix for tensile capacity. This development shall reduce the amount of reinforcement required in SFRC segments.

8 CONCLUSION

The SFRC segments are designed and related tests were performed successfully for Mumbai metro line 3, contract UGC02 with three different concrete mix. Additional reinforcement bars are required at radial joints for bursting stresses developed near curved radial joints. The design approach adopted has prospects for improvement by considering the tensile capacity of SFRC matrix while deigning bursting reinforcement for radial joints.

REFERENCES

Young, W. C. (6th ed.) 1989. *Roark's formula for stress and strain*. New York: Mcgraw-Hill.
ITAtech, 2016. *Guidance for precast fibre reinforced concrete segments - volume 1: design aspects*, Avignon: The International Tunnelling and Underground Space Association.
SFRC Consortium 2014. *Design guideline for structural applications of steel fibre reinforced concrete*. Taastrup: SFRC Consortium.
FIB 2013, *Model code for concrete structures 2010*. Berlin: Ernst & Sohn
DAUB 2013, *Recommendations for the design, production and installation of segmental rings*. Colonge: German Tunnelling Committee (DAUB)

Tunnels and Underground Cities: Engineering and Innovation meet Archaeology,
Architecture and Art, Volume 6: Innovation in underground engineering,
materials and equipment - Part 2 – Peila, Viggiani & Celestino (Eds)
© 2020 Taylor & Francis Group, London, ISBN 978-0-367-46871-2

Development and application of the next generation TBM equipped with conventional tunneling mode

K. Nishioka, A. Shigenaga, A. Yokoo & Y. Kitamura
Kajima Corporation, Tokyo, Japan

ABSTRACT: When a hard rock TBM encounters difficult ground conditions, excavation must be stopped until the completion of ground stabilization work. Such trouble obstructs to make full use of the merit of TBM's high-speed excavation. Whereas conventional tunneling is inferior to TBM in drilling speed, conventional tunneling can be adopted flexibly to complex geology by installing appropriate tunnel supports. Authors have developed the world's first hybrid excavator which can convert TBM mode into conventional tunneling mode according to the geological change. This excavator performs high-speed drilling in TBM mode for hard rock excavation and this can convert to conventional tunneling mode for weak ground excavation. Weak ground can be excavated by the operation of the bucket equipped inside the machine and tunnel supports can be installed immediately afterwards. In this paper, we will explain the detailed mechanism of the excavator and introduce the plan of adaptation to an actual tunnel.

1 INTRODUCTION

The biggest benefit of tunnel construction using a TBM is the capability of high-speed excavation of hard rock. In Japan, many TBMs were used for excavation of pilot tunnels of the Shin-Tomei and Shin-Meishin Expressways in the 1990s but, as shown in Figure 1, the number of tunnel construction projects using TBMs are on a decline.

TBMs are capable of high-speed excavation when the strata are homogeneous as those in Europe. On the other hand, cases of excavation stoppages due to ground stabilization works become greater with complex and weak geology which is peculiar to Japan, The number of tunnel construction projects using TBMs is assumed to have decreased in Japan because the benefit of high-speed excavation of TBMs was not able to be fully utilized.

We studied the number of problems involving TBM excavation stoppages for three days or longer that occurred in Japan. Figure 2 shows the number of problems during construction using TBMs classified by factors relating to the ground properties. Figure 3 shows the number of problems during construction using a TBM which are classified by machine-related factors.

Figure 2 suggests that stoppages of construction works using TBMs are mainly caused by collapse or rocks falling at the tunnel face and many problems resulting from a large amount of inflow have also occurred. In addition, Figure 3 confirms that, in many cases, problems occurred as a result of encountering such ground, including the cutter head failing to rotate and inability to get a reaction force from the gripper.

To compare the TBM operating rate between Japan and overseas, the average operating rate is 20 to 30% in Japan as compared with 35 to 50% overseas. Accordingly, for promoting utilization of TBMs in Japan, development of excavation methods for weak ground is needed to improve the operating rate.

Meanwhile, although conventional tunneling is inferior to TBM in terms of excavation speed, it can be excavated by using a construction machine suited for the ground, by visually checking the tunnel face directly, as shown in Figure 4. Tunnel supports and auxiliary methods can also be immediately selected and installed at the face in accordance with conditions of the ground.

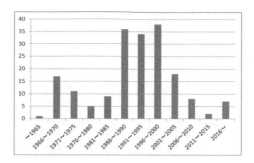

Figure 1. Change in the number of tunnel construction projects using TBMs in Japan.

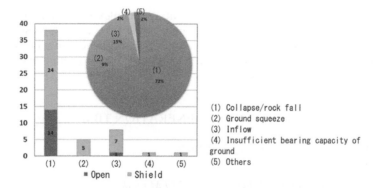

(1) Collapse/rock fall
(2) Ground squeeze
(3) Inflow
(4) Insufficient bearing capacity of ground
(5) Others

Figure 2. Number of TBM excavation problems classified by factors relating to ground properties.

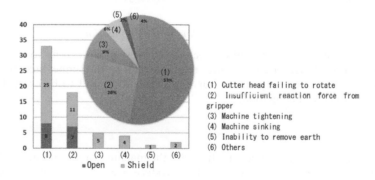

(1) Cutter head failing to rotate
(2) Insufficient reaction force from gripper
(3) Machine tightening
(4) Machine sinking
(5) Inability to remove earth
(6) Others

Figure 3. Number of TBM excavation problems classified by machine-related factors.

With these circumstances, Kajima worked jointly with Komatsu on the development of a new excavator that combines the "features of TBM capable of high-speed tunneling" and the "features of conventional tunneling capable of flexibly handling weak ground."

2 CONCEPT OF THE NEXT GENERATION TBM

The next generation TBM has two features:

(1) Uses the TBM mode for high-speed tunneling for hard rock excavation.
(2) Converted to the conventional tunneling mode when excavating weak ground and continues excavation by using the bucket excavator provided in the machine main unit.

Figure 4.   Tunnel construction by conventional tunneling.

The next generation TBM is capable of excavating complex and fragile ground which is peculiar to Japan in a conventional tunneling mode while fully using high-speed excavation capability of conventional TBM. With these features, the next generation TBM can avoid the problems and constraints on the excavator when excavating fragile ground, which will shorten the construction period and reduce costs.

3   CHARACTERISTICS OF THE NEXT GENERATION TBM

The next generation TBM has four major features.

3.1   *Prompt conversion from the TBM mode to NATM mode*

The next generation TBM can change from TBM mode to conventional tunneling mode within 1.5 days, taking into account the changes in geology.

3.2   *Stepwise opening in the cutter head according to the application*

As shown in Figure 6, the cutter head can be opened in steps according to the state of construction. This allows, when requiring minor reinforcement to the tunnel face, the cutter head to be narrowly opened at the top during forward exploration and to be opened wide at the center in conventional tunneling mode.

3.3   *TBM's function for retraction*

The TBM main unit has a telescopic structure and the machine can be retracted. As shown in Figure 7, the body of TBM can be retracted to produce a work space of 3.0 meters max in front of the machine.

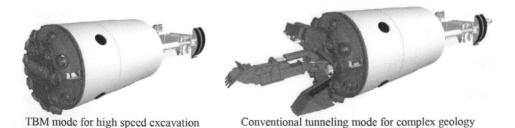

TBM mode for high speed excavation   Conventional tunneling mode for complex geology

Figure 5.   Concept of the next generation TBM.

Figure 6.   Stepwise opening of the cutter head.

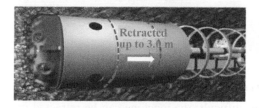

Figure 7.   Retracting function of the TBM.

Figure 8.   Tunneling in weak ground.

3.4   *Safe and secure tunneling in weak ground*

Figure 8 shows how the central part of the cutter head opens to protrude a bucket excavator which has been retracted, and starts excavation of the ground. When excavation has been completed, tunnel supports can be quickly installed to prevent loosening or collapse of the weak ground.

4   EXCAVATION PROCEDURE

The excavation steps of the next generation TBM are as follows.

4.1   *Excavation in TBM mode*

As with conventional TBMs, the gripper shoes provided on the left and the right sides of the rear part of the unit are pressed against the side walls using hydraulic jacks to secure the machine. Then, the thrust jacks on the right and left sides are extended with the cutter head in a rotating state to attack hard rock, and the supports are installed at the rear of the excavator. (Figure 9)

The next generation TBM is equipped with a function of promptly finding weak ground and drainage of ground water ahead of the tunnel face by forward exploration drilling.

Tunneling in TBM mode

Tunnel support installation in the rear of TBM

Figure 9. Excavation in TBM mode.

4.2 Retraction of the TBM and opening of the cutter head

When excavation with the cutter head becomes difficult, the machine can be retracted up to 3.0 m to obtain excavation space in front. Then, the cutter head is opened to be put in conventional tunneling mode. For retracting the machine, the TBM main unit is equipped with a retractable telescopic structure. (Figure 10)

4.3 Excavation in Conventional Tunneling mode

The cutter head is opened and a bucket excavator $(0.3\ m^3)$ inside the boring machine is sent forward. The muck initially produced when the machine is retracted is transported by a belt conveyer backward. Then, a bucket excavator continues excavation and mucking. (Figure 11)

4.4 Installation of tunnel supports

After the excavation of the portion of one ring is finished, shotcrete is placed to the side walls of the tunnel face and steel support ring is erected afterwards. Face stability can also be secured by forepoling and/or the shotcrete placement on the tunnel face. Shotcrete placement is conducted again there to fasten the steel support. (Figure 12)

4.5 Tunneling

When installation of tunnel support has been completed, the machine is advanced by 1 m. Subsequently, the poor ground can be tunneled by repeating a series of processes of excavation, shotcrete placement, tunnel support installation and forward excavation. The conventional tunneling mode is returned to the TBM mode to resume high speed boring after passing through the weak ground. (Figure 13)

Encounter with poor ground

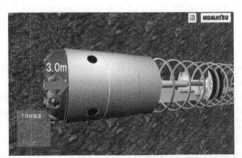

Retraction of TBM

Figure 10. Conversion to Conventional Tunneling mode.

Excavation by bucket excavator | Operation of bucket excavator

Figure 11.   Excavation in Conventional Tunneling mode.

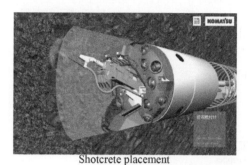

Shotcrete placement | Installation of a steel support

Figure 12.   Installation of tunnel supports in conventional tunneling mode.

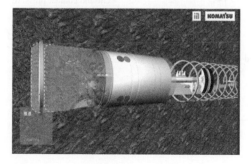

Figure 13.   Returning to TBM mode.

5   TBM RISK MANAGEMENT SYSTEM

The next generation TBM is equipped with a forward exploration function of promptly finding any weak ground ahead of the tunnel face. The machine data (volume specific energy in excavation) during tunneling in the TBM mode and the drilling forward exploration data (fracture energy coefficient) are automatically taken into a 3D BIM software to predict and visualize the geological condition 30 m ahead of the tunnel face. This makes it possible to decide on the TBM mode or the conventional mode to bore a 30 m segment in front. (Figure 14)

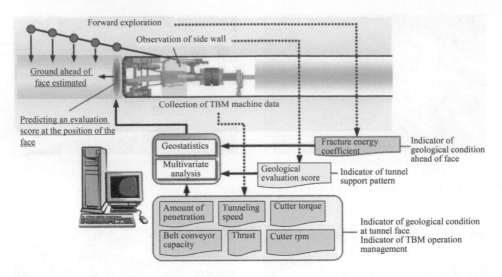

Figure 14.   Schema of TBM risk management system.

Table 1.   Specifications of the next generation TBM.

Item	Major specifications	Item	Major specifications
TBM type	Open type	Gripper	
Boring diameter	4.75 m in diameter	• Total pressing force	20,800 kN
Total length	Approx. 102 m		+265 mm to -335 mm
Total weight	Approx. 335 t	• Shoe projection amount	on right and left sides each
Cutter		• Ground contact pressure	1.1 MPa
• Cutter diameter	17 in.	Belt conveyor	
• Number of cutters	32	• Belt width x capacity	600 mm × 312 m³/h
Cutter drive unit		• Number of conveyors	3
• Motor output	200 kW × 6 units	Number of following trucks coming behind	7
• Cutter head torque	2,546 kNm max.		
• Cutter rotational speed	2.0 to 10.0 rpm		
Thruster			
• Total thrust	9,800 kN		
• Thrust jack	4,900 kN × 2		
• Thrust stroke	1,500 mm		
• Extension rate	15 cm/min max.		

6   BUILDING OF A REAL MACHINE AND ITS APPLICATION TO AN ACTUAL SITE

Currently, we have a project of constructing a headrace tunnel for a hydroelectric power plant in the geology including sepentinite and slate. For this project, we will launch the next generation TBM, and tunnel boring is scheduled to start in July 2019. At present, we are building a TBM for this tunnel. Table 1 summarizes the specifications of the next generation TBM and Figure 15 shows its structural drawing.

7   CONCLUSION

This paper has introduced the boring mechanism of a next generation TBM which features both the function of a high speed drilling of TBM and that of conventional tunneling mode which features the flexibility in dealing with weak ground, and summarized the structure of

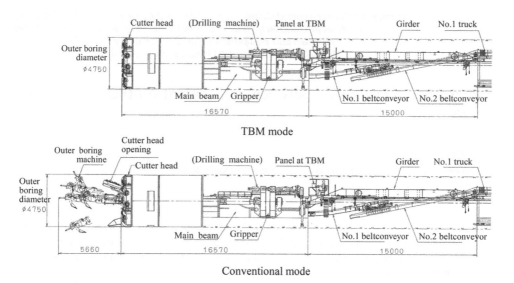

Figure 15. Structural drawing of the next generation TBM.

the machine currently being manufactured. This new boring machine is expected to minimize the lifecycle cost from construction to maintenance, thanks to the advantages over conventional TBMs.

(1) Capable of avoiding delays in construction when encountering poor ground.
(2) Capable of minimizing the potential extra cost at sites with poor ground.
(3) Since support is placed at the site before the ground ahead of the face starts loosening, it is possible to control the loosening of nearby ground, which may produce a higher endurance for the tunnel over a long span of time.

REFERENCES

Japan Tunneling Association. 2000, *TBM Handbook*: 180-182 (In Japanese)

Tunnels and Underground Cities: Engineering and Innovation meet Archaeology, Architecture and Art, Volume 6: Innovation in underground engineering, materials and equipment - Part 2 – Peila, Viggiani & Celestino (Eds)
© *2020 Taylor & Francis Group, London, ISBN 978-0-367-46871-2*

The presumed relationship between an accident on the surface and the NSS tunnel

M.H.B. Oliveira
PCE Projetos e Consultoria de Engenharia Ltda., Rio de Janeiro, Brazil

ABSTRACT: The NSS is a 4620-m long tunnel under construction in Rio de Janeiro state, Brazil. The excavation works started in November 2014 and, due to administrative issues, were interrupted in July 2016, when 3200 m had already been excavated in rock mass predominantly Classes I and II, with cover up to 260 m.

In November 2017, 14 months after the interruption of the excavation, a sinkhole occurred on the surface, 70 m exactly above the tunnel gallery. At first, it was believed that the tunnel had collapsed and caused the event at the surface. However, the investigation proved otherwise.

The investigations began immediately, to find technical elements necessary to support the studies in order to determine the eventual relation between the tunnel and the sinkhole. At the site, SPT and rotary borehole surveys were performed, as well as GPR investigation, among many others. At the office, the geological mapping and worksite staff follow-up reports issued by the constructor were analyzed.

This paper aims at showing the process and results of field surveys, the technical analysis of the data produced in the site and the analysis made to determine the cause of the sinkhole.

The conclusion was that the sinkhole is just a doline, formed by infiltration of water into sub soil and thus not related to the tunnel. It also concludes that tunnel did not cause the sinkhole but, instead, a collapse in the tunnel crown could have been caused by the sinkhole.

1   INTRODUCTION

The NSS Tunnel was under construction in Rio de Janeiro state, as a part of duplication of the road connecting the cities of Rio de Janeiro and Petropolis. It is a 4,620-m long tunnel, excavated in rock mass predominantly Classes I and II, with overburden up to 260 m. The construction began in November 2014 and was interrupted in July 2016, when 3,012-m had been excavated.

In November 2017 a sinkhole occurred on the surface and due to it many actions were performed to find the cause of the accident and if there was any relation between the sinkhole and the tunnel.

This paper aims at presenting the investigation made to find the cause of the sinkhole and describe the conclusion taken afterwards.

2   TUNNEL CHARACTERISTICS

The tunnel is a 4,620-m long road tunnel, with 2 traffic lanes, a shoulder lane and a 3.0-m width sidewalk. Under the sidewalk there is a concrete structure which constitute the safety gallery, connected to the open-air area by an auxiliary tunnel.

Figure 1 below shows the NSS tunnel cross section.

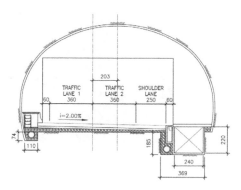

Figure 1.   NSS Tunnel cross section.

The NSS Tunnel is a single, unidirectional tube, with a severe ascending slope of 6%, and is being excavated in a mass of biotite-gneiss predominantly classes I and II (Bieniawsky). Figure 2 below shows the tunnel, the terrain and the classes of rock, as well as the slope and traffic direction.

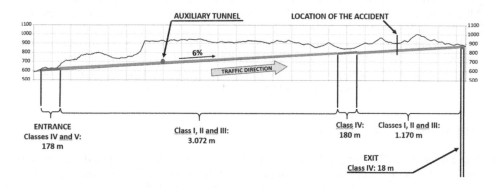

Figure 2.   NSS Tunnel longitudinal profile.

3 TUNNEL CONSTRUCTION

The excavation works began in November 2014 with two fronts, one from the entrance portal and other from exit portal. Later, as the excavation of the auxiliary tunnel was concluded, two new excavation fronts were created to accelerate the construction.

The diagram of the excavation fronts is shown on the Figure 3 below.

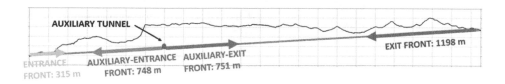

Figure 3.   Excavation services diagram.

Figure 4. Gallery during construction directly below the sinkhole.

It is important to mention that, because of the slope, the water that flows from the excavation front of the exit portal to the gallery needs to be pumped out of the tunnel. When the construction was interrupted, the pumping services was also stopped, thus accumulating water the gallery. The intake flow of water to the interior of the tunnel was small, about 156 l/min (13,2 liters/minute/100m), due to this, after months without pumping, the exit front excavation (right of Figure 3) was fully flooded with water.

Another point important to mention is the quality of the rock below the surface incident. In this segment, the rock was registered as Class II, with only 8 m in class III, and with little water flow. Figure 4 shows the gallery during excavation in the section just below the sinkhole.

As can be seen in the figure, the rock mass presents good quality and is dry.

4 THE SURFACE ACCIDENT

On November 7, 2017, a soil collapse occurred and a sinkhole was formed on the surface. Approximately 1 hour later, part of the slope around the sinkhole slid inwards. Figure 5 shows the first moment of the accident and the Figure 6, shows the sinkhole after the second moment.

As seen in Figure 7, the location of the sinkhole coincides with the excavated tunnel directly below. Because of this, the hypothesis that the tunnel could have caused the accident was considered.

Figure 5. Sinkhole at the first moment.

Figure 6. Sinkhole after the landslide of the soil around.

Figure 7. Road, sinkhole and the tunnel.

5   INVESTIGATIONS AND STUDIES

5.1   Subaquatic Inspection

Immediately after the accident and considering the relative position of the sinkhole just above the tunnel, it was necessary to investigate the situation inside the gallery. Since the tunnel was flooded, a Subaquatic Remote Operated Vehicle was provided to make the inspection in this segment of the gallery.

This inspection showed an obstruction in the gallery, almost below the sinkhole. This fact tended to corroborate the hypothesis of a relationship between the events inside the tunnel and in the surface.

Later, when the water was pumped out of the tunnel, making it possible to enter the gallery and reach the local of the obstruction, it was possible to see and confirm the obstruction, as well as state that it was constituted by sand resulting from rock decomposition.

5.2   Rotary Borehole Surveys

To better identify the rock mass above the tunnel, a geotechnical investigation campaign was performed, consisted by SPT and rotary borehole surveys. The total amount of perforation was:

- Soil: 461 m
- Rock: 691 m

It is important to mention that in all holes sound rock was found, in layers of about 45 meters, between the tunnel crown and the soil at surface. This situation denies the previous hypothesis. Besides that, borehole SR-CTN-06 made in the center of the sinkhole identified the material used to fill the sinkhole, followed by weathered rock, sound rock and, at the end, a 20-cm layer of shotcrete. This confirms the integrity of the tunnel crown in a position vertically below the sinkhole.

Although most of the holes were drilled in sound rock, some of them required special attention. Boreholes SR-CTN-02 and SR-CTN-04 found voids in the rock, that could be natural caverns, not common in in this kind of rock mass (biotite gneiss), but possible.

Another point to notice is the difference between boreholes TS-05 and MNA-04. TS-05 shows void with 59 meters depth while MNA-04, distant only 1,10 m from TS-05 reaches the tunnel crown with 70 m depth.

The results of the rotary borehole surveys campaign can be seen in the Figure 8 below.

5.3 *Ground Penetrating Radar*

As an important part of the surveys, a geophysical research campaign took part in the actions and the result of this studies showed the integrity of the rock mass above the tunnel. Figure 9 below shows an example of the GPR diagrams, where the tunnel crown can be seen in the bottom of the figure.

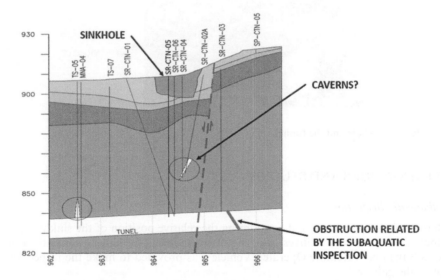

Figure 8.   Profile of terrain along the tunnel.

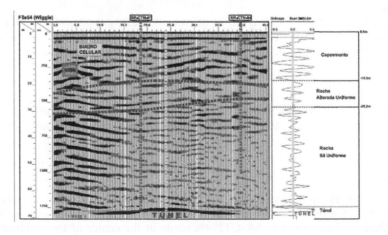

Figure 9.   Radargram from GPR survey.

The conclusion presented in the report issued by the GPR technicians is very strong in affirming that the facts indicate by themselves that there is no influence of the tunnel in the formation of the sinkhole. They based the conclusion on the outcome of GPR studies and the two other considerations shown in the following items.

5.4 Analysis of the surface of the hole

Besides the GPR analysis, a series of analysis on the surface of the hole were made. These analysis regarded the behavior of rock fractures and the geometrical form of the hole.

The fractures observed were old and stable. There were no new fractures and the old ones did not reopen. Regarding the geometrical form of the sinkhole, the specialists said that if the hole had been formed from bottom to top, meaning starting in the tunnel and developing upwards, the final form of the hole would be of a regular cone. Instead, the hole was in the shape of an upside-down cone-trunk, what means, according to them, that it was formed from the surface downwards.

5.5 Analysis of water flow at the surface above the sinkhole

The sinkhole, also called as subsidence, was formed in a middle of a valley occupied by impoverished inhabitants, in a region with no drainage or sewage systems. Due to this, rainwater and domestic sewage flows freely in the surface. When these waters reach the point where the hole was formed, they find a collector box. The pipe that flows from this box to the manhole of the main system was blocked, provoking an overflow that infiltrated the soil.

Once the water penetrated the soil, a descending flow finds a layer of micaceous sandy silt. This material is very weak and was carried by the water, along the with rock-soil interface, forming a void at the subsoil, like a cavern. When the roof of this cavern collapsed, the sinkhole appeared on the surface.

5.6 Analysis of water flow inside the tunnel gallery

During excavation works, water flow measurements inside the tunnel were made to check the infiltration of water from the rock mass to the tunnel gallery. The results of these measurements showed an average flow of water of 13,2 liters per minute for each 100 m of tunnel. These measurements were made from December 2015 to March 2016.

Months later, during a study of tunnel drainage, a new measurement was made. This time, in January 2017, the total volume of water accumulated inside the tunnel was calculated and dividing this value by the total time without pumping it was possible to determine the flow. The value obtained was 10,2 liters per second per 100 m.

When the accident occurred, a new measure of water flow was made, using the same method for the studies of January. The results of water flow along time are shown on the diagram of the Figure 10, below.

PERIOD	TIME LINE			SITUAÇÃO	MEDIUM FLOW (l/min/100m)
	2015	2016	2017		
	OND	JFMAMJJASOND	JFMAMJJASOND		
DECEMBER, 2015 to MARCH, 2016	▬▬			DURING CONSTRUCTION	13,2
JUNE, 2016 to JANUARY, 2017		▬▬		MESUREMENT FOR A STUDY IN JANUARY, 2017	10,2
JUNE, 2016 to NOVEMBER, 2017		▬▬▬	▬▬	MEASUREMENT AFTER THE SINKHOLE IN NOVEMBER, 2017	15,2

Figure 10. Water flow inside the tunnel.

As seen in the diagram above, comparing the time of excavation works with the time when the tunnel was flooded, there was no significant change in the water flow inside the tunnel. This indicates that there were no changes in the rock mass, like the initial presumed connection between the tunnel and the sinkhole on the surface. If this connection had occurred, the water flow from the rock mass (and surface) towards the tunnel was supposed to increase, which didn't happen.

5.7 *Analysis of the water flow along the valley*

Two days before the sinkhole appeared, one resident reported that the water in his well became turbid. This fact indicates that there was an underground water flow along the valley, carrying soil particles from the region of the sinkhole to the lower parts of the valley.

In Figure 11 below, the well in which the water became turbid when sinkhole occurred can be seen downstream from the subsidence site. This fact indicates the direction of the soil particles flow along the valley and not vertically to the tunnel.

5.8 *Analysis of the time line*

The excavation works began in November 2014. In January 2015 the excavation works passed directly below the point of the tunnel just under the local where the sinkhole was formed. The excavation works continued for 9 months, until October 2015, when it was interrupted. The pumping was continued, keeping the gallery dry. Only in June 2016 the pumping was interrupted and water began to accumulate in the bottom of excavation and the tunnel began to flood.

In October 2016 an inspection was made in the tunnel and the reports issued the integrity of the tunnel gallery in the point under the sinkhole. The sinkhole occurred in November 2017, 13 months after the inspection.

By calculating the speed of flooding, it was possible determine that, in the site exactly below the sinkhole, the water reached the tunnel crown in May 2017. Meaning that only 6 months passed from this event to the occurrence of the sinkhole.

Figure 11.   Water flow down the valley.

These facts are represented in the timeline of the Figure 12 below.

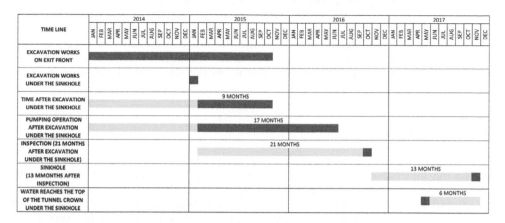

Figure 12. Time line.

The analysis of the time line shows that the rock mass maintained its stability and integrity during 21 months after excavation, even with the continuation of excavation blasting. But, as the subaquatic inspection showed, there was a collapse of the tunnel. So, two questions must be made. First: are the 13 months after inspection (that showed integrity of the tunnel) enough time to degrade the rock? The second question is: in case the collapse was caused by water, are the 6 months after water reaches the roof of the tunnel enough time to degrade the rock? In both questions must be considered that the thickness of the sound rock layer is 45 m.

6 ANALYSIS OF INVESTIGATIONS AND STUDIES

The fact that the tunnel is exactly below the sinkhole and the obstruction inside the gallery strongly suggest that there was a cause and effect relationship between the tunnel and the sinkhole. That is what was believed at first.

However, as the inspections advanced, new elements were revealed, while the studies carried out showing complementary elements to the data obtained by the investigations. The first results that changes the hypothesis that the tunnel caused the sinkhole came from the GPR inspection report, affirming the integrity of the rock mass. This information matches with the result of the rotary borehole surveys, which identified rock layer of about 45 m thickness in every hole, even the one executed vertically from the sinkhole to the tunnel. Besides these strong factors, the analysis of the sinkhole geometry and of the constitution of its walls showed that the hole was not formed in the tunnel and raised until the surface, but it was an independent event caused by sewage water and rain water coming from the neighborhood in the surface.

Complementing, the water flow studies showed that there were no changes in the intake flow to the tunnel, eliminating the possibility to have a connection from the surface to the tunnel, while the presence of residues in the water of a well downstream the valley indicates the real direction of the material that was carried from the sinkhole.

Among many other studies that were made, one specific study must be mentioned.

The rotary borehole surveys showed some fractures in the rock, many of them with presence of smectite and water. Along the years, smectite and water weakened the rock fractures.

When the sinkhole collapsed, an elevation of the water level could be registered by an instrument. The study referred in this chapter considered the possibility that this elevation of water level caused by the collapse of the sinkhole roof could have created by an over pressure in the existing water in the fractures strong enough to turn unstable one big block of rock inside the tunnel, forming the void seen in borehole SR-05, approximately 60 m distant from the obstruction (see Figure 8). To confirm this hypothesis, complementary studies need to be conducted. New rotary borehole surveys were programmed in this area, while the works on the site are concentrated in injections to fix the weathered rock in the tunnel crown. Only when the injections are finished there will be conditions to overpass the obstruction and reach the site in the interior of the tunnel below SR-05 to verify this hypothesis.

7 TUNNEL SERVICES

As the drainage of the tunnel proceeded, the water level was decreasing until it allowed access to the material of the obstruction. Then, 3 sub-horizontal rotary boreholes were made to identify the area were the rock was altered, and they found an approximate 10-meter diameter layer of altered rock. Next, a new borehole was made (SR-CTN-07) from the surface, which identified a layer of sound rock, confirming other surveys, and the height of the weathered rock about 13 meters above the tunnel crown, as shown on Figure 13.

In sequence, 15 more injections are provisioned (primary) to consolidate the material inside the tunnel and the weathered rock above.

Now, while this paper is being written (September 2018), the rotary borehole SR-CTN-07 is concluded and the injection is in course.

When these injections are made (primary, secondary and so on), there will be conditions to excavate the material of the obstruction, now consolidated, and access the part of the tunnel behind the obstruction. Only then it will be possible to proceed and finish the pumping phase, inspect the rest of the gallery and determine the services eventually necessary to complete the recovery of the tunnel.

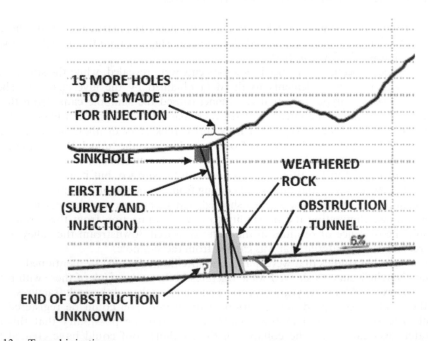

Figure 13.   Tunnel injections.

8 CONCLUSION

The occurrence of a sinkhole exactly above an obstruction inside the tunnel led to the believe that there was a relationship between events. In this case, the tunnel would be responsible for the accident on the surface.

However, studies have shown that the thick sound rock layer between tunnel and surface was intact, indicating that the two events are independent. Some studies say that the subsidence was caused by surface sewage water and rain water flow, while the tunnel roof collapsed probably due to decomposition of the rock, possibly associated with the presence of water and expansive minerals.

Tunnels and Underground Cities: Engineering and Innovation meet Archaeology,
Architecture and Art, Volume 6: Innovation in underground engineering,
materials and equipment - Part 2 – Peila, Viggiani & Celestino (Eds)
© 2020 Taylor & Francis Group, London, ISBN 978-0-367-46871-2

Fully integrated BIM in Norwegian Railway project

S.R. Olsen
Norconsult AS, Norway

ABSTRACT: The development of the technology in the tunnel industry is going faster than ever. Words like 3D, digitalization and BIM is more and more common. There is a difference between 3D and BIM, but these words are often mixed up. A BIM-model and a 3D-model looks the same, but the big difference is the "I". This paper will look at some key benefits regarding the "I" in BIM. This paper will focus on how the project have used the "I" in BIM, and what benefits we have managed to utilize. For instance, the Bill of quantities and progress planning is created from the BIM-model. Other information such as cost, construction, geological interpretation, maintenance is also included in the model. This have become possible because we are using IFC and other open formats that carries information instead of formats like DWG.

1 INTRODUCTION

The Norwegian National Railway Administration is planning and constructing in total 230 kilometres of new railway in Norway. The project is called InterCity and goal is to reduce travel time to Oslo and increasing the capacity. Our project is a part of the InterCity and containing 10 kilometres of double track from Drammen to Kobbervikdalen and between Drammen station and Gulskogen station, as part of the ongoing modernization of the Vestfold line.

Drammen station and Gulskogen station will be upgraded with both new platforms and tracks. The project is complex, with extensive earthworks and a multidisciplinary technical railway. The stations will also be raised by approximately 40 cm without disturbing the railway traffic. This force the project be carried out through a number of demanding construction phases. The project includes 10 kilometres of double track from Drammen to Kobbervikdalen, a 7-kilometre rock tunnel and a 275-metre long mined soil tunnel.

This project is said to be one of the most complex rebuilds in the history of railway works in Norway. Therefore, we decided to try to make it drawing less, so it would be the first railway project without drawings as well. There are still some drawings left, but we have already reduced the amount with approximately 85 %.

2 BIM

The utilization of Building Information Modelling (BIM) has in recent years become more common in connection with infrastructure projects involving large structures (bridges, culverts etc.), roads, earthworks, tunnels and underground construction. Although the client and consultants have utilized BIM extensively in design, the interface towards the contractors have to a large extent been based on publishing 2D drawings.

The development of the technology in the tunnel industry is going faster than ever. Words like 3D, digitalization and BIM has been used for some years, but what do they mean and what are the benefits of this new tools and technology. The first thing we should do is to understand the difference between 3D and BIM. These words are often mixed up, and nowadays a model can be called both BIM-model and 3D-model. A BIM-model and a 3D-model

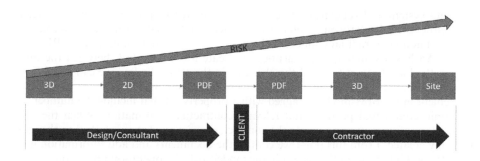

Figure 1.   Risk increasing with number of operations.

looks the same, but the big difference is the "I". This paper will look at some key benefits regarding the "I" in BIM. We have created 3D-models for many years, but we are finally starting to create real BIM-models.

By utilizing the "I" in BIM we can move the model to the center of the projects. We can give each object the necessary information for pricing, manufacturing and construction. Furthermore, we can add information such as cost, time, reports and more. With all this information, the need for drawings is reduced, and everything can be generated from or implemented in the model. The main benefit is that every piece of information is gathered at one place, and everything is connected to each other. If you change the model, you change the project. A good example of this is a change in the 3D-model leads to a change in one or several drawings, that might lead to a change in the contract or a report and so on, but to this point all the changes have been done manually and separately. This is not necessary a problem when isolated, but during a project this generates many opportunities for mistakes.

The digitalization of the industry has given us some benefits, but also some disadvantages. In Norway, most of the consultants are producing 3D-models. Often, the delivery to the client and further on to the contractor is the drawings, that might be supported by a 3D-model. To create these drawings there is usually necessary to create a 2D-model. This means that there are already three stages on the consultant side, with two manual operations involving some risk of error. Then the contractor receives the drawings and creates his own 3D-model, so that he can extract the necessary data. This is another manual operation, increasing the risk. In other words, the start of the process and the result of the process is the same. Though a bit exaggerated, in this way the client is paying more money to increase the risk in the project. The process is illustrated in Figure 1.

We concluded that to take full advantage of the model, it had to be placed in the center of the project. The one question we had to prove was to find out if it was possible to place the model in the center of the project. The quick answer is yes. With some guidelines everyone involved have agreed to follow. For instance, every model that is delivered should be on an open format. It is also necessary to avoid being tied up by these guidelines, because the most important thing is to keep this as simple as possible for the contractor.

3   OBJECT INFORMATION

3.1   *Bill of quantities*

The object information can be added as a parameter or a property to an object. You can basically add anything you want. For instance, metrics like length, area, volume and so on, is a property to an object. Combining the metrics and the contractual information, make a direct link between the contract and the model. In Norway we have a standard for building up bill of quantities (BoQs) to form the contractual basis in unit price contracts. This standard has a coding system and regulates all requirements (material, process, quantity calculation rules etc.) either directly or through reference to other standards (national and international).

To secure a good contractual connection between the bill of quantities and the BIM-model, Norconsult AS has developed an application to create a dynamic link between the different positions in the bill of quantities and the elements in the BIM. Today, we can create most of the contract through the modelling of every element and discipline.

Each element in the BIM-model is tagged with a property set that include the number, name, quantity, unit, contractual position and relevant contractual information. When the files are imported in the software were we create the contract, every part of that is linked to the model is automatically created. From this point we control the documents and add information if necessary. We also add information that is not in the model such rig, operation and so on.

The first contracts with this links, or IFC-contracts as we call them, are already in the market. Until now, we have based our bill of quantities on 3D-models, supported with drawings and manual information take out. The linking between the bill of quantities and the BIM-model give everyone a clear understanding of what and how each element is built, and what the contract element includes. The question is why having a big paper document and a set of drawings and models, when you can have everything at the same place. Figures like amount, volume, material, etc. is equal in the model and the bill of quantities, because they have the same origin.

3.2 Cost estimation

With the same properties as used to create the bill of quantities, we can also create the cost estimate for the project. The IFC-files can be imported into a software with an overview of the historical prices for the element. After importing the files, these prices are automatically linked to the different elements, and we get the total cost for the construction. After the import we can do the control of the prices and do changes if we find it necessary. This gives a better and more accurate estimation because it is based on the exact amount and not an estimated amount and everyone use the same prices.

3.3 Construction schedule

During the different phases, a construction schedule is created. The plan usually contains text elements, a lot of text elements and a timeline. It is not always given what the different elements include. By tagging the objects with a code from the schedule, the schedule and the objects can be linked. This help us create a 4D-simulation which is visual, and the experience is that more conflicts are detected. This way the progression plan gets more accurate. Another learning with the 4D-simulation is a reduce in rework. With the linking, the impact of a change in the schedule is shown in the model when the change is made, and it is easy to test different ways to carry out the project.

3.4 Requirements for execution

One of the benefits by using drawings is that we can show all the necessary information and requirements for execution of the construction. We have made properties for each object, which help us move all the information that earlier was given in the drawing to the model. If the contractor needs the information, it is shown by selecting the object in the model.

3.5 Weakness zones

Now we model the weakness zones and the different layers in the ground in the geological model. Adding properties, we can combine every known data into the model. During the construction phase this model and properties can easily be updated when new information is available, which give the users a good insight of the ground conditions based on available investigations and geological interpretation. Soon there will be released a software for geo-technical site investigation in 3D, combining this with the use of MWD data and surface scanning, we will have a complete history of the tunnel excavation.

3.6 *Revisions*

We have also created a property set that manages the status of the model. Included in this property set is among others the revision of the model, the revision of the object and the LOD. The LOD we are using to tell the viewer of the model if the object is planned, ready to be build or if it is built. This gives everyone a chance to now the status in the project and at the site. Furthermore, the revision of the model gives information of the correct version of the model, but the main benefit is the revision of the object. The revision of the object is telling the viewer of the model when the object was changed. If the model has revision 04 and the object has 03, it tells the viewer that this object was not changed in the last revision. By applying a filter, the viewer can filter out all the object revised in 04, this works as our revision marking in the model (in Norway we used to apply revision clouds on the drawings to mark the revisions).

3.7 *Surveying*

The same file is now also used surveyors. We have taken advantage of a new software, that allows the surveyor to use the IFC-file directly in the field, instead of information takeout like points and lines at the office. The surveyor now creates the points directly in the field and choose how to use the model at site. Before we started using this, the surveyor had to go back to the office if he needed more data.

4 PARAMETIC DESIGN

Parametric design has given us the possibility to streamline the design process of tunnels with Revit and Dynamo. By making advanced scripts, we can place most tunnel installations like concrete elements, signs, cable trays and so on automatically, with the properties mentioned in chapter 3. This has been used on our project and we have applied the same method at the Rogfast project. Rogfast will be, when created, the world longest and deepest subsea tunnel. The big benefit is that if the tunnel line changes, the objects and its information will follow. In the Rogfast project there is almost 600 different constructions and several thousand installations, and it would take a while to place everything. With parametric design we can place every object in 3D, with the correct rotation, distance from the roadway and we can also create cutouts and voids for them.

5 DOCUMENTATION

The last item to be addressed is the documentation of the construction phase. For instance, instead of or as an addition to writing a report of the excavation of a tunnel portal, scanning of the surface can be done. This way the right amount could be found by comparing the original surface and the excavated surface. In addition to this, the concrete structures can be design based on the actual conditions rather than the theoretical situation, giving the exact concrete volume.

6 CONCLUSION

BIM gives us a better understanding of our projects and provides several possibilities. The conflicts are easily spotted and corrected, the data can be used directly in the machines at the site and it also force the people involved to look at the processes they are used to in a new way.

In our experience, the 3D-models alone are creating a better picture of the project and the prizing gets more even. Another big advantage is the follow-up of the contract. Through the linking of contract items and the various objects in the model, the conflict level should be lower in the project. We have by giving the model properties been able to place the model in the center of the project. Now, the model is the base of creating construction schedules, bill of quantities, cost estimation and the key part of carrying out the construction work. It is important to mention that we do not need to adjust the files for the different users and software, it is the same file that is used everywhere. We can say the one model rule them all.

In our opinion the creativity and enthusiasm increase in projects that push the technology forwards. There is still a lot of work to do, but we are continuously developing the path. This paper has shed light on some of the opportunities, but there is still so much in front of us. By willingness to change and a curiosity to find the answers, anything is possible.

Tunnels and Underground Cities: Engineering and Innovation meet Archaeology, Architecture and Art, Volume 6: Innovation in underground engineering, materials and equipment - Part 2 – Peila, Viggiani & Celestino (Eds)
© 2020 Taylor & Francis Group, London, ISBN 978-0-367-46871-2

Real – Time displacement control using expansive geopolymers with Fibre Optic Sensing monitoring: Field trial at Thames Tideway West, London

M. Palaiologou & S. Psomas
Morgan Sindall Engineering Solutions Limited, Rugby, Warwickshire, UK

C. Kechavarzi & X. Xu
Cambridge Centre for Smart Infrastructure and Construction, Cambridge, Cambridgeshire, UK

ABSTRACT: Expansive geopolymers, such as Geobear Geopolymers (GPs), are usually injected post damage in weak surficial layers to re-level shallow formations. In contrast to conventional applications, GPs could also be used in a proactive real time environment to provide asset protection during deep excavation/tunneling. A field trial comprising GP injections at different depths was carried out in one of Thames Tideway West main worksites to assess their actual interaction with competent ground and especially London Clay. Fibre Bragg Grating (FBG) Fibre Optic Sensing (FOS) cables were installed in collaboration with Cambridge Centre for Smart Infrastructure and Construction to monitor the injections. The GPs expanded approximately 2 times their initial volume in London Clay, nevertheless further trials are required for the development of real time deep injections applications. The FOS system captured satisfactorily the induced strains and can therefore be considered a reliable tool for underground real time applications.

1 INTRODUCTION

As underground infrastructure will continue to develop, the need for efficient asset protection during the execution of underground works becomes even more crucial: Conventional asset protection methods comprise extensive protection measures, and can introduce significant economic cost to underground projects.

Conventionally, expansive polyurethane resins; such as Geobear geopolymers (GPs) are injected in weak surficial soils (usually post-damage) from the surface in a liquid state with the use of small diameter pipes (10mm. diameter). The injection pressure required is very small (1–2 bars), as the liquid geopolymer (GP) propagation in the ground occurs thanks to the expansion pressure (>10 MPa) developed during the state change of the material from liquid to solid. The state change is accompanied by a significant volume expansion (up to 30 times in the initial volume). After the end of the reaction, their volume remains steady in the long term. The reaction is immediate and can vary from seconds to minutes to suit different applications and ground conditions; because of the expansive GP properties; the injection process can be controlled with accuracy.

Expansive GP combined with real-time monitoring can enable immediate and efficient action depending on the ground behaviour to ensure the integrity and functionality of assets around the construction area. The properties and the behaviour of the material have been assessed before by various authors, such as Buzzi et al. (2008). The variety of the conventional applications and the control process of the injections is thoroughly illustrated by Dominijanni & Manassero (2015) and dei Svaldi et al. (2005), while the beneficial impact of the use of GPs in the restoration and strengthening of historical buildings especially in Italy has been illustrated by Berengo (2013), Dei Svaldi et al. (2005), Gabassi et al. (2011), Fischanger et al. (2013) and Mansueto et al. (2009). With the exception of Fischanger et al. (2013), and Apuani et al. (2015), very little literature is available in full scale testing and monitoring of the GPs propagation and interaction with the ground, especially at depths greater than 7m. bgl.

For major underground projects, such as Thames Tideway Tunnel, Geobear proposed for the first time to use their GPs in a pro-active real-time environment to compensate for excavation-induced ground displacements. For example, where utility pipes in the vicinity of a shaft under construction are subjected to displacements, a sentinel line of boreholes just inside the site boundary (next to the shaft) can be installed, as shown in Figure 1. These boreholes would be injected during construction of the works; and -accompanied by efficient monitoring- will provide assured protection to the services next to the excavation. Another potential application during tunneling is shown in Figure 2, where injections could be applied to protect operating tunnel against adjacent tunnel construction. This approach - upon finalisation- could lead to substantial material and cost savings thanks to the lighter equipment, space and preparatory work required; works duration; properties and behaviour of the material and the avoidance of construction or asset operation disruption.

As main designers for shafts and tunnels for the west section of the Thames Tideway Tunnel, Morgan Sindall Engineering Solutions (MSES) carried out a field trial in collaboration with Geobear; in order to assess the actual interaction of the geopolymers with competent ground at greater depths and especially London Clay (14m deep), something that has not been attempted before. Furthermore, the trial assessed the feasibility of using smart fibre optics sensing (FOS) systems to provide real time control in order to compensate for excavation/tunneling-induced displacements. This paper presents the main trial results and will provide evidence of the excavation findings.

Figure 1.   Potential real time displacement asset protection in the vicinity of deep excavation (shaft) with Geobear geopolymers.

Figure 2.   Potential real time displacement asset protection in the vicinity of a tunnel with Geobear geopolymers.

2 FIELD TRIAL DESCRIPTION AND RESULTS

The trial took place within the footprint of the Carnwath Road Riverside Shaft on the 12–14 May 2017. 4No. Boreholes were drilled in the middle of the shaft with the use of a 150mm ID cable percussive drilling rig and a temporary steel casing. The boreholes were drilled at 1m distance, one next to the other. Their relative location, cross section and ground stratification can be seen in Figure 3 and Figure 4. The soil stratification comprised 3m of made ground, followed by 1m of alluvia, a 6m thick terrace gravel layer and competent London Clay. The surface ground level at the time of injections was at 104.8m above tunnel datum (ATD).

Two of these boreholes (BH1 and BH2–15.0m deep) were used for injecting geopolymers with fast (named 1735) and slow (named 1945) reacting Geobear geopolymers at different depths (4m, 9m, and 14m below ground level -bgl.). Injection tubes were installed in the boreholes and then the boreholes were filled with bentonite and cement bentonite material. The injections would target all above mentioned soil layers. The reaction time for the 1735 geopolymer is approximately 20 sec, while for the 1945 one the reaction time can be delayed to 1–2 min, thus enabling injections at greater depths. Each injection round was carried out at only one level. The geopolymer injection sequence, type and relevant quantities are shown in Table 1. During the injections, the FOS which was installed in the intermediate boreholes FBGM1 and FBGM2 was capturing the geopolymer-soil interaction in real time.

The Fibre Bragg Grating (FBG) FOS system was developed in collaboration with Cambridge Centre for Smart Infrastructure and Construction (CSIC). The FOS consisted of a single 2mm perpendicular fibre optic cable with 16 sensors at 1m intervals, which was threaded and anchored down the FBGM1 and FBGM2 boreholes, as shown in Figure 5 and Figure 6. The sensing cable was connected to an FBG interrogator and a laptop (Figure 7) and was capturing in real time the strain developed during injections. A total station was used to capture surface displacements.

2.1 Injections in surficial layers (Made Ground-Alluvia)

FOS readings for the last injection BH2 4A are presented in Figure 8 and Figure 9. These readings are indicative of soil-geopolymer interaction in surficial layers. The fast reacting 1735 geopolymer was used for these injections. The tensile strain distribution of the cable developed around 101.0 m ATD (app. 4m bgl.) and suggested the formation of a bulb influencing the upper soil layers. The strains dissipate, as the distance from the injection point increases, enabling the estimation of an influence zone of the geopolymer equal to approximately 2m. Regarding BH2A, the total station readings provided for a 2.8mm vertical displacement at FBGM2 location (1m from the injection point) and a 1.4mm at FBGM1 location (2m away

Figure 3. Injection boreholes and monitoring installa-tion layout. From left to right: BH1, FBGM1, FBGM2, BH2.

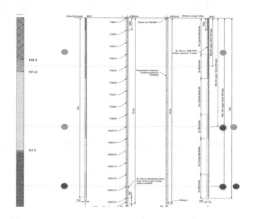

Figure 4. Boreholes cross-section, soil stratification and injection depths.

Table 1. Injection location; sequence, soil properties, injection material and relevant volumes per injection round.

Borehole location	Soil type	Modulus of Elasicity*	Injection depth	Sequece	Geopolymer type	Liquid volume
		MPa				litres
BH1	Alluvia	10	4A	1	1735	46.30
BH1	River Terrace Deposits	35	8A	2	1735	49.60
BH1	London Clay	45	14A	3	1945	92.50
BH2	London Clay	45	14A	4	1945	92.8
BH2	River Terrace Deposits	35	9A	5	1945	93.00
BH2	London Clay	45	14B	6	1945	87.90
BH2	River Terrace Deposits	35	9B	7	1735	93.00
BH2	Made Ground	15	4B	8	1735	171.10

* The given modulus of elasticity refers to drained conditions for all soil layers, apart from London Clay

Figure 5. FOS cables threaded down the borehole with the use of a reel holder and a weight tied at the free edge of the cable.

Figure 6. 2mm thick fibre optic cable threaded into the borehole (white cable coiled around the reel). Extension cable 4mm thick (Red). Temperature sensor cable next to the other two ca-bles (black.)

Figure 7. Interrogator and laptop in the cabin next to the trial lo-cation. The cable is con-nected to the interrogator via the ports.

from injection) (Figure 10). At the same locations the fibre optics gave a 2.6 mm and a 1.3mm total displacement value, which is a very satisfactory result: Another interesting finding is that after a point (14.54 line in Figure 8, corresponding to 86 litres injected) the strains captured by the cable do not increase proportionally to the injected quantity. This suggests that injection quantities might have to be limited at each round to enhance achieved results.

On the 22.11.2017 a trial pit, 4m x 6m wide and 5m deep, was excavated in order to retrieve the geopolymer material previously injected in the upper layers. Geopolymer material was recovered in the vicinity of both injection locations (BH1 4A and BH2 4A). Around BH1 position, at 100.0 m ATD (app. 5m bgl.) a horizontal geopolymer flat 'slab' was recovered from alluvia soil This level coincided with the level identified by the FOS as the geopolymer initial propagation level. The recovered sample shown in Figure 10 had dimensions 1.5m x 0.9m x

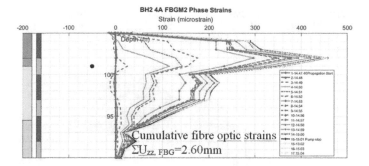

Figure 8. BH2 4A FBGM2 Phase strains development along injection at 4m bgl. (101m ATD), as captured by FBGM2 cable (which is 1m away from the injection).

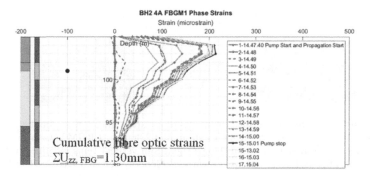

Figure 9. BH2A FBGM1 Phase strains development along injection at 4m bgl (101m ATD), as captured by FBGM1cable (which is 2m away from the injection). The impact of the geopolymer at this location is smoother. Strains are uniformly distributed around the injection point.

0.07m (thickness) and was lightweight. A smaller geopolymer fragment with a diameter less than 0.5m was recovered between BH1 and BH2. Based on the recovered samples, a rough estimated solid geopolymer volume was computed equal to 80 litres. The initial liquid volume injected was 46 litres. Defining an expansion ratio index ERI=Solid Volume Injected Liquid Volume, the ERI in alluvial soil can be estimated to 175%. Around BH2, the geopolymer interacted with made ground and created multiple sub vertical fractures propagating to the surface along with a bulb formation at approximately 101.0 m ATD (4m bgl.); validating the FOS data- at the point of injection. The sub vertical veins were lightweight and had a thickness of approximately 3 cm (Figure 10). The geopolymer veins were located exactly where the total station captured the greatest displacements (green area in Figure 10). The bulb formation was heavy and recovered in fragments. An indicative fragment, with dimensions 85cm x 45 cm x 15 cm is shown in Figure 3. Based on the recovered materials, the total estimated volume (both lightweight geopolymer veins and heavy bulbs could reach 533 litres. Taking into account that the initially injected liquid volume was 171 litres, the estimated ERI in made ground is ERI=Solid Volume Injected Liquid Volume=533/171=312%. As the upper soil layers were characterised as contaminated, no further laboratory testing could be carried out at the recovered material. Finally, it should be mentioned that the trial pit dimensions enabled the teal to verify that -at least macroscopically- no geopolymer material was found away from the injection point; allowing to draw a conclusion that there is no material loss in the upper layers.

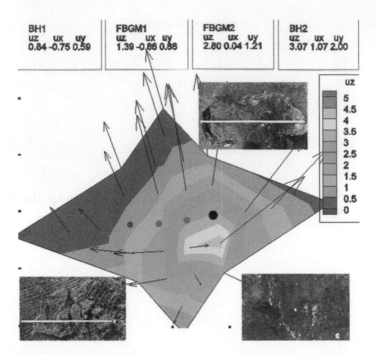

Figure 10. Total Station Surface displacement contours after injection at 4m bgl in BH2. 1.4mm recorded at FBGM1 location (2m away from injection) and 2.8mm at FBGM2 (1m away from injection). Top right: Bulb recovered at BH2 location. Down right: Sub vertical veins identified in the yellow area shown in the graph after BH2A injections. Down left: Geopolymer slab recovered in BH1; related to BH1A injections.

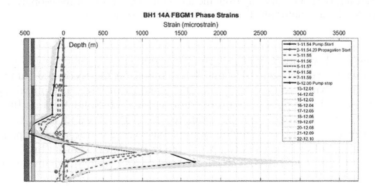

Figure 11. BH14A FBGM1 readings during injections in London Clay.

2.2 Injections in Gravel

In gravel; both the fast and slow reacting geopolymer were injected in different injection rounds. The interaction of the slow reacting geopolymer with the gravel did not cause any significant strains. The material was unfortunately not recovered during shaft excavation. Hence, further testing is required to verify the soil-geopolymer interaction.

2.3 Injections in London Clay

FOS results during injections with the slow 1945 geopolymer at 14m bgl. Indicate that the material impressively fractured and propagate in deep, competent London Clay. The

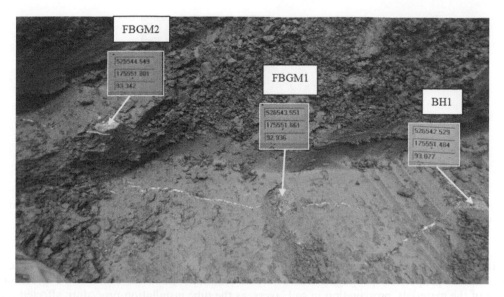

Figure 12. London Clay fractured 12m. bgl. The geopolymer propagated for more than 2m away from the injection point. The thickness of the horizontal vein varies from app.15mm close to the injection point (0.5m) to 5mm away from the injection point (2–3m away).

Figure 13. Geopolymer propagation in London Clay. Geopolymer vein thickness equal to approximately 10mm – 15mm close to the injection point.

geopolymer propagation, as shown in Figure starts at 93.0m ATD (12m bgl.), 2m higher than the targeted injection level. This observation could be related to inefficiencies during the tube installation and will be assessed through further field testing. Substantial tensile strain values were recorded, implying a localised fracture and propagation along a plane, rather than bulb formation (Fig. 11). The excavation findings were compliant with the FOS readings. confirms that the geopolymer fractured London clay and propagated through the borehole along a relatively thin plane. The shift manager, supervising the excavation and the recovery of the geopolymer described that the material created a vertical plane more than 0.5m deep and this vertical plane propagated towards FBGM1 and FBGM2 boreholes and continued its propagation for further than 2m away from the injection location. The thickness of the geopolymer plane varied from 0.010–0.015m near the injection location and decreased to 0.005mm 2m away from the injection point. A recovered fragment is shown in Figure 14. It should also be noted that the geopolymer did not propagate radially but there was a preferred propagation path, parallel to the existing river walls at the site and towards the drilled boreholes.

Figure 14.   Geopolymer fragment recovered from app. 93.0m ATD. Dimensions: 150mm (length) and 15 mm (thickness).

Subsequently, it could be suggested that the FO cables were measuring lateral and not vertical strain. The geopolymer propagation at a vertical plane should not be considered representative of the materials' propagation in soil layers, as the tube installation procedure affected the injection process. The use of wide boreholes enabled the formation of a weak path along the borehole core. Density testing of the recovered material carried out at University of Warwick and estimated an expansion of the geopolymer 2–4 times its initial volume.

3   CONCLUSIONS

The successful execution and completion of the trial demonstrated that:

- The simple FOS configuration can describe satisfactorily the subsurface deformation, capture actual injection level, geopolymer propagation and therefore enable real time control and monitoring. As the trial configuration was very simple, bespoke systems can be developed for future real time applications.
- As to be expected, the nature of the propagation pattern will depend upon several factors including soil density and grading, insitu strain and type of geopolymer applied. There is indication that in cohesive surficial alluvia layers, a horizontal geopolymer propagation is likely to occur, in contrast to non-cohesive made ground and gravel, where the bulb formation is more probable.
- Repeated injections can be carried out at the same point. Consecutive or adjacent injections interact with one another, for the actual application a conditioning round of injection is useful. Furthermore, injection quantities per round might have to be limited to enhance achieved results.
- London Clay can be fractured at a depth of 12.0–14.0m
- The geopolymer material expands 2–4 times its initial volume at depths 4m bgl. to 12m bgl. It stays in the vicinity of the injection; is not lost and can result in controllable displacements. The influence zone (in width) is approximately 1.5–2.0m.
- Further field testing is required to confirm the actual interaction with the deep gravel and London Clay layers.

The above findings are encouraging for the control of the material in deep, competent ground and the predictability of its behaviour. The presented application - upon finalisation- has the potential to improve real time displacement control. Geobear predicts substantial cost reduction thanks to reduction in material use; lighter equipment; less complicated installation processes, short duration of activities and minimisation of construction or operation disruptions. The combination of these smart materials with efficient monitoring equipment; such as fibre optics, offers assurance to the stakeholders that their assets can be well protected and remain operation at all times.

ACKNOWLEDGEMENTS

The authors would like to thank Morgan Sindall Engineering Solutions and Cambridge Centre for Smart Infrastructure and Construction for the funding of M.Palaiologou's secondment and research support, GEOBEAR and especially Mr.Roland Caldbeck for all the information provided on their methodology, Tideway and the Joint Venture Partners BAM Nuttal, Morgan Sindall and Balfour Beatty and Mr. Peter Knott for his contribution and assistance,

Special acknowledgements have to be made for Prof. Colin Eddie, who had a catalytic role in this effort.

REFERENCES

Apuani, T., Giani, G.P., d'Attoli, M., Fischanger, F., Morelli, G., Ranieri, G. and Santarato, G., 2015. Assessment of the efficiency of consolidation treatment through injections of expanding resins by geotechnical tests and 3D electrical resistivity tomography. *The Scientific World Journal, 2015.*

Berengo, V., 2013. Ground improving injections underneath historical buildings: Five case histories as an overview of a technique. In Emilio Bilotta (ed.), *Geotechnical Engineering for the Preservation of Monuments and Historic Sites, Proc. intern. symp., Napoli, 30–31 May 2013.* London: CRC Press: 165–174.

Buzzi, O., Fityus, S. and Sloan, S.W., 2010. Use of expanding polyurethane resin to remediate expansive soil foundations. *Canadian Geotechnical Journal, 47*(6):623–634.

Dei Svaldi, A., Favaretti, M., Pasquetto, A. and Vinco, G., 2005. Analytical modelling of the soil improvement by injections of high expansion pressure resin. *Bulletin für Angewandte Geologie, 10*(2):71–81.

Dominijanni, A. and Manassero, M., 2015. *Amélioration des sols par injections de résine expansive: Guide de conception.* Paris: Editions Eyrolles.

Fischanger, F., Morelli, G., Ranieri, G., Santarato, G. and Occhi, M., 2013. 4D cross-borehole electrical resistivity tomography to control resin injection for ground stabilization: a case history in Venice (Italy). *Near Surface Geophysics, 11*(1):41–50.

Gabassi, M., Pasquetto, A., Vinco, G. and Masella, A., 2011, June. Consolidamento del terreno di fondazione del palazzo di Punta della Dogana in Venezia realizzato con iniezioni di resina poliuretanica. In *XXIV Convegno Nazionale di Geotecnica; Congress proceedings, Naples* (Vol. 22424).

Mansueto, F., Gabassi, M., Pasquetto, A. and Vinco, G., 2009. Modellazione numerica di un intervento di consolidamento del terreno di fondazione di un palazzo storico sito in rue Joseph de Maistre sulla collina di Montmartre in Parigi realizzato con iniezioni di resina poliuretanica ad alta pressione di espansione. *Comunicazione personale.*

Tunnels and Underground Cities: Engineering and Innovation meet Archaeology,
Architecture and Art, Volume 6: Innovation in underground engineering,
materials and equipment - Part 2 – Peila, Viggiani & Celestino (Eds)
© 2020 Taylor & Francis Group, London, ISBN 978-0-367-46871-2

New methodology to safely install steel ribs in conventional tunnel – A14 Motorway

F. Palchetti, P. Mazzocchi, G. Giacomin, M. Maffucci & P. Bernardini
Ghella S.p.A., Rome, Italy

ABSTRACT: In recent years, conventional excavation, although making way for techniques that guarantee higher levels of safety, has been proved irreplaceable in all cases where logistics or high investment costs do not permit the use of alternative solutions. Therefore, studies and research have been promoted for solutions capable of reducing the risks associated with the activities closest to the excavation face, thus bringing about an evolution in conventional excavation, especially focussing on pre-lining with steel ribs and shotcrete. In this paper, a new methodology will be described. One which will install the rib, with different radius and types using a single operator and without any personnel below the excavation face. The new methodology allows to reduce the time for the rib's installation and the excavation's cycle. The initial testing of the "Automatic Rib" system was performed at the impetus of the client Autostrade per l'Italia S.p.A. at the Sappanico tunnel, for the works to widen the A14 motorway, between "Ancona nord and Ancona sud". Works were directed by Spea Ingegneria Europea and carried out by Ghella S.p.A.

1 INTRODUCTION

In recent years, the Technical Department of Ghella S.p.A. has tried to develop new solutions to reduce the risk and improve the safety of conventional excavation in a tunnel; therefore, studies and research have been promoted for solutions capable of reducing the risks associated with the activities closest to the excavation face. Bringing an evolution in conventional excavation .Some risks associated to the conventional excavation method are mainly due to:

- personnel close to the tunnel face, potentially unstable;
- personnel close to the tunnel face in high activity;
- potential rock spalling, also of considerable size;
- imperfect tightening of bolts due to carelessness or haste;
- incorrect plate welded of the rib;
- incorrect placed of the shotcrete;
- human error.

   In recent years, these risks have been mitigated using special equipment and an increase in the use of specific materials which prove to be safest and allow for "best practice". The improvement of safety in the workplace also brings improvement to production efficiency. Following the concept recently mentioned, the new methodology also allows to reduce the time for the installation of the rib and therefore the cycle of production.

2 INNOVATIVE TECHNOLOGIES DEVELOPED

The first target of the studies was the realization of the rapid joints between the crown and the axials of the ribs, in lieu of traditional coupling with plates and bolts. The rapid joint is ensured by a piston-cylinder system: the "cylinders" are placed on the vertical axials while the "pistons" are placed on the crown element of the rib. The release of the axials automatically activates the piston movement and consequently couple the axials to the crown, avoiding the use of plates and bolts.

Over to guarantee a safe and correct junction, this device allows the rib to assume the final configuration in a few seconds and in absence of personnel close to the tunnel face.

The structural continuity of the elements of the rib has been verified through laboratory load tests and numerical simulations (Figure 2 and Figure 3).

To avoid the risks associated with the positioning of the rib, the application of the chains and the wire mesh, it has been developed a particular system that involves the installation of two profiles with pre-assembled wire mesh (Figure 4).

The rib is assembled outside the tunnel, out of the excavation cycle's critical path, and transported close to the tunnel face, once completed the excavation phase, using a forklift designed "ad hoc" to orient the rib in each direction and lift it up (Figure 5).

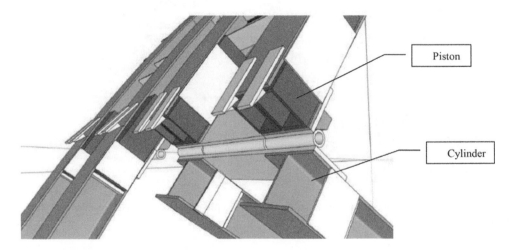

Figure 1.   Rapid joints of the automatic rib.

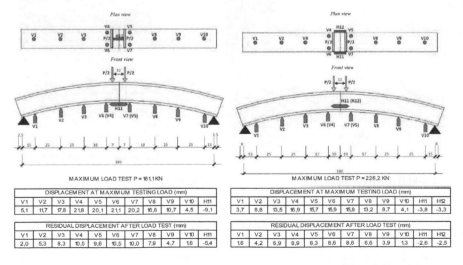

Figure 2.   Comparison between two different type of joint. Left side traditional rib. Right side automatic rib.

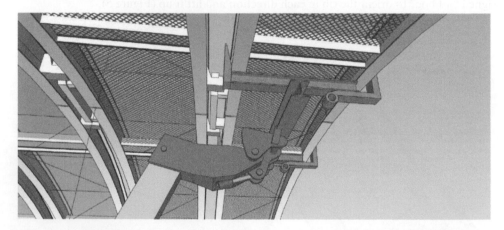

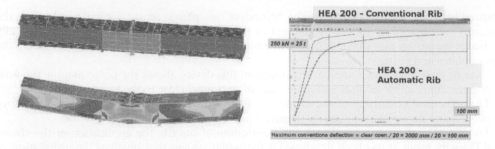

Figure 3.   Structural continuity of the elements of the rib. Modelling of automatic rib deformed (on the top-left) and non-deformed (bottom-left). Chart load and deformation (on the right).

Figure 4.   System that involves the installation of two ribs pre-coupled with HEB profiles on which the wire mesh is assembled.

Figure 5.   Forklift designed for the rib.   Figure 6.   Release claps.

Together with Professor Berry, professor of the School of Engineering and Architecture of the University of Bologna, and of Eng. Calzolari of the Safety Company of Bologna, the system of installation of the rib was sensitively improved. The improvements were made by adding headlights, video cameras with on-screen display in the cabin and an automatic release

Figure 7.   headlight and camera in the forklift.

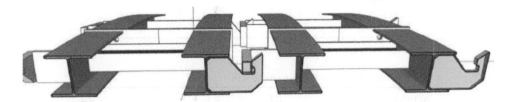

Figure 8.   Joint system of the installed ribs.

command of the claps that support the axials during transport which is operable by the operator in forklift (Figure 6 and Figure 7).

Once at the front, the rib is raised up to the lock in key (Figure 4) allowing, during the ascent, the opening of the two axials which by actuating the release tools of the element junction, allow coupling in the crown avoiding the use of plates and bolts (Figure 1).

The connection to previous ribs using fixed hooks, male-female junctions, and elimination of chains. The hook-up to the final installed rib takes place on six different points: two with saddle configuration, at the crown; two quick-coupling ones at the axials; and two directional ones at the base with appropriately sized male/female supports (Figure 8).

Finally, in the absence of workers close to the tunnel face, it is possible to follow with the shotcrete lining, starting by filling the empty space between the rib's base and the excavated ground, and then continuing from bottom to top.

3   TRIAL TEST ON SITE

To verify the effectiveness of the system, numerous test were performed on different types of joints between the ribs and looking for the most suitable for handling. The test field for the "automatic rib" was executed in the southern bound of the Sappanico tunnel, complying with the Italian standard and according to the instructions of the Engineer of SPEA – Ingegneria Europea.

The following data was collected and processed for investigations:

- investigations on shotcrete;
- core drilling on shotcrete;
- survey with georadar;
- methods and times for setting up the ribs;
- personnel and equipment used (Table 1);
- time and durations (Table 2);

- investigations on the rib and shotcrete's static behaviour (Figure 9);
- monitoring performed continuously on two instrumented ribs with strain gages a vibrating rope and load cells;
- deformation measurements performed continuously on two sections equipped with topographic sensors;
- full-scale laboratory tests on behaviour of the "guillotine" joint;

Table 1. Personnel and Equipment.

Tunnel Crew	Traditional Rib	Automatic Rib
Shift boss	1	1
Operator	2	1
Mining Equipment	2	0
Forklift	1	1
Crane	1	0

Table 2. Cycle Duration Automatic Rib.

Activity	R1	R2	R3	R4
Excavation	230	240	240	235
Handling face	16	13	18	15
Rapid Attachment	4	4	4	5
Change tools	4	5	4	4
Rapid Joint axil left	3	4	6	4
Rapid Joint feet left	5	4	5	4
Rapid Joint axil right	5	5	4	4
Rapid Joint feet right	4	3	5	4
Waiting time	9	11	10	10
Shotcrete	65	60	85	70

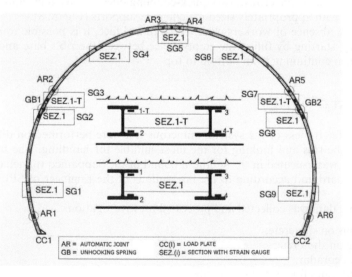

Figure 9. Investigations on the static behaviour of the rib and of the shotcrete.

4 AUTOMATIC INVERT STRUT

Along with the automatic rib, an automatic invert strut has been introduced and tested by Ghella Spa. In accordance with the same principle of the automatic rib, no workers are needed in front of the tunnel face except for the one driving the forklift. It is well known that placing the invert struts is one of the most critical operations in tunnel excavation in terms of safety and time: these problems can be avoided by using an automatic invert strut. By comparison the operations are safer and faster than the conventional method.

The automatic invert strut is brought to the front of the tunnel face with the same forklift used for the automatic rib and is actually made by a couple of struts pre-coupled with steel profiles, on which a wire mesh is assembled.

Figure 10. Automatic invert strut carried by a forklift.

At each end of the strut, the connection between the rib and the automatic invert strut is done essentially by a couple of metallic boxes which work as a piston-cylinder system: the "cylinders" are place on the strut while the "pistons" are placed on the rib.

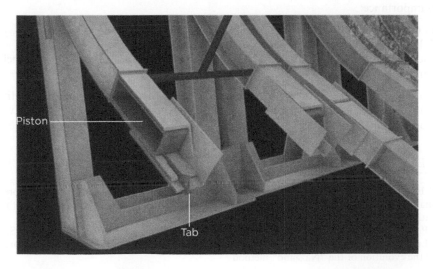

Figure 11. Connection system between rib and automatic invert strut in open configuration.

The piston is held in position by a metallic tab, designed and shaped in order to release the piston while the strut is laid on the support, automatically allowing the piston to slide into the cylinder. After that, the empty space inside the cylinder is filled with cement mortar by using an injection tube previously placed on the top of the cylinder. This operation makes the connection stable and able to adsorb compression stresses, ensuring the strut an effective structural capability.

Figure 12. Connection system between rib and automatic invert strut in closed position and with cement mortar.

5 CONCLUSION

With its long experience, despite some prejudices Ghella S.p.A's Technical Department has developed many technical solutions to improve work safety. Those measures automatically entail consistent benefits to the production. The system described in the previous chapters represents the best example of the positive correlation between safety and organizational and economic benefits.

The technological solutions developed and implemented allow to meet different aspects of primary importance:

- safety (absence of personnel at the front);
- flexibility (adaptable to all types of sections, also with variable radius);
- simplicity of application (use of standard equipment operated by qualified personnel);
- easy and quick application between 10 and 20 minutes.

The automatic rib and invert strut represent an innovative solution and an effective technique that cancels the risks connected to one of the most critical activities in the conventional excavation of tunnels.

REFERENCES

Palchetti F., (Ghella S.p.A.), Nuovo Sistema per il montaggio di centine in galleria senza la presenza di personale al fronte – NIR 2013 Notes interregional of Engineering of the safety in the tunnel excavation.
Lunardi P. (Studio Lunardi), Tozzi G. (Autostrade per l'Italia S.p.A.), Selleri A., Belfiore A. (Spea Ingeneria Europea), Ghella G. (Ghella S.p.A.), Widening the "Montedomini" tunnel in the presence of traffic: the evolution of the "Nazzano" method.

Tunnels and Underground Cities: Engineering and Innovation meet Archaeology,
Architecture and Art, Volume 6: Innovation in underground engineering,
materials and equipment - Part 2 – Peila, Viggiani & Celestino (Eds)
© 2020 Taylor & Francis Group, London, ISBN 978-0-367-46871-2

Recent developments towards autonomous tunneling and mining machinery

T. Peinsitt, H. Haubmann & H. Kargl
Sandvik Mining and Construction G.m.b.H., Zeltweg, Austria

C. Kary
Geodata ZiviltechnikgesellschaftsmbH, Leoben, Austria

ABSTRACT: Today's trend in tunneling business is clearly moving towards automation and autonomous machineries. This trend entered into mechanical excavation machineries as well and Sandvik developed a semi-autonomous excavation mode for their Roadheaders. Together with a geodetic guidance system it is possible to tele-operate the Roadheader from surface with a maximum on safety by removing operators from potentially hazardous environments. This semi-autonomous operation of the equipment requests certain requirements like a higher level of safety integration, stable tunnel network connections, reliable technologies and clear interfaces to other tunneling processes. Overcoming these challenges, a machine equipped with the automated excavation mode will provide tremendous benefits like increased excavation productivity, safety, serviceability, reduces human error, decrease costs and overcomes the lack of skilled machine operators. Further developments in Roadheader automation will continue to expand within tunneling industry to increase the productivity of tunneling sites– particularly as the infrastructure development moves underground.

1 INTRODUCTION

Automation in mining and civil tunneling have been increasingly utilized in recent decades to enhance health and safety, to improve economic profitability and to ensure sustainable surface and underground operations (Torlach, 1998). These technologies can provide more energy-efficient methods, which obviously can facilitate successful mining and tunneling projects. Moreover, mining and tunneling operations require less number of higher-skilled labor forces in tough and dangerous conditions. Therefore, risks of accidents can be reduced significantly by reduced exposure of miners to potential hazards. Although automation could be a threat to the continued employment, considering the aging and retirement of experienced mining practitioners and lack of mentorship for younger miners, automation and potentially robotics will be a unique opportunity to get training and experience.

Regardless of their sizes and types, due to technical and economic reasons, mines should be equipped with smart mining systems such as automated machines, satellite communications, and smart sensors for accomplishing sustainable mining objectives. The use of advanced technologies provides more competitive working environments in a safe and environmentally sound manner through efficient and reliable operations (Arvind, Bandopadhyay, & Kumar, 2002).

As technology advances in many industries around the world, autonomous machinery has started to replace human labor force. For the last two decades, automation and robotics have been on the agenda of the mining and tunneling industry as well. Although automation has entirely and successfully been implemented in other industries, the mining and tunneling industry does not fully benefit from a complete automation solution so far. Therefore, advancements of mining and tunneling automation are emerging, especially for underground operations where increased efficiency and productivity and reduced risk of accidents are of an utmost concern. (Entezari, 2014)

The more widespread use of the mechanical excavation systems is a trend set by increasing pressure on the mining and civil construction industries to move away from the conventional drill and blast methods to improve productivity and reduce costs. The additional benefits of mechanical cutting include significantly improved safety, reduced ground support requirements and fewer personnel. These advantages linked with recent enhancements in machine performance and reliability have resulted in mechanical cutting Roadheaders taking a larger share of the rock excavation market.

2 ROADHEADERS

Roadheaders are the most widely used underground partial-face excavation machines for soft to medium hard rocks. They are used for both development and production in soft rock mining industry (i.e. main haulage drifts, roadways, crosscuts, etc.) particularly in coal, industrial minerals and in various other rock types. In civil construction, they find extensive use for excavation of tunnels (railway, roadway, sewer, diversion tunnels, etc.) in a wide range of different ground conditions, as well as for enlargement and rehabilitation of various underground structures. Their ability to excavate almost any profile and cross-section makes them very attractive to those mining and civil construction projects where various opening sizes and profiles need to be constructed.

In addition to their high mobility and versatility, Roadheaders are generally low capital cost systems compared to the most other mechanical excavators. Because of higher cutting power density due to a smaller cutting unit, they offer the capability to excavate rocks harder and more abrasive than their counterparts, such as the continuous miners and the borers. (Copur, Ozdemir, & Rostami, 2000). Roadheaders excavate the rock by means of a cutter head mounted on a cutter boom. The cutter boom can be independently moved in horizontal and vertical direction. In addition, many Roadheaders have the ability to extend the cutter boom by a telescope unit.

A Roadheader in standard design covers the following functions:

- Cutting the rock face
- Loading the cut material on the loading table with loading arms or spinners
- Muck transfer onto consecutive haulage systems (vehicles or belt conveyor), this is done by a chain conveyor located in the center of the machine.

3 SCOPE FOR AUTOMATION

3.1 *Motivation for automation*

Development and utilization of automation in industries is motivated due to following potential improvements (Nof, 2009):

- Feasibility
- Implementation of some micro-scale operation requires high speed and accuracy, which humans cannot handle.
- Productivity
- Automatic devices can continuously operate with high speed and large capacity, rising overall efficiency and productivity.
- Safety
- Automated machines can work in environments that are not safe for humans,
- Quality and Economy

Automation allows the organized operations to be handled with high quality and it reduces economic losses due to labor salaries and insurance, safety and maintenance expenses. Besides the advantages, automation holds some limitations such as, high initial investment, labor resistance, and requirement of skilled labor. (Rajput, 2008) Suitability of automated devices for operations and payback period of the expenses for automation should be analyzed carefully. Technology without suitable integration with production can result in loss of money and time. In addition, although utilization of skilled machines raises productivity, presence of these machines in a work environment can cause jobs that are more competitive and leads to decrease in labor intensity. It can induce unemployment inside the sector. Therefore, all factors with negative and positive sides should be analyzed both socially and economically before adapting automation to production cycle.

Productivity is one of the most motivating items for automation in a cyclic production system. Companies in every sector are under pressure to achieve high quantity of production at low-level costs to raise profitability (Humbert, 2007). In general, this pressure eventuates in increasing numbers of workplaces with new improved technologies. Profitability is critical since economical implementation of sectors depends on production that can be traded above limit profits where total cash outflow for capital, operation expenses as well as environmental, and social cost is lower than cash inflow. Therefore, companies must monitor their expenses to ensure their long-term profitability. In this sense, personnel expenses such as, salary, insurance, transportation, and food, one of the largest variable cost items, can be reduced due to the implementation of automation. Automation allows the control and operation of more machines by less number of operators. In addition, unmanned machines controlled at safe distances from the face area allow operation with high accuracy in risky and tough working environments. Furthermore, manually controlled machines work with lower availability compared with automated ones since responsible personnel for manual systems cause system downtimes due to the elimination of human needs. Automated systems can operate continuously without any break except for compulsory and planned maintenance, breakdowns or other machine failures.

Another motivation for automation is clean working environment with less health and safety problems. The introduction of remotely managed systems provides direct benefits concerning occupational health and safety since automation allows the same task management of processes physically away from potentially dangerous working places. In addition to health and safety benefits, compensations paid to labors due to accidents and workforce loss are eliminated with minimizing human factor in excavation areas with unmanned autonomous systems.

Reduction in human errors and growth in production quality is another item in automation motivation. Task complexity and stress on personnel for increased performance may cause human errors in production. Sensors and diagnostic tools with programmable monitoring services mounted on automated systems support production with high accuracy as well as minimized maintenance cost and downtime periods. (Entezari, 2014)

3.2 Areas for automation

As Roadheader provide many different functions, it is necessary to automate any single functionality. The challenges lies in the fact that the mining system has to fulfil a lot of different features, like cutting, loading, roof supporting, following a defined alignment or profile and keeping communication to many other underground components. A lot of information and many degrees of freedom have to be managed at the same time. The aim for the performance of the new automated system is to reach the operational cutting performance of an average machine operator, but to deliver stable performance without significant overloading of the system. This results in smoother machine loading and consequently in higher system reliability, less downtime and reduced operational costs as well as more comfort for the underground team.

Therefore, the technical improvement and automation of such machines is mainly focusing on the integration of features, which support the operator to make the mining and excavation process safer, more reliable and easier to be handled in order to ensure good and constant operating performance.

Some examples to improve the excavation on Roadheaders are:

Positioning support

Positioning support to keep the desired orientation of the advance and to follow the predefined tunnel alignment or to keep the angle of inclination or declination for opening up the access tunnel. All these positioning activities are currently done by manual measurements of the mining and surveyor team. The machine operator currently defines the size, position and individual shape of the profile to be cut by means of manual methods. An integrated guidance system that takes care about tunnel direction and shape of profile to be cut is beneficial.

Automation of the cutting sequence

Currently, when the machine operator is cutting the face, he is controlling all the individual movements of the boom and the complete machine inclusive tracks, stabilizing cylinder, conveying system etc. This gives a lot of flexibility but it requests well skilled and experienced operator. The automation at least of some parts of the complete cutting cycle would provide a more reliable and more user-friendly system.

Maintenance planner and diagnosis system

Automated and integrated planning and controlling of maintenance activities as well as an automatic check of the conditions of wearing components at such machines would result in an increase of the reliability and reduction of the downtime of the mining system. A diagnosis system would combine existing fault messages in order to find their original roots and causes and would give the machine operator some useful hints and information how to manage and prevent faults. (Kargl, Gimpel, & Preimesberger, 2010)

Most of the automation trends are coming out of the mining industry. However, some of the new technologies could be transferred to tunneling applications as well.

3.3 State-of-the-art automation

3.3.1 Machine positioning and (coal) seam recognition

Before the machine operator is going to cut a face, he needs to define the position and shape of the area to be excavated in relation to the machine's position in order to perform suitable navigation. Therefore, one needs to know the expected layout of the tunnel and the position and orientation of the mining machine inside the tunnel. The tunnel layout is usually defined by design of the underground construction and, especially in mining, influenced by the level of the mineral seam. To evaluate the exact position and orientation of a mining machine inside the tunnel separate equipment is needed, for instance theodolite based navigation systems, which are quite commonly used at construction sites. In typical coal mining, such systems are not used for continuous position measurement and so there is much less online information available: typically, a single laser beam, which is positioned by the underground surveyor along the desired tunnel direction, provides information on heading orientation only. Normally there is no online information about the machine position along the tunnel chainage available. Vertical navigation is usually defined by the actual level of the mineral seam in front of the miner rather than predetermined in detail by a planning process. (Kargl, Gimpel, & Preimesberger, 2010)

3.3.2 Roadheader guidance system

Today's standard technology on Roadheaders include guidance systems, data logging and remote monitoring capabilities for networked machines.

The system is designed to periodically track the actual 3D position and orientation of a road header's cutter head during operation and to visualize this data together with the actual centerline and profile geometry on the on-board monitor of the operator cabin. The core component is a robotic total station serviced by the on-site surveying personnel communicating with the machine's control system.

The main features of the system are:

- Continuous determination of the absolute 3D machine position and orientation in the project coordinate system by automatic geodetic observations from the robotic total station
- Continuous determination of the absolute 3D cutter head position and orientation by means of above data and data of additional on-board sensors (2 inclinometers, 2 angular sensors, boom sensor for boom telescope position)
- 3D visualization of the actual cutter head position and orientation within the designed profile geometry and numerical display of all relevant guidance parameters (station of tunnel face, horizontal and vertical distance from cutter head to profile line, etc.)
- Due to the integration of machine data with geodetic machine position information georeferenced machine data reporting becomes available. This in turn enables the following advanced features:
  - Accurate cut volume reporting
  - Calculation of specific energy requirements
  - Specific pick consumption monitoring (manual input of pick changed necessary)

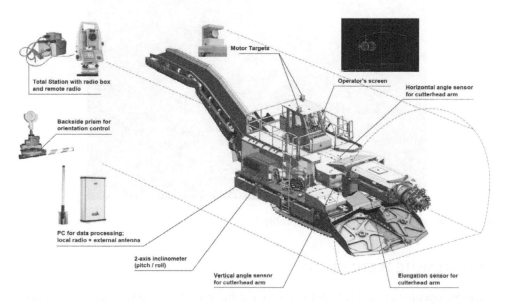

Figure 1.   Layout of Sandvik Geodata Roadheader guidance system.

3.3.2.1 AUTOMATIC CUTTING CYCLE

The development of the automated cutting cycle is a complete new approach aiming to improving the central task of the machine. Similar to the machine concept itself a balance has to be found between several key requirements on the system. On the one hand, a high amount of flexibility is necessary in order to cope with various and often changing boundary conditions such as excavation geometries, rock hardness and support requirements. On the other hand, the ultimate demand of the customer obviously is production and every automated system that targets the core process has to perform at least as well as an averagely skilled machine operator solely on his own hand. Here it has to be pointed out that the aim of such a system has never been to replace the operator. In a similar way that a cruise control of a car is not designed to replace its driver, this system aims at relieving the operator of the normally rather dull task of controlling the individual back and forth movements of the arm but still keep him in control of the governing parameters. Therefore, the machine operator still has the full control and responsibility over the entire machine operation.

One challenge of the design was to minimize this number of governing cutting parameters the operator has to keep track of.

For cutting automation of a Roadheader in principle, three degrees of freedom have to be taken into account: one horizontal, one vertical swing axis and the telescoping boom.

One further aspect has been to introduce a certain kind of learning behavior into the system, by keeping track of and analyzing key feedback parameters, such as the resulting cutter motor current, and thereby optimizing cutting performance for instance by calculating the optimum cutting height for every single cut.

The automatic cutting cycle is handled in the way that the machine operator plans the path by the online visualization component, whereby minimal configuration is necessary and the machine operator can define the cutting parameters if necessary. The planned cutting paths are visually checked by the operator and it is finally transmitted to the PLC system before executed by the equipment.

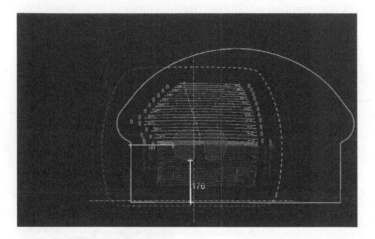

Figure 2.   Automated cutting cycle in large profile.

4   TELE-REMOTE TECHNOLOGY

Tele-remote operation is a new way to explore the full potential of automated mechanical cutting equipment while achieving the benefits of increased productivity, safety, and cost-efficiency in mining and tunneling operations. Tele-Remote is the entry-level solution from Sandvik to its industry-leading AutoMine™ offering.

Currently, machines are operated from a cabin or via radio remote control from a nearby position. Operators are exposed to hazards such as dust, noise and to moving equipment in confined space. Furthermore, immediate working area close to unsupported ground. The target of the tele-remote operation is to operate machines from a secured area in a remote position, e.g. from surface. The ultimate aim at the end of the day is, to operate machines completely autonomous.

4.1   Guidelines and standards

However, it is necessary to consider certain prerequisites for a tele-remote operation of a Roadheader via a distributed control system (DCS). Standards and guidelines to consider in Europe are the "Machine guideline 2006/42/EG (control system and actuators)" and "Standard for Safety for Machines DIN EN ISO 13849 (machine safety)". One part or the machine guideline EN13849 is the risk assessment to evaluate the safety risks ensuing from a machine.

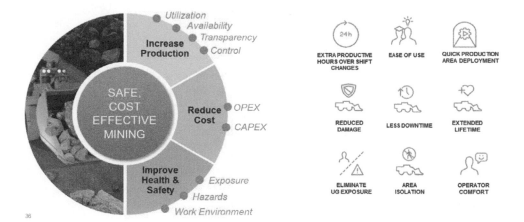

Figure 3. Value proposition of tele-remote operation.

If a risk is evaluated as too high, design actions need to reduce the risk to an acceptable residual risk.

After the execution of a risk assessment, safety system with according performance levels need to be implemented. In total, five performance levels a, b, c, d and e for safety systems describing the failure probability of a safety component related to the operating hours. (Haubmann, 2013)

4.2 *Communication and transmission technology*

The communication between the operator's location with the automated and tele-remote controlled machine is an aspect of utmost importance. The communication system layout is structured into a remote operator station with supervisory system functionality, a machine-onboard automation package including an integrated navigation system and control and monitoring capabilities. The installation of access control system barriers at the application area is used for safety isolation. Finally, the communication between remote operator station and machine is essential.

Safety is an important topic in terms of the process data communication, which means, it needs to be determined if the communication line is interrupted. It is clearly defined that any dangerous operating condition must be avoided due to communication related situations.

Thus, it is clear that there are certain functional requirement demands to the field control system. Those requirements are, for example, real-time abilities, reliability, flexibility, high bit rate and finally yet importantly cost efficiency. Such control systems are process field bus (Profibus) or CAN bus (control area network).

A typical communication system is Ethernet or WLAN communication with fast bit rates of 10 to 100 Mbit/s. Also to the Ethernet network, certain requirements like reliability under industrial or even mining or tunneling environment or real-time ability are demanded.

4.3 *Machine concept*

A potential machine concept for a tele-remote controlled Roadheader in a tunneling application requires certain prerequisites, especially safety relevant installations and protective functions defined by any applicable guideline or standard. Additionally, it is necessary to establish a reliable communication network between control station and machine. The most important aspect is to have a safe operation of the mine mitigating any kind of risk for humans and machines.

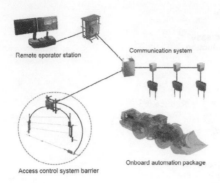

Figure 4. Communication layout with machine safety concept.

To have a distinct operation mode it is necessary to implement an "operation-mode-selector-switch" to release the tele-operation modus. On the one hand, maintenance and repairs works can be conducted at the machine; on the other hand, a tele-remote modus can only be achieved with the right position of the mode selector switch.

In the so-called maintenance mode, the machine can be operated by a line-by-sight radio remote control, if necessary. This is already a passive safety mechanism for a safe operation. The operation mode selection is an additional feature to clear danger zone.

Especially, for dangerous equipment like tele-remotely controlled machines, it is appropriate to install additional safety barriers in the danger zones. If a person or another equipment approximates the danger zone, a dangerous movement of the machine must stop. A big challenge is to detect a movement as harmless or potentially dangerous for the operational action. The system needs to decide automatically, if the machine operation must stop or proceed. However, it necessary to have other equipment or machines entering the danger zone for haulage, rock support and other necessary operational activities. This is already an evidence how difficult it is to have an operation automated.

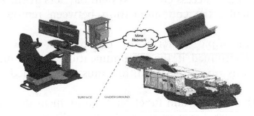

Figure 5. Machine concept for a tele-remote operation.

It is necessary to implement a safety barrier in dangerous operating areas. There are few different technologies for safety barriers like contactless safety concepts. It is necessary to not exceed the required distances between the sensors defining a working area. The concept design with minimum distances of contactless sensors is to protect humans against risky approximation to a danger zone. Potential threats like, crushing, cutting, shearing, collision, puncture hazard etc. are considered according to international standards. (Gräf, 2003)

The protection of danger zones can be established also with emergency stops. However, an emergency stop is not a substitute to safety design requirements of the machine and an emergency stop is not allowed to cause another unsafe condition during activation. When activated a machine must stop and an independent restart must be avoided. Those emergency stops must be marked with bright colors on a Roadheader. (Gräf, 2003)

Laser scanner are some other contactless safety barriers applied in tunneling and mining applications. This optical control system is scanning relevant danger zones. If an external individual or a machine not allocated to the operating area enters, the tele-remotely controlled machine must stop. A big advantage of the scanner is the potential to define three different ranges – measurement field, warning field and safety field. This gives the opportunity of a pre-warning before a deactivation takes place. (Gräf, 2003)

Light barrier and light grids are the next level of safety barriers. Those safety systems must be installed neither in the way that it is not possible to bypass them laterally nor underneath. Those systems must avoid the stay between safety field and danger zone. (Haubmann, 2013)

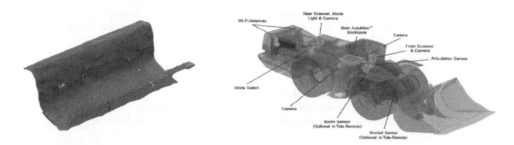

Figure 6.   Safety barriers and machine sensor for tele-remote operation of a loader.

5   CONCLUSION

Considering today's developments in automation, it can be stated that a tele-remote and semi-automated operation of a tunneling machinery is possible. Automated equipment is already state-of-the art technology and applied in many applications. The safety concept, which needs to be implemented into the machine control system, is essential for securing the operational areas and to protect miners from hazards arising from machine operations. At any time, no harm to humans and to machines are acceptable. Corresponding monitoring and emergency stop mechanisms must be implemented in the conceptual design. A reliable data connection together with a safe data protocol between machine and control station is very important.

Such a tele-remote operation protects miners as they control equipment from remote and safe places without hazardous and tough working conditions like dust, heat, noise, vibration, humidity, etc. The ergonomic situation of the operator is another advantage of the tele-remote operation.

As personal is present only during maintenance and repair works at the dangerous underground location, the risk for geological and machine related hazards is minimized. The huge advantage of an autonomous machine application is in achieving much higher performances by eliminating the human influence on the operation. Furthermore, a new level of accuracy in tunneling with nearly no overprofile can be achieve which helps to save costs for rock support, like shotcrete. The lifetime of components will be increased due to the optimized utilization of the Roadheader.

Due to the technical implementation of tele-remote controlled Roadheaders, future mining and tunneling applications will be further optimized. A big advantage will be that from one control station several machines can be controlled and operated. This will increase productivity, decrease the risk for injuries and counterbalance the lack of well-trained miners in mining and tunneling industry. However, today's technology is heading towards autonomous equipment working in underground applications with the ultimate goal to fully automate the complete mining process.

REFERENCES

Arvind, A., Bandopadhyay, L. K., & Kumar, H. (2002). *Mining Automation - Requirements and worldwide Implementations*: Indian Mining & Engineering Journal, 29–33.

Copur, H., Ozdemir, H., & Rostami, J. (2000). *Roadheader applications in mining and tunneling industries:* Golden, Colorado, United States.

Entezari, R. (2014, September). *Development of a remotely-controlled Roadheader robot*: Middle East Technical University, Mining Engineering Department.

Fuentes-Cantillana, J. L., Catalina, J., & Rodriguez, A. (2014). *Use of Computer Vision for Automation of a Roadheader in Selective Cutting Operation:* https://hal-ineris.archives-ouvertes.fr/ineris-00971823.

Gräf, W. (2003). *Maschinensicherheit: Auf der Grundlage der eurpäischen Sicherheitsnorm:* Heidelberg: Hüthig Verlag.

Haubmann, H. (2013, March 13). *Teleoperation einer Bergbaumachine:* Graz, Austria.

Humbert, M. (2007). *Technology and Workforce: Comparison between the Information Revolution and the Industrial Revolution:* Berkley: university of California.

Kargl, H., Gimpel, M., & Preimesberger, T. (2010). *Development of an automatic cutting cycle for part face mining machines*: Glückauf Mining Reporter, International Journal for Mining, Equipment and Technology – Made in Germany.

Nof, Y. S. (2009). *Handbook of Automation*: Berlin: Springer Verlag.

Rajput, R. K. (2008). *Robotics and Industrial Automation*: S.Chand Publishing.

Torlach, J. (1998). *Regulating the Mining Industry in the 2st Century*: Mine Safe International.

*Tunnels and Underground Cities: Engineering and Innovation meet Archaeology,
Architecture and Art, Volume 6: Innovation in underground engineering,
materials and equipment - Part 2 – Peila, Viggiani & Celestino (Eds)
© 2020 Taylor & Francis Group, London, ISBN 978-0-367-46871-2*

Interoperability between BIM models and 4.0 approach: Theoretical models and practical cases

K. Pini, G. Cataldi & G. Faini
CP Technology, Cinisello Balsamo, Italy

E. Piantelli
Trimble Italia, Vimercate, Italy

G. Lunardi
Rocksoil, Milan, Italy

N. Faccioli
Maccaferri, Bologna, Italy

I. Menegola
Pavimental, Barberino del Mugello, Italy

P. Sattamino
Harpaceas, Milan, Italy

ABSTRACT: Building Information Modeling (BIM) and 4.0 represent the hottest topics in construction and Industry. BIM is a process to create and manage all of the information on a project across the infrastructure lifecycle. The output of this process is a digital model that contains all the information of the infrastructure. 4.0 is a term used to refer to the developmental project in the management of manufacturing and chain production. It includes cyber-physical systems, the Internet of Things, cloud computing and cognitive computing. But how can the explosive mixture of these two approaches generate a revolution and a successful result in the construction field? Our paper wants to focus on how the 4.0 interoperability and information transparency can support and provide information to the BIM process during construction and moreover during maintenance phase of the infrastructures. Furthermore, we illustrate some of the best practises on machines carrying out aspects of production on digitalized model where BIM represents a repository for machine output information. We show the related benefit for an intelligent supply chain in real time. The result will demonstrate how the fourth industrial revolution affects the construction world and why the BIM-4.0 strategy represents a time and cost efficiency, a unique opportunity.

1 INTRODUCTION

BIM and Industry 4.0 represent the hottest topics in construction and industry. Building Information Modeling is a process for creating and managing all of the information on a project across the infrastructure lifecycle (see Figure 1). The output is a digital model that contains all the information of the infrastructure.

Industry 4.0 is a term used to refer to the developmental project in the management of manufacturing and chain production. It includes cyber-physical systems, the Internet of Things, cloud computing and cognitive computing.

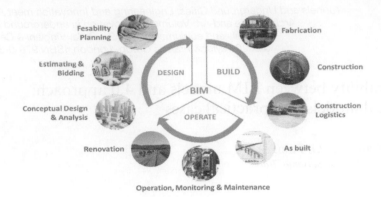

Figure 1.   Continuum Construction Cycle.

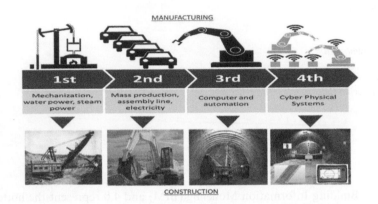

Figure 2.   Industrial revolution vs. Construction revolution.

But how can the explosive mixture of these two approaches generate a revolution and a successful result in the construction field? The presentation wants to focus on how the 4.0 interoperability and information transparency can support and provide information to the BIM Model during construction and moreover during the maintenance phase of the infrastructure.

Along the present article, we report some successful case histories on machinery which help to automatically export information functional to data collection. Therefore, we illustrate benefits related to an intelligent production chain that can be queried in real time. The result confirmed how the fourth industrial Revolution influences the world of construction and why a BIM - 4.0 strategy represents an efficiency in terms of cost and time of realization.

2   DEVELOPMENT

Here we illustrate, how interoperability BIM INDUSTRY - 4.0 positively affects all construction and maintenance processes:

- Carousel (Fabrication)
- Steel Rib Erector (Construction, Construction Logistics)
- IoT Sensors, Maintenance System (Construction Logistics)
- Maintenance Software G-Safe (Operation)
- Predictive Program (Maintenance)

The correlation between BIM and Industry 4.0 offers operational advantages in terms of information retrieval and improvement of production cycles during all phases of infrastructure implementation. A methodical approach, therefore, will not only offer a scalable digital platform at the end of construction of the infrastructure, but improve the management process in the pipeline.

Follow up reports of case studies for each of the phases of the construction continuum are shown. Some of them are completed processes, others are projects that are being developed. Another fundamental aspect for the success of any interoperability project is the correct management of organizational aspects, since the introduction of this new working methodology must be accompanied by a correct preparation of the company structure and from correct tracking of the processes inside the company, especially regarding the aspects concerning reading of the data and the operating procedures to be carried out.

2.1 *Carousel TBM segments prefabrication system*

2.1.1 *Context*

The realization of infrastructure with mechanized excavation in tunnels requires the installation of prefabricated segments. Pre-fabrication systems, whether carousel or stationary, already provide for the traceability of information related both to materials and production processes (i.e. certificates of steel related to re-bar cages, et cetera), but at present, such information is not automatically associated with a specific digital model. As a matter of fact, once the infrastructure is completed, the contractor will not have a digital model populated with automatically generated as-Built information at his disposal.

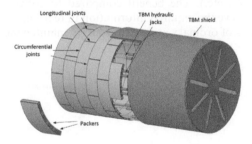

Figure 3.   Tunnel Boring Machine Method.

2.1.2 *Objectives*

The objectives of the project are:

- Implementation of an Industry 4.0 model related to the production processes of prefab elements;
- Systematic and transparent data collection related to each individual prefab element as well as of the associated production process and subsequent insertion of said data into a digital model (i.e. certificates of steel related to re-bar cages, certificates of the related concrete batch, etc.);
- Systematic and transparent data collection related to the quality compliance of the prefab elements along the whole manufacturing process and the subsequent insertion of said data into a digital model (i.e. curing curve of the specific segment);
- Generation of a production dashboard able to synthesize the data related to the production performances.

In the specific case of the Segment Prefabrication Plant, for the realization of the Fast Rail Link Tunnel between Genoa and Milan, the main purpose of the implementation of such a system is the creation of a 3D Tunnel Model containing all the construction process information. Once the infrastructure is completed, the Contractor will have a digital model generated according to BIM; this model will include all the as-built information suitable to be provided to the Project Owner, for his record and retrieval for monitoring and maintenance purposes.

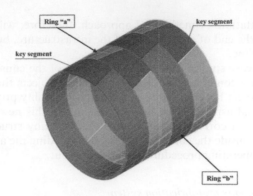

Figure 4. Configuration of segmental tunnel lining.

For the completeness of the model, the system involves the transmission of the model populated to the TBM driving system which will complete the model by adding the exact position of each prefab element along the Lined Tunnel.

2.1.3 *Advantages*

The main advantage of the implementation of such a system is the immediate retrieval of the as-built information (production process data, maturation curve data, data on the materials used, Dimensional controls, etc.). The benefit compared to conventional systems is represented by the ability to retrieve the required information in real time, thus carrying out preliminary analysis and forecasts of ordinary and extraordinary maintenance of the infrastructure.

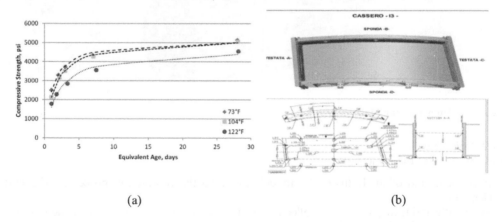

(a)   (b)

Figure 5. (a) Correlation compressive strength – maturity age – temperature. (b) Precasting mould and mould tolerances.

In addition, the programming of specific software allows the data collection and recording, including reporting of the production parameters. The risk of data loss is then eliminated.

2.1.4 *Development issues*

Obviously, to collect data in a transparent and efficient manner, the software shall receive the information directly from the Prefabrication Plant, thus in order to avoid use of human resources for the repetitive data collection activities. These resources will be used in more profitable ways for the data analysis thus in order to reveal any criticalities and to improve the quality of the processes.

The specific case showed that the interoperability between the production plant and the BIM – Industry 4.0 oriented, data collection software is fundamental. During the project's development the following criticalities have been revealed:

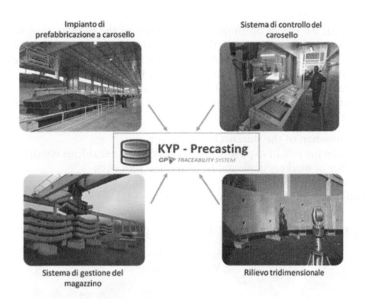

Figure 6.   KYP – CP Traceability system flow chart.

- Dialogue between sensors applied to machineries and related software;
- Dialogue between data collection systems of different machineries (TBM – Prefabrication Plant – Accounting management systems, . . .);
- Selection of essential and useful data during the construction processes of the infrastructure (not all data collected are essential, a selection of these data is required in order to let the system be simpler and durable).

2.2   Tubular steel RIB erector

2.2.1   Context

Conventional Tunnelling can be defined as the construction of underground openings of any shape with a cyclic construction process composed of the following steps:

- excavation, by using the drill and blast methods (explosives) or very basic mechanical excavators
- mucking
- placement of the primary support elements such as:
- steel ribs or lattice girders
- soil or rock bolts
- sprayed or cast in situ concrete

Ground-lining interaction control is one of the most critical processes during a tunneling project implementation. Some of the design and construction decisions during a tunnel project are very critical to reduce the ground movement around the excavated tunnel. These movements have a direct effect on the tunnel stability and the design load of the primary lining. Tunnel linings are structural systems installed during and/or after excavation to provide ground support, maintain the tunnel opening, limit the groundwater inflow, support appurtenances, and provide a base for the final finished exposed surface of the tunnel. Tunnel linings can be used for initial stabilization of the excavation, permanent ground support, or a combination of both.

During the past few years considerable advances have been taking place worldwide in the design and construction of tunnel primary linings. The trend set is a tendency to move away from traditional support, including heavy open profiles made of cold rolled steel arches, to a lighter solution with the use of optimized profiles such as the tubular steel rib, and shotcrete reinforced with wire mesh and/or steel fibers, providing a continuous and durable support. The development of tubular ribs by MACCAFERRI Tunneling, and the development of a Robotized Erector for the semi-automatization Steel Rib installation, designed and manufactured by CP-Technology, has provided engineers and contractors greater design options; increased flexibility; and brought efficient, safer, and cost-effective construction methods. As a matter of fact, the system is designed to:

- enable the automation of the steel arch installation;
- tackle and reduce the risk of exposure of the workers to hazardous conditions;
- limit the number of workers deployed to install the supports;
- reduce the installation time;
- implement effective and actual work coordination.

(a)   (b)

Figure 7.   (a) Traditional steel rib installation method. (b) Semi-automated steel rib installation method.

2.2.2 Objectives

Once the design phase of the first tubular steel rib erector prototype, currently used for the construction of the Boscaccio Tunnel by Pavimental, has been completed, a decision has been made to produce a second prototype that combines current safety ideas with the concepts of BIM modeling and Industry 4.0. So as to provide during the construction phase an advantage in terms of quality and to support management activities.

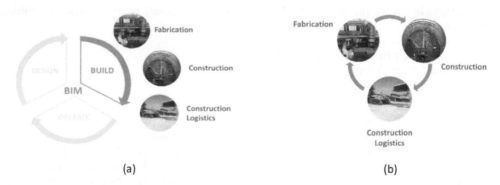

(a)   (b)

Figure 8.   (a) Continuum construction – building phase. (b) Transforming the continuum construction to a 4.0 lean supply chain.

Referring to the concept of continuum construction, we wanted to link the manufacturing, construction and management phases of the jobsite logistics working on a single digital model. The first step was the generation of a 3D model locating the ribs in the theoretical position using Tekla Structures Software.

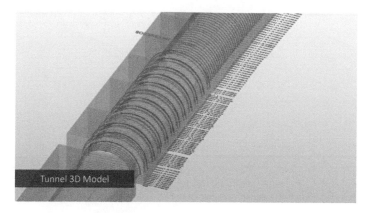

Figure 9. Tunnel 3D model.

Then, the digital model is transferred to the machine, thanks to a tool developed by Harpaceas, in order to display the correct positioning of the rib in the space, in accordance to the mathematical model, providing an effective support to the operator during the installation phase. Thanks to this solution, the correct positioning of the rib in the space is ensured, avoiding the risk connected to the lining under-thickness and therefore the corresponding risk associated with non-conformity of the primary tunnel lining works.

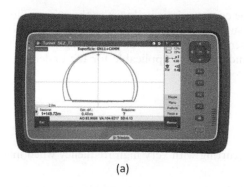
(a)

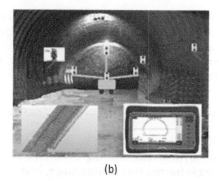
(b)

Figure 10. (a) Ekip steel rib placer topographic assistant. (b) Ekip steel rib placer topographic hardware.

The semiautomatic driving system has provided a dialogue between drawing software, machine guidance protocol and the machine positioning systems. In order to guarantee the maximum performance and safety conditions, typical automotive features have been adopted (command response times less than 500 milliseconds, complete driving system redundancy, . . .). Once the positioning phase is completed, the machine returns an as-built report with the correct position of the rib in the space in accordance to the digital model.

2.2.3 Advantages
The digital model is updated every time required with the following advantages:

- Progress Control
- Data availability for financial report
- Docs availability associated with constructed elements
- Work progress Data Availability for managin the logistics of the construction site (construction logistic)

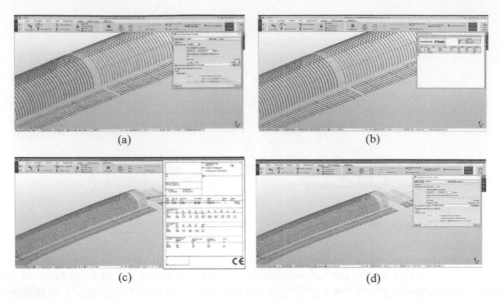

(a)　　　　　　　　　　　　　　　　　(b)

(c)　　　　　　　　　　　　　　　　　(d)

Figure 11. (a) 4D Tunnel model – Progress report. (b) 4D Tunnel model – Construction accounting support. (c) 4D Tunnel – Certification report. (d) 4D Tunnel model – Stock Report.

With reference made to the management of the construction site logistics the system allows to automatically raise an alert to the Steel Rib supplier/manufacture when the job site stock reaches the minimum threshold. The case represents a unique case in the management of the supply chain compliant to the Industry 4.0 Standards, now applicable for the underground infrastructure industry.

2.3 Maintenance and operation – G-Safe

2.3.1 Context
We may start highlighting some critical issues recurring with very high frequency when managing infrastructure maintenance, in detail:

- Difficulties for operators in collecting all of the data in a single archive for the works managed
- Difficulties in archiving and retrieving information and documentation related to the infrastructure works being managed (new and old)
- Difficulties in implementing well-structured and efficient inspection and maintenance plans
- Difficulties in verifying and certifying that inspections and maintenance activities are carried out as required
- Difficulties in forecasting budgets for routine maintenance
- Difficulties in having a real time picture of the maintenance activities for the managed infrastructure works.

On top of this the Italian Ministry of Infrastructures and Transport have enacted the ministerial decree no. 560 on 1 December 2017, which defines how to gradually integrate, by the contracting authority and operators, compulsory methods and related instruments, such as infrastructure modelling both during the design (BIM), the construction and management phases as well as the related maintenance activities. The obligation for the contracting authority will start from January 2019 for infrastructure works above 100 million euros and shall be progressively extended to contracts with a lower amount to be introduced throughout the public works system in 2025. In this context, G-Safe would be defined as: the digital system to manage the inspection and maintenance of the road infrastructures BIM-4.0 oriented.

2.3.2 *Objectives and advantages*

Which are the distinctive features of this platform?

- is able to integrate as built data in a BIM environment (Building information Modeling);
- is able to associate as built data within a unique inspection and maintenance software application;
- is able to schedule activities and draft budget;
- is able to instantly verify ongoing maintenance activities;
- is able to legally certify that inspections and maintenance activities have been carried out;

Furthermore, a GIS (Information georeferenced system) is integrated into the system with the aim of supporting the work of inspectors and maintainers. G-Safe allows for the collection oh monitoring data for road infrastructures 24/7 by means of IoTs (Internet of Things). The platform is designed to be used by infrastructure managers who can grant access to operators (surveyors, inspectors, builders, maintainers, etc.) in order to ease the sharing of information related to ongoing activities in a transparent manner.

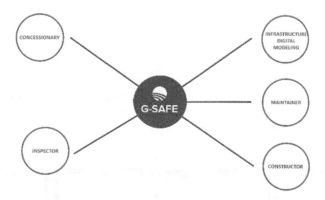

Figure 12.   G-Safe – Flow Chart.

The software platform is a subscription service and it can be accessed by means of a mobile device depending on the type of user and level of access. The opportunity of being able to consult easily and in real-time as built documentation related to the selected portion of the infrastructure being reviewed, it helps both to handle inspections and to schedule the maintenance program (considering the ease of retrieving information related to a specific supplier or to the installation date of a specific construction element or retrieving documents).

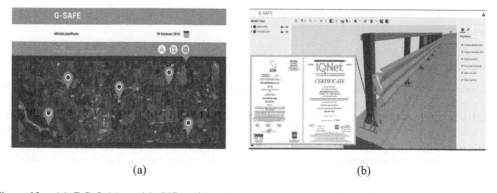

(a)   (b)

Figure 13.   (a) G-Safe Map with GIS position for maintenance plan. (b) 3DModel.

The opportunity to integrate conventional inspection activities with a sensor network able to carry out 24/7 monitoring on the most critical part of the infrastructure, constitutes a great added value.

Figure 14.   Iot Sensor.

Furthermore, the activities of field operators will change considerably by using tools which facilitate mobility tasks (tablets and smartphones) and allowing at the same time the generation of more punctual reports.

Figure 15.   G-Safe: User interface.

2.3.3 *Development issues*
Up to now the main issues faced during the development of this platform were:

- Processes Restructuring. Company processes have to be improved so as to incorporate training and the design and management of new processes.
- Information enhancement. Retrieval of historical information of existing infrastructure and related inspection and maintenance activities is an important matter to be conducted methodologically. In order to take full advantage of the platform's potential the Digitalization of the existing infrastructure will be required, at least of the most sensitive part (i.e. bridges, viaducts . . .)

The development of the platform is continually on going.

3 CONCLUSIONS

Demo application of the Platform reveals that the use of suitably designed sensors built into the equipment may be the only way for a transparent and effective information transmission to digital models according to BIM methodology in a 4.0 oriented way. As a matter of fact, managing the infrastructure according to BIM methodology requires specific skills and extensive deployment of resources, if the digital model data is transferred automatically from the sensors, there is an advantage in terms of information transparency, less data loss and lower cost in collecting data.

The data returned by sensors built into the equipment may be a valuable resource in order to manage procurement, progress control and financial information. By the way, as already mentioned, this also involves a re-design of workflow and all the associated roles.

The first, owners, managers or constructor of infrastructure who will adopt in a functional way such methods and technologies will have a huge competitive advantage both during the construction phase or the maintenance phase.

Tunnels and Underground Cities: Engineering and Innovation meet Archaeology, Architecture and Art, Volume 6: Innovation in underground engineering, materials and equipment - Part 2 – Peila, Viggiani & Celestino (Eds)
© 2020 Taylor & Francis Group, London, ISBN 978-0-367-46871-2

The management of the soil conditioning process for the excavation of the Rome Metro C line

M. Pirone, F. Carriero & R. Sorge
Astaldi, Rome, Italy

D. Sebastiani & S. Miliziano
Università degli Studi di Roma "La Sapienza", Rome, Italy

V. Foti
Metro C, Rome, Italy

ABSTRACT: Tunnel excavations in urban areas are often performed with TBM and Earth Pressure Balance technology which requires the continuous injection of chemicals in the soil conditioning process. Several advantages can be achieved by performing laboratory activities prior to starting the excavation in order to predict risks such as clogging in fine grained soils or severe cutter-head wear in coarse soils and to define an optimal range of values for injection parameters. Since the in-lab reproduction of all the conditions occurring during the excavation is hardly possible, measurement performed directly on site and TBMs monitoring data analysis are relevant tools to improve the knowledge on soil conditioning process and chemicals management. For the project of the Rome Metro C line, laboratory tests and on-site measurement were performed and useful information on excavation were drawn from the analysis of the recorded data. The results provided insights on advantages, limits and differences of results between laboratory-scale and on-site soil conditioning.

1 INTRODUCTION

Starting from the first applications of EPB-TBM in Japan, described by Fujita (1981) and in Europe with the projects of Passante Ferroviario in Milan (Peron & Marcheselli, 1994) and the subway line in Valencia (Herrenknecht & Maidl, 1995), cases of applications of the EPB technology and foaming agents, currently used in most tunnel excavation projects, especially in urban environments, are well known in literature.

Several studies and practical excavation experiences were performed by Nishitake (1990), Babendererde (1998), Babenderende (1991), Jancsecz et al. (1999) and Langmaack (2000) and provide important information and best practices on the management of the soil conditioning process.

Concurrently with said excavation experiences, several studies were developed starting from Bezuijen et al. (1999) and with the joint research activity between Cambridge and Oxford which led to the works by Milligan (2000), Psomas (2001) and Mair et al. (2003).

The evolution of the tunnelling world led us, on one hand, to more audacious projects and more and more complex excavation conditions but, on the other hand, to new, powerful and technologically advanced TBMs, new and more effective chemicals (foaming agents, additives and polymers).

Some general suggestions can even be found in literature, such as EFNARC (2005), illustrating the fast advancements of technology, geotechnical and chemistry and providing precise information which are of a considerable importance to properly manage the conditioning process during the excavation, although specific experimental studies should be performed in specifically equipped laboratories.

This work shows the results of the experimental activity performed at Sapienza University of Rome in cooperation with Astaldi and Metro C, aimed at supporting the soil conditioning process for the excavation of Rome Metro C line, from the selection of products to the management of injection parameters during excavation.

2 THE ROME METRO C PROJECT

2.1 *Metro C line project overview*

Line C is Rome's third subway line. Once completed, it will cross the city from the Northwest to the Southeast, for a total length of 25.6 km and 30 stations, almost doubling the extent of the currently existing underground network. It is also the first fully automated underground line in Rome.

Metro C S.p.A. is the General Contractor, formed of Astaldi (Leader), Vianini Lavori, Ansaldo STS and CMB, entrusted with the construction of Line C in all of its phases: from the design to the archaeological surveys, from building the tunnels to the stations and trains, up to the management of the start-up phase.

The activities started in 2006 with the archaeological surveys and the final design; currently, there are 22 stations and about 19 km of line in service.

For the construction of currently serviceable tunnels, the General Contractor used 4 EPB TBMs, having a cutter head of a diameter of 6.70 m.

Section T3 (Colosseo - San Giovanni), currently under construction, extends across the city's historical centre and very near (sometimes underground crossing) monuments dating back to the age of the ancient Romans, such as the Colosseum, the Aurelian Walls and Basilica of Massenzio. For this reason, extremely effective and innovative technical solutions were devised to reduce settlements.

In order to safeguard all the monuments, Metro C set up a specific Scientific Technical Committee (STC) the members of which include world-famous professors. The purpose of this Committee was to ensure high-quality research methods and to analyze potential interactions between the new line and the historical monumental heritage. The Committee coordinates and supervises the activities of working groups of specialists. The 5 work teams - consisting of university professors and specialists - operate in the following fields: Geology,

Figure 1.   General overview of the stretch T3 of the Rome Metro C line.

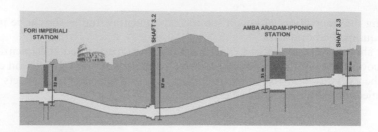

Figure 2. Longitudinal section of the tunnels excavation layout between S. Giovanni shaft and Fori Imperiali stations.

Geotechnical Engineering, Restoration and Preservation, Structural Engineering, Monitoring Systems.

The T3 section starts at San Giovanni station and is about 3 km long. It includes two stations, Amba Aradam station and Fori Imperiali station, and two shafts provided with ventilation systems (TBM shaft 3.3, and shaft 3.2). The T3-section tunnels, excavated by two EPB TBMs, have a variable depth from 30 to 60 m.

The first part of tunnels extends beneath the Aurelian Walls at Porta Metronia, while the other end of the tunnels extends along Via dei Fori Imperiali, very close to the Colosseum and the Basilica of Massenzio.

Due to the context, one of the first requirements is the control of subsidence, in order to prevent any damage to the archaeological structures above ground level. To achieve this goal during the excavation by the TBMs, good soil conditioning is essential but not trivial due to on-site geotechnical conditions.

The excavation layout is characterized by the presence of various litho-stratigraphic units; starting from the surface level downwards, the following formations are expected to be found:

– backfill material (R), with heterometric and heterogeneous elements in a sandy-silty matrix, of varying density and/or consistency;
– recent alluvial deposits of the Tiber river and its tributaries (LSO), consisting of silty clays, clayey and sandy silts, sand and silty sands with variable organic content;
– pre-volcanic fluvial-lacustrine deposits, including two sub-units: an upper unit (AR) formed of clayey silts and silty clays with various sandy interbeddings (ARS), and a lower unit (SG) of medium-coarse grain sands with gravel;
– marine sediments (APL), formed of silty clays and clayey silts of a grey - light blue colour, with alternating fine-grain sand levels.

Moreover, the excavation has to be carried out below groundwater level.

3 THE PRELIMINARY LABORATORY TESTS

3.1 Generalities

In a fruitful cooperation between Metro C, Astaldi S.p.A and Sapienza University of Rome, an intense experimental activity, to be developed preliminarily and concurrently with tunnel excavation activities, was planned to deal with the management of soil conditioning.

The general approach includes the division of the excavation layout into three successive phases: the first one characterized by AR, ARS and LSO lithotypes, the second one by the medium-coarse grain (SG) and the last one by the pliocenic grey-light blue clays (APL), encountered by the TBM in this sequence.

Before the beginning of the excavation, samples of soil belonging to the various lithotypes were taken from the vertical shaft 3.2 showed in Figure 2 and were examined at Sapienza geotechnical laboratory.

Based on the soil samples, several different products suitable for the excavation were selected from major European suppliers; different samples of the products were placed in white tanks marked with a code, so all the tests performed in the laboratory and described below should be considered as "blind tests", performed without being aware of the commercial products tested.

Laboratory activities included:

– identification and classification tests on the soil samples, including grain size distribution, water content and Atterberg limits;
– tests on each chemical product, including specific weight, viscosity, pH and Total Organic Carbon (TOC);
– stability (half-life) tests of the foam generated by using the different foaming agents;
– clogging measurements (mixing test and pull-out tests) on soil samples before and after the conditioning process performed at the laboratory by using particularly developed apparatuses, described in Di Giulio et al. (2018);
– the same tests during the excavation progress, taking samples at regular intervals and comparing results with those obtained in the preliminary tests.

The aim of the experimental activity was, on one hand, comparing the performance of different products and different dosages in order to provide preliminary indications for the selection and the dosage of the foaming agents to be used during the excavation and, on the other hand, acquiring information on the behavior of the different lithotypes described before conditioning, so as to foresee possible risks, as clogging or uncontrolled wear on the cutter-head. Moreover, the experimental activity allows to create a large database of results of the tests performed on samples taken directly from the jobsite, necessary to fill the gap between the experimental studies performed at the laboratories and actual tunnelling projects.

At present, only the first phase of the excavation has been completed and, consequently, only the preliminary activity data and the results of the samples taken on site belonging to the *AR/ARS* and *LSO* lithotypes (whose characteristics are reported in the Table 1) are available.

3.2 *Tests on foam samples*

The most commonly performed test to verify the stability of the foam over time is the half-life test or drainage time measurement. Even if, as suggested by Mori *et al.* (2018), there are significant differences between the laboratory test, carried out at atmospheric pressure and controlled temperature, and the excavation chamber, in which the foam is mixed with the ground under pressure and relevant pressure variations occur, the test results can be an important element for the assessment of the properties of a foam.

The half-life test is one of the tests proposed by the EFNARC guidelines (2005) and is essentially performed by filling a glass cylinder with 80 g of foam and measuring the time necessary to drain 40 ml of liquid into a graduated cylinder placed underneath a funnel, defined half-life time, *hlt* (Figure 3).

The results of the tests carried out on the four tested products at a standard concentration of 2.0% and at different values of *FER* are shown in Figure 4 according to the classification system developed in the Geotechnical Laboratory of Sapienza based on more than 650 tests performed and currently used to provide comparative evaluations on the stability of the foam.

Table 1. Grain size distribution and Atterberg limits of soil samples.

| Sample | Grain size distribution | | | | Atterberg limits | | |
	gravel (%)	sand (%)	silt (%)	clay (%)	LL (%)	LP (%)	IP (%)
LSO	2	21	31	46			
AR	0	4	61	35	45.1	21.4	23.7
ARS	0	40	50	10	24.6	18.6	6.0

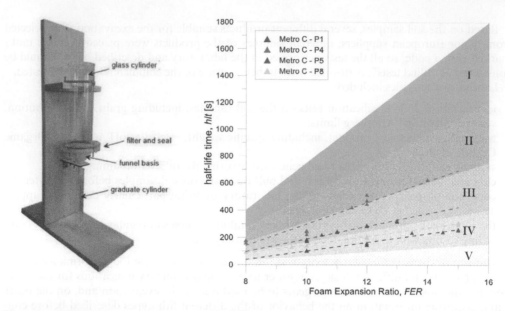

Figure 3. Half-life time test apparatus and half-life times measured.

Appreciable differences were recognized between the stability of the foam generated by the various products at the same concentration and in the range of tested *FER* values: the product P4 is always more stable than the others, positioning in the field of stability II, the product P1 has a stability corresponding to the field III while P5 and P8 have lower stability.

3.3 *Mixing tests*

After the foam characterization, tests were carried out on soil mixed with water and conditioned soil samples using injected water amount and conditioning parameters variable within commonly used ranges for the excavation of fine grained soils.

One of the most commonly performed laboratory tests to verify the achievement of the optimal consistency for the excavation and the reduction of the clogging risk is the mixing test, suggested by Mori *et al.* (2018).

The test consists in letting the soil rotate inside the Hobart mixer (Figure 4a) and in measuring the amount of soil remaining attached to the tool, expressed by the #x03BB; parameter, defined as the ratio between the stuck soil and the total amount of soil used for the test.

Tests were initially performed on the soil mixed with only with water w_{nat}, so as to have a basis for comparison with the subsequent tests performed on samples of the same soil mixed with water (the naural water content w_{nat} and the water content added, w_{add}) and foam generated using the four selected products. Figure 4b shows the results of the tests performed on *AR* and *ARS* soil samples; typical behaviors were recorded with curves having low clogging risk for low and high water contents ($w_{ant} + w_{add}$) and a peak, in the field of high clogging risk, in the middle.

As expected, due to the combined effect of grain size distribution and Atterberg Limits variation, the curves related to *AR* and *ARS* have a peak for very different water content values; tests performed on a soil sample composed of 50% of *AR* and 50% of *ARS* artificially prepared into the laboratory and representative of a mixed excavation section, resulted in a curve between the first two, establishing a relationship between the water content of the sample with an higher clogging risk and the grain size distribution.The same tests were performed on the same soil samples after a soil conditioning process performed by using foam generated at a laboratory foam generator plant equipped with a foam lance taken directly from the TBM actually used for the tunnel excavation. The undrained strength Cu on the same soil samples was measured by using a fall-cone test apparatus.

Following the EFNARC standards and on the basis of previous experiences, tests were performed by using a fixed value of Cf of 2.0%, variable values of FER in the range of 8-12 and variable values of FIR and water added apart with the purpose of making the soil reach the proper consistency (measured using standard slump and flow table tests according with EFNARC standards) required for EPB-TBM excavation.

The tested parameter combinations for the four products are listed in Table 2 below. The same Table 2 also sets forth a list of the average values (3 measurements) recorded when performing the mixing tests, as also shown in Figure 5.

The measurement obtained from the mixing tests and the fall-cone test revealed meaningful differences in the use and in the resulting effectiveness of the four tested products. The decrease in #x03BB; values related to the use of the products P4 and P5 if compared to the #x03BB; values recorded for products P1 and P8 is particularly clear. For all the products, relative better results were recorded using FER values of 8 and the required FIR values are lower for P4 and P5 (between 30% and 50%) than for P1 and P8 (between 60% - 100%). Starting from the natural content of 27%, the amount of water added for the tests was about 10% for products P4 and P5 and about 15% for products P1 and P8. The same effect is visible in the differences in the results obtained from the fall-cone test performed on the same soil samples and represented in terms of undrained strength Cu, even if, considering the variation of the saturation of soil samples due to the foam injection, it is not considered, in theory, correct to correlate the differences in results obtained to the undrained strength.

On the basis of the results of the comparative tests on the AR soil showed in Figure 6, it is deemed that the Metro C-1 and Metro C-8 products are able to effectively condition the soil only if injected in relatively large quantities and injected together with relatively high amount of water; vice versa, the results of the tests show that the Metro C-4 and Metro C-5 products are much more effective in reducing the natural tendency of the AR soil to adhere to metal elements and its undrained strength while achieving, at the same time, the right consistency in the soil.

Finally, it was noted that to obtain slightly better results a lower volume of foam generated with the Metro C-4 product was required to be used if compared with the case of the Metro C-5, which, however, is the most stable, as it appears from the half-life time described in Figure 3 and it was considered the most suitable for the conditioning of the AR lithotype.

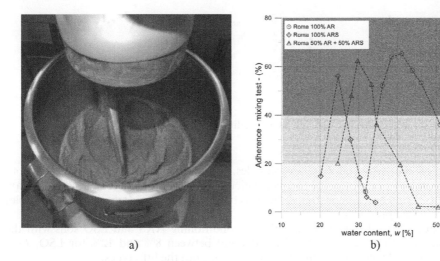

a)   b)

Figure 4.   a) Hobart mixer test apparatuses and b) results obtained performing the mixing tests on AR and ARS soil samples on the classification chart proposed by Thewes.

Table 2. Materials used and features for test tools.

Product 1	Cf (%)	FER (%)	FIR (%)	w_{add} (%)	λ (%)
	2.0	10	65	8.5	61.64
	2.0	10	85	8.5	62.41
	2.0	10	85	12.0	47.54
	2.0	10	65	15.0	47.43
	2.0	10	85	15.0	42.94
	2.0	10	100	15.0	31.39
	2.0	10	65	18.0	30.10
	2.0	8	85	15.0	5.11
	2.0	8	65	18.0	6.07

roduct 4	Cf (%)	FER (%)	FIR (%)	w_{add} (%)	λ (%)
	2.0	12	85	15.0	8.03
	2.0	12	50	15.0	11.01
	2.0	10	50	12.0	11.31
	2.0	10	40	10.0	23.81
	2.0	8	50	12.0	8.61
	2.0	8	40	10.0	19.57
	2.0	8	30	10.0	22.36
	2.0	8	40	8.5	30.74
	2.0	8	50	8.5	19.77

Product 5	Cf (%)	FER (%)	FIR (%)	w_{add} (%)	λ (%)
	2.0	10	65	12.0	14.55
	2.0	10	85	12.0	17.43
	2.0	10	85	10.0	26.13
	2.0	10	65	10.0	23.14
	2.0	8	85	10.0	15.08
	2.0	8	100	10.0	22.12
	2.0	12	50	10.0	23.46

roduct 8	Cf (%)	FER (%)	FIR (%)	w_{add} (%)	λ (%)
	2.0	12	85	18.0	37.49
	2.0	10	85	18.0	25.69
	2.0	10	100	15.0	33.80
	2.0	10	100	18.0	30.50
	2.0	10	85	20.0	26.20
	2.0	8	65	12.5	49.59
	2.0	8	65	15.0	50.42
	2.0	8	85	15.0	42.99
	2.0	8	85	18.0	16.54

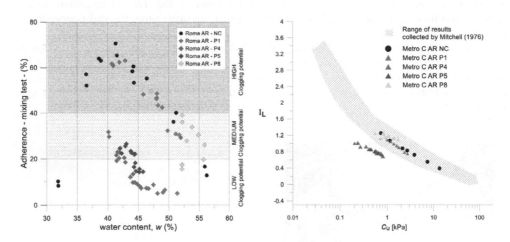

Figure 5. Mixing test and fall-cone tests results on conditioned soil samples.

The results on the *ARS* and *LSO* soil samples, not shown, confirm the relative ease to achieve the optimum consistency due to low plasticity and high percentage of sand (and gravels in the case of *LSO*) and demonstrated the possibility of properly conditioning soil samples with any of the tested products.

All the products turned out to be suitable for conditioning ARS and LSO soils with the addition of water between 2.5 and 5% for ARS and between 8% and 12% for LSO, *FER* values of 8-10 and *FIR* values between 50% and 60% for both the lithotypes.

4 THE MANAGEMENT OF CONDITIONING PROCESS DURING EXCAVATION

4.1 Generalities

The results presented in this paper are based on the database of parameters recorded during the excavation of the tunnels of the Metro C line in the section between Shaft 3.3 and Amba Aradam – Ipponio station. The excavation started in March 2018 and was completed in approximately 4 months. The data collected during the excavation of the first of the two tunnels have been processed to obtain the correlations set forth herein. Considering that the TBM launching shaft 3.3 is smaller than the TBM machine, the launching operations has to be executed in a plurality of steps. Moreover, the first excavation steps are generally used to calibrate all the machine's parameters. These are the reasons why, in order to appreciate the overall tendency in the computational phases, the first 9-10 advancements have been neglected.

4.2 TBM features

The two TBMs used, Herrenknecht S409 and S410, for the excavation are of an EPB (Earth Pressure Balance) type. The excavation diameter is 6.69 m and the machines are designed to perform excavation through various lithotypes (tuffs, pozzolanic soil, silts and clays). The opening percentage of the machines' face is 40%. They can perform a minimum radius on the alignment of 190m. The lining installed by the machine is a ring consisting of 6 segments +1 key, with a thickness of 30 cm and a length of 1.4 m. The muck can be removed by either a belt conveyor and or a track system.

4.3 The management of soil conditioning

During the excavation for the AR/ARS lithotypes, the Cf value used was 2.5% during all the excavation, the average FER value was between 6 and 8, and the FIR between 70% and 75% as shown in the charts in Figure 6 and Figure 7 for the two TBMs S409 and S410.

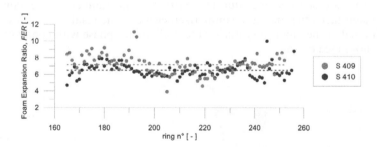

Figure 6. Foam Expansion Ratio (*FER*) values actually used during the excavation.

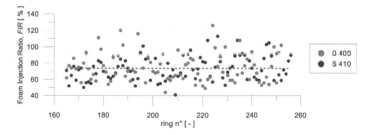

Figure 7. Foam Injection Ratio (*FIR*) values actually used during the excavation.

When compared, the soil conditioning parameter actually used and the results of the preliminary laboratory tests turned out to be generally in agreement and some general remarks can be expressed:

– the *FER* values used were similar to those used for preliminary tests; differences are due to the need, on site, to keep the excavation chamber at 3 bars of pressure at least consequently to the necessity of reducing the risk of dispersed air inside the chamber, reducing the injected air and, finally, reducing the *FER* values;
– the amount of foam injected (*FIR*) was extremely variable during the excavation; the overall average value is slightly higher if compared to the 60% values obtained from preliminary laboratory tests. The reasons therefor are likely attributable to, inter alia, the dispersion of foam during cutter-head rotation and the differences in the mixing process, apparently very effective at the laboratory, while negatively affected by the movement of the cutter-head and by the presence of blades within the excavation chamber when used with the TBM;
– the *Cf* value actually used was 0.5% higher: even in this case, considering that the foam generation is affected by the pressure in the working chamber, probably a higher surfactant concentration could compensate the lower accuracy in foam generation; in any case, it is believed that the recorded differences in *Cf* values do not affect the conditioned soil properties.

5   TESTS DURING THE EXCAVATION PROCESS

During the excavation of the tunnels, performed using Metro C-4 product, conditioned soil samples were taken from the site at regular time intervals. General characterization tests (water content, grain size distribution and Atterberg limits) were performed on this samples and the mixing and fall cone tests were performed in the preliminary tests on untreated soil samples and on soil samples conditioned at the laboratory.

Figure 8 shows the results of the tests performed on samples of soil taken from the site which, according to the grain size distribution, are attributable to a mixed AR/ARS excavation section.

Reportedly, they are samples taken during the first learning (launch) phase, samples of conditioned soil taken thereafter and, as a comparison element, the results of the tests performed on soil conditioned at the laboratory during the preliminary phase with conditioning parameters close to those used on site.

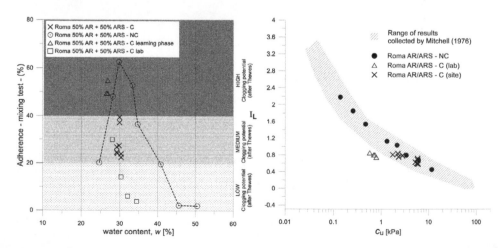

Figure 8.   Mixing test and fall-cone tests results on conditioned soil samples taken from the TBM site.

Considering the results of the mixing test performed on the soil samples taken from the construction site relative to the *AR/ARS* sample, we can preliminarily notice a series of issues:

- the samples taken during the "learning phase" in the first 10 rings showed lower values than the maximum values recorded on the not-conditioned soil sample, but still within the field corresponding to an "high clogging risk";
- the samples taken during the further advancement, where the conditioning parameters have been optimized, are positioned in the "medium clogging risk" field;
- the variability in the grain size distribution and in the conditioning process effectiveness provides samples having adherence values of about 25% but also samples having values of 40%;
- in no case, during excavation, has been reached adherence values within the "low clogging risk" field, demonstrating the fact that on site is not as easy to achieve an optimal conditioning as the one obtained in a specific developed laboratory.

Similar conclusions can be obtained by reading the chart in figure 8, in which the undrained strength values are shown in relation to the Liquidity Index; it is clear that:

- the soil samples taken during the learning phase show an undrained resistance coincident with that of the ground conditioned with only water, a clear sign of non-optimal conditioning;
- by setting the conditioning parameters correctly, it has been possible to obtain lower undrained resistance values but still within the range of typical clay values (Mitchell, 1976);
- as in the case of the mixing test, on site, it was never possible to reach the *Cu* reduction values obtained by conditioning the soil in the laboratory.

6 CONCLUSIONS AND FUTURE DEVELOPMENTS

An extensive experimental activity was performed at "Sapienza" University in cooperation with Astaldi and Metro C in order to support the management of soil conditioning prior to and during the excavation of the Metro C line in Rome.

The laboratory tests have proved to be, as a whole, a reliable tool for the optimization of several operations, from the selection of the most suitable conditioning product to the management of the parameters during the excavation. The satisfactory correspondence between the parameters actually used for the excavation and the parameters tuned in the preliminary laboratory phase lead to conclude that, for the future, similar experimental activities can be helpful to predict the proper soil conditioning parameters (FER, FIR and added water).

In no case, the laboratory activities carried out in the preliminary phase will be able to replace the observation of the excavation parameters and the variation of the conditioning during the excavation phases based on the experience and know-how acquired in decades of tunnelling projects.

The ability to foretell the soil conditioning parameters with acceptable accuracy could be an useful instrument to develop projects including soil excavation works based on an "*ahead-of-excavation knowledge*" of the geotechnical and chemical features of the debris.

In the future, since the Metro C line excavation will face sandy gravel (SG) and clay (APL) lithotypes, further preliminary tests will be performed, additional samples will be taken directly from the site and further comparison elements will be recorded to enrich the present work.

REFERENCES

Babendererde, S. 1991. Tunnelling machines in soft ground: a comparison of slurry and EPB shield systems. *Tunnelling and underground space technology*, 6(2),pp.169–174.

Babendererde, L.H. 1998. Developments in polymer application for soil conditioning in EPB-TBMs. *Tunnels and* Metropolises, Negro Jr. and Ferreira (eds.), Balkema, Rotterdam, 2, pp.691–695.

Bezuijen, A., Schaminee, P.E.L. & Kleinjan, J.A. 1999. Additive testing for EPB shields. *In Proc. of 12th European Conf. on Soil Mechanics and Geotechnical Engineering, ECSMGE* (Vol. 3, pp. 7–10).

Di Giulio, A., Sebastiani, D. & Miliziano, S. 2018. Effect of Chemicals in Clogging Risk Reduction for TBM-EPB Application. *In Proceedings of the World Tunnel Congress 2018 - The Role of Underground Space in Building Future Sustainable Cities*. Dubai, EAU.

EFNARC, A. 2005. Specification and guidelines for the use of specialist products for mechanized tunnelling (TBM) in soft ground and hard rock. *Recommendation of European Federation of Producers and Contractors of Specialist Products for Structures*.

Fujita, K. 1981. Use of Slurry and Earth-Pressure-Balance Shield in Japan. In: *Int. Congress on Tunnelling, Tunnel* Vol. 81, Bd.1, 383–406. Düsseldorf.

Herrenknecht M. & Maidl U. 1995. "Applying foam for an EPB shield drive in Valencia" *Tunnel* 5/95 pp. 10–19.

Langmaack, L. 2000. Advanced technology of soil conditioning in EPB shield tunnelling. *In Proceedings of North American tunneling*, 525–542.

Milligan, G. 2000. Lubrication and soil conditioning in tunnelling, pipe jacking and microtunnelling: A state-of-the-art review. *Geotechnical Consulting Group*, London, UK.

Mair, R., Merritt, A., Borghi, X., Yamazaki, H. & Minami, T. 2003. Soil conditioning for clay soils. *Tunnels and Tunnelling International*, 35(4),29–33.

Mori, L., Mooney, M. & Cha, M. 2018. Characterizing the influence of stress on foam conditioned sand for EPB tunneling. *Tunnelling and Underground Space Technology*, 71, 454–465.

Nishitake, S. 1990. Advanced Technology realize high performance earth pressure balanced shield. In: Franch issements souterrains pour l' Europe, 291–302, (Legrand, ed.), Balkema, Rotterdam.

Peron, J.Y. and Marcheselli, P. 1994. Construction of the 'Passante Ferroviario' link in Milan, Italy, lots 3P, 5P and 6P, excavation by large earth pressure balanced shield with chemical foam injection. *In Proc. of Tunnelling 94 conf*erence, 679–707. Springer, Boston, MA.

Psomas, S. 2001. *Properties of foam/sand mixtures for tunnelling applications*. Doctoral dissertation, University of Oxford.

Sebastiani, D., Ramezanshirazi, M., Miliziano, S. & Di Giulio A. 2017. Study on short and "long term" effects of chemicals on fine grained soils for mechanized tunnelling conditioning. *In Proc. of the World Tunnel Congress 2017 – Surface challenges – Underground solutions*. Bergen, Norway.

Tunnels and Underground Cities: Engineering and Innovation meet Archaeology, Architecture and Art, Volume 6: Innovation in underground engineering, materials and equipment - Part 2 – Peila, Viggiani & Celestino (Eds)
© 2020 Taylor & Francis Group, London, ISBN 978-0-367-46871-2

Hybrid lining segments – bearing and fracture behavior of longitudinal joints

S. Plückelmann & R. Breitenbücher
Ruhr University Bochum, Institute for Building Materials, Germany

ABSTRACT: In order to utilize the benefits of steel fiber reinforcement and ensure an economic design at the same time, a new approach to produce hybrid lining segments was developed and is presented. The basic principle is to strengthen the vulnerable zones of segments with a high performance steel fiber reinforced concrete (HPSFRC), while for less critical zones an ordinary lining concrete is used. Focusing on the longitudinal joints, an experimental study on the bearing and fracture behavior of hybrid specimens, produced with a so called "wet-on-wet" production procedure, has been conducted under concentrated load. It was proven that the installation of a HPSFRC layer only in zones where critical splitting stresses occur leads to a remarkable increase in the bearable loads. The results showed that the use of vertical instead of horizontal formworks had a positive impact on the fiber orientation and thus consequently on the bearing behavior.

1 INTRODUCTION

The use of steel fiber reinforced concrete (SFRC) for the production of segmental lining is increasingly gaining importance due to its advantageous material properties especially in terms of an enhanced spalling behavior, reduction of crack widths and capability of stress redistribution. A further advantage compared to conventional reinforcement with steel bars is that steel fibers allow to reinforce the peripheral zones of segments, especially their edges and corners, which are particularly vulnerable to damage because of for example imperfect segment installation scenarios.

For the production of steel fiber reinforced segmental lining, the steel fiber content usually ranges between 30 and 60 kg/m³ in order to obtain a sufficient workability and homogenous distribution of steel fibers in the concrete matrix (Breitenbücher 2014). However, not only due to technical circumstances but also from an economic point of view, a higher steel fiber content is not feasible in most cases. Therefore, in particular with regard to high concentrated loads as they frequently develop in the circumferential and longitudinal joints of the segments, the reinforcement only with steel fibers is often insufficient to withstand the resulting tensile splitting stresses. In such cases, additional steel bar reinforcement is necessary.

In order to utilize the mentioned benefits of SFRC and ensure an economic design at the same time, a new approach was developed in the Subproject "Optimized structural segments for durable and robust tunnel lining systems" within the Collaborative Research Center (SFB) 837 to produce hybrid lining segments. Instead of a conventional full range fiber reinforcement and, if necessary, additional steel bar reinforcement, the basic principle of the hybrid concept is to partially strengthen the segments with a high performance steel fiber reinforced concrete (HPSFRC) in vulnerable zones, while for less critical zones an ordinary lining concrete is used.

In this paper, a so called "wet-on-wet" procedure for the production of hybrid lining segments is presented. Focusing on the longitudinal joints of segments, an experimental study on the bearing and fracture behavior of hybrid specimens cast with the before mentioned production procedure has been conducted under concentrated load in order to investigate the potential of the hybrid concept. Particular attention was attributed to the special circumstances of the "wet-on-wet" production procedure with respect to the casting direction, which is influencing the steel fiber orientation and consequently the bearing behavior of longitudinal joints.

2 PRODUCTION PROCEDURE FOR HYBRID LINING SEGMENTS

Since the peripheral zones of lining segments (i.e. corners, edges, joints) are significantly more vulnerable to damage than the interior zones a new hybrid concept is pursued, which is characterized by the combined use of HPSFRC and ordinary lining (LC) concrete. However, for the production of hybrid lining segments a modification of the conventional production procedure is required.

One approach is to install tailor-cut precast components made of HPSFRC in the critical zones in segment formwork, followed by the casting of a freshly mixed lining concrete ("wet-on-solid"). However, the bonding between the precast components and the subsequently added concrete is a weak point of this procedure and has to be strengthened by suitable measures. A detailed description of the "wet-on-solid" production procedure is given in (Plückelmann et. al. 2017a). As an alternative approach, a so called "wet-on-wet" procedure was developed, which is illustrated schematically in Figure 1.

Contrary to the conventional production procedure for lining segments, the "wet-on-wet" procedure is characterized by the use of a vertical instead of a horizontal formwork. In a first step, the HPSFRC is cast in the vertical formwork and compacted to a certain predefined layer thickness (Figure 1, left). Directly afterwards, the two different concretes (HPSFRC and LC) are simultaneously cast in the formwork. However, the two concretes are firstly separated by moveable partition plates serving as an interior formwork. In this way, a layer of HPSFRC is molded along the longitudinal joints, whereby the thickness is adjustable by the appropriate positioning of the partition plates. During the compacting through vibrating, the partition plates are lifted up so that the concretes get in contact with each other in their fresh state without becoming mixed (Figure 1, middle). Finally, the upper layer is added so that all the outer edges of the segment are covered with HPSFRC (Figure 1, right). In contrast to the "wet-on-solid" procedure, the "wet-on-wet" procedure ensures a monolithic bonding due to a more or less simultaneous start of the hydration process of the two concretes.

3 LOAD SITUATION IN LOGITUDINAL JOINTS

Compressive forces in ring direction are transmitted through contact surfaces in the longitudinal joint between two adjacent segments. Figure 2 (left) shows the typical constructive design of plane longitudinal joints for segments with all-round gasket (Schmid & Colombi 2002). The contact surface (i.e. loaded area) is limited in order to prevent the introduction of compressive forces into the outer edges of the segment's cross section. The width of contact surfaces usually ranges between 1/2 and 1/3 of the total segment thickness (Maidl et. al. 2011). This situation of concentrated load introduction leads to a load diffusion along with a deviation of compressive stresses. As a result, tensile splitting stresses are generated, which can be

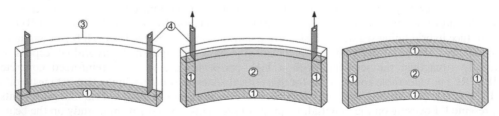

Figure 1. Schematic description of the "wet-on-wet" production procedure for hybrid lining segments (1: HPSFRC, 2: LC, 3: vertical segment formwork, 4: partition plate/interior formwork).

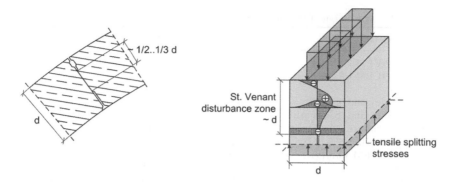

Figure 2. Typical constructive design of plane longitudinal joints (left) and simplified illustration of the load situation and resulting stress distribution (right).

decisive for the reinforcement design of the segment's longitudinal joints and thus have to be taken into account accordingly. A simplified illustration of the described load situation and the resulting characteristic state of stresses occurring at longitudinal joints is given in Figure 2 (right). As it is shown, zones directly adjacent to the loaded area are subjected to compressive stresses. However, due to the load diffusion, tensile splitting stresses are generated, which reach their maximum at some distance to the loaded area and spread along the through-thickness direction perpendicular to the load. With increasing distance to the loaded area, the tensile splitting stresses tend to diminish gradually. From a distance approximately equal to the thickness of the cross section, the state of stresses exhibits a uniform distribution. The zone within this distance is referred to as "St. Venant disturbance zone" (Wichers 2013).

Based on this load situation, it is not appropriate to reinforce the entire segment with an adequate high amount of steel fibers, rather only those zones where high splitting stresses are expected. Therefore, with regard to the design of hybrid segments, the thickness and position of the HPSFRC layer should be aligned to the shown distribution of tensile splitting stresses.

4 EXPERIMENTS

4.1 Materials

In this experimental study, hybrid specimens were produced using both plain concrete (PC) and HPSFRC. The mix designs (Table 1) are similar to (Plückelmann et. al. 2017b). The HPSFRC was produced using a steel fiber cocktail consisting of 60 kg/m³ hooked-end macro fibers (3D 80/60 BG) and 60 kg/m³ straight micro fibers (FM 13/0.19). The properties of the steel fibers are listed in Table 2. In Table 3 the fresh and mechanical properties of PC and HPSFRC are shown.

With regard to the bearing behavior of hybrid specimens under concentrated load, the tensile behavior of the HPSFRC is of major importance. However, as usual for SFRC the tensile behavior of the HPSFRC used in this study was characterized by performing four-point bending tests on beam specimens (150×150×700 mm³) according to (DAfStb 2012). Figure 3 shows the stress-deflection curves of the HPSFRC and PC. Furthermore, the stress-deflection curve of a conventional SFRC with a moderate fiber content of 50 kg/m³ (otherwise same mix design as the PC) is illustrated for comparison only.

4.2 Specimens

Hybrid concrete specimens with dimensions of 25×50×50 cm³ were produced ("wet-on-wet") in three different ways using either vertical or horizontal wooden formworks. In the first case (case A), the freshly mixed PC and HPSFRC were cast and compacted in layers one after the other. In the second case (case B), a vertical formwork was also used but here the two different

Table 1. Mix design of PC and HPSFRC.

Component		PC	HPSFRC
Cement (CEM I 42.5R)	[kg/m³]	320	-
Cement (CEM I 52.5R)	[kg/m³]	-	400
Fly ash	[kg/m³]	60	100
Aggregate (sand, gravel)	[kg/m³]	1880	-
Aggregate (sand, basalt grit)	[kg/m³]	-	1860
max. grain size	[mm]	16	8
Steel fiber (3D 80/60 BG)	[kg/m³]	-	60
Steel fiber (FM 13/0.19)	[kg/m³]	-	60
Water	[kg/m³]	155	154
$(w/c)_{eq}$	[-]	0.45	0.35
Superplasticizer*	[%]	0.4	1.5

Table 2. Properties of steel fibers.

Fiber	Shape	Length [mm]	Diameter [mm]	Aspect-ratio	Tensile strength
3D 80/60 BG	end-hooked	60	0.75	80	1225
FM 13/0.19	straight	13	0.19	68	2000

Table 3. Fresh and mechanical properties of PC and HPSFRC.

Properties		PC	HPSFRC
Consistency[a]	[cm]	42	40
Bulk density[c]	[kg/m³]	2.3	2.6
Air void content[b]	[vol.-%]	1.6	1.0
Compressive strength[d]	[MPa]	70	120
Splitting tensile strength[e]	[MPa]	4.0	7.8
Young's Modulus[f]	[GPa]	35	50

a according to EN 12350-5
b according to EN 12350-6
c according to EN 12350-7
d according to EN 12390-3
e according to EN 12390-6
f according to EN 12390-13

concretes were simultaneously cast in the formwork. However, they were firstly separated by a movable steel plate. During the compacting this steel plate was lifted up so that the concretes got in contact with each other in their fresh state. The third case (case C) is also characterized by the initial separation of the two concretes by a steel plate, however, in this case a horizontal formwork was used. Figure 4 (left) shows the horizontal and vertical wooden formworks with the moveable steel plate used for case B and C. It was found, that even for case B and C an accurate incorporation of a thin HPSFRC layer and adjacent PC can be achieved. As an example, Figure 4 (right) shows a sectional image of a hybrid specimen produced in a vertical formwork (case B) with a HPSFRC layer of approximately 10 cm.

4.3 *Scope of test program*

With regard to the load situation and the resulting distribution of tensile splitting stresses occurring at the longitudinal joints of segments (cf. Section 3), the thickness t of the HPSFRC

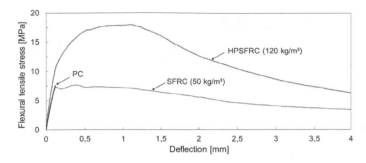

Figure 3. Stress-deflection curves of PC, SFRC and HPSFRC according to DAfStb (2010).

Figure 4. Formworks (vertical/horizontal) for the "wet-on-wet" production of hybrid specimens (left) and sectional image of a hybrid specimen with a HPSFRC layer thickness of approx. 10 cm (right).

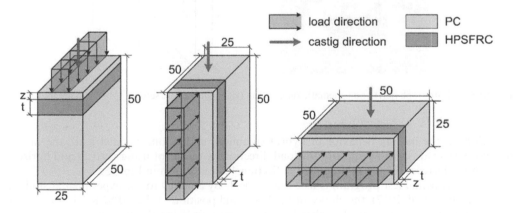

Figure 5. Casting and load direction for hybrid specimens with varying thickness and position of the HPFRC layer produced according to case A (left), case B (middle) and case C (right).

layer was varied in order to determine the optimal configuration of the hybrid fiber reinforcement. Due to the fact that zones directly beneath the loaded area are mainly exposed to compressive stresses, also the position z (i.e. distance to the loaded area) of the HPSFRC layer was varied. For hybrid specimens produced according to case B and C (cf. Section 4.2) where the HPSFRC layer was incorporated in between the PC ($z > 0$), the position and thickness of the HPSFRC was adjusted by using a second steel plate.

Experimental investigations by Breitenbücher et. al. (2014) and Song (2017) showed that the casting direction is strongly influencing the fiber orientation and thus consequently the structural performance of SFRC. As reported by the authors, fibers are preferably oriented perpendicular to the casting direction. In order to investigate the impact of the casting

Table 4. Experimental test program.

Series index	production according to	thickness t [cm] of HPSFRC layer	position z [cm] of HPSFRC layer
REF	-	-	-
A_t10_z0	case A*	10	0
A_t10_z5	case A*	10	5
A_t20_z0	case A*	20	0
B_t10_z0	case B*	10	0
B_t10_z5	case B*	10	5
B_t20_z0	case B*	20	0
C_t10_z0	case C*	10	0
C_t10_z5	case C*	10	5
C_t20_z0	case C*	20	0

* case A – C: cf. Figure 5 for corresponding casting and load direction

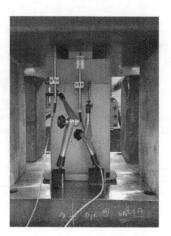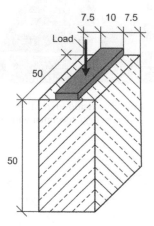

Figure 5. Photograph (left) and schematic description (right) of the test setup.

direction on the bearing behavior of hybrid specimens under concentrated load, the casting direction was varied with respect to the load direction by means of using vertical and horizontal formworks (case A - C), as described in Section 4.2 and shown in Figure 5.

The experimental test program is listed in Table 4. By analogy to the experimental study in (Plückelmann et. al. 2017), the choice of thickness and position of the HPSFRC layer is based on the analytical solution by Leonhardt & Mönning (1986) for the distribution of splitting stresses within a concrete element under concentrated load. However, it has to be noted that this solution is not quite exact for hybrid specimens due to different stiffness of PC and HPSFRC (cf. Table 3).

For comparison purpose and in addition to the hybrid specimens, also reference specimens containing only PC were produced and tested under the same testing conditions. For each test series listed in Table 4, three specimens were produced and tested.

4.4 Test setup

The tests were performed by using a universal testing machine with a maximum load capacity of 20 MN. Figure 6 (left) shows the test setup. The concentrated load (strip load) was transmitted onto the upper surface of the specimens through a steel plate. With respect to typical ratio between the width of the contact surface of longitudinal joints and the total segment thickness (cf. Figure 2, left), the width of the steel plate (i.e. loaded area) was chosen to 10 cm, which is corresponding to a ratio of 0.4. The length of the loaded area was equal to the length of the specimens (Figure 6, right). The load was applied continuously at a loading rate of 0.5 mm/min. In order to avoid stress concentrations due to surface roughness of specimens, hardboard strips were placed between steel plate and loaded area. Furthermore, a fast-curing, high-strength grouting mortar was poured under the specimens in order to prevent any inclinations. During testing, the vertical displacements of the machine and the specimens were measured using LVDTs.

5 RESULTS

5.1 *Ultimate local compressive stresses*

In Table 5 the mean values (M) and variation coefficients (CV) of the ultimate local compressive stresses (σ_{max}, defined as the ultimate load divided by the loaded area) are summarized. For comparison purpose, the increase in σ_{max} in relation to the reference specimens is indicated for each test series as $\Delta\sigma$.

Regarding the different cases for the production of hybrid specimens (case A – C), a considerable increase in the ultimate local compressive stresses was observed compared to the reference specimens (REF) with increasing thickness of the HPSFRC layer. For hybrid specimens produced in vertical formworks (case A and B) with a HPSFRC layer thickness of 20 cm the increase in σ_{max} was up to 90.3 % (A_t20_z0) and 87.5 % (B_t20_z0), respectively.

The downward relocation of the HPSFRC layer with a thickness of 10 cm (z = 5 cm) led also to a moderate increase in σ_{max} for all cases (8.1 % for case A, 10.7 % for case B and 2.8 %

Table 5. Results of concentrated loading tests.

Series index	M(σ_{max}) [MPa]	CV(σ_{max}) [%]	$\Delta\sigma$ [%]
REF	69.6	7.2	-
A_t10_z0	105.6	4.3	51.8
A_t10_z5	111.3	4.7	59.9
A_t20_z0	132.4	3.3	90.3
B_t10_z0	103.0	4.8	48.0
B_t10_z5	110.4	4.0	58.7
B_t20_z0	130.5	3.1	87.5
C_t10_z0	92.2	2.4	32.5
C_t10_z5	94.1	5.4	35.3
C_t20_z0	100.7	4.5	44.7

Figure 6. Fracture patterns of PC specimens (left), B_t10_z0 (middle left), B_t10_z5 (middle right) and B_t20_0 (right).

for case C). This can be explained by the fact that zones of the specimens directly beneath the loaded area are mainly exposed to compressive stresses (cf. Figure 2, right). According to the analytical solution by Leonhardt & Mönning (1986) for the distribution of tensile splitting stresses, the location of the maximum tensile splitting stresses has to be expected at a distance of approximately 10 cm to the loaded area. Based on this assumption, for specimens with a HPSFRC layer of 10 cm located directly beneath the loaded area ($z = 0$ cm), the maximum splitting stresses occur close to the transitional area from PC to HPSFRC. As a result, splitting tensile stresses which lead to a cracking in the PC cannot be prevented due to the absence of steel fibers. In contrast, for specimens with a distance of the HPSFRC layer of 5 cm to the loaded area, the maximum tensile splitting stresses should occur close to the middle of the HPSFRC layer so that the tensile splitting stresses generated in the adjacent zones of PC are comparatively smaller. This results in a higher utilization of steel fibers in the HPSFRC layer and thus to a higher effectiveness of splitting fiber reinforcement.

As shown in Table 5, specimens produced in horizontal formworks (case C) consistently exhibited significantly smaller ultimate local compressive stresses compared to specimens produced in horizontal formworks (case A and B) with otherwise same configuration of the HPSFRC layer. Compared to the reference specimens, the increase in σ_{max} for specimens produced in vertical formworks with a HPSFRC layer thickness of only 10 cm was 51.8 % for A_t10_0 and 48.0 % for B_t10_0 and therefore higher than the increase in σ_{max} of 32.5 % for C_t20_0, even though the HPSFRC layer for C_t20_0 was twice as thick. This clearly shows that the fiber orientation in the HPSFRC layer and thus consequently the bearing capacity of hybrid specimens is considerably affected by the casting direction with respect to the load direction.

As mentioned in Section 4.3 and reported by Breitenbücher et. al. (2014) and Song (2017), steel fibers are preferably oriented perpendicular to the casting direction. Because the tensile splitting stresses spread along the through-thickness direction of the specimens perpendicular to the load (cf. Section 3), a fewer number of fibers are aligned towards the direction of tensile splitting stresses for specimens produced in horizontal formworks (case C) compared to specimens produced in vertical formworks (case A and B). As is well known, fibers aligned to the acting direction of tensile stresses have the best crack-bridging capacity. As a result, the effectiveness of the HPSFRC layer served as splitting reinforcement was significantly lower for specimens produced according to case C. The ultimate compressive stresses of specimens produced according to case A and specimens produced according to case B (with otherwise same configuration of the HPSFRC layer) differ only marginally. In these cases, a similar amount of fibers was aligned in the direction of tensile splitting stresses.

5.2 *Fracture patterns*

For all specimens, no visible cracking or spalling occurred until shortly before reaching the ultimate local compressive stresses. When reaching the ultimate stresses, all specimens showed a more or less abrupt and brittle failure. In this study, a distinct ductile fracture behavior as it was reported for hybrid specimens subjected to point loading in (Song 2017) could not be observed. This is mainly attributed to the special case of concentrated loading investigated here (i.e. strip loading), which results in an incomplete effect of confinement.

However, the fracture patterns are clearly different depending on the presence and configuration of the HPSFRC layer. Figure 6 shows the fracture patterns of selected specimens after testing. The PC specimens (REF) are characterized by the formation of a splitting wedge directly beneath the loaded area and the complete splitting through the central axis in longitudinal direction (Figure 6, left). In contrast, all hybrid specimens stuck together even after failure occurred. Due to the crack bridging fibers in the HPSFRC layer, no complete splitting has occurred as it was observed for PC specimens.

Hybrid specimens with a HPSFRC layer thickness of 10 cm share a similar crack pattern independent of the position of HPSFRC layer. This crack pattern is also characterized by the formation of a splitting wedge. Two main cracks starting from the edges of the loaded area propagate diagonally downwards, until they merge into one vertical crack, which spreads down to the bottom of the specimens (Figure 6, middle left/middle right). While in zones of

HPSFRC no concrete spalling could be observed, zones of PC adjacent to the HPSFRC layer were characterized by major concrete spalling. However, resulting from the position of the HPSFRC layer, concrete spalling in the peripheral zones directly beneath the loaded area only occurred for specimens where the HPSFRC was installed as an intermediate layer between the PC.

Specimens with a HPSFRC layer thickness of 20 cm showed a significantly different fracture pattern compared to the other specimens. As shown in Figure 6 (right) no cracking or spalling occurred in the peripheral zones close to upper edge of the specimens. In this case, one main crack starting with a certain distance to the upper edge propagates vertically downwards to the borderline of PC and HPSFRC. From there the crack splits up and the two cracks spread out to the bottom of the specimens along with major concrete spalling.

Specimens with the same thickness and position of the HPSFRC layer showed a very similar fracture and crack pattern independent of the casting direction.

6 CONCLUSIONS

Focusing on the longitudinal joints, an experimental study on the bearing and fracture behavior of hybrid specimens, produced with a so called "wet-on-wet" production procedure, has been conducted under concentrated load (strip loading). The results of the experimental study presented in the present paper allow to draw the main concluding remarks listed in the following.

1. It was proven that the installation of a HPSFRC layer only in zones where critical splitting stresses occur leads to a remarkable increase in the bearable loads or ultimate local compressive stresses, respectively.
2. The results showed that the position of the HPSFRC layer with respect to the distribution of tensile splitting stresses has an impact on the bearing behavior of hybrid specimens. The downward relocation of the HPSFRC layer leads to an increase in the ultimate local compressive stresses. However, this improvement was more or less inconsiderable, especially because in this case major concrete spalling was observed in the peripheral zones of the specimens adjacent to the loaded area. As concrete spalling in the peripheral zones of segments should be absolutely prevented, the installation of HPSFRC as an intermediate layer is not to be recommended. Moreover, this would complicate the production process additionally.
3. The casting direction has a significant influence on the fiber orientation and thus consequently on the bearing behavior of hybrid specimens. The results showed that the use of vertical instead of horizontal formworks lead to a remarkable increase in the ultimate local compressive stresses. This finding is of great importance for further development of the "wet-on-wet" production procedure for hybrid lining segments because the use of a vertical segment formwork is indispensable for this procedure.

7 OUTLOOK

In order to further develop the "wet-on-wet" production procedure for hybrid lining segments and to investigate the practical feasibility in a realistic scale, the Ruhr-University Bochum recently purchased a vertical segment formwork. This formwork is shown in Figure 7 and allows to cast rectangular segments with an external ring diameter of 7.80 m, a thickness of 0.40 m and a width/height (length of segment in longitudinal direction of the tunnel) of 1.00 m in compliance with standard dimensional tolerances. The formwork is equipped with compressed air vibrators for the compaction of concrete. For the demolding of the segment, the formwork can be folded open on three of its four outer sides. In a next step, the formwork will be modified through the installation of moveable partition plates as shown in Figure 1 to enable the "wet-on-wet" production of hybrid lining segments on a large-scale.

Figure 7.   Vertical segment formwork available in the laboratory of Ruhr University Bochum.

ACKNOWLEDGEMENTS

The Authors would like to give their appreciation to the German Research Foundation (DFG) for the financial support of the subproject B1 "Optimized Structural Segments for Durable and Robust Tunnel Lining Systems" within the Collaborative Research Center (SFB) 837 "Interaction Modeling in Mechanized Tunneling". Furthermore, the authors want to thank Herrenknecht Formwork Technology GmbH for the design and manufacturing of the vertical segment formwork (cf. Figure 7) and their support.

REFERENCES

Breitenbücher, R. 2014. Spezielle Anforderungen an Beton im Tunnelbau. In Bergmeister, K, Fingerloos, F & Wörner, J.-D. (ed.), *Beton-Kalender 2014: Unterirdisches Bauen, Grundbau, Eurocode 7*: 393–422. Berlin: Ernst & Sohn

Breitenbücher, R., Meschke, G., Song, F., Hofmann M., & Zhan, Y. 2014. Experimental and numerical study on the load-bearing behavior of steel fiber reinforced concrete for precast tunnel lining segments under concentrated loads. *Fibre Reinforced Concrete: from Design to Structural Applications, proc. FRC 2014 Joint ACI-fib International Workshop, 24–25 July 2014, Montreal, Canada*: 431–443

Deutscher Ausschuss für Stahlbeton (DAfStb). 2012. DAfStb-Richtilie – Stahlfaserbeton: Berlin: Deutscher Ausschuss für Stahlbeton e. V. – DAfStb

Leonhardt, F & Mönning, E. 1986. Einleitung konzentrierter Lasten oder Kräfte. In Leonhardt, F & Mönning, E. (ed.), *Vorlesungen über Massivbau – Teil 2: Sonderfälle der Bemessung im Stahlbetonbau*: 67–121. Berlin: Springer

Maidl, B., Herrenknecht, M., Ulrich, M. & Wehrmeyer, G. (ed.) 2011. *Maschineller Tunnelbau im Schild-vortireb. 2nd Edition.* Berlin: Ernst & Sohn

Plückelmann, S., Song, F. & Breitenbücher, R. 2017a. Load-bearing and Bonding Behavior of Hybrid Concrete Tunnel Lining Segments. *Surface challenges – Underground solutions, proc. ITA-AITED World Tunnel Congress, 9–15 June 2017, Bergen, Norway*: 1311–1318

Plückelmann, S., Song, F. & Breitenbücher, R. 2017b. Hybrid Concrete Elements with Splitting Fiber Reinforcement under Two-Dimensional Partial-Area Loading. *High Tech Concrete: Where Technology and Engineering Meet; proc. fib symposium 2017, Maastricht, Netherlands, 12–14 June 2017.* Springer International Publishing AG: 347–355

Schmid, L. & Colombi, L. 2002. Designing segmental Joints in extreme Conditions. *Tunnel 3/2002*: 28–36.

Song, F. 2017. Steel fiber reinforced concrete under concentrated load. *Dissertation, Institute for Building Materials, Department of Civil and Environmental Engineering*, Bochum: Ruhr University Bochum

Wichers, M. 2013. Bemessung von bewehrten Betonbauteilen bei Teilflächenbelastung unter Berücksichtigung der Rissbildung. *Dissertation, Institute of Building Materials, Concrete Constructions and Fire Protection*, Brunswick: University of Brunswick

Tunnels and Underground Cities: Engineering and Innovation meet Archaeology, Architecture and Art, Volume 6: Innovation in underground engineering, materials and equipment - Part 2 – Peila, Viggiani & Celestino (Eds)
© 2020 Taylor & Francis Group, London, ISBN 978-0-367-46871-2

Relieving shotcrete and other linings from pressure with innovative drainage solution

J.R. Poulsen
Dolenco Tunnel Systems ApS, Copenhagen, Denmark

ABSTRACT: A tunnel lining is often submitted to hydrostatic pressure. This solution can be installed behind the tunnel lining, for example a fiber reinforced shotcrete layer or spray membrane. The patented network drain module is encapsulated in the concrete, so the lining will be monolithic. This makes it possible to design a slim structure, with a network of drain channels inside. Once installed, it does not have a limited life span. Low LCC makes it very cost-effective. Quick and safe repair after physical damage requires minimal downtime of tunnel. The solution is installed without the use of anchorage and constitutes no fire hazard. The design is made from recycle material or sustainable biodegradable material and is therefore a solution for the future.

1 INTRODUCTION

Constructing a tunnel is seen as one of the most complex challenges in the field of civil engineering. Many tunnels are recognized as technological masterpieces. More than 20 years in the business of waterproofing of concrete structures made us aware of a demand for a solution, which addresses the many difficulties concerning water pressure on the tunnel structures.

It is broadly understood that water can cause major damages, because of its ability to find its way through concrete. Water is pressured into and through the concrete, which has the following consequences: The water pressured through the concrete structure can cause dripping onto trains, vehicles, and installations. Also, dripping onto roads poses a risk to the traffic — especially formation of ice on the road and icicles falling from the ceiling. Furthermore, the water pressured into, and accumulating inside, the concrete, causes deterioration by corrosion and freeze/thaw. Also, damages caused by fire leads to spalling and decreased structural strength and in worst case the collapse of the tunnel.

Figure 1. Icicles damage to concrete structure (photo: Washington State Dept. of Transportation).

The need for injection work, more shotcrete layers and new anchorage, and difficult working conditions are some of the common problems, which can be minimized or avoided.

Repairs of damaged tunnel structures and maintenance are complicated and costly, including the socioeconomic effects. Closing the tunnel, locating all the damages and fully repairing the tunnel with today's solutions, can take up to a year. This is expensive, time-consuming and completely unnecessary with the innovative Dolenco Tunnel Drainage System, which can provide a safe repair within days of an existing Dolenco solution.

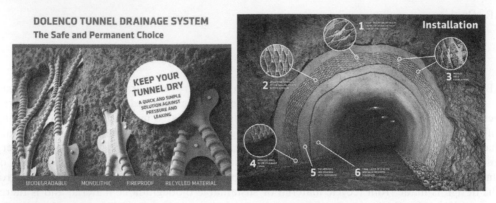

Figure 2.   Dolenco Tunnel Drainage System (DTDS) and Installation Guidance of DTDS.

The present paper is organized as follows. Section 2 outlines examples of solutions available to cope with water pressure on structures in tunnels today and the individual challenges of the solutions. Section 3 provides information on how to implement DTDS into the existing solutions and the effects on the tunnel solution. The advantages are portrayed from a technical, economical and environmental point of view. Finally, Section 4 summarizes the paper with conclusions.

2   THE STANDARD SOLUTIONS AVAILABLE

2.1   Examples of standard solutions

As with all underground structures it is important to drain tunnels. The solutions today are complicated, costly, time-consuming, and they cause significant disturbances to traffic. This, amongst other challenges, has been reason to innovate the standard drainage method and develop alternative systems. Presented below are some of today's standard solutions with individual challenges compared to the Dolenco Tunnel Drainage System.

Solution 1a. Blasting/boring/digging the tunnel, pre-injection, adding about 20–30 mm shotcrete for stabilizing the surface, and then a final layer of 80 mm shotcrete (e.g. fiber reinforced).

Challenges: High risk of water pressure damaging the structure. Even though it is hard to tell if there is any pressure, it is still the case that pressure builds up over time enlarging the risk of damage. The shotcrete does not remove the water pressure, which is accumulating over time.

Solution 1b. Blasting/boring/digging the tunnel, pre-injection, anchorage, adding about 20–30 mm shotcrete for stabilizing the surface, covering the surface with a membrane, and then a final layer of 80 mm shotcrete (e.g. fiber reinforced).

Challenges: High risk of water pressure damaging the structure. Even though it is hard to tell if there is any pressure, pressure will build up over time enlarging the risk of damage. The membrane is meant to ensure that no water comes through, but that only addresses the issue of water entering the tunnel and not the water pressure problem.

Solution 2. Blasting/boring/digging the tunnel, pre-injection, anchorage, about 20–30 mm shotcrete for stabilizing the surface, mounting of membrane sheets bolted to the surface of outer shotcrete lining, and then a final lining of 80–300 mm reinforced shotcrete.

Challenges: It is a complicated solution resulting in a medium to long-term risk of damages through errors. Also, the membrane sheets have a limited lifespan and are made of flammable material increasing the risk of severe damage. The typical damages from fire are spalling and scaling. In case of fire, the air pockets between the structure and the membrane sheets work as canals so the fire spreads faster. Finally, the risk of damage from water pressure is still enlarged over time since the solution does not solve this challenge.

Solution 3. During boring of tunnels (e.g. TBM) unforeseen water ingress is seen from the ceiling and the sides. Usually, the solution is that an extra layer of shotcrete is applied for better working conditions. This does not necessarily solve the problem entirely.

Challenges: The extreme working conditions sometimes makes it difficult or impossible to manage the water seeping and flowing into the tunnel during construction.

Solution 4. Sheet piling for shafts and retaining walls for stations etc.

Challenges: The water can find a way between the sheet piles. The pressure prevents the concrete from hardening properly, and water continues to work its way into and through the concrete. Injection of the cracks is time-consuming, technically challenging and can be costly.

Solution 5. Secant piling for shafts and retaining walls for stations etc.

Challenges: The water can find a way through and between the secant piles. The pressure prevents the concrete from hardening properly, and water continues into and through the concrete. Injection of the cracks is time-consuming, technically challenging and can be costly.

Figure 3. Leaking secant piling and sheet piling (photo: Aarsleff).

3 APPLICATION AND ADVANTAGES

3.1 *Specifications*

Dolenco Tunnel Drainage System is used for draining structures. It leads water out of the structure with a minimum of space required. It prevents water build-up and pressure on the structure and can drain up to 8 qbm water per linear meter per hour. The module is made of HDPE, which is recyclable and known for its durability. It measures 800 x 1200 x 14 mm. The modules are mounted in extension of one another with an easy click system, nailed to the surface with a nail gun, covered with a layer of shotcrete, and then covered with a layer of specially designed shotcrete (e.g. fiber reinforced).

3.2 *Examples of implementing the Dolenco Tunnel Drainage System (DTDS)*

In Section 2, a few examples of standard solutions were discussed. To solve the challenges of the standard solutions, DTDS is the safe and permanent choice.

Solution 1a and 1b. To drain the water away from the structure DTDS is mounted on the stabilizing layer of shotcrete with a nail gun from top to bottom. The modules are covered with shotcrete, and then finally a lining of 6–10 cm specially designed shotcrete is added. As a result of installing the modules, the water will be drained to the bottom of the tunnel, reaching the drainage canals, and hereby permanently preventing pressure build-up on the structure.

Advantages: The water is not pressured into and through the concrete lining, which prevents damages caused by the water. This makes it an economical and technical sustainable solution.

Solution 2. As in Solution 1a and 1b the water is drained away from the structure by mounting the DTDS on the stabilizing lining of shotcrete with a nail gun from top to bottom. Then it is possible to use DTDS both as an addition as well as a substitute to the membrane sheet to minimize the risk of damages to the structure.

Advantages: Reducing risk of medium to long-term deterioration, damage and repair. Adding DTDS will relieve the membrane from the water pressure ensuring a longer lifespan. The DTDS solution is a simpler solution, decreasing the risk of errors. This leads to longer lifespan and less need for repairs during maintenance.

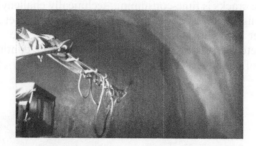

Figure 4.   Shotcrete used to cover drainage mats of foamed polyethylene.

Solution 3. To temporarily drain the water away from a TBM tunnel during construction, shotcrete is added, the DTDS is mounted from top to bottom with a nail gun, and a layer of shotcrete is applied to cover the modules. This will divert water to the sides and down.

Advantages: The water is drained away, allowing construction work to be continued.

Solution 4. When the sheet-piling wall is installed, DTDS is mounted directly onto the surface with a nail gun. Then either shotcrete is sprayed onto the surface or a geotextile is mounted before casting concrete.

Advantages: This is a very simple solution that ensures no water pressure as the water is drained away behind the concrete wall.

Solution 5. When the secant-piling wall has been cast, DTDS is mounted directly onto the wall and a layer of specially designed shotcrete is applied.

Advantages: This is a very simple solution that ensures no water pressure as the water is drained away behind the concrete wall.

Avoiding or minimizing post-injections and repairing, Dolenco Tunnel Drainage System is a safe and permanent choice. It is a quick and simple solution against pressure and leaking, resulting in keeping the tunnels dry. Once installed, the Dolenco Tunnel Drainage System does not have a limited lifespan, and it improves the tunnel durability. This is compared to the standard solutions that have a lifespan of 20–60 years. DTDS is suitable for both new and existing structures, and the slim and monolithic design can be installed with very little space available. The solution requires minimal downtime of tunnel and is a quick and safe repair of concrete after physical damage.

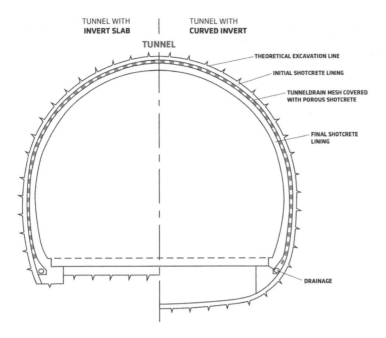

Figure 5.   Technical drawing of DTDS implemented in tunnel with invert slab and curved invert.

The modules can be made from HDPE as well as from a recycled material, which is 100 percent biodegradable over time. There is no need for waste separation, making it easy to recycle the concrete and a part of a true circular economy solution. Not only does it have a low LCC making it very cost-effective, it is also CO_2-neutral.

If blockages occur, On Demand Tunnel Servicing is available. Firstly, the tunnel is scanned systematically to identify any blockages and standing water. Secondly, blockages are safely removed by learning the channel mechanically without applying pressure. Finally, the reinstated function of the drain can be checked visually.

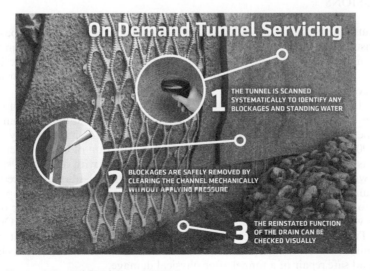

Figure 6.   On Demand Tunnel Servicing of DTDS.

Comparing Dolenco Tunnel Drainage System with conventional drainage systems of today is complicated, because it does not directly substitute the solutions available today. It is an addition to existing solutions, because it improves the overall tunnel solution in a variety of ways. A comparison has been made to a standard solution, and it is expected that the cost reduction of construction is 40 percent and the overall cost reduction 55 percent. Also, there is an energy reduction of 40 percent when switching from a conventional drainage system to DTDS. The life expectancy of the drains is estimated to 120 years compared to conventional drainage systems.

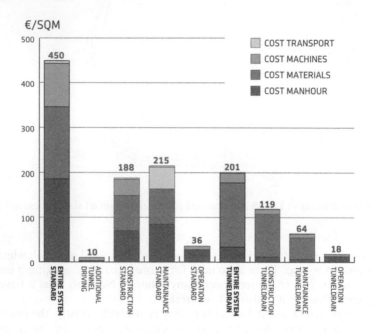

Figure 7.   Standard vs. TunnelDrain (LCC).

4   CONCLUSIONS

Finding a solution to pressure and leaking is important. Today's solutions can be complicated, costly and insufficient or non-existing. Most common damages in tunnels include: Water inflow, accumulation of water, and cracking and scaling/spalling of the surface. The solutions available today do not address all the challenges of water pressured into and through the concrete by the accumulated water pressure.

To counter these challenges, the Dolenco Tunnel Drainage System has been developed. It prevents water build-up and pressure on the structure and can drain up to 8-qbm of water per linear meter per hour. The module is made of HDPE, known for its durability, and it measures 800 x 1200 x 14 mm. The modules are mounted in extension of one another with an easy click system, is nailed to the surface with a nail gun, covered with a layer of shotcrete, and then covered with a final layer of specially designed shotcrete (e.g. fiber reinforced).

Dolenco Tunnel Drainage System provides a solution that is tailored to the specific project. It is suitable for both new and existing structures, and the slim and monolithic design can be installed with very little space available — even on wet and seeping structures. Once installed, the Dolenco Tunnel Drainage System does not have a limited lifespan, and it improves the tunnel durability. The solution requires minimal downtime when repairing tunnels, and therefore a quick and safe repair of a tunnel after physical damage.

The Dolenco Tunnel Drainage System does not directly substitute the standard solutions available today. It is an addition to existing solutions, because it improves the overall tunnel solution in a variety of ways. It is expected that the cost reduction of construction is 40 percent and the overall cost reduction is 55 percent compared to standard solutions available today. Also, there is an energy reduction of 40 percent when switching from a conventional drainage system to Dolenco Tunnel Drainage System.

Besides the HDPE version, the modules can be made from a recycled and biodegradable material, which degrades 100 percent over time. There is no need for waste separation, making it easy to recycle the concrete and a part of a true circular economy solution. Not only does it have a low LCC making it very cost-effective, it is also CO_2-neutral. If blockages occur, On Demand Tunnel Servicing is available.

Constructing a tunnel is seen as one of the most complex challenges in the field of civil engineering, and Dolenco Tunnel Drainage System is an innovative solution, which addresses the many difficulties concerning water pressure in an easy, safe, permanent and cost-effective manner. This is the solution that will help to create true technological masterpieces.

REFERENCES

Aram, M. 2016. Armenian and European Methods of Tunnel Waterproofing. *International Journal of Research in Chemical, Metallurgical and Civil Engineering* (ISSN 2349-1442 EISSN 2349-1450 IJRCMCE) Volume 3, Issue 1.

Funahashi, M. PE. 2013. Corrosion of Underwater Reinforced Concrete Tunnel Structures. *MUI International Co. LCC.*

Poulsen, J.R. 2018. Dolenco Tunnel Drainage System. *www.TunnelDrain.com*

Russell, H.A. 2008. Guidelines for Waterproofing of Underground Structures. *Parson Brinckerhoff.*

Verya, N., PhD, PE. 2016. Waterproofing and Final Tunnel Lining. *AECOM/University of Colorado*: 46–66.

Tunnels and Underground Cities: Engineering and Innovation meet Archaeology,
Architecture and Art, Volume 6: Innovation in underground engineering,
materials and equipment - Part 2 – Peila, Viggiani & Celestino (Eds)
© 2020 Taylor & Francis Group, London, ISBN 978-0-367-46871-2

Service limit state design for pressurised steel fibre reinforced concrete tunnel linings

S. Psomas

Morgan Sindall Engineering Solutions Limited, Rugby, UK

ABSTRACT: The use of steel fibre reinforced concrete (SFRC) for tunnel linings has been established in the UK for more than 20 years. Although the Ultimate Limit State (ULS) of steel fibre reinforced concrete (SFRC) can be verified in a number of ways, the design analysis for Service Limit States (SLhS) lacks a widely-agreed set of design rules due to the difficulty of estimating crack widths. A method of analysis is presented based on design principles established by EN 1990 and the fib Model Code 2010. The analysis presented here relates mean crack width, curvature and tensile strain to circumferential hoop force and bending moment, for a given crack width limit. The latter is an important parameter to satisfy the durability performance. The development of post-cracking inelastic behaviour is described, and its use for prediction of redistribution of moments prior to plastic hinge formation is discussed. This is illustrated by an outline of the current design practice for cast in-place (CIP) SFRC tunnel lining (pressurised sewer tunnels) for two sewer projects in London (Lee Tunnel and Thames Tideway West).

1 INTRODUCTION

The use of steel fibre reinforced concrete (SFRC) for tunnel linings has been established in the UK the last 25 years. One of the main benefits of using steel fibre-only reinforcement instead of traditional steel rebars, is the elimination of the risk of long-term corrosion resulting in an enhancement of structure durability. However, there is no established method of design analysis under the Service Limit State format that has ever been incorporated in Design Standards and the current design rules for SFRC tunnel linings are generally absent in the main international structural design codes, even though Model Code 2010 (fib, 2013) include procedures that can be adopted (with some interpretation) for the design of tunnel linings. Furthermore, there has been - for some time now - a number of European Guidelines, all of which although useful for SFRC design, are not intended for use by tunnel lining designers. The work exposes that stronger evidence from tests and site measurements is needed, to streamline the design of SFRC tunnel linings and reduce the embedded over-conservatism.

1.1 *Problem definition*

It is often assumed that the behaviour of a structure can be approximated by the behaviour of a test beam or slab. This assumption is markedly restrictive for SFRC tunnel linings and leads to a significant underestimation of the actual resistance of the lining structure –which is an indeterminate structure - as opposed to the resistance of a beam. In fact, a linear elastic approach cannot properly take into account the beneficial effects of fibre reinforcement which become effective only after cracking of the concrete matrix when SFRC behaviour is significantly non-linear. Additionally, very little can be found in technical literature relevant to the Service Limit State design and particularly crack width control for SFRC structures. Fibre reinforcement is known for improving crack propagation resistance, promoting multiple-crack development and therefore reducing the permeability of concrete. This is of particular significance for tunnel lining applications in hydraulic tunnels where the linings are potentially in moderate tension. Then the durability of SFRC concrete depends upon the crack width and the ability to self –heal, preventing the ingress of water.

2 SECONDARY LININGS IN PRESSURE TUNNELS

2.1 Project information and requirements

The design methodology presented here has been adopted for the Main Tunnel Secondary Lining for Lee Tunnel Project and Thames Tideway West Contract. The 7.2m ID Lee Tunnel Secondary Lining is discussed as being the first SFRC tunnel lining designed to satisfy SLS (crack control) using a steel fibre-only reinforced concrete mix. Thames Tideway West contract adopts the same solution and is to start construction in 2019. The Thames Tideway Lee Tunnel was a £635m project situated in East London and constitutes the first phase of the £4bn Thames Tideway Tunnel, which ultimately aims to capture the untreated sewage discharged into the River Thames and also to serve as storage and transfer tunnel from Acton Storage Tanks (west) to Beckton Sewage Works (east). Lee Tunnel Underground Civil Works comprised of a 6.9km long, 7.2m ID tunnel and 5no. deep (up to 80m) shafts. The Tunnel is constructed at a mean depth of 75m from ground surface, predominately in water bearing Chalk. The tunnel lining system consists of the 7.8m ID, 350mm thick, primary bolted segmental precast SFRC lining and the 300mm thick secondary cast-in-place (CIP) SFRC lining. The Thames Tideway West project features a 7km 6.5m ID tunnel and 7No. shafts up to 25mID and 50m deep.

2.2 Design Approach and requirements

The Tunnel Primary and Secondary Linings (TPL and TSL respectively) performance requirements - set out in the Works Information (WI) - include a design life of 120 years and a limited infiltration criterion set to 0.1lt per square metre of tunnel lining per day (without running water). Thames Water's deep Sewer Tunnels in London are characterised by the relative high internal surge pressures (up to 8bar) and external groundwater pressures (up to a depth of 75m). If a steel bar RC had been adopted for the TSL, very onerous crack widths limits (up to 50microns) would be required to comply with EN 1992-3.

Performance based specification allowed the Designer to adopt the SFRC solution allowing 'controlled' cracking for the secondary linings, avoiding thousands of tonnes of steel rebar. The Design Verification was carried out according to EN 1990 Principles, 'Design assisted by testing' (EN 1990 Section 5.2) and the partial factor of safety format.

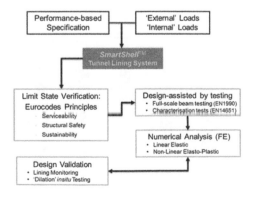

Figure 1.   Adopted Design Approach ('design –assisted by testing').

The dominant design situation is the serviceability during the surge event. The tensile strains in the lining will exceed the elastic tensile strain capacity of the lining resulting in the development of a number of cracks of limited width. In these tunnels the Primary lining is always in compression (confined by the geostatic compressive stress) whereas the Secondary lining during surge events goes into tension and expands. Provided that cracks are controlled and there is no admissible mechanism of failure, the verification of the design is governed by the SLS. The design philosophy requires the primary and secondary linings to work as a combined lining system to ensure compliance with the requirements of the WI. The Primary Lining provides the watertightness of the system and the secondary lining is required to ensure that if when cracked during transient events (surge pressure), the cracks remain within the self-healing limit (less than 0.15mm). The verification required therefore is full 'thickness' structural testing to be performed. This type of testing ensures the ductility of SFRC at full thickness (deflection-hardening), a sound representation of fibre distribution and the elimination of scale effects. Testing was carried out at Building Research Establishment (BRE) at Watford, UK.

2.3 SFRC Characterisation - Small Beams

Analytical Verification of full section tests was carried out to assess the ductility in relation to the standard small scale beam tests. The results of tests of the SFRC material on notched beams to BS EN 14651 are given in Table 1.

The strength indices are calculated from the bending moment at the notch using $f = M/bh^2$ with $h = 125$ mm. This ignores the effect of stress concentration at the notch. It is allowed for later in the design method of the Model Code 2010, which defines the tensile strength for crack widths up to 0.5 mm by $f_t = 0.45 f_{R1}$. This ratio is examined elsewhere [Johnson et al, 2017]. An isotropic distribution of fibres is assumed reasonable for cast linings, so the orientation factor of clause 5.6.7 of the Model Code is taken as 1.0. It is evident that the ductility ratio (fib, 2013) as defined through the ratio of f_{R3} to f_{R1} (~1.25) demonstrated that deflection-hardening behaviour can be achieved:

The strength indices can be classified as C60-6d to fib, 2013. They have only been used to characterise the SFRC material (Ductility class) but not for the derivation of the design strength of the lining. This compares well with the yield strength derived from the large scale tests as it is demonstrated hereinafter.

Table 1. SFRC notched beam test results in accordance to BS EN14651.

Beam specimen origin	f_{R1k}/f_{Lk}	f_{R3k}/f_{R1k}
Trials (test batches)	0.8	1.4
Works (shutter)	0.9	1.3

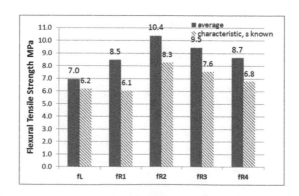

Figure 2.   Strength indices to EN14651 from 19no. tests at 56 days (Works).

2.4 Full section thickness testing (BRE)

In order to accurately measure crack widths and strains developing in concrete and steel fibre-reinforced concrete beams under load, digital photography and Particle Image Velocimetry (PIV) techniques were used to estimate crack widths and total strain. The experimental set-up is shown in Figure 3:

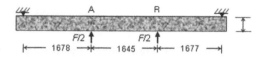

Each beam was supported on two jacks, 1645 mm apart. It was held down at points about 0.2 m from each end, spaced 5.0 m apart. It thus had a simply-supported span of 5 m, tested upside-down to enable cracking in the almost constant-moment region AB to be observed.

Figure 3. Experimental set-up & Side View sketch.

Each beam was supported on two jacks, 1645 mm apart. It was held down at points about 0.2 m from each end, spaced 5.0 m apart. It thus had a simply-supported span of 5 m, tested upside-down to enable cracking in the almost constant-moment region AB to be observed.

A row of 120-mm fibre-optic Bragg gauges (FOBG) was attached in a line 1.20 m long along the centre-line of the top surface. Each recorded the extension of a 120-mm length. The mean tensile strain over length AB (Figure 2), ε_{av}, was found from the sum of their readings.

During PIV analysis, the side surface of the beam would be divided into a grid of squares – or Interrogation Areas (IA) – each of which would be tracked by the software over the series of photographs to determine its location. Texture in each IA was created on the surface of the beam by adding patterns (colours and/or shapes). One plain concrete (specimen 1) and three SFR concrete beams (specimen 2.1, 2.2 and 2.3) were tested to failure. Photographs (16Mp) were taken at an uneven rate throughout the tests using remote control at load increments until 80% of the predicted failure load, after which continuous shooting at a rate of 2 frames per second was used. The load was applied by way of four load cells situated underneath the beam, at the edges of the field of view. As the load was increased, the beam was held in place by mild steel restraints placed above the beam at either end and held in place with screws. It must be noted that strains were calculated at points in time that corresponded to each crack. These points in time were found by identifying when the load on the beam (in the load-displacement curve) dropped suddenly. More details about PIV processing and analysis of these particular test results can be found in Hover et al, 2015.

2.5 Design assisted by testing

The RILEM TDF162 (RILEM, 2003) and Concrete Society TR 63 (CS, 2007) define different 'depth factors' for reducing tensile strengths in beams more than 0.125 m deep. These presumably allow for the less steep strain gradient in a deeper beam. Considering f_{R1} as the strength related to SLS then it was found that the fit to the test M-ε_{av} curve (see Figure 8 below) had a good fit. The results given later adopt the factor $k=0.88$, which relates to size effects and is explained elsewhere (Johnson et al, 2017)

$$f_t = 0.45 f_{R1}k \qquad (1)$$

It was assumed in calculations that for concrete in compression, linear-elastic behaviour up to 25 MPa was followed by plastic deformation at constant stress.

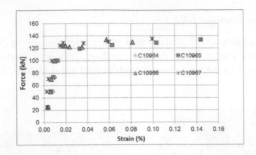

Figure 4. Full scale test results force vs total average strain.

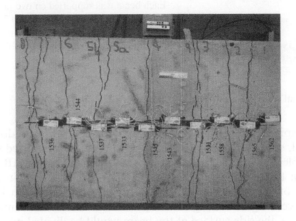

Figure 5. Crack pattern on top surface of beam 2.3 (#C10967) – Note the FOBG array on top.

Table 2. Test Results when the first crack was observed.

Test No	M-kNm	$f_{t,e}$-MPa	ε_{av}-%
Specimen1: #C10964 (plain) -	62.0	4.30	0.0116
Specimen2: #C10964 2.1 (SFRC)	82.4	5.72	0.0173
Specimen3: #C10964 2.2 (SFRC)	82.3	5.71	0.0198
Specimen4: #C10964 2.3 (SFRC)	82.8	5.75	0.0149
Mean(SFRC)	82.5	5.73	0.0173

Before the jacks had lifted the beam into contact with its end restraints, the dead weight F was 38.1 kN, the weight of the beam and its attachments, and the bending moment at A and B (Figure 3) was calculated. Figure 5 shows the constant-moment region of beam 2.3, the FOB gauges, and the cracks at the ultimate strain – maximum bending moment. The sequence in which the cracks appeared is given by the sequences of times at which a crack reached a given width and was determined by the PIV method. The sequences (Figure 5) for example at w=0.15mm was 7 8 4 1 6 2 3 5 whereas at w=0.25mm: 1 2 7 3 6 8 5 4. This sequence suggested that the cracks opened at different rates; for example, much faster for crack 1 than for crack 4. The same effect was observed in the other two specimens.

From the test results in Table 2 for the three SFRC beams the mean load at first crack, F = 123.9 kN, corresponds to a mean bending moment of 82.48 kNm at loading points. From this and the dead weight reading at F = 38.1 kN, the relation between force F and the bending moment per unit width, M, was found in kN, m units to be:

$$M = 0.845 \, (F - 22.3) \tag{2}$$

Elastic moduli were found from the results in Table 2, as

$$E = f_{t,e}/\varepsilon_{av}. \tag{3}$$

From the mean strength from tests 2.1 to 2.3, the secant tensile modulus is

$$E_t = 5.73\text{MPa}/0.173e - 2 \; = 33\text{GPa}. \tag{4}$$

The ratio of flexural strength to direct tensile strength is a function of the section thickness (fib, 2013):

$$A = \left(f_{ct.}/f_{ct,fl}\right) = \left(0.06h^{0.7}\right)/\left(1 + 0.06h^{0.7}\right) \tag{5}$$

Therefore for structural thickness of 250mm (as assumed in this case) the ratio is about 0.74 and the mean tensile strength is:

$$f_{ctm} = f_{t,e}A = 5.73\text{MPa}^*0.74 = 4.2\text{MPa} \tag{6}$$

This is similar to what **BS EN 1992-1.1** predicts for a concrete C50/60. The important aspect is that this corresponds to full scale bending tests. However, in order to carry out any plastic FEA a 'yield' residual strength needs to be derived. Therefore, for 3No. tests a coefficient $k_n = 3.37$ (as per BS EN 1990, assuming a coefficient of variability Vx unknown) is used and the strength is calculated, based on the mean strength (variability is less than 5% up to strains of 0.1%), as:

$$f_{ctd,s} = f_{t,e}(1 - k_n V_x)/\gamma_{m,\text{sls}} = 5.73^*0.45^*(1 - 3.37^*0.05)\text{MPa}/1.0 = 2.1\text{MPa} \tag{7}$$

ULS design strength assuming a material factor of $\gamma_{m,uls} = 1.50$ is:

$$f_{ctd,u} = f_{ctd,s} / g_{m,uls} = 1.4\,\text{MPa} \tag{8}$$

2.6 Design analysis

The design analysis of the lining was facilitated by the adoption of non-linear FE analysis. The purpose of this was to capture the ductility and the residual tensile capacity of SFRC, which in this case is necessary to verify SLS during surge events. The stress-strain curve of the SFRC material (based on test response see Figure 4) was adopted so that the critical strain could be derived. The methodology of FEA and the numerical modelling is not part of this Paper. The important aspect to be noted is that this numerical analysis enabled the derivation of the critical (mean) cracking strain after the stress redistribution. This is a necessary stage in order to determine the critical strain required for the crack width estimation. The design analysis methodology can be summarised as follows:

1. Derive load deflection vs time and load vs deformation curves from PIV analysis;
2. Deduce from data post-processing analyses the total tensile strains;
3. Determine, from the FEA analysis of the lining, the critical strain distribution;
4. Produce the strain versus time graph and identify the critical strain limits (Figure 6);
5. Calculate the strains (PIV analysis) and estimate the average crack widths (PIV and FOBG).

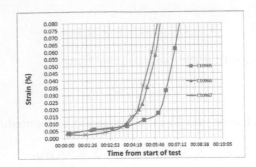

Figure 6. Strain development with time from the full scale tests.

The large scale tests assisted in designing the tunnel lining for SLS, limiting crack widths to 0.15mm. This design methodology enables the development of an analytical design approach for SLS verification (see below), which can be used for future applications.

3 SERVICE LIMIT STATE VALIDATION

A relatively simple model was developed to validate this design (Johnson, 2014): Figure 7 shows a length s of a member of rectangular cross-section of unit breadth and depth h, subjected to bending moment M. There is a single crack, of width w, shown at the bottom surface. The assumed crack spacing is s. At distance $s/2$ from the crack the behaviour is elastic, with stresses as in Figure 7 (a), with:

$$6M = f_{t,e}h^2 = E h^2 \varepsilon_{av}, (b = 1\text{m})$$ (9)

For serviceability verification, crack width is always less than 0.5 mm. If the 'linear model' of the Model Code (fib, 2013) is used, it gives the stresses in Figure 7(c), with f_t from Eq. (1). Here, for prediction of test results, f_t is based on the mean value of f_{R1}. Its characteristic value is used in design. For simplicity the tensile region of the uncracked concrete up to its cracking stress (the dashed lines) has been ignored as this complicates the analysis and barely affects the results. The objective is to obtain curves of bending moment in terms of mean tensile strain, ε_{av}, for several assumed numbers of cracks in the almost constant-moment region in the tests, of length 1.645 m; that is, for given ratios s/h, where s is the crack spacing. This is done for one, three and six cracks, so the ratios s/h are 5.48, 1.83 and 0.91. It is recommended (Johnson et al, 2017) that for fully-developed cracking, crack spacing can be taken as the crack depth, which will be less than 0.91 h. The ratios here represent progress in the tests towards fully-developed cracking, but do not reach it.

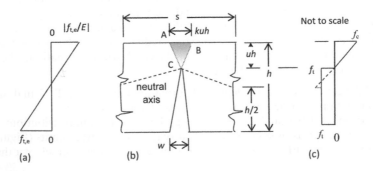

Figure 7. Rectangular cross section subjected to bending moment M (Johnson, 2014).

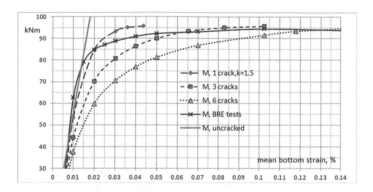

Figure 8. Mean test results, and predicted bending moments as a function of mean tensile strain.

The key assumption of this 'hinge' theory is that the maximum compressive stress is constant at f_t over some length $[k\,u\,h]$, where k is to be determined, and $u\,(= x/h)$ is the non-dimensional depth in compression at the crack. The compressive stress falls to $|f_{t,e}|$ at the ends of length s.

The extension of length CD in Figure 7(b) is:

$$k\,u\,h\,f_c/E,\tag{10}$$

so the hinge rotation is

$$(k\,u\,h\,f_c)/(E\,u\,h)\tag{11}$$

It is also the angle between the faces of the crack,

$$w/[h(1-\,u)].\tag{12}$$

Equating these rotations gives:

$$w = k\,h(1-u)f_c/E\tag{13}$$

The steel fibres are closely spaced, so at the face of the crack, the stress f_t is transferred to the concrete within a short distance of the crack. The surface tensile strain is assumed to be f_t/E at the crack and $f_{t,e}$ at the ends of length s. These strains are similar (in contrast with f_c/E), so a linear variation can be assumed between them, over length $(s-w/2)$.

Assuming that crack width w is much smaller than s, and using Eq. (13), the mean tensile strain over length s is given by:

$$\varepsilon_{av} = (w/s) + \left[(6M/h^2 + f_t)/2E\right]\tag{14}$$

For longitudinal equilibrium at the crack,

$$f_c\,u\,h/2 = f_t\,h(1-\,u)\tag{15}$$

At the crack, the bending moment is given by

$$12M/h^2 = f_c u(3-2u) + 6f_t u(1-u)\tag{16}$$

2905

Elimination of w, M and f_c from eqs (13) to (16) leads to

$$u^3 + 2u^2 - 4K_1(1-u)^2 + 2K_2\varepsilon_{av}u - 4u = 0 \qquad (17)$$

where

$$K_1 = kh/s \qquad (18)$$

$$K_2 = E/f_t \qquad (19)$$

This is solved for assumed values for ε_{av}; then w, M and f_t are found. This gives $M(\varepsilon_{av})$ for any assumed number of cracks (s/h) and ratio k. The best match with the curve from the tests was found with $k = 1.5$. Elimination of u and f_c from Eqs (13), (15) and (16) gives a curve M (w) which is independent of the crack spacing, as it is based only on the model at the crack. It is given in Figure 9.

The mean curvature φ_{av} can be found by assuming that the depth of member in tension, d, varies linearly from $h(1-u)$ at the crack to $h/2$ at the ends of length s. Hence,

$$d_{av} = h(3 - 2u)/4 \qquad (20)$$

and from Eq. (14):

$$\varphi_{av} = \varepsilon_{av}/d_{av} \qquad (21)$$

The rotation θ of length s is then:

$$\theta = \varphi_{av}s + w/[h(1-u)] \qquad (22)$$

The curves shown in Figure 8 (Johnson, 2014) were calculated for a beam 1 m wide using these values: $h = 0.300$ m, $f_t = 1.54$ MPa, $E = 33$ GPa, $k = 1.5$ and ratio of crack spacing s to thickness h as 5.48, 1.83 and 0.914. Interpolation between them can be used. The 'uncracked' line in Figure 8 is based on the secant modulus at 'first cracking', that is, when a crack becomes wide enough to be noticed. The predicted curve, 'one crack', meets the test curve at $\varepsilon_{av} = 0.02\%$, or 200$\mu\varepsilon$, slightly above the expected cracking strain for normal concrete of this strength, but very close to the mean (173$\mu\varepsilon$) observed in BRE tests.

4   DISCUSSION

The curve of Figure 9 enables crack widths to be predicted based on the total strain. It shows that the bending moment continues to increase until the mean width reaches about 0.5 mm; an important result because up to the serviceability limit, likely to be 0.15-0.20mm or less,

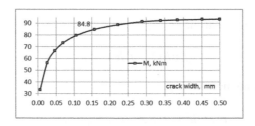

Figure 9.   Predicted bending moment as a function of mean crack width, for any number of cracks.

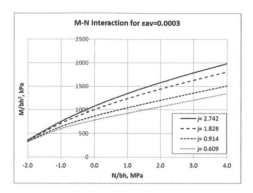

Figure 10. M-N interaction diagram for a different number of cracks (-ve is tension).

behaviour is clearly deflection-hardening and therefore stable. The analysis above can be extended to include circumferential tension or compression, and for crack widths up to 2.5 mm (ULS). The stress fc reaches the assumed 'yield' value of 25 MPa, when the bending moment reaches 91.3 kNm, after which yielding has been allowed for.

Yielding begins to affect the curves shown, but it is still small at 0.5 mm. At a crack width of 0.5 mm, the predicted mean strain is 0.0019, an unserviceable situation, which this model finds to be far short of an ultimate limit state. It is interesting to note that one 0.5-mm crack in a length of 270 mm increases its mean curvature to 3.7 km^{-1}. In a tunnel lining this great increase would cause redistribution away from the hinge region. The model predicts that the bending moment is still rising very slowly, at 93.5 kNm.

Further expansion of this approach can be achieved by introducing the axial force N and limiting the strain to the chosen critical value (0.03% for this application based on FEA). Then different response curves can be derived for different number of cracks (Where $j = s/h$).

Notwithstanding this it is encouraging that the methodology above can be used to predict the M-N response for an assumed number of cracks. This model, using mean strains and uniform widths of uniformly-spaced cracks, cannot represent the complexity and variability of cracking in a set of three beams. However, it predicts their average behaviour. These findings are based on a single series of tests, with one fibre concrete mix and one type of fibre (40kg/m^3, Bekaert 5D - 2300MPa yield strength). Further analysis on the statistical distributions of crack width or spacing was not undertaken.

5 CONCLUSIONS

Hydraulically pressurized tunnels can be designed as steel fibre reinforced structures. The principle of design assisted by testing methodology was adopted for the Thames Lee Tunnel Project and currently for the Thames Tideway West Project. The design verification for the SLS required full section depth testing and based on the test results, the average structural crack width is expected to not exceed the 0.2mm limit. These tests demonstrated that SFRC can yield deflection-hardening behaviour within the anticipated working strain and measurable ductility (multi-cracking) at ultimate strain. Additionally, an analytical model is presented to validate the tests and attempts to predict behaviour . In view of the extensive use of this material, the scarcity of reported research on these matters is remarkable. The model presented here fits the test data well, but needs to be tried out in tests for a variety of concrete mixes and fibers.

ACKNOWLEDGEMENTS

The Author would like to thank Prof. R.P. Johnson for his fruitful contribution in developing the validation model as a result of a number of stimulating discussions. It is important also to acknowledge MVBJV and Project Management Team at Lee Tunnel for their support all along Lee Tunnel project as well as the structures lab staff at BRE for their excellent execution of a challenging set-up. Special thanks also to certain individuals whose contributions were substantial and essential to the success: Dr Eyre Hover, who carried out the PIV analysis, Prof C. M. Eddie for initiating and promoting the design of SFRC linings, Mr M. Rimes (Materials Manager) and Dr V. Astrinidis (FOBG measurements).

REFERENCES

British Standards Institution 2005. EN 14651: 2005. 'Test method for metallic fibered concrete- Measuring flexural tensile strength'. London, BSI

British Standards Institution. 2004. EN 1992-1.1: 2004. 'Eurocode 2: Design of concrete structures' & 'National Annex for EC2'. London. BSI

British Standards Institution. 2005. EN 1990 (A1): 2005. 'Eurocode – Basis of structural design' & 'National Annex for EC0'. London, BSI

British Standards Institution. 2006. EN 1992-3: 2006. 'Eurocode 2: Liquid retaining structures'. London, BSI

CS, 2007. Concrete *Society Report TR63 - Guidance for the Design of steel fibre reinforced concrete*, CS;

fib, 2013. *Model Code for Concrete structures 2010* - Report FIB

Hover, E., Psomas, S. and Eddie, C.M., 2015. Estimating crack widths in steel fibre-reinforced concrete, *Proceedings of the Institution of Civil Engineers-Construction Materials*:1–12

Johnson R.P., 2014. Review of PN48 – unpublished Technical Notes, August

Johnson R.P., Psomas S., Eddie C.M., 2017. Design of steel fibre reinforced concrete tunnel linings, *Proceedings of the Institution of Civil Engineers-Structures & Buildings*, Volume 170 Issue 2, February:. 115–130

Psomas S., 2016. *SFRC segmental lining design for a pressure tunnel* in WTC 2016, San Francsisco

Rilem, 2002. 162-TDF. Test and design methods for steel fibre reinforced concrete. *Materials and structures*, 35: 579–582.

Notation (based on EN 1992-1.1 and the Model Code 2010)

E effective Young's modulus for concrete in compression

F total force in the two loading jacks

L length of region of almost constant bending moment

M bending moment in a member of unit width at a cracked cross-section

b breadth of member, taken as unity and omitted from most equations

f_c maximum compressive stress in the concrete at a cracked cross-section

f_{cy} compressive 'yield' strength of concrete

f_{R1} stress calculated from $f = 6 \, M/bh^2$ from a standard test on a notched beam 0.15 m deep

f_t assumed flexural tensile stress across a crack at the neutral axis

$f_{t,e}$ extreme fibre stress for an elastic uncracked cross-section

h depth of a member or test specimen

k hinge length factor

s length of beam affected by a crack, and assumed final crack spacing

u non-dimensional ratio x/h

w width of a crack at the surface in tension

x depth of compression zone at a crack

ε_{av} mean strain at the surface in tension over length L

φ_{av} mean curvature of length L of the beam

Tunnels and Underground Cities: Engineering and Innovation meet Archaeology,
Architecture and Art, Volume 6: Innovation in underground engineering,
materials and equipment - Part 2 – Peila, Viggiani & Celestino (Eds)
© 2020 Taylor & Francis Group, London, ISBN 978-0-367-46871-2

Design and construction of permanent steel fibre reinforced sprayed concrete lining shafts for Thames Tideway West Project UK

S. Psomas, P. Coppenhall & M. Rimes
Morgan Sindall Engineering Solutions Limited, Rugby, UK

D. Brown & E. Cheevers
BMB JV, London, UK

ABSTRACT: Thames Tideway West Project involves the construction of 7No deep shafts, one 7km Main Tunnel and 4No. connecting Tunnels through London Strata. Of particular interest are the Shafts, which are designed as composite linings with Steel Fibre Reinforced (SFR) Sprayed Concrete Lining (SCL) as Primary Lining and Cast In Place (CIP) SFR Concrete as Secondary Lining. Emphasis is given to the design and construction of the largest SCL shaft of the West contract (also the largest SCL Shaft ever constructed in the UK), at Carnwath Road Riverside (up to 25m ID, 50m in depth), from which the Launching of the Tunnel Boring Machine (TBM). The design philosophy required the development of a special SFR SCL mix with enhanced water resisting capacity to enable the omission of waterproofing membrane from the works. This Paper discusses the design philosophy, the extensive SCL trials, the construction challenges and the monitoring results thus far.

1 INTRODUCTION

1.1 *General*

The quality and performance of critical city infrastructure, such as Thames Tideway is essential for supporting economic growth and productivity in a city, such as London, in the 21st century. The construction of the new 'super' sewer aims to enhance city infrastructure and make it resilient to changing patterns. It also needs to be constructed efficiently in terms of cost, low carbon footprint and service performance.

The West Contract of Thames Tideway Project includes the 6.5m ID, 7km long Main Tunnel alignment and a number of associated structures between Acton Storm Tanks (ACTST) and Carnwath Road River side (CARRR) worksites. This also includes the Frogmore Connection Tunnel (FRGCT) that links the King Georges Park (KNGGP) and Dormay Street (DRMST) worksites to the Main Tunnel. The Main Works Contractor, BMBJV (Balfour Beatty-Morgan Sindall-BAM Nutall Joint Venture), has appointed Morgan Sindall Engineering Solutions Limited (MSES) for the design of the Tunnels and Shafts structures. The majority of the shafts feature double linings, a Primary Steel Fibre Reinforced (SFR) Sprayed Concrete Lining (SCL) and a Secondary Cast In Place (CIP) Steel Fibre Reinforced (SFR)/Steel Bar Reinforced Concrete (RC) Lining.

This Paper focuses on the Shaft Lining design philosophy and construction aspects of the largest Shaft (CARRR) in the Tideway West Contract, which is also the largest shaft ever constructed in the UK in 'soft' ground. The CARRR shaft (as per all shafts) includes internal Vortex structures, high level platforms and other features to facilitate their hydraulic function. A Building Information Modelling (BIM) model has been created to enable effective design

Table 1. Thames Tideway West Shafts – general information.

Shaft (acronym)	Internal Diameter (m)	Excavation Depth (below GL) (m)	Primary Lining Thickness (mm)	Secondary Lining Thickness (mm)
CARRR	25	49.5	SCL: 575 to 1050	SFR/RC: 600
HAMPS	11	35.5	SCL: 425	SFR/RC: 325
BAREL	6	36.0	SCL: 325	SFR/RC: 250
ACTST	15	35.7	SCL: 400	SFR/RC: 400
DRMST	12	26.5	SCL: 325 to 750	SFR/RC: 300
KNGGP	9	23.3	SCL: 325	SFR/RC: 500
PUTEF	6	38.3	RC (caisson): 725	

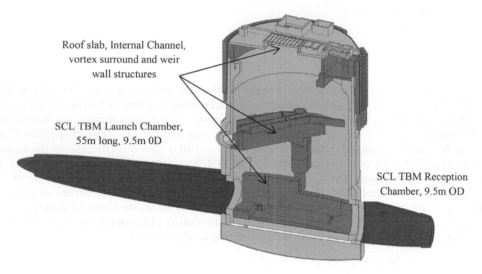

Roof slab, Internal Channel, vortex surround and weir wall structures

SCL TBM Launch Chamber, 55m long, 9.5m 0D

SCL TBM Reception Chamber, 9.5m OD

Figure 1. A detailed BIM model was created including all permanent works elements.

delivery, updated construction information and provides a framework for efficient asset management and maintenance.

1.2 Carnwath Road Riverside (CARRR)

The Carnwath Road Riverside Worksite site located adjacent to the tidal River Thames in the London Borough of Hammersmith and Fulham. The CARRR shaft is a tunnel drive and reception shaft for the main alignment tunnel and shall receive flow from the main tunnel as well as from Frogmore Connection Tunnel.

1.3 Ground and groundwater conditions

Ground conditions are typical of the London Basin with made ground and soft deposits overlain the medium dense Terrace Gravel and London Clay Formation, in which the majority of the shaft excavation took place. The shaft base slab construction is in the variable Lambeth Group, which is characterized by mixed ground conditions.

Thanet Sand is underlain by the Chalk formation (+32mATD). There are three distinct aquifers; Upper (Terrace Gravels), Medium (London Clay and Upper Lambeth Group) and Lower (Upnor, Thanet Sand and Chalk). See Figure 6 for a geological long-section.

Table 2. CARRR Geological and Groundwater Conditions.

Soil layer	Top Level (mATD)	Groundwater (kPa)	Soil Shear Strength (phi)	Soil Stiffness (MPa)	Permeability (m/day)
Made Ground	105.5	0	30	10	0.864
Alluvium	102.0	-5	25	5	0.864
Terrace Gravel	101.0	-10	35	35	8.64
London Clay	94.5	-25	25	72	8.64 x10-6
Harwich	59.5	-320	27	72	0.864 x 10-3
USB/UMB	59.0	-380	27	80	0.864 x 10-3
LMB Silt	52.5	-445	35	80	8.64x10-3
LMB clay	49.0	-480	27	80	0.864 x 10-3
Upnor	45.0	-410	40	80	0.363
Thanet Sand	39.0	-460	40	200	0.950

2 DESIGN

2.1 Design philosophy

The design philosophy for the SCL tunnels and shafts encompasses all the necessary provisions to satisfy the Works Information requirements and also minimises construction, operation, maintenance and decommissioning Health and Safety risks. In order to fulfil the above all lining works are designed in accordance with the principle that the Primary and the Secondary Lining work compositely.

The Primary Lining is installed taking the initial ground and water loading, therefore ensuring the Secondary Lining installation is carried out stress-free. The inevitable additional external pressure resulting from clay consolidation, variations in the underground water, seepage, surcharge changes, creep etc., lead to the development of an additional hoop force. Under this load, the Primary Lining deforms and activates the Secondary Lining, resulting in the two linings working together in a combined manner. The two linings are bonded with friction, adhesion and interlocking at the concrete interface. See Figure 2 below:

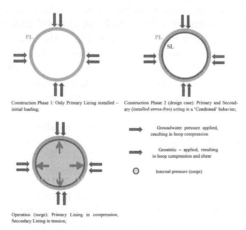

Construction Phase 1: Only Primary Lining installed – initial loading;

Construction Phase 2 (design case): Primary and Secondary (installed stress-free) acting in a 'Combined' behavior;

→ Groundwater pressure applied, resulting in hoop compression

→ Geostatic – applied, resulting in hoop compression and shear

○ Internal pressure (surge)

Operation (surge): Primary Lining in compression, Secondary Lining in tension;

Figure 2. The composite behaviour of the lining in the temporary and permanent condition.

2.2 Performance requirements

The Works Information specifies the performance requirements which form the key criteria for the SCL and can be summarized as follows:

1. Design Life 120 years
2. Limited water ingress/egress (damp patches no running water), on average 0.1 l/m²/day ingress permitted
3. Design for a maximum internal grade line of 104mATD.

2.3 Design analysis

2.3.1 Design cases

During operation, the shafts will be filled with sewage at regular intervals and then emptied, according to the operational requirements. Consequently, based on the above description, two principal Design Cases determine the design limit states for the shaft:

SHAFT EMPTY: Highest external groundwater level with no internal pressure. The external ground water and earth pressures oppose the internal pressure, and therefore the worst loading case is when the shaft is empty. This state of stress represents the worst load condition for the primary lining and the portal structures, the maximum bending moments and stresses are anticipated to develop in the structure, which govern the shaft design. The ground water level is taken as the maximum level.

SURGE CONDITION: Lowest characteristic external groundwater level, highest internal water level (104mATD). That means the internal pressure at base slab level will be a 45m head.

The design has been facilitated by the use of Non-linear Finite Element (FE) analysis. To that end PLAXIS 2D and Strand7 have been used in 2D and 3D analyses respectively.

Design of SCL has been carried out considering limited relaxation in soft ground resulting in early loading of the lining. Therefore, the SCL will provide early age support of the ground, minimising the settlement and optimising safety. Furthermore, critical intermediate stages of construction are modelled and their structural stability is checked.

For the STR/GEO Ultimate Limit State verification of the shaft, the BS EN 1997-1:2004 Design Approach 1 has been adopted in line with UK National Annex. Two calculations (Combination 1 for STR limit state and Combination 2 for GEO limit state) are hence performed for both the empty shaft and the surge design conditions. Within the context of the FE design analysis, the partial factors in Combination 1 are applied to the effects of actions (forces/bending moments are multiplied by a partial factor – greater than 1.0 - to yield design values). It is however difficult to apply Combination 2 within the framework of FE design analysis, as factoring the soil strength often results in the soil exhibiting unrealistic yielding behaviour. Therefore, Combination 2 is used to explore the sensitivity of the structure to extreme situations such as low soil softening and seepage analysis through the shaft lining. In these cases, the partial factors for actions in Combination 2 are applied directly to material properties and actions. For the excavation stability, the excavation stages were checked carrying out a C-Phi reduction analysis and couple stress-flow analysis using PLAXIS (FE modelling).

Lining thicknesses were verified for Ultimate and Serviceability conditions according to BS EN 1992-1-1:2004+A1:2014 and where SFR Concrete elements were included according to *fib* Model Code 2010 (fib, 2013).

2.3.2 Shaft watertightness

The Primary Lining and Secondary Lining will work in a combined manner in order to provide the necessary watertightness. The Primary Lining will operate in full hoop compression and hence no tension is expected to develop within it, whilst the Secondary Lining is partially in tension under surge conditions. The required degree of watertightness will be achieved by adopting a combination of measures, including the combined action of Primary and Secondary Lining. The 'Primary' measures include the Primary SCL acting compositely with the CIP

Secondary Lining along with selection of a suitable concrete mix. The concrete mixes specified ensure the structural elements are of sufficiently low permeability, which include the use of high performance steel fibres to control crack widths, and with the adoption of staggered joint construction for the Primary Lining further help control the watertightness. To enhance the watertightness of the concrete, in particular at joints, the Primary Lining SCL mix will incorporate a waterproofing admixture where a crystallisation agent (additive) is used to promote self-healing within the concrete after placement. Compliance with infiltration and exfiltration through the shaft lining will be confirmed by numerical seepage (flow) analysis.

2.4 Durability assessment

The durability assessment of the structural elements (permanent works) required extensive analysis, especially for the Secondary Linings durability modelling according to the *fib* standards. For SCL all potential deterioration mechanisms have been considered in relation to the exposure conditions in the ground. Prevailing ground and groundwater conditions were assessed and the ground classified as DC3 in all strata and DC2 in the London Clay Formation. For the base slab the exposure condition warrants Concrete Class DC3 plus the use of an additional sacrificial cover, otherwise known as an Additional Protection Measure (APM).

The maximum crack width requirement is primarily related to the exposure conditions, magnitude of tensile stresses and the self-healing capacity of concrete. To meet durability requirements a crack width of 0.3mm is adequate for the exposure conditions assessed. Corrosion is not a consideration of SFR Concrete Linings, because they are in compression during the operational life of the structure. For Secondary Linings which contain fluid pressure, the structural crack widths have been limited to 0.2mm at the surface. This ensures the risk of reinforcement (where it is specified) corrosion with adequate cover will be minimal. At this level, crack widths enable the self-healing of concrete in a wet environment, albeit this is not relied upon in the design.

In order to enhance durability a sacrificial layer (APM) of 75mm for the Primary Lining and 150mm for base slab is provided. This layer is not considered within the structural thickness during design calculations. Additives are also included in the SCL mix in order to enhance interlayer and concrete matrix impermeability.

2.5 SCL development & trials

Significant effort was invested by MSES and BMBJV to design and develop the appropriate SCL mix specifically for the Tideway project. To this end early trials have been carried out to design the SCL Mix. Experience from previous infrastructure projects proved invaluable when designing the mix.

Pre-trials carried out at the supplier's facilities established a C35/45 mix (CII-D+SR to BS 8500, with chemical class DC-2 '+' APM) available for use on the Tideway West Contract. The mix was designed to satisfy the design specification (compressive strength, residual flexural strength, permeability, workablity). The target water cement ratio was set to 0.48 and the cementious content to 450 kg/m^3. The maximum coarse aggregate size was 6mm (granite).

High performance hooked end steel fibres (1800MPa yield strength, 65 aspect ratio) were used to enhance the short-term ductility and crack control capacity of the mix. Subsequent pre-commencement trials were also carried out to test the spraying equipment and the SCL performance (encapsulation) when applied against reinforcement.

It is common knowledge that the mass permeability of the SCL is very often controlled by the permeability of their construction joints; hence this is why a crystallisation agent (Xypex C500NF) admixture has been used to enhance the impermeability of the SCL joints. Although, the addition of this agent is not a substitute for a good quality dense concrete mix, the lab tests demonstrated that the addition of crystallisation agent greatly enhances the mass and the joint impermeability after water penetration tests were carried out.

Figure 3. Sprayed Beams and panels prepared for testing.

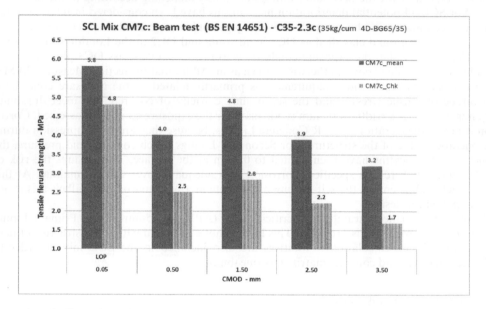

Figure 4. The flexural tensile strength of the SCL Mix (from the works).

Tensile Strength (fib, 2013) is based on residual strength indices from the BS EN 14651 test. The Characteristic tensile flexural strength at $CMOD$ = 0.5mm and $CMOD$ = 2.5mm specified f_{R1k} = 2.3MPa and f_{R3k} = 2.1MPa, respectively. The benefit of enhanced tensile capacity of the SCL is only considered during the early age and short-term conditions i.e. during construction. Therefore, SCL linings are assumed to have zero tensile strength in the persistent design situations. Where reinforcement is required to resist induced tensile stresses (e.g. at portal openings), steel bar reinforcement has been provided.

3 SHAFT SCL CONSTRUCTION

3.1 *Shaft basal depressurisation*

Depressurisation was required in the Lambeth Group for CARRR Shaft during excavation and SCL works. The purpose of the depressurisation is to ensure that pore pressures in the permeable layers are reduced sufficiently to avoid risk of basal heave.

At CARRR the standing groundwater level in the Clay within the UMB (Upper Mottled Beds) of the Lambeth Group (95.0mATD head recorded at a depth of 53-

55.0mATD) has been shown to be above the lowest excavation level (56.05mATD). Measurements from additional boreholes revealed that the standing water level at the top of coarse-grained Upnor Formation (42mATD) was at approximately 83.0mATD and therefore would generate an uplift pressure greater than the resistance of the fine-grained Lambeth Group 'plug'. This difference in pressure suggested that there was no certain hydraulic connectivity between the strata where the piezometer readings were taken and that the low permeability Clay in the LMB (Lower Mottled Beds) of the Lambeth Group in between was acting as an aquiclude.

A depressurisation plan was created in order to reduce the pore water pressure in the lower aquifer (in the confirmed coarse-grained strata below 45.0mATD). There was also potential water bearing soil layers suspected from Site Investigation data at the top of the Harwich Formation (59.5mATD - 59.0mATD) and a Silt layer at the top of the Lower Mottled Beds of the Lambeth Group (53.0mATD - 52.16mATD) which required investigation to confirm the pore water pressure and any hydraulic connectivity to the aquifer. The ground conditions and water pressures were verified from additional boreholes as part of the depressurisation works.

Figure 5.   Passive wells layout within the shaft.

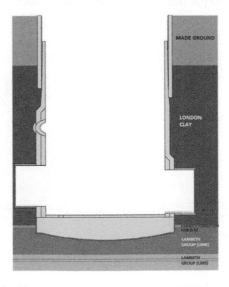

Figure 6.   Geological Long-Section.

The depressurisation plan required deep wells in Thanet Sand, avoiding disturbing the Chalk, and was carried out in two stages:

Stage 1: 5No. Active External wells installed from ground level:

4No Thanet Sand Wells were installed and had a response zone from 44.5 to 34.5mATD and were actively pumping throughout excavation with the use of a pump installed at 36mATD. The flow for each well did not exceed 1.5l/s. 1No Well in Upnor Formation was also installed, which comprised a narrower response zone between 45.0-40.0mATD. The Upnor Well had a pump installed at 41mATD and yielded 0.7 l/s. All Wells at this stage were drilled from ground level and activated early on during construction; this enabled a drawdown in the shaft to a GWL of less than 75mATD to be achieved and subsequently allowed the drilling of internal wells.

Stage 2: Development of 7No. internal relief wells to depressurise the Upnor Formation and Thanet Sands.

These wells were developed at the locations shown in Figure 5 from a ground level of 78.5mATD and had a response zone between and 45.0mATD and 37.0mATD. During the initial pumping test phase the internal wells were flow tested and two wells, PRW4 and PRW6, were selected to be actively pumped with a pump installed at 37.5mATD and 38.5mATD respectively. During the course of the excavation flow from these wells did not exceed 2l/s. 5No of the remaining Wells acted as passive relief Wells, PRW1,3,5,8,9. Wells PRW2 and PRW7 were unsuccessful and subsequently were abandoned.

Furthermore, 2No. internal relief wells were installed, ILG1 and ILG2, with a response zone between 63.0mATD and 52.0mATD. These wells intended to relieve the potential excessive flows in the Harwich Formation and Upper Mottled Beds.

Overall the scheme was largely successful and enabled the construction of the base slab.

3.2 Shaft excavation

The shaft was constructed using the sprayed concrete lining (SCL) method. This well-established method comprises of staged excavation of the shaft using mechanical excavation and the application of sprayed concrete to the excavated surfaces to provide ground support and create the lining thicknesses.

Due to the size of the shaft the thickness of the Primary Lining tapered up to a maximum of 1050mm. In order to construct these abnormally large thicknesses the sprayed concrete had to be sprayed in 3No. separate layers with each layer being 350mm thick. A Sealing Layer (also acting as the APM) was incorporated in to the first Layer, which helped provide the immediate ground support.

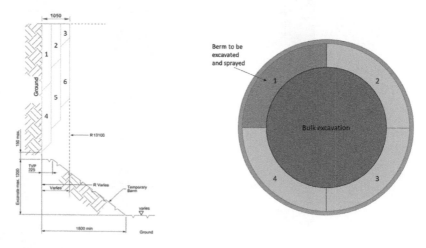

Figure 7.   Section View of Shaft Lining and Plan View of Construction Sequence.

In order to help control the ground movements the amount of the ground excavated was limited by two main criteria:

The first was the advance depth; this was limited to a maximum of 1.2m high. This ensured a reasonable advance height and controlled the amount of open ground.

The second one was the excavation and spray sequence. The sequence started by bulk-excavating the centre of the shaft and leaving a berm in place to provide temporary support and prevent softening of clay. Each Berm was then removed and subsequently sprayed (the first layer only) for a section of the shaft (such as a quadrant). Once the first layer was complete the second and third layers were individually sprayed fully round the circumference of the shaft in order to minimise the number of vertical joints. This methodology was developed over years of experience of SCL construction in soft ground. In all cases staggered joints were formed.

Figure 8.   Spraying operation - building the second layer.

3.3   Base slab

The 2800m³ base slab was designed with a flat top and domed base. It was designed as a slab of varying cross section to resist the uplift pressures generated from the groundwater and long term geostatic pressures. The slab was designed assuming no structural rotational or horizontal connectivity with the Primary and Secondary Linings. This consequently removed the need to provide reinforcement across the joint and hence provided benefits in terms of simplifying the connection details, improving the durability and watertightness. The joint is in compression and the rotations being transferred into the lining wall are limited; this removes restraint to horizontal movements (expansion/contraction) within the base slab. FE thermal modelling was carried out to capture the heat of hydration development.

3.4   Launch and reception chambers

The shaft consists of three chambers, one chamber is for the launch of the main Tideway West Tunnel Boring Machine (TBM) and the other two chambers allow for the reception of the TBMs (one smaller TBM from the Frogmore Connection Tunnel at a high level and one from the Tideway Central Contract). The chambers are up to 10m wide and 55m long, plus each chamber will have a sprayed concrete steel bar reinforced portal constructed prior to the chamber construction. The bar sizes, spacing and layering takes into account constructability considerations for encapsulation of reinforcement using the SCL method. The use of reinforcement couplers are employed where appropriate to assist with constructability of the larger portals during staged construction.

3.5 *Ground movements and monitoring*

Extensive monitoring has been carried out at surface. The overall movement due to excavations was within the predictions, which were derived from empirical and FE Analysis capturing the effects of construction activities, excavation, soil consolidation and dewatering. The SCL deformations were within the 'Green' Trigger value (radial - 5mm) with one exception at the south side at mid shaft level, where 'Amber' Trigger levels were reached (radial - 8mm).

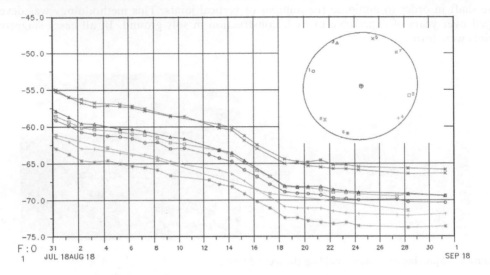

Figure 9.   Surface settlement variation with time.

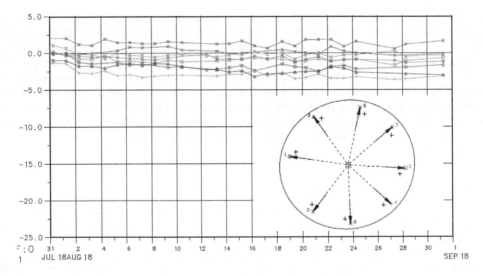

Figure 10.   Radial lining displacement variation with time, at lower level.

4 CONCLUSIONS

The Thames Tideway West project has constructed the largest ever built SCL shaft in the UK in soft ground. A new SFR mix with enhanced watertight features has been adopted and successfully applied by controlling the construction sequence. Ground movements and lining deformations were within the predicted values.

ACKNOWLEDGEMENT

The Authors would like to thank BMBJV for their support and implementation of an innovative design and challenging construction, as well for their permission to publish this Paper. Furthermore, the Authors are also grateful to our current and ex-colleagues in Morgan Sindall Engineering Solutions Limited involved in the design development

REFERENCES

BS EN 1990 (A1): 2005. *Eurocode 0 – Basis of structural design*, BSI
BS EN 1992-1.1: 2004 (2014). *Eurocode 2: Design of concrete structures*, BSI
BS EN 8500-1 & 2: 2015. *Concrete- Complimentary to BS EN 206-1*, BSI
BS EN 14651: 2005. '*Test method for metallic fibered concrete- Measuring flexural tensile strength*', BSI
fib, 2013. *FIB Model Code for Concrete Structures 2010*. (2013). Berlin: Ernst & Sohn.

Tunnels and Underground Cities: Engineering and Innovation meet Archaeology,
Architecture and Art, Volume 6: Innovation in underground engineering,
materials and equipment - Part 2 – Peila, Viggiani & Celestino (Eds)
© 2020 Taylor & Francis Group, London, ISBN 978-0-367-46871-2

De-bonding of shotcrete linings at the substrate boundary

S.G. Reid
University of Sydney, Australia

E.S. Bernard
BG&E, Sydney, Australia

ABSTRACT: This paper examines a series of scenarios in which possible failure mechanisms involving shrinkage-induced cracking, delamination, and creep of a shotcrete lining are assessed to identify the possible causes of a recent shotcrete lining collapse observed in a Sydney tunnel. Derivations of structural behaviour based on shrinkage, creep, cracking and de-bonding, have been developed from basic engineering principals. Typical geometries for recently built linings in Sydney have been used to determine which of several postulated scenarios are likely to actually occur. This has been done by comparing tensile stresses generated within the lining and at the boundary with the substrate against estimated tensile and adhesive bond strengths. The results indicate that drying shrinkage cracking is highly likely at the exposed surface of a lining, and that these cracks are likely to induce de-bonding of the lining from the substrate in the immediate vicinity.

1 INTRODUCTION

In 2014 a section of the Lane Cove Western Vent Tunnel in Sydney suffered a failure in which a portion of the Steel Fibre Reinforced Shotcrete lining collapsed (Parsons Brinckerhoff, 2014). The area of fall-out extended to several square metres of lining in the crown in an area with a lining thickness of 50–125 mm. This collapse occurred despite the lining being less than 10 years old in ground considered to be stable at the time of excavation. The steel fibres within the shotcrete matrix proved to be ineffective in holding the lining in place. Limited ground instability was noted, but some degradation in rock mass class was observed due to an apparent increase in weathering beyond expectations. Nevertheless, it appeared that the instability in the lining was not related to ground movement, so several alternative scenarios have been considered to explain this early-age failure. These have centred on shrinkage cracking and delamination of the Steel Fibre Reinforced Shotcrete (SFRS) lining from the ground and potential instability of the lining. Detailed analyses of this problem were presented by Reid and Bernard (2016).

1.1 *The structure*

The structure is an arched Steel Fibre Reinforced Shotcrete (SFRS) lining of uniform radius and thickness supposedly bonded to the excavated rock at all points across the span. The 'abutments' of the arch comprised slightly enlarged 'elephants feet' that were intended to transfer axial compression from the arch to the adjoining ground. The radius was quite large (12230 mm) giving rise to a shallow arch as is common in most Sydney road tunnels. The SFRS lining is a permanent lining, sprayed in one pass, with additional shotcrete sprayed over bolt heads resulting in an irregular surface profile. Rock bolts were located at regular intervals of 1.5 m across the span of the arch. The normal practice in Sydney is to spray the lining over the bolts such that most bolt end plates are attached to the rock face and do not brace the lining. This is done for reasons of durability and safety during construction, but is likely to result in limited shear transfer between a shotcrete lining and the end of a bolt should substantial ground movement occur. Regardless of the efficacy of shear transfer, the bolts have been excluded from the current analyses since their only contribution is to restrain the potential extent of delamination.

All the analyses undertaken as part of this research have been two-dimensional and have focused on beam-based mechanisms between hinges at cracks; fan-shaped yield line failures have not been considered. Geometric imperfections in either the rock surface or the lining thickness have been ignored for the present, as have uneven loads. All potential failure mechanisms are assumed to occur near the crown of the arch.

1.2 *Shotcrete lining performance expectations*

Many of the presumed failure modes are assumed to be preceded by de-bonding and progressive deformation of the lining due to the effects of drying shrinkage and creep. These phenomena are known to proceed slowly, so initiation commences slowly during the shotcrete curing process and it will take several years before delamination and shrinkage cracking are sufficiently advanced to result in failure in the shotcrete lining. Drying shrinkage and creep will, by this stage, have had sufficient opportunity to significantly influence stresses and strains within the lining.

Drying shrinkage and creep are related phenomena since both are caused by movement of moisture under an external stress or migration of moisture between CSH crystals due to a hydraulic gradient within the cementitious paste phase of concrete (Wittman, 1982). Shotcrete has a large amount of cementitious paste compared to other types of concrete, and additionally contains a relatively small amount of coarse aggregate. Both these attributes lead to an increased amount of drying shrinkage and creep in shotcrete compared to conventional concrete.

Fibres do not alter the unrestrained shrinkage of concrete (that is, shrinkage in which strains are free to occur due to an absence of restraint) because unrestrained shrinkage does not lead to cracking and fibres are only effective once a crack has been initated. However, fibres can have some effect on restrained drying shrinkage (that is, shrinkage in which restraint can give rise to a tensile stress resulting in crack formation) because fibres can act to limit the development of crack widths by promoting tensile creep of the concrete matrix. The challenge in designing for restrained drying shrinkage is to estimate the degree of restraint and its effect on tensile creep and thus crack widths.

Creep of shotcrete is known to be greater than that of conventional concrete containing normal amounts of 20 mm aggregate (Collota et al, 2010), but there appears to be limited long-term data on creep (or drying shrinkage) of shotcrete used in Sydney in either compression or tension. Unpublished test data indicate that for shotcrete loaded at an age of 56 days the creep strain is approximately 90–100 $\mu\varepsilon$/MPa at 180 days, while the unrestrained drying shrinkage is 1200–1400 $\mu\varepsilon$ at 180 days (for drying commencing at an age of 7 days). It is also known that set accelerators increase drying shrinkage (Lagerblad et al, 2010), but it is not known precisely how much set accelerator is used in any particular region of shotcrete in a given tunnel. It is not known how set accelerators may potentially affect the creep characteristics of shotcrete.

Despite these partially-understood characteristics of shotcrete, it can be estimated that the long-term unrestrained drying shrinkage potential of shotcrete (measured under standard test conditions as per AS1012 over at least one year) is about 1400 $\mu\varepsilon$, and the long-term creep strain is estimated to be about 200 $\mu\varepsilon$/MPa. The creep factor ϕ (taken as a multiple of short-term elastic strain) under moderate stresses is therefore about 5–6. These estimates need to be verified by experimental means, and should ideally be confirmed over a period of time much greater than the 56–91 days normally used in tests. Moreover, these estimates will need some modification to account for the drying characteristics of a typical shotcrete lining, which could consist of a layer as thin as 50 mm with a single side exposed to a ventilated atmosphere and the other side sealed but possibly exposed if delamination occurs.

The unconfined compressive strength (UCS) of shotcrete used in most Sydney tunnels is nominally 40 MPa at 28 days, but the actual long-term UCS is about 55 MPa (eg. Hanke et al, 2004; Asche & Bernard, 2004; Clarke et al, 2014). The flexural tensile strength under quasi-static loading is about 6 MPa, but evidence from Zhou (1992) indicates that sustained flexural-tensile loading of concrete can lead to rupture at a stress up to 50% lower than indicated by short-term tests. It is therefore estimated that the long-term flexural-tensile strength of shotcrete under sustained loading is no more than 3 MPa. Size effects (Zi et al, 2014) and geometric imperfections such as pits in the surface and voids inside the shotcrete can diminish this further such that a lower bound to the long-term in situ flexural-tensile strength can be estimated at about 2 MPa.

2 POSTULATED SCENARIOS

Fibre Reinforced Shotcrete tunnel linings have presently been examined using analytical models based on several scenarios involving shrinkage cracking and possible delamination. The models involve cracks arising from drying shrinkage and gravity loading. The mechanisms involved in drying shrinkage cracking and delamination of a lining from the rock substrate have only been assessed in the published literature for a limited number of cases (Ansell, 2010; Bryne et al, 2014), but shrinkage cracking and delamination have been extensively investigated for the case of concrete overlays on roadways (RILEM, 2011). There are many similarities in the structural behaviour of concrete overlays and shotcrete linings, thus some of the findings obtained from research on overlays has been incorporated into the models described below.

The primary assumptions underlying the current analyses are:

1. Drying shrinkage and creep are the primary factors that affect the stress development in a lining in the long term, not external loads.
2. Delamination is initiated at flaws in the shotcrete/ground boundary and spreads by various mechanisms,
3. The ground is essentially stable during the period of delamination and potential lining instability,
4. Peeling strength at the shotcrete/ground boundary is lower than the direct tensile bond strength
5. The long term bond strength at the shotcrete/ground boundary is lower than the short-term bond strength.

For each of the models described, the ground is assumed to be more stable than the lining. This means that the lining does not support a compressive arching stress even if arch-shaped, primarily because straining of the ground is insufficient to overcome the dissipation of stress attributable to creep of the shotcrete (Chen and Small, 2008). Moreover, in Sydney tunnels it has been the practice until recently that shotcrete is typically applied only after the ground is assessed to be 'stable' (which can be 100–200 metres behind the heading), thus it can be argued that movement of the ground has largely ceased by the time the FRS lining is installed. In more recent projects the excavation cycle has been altered to a more conventional in-cycle application of shotcrete immediately behind the face. The ground in the majority of cases consists of Hawkesbury Sandstone, and in a minority of cases shale or mudstone/siltstone. Hawkesbury Sandstone is usually very competent (Pells, 2002). Weathering and weak intersections occur in a few limited regions, and these have typically been the cause of unstable ground in the past. However, the present investigation has focussed on lining behaviour in competent stable ground in which poor bond, drying shrinkage, and creep are taken to be the drivers of lining failure, not ground instability. The lining is therefore assumed to consist of a delaminated region (except in Scenario 1) which is restrained by the adjacent lining which is bonded to ground that is assumed to be both rigid and stable.

The causes of initial de-bonding are difficult to identify with certainty. Possible causes include: changes in hydrology behind the shotcrete lining leading to weakening of the substrate and loss of tensile bond capacity; in situ stress changes due to hydromechanical coupling (eg. as pore pressures increase, stress also increases); stress changes arising from excavation of the tunnel void which can lead to an indirect change in hydraulic properties; de-bonding along planes of weakness in the rock (such as joints and bedding planes in shale); vibration due to excavation; initial imperfections such as regions of excessive set accelerator sprayed on the substrate; or areas of accumulated rebound in hollows in the surface during spraying. It is believed that the zones of debonding need only be small to act as an initial weakness that can grow with the passage of time and the progress of drying shrinkage in the lining. Since Sydney tunnels are almost all excavated using road headers, the surface has tended to be quite smooth and the geometry uniform. This means that potential peeling action at the boundary between debonded and bonded regions is relatively unimpeded by geometric obstructions. However, irregular areas may arise from time to time, and the finished surface of the lining can appear quite bumpy due to the requirement to spray additional shotcrete over the heads of bolts.

Contractors responsible for tunnel linings constructed in Sydney usually claim that 'best practice' was followed in preparing the ground for spraying and during application of shotcrete. In reality, this has not always been true, so the potential can arise for bonding problems even in the short term. Bond tests are regularly undertaken to ascertain the minimum levels of bond achieved soon after construction. However, published results of bond tests by Bernard (2009), Clements et al (2004), and Bryne et al (2013) on a variety of rock types, and anecdotal evidence from QC testing in Sydney, all indicate a wide range in the magnitude of bond strength developed between shotcrete and rock with numerous 'poor results' in each set of data which are usually assumed to be caused by eccentric load application during tests. In reality, it is not known whether the 'poor results' which so frequently occur in testing are actually poorly bonded shotcrete or experimental errors. Nevertheless, even the best results indicate a tensile bond strength of only 0.5–1.5 MPa in direct short-term tension (that is, measured over several minutes). If peeling (that is, eccentric) tensile stress is applied, the effective tensile strength is likely to be much lower since crack growth at a peeling front is known to take place under a lower tensile stress than would occur under a uniform tensile stress. Peeling mechanisms are known to occur at the edges of curling concrete membranes and on the sides of cracks (RILEM, 2011). Moreover, tensile strength under a sustained load is known to be lower than under short-term loading (Zhou, 1992) due to progressive linking up of otherwise dispersed micro-cracks. For these reasons, plus potential deterioration of the substrate and shotcrete at their common boundary as a result of water migration, it can be assumed that the tensile bond strength or adhesion between shotcrete lining and ground is lower in the long term than in the short term and is independent of the gradual increase in compressive strength of the shotcrete.

The basic drying shrinkage strain $\varepsilon_{cs,b}$ has been taken to be 1200 µε, with the hypothetical drying thickness t_h equal to 150 mm (drying on the exposed face of the tunnel lining only). Hence, in accordance with AS3600 (2018), the average cross-sectional drying shrinkage strain ε_{cs} for the tunnel lining (for an inland temperate region) is estimated to be 603 µε at 90 days, 819 µε at one year, and 932 µε at 5 years. For the case of non-uniform shrinkage, it is assumed that the shrinkage on the boundary face of the lining is half the shrinkage on the exposed face.

For all the potential scenarios, appropriate concrete properties must be estimated. Due to the absence of experimental data for shotcrete used in Sydney tunnels, we have used estimates of drying shrinkage strain published by Goodier et al (2008), creep factors extrapolated from data by Collota et al (2010), and concrete properties estimated using AS3600 (2018 version, with the elastic modulus for shotcrete taken to be 10% less than for cast concrete). The concrete strength in compression has been taken to be 20 MPa at 3 days, 40 MPa at 28 days, 45 MPa at 90 days, and 50 MPa after one year. Thus, E_c for the shotcrete is estimated to be 20,300 MPa at 3 days, 29,700 at 90 days, and 30,700 MPa after 1 year.

A time-dependent age-adjusted effective modulus is used to assess the stresses arising from restrained shrinkage, taking

$$E_{c,\,eff} = \frac{E_c}{1 + 0.85\varphi_{cc}} \tag{1}$$

where E_c is the initial elastic modulus at the commencement of loading (at 3 days) and ϕ_{cc} is the creep factor that would be applicable for a constant sustained load. The creep factor ϕ_{cc} is estimated as for normal concrete in accordance with AS3600 (2018), assuming that the basic creep factor $\phi_{cc,b}$ is equal to 4. Accordingly, for a constant stress sustained from an age of 3 days, the creep factor ϕ_{cc} would be 3.6 at 90 days, 4.9 at 1 year, and 5.6 at 5 years. Hence, the age-adjusted effective modulus $E_{c,\,eff}$ for the assessment of shrinkage effects is 5020 MPa at 90 days,

3950 MPa at 1 year and 3550 MPa at 5 years. These estimates of shotcrete properties have been used to estimate the conditions associated with potential failure modes as described below.

2.1 *Shrinkage cracking in the absence of delamination*

The first mode of failure considered in this investigation is the case of cracking induced by drying shrinkage in the absence of delamination. Inspections of aged shotcrete linings in tunnels such as the Norfolk tunnels on the M2 Motorway in northern Sydney have indicated widespread cracking in the absence of ground instability. This pattern of extensive shrinkage cracking has also been observed in other countries (Ansell, 2010; Bryne et al, 2014; Grasberger and Meschke, 2003). These cracks appear to be induced by drying shrinkage of the shotcrete lining as they are not associated with deformation of the ground. Indeed, research has shown that drying shrinkage stresses may be sufficient to overcoming ground-induced compressive stress in a lining leading to crack formation (Grasberger and Meschke, 2003). In this scenario (Figure 1), the lining is fully bonded to the ground, and prior to cracking, shrinkage of the curved lining is resisted by adhesive stresses acting normal to the bonded surface of the lining. These stresses prevent the lining from pulling away from the rock face and associated circumferential (hoop) stresses are induced by the restrained shrinkage (analogous to the hoop stresses in a pressurised pipe). Over time, the hoop stresses associated with the restrained shrinkage may become large enough to induce cracks which will extend through the thickness of the tunnel lining, but if the lining remains fully bonded to the rock face then shear stresses and normal tensile stresses will act on the bonded boundary adjacent to the crack to prevent movement of the lining at that surface. The hoop stresses associated with restrained shrinkage in the lining are relieved at the crack face, but this relief dissipates each side of the crack over a length that is similar to the depth of the crack.

For a bonded lining of 14850 mm radius and thickness 75 mm, the stresses estimated to occur in the lining prior to cracking are listed in Table 1 (Reid and Bernard, 2016). The normal stress to be resisted by adhesion at the lining-rock interface is approximately 20 kPa (which is significantly greater than the loading due to self-weight of the tunnel lining) and the tensile hoop stress in the lining is approximately 5 MPa. Accordingly, a crack due to shrinkage is very likely to occur on the exposed face, and this is consistent with observations in the field that cracking is commonplace in relatively thin linings. Delamination after cracking (due to peeling stresses) is also very likely immediately adjacent to each crack but the extent of this delamination is unclear. It can therefore be surmised that cracking and delamination will be likely in FRS linings. These results are similar to the results presented in the RILEM Report 193 (2011), which indicate that drying shrinkage cracking is 'inevitable' for concrete overlays.

2.2 *Delamination of a lining in the absence of cracking*

The second possible scenario considered in this investigation is the case of lining delamination in the absence of drying shrinkage cracks. This scenario has been considered because anecdotal evidence exists of delamination in some shotcrete linings, both in civil tunnels and in mines, in the apparent absence of any cracks. However, it is generally not known what the cause of this delamination is. In this scenario, the delamination is assumed to originate from insufficient adhesion (cf. Table 1) to prevent the curved lining from pulling away from the rock face as

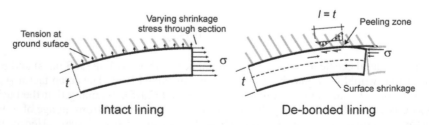

Figure 1.   Scenario 1 - involving cracking of fully bonded shotcrete lining due to drying shrinkage.

shrinkage stresses develop in the lining. The incremental changes associated with a loss of adhesion over the debonded section of an intact (uncracked) lining can be modelled by applying an equal and opposite action (comprising a surface pressure). The debonded lining then acts like an end-restrained beam subject to a uniformly distributed load w (Figure 2). This load increment induces bending moments at the ends of the debonded region which are resisted by the flexural rigidity of the bonded lining and peeling stresses at the lining-ground boundary.

Taking a lining of 14850 mm radius and 75 mm thickness, subject to shrinkage and gravity loading due to self-weight, the results are listed in Table 2. The results indicate that if an uncracked section of lining becomes unbonded, shrinkage will generate high debonding stresses at the edges of the debonded region (leading to propagation of the area of delamination) and high tensile stresses for relatively small debonded segments. As the debonded region grows in size, the internal tensile stresses due to shrinkage are reduced significantly. If debonding is arrested before the unbonded section becomes large, then cracking is likely to occur at the crown or ends.

2.3 Cracking at the crown of a delaminated lining

The third scenario considered in this investigation is the case of cracking induced by drying shrinkage and gravity loading of a delaminated segment of lining. In this scenario, the initial debonding may have been caused by poor bond to the ground, delamination within the ground, or

Table 1. Stresses in lining prior to cracking, and stresses at lining-rock interface after cracking.

Stresses Prior to Cracking	90 days	1 year	5 year
Circumferential Tension, Max tensile stress σ_i (MPa)	4.5	4.8	4.9
Radial Stress at lining-rock Boundary (MPa)	0.017	0.018	0.019
Stresses After Cracking	90 days	1 year	5 year
Normal peeling stress at boundary (MPa)	16	17	17
Interface shear stress (MPa)	6.7	7.2	7.4
Principal tensile stress (MPa)	18	20	20

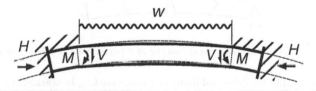

Figure 2. Scenario 2 - Debonding of shotcrete lining in the absence of cracking. M is a moment induced at the ends of the delaminated region due to shrinkage and self-weight of the lining, H is a lateral restraining force and V is the corresponding shear force.

Table 2. Stress and load ratios for uncracked debonded lining of various arc lengths.

Segment with delaminated length = 4158 mm	90 days	1 year	5 year
M/M_{cr} at crown	1.1	1.2	1.2
M/M_{cr} at supports	1.2	1.2	1.2
Tensile stress/strength at support interface with rock	2.0	2.0	2.1
Segment with delaminated length = 5346 mm	90 days	1 year	5 year
M/M_{cr} at crown	0.68	0.69	0.72
M/M_{cr} at supports	0.75	0.76	0.77
Tensile stress/strength at support interface with rock	1.8	1.8	1.8
Segment with delaminated length = 6534 mm	90 days	1 year	5 year
M/M_{cr} at crown	0.45	0.46	0.48
M/M_{cr} at supports	0.50	0.51	0.51
Tensile stress/strength at support interface with rock	1.6	1.6	1.6

shrinkage-induced stresses associated with the first scenario considered in Section 2.1. Regardless of the cause, a shrinkage crack located over a delaminated region will cause the two halves of the delaminated lining to act as cantilevers subject to a uniformly distributed load w comprised of self-weight plus a possible surcharge associated with delaminated ground. Cracking (partially) relieves the shrinkage stresses in the delaminated region, but curvature associated with differential shrinkage of the outer and inner faces of the lining may persist. Depending on shrinkage, creep, thickness, and the extent of debonding, deformations associated with loading might bring the cantilever tips into contact and arching action may subsequently develop. Actions at the supported ends of the delaminated segment include peeling stresses at the periphery of the debonded region, which possibly lead to an increase in the size of the debonded region (Figure 3).

For this analysis, we take a lining of 14850 mm radius and 75 mm thickness, and impose gravity loading equal to self-weight of the lining plus an external surcharge of 5 kPa corresponding to a thin layer of debonded ground. The results indicate that a delaminated and cracked section of lining will remain relatively stable due to the development of arching as the size of the debonded region increases. However, crack widths will be large and corrosion of steel fibres is very likely leaving the lining unable to support a moment at these cracks.

2.4 *Cracking at the support boundaries of a delaminated lining*

The fourth scenario considered in this investigation is the case of lining delamination with cracking at the boundaries due to drying shrinkage (Figure 4). This scenario represents a progression from Scenario 2 for an uncracked lining in which the bending moments at the ends (due to gravity loading and the partial loss of shrinkage restraint) have reached a sufficient magnitude to crack the lining in bending at each end. The de-bonded portion of the lining between the cracks acts as a partially restrained beam subject to a uniformly distributed load w in which the restraint is provided by the bonded sections of lining located either side of the delaminated

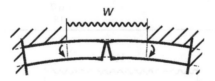

Figure 3.   Scenario 3 - cracking due to drying shrinkage of a partially de-bonded shotcrete lining.

Table 3.   Stress and load ratios for debonded lining of various arc lengths cracked at crown.

Segment with delaminated length = 2970 mm	90 days	1 year	5 year
M/M_{cr} at supports	0.21	0.24	0.25
Tensile stress/strength at support interface with rock	0.56	0.58	0.60
Segment with delaminated length = 4456 mm	90 days	1 year	5 year
M/M_{cr} at supports	-0.38	-0.36	-0.35
Tensile stress/strength at support interface with rock	0.22	0.21	0.21
Segment with delaminated length = 5346 mm	90 days	1 year	5 year
M/M_{cr} at supports	-0.56	-0.55	-0.55
Tensile stress/strength at support interface with rock	0.18	0.18	0.18

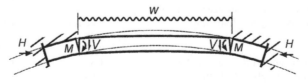

Figure 4.   Scenario 4 - involving cracking due to drying shrinkage of a partially de-bonded shotcrete lining. M is the moment associated with the eccentric compression block at each end, H is a lateral restraining force and V is the associated shear.

Table 4. Stress and load ratios for at boundaries of a debonded lining of various arc lengths.

Segment with delaminated length = 2970 mm	90 days	1 year	5 year
M/M_{cr} at crown	0.46	0.46	0.46
Tensile stress/strength at support interface with rock	1.55	1.51	1.53
Segment with delaminated length = 4456 mm	90 days	1 year	5 year
M/M_{cr} at crown	0.43	0.42	0.43
Tensile stress/strength at support interface with rock	1.64	1.59	1.60
Segment with delaminated length = 5346 mm	90 days	1 year	5 year
M/M_{cr} at crown	0.42	0.41	0.41
Tensile stress/strength at support interface with rock	1.60	1.60	1.61

Table 5. Response for 50 mm thick lining (for delaminated length 3669 mm, self-weight only).

Stress ratio under Self-weight only	$\varepsilon_{cs,\,b}$	90 days	1 year	5 year
M/M_{cr} at ends	600	0.09	0.08	0.08
	1200	0.06	0.05	0.05
	1600	0.04	0.04	0.04
Stress ratio under Self-weight only	$\varepsilon_{cs,\,b}$	90 days	1 year	5 year
Tensile stress/strength at support interface	600	0.04	0.04	0.04
	1200	0.05	0.05	0.04
	1600	0.06	0.05	0.05
Crack widths under Self-weight only	$\varepsilon_{cs,\,b}$	90 days	1 year	5 year
Crack widths (mm)	600	0.09	0.08	0.08
	1200	0.39	0.38	0.36
	1600	0.60	0.57	0.54

Table 6. Response for 125 mm thick lining (for delaminated length 3669 mm, self-weight only).

Stress ratio under Self-weight only	$\varepsilon_{cs,\,b}$	90 days	1 year	5 year
M/M_{cr} at ends	600	0.24	0.28	0.25
	1200	0.58	0.66	0.59
	1600	0.81	0.92	0.83
Stress ratio under Self-weight only	$\varepsilon_{cs,\,b}$	90 days	1 year	5 year
Tensile stress/strength at support interface	600	0.38	0.44	0.39
	1200	0.87	0.98	0.89
	1600	1.2	1.3	1.2
Crack widths under Self-weight only	$\varepsilon_{cs,\,b}$	90 days	1 year	5 year
Crack widths (mm)	600	0.27	0.35	0.32
	1200	0.70	0.86	0.80
	1600	0.99	1.2	1.1

region. A moment is taken to act at each end of the delaminated region, partly due to the presence of fibres but also due to the eccentric compression block generated at the points of contact.

For this analysis, we take a lining of 14850 mm radius and 75 mm thickness, and impose gravity loading equal to self-weight of the lining plus an external surcharge of 5 kPa corresponding to a thin layer of debonded ground. These results indicate that a delaminated segment with end cracks acts as a relatively stable arch and is unlikely to break into smaller segments, but delamination is likely to continue to propagate at each end.

2.5 Sensitivity to drying shrinkage

A parametric analysis was undertaken to examine the sensitivity of stresses, deformations, and crack widths to the magnitude of the drying shrinkage strain. The scenario considered was that of a delaminated region with a central crack (Scenario 3). The lining radius was taken to be 12230 mm, and the thickness 50 or 125 mm. The basic drying shrinkage strain $\varepsilon_{cs,b}$ was varied across the range 600, 1200, and 1600 microstrain. The results presented below include an allowance for concrete creep following the initiation of cracking at the ages listed.

These results indicate that the behaviour of a cracked lining is very sensitive to the magnitude of drying shrinkage strain, especially in the range 600–1200 $\mu\varepsilon$. The thinner 50 mm thick lining is unlikely to crack at the ends once a central crack has occurred, but the thicker lining may possibly crack (due to gravity loading) giving rise to a three-hinge mechanism. For both thicknesses the width of crack generated at the centre is very large and far exceeds normal crack width limits. The estimates of crack width at the centre of the delaminated region agree quite well with measured crack widths reported by Bryne et al (2014) for the case of shotcrete sprayed over a strip drain. This is because a delaminated lining is similar in concept and behaviour to a lining covering a wide strip drain (although the drains were primarily on the walls of tunnels and thus gravity does not exacerbate deformations). Bryne et al found that 60% of cracks in 1500 mm wide drains exceeded 0.50 mm in width, and that some cracks were up to 1.2 mm wide. These cracks were assessed to be caused by drying shrinkage of the shotcrete and were inadequately controlled by the inclusion of steel fibres in the shotcrete mixture.

3 DISCUSSION

The derivations and analyses described above indicate that the mechanisms involved in cracking and delamination of the shotcrete lining from its substrate are very complex and involve numerous parameters. Unfortunately, the magnitude of drying shrinkage strains and creep factors pertinent to shotcrete used in Sydney tunnels is poorly known and subject to large margins of error. It is therefore difficult to make specific recommendations as to minimum lining radii, thicknesses, or bolt spacings that should be maintained in order to avoid possible instability or collapse. Only when long-term drying shrinkage strains, creep factors, strength, and ductility in FRS used in Sydney tunnels have been identified can an analysis be undertaken and used to arrive at sensible design limits.

Despite the present lack of clarity on design limits for lining stability, one outcome that can be identified is the sensitivity of the present results to the magnitude of drying shrinkage strain for the shotcrete. The results for a lining conforming with the geometry of the Lane Cove Western Vent Tunnel (50–125 mm thickness and 12230 mm radius) indicates that cracking of the lining as a result of drying shrinkage is highly likely (even for relative low estimates of the long-term drying shrinkage), and that cracks are very likely to be associated with regions of delamination. Depending on the thickness and extent of delamination, the initial cracking can progress to the stage where secondary cracks occur at the supports giving rise to a potentially unstable mechanism.

One of the additional outcomes of the analyses is that crack widths in a delaminated FRS lining can become very wide with the passage of time, even for conservative (that is, low) estimates of long-term drying shrinkage strains. This is supported by observations in the field in Sydney, and by observations in other countries (Ansell, 2010; Bryne et al, 2014; Grasberger and Meschke, 2003). The maximum permissible width of cracks in steel FRC is normally limited in underground environments to 0.15–0.20 mm by code requirements intended to preclude the onset of corrosion (AFTES, 2013). In-field research on the rate of corrosion of steel fibres in shotcrete in Swedish road tunnels by Nordström (2001) indicated that the maximum acceptable crack width for long term exposure of SFRS is only 0.10 mm. More recent work by Kaufmann (2014) in Swiss tunnels indicated that cracks of 0.50 mm width can lead to rapid loss of structural performance in steel FRS (within one year), but negligible to moderate rates of degradation in macro-synthetic FRS. Marcos-Meson et al (2017) summarised the results of numerous investigations and found that crack widths in excess of 0.15 mm in SFRC generally led to steady reductions in post-crack performance with the passage of time. The present investigation has indicated that crack widths in excess of 0.5 mm can be encountered in as little as 1–2 years after delamination of a lining, due principally to shrinkage and creep of the shotcrete, thus the consequences for durability of steel fibres at these cracks are likely to be severe.

REFERENCES

AFTES, 2013. *Design, dimensioning, and execution of precast steel fibre reinforced concrete arch segments*, French Tunnelling and Underground Space Association.

Ansell, A. 2010. Investigation of shrinkage cracking in shotcrete on tunnel drains, *Tunnelling and Underground Space Technology* Vol. 25(5),pp 607–613.

Asche, H.R. and Bernard, E.S., 2004. Shotcrete design and specification for the Cross City Tunnel, Sydney, Shotcrete: More Engineering Developments, Bernard (ed.), pp 27–38, Taylor & Francis, London.

Australian Standard AS1012 *Testing Concrete*, Standards Australia, Sydney.

Australian Standard AS3600 (2018) *Concrete Structures*, Standards Australia, Sydney.

Bernard, E.S., 2009. Load Capacity of Early-age Fibre Reinforced Shotcrete Linings, *TSE Report 197*, Technologies in Structural Engineering P/L, Sydney.

Bryne, L.E., Ansell, A., and Holmgren, J., 2013. Laboratory testing of early age bond strength between concrete for shotcrete use and rock, *Nordic Concrete Research* 47(1) pp81–100.

Bryne, L.E., Ansell, A., and Holmgren, J. 2014. Investigation of restrained shrinkage cracking in partially fixed shotcrete linings, *Tunnelling and Underground Space Technology* Vol. 42, pp 136–143.

Chen, R., Small, J., 2008. Analysis of the Creep Behaviour of Tunnels in Sandstone/Shale. *13th Australian Tunnelling Conference 2008*, Carlton, Victoria, Australia.

Clarke, S.J., de Ambrosis, A., Bertuzzi, R., and Redelinghuys, J., 2014. Design and construction for the widening of the M2 Norfolk Twin Tunnel, *15th Australasian Tunnelling Conference*, Sydney, 17–19 Sept, pp 1–10.

Clements, M.J.K., Jenkins, P.A., and Malmgren, L., 2004. Hydro-scaling – An overview of a young technology, *More Engineering Developments in Shotcrete*, E.S. Bernard (ed.), pp 89–96, Taylor & Francis, London.

Collotta, T., Barbieri, T., and Mapelli, M., 2010. Shotcrete tunnel linings with steel ribs: stress redistribution due to creep and shrinkage effects, *Shotcrete: Elements of a System*, Taylor and Francis, London.

Goodier, C.I., Austin, S.A. and Robins, P.J., 2008. Low-volume wet-process sprayed concrete: hardened properties, *Materials and Structures*, 41(1), pp. 99–111.

Grasberger, S., and Meschke, G., 2003. Drying shrinkage, creep and cracking of concrete: From coupled material modelling to multifield structural analyses, *Computational Modelling of Concrete Structures*, EURO-C 2003, pp. 433 442.

Hanke, S.A., Collis, A. and Bernard, E.S., 2001. The M5 motorway: an education in Quality Assurance for fibre reinforced shotcrete, *Shotcrete: Engineering Developments*, Bernard (ed.), pp 145–156, Swets & Zeitlinger, Lisse.

Kaufmann, J.P., 2014. Durability performance of fiber reinforced shotcrete in aggressive environment, *World Tunnelling Congress 2014*, (Ed. Negro, Cecilio and Bilfinger), Iguassu Falls Brazil, p 279.

Lagerblad, B., Fjällberg, B, Vogt, C., 2010. Shrinkage and durability of shotcrete, *Proceedings of the 3rd International Conference on Engineering Developments in Shotcrete*, Queenstown, pp. 173–180.

Marcos-Meson, V., Michel, A., Solgaard, A., Fischer, G., Edvardsen, C., Skovhus, T.L., 2017. Corrosion resistance of steel fibre reinforced concrete - A literature review, *Cement and Concrete Research*, Vol. 103, 2017, p. 1–20.

Nordström, E., 2001. Durability of steel fibre reinforced shotcrete with regard to corrosion, *Shotcrete: Engineering Developments*, Bernard (ed.), pp 213–217, Swets & Zeitlinger, Lisse.

Parsons Brinckerhoff, 2014, Memo regarding Lane Cover Tunnel - Sirius Road and Western Ventilation Tunnel - safety issues associated with shotcrete inspection, Ref 2152190B-TPT-MEM-244A Rev A

Pells, P.J.N., 2002. Developments in the Design of Tunnels and Caverns in the Triassic Rocks of the Sydney Region, *Int. J. of Rock Mechanics and Mining Sciences*, 39, pp569–587.

Reid, S.G. and Bernard, E.S., 2016. Investigation of the Potential for Shotcrete Lining Delamination in Sydney Tunnels and Its Consequences, *TSE Report 257*, January 2016.

RILEM Technical Committee 193-*RLS Report*, 2011. Bonded Cement-based Material Overlays for the Repair, the Lining, or the Strengthening of Concrete Slabs and Pavements, RILEM, Paris.

Wittman, F.H., 1982. Creep and shrinkage mechanisms, *Creep and Shrinkage of Concrete Structures*, Bažant & Wittman (eds), Wiley & Sons, pp 129–161.

Zhou, F.P., 1992. Time-Dependent Crack Growth and Fracture in Concrete, Doctoral Thesis TVBM-1011, Division of Building Materials, Lund Institute of Technology.

Zi, G., Kim, J., and Bažant, Z.P., 2014. Size Effect on Biaxial Flexural Strength of Concrete, *ACI Materials Journal*, V. 111, No. 1–6, pp 1–8.

Tunnels and Underground Cities: Engineering and Innovation meet Archaeology,
Architecture and Art, Volume 6: Innovation in underground engineering,
materials and equipment - Part 2 – Peila, Viggiani & Celestino (Eds)
© 2020 Taylor & Francis Group, London, ISBN 978-0-367-46871-2

Eurasia Tunnel – Construction features of a tunnel under severe pressure conditions in high seismic area

G. Rinaldi, P. Cuppone & A. Turi
Italferr SpA, Rome, Italy

ABSTRACT: Eurasia Tunnel is a submarine double-decker road tunnel underneath the Bosphorus. It is part of the Istanbul Strait Road Tube Crossing Project: it connects the Asian side and the European one of Istanbul and it is 14.6 km long. The underground section is about 5.4 km long: 3.3 km have been excavated by Mixshield TBM. The tunnel is to be considered unique and problematic due to the large excavation diameter (13.7 m), the high pressures (11 bar), the heterogeneity of the soil (soft and hard rock basement and marine sediments with erratic boulders), the high permeability and the seismic conditions. In fact, the tunnel lies in a geological zone of high seismicity, located parallel to the Anatolia plate-boundary fault in the Marmara Sea (17 km offshore). Therefore, the project implementation required specific seismic joints located inside special metallic rings, allowing structural deformation under the action caused by an eventual earthquake.

1   INTRODUCTION

Like many metropolises, Istanbul suffers from large-scale traffic congestion. The situation here is made even worse by the fact that the city is divided into two parts (European – Asian) by the Bosphorus strait, which represents a huge obstacle for road connections.

Until the end of 2016, in fact, there were only two bridges (not considering the insufficient ferry service): the '15 July Martyrs' Bridge (previously called the Bosphorus Bridge, built in 1973) and the 'Fatih Mehmet Sultan' Bridge (opened to traffic in 1988), which are often congested and obviously insufficient for a city of about 16 million inhabitants and continuously growing.

A third bridge, the 'Yavuz Selim' Bridge, was inaugurated on 26 August 2016, in the north of Istanbul, but close to the Black Sea.

In order to improve road connections between the Asian and European parts of Istanbul, the Istanbul Strait Road Tube Crossing Project was developed. The project allows the crossing of the Bosphorus, this time, through an undersea tunnel also known as the Eurasia Tunnel (Avrasya Tüneli in Turkish). The same name also indicates all the tunnel-related underground works, which will be illustrated further on in this paper (Figure 1).

2   THE PROJECT

The first feasibility study was commissioned by the Turkish Ministry of Transport, Maritime Affairs and Communications to Nippon KOEI – NCC, in 2005. In June 2008, the General Directorate of Infrastructure Investments (AYGM which is an agency of the above mentioned Ministry) tendered the design, construction, financing, operation and subsequent transfer to the State of the infrastructure, on the basis of a BOT (Build Operate and Transfer) contract.

The evaluation of the tenders was completed in December 2008 and the BOT contract was awarded to the ATAŞ (Avrasya Tüneli İşletme İnşaat ve Yatırım A.Ş.).

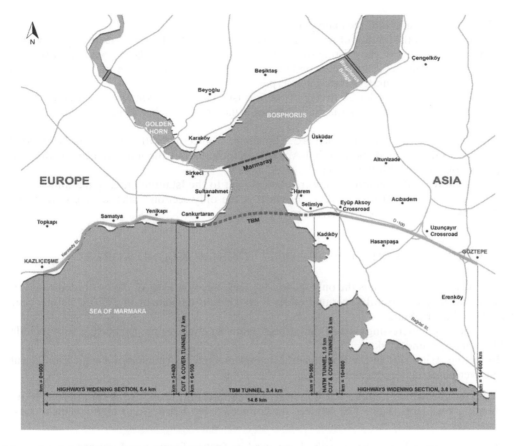

Figure 1. Istanbul Strait Road Tube Crossing Project.

The construction work was carried out by YMSK JV, a joint venture established between the Turkish company Yapı Merkezi and the South Korean SK E&C (these companies are also members of the ATAŞ.

After an initial stage (2011), required for the preliminary works in the areas crossed by the infrastructure, and following the supplementary agreement entered into in 2012, the works began in earnest in January 2013. The forecasted construction time was 4 years and 7 months, so the completion date was anticipated to occur in August 2017, but the works were completed 8 months earlier than planned, in December 2016.

The total price amounted to approx. 1.25 billion dollars.

In May 2014, Italferr SpA, as part of a joint venture with the Turkish company Altınınok CE LLC, was awarded for works supervision, as the Employer's Representative (ER) of the Client AYGM, for the stages of the development of detailed design execution of works, acceptance testings and the commissioning of the infrastructure as a whole. The construction required approx. 14 million hours of work of the engineers, technicians and construction workers.

3 THE PROJECT PARTS

The project consists of an intercontinental road infrastructure connecting the district of Kazlıçeşme, on the European side of the Bosphorus, and Göztepe, on the Asian side, crossing underneath the Bosphorus strait.

In particular, the road links Kennedy Street (leading to Atatürk Airport) and the D100 road to the Istanbul-Ankara motorway.

The project consists of three main parts:

Part 1) includes the section of road on the European side, 5.4 km long, featuring the widening and improvement of the existing road (increasing the number of lanes by adding two lanes leading into the tunnel), and the secondary roads, and the construction of junctions and of vehicle and pedestrian underpasses and overpasses.

Part 2) includes the undersea stretch of the Eurasia Tunnel, 5.4 km long (of which 3.34 km excavated with a TBM), underneath the Bosphorus strait, which will be described in more details further on.

Part 3) is the Asian section which is 3.8 km long, connecting with the existing D100 road, which links Göztepe and the Istanbul-Ankara motorway, and the improvement of the road network connecting with the tunnel.

Parts 1) and 3), after completion, were handed over to the Istanbul municipal authorities.

The operation period for the Part 2) (which is the only section under BOT contract) will end in 2043, when it will be transferred to the State.

4   DETAILED DESCRIPTION OF PART 2 – UNDERGROUND WORKS

This part of infrastructure is the only section which usage is subject of payment, dedicated for vehicle traffic and currently operated under a concession agreement by ATAŞ, before being transferred to the State.

It is the most interesting part of the project, from an engineering perspective, and includes all the underground works.

From the Asian side of the Bosphorus, after short stretches running in a trench and a cut-and-cover construction, there are:

– Two parallel tunnels, approx. 950 m long, with an average cross-section of approx. 85 m^2, hosting two lanes each. The tunnels are connected by 4 pedestrian and 1 vehicle cross-passages.

The conventional excavation method was used in Trakya formation. Four excavation faces were created, at different times, from both ends, to reduce overall excavation times.

– The Asian Transition Box – ATB – is a huge chamber 170 m long, between 25 and 36 m wide and 37 m deep, located at about 80 m from the sea. It was built by means of reinforced concrete secant piles wall with prestressed tie-backs (ground anchor) and struts in the upper sections. The twin tunnels have been conventionally excavated from one side of the chamber, at different heights, to allow the proper full connection, inside the ATB, with the twin-deck arrangement of the TBM tunnel – one of them leading towards Asia (URD) and the another toward Europe (LRD).

Excavations for this structure was also used to assemble and launch the TBM.

– The 3340 m long TBM section crossing under the Bosphorus, which will be described in more details further on.
– The European Transition Box – ETB – is a chamber 70 m long, between 25 m wide and 31 m deep, located at about 30 m from the sea. It was built by means of reinforced concrete diaphragm walls and struts and is the final destination and disassembly point of the TBM. In the ETB the carriageways are arranged in parallel, and do not feature the twin-deck arrangement, connecting with the surface road network by means of cut-and-cover constructions and trenches.

Part 2 also includes the toll areas, two ventilation buildings, three electricity substations and all the necessary systems and installations for the safe operation of the infrastructure.

The project also includes a building housing the operation offices and a control room, active on a 24/7 basis, which receives all the monitoring and operating data regarding the road section, in real time, via a SCADA (Supervisory Control and Data Acquisition) system.

5 THE EURASIA TUNNEL

The following paragraphs describe the tunnel section of the road, which represents the most complex and significant part of the entire project.

5.1 Geology

The infrastructure crosses through two sedimentary formations dating back to the Carboniferous period, with completely different characteristics. 70% of the tunnel was excavated in an intensely fractured formation called 'Trakya', consisting of alternating layers of sandstone, mudstone and siltstone (with uniaxial compressive strength of between 5 and 120 MPa), generally dark grey or brownish-grey in colour. This formation is intensely fractured due to the region's considerable seismicity. The Trakya formation has also been intruded by dikes of andesite and diabase.

23% of the tunnel alignment was excavated through 'marine sediments', consisting of interposed mixed layers with cobbles, gravel, sand, silt and clay. The remaining 7% of the tunnel has been excavated through two transition areas consisting of a combination of Trakya and marine sediments, which is where the two seismic joints of the Tunnel have been installed, as illustrated further on (Figure 2).

Frequent erratic boulders were also found, causing problems during the tunneling, also in terms of the damage to the cutterhead.

5.2 Seismicity of the area

The region of Istanbul is seismically very active and, given that it has been continuously inhabited, there are many historical evidences of earthquakes occurring over the centuries.

The southern part of Istanbul faces the Sea of Marmara and below this sea, running parallel to the shoreline and the Eurasia Tunnel (in an east-west direction), at a short distance (about 17 km), is a section of the North Anatolian Fault, which separates the Eurasian and the Anatolian plates and runs across Turkey for over 1600 km.

The area of the Sea of Marmara closest to the fault line is affected by repeated and periodical earthquakes, with at least one of medium intensity every 50 years.

5.3 The TBM used

The TBM would have to be capable of boring through both rock (Trakya) and very loose and permeable soils, in conditions of high hydrostatic pressure (11 bars).

Therefore, the choice fell on a Hydroshield TBM, which has turned out to be particularly suited to this type of excavation because it can provide very high pressures, and enable a greater control of the pressure, at the excavation face.

So, it was decided to employ a Mixshield Slurry TBM (manufactured by Herrenknecht), having a diameter of 13.71 m and a capacity to provide a maximum of 12 bar of hydrostatic pressure. The cutterhead was capable of boring through the different types of soils, being equipped with scrapers for the softer soil and disc cutters for the harder rock (Figure 3).

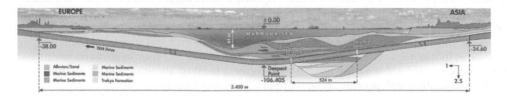

Figure 2.   Geology of the TBM subsea tunnel – Seismic joints.

TBM Type	Mixshield - Slurry
TBM Bore Diameter, D	φ 13.7 m
Disc Cutters	35 No X 19" (twin monoblock) -changeable under atmospheric condition-
Cutting Tools (scrapers)	* Atmospheric accessible (48 pieces) * Hyperbaric accessible (144 pieces) * Total (192 pieces)
TBM Total Length, L_{TBM}	120 m
TBM Shield Length, L_{SHIELD}	13.5 m
TBM Weight, W_{TBM}	~3300 t
TBM's Design Face Pressure, P_f	12 bar
TBM's Total Installed Power, P_i	10,330 kW
TBM's Cutterhead Power, P_c	14 X 350 kW = 4,900 kW
Nominal Torque, TQ_n	23,289 kNm
Overload Torque, TQ_o	34,933 kNm
Total Thrust Force, T_t	247,300 kN
Thrust Force per unit Excavation Area	1,678 kN/m²
Thrust Cylinders	Triple cylinder arrangement 17 X 3 No with 3000 mm stroke
Ring Diameter	* Outer (13.2 m) * Inner (12.0 m)
Ring Arrangement	* Universal * 8 + 1 keystone
Segment Durations, Weight and Concrete Strength Grade	* Segment thickness (600 mm) * Segment Length (2,000 mm) * Total ring weight (~127 t) * ERQ requirement is 50 MPa -28 days-

Figure 3.   TBM and its main parameters. (Valenza et al. 2015).

In particular, 35 x 19′ double ring monoblock disc cutters were used (all atmospherically replaceable from outside the chamber) and 192 cutting knives (of which 48 atmospherically replaceable).

During the tunneling about 400 elements were replaced, on the basis of individual wearing sensor system.

Of course, by using Hydroshield TBM, it was necessary to build a slurry treatment plant near the ATB, for regenerating the mix of water and bentonite.

The plant enabled the treatment of 2,800 m³/h of bentonite slurry.

5.4   Excavation

Considering the great size of the machine, the high pressure at the excavation face, and the variable nature of the materials crossed by the tunneling process, the project can be considered as unprecedented and extremely complex.

The TBM was launched on the Asian side (5% gradient), emerging on the European side with the same gradient.

The pressure at the excavation front reached 11 bars and the injection pressure of the two-component back-fill grout used behind the concrete lining was several bars higher.

The deepest point reached, in relation to the tunnel axis, was -106 m below sea level.

The tunneling process began in April 2014 and ended 16 months later, on 22 August 2015, the day on which the TBM, after having tackled a last stretch consisting of secant piles made of concrete reinforced with fiberglass, cut through the last diaphragm separating it from the ETB.

The excavated section is approx. 3340 m long.

On average, the advance rate (including standstills for maintenance/repairs/religious holi-day) was about 7 m/day (about 9 m/day without standstills), with maximum peaks of 18 m/day. Obviously, the work was carried out 24 hours a day, 7 days a week.

The difficult excavation conditions and the stress affecting the components of the TBM, including the presence of erratic boulders, required both ordinary and extraordinary mainten-ance activities, which also entailed, inter alia, the replacement of hundreds of cutters, the replacement of the innermost ring of the brushes (the only one of the three rings of brushes accessible during the excavation) and repairs to the suction grid and stone crusher.

Some activities required working in hyperbaric conditions and four hyperbaric maintenance events were carried out over the project period (for a total of about 6 weeks), at considerable depths and dealing with pressure ranges of between 8 and 10 bars.

For these purposes, specially trained divers were engaged, together with the TBM assistance personnel. The most significant operation was to repair a stone crusher, which took ca. 3 weeks between April and May 2015. To reduce the decompression time the divers were kept constantly in a hyperbaric environment, working inside the machine, and then transported, in the rest phase, by means of a pressurized shuttle, to a special hyperbaric chamber outside the tunnel.

This was significant standstill of the tunnel face advancement, which otherwise would have been much greater.

In any case, the preparation of all the equipment required for this type of operation, and the professional competence of the personnel involved, made it possible to solve several situations which, at the time, had given rise to a deal of concern.

Figure 4 shows the monthly average advance speed (expressed in m/day). The months featuring the lowest average speed are January and April 2015 (excluding June 2014, when the low advance rate was due to the initial machine cutterhead maintenance). The reason for this is that most of the hyperbaric activities, including the longest, were concentrated in precisely these two months (resulting in TBM standstill of 15 and 19 days, respectively).

The considerable differences in the advance rates may pose data interpretation problems. However, if the calculation is made by removing the days of the main standstills for religious holidays, starting operations, cutterhead maintenance, changing of brushes, hyperbaric repair works carried out by the divers, we will see that the values obtained are more uniform.

Therefore, with reference to Figure 5. we may see that the advance rate increases initially in proportion to the improved knowledge of how to use the TBM (learning curve), and then decreases in the middle of the diagram, in connection with the tunneling work at higher hydrostatic pressure, increasing once again over the final months, in the stretches with decreasing hydrostatic pressure.

5.5 *Tunnel*

Two horizontal structures have been built inside the tunnel, one for the Upper Road Deck (URD), for traffic travelling towards Asia, and one for the Lower Road Deck (LRD), for traffic travelling towards Europe.

These horizontal structures are supported by corbels anchored to the lining by means of embedded reinforcement rebars.

The URD was the first deck to be built, consisting of a reinforced concrete slab casted in situ over a bridge-like formwork structure, to enable the movements of the work vehicles on the deck below.

Instead, the LRD was carried out after the completion of the upper deck, by means of pre-fabricated upturned T shaped elements fastened together by means of reinforcement and concrete casting.

The invert space (beneath the LRD) houses the pipelines and cable routes for plants and systems, including the water pumps, which are draining out collected waste-water from the deepest part of the tunnel outside to the European side (Figure 6).

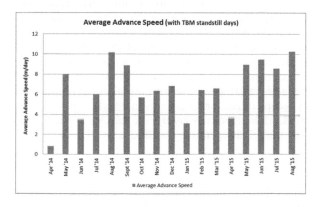

Figure 4.   Average advance speed including TBM standstill days.

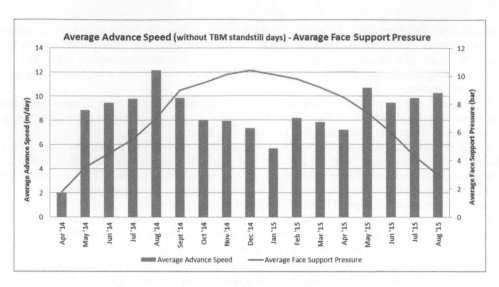

Figure 5.   Average advance speed without TBM standstill days and average face support pressure.

Figure 6.   Tunnel full-scale model.

The concrete rings forming the tunnel lining have been entirely coated with fire-resistant plaster (30 mm thick), which ensures protection against fire, according to the RABT-ZTV (EUREKA) fire time-temperature curve (exposure to a temperature of 1200° C for 60 minutes and 110 minutes cooling). The protective layer must guarantee the absence of concrete spalling and of any deterioration of the structural resistance of the rings.

The temperature of the concrete, in fact, in the event of a fire, must remain below 380° C, while that of the reinforcement rebars must remain below 250° C (as verified by means of ad hoc laboratory tests, on a real scale, conducted on several of the reinforced concrete segments).

There are emergency escape routes for pedestrians that vertically connect the two decks through a 'filter zone' every 200 m, while emergency lay-bys for vehicles can be found every 600 m.

The speed limit in the tunnel, in operating conditions, is 70 km/h.

5.6   Seismic joints

As previously mentioned, the area of Istanbul is intensely seismic, with the 'North Anatolian' fault line running near the tunnel alignment, at a distance of approx. 17 m.

Geologically, the area crossed by the tunnel can be roughly divided into two different structures: a rock area (the so-called Trakya formation) and an area of marine sediments with extremely heterogeneous stratifications.

Normally, tunnels do not require the construction of structures for minimizing the effects of an earthquake, but in this particular case, to take into account the different behavior of the two geological structures during an earthquake (due to their different degrees of rigidity), two identical seismic joints have been installed.

In the event of a large earthquake, in fact, it is expected that the section of tunnel built in the soft alluvial deposits will be subject to longitudinal and transverse displacement, unlike the section in the Trakya Formation. Therefore, different types of displacement will be produced between the tunnel sections in the soft soil and in the rock.

To prevent these differential movements to produce concentration of strains and to enable the dissipation of such effects, the tunnel has been equipped with two flexible joints (Figure 7) installed in two contact areas between the Trakya formation and the marine sediments.

These two areas can be found at approx. 1/4 and 2/5 of the length of the TBM excavation (from the ATB on the Asian side).

However, it should be taken into account that these two joints, by allowing the adjacent sections to move longitudinally and laterally, relative to each other, also reduce the longitudinal confinement in the other parts of the tunnel, as a result of which water could infiltrate into the tunnel due to the decompression of the gaskets around the circumferential joints of the rings.

To prevent this, the rings closest to the seismic joints feature special bolted connections, in order to ensure that the gaskets remain confined and compressed.

The transition from the precast reinforced concrete standard rings (the standard tunnel lining) to the metal ring housing the seismic joint occurs gradually (Figure 8).

Initially, the standard concrete ring is coupled to the first transition concrete ring capable of receiving, on one face, the usual inclined circumferential bolt and featuring on the other face special pockets for enabling its connection to the adjacent second transition concrete ring, by means of short circumferential bolts (130 mm), which ensure effective coupling even in seismic conditions and the suitable compression of the gaskets. The second adjustment/transition concrete ring features pockets, on both faces, for housing the short circumferential bolts connecting it to the first metal ring. This metal ring, which is as thick as the concrete ring, is finally coupled, on the other face, to the metal ring housing the seismic joint.

The ring with the seismic joint is made of specific metal segments which lack, in the middle, the bracing structures of the external sheet lining and, therefore, allow deformation during the earthquake.

Figure 7.   Seismic joint segment model (ATAŞ et al. 2015).

Standard concrete ring	1th Transition concrete ring	2nd Transition concrete ring	Transition steel ring	Steel ring with Seismic Joint	Transition steel ring	2nd Transition concrete ring	1th Transition concrete ring	Standard concrete ring

Figure 8.   Rings assembly sequence for the installation of the seismic joint.

According to the assembly order, the ring housing the seismic joint is followed by a steel ring and a further two transition concrete rings and, finally, by the standard concrete rings.

At the moment of installation of a ring containing the seismic joint, of course, it is still not suitably configured for performing its function, because it still contains metal thrust members that are blocking its movements, to enable TBM to advance, by applying its thrust, without causing any deformation to the joint itself, during construction (Figures 9, 10).

In this initial stage, after the assembly of all the segments of the seismic ring, the primary water sealing rubber is also put into place towards the external section of the ring.

Later on, when the advancement of the TBM can no longer affect the joint in any way, the thrust members can be removed and the secondary water sealing rubber installed.

The two water sealing rubber bands are necessary to ensure the water tightness of the joint, in the event an earthquake damages the part of the extrados lining of the metal ring in contact with the ground. This is particularly important for ensuring tunnel safety, if one considers that the seismic joints are located in points where the hydrostatic pressure is particularly high.

Finally, the joint is completed with the putting into place of the load-bearing bar and sleeve, which purpose is to limit movements during the earthquake.

Figure 9.   Assembly of the metal ring containing the seismic joint.

Figure 10.   The second transition concrete ring, metal transition ring and metal ring with the seismic joint.

The seismic joints allow displacement limits of +/- 50 mm for shear and +/- 75 mm for extension/contraction.

6 CONCLUSIONS

The construction of the Istanbul Strait Road Tube Crossing Project, and of the Eurasia Tunnel in particular, has proved a resounding global success, from both an engineering point of view, and with regard to the management of the entire project, including its financing.

The BOT (Build Operate and Transfer) arrangement, in fact, has attracted financing from investors without draining public resources.

After the full definition of the financing stage and the conclusion of the final agreements, the project was completed eight months earlier than the anticipated 4 years and 7 months.

From an engineering point of view, the tunnel crossing underneath the Bosphorus strait represents an encouragement and a benchmark for other challenging projects, which may learn from its ability to successfully tackle and overcome particularly critical and unique environmental conditions, also in consideration of the large excavation diameter (13.71 m), the intense seismicity of the region, the high water pressures at the excavation face (11 bar) and the particular nature of the surrounding ground.

ACKNOWLEDGEMENTS

The authors would like to thank:

AYGM Istanbul - Altyapı Yatırımları Genel Müdürlüğü İstanbul - General Directorate of Infrastructure Investments Istanbul.

ATAŞ - Avrasya Tüneli İşletme İnşaat ve Yatırım A.Ş.

YMSK JV - Yapı Merkezi İnşaat ve Sanayi A.Ş. e SK Engineering & Construction Co. Ltd.

REFERENCES

Arioğlu, E, Arioğlu, B. 2015. The most challenging phase has been completed. *Announcement from Yapı Merkezi*. www.ym.com.tr.

Burger, W. & Bäppler, K. & Schleer, G. 2017. La costruzione del tunnel Eurasia. *Strade & Autostrade* n.2.

Chiara, S. 2014. Ancorato al Bosforo. *leStrade* n.4.

Pigorini, A. 2009. The New High Speed Turin-Salerno line: design and construction choices for sustainable infrastructures. *Proceedings of the Tunnels workshop for high-speed railways*. Porto, Portugal.

Rinaldi, G. & Cuppone, P. 2017. Il sottoattraversamento stradale del Bosforo – Eurasia Tunnel. *Gallerie e Grandi Opere Sotterranee* n.123.

Valenza, P. & Arioğlu, B. & Burak Gökçe, H. & Arioğlu, E. 2015. A Challenging Project under Bosphorus: Eurasia Crossing. *Proceedings of the International Conference The great infrastructures and the strategic function of alpine tunnels*. Rome, Italy.

Tunnels and Underground Cities: Engineering and Innovation meet Archaeology,
Architecture and Art, Volume 6: Innovation in underground engineering,
materials and equipment - Part 2 – Peila, Viggiani & Celestino (Eds)
© 2020 Taylor & Francis Group, London, ISBN 978-0-367-46871-2

TBM data processing for performance assessment and prediction in hard rock

A. Rispoli & A.M. Ferrero
Università degli Studi di Torino, Turin, Italy

M. Cardu
Politecnico di Torino, Turin, Italy

ABSTRACT: TBM data processing is an important aspect that affects several phases of the construction process of a tunnel excavated by TBM. The real-time machine monitoring is required during the excavation phase, and the TBM data are used after the tunnel completion to gather information for predictions in new tunnel projects. The main issues typically involved in TBM data processing include the raw data management and filtering, the implementation of the rock mass data, and the machine performance evaluation. The paper presents a series of procedures for dealing with the TBM data processing in hard rock context, starting from the raw data provided by the machine acquisition system. Guidelines for an efficient data filtering are provided, together with techniques to improve the representativeness of the data sample. Several techniques for analysing and modelling the datasets are reported, including descriptive statistics, regression analysis and neural net fitting. An approach to carry out probabilistic analysis on the TBM parameters is also described.

1 INTRODUCTION

The modern TBMs are equipped to record a high amount of data during the excavation, providing useful information for dealing with important aspects, such as the real-time monitoring and the machine performance assessment and prediction. However, the conventional data acquisition systems provided by TBM manufacturers are often not able to supply data that are directly usable to solve all the specific issues required, thus making the data processing an important factor both during and after the tunnel construction process. In this regard, some works made over the last years are focused on the machine data management and monitoring (e.g. Mooney et al. 2012; Marchionni et al. 2013; Moreno et al. 2015; Wang et al. 2018). Many other studies are devoted to machine performance prediction models on the basis of the TBM data analysis of past projects (e.g. Bruland 1998; Barton 1999; Alber 2000; Bieniawski et al. 2006; Gong and Zhao 2009; Hassanpour et al. 2010; Khademi Hamidi et al. 2010; Farrokh et al. 2012; Delisio and Zhao 2014; Benato and Oreste 2015; Salimi et al. 2016; Zare Naghadehi and Ramezanzadeh 2017; Avunduk and Copur 2018). Nevertheless, the procedures employed for TBM data processing are often not detailed in the literature, including aspects such as the database construction, data filtering, rock mass data implementation, datasets creation and analysis.

This paper describes the data processing used to carry out the TBM performance analysis and prediction in Rispoli (2018), which refers to the 7 km Maddalena exploratory tunnel, excavated by a 6.3 m diameter gripper TBM in hard rock (e.g. Rispoli et al. 2016; Parisi et al. 2017; Rispoli et al. 2018). The tunnel crossed two main lithological units: the first unit extends for about 1 km and is characterized by aplitic gneiss (RMR~75), whereas the second unit has a length of around 5.7 km and mainly consists of mica-schists (RMR~59).

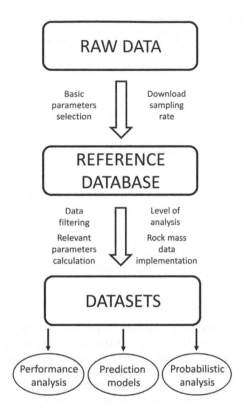

Figure 1.   The main steps of the TBM data processing described in this study.

Figure 1 shows the main steps of the data processing performed. The first step was the reference database creation, which involved two main issues: the selection of the basic instantaneous parameters and the choice of the sampling rate of the download from the TBM acquisition system. The reference database includes the raw data used as a basis for the datasets generation. Each dataset provides the relevant parameters calculated for a given interval sampled from the reference database. The datasets creation involved the selection of the level of analysis, the calculation of the TBM parameters and the data filtering, as well as the implementation of the rock mass data, which were provided by the geological surveys carried out during the excavation. The datasets generated were employed to perform several studies, including machine performance analysis, creation of prediction models and probabilistic analysis.

2   REFERENCE DATABASE

2.1   *Selection of the basic parameters*

At present, the TBM acquisition systems are able to store many parameters, providing information about several aspects of the tunnel excavation. However, only some of them are actually essential for TBM performance assessment and prediction, whereas others do not provide useful information for that (e.g. belt data, water system status, cutterhead brake pressure). Moreover, some potentially relevant parameters may not be useful because of their implementing rules, e.g. the thrust cylinder stroke was not used to define the cutterhead advancement, since the direction of movement of the cylinders was tilted compared to the tunnel axis. The parameters accuracy, according to the equipment employed for their measurement, is

Table 1. Basic instantaneous parameters included in the reference database.

Parameter	Unit
Date	dd-mm-yyyy hh:mm:ss
Total thrust	kN
TBM excavation	yes or not
Cutterhead speed	rev/min
Torque	kN.m
Chainage	m

another important factor to consider. In this regard, an efficient synergy between the TBM manufactures and the on-site technical staff is not always possible, and the exact nature of some TBM parameters provided by the machine acquisition system is sometimes unknown.

On this basis, only a limited number of parameters directly acquired by the TBM were included in the reference database (Table 1), whereas the other relevant instantaneous parameters were obtained by means of user-selected formulas.

2.2 Sampling rate of the download

The sampling rate of the download was selected on the basis of the analysis to perform and of the data management software available. In order to increase the representativeness of the database, the maximum sampling rate allowed by the machine acquisition system was used (i.e. one data every 5 s). However, this solution involved managing a high volume of data, namely more than 17,000 acquisitions for each of the more than 1000 working days sampled, requiring a high capacity of storage by the software and the development of specific calculation codes.

3 DATASETS CREATION

3.1 Level of analysis

The level of analysis defines the sampling intervals of the reference database in which each dataset is evaluated. This issue is particularly relevant, since different sampling procedures may involve variations in the results produced.

Several factors were taken into account to select the level of analysis, including database characteristics, purpose of the analysis, and accuracy of the parameters involved. According to the reference database, the level of analysis may be defined by two different parameters: the acquisition date or the tunnel chainage. Using the date allows evaluating each dataset on a time basis, which is useful for assessing the machine performance in pre-determined temporal intervals; using the chainage, the tunnel alignment can be divided into sections with specific mechanical properties. On this basis, two different levels of analysis were selected for the TBM performance assessment:

– Daily level, where each dataset refers to a working day, according to the work schedule;
– Survey level, where each dataset is related to a tunnel section that was subjected to a geo-structural survey.

3.2 Calculation of the relevant TBM parameters

The relevant parameters in TBM performance assessment include overall performance, net performance, rock mass boreability and cutting efficiency (Rispoli et al. 2018). Moreover, an accurate performance evaluation cannot disregard the excavation parameters which define the machine operating level, such as total thrust (FT), rotational speed (RPM), and torque (T).

Table 2. Calculation of the relevant TBM parameters for each dataset. FT is the total thrust, RPM is the rotational speed, T is the torque, and A_{face} is the cross-sectional area of the tunnel.

Parameter		Equation	Level of analysis
OVERALL PERFORMANCE	Advance rate AR [m/d]	$\dfrac{Boringlength[m]}{Workingtime[d]}$	DAILY LEVEL
	Utilization factor UF [%]	$\dfrac{Boringtime[h]}{Workingtime[h]} \cdot 100$	
NET PERFORMANCE	Penetration rate PR [m/h]	$\dfrac{Boringlength[m]}{Boringtime[h]}$	SURVEY LEVEL
	Rate of penetration ROP [mm/rev]	$\dfrac{PR[m/h] \cdot 100}{RPM[rev/min] \cdot 60}$	
ROCK MASS BOREABILITY	Field penetration index FPI [kN/mm]	$\dfrac{FT[kN]}{No.ofcutters \cdot ROP[mm/rev]}$	
CUTTING EFFICIENCY	Field specific energy SE [MJ/m³]	$\dfrac{2 \cdot \pi \cdot RPM[rev/min] \cdot T[MN \cdot m]}{PR[m/s] \cdot A_{face}[m^2]}$	

The overall performance are usually evaluated by the advance rate (AR) and the machine utilization factor (UF), whereas the net performance are often defined by the penetration rate (PR or ROP). The field penetration index (FPI) and field specific energy (SE) can also be used for assessing the rock mass boreability and the cutting efficiency, respectively.

To choose a proper level of analysis in the calculation of the TBM parameters, the factors that mainly affect the machine performance were taken into account. In particular, the overall performance are heavily influenced by the operational factors, whereas the net performance are most affected by the mechanical and geological factors (Bilgin et al. 2014). For these reasons, the daily level was used to assess the overall performance in uniform time ranges, allowing a better evaluation of the effect of the operational variations, including those related to the work schedule. On the other hand, the survey level was more consistent in the evaluation of the net performance, rock mass boreability and cutting efficiency, since each tunnel portions surveyed showed uniform mechanical properties. As a result, two different groups of datasets were generated.

For each dataset, the relevant TBM parameters were obtained from the basic instantaneous parameters included in the reference database, as reported in Table 2. The equations used are based on the following parameters:

- Working time, calculated as the difference between the date parameter at the end and at the beginning of the interval sampled (the contribution of the non-working days/festivities was subtracted);
- Boring time, obtained from the number of acquisitions when TBM is cutting within the interval sampled;
- Boring length, calculated as the difference between the chainages at the end and at the beginning of the interval sampled;
- Excavation parameters (i.e. FT, RPM, T), obtained as the average value of the instantaneous parameters within the interval sampled.

3.3 Implementation of the rock mass data

The implementation of the rock mass data in the datasets is one of the main issues to solve in TBM data processing (Moreno et al. 2015). The combination between TBM and rock mass parameters is made difficult by the different acquisition rate. Moreover, in order to perform a reliable comparison between machine performance and rock mass properties, each TBM dataset must be representative of the mechanical properties provided by the corresponding rock mass dataset.

In the case under study, the rock mass datasets refer to the information provided by the geo-structural surveys performed on the tunnel walls during the excavation, including RMR, GSI, uniaxial compressive strength (UCS), degree of fracturing, joints orientation and conditions. Therefore, the TBM datasets generated according to the survey level showed a perfect combination with the rock mass datasets, resulting in the creation of complete datasets, which were used to investigate the relationship between net machine performance and rock mass properties, and to create the prediction models.

3.4 Data filtering

The data filtering is an important aspect in TBM data processing, in order to ensure the representativeness of the data sample, improving the reliability of the results. In this study, different filtering conditions were used depending on the parameters involved and on the analysis to perform. They can be divided into three main groups:

1) Anomalies in the reference database;
2) Outliers in the excavation parameters distribution;
3) Reliability tests of the datasets.

The first group includes the early filtering operations to remove from the reference database the data related to values beyond the machine capacity, which are usually due to acquisition system failures.

The second group of filtering conditions was used in the calculation of the excavation parameters of each dataset. In particular, the average value was calculated by not taking account of the outliers, defined by a quantile-based approach (Figure 2) originally proposed by Tukey (1977). This approach was applied to each interval sampled and allowed to assess the outliers on the basis of the characteristics of the parameter distribution, improving the reliability of the average value estimated, which is quite affected by extreme data (limited robustness).

The last filtering operations were applied to the datasets. Several calculation codes were developed to identify the datasets that are not reliable, due to the machine acquisition system failures that were not detected in the previous filtering operations. In particular, a dataset was removed when at least one of the following conditions was observed in the database interval sampled:

– There is a lack of data greater than 10%;
– The chainage is not updated at the beginning or at the end of the interval;
– The chainage shows misleading variations while TBM is cutting;
– The excavation parameters that typically vary during the boring time (e.g. FT, RPM, T) are frozen for an extended range.

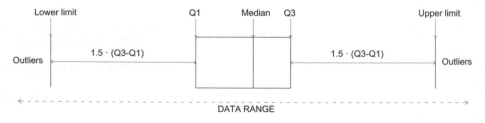

Figure 2. Box plot representation of the quantile-based approach adopted for data filtering. Q1 and Q3 are the first and third quartile of the distribution in the interval sampled (from Rispoli 2018).

4 DATASETS PROCESSING

Three main types of processing were carried out on the datasets created: performance analysis, creation of prediction models and probabilistic analysis. The performance analysis was used to assess the machine performance achieved along the tunnel alignment (Rispoli et al. 2018), and to investigate the relationship between the main parameters analysed. The prediction models were developed with the goal of predicting the TBM performance in new tunnel projects. In this regard, several modelling techniques were employed, and the two main lithological units crossed during the excavation were analysed separately, considering the significant differences in terms of rock mass properties. The most reliable models were obtained through standard regression analysis for the first tunnel portion and by means of neural net fitting for the second unit. Finally, the probabilistic analysis was used for assessing the distribution of the relevant TBM parameters. In this case, the datasets consisted of the whole reference database, properly filtered according to the Section 3.4. Table 3 summarises the characteristics of the different datasets involved in the data processing, including their size, the tunnel length considered and the boring hours analysed.

4.1 *Performance analysis*

Several techniques of descriptive statistics were employed to analyse the TBM performance, including the use of statistical indexes, scatter plots, and matrix of scatter plot and correlation. Statistical indexes such as mean, median, standard deviation, minimum and maximum value were used to summarise the distribution of the parameters for each lithological unit crossed. In this regard, since the length of the tunnel portions surveyed was not constant (typically between 6 and 14 m), the statistical indexes related to parameters from the survey level analysis were weighted in respect of each survey length. The scatter plots were useful to examine the trend of the parameters along the tunnel alignment, highlighting aspects such as the driving philosophy and the tunnel portions where a trend variation occurred. The scatter plot matrix (Figure 3) allowed a qualitative evaluation of the shape and direction of the relationships between the parameters, including those between machine performance and rock mass properties. The correlation matrix provided the measure of the strength and direction of the linear relationship, by means of the correlation coefficient (Pearson's ρ).

4.2 *Prediction models*

One of the crucial issues in the prediction models development is the selection of the variables involved in the model, which are the output variable and its predictors. As for the output variable, PR and ROP are the two most common indexes employed to define the net performance; however, both are affected by the machine operating level. For this reason, reference is often made to FPI (e.g. Gong and Zhao 2009; Hassanpour et al. 2010; Khademi Hamidi et al. 2010; Salimi et al. 2016), which is not directly dependent on mechanical and operational variations, and allows defining the PR if FT and RPM are known. Nevertheless, as noted by Farrokh et al. (2012), FPI models are reliable in new tunnel performance predictions only if the thrust level applied is similar to that of the original project.

Table 3. Characteristics of the datasets involved in the data processing.

Type of processing	Tunnel length [m]	Boring time [h]	No. of datasets
Performance analysis	4075	2819	295
Regression analysis	299	214	33
Neural net fitting	3776	2605	262
Probabilistic analysis	4075	2819	2,029,391

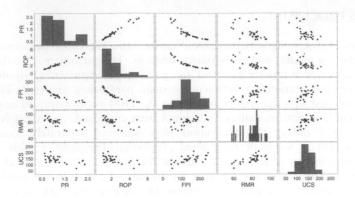

Figure 3.   Example of scatter plot matrix used for performance analysis.

With regard to the predictors, the choice falls on the parameters with the greatest impact on the output variable, and the performance analysis reported in Section 4.1 may provide useful information in this respect. Some modelling techniques like the stepwise regression are able to automatically select the best predictors depending on statistical criteria. The use of parameters that are usually predictable in the early stage of a tunnelling project is preferred, in order to ensure the applicability of the model in new project predictions.

4.2.1 *Regression analysis*
Simple and multiple regression analysis were performed through standard and stepwise regression procedures. The impact of each predictor on the regression model was checked by means of the standard error of the coefficients, p-value and variance inflation factor. Several regression models were developed by testing different regression types, fitting methods and combinations of the predictors. The goodness of fit was tested by statistical indexes such as root mean square error (RMSE), standard and adjusted coefficient of determination (R^2 and AR^2). The data estimated by the regression equations were also compared with the data points, in order to test the reliability of the models.

The most significant model developed is given in Equation 1, obtained by power regression and least square fitting method. It includes the natural logarithm of FPI as output variable, and the compressive strength of the rock mass (UCS_{rm}) as single predictor, which was already used in other TBM performance prediction models (e.g. Alber 2000; Hassanpour et al. 2010; Salimi et al. 2016). The goodness of fit is quite limited: less than 50% of the total variance of the observed data is explained by the regression model. However, an AR^2 of 0.7 can be obtained by removing the two more extreme data points. The reliability of the model for TBM performance prediction is likely limited as well, since the excavation of the first lithological unit crossed did not occur in optimum cutting conditions (Rispoli et al. 2018), and this may result in an underestimation of the machine performance in a new project.

$$LN(FPI) = 3.2826 \cdot UCS_{rm}^{0.10302} \tag{1}$$

where *FPI* = field penetration index in kN/mm; and UCS_{rm} = compressive strength of the rock mass in MPa.

4.2.2 *Neural net fitting*
The neural net fitting uses the artificial neural networks (ANN) logic for fitting a model to the experimental data and is particularly effective to solve nonlinear fitting problems that involve a high volume of data. Several ANN models were tested by varying the input/output variables, number of hidden layers and training algorithm. The most effective model developed involves

PR as output variable, and eight input variables, including FT, RPM, UCS, some RMR partial ratings and the tunnel depth. The optimum number of hidden nodes was 20. The 80% of the datasets were employed for training the algorithm, whereas the 20% were used for testing the model. The training algorithm employed was the Bayesian Regularization algorithm, which is generally more robust compared to the standard back-propagation algorithms (Burden and Winkler, 2008). During the training, the performance of the model were measured by the mean squared error (MSE).

The model developed shows a good fit with the data points (R^2=0.8). Nevertheless, other aspects should also be taken into account to test the actual reliability of the model. First, an explicit equation of the model is not available, as usual in ANN modelling; therefore, the influence of each predictor cannot be properly evaluated. The large number of the predictors is another limiting factor; moreover, some of them may not be available in the early stage of a tunnel project.

4.3 *Probabilistic analysis*

The prediction models developed do not allow a detailed assessment of the output variable distribution. Moreover, the datasets described in Section 3 represent a summary of the reference database within given intervals, resulting in a loss of information. In order to manage the whole reference database and perform probabilistic analysis on the relevant TBM parameters, a user interface in MATLAB (MathWorks) environment was created and it is described in detail in Rispoli (2018). It allows assessing the distribution of a TBM parameter depending on specific conditions defined by the user, including the machine operating level and the rock mass properties. The structure of the user interface consists of a series of blocks, which include commands, editable spaces and the distribution plot, where the histogram of the output variable analysed is shown. Figure 4 shows the screenshot of the user interface during the evaluation of the FPI distribution for a given RMR range; in addition to the distribution plot, information on the interval sampled are included and a particular portion of the variable distribution (i.e. FPI between 15 and 38 kN/mm) is investigated in detail.

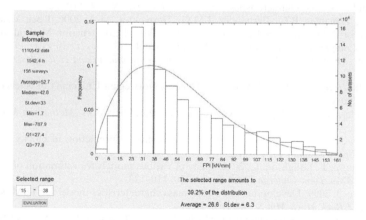

Figure 4. Screenshot of the user interface during the assessment of the FPI distribution for RMR between 40 and 60.

5 CONCLUSIONS

This study provided an approach to perform the TBM data processing for performance assessment and prediction in hard rock. For this purpose, the procedures used to process the raw data from a 7 km tunnel excavated by a 6.3 m diameter Gripper TBM were presented.

The reference database consisting of more than 2 million acquisitions was created by downloading the basic instantaneous parameters from the machine acquisition system, according to the maximum sampling rate.

Starting from the reference database, the TBM datasets were created on the basis of two different levels of analysis/sampling intervals: the daily level was useful for assessing the overall machine performance for each working day, whereas the survey level was used to evaluate the other relevant TBM parameters for each tunnel portion that was subjected to geo-structural survey. This second level of analysis also allowed an effective combination between the TBM and rock mass datasets. Different data filtering procedures were employed to improve the representativeness of the datasets.

Various techniques of descriptive statistics were applied to assess the machine performance and to investigate the relationship between TBM parameters and rock mass properties. Regression analysis and neural net fitting were found to be the most effective techniques in the development of prediction models for the two main lithological units crossed by the tunnel. The creation of a user interface allowed to perform probabilistic analysis on the relevant TBM parameters depending on conditions such as the machine operating level and the rock mass properties.

REFERENCES

Alber, M. 2000. Advance rates of hard rock TBMs and their effects on projects economics. *Tunn Undergr Space Technol* 15(1): 55–84.

Avunduk, E., Copur, H. 2018. Empirical modeling for predicting excavation performance of EPB TBM based on soil properties. *Tunn Undergr Space Technol* 71: 340–353.

Barton, N. 1999. TBM performance in rock using QTBM. *Tunnels & Tunnelling International* 31: 41–48.

Benato, A., Oreste, P., 2015. Prediction of penetration per revolution in TBM tunneling as a function of intact rock and rock mass characteristics. *Int J Rock Mech Min Sci* 74: 119–127.

Bieniawski, Z.T., Tamames, B.C., Fernandez, J.M.G., Hernandez, M.A. 2006. Rock Mass Excavability (RME) Indicator: New Way to Selecting the Optimum Tunnel Construction Method. In: *ITA-AITES World Tunnel Congress & 32nd ITA General Assembly*, Seoul.

Bilgin, N., Copur, H., Balci, C. 2014. *Mechanical excavation in mining and civil industries*. Boca Raton: CRC Press.

Bruland, A., 1998. *Hard Rock Tunnel Boring*. PhD Thesis, Norwegian University of Science and Technology (NTNU), Norway.

Burden, F., Winkler, D., 2008. Bayesian regularization of neural networks. *Methods in Molecular Biology* 458: 23–42.

Delisio, A., Zhao, J., 2014. A new model for TBM performance prediction in blocky rock conditions. *Tunn Undergr Space Technol* 43: 440–452.

Farrokh, E., Rostami, J., Laughton, C. 2012. Study of various models for estimation of Penetration rate of hard rock TBMs. *Tunn Undergr Space Technol* 30: 110–123.

Gong, Q.M., Zhao, J. 2009. Development of a rock mass characteristics model for TBM penetration rate prediction. *Int J Rock Mech Min Sci* 46(1): 8–18.

Hassanpour, J., Rostami, J., Khamehchiyan, M., Bruland, A., Tavakoli, H.R. 2010. TBM performance analysis in pyroclastic rocks, a case history of Karaj Water Conveyance Tunnel (KWCT). *Journal of Rock mechanics and Rock Engineering* 4: 427–445.

Khademi Hamidi, J., Shahriar, K., Rezai, B., Rostami, J. 2010. Performance prediction of hard rock TBM using Rock Mass Rating (RMR) system. *Tunn Undergr Space Technol* 25(4): 333–345.

Marchionni, M., Selleri, A., Stahl, F., Messina, L. 2013. Intensive application of the TBM data management system for the work supervisor of the largest worldwide TBM-EPB project. In: Anagnostou, G., Ehrbar, H. (ed.), *World Tunnel Congress 2013 Geneva, Underground – the way to the future*: 1312–1319. London: Taylor & Francis Group.

Mooney, M.A., Walter, B., Frenzel, C. 2012. Real-time tunnel boring machine monitoring: A state of the art review. *North American Tunnelling*: 73–81.

Moreno, P., Ruiz, M., Gorines, F. J., 2015. TBM Process Data Management System. *International Journal of Engineering and Technology* 7(5): 431–437.

Parisi, M. E., Brino, L., Gilli, P., Fornari, E., Martinotti, G., Lo Russo, S. 2017. La Maddalena exploratory tunnel. *Geomechanics and Tunnelling* 10(3): 265–274.

Rispoli, A., Ferrero, A.M., Cardu, M, Brino, L., Farinetti, A. 2016. Hard rock TBM performance: preliminary study based on an exploratory tunnel in the Alps. In: Ulusay et al. (eds). *Rock mechanics and rock engineering: from the past to the future*. Taylor & Francis Group, London, pp 469–474.

Rispoli, A. 2018. *Hard rock TBM excavation: performance analysis and prediction*. PhD Thesis, University of Turin, Italy.

Rispoli, A., Ferrero, A.M., Cardu, L., Farinetti, A. 2018. TBM performance assessment of an exploratory tunnel in hard rock, in: Litvinenko (ed), *Geomechanics and Geodynamics of Rock Masses*, Taylor & Francis Group, London, pp 1287–1296.

Salimi, A., Rostami, J., Moormann, C., Delisio, A. 2016. Application of nonlinear regression analysis and artificial intelligence algorithms for performance prediction of hard rock TBMs. *Tunn Undergr Space Technol* 58: 236–246.

Tukey J.W. 1977. Box-and-Whisker Plots. §2C. In: *Exploratory Data Analysis*: 39–43. Reading: Addison-Wesley.

Wang, Q., Xie, X., Zhou, B., Huang, Z. 2018. Intelligent Sensing of Shield Tunneling Information in Real-Time via Mobile Application. *World Tunnel Congress 2018 Dubai, The Role of Underground Space in Building Future Sustainable Cities*.

Zare Naghadehi, M., Ramezanzadeh, A. 2017. Models for estimation of TBM performance in granitic and mica gneiss hard rocks in a hydropower tunnel. *Bulletin of Engineering Geology and the Environment* 76(4): 1627–1641.

Tunnels and Underground Cities: Engineering and Innovation meet Archaeology,
Architecture and Art, Volume 6: Innovation in underground engineering,
materials and equipment - Part 2 – Peila, Viggiani & Celestino (Eds)
© 2020 Taylor & Francis Group, London, ISBN 978-0-367-46871-2

Building information management for tunneling

F. Robert & A. Rallu
CETU (Tunnel Study Centre), Bron, France

C. Dumoulin
French national research project MIN^{n}D, Paris, France

N. Delrieu
ANDRA (French National Radioactive Waste Management Agency), Châtenay-Malabry, France

M. Rives
Vianova Systems, Boulogne-Billancourt, France

M. Beaufils
BRGM (French Geological Survey), Orléans, France

ABSTRACT: MIN^{n}D (Modeling Interoperable Information for sustainable Infrastructures) is a French national research project dedicated to BIM for infrastructures. Many working groups are in progress, treating roadways, bridges, environment, etc. On the initiative of ANDRA, the French National Radioactive Waste Management Agency, a major effort has been undertaken for underground works. In 2017, a first step was to build the Data Dictionary, proposing an organic decomposition of the systems and subsystems, as well as the necessary data flow between the different stakeholders (Information Delivery Manual). This paper presents the main results of this work. This concerns the definition of relevant data concerning the geometry, geology, hydrogeology, geotechnics, design, construction of the structure itself, including ventilation, electrical and management equipment.

1 INTRODUCTION

The French national research project MIN^{n}D (French acronym meaning Modeling Interoperable Information for sustainable Infrastructures) was born from the following observation: the digital model that was developed for industrial products (cars, aircrafts, etc.) that have the distinction of being modeled independently their environment, is not satisfactory for infrastructures that interact with the ground, and whose construction phases must be able to be modeled (earthworks, pouring, construction tools, etc.). If things are changing for the building, the MIN^{n}D national research project was launched to deal with the problem of infrastructures. Several use cases have been identified. For example, use case No. 3 "Bridge" for bridges, was developed in MIN^{n}D.

The use case No. 8 relating to the underground infrastructures is emblematic of the interaction between the structure and its environment insofar as the structure is in permanent interaction with it (soil, networks, and neighboring works) as well during construction phase to the operation phase. In 2017, under the leadership of ANDRA (French National Radioactive Waste Management Agency), an important work was led to work on the modeling of data relating to underground structures. This work was carried out on two levels: on the one hand the modeling of the civil engineering structure and its equipment (led by the working group named GC – cf. Section 5) and on the other hand the modeling of the environment of the structure (led by the working group named GT – cf. Section 6), namely the soil and its various components. This paper presents the main results achieved and future prospects.

2 MIN^{N}D PROJECT

MIN^{n}D is a French national research project involving 71 partners and gathering stakeholders of the construction sector including owners, engineering companies, contractors, industrial companies, public and private laboratories, universities, *Grandes Ecoles*. Cf. www.minnd.fr.

Controlling and sharing information are key issues for the construction industry which must nowadays cope with major changes in its activity, such as project complexity, eco-design development, new types of partnership between actors (Public Private Partnerships, Concessions), obligation to manage risks (anticipation, evaluation, distribution) or development of Building Information Modeling (BIM). Drawings, notes and records, files have shown their limits. Moving to information modeling applied to infrastructure is already in progress. The first challenge is therefore to move to the item that is the finest information by establishing a structure and a recognized internationally standard for information exchange as well as to use adapted tools, either transversal, such as digital models, or specialized and developed internally by each actor.

The digital mockup of a manufactured product focuses on a "new" product. Regarding buildings, the BIM describes only the building itself, often neglecting the surrounding environment. Foundations and connections to the external networks are modeled, but the surrounding ground is often omitted, or described only at its final stage. The focus is clearly put on the definition of the delivered product, mainly on the design. The construction steps including stop pouring concrete, the necessary tools, logistics and temporary elements are not modeled.

Regarding tunnels, the focus cannot be limited to the supporting structure and functional systems. On one hand, the geometric design of the supported road or railway is a key driver of the tunnel design. And on the other hand, the surrounding ground has to be taken into account as well as the impacted and impacting environment.

Some considerations imposed by the road must be properly addressed during the design process, including design speed, design traffic volume, number of lanes, level of service, sight distance, alignment, super-elevation and grades, cross section, lane width, horizontal and vertical clearance. These considerations are also mandatory for bridge design. MIN^{n}D is currently finishing a research work regarding BIM applied to bridges. The results of this research have been shared with the Infrastructure Room of buildingSMART International (bSI) and will contribute to the extension of the international standard *"ISO 16739:2013 Industry Foundation Classes (IFC) for data sharing in the construction and facility management industries"* for application to bridges. This work will provide a good input for supporting road geometry applied to tunneling.

What is really new regarding tunneling is geo-modeling Instead of designing what has to be built and how to do it, only observations can provide some information locally about what is in place and interpretations try to fill the gap and imagine the missing information. When the tunnel is bored or the ground cut, the characteristics of the extracted ground are finally known, but only at the end when analyzing the muck. But the influence of the surrounding environment and the impact of the tunnel construction on the environment will remain estimations and therefore induce risk management.

Except the geotechnical aspect, a tunnel looks like a "long" building following the alignment of the hosted road or railway. Several functional systems describe the operational systems, such as drainage, ventilation, electricity… But risk management, construction management, ground boring including replacement of the ground in place by the built structure, temporary works, logistics require a 4D modeling, which means the capability of modeling the progress of the construction versus time.

Finally, Building Information Modeling should offer a 3D "architectural" model defining the tunnel project as it will be delivered to the client, including the tunnel structure, the operational systems, the environment (ground, terrain, external networks) modified by the project.

This model will be associated to a 4D model able to manage the construction sequences, and to a structural analysis model. In addition, all the knowledge related to the existing environment that will not be modified directly by the project but could impact or be impacted by the project will be described in a geo-model. Links between the geo-model and the project models will be established to manage the changes that can occur in the different models during the project lifecycle.

All this activity is developed in a specific sub-project called MIN^nD UC8 (Use Case n° 8) Tunnel.

3 THE ORIGIN OF THE USE CASE N°8 DEDICATED TO UNDERGROUND INFRASTRUCTURES

3.1 *Context*

ANDRA (French National Radioactive Waste Management Agency) is a public industrial and commercial establishment (EPIC), in charge of long-term radioactive waste management operations.

In order to manage the ultimate radioactive waste that cannot be stored on the surface or on sub-surface (at a shallow depth), ANDRA is designing a deep geological underground radioactive waste storage facility: Geologic Industrial Storage Center (Cigéo). This project combines all difficulties related to data management and collaborative work of complex projects, such as (i) industrial installations, (ii) large infrastructures, (iii) complex underground work and (iv) basic nuclear installations. This kind of project involves a large panel of professions, each of them using different methods of communications. So data exchange can be hard.

Consequently, ANDRA identified quickly the challenge of a single model shared database and making easier collaborative work and data management. This latter must be accessible by all stakeholders and during all phases of the life cycle of a work via an interoperable format.

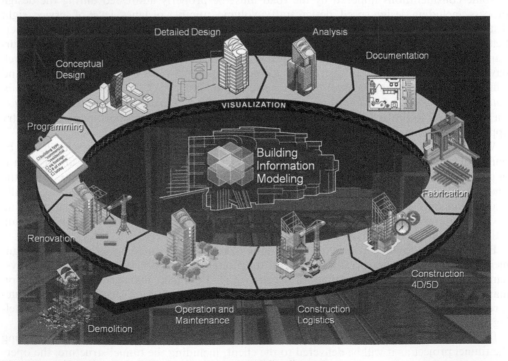

Figure 1.   Life cycle of a building project (source: buildipedia.com).

3.2 *PLM and BIM for information sharing and data management*

This project involves many works of different nature:

- large-scale ground-level infrastructure integrated in its surrounding environment;
- complex underground structures with shafts and galleries;
- industrial equipment's and facilities;
- multiple working and living quarters for several hundred people;
- operating sites.

Indeed, a key point is to share the information between the various stakeholders throughout the project lifecycle, which is designed for secular times. The challenges in terms of coherence and technical, geometric and geographical coordination are numerous.

A tunnel project is therefore part of an approach of more than a century of multi-sector, multi-stakeholder cooperation and co-activity, working in parallel across various stages of the project: (i) design, (ii) construction, (iii) operation and maintenance, and (iv) post-closure monitoring.

This therefore represents a challenge in terms of secure data management and traceability, and one that ANDRA is anticipating through the use of two data management tools that are tried and tested in both industry and construction: PLM and BIM.

A first tool is essential to manage in real time the integral memory of the project: the Product Lifecycle Management (PLM), i.e. the management of the life cycle of the infrastructure project. It supports, structures and archives all project information in a secure database. This includes a digital biography that incorporates supplementary data to BIM (knowledge of waste producers' packages, operating history, procedures, etc.) and which facilitates the monitoring, traceability, exchange and archiving of any information. PLM is configured to provide tailored solutions to the specific needs of each stakeholder through industry-oriented views. Responsible for all technical data, PLM also supplies the corporate information system.

The second data and project management tool is BIM: Building Information Model/Modelling/Management, all in one. This is a collaborative working method for generating and working with construction data across the project lifecycle. BIM draws on information exchanged via multi-sector and multi-stakeholder digital models. It is shared by design offices, technical staff, service providers and operators, who all interact around this same dynamic representation and model. Each change is reflected throughout the overall model, which enables an accurate assessment of the future design and makes it possible to carry out simulations. BIM thus provides a shared vision of the project and allows access to the basic data for each component: geometry, materials, mass, costs, timeframes, suppliers, etc.

For his infrastructure project, ANDRA wants to meet the challenge of marrying

- BIM, which manages 3D geometry and basic information for the project's components;
- and PLM, which organizes and combines this information with the records of all project data.

It is a digital innovation that combines the benefits of the two data management tools and presents the complete information of PLM in BIM's highly visual digital model. This synergy has to be able to ensure more streamlined collaboration between the various sectors, driving operational performance. To achieve this goal, ANDRA's teams work with the MIN^{n}D UC8 working group to identify and harmonize the standards and requirements of each stakeholder to integrate them into this new collaborative interface, able to evolve over time.

4 BIM STANDARDIZATION PANORAMA, MIN^{n}D POSITIONING

4.1 *Standardization through IFC*

All what is created or modified by the project has to be defined in a 3D "architectural" model for approval by all project stakeholders. The model description could be based on the international standard "*ISO 16739:2013 Industry Foundation Classes (IFC) for data sharing in the construction and facility management industries*" including the infrastructure extension foreseen by buildingSMART International. Standardized description of the environment is also critical for infrastructure design, especially for engineering structure such as tunnel. This will include the terrain model and the ground layers included in the project limits.

To cover this need, one approach is to extend IFC and define new classes. The China Railway BIM Alliance, for example, proposes definition of concepts such as IfcRockSoilMass, IfcDrillHole and IfcDrillHoleLayer to respectively describe geological units, boreholes, and layers identified in the geology core. The project has to manage and track the construction of the tunnel, and therefore the associated earthworks and tunnel boring activities.

But the aim is not to address geosciences features, but to identify existing standards outside of the BIM/IFC community and define links with them. For a given project, geosciences have to cover a larger space than the project limits in order to identify the ground characteristics related to the tunnel.

4.2 *Alternatives to IFC*

In terms of standardization of Geographic Information Systems (GIS) data in general, several relevant initiatives and standards can be mentioned.

In 2007, INSPIRE, the European Directive introduce several themes and models for data that should be shared in the European Community. Among those topics, most relevant to mention are Building (BU), Geology (GE), Soil (SO), Environmental Monitoring Facility (EF), Natural Risk Zones (NZ).

Most models from INSPIRE rely on standards developed by the Open Geospatial Consortium (OGC), an international non for profit organization dedicated to support and development of standards for geographical information.

One regularly mentioned standard from OGC for bridging BIM and GIS is CityGML that aims at describing city models. Reciprocal conversion from IFC to CityGML is a topic addressed by several working groups, especially one dedicated joint working group between bSI and OGC named Integrated Data Built Environment Domain Working Group (IDBE DWG).

Concerning geology, OGC launched in 2017 a Domain Working Group (DWG) to cover this topic: the GeoScienceDWG. Main objective of this group is to harmonize practices, support and enhance existing standards in that domain inside and outside OGC. This task particularly influences the development and refinements of OGC geoscience standards GeoSciML and GroundWaterML2, which have been designed to describe both geological and hydrogeological features, introducing concepts such as GeologicUnit, Faults, Boreholes, HydrogeoUnits, Voids... but also their associated properties such as EarthMaterial. Those standards have been designed by geoscientists, and are widely implemented by providers of this kind of data, especially geological surveys.

5 MAIN RESULTS OF GC WORKING GROUP

The GC (Civil Engineering and Equipments) working group of the MIN^nD-UC8 National Project was set up to gather the required expertise in order to be able to address the various domains that compose an underground infrastructure, being its civil works parts or its equipments.

This comprehensive knowledge was carried out, after a formal request for proposal process, with more than 25 experts from underground infrastructure owners, consultants and contractors, as well as with IFC experts; all with a solid experience of underground infrastructure projects, in France and internationally.

The goal of the GC working group is to provide specifications to the independent bSI organization for the production of extensions to the current IFC4.x (x being the latest version) format in order to keep fruitful exchange mechanisms of design/build/operate information representing underground infrastructures data; such information being organic, functional or spatial related.

Ultimately, such IFC extensions for underground infrastructures are to be proposed to the ISO organization for deployment as internationally applicable industry standards.

As a first step, the GC working group scoped what underground infrastructures might be: tubes, shafts and storing cells, with a view to produce a breakdown of all of their civil works structures and equipments, their interdependences (relationships) and their characteristics (properties).

The approach used to conduct this work followed a 'why'/'what'/'how' systemic analysis to outline an exhaustive perspective of the programmatic/functional/organic aspects of the infrastructures concerned. This led us to identify a series of sub-systems, as featured below:

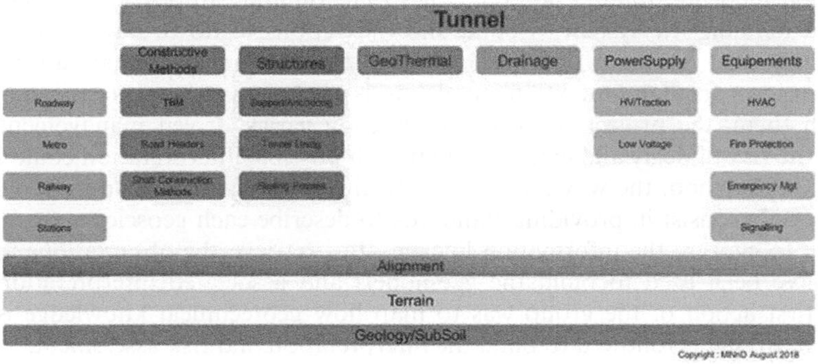

Figure 2.   Decomposition into sub-systems.

In parallel, we focused on where data exchanges take place in a design process (between industry domains experts/at what phases), in a construction process (including between detailed design experts and constructive methods experts) as well as in an operate/maintain context.

As a second step, GC working group started an analysis of the possible implementation of the new objects classes (in IFC terms) needed to provide the functional, organic and spatial representations of the various components of the sub-systems, their relationships and their properties.

This work required first to leverage the existing features classes and their hierarchies available in IFC4.1, and second to propose an enrichment of these, while aiming at introducing as minimum extra complexity as possible into the existing IFC conceptual model.

Then, the GC working group will develop a proposition for an extended conceptual model and the corresponding IFC schema (based on the IFC4.1), with a view to provide definitions for the IFC5 development roadmap, in capitalizing the IFC Alignment 1.1 and IFC Overall Architecture projects led by bSI. Ultimately, this work shall help the bSI organization in:

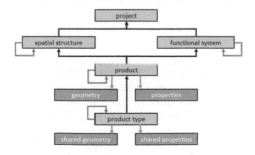

Figure 3.   IFC Tunnel conceptual model (simplified view).

- The development of a semantic description of underground infrastructures in a language implementing the concepts and logic used by underground construction experts;
- The set up and use a common data dictionary for underground infrastructures;
- The development of subsoil modeling (based upon industry practices and input from OGC);
- The resolution of the appropriate objects breakdown into the global IFC schema.

This initiative will also be fed by other national projects shooting for open, international standards and identify use-cases for underground infrastructures.

6 MAIN RESULTS OF GT WORKING GROUP

Environmental modeling follows different rules than Building Information Modeling. While BIM aims at defining the system to build and the actions to do to have it, geomodeling is about exposing what has been observed and possible interpretation that can be built from those observations. In that context, geomodeling is always associated to uncertainties. Then to cope with them, the project coordinator relies on a risk driven management approach, focusing on the risk of delay and extra cost that each possible interpretation could lead to.

Based on that method, the way that geotechnical standardization is addressed in that project does not only consist in providing standards to describe each geoscience feature, but also in being able to capture the information lineage, thus retrieve the observations and assumptions that have been used to build the geomodels and associated interpretations. For this reason, the first action of the group was to map how geotechnical knowledge is built, and describe the path that is followed to build the interpretation and risk assessment.

6.1 *Definition of geotechnical work*

In France, the geotechnical activities are described by the standard NF P 94–500 from the French Association of Normalization (AFNOR). This standard defines the goals that should be reached at each time of the project in several domains (geology, hydrogeology…). In addition, the French Tunneling and Underground Space Association (AFTES) also propose guidelines for interpretation and illustration of the standard to facilitate its application (GT43R1F1).

Lecture and analysis of that standard lead to the proposition of dividing geotechnical activity in ten main topics: RECO deals with the collection of observations, measures and information on the field, test, laboratory analysis and survey. GEOL, HYDR and GTCH respectively deals with the creation of geological, hydrogeological and geotechnical model, including the research of coherence between them. CALC and MECO respectively deals with infrastructure sizing, determination of interaction with the soil, definition of the geotechnical influenced zone (ZIG) and the proposition of appropriate constructions. ENVI and AVOI respectively discuss the management of the impacts on both environment and built areas. Finally, RISK is about the assessment and management of risks and uncertainties associated to the data and proposed models and MRCH addresses the collection of useful data and deliverable that shall be available for tenders.

6.2 *Information Delivery Manual*

For this project, ten Information Delivery Manuals (IDM) were developed, one per topic. Each IDM consists in a process map and a glossary. The process map provides a graphical representation of the sequence of activities in the topic and the involved data. The glossary acts as a legend for the process map, and enable to provide more detailed description of both data and actions.

Process maps from UC8-GT all have the same design: in a middle part, actors and processes sequence. On the left, data used in the processes. On the right, data updated or created by the processes. Each process and datum have an identifier in order to ensure that same data does not have different names in different IDMs and facilitate links between IDM (eg. CALC.D3 stands for the datum "D3" of the "CALC" IDM). Such IDM design also enables to identify easily the key concepts and properties to consider in the Conceptual Model.

6.3 *Conceptual Model inspiration and first propositions*

The IDM clearly highlight that geotechnical activities consist in a chain of tasks including collection of information and formulation of hypothesis that once linked, enable to model the subsurface and determine important information for construction. Discussions in the group also emphasize that results from modeling can be very different from a project to another. The context of the project often constrains and centers the modeling tasks, and necessitates use of specific methods and tools. This lead to a wide range of possible model outputs: from 3D voxels to 2D cross-sections for geometrical modeling, to more numerous and original results for analytical modeling.

Instead of trying to provide standard description for all of those possible outputs, the MIN^{n}D UC8-GT decided to focus on two objectives: 1- describing and associating properties to identified "components" of the subsurface, 2 – describing the geotechnical processes activities and context to facilitate data reuse.

Regarding the first objective, OGC standards GeoSciML and GroundWaterML2 already introduce concepts to describe components of the subsurface. This includes geologic units, geologic structures, aquifers, water bodies, voids. As several international organizations, especially geological surveys adopted them and ensure the maintenance of those standards, relying on them, in opposition to extending IFC for example, appears to be a sustainable and long term solution.

Regarding the second objective, the MIN^{n}D UC8-GT focuses on the ISO: 19156 standards, also known as OGC Observation & Measurements. Basic purpose of this standard is to facilitate exchange of information describing observation acts and their results. The key concept is called "Observation" that aims at determining the value of a property (cf. Figure 4). The observation result – the value – describes a phenomenon or property of a feature, the feature-of-interest of the observation. Observations can be realized by an instrument or a sensor, but also by a process chain, human observer, an algorithm, a computation or a simulator. Observation properties provide context or metadata to support evaluation, interpretation and use of the observation results. Finally observation can be linked to another through the relatedObservation association.

In that context, Observations & Measurements (O&M) can be used both for describing acts relative to data collection (observed by humans or sensors), modeling (human interpretation

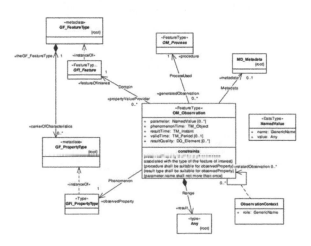

Figure 4.   Observation schema from ISO19156/OGC Observations & Measurements standard.

assisted by software), but could also by extension be used for preconisation (proposition based on model run). The relatedObservation association can be used for lineage and data discovery from results. Finally, the use of such standard would also facilitate data provision and access, as OGC also proposes several interfaces to handle O&M structured data: e.g. Sensor Observation Service (SOS) and SensorThings API. Model results and associated data would then be retrieval through standardized protocols.

7 BUSINESS APPROACH, CONSIDERATION OF AFTES'S GUIDELINES

The work to be carried out had to take into account the rules of the art and the uses recommended by the profession of the underground works. This is why the CETU, as a government institution, joined the partners of the MIN^{n}D project in trying to bring its neutral vision of things and by eliciting the AFTES (The French Tunneling and Underground Space Association, member of ITA) recommendations in the process of data exchange between the different actors, and the formalism of the usual deliverables.

The French Tunneling and Underground Space Association (AFTES) aims to unite and mobilize all players in the profession: investors, project owners, project managers, design firms and inspection bodies, research and training institutes, contractors, consultants, architects, town planners, equipment manufacturers, academics...

Thus, for the constitution of the data dictionaries and the organic decomposition of the constituent elements of an underground infrastructure, the bilingual French-English dictionary, used in the framework of the CETU international actions (PIARC, ITA) was used to dispose of an unambiguous basis for working directly in English in order to have a bilingual taxonomy of systems and subsystems.

From this organic decomposition into systems and subsystems, the information delivery manuals (IDM) were produced using specific schemes from the AFTES recommendation GT32R2F1 dealing with the characterization of geological, hydrogeological and geotechnical uncertainties and risks, and the recommendation GT32R3F1 dealing with the consideration of technical risks in projects. Moreover, some parts of the Information Document on the Tunnels Prices, published by CETU in 2016, were used in order to structure some IDM according to a representative decomposition in underground works.

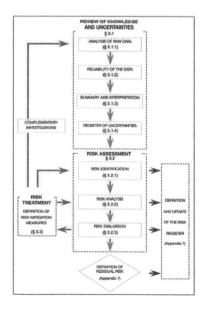

Figure 5. Risk Management Methodology Summary Flowchart (AFTES recommendation GT32R2F1).

These recommendations describe the iterative process of risk management to be implemented during the different stages of the design process (cf. Figure 5), as well as the documents to be produced (the deliverables) for tendering. We can quote the following deliverables:

- Book A1, which includes geotechnical input raw data;
- Book A2, which contains input raw data for nearby buildings;
- Book A3, which contains input raw data for the natural and human environment;
- Book B1, which is the geotechnical summary report;
- Book B2, which is the sensitivity report of neighboring works;
- Book B3, which contains comments on environmental constraints;
- Book C, which is the design report;
- Notice of respect of the environment;
- Risk management plan and the residual risk register.

Thus, the different deliverables, their descriptive content and the flow of information between the various project stakeholders were described in accordance with AFTES recommendations, and therefore according to the business approach advocated by the entire profession.

8 CONCLUSION

The work of both working groups (GC and GT) goes on in order to have a panorama covering the entire life cycle. The next step is to manage the information related to the operation, maintenance, evolution, and, should they occur, damages, repairs, renovations. Data related to the neighboring structures may also be added in so far as they are available. In order to ultimately lead to ISO standardization, the work continues on with the extension of standard proposals from bSI (building SMART International) and OGC (Open Geospatial Consortium), such as IFC-Tunnel, GeoSciML and GroundWaterML. The MINnD project welcomes any partner who would like to get involved.

REFERENCES

www.minnd.fr

Guide d'application au domaine des ouvrages souterrains de la norme NF P 94-500 (version 2013) relative aux missions d'ingénierie géotechnique – AFTES Recommendation n°GT43R1F1 - November/December 2015 – n°252.

Caractérisation des incertitudes et des risques géologiques, hydrogéologiques et géotechniques – AFTES Recommendation n°GT32R2F1-2012 - n°232.

Prise en compte des risques techniques dans les projets d'ouvrages souterrains en vue de la consultation des entreprises – AFTES Recommendation n°GT32R3F1 - November/December 2016 – n°258.

CETU - *Document d'information – Prix des tunnels* – March 2016 - (http://www.cetu.developpement-durable.gouv.fr/genie-civil-a572.html).

ISO 16739:2013 Industry Foundation Classes (IFC) for data sharing in the construction and facility management industries.

AFNOR Standard NF P 94-500 *Missions d'ingénierie géotechnique – classification et spécifications* - November 2013.

Stascheit J., Meschke G., Koch C., Hegeman F. & König M., 2013, Process oriented numerical simulation of mechanized tunneling using an IFC-based tunnel product model, *13th International Conference on Construction Applications of Virtual Reality, 30–31 October 2013*, London.

Mao S., Lebrun JL, Doukari Ö., Aguejdad R. & Yuan Y, Modélisation 3D BIM multi-échelle d'un projet BTP Tunnel, *Spatial Analysis and GEOmatics conference, SAGEO 2015*.

Yabuki N., Representation of caves in a shield tunnel product model. In book: *eWork and eBusiness in Architecture, Engineering and Construction*, pp.545–550, *September 2008*.

*Tunnels and Underground Cities: Engineering and Innovation meet Archaeology,
Architecture and Art, Volume 6: Innovation in underground engineering,
materials and equipment - Part 2 – Peila, Viggiani & Celestino (Eds)*
© *2020 Taylor & Francis Group, London, ISBN 978-0-367-46871-2*

Rehabilitation of the slab under the Mont Blanc Tunnel roadway

N. Rogès
Spie batignolles GC, Boulogne –Billancourt, France

F. Bertino
Cogeis SpA, Quincinetto, Italie

E. Clayton
Setec Als, Paris, France

G. Rakoczy, G. Schwarshaupt & M. Cipollone
GEIE Tunnel du Mont Blanc, Courmayeur, Italie

ABSTRACT: The Mont-Blanc tunnel is operated by a consortium, the GEIE-TMB. It has been operating, maintaining and making the necessary investments since its creation in 2002. It provides all continuous monitoring and inspection operations of civil engineering structures inside the tunnel. This regular monitoring made it possible to detect the need to renew the concrete slab over a 555 metre section in the centre of the tunnel. This contract, attributed to our consortium (Spie batignolles Génie civil/Cogeis/Setec Als) in June 2017, is part of the ongoing safety operation of the structure, a major challenge for the GEIE-TMB. Our group proposed a solution based on an innovative technology: a design of the new prefabricated slabs and the use of two gantries developed for the site. This system enabled us to work mainly at night, during 32 traffic interruptions, while ensuring that the tunnel reopened every morning.

1 INTRODUCTION

During its inspections of the slab under the roadway, the GEIE-TMB noted a gradual deterioration in its condition, particularly in the central zone of the tunnel.

A design-build call for tender was therefore launched for the replacement of 555 linear metres of slabs in the middle of the tunnel (PM6125 to 6680) during short closure periods.

The technical solution proposed by the consortium composed of Spie Batignolles civil engineering and Cogeis, assisted by Setec engineering, and selected by the GEIE-TMB, consists of a multi-stage process:

- the cutting of the existing slab at the level of the 4 supports (2 vertical supports and 2 central partitions) then in the width direction of the tunnel in order to form lifting and transportable elements;
- the removal of the slab pieces by means of a custom made lifting gantry designed for the operation, then the evacuation of these;
- the conveying of the new prefabricated slabs 8.6m long, 2.5m wide and 20cm thick;
- the laying of these slabs still using the lifting gantry by interlocking them together;
- the installation of a temporary connection at the end of each night between the existing slab and the newly laid slabs;
- the reopening of the tunnel with the traffic driving over the concrete slab.

From conception to completion, the project will have lasted a total of 12 months, which represents a major achievement in the framework of a design-realisation of this magnitude.

Right from the bid phase, our solution provided for the cutting and removal of the existing structure with the laying of new prefabricated slabs. We called on our group's factory, located in Aosta Valley, to prefabricate the 222 slabs and 425 pavement elements. The 555 metres of slab were replaced between the end of March and June 2018, totalling 350 hours. To meet this deadline, two work zones were held simultaneously on each side of the site at each night of closure, thus ensuring an average of 20 metres of work.

This system also guaranteed the safety of the tunnel crews and users as well as that of operating conditions.

Figure 1.   3D view of the slab removal stage.

2   THE DESIGN: PRECAUTIONS AND ANTICIPATION

The 6 months of the technical solution design (3 months of PRO studies and 3 months of EXE studies) resulted in choices which formed the basis of the success of the project, some of which are detailed below.

Like any new project, this one had to be carried out according to the latest industry standards, whatever the age of the structure. From the first design stages, it was therefore necessary to calculate the less common elements in conventional reinforced concrete, such as fire resistance:

– modelling on specialized software such as Safir or Phytagore made it possible to define the first slab settings:
  o a thickness of 20cm (22cm above the exhaust air duct which is the most exposed in the event of fire)
  o substantial reinforcement (276 kg/m^3) to resist the colossal forces caused by the rise in temperature under standard fire Increased Hydrocarbon tests on the slab (1300°C for 2 hours)
  o a concrete formula with 2kg/m^3 of polypropylene fibres. This option allows concrete water to escape by means of evaporation during a fire, thus limiting the occurrence of flaking and the ruining of the structure by the destruction of the steel's concrete coating. In addition, a balance had to be found in the cement proportioning, the choices of aggregate granulometry and the fibre quality.

– a flaking test under HCM fire for 120 min was therefore carried out in order to validate the slab's resistance and the assumptions that had been made on the depth of flaking of the concrete. For this purpose, thermocouple components were embedded in a test slab in order to monitor the temperature evolution depending on the depth and to check that the steel did not exceed 500°C (cf. Figure 3 below)

This laboratory test was also carried out by reproducing the calculation of the stress state of the concrete and that expected in a fire situation, thus making it possible to confirm the fire resistance of the slab. The deadlines to carry out these tests and studies were greatly reduced in order to start the prefabrication production only 6 months after the starting of the design.

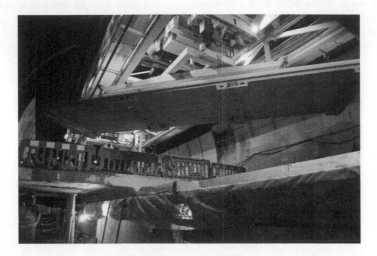

Figure 2.   Installation of a new slab – view from the ventilation ducts.

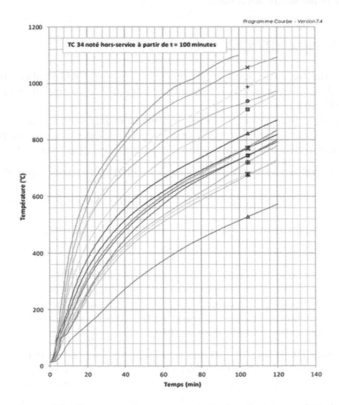

Figure 3.   Temperatures of the thermocouple components during the increased Hydrocarbon testing in relation to the depth.

To successfully carry out rehabilitation work on an old structure, it is also necessary, from the beginning of the design of the project, to anticipate the elements related to the state of the existing one:

– it is impossible to assess, by calculation, the condition of the slab in every respect and how it reacts to lifting and handling. For this reason all measures must be taken to manage all scenarios that may occur.

Figure 4. Removal of very degraded slabs.

These resources must be available in tunnels or in external zones, as, at night and at week-ends it is not possible to rely on outside help;

– the geometry of the tunnel in which the new slab must be replaced is also to be queried: it is impossible to know every aspect pertaining to the width between the vertical supports before removal of the existing slab.

The geometry of the new prefabricated slab must therefore be chosen with care, i.e. neither too short, in order to have sufficient supporting force, nor too long to enable insertion into the existing one. In light of ad hoc surveys, a statistical analysis of the predicted situation was carried out in order to choose the slab length and the margins to be taken into account at the supports and vertical supporting walls for the final adjustment, which took place 9 months later.

Methods are another major element to be integrated into the project as early as possible:

– concerning the handling means, the designed slab must be easy to handle/turn inside the tunnel and in line with the characteristics of the lifting gantry (weight, height of rotation,...);
– the installation and adjustment of the slabs must be done quickly and simply. To achieve this, the adopted principle is based on a concrete male/female shear key on the slab edge, supporting the interlocking of the slabs, followed by bolting of the components and the cementing of the shear key with mortar. This renders them structurally sound, as with the tunnel boring machine segments;
– the solution must be as adaptable as possible by way of safety margins on all components according to the expected degree of risk: dimensions, implementation time, accessibility of the work zone, supply means, amount of human resources...

Figure 5.   The gantry in working position in the tunnel.

3   LOGISTICS: TIME-BASED OPERATIONS

Closely linked at the design stage, the logistics were a key issue in finalising the replacement of 222 slabs over 32 closure slots. Bearing in mind that each night 25% of the available time is devoted to setting up the work zones at the beginning of the shift and then taking them back down at the end.

This key concept of optimising the nights of closure and reducing the unit time of each task was dealt with largely through the choice of equipment and the organisation of men's work.

3.1   The choice of "made-to-measure" material

Several innovative solutions had been developed specifically for this site. The most spectacular concerns custom-designed lifting gantries, in order to respect numerous constraints:

– rapid transport within the tunnel despite their large dimensions (25m long, 6m wide). Each night they were towed over 6km at 30km/h, which corresponds to a total of approximately 400km travelled during this site. The coupling system between the gantry and its traction represents one of the strong points of the solution that was put in place;
– easy installation in the tunnel's working zone, via an electrical networks connection and transferral from the transportation position to the lifting position by lengthening the 4 hydraulic posts in 5 minutes;
– a guarantee of high performance in the handling phases:
  o a lifting capacity of 20T distributed over 4 hoists which enables the lifted load to be moved in all directions and in a cramped environment offering little room for error, at the risk of damaging equipment, critical to the tunnel operations;
  o the loading or unloading of the slab on the trailers in less than 15 minutes.

The second key element in this site's logistics concerns the supply or removal of slabs and other tunnel equipment. To achieve this, the group chose 12 port trailers and port tractors offering considerable manoeuvrability, in particular for the machines to turn around in the garage area.

The rigorous organization of supplies and the anticipation of the following day's needs represented one of the priorities throughout the project, any flaw or error may result in the loss of a night's work.

Figure 6.   The transfer of a gantry from the external zone to the tunnel.

Figure 7.   New slab installation. View from the roadway.

3.2   *The tunnel crews: an optimized arrangement*

The crews and their safety constitute the other predominant factor in the logistical organisation of the site. The operating methods therefore ensured that:

– the men's positioning was optimised, by dividing them up between the ventilation ducts, the pavement and the external zones in order to limit their movements to a minimum during the night. This organisation increased work efficiency by concentrating as many resources as possible on the critical path's tasks (which does not include the movement of individuals);
– the same person was assigned the same task every day for each slab. Each person was thus responsible for his own activity at his level and for making sure he took his small equipment that he would need that night with him;

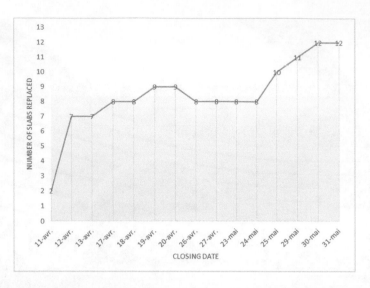

Figure 8.   The number of slabs replaced per night.

These organisational principles produced a rapid rise in pace. The objectives were quickly achieved and even exceeded on the last closures. This is explained by the design choices of a highly industrialised solution, through task repetition and their improvements.

4   WORKING CONDITION PARTICULARITIES IN AN OPERATING STRUCTURE

Working in a structure that is put back into service every morning increases some additional restrictions.

– The proximity of electrical/fire equipment in use requires moving, monitoring or simply protecting these networks in preparation for works to be carried out;

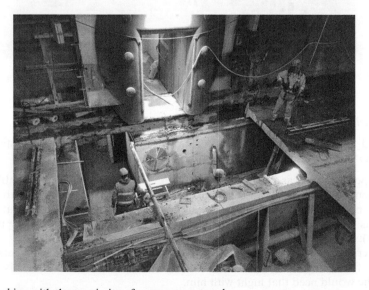

Figure 9.   Working with the proximity of numerous networks.

– In order to maintain ventilation performance during the provisional phases, maximum duct dimension requirements were calculated during the design phase;
– Support instrumentation using movement sensors was used to monitor the way the new slabs reacted in the work area after reopening;

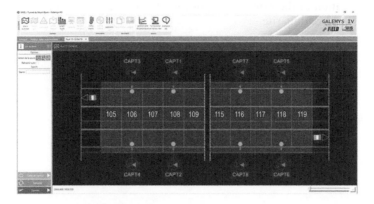

Figure 10.   Movement monitoring software.

– Restrictions related to the safety of the temporary phases had repercussions on the work:
  ○ each slab was bench-tested at the prefabrication plant, recreating stress variations in accordance with operating specifications;
  ○ systematic stop point surveys at the end of each night enabled us to guarantee the strength of the slab and the efficient operating of the equipment.

5   CONCLUSION

It is certainly more complicated for companies to carry out heavy work in an existing tunnel which is put back into service after each intervention period. However, it represents a considerable benefit by ensuring that project owners limit operating losses and reduce the impact the works have on users.

In the Mont Blanc tunnel, the rigorous organisation of tasks in a limited space and short closing slots led to successful multi-tasking management while guaranteeing the teams' safety. The choice of custom-designed equipment and the timing of interventions also contributed to meeting deadlines, the rate at which the slabs were replaced improved as the work progressed. Using innovative solutions accompanied by appropriate phases and meticulous preparation of the works incorporating the risks, enabled our companies to meet the challenge of this building site.

Tunnels and Underground Cities: Engineering and Innovation meet Archaeology,
Architecture and Art, Volume 6: Innovation in underground engineering,
materials and equipment - Part 2 – Peila, Viggiani & Celestino (Eds)
© 2020 Taylor & Francis Group, London, ISBN 978-0-367-46871-2

Load and resistance factor design coefficient calibration for tunnel design in Buenos Aires

L.A. Roldan, G. Dankert, J.M. Santa Cruz & L. Sambataro
Transport Planification, Federal Department of Transportation, Buenos Aires, Argentina

R. Bertero
University of Buenos Aires, Buenos Aires, Argentina

ABSTRACT: A Load and Resistance Factor Design approach is intended to be used to calculate the tunnels of the Buenos Aires Network of Regional Express Trains (RER) project. AASHTO's Tunnel Design and Construction Guide Specifications (2017) is used as reference and a recalibration of the load/resistance factors is proposed, to analyze the influence of the local ground properties and their statistical distribution. To do so, a beam and spring finite element model was coupled with a structural reliability software to perform a FORM analysis. From this study importance factors were obtained for the involved parameters and load/resistance factors were calibrated for different limit state cases with an objective reliability. Finally, the RER definitive linings were analyzed using the classical Allowable Stress Design method, the latest AASHTO's Tunnel Specification and the resulting limit state functions obtained from this study. A performance comparison was carried out for the three methods.

1 INTRODUCTION

The current regional railroad network of the City of Buenos Aires consists of 6 train lines adding up 790km of existing rails and 241 train stations. All of them reach the city center at either one of three central stations, located around 4km from each other. The goal of the new Network of Regional Express Trains project (RER, for its Spanish acronym), is to link these lines through 20km of new tunnels (Ministerio de Transporte, 2018). This interconnection, would allow the reduction of travelling times due to the increase of the trains frequency and reduction of passenger transfers between train-subway-bus lines.

Due to the magnitude and complexity of the project, it is divided in several sections for its construction and it is expected that several local and international contractors will work simultaneously on it. Having no local specifications or written recommendations for the structural design criteria, there is no guarantee that the different contractors will submit designs and calculations with a minimum/uniform quality. In order to achieve this, the Planification division of Department of Transportation (DOT) decided to publish a document establishing the design criteria that should be used for the project.

The mentioned document uses as main reference the newly published LRFD Tunnel Design and Construction Guide Specifications (AASHTO, 2017), where LRFD stands for Load and Resistance Factor Design. Although AASHTO is a well-known organization, the local geotechnical engineers are not used to work using the methodology proposed by the LRFD Specifications. Given that, and the fact the tunnel section geometry is not circular and the geological conditions of the region have some particular characteristics not taken into account by AASHTO, it was decided to perform a calibration of the load and resistance factors and compare the resulting structures from AASHTO's Tunnel Specification, the newly-calibrated LRFD and the current most common practice.

2 THE CURRENT PRACTICE

Since the 1980's until 2012, structural calculations in Argentina were carried out using an Allowable Stress Design (ASD) approach, based on the old DIN standards. In 2013, the government approved a new series of standards for buildings and surface structures using a LRFD approach. Specifically CIRSOC 201:2005, based on ACI 318, regulates the design methodology for concrete structures.

As there are no local specifications under LRFD paradigm for underground concrete structures, the most popular approach consists on a blend of both methodologies, which is a widely known fundamental error. The result of mixing classical loads for tunnel linings with LRFD load combinations that where calibrated for buildings are structures of unknown reliability and reinforced to withstand highly unlikely or even physically impossible scenarios.

3 UNCERTANTIES AND THE LRFD APPROACH

Soil is a complex material formed by combination of geological, ambient and chemical processes, producing vertical and horizontal variations of its properties (Peck, 1969). That is why, adding up the varying properties of concrete and the uncertainty due to systematic errors in the estimation of geotechnical characteristics, a great dispersion of loads (Q) and resistances (R) in soil-structure interacting systems is obtained.

The limit state function is defined as $G = R - Q = 0$, being the difference between the resistance and the loads effects. If their probability density functions are considered, the probability of failure is the probability of G being less than 0.0 (shaded area in Figure 1).

Another common way of accounting for this probability of failure is the reliability index β, which normalizes G with respect to its standard deviation (Baecher & Christian, 2002).

The LRFD method ensures the reliability of a structure (given an objective β), by amplifying or reducing each of the loads and resistances involved in the structural calculations, using factors that take into account the variability, the probability of occurrence and the sensitivity of the loads and resistances involved in the mathematical model.

Note that 'loads' and 'resistances' should be understood in a broader sense than forces and stresses. It is only a standard nomenclature used in structural engineering for differentiating parameters that are affected by a coefficient higher or lower than 1.0.

In the calibration of AASHTO's Tunnel Specifications (Wisniewski, Nowak, & Hung, 2016) the authors make certain hypothesis such as the shape of the tunnel, the loads involved and the parameter distributions that not necessarily coincide with those present in the RER project and the geological conditions of Buenos Aires. This differences encouraged this study considering our particular conditions.

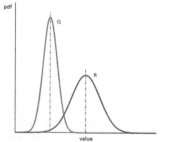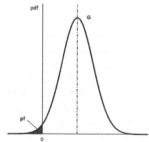

Figure 1.   Left: typical probability density functions for loads (Q) and resistance (R). Right: probability density function for margin (G) and probability of failure (pf).

4 COEFFICIENT CALIBRATION

4.1 Finite Element Model

A beam & spring Finite Element Method (*FEM*) model was used to perform the calculations, in an analog way to the calibration of the AASHTO's Tunnel Specification (Wisniewski, Nowak, & Hung, 2016), using the finite element framework *OpenSees* (University of California Berkley, 2018). The finite element formulation used consisted on linear Euler beam elements with two integration points. Radial springs (only-compression) and tangential springs were used, with stiffness calculated, depending on the soil stiffness and tunnel curvature, as stated in section 7.7.5.4 from AASHTO's Tunnel Specification. Loads were applied in the nodes of the elements. Non-linear calculations using an iterative Newton-Rhapson method to consider non-linarites from the multiple spring support.

4.2 Load Model

The following load conditions were considered and calculated as recommended in AASHTO's Tunnel Specification and AASHTO Bridge Design Specification (AASHTO, 2012), integrated over the length of the elements and applied on the nodes.

- Dead loads of structural components
- Vertical earth pressure
- Lateral earth pressure
- Hydrostatic pressure
- Surcharge due to buildings
- Surcharge due to internal vehicles
- Surcharge due to surface vehicles

4.3 Limit state function

Different limit state functions were defined depending on the position of the section analyzed and whether the lining element was considered as plain concrete or reinforced concrete. For simple concrete elements, the nominal resistance for the combination of flexure and axial compression was calculated following the ACI-318 equations. For symmetrical and non-symmetrical concrete elements, the nominal resistance was calculated as a simplification of the flexural and axial compression diagrams developed in ACI-318. The limit state function is defined as $G = R - Q$, where R is the resistance of the structure and Q are the moments and axial forces acting on the structure.

4.4 Reliability analysis method

In order to perform a reliability analysis for the later calibration of the load/resistance coefficients a First Order Reliability Method (*FORM*) is used. To find the design point u^* (most probable point of failure), the following optimization problem is solved (Bourinet, 2010):

$$\mu^* = \arg\min\{\|u\| \,|\, g(x(u), \theta_g) = G(u, \theta_g) = 0\} \tag{1}$$

where g or G are the limit state function, x is a realization of a n-dimensional vector for the random variables, u is a vector of those variables expressed in the normalized space, θ_g denotes a vector of deterministic limit-state function parameters.

Once the design point is found the reliability index β is computed as $\beta = \alpha^T u$ where $\alpha = -\nabla_u G(u^*)/\|\nabla_u G(u^*)\|$ is the negative normalized gradient vector at the design point, see Figure 2. The gradient ∇_u is obtained by finite differences. The probability of failure (p_f) is obtained by a first order approximation such that $p_f = \phi(-\beta)$, where ϕ is the standard normal

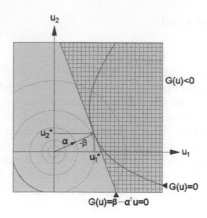

Figure 2.   First-Order Reliability Method (FORM) (Bourinet, 2010).

Table 1.   Statistical distribution of parameters used.

Parameter	Type	μ	σ	L1	L2	Description
K0 [-]	Beta	0.65	0.09	0.40	1.20	Lateral earth pressure coefficient
Es [MPa]	Log-normal	500	100	-	-	Modulus of elasticity of the soil
γs [kN/m3]	Beta	18.5	0.27	17.0	20.0	Unit weight of the soil
zw [m]	Normal	16.5	1.00	-	-	Depth of the groundwater level
fc [MPa]	Log-Normal	43	6.45	-	-	Concrete compression resistance
t [cm]	Normal	40	0.40	-	-	Concrete thickness

Where μ= mean, σ=standard deviation, L1=inferior limit, L2=superior limit.

cumulative distribution function. This analysis is performed by the open source code *FERUM* (Haukaas & Der Kiureghian, 1999).

4.5   *Statistical distribution of parameters*

In Table 1, parameters that describe the statistical distribution used for the analysis are shown. There are several other parameters with uncertainties in its determination (concrete unit weight, internal surcharges, etc.) that proved to have despicable influence in the results and adding only computational cost to the analysis. Those parameters, after prelaminar analyses, where accounted as deterministic ones, using the mean value as design criteria. Note that the analysis was carried out for a deep tunnel, therefore superficial load variability such as building or road surcharges become irrelevant.

The values shown in Table 1, are the result of known distributions used in local normative (fc) (INTI, 2005), unpublished hydrological studies done specifically for the RER project (zw) and advisors experience in the construction of the Buenos Aires subway system (K0,Es,γs,t).

It is important to note the lack of soil test studies done in Buenos Aires to directly measure K0 and Es (Santa Cruz, et al., Unpublished). A better understanding of the variability of the analyzed parameters would result in more optimal and safer tunnels from a structural perspective.

4.6   *Algorithm description*

An in-house code was written in *Matlab* to link both programs *FERUM* and *OpenSees*, pre-process loads and properties for the structural model and post-process the results from the finite element analysis. As shown in Figure 3, the statistical distribution for the values of the properties of interest (groundwater level, soil stiffness, lateral pressure coefficient, etc.) is given

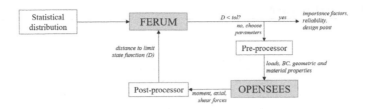

Figure 3. Flow diagram for the *FERUM/OpenSees* coupling algorithm.

to *FERUM*, which sends a single value for each properties to the pre-processor. The pre-processor calculates the boundary conditions and model properties and writes the input file for the *OpenSees FEM* model, which performs the structural analysis and returns the resulting forces (moment, shear and axial) at every element. The post-processor compares the forces at one section with the concrete limit resistance and returns the "distance" between them. In the *FORM* analysis, *FERUM* calculates the derivatives numerically by performing small perturbations to the *OpenSees* inputs. The loop is repeated as *FERUM* changes the properties values until the most probable failure is found (the distance to the limit state function is practically zero), and the importance factors, reliability index and design points are returned.

4.7 *Analysis results and calibrated coefficients*

A deep tunnel was subjected to the reliability analysis considering a 21m cover. Three sections were studied, considering simple concrete for the top heading and reinforced concrete for the bench and bottom. Two failure modes were identified: flexural and axial yielding, resulting in different failure probabilities depending on the concrete thickness. In Table 2 the sensitivity analysis and resulting amplification/reduction coefficients can be observed, being $\gamma_r = 1 - \alpha_i \delta_i \beta$.

It can be observed that the two most important parameters (higher α), which have a great impact on the calculations, are K0 and Es. It is no coincidence that they present the bigger statistical variation, as they have proven to be difficult to test in geotechnical studies. In one hand, the young modulus acts always as a resistance parameter (its importance factor is always positive), consequently needing to be reduced by factor γ in every case. On the other hand the lateral earth pressure coefficient can have a negative or positive effect on the system, resulting on amplification or reduction coefficients depending on the section and failure mode (reduction for the heading and amplification in the bottom).

The obtained amplification/reduction coefficients for K0 can be compared against AASHTO's Tunnel Specification coefficients by analyzing the effect over the vertical and horizontal earth load relationship. Even though, AASHTO does not affect K0 by a coefficient, it does proposes several combinations of majored/minored earth pressure values (lateral earth pressure Hs and vertical earth pressure Vs). For the case where mean K0=0.65, $0.43 < H_s/V_s < 0.77$ for AASHTO where $0.57 < H_s/V_s < 0.78$ for our calibration.

5 CASE STUDY

5.1 *Problem description*

In order to test the performance of the three different methods (current, AASHTO's Tunnel Specification for tunnels and the newly calibrated LRFD), the final lining of a typical tunnel sections were analyzed and the resulting concrete reinforcement need compared. The geometry and considered loads can be observed in Figure 4.

In Table 3 the parameters and considerations used in the calculations are listed.

Table 2. Reliability analysis results.

Section	Yield mode	Parameter	α	δ	γ β=3	γ β=4	γ β=5
Heading	Comp.	K0	0.28	0.15	0.87	0.83	0.79
"	"	Es	0.15	0.20	0.91	0.88	0.85
"	"	γs	-0.12	0.02	1.01	1.01	1.01
"	"	zw	-0.03	0.06	1.00	1.01	1.01
"	"	fc	0.94	0.15	0.58	0.44	0.30
Heading	Flex.	K0	0.39	0.15	1.03	1.04	1.05
"	"	Es	0.89	0.20	0.47	0.29	0.11
"	"	γs	-0.12	0.02	1.01	1.01	1.01
"	"	zw	-0.17	0.06	1.03	1.04	1.05
"	"	fc	-0.12	0.15	1.06	1.07	1.09
Bench	Comp.	K0	-0.32	0.15	1.14	1.19	1.24
"	"	Es	0.15	0.20	0.91	0.88	0.85
"	"	γs	-0.11	0.02	1.01	1.01	1.01
"	"	zw	-0.07	0.06	1.01	1.02	1.02
"	"	fc	0.92	0.15	0.59	0.45	0.31
Bench	Flex.	K0	0.25	0.15	0.89	0.85	0.81
"	"	Es	0.90	0.20	0.46	0.28	0.10
"	"	γs	-0.03	0.02	1.00	1.00	1.00
"	"	zw	0.05	0.06	0.99	0.99	0.99
"	"	fc	-0.33	0.15	1.15	1.20	1.25
Bottom	Comp.	K0	-0.38	0.15	1.17	1.23	1.29
"	"	Es	0.31	0.20	0.81	0.75	0.69
"	"	γs	-0.10	0.02	1.00	1.01	1.01
"	"	zw	0.00	0.06	1.00	1.00	1.00
"	"	fc	0.86	0.15	0.61	0.48	0.35
Bottom	Flex.	K0	-0.46	0.15	1.21	1.28	1.35
"	"	Es	0.84	0.20	0.50	0.33	0.16
"	"	γs	-0.02	0.02	1.00	1.00	1.00
"	"	zw	0.05	0.06	0.99	0.99	0.99
"	"	fc	-0.26	0.15	1.12	1.16	1.20

Where α = importance factor, δ =variation coefficient (σ/μ), β =objective reliability index, γ = amplification/ reduction factor. Parameter "t" is not included in the table as it's importance is low enough to yield in amplification factors of 1.00 in every case.

5.2 *Application of different methods: results comparison*

As different methodologies of work are being contrasted, it is irrelevant to compare the resistances and loads effects obtained over the tunnel concrete lining. Instead, the reinforcement requirement for of the concrete structure is analyzed for every approach. In Figure 5 the regions of the lining where a simple concrete analysis does not verify is plotted.

It can be noted that the newly calibrated LRFD method requires less steel reinforcement than AASHTO Tunnel Specification and the current method. It can be observed that using this last method reinforcement is required in the bottom center, which comes from a load combination with self-weight and water pressure without earth pressure.

6 DISCUSSION

The calibration of the parameters was carried out using a typical geometry of the RER project, resulting in great performance when compared with AASTHO's LRFD and the local current approach. Nevertheless, certain hypothesis were considered, which does not account for several possible conditions:

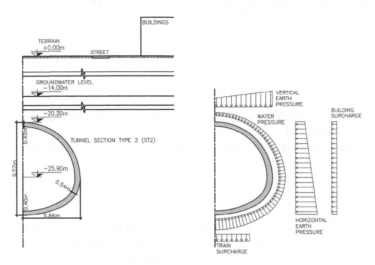

Figure 4. Case study geometry and load diagrams.

Table 3. Parameters used for the performance analysis.

Parameter	Value	Description
γs [kN/m3]	18.5	Mean soil self-weight
Es [MPa]	500	Mean soil elastic modulus
v [-]	0.3	Mean soil Poisson coefficient
K0 [-]	0.65	Mean lateral earth pressure coefficient
γh [kN/m3]	24	Mean concrete self-weight
v [-]	0.2	Mean concrete Poisson coefficient
fc [MPa]	43	Mean concrete compression resistance
Lti [kPa]	20	Mean distributed pressure due to trains
Le [kPa]	105	Mean pressure at terrain level due to buildings (15m from the center of the tunnel and 35m wide)

Concrete elastic modulus is estimated as $E_c = 4700\sqrt{fc}$.

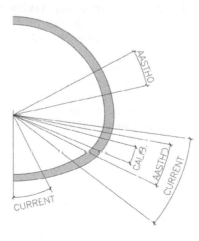

Figure 5. Comparison for reinforcement need.

- The parameters were calibrated for a variable groundwater level, but not for a totally drained condition, which is a load state that can happen immediately after the construction of the tunnel but before the restoration of the natural groundwater level.
- The influence of superficial overloads such as railways, roads and buildings were not considered as probabilistic variables as in deep tunnels they become less relevant to the structure design. No reliability analysis was carried out for shallow tunnels.
- Only one typical geometry was studied in this case, when at least 5 other typical ones are involved in the project. Different geometries could result in different sensitivities for the different parameters involved.
- Only axial-flexion yielding was analyzed. A similar analysis should be done for shear forces.

It is important to remark that a total independence between the probabilistic parameters was assumed for this study, where soil properties typically present some correlation between their values.

Additionally, neither in our calibration and in AASHTO recommendation the load condition where only the self-weight of the concrete structure and the groundwater effect is considered. This state can occur when the final lining is already constructed but the soil surrounding does not have had time to relax and water had entered the space between the final lining and initial support shotcrete. It is important to note that the current method does consider that case. That particular case should be studied in detail, performing a reliability analysis to find the amplification/reduction coefficients needed for that load combinations.

Even though reliability analysis is associated with time consuming calculation, the design point of each of the models run in Table 2 took a couple of hours, allowing the team to plan several other analysis in order to account for the hypothesis not considered in this work. This simple, yet powerful analysis, shows where the modeler should strengthen its input data and the value of simple models to understand the physics of the problem.

7 CONCLUSION

Given the importance and size of the Buenos Aires RER project and the inconsistencies in the local industries current method for designing the final lining for tunnels, it was decided to use a full LRFD approach in the calculations. In order analyse and test the applicability of AASHTO's Tunel design guide, an in-house calibration of parameters for a reliability based methodology was carried out, taking into account some geometrical an geothechnical considerations particular to this project. To do so, a program was written to link a finite element code (OpenSees) and a reliability analysis software (FERUM).

The obtained coefficients were tested, and the results compared with AASHTO's LRFD and the local practice. In spite of some limitations of the analysis, the similarities between the relationships of vertical and horizontal soil pressure and the resulting Moment/Axial relationship that led to similar reinforcement regions, strengthen the confidence in the applicability of AASTHO's LRFD to tunnel geometries slightly different than circular considering the local geology.

8 ACKNOWLEDGMENT

We thank our colleagues E. Zielonka, H. Tymkiw, D. Correa, C. Cogliati, I. López Godoy and S. Wolkomirky from the DOT, F. Bissio from National University of La Plata and A. Sfriso and P. Fernandez from SRK Consulting, who provided insight and expertise that greatly assisted the research, although they may not agree with all of the interpretations/conclusions of this paper.

REFERENCES

AASHTO. 2012. LRFD Bridge Design Specification, customary U.S. units. In American Association of State Highway and Transportation officials (ed.). Washington, D.C, United States of America.

AASHTO. 2017. LRFD Road Tunnel Design and Construction Guide Specification, First edition. In American Association of State Highway and Transportation officials (ed.). Washington, D.C, United States of America.

Beacher, G. & Christian, J. 2002. Reliability and Statistics in Geotechnical Engineering. In John Wiley & Sons (ed.). Chichester West Sussex, England.

Bourinet, J.M. 2010. FERUM 4.1 User's Guide. In Institute Français de Mécanique Avancée (ed.). Clemont-Ferrand, France.

Haukaas, T. & Der Kiureghian, A. 1999 FERUM University of California Berkley. Retrieved July 2018, from http://projects.ce.berkeley.edu/ferum/index.html

INTI. 2005. CIRSOC 201 Reglamento Argentino de Estructuras de Hormigón. In Instituto Nacional de Tecnología Industrial (ed.). Buenos Aires, Argentina.

Ministerio de Transporte. 2018 Red de Expresos Regionales. Retrieved July 2018, from https://www.argentina.gob.ar/transporte/rer

Peck R. 1969. Advantages and Limitations of the Observational Method in Applied Soil Mechanics. In ICE Publishing Ltd (ed.), *Geotechnique, 2, 171–187.*

Santa Cruz, J, Roldan, L, Dankert, G, Sambataro, L, Sfrisso, A. & Bertero, R. 2018. How to strengthen a geotechnical campaign for tunnel's secondary lining. Unpublished.

University of California Berlkey. 2018. OpenSees, 2.5.0. Retrieved July 2018, from http://opensees.berkeley.edu/index.php

Wisniewski, J, Nowak, A. & Hung, J. 2016. Recommended AASHTO LRFD Tunnel design and construction specifications (Draft Final Report). In Transportation Research Board (ed.). Washington, DC, United States of America.

Tunnels and Underground Cities: Engineering and Innovation meet Archaeology,
Architecture and Art, Volume 6: Innovation in underground engineering,
materials and equipment - Part 2 – Peila, Viggiani & Celestino (Eds)
© 2020 Taylor & Francis Group, London, ISBN 978-0-367-46871-2

Steel fibre reinforced concrete for the future of tunnel lining segments – A durable solution

B. Rossi & S. Wolf
ArcelorMittal Fibres, Bissen, Luxembourg

ABSTRACT: The application of steel fibre reinforced concrete is more and more common due to a lot of improvements, mainly related to durability and sustainability. The production process is easier, minimizing edge cracks during handling, the crack width and depth are greatly reduced. Due to a higher resistance against corrosion and impact toughness, the segments are more durable. The service life is longer according to the very low need for repair or maintenance and, on top, the production costs are also lower than for conventional steel reinforced tunnel segments. Nowadays, many standards have been published for the definition of the fibres characteristics, for the measuring of the SFRC mechanical properties and for the design, production and control of the SFRC structures. The aim of this paper is to provide some useful indications for the design of SFRC Tunnel Lining Segments and to improve the durability in aggressive environment.

1 INTRODUCTION

The application and the use of SFRC in precast tunnel lining design is a growing trend due to its advantages in performance, durability and ease of manufacture compared to traditionally reinforced concrete.

From the early 90's, many projects have been realized with steel fibres only. Some of them are listed hereunder:

Table 1.   Steel fibre reinforced only tunnel lining segments – Some projects realized worldwide.

Tunnel name	Year	Country	Usage	Diam. (Int.) (m)	Thickness (mm)	Dosage (kg/cm)
Metrosud	1982	Italy	Subway	5.80	300	n.a.
Heathrow Baggage Handling	1993	UK	Service	4.50	150	30
Essen	2001	Germany	Subway	6.30	350	n.a.
Trasvases Manabi (La Esp.)	2001	Ecuador	Water supply	3.50	200	30
Lotschberg	2007	Switzerland	Temporary p.	4.50	220	n.a.
Hobson Bay	2009	New Zealand	Wastewater	3.70	230	35
Copenhagen District Heating	2009	Denmark	Water supply	4.20	300	35
FGC Terrassa	2010	Spain	Railway	6.00	300	25
Brightwater	2011	USA	Wastewater	5.10	260	35
Pando	2012	Panama	Water supply	3.00	250	40

(Continued)

Table 1. (*Continued*)

STEP Abu Dhabi Lot T-02	2014	UAE	Wastewater	6.30	280	30
London CrossRail	2015	UK	Railway	6.20	300	35-45
Legacy Way	2015	Australia	Road	11.30	350	40
Ejpovice	2016	Czech Rep.	Railway	8.70	400	40
Doha Metro (Red & Gold lines)	2016	Qatar	Subway	6.00	300	40
Downtown Line 3	u.c.	Singapore	Subway	5.80	280	40

The SFRC is a composite material made of basic concrete in which the fibre reinforcement is incorporated and homogeneously distributed. Steel fibre's addition in concrete controls plastic and hydraulic shrinkage cracking, considerably improves the concrete post-cracking behavior. SFRC behavior depends on:

- Fibre characteristics (geometry, aspect ratio, hooked end shape, tensile strength, etc.)
- Concrete matrix properties (compressive strength at early age and long-term)
- Fibre dosage rate (kilos per concrete cubic meter)

The fibres sew the cracks and redistribute the tensile stresses in a larger concrete area, reduce the crack width and spread the cracking due to loads or constraints, prevent spalling due to load or impacts, stop water paths; it will result in a better aesthetic presentation and increase of durability.

2 TUNNEL LINING SEGMENTS DESIGN

2.1 *Material mechanical properties*

Nowadays, many standards have been published for the definition of the fibres characteristics, for the measuring of the SFRC mechanical properties and for the design, production and control of the SFRC structures.

Fibre's geometrical and mechanical characteristics are defined by the following standards:

- EN 14889_2006: Fibres for concrete, specifies the requirements of fibres for structural or non-structural use in concrete, mortar and grout. This standard comprises two parts:
  - Part 1 dealing with steel fibres for concrete;
  - Part 2 dealing with polymer fibres.
- ASTM A820/A820M – 11, defines the minimum requirements for steel fibres for use in fibre-reinforced concrete;
- ISO 13270_2013: Steel fibres for concrete. This International Standard explains the main reason for which steel fibres can be used as a concrete reinforcement: "steel fibres are suitable reinforcement material for concrete because they possess a thermal expansion coefficient equal to that of concrete, their Young's Modulus is at least 5 times higher than that of concrete and the creep of regular carbon steel fibres can only occur above 370 °C".

The Flexural behaviour of Fibre Reinforced Concrete has been regulated as follows:

- EN 14651_2005: Test method for metallic fibre concrete. Measuring the flexural tensile strength (limit of prop. (LOP), residual); this is a Three Points Bending test, correlating centrally applied loads versus crack openings;
- ASTM C1609/C1609M_2012: Standard Test Method for Flexural Performance of Fiber-Reinforced Concrete;
- JSCE-SF4: Method of tests for flexural strength and flexural toughness of steel fibre reinforced concrete.

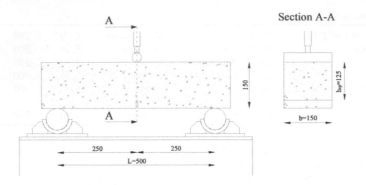

Figure 1. Flexural bending test set up.

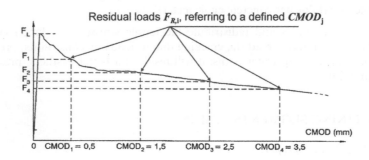

Figure 2. F-CMOD diagram.

2.2 *Design guidelines*

Several standards and guidelines are available for the design at serviceability and ultimate limit states of SFRC structures:

- EN 1992-1-1-2004: Eurocode 2: Design of concrete structures - Part 1-1: General rules and rules for buildings;
- RILEM TC 162-TDF - 2003: Test and design methods for steel fibre reinforced concrete;
- fib CEB - FIP Model Code 2010 (fib Bulletins 65-66):
  - Chap. 5.6 Fibers/Fiber Reinforced Concrete;
  - Chap. 7.7 Verification of safety and serviceability of FRC structures;

Fibre materials with a Young's-Modulus which is significantly affected by time and/or thermo-hygrometrical phenomenon are not covered by this Model Code. Structural design of FRC elements is based on the post-cracking residual strength provided by fibre reinforcement. For structural use, a minimum mechanical performance of FRC must be guaranteed. Fibre reinforcement can substitute (also partially) conventional reinforcement at ultimate limit state, if the following relationships are fulfilled:

- $f_{R1k}/f_{Lk} > 0.4$
- $f_{R3k}/f_{R1k} > 0.5$

For the fibre reinforced concrete tunnel lining segments design, some guidelines have been recently published all over the world from fib, ITA-AITES, France, Germany, USA and UK:

- AFTES Recommendations GT38R1F1 (2013): La conception, le dimensionnement et la realisation de voussoirs prefabriques en beton de fibres metalliques;

- DAUB (Deutscher Ausschuss für unterirdisches Bauen e.V.) (2013): Empfehlungen für den Entwurf, die Herstellung und den Einbau von Tübbingringen;
- ACI 544.7R-16 (2006): Report on Design and Construction of Fibre-Reinforced Precast Concrete Tunnel Segments;
- ITA-TECH (2016): Design Guidance for Precast Fibre Reinforced Concrete Precast Segments - Draft Report;
- ITA-Working Group 02 (2016): Twenty years of tunnel segments practice: lessons learnt and proposed design procedure;
- PAS 8810 (2016): Design of concrete segmental tunnel linings - Code of practice
- fib Bulletin 83 (Oct. 2017): Precast tunnel segments in fibre reinforced concrete. State of the art report;

The ACI 544.7R-16 is the most comprehensive document, as it lists eleven different load cases, divided in three main categories: production and transient stages, construction stages and service stages. For each of them, useful and simple closed-form expressions are available, easy to implement in a program or spreadsheet.

The most challenging situations are demolding, handling, storage, installation (erection and TBM thrust applied loads during the tunnel excavation), localized back-grouting pressure and permanent ground loading in service. The design values are derived from residual flexural strengths of FRC, evaluated with bending tests ASTM C1609 or EN 14651.

2.3 *Input data*

For a preliminary segmental lining design, the following information are required:

- Geometry of the Tunnel Segments: thickness, radius, width, segment angle, number of the segments per ring, RAM shoes dimensions;
- Material properties: compressive and residual flexural strengths, at early and long-term age;
- Material safety factors: concrete, FRC and steel rebars depending on the loading conditions (EC2);
- Loading safety factors: for demoulding, stacking and for TBM thrust loads;
- Demoulding and handling set-up: type of lifting device (mechanical or vacuum), lifting points spacing or vacuum width;
- Segments storage: n° of segments stacked, stacking spacing, misalignment of the supports;
- TBM thrust loads: max. TBM load, n° of rams, ram shoe dimensions (width and length, eccentricity);
- Sectional forces in service: Axial forces N, bending moments M, shear forces V (ground loading).

The most critical load case is when Tunnel Boring Machine thrust jacks are pushing on circumferential joints, causing tensile forces under jack shoes (bursting, also known as splitting) and between two adjacent pairs of them (spalling), schematically represented by Groeneweg 2007. These forces, computed adopting strut-and-tie model (Morsch), are evaluated in tangential and radial directions of the segment. The respective areas on which these forces are distributed, have been estimated, at Ultimate Limit State, following the DAfStb 2012. Finally, the related acting stresses must be compared with the factored 28-days residual tensile strength of the SFRC. Radially, the maximum value of the bursting stress occurs when the load is centrally applied, while the spalling stress increases with the eccentricity (DAUB 2013). Tangentially, the most critical case is the spalling stress between two adjacent pairs of jack shoes (Plizzari & Tiberti). The addition of steel fibres results in a larger depth of the zone under the jack loads involved in the cracking process (B. Schnütgen & E. Erdem).

Actually, the material specifications for the Tunnel Lining Segments require a 4c material (Model Code 2010 class, where $f_{R1k} > 4.0$ MPa and $f_{R3k} > 3.6$ MPa) but, due to the continuously increase of the concrete strength, C50/60 and more, the steel wire strength used for the fibre's production will be higher.

According to our experience, to fulfill all the verifications at ULS, for an only steel fibre reinforced tunnel lining segment, a material 5d ($f_{R1k} > 5.0$ MPa, $f_{R3k} > 5.5$ MPa) is recommended.

Figure 3. Tunnel Boring Machine thrust jacks pushing forward.

3 SOME CASE STUDIES – DOHA METRO RED LINE SOUTH IN QATAR AND CROSSRAIL LINK IN LONDON

3.1 *Doha Metro Red Line South in Qatar*

The Metro in Doha, Qatar's capital, will start operation at the end of 2019 and will be ready not only to ease travelling times and distances to the different stadiums for the fans of the Football World Championship (FIFA World Cup) in the year 2022. Beside the well-established road network in Qatar it is the objective of the transport master plan for Qatar to develop an attractive, efficient and reliable public transport system in the country. The Red Line is one of four Metro lines of the new and modern Doha railway system, which will be altogether 300 km long at least. A joint venture of Qatari Diar and Vinci Construction Grands Projects, France, (QDVC) realizes the project for the Qatar Railways Company. The Red Line South is a 42.8 km long tunnel with 12 stations connecting the cities Mesaieed and Msheireb. The construction time was only two years, from 2014 to 2016, with costs of approx. 1.8 to 2.0 Billion USD. To meet the strong requirements of the aggressive groundwater and the durability of 120 years, the decision was made to use steel fibre reinforced concrete for the tunnel segments. The inner diameter of the tunnel is 6.00 meters and the maximum depth is 60 m below surface. To produce these tunnel segments, 8,088 tons of ArcelorMittal steel fibres type HE ++ 90/60 were used with a dosage of 40 kg/m3.

3.2 *Crossrail Link in London*

A rapidly growing population in Great Britain's capital London means for Transport for London (TFL) to do a challenging job to ensure and to offer a working urban transport system for the approx. 8.4 million people living in the heart of London. With the forecast for the population in London to grow up to 10 million people until 2030, especially the public transport system needs to be upgraded. From 2018, Crossrail trains will transport up to 72,000 passengers per hour through the new tunnels with a parallel increase of rail capacity and a reduction of journey times for every passenger. The largest of the Crossrail tunneling projects, known as the Eastern Running Tunnels (project C305) comprises a 12 km stretch of twin tunnels under the East End of London. For the tunnel segments here, steel fibres were used as a sole reinforcement, as opposed to a combination of steel fibres and rebars. Specifically for the London Crossrail project, two new premium quality fibres have been developed: the HE++ 55/35 (for shotcrete) and the HE++ 90/60 (for the tunnel lining segments). 5,100 t of HE++ 90/60 steel fibres were supplied for the tunnel segments (35 kg/m3 dosage) for the running tunnels between January 2012 and May 2014. Approx. 1,100 t of HE++ 90/60 steel fibres were supplied to produce tunnel lining segments for the 2.6 km of twin tunnels and portal structures of the Thames Tunnels situated in southeast London (project C310). At these both projects around 21 km of steel fibre reinforced tunnel lining segments were built in the tunnels. In maximum, the tunnel was 42 m below the surface and under the river Thames in London. The internal tunnel diameter of the tunnel was 6.20 m and each ring was built with 6 segments plus 1 key.

Figure 4.   Crossrail – London Dragados-Sisk JV – C305.

4   DURABILITY OF STEEL FIBRE REINFORCED CONCRETE TUNNEL LINING SEGMENTS

The advantages of steel fibres in concrete for tunnel segments are great, but nevertheless there are some parameters like exposure conditions, concrete quality and more, which affects the durability of SFRC segments. If there is a corrosion of steel fibres possible, mostly one or a combination of the following exposure conditions are responsible:

– Chloride exposure
– Carbonation

Also, there is an influence, if cracks occur or not, because with these cracks there is an ingress of chlorides possible with an influence on the durability.

4.1   Chloride Exposure in Concrete without and with Cracks

Basically, the resistance against chloride-induced corrosion of SFRC in concrete without cracks is higher than in concrete which is traditionally reinforced with steel bars and stirrups. The reasons for that performance are due to the following conditions:

– Dimensions of the steel fibres
– Casting conditions of the SFRC
– Chloride threshold of the steel fibres

By comparing traditional reinforced concrete with SFRC, the chloride threshold for carbon steel in concrete without cracks is up to ten times lower than for steel fibres. According to a study, the measured chloride threshold of steel fibres in concrete is greater than 5.2 %/wt (water tank) of cement. The fib Model Code for Service Life Design specifies a mean value for the chloride threshold of reinforcement bars of 0.6 %/wt of cement. The interface of concrete and steel is one of the most important controlling factors for chloride threshold of in concrete embedded steel. If there are voids or defects, the protection of the steel against chlorides will be reduced. Due to an easier and denser mixing process with the steel fibres, the connection with the concrete is much more better than with the traditional reinforcement. The casting conditions of the steel fibres additionally are favoring the dense concrete/steel fibre situation. This improves the protection of the embedded steel fibres resulting in an increased chloride threshold. Additionally, the difference of the length of the steel elements is a reason for the improved corrosion resistance of the steel fibres compared with reinforcement bars. Due to the limited length of the steel fibres, the possibility of building up the required electrochemical potential difference, the anode and cathode areas are therefore reduced compared to the scenario with reinforcement bars, which may be several metres long. In summary, it is concluded that the resistance to chloride-induced fibre corrosion in uncracked concrete is significantly higher for SFRC than for traditional reinforced concrete. If there are cracks, the situation is different. To avoid the ingress of substances like chlorides, there should be no cracks or only controlled cracks allowed. Depending the exposure conditions, normally the crack width is limited in traditional reinforced concrete. For SFRC, the mechanisms controlling the fibre corrosion are not fully understood. Results in the literature concern experimental investigations on factors such as exposure conditions, fibre type, crack width and for obvious reasons the crack width limitations proposed vary significantly. However, there appears to be a consensus that the long-term durability of the fibres is not affected by crack widths of less low than 0.2 mm. Due to ongoing discussions about cracks of SFRC, cracks are often not allowed during the permanent loading conditions.

4.2 *Carbonation*

The highest availability of oxygen which is affecting the carbonation-induced corrosion of steel fibres in uncracked concrete is typically limited to the outer regions of the SFRC. Here is the thickness of the outer region important where fibres corrode due to carbonation. The process is controlled by various parameters like concrete quality, type of cementitious materials used, exposure conditions and time and some more. For dense low permeability concrete with a w/c < 0.40, which is used for permanent SFRC structures with high requirements for the service life, the thickness of this region is a couple of millimeters. If steel fibres are exposed at the concrete surface, it is possible to see some rust stains on the surface which are not causing cracks or spalling of the concrete cover.

4.3 *Corrosion resistance of steel fibre reinforced – Available literature*

There is a huge literature about corrosion resistance of SFRC to chloride and carbonation exposure. Victor Marcos-Meson et all published recently, Cement and Concrete Research 103 (2018), a literature review of the available experimental experiences and standards. The main conclusion of this work is the following: "The durability of cracked SFRC exposed to chlorides and carbonation is under discussion at the technical and scientific level. There is substantial insight among academics regarding the existence of a critical crack width, below 0.20 mm, where fibre corrosion is limited and the structural integrity of SFRC can be ensured for long-term exposures. However, the mechanisms governing corrosion of carbon-steel fibres in cracked SFRC subject to chloride and carbonation exposure are still unclear. In particular, the influence of fibre corrosion on the residual strength of SFRC is in focus and under discussion".

5 CONCLUSIONS

The application and the use of SFRC in precast tunnel lining design is a growing trend due to its advantages in performance, durability and ease of manufacture.

Many standards have been published for the design, production and control of the SFRC structures.

The most critical load case is when Tunnel Boring Machine thrust jacks are pushing on circumferential joints.

For a TLS reinforced with steel fibre only, a material 5d ($f_{R1k} > 5.0$ MPa, $f_{R3k} > 5.5$ MPa) is recommended.

The fibres reduce the crack width and depth, it will result in a better aesthetic presentation and increase of durability.

REFERENCES

ACI 544.5R-10. 2010. Report on the Physical Properties and Durability of Fiber-Reinforced Concrete, reported by ACI Committee 544, Farmington Hills, U.S.A.

ACI 544.7R-16. 2016. Report on Design and Construction of Fiber-Reinforced Precast Concrete Tunnel Segments. Reported by ACI Committee 544, Farmington Hills, U.S.A.

AFTES Recommendations GT38R1F1. 2013. La conception, le dimensionnement et la realisation de voussoirs préfabriqués en béton de fibres métalliques, *Tunnels et espace souterrain*, France.

ASTM A820/A820M. 2011. Standard Specification for Steel Fibers for Fiber-Reinforced Concrete, West Conshohocken, U.S.A.

ASTM C1609/C1609M. 2012. Standard Test Method for Flexural Performance of Fibre-Reinforced Concrete (Using Beam with Third-Point Loading), West Conshohocken, U.S.A.

Bakhshi, M. & Nasri, V. 2014 AECOM, New York. Design considerations for precast tunnel segments according to international recommendations, guidelines and standards, *Tunneling in a Resource Driven World*, Vancouver TAC 2014, NY, USA.

DAfStb. 2012. DAfStb Richtlinie Stahlfaserbeton. Berlin. Deutschland

DAUB. 2013. Empfehlungen für den Entwurf, die Herstellung und den Einbau von Tübbingringen. Deutscher Ausschuss für unterirdisches Bauen e. V. Köln, Deutschland.

EHE-2008 ANEJO 14. 2008. Recomendaciones para la utilización de hormigón con fibras. Spain.

EN 1992-1-1-2004, Eurocode 2. 2004. Design of concrete structures — Part 1-1: General rules and rules for buildings. CEN.

EN 14651. 2005. Test method for metallic fibered concrete - Measuring the flexural tensile strength (limit of proportionality (LOP), residual). CEN.

EN 14889-1. 2006. Fibres for concrete - Part 1: Steel fibres - Definitions, specifications and conformity; CEN.

fib Bulletin 83 (Oct. 2017): Precast tunnel segments in fibre-reinforced concrete. State of the art report. Federation International du Beton. Lausanne, Switzerland.

fib CEB – FIP Model Code 2010 (fib Bulletins 65-66). 2012. Chap. 5.6 Fibres/Fibre Reinforced Concrete; Chap. 7.7 Verification of safety and serviceability of FRC structures. Federation International du Beton. Lausanne, Switzerland.

Groeneweg, T.W. 2007. Shield driven tunnels in ultra-high strength concrete. Reduction of the tunnel lining thickness. Technische Universiteit, Delft. Netherlands.

ITAtech - 2016 - Design Guidance for Precast Fibre Reinforced Concrete Precast Segments. ITA-AITES. Châtelaine. Switzerland.

ITA Working Group 2-2016 – Twenty years of FRC tunnel segments practice: lessons learnt and proposed design principles. ITA-AITES. Châtelaine. Switzerland.

JSCE-SF4: Method of tests for flexural strength and flexural toughness of steel fibre reinforced concrete. Japan.

ISO 13270. 2013. Steel fibres for concrete – Definitions and specifications, Genève, Switzerland.

PAS 8810. 2016. Design of concrete segmental tunnel linings - Code of practice. BTS. United Kingdom

RILEM TC 162-TDF. 2003. Test and design methods for SFRC. *Materials and Structures, Vol. 36, October 2003, pp. 560-567*. RILEM - International Union of Laboratories and Experts in Construction Materials, Systems and Structures, Paris. France.

Schnütgen, B. & Erdem, E. 2003. Annex 1 to Final Report Sub-task 4.4, Splitting of SFRC induced by local forces, BRITE-EURAM BRPR-CT98-0813, Ruhr-University, Bochum. Deutschland.

Tiberti, G. & Plizzari, G.A. 2014. Structural behavior of precast tunnel segments under TBM thrust actions. *Proceedings of the World Tunnel Congress 2014*. Foz do Iguaçu, Brazil.

Marcos-Meson V., Alexander M., Anders S., Fischer G., Edvardsen C. & Torben L. S., *Cement and Concrete Research 103 (2018)*

*Tunnels and Underground Cities: Engineering and Innovation meet Archaeology,
Architecture and Art, Volume 6: Innovation in underground engineering,
materials and equipment - Part 2 – Peila, Viggiani & Celestino (Eds)*
© 2020 Taylor & Francis Group, London, ISBN 978-0-367-46871-2

Logistic challenges for TBM operation during Sofia Metro Line 3 extension

N. Rubiralta & Á. Hernández
TunnelConsult Engineering SL, Spain

S. Ergut & A. Shaban
DVU JV, Bulgaria

ABSTRACT: The Project Extension of the Sofia metro (Bulgaria), Line 3, phase I, aims to build almost 8 km of new line and 7 new stations. Main tunnel consists in a metro single tube configuration excavated by an EPB Machine. In order to minimize interferences with other contractors involved in the project, new intermediate site establishment to operate the TBM was planned. To maintain original TBM logistic concept, 2 temporary connection from surface to the tunnel where designed: a vertical shaft over the tunnel section, to proceed with the precast segments and other tunnel material supply; Lateral shaft with a ramp to the surface including 2 lateral connections to the main tunnel in order to manage the TBM belt conveyor and transfer out the excavated material. Planning the overall tunnel execution, logistic TBM shaft concept design and TBM performance analysis from each site establishments are presented.

1   PROJECT DESCRIPTION

"Extension of the Sofia metro, Line 3, phase I, Boulevard Valdimir Vazov-Centre-Zhitnitsa Street" aims to build almost 7.8 km of new line and seven new stations linking several residential districts with the centre of Sofia. This is the main central section of Line 3.

Figure 1.   New Metro Line 3, stage I, route through Sofia.

Considering existing geological and hydrogeological conditions, including passing right under Perlovska River and other relevant pre-existing structure, tunnel section along the larger part of the route is being constructed by tunnel boring machine.

Main tunnel consist in a single tube configuration of 8.43 meters internal diameter excavated using an EPB Machine of 9.4 boring diameter.

A consortium of Doğuş Construction, Via Construct and Ultrastroy is the lead contractor for tunneling works. TunnelConsult Engineering has been in charge of detailed design related to the underground works (permanent and temporary structures) and technical assistance during construction.

2 TBM TUNNEL LOGISTICS AS A CHALLENGE

For Stage I of Line 3, Metropolitan (Project's owner) considered for the overall project a procurement and delivery strategy considering the splitting of major works in different contracts as follows:

- Underground stations
- Main tunnel civil works
- Rail infrastructure
- Electromechanical systems

The simultaneous presence of several contractors usually generates interferences that must be considered and analyzed during the previous phases of the project.

Regarding tunneling works, contractor (DVU JV) had to deal with every station contractors in terms of coordination for TBM arrival, crossing and launching, and on the other hand the need to release the main tunnel already constructed to the track work contractor in a very tight schedule (even with a partial delivery).

Due to this kind of project constrains, logistics for TBM operation during tunneling works was considered very challenging from early stages.

Once analyzed these factors and their influence to the overall program, Contractor decided to consider an intermediate site establishment to operate the TBM (independent of any station) in order to minimize affections to station works progress and rail infrastructure.

Initial TBM launching site located on eastern extreme of the alignment should be changed to a new location close to the central part of the alignment to release a first stretch of tunnel (including stations in between) of any activity related to TBM tunneling operations.

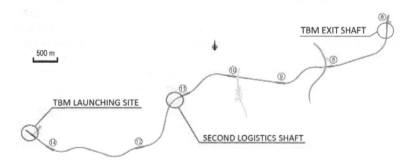

Figure 2. Project alignment showing TBM working sites location.

This new location needed to be able to supply the TBM with all materials and utilities for the boring process, and allow the muck out of the excavated material, managing the tunnel belt conveyor to the surface.

Location options for requested intermediate shafts were very limited due to the population density close to Sofia city center, with all the typical constrains related to land availability, tight working space, narrow right of way and interference with existing structures and utilities implications of dealing with them.

Finally a suitable area for the planned second stage was located just before the MC11 station, with an available area of 350 square meters close to the Medical University Complex of Sofia.

To be able to keep all the functionalities requested for tunnel boring machine operation, 2 temporary connection from surface to the tunnel where designed at the second logistic site.

A vertical round shaft over the tunnel section (6.75 meters diameter and 12 m depth) in order to proceed with the segments, backfill mortar and other material supply to the tunnel MSVs and the rest of TBM linear utilities.

Vertical shaft beside the main tunnel with a ramp up to the surface with a lateral connection to the TBM tunnel in order to facilitate the exit of the belt conveyor to the surface.

Detailed engineering for conveyor belt configuration to muck out through the opening and achieve the belt conveyor storage unit on surface, was defined by H+E logistics according to DVU JV indications.

3 TUNNEL CONNECTIONS CONCEPT DESIGN

DVU JV (Tunnel contractor) main requirement was to design a final solution for each connection minimizing the works inside the main tunnel, to not interfere with the TBM operations and be able to advance most of the works regardless the TBM position.

Another major requirement was to avoid the use of any temporary heavy steel frame inside the tunnel to hold the segments during ring opening and connection works, very common in usual openings associated to cross passage construction or ventilation connections.

One of the main challenges was the unforeseen use of any special heavy segment or any use of shear-cones between ring to ring, to redistribute loads from the opened rings to the adjacent ones prior to the construction of the permanent structure.

All these limiting factors were key-points in order to find out a proper technical solution.

Proposed design considered all the construction stages from surface with special focus on reinforcement connecting new structures to the main tunnel prior the saw cut of the segmental lining from inside.

Two concept connections were developed:

- One upper connection for TBM logistic porpoises, basically to feed MSVs with segments by a tower crane.
- One lateral connection, to manage tunnel belt conveyor to muck out the excavated TBM material.

Required dimensions for segments logistic shaft were already fixed, showing a minimum area for lift operations about 35 square meters in the upper part of the main tunnel.

For belt conveyor connection, the early requests were to allow a minimum lateral opening about 40 square meters, for and easy fit of the overall conveyor diversion concept.

This preliminary request implies an opening of a minimum of 4 contiguous rings.

This situation was impossible to handle without the installation of complementary temporary steel frames inside the tunnel or the use of heavy segments in combination with shear cones between rings, as was discarded from the beginning.

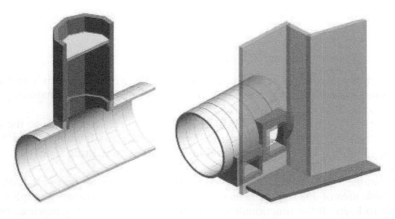

Figure 3. Connection concept designs for each option.

Due to these mandatory constrains, a more adjusted option was developed including 2 small openings (10.5 and 6.7 m^2 each one) adapted to the minimum space required for H+E for each conveyor belt included in the design (main tunnel conveyor and transverse conveyor).

With this option, even a ground improvement was required for the lateral connections, the openings were feasible in structural terms without any complementary temporary structure inside the tunnel.

In order to evaluate the influence of the connections construction foreseen in each location and to analyze the forces acting on the opened rings, different 3D FEM model of the tunnel aperture were performed using Sap2000 and MIDAS GTS software.

In each developed model, the ground is modeled through three-dimensional elements, which behavior for this particular case has been defined according the geotechnical parameters obtained from boreholes on site and laboratory tests results.

3D model was used to simulate the whole process of connection construction:

- Main tunnel excavation using the deconfinement method.
- Ground treatments execution by modifying soil properties.
- Excavation to the main tunnel.
- Opening the segmental lining.

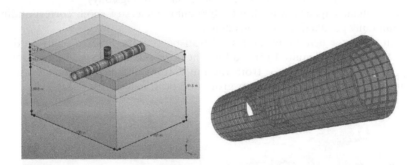

Figure 4.   3D models used showing geology scheme considered for vertical shaft and lateral openings geometry for conveyor connection.

New forces redistribution (axial forces N, moments M and shear forces T), on the segmental lining after ground excavation phase and after the rings opening was pointed out. In particular, shear forces that are transferred to adjacent rings, after the opening.

Using these results, connections between structures by collar beams were defined in order to achieve a successful design.

4   PLANNED TBM LOGISTICS

4.1   *Initial portal logistic concept*

Located at Zithnitsa depot area, 510 square meters were used to launch and perform the initial TBM works.

Ancillary equipment included were:

- Conveyor belt for the muck
- Muck pit
- Batching plant
- Pumps and piping for air and water supply
- MSV vehicles for the segments and grout transport
- Warehouse
- Ring segments storage area...

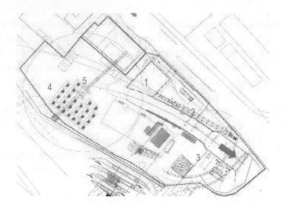

1. Muck pit
2. Belt conveyor
3. Batching plant
4. Segment rings storage
5. Tower crane

Figure 5.   TBM launching site layout.

All the logistics to transport materials in and out of the tunnel were originally designed to enable the EPBM to work in a full capacity of 40 m/day (26 rings/day).

Mucking pit volume capacity was about 3.000 cubic meters of spoil, corresponding to 1,5 times maximum daily production rate considered.

With a main conveyor belt along the tunnel transporting the muck directly to the surface, only one tower crane was required for loading the segments on MSVs, in addition, the segment transport gained independence from the muck transport, permitting the use of only 3 MSVs per tunnel (2 for segments and 1 for grout and personnel).

Figure 6.   Main tunnel conveyor on launching site, and MSVs used in the Project.

For MSV circulation along the tunnel, concrete backfill was planned after each Station arrival, in order to allow the MSV crossing inside the tunnel during TBM advance and minimize possible stops due to lack of materials provided by MSVs.

4.2   Intermediate TBM logistics shaft

New location at Sveti Georgi Sofiyski Street with an area of 350 square meters was performed, just beside the Medical University complex of Sofia.

Note than the area available was roughly a 40% less than the initial launching site, were no space limitation was considered.

This site configuration includes two different shafts, one located on the left side of the tunnel for belt conveyor system and one located over the tunnel for TBM logistics.

At the lateral shaft, a transverse conveyor belt transported the muck through the tunnel connection to another conveyor belt to transport the muck from the pit bottom to the surface in an inclination of 17°.

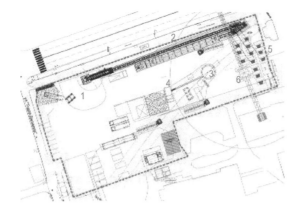

1. Muck pit
2. Belt conveyor
3. Logistic shaft
4. Batching plant
5. Segment rings storage
6. Tower crane

Figure 7.   Intermediate TBM logistic shaft layout.

Figure 8.   Belt conveyor exiting from the main tunnel and ancillary inclined belt conveyor mucking out.

Mucking pit capacity on the site was adjusted up to 2.135 cubic meters, corresponding to 18,5 excavated rings, lower than original site establishment.

In that case, the absence of a reasonable buffer in accordance with the maximum planned TBM rate, made the handling of the muck specially challenging due to the new location close to the city center.

Time constraints for trucks circulation to the landfill in order to minimize discomfort in the neighborhood, as well as the greater intensity of traffic around the work site during peak hours, were limiting factors during the advance of the tunnel boring machine.

In order to facilitate the main conveyor extension operations, horizontal storage unit was placed on surface. Two 90 degrees turnings on the belt conveyor (including vertical disposal) were used to achieve the surface on the new logistic site.

Figure 9.   Vertical disposal of the main tunnel conveyor belt to reach the horizontal storage unit on the surface.

Through the shaft over the main tunnel, segmental lining and other complementary materials were loaded to the MSVs, and the rest of linear items as well.

Figure 10. Loading materials on an MSV through the logistic shaft.

Related to the new site relocation, defined deadlines were met, including the belt conveyor set-up process and the rest of ancillary equipment, allowing to restart TBM operations in 2 months' time.

Moving of logistics from initial construction site to new intermediate TBM logistic site was carried out following the break through into MC11 station, the third one after launching the TBM.

About 80 per cent of structures and equipment from initial site were relocated at intermediate shaft.

During the course of the logistics moving operation, TBM was being pulled through MC11 station and a maintenance program for the TBM was performed, not interfering any of scheduled activities for connections.

Table 1. Main activities durations related to the new site set-up.

	Start date	Finishing date	Duration
D-Walls and shafts excavation	02.06.2017	25.11.2017	176 days
Ground improvement works	14.12.2017	27.01.2018	41 days
Conveyor belt/vertical shaft set-up	28.01.2018	31.03.2018	62 days
Site mobilization	03.02.2018	12.04.2018	69 days

5 TBM PERFORMANCE FROM EACH LOGISTIC SITE

TBM launching was performed on March 2.017, and after a first stretch to the Station MC14 considered as a learning curve process, following sections achieved expected advance rates.

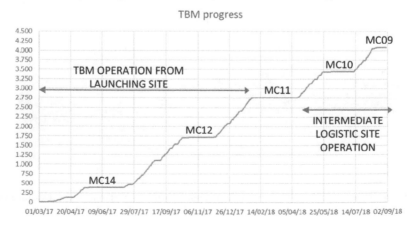

Figure 11. Frequency analysis regarding rings per day production.

A good daily maintenance plan, combined with a highly skilled staff in charge of TBM operations, had allowed to achieve really high and maintained performance in time.

As can be seen in the following table, once the learning curve was overcome after TBM launching, performance achieved was very regular all along the alignment around 10–11 rings per day as an average.

Table 2. Main TBM performance data during tunnel excavation.

	Length	m/day	ring/day	Best day (rings)
LS to MC14 (*)	375	5.21	3.47	15
MC14 to MC12 (*)	1323	14.07	9.38	21
MC12 to MC11 (*)	1032	16.65	11.10	21
MC11 to MC10 (**)	678	16.14	10.76	21
MC10 to MC09 (**)	634.5	17.15	11.43	25

* Logistics from initial launching site
** Logistics from intermediate shaft

If the frequency of rings per working day is analyzed, it can be noted that there is no difference in terms of TBM performance capacity, despite from which logistic site was operating.

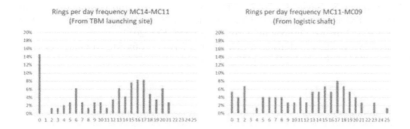

Figure 12. Frequency analysis regarding rings per day production.

These statistics shows that even original launching shaft configuration was planned for a full capacity of 26 ring/day, intermediate site with greater restrictions in terms of space and general availability, tunneling performance was not been affected thanks to the previous work of sizing and planning on this option.

Rest of stretches from MC09 to the exit shaft are still ongoing (September 2018), and expected TBM rates for the rest of the sections seem that can be maintained without major problems until the end of the project.

End of tunneling works included in Phase I of Line 3, are planned to be finished for first quarter of 2019.

6 CONCLUSIONS

Most of challenges in urban tunneling are related to tight working space, narrow right of way and interference with existing structures and utilities implications of dealing with them.

In the top of this, if the overall project involves due to the planned delivery strategy, different contractors in charge of different parts of the same project, inevitable interferences are expected and some challenges can achieve higher levels of difficulty (as the example of Sofia metro project).

Dealing with these challenges starts in the conceptual design and continue through the preliminary and detailed design when constrains and solutions are detected and defined during the early stages of the project.

During planning and construction stage, Contractor may find some answers to face design challenges and cost savings through innovative solutions and methods.

Sofia Metro Line III can be considered a good example about this kind of planning in early stages performed by Doğuş Construction, Via Construct and Ultrastroy JV.

Presented temporary logistic shaft for TBM operation has been successfully constructed during the project execution, fulfilling all the considered requirements during planning stage, allowing a high TBM performance until the end of the Project.

ACKNOWLEDGEMENTS

The authors would like to thank the consortium of Doğuş Construction, Via Construct and Ultrastroy, and TunnelConsult Engineering for the source of data and support during this paper preparation.

A special thanks also to Eng. Stoyan Bratoev managing director of Sofia metro operator "Metropolitan", in charge of the overall metro extension.

REFERENCES

Guglielmetti V., Grasso P., Mathab A., Xu S. 2007 "Mechanized Tunneling in Urban Area. Design meth-odology and construction control". Ed. Taylor and Francis.
Della Valle, N., Babucci, S., Savin, E., 2012. "EPB pressure control and volume loss resulting in tunnelling Sofia Metro Extension". UNDER CITY Colloquium on Using Underground Space in Urban Areas in South-East Europe. Dubrovnik, Croatia, pp.526–537, 2012.
Úlehla, J., Pekárek, V., Seidl, V., Mensík, A., Hasík, O., 2013. "Sofia Metro Diameter III". Published in Tunel Magazine. 22.ročník - č. 4/2013, pp.47–53, 2013.

Tunnels and Underground Cities: Engineering and Innovation meet Archaeology,
Architecture and Art, Volume 6: Innovation in underground engineering,
materials and equipment - Part 2 – Peila, Viggiani & Celestino (Eds)
© 2020 Taylor & Francis Group, London, ISBN 978-0-367-46871-2

An innovative method of large space underground construction in soft and shallow ground using concrete arch pre-supporting system

M.H. Sadaghiani
Sharif University of Technology, Tehran, Iran

ABSTRACT: Construction of underground structures in urban areas is a very challenging work. There are generally two methods of their construction, an open cut-and-cover and underground method. To eliminate the subsurface and surface disturbance and street traffic problems, underground method is preferred. The stress redistribution caused by underground excavation induces movements in the earth mass and ultimately at the ground surface. This is more pronounced in large space excavations. Generally pre-supporting system is used to control ground deformation and thus enhance its stability. In this paper an innovative pre-supporting system is presented. Concrete Arch Pre-supporting System (CAPS) is introduced in Tehran Metro in 2002. This method has roots in construction method of an old Iranian small water tunnels, called *Quanat*. CAPS is an efficient method for stabilizing large span underground spaces constructed in shallow and soft ground. In this technique underground reinforced concrete elements are constructed around the proposed underground space prior to main excavation. This method is used successfully in over 50 large span underground structures in Tehran Metro. CAPS has a potential to be used to pre-support the large span underground spaces at any weak ground condition in an urban area.

1 INTRODUCTION

Underground subway stations can be constructed either by open cut-and-cover or underground construction. Due to the presence of subsurface utilities and street traffics in urban areas, underground are preferred to eliminate relocation of the subsurface utilities, and other city disturbance. Underground construction in an urban area is a very challenging effort. In a weak and shallow ground and large span underground structures, the complexity is more pronounced. The stress redistribution caused by underground excavation induces movements in the earth mass and ultimately at the ground surface causing damage to adjacent structures.

The need to control ground surface settlements in urban area is widely recognized and new construction methods are continuously developed. Settlements induced by underground excavation may cause serious damages to nearby structures and subsurface underground utilities, (Sekimoto et. al. 2001). Various ground treatment techniques to improve stability of underground excavation and reduce settlements are presented in recent literature, Carrieri, et.al. (2002), and Ocak (2008). Different forepoling, grouting methods and pre-supporting the main tunnel by concreted small horizontal tunnels were introduced, (Johnson, et.al., 1983).

Concrete Arch Pre-supporting System (CAPS) is introduced in Tehran Metro in 2002, (Sadaghiani, Gheysar, 2003). This pre-supporting system is a very effective method for stabilizing large span underground spaces in shallow and soft ground and reducing the ground deformation and subsequently the surface settlement. In this technique underground reinforced concrete elements, including piles and curved beams (arches), are constructed around the proposed underground space prior to main excavation in order to support the ground during the excavation of the underground section, Figure 1. Subsequent to pre-supporting, excavation can be executed in variety of methods. Generally multi-stage excavation and supporting is used to finalize the construction of the desired section.

2   CONSTRUCTION OF CAPS

Concrete Arch Pre-supporting System (CAPS) method is a pre-supporting system that is constructed prior to main large span underground excavation. This method has roots in construction method of an Old Iranian small water tunnels, called Quanat. Quanat consists of wells and small semi-horizontal continuous adits which connects wells at the bottom level. They are all hand excavated in an alluvium ground to collect water, distribute and convey it to desired location in downstream. In a very weak ground small perforated concrete segments, KAVAL, are installed to maintain stability as the excavation proceeds. Hand underground excavation method in Iran, generally cost less and takes much less time compared to forepoling and grouting pre-supporting systems. In CAPS, small adits in vertical direction as piles, in semi-horizontal and horizontal direction as arch beams and connecting galleries in horizontal longitudinal direction are hand excavated around the future large underground space. The rate of hand excavation of an adit is about 3 m daily. By increasing man power, the adits can be excavated from number of faces, and several adits can be excavated simultaneously, thus generally it takes very short time to excavate the galleries, piles and arch beams. Using new mechanized technologies such as micro-tunneling as an innovative mechanized method of constructing CAPS can be developed.

In order to use this method in construction of large span underground subway stations or transition tunnels, initially either heading of main tunnel in NATM tunneling or the TBM bored tunnel should be excavated along the station line. In the cases that the construction of large underground space is prior to the main tunneling, small adit or adits along the underground space is excavated. From this opening, small access adits in both transverse directions are excavated at desired spaces. The transverse access adits are excavated to the length beyond the transverse excavation line of main station opening. From the end of transverse access adits, side adits (galleries) in longitudinal direction are excavated with total length equal to the length of station at both side of opening. The location of longitudinal adits is beyond the excavation lines of main station opening. Inside both longitudinal adits, series of piles are constructed at designed intervals. Small adits for arch beams, with horseshoe-shape section are excavated from the top of piles in two sides connecting piles and creatig an arch shape as shown in Figure 1. The construction stages of CAPS are shown in Figure 2.

Another longitudinal adit is excavated over the crown for concreting the excavated arch frames. The excavated pile and arch beam adits are filled with reinforced concrete to make a rib shape underground structural frames prior to main large excavation. The longitudinal crown and side adits sometimes are filled with concrete to produce the continuous 3-D underground frame structure around desired section of main station opening. The overall system

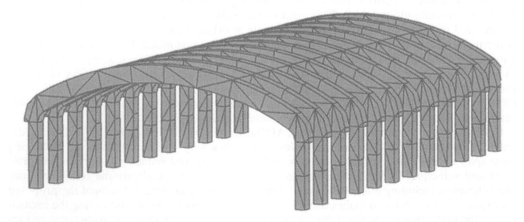

Figure 1.   Rib-shape Concrete Arch Pre-supporting Structure (CAPS) constructed prior to main excavation.

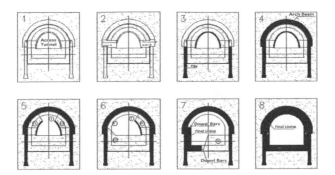

Figure 2. Construction stages of *CAPS*.

make series of parallel reinforced concrete arch frames as shown in Figure 1. CAPS supports the surrounding ground during the main excavation and requires minimal initial supporting system.

Practically, CAPS is introduced in construction of Mellat Station in Line 2 of Tehran Metro in 2002. The ground settlement from numerical investigation was 24 mm and from field monitoring measurement was 19mm (Sadaghiani, Gheysar, 2003). Due to its success, CAPS has been the dominant construction method of underground stations and some other large span spaces such as Transitional tunnel in Tehran Metro Lines (Sadaghiani, Ebrahimi, 2006 and Sadaghiani, Taheri, 2008).

The main advantages of CAPS in large underground spaces are that by pre-supporting the soil around the underground space prior to excavation, it reduces soil deformation and thus increases the stability of the large excavation with low overburden in soft grounds. This system constrains the ground settlement, thus enhances the general stability. Another advantage is that, it reduces the construction time due to the fact that the piles and arch beams in CAPS can be constructed simultaneously from several faces and the main core excavation can be done in large sections all in short period of time.

After CAPS construction, the station construction is preceded by the multi-stage excavation and initial supporting of the main opening. The main excavation proceeds in large sections followed by installing very light initial supporting system such as shotcrete and welded wire mesh over the excavated surface. Due to simultaneous sequential construction, the rate of construction advance increases dramatically.

3 EXAMPLES OF CAPS SYSTEM

Mellat Station was one level underground station. Following its construction in 2003, over 30 similar one level stations in all Lines of Tehran Metro are being designed and constructed. Similar large span underground spaces such as transitional tunnels for connecting intersecting between Metro Lines are also designed and constructed. This method is also used in road tunnels in Tehran. Followings are some other examples of using CAPS method in underground structures in Tehran.

3.1 *Transition tunnel between stations A4-3 and A4-4*

At the west part of Line 4 of Tehran Metro, passing A4-3 Station in Ekbatan Complex, there is Ekbatan West Train Parking for Metro. In order for the train to either go to A4-4 Station or to parking, after A4-3 Station, a transition tunnel (TT) is required for this purpose. Figure 3 shows the plan view of this TT, Figure 4 illustrates sample construction stages 7 - 12 of part A, and Figure 5 shows cross sections of A, B and C parts of TT. The largest section of

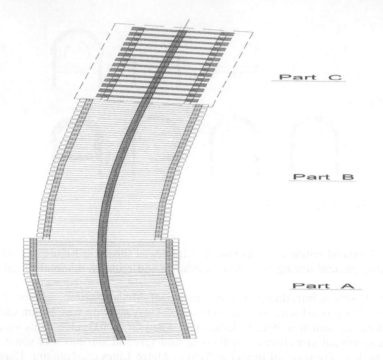

Figure 3.   Plan view of Transition Tunnel *(TT)* between A4-3 Station and Ekbatan Parking.

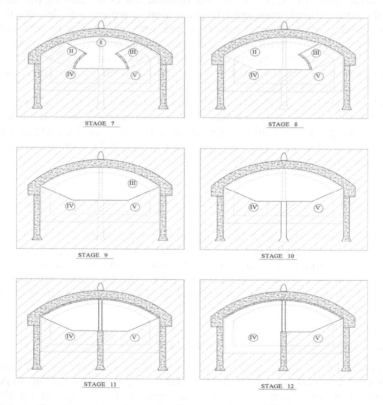

Figure 4.   Construction stages 7–12 of part *A* of Transition Tunnel (TT).

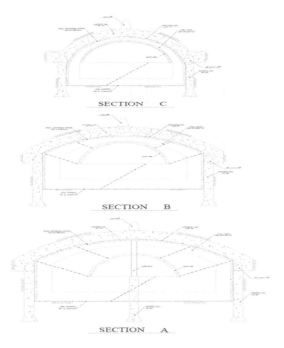

Figure 5.   Cross-section of Parts *A, B* and *C* of TT.

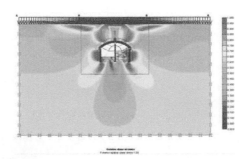

Figure 6.   Shear stress distribution in stage 14, Part A, *TT.*

this TT is part A which has over 24 m span and less than 10 m overburden. There are central piles and structural wall at the middle of this part which divide the path to parking or A4-4 Station. All construction stages of these sections are numerically modeled and based on the results, they are designed and constructed. Figure 6 shows the shear stress distribution in the ground and the supporting system for one of the last stages.

3.2   *CAPS for a twin tunnel and Island stations*

Subway system in Tabriz, Esfahan and Shiraz has twin tunnels and the island stations. For these cases, CAPS also can be applied. A research had been conducted at SUT (Sadaghiani, Dadizadeh, 2010). An example of a station in Tabriz Metro is investigated. The construction stages of this type of CAPS are shown in Figures 7 and 8. Figure 9 illustrates 3-D numerical modeling for stability analyses and determination of deformation and stress distribution for variation of construction stages.

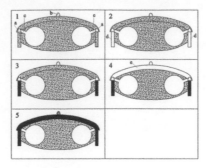

Figure 7. Construction sequence of CAPS in Island Stations with Twin Tunnels; a: Longitudinal side adits, b: Longitudinal crown adit, c: Transverse adits, d: Side piles, e: Arch beams.

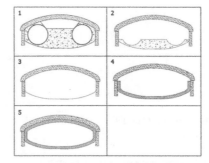

Figure 8. Construction stages of excavation and main structure in Island station after CAPS.

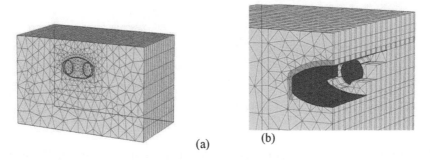

Figure 9. (a) General 3-D FEM model, (b) Part of the half of cross section and stage construction.

3.3 *Two-level stations*

In narrow locations where there is not enough space for side structures for ticket hall and utility rooms, the underground station should be in two levels. Therefore there is a need for construction of an underground structure with deeper section. The construction stages of these types of structures are similar to one story spaces. In relatively competent ground, the side piles are deeper.

3.3.1 *Ghaem Station, last station in Line 3 of Tehran Metro*

Ghaem Station is the last station in north part of Line 3 of Tehran Metro. This station has two levels. Figure 10 illustrates two Level underground Ghaem Station. Stability analyses of all the construction stages are conducted by numerical modeling, Figure 11. The results

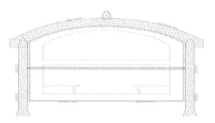

Figure 10. Two level underground station, *Ghaem* Station.

Figure 11. Part of numerical model of two level *CAPS* for *Ghaem* Station.

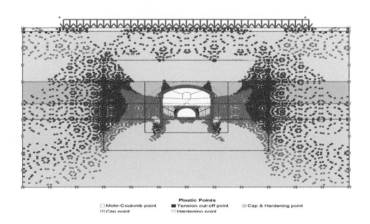

Figure 12. Plastic zone around underground space at construction stage 13 For Ghaem Station.

include ground deformation, surface settlement and stresses in the ground and structural elements. Figure 12 illustrates the plastic zone around the excavated section.

3.3.2 *Satari Station, Line 6 of Tehran Metro*

A new method of CAPS is recently applied in Satari Station in Line 6 of Tehran Metro. In this method there are two levels of piling. The first level of CAPS is slightly wider to make space for the first level of piles. Piles of the lower level are constructed from inside of the upper CAPS level. Construction stages of this method are different. The main tunnel of Line 6 is constructed by TBM. Figure 13 illustrates overall section of Satari Station. Sample of construction stages of Satari Station is shown in Figure 14. In Figure 15 shear stress distribution after excavation of arch Beam in Satari Station is shown.

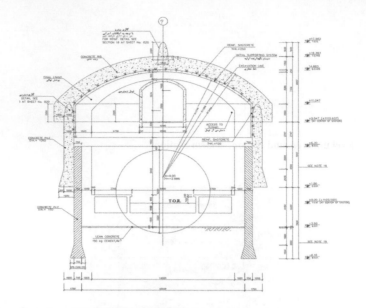

Figure 13. Overall section of CAPS in Satari Station.

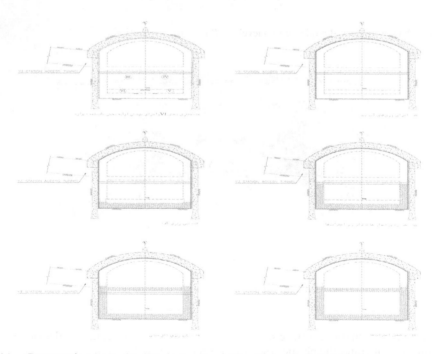

Figure 14. Construction stages 12–17 (construction of final lining) in Satari Station.

4 CONCLUSION

To eliminate surface disturbance and relocation of subsurface utilities in urban area, subway stations are constructed using underground methods. Construction of large span underground Structure in an urban area where the overburden is very low and the ground is generally weak is a very challenging work. To stabilize the underground space, usually pre-supporting system

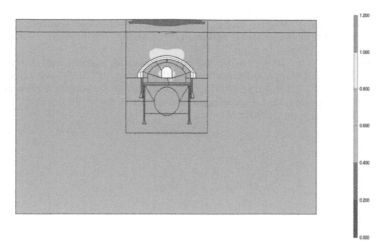

Figure 15. Shear stress distribution after excavation of arch beam in Satari Station.

is used. In this paper an innovative pre-supporting method, Concrete Arch Pre-supporting System, CAPS is presented. In this method a reinforced concrete underground arch frame is constructed around the proposed underground structure prior to the main excavation. This structure reduces the ground deformation and stress concentration in the ground and main structure. The work including CAPS and overall ground excavation and construction of main structure is conducted by logical stages. Numerical modeling can be used to determine the logical construction stages and structural design for controlling the ground deformation, surface settlement and stress concentration.

CAPS has been used successfully in over 50 underground structures in Tehran Metro. In Tehran Metro it is used for one and two level underground stations and transition tunnels. The main method of excavation is an Old Iranian adit excavation, Quanat. Mechanized method of CAPS excavation can be developed using micro-tunneling concept. CAPS can be utilized in any urban large underground space development.

ACKNOWLEDGMENTS

Author would like to thank Sharif University of Technology for providing facility for conducting research on subject of CAPS and Sazbon Consulting Engineering Co. for providing information about CAPS in construction of underground Stations and Transition Tunnels in Tehran Metro.

REFERENCES

Carrieri, G., Fiorotto, R., Grasso, P.G., Pelizza, S. (2002), "Twenty years of experience in the use of the umbrella-arch method of support for tunneling", *International Workshop on Micropiles*, Venice.
Johnson, E.B., Holloway, L.J., Kjerbol, G. (1983), "Design of Mt. Baker Ridge freeway Tunnel in Seattle", *Proceeding, Rapid Excavation and Tunneling Conference*, Vol. 1, pp. 439–458.
Ocak, I. (2008), "Control of surface settlements with umbrella arch method in second stage excavations of Istanbul Metro", *Tunneling and Underground Space Technology, Elsevier*, pp. 674–681.
Sadaghiani, M.H., Dadizadeh, S. (2010), "Study on the Effect of a New Construction Method for a large Span Metro Underground Station in Tabriz-Iran", *Tunneling and Underground Space Technology*, 25, PP. 63–69.
Sadaghiani, M.H., Ebrahimi, A. (2006), "Stability analysis of construction sequences of a large underground metro station using Concrete Arc Pre-Supporting System (CAPS)", *Proceedings, 7th Iranian Tunneling Conference*, Tehran, Iran. pp. 417–426.

Sadaghiani, M.H., Gheysar, Y. (2003), "Study on the effect of construction sequence of a large underground on the ground behavior in Mellat Station, Tehran Metro", *Proceedings, Fourth National Civil Engineering Conference*, Tehran, Iran, pp. 287–294.

Sadaghiani, M.H., Taheri, S.R. (2008), "Numerical investigations for stability analysis of a large underground station, Mirdamad Station in Tehran Metro", *Proceedings, 34th WTC*, Agra, India.

Sekimoto, H., Kameoka, Y., Takehara, H. (2001), "Countermeasures for surface settlement in constructing shallow overburden urban tunnels penetrated through active fault", *Proceedings, Modern Tunneling Science and Technology*, pp. 711–716.

Tunnels and Underground Cities: Engineering and Innovation meet Archaeology,
Architecture and Art, Volume 6: Innovation in underground engineering,
materials and equipment - Part 2 – Peila, Viggiani & Celestino (Eds)
© 2020 Taylor & Francis Group, London, ISBN 978-0-367-46871-2

Passive fire protection for road tunnel structures

K.M. Sakkas, N. Vagiokas & M. Panias
Geomaterials Valorization Innovation PC, Athens, Greece

D. Panias
National Technical University of Athens, Athens, Greece

ABSTRACT: An essential element of a tunnel lining design is to ensure that the final lining will not fail due to the high temperatures, which may be developed during a fire. For testing the fire resistance of the passive fire protection materials, a series of standard time – temperature curves have been developed which last maximum three hours. However, there are plenty of real tunnel fires which lasted more than three hours as in the fires of Mont Blanc, Tauren and Gotthard tunnels. On this purpose an innovative fire protection material has been developed based on the technology of geopolymerization. In the specific research the properties of this material are presented (mechanical, physical, thermal) as well as the behavior of the material in fire scenarios which last for more than 3 hours and for different thicknesses of the material. The results of the test show that a) the material is not damaged after an extended time exposure to fire and retains its structural integrity b) with a thickness of 4 cm the material can achieve the RWS fire scenario limits (380 °C in the interface) for substantial more time than the required time of 3 hours. The results of the laboratory scale tests proved that the material has a good thermal insulating capacity which may render it a promising material for passive fire protection of underground constructions.

1 INTRODUCTION

1.1 *General*

A number of serious tunnel fire incidents have been reported worldwide that have led to injuries and life losses, heavy damage in the concrete lining, excess material damage, and significant time periods of tunnel restoration during which the tunnels were unavailable for traffic (Jönsson J. and Herrera F. 2010). Fires in tunnels can seriously damage their concrete lining rendering it to collapse. The damage is caused particularly by the spontaneous release of great amounts of heat and aggressive fire gases, resulting to spalling of concrete. The spalling phenomena are expected at several temperatures depending on the strength of the concrete. It is generally accepted that concrete exposed at temperatures higher than 380 °C is considered as damaged and should be removed and repaired (Boström and Jansson R., 2004). In addition to the damage caused by fire to concrete, special attention has to be paid to the damage caused to the structural steel rebars that normally reinforce the concrete structures since are considered to lose their strength at temperatures between 500 and 600 °C. Therefore, steel and concrete are both fire sensitive construction elements requiring passive protection against fire in order to be capable of withstanding of fire for an appropriate period of time without loss of stability (Fletcher, I. 2007). Passive fire protection methods are generally divided in two categories: external (insulation) and internal (concrete design). The former are more advantageous being applied in new as well as in existing tunnels and consist of the cladding of the concrete by a fire resistant material which creates a protective external insulation envelope. However, even this kind of protection necessitates renovation of the tunnel lining after the fire. Most of the passive fire protection materials either lose their structural integrity or undergo severe damage which necessitates replacement of them. As a result they offer protection to the tunnel in case of one only fire. Apart from that, most of them necessitate long period of time for restoration and as a result tunnels remain closed with obvious negative economic effects. The RWS curve considers a 50 m^3 petrol tanker fire with a fire load of 300 MW lasting up to 120 min or 180 min depending on the fire scenario. The initial rapid fire growth is simulated with temperature rise up to 1200 °C within the first 10 min. Then, the temperature increases slowly and reaches the highest achieved value of 1350 °C within the next 50 min, followed by a gradual drop in temperatures to 1200 °C in the next hour as the fuel load is burnt off. The RWS temperature-time curve is able to describe very severe fire incidents in tunnels that are accidentally created by large tanker vehicles transporting oil or petrol. The performance requirements for an efficient fire resistant material are that the temperature of the interface between the concrete and the fire protective lining should not exceed 380 °C and the temperature on the reinforcement should not exceed 250 °C (Khoury 2000, Phan 2008, Sakkas 2013, PIARC 2004) . In this paper, the behavior of an innovative fire resistant material under a 4 hour RWS fire incidents will be evaluated, in order to test the possibility of leaving the material on a tunnel lining for more time than the typical of the proposed by the standards.

This work aims at the production and testing of an innovative fire resistant material for external passive fire protection of concrete tunnels linings under a 4 hour RijksWaterStaat (RWS) fire incidents. Following the European Federation of National Associations Representing producers and applicators of specialist building products for Concrete (EFNARC) guidelines the material was subjected under the RWS curve for substantial more time than the required time. The total exposed time was 4 hours, which is almost the double time of the ordinary standards. The results of the laboratory scale tests proved that the material has a good thermal insulating capacity remaining lower than the limit of the specific curve for 4 hours without damage and yielding. It should be noticed that all the commercial available materials are seriously damaged after the fire and should be fully renovated.

2 EXPERIMENTAL

2.1 Raw Materials

The slag used in the present study was provided by a large metallurgical plant. The aforementioned slag was used as the main ingredient for the synthesis of a fire resistance mater al, utilizing the technology of geopolymerisation.

2.2 Analysis and Tests

The mechanical properties of the material investigated in the specific research include uniaxial compressive strength and flexural strength. The mechanical properties were measured in triplicate and the described values comprise the mean values of the measurement. Compressive strength was measured according to ASTM C109 using cubic specimens of 50 mm edge. The flexural strength was measured according to ASTM C348 using prismatic beam specimens of 40 x 40 cm^2 cross section and 160 mm length. The properties were measured at 28 days and three months after the curing period. The thermal conductivity of the material was measured according to ASTM C518 standard test using a 15 cm x 15 cm x 2 cm specimen.

The passive fire protection test was performed in the laboratory by using a test furnace which was designed according to European Federation of National Associations representing producers and applicators of specialist building products for Concrete (EFNARC) guidelines. The furnace has the ability to simulate the temperature-time curves employed in several international standards such as the RWS fire load curve. For this test a 15 cm x 15 cm x 14 cm specimen was prepared, consisting of four cm thick geopolymer material and 10 cm thick concrete slab. The latter was prepared by using CEM II cement type (>350 kg/m^3) and coarse aggregates from crushed limestone with maximum particle size between 16 and 20 mm as well as a water to cement ratio less than 0.48 according to the EFNARC guidelines (EFNARC, 2009). The test was performed 28 days after the production of the specimens. The adhesion of the fire resistant material on the concrete slab was conducted through special anchors provided by Fisher. The specific anchors have the ability to withstand at all the fire scenarios. During the test the free surface of the geopolymer material was exposed to a heat flux simulating the RWS fire load curve. The temperature at the geopolymer/concrete interface was measured by using a "K"-type thermocouple, while the temperature of the back surface of the specimen was measured with a high performance infrared thermometer. The 4 hour test was performed with a specimen following the above standards, and simulating the RWS curveThe complete abstract will fall in the abstract frame, the settings of which should also not be changed (width: Exactly 15.0 cm or 5.91"; height: Automatic; vertical: 7.2 cm or 2.83" from margin; Lock anchor).

3 RESULTS

3.1 *Properties*

The compressive strength and flexural strength of the material was measured equal to 4,5 MPa and 1,2 MPa respectively at 28 days after the production. The thermal conductivity of the material was measured to be 0.09 W/(m·*K) at 300 K with a density of 980 kg/m^3. The thermal conductivity of the material measured is substantially lower than the corresponding ones of the commonly used structural building materials, such as concrete blocks (0.5–0.6 W/m·K) and cement or gypsum plasters (0.2–0.8 W/m·K). This property is very crucial for the performance of fire resistant materials because it determines the ability of materials to operate as efficient heat flux barriers. In general, for a given heat flux the lowest the thermal conductivity, the highest the established temperature gradient across the fire resistant material. In addition, for given heat flux and predetermined temperature drop inside the fire resistant material the lowest the thermal conductivity, the lowest the thickness of the superficial fire resistant material.

3.2 *Passive fire protection testing*

The behavior of the material under the four hour fire loading test is shown in Figure 1, where the attained temperature at the concrete/geopolymer interface, the real furnace temperature simulating RWS fire load curve, the temperature of the back surface of the concrete are shown as a function of time. As mentioned above the total time of the test was 4 hours. As it is observed, the temperature at the geopolymer/concrete interface was lower than 380°C during the whole duration of the 4 hour fire test which is the limit of the RWS curve according the EFNARC guidelines. The temperature at the 2 hour and 3 hour test was 158°C and 270°C which extremely lower comparing to the limits of the standard set by the E.F.N.A.R.C for a passive fire protection test with the RWS fire loading curve and the behavior of the commercial available materials. At the first 50 minutes, where the temperature in the furnace was increased rapidly from the ambient temperature to almost 1350°C, the interface temperature was remained lower than 50°C establishing a temperature gradient across the fire resistant material equal to 25°C/mm. Then, the interface temperature started increasing, with a low rate of 16 C/min reaching almost 100°C at about 85–90 minutes of test duration. During this time period the furnace temperature reached the maximum value of the test (=1350°C),while the temperature at the interface was just 90–100°C, establishing a high temperature gradient which is attributed to its low thermal conductivity value (0.09 W/m·k).

At that time of the test, the temperature remained for some minutes (8–10) constant at around 100°C with a temperature gradient at.The temperature plateau at 100°C was attributed to the removal of geopolymeric water through an endothermic water evaporation process consuming large amount of the incoming heat due to the large latent heat of water

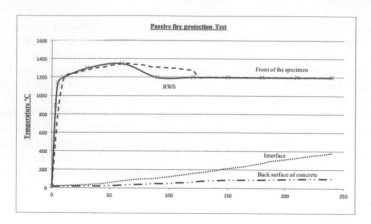

Figure 1. Performance of the fire resistant material under the 4 hour RWS fire load.

Figure 2. Performance of the fire resistant material under the 4 hour RWS fire load.

evaporation and keeping the interface temperature more or less constant at around $100°C$. From this point onwards, the temperature at the geopolymer/concrete interface increased with an average rate of $1.5°C/min$ while the furnace temperature decreased to $1200°C$ reaching the 2 hour of the test. At the end of the 4 hour fire test the interface temperature was $376°C$ and the temperature gradient was $19.7°C/mm$. Finally, the temperature in the back surface of the concrete slab did not exceed $100°C$ during the whole duration of the fire test as is seen in Figure 1, which means that across the concrete slab the temperature was varied in-between $100°C$ and $376°C$, temperatures that do not seem to create damage to the concrete and the structural steel.

Also as it is seen in Figure 2, the fire resistant material did not appear any macroscopically visible mechanical damage after the end of the test. It remained on top of the concrete slab without any change of its geometry as well as without the appearance of any cracks. The adhesion of the material with concrete through the steel anchors proved to be excellent and no serious detachment at the materials interface was observed. Also the concrete slab protected by the material did not appear any form of spalling or other mechanical damage and remained as it was initially before the fire test (Figure 2).

4   CONCLUSIONS

The fire resistant material that was produced demonstrated through a series of tests the following attributes:

a) Its 28 days compressive strength reached the value of 4.5 MPa with a flexural strength of 1.2 MPa The mechanical properties of the material are inferior to the ones of structural concrete and thus, it is not capable to be used as a load bearing construction element. On the other hand, the material has comparable mechanical properties with the existing fire proofing materials and therefore, has the potential to be used as a superficial material for passive fire protection of underground constructions.

b) The thermal conductivity of the material is 0.09 W/m.K at 300 °K which is substantially lower from the thermal conductivities of the commonly used structural building materials. In comparison to the marketable fire resistant materials, the thermal conductivity of the material is also lower indicating that it can perform as an insulating superficial material setting a thermal barrier in front of the fire sensitive tunnel concrete lining.

c) The material has a density of 980 kg/m^3 which is also similar to the commercial available fire resistant materials

d) The behavior of the fire resistant material during the 4 hour fire test with the most severe RWS fire load curve was excellent. The maximum temperature at the concrete/geopolymer interface was 374 °C which is lower from the required temperature set by EFNARC for protection of tunnels concrete linings ($\leq$ 380°C) while the temperature at the 2 hour and 3 hour test was 158° C and 270° C respectively which is also substantially lower from the required temperature. The material after the four hour fire test did not appear damages as well as visible deformations and its adhesion with the concrete through the steel anchors proved to be excellent. The passive fire protection test proved that the material can put an effective heat flux barrier offering very successful passive fire protection to the tunnels concrete linings as well as the steel reinforcement from the most severe fire incidents that can happen in underground constructions.

e) The material after the end of a fire incident remains on top of the concrete lining without any mechanical damage or deformation. Therefore, there is no need for serious repairing works but only for minor interventions which will not stop the traffic.

REFERENCES

Boström, L. & Jansson, R. 2004. Spalling of Concrete for Tunnels, *Proceedings from the 10th international Fire sience & Engineering Conference (Interflam '04), Edinburgh, Scotland.*

European Federation of National Associations Representing producers and applicators of specialist building products for Concrete 2009. Specification and Guidelines for Testing of Passive Fire Protection for Concrete Tunnels Lining. *Norfolk, UK.*

Fletcher, I., Welch, S., Torero, J.L., Carvel, R.O. & Usman, A. 2007. Behaviour of concrete structures in fires. *Journal of Thermal Science, 11*: 37–52.

Jönsson, J. & Herrera, F. 2010. HGV traffic – Consequences in case of a tunnel fire. *Fourth International Symposium on Tunnel Safety and Security, Frankfurt am Main, Germany, March 17–19.*

Khoury, A. 2000. Effect of fire on concrete and concrete structures. *Progress in Structural Engineering and Materials 2*: 429–447.

Phan, L.T. 2008. Pore pressure and explosive spalling in concrete. *Materials and Structures* 41:1623–1632

PIARC, 2004. ITA Guidelines for Structural Fire Resistance for Road Tunnels. *Fire Safety Engineering: Design of Structures*, Butterworth-Heinemann.

Sakkas, K., Nomikos, P., Sofianos, A. & Panias, D. 2013. Inorganic polymeric materials for passive fire protection of underground constructions. *Fire and Materials.* 37:140–150.

Tunnels and Underground Cities: Engineering and Innovation meet Archaeology,
Architecture and Art, Volume 6: Innovation in underground engineering,
materials and equipment - Part 2 – Peila, Viggiani & Celestino (Eds)
© 2020 Taylor & Francis Group, London, ISBN 978-0-367-46871-2

Comparative study on shotcrete performance in tunnels based on different constitutive approaches

G.G. Saldivar
National University of Mexico, Mexico

F.A. Sánchez
President of the Mexican Association for Tunnelling and Underground Works (AMITOS), Mexico

ABSTRACT: For decades, shotcrete in underground works has been one of the most commonly used support systems due to the benefits that provides in terms of efficiency, ground stabilization and reinforcement. Thus, technology applied to its implementation, mix design, reinforcement elements (as fibers) as well as test and calculation methodologies to verify its structural performance, have experienced great progresses. In this paper some advanced techniques for its better representation and design will be presented.

1 INTRODUCTION

The need for economy and safety in underground works requires the use of increasingly advanced techniques for calculation. Nowadays, software used in boundary value problems allows performing large scale simulations that, in the case of shotcrete analysis and design, go beyond classical structural theory.

To perform advanced designs it is necessary to understand the basic aspects of shotcrete behavior and represent them adequately, i.e., phenomena associated with stiffening and strengthening in time, as well as with its quasi-brittle nature once hardened (stress-strain levels associated with the onset of damage and post-peak softening where strain localization appears).

While modelling, strain localization triggers a strong mesh dependency. In this work, two constitutive models are calibrated with real laboratory tests at structural level; some adaptations are made in order to minimize mesh dependency with the aim to, later, extend the approach to real shotcrete structures.

The aim of this research is proposing mathematical approaches useful to better represent the behavior of shotcrete during tunnel excavations while creating realistic numerical geotechnical scenarios that can lead to optimizing designs.

2 CONSTITUTIVE MODELLING

In 2016, the firm Plaxis introduced a constitutive model for its use in the simulation of shotcrete called the *Shotcrete Model* (SM), which considers a Mohr-Coulomb type yield surface for the compression branch (F_c) and a Rankine type surface for the tension branch (F_t); when combined, they constitute the well-known modified Coulomb criterion.

The Shotcrete Model has a formulation based on classic elastoplasticity with hardening and isotropic softening; In addition, it considers the evolution of time-dependent parameters (modulus of elasticity, uniaxial compressive strength, etc.), shrinkage and creep of the material. Likewise, the model uses the Smeared Crack theory as regularization technique, which

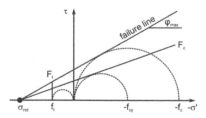

Figure 1.   Yield surfaces and failure envelope. Taken from Schädlich et al. (2014).

allows avoiding in a certain way the pathological mesh dependency problem, commonly presented in the modelling of boundary value problems when working under strain states beyond the elastic branch.

On the other hand, Itasca's FLAC^{3D} is one of the most powerful commercial codes for geotechnical applications, which includes a constitutive model, also based on classical elastoplasticity with hardening and softening called *Strain Hardening/Softening Mohr-Coulomb Model* (SHSMCM). This constitutive model does not have time-dependent laws that allow to represent the phenomena associated with aging; in addition, it lacks a regularization technique to reduce the grid dependency problem; However, there is the possibility of introducing evolution laws for parameters in time and providing the model with a simple methodology that mitigates grid dependency using the in-built programming language (FISH).

3   MESH DEPENDENCY PROBLEM

Steel fiber reinforced shotcrete (SFRS) is a quasi-brittle material that, when subjected to considerable levels of deformation, experiences the so-called localization phenomenon, manifesting itself in thin bands of intense plastic flow known as shear bands.

Regarding the modelling stage, the presence of shear bands causes certain kinematic fields to become discontinuous, therefore, numerical codes formulated under continuum mechanics theory, yield solutions showing a strong dependency to the degree of mesh or grid refinement, forcing to use small elements or zones, which implies a great computational cost, being impractical in many occasions; This is known as the mesh dependency problem, so, in order to use coarser meshes or grids, lowering calculation time and at the same time providing reliable results, this paper presents mathematical expressions that can give a practical solution to the mesh dependency problem.

4   LABORATORY TEST MODELLING

The aim of the present research begun with the need for a review of the geotechnical and constructive aspects of the excavation of a metro tunnel in Mexico City with the objective of determining the excavation's safety conditions according to some changes proposed by the contractor in order to improve the excavation procedures and the support systems. One of the most important of these changes was the replacement of double-welded wire mesh within shotcrete layers, included in the original project, by a mixture of steel fiber reinforced concrete of a type and characteristics such that it complied with the common specifications used in underground works.

To justify this change, 2D and 3D numerical analysis where carried out using the Plaxis Shotcrete Model, which adapts very well to the behavior exhibited by this type of mixtures, taking into account the stiffening and strengthening processes in the time (same that adjusts to the constructive process simulated in the model), as well as several of the particularities of the stress-strain relationships of shotcrete.

Subsequently, these models were used to determine if the FLAC^{3D} Strain Hardening/Softening Mohr-Coulomb constitutive law was able, with a certain mathematical treatment, to represent in a realistic way the behavior of SFRS. The adaptation of the model was achieved by calibrating the behavior of the material in response to different structural demands using mathematical expressions that, in addition to adequately describing the behavior of SFRS, mitigate the problem of grid dependency.

4.1 Time dependent Young's modulus

The increase of Young's modulus in time implicit in the SM and the one programmed for the Strain Hardening/Softening Mohr Coulomb Model follows the recommendations of the European standard CEB-FIP Model Code (2010).

$$E_{ci}(t) = \beta_E(t)E_{ci} \text{ with } \beta_E(t) = [\beta_{cc}(t)]^{0.5} \text{ and } \beta_{cc}(t) = \exp\left\{ s \left[1 - \left(\frac{28}{t}\right)^{0.5} \right] \right\} \qquad (1)$$

where $E_{ci}(t)$ is the Young's modulus at an age of t days; E_{ci} the Young's modulus at 28 days; $\beta_E(t)$ a coefficient depending on the age of concrete (t in days); $\beta_{cc}(t)$ a function describing the evolution of concrete in time; s a coefficient that depends on the type of cement or accelerating additives (s=0.38 for normal gain, s=0.25 for rapid gain and s=0.20 for very rapid gain).

4.2 Time dependent uniaxial compressive strength

To calculate the evolution of shotcrete's uniaxial compressive strength f_c over time, the SM uses the same approach as the CEB-FIP Model Code-2010 with a lower limit of $f_c = 0.005 \times f_{c,28}$, which is used at very early ages; alternatively, the evolution of strength can be simulated according to the early ages of shotcrete classes J1, J2 and J3 of the standard EN 14487, 2006.

On the other hand, the ability of shotcrete to withstand large deformations is not only the result of the low modulus of elasticity at early ages, but also a consequence of its great ductility. When concrete ages, ductility decreases. In the SM, this behavior is represented by a time dependent plastic deformation peak ε_{cp}^p. Using an approach similar to that proposed by Meschke (1996), a tri-linear function is adopted over time. The parameters introduced are the peak plastic deformations for $t = 1h, 8h, 24h$. After 24 hours it is assumed to be constant. The evolution of the uniaxial compressive strength in the first 24 hours is calculated as:

$$f_c(t) = f_{c,28} \cdot (f_{c,1}/f_{c,28}) \frac{t_{hydr} - t}{(t_{hydr} - 1d) \cdot t} \qquad (2)$$

where $f_{c,28}$ is the strength at 28 days, $f_{c,1}$, the strength in 1 day, t the time and t_{hydr} the hydration time which is usually equal to 28 days.

In the case of FLAC^{3D}, an evolution law for the SHSMCM was implemented (Sánchez 2016), which consists of expressions allowing to calculate the change of strength properties over time and, from this, such evolution can be programmed within construction stages.

$$f_t = \frac{t \cdot t_{hydr}}{t_{hydr}(a + t) - a \cdot t}; \quad \tan \phi(t) = \frac{f_t \frac{\sin \phi^{peak}}{1 - \sin \phi^{peak}}}{\sqrt{1 + 2f_t \frac{\sin^{peak}}{1 - \sin \phi^{peak}}}}; \quad c(t) = c^{peak} \frac{\tan \phi(t)}{\tan \phi^{peak}} \qquad (3)$$

where a is a parameter that controls the strength gain ratio and its values vary according to the concrete class: $a = 1.9$ for concrete with very rapid strength gain, $a = 2.5$ for rapid gain and $a = 4.5$ for normal gain; f_t is the strength factor as a function of time; c^{peak} and ϕ^{peak} are the cohesion and friction of the concrete at 28 days, respectively. If the value of the strength

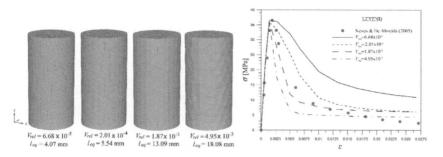

$V_{rel} = 6.68 \times 10^{-5}$
$l_{eq} = 4.07$ mm

$V_{rel} = 2.01 \times 10^{-4}$
$l_{eq} = 5.54$ mm

$V_{rel} = 1.87 \times 10^{-3}$
$l_{eq} = 13.09$ mm

$V_{rel} = 4.95 \times 10^{-3}$
$l_{eq} = 18.08$ mm

Figure 2.   Presence of the mesh dependency problem when using the Shotcrete Model.

parameters is unknown, it can be assumed that $37° \leq \phi^{peak} \leq 39$ ° and then calculate the peak cohesion from the Mohr-Coulomb criterion.

4.3   Avoiding mesh dependency with the finite element method

As mentioned above, Plaxis SM model has a regularization technique based on the Smeared Crack theory that supposedly would help mitigating mesh dependency; nevertheless, while carrying out a series of calculations with meshes having different degrees of refinement, it could be observed that such treatment in the model does not provide satisfactory results, that is, the problem of dependence of the mesh prevails. Figure 2 shows some different meshes used for the simulation of a test carried out by Neves & De Almeida (2005) on a shotcrete sample reinforced with 1.13% steel fibers.

To mitigate the problem a methodology proposed in García (2017) was implemented, which consists of using mathematical expressions that allow calculating mesh-dependent parameters based on the relative volume concept (V_{rel}). It is calculated as the quotient between the average volume of the elements in the mesh and the total volume of the same; this value turns out to be an indicator of the degree of refinement: the lower the value, the greater the degree of refinement of the mesh.

The equations developed to reduce mesh dependency are presented below:

$$G_c = G_{c_r} - \left(G_{c_r} - G_{c_0}\right)e^{-\frac{V_{rel}}{b}} \tag{4}$$

$$G_t = 133387(V_{rel})^2 - 35.27(V_{rel}) + 1.07 \tag{5}$$

$$f_{t,28} = 82367(V_{rel}) + 1311.4 \tag{6}$$

$$f_{t,28} = 7.251 \times 10^9 (V_{rel})^2 - 4.294 \times 10^6 (V_{rel}) + 2013 \tag{7}$$

$$f_{tun} = -4.452 \times 10^6 (V_{rel})^2 + 1515(V_{rel}) + 0.09 \tag{8}$$

$$G_t = -172442(V_{rel})^2 + 540.64 V_{rel} + 3.3502 \tag{9}$$

$$f_{tun} = -6923.9(V_{rel})^2 + 53654 V_{rel} - 0.005 \tag{10}$$

Where G_c is the fracture energy, G_{cr} is the ultimate value of the G_c-V_{rel} curve (in this case 0.95); G_{c_0} the initial value of the G_c-V_{rel} curve (in this case 1.0) and b an adjustment parameter. Table 1 summarizes the application of each equation and indicates the parameter to be set as a function of the relative volume in order to avoid mesh dependency.

a. *Uniaxial compression test*

The fracture energy in compression (G_c), which represents the area under the stress-strain curve of the element, is defined as the amount of energy needed to produce a unit of continuous crack.

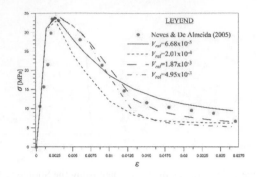

Figure 3. Stress-strain curve corresponding to a SFRS sample with 1.13% fibers using the mathematical expression (9).

Table 1. Summary of the application of each equation and its corresponding mesh-dependent parameter.

Equation	Test	Mesh-dependent parameter
(4)	Unconfined compression test	Fracture energy in compression, G_c
(5)	Energy absorption test	Fracture energy in tension, G_t
(6)	Energy absorption test	28 day tensile strength, $f_{t,\,28}$
(7)	Three point bending test	28 day tensile strength, $f_{t,\,28}$
(8)	Three point bending test	Ultimate residual tensile strength, f_{run}
(9)	Direct tension test	Fracture energy in tension, G_t
(10)	Direct tension test	Ultimate residual tensile strength, f_{run}

Figure 3 presents the results obtained by means of this methodology and the results published by Neves & De Almeida (2005). Note that, compared to the great dependency shown by the simulations in Figure 1, the problem is somehow mitigated.

b. *Energy absorption test*

In the case of the square panel test (EFNARC 14488-5: 2006), two mathematical expressions where developed (5 and 6) for the tensile fracture energy, G_t and tensile strength at 28 days, $f_{t,\,28}$.

Figure 4 shows the final results obtained from the simulations once the expressions 5 and 6 are introduced, comparing them with real laboratory tests; it can be seen that the problem of mesh dependency has been almost completely eliminated. Figure 5 shows the incremental deviatoric shear strain contours generated in the panel once the test is finished. Note the similarity it has with a real panel tested in the laboratory.

c. *Three point bending test on square panel*

In 2011 EFNARC published the ENC 371 FTC V1.1_18-06-11 test, which makes use of a panel with the same dimensions of the one for the energy absorption test but has a 5 mm width by 10 mm height notch in the lower central part.

The panel is supported on two, free to rotate, steel cylinders located 5 cm from each edge. The load is transmitted by another cylinder placed in the upper central part of the panel; in this way, the panel is subjected to bending and, therefore, undergoes tension in the lower part.

Figure 6 shows the resulting curves when using the adjustment expressions developed by García (2017) and compare them with actual test results published by de Rivaz (2011); also, it presents the failure mechanism of the panel at the end of the test. Table 2 presents the results of residual strengths obtained by the model and the experiment.

d. *Direct tension test*

Figure 6 shows the results obtained from one of the series of simulations carried out and compares them with the results of real tests published in Carnovale (2013) for shotcrete samples reinforced with 1% of steel fibers (sample DC-P2). These results already include the mathematical treatment to mitigate mesh dependency.

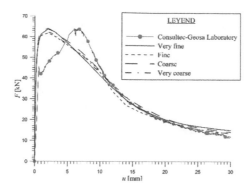

Figure 4.   Force deflection curve corresponding to the energy absorption test.

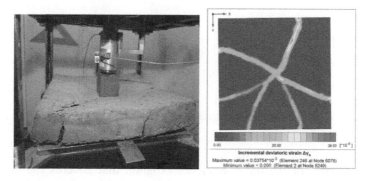

Figure 5.   Cracks and incremental shear strain contours over the width of the sample.

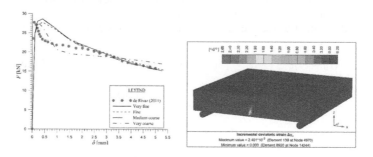

Figure 6.   Comparison between load-deflection curves obtained from simulations and incremental shear strength.

4.3.1 Avoiding mesh dependency with the finite difference method

As mentioned above, the model used in the Finite Differences program is the Strain Hardening/Softening Mohr Coulomb Model, which does not have a mathematical treatment to mitigate mesh dependency. In García (2017) a practical methodology to achieve this goal is presented; it is based on certain equations that describe the evolution of the parameters which depend on the refinement of the grid as a function of a relation of the evolution of strength (χ) and the relative volume of the mesh to be used (V_{rel}).

It should be mentioned that each mechanical application has a set of particular expressions that allow it to adequately describe the behavior of fiber-reinforced shotcrete, both in the peak and post-peak branches, while reasonably mitigating the problem of grid dependency.

3015

Table 2. Residual stresses and limit of proportionality of each of the simulations and of the tests published by de Rivaz (2011).

Mesh coarseness	Very coarse	Medium coarse	Fine	Very fine	Test (de Rivaz, 2011)
$f_{R,1}$ [N/mm^2]	26.8	30.3	30.1	30.5	25.0
$f_{R,2}$ [N/mm^2]	21.7	28.1	24.7	24.7	23.4
$f_{R,3}$ [N/mm^2]	21.0	21.6	21.6	21.2	21.0
$f_{R,4}$ [N/mm^2]	19.7	19.2	18.7	18.3	18.5
LOP [N/mm^2]	12.4	12.4	12.4	12.4	29.2

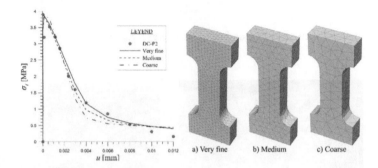

Figure 7. Stress-displacement curve of the direct tension test and results from different meshes used for the simulation.

The internal variable of the finite difference program that governs the evolution of the model's strength parameters, Δk^s, is a measure of the second invariant of the plastic deformation tensor (see Flac^{3D} manual). Once obtained the value of Δk^s in each of the calculation stage, it is possible to establish a relationship with the evolution of the Mohr-Coulomb criterion parameters and, in this way, establish it as a function.

The equations developed to reduce mesh dependency in this case are listed below:

$$c = \left(c_r + (c_0 - c_r)e^{-\left(\frac{\Delta k^s}{\chi}\right)}\right)c^{peak}; \quad \phi = \tan^{-1}\left[\left(\phi_r + (\phi_0 - \phi_r)e^{-\left(\frac{\Delta k^s}{\chi}\right)}\right)\tan\phi^{peak}\right] \qquad (11)$$

$$\chi = \chi_{ult} + (\chi_0 - \chi_{ult})e^{-\left(\frac{v_{rel}}{b}\right)} \qquad (12)$$

$$\sigma_t = \sigma_t^{peak} + m\,\Delta k^t \qquad (13)$$

$$m = m_0 - (m_0 - m_r)e^{-\left(\frac{v_{rel}}{b}\right)} \qquad (14)$$

$$\sigma_t = \begin{cases} \left(\sigma_r^t + (\sigma_0^t - \sigma_r^t)e^{\left(\frac{-\Delta k^t}{\chi_1}\right)}\right)\sigma_t^{peak} & for\,\Delta k^t \leq 0.05 \\[2ex] \left(\sigma_0^t - (\sigma_0^t - \sigma_r^t)e^{\left(\frac{\Delta k^t}{\chi_2}\right)}\right)\sigma_t^{peak} & for\,\Delta k^t > 0.05 \end{cases} \qquad (15)$$

$$\chi_1 = \chi_1^{ult} + (\chi_1^0 - \chi_1^{ult})e^{-\left(\frac{v_{rel}}{b_1}\right)}; \quad \chi_2 = \chi_2^{ult} + (\chi_2^0 - \chi_2^{ult})e^{-\left(\frac{v_{rel}}{b_2}\right)} \qquad (16)$$

$$\sigma_t = \left(\sigma_{tr} + (\sigma_{t0} - \sigma_{tr})e^{-\left(\frac{\Delta k^t}{\chi}\right)}\right)\sigma_t^{peak} \qquad (17)$$

Table 3. Summary of the application of each equation and its corresponding mesh dependent parameter.

Equation	Test	Mesh-dependent parameter
(11)	Unconfined compression test	Cohesion, c
(11)	Unconfined compression test	Friction angle, ϕ
(12)	Unconfined compression test	Hardening parameter, χ
(13)	Energy absorption test	Tensile strength, σ_t
(14)	Energy absorption test	Slope of the softening law, m
(15)	Three point bending test	Tensile strength, σ_t
(16)	Three point bending test	Hardening parameter, χ
(17)	Direct tension test	Tensile strength, σ_t

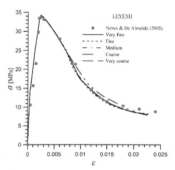 

Figure 8. Stress-strain curve of the UCS test after using the mathematical treatment to mitigate the problem of grid dependency and shear band localization in the grid.

where subscripts r and 0 correspond to residual and initial values on the normalized softening curves, respectively; χ_{ult} and χ_0, the residual and initial values of the curve χ-V_{rel} and b, an adjustment parameter. Table 3 summarizes the application of each equation and indicates the parameter to be set as a function of the relative volume in order to avoid mesh dependency.

a. *Uniaxial compression test*
   In the case of UCS, parameters governing the behavior of the material are cohesion and the friction angle. For a sample reinforced with 1.13% steel fibers, the mathematical treatment resulted in the curves shown in Figure 8, in which they correspond to grids with different degrees of refinement, compared with the results published by Neves & De Almeida (2005).

b. *Energy absorption test*
   In this test the parameter that governs the behavior of the panel is the tensile strength, however, unlike the previous test, the evolution proposed is linear (eq. 13). The results that it provides are quite adequate, i.e. compared with laboratory tests, a good fit is observed (Figure 9) and the failure mechanism is very similar to that presented in real panels.

c. *Three point bending test on square panel*
   The proposed law for the evolution of the tensile strength during the execution of this test corresponds to a discontinuous function with two intervals. The results provided are shown in Figure 10 where, in addition, they are compared with results published by de Rivaz (2011); note the good fit between results.
   Table 4 presents the limit of proportionality and the residual stresses calculated from the curves; the similarity that exists between the simulations carried out and the actual tests is clear. Figure 11 shows the deformed grid and the incremental shear strain contours at the end of the test for the finest grid.

d. *Direct tension test*
   Figure 12 shows the performance of these expressions for samples reinforced with 1.0% (sample DC-P2) of steel fibers comparing them with results published in Carnovale (2013) and the actual failure mechanism of a real sample and from one of the simulations.

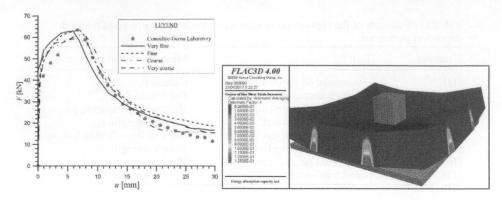

Figure 9. Stress-strain curves obtained from the energy absorption capacity test (experimental and numerical) and failure mechanisms of the model.

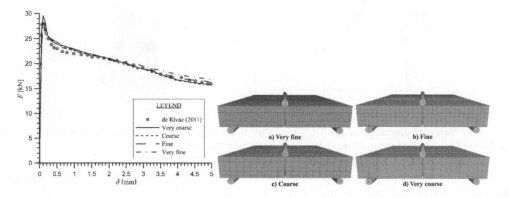

Figure 10. Comparison between the force-deflection curve of de Rivaz (2011) and those obtained from the simulations and different grids used.

Table 4. Residual stresses and limit of proportionality of each of the simulations and of the tests published by de Rivaz (2011).

Coarseness	Very coarse	Coarse	Fine	Very fine	Test (de Rivaz, 2011)
$f_{R,1}$ [N/mm^2]	26.3	26.0	26.4	25.9	25.0
$f_{R,2}$ [N/mm^2]	23.5	23.2	23.3	23.5	23.4
$f_{R,3}$ [N/mm^2]	20.7	20.4	20.9	21.4	21.0
$f_{R,4}$ [N/mm^2]	17.9	17.9	18.6	19.3	18.5
LOP [N/mm^2]	26.4	25.8	25.0	24.2	29.2

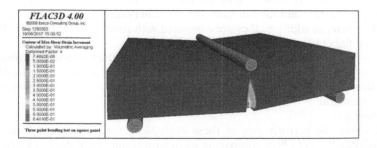

Figure 11. Incremental shear strain contours at the end of the test.

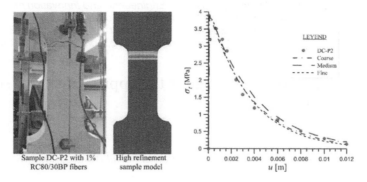

Figure 12. Failure mechanism of a real sample and the model; Stress-displacement curves of the direct tension stress and from the model.

5 CONCLUSIONS

Both FLAC^{3D} and Plaxis 3D codes, are very powerful calculation tools that allow several considerations to be made during the modelling of a geomechanical problem; This is quite practical when the models are fed properly since the results obtained can be very realistic; however, it should not be expected that both constitutive approximations provide exactly the same results because each one has its own formulation and its own methodology: Plaxis solves the problems implicitly, that is, carries out iterations to solve the differential equations that govern the problem, while FLAC approximates these equations in difference equations.

It should be noted that the methodology shown in this article to mitigate the problem of mesh and grid dependency is practical and provides reasonable results. It should be also noted that in the case of finite differences, better results are generated than with the SM model for the finite element program.

Finally, this methodology can be easily extended to more complex boundary value problems like a tunnel excavation, allowing to extend the calculation potential for design purposes. That will be the subject of the next paper in these same proceedings.

REFERENCES

Carnovale, D. J. (2013). Behaviour and analysis of steel and macro-synthetic fibre reinforced concrete subjected to reversed cyclic loading: a pilot investigation. *Doctoral dissertation.*

de Rivaz, B. (2011). Fibre reinforced spray concrete for compliance with site safety requirement. *Concreto y cemento.* Investigación y desarrollo, 2(2),48–58.

EFNARC Sprayed Concrete Technical Committee. (2011) Three Point Bending Test on Square Panel with Notch. ENC 371 FTC V1.1_18-06-11

EN, C. (2006). 14488-5: Testing sprayed concrete. *Determination of energy absorption capacity of fiber reinforced slab specimens.* ICS 91.100.30

García, G. (2017). Estudio comparativo del comportamiento del concreto lanzado en túneles a partir de distintas aproximaciones constitutivas; *Tesis de maestría, UNAM,* México.

Neves, R. D., & De Almeida, J. F. (2005). Compressive behaviour of steel fibre reinforced concrete. *Structural concrete,* 6(1),1–8.

Sánchez, F. (2016). El colapso del túnel Xicotepec I Una investigación sobre sus causas y un estudio para su reconstrucción. *Revista Obras Públicas 2016/Número 3579: Túneles (Monográfico).* España.

Schädlich, B., Schweiger, H. F., Marcher, T., & Saurer, E. (2014, January). Application of a novel constitutive shotcrete model to tunneling. *ISRM Regional Symposium-EUROCK 2014.* International Society for Rock Mechanics.

Tunnels and Underground Cities: Engineering and Innovation meet Archaeology,
Architecture and Art, Volume 6: Innovation in underground engineering,
materials and equipment - Part 2 – Peila, Viggiani & Celestino (Eds)
© 2020 Taylor & Francis Group, London, ISBN 978-0-367-46871-2

Advanced constitutive modelling for the approach of real shotcrete performance in tunnels

F.A. Sánchez
President of the Mexican Association for Tunnelling and Underground Works (AMITOS), Mexico

G.G. Saldivar
National University of Mexico, Mexico

ABSTRACT: The use of advanced constitutive approaches to better represent the behavior of shotcrete during tunnel excavations is a common practice nowadays. Moreover, with some special mathematical treatments focused on the mitigation of the well-known mesh dependency problem, they can to contribute in creating realistic numerical and geotechnical scenarios that can lead to optimize designs.

1 INTRODUCTION

In this paper, a case-study on the excavation of a metro line in Mexico City is presented, where a real optimization of excavation phases and shotcrete support was achieved. The mathematical functions developed in the first part of this research (García, 2017, García & Sánchez, 2019) were implemented as part of the constitutive behavior of the solid elements and zones representing shotcrete layers at the excavation boundary. Finite element and finite difference models of the tunnel where calibrated to reproduce as accurately as possible the monitored ground. Once the full models where considered as representative of the global response during construction, stresses and forces at the shotcrete lining where used to verify its structural performance.

From the review of the geotechnical and constructive aspects during the construction of this particular metro tunnel in Mexico City, it was possible to determine the safety conditions prevailing at the excavation before and during the implementation of some changes proposed by the contractor focused on the improvement in the construction procedures and the support systems. With the results of this study the authorities approved such changes, in particular the replacement of double-welded wire mesh within shotcrete layers, included in the original project, by a mixture of steel fiber reinforced concrete of a type and characteristics such that it complied with the common specifications used in underground works.

1.1 *Grid/mesh dependency problem*

During the modelling of the post-peak behavior of materials, the development and width of shear bands depend strongly on the size of the elements or zones; when strong shear deformation fields start to localize, the amount of softening represented by the model will be inevitably related to the mesh or grid refinement if any regularization technique is utilized.

To better understand this problem, a simple compression test was modeled on a cubic sample of steel fiber reinforced shotcrete. Figure 1 shows the 6 meshes used and the results obtained; it can be observed that before reaching the peak, strength at all graphs is equal, however, even though the 6 models have exactly the same input parameters, immediately after this point the dependence on the mesh is manifested and all the models provide different amounts of softening, being the results from the thinner mesh the most reliable.

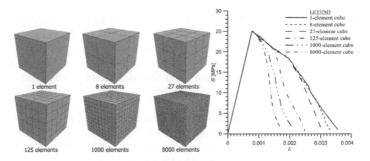

Figure 1. Finite difference grids used for the modelling of a cubic sample subjected to uniaxial compression and stress-strain curves from the different models.

2 CALIBRATION OF THE CONSTITUTIVE MODELS

After having developed the necessary relations to control the evolution of strength parameters during the post-peak stage of the shotcrete elements behavior in the finite element mesh and the shotcrete zones in the finite element grid, the initial, peak and residual values of such variables where calibrated according to laboratory test over shotcrete samples from the construction site as well as from in-time stiffening and strengthening, energy absorption capacity and residual strength specifications given to the contractor in order to guarantee the adequate performance of the mixture to use in the construction site.

As explained in García & Sánchez (2019), shotcrete elements models in Plaxis where simulated with the Shotcrete Model (SM) constitutive law, while for shotcrete zones the Strain Hardening/Softening Mohr Coulomb model was utilized.

2.1 Time dependent parameters

2.1.1 Young's modulus
Stiffening in time of the shotcrete material in the models was adapted to the laboratory results over samples from the construction site (Figure 2).

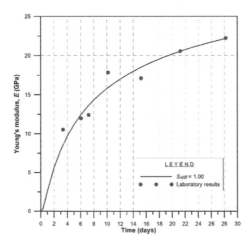

Figure 2. Laboratory results and the evolution of Young's modulus introduced in the model, according to CEB-FIP Model Code (2010) with $s=1.0$.

In Figure 2 one can see that the relation E_c vs time is quite low in speed, compared to what is usually established by other laws in literature and in Model Code, however, since the aim was to adapt the models to real data from the construction site, such an evolution in time was implemented during the simulation. One important fact is that the value E_{28}, matches almost exactly to the one calculated with the formula from the Mexican Construction Code.

2.1.2 *Uniaxial compressive strength*

Evolution of strength was also determined from laboratory tests and programmed in both finite element and finite difference models; In Plaxis Shotcrete Model the corresponding parameters for the in-built evolution law where specified, while in FLAC, an evolution law according to Sánchez (2016) was programmed in Fish language.

Also, constitutive models where calibrated to represent as better as possible the post peak behavior shown by the laboratory samples from the project; Figure 3 shows a sample modeled with FLAC, and a plot with the laboratory results compared with the ones obtained with the simulation.

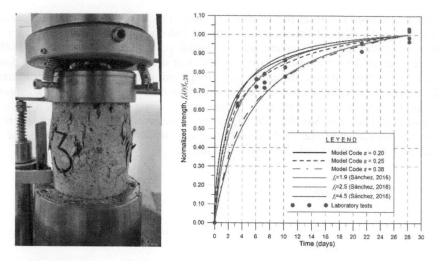

Figure 3. Sample tested, laboratory results and time-evolution laws introduced in the model, according to CEB-FIP Model Code (2010), and Sánchez (2016).

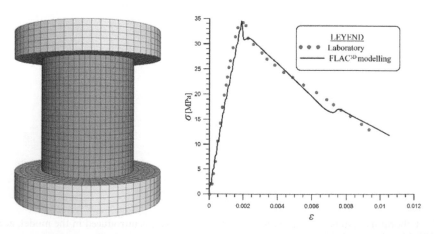

Figure 4. Sample finite difference model and laboratory and simulation results.

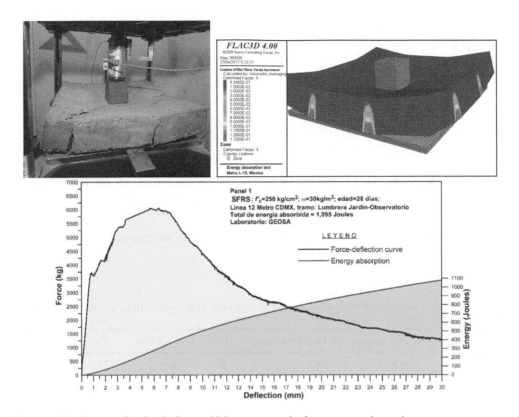

Figure 5.   Broken sample, simulation and laboratory results from energy absorption test.

2.1.3 *Energy absorption test*

Shotcrete for this project was specified to have an energy absorption capacity of over 900 Joules, corresponding to difficult ground conditions. Laboratory tests were performed and the models calibrated to represent such toughness. Figure 5 shows lab results from one of the tested panels. Results from the calibration of these tests are presented in García & Sánchez (2019).

Figure 5 shows one sample already tested, the deformed grid and incremental shear contours of one finite difference model and the results from the real test.

2.1.4 *Residual flexural strength*

According to the specifications of this particular project, shotcrete should have a pair of residual flexural strengths of $= f_{R,1} = 2.5$ and $f_{R,4} = 2.0$ (MPa).

Figure 6 shows the resulting finite element curves when using the adjustment expressions developed by García (2017) and compares them with actual test results published by de Rivaz (2011); also, it presents the failure mechanism of the panel at the end of the test. Table 1 presents the results of residual strengths obtained by the models and the experiment.

3   CASE STUDY. METRO TUNNEL IN MEXICO CITY

In the following models the evolution-in-time laws for stiffness and strength, as well as the mesh dependency treatments for hardening-softening (García. 2017; García & Sánchez, 2019), where implemented.

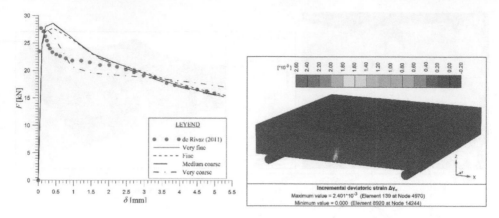

Figure 6. Comparison between load-deflection curves obtained from simulations and incremental shear strength.

Table 1. Residual stresses and limit of proportionality of each of the simulations and of the tests published by de Rivaz (2011).

Grid coarseness	Very coarse	Medium coarse	Fine	Very fine	Test (de Rivaz, 2011)
$f_{R,1}$ [N/mm²]	26.8	30.3	30.1	30.5	25.0
$f_{R,2}$ [N/mm²]	21.7	28.1	24.7	24.7	23.4
$f_{R,3}$ [N/mm²]	21.0	21.6	21.6	21.2	21.0
$f_{R,4}$ [N/mm²]	19.7	19.2	18.7	18.3	18.5
LOP [N/mm²]	12.4	12.4	12.4	12.4	29.2

3.1 Finite element modelling

This part of the study consisted in a review of the geotechnical and construction aspects of the excavation of a metro tunnel in order to verify the safety conditions after suggested changes in the excavation procedures and support systems during the construction. Numerical modelling took into account monitoring during construction for back analyzing, in order to approximate the real behavior.

In a first stage, modelling was carried out with the finite element program Plaxis 3D and, in order to speed up calculations and to easily obtain design forces in shotcrete elements, Plaxis 2D was also utilized; that is, although the three-dimensional mesh constructed for this analysis is simple (Figure 7-left), a complete calculation can take several days, while a well-calibrated two-dimensional mesh allows to obtain reasonable results in just minutes. The two-dimensional model coincides with one of the monitored sections (control station) during construction works.

To validate the calibration in two and three dimensions, control points were placed just in the same positions as in the construction site (Figure 7-right) and related to the progress of the different construction stages, the age of shotcrete and with the distance from the excavation face.

The plot in Figure 8 shows the results of convergence measurement on line A-B of the control station with respect to the distance of the excavation front. The cumulative shortening of this line in the 3D model is added to the instrumentation data at the time when the front of the upper half section is 4.5 m ahead of the control station (distance to the position of the front when monitoring devices were installed); thus the graph represents the complete deformational history obtained from the calibration. Once the three-dimensional model was calibrated, the results of the 2D model were adjusted so that they coincided with each other (Figure 9).

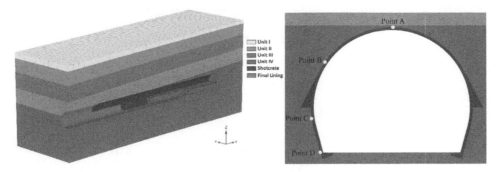

Figure 7. 3D mesh in an a certain excavation stage with an advance of 28.5 m of the front (left) and control points for the calibration of the 2D model (right).

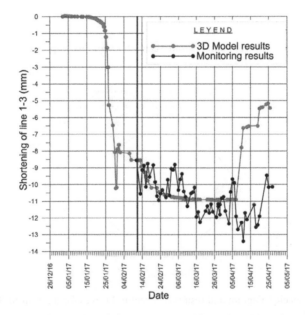

Figure 8. Extrapolation of the results of the modelling with PLAXIS 3D, from the beginning of the excavation to a distance of 85 m from the excavation front and equivalent positioning of the convergence measurements.

3.1.1 Structural verification of shotcrete support

Once the models were calibrated, it was assumed that the stress-strain states obtained represent a sufficiently realistic geotechnical scenario to be used in the structural review. This verification was carried out based on the results of structural forces obtained from the 2D model, corresponding to the last stage of analysis and for a concrete age of 28 days; axial forces, shear forces and bending moments are obtained directly from the solid elements modeled with the shotcrete model (SM).

From the results of axial forces and bending moments, a structural verification was made introducing them into the bearing capacity diagram (Figure 10). The latter was developed using the RILEM committee methodology, known as the σ-method (Vandewalle, 2003), taking as the main design parameters $f_{R,1} = 2.5$ and $f_{R,4} = 2.0$ (MPa) from specifications and using the previously calibration of the constitutive model (SM), which is based on the results

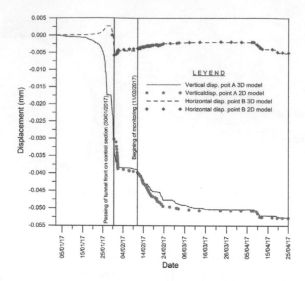

Figure 9.   Displacements in control point A from 3D analysis and its respective calibration in 2D.

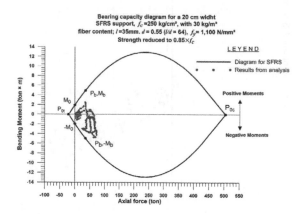

Figure 10.   Bearing capacity diagram and results of structural forces for a 20 cm SFRS support. Diagram with strength reduction.

obtained from the three-point bending test. Note that the points fall inside the diagram with enough safety margin even though the diagram is affected by the strength reduction coefficients proposed by the Mexican construction code.

A very important characteristic of this type of analysis is that, being an elastoplastic continuous model, no matter how big the structural demand is, forces can reach the bearing diagram but they can never get out, as would happen if an elastic analysis with typical structural elements (plates, shells, beams, etc.) was performed.

3.2   *Finite difference modelling*

Finite Difference modelling (Figure 11) was carried out with FLAC[3D] using the Strain Hardening/Softening Mohr Coulomb Model. As mentioned above, this constitutive law is not exclusive for the modelling of concrete structures, therefore, various adaptations were made using the Flac's in-built FISH language, among which the following can be listed:

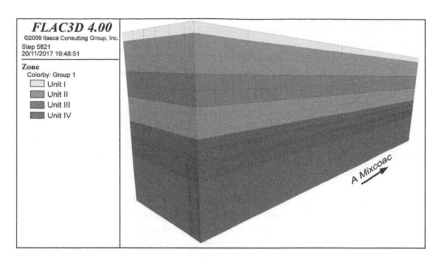

Figure 11.   Finite difference grid in FLAC^{3D}.

a) Implementation of a time-dependent law to describe the evolution of the shotcrete's UCS (Sánchez, 2016)
b) A time-dependent law to describe the evolution of the tensile strength of shotcrete (García, 2017)
c) A time dependent evolution law for Young's modulus and Poisson ratio (Sánchez, 2016, García, 2017).
d) Treatments for avoiding grid dependency for softening behavior (García, 2017)

The excavation-support processes were programmed in such a way that it approximated as much as possible the Plaxis model. This was due to the fact that in FLAC^{3D} the programmed FISH routine considers discrete increments of time of one day, whereas in Plaxis it is possible to define increments of hours. Likewise, the same control points were placed in the analyzed section to be able to perform the calibration between both calculation techniques.

Figure 12 shows the calibration between the Plaxis 3D models (previously calibrated with monitoring results) and FLAC^{3D}, corresponding to the vertical displacements measured in the crown and in the tunnel wall; it shows that both the magnitude and the behavior of the

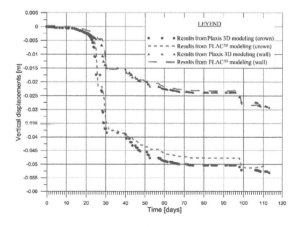

Figure 12.   Comparison of Plaxis 3D and FLAC^{3D} results for vertical displacements at the crown and the side of the tunnel.

displacements calculated using the finite difference technique are quite similar to the results obtained by finite elements.

The previous figure show that the adaptations made in the Strain Hardening/Softening Mohr Coulomb Model allow a realistic representation of the tunnel behavior, at the time the problem of grid dependency is significantly mitigated.

3.2.1 Bearing capacity of the structural system under loosening pressure

Finally, after calibrating the finite difference model, a series of verifications where carried out to estimate which would be the maximum loosening pressure the system would be able to withstand. A simple model was built in which a vertical pressure was applied incrementally over the solid shotcrete arch in a length of 10 tunnel meters, until failure was produced (Figure 13).

A 2D finite element model was again calibrated with the 3D finite difference model in order to get the same behavior in terms of stress, strain, displacements and failure; shotcrete internal forces obtained from the 2D model where then compared with the bearing capacity diagram, being possible to verify that in fact, these forces at the failure state coincide very well with the strength envelope.

Figure 14 shows the bearing capacity diagram of the 20 cm SFRS support and the results from the simulation of an incremental loading until failure occurred (q_{ult}).

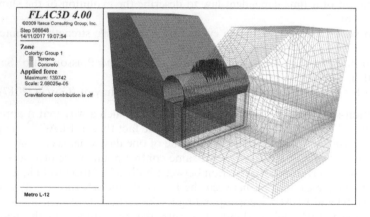

Figure 13.   Finite difference grid for analyzing bearing capacity of the shotcrete system under loosening pressure.

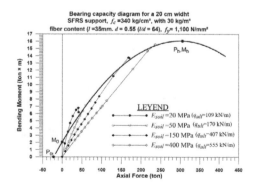

Figure 14.   Bearing capacity diagram of the 20 cm SFRS support and the results from the simulation of an incremental loading until failure occurred (q_{ult}).

4 CONCLUSIONS

The methodology presented in this paper for the design and performance verification of shot-crete support systems in tunnels corresponds to the current state of the art. It is important to consider that in many countries such as Mexico, the use of this type of technologies is not a common practice and, therefore, it should be promoted, since it allows for safer, more efficient and more economical designs. The changes achieved for the metro tunnel under study are a good example of the above: a good amount of money and time were saved by the suppression of the double welded wire mesh.

REFERENCES

García, G. (2017). Estudio comparativo del comportamiento del concreto lanzado en túneles a partir de distintas aproximaciones constitutivas; *Tesis de maestría*, UNAM, México.
Neves, R. D., & De Almeida, J. F. (2005). Compressive behaviour of steel fibre reinforced concrete. *Structural concrete*, 6(1),1–8.
Sánchez, F. (2016). El colapso del túnel Xicotepec I Una investigación sobre sus causas y un estudio para su reconstrucción. *Revista Obras Públicas 2016 / Número 3579: Túneles (Monográfico)*. España.
Vandewalle, L. 2003: Design with σ-ε method; Proceedings of the RILEM TC 162-TDF Workshop: *Test and Design Methods for Steel Fibre Reinforced Concrete*. Schnütgen, B. & Vandewalle, L. Editors.

*Tunnels and Underground Cities: Engineering and Innovation meet Archaeology,
Architecture and Art, Volume 6: Innovation in underground engineering,
materials and equipment - Part 2 – Peila, Viggiani & Celestino (Eds)
© 2020 Taylor & Francis Group, London, ISBN 978-0-367-46871-2*

Tunnel inspection with high-performance devices. Methodology

F. Sánchez-Domínguez, J.A. Ramos-García, Álvaro Barbolla-Díaz & E. Calvo-Haro
Euroconsult NT, Spain

T. Vitti & Q. Vitti
Euroconsult NT, Italy

ABSTRACT: Management and maintenance of tunnels requires the application of high performance lining inspection techniques. Many infrastructure maintenance contracts, including those of tunnels, are managed through quality indicators. In order to assess these indicators, objective inspection tools and devices are necessary. This paper will present different automated techniques; some brand new, while others well-known like the Lidar technique. We will describe how the results obtained in different tunnels through visual inspections performed by specialized surveyors, can be completed with the information obtained from a combination of techniques, mainly based on imaging and laser. The automated inspection offers a better knowledge of the state of the tunnel and avoids subjectivity problems. These techniques may also function as a basis for the acceptance of new tunnels. Examples of the use of these devices for tunnel lining inspections on different infrastructures (metro, railway, roads) in Italy, UK, Spain and Japan will be presented.

1 TUNNEL INSPECTION: INTRODUCTION

Over the last years, the number and length of different types of tunnels (road, railway, and utilities) have increased significantly in developed countries. Furthermore, some of the oldest infrastructures have had an important increase of passengers, traffic, occupancy, and operation time. However, the time for maintenance has been reduced. In this context, it is necessary to implement new techniques to inspect the condition of infrastructures (which are critical for city mobility), allowing a preventive maintenance of the structural stability without affecting the traffic. The maintenance inspections allow to analyze the preservation and deterioration condition of tunnels on a routine basis in terms of cracks, breaks, significant bulges of the lining affecting the clearance, damp areas, etc. We must keep in mind that all these signs of deterioration are indicative of possible structural problems, hence the importance of these inspections.

Scheduled routine inspections are key to detect safety issues and to prevent structural, geotechnical or infrastructure/operation issues or failures. Moreover, if these inspections are conducted with high-performance equipment, which allow an objective and automated analysis of the real state of the infrastructure, it is possible to accurately assess the key performance indicators from a preventive maintenance perspective and to consider the most appropriate investments for a safe tunnel operation.

2 AVAILABLE TECHNIQUES

Detailed inspections of tunnels, conducted with non-destructive methods (radar, impact-echo, ultrasonic technologies…) present the drawback of being slow, entailing complex tests and demanding many material resources. On the other hand, visual inspections tunnels conducted by specialized staff do not ensure the most comprehensive and swift information, as they are subjective and time-consuming. In order to solve both issues, different new inspection systems have been under development during the last years in order to get fast and most importantly, accurate inspections of the tunnel condition focusing on non-destructive and detailed visual techniques.

The techniques currently used in the market for high-performance inspections in new-built tunnels and particularly, in currently-built tunnels are the following:

- 3D linear laser cameras
- Lidar
- Thermography
- Cameras with near infrared illumination

Among all these techniques, the 3D linear laser cameras have been the last ones to enter the market and they are currently in constant improvement. These cameras are changing the methodology of damage inspection and management for different infrastructures (pavements, railways, tunnels) due to their high productivity, performance, and their ability to incorporate automatic algorithms to the acquired images analysis. The inspection systems incorporate these 3D cameras combined with the thermography and Lidar information in order to ensure a more comprehensive inspection. In addition to this, techniques such as near infrared cameras are capable of offering the user a 360° field of view inside the tunnel allowing the surveyors to conduct an in-situ inspection.

These techniques are summarized below, where we will see the 3D cameras in closer detail, as these are new systems in the domain of tunnel inspection.

2.1 3D Laser cameras

The 3D laser cameras apply the triangulation principle to extract 3D information from the sensors, as shown in Figure 1. A pattern of known light (a line in this case) is projected from the laser onto the object to be inspected. The line is recorded by an area-scan camera placed at a fixed distance with an oblique angle relative to the projected light. The relative position (rotation + translation) between camera and laser is obtained by means of a precise calibration process.

The 3D sensor cameras used in tunnel applications (Figure 2) are the LCMS camera-laser units originally developed for road pavement surface inspection tasks by INO (National Optics Institute, Canada), and marketed by Pavemetrics. These sensors have been adapted and integrated for tunnel inspection within the Tunnelings system [1].

The 3D laser cameras are triggered by a Distance Measurement Instrument (DMI) to acquire profiles of 3D points, creating images by means of profile aggregation.

Each camera of the Tunnelings system captures an array of 2352 points per profile in a width of approx. 2m with three-dimensional information. The profiles are acquired every time the system receives a signal from the Distance Measurement Instrument (DMI) of the vehicle. The DMI measures the displacement and distance in the driving direction of the survey with millimeter resolution. This allows, for every 1000 consecutive

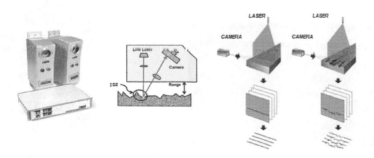

Figure 1. 3D camera technology.

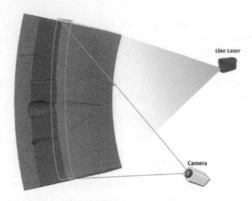

Figure 2.   Triangulation measurement principle for high-speed acquisition of 3D information.

merged profiles, a 2352mm x 1000mm resolution image from each camera, to be saved in a common file.

With this technology, luminance images (Figure 3a) that any camera can offer is obtained, but at the same time the distortions of the laser line are obtained, which can be represented on a range image (Figure 3b) showing the distances between sensor and surface. Using both images, it is possible to carry out a later 3D reconstruction (Figure 3c) and develop numerical algorithms of damage analysis. The three described images are shown below.

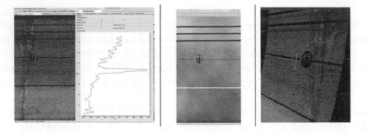

Figure 3.   a. Illumination image. b. Range image. c. 3D reconstruction.

The high resolution images allow to perform surveys as well as programming deterioration analysis algorithms. Figure 4 shows a 3D high resolution image.

Figure 4.   High resolution 3D reconstruction.

Figure 5.   Image of 2 Lidar sensors mounted in parallel.

2.2 Lidar

Lidar is a technique used currently in mobile mapping and topography for 3D reconstruction in a wide range of infrastructures. This technology measures the distance to the object to be analyzed through an emitted pulsed laser, measuring different physical principles of the reflected light from the object.

Depending on the technology used, the Lidar system is based on the physical principle of 'Time Of Flight (TOF)' or 'Phase-Shift' to take the measurements. The principle of 'Time Of Flight (TOF)' measures the time that a punctual light beam takes to be reflected on the surface, while the principle of 'Phase-Shift' measures the difference of the emitted wavelength phase. The use of one principle or the other affects the precision of the survey and its operation in dynamic conditions.

The measurement range of Lidar systems usually covers distances from approx. 0.5m up to (or greater than) 80m, with a measurement accuracy greater than 5mm in depth even in the most accurate systems. Considering that 360° are covered, the transverse resolution depends on the distance, as the system performs a circular scanning and records data at certain angles. The resolution then varies in proportion to the distance covered by the laser. Below is included an image with two Lidar sensors installed on the tunnel inspection vehicle to scan 190° x 2.

The Lidar system is useful in allowing the generation of a 'map' of the inspection configuration, but it is difficult to use it for defect analysis purposes due to the acquisition speed, the resolution and the accuracy in comparison with 3D vision cameras. However, it offers the advantage of being able to detect obstacles at great distances, such as slopes in portals, tunnel headwalls and clearance (Figure 6).

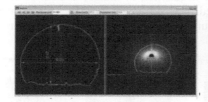

Figure 6.   Image of Lidar sensor within a tunnel, obtaining cross section clearance.

2.3 Thermography

Infrared thermography allows the acquisition of high resolution surface temperatures from the lining, while measurements are taken with the laser sensors (3D cameras and Lidar). The measurement of the surface temperature contributes to a better analysis of the dampness present on the lining. It also allows for surveying hidden damages, for instance leakages, delamination and concrete hollows or cavities that might be present in the building materials or within the backfill. The thermography is recorded in synchrony with the rest of the sensors, allowing the combined usage of these techniques for the detection of eventual damages.

Thanks to the thermography technique and through post-processing, it is possible to show temperature images with their pixels in different colours, in accordance with the selected temperature interval, and with a resolution of a tenth of a degree. The thermography cameras installed in the tunnel inspection vehicle described in this document have the following features:

- Temperature range: 0–250°C
- Spectral range: 7.5 to 13μm
- Detector: 120Hz
- Pixel size: 4cm
- Accuracy in absolute temperature measurements: ±2°C or ±2%

Below is included an image (Figure 7) acquired with this type of sensor in an underground tunnel.

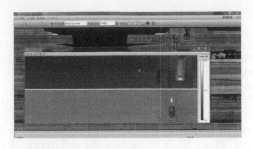

Figure 7. Thermographic images synchronized with the rest of the sensors.

The camera has a software that makes it possible to make punctual measurements on the acquired image, to escalate the temperature gradient and to generate temperature histograms so as to identify elements with a different temperature in the tunnel.

With the infrared images it is also possible to inspect all the electrical equipment installed in the tunnel, including cables, connections or splices that are, for example, either hidden or located in exposed areas of the tracks. In this case, the goal is to detect potential problems that could result in a future failure or even fire risk.

2.4 Cameras with visible infrared illumination

These types of cameras (Figure 8) allow the acquisition of images in unlit environments from which a global vision of the tunnel is required. The equipment includes a set of infrared illuminators and cameras able to acquire high-resolution images in completely dark environments.

Figure 8. Near infrared cameras.

Figure 9. Images of a very dark tunnel with visible infrared cameras.

The cameras have a CMOS sensor of e2v which features high quality and sensitivity in the near-infrared part of the spectrum. Furthermore, these cameras include a global shutter allowing for accurate moving snapshots and reproduction of sharp contrasts. Figure 9 shows an example of images captured with these cameras in a completely dark tunnel.

3 STRUCTURAL DAMAGES TO BE ANALYZED IN DIFFERENT LININGS

Through the combination of the techniques explained in this paper, it is possible to inspect tunnels ranging from large-diameter to small-diameter, with a reasonable measurement speed and ensuring maximum resolution for each sensor.

The following images shows how the scanning system has been installed on the inspection vehicle to be used in different types of infrastructures and in tunnels with different diameters and with different lining materials.

Even if all of the described techniques are combined for the analysis of the structural condition of the tunnel, the analysis of damages is mainly based on 3D cameras, due to their high

Figure 10. Different installations for road, railway and water tunnels.

resolution and their capacity to reconstruct three-dimensional images that may be automatically processed for detection.

The analysis of the results will vary according to the type of lining of the tunnel. Obviously, the types of visible deterioration in a brick tunnel or a masonry tunnel will not be the same as the types observed in a segment tunnel or a metal frame tunnel.

In general, we can identify four main types of tunnel linings [2] if the classification criterion is that of homogeneity in the visible results after an automatic analysis.

- Metal frame tunnels.
- Concrete tunnels with precast segments.
- In situ concrete tunnels and shotcrete tunnels.
- Brick and masonry tunnels.

These four groups evidently present defects that could be present in all tunnel types –for instance, dampness and cracking. As for the cracking, it might be classified according to its location (at the keystone or at the arches), or its development and orientation (longitudinal, transverse, oblique). The severity of cracks may be later considered in terms of extent, width and depth. Other frequent defects in all of these different tunnels –although with their own specificities depending on the mechanical behavior of the lining material– are spalling or bird-smouthing, bulging or clearance deformations [3].

The following Figures show some cracking and spalling images in different linings.

Figure 11. Longitudinal cracking in a brick tunnel.

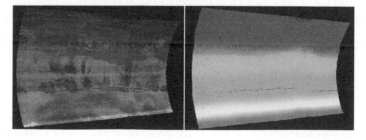

Figure 12. Longitudinal cracking in a concrete tunnel.

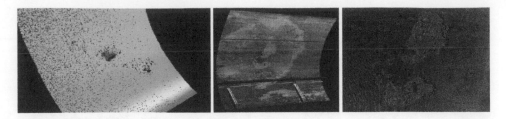

Figure 13.   Spalling of in situ concrete tunnels.

In addition to these defects which are common to all these linings, there are other distresses that are particular to metal frame tunnels, to concrete tunnels with precast segments, metal structures and to brick and masonry tunnels.

3.1   Metal Frame Tunnels

For metal frame tunnels, attention must be paid to the areas presenting corrosion and surface calcareous deposits, either at the joints or the center of the metallic segments, and to the condition of the bolts joining such segments (if they are missing, deflected, bent or the nuts are missing). Some examples of deterioration in metal frame tunnels are shown below. They are analyzed according to the image and the reconstruction obtained thanks to the 3D cameras.

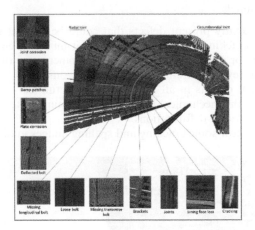

Figure 14.   Metal frame tunnels.

3.2   Concrete Segment Tunnels

In concrete segment tunnels, the monitoring of faulting and of the openings between segments, may be as accurate (or better) than 1mm thanks to the 3D camera system [4]. Thus, it is possible to measure the faulting and the openings along the whole joint and compare it with a second inspection to check whether any changes have taken place.

Figure 15 shows examples of the measurements of such faulting and their representation along the tunnel.

3.3   Brick Tunnels

In brick and masonry tunnels, the loss or fall of bricks and the deterioration of joints are treated specifically. To do so, the algorithms calculate the presence of joints with mortar loss and with a depth greater than a certain threshold; they are also used to calculate the affected volume and to

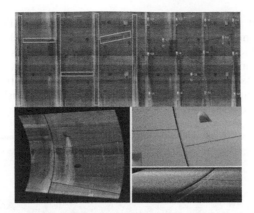

Figure 15. Faulting and opening analysis along a TBM tunnel.

identify which elements have fallen off the lining. Some examples of brick loss and mortar loss detection using 3D cameras are shown below. As can be noted, this information makes it possible to obtain measurements and to generate sections of deterioration.

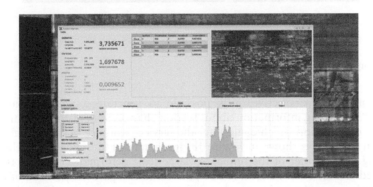

Figure 16. Brick loss and mortar loss analysis.

4   GLOBAL ASSESSMENT OF THE TUNNEL CONDITION KEY PERFORMANCE INDICATORS

4.1   Calculation of indicators

Regarding the cracking, it is possible to calculate the extent and severity per rings or per sections of fixed distances (5m, 25m, etc.), with identification of their location in the transverse section of the tunnel (keystone, shoulders or arches), and to determine whether the cracking is longitudinal, transverse or oblique. As for the dampness, the presence of water on the surface is analyzed to obtain the percentage of area affected. Regarding breaks and chipping, it is possible to calculate the volume of material affected.

All this information, together with more specific data in cases such as those of brick tunnels, segment tunnels or metal frame tunnels, is statistically processed to obtain adapted key performance indicators allowing to assess the global condition of the tunnel [5].

4.2   Monitoring

The use of a high-performance system to record deterioration and defects, with automatic analysis and storage of every single element in a database, offers a more interesting assessment, as it makes it possible to compare different inspections. This comparison is necessary for the monitoring and supervision of the evolution of the tunnel.

If the system is capable of storing automatically detected cracks (with all characteristics such as location, width and depth) and to compare with successive inspections conducted under the same conditions, then it will be able to evaluate the evolution of defects and determinate how the deterioration is progressing to prevent potential risks. Figure 17 shows an example of a comparison of two surveys.

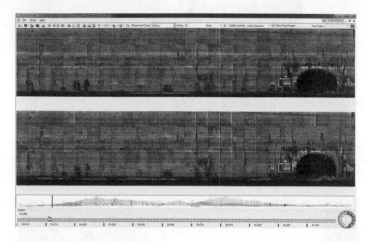

Figure 17.   Comparison of two surveys.

From a geometrical perspective, it is possible to generate a relative mapping of the deformations[6], providing information about section changes and, at the same time, revealing any bulging or deformation which could alter the service clearance of the tunnel. An example of deformation mapping is shown in Figure 18. It is intended to indicate the relative distance from the lining to the center of the tunnel, so large-scale deformations of the tunnel section can be observed. On the other hand, it is also possible to compare the position of the segments inside the tunnel so as to determine whether millimeter movements are taking place.

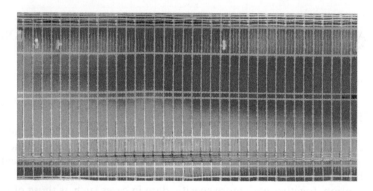

Figure 18.   Deformation mapping in a metallic tunnel.

5   CONCLUSIONS

High-performance systems incorporating 3D technology, Lidar and thermography cameras allow the calculation of objective key performance indicators that make it possible to assess the condition of any type of lining in different types of tunnels.

Thanks to these indicators, we are able to calculate the severity and the extent of the most important damages and to prioritize maintenance activities. For this reason we categorize the intervention areas, separating the areas which are in good condition and those with minor defects (which would be simply subject to monitoring), from those areas where repair works need to be done without delay due to stability risks.

REFERENCES

Sánchez-Domínguez, García Ramos & et al 2017 "Tunnel inspection with high-performance devices. Performance Indicators". *The 8th International Conference on Structural Health Monitoring of Intelligent Infrastructure Brisbane, Australia, 5–8 December 2017.*

"Pavemetrics LCMS system". Retrieved from http://www.pavemetrics.com/.

Sánchez- Domínguez, García- Ramos & et al 2015 "Automated Structural Evaluation of Tunnels in Service". *ITA WTC 2015 Congress and 41ˢᵗ General Assembly, Lacroma Valamar Congress Center,* Dubrovnik, Croatia.

Sánchez- Domínguez, Duval A. & et al 2014 "Use of 3D Scanning Technology for Automated Inspection of Tunnels". *Proceedings of the World Tunnel Congress (WTC) 2014,* Foz do Iguaçu – Brazil.

Gavilán, Sanchez-Domínguez, & et al 2013 "Mobile inspection system for high-resolution assessment of tunnels". *The 6ᵗʰ International Conference on Structural Health Monitoring of Intelligent Infrastructure (SHMII-6),* Hong Kong – China.

Sánchez- Domínguez, García- Ramos & et al 2016 "Euroconsult Tunnelings" http://www.euroconsult.es/pdf/tunnelingspliegos-baja.pdf

Tunnels and Underground Cities: Engineering and Innovation meet Archaeology, Architecture and Art, Volume 6: Innovation in underground engineering, materials and equipment - Part 2 – Peila, Viggiani & Celestino (Eds)
© 2020 Taylor & Francis Group, London, ISBN 978-0-367-46871-2

Preliminary test of an innovative pre-convergence monitoring tool for traditional tunnels

R. Savi, A. Valletta, E. Cavalca, A. Carri & A. Segalini
University of Parma, Parma, Italy

ABSTRACT: During a traditional tunnel excavation, the behaviour of the surrounding rock mass is a meaningful evidence of the appropriate design assumptions. Various studies have highlighted the relationship between tunnel convergence and the pre-convergence, which is recognized as the key parameter to understand and control this behaviour. Up to now, the pre-convergence has been assessed with an indirect approach, deriving its values from other parameters. This paper presents a series of laboratory tests performed on a new instrument designed for the direct monitoring of pre-convergence. The tool (called PreConv Array) is based on 3D MEMS sensors and it is designed to be inserted inside an advancing borehole, providing information about the deformation of rock mass ahead the excavation face. The tests were aimed to evaluate the sensitivity, accuracy and uncertainty of the instrumentation, by imposing various deformations to a PreConv Array prototype and comparing the readings with those obtained by a topographic survey.

1 INTRODUCTION

Tunnel design and construction are particularly complex problems, requiring the acquisition of a wide number of data and information. In this context, an in-depth knowledge of the rock mass behaviour and its interaction with the structure is mandatory to guarantee the correct development of tunnel design. In particular, it is especially relevant to study the behaviour of the material near the excavation front, where instability phenomena and deformation problems could occur (Tanzini, 2015).

For these reasons, starting form the second half of the 20th century, monitoring instrumentations have assumed a key role within the tunnel realization procedures. The importance of on-site measurements derives from the formulation of the "observational method" proposed by Peck (1969). Terzaghi firstly theorized this approach in 1940s as a "learn-as-you-go" philosophy to overcome the setbacks encountered. Nowadays, instrumentation and monitoring are an essential part of current tunnelling practice, allowing for a continuous adaption of excavation and support design (AITES, 2011). In particular, the monitoring activity aims to:

- Obtain information on ground response to tunnelling;
- Provide construction control;
- Verify design parameters and models;
- Measure performance of the lining during and after construction;
- Monitor the impact on the surrounding environments;
- Give warning of any safety-critical trend;
- Make predictions on the performance and management of the completed tunnel.

2 TUNNELLING METHODS AND MONITORING

The monitoring procedure and the results deriving from this activity play a fundamental part in several tunnel design methods. For example, the New Austrian Tunnelling Method (NATM) proposed by Rabcewicz (1964) includes monitoring activities to ensure safety of design and construction. In particular, monitoring and optimizing the ring closure time is of crucial importance for the correct application of the NATM (Karakuş & Fowell, 2004).

More recently, Lunardi (1988) proposed the ADECO-RS method, focusing the attention on the response of the rock mass to the excavation process. In fact, the study and analysis of the medium behaviour near the excavation face should allow the identification of stability problems (Lunardi, 2008). According to the ADECO-RS method, it is possible to observe three different components of deformation process at the front, as represented in Figure 1 (Lunardi, 2000):

- Extrusion: it is the primary component of the deformation response of the medium to the action of excavation. Extrusion depends on the strength and deformation properties of the core together with the undisturbed original stress field. This deformation develops along the longitudinal axis of the tunnel;
- Convergence: this phenomenon defines a decrease in the size of the cross section of the tunnel cavity after the passage of the face. This deformation follows a radial trend and it is caused by the stress redistribution around the excavation free surface;
- Pre-convergence: it represents the convergence of the theoretical tunnel profile ahead of the excavation face, dependent on the relationship between the strength and deformations properties of the advance core and its original stress state.

During the years, several monitoring instruments have been used to measure the tunnel deformations. In particular, monitoring of extrusion and convergence is a primary task when approaching tunnel construction phase.

Unlike these two quantities, nowadays it is possible to directly measure pre-convergence only in rare conditions, specifically when the tunnel section can be accessed from the surface. In these particular cases, multi-point extensometers are inserted vertically into the ground above the crown and the springline of the tunnel to be constructed (Lunardi, 2008). Alternatively, this parameter can be derived from measurements of advance core extrusion by using simple volumetric calculations (Lunardi, 2000). As evident from Figure 2, these relationships depend on the extrusion shape, which can be detected thanks to topographical measurements of the excavation face.

Nonetheless, since this phenomenon plays a key role in the total radial deformation, a specific tool able to directly monitor convergence ahead of tunnel excavation face could be extremely useful to improve the measurements accuracy.

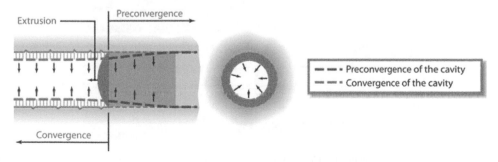

Figure 1.   Typologies of deformation (modified after Lunardi, 2000).

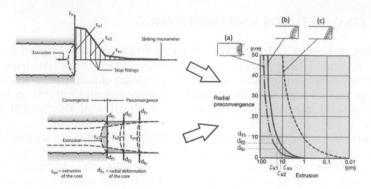

Figure 2. Assessment of pre-convergence using experimental measures of extrusion: (a) spherical dome-like extrusion, (b) combined cylindrical dome-like extrusion, (c) cylindrical extrusion (modified after Lunardi, 2000).

3 MATERIAL & METHODS

In order to achieve the objective of developing an innovative tool able to measure directly the tunnel pre-convergence, the authors applied the MUMS technology (Segalini & Carini, 2011; Segalini et al., 2015) to the monitoring of rock mass deformations ahead of excavation front. This system, originally intended to be applied on slope stability problems (Carri et al., 2015), has laid the basis for the production of new instruments for tunnel monitoring purposes (Carri et al., 2017a).

3.1 *PreConv Array*

The PreConv Array is a chain of sensors (defined PreConv Link) located at determined and fixed distances, equipped with a 3D accelerometer and a thermometer. A fiberglass rod links these nodes, preserving the axis alignment during installation, while the recorded data are firstly transmitted through a single quadrupole cable to a radio device located at the onset of the drilled hole and connected wireless to the control unit. The instrument is designed to be inserted and cemented inside an advancing borehole, or directly in one of the forepole reinforcement (Figure 3).

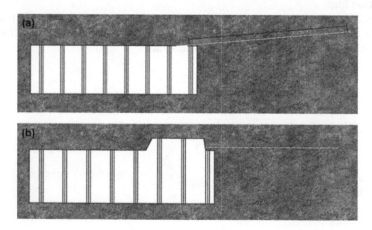

Figure 3. Examples of possible PreConv Array installations: (a) tilted configuration (b) horizontal configuration.

By taking as a reference position the farthest point from the excavation front, every sensor provides its own position in the space with respect to the previous one. Starting from a reference reading, the tool provides the differential displacements over time. The data logger automatically queries each node at defined time intervals. Thanks to this automated procedure, it is possible to gather a considerable amount of data with a high sampling frequency. Subsequently, the collected dataset is sent offsite and processed with self-check controls, together with statistical and reliability analyses, in order to improve its quality.

The PreConv Array is designed for quick installation and deployment, starting the monitoring process just after the installation. The automated data recording is intended to be carried out without interfering with excavation works. Moreover, the tool could remain embedded in the final lining, providing a continuous control of the tunnel stability during its whole lifetime.

3.1.1 Calculation algorithm

The elaboration procedure of the PreConv Array measures derives from the CSC algorithm, originally developed to process data collected by other MUMS-based structured system such as Cir Array and Rad Array (Carri et al., 2017a).

Once a new raw data is received, the software elaborates it by converting the electrical units into physical units. To achieve this objective, the algorithm applies a conversion equation that includes calibration and thermal correction parameters, obtained at the system production factory:

$$a_{xi} = g_{xi}d'_{xi} + g_{xti}T'_i + o_{xi} \tag{1}$$

Where:

- a_{xi} is the acceleration, expressed in physical units, on the x-axis (according to the instrumental reference system) of node i;
- g_{xi} is the x-axis acceleration gain of node i;
- d'_{xi} is the raw acceleration value on the x-axis of node i;
- g_{xi} is the gain of thermal correction along the x-axis of node i;
- T'_{xi} is the raw temperature value of node i;
- o_{xi} is the thermal offset along the x-axis of node i.

Then, the algorithm calculates the position of every node related to the reference system. This value depends on the distance between two consecutive Links, according to the following equation:

$$x_i = SoR_i a_{xi} \tag{2}$$

Where:

- x_i is the local displacement along the x-axis recorded by the node i;
- SoR_i is the segment of relevance of node i;
- a_{xi} is the acceleration expressed as physical units on the x-axis of node i.

Additionally, the algorithm provides the cumulated results in the Calculation Point, defined as the intermediate location between a node and the following one, as in the following Equation:

$$X_i = X_{i-1} + x_i \tag{3}$$

Where:

- X_i is the cumulated displacement along the x-axis recorded by the node i;
- x_i is the local displacement along the x-axis recorded by the node i.

It is possible to evaluate also the temperature measured by each node by applying the following conversion equation:

$$T_i = c_i T_i' + d_i \qquad (4)$$

Where:

- T_i is the value of temperature, expressed in physical units, of the node i;
- c_i is the temperature gain of node i;
- T_i' is the raw temperature value recorded by the node i;
- d_i is the offset of node i.

3.1.2 Calibration procedure

The calibration of PreConv Array requires a thermal chamber and a PLC (Programmable Logic Controller). The instrumentation is read with at least 24 different positions for each temperature and a linear regression is applied. The norm of the three components of the instrumental axes have to result equal to the acceleration of gravity. Therefore, it is possible to evaluate the $g_{xi}, i_{xi}, g_{yi}, i_{yi}, g_{zi}$ and i_{zi} parameters by minimizing the value of error ε, according to the following equation:

$$\varepsilon = \sum_{i=1}^{24} \left(G - \sqrt{\left(g_{xi} a'_{xi} + i_{xi}\right)^2 + \left(g_{yi} a'_{yi} + i_{yi}\right)^2 + \left(g_{zi} a'_{zi} + i_{zi}\right)^2} \right) \qquad (5)$$

Where:

- G is the acceleration of gravity
- ε is the error minimized with the linear regression

It is possible to derive the parameters for the thermal effects compensation and the calibration values for thermometers by repeating the procedure at different temperatures. At the end of the calibration procedure, every node features eleven values (three for each instrumental axis and two for the temperature conversion).

3.2 Laboratory test

A series of laboratory tests was carried out on a PreConv Array prototype, in order to verify the possible applications of the proposed system. Each node composing the Array was placed in the middle point of a fibreglass rod 1-metre long. Six fibreglass rods were connected to each other, thus resulting in a six metres long PreConv Array composed of six sensors. The system was anchored to threaded metal rods, located at 0.5 metres from each sensor and fixed to a wall. It was possible to impose a predetermined displacement in the calculation point of each node by using an L-shaped connector able to slide along the vertical direction. Additionally, a topographic target was placed on this particular point to measure the imposed displacement (Figure 4).

The authors derived a pre-convergence dataset by using the empirical curves presented in Figure 2, assuming a spherical dome-like extrusion shape, in order to have a numerical reference for what concern displacement values to be applied on the tested tool. In particular, the input data were taken from extrusion measures related to a full-scale experiment presented in Lunardi (2008), thus obtaining the values reported in Table 1. It should be specified that these values were imposed manually on the PreConv Array prototype, while the real displacements were measured with a topographic survey during every step.

Displacement values were applied gradually to the monitoring tool by following a three-step approach as displayed in Table 1, in order to simulate a gradual deformation of the material ahead of excavation front. Five measurements were taken for each step, thus allowing to calculate a mean value of the assigned deformation.

Figure 4. Detail of the mechanical structure developed to connect the PreConv Array to the threaded metal rod, the locking system and the topographic target placed at the calculation point.

Table 1. Pre-convergence reference values derived from extrusion monitored data.

Distance from excavation front [m]	0	1	2	3	4	5	6
Zero reference	0	0	0	0	0	0	0
Step 1 [cm]		-2.38	-2.01	-1.55	0	0	0	0
Step 2 [cm]		-4.75	-4.03	-3.10	-1.35	-0.85	0	0
Step 3 [cm]		-7.20	-6.10	-4.70	-2.70	-1.70	-0.90	0

This procedure was repeated for two different initial conditions: the first test was performed starting from a horizontal configuration, while in the second case the PreConv Array was inclined of 4 degrees. A percentage error is evaluated for each step, in order to evidence the difference of measurements recorded by the theodolite x_t and the tested monitoring tool x_{PC}:

$$E[\%] = \left(1 - \frac{x_{PC}}{x_t}\right) * 100 \qquad (6)$$

The topographic survey was carried out using a Topcon IS200 total station, with a focus length of 165 mm and a focal lens of 45 mm. The sensibility in the measure of the distance is ±2mm while the accuracy on the angle is 0.3 mgon (source: Topcon website). The zero for the azimuth was assessed using a referral point P_0, while the roto-translation of theodolite reference system to PreConv Array reference system was performed using known points P_1, P_3 and R_7. The physical apparatus and topographic targets position are displayed in Figure 5. In this structure, the rod R1 represents the excavation front, while R7 is taken as a fixed reference point for the tool installation.

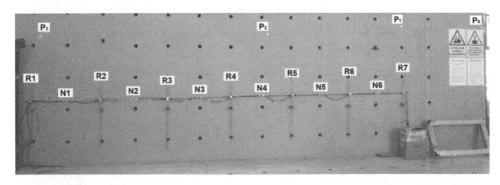

Figure 5. Physical apparatus developed to test the PreConv Array prototype (Rods R1-R7; Nodes N1-N6; Fixed points P0-P3).

4 RESULTS AND DISCUSSION

4.1 *First test – horizontal configuration*

Figure 6 and Table 2 report the results obtained from three deformation steps with respect to a horizontal starting position. It can be noted that the monitoring tool and the topographic survey display comparable trends, properly representing the assigned displacements.

By analysing displacement data reported in Table 2, it appears that the monitoring tool has a slight tendency to underestimate deformations further from the fixed point. This behaviour is probably due to the cumulated nature of displacements presented in the chart, where the presence of a minor difference in a specific node can propagate through the following calculation points. In the case here exposed, this difference reaches a maximum value of 0.51 cm.

Nonetheless, as can be observed in Table 3, this difference is often a small percentage of the total displacement. In fact, with the exception of a single case, all measures recorded by the PreConv Array feature an error less than 10% with respect to the topographic survey results.

It should be taken into account that results here presented refer to almost undisturbed laboratory conditions. For this reason, a high sampling frequency is suggested in order to obtain similar results in an on-site application, where several external actions are usually present.

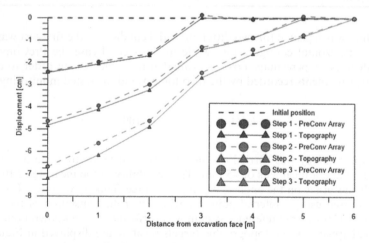

Figure 6. Displacement recorded by PreConv Array and topographic survey, starting from a horizontal configuration.

Table 2. Displacement comparison between topographic and monitoring tool measurements – horizontal starting configuration.

Distance from excavation face [m]	Topographic survey [cm]			PreConv Array [cm]		
	Step 1	Step 2	Step 3	Step 1	Step 2	Step 3
0	-2.45	-4.83	-7.19	-2.43	-4.65	-6.68
1	-2.07	-4.10	-6.15	-1.97	-3.95	-5.62
2	-1.69	-3.22	-4.87	-1.61	-2.97	-4.61
3	0.00	-1.42	-2.68	0.13	-1.31	-2.45
4	0.01	-0.86	-1.59	-0.05	-0.87	-1.38
5	0.02	0.00	-0.80	0.03	0.11	-0.74
6	0.00	0.00	0.00	0.00	0.00	0.00

Table 3. Difference between displacements recorded by theodolite and PreConv Array, and corresponding percentage error - horizontal starting configuration.

Distance from excavation face [m]	Measure difference [cm]			Error [%]		
	Step 1	Step 2	Step 3	Step 1	Step 2	Step 3
0	0.02	0.18	0.51	1%	4%	7%
1	0.10	0.15	0.53	5%	4%	9%
2	0.08	0.25	0.26	5%	8%	5%
3	0.13	0.11	0.23	-	8%	9%
4	-0.06	0.01	0.21	-	1%	13%
5	0.01	0.11	0.06	-	-	7%
6	0	0	0	-	-	-

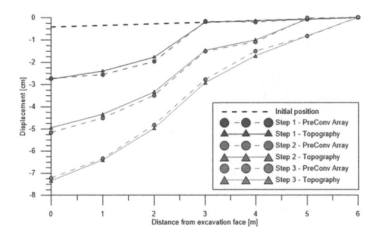

Figure 7. Displacement recorded by PreConv Array and topographic survey, starting from a 4-degress tilted configuration.

4.2 Second test – tilted configuration

In a similar way, Figure 7 displays the results obtained from the test performed with a different initial position, featuring a PreConv Array prototype tilted by 4 degrees with respect to the horizontal reference. In this particular setup, the fixed point at R7 is taken as a zero-vertical reference, while all the other calculation points are located according to the chosen tilt angle. Table 4 reports the differential numerical resulting displacements.

The data obtained from this test confirm the consideration expressed in the previous paragraph, demonstrating a good correspondence between PreConv Array and topographic survey outcomes.

Moreover, it should be noted that the slight underestimation highlighted in the previous test is not present in this case. In fact, the first two steps display an opposite behaviour, while its entity is almost negligible compared with the assigned displacement. The third step apparently shows a trend similar to the horizontal test, but the difference appears to be less important compared to the previous configuration results.

The error values reported in Table 5 confirm this statement, underlining the good correspondence between data recorded by PreConv Array and the topographic survey. In particular, in this specific case 10 out of 14 measurements feature a difference equal or less than 5%, with a maximum difference of 0.21 cm.

Sensibility analysis performed during both tests returned a value of 0.15 mm/metre, which represent an improvement if compared to other MUMS applications where sensibility is equal to 0.2 mm/metre, due to the mono-dimensional approach of the PreConv Array data elaboration.

Table 4. Displacements recorded by theodolite and PreConv Array, and corresponding percentage errors - 4-degrees tilted starting configuration.

Distance from excavation face [m]	Topographic survey [cm]			PreConv Array [cm]		
	Step 1	Step 2	Step 3	Step 1	Step 2	Step 3
0	-2.75	-4.95	-7.34	-2.72	-5.16	-7.22
1	-2.39	-4.34	-6.40	-2.56	-4.50	-6.34
2	-1.78	-3.34	-4.98	-1.97	-3.49	-4.82
3	-0.21	-1.47	-2.92	-0.18	-1.51	-2.78
4	-0.14	-1.00	-1.71	-0.20	-1.09	-1.50
5	-0.07	-0.07	-0.82	-0.08	-0.02	-0.84
6	0.00	0.00	0.00	0.00	0.00	0.00

Table 5. Difference between displacements recorded by theodolite and PreConv Array, and corresponding percentage error - 4-degrees tilted starting configuration.

Distance from excavation face [m]	Measure difference [cm]			Error [%]		
	Step 1	Step 2	Step 3	Step 1	Step 2	Step 3
0	0.03	-0.21	0.12	1%	5%	2%
1	-0.15	-0.16	0.06	8%	4%	1%
2	-0.19	-0.15	0.16	13%	5%	3%
3	0.03	-0.04	0.14	-	3%	5%
4	-0.06	-0.09	0.21	-	10%	14%
5	-0.01	0.05	-0.02	-	-	3%
6	0	0	0	-	-	-

5 CONCLUSIONS

Tunnel design and excavation are significantly complex problems, where it is of capital importance to implement a reliable and accurate monitoring system. Several instruments are available to identify and control rock mass deformation, since it is one of the most important parameters to keep under control in tunnelling activities. In this paper is presented the development and testing of an innovative MEMS-based monitoring tool (called PreConv Array) designed to record pre-convergence data ahead of the excavation face with a direct approach. This parameter is a relevant component of deformation processes, but nowadays a direct measure is rarely viable and its value is usually derived from measurements of other quantities.

A PreConv Array prototype and a physical apparatus were realized in order to test accuracy and applicability of the monitoring tool. In particular, a series of predetermined displacements were applied to the instrumentation, starting from a horizontal and 4-degrees tilted configuration, and the results acquired by the sensors were compared with topographic survey measures. Both tests displayed satisfying results, highlighting a good match between displacement data recorded by the two different tools. For the majority of cases analysed, the resulting error is less than 10% compared to the corresponding topographic measure, with maximum difference observed at the excavation front and sensibility of 0.15 mm/metre.

While the prototype tool proved to be an effective way to test the PreConv Array applicability, a series of future developments has been already planned to further improve the efficacy and reliability of the mechanical apparatus. In particular, a larger setup would allow the test of a longer array of sensors. Moreover, by imposing different deformation values and steps it would be possible to analyse the tool applicability to a wider range of theoretical cases, in the interest of a following installation in a real-case scenario

Furthermore, the PreConv Array prototype presented in this paper is equipped with MEMS sensors only. It would be possible to test a new tool featuring also electrolytic cells, in a similar way to other MUMS-based monitoring tools (Carri et al. 2017b), in order to improve considerably the instrumentation accuracy and sensibility.

REFERENCES

AITES 2011. Monitoring and control in tunnel construction. *AITES/ITA WG2-Research, ITA Report #009/Nov. 2011*

Carri, A., Chiapponi, L., Giovannelli, R., Segalini, A. & Spaggiari, L. 2015. Improving landslide displacement measurement through automatic recording and statistical analysis. *Proceedings of World Multidisciplinary Earth Sciences Symposium, WMESS 2015, Prague*

Carri, A., Chiapponi, L., Savi, R. & Segalini, A. 2017a. Innovative technologies for monitoring underground excavations during construction and usage. *International Society for Rock Mechanics and Rock Engineering – AfriRock "Rock Mechanics for Africa", 3-5 October, Cape Town, South Africa, Volume: Symposium Series S93*

Carri, A., Grignaffini, C., Segalini, A., Capparelli, G., Versace, P. & Spolverino G. 2017b. Study of an active landslide of A16 Highway (Italy): modelling, monitoring and triggering alarm. *In*: Mikoš M., Arbanas Ž., Yin Y., Sassa K. *(eds) Advancing Culture of Living with Landslides. WLF 2017. Springer, Cham*

Karakuş, M. & Fowell R.J. 2004. An insight into the New Austrian Tunnelling Method (NATM). *ROCKMEC'2004-VIIth Regional Rock Mechanics Symposium, 2004, Sivas, Turkey.*

Lunardi, P. 1988. ADECO-RS Analisi delle deformazioni controllate nelle rocce e nei suoli. *Seminar on "Design and Construction of Tunnels", ISMES, Bergamo*

Lunardi, P. 2000. The design and construction of tunnels using the approach based on the analysis of controlled deformation in rocks and soils. *T&T International ADECO-RS Approach, May 2000*, pp. 3–30

Lunardi, P. 2008. Design and construction of tunnels – Analysis of controlled deformations in rocks and soils (ADECO-RS). © *Springer – Verlag Berlin Heidelberg, 2008*

Peck, R.B. 1969. Advantages and limitations of the observational method in applied soil mechanics. *Geotechnique 19, No. 2*

Rabcewicz, L. 1964. The New Austrian Tunnelling Method. *Part one, Water Power, November 1964, 453-457. Part two, Water Power, December 1964, 511-515*

Segalini, A. & Carini, C. 2011. Underground landslide displacement monitoring: a new MMES based device. *In*: Margottini, C., Canuti, P. & Sassa, K. *(eds) Landslide Science and Practice. Springer, Berlin, Heidelberg*

Segalini, A., Chiapponi, L. & Pastarini, B. 2015. Application of Modular Underground Monitoring System (MUMS) to landslides monitoring: evaluation and new insights. *Engineering Geology for Society and Territory. Vol. 2: Landslide Processes. Proceedings of the XII IAEG Congress*, pp. 121–124.

Tanzini, M. 2015. Gallerie – Aspetti geotecnici nella progettazione e costruzione, II edizione aggiornata. *Palermo: Dario Flaccovio Editore*

Topcon IS200 Total Station – Specifications and technical information available at http://www.topcon.co.jp/en/positioning/products/product/ts00/IS_E.html

Tunnels and Underground Cities: Engineering and Innovation meet Archaeology,
Architecture and Art, Volume 6: Innovation in underground engineering,
materials and equipment - Part 2 – Peila, Viggiani & Celestino (Eds)
© 2020 Taylor & Francis Group, London, ISBN 978-0-367-46871-2

The use of artificial intelligence for a cost-effective tunnel maintenance

O. Schneider & A. Prokopová
Amberg Technologies AG, Regensdorf-Watt, Switzerland

F. Modetta & V. Petschen
Amberg Engineering AG, Regensdorf-Watt, Switzerland

ABSTRACT: Every year 4700km of new tunnels are built with an annual growth value of 7%. This results in the fact, that the total amount of tunnels which must be inspected will increase in the future. Furthermore, their operation reliability must be guaranteed with safe and cost-efficient means. Building upon Amberg Technologies' and Amberg Engineering's experience from several international tunnel projects, we have come up with a new platform which allow us to optimize the tunnel inspection process. Today, tunnel assessment is mostly based on a slow and subjective human inspection process. Automatic defect detection will provide a more objective and quantifiable approach to the task of tunnel inspection. By manual inspection it is difficult to assess anomalies objectively, especially cracks at an accuracy of 0.2 mm with random size and shape. It is also difficult to compare them to the historic state of previous assessment campaigns. Thus, it makes sense to aim at an automated detection procedure and a fully digitized workflow for tracking the defects over time. Recent developments in the field of big data and artificial intelligence can be applied to the field of tunnel inspection and bring it to a higher degree of automatization. Amberg is focusing on developing a computer-based and BIM-compatible implementation of the new methods for damage classification and tunnel assessment. This paper will explain the principle of the new platform and show some initial test projects.

1 INTRODUCTION/MOTIVATION

Currently tunnel inspections are carried out mainly by (visual) walking through inspections based on some input data available. This allows to focus on certain phenomena and locations which need special attention. These walk through inspections, which may also include some visual close-up inspections are carried out by experienced but sometimes not specially trained and educated staff. In often cases knowledge is missing on the severity of certain damages as such and of the relevance of their development.

During these walk through inspections notes, images and data in general are taken. After the inspection, this data is integrated into some kind of database, partly in digital form, partly still on paper. This data serves as an input for further inspection at regular intervals or as a trigger for additional inspections and measurements. As well as for immediate necessary maintenance and rehabilitation activities. Evaluation of the data for a holistic state assessment, damage development, maintenance etc. is done only to a limited extend and not in a digitalized manner.

The Project owner needs answers in real time to the following questions:

– What is the condition of my structure?
– What types of damages are present?
– Is there a threat to structural and operational safety?
– How much money must be budgeted for maintenance?

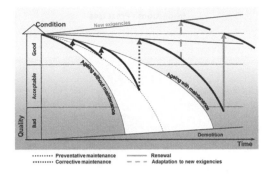

Figure 1. Aging process of an infrastructures.

To get answers to the questions above, project owners deploy systematic tunnel inspections. Periodical inspections should help to keep the overall costs low. It helps to identify the critical areas early and start with refurbishment work when the areas of damages are still small. Without regular inspections you can increase the risk of a big incident which ends up in bigger refurbishment work and also costs (Figure 1).

2 METHODOLOGY OF TUNNEL MAINTENANCE WORK

2.1 State of the art today

To maintain operational safety, serviceability and structural safety of tunnel infrastructure as economically as possible, a mature value preservation concept must be implemented.

The basis of the value preservation concept is regular monitoring, which means regular surveying and inspections each 5th year in most of cases (number of years can be different in each country and for each tunnel).

Preservation of tunnel infrastructures is divided into surveying measures and maintenance measures. Surveying measures consist of inspections, visual and measurement control, function and material tests. Maintenance measures are further divided into preventive maintenance, corrective maintenance and renewal measures. Maintenance measures are defined as the result of the condition assessment based on the surveying measures.

Maintenance management starts as soon as the tunnel is finished by the first inspection to document the as-built state. The survey measures serve both as initial documentation of the condition and as continuous or periodic documentation of the condition development.

Then each 5 year inspections are scheduled to register the tendencies of the state of the construction and to be able to define when a maintenance work is needed. After an intervention, the new state is also inspected.

Inspections are fundamental for all focused management processes for structural value preservation that must be based on detailed knowledge about the condition of the tunnels. During the surveying measures, technical data about the properties, characteristics and the condition of the tunnel must be collected:

– Construction materials and their properties.
– Geometry and profile of the cavity.
– Installations for operation and clearance profile.
– Damages and deficiencies.
– Additional data: Construction method and tunnel age, geology and geologically induced forces, hydrogeology and chemical characteristics of the rock water, climatic condition, rolling stock and operation types.

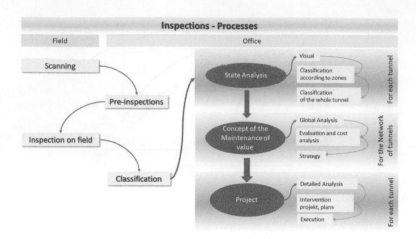

Figure 2.   Inspection processes.

Inspection or surveying work is partly done in Field (Figure 2). For safety reasons, work in Field must be carried out in closure periods. These can be done usually only during night shifts and therefore it is very expensive. Therefore it is important to minimize time in the field as much as possible. This is facilitated by computer-aided tools for survey activities and data analysis. Carrying out a pre-inspection in the office using the scanned images as a background for the inspection software reduce the time needed in field. Accordingly, the tunnel structure, materials, installations and damages that have been already registered only needs to be checked, corrected or added, once the inspection is lead through the tunnel. The data collected and analyzed are the basis for the tunnel state assessment and future planning. Damage categories are for example deformations, cavities, cracks, water inflow, corrosion etc. or combination of these phenomena.

After a complete acquisition of the damages data and deficiencies of all tunnels on the network, it needs to be defined to which condition class each zone belongs to.

Each zone is assigned to a condition class. Classes are defined as follows:

– 1: Good.
– 2: Sufficient.
– 3: Insufficient.
– 4: Aggravating.
– 5: Alarming.

After classification the management approach has to be defined (preventive or corrective maintenance or renewal). This is going to be applied to the different condition classes 2 to 5.

Then this process needs to be done for the whole tunnel, then for the whole network in order to make a prioritization among the tunnels in function of the seriousness of the damages and the construction costs. The owner can organise finance plans and intervention schedules. Finally, a maintenance intervention project of the tunnel (or zone) is made according to the priorities defined for the whole network.

2.2   Pros & Cons of used approach

The assessment is carried out by experienced engineers but it is naturally subjective. The influence of subjectivity must be reduced without relieving the engineer from responsibility. Digitalization and working with artificial intelligence pushes the industry to work out an increasing amount of standards that makes the evaluation of the different phenomena easier, faster and more unified. This development is already an important factor to simplify, uniform, clarify the moreobvious results and to accelerate the communication between the different possible participants of the whole maintenance process.

Beside the fact that digitalization pushes the development of the standards, an important acceleration happens in each step of the inspection project. For example, scanning data can be transformed to action data of the maintenance project execution and then it's going to be the part of the as-built model. Later it can be used as a basis for further inspections/maintenance management tools. Firstly, the image data from the scanning is automatically analyzed by the software with image and object recognition tools. It accelerates the pre-inspection because it recognizes (semi-) automatically most of the defects. Due to its neural learning capacity, as the system get used, it detects more incidents more precisely and marks them on a layer over the scanning picture. This data must be checked and corrected if needed which implies much less work than drawing all the incidents by hand on the layer. The expectation is that the more this tool is used, the less corrections will be needed.

Second, with a better quality of pre-inspection documents, the inspection can be done faster because fewer new damages have to be discovered on site. The inspection data can be registered in the cloud and be followed real-time.

On the other hand, the synergies of differing damages are quite complex. The same phenomena can have different importance according to:

– The geology.
– The building materials.
– Geometric position.
– The age of the damage, and the speed of change.

Even though the combination of most of these factors with the damage or deficiency types can be programmed into a software to support decisions, the human control factor cannot be excluded. For example, the possible complexity of the geology or a lack of information can cause individual cases.

Even though some decisions in the process of value preservation can't be 100% unmanned, still many steps can be standardized, digitalized, supported, optimized and finally accelerated by artificial intelligence. This kind of evaluation of data capturing, processing and analysis is a necessary must for integrating tunnel maintenance workflows in the world of the BIM. This is going to be more and more common in underground construction.

2.3 Metric of today's solution, time needed for each step is compared later with the new system in the conclusion

Based on the experience of 450 km of tunnels inspected by Amberg Engineering, we calculated the time needed when using a standard methodology used in the past 10 years in all instances, i.e. "Trolley and TunnelMap". Comparing this already computer-based methodology with the analog 2-dimenstional CAD support indicates already significant improvements as shown in the first column of Table 1. The third column shows the time-saving effect of the use of artificial intelligence compared to Trolley and Tunnel Map.

Table 1. Potential time-saving on each steps of surveying/inspection, if by each step 100% is equal to the inspection method with TunnelMap and trolley.

Time needed for each step	Analog surveying-methods + CAD [%]	Trolley + Tunnel-Map [%]	Use of artificial intelligence [%]
Data capture (closing the tunnel)	0	100	100
Pre-inspections (analysis)	70	100	40
Field (closing the tunnel)	400	100	80
Field (Engineers work)	300	100	75
Documentation	300	100	60
Preparing the next insp.	200	100	70

While below shows that data capturing itself is not much faster, it could be further accelerated if combined also with GPR (Ground Penetrating Radar) data capture that would mean getting more information at the same time. The data capturing is going to be more detailed which contributes a lot to the clarity and precision of pre-inspection reports. A special case where digitally supported data capture is beneficial would be the use of drones for scanning in tunnels which are difficult or dangerous to enter. Related benefits depend on the individual circumstances.

The pre-inspections are going to be up to 60% more efficient thanks to the reasons described in detail in chapter 2.2 such as e.g. the fact that (semi-)automatic change and defect detection helps saving an encouraging number of working hours of engineers.

As for the inspection itself, working in the field is going to need less time as more detailed pre-inspections save detection work in the tunnel. Due to this fact less personnel is needed and the tunnel can be opened sooner. We consider that the time saving would be 25–30%.

An important aspect of the use of new technologies is that a part of the documentation of the inspection can be directly generated from the data capture itself. The data is going to be accessible in a cloud for everybody with a permission, so the owner of the tunnel can see the inspection results online in real time.

Finally, preparing the data for the next inspection happens automatically in the cloud as the previous inspection which helps accelerating the preparation process.

3 METHODOLOGY OF DATA CAPTURE AND DATA PROCESSING

3.1 *Data capture*

Data capture in the tunnel is very demanding due to low light, harsh conditions, usually very limited time for measurement, high expected accuracy and safety regulations. Data capture techniques have developed from visual inspection on sight to more automated technologies. Most common data acquisition method is laser scanning. The advantage is that light conditions do not influence the quality of results. On the other hand, the color information is lost. The opposite characteristic applies for photogrammetry.

Another parameter is the way how the measuring equipment is moved in tunnel. This varies from manually pushed trolleys, small robots up to big cars or locomotives. Obviously, the speed and purchasing cost vary. Lately drones are being tested for use in tunnels. The data capture methods are summarized in Table 2.

3.2 *Data processing*

Raw data processing starts with data import from measuring instrument and continues with positioning and exporting of files in formats needed for data analytics. These are usually pictures referenced in coordinate system of a tunnel axis.

Table 2. Evaluation of data capture methods.

Method	Speed	Data capture	Accuracy	Personnel
Manual inspection	Walking speed	Camera images, Test results of material behavior	0.1m	2
Trolley based inspection	Kinematic laser scanning (profiler), 3 - 5 km/h	Gray scale image, RGB image,	0.005m	1
Train based inspection	Kinematic laser scanning (profiler), operating speed (40 – 60km/h)	Gray scale image, RGB image, Ground penetrating radar information, thermal image, ultra-band width	0.01m	-
Unmanned robot	Kinematic data capture, autonomous inspection in walking speed 3 – 5 km/h	Kinematic scanning (profiler), camera image, thermos image, GPR, ...	0.1m	-
UAV	Kinematic data capture, autonomous inspection	Kinematic scanning (profiler), camera image, thermos image, GPR,...	0.5m	-

As the table in previous chapter shows, there are many ways how to capture data. What matters in the end is the quality of data. When it comes to data processing, low quality data is inefficient because the inspected phenomena are usually not visible and the repetitive tunnel visits are needed.

When there is good quality data the second important factor is to be able to effectively use and reuse this.

3.3 Data analytics

Having good quality data is a promising start. But the real value is still missing. Equally important as the quality of the data is good data processing. Just the grayscale images are futile. Grayscale images with the ability to add features drawing is proving advantageous. But the aim is to have an intelligent system where each picture and phenomena relates to database where all the additional information is organized. This connection makes it easy to analyze the data and reuse them in the future.

To analyze the collected data, referenced grayscale pictures are used as a background and tunnel features are drawn on it. All the feature drawings are stored in a database. The common coordinate system provides connection between drawings, pictures and the real tunnel. Experienced engineers evaluate state of the tunnel based on feature drawings and determine condition class of each block.

3.4 Motivation of our development project

Currently used software TunnelMap (Figure 3) provides the functions to create database from phenomena drawing. But the whole software has its limitations caused mainly by the age of the software. It was developed roughly two decades ago when data synchronization, ideas about BIM and possibilities of smart construction were on a completely different level (if existed at all).

Nowadays the gap between the old software and modern inspection demands reached the size when using it is not effective and therefore unsustainable any more. It is time for modern web based platform that will keep pace with today technology and market expectations.

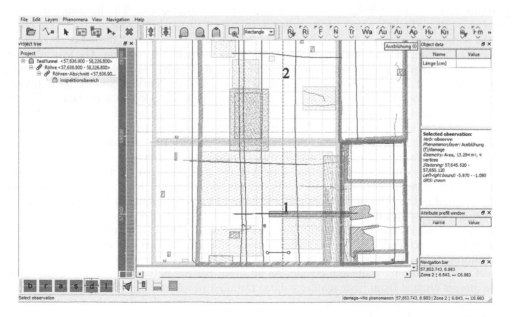

Figure 3.   TunnelMap software - drawing page.

4 THE SYSTEM

4.1 *Workflow*

The paper drawings are rather old fashioned. Working on the computer is a main stream nowadays. Looking in the future we need to go further. Move the data to the cloud to make these available everywhere, every time, from all types of devices and for all users.

In this chapter the new platform called Amberg Inspection Cloud will be described. Mainly the new methodology and advantages for tunnel inspection will be discussed.

4.2 *Workflow optimization*

The basic workflow remains identical as described in the *Chapter 3*. The main goal is to make the whole inspection process easier, more automated and therefore less time consuming and cheaper.

Our focus is on the data processing and data analysis itself. There are many ways how to collect data efficiently in automatic or semi-automatic way. But the inspection and damage drawing itself has always been tedious and time consuming manual work.

The workflow optimization can be done in two steps. The first step is helping users with manual drawing by highlighting areas with high crack (damage) probability. This way users do not have to see the whole tunnel in detail and can focus only on problematic areas.

The second step is an automatic phenomena drawing. It means that inspected damages or utilities are found and drawn fully automatically. Users can then perform a quick visual check. This second level of the workflow optimization is expected to save ¾ of the inspection data processing time.

In both above mentioned steps an artificial intelligence (deep neural network) is used for damage highlighting and automatic drawing. The deep neural network will be trained either by your own work (manually inspecting tunnels) or by the community (all users of the platform).

4.3 *The new platform – Why artificial intelligence?*

The evolution of the artificial intelligence and machine learning started already in the middle of the last century. Recent findings in this field and powerful hardware enabled us in the last years to transfer this technique from academic field to business use. The hardware development was especially important for so called deep neural networks. These networks are based on the same principle as a human brain. The power and weakness of these networks lie in millions of parameters that need to be adjusted during the network training. The learning process may take from a few minutes up to several days. But with a suitable network structure, very complicated phenomena can be extracted from pictures. Various studies have already been conducted on this topic. The biggest effort was probably done in a medical field. But some investigations were already done in road, rail and concrete damage detection [Yokoyama et al. 2017, Cha et al. 2017, Faghih-Roohi et al. 2016, Yang et al. 2017, Eisenbach et al. 2017]. In these studies, we can find many similarities with tunnel damages. But up to this date we are not aware of any work on this topic in tunnel environment.

Using the previous research dealing with damage detection and available tools (Keras, TensorFlow™) we tested LeNet and VGG-16 network. The tools Keras and Tensorflow are common used in this field of development and have a huge community of users. The networks (LeNet & VGG-16) promised good results in reasonable computation time.

The main limitation in the first development stages was a limited amount of a training data. Therefore, we excluded very deep networks from testing. Even VGG-16 ended up overfitting the training data. After this observation we fine-tuned the LeNet network for searching cracks or wet areas on tunnel walls and ceiling. As an input for the network training we used gray scale images created from laser scanning data. Detailed results are described in more detail in *Chapter 5*. The visual results (Figure 4) are presented in more detail on the new web-based platform Amberg Inspection Cloud.

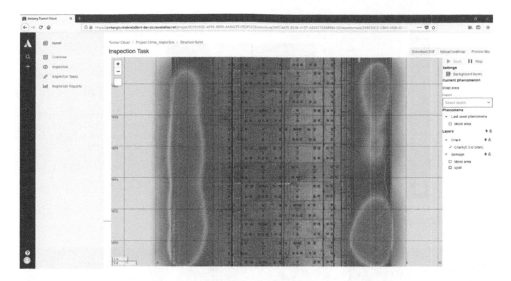

Figure 4.   Color map showing probability of wet area in tunnel.

4.4   *The new platform – From tunnel to cloud*

Amberg Inspection Cloud is a brand-new web-based platform specialized on tunnel inspections. It is open for data from different sources and measurement methods. As long as the inputs are good-quality, high-resolution images the measurement method is not important.

Tunnel design and the high-resolution images are uploaded to a cloud. Pre-trained neural network can automatically detect damage probability in each area and draw a damage probability color map. Experienced operator then just looks at the highlighted areas and draw the exact tunnel damages. In case of bigger project more operators can work independently without tedious data transfer and synchronization.

When the drawing is finished the analysis can be accessed directly in Amberg Inspection Cloud. That includes comparison of the tunnel state in different years, visualization of damages along the tunnel and the standard DXF export.

In the future, automatic phenomena drawing will be introduced. When Amberg Inspection Cloud is fully developed, the inspection engineer only checks the automatic drawings and potentially make some small changes if needed. That will save enormous amount of time compared with manual drawing.

4.5   *Outlook*

The final goal is to eliminate manual work from tunnel inspection as much as possible. Nowadays it does not seem possible to skip manual input and expert knowledge completely. In the future experts will still need to decide on some unclear cases. But the tedious manual marking of all the clear cases will be replaced by the artificial intelligence. Automatic crack detection and drawing is on the horizon.

Not only is the inspection evaluation itself an issue. Other time-consuming aspects are the result evaluation and data transfer. The goal in this area is to be fully BIM compatible. That is somehow restrained by the fact that BIM standard for tunnels is not yet defined. The Amberg Inspection Cloud already combine position information (3D), feature info (1D), time (1D) and powerful data analysis. The missing part of the chain is a BIM format that allow easy transfer to other programs.

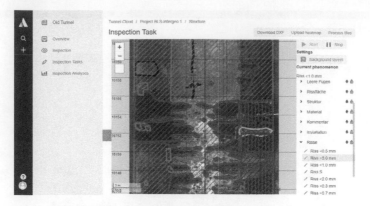

Figure 5. Finished phenomena drawing of masonry tunnel.

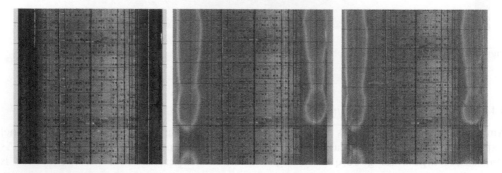

Figure 6. Workflow with wet area probability predicted by neural network.

5 TEST PROJECTS

As test data we have some old masonry tunnels in Switzerland where the inspection is done regularly. The second type of the tunnel that was used for testing is significantly newer rail segmental lining tunnel.

We realized that it is important to distinguish between different types of the tunnels. Obviously, the tunnels and the damages in them are very different. Therefore, it seems impossible to train one neural network that is applicable for all tunnel types.

5.1 Old masonry tunnel

Obviously, the inspection of old tunnel (Figure 5) is much more demanding because significantly bigger amount of damages is found here. Moreover, different types of materials (for example: bricks, shotcrete) makes automatic recognitions even more difficult.

5.2 Segmental lining tunnel

Segmental lining tunnels are in general not so old. Therefore, less damages can be found in these tunnels. The most frequent are wet areas. Sometimes small cracks can be found.

That means on one hand that the damage data are rarer and it is more difficult to gather enough pictures with damages to train the network. On the other hand, the damages are much more uniform so it is easier to recognize them automatically even with smaller training dataset.

In detecting the wet zones in segmental tunnel, we achieved accuracy over 90% (Figure 6). This can be further improved by adding more training data and tweaking the network model.

6 CONCLUSION

Artificial intelligence is not a new technology, but the change of access to powerful processing engines changed the usage of these technologies dramatically. The transfer from academic field to business area started in the last few years. The construction industry is nowadays also transferring many operations and applications with neural networks approach. Our industry can now profit from applications and learnings from different industries - for example the medical industry.

Our challenge for the future requires safe, predictable and adaptable tunnels which help us to optimize the costs for the entire life-cycle of a tunnel infrastructure. Thanks to the digitalization of our industry we will collect more data from our infrastructure. However, the challenge will be to increase the data analytics. Artificial intelligence helps to automate this process. The usage in this paper for artificial intelligence for defect detection has shown that these technologies will change the way tunnel inspection will be done in the future. However, for the time being there is still some work to do and to collect some more reliable training data. First test project shown that the time saving is up to 60% of the initial time. However currently we reached with the actual state of the software around 30%. We're optimistic that with improving our neural network and training the system with existing project, which has been already carried out with our previous software TunnelMap, we can increase the degree of automation.

Once we can automate the data processing for defect detection we can also increase the data capturing cycle. The data capture process is currently also in a big change. New sensors and platforms are developed which allow an automated data capture of the entire tunnel infrastructure. Combining this data, we can then build up a digital tunnel twin. With these technologies, it then should be possible to start changing the periodically inspections to event driven inspections. Which means that the owner of an infrastructure can plan their inspections based on the real conditions and the changes in certain sectors of the tunnel.

REFERENCES

Cha, Young-Jin, et al. 2017. "Deep Learning-Based Crack Damage Detection Using Convolutional Neural Networks." *Computer-Aided Civil and Infrastructure Engineering*, vol. 32, no. 5, pp. 361–378, doi:10.1111/mice.12263.

Eisenbach, Markus, et al. 2017. "How to Get Pavement Distress Detection Ready for Deep Learning? A Systematic Approach." *2017 International Joint Conference on Neural Networks (IJCNN)*, doi:10.1109/ijcnn.2017.7966101.

Faghih-Roohi, Shahrzad, et al. 2016. "Deep Convolutional Neural Networks for Detection of Rail Surface Defects." *2016 International Joint Conference on Neural Networks (IJCNN)*, doi:10.1109/ijcnn.2016.7727522.

Winkler N. & Ackermann A.W., 2011, "Value Preservation of underground infrastructure through focused conceptional planning", *Amberg Engineering Ltd., Switzerland*

Yang, Liang & Li, Bing & Li, Wei & Zhaoming, Liu & Yang, Guoyong & Xiao, Jizhong. 2017. "Deep Concrete Inspection Using Unmanned Aerial Vehicle Towards CSSC Database." *Conference: 2017 IEEE/RSJ International Conference on Intelligent Robots and Systems*

Yokoyama, Suguru, and Takashi Matsumoto 2017. "Development of an Automatic Detector of Cracks in Concrete Using Machine Learning." *Procedia Engineering*, vol. 171, pp. 1250–1255, doi:10.1016/j.proeng.2017.01.418.

Tunnels and Underground Cities: Engineering and Innovation meet Archaeology,
Architecture and Art, Volume 6: Innovation in underground engineering,
materials and equipment - Part 2 – Peila, Viggiani & Celestino (Eds)
© 2020 Taylor & Francis Group, London, ISBN 978-0-367-46871-2

Design of the support medium for Slurry Pressure Balance (SPB) shield tunnelling in demanding ground conditions

B. Schoesser
Ruhr University Bochum, Bochum, Germany

M. Straesser
Herrenknecht AG, Schwanau, Germany

M. Thewes
Ruhr University Bochum, Bochum, Germany

ABSTRACT: The application range of slurry shield machines aims primarily non-cohesive soils. In addition, slurry shields are preferably utilized in tunnelling projects with challenging ground conditions, e.g. coarse and highly permeable soils, extremely heterogeneous geology as well as very high ground water table. These boundaries impose particularly high requirements on the support fluid. Here, bentonite suspensions are applied, which are characterised by the rheological parameters yield point and viscosity. For special applications, further characteristics and quality standards are important, e.g. density and the interaction with the in-situ soil at the tunnel face. In this paper, the fundamental rheological parameters of bentonite suspensions are discussed, including widely used measuring devices such as the Marsh-funnel, ball-harp rheometer and rotational viscometer. The parameters gained from these devices are supplemented by advanced rheometric tests. The measurement profiles provide further information on the flow behaviour. As an example, the flow curve is a valuable tool to constitute the yield point and the fluid flow under very low and high shear rates. This extended analysis allows a detailed discussion within the suspensions design phase. Finally, the suspension requirements are summarized and related to practical experience for the needs of a construction site. These settings are transferred to suitable suspension design supported by the findings of research work and laboratory test.

1 FACE SUPPORT OF SLURRY PRESSURE BALANCE (SBP) SHIELDS

Among the mechanized tunneling machines, slurry pressure balance (SPB) shields are known as a reliable excavation technology in non-cohesive soils under groundwater table and unstable tunnel face conditions (Fig. 1). Utilizing pressurized bentonite suspensions as support medium in the excavation chamber, the main advantage of a SPB shield is the exact control of the support pressure. In such conditions, the tunnel face is actively supported preventing surface deformations and other ground movements.

To achieve a stabilised tunnel face, two key conditions must be fulfilled. First, a sufficiently large slurry pressure in the excavation chamber of the shield has to be applied with reference to the acting boundary conditions. The required pressure in the excavation chamber can be determined by various methods, e.g. Anagnostou & Kovari (1994), Jancsecz & Steiner (1994) or following the recommendation of the German Tunnelling Committee for the calculation of face support pressure (DAUB 2016). Second, the slurry excess pressure in terms of the difference between the slurry pressure and the groundwater pressure must be transferred at the soil

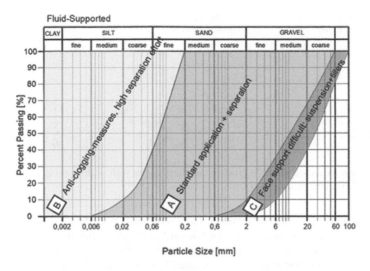

Figure 1. Application range of slurry pressure balance (SPB) shields (Thewes 2009).

skeleton to counteract the earth pressure. In practice, the German standard DIN 4126 (2013) is often used to predict the support pressure transfer at the soil skeleton.

As summarized by Zizka et al. (2018), Morgenstern & Amir-Tahmasseb (1965) conducted one of the first attempts to analyse the interaction between bentonite slurry and soil for the purposes of open trench stabilization. The authors pointed out that at the time of publishing their paper, several mechanisms such as hydrostatic pressure, arching of the soil and electro-osmotic forces were discussed as the main mechanisms responsible for the slurry support of non-cohesive open trenches. Weiss (1967) performed laboratory experiments dealing with stabilization of non-cohesive soils by bentonite slurry. In his experiments, he visualized the penetration behaviour of slurry into the pores of soil and suggested that the stabilisation behaviour depends on the equivalent pore diameter of soil and the yield point of the slurry. Weiss (1967) summarizes the stability condition of a trench in non-cohesive soils as three complementary conditions.

- yield point of slurry is required to achieve the equilibrium of forces on a single soil grain.
- slurry pressure must exceed the groundwater pressure.
- slurry excess pressure must counterbalance the earth pressure.

Further research on the stability of slurry-stabilized trenches was conducted by Müller-Kirchenbauer (1972). The author performed both, experiments and theoretical analysis. He stated that penetration of slurry in the pores of soil influences the slurry pressure transfer on the soil skeleton. The penetration behaviour of slurry is determined by the grain size distribution of soil and the stagnation gradient of slurry. He distinguishes two cases of slurry-soil interaction (Fig. 2). The first is the formation of a filter cake at the soil's surface (Type I) and the second is slurry penetration inside soil's pores without any aggregation of slurry particles at the surface (Type II). In Type II, the shear resistance of the pore channel wall is activated. The shear resistance is in equilibrium with the excess hydraulic head of the slurry in the trench at the end of the penetration process. The pressure transfer Type III consist of both, penetration zone and the filter cake (Zizka et. al. 2018).

For tunnelling in coarse and permeable soils, the pressure transfer according to type II is the applicable principle. Here, the bentonite suspension penetrates into the ground. The movement is induced by the acting pressure and leads to the interaction between the suspension and the surface of the soil particles. Depending on the flow velocity, a certain shear rate acts within the suspension. At the contact areas, shear stresses of the magnitude of the yield point of the suspension are transferred to the soil particles. When the penetration depth has become so large that the integral of the transferred shear stresses is in equilibrium with the difference between the suspension pressure and the prevailing earth pressure, the penetration process stagnates.

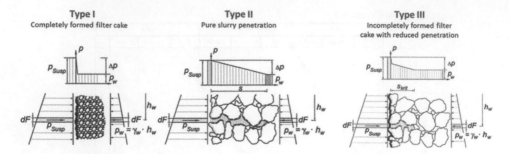

Figure 2. Different types of pressure transfer in diaphragm wall stabilization according to Müller-Kirchenbauer (1972) (Zizka et. al. 2018).

For the design of a bentonite suspension, the rheological parameters need to be adapted to the boundaries of the in-situ soil. Here, the relevant geological characteristics of non-cohesive soils cover the description of the pore space in terms of porosity, permeability, compactness and effective or characteristic grain size d_{10} gained from the particle size distribution.

2 SUSPENSION RHEOLOGY AND RHEOMETRY

Bentonite suspensions are classified as non-Newtonian fluids. They inhibit the rheological parameters yield point, viscosity and thixotropy. The parameters can be determined constituently but they interact with each other. For practical application, the flow model according to *Bingham* is used for the rheological description of bentonite suspensions.

For the determination of the rheological parameters, different technical devices are available. Marsh funnel, ball-harp rheometer and rotational viscometer belong to the proven rheometric technologies utilised on construction sites.

2.1 *Rheology*

Rheology describes the deformation and flow behaviour of materials and fluids in general. The flow behaviour of a fluid is caused by the introduction of an external force. The flow starts when the outer energy is bigger than the inner energy. Overcoming the inner physical structure allows the fluid to flow. This inner resistance is described as *viscosity*. Water and oil show varying viscosity values indicating different flow velocities at ident boundary conditions. In addition, the specific parameter of a bentonite suspension is the *yield point*. Below the yield point, the bentonite suspension acts as a solid (including a thixotropic effect); above the yield point it acts as a fluid. The transition between solid and liquid behaviour is marked by the yield point.

2.1.1 *Viscosity*

As mentioned before, viscosity is the internal resistance of the fluid to flow. Temperature as well as type, amount and duration of the shear exposure influences the behaviour of the fluid flow. The smaller the viscosity, the lower the resistance and the faster the flowing motion. Viscosity values are gained directly using the viscometer (Chap. 2.2.3) or indirectly using the Marsh funnel (Chap. 2.2.1)

The general law describes the relation between viscosity η [Pa s]; shear stress τ [Pa] and shear rate $\dot{\gamma}$ [1/s] as follows:

$$\eta = \frac{\tau}{\dot{\gamma}} \tag{1}$$

2.1.2 Yield point

After reaching the yield point, fluids that show a linear flow relation are categorized as *Bingham* plastic fluids. The following equation represents the mathematical model of the *Bingham* type. Figure 3 shows the theoretical context.

$$\tau = \tau_0 + \eta\dot{\gamma} \qquad (2)$$

where τ = shear stress [Pa]; η = viscosity [Pa s]; $\dot{\gamma}$ = shear rate [1/s]; and τ_0 = yield point [Pa]

The transferable shear stress τ at a shear rate $\dot{\gamma}$ of 0 provides the yield point of the *Bingham* fluid. Overcoming the yield point, the shear stress τ increases linearly with increasing shear rates $\dot{\gamma}$. The slope of the straight line describes the viscosity η of the fluid (Fig. 3).

2.2 Rheometry

2.2.1 Marsh funnel

The *Marsh time* t_M is measured according to DIN 4127 (2014) for 1,000 cm³ and according to API 13B (1997) for 948 cm³ with a standardised Marsh funnel (Fig. 4 left). The Marsh funnel is easy to use on site and provides rapid detection of any change in suspension properties relative to an initial value. The Marsh funnel reaches its limits with very viscous suspensions, since the run-out times become impracticably long or the suspension does no longer flow out of the Marsh funnel (Praetorius & Schösser 2017).

The functionality of a certain Marsh funnel can be examined by determining the run-out time of water. When the funnel contains 1,500 cm³ of water, the amount of 1,000 cm³ should spill in 28 s (DIN 4127, 2014) and in 26 s for an amount of 948 cm³ (API 13B 1997).

2.2.2 Ball-harp rheometer

The *static yield point stat* τ_f is determined by the process of *von Soos* as laid down in DIN 4127 (2014) with a ball-harp rheometer. Ten balls (Fig. 4 middle) hang from a disc on nylon threads.

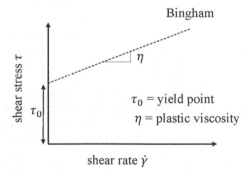

Figure 3.   Flow model of *Bingham*.

Figure 4.   Rheometric devices for the determination of suspension parameters: Marsh funnel (left), ball-harp rheometer (middle), and rotational viscometer (right).

Figure 5. Determination of the static yield point with the ball-harp rheometer.

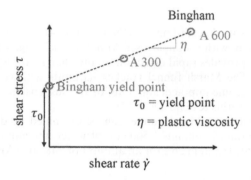

Figure 6. Determination of *Bingham* parameters using rotational viscometer.

The balls have different diameters and consist of either glass or steel. Each of these balls corresponds (depending on their specific weight and the density of the suspension) to a yield point.

The balls are arranged round the disc with an equivalently increasing yield point and each is marked with a number. The disc is fixed to an apparatus, which enables vertical dipping of the disc and thereby the balls into the surface of the suspension. When the disc is lowered, all balls corresponding to a lower yield point than that of the suspension float on the surface, and all balls corresponding to a higher yield point sink into the suspension (Praetorius & Schösser 2014). When the balls are dipped, the weight of the balls is opposed by the floating force and the yield point of the suspension. Balls that sink into the suspension can be recognised by their tight threads, and the balls that float can be recognised by their slack threads (Fig. 5).

Since the balls are arranged in a sequence of increasingly assigned yield points, the actual yield point of the suspension is located between the values of the last floating ball and the first sunken ball (Schulze et. al. 1991).

2.2.3 *Rotational viscometer*

To determine the *plastic viscosity* η_p and the *Bingham yield point* τ_B, the rotational viscometer according to API 13B (1997) is used. In this apparatus, the suspension is filled into the annular gap between two rotationally symmetrical and coaxially mounted cylinders (Fig. 4 right). One of the cylinders rotates with the angular velocity Ω and the other remains stationary. The required force to overcome the flow resistance of the suspension in the annular gap can be determined as a function of torque and rotation velocity (DIN 53019 2008).

In the case of apparent viscosity, the measurement of shear stress is performed at a shear rate of 600 rpm (A 600) (Fig. 6). In the case of plastic viscosity, the shear stresses are determined at shear rates of 600 rpm (A 600) and 300 rpm (A 300). Subtracting the A 300 value from the A 600 result in the plastic viscosity value. The *Bingham* yield point is calculated by subtracting the plastic viscosity from the value of A 300 (API 13B 1997).

Addressing the *Bingham* model, further analysis can be derived by means of the flow curve, as it represents the relation between the shear stress and the shear rate. Here, systematic derivations can be drawn for very low and very high flow velocities. They help to assess the behaviour of the bentonite suspensions within the complete slurry circuit including excavation chamber as well as feeding and transportation lines during tunnelling works.

3 RELATION BETWEEN SUSPENSION RHEOLOGY AND SOIL MECHANICAL PARAMETERS

For the selection of a suitable bentonite suspension, the relevant geotechnical characteristics of the ground must be known. The objective is to find the optimal bentonite suspension depending on the geotechnical conditions of compactness, permeability and pore space to fulfil all requirements regarding the face support.

In the following, formulae are summarized to calculate the required value of the yield point, which suits the certain geological condition. On this basis, a target value for different bentonite products and varying solid contents can be derived (Praetorius & Schösser 2017).

The aim of slurry assessment is to specify the requirements concerning the suspension rheology (yield point τ_0) depending on the geotechnical conditions (characteristic grain size d_{10}). The standardised approaches for determination of slurry properties can be found in DIN 4127 (2014) and are summarized in the DAUB recommendation (DAUB 2016).

3.1 *Required yield point of the suspension depending on the pressure gradient fso*

The interaction of the bentonite suspension with the ground can be described with the help of an existing pressure gradient f_{so}, which represents the decrease in slurry excess pressure over a meter of the penetration distance into subsoil. The pressure gradient f_{so} is a theoretical variable, since the penetration depth is usually much smaller than 1 m. According to DIN 4126 (2013), the existing pressure gradient can be calculated from Equation (3).

$$f_{So} = \frac{3,5 \cdot \tau_f}{d_{10}} \tag{3}$$

where f_{so} = pressure gradient [kN/m³]; τ_F = yield point of the suspension [kN/m²]; and d_{10} = characteristic grain size of the soil [m].

According to DIN 4126 (2013), Equation (3) is valid for a pure suspension without sand if the pressure gradient f_{so} > 200 kN/m³. Substituting f_{so} = 200 kN/m³, the required yield point τ_F of the pure suspension can be derived from Equation (4).

$$required \; \tau_f = \frac{200.000 \cdot d_{10}}{3,5} N/m^2 \tag{4}$$

where τ_f = yield point of the slurry [kN/m²]; and d_{10} = characteristic grain size of the soil [m].

The calculated values of the "required yield point τ_f" according to Equation (4) are listed in Table 1. Even with a reduced particle size of the soil d_{10}, the value of the required yield point increases rapidly, e.g. it is a challenging task to put a yield point of 40.0 N/m² into practice

Table 1. Calculation of the "required yield point τ_f" [N/m²] of the bentonite suspension depending on the pressure gradient f_{so} = 200 kN/m² (Eq. 4).

d_{10} [mm]	0.1	0.2	0.3	0.4	0.5	0.6	0.7	0.8	0.9	1.0
τ_f [N/m²]	5.7	11.4	17.1	22.8	28.6	34.3	40.0	45.7	51.4	58.3

Table 2. Calculation of the "required yield point τ_f" [N/m²] of the bentonite suspension depending on the micro-stability at the tunnel face (Eq. 5).

d_{10} [mm]	0.1	0.2	0.3	0.4	0.5	0.6	0.7	0.8	0.9	1.0
τ_f [N/m²]	2.4	4.8	7.2	9.6	12.0	14.4	16.8	19.2	21.6	24.0

within the slurry circuit. For highly permeable or heterogeneous soils, additional requirements concerning the support medium need to be defined.

3.2 Required yield point of the suspension depending on micro-stability

From a local pressure transfer perspective, face stability is considered at the soil particle scale (micro-stability). Micro-stability is the stability of a single grain or group of grains that prevents them from falling out from the soil skeleton under gravity. According to DIN 4126 (2013), Equation (5) defines the "required yield point τ_F" to satisfy the criterion for micro-stability (fresh suspension).

$$\frac{d_{10}}{2 \cdot \eta_F} \cdot \frac{\gamma\varphi}{\tan \varphi'} \cdot (1 - n) \cdot (\gamma_B - \gamma_F) \cdot \gamma_G \leq \tau_f \tag{5}$$

where d_{10} = characteristic grain size of the soil [m]; n = soil porosity [-],γ_B = unit weight of soil grains [kN/m³], γ_G = partial safety factor for permanent load case in GZ1C acc. to DIN 1054 (=1.00) [-], γ_F = unit weight of fresh slurry [kN/m³], γ_φ = partial safety coefficient for drained soil within the status GZ1C in load case LF2 acc. to DIN 1054 (=1.15) [-]; φ = characteristic drained friction angle of the soil [°]; η_F = safety factor accounting for deviations in the yield point of suspensions (= 0,6) [-]; and τ_F = yield point of the slurry [kN/m²].

The calculated values of the "required yield point τ_f" according to Equation (5) are listed in Table 2 applying a soil porosity of n = 0.26 [-]; unit weight of the soil γ_B = 2.65 [kN/m³]; unit weight of the fresh slurry γ_F = 1.025 [kN/m³]; and a characteristic drained friction angle of the soil φ' = 37 [°].

Comparison of the calculated values of the yield point depending on the pressure gradient f_{so} (Tab. 1) and on the micro-stability (Tab. 2) shows that the criterion of micro-stability leads to significantly lower yield point values (< 50%). Based on micro-stability, they appoint lower limit values, which must be kept as an absolute minimum during the tunnelling works. In the practical use on site, a warning range for the yield point should be implemented.

4 SLURRY ASSESSMENT FOR DEMANDING GROUND CONDITIONS

In slurry pressure balance (SPB) shields, temporary support to the tunnel face is provided by the bentonite suspension under pressure. The viscosity of the suspension and the associated reduced flow capability lower the risk that support medium is lost through sudden uncontrolled escape into the surrounding ground. The slightly higher density of the regular bentonite suspension compared to that of water means that the overpressure of the support medium at the top of the face is low and blowouts are thus avoided (Anagnostou & Kovari 1994).

The stability of the tunnel face is maintained by a suspension with a proper rheology (yield point, viscosity) adapted to the prevailing geotechnical conditions and a sufficiently high support pressure in the excavation chamber. The latter is achieved through the capability of the support medium to develop a mechanism in the ground, which enables the support pressure to be transferred into the grain structure of the soil in the form of effective stresses.

For demanding ground conditions e.g. higher permeability, coarse to very coarse soil, heterogeneous particle size distribution, a support medium is required that develops a suitable pressure transfer mechanism reliably. For these in-situ ground, the presented types of pressure transfer (Fig. 2) are only applicable to a limited extend.

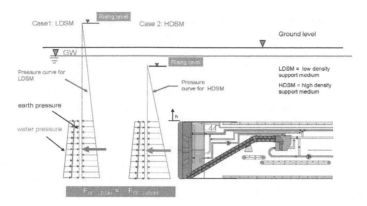

Figure 7. Pressure gradients of the support pressure in the excavation chamber for LDSM (low density support medium) and HDSM (high density support medium) according to (Straesser et. al. 2016).

Facing that challenge, the concepts of Low Density Support Medium (LDSM) and High Density Support Medium (HDSM) was developed in the laboratory of the Ruhr University Bochum (RUB) in cooperation with Herrenknecht AG for the world's first use of the VD-TBM (variable density) at Klang Valley MRT Project in Kuala Lumpur, Malaysia. This concept was also proven in slurry shield tunnelling for undercrossing the river Spree with very small overburden at project Metro U5 in Berlin, Germany.

The basic idea is to increase the density of the basic bentonite suspension up to a certain value. With increasing density, the viscosity and yield point of the support medium also increase. This enables or improves the support pressure transfer at the tunnel face. The theory is shown in Figure 7.

The pressure distribution of the earth and groundwater pressure increases linearly with depth (Fig. 7). This distribution is reproduced by the pressure gradients in the excavation chamber. Via compressible air cushion in the working chamber, the support pressure is adjusted slightly above the earth and groundwater pressure, including a safety factor. This positive pressure is also acting in vertical direction and must be compensated by the static imposed load from the cover depth. If this imposed load is raised by decreasing density or increased permeability of the ground, the risk of blowout or heave arises (Straesser et. al. 2016). According to the DAUB recommendation (DAUB 2016) the safety factor for the blowout verification should be > 1,1.

Applying a suspension with high density, the pressure gradient curve is flatter with altogether higher values (Fig. 7). The pressure ordinates in the crown are slightly shorter compared to LDSM. This improves the safety factor concerning incidents as blowout and soil heave. For a constant support pressure, the high density of the support fluid results in a smaller pressure ordinate in the crown. This value has a direct effect on the verification of safety against heave. A low value here increases the factor of safety or enables passing through a zone with shallow cover without additional safety measures such as ballasting (Straesser et. al. 2016).

Increasing the density through the addition of inert solids, the fines content in the suspension raises and the HDSM becomes altogether more viscous. As mentioned before, this viscosity is caused by internal friction between adjacent layers (Praetorius & Schösser 2017). The higher the viscosity, the stronger is the bonding between the molecules and the lower is the flow capability. Through the reduced flow ability of HDSM, the risk of a sudden loss of support medium through uncontrolled escape into an extensive network of pores or fissures is considerably reduced. At the same time, the fine particles in the HDSM help to mechanically block or clog smaller pores or fissures through successive accumulation (Praetorius & Schösser 2017).

With the increased viscosity of the HDSM due to the addition of fines, the yield point of the HDSM also rises. The higher the yield point of a suspension, the higher is the magnitude of the shear stresses that can be transferred to the grain structure of the soil (Praetorius & Schösser 2017). This achieves a stagnation of the HDSM in highly permeable ground, such as coarse soil (gravel) and highly karstic rock, at a much smaller penetration depth.

Beside the density and the yield point of the suspension as well as the characteristic grain size d_{10} of the soil, the slurry assessment considers also the actual flowing properties of the support medium. These include the penetration behaviour into the surrounding ground and the capability of transferring the support pressure applied in the excavation chamber to the tunnel face. For investigating the interaction of HSDM with certain ground condition, injection tests were developed and performed at the laboratory of Ruhr University Bochum. Aim of the test series was to evaluate the capability of the support medium to develop a mechanism with the ground for reliable support pressure transfer.

For the injection tests, a basic bentonite suspension (LDSM) with a solid content of 50 kg/ m^3 was modified. The density was increased from 1.025 t/m³ up to 1.7/1.8/1.9 t/m³. To intentionally enlarge the scope of application of the HDSM, special additional aggregates (sand with particle size 2-4 mm) were added to selected suspensions and then injected into coarse-grained soil. During the injection tests, the development of the support pressure and the penetration depth were monitored.

A supporting effect in the subsoil can only be achieved if the difference between the supporting pressure in the excavation chamber and the sum of the earth pressure and the groundwater pressure at the working face can be transferred as effective stress to the grain skeleton. For this purpose, the suspension must be able to build up a zone of lower permeability in the soil. The penetration behaviour of the suspension into the soil plays a decisive role here. Within the framework of the injection tests carried out, the coarse soils contain particle sizes of 16-25 mm and 25-63 mm with high permeability values.

Figure 8 shows an overview of the results of the injection tests in reference to HDSM density, the size and amount of additional sand and the particle size of the injected soil including the permeability coefficient k_f.

For the evaluation of the injection test series, three categories were laid down (Fig. 8):

(1) Small penetration depth < 10 cm transferring support pressure ≥ 2.0 bar
(2) Limited penetration depth < 50 cm transferring moderate support (≥ 0.8 bar)
(3) Theoretically infinite penetration depth without any support pressure transfer

Figure 8 shows that the density of the HDSM is the basic need for face support pressure, In the soil with particle size of 16-25 mm, a minimum density of 1,700 kg/m³ is necessary. With increasing amount of additional sand, the transferred support pressure is increased (HDSM 1,800 kg/m³ + additional sand 125g and 225g). Furthermore, the same modified suspension (HDSM 1,800 kg/m³ + additional sand 225g) transfers high support pressure (cat. 1) in soil 16-25 mm, in the soil 25-63 mm the support pressure decreases to cat. 2.

In summary, the challenge of the injection test is the adjustment of the suspension design in terms of optimal density of the HDSM and the right amount of additional sand to improve the support pressure transfer at the tunnel face. The extreme soil conditions were tested to investigate the potential of the HDSM slurry. For application on the construction site the HDSM should be assessed in the individual case whether, having regard to the financial and technical feasibility.

HDSM		Soil (Particle size acc. DIN 18123)	
	Density [kg/m³]	16-25 mm $k_f \gg 10\text{-}3$ m/s	25-63 mm $k_f \gg 10\text{-}3$ m/s
*additional Sand = 2-4 mm			
HDSM + additional Sand 225g	1,600	3	
HDSM + additional Sand 275g	1,600	3	
HDSM + additional Sand 175g	1,700	1	
HDSM + additional Sand 275g	1,700	1	
HDSM + additional Sand 275g	1,700		3
HDSM + additional Sand 125g	1,800	2	
HDSM + additional Sand 175g	1,800		2
HDSM + additional Sand 225g	1,800	1	
HDSM + additional Sand 225g	1,800		2

Figure 8. Overview of the results of injection tests in reference to HDSM density and particle size of the injected soil – classification in categories 1, 2, 3 (Straesser et. al. 2016).

5 CONCLUSION

With the introduction of the Variable Density-TBM, the terms Low Density Support Medium (LDSM) and High Density Support Medium (HDSM) are used to differentiate the support medium used. LDSM is a conventional bentonite suspension as used in slurry pressure balance (SPB) shield tunnelling. The properties of LDSM are defined through the physical parameter *density* and the rheological parameters *yield point* and *viscosity*. A HDSM is a support medium with a significantly increased density. It is based on a standard bentonite suspension, whose density has been intentionally increased by the addition of inert, chemically inactive solid materials. The quality of a HDSM is also evaluated with the parameters density, yield point and viscosity. The interaction with the soil needs further investigations to evaluate the transferable support pressure and the in-situ penetration depth.

REFERENCES

American Petroleum Institute: *API Recommended Practice 13B-1 (1997) Standard Procedure for Field testing Water based Drilling Fluids*. September.

Anagnostou G. & Kovari, K. (1994), The Face stability in Slurry-shield-driven Tunnels. *Tunnelling and Underground Space Technology* No. 2 1994, pp. 165–174.

DAUB (2016) Recommendations for Face Support Pressure Calculations for Shield Tunnelling in Soft Ground. *DAUB - Deutscher Ausschuss für unterirdisches Bauen e. V. (German Tunnelling Committee)*, October 2016, download at: www.daub-ita.de

DIN 4126 (2013) *German Standard Nachweis der Standsicherheit von Schlitzwänden (Stability analysis of diaphragm walls)*, September.

DIN 4127 (2014) *German Standard: Erd- und Grundbau – Prüfverfahren für Stützflüssigkeiten im Schlitzwandbau und für deren Ausgangsstoffe (Earthworks and foundation engineering – Test methods for supporting fluids used in the construction of diaphragm walls and their constituent products)*, Februar.

DIN 53019 (2008) *German Standard: Viskosimetrie – Messung der dynamischen Viskosität newtonscher Flüssigkeiten mit Rotationsviskosimeter, Teil 1: Grundlagen (Viscometry – Measurement of viscosities and flow curves by means of rotational viscometers – Part 1: Principles)*. September.

Jancsecz S., Steiner W. (1994), Face support for a large mix-shield in heterogeneous ground conditions, in: *Proc. Tunnelling '94*, pp.531–550, Chapman and Hall, London.

Kilchert, M.; Karstedt, J. (1984), Schlitzwände als Trag- und Dichtungswände, *Band 2: Standsicherheitsberechnungen von Schlitzwänden nach DIN 4126*. Bauverlag, Wiesbaden.

Morgenstern N., Amir-Tahmasseb I. (1965), The stability of a slurry trench in cohesionless soils, *Geotechnique*, 1965; Vol. 15 No. 4, 387–395.

Müller-Kirchenbauer H. (1972) Stability of slurry trenches, *Proceedings of 5th European conference on soil mechanics and foundation engineering*, Madrid, 1972, 543–553.

Praetorius, S.; Schösser, B. (2017) *Bentonite handbook – lubrication in pipe jacking*. Verlag Ernst & Sohn, Berlin, ISBN: 978-3-433-03137-7.

Straesser, M.; Klados, G.; Thewes, M.; Schoesser, B. (2016) Entwicklung des LDSM und HDSM Konzepts für die Variable Density-TBM/Development of LDSM and HDSM Concept for Variable Density TBMs, *Tunnel* 7, pp 18–37.

Schoesser, B. (2004) The transfer of tangential stress at circumference of pipes which have been laid by Pipe-Jacking in non-cohesive soil, in: *Proceedings 22nd International No-Dig Conference and Exhibition*, Hamburg, Nov. 15–17.

Schulze, B.; Brauns, J.; Schwalm, I. (1991): Neuartiges Baustellenmessgerät zur Bestimmung der Fließgrenze von Suspensionen. *Geotechnik Zeitschrift*.

Thewes, M. (2009) Bentonite slurry shield machines *In:* Pérez de Ágreda, Eduardo Alonso; Álvarez de Toledo, Marcos Arroyo *(Eds.)*: *Operación y mantenimiento de escudos: presente y futuro*. Barcelona, Universitat politécnica de Catalunya, 2009, pp. 41–68.

Weiss F., (1967) Die Standfestigkeit flüssigkeitsgestützter Erdwände, PhD thesis, Technische Universität, Berlin, 1967.

Zizka, Z., Schoesser, B., Thewes, M., Schanz, T. (2018) Slurry shield tunneling: new methodology for simplified prediction of increased pore pressures resulting from slurry infiltration at the tunnel face under cyclic excavation processes, *International Journal of Civil Engineering, Special Issue "Current trends and challenges in subsurface engineering"*, 15 (4), p. 387. DOI: 10.1007/s40999-018-0303-2

*Tunnels and Underground Cities: Engineering and Innovation meet Archaeology,
Architecture and Art, Volume 6: Innovation in underground engineering,
materials and equipment - Part 2 – Peila, Viggiani & Celestino (Eds)
© 2020 Taylor & Francis Group, London, ISBN 978-0-367-46871-2*

Development of a test setup for the simulation of the annular gap grouting on a semi technical scale

C. Schulte-Schrepping & R. Breitenbücher
Institute for Building Materials, Ruhr-University Bochum, Bochum, Germany

ABSTRACT: In mechanized tunneling, an early bedding of the recently installed segment ring is a constructive necessity and of considerable importance for the durability of the whole structure. Alongside the actual material technology of the grout, procedural aspects like advance rate of the tunnel boring machine, therefore the interacting parameters like grout volume per unit of time and gelation and hardening of the grout also play an important role for a successful injection process. Due to adjustments of the grout composition during the construction the flow properties before and especially after activation will change. In order to observe the flow behavior and the interactions between the grout and soil a large-scale test setup was developed to simulate the annular gap grouting under realistic boundary conditions. Herein, it is possible to observe the flow behavior and to examine the material for its homogeneity and strength characteristics after grouting.

1 INTRODUCTION

The complete and homogeneous filling of the annular gap ensures the position stability of the tunnel tube, prevents it from floating and reduces surface settlements. The development of a suitable grout currently is carried out on an empirical basis and the requirements of the fresh and solid grout properties are basically project-specific and sometimes vary in large ranges. This affects in particular the required strength and stiffness development. At the material-technological level, in the case of two-component grouts, these properties are controlled by the content and composition of the binder and by the activator content. As part of the formulation, the so-called gel time of the material changes along with it. This is defined as the time after activation until the material reaches the gel state, which should take place after about 5 to 15 seconds after mixing the two components (Hashimoto, 2004; Pelizza, Peila, Sorge, Cigniti, 2012). In the event that a rapid stiffness development of the material is required in the annular gap and this is achieved by increasing the binder content or its reactivity, this also accelerates the gel time. In the worst case, this could lead to a clogging of the pilaster strip or else to a blockage at the injection opening in the annular gap. A homogeneous backfilling is then not guaranteed. When using one-component grouts the stiffness development is regulated next to the material composition by the grouting pressure. Due to the pressure induced dewatering of the material in the surrounding soil and therefore the consolidation, the material develops shear strength and stiffness. The main part here is next to the composition of the grout the permeability of the surrounding soil.

To evaluate the flow behavior, to examine the material parameters like shear strength and to investigate the interaction between soil and grout a large scale test setup was developed within the collaborative research center 837 at the Institute for Building Materials at the Ruhr-University Bochum. In addition, the test setup should prove the suitability for a grouting process of innovative more-component grouts, like a solidification by superabsorbent polymers. In this paper, the development of the test setup and also the results of grouting tests of typical and modified grouts are shown.

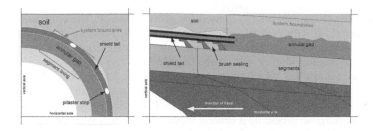

Figure 1.   Cross section of a segmental lining and differentiation of the system boundaries of the test setup.

2   CONCEPT OF THE TEST SETUP AND SYSTEM BOUNDARIES

The test setup forms in cross section the shield tail with one pilaster strip, the shield tail brush sealing, the annular gap and the surrounding soil (Figure 1). For simulation of the annular gap grouting as realistic as possible, the system sizes of actual tunnel construction projects were adapted within the framework of the planning and development. The average annular gap height is about 13 to 18 cm (Thewes & Budach, 2009), 15 cm were chosen. A typical segment width in the longitudinal direction of the tunnel is approximately 1.5 m. These geometric parameters were adapted and the annular gap width was chosen to 0.6 m. Thus, the annulus has the shape of a cuboid and the maximum grout volume is approx. 135 liters. The height of the simulated soil was chosen to 30 cm in order to be able to record the dewatering over the depth as part of the pressurized grouting of one-component grouts and to observe in general the interaction between grout and soil.

The grouting time was derived from known advance rates of real tunnel projects. The maximum advance speed was chosen to 90 mmm/min in compliance with the Sao Paulo Metro Project, in which an advance speed of 80 mm/min was achievable (Pellegrini & Peruzza, 2009). As lower limit value, a minimum advance speed of the tunnel boring machine was chosen to 30 mm/min.

3   CONSTRUCTIVE DESIGN

The whole test setup can be divided into six subassemblies (Figure 2): bucket (1, soil), slide (2, simulated TVM or its shield tail), the annular gap (3), the nozzle (4, mixing of component A and B), the hydraulic system including the abutment (5) and the cover (6). In Figure 2 an overall view on the left and a longitudinal cut through the design drawing is shown on the right.

The slide (Figure 2, pos. 2) simulates the tunnel boring machine or the shield tail with one pilaster strip. This is divided into a front and rear subassembly. In addition, at the rear of the slide the connection to the hydraulic cylinder is mounted. The front slide fulfills several tasks.

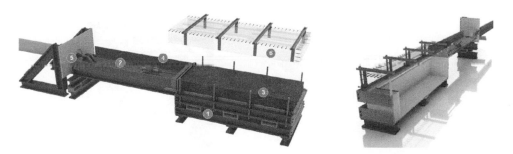

Figure 2.   Left: Overall view of the grouting device. Right: Longitudinal cut through the grouting device.

The central opening (Figure 3, pos. 1) is used for grout injection, the remaining four openings (pos. 2) can be provided with pressure sensors to control the pressure during the grouting process. The sensors are screw-in transmitters, which are specially designed for pressure detection of viscous or pasty media. One of the left-hand positions can also be provided with a safety valve when pressure is applied, which opens automatically at 2.5 bar. In order to ensure that the grouting material is not pressed out of the simulated annulus gap during the pressurized filling, two brush seals (Figure 3, right, pos. 1) are arranged. The space between (Figure 3, right, pos. 2) is filled with multi-purpose grease by a central lubrication system over nine ports, which creates an additional sealing level and also reduces the friction to the cover on all sides.

The mixing of two-component grouts is done by a specially made nozzle, which consists of components of a shotcrete nozzle. The nozzle can be divided into four components. The centerpiece is the injection unit for the activator (Figure 4, left, pos. 1). This unit has two quick couplings. One of these couplings is used for component B, the other one serves for the water connection for cleaning the injection channels and the entire nozzle. In addition to this unit, there are two coupling pieces (pos. 2), which allow the connection of a DN 50 coupling and a two inch screw thread. Furthermore, the nozzle has a reduction unit (pos. 3), which allows the attachment of static mixers with different diameters, lengths and variable arranged mixing tools. Finally, the static mixer (pos. 4) ensures homogeneous mixing of the two components. In Figure 4, right the actual injection of the activator in the nozzle is shown.

The arrangement of the grout pipe, the grease channels and the sensor cables in the slide is shown in Figure 5. Component A is transported via a DN 50 pipe (pos.1). There are two alternative grout delivery options available. On the one hand, the component A can be conveyed by an eccentric screw pump. Alternatively, in particular for grouts with aggregates, a pressure vessel with a capacity of 300 liters is used. Here, the component A is impinged by air pressure. Component B reaches the nozzle via a 20 mm pipe (pos. 2). The activator is conveyed by a horizontally arranged eccentric screw pump. The grease required for the seal is conveyed to the front slide through nine channels (pos. 3). The grease channels are connected to a central lubrication system via quick-release couplings so that there is a fixed line system in the slide. The data cables (pos. 4) of the sensors are arranged parallel to the grease channels. In Figure 5 in the middle the fully mounted test setup and on the right a view into the simulated annular gap is shown.

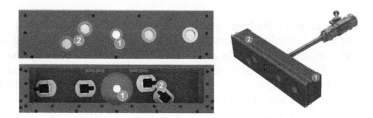

Figure 3. Detailed view of the front slide.

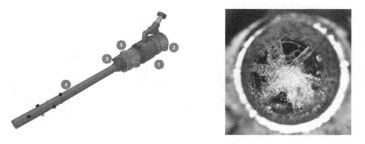

Figure 4. Left: Fully assembled nozzle. Right: Activator injection.

Figure 5.   Left: Cable/Pipe management, Middle: Built-up device, Right: View into the annular gap.

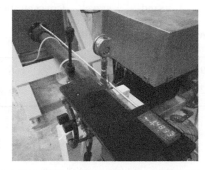

Figure 6.   Left: Coupling slide-cylinder. Middle: Control unit and process meter. Right: Tackle way receiver.

The movement of the slide takes place by a hydraulic cylinder. This cylinder is connected to the carriage by a joint head, which ensures a torque-free coupling. By appropriate valves, the advance rate can be controlled continuously. The current speed is displayed by a process meter, which is connected to a speed sensor based on a tackle way receiver. The cylinder is designed for a maximum speed of 2.0 mm/s, which corresponds to a maximal advance rate of 120 mm/min. The connection between hydraulic cylinder and slide, the process meter and the tackle way receiver are shown in Figure 6

4   ANNULAR GAP GROUTING – RESULTS

4.1   *Used annular gap grouts*

In the following chapter, excerpts of the grouting tests of a modified one- and a typical two-component grout are shown. In Table 1 the used two-component grout composition and practice-orientated minimum and maximum levels (Pelizza, et al., 2010; Bäppler, 2006) of typical composites are given. The two-component grout is an in-house development in order to function as a basic formulation for further material variations and as comparison regarding the hardening and stiffening behavior after activation (Schulte-Schrepping, Youn-Cale, & Breiten bücher, 2018).

In Table 2 an adapted one-component grout "mod. B-60" compared to a typical composition of semi-inert grout (Thewes & Budach, 2009) is shown. The modification was done, because this grout forms the basis for a study within the collaborative research center 837, in which the stiffening of the grout is forced by an internal dewatering due to the addition of superabsorbent polymers. Therefore, the water content was reduced and by the addition of superplasticizer the fluidity and pumpability was ensured.

Table 1. Two-component grout composition and practice-orientated minima and maxima.

| | Composites in kg/m³ | | | | | |
	water kg	bentonite kg	cement kg	fly ash kg	retarder m.-%/c.	activator l
Practice-orientated minimum	730	25	85	0	0,2	37
Used two-component grout	836	30	250	0	2,5	66
Practice-orientated minimum	870	60	482	150	5,2	100

Table 2. Modified one-component grout compared to a typical semi inert grout.

| | | | aggregates | | | bentonite-slurry | | |
| | Cement | water | sand | sand | gravel | (concentration 6 %) | fly ash | Super plasticizer |
	kg/m³	kg/m³	0-1 mm kg/m³	0-2 mm kg/m³	2-8 mm kg/m³	kg/m³	kg/m³	kg/m³
B-60	60	164	169	674	454	166	328	0
mod. B-60	60	68	272	735	598	166	328	2,4

4.2 One-component grout

In the tests with the modified one component grout different grouting pressures of about 0.5, 1.0 and 1.5 bar were investigated. For the tests under pressure, a closed cover made of steel is used. The advance rate of the slide was set in the test with 0.5 bar approx. 0.5 mm/s which corresponds to a typical advance rate of a TBM of 30 mm/min. When using 1.0 and 1.5 bar as grouting pressure, the advance rate was set to 1.0 mm/sec e.g. 60 mm/min. Due to the use of aggregates in this grout a nozzle with a diameter of two inches without any mixing tools was used. To observe the flow behavior after grouting, approx. one third of the grout volume was colored with high intensity color. The surrounding soil had a permeability coefficient of $k_f = 2.3 * 10^{-4}$ m/s (permeable). The demolded grout, which has been pressed with 1.0 bar, is shown in Figure 7.

The color differences show, that the grout is distributed annularly from the nozzle (Figure 7, right). Next to the observation of the flow behavior, the shear strength was tested with a shear vane according to DIN 4094-4:2002-01 for all three tested grouting pressures. The average shear strengths (evenly distributed over the total samples, per sample 5 measuring points) for the grout after the grouting process at an age of two hours for the three different grouting pressures are shown in Figure 8.

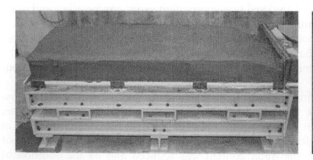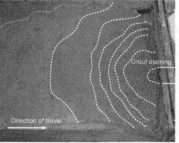

Figure 7. Left: Pressed one component grout, Right: Flow behavior of the grout.

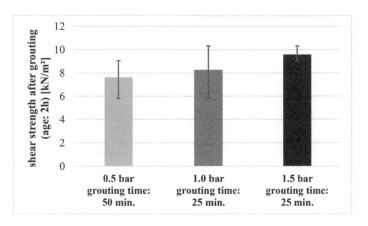

Figure 8.   Shear strength according to the grouting pressure.

It can be seen, that the shear strengths correlates non-linearly with an increasing grouting pressure. A three-fold increase in the injection pressure causes an increase in shear strength of approx. 25% based on the averaged values. The results are lower than those of the investigations in (Youn, 2016). Herein, shear strengths for the classical semi-active one-component grout B-60 (see Table 2) after pressurized dewatering for 30 minutes at a pressure of 2.0 bar in the amount of approx. 30 kN/m² were obtained. The fact, that the modified grout "mod. B60" has not dewatered in the surrounding soil and therefore does not consolidate in the same manner as the grouts in (Youn, 2016), can be assumed as the reason for this discrepancy. Also, in (Youn, 2016), there was no soil arranged under the grout in the test device. In Figure 9, a cut surface of the injected grout (age: 2 hours) is shown on the left and a detail of the boundary layer between the grout and the soil on the right.

It gets obvious from Figure 9, that due to the applied injection pressure a water transport to the bottom within the material takes place and not into the surrounding soil. The mortar "mod B-60" shown here has been modified to such an extent, that for the aspired activation by superabsorbent polymers (component B) an optimized water content is present in the over-

Figure 9.   Left: Pressure induced water transport within the grout, Right: Detail view of the boundary layer between grout and soil.

Figure 10.   Left: Contact zone of grout and sealing during an injection process. Right: Demolded grout.

Figure 11.   Injection of high intensity color in a running grouting process of a two-component grout and visualization of the flow directions.

all system, taking into account a sufficient workability of the grout (component A).The experiments on the grouting behavior of the grout "mod. B-60" show, that with the soil chosen here in combination with the mix design of the grout the desired boundary conditions for the solidification with superabsorbent polymers, primarily no pressure induced dewatering in the surrounding soil, are given. The field of application for this annular gap grout concept are less permeable soils, which do not allow a dewatering of the grout, which in turn does not allow the use of classical one-component grouts.

4.3   Two-component grout

As a basis for further material modification a typical two-component grout (Table 1) was chosen for first grouting test. The surrounding soil had a permeability coefficient of $k_f = 2.3 * 10^{-4}$ m/s (permeable). In this case, the grouting was done without pressure. This was made possible by elongated openings in the cover. The cover is made of Perspex, so that the flow behavior of the grout can be observed in live view. Figure 10 shows the contact zone of the grout and the brush sealing (left) and the completely pressed two-component grout after demolding (right).

The shear strength was then determined on the pressed grout. Here a very good agreement with the laboratory samples could be determined. The shear strengths of the compressed grout were about approx. 23 kN/m^2 for between 40 and 50 minutes and 52 kN/m^2 90 minutes after activation. The shear strength of the laboratory samples after 50 minutes were about 20 kN/m^2 and after 90 minutes approx. 48 kN/m^2. In addition, the pressed material showed no inhomogeneity's in form of insufficiently solidified areas. In a next step, through a separate opening in the grout pipe, during another grouting test at two different points in time a high intensity color was injected to observe the flow behavior of the grout while grouting. The results are shown in Figure 11. It can be seen that the grout spreads homogeneously in both, the transverse and longitudinal direction.

5 CONCLUSION

The presented test setup was built to prove the suitable of grouts for the annular gap grouting in general but also for new developed grouts on an innovative basis, like the activation or stiffening due to the use of superabsorbent polymers. Therefore, different control parameter of a real tunnel construction can be simulated and typical surrounding soils with different permeability's can be arranged. In this paper, the development of the device and grouting tests of a typical two-component grout and of a slightly modified one-component grout are shown.

By now these tests show, that the grouting process with the developed device in combination with the in each case adapted nozzle can be simulated realistically in the laboratory. The flow behavior, as an important characteristic of such grouts can be detected and the material properties like shear strength or uniformity of the pressed material can be examined shortly after grouting.

In conclusion, with this test device, a possibility is created, which allows, above the known test methods, to prove the suitability of different kinds of grouts within the development process of annular gap grouts.

ACKNOWLEDGEMENT

Financial support was provided by the German Science Foundation (DFG) in the framework of project B3 of the Collaborative Research Center SFB 837. This support is gratefully acknowledged.

REFERENCES

Bäppler, K. 2006. *Development of a two-component dynamic grouting system for tailskin injected backfilling of the annular gap for segmental concrete lining in shielded tunnel boring machines.* Golden, Colorado: Dissertation.

Hashimoto, T. B. 2004. Simultaneous Backfill Grouting, Pressure Development in. *Tunnelling and Underground Space Technology. Underground Space For Suitainable Urban Development. Proceedings Of The 30th ITA-AITES World Tunnel Congress, SIngapore, 22 - 27 MAY 2004.*

Pelizza, S., Peila, D., Borio, L., Dal Negro, E., Schulkins, R.,Boscaro, A. 2010. Analysis of the Performance of Two Component Back-filling Grout in Tunnel Boring Maschines Operating under Face Pressure. *ITA-AITES World Tunnel Congress 2010, Vancouver (CA), 14 to 20 May, 2010.*

Pelizza, S., Peila, D., Sorge, R.,Cigniti, F. 2012. Back-fill grout with two component mix in EPB tunneling to minimize surface settlements: Rome Metro—Line C case history. In G. Viggiani, *Geotechnical Aspects of Underground Construction in Soft Ground – Viggiani (ed)* (S. 291–299). London: Taylor & Francis Group.

Pellegrini, L.,Peruzza, P. 2009. *Sao Paulo Metro Project – Control Of Settlements In Variable Soil Conditions Through EPB Pressure And Bicomponent Backfill Grout.* Las Vegas, Nevada.

Schulte-Schrepping, C., Youn-Cale, B.-Y.Breitenbücher, R. 2018. Strength Development of Two-Component Grouts for Annular Gap Grouting. *Tunnel 3/2018*, 24–33.

Thewes, M.,Budach, C. (2009). Mörtel imTunnelbau - Stand der Technik und aktuelle Entwicklungen. *BauPortal* (12/2009).

Youn, B.-Y. 2016. Untersuchungen zum Entwässerungsverhalten und zur Scherfestigkeitsentwicklung von einkomponentigen Ringspaltmörteln im Tunnelbau. Bochum: Shaker Verlag, Schriftenreihe des Instituts für Konstruktiven Ingenieurbau. Heft 2016–04.

Tunnels and Underground Cities: Engineering and Innovation meet Archaeology,
Architecture and Art, Volume 6: Innovation in underground engineering,
materials and equipment - Part 2 – Peila, Viggiani & Celestino (Eds)
© 2020 Taylor & Francis Group, London, ISBN 978-0-367-46871-2

Two-component grouts with alkali-activated binders

C. Schulte-Schrepping & R. Breitenbücher
Institute for Building Materials, Ruhr-University Bochum, Bochum, Germany

ABSTRACT: In mechanized tunneling, the annular gap between the segmental lining and the surrounding soil must be filled with a suitable grouting material simultaneously. When using two-component grouts as grouting material with a composition adapted to the geological and structural conditions, the early bedding of the recently installed segment ring but also a permanent position stability of the tunnel can be reliably achieved. In the compilation of two-component grouts, by now exclusively on an empirical basis, primarily hydraulic binders are used. In order to increase the economics of such systems, it is obvious to replace the hydraulic fraction in the binder by suitable substituents without adverse consequences. As a result of systematic investigations, the use of latent-hydraulic blast furnace slag has proved to be suitable in the system of activated annular gap grouts. In combination with different activator configurations, exchange rates of up to 70% of the cement are possible.

1 INTRODUCTION

The complete and homogeneous filling of the annular gap ensures the position stability of the tunnel tube, prevents it from floating and reduces surface settlements. The development of a suitable grout currently is carried out on an empirical basis and the requirements of the fresh and solid grout properties are basically project-specific and sometimes vary in large ranges. This affects in particular the required strength and stiffness development. At the material-technological level, in the case of two-component grouts, these properties are primarily controlled by the content and composition of the binder and by the activator content. Herein the knowledge about the strength development, in particular in the young age of the grout is significant, to ensure a permanent position stability of the tunnel. In practice, however, there are occasional damages to the segmental lining, which show up in the form of offsets in the segment joints or as bending cracks on the tunnel lining. Usually, the back filling of the annular gap (Figure 1) takes place simultaneous to the advance of the tunnel boring machine through pilaster strips in the shield tail. In the case of two-component grouts, the so-called TAC system is used, in which the two components are mixed together shortly before they enter the annular gap (Tac Corporation, 2018) (Figure 1, left). In the case of typical cementitious two-component grouts, after the two components are mixed, the material gels within a few seconds and develops a mechanical strength within the first hours. The gelling of the two-component grout is necessary in water-bearing soils to prevent excessive washing out of the grout. The required strength development is project-specific and depends on the structural boundary conditions such as the advance speed and the ring construction time. Possible control variables in order to be able to adapt the grout to varying boundary conditions as well as fundamental systematic factors influencing the strength development but also their interactions with the workability of component are not available by now.

As part of the Collaborative Research Center 837 "Interaction Modeling in Mechanized Tunneling" funded by the German Research Foundation (DFG), the Institute for Building Materials at the Ruhr University Bochum is developing innovative multi-component grouts. In particular, modifications of the used binders and interactions with different accelerators are

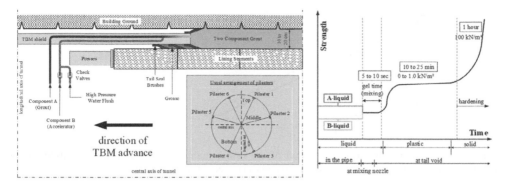

Figure 1. Left: Schematic illustration of the annular gap grouting and typical pilaster strip arrangement; Right: Required strength development of two-component grouts (Hashimoto, 2004).

considered and also innovative approaches such as the use of superabsorbent polymers as component B and, as a result of their water absorption capacity, induced gelation and solidification of the mortar are investigated. Based on a typical hydraulic reference mixture, the focused alkaline activation of primarily latent-hydraulic, but also pozzolanic additives and their applicability in the system "two-component grout" and their influence on the strength development is evaluated. Excerpts from the results are presented in this publication

2 ALKALINE ACTIVATION - BINDER AND ACTIVATOR

For the alkaline activation, a wide range of composite materials are suitable (Table 1). Due to the high amounts of amorphous SiO_2 and Al_2O_3 as activating main components for alkali activated binders mainly secondary materials from the steel production (blast furnace slag) or coal combustion (fly ash) and calcined kaolins (metakaolin) are used. Similarly, mixtures of the mentioned materials are suitable for the alkaline activation. The reactive solids can be combined with a wide range of suitable alkaline activators, which differ in the level of their pH value (see Table 1). These include alkali hydroxides, alkali silicates (water glass) and alkali salts (e.g., carbonates and sulfates). Usually, aqueous solutions of water glass or a mixture of alkaline solutions and water glass are used (Buchwald, 2012; Herrman, König, & Dehn, 2015).

Starting with Portland cement clinker to blast furnace slag and fly ash to metakaolin, the calcium oxide (CaO) content in the material is steadily reduced (Figure 2). In order to be able to achieve sufficient strengths with decreasing CaO content, the alkalinity of the activator, which is characterized by the M2O content must be increased (Buchwald, 2012; Tänzer & Stephan, 2011).

In general, a distinction is made between calcium scarce alkaline activated binders with a CaO-content of at most 10% by mass and calcium-rich binders with a CaO-content of at least 10% by mass (Table 2). Low-calcium binders, which are also called geopolymers, have a different reaction mechanism than calcium-rich binders. At a CaO-content below 10% by mass, a

Table 1. Combinable mineral materials and suitable activators.

Mineral materials	Suitable activators
Blast furnace slag	Alkali hydroxide (MOH)
Fly ash	Alkali silicate (M2O*rSiO2)
Metakaolin	Alkali carbonate (M2CO3)
Pozzolan	Alkali sulfate (M2SO4)

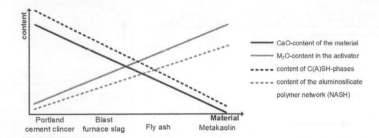

Figure 2. Correlation between CaO-content of the source materials, activator concentration and the hydration products of alkaline-activated binders modified according to (Škvára, 2007).

Table 2. Classification of alkaline activated binders according to (Herrman, König, & Dehn, 2015).

characteristics	CaO-scarce binders	CaO-rich binders	
CaO-content [M.-%/Binder]	0 - 10	10 - 40	> 40
Dominating reaction product	NASH	Mixed phases of NASH and C (A)SH	C(A)SH
Typical materials	Calcium scarce fly ash, metakaolin	High calcium fly ash	Blast furnace slag

three-dimensional alumino silicate polymer network (NASH) is formed as a result of alkaline activation. As the CaO-content (10 to 40% by mass) increases, mixing phases from this alumino silicate polymer network and hydrate phases in the form of CSH phases with aluminum incorporation (C(A)SH phases) are formed. From a CaO-content above 40 m.-%, the C(A)SH phases dominate the strength-forming structure as a result of an alkaline activation. Therefore, in the case of calcium-rich binders, due to the different dominant reaction products, further differentiation into the two subgroups occurs with the CaO-contents of 10 to 40% by mass and over 40% by mass (Table 2) (Buchwald, 2012; Herrman, König, & Dehn, 2015).

The reaction products and the associated performance of alkali activated binders are highly dependent on the nature and concentration of the used activator. In the concrete industry, aqueous solutions of alkali silicates or a mixture of alkali silicates and alkali hydroxides are usually used (Buchwald, 2012). In most cases, for reasons of economy, sodium or potassium alkalis are used as the starting material for these activator types (Provis & van Deventer, 2014)

From a purely chemical point of view, the two activator types differ based on their SiO_2 content. While alkali hydroxides consist only of H_2O and M_2O (M as placeholders for a cation such as Na or K), alkali silicates in addition to H_2O and M_2O have a certain amount of amorphous SiO_2. Depending on the chemical composition, the pH value also varies. The pH value is generally between 13 and 14 for alkali hydroxides and between 11 and 13.5 for alkali silicates (Provis & van Deventer, 2014). The composition of alkali silicates can generally be described by the formula "$M_2O * n SiO_2$". In this case, n corresponds to the molar ratio which represents the number of silicon dioxide moles per alkali metal oxide mol. This molar ratio (MR) is to be determined by means of the weight ratio (WR), which corresponds to the quotient of SiO_2 to M_2O [both in %], and a subsequent multiplication by a factor (Sodium silicate: $MR_{Na} = 1{,}032 * WR$, Potassium silicate: $MR_K = 1{,}566 * WR$) for the molar conversion.

By mixing an alkali metal hydroxide solution and a water glass, the MR of the mixture can be adjusted as needed depending on the proportions of the two activators. With a decrease of the molar ratio the alkalinity of the activator increases, since more alkalis are present. But at the same time the content of SiO_2 in the activator decreases as well. Thus, for an activator consisting only of an alkaline solution, the MR is 0.0, because there is no amorphous silica in the solution.

3 USE OF THE ALKALINE ACTIVATION FOR TWO-COMPONENT GROUTS

To function as a basic formulation for the alkaline activation and to ensure the comparability of the new developments a usual two-component grouts was developed. For further details see (Schulte-Schrepping, Youn-Cale, & Breitenbücher, 2018). In Table 3 this two-component grout composition and practice-orientated minimum and maximum levels (Pelizza, et al., 2010; Bäppler, 2006) of typical composites are given. In the further shown investigations the water content, the bentonite content and the retarding agent correlated to the cement content and also the volume of the activator were kept constant.

Within the investigations, different activator configurations, composed of sodium and potassium water glasses with the ratios 100/0 (MR 3.7), 70/30 (MR: 3.0) and 50/50 (MR: 2.6) were used. The application of the targeted alkaline activation of latent hydraulic or pozzolanic additives is limited by the central requirement of a rapid gelation of the two-component grout and a sufficient early strength development after addition of the activator (component B). A typical requirement of the compressive strength after 24 hours of two-component grouts is about 0.5 MPa (Pellegrini & Peruzza, 2009). In addition, two-component-grouts have to gel within a few seconds after mixing of the two components in the annular gap. Pretests have shown, that these requirements could not be achieved with only low-calcium binders, so that the focus in the further course of the investigations was set on hybrid binders, as described in (Provis & van Deventer, 2014) with a small amount of cement combined with binders like granulated blast furnace slag or fly ash (Figure 3).

It can be seen from the results in Figure 3, that a higher CaO content in the binder causes an increase in the compressive strength after 24 hours. Next to this, the chemical composition of the activator influences significantly the strength development. Therefore, an activator with an MR of 3.7 (clear sodium water glass) leads to a higher compressive strength. This is caused due to the immediately reaction between the water glass and the CaO – primarily of the cement – and the presence of free silicon dioxide in the pore solution. With water in the system, a huge amount of hydrate phases can be built in a short time period. With lower CaO-content in the system, the amount of reaction products within 24 hours decreases. The decreasing amount of SiO_2 when using activators with an MR of 3.0 and 2.6 is also the reason for the decreased strength development.

For the further investigations, the granulated slag 2 was used, as its physical properties correspond to typical available granulated blast furnace slags. The variations in the composition of the activator or the proportions of sodium and potassium water glasses only have a strengthening effect on a cement-slag mixture at a later time, as shown in Figure 4. The first number in the sample designations stands for the cement content in the binder (total binder quantity: 319 kg/m³).

The proportional use of blast furnace slag in the binder in combination with pure sodium water glass compared to the mix design M0 (Table 3) shows, that higher strengths are achieved with a cement content of 160 kg combined with 239 kg blast furnace slag from a time

Table 3. Two-component grout composition and practice-orientated minima and maxima.

	Composites in kg/m³					
	water [kg]	bentonite [kg]	cement [kg]	fly ash [kg]	retarder [M.-%/Z.]	activator [l]
Practice-orientated minimum	730	25	85	0	0,2	37
Used two-component grout - M0	836	30	250	0	2,5	66
Practice-orientated minimum	870	60	482	150	5,2	100

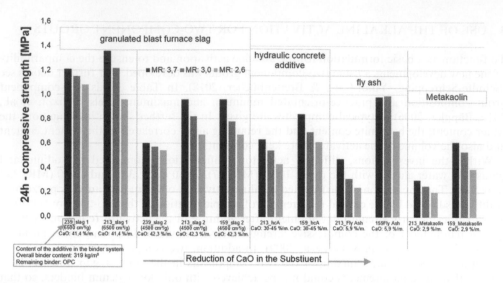

Figure 3. 24h-compressive strength as a function of the additive and chemical composition of the activator.

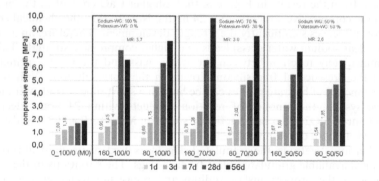

Figure 4. Influence of the activator composition on the strength development of modified binders.

of just one day after activation. This suggests that within the first 24 hours the blast furnace slag already contributes to the strength development due to its incipient hydration.

After 7 days, the use of granulated slag in the binder leads to a significant increase in compressive strength. The proportionate use of potassium water glass also causes an increase in strength in grouts with blast furnace slag. This continues until 28 days, with increasing age, the potassium water glass leads to no strength increase compared to the pure use of sodium water glass (100/0). This increase from 7 days is in part due to the fact, that potassium water glasses solves the reactive constituents of the granulated slag due to their higher alkalinity more effectively compared to sodium water glass. At the same time, and this factor limits the substitution rates, the 24 hours compressive strength decreases, when the sodium water glass was replaced by potassium water glass. Within the framework of the binder configurations shown here, the cement exchange by blast furnace slag was up to 50% to 75%. From this, it can be concluded that the potassium water glass accelerates the formation of hydrate phases of the slag within a limited time frame, however, this effect decreases with increasing age (> 28 days).

The higher strengths of the slag-containing grouts can also be clearly identified on the basis of scanning electron microscope images (Figure 5). The structure of the alkaline activated mortar is substantially denser, which in addition to the reactivity and the geometries of the hydrate phases formed also may be due to a higher packing density.

In addition, the shear strengths in the structurally relevant time window over a period of 30 to 360 minutes and additionally after 24 hours after mixing of component A and B and the static Modulus of Elasticity were determined for appropriate grouts.

Analogous to the compressive strengths, the results of the shear strength development show that sodium silicates in combination with a certain amount of cement in the binder is of considerable importance for ensuring the requirements of early shear strength, in this case 30 minutes after back filling (Figure 6, left). Thus, the mortar 80_100/0 (80 kg OPC, 239 kg slag, pure sodium water glass) results in comparison to the mortar with the same component A, but a certain proportion of potassium water glass in the activator approx. 70% higher shear strengths after 30 minutes (80_100/0: 3.2 kN/m², 80_70/30: 1.9 kN/m²). After 4 hours, the grout 80_100/0 shows an increase in shear strength, whereas mortar 80_70/30 still stagnates until 360 minutes after activation. By doubling the cement content and using pure sodium water glass (MR: 3.7), shear strengths of around 5.8 kN/m² are achieved after just 30 minutes. It can therefore be

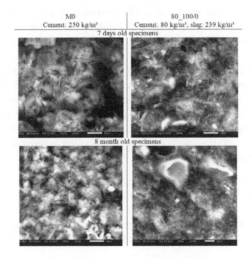

Figure 5.   Scanning electron microscope images of two different grouts after 7 days and 8 month.

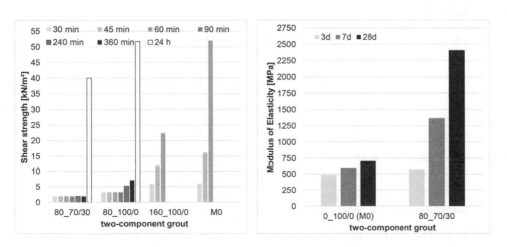

Figure 6.   Left: Shear strength development, Right: Development of the modulus of elasticity of a slag containing grout compared to the basis formulation M0.

deduced from the sum of the strength tests that with a maximum cement content of 80 kg (about 25% of the total binder) in combination with blast furnace slag and an activator based on pure sodium water glass, a sufficient and rapid development of strength occurs. This meets the requirements for a material, which is suitable for annular gap grouting.

The development of the stiffness shows the same behavior as the compressive strengths. The combination of slag and cement in the binder leads to a significant increase (+ 200%) of the stiffness of the grout proceeding from an age of the specimens of 7 days.

4 CONCLUSION

The results presented in this paper show, that the use of granulated blast furnace slag and the alkaline activation of these can be considered advantageous in the field of annular gap grouts for mechanized tunneling. In addition to the cost-effectiveness of this material compared to cement, this also concerns the material-technical optimization potentials. Because of the required gelation of the grout immediately after backfilling, the use of comparatively high reactive composites is necessary. Therefore, when combined with water glass, the reactant has to own a certain amount of calcium oxide. The combination of hydraulic cement and latent hydraulic blast furnace slag, forming a hybrid binder, meets the requirements for a suitable binder for annular gap grouts. The cement ensures a rapid gelation (<60 seconds) and the first strength development.

Due to the proportionally used slag an enormous increase in strength and stiffness compared to conventional pure cementitious two-component grouts can be generated. Due to the developing dense microstructure of slag-containing systems, such binder configurations also offer advantages in the case of a chemical attack. In addition, the use of blast furnace slag also has an advantageous effect on the workability of the material, since the slow-acting hydration in water in combination with cement when using a retarder cannot form any structures which negatively affect the workability.

ACKNOWLEDGEMENT

Financial support was provided by the German Science Foundation (DFG) in the framework of subproject B3 of the Collaborative Research Center SFB 837. This support is gratefully acknowledged.

REFERENCES

Bäppler, K. 2006. *Development of a two-component dynamic grouting system for tailskin injected backfilling of the annular gap for segmental concrete lining in shielded tunnel boring machines*. Golden, Colorado: Dissertation.

Buchwald, A. 2012. *Der Einfluss des Kalziums auf die Kondensation von Silikaten in alkali-aktivierten Bindern. Habitilationsschrift*. Bauhaus-Universität Weimar.

Hashimoto, T. B. 2004. Simultaneous backfill grouting, pressure development in construction phase and in the longterm. *Tunnelling and Underground Space Technology. Underground Space For Suitainable Urban Development. Proceedings Of The 30th ITA-AITES World Tunnel Congress, SIngapore, 22-27 MAY 2004.*

Herrman, A., König, A., & Dehn, F. 2015. Vorschlag zur Klassifizierung von alkalisch aktiviwerwten Bindemitteln und Geopolymer-Bindemitteln. *Cement International*, 62–69.

Pelizza, S., Peila, D., Borio, L., Dal Negro, E., Schulkins, R., & Boscaro, A. 2010. Analysis of the Performance of Two Component Back-filling Grout in Tunnel Boring Maschines Operating under Face Pressure. *ITA-AITES World Tunnel Congress 2010, Vancouver (CA), 14 to 20 May, 2010.*

Pellegrini, L., & Peruzza, P. 2009. Sao Paulo Metro Project – Control Of Settlements In Variable Soil Conditions Through EPB Pressure And Bicomponent Backfill Grout. *RETC June, 14-17 2009; Las Vegas, Nevada.*

Provis, J. L., & van Deventer, J. S. 2014. *Alkali Activated Materials. RILEM State-of-the-Art Report.* Dordrecht, Netherlands: Springer.

Schulte-Schrepping, C., Youn-Cale, B.-Y., & Breitenbücher, R. 2018. Strength Development of Two-Component Grouts for Annular Gap Grouting. *Tunnel 3/2018*, 24–33.

Škvára, F. 2007. www.geopolymery.com.

Von http://www.geopolymery.eu/aitom/upload/documents/publikace/2007/2007_03_173.pdf (20.07.2018 retrieved)

Tac Corporation. 2018. Von TAC Backfill Grout Injection-Two Component Backfill Gout Injection: http://www.tac-co.com/enoutline/TAC%20Presentation.pdf (20.07.2018 retrieved)

Tänzer, R., & Stephan, D. 2011. Portlandzementfreie Bindemittel. *Die Aktuelle-Wochenschau der GDCh - Bauen und Chemie*, 1–5.

Tunnels and Underground Cities: Engineering and Innovation meet Archaeology,
Architecture and Art, Volume 6: Innovation in underground engineering,
materials and equipment - Part 2 – Peila, Viggiani & Celestino (Eds)
© 2020 Taylor & Francis Group, London, ISBN 978-0-367-46871-2

Tuen Mun – Chek Lap Kok Link in Hong Kong – Innovative solutions for construction of an outstanding Subsea Tunnel

A. Schwob & F. Guedon
Dragages Hong Kong Ltd, Hong Kong SAR, China

B. Combe & T. Lockhart
Bouygues Travaux Publics, France

ABSTRACT: For Tuen Mun Chek Lap Kok Link subsea tunnels construction, numerous innovative solutions were implemented to overcome the challenging ground conditions and the surrounding pressure higher than 5 bars at tunnel face. Two 14m diameter TBMs were used along the 4.5km long drive below the sea. One of them had started its drive with the world's largest diameter (17.63m) along the first 630m before being reconfigured into a 14m diameter TBM. For cutting tools replacement, saturation diving system and full automatic robotic arms were developed to change a total of about 2000 disc-cutters on both TBM cutterheads. Another major challenge was the construction of 41 subsea cross passages between tunnels for which the pipe jacking methodology was adapted for the project. Apart from TBM drives, innovative design solutions were developed for the construction of Shafts and Cut and Cover sections including a 500m-long and 43m-deep Caterpillar-shaped cofferdam with 15 cells.

1 INTRODUCTION

In August 2013, Dragages Hong Kong Ltd and Bouygues Travaux Publics were awarded the largest construction contract ever awarded from Hong Kong Government: Design and Built of the Tuen Mun – Chek Lap Kok Link, Northern Connection Subsea Tunnel Section.

This New Road will provide alternative access route to the Hong Kong International Airport, located on Chek Lap Kok Island, which, until now, is accessible only via the Tsing Ma suspension bridge. TM-CLKL is also part of a cross-border project connecting the Northwest New territories (few kilometers away from the mega-city of Shenzhen, entrance gate to Mainland China) to Zuhai and Macau via the new Hong Kong – Zuhai – Macau Bridge. The artificial island of Hong Kong Boundary Crossing Facilities (HKBCF) on which the tunnel lands on its south end will be the heart of this new cross-border route.

The 4.5km long subsea section of the tunnel is made of two tubes built with two Slurry Tunnel Boring Machines (14m excavation diameter). Each tunnel has a 2-lane carriageway. On both sides, the North (630m) and South (670m) approach tunnels were built in freshly reclaimed land (Figure 1). While the construction of the HKBCF artificial island was part of a separate contract, the completion of the North reclamation in a very short time-frame was the first challenge faced by Bouygues Construction's teams.

2 NORTH APPROACH TUNNELS – MAIN CHALLENGES

2.1 Northern Landfall

After less than a year of intensive day and night work, a new piece of land emerged from the sea constituting phase 1 of the north reclamation. With an area of about 16.5 hectares for a

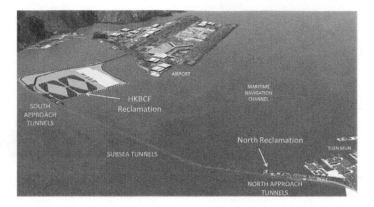

Figure 1.   TM-CLKL Project Layout.

Figure 2.   North Reclamation.

Figure 3.   North Reclamation.

total length of 1.2km and width of 140m (Figures 2 & 3), the north reclamation allows the tunnels to deepen and pass below the existing seabed with a gradient of 5%.

Two large shafts were built within the North reclamation:

– The 3-cells caterpillar-shaped Launching shaft is located at the northern end. Relatively shallow (20m) but large (80m long, 40m wide), its main function was to give space to assemble the two TBMs used to build the North Approach tunnels. A short section of Cut and Cover Tunnels, forming the North Entrance of the Tunnels, is also located there.
– The circular-shaped Ventilation Shaft (56m in diameter), is located at the other end of the reclamation. Much deeper (45m), it forms the transition between the North Approach tunnels and the subsea tunnels. During construction stage, it was used for receiving and reconfiguring the TBMs for them to start the subsea drive. In permanent stage, it is filled with many large concrete ventilation ducts connecting the tunnels to the North Ventilation Building constructed right above.

2.2   *The World's largest Tunnel Boring Machine*

At the early stage of the Project, the Reference Design for the North Approach Tunnels was a Cut and Cover structure. The construction of such a structure (630m-long, 45m-deep and 40m-wide) in a freshly-reclaimed land was one of the major challenges faced initially. The construction programme for this Cut and Cover section was very tight as the land above the North Approach tunnels had to be available partly for handover to third parties but also for site installations required for a timely start of the subsea TBM drives.

Figure 4.   North Approach Tunnel (Ramp up).   Figure 5.   World's Largest TBM (S-880).

The teams of Bouygues Construction therefore proposed to replace this Cut and Cover by bored tunnels with slurry TBMs. This could reduce the risks during construction, minimize earth movements and secure the programme. It was approved by the Employer's Supervising Officer as it satisfied all the functionalities of the structure defined in the contract while reducing the risk profile of the project.

Starting from the surface at the northern end of the reclamation and in order to reach sufficient cover underneath the seabed at the southern end, the tunnels have to follow a steep gradient of 5%. With such a gradient, the regulation in Hong Kong imposes a 3-lane carriageway for the ramp up. To accommodate this third lane and the required ventilation duct area, a tunnel with an internal diameter of 15.6 meters is needed (Figure 4) leading to the design and construction of the largest TBM ever used in the World with a diameter of 17.63 meters (Figure 5). The transition from 2 to 3 lanes takes place at the Ventilation Shaft. The other tunnel (ramp down) remains identical to the subsea section with an excavation diameter of 14 meters and an internal diameter of 12.4 meters.

The high variability of the ground encountered by the TBMs, from sand and clay to the hardest granite, was taken into account in the TBMs specifications. On site, the engineers decided to prioritize power and robustness for the cutter head and imposed to the TBM Manufacturer, Herrenknecht, a large number of disc cutters to cross the rock without any difficulty. This allowed the completion of the North Approach tunnels and associated works a few weeks ahead of the initial programme, 30 months after Project award.

2.3   Enhancement to the North Reclamation

Driving TBMs in a freshly reclaimed land was a major challenge that required some enhancement to the North reclaimed land.

In addition to the geotechnical data available from earlier investigation campaigns, a thorough survey of 157 Cone Penetration Tests (CPTs) was conducted along the tunnel alignment to map the geological conditions under the reclaimed land. The Marine deposits and the clayey Alluvium layers are fine-grained soils. These layers would be disturbed by the backfilling works of the North reclamation, then by the tunneling activity, and would consolidate under the new loads. The design of the reclamation had therefore to take into account the pre-existing state, its own requirements, but also prepare the ground for the subsequent boring of the tunnels.

In order to bring strength parameters of the compressible soils – namely, the Marine Deposits and the clayey Alluvium – to the minima given by the design of the tunnels, a classical scheme of consolidation under surcharge was chosen.

Due to the surcharge imposed on the compressible soils, in which drains of high permeability were placed, the water was squeezed out of the soil matrix and the grains were packed more tightly against each other, leading to an increase in the shear strength of the material. The height of surcharge (up to 12.5m) was defined in order to reach the required shear

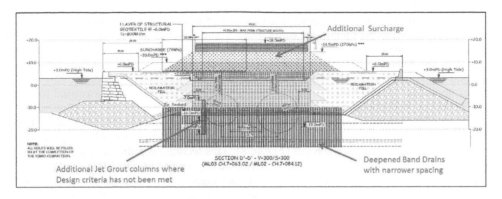

Figure 6. Extra measures for Consolidation of North Reclamation.

strength for tunneling. From the schedule of works that allowed for a consolidation period of 4 to 6 months, the pattern and spacing of the band drains were defined using the radial consolidation theory (Barron, 1948), to reach the final efficiency of 90% consolidation under the weight of the surcharge at the end of the consolidation period (Figure 6).

Plastic band drains were installed by a derrick lighter using a hydraulic hammer. After the installation of the band drains, the reclamation was backfilled up to the levels defined for the consolidation, and monitoring of the consolidation begun.

The monitoring of the consolidation was of paramount importance: what was at stake is to guarantee that the target degree of consolidation of 90% under surcharge weight was achieved, which in turn is a *sine qua non* condition to the success of the tunneling works.

In order to check the efficiency of the consolidation on the strength parameters of the ground, a post-consolidation CPT campaign was conducted.

Where the success conditions were not met, a local ground improvement by jet grouting was performed. The reason for this lack of performance was investigated, and was often correlated to difficulties during the installation of the band drains.

After the end of the consolidation and surcharge period, the twin TBM tunnels under the newly constructed Northern landfall were excavated.

2.4 Crossing the North Ventilation Shaft

After a 630m long journey below the North reclamation, TBM S-880 (17.63m diameter) reached the North Ventilation Shaft. To secure the break-out of this mega machine, the shaft

Figure 7. S-880 Break-out in flooded shaft (Photo taken after shaft dewatering).

Figure 8. Change Diameter Operation and Crossing Steel Bell.

was flooded to balance the hydrostatic pressure inside the shaft with the outside ground pressure (Figure 7). After pumping the water out, TBM S-880 was reconfigured into S-881 (Figure 8). Its diameter had to be reduced to 14m to bore the 2-lane carriageway subsea tunnel. This modification was thoroughly prepared in order to re-use a maximum of noble pieces and to limit the changes to the cutterhead and the shield elements only. This change-diameter operation required the use of the 580t capacity gantry crane that was used to assemble the TBMs in the launching shaft 8 months earlier.

The second TBM (S-882) crossed the North Ventilation Shaft with no interruption to its production cycle. For this purpose, a crossing Steel bell weighing 1500 tons, filled with concrete, was installed at the bottom of the Shaft (Figure 8). This technique eliminated the need for any ground treatment outside the shaft as the confinement pressure of the TBM could be maintained throughout the crossing. In addition, this arrangement avoided any disruption to the adjacent on-going change diameter operation while crossing with S-882. Furthermore, it saved significant time as S-882 was not slowed down by the Break-out/Break-in operations that are always heavy and time consuming for a TBM of this size. It allowed an early start for the main section of the project: the subsea tunnels and associated Internal structures and cross passages.

3 SUBSEA TUNNELS

3.1 *Adverse Geology*

An extensive geotechnical campaign was carried out immediately after the project award in order to get an accurate knowledge of the geology, starting point for the design of the structures and equipment. The subsoil of this region of Hong Kong is extremely disturbed and shows a high variability (Figure 9). Along the Northern part of the alignment, the Granite gives way progressively to CDG (Completely Decomposed Granite). Through this section, the TBMs had to encounter mixed face conditions which was a challenge for cutterhead

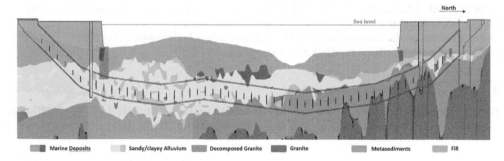

Figure 9. Geological Profile along the Tunnels.

Figure 10. Subsea Tunnels driven from Northern landfall.

Figure 11. TBM Excavation.

maintenance and hyperbaric interventions. Along the southern half of the alignment, the ground conditions change to alluvium with interbedded layers of sand and clay. While driving within alluvium, the wear of the disc cutters was much reduced. However, ensuring the face stability when hyperbaric intervention was required became an issue. The seabed is composed of a thick layer of Marine Deposits, very soft and of low density.

3.2 Subsea Tunnels Drive and TBM Maintenance

In this adverse geology, the two Tunnel Boring Machines specifically designed for the project were driven in parallel. With an outside diameter of 14m (Figures 10 and 11), the face was maintained by a Slurry Pressure system ensuring a confinement pressure up to 5.8bar at the deepest sections along the alignment.

In order to be able to complete the subsea section on time, barely more than a year was allocated to the excavation of the twin 4.5km long subsea tunnels. The TBMs had therefore to progress at a steady pace allowing the progress of various work fronts immediately at their back: installation of Tunnel internal structures, Cross passages construction, etc... The production rates required for TBMs were very demanding and required a well mastered logistics. Along the last section of the alignment, while the two TBMs were progressing in alluvial soils, each TBM could erect 7 to 10 rings per day. A total of approx. 35m of tunnel were erected daily.

In order to avoid breakdowns and maintain these rates of TBM excavation, ensuring an efficient maintenance of the TBMs cutter tools was crucial. The geology along the first part of the alignment was particularly demanding for the machines: the disc cutters worn out quickly in the abrasive granite and CDG and had to be regularly replaced. It was therefore required to have daily intervention under hyperbaric conditions, in the excavation chambers of the TBMs allowing access to the disc cutters and tools. Working in a hyperbaric environment reaching a pressure up to 5.8bar was certainly the most difficult challenge faced by the teams of Bouygues Construction.

3.3 Saturation Technique

In order to avoid exposing workers to the risks associated to daily compression/decompression cycles and to increase the effective time of interventions, the saturation technique was developed and implemented for the first time at such a scale on an underground project. It consists in lengthening the cycles by maintaining in a pressurized environment a team composed of 4 workers, specialized in hyperbaric works. The duration of the cycles is 28 days. For that purpose, a living habitat was built on surface. It included, among other things, the main hyperbaric chambers in which the workers were living, compression/decompression chambers and medical caissons equipped with the necessary equipment to provide care in case of injury (Figure 12).

To ensure the proper functioning of the installation, many parameters had to be constantly monitored: the pressure of the various chambers but also the temperature, humidity and the mixture of gas including helium that the workers were breathing. From a fully equipped control room, a surface team was constantly checking the proper functioning of the installation. A doctor certified for hyperbaric works was also full-time available to intervene when needed.

Every day, the teams were transferred via a pressurized shuttle from the living habitat to the TBMs. This shuttle was then connected directly to the hyperbaric chambers of the TBMs (Figure 13). Replacement of the disc cutters could then start.

3.4 TBM Monitoring System

In order to make full benefit of the time allowed for each intervention, it was essential to have in advance a good knowledge of the condition of the cutter head. Hyperbaric workers could

Figure 12.   Living Habitat for Saturation Diving.   Figure 13.   Shuttle clamped to TBM Chamber.

then prioritize the most critical areas. For this reason, the cutterheads of the TBMs were equipped with innovative technologies developed by Bouygues Travaux Publics:

– The Mobydic system performs a real-time mapping of the cutterhead. Each disc cutter is identified and any abnormal behavior caused by excessive wear is tracked and immediately reported to the system. Mobydic is also able to assess the type of ground encountered based on the pressure applied to the disc cutters. It allows getting an accurate view of the environment in which the TBM operates.
– The Snake system is a remote-controlled, poly-articulated arm equipped with a camera and a high pressure water jet to clean the cutterhead and carry out its inspection.

These systems are proven to be crucial to ensure the optimal operation of Tunnel Boring Machines while minimizing human interventions. They now equip most of the Company's TBMs.

3.5 *Telemach Robot*

The TM-CLKL project made a step further in the automation of maintenance systems with Telemach robot, developed by Bouygues Travaux Publics. Telemach is an articulated robot located in a dedicated air locked chamber in the TBM Shield (Figure 14). It is able to access the excavation chamber under hyperbaric conditions, deposit a worn or damaged disc cutter and replace it by a new one. Telemach is similar to robots used in the car industry but with greater difficulty. On a car assembly line, the car frames always stop precisely at the same place, facing the robots. Here, the disc cutter to be replaced can be anywhere. It was therefore necessary to develop a system able to locate the disc cutter and adapt the movements of the robot accordingly in a compressed air environment in the presence of spoil. In January 2016, for the first time, Telemach performed a complete cycle of disc cutter replacement in a fully

Figure 15. Telemach Robot changing Disc cutters.

Figure 14.   Telemach Chamber in TBM Shield.

automated way (Figure 15). This system proved its effectiveness on TM-CLKL Project and has, since then, been used on several projects by Bouygues Construction.

4 CROSS PASSAGES WITH PIPE JACKING TECHNOLOGY

4.1 *Context*

To provide means of egress for pedestrians in case of emergency, cross-passages linking the two tunnels are provided every 100m. The typical length of a cross-passage is approximately 13m. Value engineering was conducted at the early stages of the project to standardize the diameters and levels, given the high number of 57 cross-passages to be built.

The initial design was based on ground freezing (using brine) and conventional excavation. However, it quickly became obvious that this option was bringing significant threats to the project, because of risks in construction and potential delays.

As a result, a decision was made, during an early optioneering & value-engineering stage, to opt for a mechanized solution, with the clear intention to industrialize as far as possible the realization of the cross-passages.

4.2 *Design and Methods of construction*

An innovative solution was developed using, for the first time in the World, small diameter (3.665m) slurry TBM for construction of Cross Passages (Figure 16). The well-known pipe jacking technology was used but this technique had to be fully revisited in order to adapt to this specific context.

The general concept of the solution is as follows:

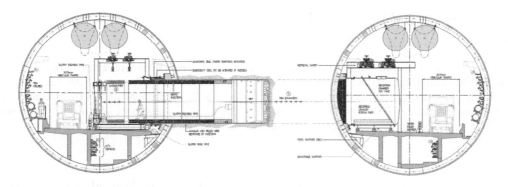

Figure 16.   General Concept of the Mechanized solution.

Figure 17.   Launching Side - Pipe Jacking Equipment.

Figure 18.   Receiving Side – Steel Bell.

- Construction of a concrete structure acting both temporarily and permanently
- On one side, installation of the pipe-jacking equipment (Figure 17)
- On the other side, installation of the receiving steel bell (Figure 18)
- Excavation using the pipe-jacking technique, with precast concrete pipe as lining.

A key element of the mechanized solution is the use of a multi-purpose concrete tympanum, cast in the main tunnels, and serving both temporary and permanent purposes. This tympanum is designed to

- stiffen the main tunnel lining against the squat around the opening.
- withstand the loads brought by the pipe-jacking machine
- act as a permanent structure on the long term, with embedded waterproofing and second-phase concrete for the door.

On top of the time constraint, the cross passages works were concurrent to all the other main tunnel activities such as the main TBMs logistics but also all the internal structures works on-going at the back of CP works areas in each tunnel. This meant that half of the tunnel section had to be kept free for traffic at all times and all the equipment were designed accordingly.

4.3 Achievement

The greatest achievement was to make 39 cross passages in less than 13.5 months, which represents more than 3 CPs per month in variable geology under high water pressure.

This groundbreaking mechanized solution allowed controlling the production schedule while minimizing exposure to geological risks. From design to production, the guiding thread was to make the solution as industrial as possible, and to use permanent elements for temporary stages to minimize the number of operations.

5 CATERPILLAR COFFERDAM FOR SOUTH CUT AND COVER TUNNELS

5.1 Southern Landfall

Upon their arrival below the HKBCF reclamation tip, the two TBMs have crossed the 2 circular South Ventilation Shafts (55m deep - future connection to the adjacent South Ventilation Building – Figure 19). Immediately southwards, the South Approach tunnels are made of a 250m long section of TBM tunnels with the same diameter as the subsea section, followed by a 420m long Cut and cover structure (Figure 20).

5.2 Context

On the South reclamation (HKBCF Island), the geology is particularly adverse, with very thick (more than 30m) and very deep (up to 50m below ground level) layers of marine deposits

Figure 19.   South Ventilation Shafts.

Figure 20.   Southern Landfall.

and alluvium clays. In these conditions, building a large cofferdam for construction of South Cut and Cover tunnel is a major challenge.

A conventional straight D-wall cofferdam would have required up to 9 layers of strutting with large steel members as well as an extensive ground treatment below the final excavation level to maintain toe stability of the D-walls (soil layers could not provide sufficient passive resistance)

A very positive experience was gained from the design and construction of the 3-cell caterpillar shaped North Launching Shaft. Therefore, the project's technical team decided to push the concept to a world's first 15-cells caterpillar-shaped cofferdam of 500m long.

5.3 Design of the Caterpillar Cofferdam

The proposed structure consists of 15 numbers of truncated circular cells ranging from approx. 25 to 37m long and approx. 44 to 57m (diameter of cell) wide each. Each cell is formed by perimeter D-walls in an arc shape, which resists lateral pressures by hoop force. The hoop forces induced on the perimeter arc D-walls are then transferred at the junction of the cells to heavy duty D-walls called "Y-Panels", which are transversally supported by reinforced concrete struts and cross walls.

The Y-Panels transmitting vast loadings (up to 44m long to 56m deep of lateral pressure), a special type of D-wall panel had to be invented. It had to be cast from a single trench of 3.6m wide by 6.5m long with varying geometry. In-situ reinforcement installation method had to be adopted for the heavy reinforcement cage (Figure 21).

The design modelling required 3 different 3D Models (Plaxis 3D, SAP2000, Strand 7) and 4 consultants to validate properly the design down to even the D-wall joints behavior (Figure 22).

Overall, the caterpillar scheme went beyond a simple cofferdam proposal and became a fully integrated solution involving innovations on the design modelling, the construction methods, and the permanent works design

5.4 Main Advantages of the Caterpillar cofferdam

The principle of the caterpillar-shaped structure is that it uses the arch effect instead of vertical bending for retaining the earth and water pressures acting on the excavated trench. This makes the D-walls much more efficient as the main forces are transferred in compression inside the D-wall arches.

A main advantage is the absence of steel struts and associated king posts and waler beams. It offers a drastic optimization to the construction programme both during:

Figure 21. Y-Panel Cage.

Figure 22. 3D Modeling of the 15-cell Caterpillar.

Figure 23.   Excavation of first 3 cells.

– Excavation stage: whilst the total amount of excavation volume is increased, the absence of steel struts allows for ease of plant movement, ease of vertical lifting leading to a much faster excavation overall.
– Permanent structure construction stage: The permanent structure can be cast as if the works were conducted in open air, instead of being confined by multiple strut layers.

The caterpillar cofferdam also removes the necessity for ground treatment below the excavation level as the arch effect still exists below the excavation level and toe stability is therefore not an issue.

6   CONCLUSION

In order to meet the numerous technical challenges for the construction of the Tuen Mun – Chek Lap Kok Link – Northern Connection Subsea tunnels in Hong Kong, a large number of innovative design and technologies were put in place over the past five years.

Launching the World's largest Tunnel Boring Machine in a freshly reclaimed land was the starting point of a 5.5km long adventure.

In adverse ground conditions, at more than 55m below sea level, state of the art technologies were used for facilitating the daily hyperbaric interventions on the TBM cutterheads including the use of a robotic arm able to change disc cutters without human interventions.

Another major milestone of the project was the construction of the cross passages for which small-size TBMs were launched more than 40 times from one tunnel to the other.

The Tunnels will then end up with a cut and cover section built within a gigantic caterpillar-shaped cofferdam (500m long and 43m deep).

These innovative technical solutions, developed through a thorough preparation of the design and construction methods, laid the foundations of recent developments in underground works, pushing further the limits of feasibility of the techniques used.

Tunnels and Underground Cities: Engineering and Innovation meet Archaeology,
Architecture and Art, Volume 6: Innovation in underground engineering,
materials and equipment - Part 2 – Peila, Viggiani & Celestino (Eds)
© 2020 Taylor & Francis Group, London, ISBN 978-0-367-46871-2

Tunnel deformation evaluation by Mobile Mapping System

Y. Shigeta, K. Maeda, H. Yamamoto & T. Yasuda
Pacific Consultants Co.,Ltd., Tokyo, Japan

S. Kaise, K. Maegawa & T. Ito
NEXCO Research Institute Japan, Tokyo, Japan

ABSTRACT: The Mobile Imaging & Mobile Mapping "MIMM" System, mounting Mobile Imaging Technology System (MIS) and Mobile Mapping System (MMS) on the same vehicle, was developed in 2010; practical use in the assessment of tunnel soundness was verified with the system, which was then applied to many tunnels at the practical level. This vehicle does not require traffic control and deformation can be accurately detected; while traveling at high speed, it continuously carries out the image measurement and laser measurement of the surveyed tunnel wall surface. Here, regarding the evaluation of deformation caused by external force, we report on the external force assessment of tunnels in which deformation due to external force is suspected and also on the verification of progressiveness in an actual size large tunnel.

1   INSTRUCTION

Topographical and geological conditions are complicated in Japan. Therefore, after the tunnel is completed, it may deform due to external force from bedrock and countermeasures may be needed. Deformations caused by external force can be hazardous to the structure of the tunnel, so it is necessary to accurately grasp these variations at an early stage.

On the other hand, the cause of the cracking of the lining concrete is not only due to external force, but also due to drying shrinkage of concrete. The necessity of countermeasures is generally low because cracks due to drying shrinkage of concrete usually stop growing in about 5 years.

Generally, the cause of deformation is first identified from the position and direction of cracks occurring on the lining surface, and then identified in more exact detail by boring survey, indoor tests and various measurements so that we judge the progressiveness of the deformation and the necessity of countermeasures. In addition, since boring survey, indoor tests and various measurements require a lot of time and cost, it is desired to develop a method that is relatively simple and low cost.

Therefore, the authors developed a system that acquires images of the surface of lined concrete surface and cross-section while traveling and estimates the cause of the change. Since the lining concrete cracks due to external force involve deformation, we focus on the change in cross-sectional shape and the occurrence of cracks to estimate the cause of deformation. In addition, we developed a system to three-dimensionally evaluate the progressiveness of deformation through the mobile measurement of cross-section.

In this paper, in regard to the applicability of mobile cross-section measurement, we show the estimation results of the cause of deformation on the road tunnel and the experimental results of the mobile cross-section measurement at the full-size road tunnel.

2   OUTLINE OF MEASUREMENT VEHICLE SYSTEM

Mobile photographic survey to determine the deformation of the lining surface, and mobile laser measurement to determine the shape and the deformation mode of the tunnel cross

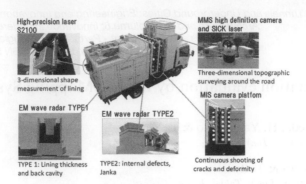

Figure 1. Outline of Measurement Vehicle (MIMM-R).

section, are some diagnostic supporting technology for the periodic inspection of tunnels. In particular, on-vehicle diagnostic imaging (mobile photographic survey) is often used by NEXCO (the Nippon Expressway Companies), the Japanese Ministry of Land, Infrastructure, Transport and Tourism, and municipalities.

MIMM – on-vehicle diagnostic imaging by the Mobile Imaging Technology System (MIS) and mobile laser measurement by the Mobile Mapping System (MMS) together on the one vehicle – is a further evolution of the measurement vehicle system. These systems, which were found useful in tunnel inspection, were the result of the 2006–2008 Tunnel Soundness Evaluation Project which was chaired by Professor Yuzo OHNISHI, then at Kyoto University, and which was part of the New Urban Technology Workshop centered on Kinki Regional Development Bureau of the Ministry of Land, Infrastructure, Transport and Tourism.

Further, MIMM-R has been developed, installing on to MIMM an on-vehicle radar probe device to determine places where the thickness of the concrete lining the tunnel is insufficient and to discover cavities behind the lining, contributing to cost savings, labor savings and efficiencies in periodic tunnel inspection and investigation (Figure 1).

MIMM-R consists of image acquisition equipment, laser survey equipment, non-contact radar sensor equipment, and GPS, odometer etc. The laser survey equipment is able to collect high precision data from one million points per second, at a resolution of 0.1mm. The imaging acquisition equipment consists of 20 CCD cameras (at 380 kilo pixels each) and 60 LED lights, and is able to detect a 0.3mm crack on the lining at speeds of 70km/h.

3 METHOD TO DETERMINE THE DEFORMATION MODE

The first step in determining the deformation of the tunnel lining is to eliminate outlier data and then categorize the remaining data into each lining span for analysis. Next, the central axis of the tunnel is established and the presumed executed cross section about that central axis is calculated. A contour plot is drawn to indicate the difference between the presumed cross section and the measured plot points resulting from the laser measurement.

3.1 Establishing the average cross section (presumed executed cross section)

The raw plot point data obtained from the laser survey is no more than 3-dimensional coordinate data. To be able to use this data meaningfully to assess the deformation of the tunnel lining, an average tunnel cross section must be established as a basis point, and the variation of the laser survey data compared to that basis point indicated on a contour plot to visualize the results.

First, the central axis of the tunnel cross section is calcu-lated. Next, the laser survey cross section data is project-ed perpendicular to this central axis, with the result be-coming a projected face. The presumed executed cross section is the result of the averaging of those projected data points, with that averaging being performed based on the angle and direction centered on the point of view of the resulting projected face (Figure 2).

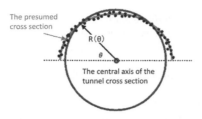

Figure 2. The presumed cross section.

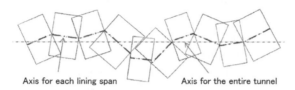

Axis for each lining span   Axis for the entire tunnel

Figure 3. Image of the central axis.

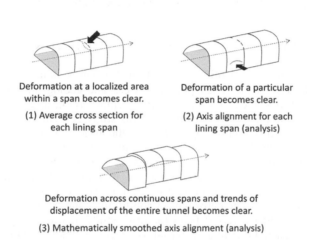

Deformation at a localized area   Deformation of a particular
within a span becomes clear.   span becomes clear.

(1) Average cross section for   (2) Axis alignment for each
each lining span   lining span (analysis)

Deformation across continuous spans and trends of
displacement of the entire tunnel becomes clear.

(3) Mathematically smoothed axis alignment (analysis)

Figure 4. Mode of deformation.

3.2 Establishing the central axis

Tunnel lining deformation may occur at a localized area within a particular lining span, or it may occur across a number of lining spans, moving from one to another. In order to sample both these deformation types, two tunnel central axes should be established, one axis for each lining span, and one axis for the entire tunnel (mathematically smoothed face) (Figure 3).

3.3 Determining the deformation mode

Irregularities in the surface of the tunnel lining (differences between the average cross section and the actual laser survey data) is shown on a colored distribution contour diagram, from which we are able to understand what type of deformation is occurring in the tunnel lining. A contour diagram is created for each type of deformation mode as shown in Figure 4, and the characteristic of each type are as follows.

3.3.1 *Average cross section for each lining span*

A contour diagram for the average cross section for each lining span is created by preparing a color varied diagram based on the degree by which the laser survey data, which is itself based on the central axis calculated for that span, varies from the presumed executed cross sec-tion calculated for that span. In this way, deformation at a localized area within a span becomes clear.

3.3.2 *Axis alignment for each lining span (analysis)*

A contour diagram for the analysis of the axis alignment for each lining span is created by preparing a color varied diagram based on the degree by which the laser survey data, which is itself based on the central axis calculated for that span, varies from the presumed executed cross section calculated as the average for the whole tunnel. In this way, deformation of a particular span becomes clear.

3.3.3 *Mathematically smoothed axis alignment (analysis)*

A contour diagram for the analysis of the mathematically smoothed axis alignment is created by preparing a color varied diagram based on the degree by which the laser survey data, which is itself based on the central axis calculated for the entire tunnel, varies from the presumed executed cross section calculated as the average for the whole tunnel. In this way, deformation across continuous spans and trends of displacement of the entire tunnel becomes clear.

4 BRIEF OVERVIEW OF THE DEFORMATION AT THE TUNNEL STUDIED

Uplift in the road surface was identified in the tunnel which is subject to this study in 1996, three years after the tunnel had been put in service. The geological condition around the tunnel consists of volcanic rock and volcanic fragmentary rock, and contains clay mineral. The reported road surface uplift area subject to this study is located about 1000m distance from a starting portal in a tunnel of approximately 3000m length, in spans A to D. The area was confirmed to be a geological boundary and a fault fracture zone in a prior study.

Uplift in the road surface, and a crack on the cross section side lining was confirmed in a section of the tunnel without an invert arch structure. In addition, numerous cracks in a longitudinal and an oblique direction on the side wall lining were confirmed, and the side trench had collapsed in some areas. The deformation condition is shown in Figure 5. Based on the deformation condition and the report from the prior study, it was thought possible that the tunnel deformation occurred due to an external force with plastic behavior from the direction of the side of the tunnel.

4.1 *Results of Measurement Vehicle System*

A mobile photographic survey and mobile laser measurement was implemented at several sites of recognized road surface uplift. A deformation mode analysis was performed using the data points from the mobile laser measurement, based on Report on Research into Implementation of Evaluation of Soundness of Road Tunnels Using Measurement Vehicle System Technology 1), and the result is shown in Figure 6.

The color of deformation contour diagram are showed to red means the inside deformation and to blue means the outside deformation. And it has possibilities to be estimated that the deformation area fit in crack area means the external force is happen to the deformation.

1. The contour diagram means the average cross section of the each span are showed deformation at a localized area within a span.
2. The contour diagram means the axis alignment for each lining span are showed deformation of a particular span.
3. The different span axis from the smoothed axis are showed horizontal difference means the cross section disconnect in horizontal direction.

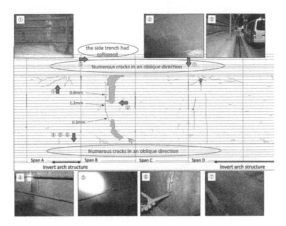

Figure 5. The deformation condition.

4. The different span axis from the smoothed axis are showed longitudinal difference means the cross section disconnect in longitudinal direction.
5. The contour diagram are showed the mathematically smoothed axis means consecutive deformation of the tunnel.

The span B which was confirmed of road surface uplift was identified the many cracks in an oblique and a longitudinal direction. Figurer 6(2) contour diagram was showed the side wall deformation and the crack area are fit in. In consequence, the deformation of this span was occurred by the external force condition.

4.2 Inference of cause of deformation

With the confirmation of distortion towards the inside of the tunnel in the side wall as a result of the analysis of deformation mode in span B, it is conceivable that the deformation of this span was caused by the plastic behavior pressure from the lateral rock. It is conceivable that at that time, the plastic pressure caused the side wall to be distorted towards the inside, together with longitudinal cracks. Further, since span B was not constructed with an inverted arch structure, if it is subject to plastic pressure from a lateral direction, it is conceivable that the buildup of stress escapes, as shown in the first example of deformation patterns in Figure 7, to the road surface which has been built without an inverted arch structure. It is possible that the road bed in span B was uplifted in consequence of this.

Across spans A to D, numerous cracks were confirmed in the vicinity of the tunnel wall, and in span B, a cross sectional crack in the arch crown and a slight misalignment was found. A span constructed without an inverted arch structure which is subject to plastic pressure from a lateral direction is structurally compromised and easily deformed, as shown in the second example of deformation patterns in Figure 7.

Once that span had experienced deformation towards the inside of the tunnel, it is conceivable that the neighboring spans were subject to tensile stress and deformed in a chain reaction, and at that time cracks developed in an oblique direction.

Further, it is conceivable that the entire span lining was deformed in an upper direction and itself uplifted by the uplift of the road bed. With this, the entire central span was subject to tensile stress and dragged by the neighboring spans, causing the cross sectional crack in the arch crown.

In this way, by implementing deformation mode analysis in addition to measuring the deformation of the tunnel, the likelihood that the deformation was caused by external forces can be determined, and this method is very effective in inferring causes of deformation.

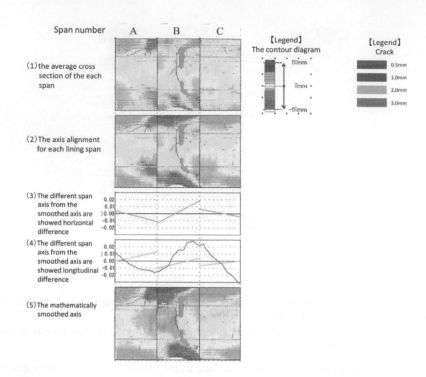

Figure 6.   Results of Measurement Vehicle System.

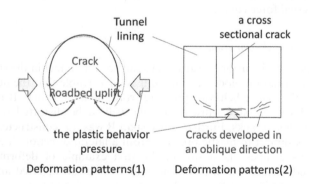

Figure 7.   Mode of deformation.

5   PROGRESSIVE EVALUATION METHOD OF DEFORMATION

5.1   *Overview*

In general, the measurements of internal displacement, cross-section, crack displacement, etc. are carried out as a method for grasping the progressiveness of the deformation of tunnel. The evaluation can be highly accurate because these methods focus on the same measurement point; however, a lot of work is required because a large number of cross-sections are measured in order to evaluate the extent of occurrence of the change. For this reason, in the evaluation, it seems necessary to combine highly efficient mobile cross-section measurement and highly accurate measurement. Therefore, in regard to the mobile cross-section measurement, from the results of the deformation mode analysis of multiple mobile cross-section measurements, we developed a method of progressive evaluation of deformation and tested the simulated full-size road tunnels.

5.2 Test method

In order to verify the validity of the deformation mode analysis, measurements were carried out at the running speed of 20 km/h in the simulated tunnel. Tests were carried out using a simulated tunnel with a length of about 80 m having the same width and height as the real road tunnel. Figure 8 shows the images of test cases and simulated displacements. The displacement was simulated by placing plywood in the full-size road tunnel. The test case was set considering actual tunnel deformation. In Case 1, the current tunnel is evaluated without simulated displacement (see Figure 8 (a)); Case 2 simulates the pressure from a sidewall (see Figure 8 (b)); and Case 3 simulates the external forces from sidewalls on both sides such as plastic pressure and water pressure (see Figure 8 (c)).

5.3 Test results

Figure 9 shows the results of deformation mode analysis of each case. Here, because the tunnel lining concrete simulates the external force deformation, we put the axis alignment in order for each span (evaluation of span deformation). The upper end and the lower end of each of Figs. (a) to (c) correspond to the foot of the tunnel while the center corresponds the tunnel crown center. The left and right edges are the start point and the end point, respectively. In the figures, the displacement and deformation towards the inner space is shown by the reddish color and the displacement toward the bedrock side (outer side) by the bluish color. Yellow indicates equivalence to the estimated completed cross-section.

First, by looking at the place where simulated displacement occurs, it can be seen that the place of simulated displacement is changed to red in Cases 2 and 3 as compared with Case 1. As a result, it is considered that changes in the cross-section can be read by comparing with the estimated completed cross-section at the location where displacement or deformation occurs.

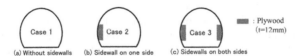

Figure 8.   Test cases and simulated displacements.

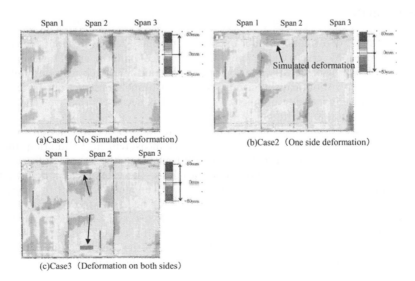

Figure 9.   Results of deformation mode analysis of each case.

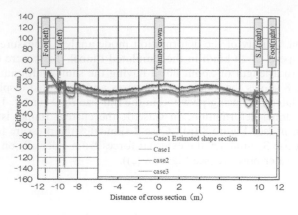

Figure 10.   Results of deformation mode analysis of each case.

Also, looking at the overall tendency, the distributions of the red and the blue based on the estimated completed cross-section are similar, though some variations are seen. In the mobile cross-section measurement, this variation occurs because the same point cannot be measured in a plurality of measurements, and that variations are seen in the obtained point groups just as observed in the accuracy verification. However, since the same contour distribution is shown in all cases, we can judge between displacement and deformation by considering these points in the evaluation of the cause of deformation and the judgment of progressiveness.

Next, the displacement detection accuracy of the simulated displacement is put in order. Figure 10 shows the distribution of displacements in the estimated completed cross-section and the simulated displacement section of Case 1. In the figure, when looking at the estimated completed cross-section and data variation in each case, a difference of about ±20.0 mm is seen to occur. It is presumed that this error affects the error in the setting of the estimated completed cross-section and the error and dispersion of the measurement system. In the simulated displacement on the right side (caliper measurement: 12.5 mm), the absolute value is 0.5 mm in Case 2 and –9.1 mm in Case 3. However, as a trend, it can be distinguished as protrusion, and we can observe the tendency of progress of deformation.

6   CONCLUSION

Evaluation of tunnel deformation under external force has been carried out focusing on the cracks characterized by crack width, length, occurrence pattern, etc. However, cracks in tunnel lining concrete occur mostly due to shrinkage strain and back restraint caused by the material of concrete, without causing a big problem of tunnel construction. In order to take appropriate measures for a large number of tunnels, it is necessary that we can estimate the cause of the deformation in the tunnel.

In this study, it was found that the existence of deformation due to external force is easily judged by the deformation mode analysis of the mobile measurement. In addition, the possibility of progressive evaluation of deformation was confirmed from the test of a full-size tunnel.

However, in the existing deformation mode analysis, the cross-section deformation and heaving of roadbed as shown are not absolute but relative values. In order to properly evaluate the variation caused by external force, it is considered necessary to determine the absolute amount, and therefore further technical development is necessary. Furthermore, in order to utilize the moving measurement technology to the fullest, it is considered necessary to improve the reproducibility of the measurement and the accuracy of the comprehension of progressibility ; therefore, we will carry out verification experiments including the accuracy verification of various surveying instruments.

Tunnels and Underground Cities: Engineering and Innovation meet Archaeology,
Architecture and Art, Volume 6: Innovation in underground engineering,
materials and equipment - Part 2 – Peila, Viggiani & Celestino (Eds)
© 2020 Taylor & Francis Group, London, ISBN 978-0-367-46871-2

Probabilistic modelling of fibre reinforced shotcrete

A. Sjölander & A. Ansell
KTH Royal Institute of Technology, Division of Concrete Structures, Stockholm, Sweden

ABSTRACT: Shotcrete is widely used as rock support and can support the load from blocks either by bonding to the rock or by bending between rock bolts. By introducing fibres, the ductility of the shotcrete will increase and the crack widths decrease. Predictions of the structural behaviour for fibre reinforced shotcrete (FRS) are however complicated due to the large scatter normally seen in test results. The reason is mainly related to the non-uniform distribution and orientation of the fibres which could lead to uncertainties regarding the quality of in-situ shotcrete. The aim with this paper is therefore to investigate if a probabilistic material model for FRS can be used to capture the scatter in the results. An isotropic damage model that combines an exponential softening curve for unreinforced shotcrete and a bi-linear curve to account for the effect of fibres were used. Suitable distributions for each of the parameters in the model have been proposed based on fitting of experimental results. Thereafter, Monte Carlo simulations were used to produce results for a large number of lab tests. Results show that the model together with the proposed distributions was able to capture the scatter in test results.

1 INTRODUCTION

Fibre reinforced shotcrete (FRS) in combination with rock bolts have since the beginning of the 1970's been the predominating support method for rock tunnels in Sweden and Norway (Franzén 1992). Shotcrete is concrete applied by spraying in which the properties of the concrete have been adjusted to simplify spraying and harden rapidly so it sticks to the wall. The post-cracking structural behaviour of FRS is governed by the fibres ability to bridge and transfer stresses across cracks that develops in the concrete. As for normal (bar) reinforced concrete, the influence of the fibres on the mechanical properties of concrete before cracking is negligible. After cracking, the major energy consumption is sliding of fibres which normally leads to a pull-ot failure of the fibres (Banthia 1992). This is desirable since it leads to a more ductile failure compared to rupture of the fibres. The behaviour of fibre reinforced concrete post-cracking depends on the magnitude, orientation and anchorage length of the fibres in relation to the cracks that occur. The random distribution and orientation of fibres is the main reason to the large scatter seen in test results and this have therefore attracted great interest and have been studied e.g. by using X-ray tomography, see Heiko et al. (2016) and Ansell et al. (2017). Numerical models to predict distributions have been presented by Soetens & Matthys (2014) and by Alberti et al. (2017). The random distribution of fibres is one aspect that makes the design of FRS complex since the large scatter in test results leads to uncertainties regarding the structural capacity and behaviour of FRS post-cracking. Typical test results are shown in Figure 1 in which results for 24 notched beams are presented. As pointed out by Cavalaro & Aguado (2015) and Bernard & Xu (2017), the scatter in post-cracking behaviour leads to that the characteristic values for residual strength will be far below the average. The consequence of this is that higher

quantities of fibres are required to reach a desired residual strength which leads to an uneconomic design. One possible way to overcome this problem is to instead use a probabilistic model and study the probability of failure.

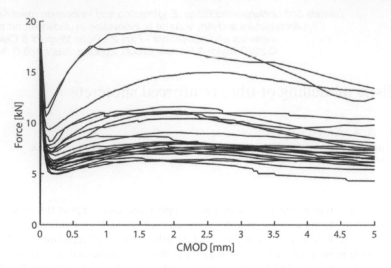

Figure 1. Force-CMOD (Crack mouth opening displacement) curve from three-point bending tests provided by Bekaert.

The aim with this paper is to present a computationally efficient material model based on damage mechanics to describe the non-linear material behaviour of FRS. Based on data from 24 tested beams, a suitable distribution will be suggested for the governing parameters in the material model and Monte Carlo simulations will then be used to investigate if the suggested model can predict the behaviour of the experimental tests.

2 MECHANICAL MODEL FOR FIBRE REINFORCED CONCRETE

Hooke´s law is used to describe the behaviour up until the tensile strength σ_t is reached. Once σ_t is reached, cracking occurs and the relation between stress σ and crack opening w is used to describe the structural behaviour. An isotropic damage model is used to allow localization of the crack:

$$\sigma = (1 - \omega)E_\kappa \tag{1}$$

Here, ω is a damage parameter that ranges from zero for the undamaged material to one for a fully damaged material, E is Young's modulus for the undamaged material and κ is the total strain. Further presentation of the numerical model is given by Oliver et al. (1990) and the implementation of the model in a finite element software by Gasch (2016). Different $\sigma(w)$ functions are presented in the literature. Model Code (FIB 2013) suggest that a bi-linear or an elastic-perfectly plastic model could be used while Yoo et al. (2015) suggest a tri-linear curve to account for a combination of softening and hardening behaviour. It should be noted that the $\sigma(w)$ function is not a constitutive law and several different shapes of the function could be used to match experimental results. The mechanical model used here is based on a combination of an exponential and a bi-linear material model presented by Sjölander et al. (2018), schematically shown in Figure 2. The first part of the post-cracking response is modelled with an exponential softening function that represents the behaviour of unreinforced concrete, see $\sigma_A(\kappa)$ in Figure 2.

$$\sigma_A(\kappa) = \sigma_t \exp\left(-\frac{\kappa - \varepsilon_0}{\varepsilon_f}\right) \text{ for } \varepsilon_0 < \kappa < \varepsilon_1 \tag{2}$$

$$\varepsilon_f = \frac{\varepsilon_0}{2} + \frac{G_f}{\sigma_t} \tag{3}$$

As can be seen in Equation 2 and 3, the softening behaviour of unreinforced concrete is governed by the fracture energy of the concrete G_f, tensile strength σ_t and strain at crack opening ε_0. At a defined crack width w_1, fibres are activated and a bi-linear model is used, see $\sigma_B(\kappa)$ and $\sigma_C(\kappa)$ in Figure 2. The stress at the intersection between the exponential and bi-linear curve is calculated as $\sigma_1 = \sigma_A(\varepsilon_1)$ where ε_1 is the strain at the crack width w_1. The conversion between crack width and strain is based on the crack band width h_f which has been set equal to the element length and hence:

$$w = (\kappa - \varepsilon_0)\, h_f \text{ for } \varepsilon_0 < \kappa \tag{4}$$

The bi-linear curve can account for hardening, perfectly plastic or softening behaviour and is defined by two linear equations.

$$\sigma_B(\kappa) = \sigma_{f1} - \kappa k_1 \text{ for } \varepsilon_1 < \kappa \le \varepsilon_2 \tag{5}$$

$$\sigma_C(\kappa) = \sigma_{f2} - \kappa k_2 \text{ for } \varepsilon_2 < \kappa \le \varepsilon_3 \tag{6}$$

Where σ_{f1} and σ_{f2} are stresses when the strain is equal to zero, see Figure 2. The slope of each curve k_i is defined as:

$$k_i = \frac{\sigma_i - \sigma_{i+1}}{w_{i+1} - w_i} \tag{7}$$

The damage function for exponential softening $\omega_A(\kappa)$ is derived from Equation 1 and 2 while the two linear damage functions $\omega_B(\kappa)$ and $\omega_C(\kappa)$ are derived from Equation 1 and 5 or 6.

$$\omega_A(\kappa) = 1 - \frac{\varepsilon_0}{\kappa} \exp\left(-\frac{\kappa - \varepsilon_0}{\varepsilon_f}\right) \text{ for } \varepsilon_0 < \kappa \le \varepsilon_1 \tag{8}$$

$$\omega_B(\kappa) = 1 - \frac{\sigma_{f1} - \kappa k_1}{E\kappa} \text{ for } \varepsilon_1 < \kappa \le \varepsilon_2 \tag{9}$$

$$\omega_C(\kappa) = 1 - \frac{\sigma_{f2} - \kappa_2}{E\kappa} \text{ for } \varepsilon_2 < \kappa \le \varepsilon_3 \tag{10}$$

An equivalent strain based on Rankine's theory was used which means that failure occurs when one of the principal stresses reaches the failure stress and hence:

$$\varepsilon_{eg} = \frac{1}{E} \max \langle \sigma_i \rangle; \quad i = 1, 2, 3 \tag{11}$$

The maximum stress in Equation 11 indicates that only tensile stresses and failure is accounted for in the model. To ensure that damage of concrete is irreversible, a history dependent variable κ was used to keep track of the maximum tensile strain in each element.

$$\kappa = \max(\varepsilon_{eq}, \kappa_{old}) \tag{12}$$

The damage criterion is defined in Equation 13 and the growth of damage was specified with a Kuhn-Tucker condition in Equation 14. This ensures that damage can only increase when the current strain in an element, ε_{eq} is larger than the previous maximum strain in that element.

$$f(\varepsilon_{eq}, \kappa) = \varepsilon_{eq} - \kappa \le 0 \tag{13}$$

$$f \le 0;\, \kappa'\, 0;\, \kappa' f = 0 \tag{14}$$

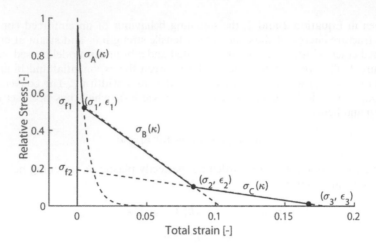

Figure 2. Schematic presentation of the material model.

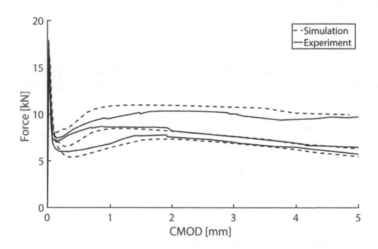

Figure 3. Comparison between simulated (dashed) and experimental (solid) results.

3 DISTRIBUTION OF MATERIAL PARAMETERS

Test results for notched fibre reinforced concrete beams according to EN 14651 (CEN 2007) were provided by Bekaert. According to the standard, the length, width and depth of the test beam is $500 \times 75 \times 125$ mm and during testing measurements of Stress - CMOD (crack mouth opening displacement) are performed. Results are shown in Figure 1 for a total of 24 beams. Using the presented numerical model, a 2D plane stress numerical model was created in the finite element software Comsol Multiphysics (2016) to simulate the experiments. The average mesh size of the model was 13 mm and a comparison between numerical and experimental results is shown in Figure 3, for three beams. Parameters for the numerical model were chosen to fit each of the experimental tests. Based on this, the suggested distributions for each parameter are presented in Table 1, along with mean values and standard deviations.

Matlab was then used to generate 2000 samplings from each of the distributions and in Figure 4, the sampled and experimental values for each parameter are presented together with the fitted distribution. The selected distributions shows a reasonable agreement with the experimental data for parameters w_1, a_3, f_t and G_f. In the distributions for w_2 and a_2, one

Table 1. Suggested distributions for parameters.

Parameter	Notation	Unit	Distribution	Mean value	Standard deviation
Young's modulus	E	[GPa]	Deterministic	33,0	-
Tensile strength	σ_t	[MPa]	Lognormal	4.230	0.142
Fracture energy	G_f	[Nm]	Lognormal	78.61	55.43
Crack width	w_1	[mm]	Normal	0.040	0.006
Crack width	w_2	[mm]	Normal	0.930	0.280
Crack width	w_3	[mm]	Deterministic	5.5	-
Stress ratio	a_2	[-]	Lognormal	0.250	0.008
Stress ratio	a_3	[-]	Lognormal	0.134	0.002

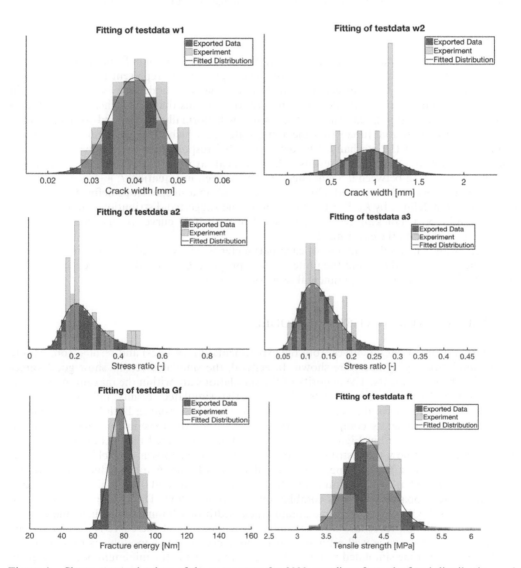

Figure 4. Shows exported values of the parameters for 2000 samplings from the fitted distribution and the experimental values for the parameters according to Table 1.

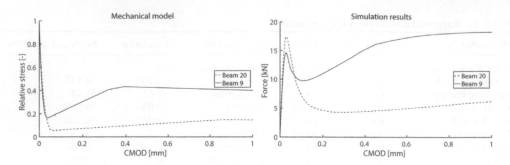

Figure 5. Comparison between material model and simulated result, i.e. global response, for beam number 9 and 20.

distinctive peak exist and no suitable distribution was found that could fit both the distinctive peak and the other values of the distribution. These parameters represent the crack opening and the stress ratio for the intersection of the bi-linear material model, see $\sigma_B(\kappa)$ and $\sigma_C(\kappa)$ in Figure 2. The suggested distribution for stress ratio a_2 means that it is rather unlikely that a ratio of 0.5 will be simulated. Such a stress ratio will normally correspond to a hardening phase (increase in external force) after the initial softening which can be seen for beam 9 in the comparison between the mechanical model and global response for two beams in Figure 5. The distribution of experimental results indicate a peak at $w_2 = 1.2$ mm and then an almost uniform distribution between 0.5 mm and 1.0 mm. As seen in Equation 7, w_2 is used to define the slope of the $\sigma_{B,C}(\kappa)$ curve and a low value will result in a steeper slope defined by k_2 and a more flat curve defined by k_3. For the simulations, the suggested distribution could result in a slightly stiffer post- cracking response, i.e. the Force-CMOD curve will increase more after cracking compared to the experiments.

The random generated samplings of each parameter were then used as input in the numerical model presented above and the finite element program Comsol Multiphysics was used to run 2000 simulations of the experimental setup of the beam.

4 RESULTS FROM STATISTICAL MODEL

In Figure 6, Force-CMOD curves from the experiments (black lines) and results from Monte Carlo simulations (gray lines) are shown. In general, the simulated results show good agreement with the experiments. The majority of the simulations are within the maximum and minimum results from the experiments. Some of the simulated results fall outside these boundaries which is natural since the suggested distributions result in both lower and higher values for the parameters compared to those used to fit the experimental data shown in Figure 4. As can be seen in Figure 5, few of the simulations exceed the maximum boundary. This is an effect of how the distributions for w_2 and a_2 were choosen. To achieve an increase in the external force after cracking, as seen for Beam 9 in Figure 6, a low value of w_2 (around 0.4 mm) is required in combination with a high value of a_2 (around 0.4). With the chosen distributions, the probability for this combination to occur is low. The right part of Figure 6 shows Force-CMOD curves for a maximum crack width of 1.0 mm, to highlight the estimation of the maximum force and how the initial softening part of the curve is simulated. The initial part of the curve is well estimated so the assumed Young's modulus is reasonable. It can be seen that some simulated results fell close to but outside the minimum boundary given by the experiment. This means that the simulated stress in the material when the fibres are activated is smaller than expected. This could be caused by a large crack width w_1 when the fibres are activated or a combination of large w_1 and small fracture energy of the concrete G_f Again, the majority of the results shows good correlation with the experimental results and it

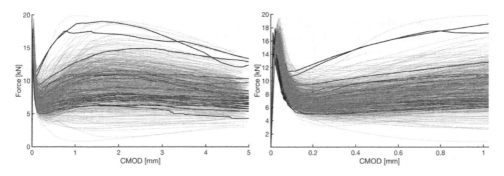

Figure 6. Force - CMOD curve from experiment (black lines) and from Monte Carlo simulations (grey lines).

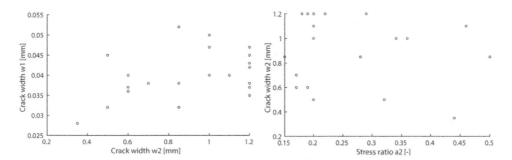

Figure 7. Relation between crack width w_1 and w_2 and between w_2 and stress ratio a_2.

can be assumed that discrepancies is a natural result of that the suggested distributions contains values both higher and lower than those suggested to fit experimental results.

The accuracy to simulate the maximum and minimum response of the fibre reinforced con crete could possibly be improved if any correlation exists between the parameters of the material model. However, the plots in Figure 7 show that no correlations exist between crack widths w_1 and w_2 or between crack width w_2 and stress ratio a_2. Other combinations of parameters showed similar results.

5 CONCLUSIONS

A numerical model that can describe the linear and non-linear structural behaviour of fibre reinforced shotcrete has been presented. Based on experimental results from 24 notched fibre reinforced concrete beams tested in a lab, suitable distributions of the parameters controlling the numerical model were presented. Random sampling of the parameters was performed to produce input to Monte Carlo simulations. A total of 2000 simulations were performed using the same set-up as the experiments to investigate if the presented probabilistic model was able to capture the typical scatter in results for fibre reinforced shotcrete.

Overall good agreements between numerical and simulated results were obtained. The majority of the simulated results are within the boundaries of the experimental results while a few results fell outside of the minimum response curve. This was expected since the suggested distributions include lower and higher values of the parameters in the material model than those used to fit the experiments. From the experiments shown in Figure 1, two beams showed a significantly larger increase in force compared to the others, up to a crack width of 1.0 mm. To simulate these beams a combination of low crack width w_2 and high stress ratio a_2

was required. With the suggested distributions shown in Figure 4 the probability that such combination occurs was low and therefore, the hardening behaviour shown in Figure 6 for two beams for a crack width up to 1.0 mm could not be entirely captured.

The probabilistic model presented here shows promising results on how the structural response of fibre reinforced shotcrete could be predicted. The benefit of using such a model is that the characteristic values for residual strength of fibre reinforced shotcrete could be over-conservative due to the large scatter in test results. A probabilistic design could therefore yield a more economic design and/or prediction of structural behaviour.

ACKNOWLEDGEMENT

The financial support for this research has been provided by the Swedish Rock Engineering Research Foundation. The authors would also like to thank Bekaert for sharing test results for the fibre reinforced concrete beams.

REFERENCES

Alberti, M.G., Enfedaque, A. and Gálvez, J.C. 2017. On the prediction of the orientation factor and fibre distribution of steel and macro-synthetic fibres for fibre-reinforced concrete. *Cement and Concrete Composites*, 77:29–48.

Ansell, A., Nordström, E. and Guarin, A. 2017. Laboratory investigation of steel fibre reinforced sprayed concrete using a computed tomography method. *In Proceedings of 8th International Symposium on Sprayed Concrete*. Trondheim.

Banthia, N. 1992. Steel Fibre-Reinforced Concrete at Sub-Zero Temperatures: From Micromechanics to Macromechanics. *Journal of Materials Science, Letters*, 11:1219–1222.

Bernard, S. 2010. *Shotcreting in Australia*. Technical report, Sydney.

Bernard, S. and Xu, G. 2017. Influence of fibre count on variability in post-cracking performance of fibre reinforced concrete. *Materials and Structures*, 50:169–177.

Cavalaro, S. H. P. and Aguado, A 2015. Intrinsic scatter of FRC: an alternative philosophy to estimate characteristic values. *Materials and Structures*, 48:3537–3555.

CEN. 2007. *Test method for metallic fibre concrete - Measuring the flexural tensile strength (limit of proportionality (LOP), residual)*. Technical report, Brussels.

Comsol. 2016. *Comsol Multiphysics version 5.2a*. Manual. Stockholm.

FIB. 2013. *fib Model Code for Concrete Structures 2010*. Technical report, Berlin.

Franzén, T. 1992. Shotcrete for underground support: a state-of-the-art report with focus on steel-fibre reinforcement. *Tunnelling and Underground Space Technology*, vol. 7, no. 4.

Gasch, T. 2016. *Concrete as a multi-physical material with applications to hydro power facilities*. Stockholm, Sweden.

Herrmann, H., Pastorelli, E., Kallonen, A. and Suuronen, J. P. 2016. Methods for fibre orientation analysis of X-ray tomography images of steel fibre reinforced concrete (SFRC). *Journal of Material Science*, 51: 3772–3783.

Oliver, J., Cervera, M., Oller, S. and Lubliner, J. 1990. Isotropic damage models and smeared crack analysis of concrete. *In Proceedings of the 2nd International Conference on Computer Adided Analysis and Design of Concrete Strucutres*, pages 945–957, Zell am See.

Sjölander, A., Hellgren, R. and Ansell, A. 2018. Modelling aspects to predict failure of a bolt- anchored fibre reinforced shotcrete lining. *In Proceedings of 8th International Symposium on Sprayed Concrete*. Trondheim.

Soetens, T and Matthys, S. 2014. Different methods to model the post-cracking behaviour of hooked-end steel fibre reinforced concrete. *Construction and Building Materials*, 73:458–471.

Yoo, D. Y., Yoon, Y. S. and Banthia, N. 2015. Predicting the post-cracking behavior of normal- and high-strength steel-fiber-reinforced concrete beams. *Construction and Building Materials*, 93:477–485.

Tunnels and Underground Cities: Engineering and Innovation meet Archaeology, Architecture and Art, Volume 6: Innovation in underground engineering, materials and equipment - Part 2 – Peila, Viggiani & Celestino (Eds)
© 2020 Taylor & Francis Group, London, ISBN 978-0-367-46871-2

The Tunnel Laser Scanner technique: Applications to the road tunnels monitoring

C. Sorce, L. Cedrone & A. Lattanzi
ANAS S.p.A., Design Department, Rome, Italy

ABSTRACT: The present paper shows some applications of the Tunnel Laser Scanner technique. Such technology, based on the laser and infrared joined employment, allows getting digital photographic image of the tunnel, its continuous geometric shape and extracting transversal sections at any progressive and the cavity thermographic image, simultaneously. Using this detection method, it is possible to control the lining thickness and the quantity of the materials used during the construction stage, to verify the actual conditions of an existing tunnel and to create a local mapping for a maintenance check of the structures. The Laser Scanner instrumentation used for the surveys of the road tunnels represents an innovative technique of investigation characterized by the significant level of details and accuracy of the measurements, as well as by the information acquisition rate and consequent reduction of the interference between construction and monitoring activities. Further point that is of particular importance it is the objectivity and reliability of the collected information that are independent from any operator's personal decision, unlike the use of the traditional topographic instrumentation. In the paper, some applications of the Laser Scanner to roads tunnels will be illustrated. The main purpose of the adoption of such technique was to ensure an efficient control of the structure and a consequent resource savings and performance optimization.

1 INTRODUCTION

In recent years the use of Terrestrial Laser Scanner technique in engineering field is gaining an increase of use due to its advantages.

In fact, underground conditions are not the most suitable for taking measurement: a dark, damp and dusty environment may create several disturbs and it is not possible to take traditional measurements during active construction work, but, at the same time, the work cannot be stopped for very long time for survey. However, there are many applications that require accurate information of the tunnel's features like the thickness of the concrete surface layer sprayed on the rock, the likely presence of structural discontinuities, possible strains and seepages, as well as the construction quality.

Terrestrial static or mobile Laser Scanning offers the fastest, safest and most accurate way of gathering data for the tasks above mentioned. Static scanners are portable and mobile scanners can be set up on a car, train or any other vehicle: they produce a dense 3D point cloud in short time that has several applications, such as construction survey of tunnels, cross-section extraction and deformation measurement. For example, the surveying of the railway or motorway tunnels can be conducted with mobile scanner without disrupting the traffic flow or placing personnel at risk in the traffic.

2   TERRESTRIAL LASER SCANNER

The core technology is the LiDAR technique, which is used to obtain the distance of each object point from the lens. The laser system produces and emits a beam of electromagnetic radiation: when the light reflected by the surface of an object is received, the system can calculate the range by the flight time and acquire the reflectivity of the surface (Figure 1).

The laser scanner consists of a transmitter/receiver of the laser beams, a scanning and a timing device. The scanner sends out laser pulses and then it receives and records the reflected signals. The timing device measures the time of flight of the signal and the scanner can compute the distance covered; knowing the direction and the angle of the light ($\cos \alpha$, $\cos \beta$, $\cos \gamma$) allows determining the relative position (x, y, z) of a reflective surface to the device.

$$\begin{cases} x = d \cos \alpha \\ y = d \cos \beta \\ z = d \cos \gamma \end{cases} \tag{1}$$

The surveys are performed by specialized technical advice on the instrumentation and application software. The scanner position is calibrated by measurements obtained by the total station. The necessary time, including the scanner positioning, in normal condition, does not overcome the 5 minutes. The time of one scanning (rotation to 360° of the scanner) is does not last more than 2 minutes. The distance between two positionings of the laser scanner must not be more than 1,5 diameter of the gallery and however not more than 25 m. The system performance allows the acquisition, under normal conditions, of approximately 120 m of tunnel for hour. To obtain a 3D model the scanner must work from different positions.

Figure 1.   An example of laser scanner.

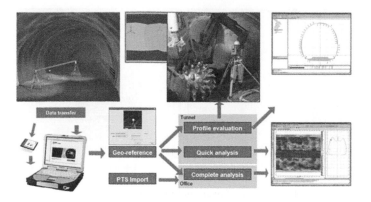

Figure 2.   Principle of Laser Scanner data acquisition.

3   THE LASER SCANNER IN TUNNEL EXCAVATION PHASE: THE 'CONIGLIO' TUNNEL

Terrestrial Laser Scanner is a high-precision measurement and analysis system for a complete and an efficient gathering of information of tunnels and caverns in all construction phases and during its entire life. The system has several advantages and potential applications of interest, which may lead to significant cost savings. In fact, it is possible to get more reliably the accuracy of the contours of the excavated profile, the thickness of the shotcrete, to control and optimize the construction activities, the occurrence of possible oversizing and to check the geological context (Figure 3). The software provides to input, managing and archiving the project and measurement data and results, to generate a digital three-dimensional surface model and to create cross sections with indication of over- and under-profiles (Figure 4). It is then possible to make a graphical comparison and calculation of the differential volumes along the profile of two construction phases, to produce a graphical and tabulated presentation of the results with the goal of concrete optimization for the inner lining, in accordance with project-specific requirements.

Since 2015 ANAS S.p.A.is monitoring by the laser scanner technique the 'SS117 - Itinerary North – South batch B4b dal Km 32+000 al Km 38+700 – Coniglio tunnel' project, with the aim to control the excavation working progress. The 'Coniglio' tunnel develops between pk 1 +610.00 and 2+598.00, with a total length of 988 m. The geological context is characterized by sand and silty soils, belonging to the Numidic flysch, typical of the Sicilian region.

Considering the large investments involved on the project realization, ANAS decided to adopt useful instrumentations to control lining thickness and quantity of materials used, during construction phase. In particular, it was adopted the following devices:

Figure 3.   Image representation of excavated surface.

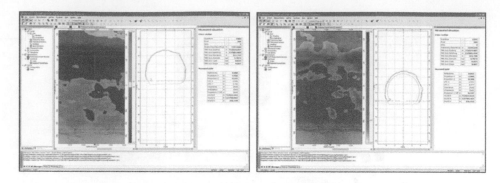

Figure 4.   Cross sections with indication of over- and under-profiles.

- totally automatic station with remote control and characterized by an angular precision of 0,5 sec;
- laser scanner with a data acquisition rate of 1.000.000 points/sec;
- dual frequency GPS for the georeferencing of external targets;
- PC 'Full rugged' and software application for the survey management and graphical comparison and calculation of the differential volumes along the profile.

Figure 5.   Laser scanner instrument in Coniglio tunnel.

Before taking the measurements, a preliminary design phase was needed to define the surfaces object of investigation, accuracy of the instrumentation and the possibility of obstacles presence (like ribs, working machines, etc.).

The first phase of the survey planned the collection and the georeferencing of the cloud of points resulting from the laser scanner survey (Figure 6).

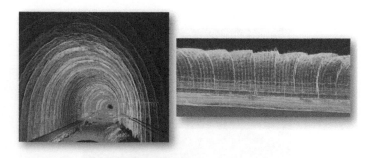

Figure 6.   Collection and georeferentiation of the cloud of points.

The investigation technology ensured to obtain an image representation on the tunnel plane and to visually compare, through a scale and a chromatic variation, the differences between the measured thickness and the designed one or between the measures in different epochs (Figure 7).

By these comparisons, it was possible to detect the problematic sections before the final lining application. Moving the cursor on the image it was possible to visualize, on the same monitor and in real time, geometry variations of the profile and the effective existing thicknesses or those to build.

In Figure 8 – Figure 9 there are two examples of comparison between design data and excavated profile during the working phase. It was possible to know, in real time, the areas characterized by under- or over-thickness problems, where it was necessary to act in order to obtain the correct design layer thickness.

The tunnel laser scanner enabled also a generation of a digital, three-dimensional surface model, useful to visualize, on 3D space, areas interested by design conditions restoring (Figure 10).

Coupling the results of the terrestrial laser scanner with Surface Penetrating Radar investigations, it was additionally possible to confirm the results about the correct layer thicknesses

Figure 7. Scaled image documentation.

Figure 8. Excavation profile comparison.

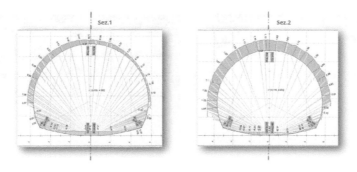

Figure 9. Cross-section extraction: comparison between design data and excavation phase.

Figure 10.   3D model reproduction.

defined in the design phase. Furthermore, the instrumentation let to check possible cavity extrados occurrences, to individuate the position and the type of installed ribs and to detect the final lining discontinuities.

Figure 11.   Surface Penetrating Radar test in 'Coniglio' tunnel.

During the excavation phases, the monitoring of eventual front extrusion, which could be characterized by different geometric shape depending on material type and tensional state (translation or rotation), is essential. The periodic analysis of the excavation face allows evaluating its geometric evolutions (Figure 12).

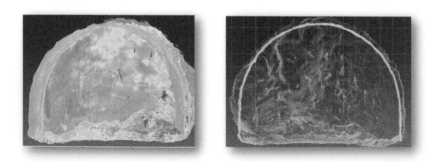

Figure 12.   Geometric control of the eventual front extrusion.

4 THE LASER SCANNER FOR EXISTING TUNNELS

The surveying and the mapping of the tunnels are crucial for the optimal use after construction and in routine inspections. The application of the laser scanner provides some geometric information of the tunnels, consisting of two kinds of indications, about strain measurements or the extraction of some features. These features, such as the cross-section, the centerline and the installations of the rails and the pipes, are necessary for the acceptance inspection and the tunnel mapping.

In July 2018 ANAS utilized this technique to survey the 'Palizzi' tunnel, located in Calabria region of the Southern Italy, which was interrupted for several problems during phase construction. The tunnel was characterized by the presence of the one provisional lining (realized by ribs and spritz-beton) and it was partially fill with infill material.

Figure 13.   Geometric shape control of the Palizzi tunnel.

Through this inspection, it was possible to know the temporary lining condition and the relative under-sized profile excavation, with the aim to foresee necessary and appropriate actions for the safety resumption of the construction works.

The traditional strains measurements in underground construction are generally performed by permanent control points installed around the excavation profile. This traditional control does not provide satisfying information about the whole tunnel surface. To overcome this drawback, it is possible to adopt laser scanning to have a full characterization of the modifications to the profile.

In this sense, the monitoring by the laser scanner ensures a further advantage because it does not imply permanent installation of control points. Finally, no damage is done during investigation and it is possible to execute the survey very quickly.

An important laser application in existing construction is the periodic control of the tunnel convergences, generally coupled with traditional topographic monitoring. ANAS is using this

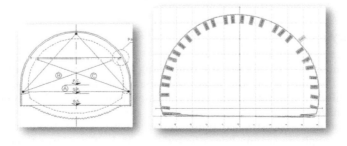

Figure 14.   Convergence monitoring using laser scanner in 'Coniglio' tunnel.

Figure 15. Laser scanner survey in 'Coreca' tunnel.

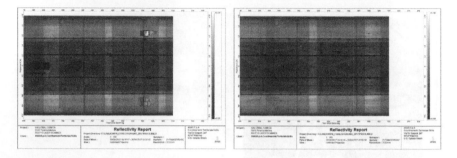

Figure 16. Evidence of cracks on the final lining of 'Coreca' tunnel.

kind of monitoring in 'Coniglio' tunnel since 2015 (Figure 14). The survey is normally executed using five control points applied to defined sections. Using this technology it is possible to integrate these measurements, obtaining precise convergence data about an unlimited number of sections and the corresponding points.

In July 2017 ANAS performed a final lining cracking control in 'Coreca' tunnel using the tunnel laser scanning analysis (Figure 15). This gallery belongs to the S.S. 18 'Tirrena Inferiore', that is an important transport route located in Calabria. The quick acquisition data was a fundamental property that allowed a reduced interruption of the intense road traffic characterizing this area.

Figure 16 shows the output analysis obtaining from the software, which allows the conformity check of the tunnel surface layer. It is evident the presence of several cracks that could have important implications on the final lining functionality.

Because of the crack damages detected by laser scanner inspection, the final lining conditions are constantly controlled through a traditional strains measurement. The conjunction of these instrumentations can provide, with a series of permanent control points installed around the profile, a more complete control and a global vision of the convergences of the tunnel.

5 CONCLUSION

Underground context are not suitable for conducting measurements, but rather they present several issues. In this framework, the Terrestrial Laser Scanning is increasingly being used as survey and monitoring technique, becoming an important tool for both construction and maintenance management. In fact, it is possible to obtain important information, like the complete detection of the constructed tunnel, the geometry shape, the conformity check of undulation before applying sealing liner, the quality check of the layer thickness of sprayed concrete and the detection of over- and under-profiles. All of these information can be obtained with a very fast, safe and accurate way of gathering data. Further point that is of particular importance it is the objectivity and reliability of the collected details that are independent from any operator's personal decision, unlike the use of the traditional topographic instrumentation. The capacity to create cross sections and digital two- and three-dimensional surface model, with these important features, is a unique peculiarity of the Tunnel Laser Scanner technique. ANAS S.p.A. has assumed this technique as a standard monitoring control in all its own projects.

REFERENCES

Amberg technologies. TMS Solution: surveying for tunneling professionals. https://ambergtechnologies. com/solutions-services/amberg-tunnel/amberg-tunnelscan.

Delaloye, D, Hutchinson, J., Diederichs, M., 2012. Using terrestrial lidar for tunnel deformation monitoring in circular tunnels and shafts. Eurock 2012, Stockholm.

Fekete, S., Diederichs, M., Lato, M., 2010. Geotechnical and operational applications for 3-dimensional laser scanning in drill and blast tunnels. Tunnelling and Underground Space Technology, 25 (5): 614–628

Gikas, V., 2012. Three-dimensional laser scanning for geometry documentation and construction management of highway tunnels during excavation. Sensors, 12 (8): 11249–11270.

Han, J. Y., Guo, J., Jiang, Y.S., 2013a. Monitoring tunnel deformations by means of multi-epoch dispersed 3D LIDAR point clouds: an improved approach. Tunnelling and Underground Space Technology, 38: 385–389.

Han, S., Cho, H., Kim, S., et al., 2013. Automated and efficient method for extraction of tunnel cross sections using terrestrial laser scanned data. Journal of Computing in Civil Engineering, 27 (3): 274–281.

Heritage, G.L., Large, A.R.G., 2009. Laser scanning for the environmental sciences. John Wiley & Sons, Hoboken.

Kong, X.L., Ou, B., 2013. The application and research of 3D laser scanning technology in finish tunnelling survey. Urban Geotechnical Investigation & Surveying, (2): 100–102.

Lam, S.Y.W., 2006. Application of terrestrial laser scanning methodology in geometric tolerances analysis of tunnel structures. Tunnelling and Underground Space Technology, 21 (3): 410–410.

Lindenbergh, R., Uchanski, L., Bucksch, A., et al., 2009. Structural monitoring of tunnels using terrestrial laser scanning. Reports of Geodesy, 87: 231–238.

Monserrat, O., Crosetto, M., 2008. Deformation measurement using terrestrial laser scanning data and least squares 3D surface matching. ISPRS Journal of Photogrammetry and Remote Sensing, 63 (1): 142–154.

Schulz, T., Ingensand, H., 2004. Terrestrial laser scanning-investigations and applications for high precision scanning. Proceedings of the FIG Working Week-The Olympic Spirit in Surveying, Athens

Seo, D. J., Lee, J.C., Lee, Y.D., et al., 2008. Development of cross section management system in tunnel using terrestrial laser scanning technique. XXIst ISPRS Technical Commission V Beijing.

Wang, W., Zhao, W., Huang, L., Vimarlund, V., Wang, Z., 2014. Applications of terrestrial laser scanning for tunnels: a review. Journal of Traffic and Transportation Engineering (English Edition). 1(5): 325–337

Tunnels and Underground Cities: Engineering and Innovation meet Archaeology,
Architecture and Art, Volume 6: Innovation in underground engineering,
materials and equipment - Part 2 – Peila, Viggiani & Celestino (Eds)
© 2020 Taylor & Francis Group, London, ISBN 978-0-367-46871-2

BIM implementation – Brenner Base Tunnel project

R. Sorge, D. Buttafoco & J. Debenedetti
BTC (Brennero Tunnel Construction), Fortezza, Bolzano, Italy

A. Menozzi, G. Cimino, F. Maltese & B. Tiberi
SWS Engineering spa, Trento, Italy

ABSTRACT: To improve quality and efficiency of all engineering services related to the detail design and construction engineering for Brenner Base Tunnel lot Mules II-III, BTC Contractor consortium and SWS have implemented BIM within the project. The tunnel consists of two main parallel single-track tubes that connect Innsbruck to Fortezza and of a central smaller test tunnel. All activities are supported by BIM adoption as tool for design, sharing and information check of the entire infrastructure facility. With this purpose, it has been possible to optimize costs and timing of design and construction work. Implementing innovative workflows, BIM utilities handled are concerning mainly following fields: advanced BIM modeling of structural and MEP parts with real constructive details, parametric segmental lining modelled, construction elements linked to planning schedule, automatic clash detection during design stage, mucking storage modelling, computational design with extensive use of programming software, automated modelling, automated procedures for drawings production.

1 INTRODUCTION

The Brenner Base Tunnel Project is a straight, flat railway tunnel between Austria and Italy. It runs from Innsbruck to Fortezza (55 km). Considering the Innsbruck railway bypass, which has already been built and which is the endpoint for the Brenner Base Tunnel, the entire tunnel system through the Alps is 64 km long. It is the longest underground rail link in the world.

Moreover, the Brenner Base Tunnel is part of the SCAN-MED corridor, one of the nine trans-European communication corridors. This network plays an extremely important role at the European level, as it will link important cities in Germany and Italy to the most important Scandinavian and Mediterranean ports. The Italian Mules II-III lot is the largest under construction, assigned to the BTC consortium with Astaldi S.p.A, Ghella S.p.A., Oberosler Cav Pietro, Cogeis S.p.A., PAC S.p.A for a total work amount of 1 billion €. The lot extends from the town of Mezzaselva in the province of Bolzano, up to the state border with Austria.

The Italian lot, whose completion is scheduled for 2023, will include the excavation of approximately 5,000,000 cubic meters of material for 65 km of tunnels divided as follows:

- 39.8 km of main tunnels;
- 14.7 km of pilot tunnel;
- 1 by-pass between the main tunnels each 333 m;
- 1 emergency stop in Trens;
- 4.5 km of access tunnel.

The first part of the excavations is planned with the drill and blast method to complete the crossing of the most important Alpine fault. The remaining part of the work will be excavated with mechanized method, with the use of three TBM Double Shield: two for the main tunnels and one for the pilot tunnel.

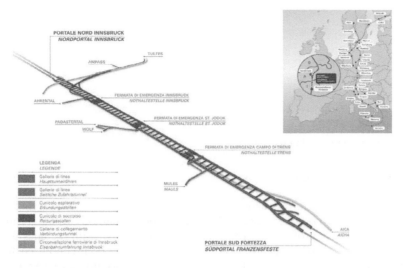

Figure 1.   Brenner Base Tunnel overview.

1.1   *Main Activities*

SWS Engineering S.p.A. is providing all engineering services related to the Detail Design and Construction Activities for lot Mules II-III, mastering geotechnical and structural design activities, MEP design (Mechanical Electrical and Plumbing), hydraulic design and supporting on site activities. All these activities are constantly supported by BIM workflow adoption. The infrastructure digitalization plays a central role as a tool during the design phase, for sharing information and data-attributes check concerning the entire project.

Here below it is shown a brief list of services highlights:

- Advanced BIM modelling of structural and MEP parts with real constructive details;
- Parametric segmental lining modelled;
- Construction elements linked to planning schedule;
- Automatic clash detection to highlight design inconsistencies;
- Improvement of design and quality check;
- Excavated material information;
- Computational design with extensive use of Dynamo, visual programming software;
- Automated modelling;
- Automated procedures for drawings production;
- Construction assistance;

1.2   *SWS BIM Approach*

A complex facility such as the Brenner Base Tunnel require powerful and centralized data management during design and construction. One of the most important results of such management is enhancing communication between the different parties working on the project.

According to the typical SWS approach for implementing BIM into a project, when the Input comes from the Client, it starts the project data management strategy. The strategy adopted is for ensuring that input data is consistently organized and shared on a common data environment. The approach strategy transforms traditional design input into digital models which can be handled in a database and visualized in 3D. Models and non-graphical data are coordinated within a unique clash detection resolution model.

Furthermore, information processing is performed on the digitalized project model, and design is validated with further postproduction activities leading to a high-quality output.

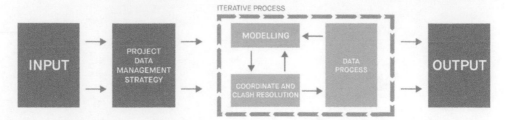

Figure 2.   Information processing scheme.

2   DESIGN SUPPORT ACTIVITIES

A BIM approach in a large infrastructure project includes a series of activities that support the development of the design process together with the traditional methods, collaborating and integrating works with other professional roles. For this project, the team includes geotechnical, structural, hydraulic and MEP engineers. All the activities have been conducted to reach the defined standard for this project, which is a Level 2 of maturity according to PAS 1192-2:2013.

2.1   *Construction design and detail drawings production*

The models have been produced with the purpose to become the unique central database of the project, repository of geometrical and non-geometrical information and data that can be reviewed for consultation during every phase: from design to construction, until management. Design solutions have been studied with the support of BIM technologies, to properly analyze design options as well as endorse the better solution.

At this time, infrastructure design has been dominated by traditional design and thus traditional representation: two-dimensional views, such as plans, elevations and cross-sections. However, the traditional representation is still an essential part of new BIM workflow which allows the direct link between 3D models and traditional 2D drawings directly produced from it, but also including in references and links to project drawings and documents, the model elements, connecting the interactive 3D BIM model to traditional documentation.

Moreover, thanks to a well-structured workflow and a strongly organized design team, the model does not require complete rework even in case of big changes of cross-sections rather than alignments. Production of construction drawings from BIM models is consequently a straightforward process that reduces elaboration time compared to traditional drawing techniques (CAD) and considerably lowers the risk of introducing errors due to complex geometries and space positioning.

2.2   *Data extraction: scheduling and quantity take-off*

The models include detailed information of construction phases, costs, and geometrical dimensions as areas, volumes and lengths that can be easily accessed and extracted through interactive schedules.

The BIM model has been structured as a series of successive groups of typological elements, reflecting the logic of operations and quantity calculations as defined by the Contractor's Accountants, and properly parametrized to automatically give results for the specific and desired computation item and, if necessary, to edit and update items with a smart workflow, with biunique correspondence between cost codes and model elements.

The BIM design team has developed tools for the automatic export of these schedules-tables in open format (xml, txt, csv) instead of Excel spreadsheet (xls). In this way, exchanging information between parties is clearly enhanced: faster and more reliable. The project's federated model becomes a tool for precise cost control and evaluation of changes during construction for a balanced planning and production of scenarios.

2.3 On-site construction team support

During the development of the project SWS is giving continuous feedback to on-site works. On this side, the BIM federated model is a fundamental tool that collects and merges data and projects, highlighting potential interferences and inconsistencies.

With specific reference to interference checking, an extensive study has been conducted regarding temporary works. Temporary works – such as conveyor belts, ventilation systems, cooling systems, etc. - are often designed by various manufacturers and therefore need coordination to minimize the impact of their construction on the work progress.

Collaboration with the construction team has been a key point in taking advantage of the BIM model and procedures. Specific activities have been developed with respect to spatial interference checking, control and planning of excavated material, spatial positioning of temporary structures and MEP systems, with the support of laser-scanning techniques and on field surveys. Laser scan data and surveys are integral part of digital models, inserted and correctly georeferenced within various models. With this purpose, coordination and control activities are simplified, due to the presence of many more information and tools accessible.

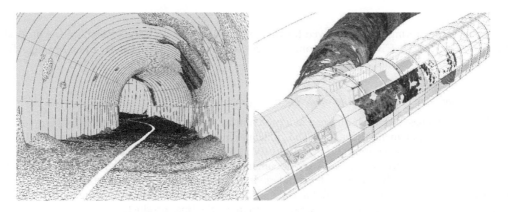

Figure 3.   Model of theoretical and as-built excavation recreated from laser scan (left) and point clouds (right).

2.4 Communication and coordination

Communication is a key component of all construction projects and to be successful, it requires many different people working together to achieve a unified goal. Implementation of BIM introduces and nearly obliges to constant collaboration between parts and sharing of information.

Collaboration among disciplines and stakeholders involved is a critical point that, when not accurately managed, can result in incomplete or incorrect data sharing or a need to fully remodel a design.

3   METHODS AND PROCEDURES

To perform the activities pictured before, different procedures and tools have been developed since pre-contract phase, according with contractor's needs, to outline the type of information that had to be created and the way they had to be accessed. In post-contract phase, the creation of contents took advantage of a rigorous planning and definition of methods, often facing the limits imposed by the state-of-the-art of the authoring software and thus leading to a large use of automation, coding, and in-house software development, thus minimizing errors in contents creation that a manual process could be prone to. R&D department played crucial role to support these activities, combining data science with civil engineering expertise.

3.1 Software and Regulations

For each activity and part of the project, a regulation reference - when available - has been defined. Because of the large number of regulations available in the construction field but the lack of specific rules in infrastructure, a fundamental preliminary work has been done, collecting international regulations and from them extracting key points and principles applicable to the Brenner Base Tunnel project thus assigning each of them to a specific part of the project (for example: guidelines, coding, LOD, model sharing, etc.).

I – rif. primario / II – rif. secondario	Linee Guida	Processi Collaborativi	Codifica dei File	Codifica degli Elaborati	Elaborati Grafici	Classificazione dei Contenuti	Formati di Scambio	LOD	CDE	Stima dei Costi	COBie	Indicazioni Cotrattuali
BS1192:2007		I	I	I	I				I			
PAS1192-2:2013	I	I	I	I				II	I			
PAS1192-3:2014	I											
PAS1192-4:2014	II										I	
PAS1192-5:2015	II											
NBS BIM Toolkit	II			II	II	II		II				
COBie-UK 2012											I	
Uniclass 2015						I						

Figure 4. Example of regulation references table of priorities.

The main authoring tools approved for this project are Autodesk Revit, Dynamo, Autodesk Inventor, Autodesk Civil 3D for content creation and Autodesk Navisworks Manage for coordination, review and sharing.

3.2 Management and control of modeling activities

All BIM guidelines have been outlined in a pre-contract agreement following the EIR (Employers Information Requirements) structure. The guidelines have then been developed in a more detailed and specific set of instructions and practical procedures in the BIM Execution Plan (BEP) of the project. The BEP represents a necessary document in a collaborative project to ensure collaboration and cooperation between all stakeholders. It evolves with the work progress and therefore is updated periodically to become a full and detailed manual that, together with the models, represents the complete package of data necessary to manage the final product.

BIM Execution Plan's main contents include: general information, key roles and people involved, regulations, project milestones, Model Breakdown Structure, model contents creation and levels of detail, outputs and data sharing system, information management, hardware and software requirements.

3.3 Model Breakdown Structure (MBS) and modeling principles

The main reference for classification of elements is the NISTIR 6389 Uniformat II. Because of its lack of information in infrastructure field though, a new classification system has been specifically developed for the Brenner Base Tunnel project.

For each new Uniformat Level, the Level of Detail (LOD) has been defined according to the BIM Forum LOD Specification 2015, together with their attributes and parameters. In the table below is shown an extract of the customized classification.

UniFormat Level					Relevant Attribute Tables	LOD	Date MEA	Notes
1	2	3	4					
T				TUNNEL				
T	10			Tradizionale				
T	10	10		Scavo			GEO	
T	10	10	.10	Terreno	A, B Materiali	400		
T	10	20		Rivestimento I fase		400	STR/GEO	
T	10	20	.10	Calcestruzzo (shotcrete)	A, B Calcestruzzo	400		
T	10	20	.10	Centine	A, B Acciao strutturale	300		
T	10	20	.10	Chiodi	A, B Acciao strutturale	300		

Figure 5. Example of the classification system based on Uniformat II.

The final model of the project is composed of several nested and linked models each one representing a significant part of the project according to the Work Breakdown Structure (WBS) and to its subcategories and thus reflecting the original project organization.

Following a multi-level approach every single model consists of parametric objects that allows to reach the desired level of detail up to – for example - the millimeter-scale detail of segments' features in the mechanized tunnels or rebars in concrete structures.

In order to obtain internal coherence between a large number of models and families, detailed templates have been developed by SWS for each discipline, defining view templates, annotation styles, schedules style, shared coordination system, etc. The model's names are defined by a code that allows to identify the fundamental part of the project by its WBS, chainage and discipline.

3.4 *Data sharing and quality check*

Data and information are managed according to BS 1192:2007 and PAS 1192-2:2013 in a shared cloud-based platform used to exchange, revise and output data. With this purpose, different phases have been managed: during design phase, between different disciplines inside the project team, and during construction phase, between project and construction team. Every part of the project is transferred in the Common Data Environment (CDE), granting the team managers to control the status of each model (approved, reviewed, delivered, pending requests), and the actors involved in the process of creation, validation, and delivery.

Every model and every information provided is subjected to a procedure of quality control. The procedure is articulated in two main steps: the first step is in charge of the designer, SWS, which takes the models under a systematic examination before every intermediate delivery, checking the adherence with the prescriptions set in the BEP and including in the BIM file a report highlighting each possible issue. This phase takes great advantage of in-house automation procedures specifically developed for the Brenner Base Tunnel project to implement automatic quality checks and to validate the model and its content. After the delivery on the Common Data Environment, a second and final quality control is held by BTC which, if positive, validates the model.

4 DIGITAL CONTENT CREATION

The digital content creation process has been regulated by defining initial standards, in order to achieve internal models' congruency and meet the Contractor's needs, such as optimizing design, costs and timing of construction works. The digital content creation is based on smart and intelligent digital objects which behaves like real-world elements, to avoid time losses and obviously improve design quality.

The infrastructural modelling is realized using not only traditional BIM tools: a fundamental role is represented by the integration of programming and visual programming tools and workflow implemented within the BIM design team for the infrastructure.

As general rules, SWS defined some fundamental modeling strategies to be adopted during the modeling process like a dual Coordinates System defined, in every model, by the Project Coordinates System and a Shared Coordinates System. Also, elements' geometric definition and modeling followed construction processes, adhering the project typological categories' subdivision, implementing references to those values in every model's element (e.g.: Sheet Code, WBS, Pricing Codes and Construction Phases).

4.1 *Input Data*

Inputs for Brenner Base Tunnel were two-dimensional and written inputs mainly consisting in:

• Bi-dimensional alignment and profile drawings;

- Bi-dimensional tunnel cross-section drawings (cast-in-place concrete lining or precast ring segmental lining);
- Contractor's Project organization, such as Lots, WBS, etc.;
- Time schedules.

Many different software have been used in an integrated workflow to achieve the desired result, and, starting from the inputs data, the entire BBT project has been modeled to obtain a 3D BIM Model.

4.2 *Element creation*

Each element composing the tunnel has been modeled as a separate component to distinguish functions and materials, such as excavation theoretical profile, primary lining and anchors or bolts as well as drains, waterproofing and secondary lining (for both drill and blast or TBM methods).

All those elements have been defined referring typological base-unit elements according to executive design specifications or Contractor's needs and have been created as Autodesk Revit adaptive parametric objects, capable of representing in the most accurate way possible their real position on the alignment, by adapting their placement or form inside the 3D environment.

For highly complex geometries, such as ring segments, Autodesk Inventor was used, developing a parametric configurator, which can generate high-fidelity ring segments' models by linking the software's engine to the parametrized geometrical ring's properties. By importing the ring geometry into a Revit adaptive family, in addition, it has been possible to create the parameters for ring's placement to adhere its real behavior along the alignment.

Figure 6.   TBM ring segments creation in Inventor and Revit.

BIM model elements have been defined with an increasing level of geometrical detail and nested information. Parameters reporting terrain, construction material and project organization related information were assigned to every specific typical cross-section and modeled component.

The overall model, and all its components, have been georeferenced according to the project coordinate system, so that all models are correctly real world oriented, with information related to chainage and to the coordinates of its base point.

4.3 *Design Automation and Programming Activities*

Due to the magnitude of the project and the high LOD required, the number of elements composing each single model easily reached significant amount. With those assumptions, a conventional workflow would have led to a potentially uncontrollable and error-leading situation, or to an excessive and inefficient resources effort. Therefore, SWS implemented innovative workflows, characterized by an extensive use of programming software (coding language such as Pyhton), computational design, automated modelling and procedures for data extraction and drawings production (by means of Dynamo), developing its own tools to reduce the risk of human errors and obtaining total control over the procedures of element creation and parameters insertion.

5 INTERFERENCE CHECKING

During the detail design phase, it is fundamental to coordinate the various designs to prevent any inconsistencies which could otherwise reveal themselves directly on site, causing delay and unexpected costs.

5.1 *Site Logistics and Construction Phases Simulation*

Construction site and its logistic aspects have been interested by BIM implementation, aiming to visualize possible interferences. Therefore, supply flows and storage or operational areas were included in the model (e.g. spoil material, ring segments stockpiles). The picture below shows Hinterrigger's construction site, for which BIM played a fundamental role in its logistic organization, regarding excavated material and the site area itself.

Figure 7.   Hinterrigger site BIM model, for stockpiles and material flows simulation.

By means of laser scan techniques, it has been possible to determine the excavated rock's bulk modulus, estimating material's volume. Then, storage and usage dates have been assigned to materials, to simulate and display phases and the corresponding volumes of use, also for a visual evaluation of the possible scenarios domino effects. BIM has been used for construction phases simulation, exploiting WBS and timing parameter associated to the various items, displaying operative programs and elementary activities throughout which the various project stages will be completed.

5.2 *MEP systems*

MEP systems play an extremely important role in an infrastructure project, because of their size and complexity, especially during the construction phase in which many activities are conducted simultaneously with a high chance of changes during works.

Therefore, temporary MEP system have been modeled to the highest LOD to check their actual fitting in the as-built primary lining and, if needed, adjust the position defined in the Detail Design. Also, in particularly complex nodes, all construction supporting plants (such as spoil conveyors structures, hoppers, etc.) have been modeled, as interferences may rise between them due to physical space lack or in case of vehicle transit (e.g. dumpers or TBM pieces transportation).

MEP system modeling also included permanent ones, for realization feasibility check and scheduling/BoQ purposes.

Figure 8.   Permanent MEP systems (left) and temporary works MEP systems (right): high complexity logistic node including topographic targets and collimation lines, TBM service watering pipes, ventilation plenum/ducts and conveyor belts structures, modeled to check interferences.

6 RESULTS AND DISCUSSIONS

BIM implementation increased data sharing, improving data accessibility and interaction between parties involved. It strongly enhanced teamworking among different departments and collaboration with the involved manufacturers. It also made possible to overcome limits of proprietary software used for creating infrastructure models, using widely diffused file formats which converged into a coordination model, which highlighted tens of inconsistencies or clashes.

Figure 9.   Cooling towers comparison: design BIM model and as-built.

BIM and computational design also led to otherwise unreachable results, such as connecting alignments and profiles to BIM geometries, during the laying-out of the elements, from drill and blast or segmental lining ones, to worksite supporting and/or final MEP systems with real constructive details, even at the end of the modeling process, establishing a biunique connection.

Post-processing of laser scan survey data, such as flat sections or point clouds, has been made thanks to in-house Dynamo scripting, through which it has been possible to recreate and physically show the as-built solid geometries.

Results lead, in example, to design optimization of particularly complex tunnel nodes, to precise excavation volume value calculation, useful to check the actual material flow of a desired excavation section, and to study/verify the actual over/under excavation in desired tunnel parts.

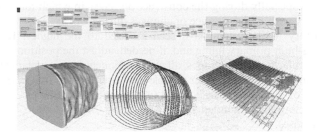

Figure 10.   Volume extraction from point cloud processing and under/over excavation study from as-built flat sections survey data.

Thanks to timing data integration, 4D simulation were made to check the models and draw attention to any clashes or other interferences between concurrent activities, comparing different scenarios of site logistics and material fluxes (supply/storage/use), highlight the effects of the choices or realization method on the storage areas utilization.

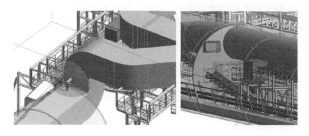

Figure 11.   Example of clashes between projects from different manufacturers.

7   CONCLUSIONS

The BIM implementation to the Brenner Base Tunnel project has been described in this article. The main conclusions are summarized as follows:

- Increase of collaboration between all parties involved from designer to contractor and client with easy and accessible data sharing;
- Interoperability between disciplines led to great working time reduction especially in project/decision making phases;
- Ability to manage different design option, finding faster solutions and with higher quality of design thanks to parametric design and spatial coordination;
- Crucial tools and workflows have been developed to face and manage such a big project, defining a standard method for future applications in large infrastructure projects;
- More than 190 models with more than 1000 elements each, result in thousands of parameter values: impossible without BIM tools and workflows.

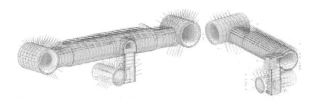

Figure 12.   Final model of a by-pass tunnel linking the two mechanized main tunnels.

REFERENCES

BIM Forum, 2015. Level of Development Specifications
British Standard, 2007. BS 1192 Collaborative production of architectural, engineering and construction information - Code of practice
https://www.bbt-se.com/
National Building Specification, 2015. Uniclass 2015
Publicly Available Specifications, 2013. PAS 1192-2 Specification for information management for the capital/delivery phase of construction projects using building information modelling
Schiavinato, L. & Colombelli, S. & Eccher G. & Nave, D. & Cucino, P. (SWS Engineering spa) & Bona, N. (University of Rome Tor Vergata), 2016. BIM Use in the Infrastructural Field: The Case of the Extension of the Railway in the Underground Track of Catania, From the Central Station F.S. to the Airport

Tunnels and Underground Cities: Engineering and Innovation meet Archaeology, Architecture and Art, Volume 6: Innovation in underground engineering, materials and equipment - Part 2 – Peila, Viggiani & Celestino (Eds)

Concrete final lining inspection procedure and case study

C. Steiakakis & S. Delmadorou
Geosysta Ltd, Athens, Greece

A. Simopoulos
Nea Odos Inc., Athens, Greece

ABSTRACT: A significant financial investment is required to construct, operate and maintain tunnels and therefore, a systematic inspection procedure in place is imperative to effectively evaluate the condition of the lining during their operation. Thus, a proper maintenance schedule can be followed preventing unexpected repairs and keeping costs down. This paper presents the implemented procedure for the concrete lining visual inspection of "Ionia Odos" highway tunnels in Western Greece. It provides a systematic approach to visually evaluate the structural integrity and record issues needing to be addressed during operational life of the tunnels. Furthermore, a case study from the first structural visual inspection of these tunnels is presented. Initial findings from the implemented procedures along with aspects of structural issues and operational challenges that occurred during the inspection are outlined in detail.

1 INTRODUCTION

Having a total length of approximately 200km, Ionia Odos is the main corridor of Western Greece connecting the region of Antirio with the city of Ioannina in the north. The Ionia Motorway route required the construction of four twin tube rock tunnels measuring 11km in total length and utilizing either reinforced or plain concrete for their final lining support (Figure 1). The considerable financial investment needed for the construction, operation and maintenance of such tunnels leads to the conclusion that a systematic inspection procedure of the tunnels is imperative to ensure their overall integrity and prevent failures.

The very first visual structural inspection of Ionia Odos' tunnels was conducted in March 2018, only a few months after the motorway was put in full operation. Makyneia, Klokova,

Figure 1. Ionia Odos Motorway – Tunnels.

Kalydona and Ampelia tunnels were visually inspected in order to record any structural issues derived right after their construction and to create an initial database for future reference. All the required information pertaining to the theoretical basis of the inspection procedure, the overall arrangements needed for the inspection to be conducted, as well as the issues recorded and the difficulties encountered during the inspection are delineated in the following sections.

2 CONCRETE LINING DEFICIENCIES

A visual structural inspection of a concrete lining refers to the detailed examination of the concrete surface in order to document any defects. There are different kind of defects or problems that may be observed on a concrete lining, which are presented below.

One set of observations may refer to surface textural defects which can be easily detected and usually are caused by unsatisfactory concrete mixture, placement and hardening. Improper vibration of concrete and the presence of larger percentage of coarse aggregate particles than needed typically leads to the formation of bugholes and honeycombs respectively (ACI 201.1R-08). Bugholes (Figure 2) appear as small surface voids resulting from entrapped air between the concrete – form interface due to inadequate vibration of the concrete during placement. These features are especially difficult to overcome if the lower part of the lining has a convex geometry and air is trapped more easily between the formwork and the lining. Honeycombs (Figure 3) appear as cavities and visibility of aggregates due to the failure of the concrete paste to insert in-between the voids generated by coarse aggregates especially in heavily reinforced sections. Poor curing of concrete combined with water saturation, excessive evaporation and freeze-thaw conditions may also result in the development of crazing and scaling of the concrete. As shown in Figure 4 crazing appears as a network of closely spaced micro cracks on the surface mortar (PCA, 2002), where scaling refers to the flaking of the finished concrete surface as presented in Figure 5 (Soudki, 2001). Additional textural feature observations may also refer to the appearance of pour lines on the concrete surface and dusting defects. Delay in concreting stages usually results in the appearance of visible pour lines, as depicted in Figure 6, which indicate the hardening of the first concrete layer before the subsequent layer was placed (PCA, 2002). In Figure 7 the surface dusting effect is displayed which appears as a powered material generated due to the combination of excessive water content and rapid drying of concrete (PCA, 2002).

Although all the aforementioned concrete observations may create the impression of significant findings requiring immediate treatment or repair, when examining a tunnel concrete lining they mostly pertain to cosmetic problems. Such issues lay exclusively onto the lining surface and do not affect the structural integrity of the tunnel. There are only two exceptions made regarding the scale of honeycombing and pour line effect and hence these two should be cautiously examined. When honeycombing extends deeper into the tunnel section, further deterioration may develop as moisture could easily penetrate and affect the section (Soudki, 2001) or reinforcement. Pour lines should also be carefully evaluated if they form a cold joint and if this joint is significant from a structural point (ACI 201.1R-08). This issue will be discussed in more detail later in this paper.

Figure 2. Bugholes.

Figure 3. Honeycombs.

Figure 4. Crazing.

| Figure 5. Scaling. | Figure 6. Pour lines. | Figure 7. Dusting. |

The development of cracks on concrete surfaces is also a frequent observed problem the significance of which can vary depending on the type, location, extent and scale of the crack. Concrete cracks can be basically divided in two main categories, surface cracks and deep cracks (Soudki, 2001). Shrinkage cracking is the most common example of surface cracks caused by the tensional failure of restrained concrete during its curing phase (Figure 8) (ACI 201.1R-08). The development of shrinkage cracks is a common observation onto tunnel final linings, especially if plain concrete is utilized. They are generated right after concrete pouring and remain inactive afterwards. As a result, most of the time they stand as insignificant cosmetic issues since such cracking is not even identified by transit vehicles and most of the time it does not continue into the section. All other types of cracks may be either detected at the surface of the concrete or may extend to a greater depth into the concrete section. Their severity depends on the reason of cracking development, the parameters affecting its propagation and their location. The most ordinary crack types in tunnel concrete final lining pertain to individual longitudinal or vertical cracks, diagonal cracks, cold joints, D shaped cracks or to a general pattern cracking. Longitudinal cracking usually develops parallel to the length of the tunnel block in the crown area, diagonal cracking is inclined to the longitudinal tunnel axis, where the D shaped cracks are located near the construction joints of the concrete structure at the crown area. Such crack types are presented in Figures 9, 10 and 11 respectively. Cold joint cracks (Figure 12), are usually formed from the delay in concrete placement where a joint is formed between the already placed concrete (bottom) layer, which has hardened substantially, before the top subsequent layer is poured. Depending on the concrete placement delay duration the cold joints may lead to severe stability effects. Pattern cracking, as shown in Figure 13, refers to a cracking form of repeated sequence (ACI 201.1R-08).

In general, the significance of cracks detected on tunnel surfaces cannot be exclusively related to their category type. The crack extend, location and the reason of formation, specify the severity of each crack and classify them as cosmetic or structural. The only exception applies to shrinkage cracking, which as described above, is a minor cosmetic concern, but if extended, it could reduce the long term durability of the concrete. However, any presence of moisture in the tunnel could render shrinkage cracking as a major concern in reinforced

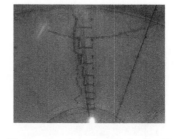

| Figure 8. Shrinkage cracks. | Figure 9. Longitudinal & D shaped crack. | Figure 10. Diagonal crack. |

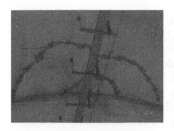

Figure 11.   D shaped crack.

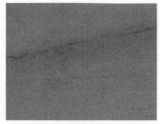

Figure 12.   Cold joints.

Figure 13.   Pattern cracking.

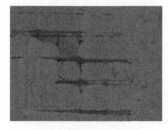

Figure 14.   Reinforcement
exposure & mortar flaking.

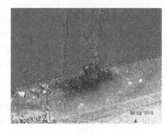

Figure 15.   Rust Staining.

Figure 16.   Leakage.

linings, since if the cracks have openings greater than 0.3mm they can allow moisture to penetrate deeper into the concrete section and adversely affect reinforcement. Thus, when examining cracking effects on a tunnel surface, all concurrent effects should be evaluated in order to reach a valid conclusion regarding the crack severity level.

In case of reinforced concrete, observations regarding reinforcement exposure, corrosion and concrete rust staining are also very important and should be documented. Inadequate reinforcement coating during construction may result to exposure of steel bars, which in combination with possible moisture effects, progressively convert the steel to rust which results in rust staining and mortar flaking on concrete surface (Figures 14, 15). The resulting rust steel expands inside the concrete and creates tensile zones which lead to local flaking (PCA, 2002).

Exposure and resultant corrosion of tunnel lining reinforcement is an issue of great importance since omission of remediation techniques may affect the functionality of the tunnel (Soudki, 2001).

Last but not least, the existence of external factors such as moisture effects, underground water level combined with lack of waterproofing measures and oxidized air also affects tunnel concrete linings. The most common impact resulting from the aforementioned effects is water leakage (Figure 16) (ACI 201.1R-08). Such impact should be cautiously examined and evaluated since it can lead to deterioration of the tunnel concrete section, corrosion of the steel reinforcement and eventually affect the structural integrity of the tunnel.

3   PRE-INSPECTION GENERAL ARRANGEMENTS

The accuracy and detail of observations and findings made during a concrete lining visual structural inspection is strongly associated with the preliminary desk study. More specifically the following items should be addressed.

It is very important that all pertinent information from the construction phase is collected. The inspector must be aware of the geotechnical profile of the area, the type and the geometry of the concrete final lining as well as of all the difficulties and problems that occurred during construction and how they were handled. For instance, the significance of a recorded crack on the concrete surface of a tunnel block may vary and is strongly related to whether the block supports a stable or a squeezing rock mass, if the section of the block is of an open or closed type, if the concrete lining is reinforced or plain, if in the current location there are stress concentrations or if any other type of geotechnical problems preexisted, etc. On account of this, all the available information for the construction of Ionia Odos' tunnels were gathered and examined by the inspectors in order to have in hand all available information regarding construction practices and encountered problems.

The observations during a tunnel visual inspection must be evaluated correctly and recorded accurately and with the appropriate detail in a difficult environment (dark, windy, noisy etc.). Thus, the appropriate equipment must be available. When examining the condition of a concrete tunnel, the use of a crackmeter, a tape measure, a torch and a high quality digital camera is imperative. The need for such items is obvious for the correct and accurate measurement and recording of defects. Chalks and water spray are two additional items that initially may seem of minor importance, however they significantly ease the work of the inspectors. More specifically, the utilization of chalks to number the tunnel blocks serves to constantly remind the inspectors their exact location within the tunnel, while numbering the cracks or painting in a non-permanent way the crack pattern. Water spray is utilized to reveal hairline shrinkage cracks after being moisturized. The inspection must also be performed in the safest possible manner, so all appropriate safety gear must be available at all times.

During tunnel inspection any defect or problem encountered must be recorded appropriately and with sufficient accuracy and detail. It seems that a general tunnel inspection template is not readily available in the international literature. Therefore, it was essential to create a template inspection sheet for Ionia Odos' tunnels (Figure 17). Such a template should be simple to understand and utilize and should include all the required information regarding the exact location into the tunnel and the type of issues recorded.

Furthermore, when an inspection of a tunnel in service is scheduled to be conducted, the traffic arrangements need to be made. At least one traffic lane must be closed to vehicular traffic and if a lifter is to be utilized the whole tunnel branch may have to be closed to traffic for safety reasons. It should be mentioned that even though tunnels are underground structures and weather conditions seem not to affect the work inside them, it is strongly recommended that tunnel inspections should not be scheduled during periods of remarkably low

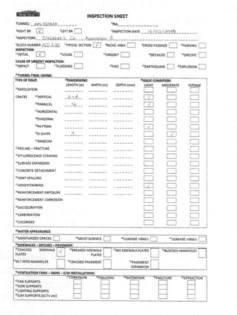
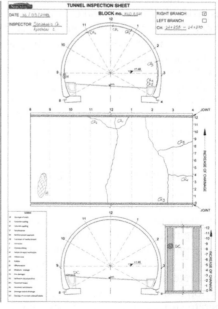

Figure 17. Tunnel lining inspection and recording sheet.

temperatures because such conditions may force inspectors to work for short time intervals due to wind currents formed in the tunnel tube.

Last but not least, the necessity of using personal protective equipment (PPE) should be emphasized. Each inspector must be equipped with the required safety gear such as hard hats, goggles, vests and safety gloves for hand protection. Thus, they would be clearly identified by passing vehicles and they would also be totally protected in case of unexpected hazards.

4 TUNNEL STRUCTURAL VISUAL INSPECTION PROCEDURE

4.1 General

As mentioned above, there are four twin tube tunnels along the Ionia Odos motorway named after the adjacent villages. Makyneia tunnel is the first and shortest tunnel having a length of approximately 500m and an overburden of 50m in flysch formation, followed by the longest and deepest Klokova tunnel measuring 3km and excavated under 600m of mostly limestone rockmass. The two remaining Kalydona and Ampelia tunnels present similar characteristics each having an approximate length of 1km and an overburden of 100m excavated in flysch and limestone rock mass respectively

After examining all the data from the construction phase of the tunnels it was verified that the final lining of Ionia Odos' tunnels had utilized either reinforced or plain concrete, consisting mostly of 12m length blocks, except for specific areas where the length of the cast blocks were shorter. More specifically Makyneia, Ampelia and Kalydona tunnels have mostly been constructed with unreinforced concrete lining with very few reinforced blocks whereas the final lining of Klokova tunnel is entirely constructed with rebar reinforced concrete. The cross section of the tunnels was generally designed as an open type section, provided that the excavation took place in good quality rockmass, except of localized geotechnically problematic areas where a closed type with invert section was used (Figure 18).

The average structural concrete thickness of the vault was 0.35m and the joint between the vault and the footings was constructed as a hinge.

4.2 Findings

First of all, it should be mentioned that the overall structural condition of the four inspected tunnels was found quite sufficient and of good quality with no significant problems. Such results were reasonably expected as Ionia Odos motorway has been in full service for a short time and was constructed with high standards. Nevertheless, some minor problems or defects were documented and are presented in the following paragraphs.

Minor textural features of the tunnel concrete surface, were identified in all tunnels. The frequency and scale of these features was different in each tunnel. Mostly bugholes, and in a much lesser extend honeycombs and pour lines were findings along the Makyneia, Klokova and Kalydona tunnels. It was impressive that although in the first three tunnels all of the above findings were recorded in regular intervals, in the Ampelia tunnel they were all detected in very localized areas. This was contributed to a much better aggregate gradation and type

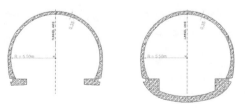

Figure 18. Geometry of typical support section at Ionia Odos' tunnels (left: open type section; right: invert section for problematic areas).

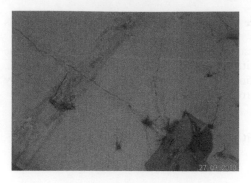

Figure 19.   Longitudinal crack near the crown.

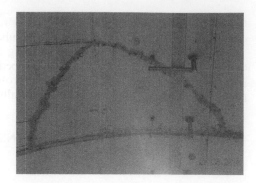

Figure 20.   Repaired D shaped crack in the crown.

available for the concrete mix in Ampelia tunnel
which is located almost 150km north from the other tunnels. Minor crazing and scaling effects were also identified in multiple areas in the Klokova tunnel. Crazing was mostly located near the pavement area of the tunnel section while scaling tracked in the vault. Such features were found almost in every tunnel block throughout its total length and were attributed to the concrete mix properties and possibly in some cases the premature form removal. Considering that such observations were recorded for the three southern tunnels, it was assumed that the aggregates available in the southern part of the motorway were not as good as the ones utilized for the northern Ampelia tunnel.

Considering the honeycombing effects, they were found in a limited extend and in areas mostly in compression near the footing of the tunnel. Due to their limited extend they were considered as minor structural problems that would not affect the behavior of the lining. Nevertheless, they should be examined and reevaluated in the future in case external conditions, such as moisture effects, air oxidization etc., affect their propagation and impact in the long term.

Concerning the cracking of concrete tunnel section, it was observed that cracks mostly developed onto sections utilizing unreinforced (plain) concrete. Consequently, Klokova was the only tunnel not to present such crack issues. Most of the recorded cracks were characterized as shrinkage cracks that could have been reduced with the utilization of better quality aggregates and more sufficient curing time. The absence of cracking onto Ampelia's tunnel blocks is justified since it reveals prime concrete quality as aforementioned. More extensively, both at Makyneia and Kalydona tunnels a pattern cracking was observed onto almost every single block. The formation of such cracks in unreinforced concrete tunnels is standard and occurs during curing of the concrete. Hairline shrinkage cracks were detected on both sides of Makyneia's tunnel walls, while the development of longitudinal (Figure 19) and D shaped

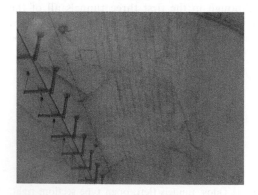

Figure 21.   Reinforcement shading.

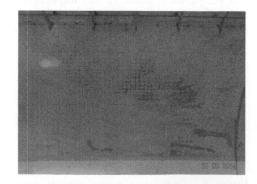

Figure 22.   Reinforcement shading and exposure.

cracks (Figure 20) at the crown were also documented. The longitudinal and D shaped cracks at Kalydona's crown was a frequent observation, but most of them had already been repaired. In the widest longitudinal cracks, crack meters had been installed and monitored and based on the data gathered, they did not produce any movement. The cracks after the initial setup time remain dormant and do not pose any structural issues unless the tunnel is not water-proofed or the crack pattern forms unconstrained geometric features such as the D shaped cracks.

D cracks were formed during construction at the time for formwork placement for the next block to be concreted. The pressure extended by the formwork onto the newly and unhardened previous block in addition to the minor geometrical curvature incompatibilities between formwork and already constructed lining lead to the creations of D shaped tensile cracks. These cracks present problems due to the free and unconstrained surfaces formed all around them and may produce fallouts during cold temperatures or high vibrations if not repaired. Pour joints due to different phase concreting were also reported at Kalydona tunnel.

The rest of the findings pertain to the shading (Figure 21) and/or the exposure (Figure 22) of the implemented reinforcement bars. The shading and exposure of reinforcement was attributed to incorrect placement of steel, resulting in shallow cover, lower than required by the design. The shallow cover in tunnels is produced mostly due to fewer reinforcement spacers installed during steel erection and formwork placement.

As a result, such phenomena were found only at the reinforced concrete Klokova tunnel and were recorded less frequently at the few reinforced tunnel blocks of the rest of the tunnels. Even though, a simple exposure of the rebars may not seem critical, their possible corrosion with time would lead to significant effects (concrete spalling, etc.), which could affect the passing vehicles and as result obstruct traffic through the tunnel, thus presenting safety issues.

The last type of observations concerns the moisture and/or leakage effects on the tunnel final lining. Such issues were extremely localized in two areas. The first area refers to a block located near the exit portals of the Klokova tunnel, where water leakage was detected at the bottom of the vault in just one place, while the second area pertains to the Ampelia tunnel, where moisture effects were recorded at a few blocks near the tunnel exit. Due to the water-proofing of the tunnels and the localization of the leakage – moisture records, they are not considered as critical.

The histogram in Figure 23 briefly presents the different types of recorded issues on the concrete surface of the tunnels and the frequency in their appearance on either plain or reinforced concrete blocks.

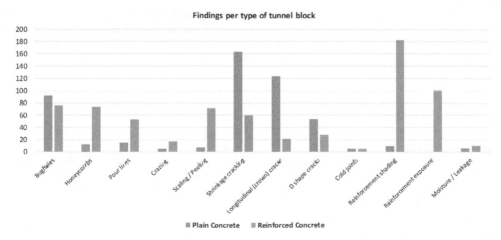

Figure 23. Different type of findings per type of tunnel block (the observation for reinforcement shading on plain concrete lining pertains to limited use in some locations of partial reinforcement).

As can be seen from the Figure 23, some problems such as bugholes and pour lines and cold joints, are present regardless whether the lining is reinforced or plain. Some defects are mostly associated with plain concrete, such as shrinkage and longitudinal (in the crown) cracking, while honeycombs are mostly related to reinforced concrete, probably due to congested reinforcement and the difficulty of concrete pour.

4.3 Prevention and repair of defects

As previously mentioned most of the findings recorded during the Ionia Odos tunnels visual inspection do not affect directly the structural integrity of the tunnels and can be described mostly as cosmetic defects, except the D shaped cracks. However, most of these could have been avoided during construction using a better concrete mix or better curing and formwork placement and removal. In the following paragraphs, recommendations on prevention and reparation techniques of the documented concrete observations are made with an emphasis on the most severe defects which threaten the tunnel structural integrity in the long term. Concisely these defects pertain to longitudinal and D shaped cracks at the crown of the tunnel vault, the development of cold joints and the extensive reinforcement exposure.

Observations with regards to honeycombing, bugholes, crazing, shrinkage cracking and pour lines are of minor cosmetic significance for tunnel sections. Their prevention is completely related to the concreting of the tunnel. Appropriate aggregate gradation, well-adjusted water content, deliberate concrete mixture placement and careful curing could significantly reduce the development of such issues.

The observations regarding leakage and moisture effects at Ionia Odos tunnels did not seem to require remediation, since they were identified in a few localized areas and their extend was insignificant. These observations usually result from failure or puncture of the waterproofing membrane and could be repaired by utilizing appropriate grouts to seal the water flow path.

With respect to D shaped cracks at the edge of the tunnel block crown, they require sealing and bonding with epoxy, cement grout or similar injections (Soudki, 2001).

Similar techniques may be applied for the sealing of the longitudinal tunnel crown cracks as well if found to generate free and unconstrained surfaces. Such cracks were mostly identified at the Kalydona tunnel and developed in the crown due to the low tensile strength of concrete and the bending stresses at this location. This issue was addressed during the design phase and it was found that a significant portion of the section is in compression and this tension crack does not produce a structural issue. Since this longitudinal crown crack is almost always present in plain concrete linings it could have been constrained inside a false joint during construction along the tunnel block crown in order to predetermine the location of the cracking (DAUB, 2007).

Last but not least, problems referring to the shading and exposure of the tunnel section reinforcement could be prevented if the required cover of the reinforcements is thoroughly applied during construction. However, if reinforcement exposure is recorded, as occurred along some blocks of the Klokova tunnel, protection techniques are crucial so as to prevent corrosion and consequent concrete spalling.

5 CONCLUSIONS

This paper presented the visual inspection procedure implemented for the very first structural inspection of the tunnels along the Ionia Odos motorway. The theoretical basis, the implementation procedure, all the required arrangements for its proper and safe execution as well as all the recorded findings in each tunnel and their possible prevention and repair techniques are briefly presented. The entire inspection procedure is considered successfully completed despite all the difficulties derived before and during its execution. The need of creating from scratch comprehensible tunnel inspection sheets, the inspection taking place concurrently to partial vehicle traffic and throughout predefined time periods, and the strenuous procedure for the inspectors of identifying and accurately evaluating issues under such conditions were effectively overcome and the final outcome was successful.

During the visual inspection and recording for the different types of concrete used (plain, reinforced) similar and different defects where found. In plain concrete linings, significant shrinkage cracks, longitudinal cracks in the crown and D shaped cracks were recorded. In reinforced concrete lining, the predominant defect was, inadequate reinforcement cover and sometimes exposure and honeycombs. In both types of concrete linings, bugholes and pour lines where observed and recorded. It was verified that small scale cracking, mostly shrinkage cracking, is an inevitable effect on the surface of unreinforced concrete tunnel lining and they develop right after construction and henceforth remain inactive as was evaluated from crack meters that were installed.

Regarding the inspection process, it is of great importance that a tunnel's first visual inspection is conducted right after it is constructed, as was currently accomplished for the tunnels along the Ionia Odos motorway. Therefore, a secure record for following inspections is created since all the issues, even from the beginning of the service life of the tunnels, are well documented and will be useful as a future reference. Furthermore, the inspection procedure could be significantly less time consuming if the inspection sheets could be filled in electronically using a mobile device. Last but not least, it should be noted that the visual inspection of a tunnel must be performed at reasonable time intervals and should be combined with an instrumented monitoring procedure.

REFERENCES

ACI Committee 201 2008. *Guide for Conducting a Visual Inspection of Concrete in Service.* Michigan, USA.
Soudki, K.A. 2001. Concrete Problems and Repair Techniques. *Professional Engineers Association,* Beirut, February 2001, Lebanon.
PCA 2002. *Types and Causes of Concrete Deterioration.* IS536, Portland Cement Association, Skokie, Illinois, 2002, 16 pages.
DAUB, German Committee for Underground Construction 2007. *Recommendations for Executing and Application of unreinforced Tunnel Inner Linings.* Tunnel (int. journal of DAUB, D-50827 Köln), 5, 19–28.

Tunnels and Underground Cities: Engineering and Innovation meet Archaeology,
Architecture and Art, Volume 6: Innovation in underground engineering,
materials and equipment - Part 2 – Peila, Viggiani & Celestino (Eds)
© 2020 Taylor & Francis Group, London, ISBN 978-0-367-46871-2

Building information modeling in Warsaw metro extension project

L.J. Szczepaniak, F. Bizzi & R. Sorge
ASTALDI SpA, Rome, Italy

ABSTRACT: The first BIM implementation in underground project in Poland was carried out by ASTALDI in the extension of Warsaw Metro (WM). The BIM uses in the design phase started from the 3D modelling of 2D design documentation, clash detection and quantity take-off support. In the construction phase, the added value arose in the field of scheduling. The site activities were monitored by providing periodical model-based progress reports. The engineers had quick access to the BIM database via mobile and desktop devices. The project team used the laser scanning to enhance surveying and quality assurance. Methodical data collection, through all phases of the project within a single BIM database, greatly aided the development of the As-Built model that will be used in preparation of facility management tool. The BIM implementation in the WM project, allowed ASTALDI to better understand the cost-effective uses that information management provides for application in forthcoming projects.

1 INTRODUCTION

The BIM use in construction projects in Poland is still in its early stage, despite this, the Polish government has already started to research the ways to make it mandatory on public-funded projects. In order to be one step ahead of competition and to benefit from the BIM processes advantages over "old school" 2D documentation approach, the Warsaw Metro Extension Project became another ASTALDI's BIM implementation pilot project. This article will present some of the benefits that can be obtained from the use of BIM processes in the construction works.

2 BRIEF PROJECT DESCRIPTION

The Warsaw Metro Extension Project agreement was signed in March 2016 and it is planned to finish in May 2019. The project consist of 3 stations, named C16, C17, C18; 3 ventilation shafts, called V16, V17, V18 and the mainline tunnels totaling in length at 4480m. The building permit design was prepared by the Client and this became a base for the executive design prepared by an external consultant.

All official contract design documentation is in 2D paper format, as there is no mandatory requirement for the BIM use in this project. In order to achieve the most of anticipated benefits the BIM is being implemented only for the most complex station with holding tracks, 449m long "Trocka" station (C18) and the affected underground utilities in the close vicinity.

3 BIM SCOPE IN THE PROJECT

3.1 *BIM objectives*

The main general objectives of BIM implementation in Metro Warsaw project are:

- To promote collaboration through systems, departments and software interoperability
- To use quantity extraction for external activities of budgeting and offers formulation;

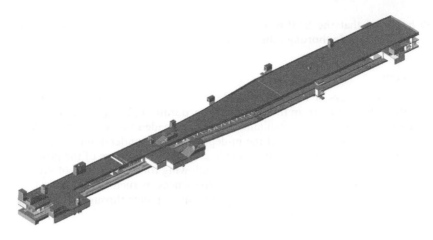

Figure 1.   C18 Trocka station overview.

- To maximize production efficiency through the adoption of a coordinated, consistent and integrated BIM workflow;
- To improve the overall quality of the provided design solution, through a coordinated and clash-free set of modelled content;
- To enhance the constructability review by referring to the BIM model, anticipating the constructability issues through digital simulation and validation of the design solution;
- To create a project file containing the information on the performed works, in support to the as-built drafting phase;
- To develop and improve BIM know-how in ASTALDI.

3.2   *BIM activities*

BIM activities are divided in two main tasks:

- Preliminary phase: modelling and BIM development of C18 Station (architectural, structural and external underground utilities networsk) Building Permit Design;
- Execution design phase: training and supporting the staff employed by ASTALDI for the implementation of BIM and BIM quality systems, implementation of project execution design in the model, development of "application tools" related to the current system of ASTALDI for planning, procurement, quantity estimation, cost control, documents archiving, site communication and work progress measurements.

3.3   *BIM Execution Plan*

A detailed BIM Project Execution Plan (BEP) has been prepared by the project team, in order to define objectives, uses, structured procedures and expected results for BIM on the project, along with a detailed design of the processes for executing BIM throughout the project lifecycle. Main standards and guidelines followed as reference for the preparation of the BEP are:

- BIM Project Execution Planning Guide by CIC Research Group, Pennsylvania State University;
- Level of Development Specification for Building Information Models (version 2015) by BIM Forum;
- BS 1192 – 2007 Collaborative production of architectural, engineering and construction information;
- National BIM Standard-United States ® (NBIM-US™)
- BEPs from previous ASTALDI project with BIM implementation.

It shall be noted that the BEP is a live document that has been dynamically adjusted to reflect the decisions made thorough the project life.

3.4 BIM Team

ASTALDI formed 3-person team responsible for the BIM implementation in the project. Along with the help coming from the ASTALDI corporate staff, the team has been supported also by an external experienced consultant, who has provided the team with his knowledge of the BIM implementation process and the in-depth software understanding. The team organization chart has been prepared, however due to experimental nature of the project, the whole team was engaged in various types of tasks, thus allowing for a quick gain of skills in different BIM-related activities and rapid exchange of experiences in such broad subject. The gained knowledge has been spread among other project team members through series of training sessions, in order to support their daily activities.

3.5 BIM implementation and uses

3.5.1 3D Modelling
The first phase of the implementation covered the preparation of the 3D models based on the building permit design documentation, provided by the Client. This phase finished at the beginning of April 2016. Since then, there is ongoing phase of the 2D executive design implementation in the 3D model. For this purpose, the model has been divided into the sub-models in accordance with their disciplines.

To assure reliability and consistency of the models, some standard quality control processes were established, allowing to ensure the level of quality required for each modeling use. Each project team member is responsible for performing of the quality control checks according to the roles and responsibilities assigned. The Contractor BIM Manager is the one who confirms final quality of the model after the revision are made. Requirements in terms of BIM elements level of definition and level of information are described in BIM Execution Plan via Level of Development (LOD) specification. Additionally, all elements have been modelled in the parametric way, wherever possible. In order to accomplish this purpose, certain types of object families have been designed and are now stored as a local library which will be developed in future projects.

The models have been divided by disciplines as shown in green circles in Figure 2, whereas 3 additional types of models have been prepared, based on their usage as listed below:

- MM - Master model – coordination model, where all the models listed below are linked in, used for coordination, quality inspection and clash detection purposes (red circle);
- LAY - Layout model – drawing sheet model where all the models listed above are linked in, used for documentation publishing (red circle);
- DRV - Driver model- input models, contain linked in 2D design input files for given disciplines (blue circle).

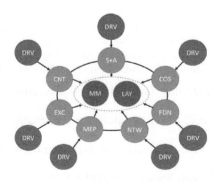

Figure 2. Model structure chart.

3.5.1.1 EXCAVATION MODEL (EXC)

The main elements that are included in the Excavation Model are ground layers to be excavated.

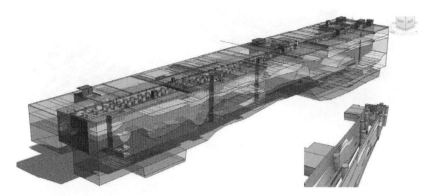

Figure 3. Excavation model view with boreholes information.

3.5.1.2 FOUNDATION MODEL (FDN)

The main elements that are included in the Foundation Model are guide walls, diaphragm walls, barrettes, struts, jet grouting and temporary steel beams for entrances execution and sheet pilings.

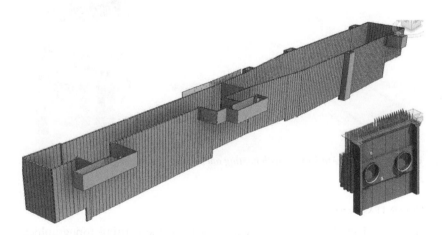

Figure 4. Foundation model view with plug-out detail.

3.5.1.3 STRUCTURAL AND ARCHITECTURAL MODEL (S+A)

The main elements that are included in the Structural and Architectural Model are concrete slabs, stairs, columns, interior walls, curtain walls, ceiling, suspended ceilings, floors and raised floors, walls, ceiling and floor finishes, doors handrails and room information.

Figure 5.   Structural and architectural model view with interior walls detail.

3.5.1.4 UTILITIES NETWORK MODEL (NTW)

The Utilities Network Model consist of the existing and proposed utilities network, i.e. electrical, gas, municipal heating, telecommunication, potable water and sewage water network.

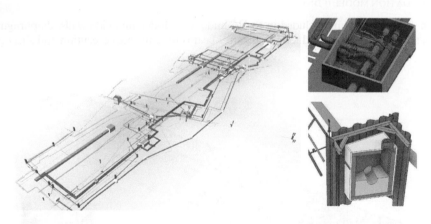

Figure 6.   Utilities *Network Model view with heating and surface water drainage chamber details.*

3.5.1.5 URBAN CONTEXT MODEL (CNT)

The main elements included in the Context Model are the existing topographical surface, building, trees, road and footways.

3.5.1.6 CONSTRUCTION MODEL (COS)

The Construction Model is based on the initial layouts produced by the Designer and since then, it is in on-going development to suit the site management needs. This model is used to run simulations and clash detection for the construction activities allowing for an easier identification of the possible safety hazards or space constraints during construction works.

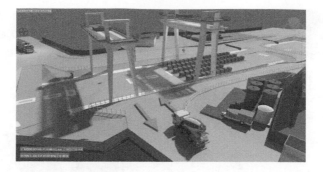

Figure 7. Construction model view.

3.5.1.7 MECHANICAL, ELECTRICAL AND PLUMBING MODEL (MEP)

The MEP model is prepared directly by the external company responsible for the executive design. This model is a base from which the 2D design drawings are created. However, as there is no contractual obligation for the BIM implementation by the designer, it is transmitted only in a read only viewer format on a monthly basis. Nevertheless, it contains most of metadata and can be successfully used for coordination and clash detection purposes as well as for the verification of the quantities. The model contains the information about HVAC, electrical MV and LV potable water installation, firefighting water and sewage installation.

3.5.2 *3D model usage: collaboration*

To achieve one of the main targets of the BIM implementation, i.e. increase the effective collaboration and information exchange, the models had to be distributed to the project team members. For this purpose, the project document sharing platform (CDE) has been used with embedded free model viewer. This allowed the team to have a direct access to the published models with no installation of additional software required and to get acquainted with this new comprehensive way to present the design information. The project team members, often with a faint initial knowledge about BIM, had the opportunity to have a hands on experience, that resulted in many ideas about the its usage in their departments. By now, the models have become more popular and are often used during coordination meetings and as a support in design review process, while the BIM team focus on the incorporation of the newest incoming executive design packages, in pursuit to provide the other project members with the most up to date and accurate information possible. The models are also accessible on-site via mobile devices, like tablets and smartphones.

Figure 8. CDE model viewer gives direct access to the models for all project staff.

Figure 9.   Engineers using mobile device to verify the model on site.

3.5.3 *3D model usage: quantities extraction, verification and specification schedules preparation*
Another area where the 3D models have become handy, is the quantity surveying field. Under the supervision of the QS department, the BIM in the process of preparation of a universal bill of quantities, tailored to Polish contract needs, that could be used in future contract. As the new design packages arrive, the bills of quantities are prepared to the specification of the ordering person.

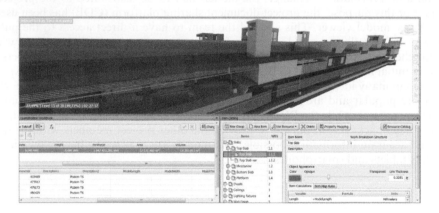

Figure 10.   Quantification takeoff work in progress.

3.5.4 *3D model usage: coordination and clash detection*
All discipline models are linked together in the Master Model which is used during coordination meeting and for clash detection purposes. The presented views can be adjusted to the project member needs, thus allowing them to see very clearly the possible forthcoming design and construction feasibility issues, while site engineers are able to assess the best delivery routes for abnormal loads. The BIM Team worked out clash detection procedures and clash detection matrix for different fields of engineering. This activity is performed every time one of the models is updated with a new design package, permitting for a fast design snag reporting. This enable the interference check to be run in a structured manner with comparable results, which is very important for such complex multidisciplinary project.

Figure 11.   Example of clash detection results.

3.5.5 *4D Modelling – time*

The 4D simulations allowed the project members to have a better control and a clear view of the impact made by decisions made during construction works planning. The availability of an early identification of the works that could interfere with each other, allows the team to prepare error free work schedules, that increase the productivity on-site and decrease the delay times and works stoppages. To perform the 4D simulations two applications have been used in parallel: one professional planning tool with BIM extension while the second one was a BIM model coordination tool with 4D simulation capabilities. The first one have been already used by the planning department, but without connection to the 3D models. This allowed to integrate the process in the project workflow seamlessly, as the planner's amendments to the schedule are automatically reflected in the 4D simulation permitting the project management to plan their decisions in a graphical interface, with a clear view of the consequences of proposed changes.

Additionally to simulations, a monthly progress report is prepared. This document uses the combined data from design authoring tool and planning tool with BIM extension, which is submitted to the project management staff in order to provide them with a summary of all site activities and their relation to the program of works.

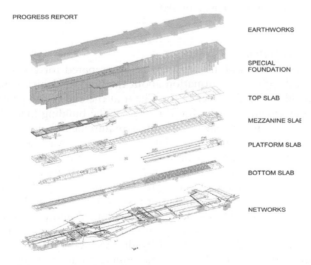

Figure 12.   Example of the summary sheet of progress report.

Not in all 4D related tasks the planning tool is a winner, it is still a software developed mainly for the scheduling and its BIM extension does not allow for a lot of personalization and changes in the model appearance. For the presentation purposes, where the good quality of the product is desired another product was used. The animation prepared by the BIM Team, explaining the sequencing of works in a simplified way, are displayed in the project information point and have been met with a good reception from the public, who are able to easily understand what is the status of the ongoing and planned works in their neighborhood.

3.5.6 *5D Modelling – cost*

Since the beginning of the project, the future cost – time analysis has been in mind during creation of the models. All items have work breakdown structure and Uniformat codes assigned, beside that there is ongoing process of bill of quantities codes assignment. The 5D cost modeling is still being under development and so far has been implemented in the simulation prepared in the coordination tool, where the current production value is displayed. It is planned to prepare a database with costs, pricing information and labor productivity rates, that will be used for future ASTALDI projects in Poland and shall help with the analysis of the correlation between program of works changes and costs.

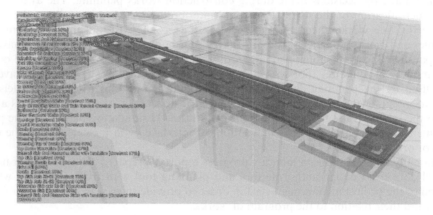

Figure 13.   5D simulation – construction of top slab.

3.5.7 *6D Modelling – as-built model*

As the project progresses, the as-built model is being prepared for every discipline. The as-built survey data and the quality information about the built in elements, material are linked to the model that could be handed over to the Client at the end of the project and will transferred to facility management software for 7D stage of the building lifecycle, in this case for educational purposes, as there is no such contractual requirement from the Client.

One of the methods to gather the as-built data in this project was the use of laser scanning, this provides an aid for the surveying team and the construction site office in recording very accurate geometrical data in form of a point cloud, as well as allows to verify the deviations of the construction works in relation to the design. This work was performed in two different phases of the construction works before and after installation of the building installations.

The first phase took approximately 4 days of measurement and processing of point clouds and spatial photo shooting of the station by an external surveying company. The working material that has been obtained was analyzed by the ASTALDI BIM Team in terms of its usability. In this way, at each station location, panoramic photos were created with 360° field of view, that allow to perform a virtual site visit and a quick visual evaluation of the areas and elements of interests.

The point cloud, instead, juxtaposed with the BIM model of the station, with the use of a modeling software add-on, helped to verify the accuracy of the constructed works, without the need for laborious on-site measurements. The add-on allows for automatic analysis of the

Figure 14. Point cloud views (top) and 360° image from the scan location.

distance between points in the surface of the elements of the design model compared to the points in the point cloud scanned on-site. Deviations are shown graphically in different color scale. Additionally, there is also a possibility to visually compare the point cloud and the design model in layouts and cross-sections.

Figure 15. Construction deviations analysis.

After testing and analysis of the point cloud method, it shows great potential for surveying and project control. In a quick way, the engineers can get an accurate comparison and detection of construction errors and potential risks for the future works.

Fortunately and above all, due to great team effort, in the Warsaw Metro Project, the verification proved the constructed element are all within the allowed range of accuracy.

4 CONCLUSIONS

From the start of the BIM implementation in the project of the Warsaw Metro extension, it has been established that such approach allows for more effective work and improves information exchange. During the project development all of the BIM implementation objectives has been achieved, which lead to substantial cost savings, even after taking into account the cost of the software and the hardware upgrades. Moreover, there has been a lot of positive feedback from employees involved in the construction of the previous stations of metro in Warsaw. The significant decrease in the number of the clashes was noted, thankfully to the greater coordination of the design. Currently the information models are prepared in the background and serve as a support for all departments. This process has allowed the project staff to improve the efficiency and the quality of the construction work. Based on the success of the ongoing pilot project, the recently acquired new Warsaw metro extension projects are developed with the BIM approach in place from the beginning and utilizing all currently developed BIM features in the daily work routines of all employees, with intention to place the BIM in the core of the project and results in future improvements.

Tunnels and Underground Cities: Engineering and Innovation meet Archaeology,
Architecture and Art, Volume 6: Innovation in underground engineering,
materials and equipment - Part 2 – Peila, Viggiani & Celestino (Eds)
© 2020 Taylor & Francis Group, London, ISBN 978-0-367-46871-2

Concrete lining covered with an outer compressible layer applied to tunnels in viscoplastic rocks

R. Taherzadeh & E. Boidy
Convergences, Tractebel, Engie, Antony, France

H. Ouffroukh
Convergences, Arcadis, Antony, France

ABSTRACT: From a geomechanical point of view, viscoplastic rock behavior is likely to create significant loads on the lining, which can considerably increase its size. Excessive rock deformations can also challenge the construction method, which affects project costs and planning. A concrete lining covered with a compressible layer is therefore of great interest for tunnel design in viscoplastic rock.

With this new lining concept, the compressible layer absorbs deformations to limit stresses in the outer concrete ring. It thus allows the use of TBMs in strongly time-dependent viscoplastic rock. In the present case, the effectiveness of this solution has been studied for two construction methods: TBM and conventional. Application of the double-layer solution has also been considered for tunnel intersections. This concept has been adapted to the basic design for the galleries in "Cigeo", the deep geological repository for Intermediate Level (ILW) and High Level (HLW) radioactive waste in France.

1 INTRODUCTION

1.1 *Definition of a new design for the lining of galleries*

The civil engineering design of an underground structure faces major difficulties when the excavation is carried out at great depths, and therefore under heavy in-situ stresses, in highly deformable grounds such as argillite, shale or phyllite. From a geomechanical point of view, the viscoplastic behavior of the ground can have a considerable impact on lining design. Moreover, in the short term, the strong deformation of the ground during excavation requires the use of complex and expensive excavation and supporting techniques.

Tunnel boring machines (TBMs) are being used more and more frequently in underground civil works because they now make it possible not only to excavate tunnels quickly and safely, but also to place prefabricated segments at the same time, which constitute the final lining of the structure. This tunnel construction technique is fast and economic but is not applicable to all ground conditions and all geometrical configurations. Indeed, concrete liners backfilled by mortar or gravel do not have sufficient resistance to withstand the high level of stresses due to time-dependent behavior of the ground.

A first approach has been applied by Andra in the underground laboratory at Meuse Haute-Marne, involving the injection of a compressible filling mortar (DeCoGrout®) on the outer side of TBM segments to partly absorb viscoplastic ground deformations and consequently limiting internal forces on the concrete rings Zghondi et al. (2015).

This approach presents the advantage of having no impact on the fabrication and setting of the segmental lining. Feedback from experience shows that fractures in argillite that occur during the excavation result in the accumulation of argillite blocks in the annulus along the TBM shield prior to injection of the filling mortar placed on the outer side of the lining. But, in the case of injection, the theoretical thickness of the filling mortar is not guaranteed. In addition, non-uniform thickness of fill material may create some rigid points, due to the presence of blocks in the annulus. These blocks induce a heterogeneity in the distribution of the pressure applied on the concrete ring, leading to the generation of high bending moments in the concrete ring.

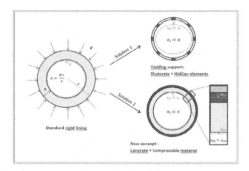

Figure 1.   Technical solutions to limit the thickness of the rigid lining.

The new concept studied by the engineering group "CONVERGENCES", involves direct placement, in the case of TBM, of double-layer segments composed of a rigid concrete layer inside and a compressible material layer outside, to ensure constant thickness of this material around the ring. Ground pressure exerted on the lining is thus reduced efficiently and the risks of heterogeneity are eliminated; the concrete thickness required can therefore be significantly reduced.

With this new type of liner, it is possible to widen the use of TBMs. This new concept is also applicable to cast-in-place concrete linings. A compressible material placed between the temporary support and the final liner absorbs some of the ground deformations over the lifetime of the structure and thus limits the stresses in the concrete lining, particularly for tunnel intersections.

1.2   *Applicability of the double-layer lining concept*

Yielding supports with lining stress controller elements incorporated in the shotcrete are traditionally used for deep tunnels in deformable grounds. In these supports, significant convergences are produced by deformations of the yielding elements. Although these deformations can be considered acceptable for road/train tunnels or hydraulic galleries, this is not the case for nuclear underground facilities such as Cigeo, in which the interior dimensions must be strictly observed over a long period. These kind of deformable support have been applied to certain Alpine tunnels in Switzerland, the Saint-Martin-La-Porte drift, and the GCS, GCR and GER galleries at the Meuse Haute-Marne Underground Research Laboratory (LSMHM).

Actually in an industrial and nuclear context, such as in the Cigeo project, it is necessary to guarantee maintenance of the internal clearance for operational equipment for at least hundreds of years. This requirement has led the engineering group "CONVERGENCES" to get involved in the study of the alternative solution for double-layer linings as described above. Thanks to its deformation characteristics, the compressible material not only makes it possible to decrease the internal forces on the lining in the short to medium term but also to neutralize deformations of the internal clearance over the lifetime of the structure (Figure 1).

2   CHARACTERIZATION OF VISCOPLASTIC GROUNDS

2.1   *Saint-Martin-La-Porte access tunnel*

As part of the international section of the Lyon-Turin rail project, the Saint-Martin-La-Porte access tunnel was performed with an excavation section of 80 m^2 and a length of 2200 m. The tunnel faced a series of extremely fractured rocks, very compressive carboniferous shales, generating very large convergences reaching 2 m under the overburden of 300 m.

After discovering the squeezing zone, sequential excavation was carried out with a temporary support composed of yielding supports with hiDCon® elements incorporated in the shotcrete lining and steel sets with yielding couplings (Figure 2).

2.2 *Case of the Meuse Haute-Marne Underground Research Laboratory (LSMHM)*

The Flexible Design Gallery (FDG), oriented in the direction of the major horizontal in-situ stress, was excavated using a conventional method with a diameter of 5.2 m. A yielding support system, consisting of an 18-cm thick shotcrete layer with 12 compressible hiDCon® elements and 12 HA25 radial bolts of 3.0 m length, was installed at intervals of 1.2 m Bonnet et *al.* (2011).

The compressible hiDCon® elements have a yielding strength of between 4 and 5 MPa with a maximum allowable strain of 40%. The measurements give an orthoradial stress varying from 1.36 to 7 MPa with an average of 4 MPa. These values show the correct operation of this support. The highest stresses are observed in the vault and in the invert.

In addition, the convergence level measured in galleries with flexible support depends on the orientation of the gallery with respect to the initial stress as shown in Figure 3.

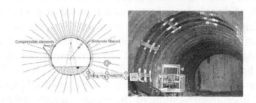

Figure 2. Saint-Martin-La-Porte access tunnel – Support systems.

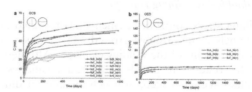

Figure 3. Convergence measurements in galleries oriented in the direction of the major horizontal stress (on the left) and the minor horizontal stress (on the right) Armand *et al. (2013)*.

3 CHARACTERIZATION OF THE COMPRESSIBLE MATERIAL

3.1 *Objectives sought in underground works*

In the case of underground structures excavated in viscoplastic rock, the characteristics of the compressible materials must meet certain objectives:

- Supporting and stabilizing the inner concrete ring, particularly during transport and its implementation,
- Absorbing ground deformations and limiting the ground pressure exerted on the lining,
- Providing a significant strain level to limit the thickness of the compressible material used, and thus the excavation diameter.

To summarize, the stress-deformation behavior for this type of material must be as follows:

- A first elastic phase with a relatively high deformation modulus, which makes it possible to stabilize the concrete ring and quickly put it under stress,
- A second plastic phase, with the lowest possible deformation modulus, ideally zero (perfectly plastic phase), starting from a stress value that is defined by the designer (generally between 1.5 and 2.5 MPa). This phase makes it possible to absorb ground deformations, retaining the constant stress transmitted to the lining to some extent.

At the end of the plastic phase, there is a third phase with a very high deformation modulus, the material becoming relatively "incompressible" (Figure 4).

Moreover, the compressible material must be strong enough to make it easy to transport, store and handle in confined environment.

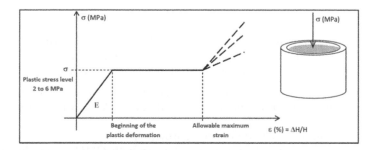

Figure 4.   Typical behavior of the compressible material under oedometric loading.

3.2   *Behavior of the double-layer lining*

From a theoretical point of view at first approach, the stresses induced in the lining are proportional to the pressure applied by the ground. Therefore, limiting this pressure reduces the concrete lining internal stress.

To better illustrate this, Figure 5 simply compares the impact of rigid and double-layer linings in terms of pressure and displacements using a convergence-confinement interaction method.

In the double-layer lining, the compressible material deforms to absorb the pressure applied on the outer side of the lining.

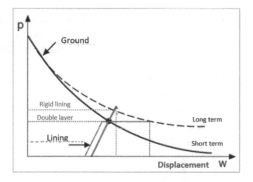

Figure 5.   Comparison of the impact of a rigid lining (without compressible material) or a double-layer lining in terms of pressure and displacements using a convergence-confinement interaction graph.

Assuming an outer diameter of the lining of 9 m and a lining thickness of 50 cm, the stress curves in the lining as a function of the ground pressure indicate that to limit the stress in the lining to a value of 30 MPa, the pressure applied by the ground should be lower than 3 MPa (Figure 6).

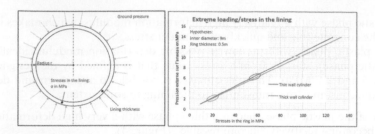

Figure 6. Relationship between ground pressure and lining stress.

3.3 *Compressible materials*

In general, the compressible materials identified at this stage of the studies still need to be developed and tested in order to guarantee the desired allowable strain.

As an example, reference can be made to the following main materials:

- Existing materials developed specifically for underground works: DeCoGrout® (Hochtief Construction AG/Schretter) or hiDCon® (SolExperts AG), which are cement-based products
- Foamglas® type materials as used in building isolation
- Material under development: Andra, in partnership with CMC, is developing a double-layer liner with a compressible clay shell layer.

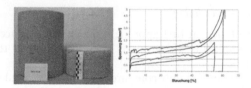

Figure 7. Sample before and after oedometric compression test (left) and example of stress-strain curves (DeCoGrout ®) (right) (Billig et al. 2007, Billig et al. 2008).

The compressibility of this prototype is related to the hollow form of the mini-cylinders, which are the main component, and to their maximum allowable stress. The elements are connected with a cement mortar. The behavior is controlled by the diameter and length of the cylinders.

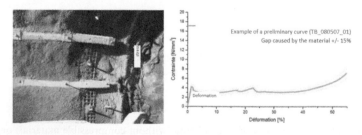

Figure 8. Example of stress-strain curves (hiDCon®) Guayacan et al. (2015).

Materials developed for other applications but with compressibility properties that can be compatible: for example, metal foams (in particular those based on aluminum oxide) or glass (Foamglas® type) can have an allowable maximum strain much higher than the other products (between 30 and 50%) and can provide high compressibility.

Figure 9.   Prototype of new lining concept (Andra-CMC) – Reference: patent No. FR3021346-29/07/ 2016.

3.4 *Packaging and implementation of the compressible material*

3.4.1 *Double-layer segmental lining*
For the Cigeo underground facilities, expected convergences over their service life at 100 or 150 years is about 20–30 cm for a flexible support, depending on the diameter and/or orientation of the galleries in relation to the major in-situ ground stress. Basic design of the Cigeo project shows that a thickness of 20–30cm for the compressible material on the outer side of the segments, with an allowable maximum strain of 50%, is sufficient to absorb the long-term ground deformation.

The double-layer lining can be manufactured as follows:

- Pouring the compressible material directly onto the outer side of the concrete segmental lining using specific removable formwork,
- Fixing the prefabricated material to the outer side of the concrete segment.

Prefabrication of the compressible layer makes it possible to have two distinct manufacturing sites: one for the concrete segments and one for the compressible layer. In this case, handling operations of the concrete segments are simplified, as the compressible layer is not in place, but require a device for fixing this element to the outer side of the concrete segment. Automatic gluing or spanning/nailing could be alternative solutions. This operation can be carried out at different stages of TBM mounting: at the concrete segment factory, the site stockyards or even on the backup train.

The formwork for the compressible material requires specific rheology to be able to perform the casting and surfacing on the outer side of the concrete segment, as well as to provide sufficient adhesion to the concrete. This method is preferable in order to reduce transport and handling operations, which would likely damage the relatively fragile compressible material.

3.4.2 *Double-layer cast-in-place linings*
The lining of tunnels excavated by conventional methods will face the same problems as with TBMs. However, the implementation of this lining after excavation by conventional method can be delayed until after excavation and support setting.

The duration of the delay is calculated based on:

- Construction planning: in the case of Cigéo, a final lining is placed between 6 months to 1 year after excavation;
- Impact on the lining design.

Implementation of the compressible layer between the temporary support and the final lining can go beyond this delay. In this case, the lining is cast in place by formwork and pouring, so it is necessary to place the compressible layer first.

The most appropriate method for underground works would be implementation of the compressible material by projection, which could be performed rapidly and in large volume. However, as things stand, it is considered that projection of a thickness of 20 – 30 cm is probably not possible without significantly modifying the original mechanical properties of this material. Indeed, the high energy required for the projection operation may densify the material, modifying the plastic level as well as the allowable maximum strain.

On the contrary, it seems more feasible to install the compressible material in the form of prefabricated plates (cement, metal foam or glass), fixed to the walls using a traditional nailing system. These plates must be light enough for manual installation.

4 LONG-TERM THEORETICAL BEHAVIOR OF TUNNELS

4.1 Numerical modeling principles

The numerical method, based on finite elements or finite difference calculations, is used in case of a complex ground-structure interaction. To justify this new concept, the finite difference model using FLAC software is applied here.

This numerical method can take account, in a sophisticated way, of:

- The actual shape of the galleries,
- The complex behavior of the ground, including its thermomechanical behavior if necessary,
- Anisotropy of in-situ stresses,
- Construction phases,
- The presence of structural elements without any simplification,
- The ground-structure interaction, continuously in real time.

In particular, 3D calculations are used to take account of the actual 3D character of the structures, particularly intersections.

In 2D analysis, the compressible material is modelled by FLAC "support" elements, oriented radially. Their stiffness is determined by a non-linear behavior defined in Figure 10. Note that this compressible material is assumed to be prefabricated, so its thickness is perfectly controlled on the outer side of the concrete ring.

In 3D modelling, the use of "support" elements have some limitations, particularly in the case of tunnel intersections. In order to solve this problem, the compressible material is modelled with volume elements, to take account of the 3D effect of stress tensors. For this purpose, the double-yield model, implemented in FLAC, is used. This model considers three yield surfaces that includes deviatoric (shear), volumetric (cap) and tension cut off. Note that to neutralize the development of an orthoradial stress in the material, the Poisson ratio is assumed to be zero. This prevents the development of a lateral deformation during the elastic phase of behavior.

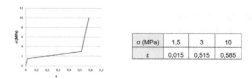

σ (MPa)	1,5	3	10
ε	0,015	0,515	0,585

Figure 10. Stress-strain behavior of the compressible material.

4.2 Results for TBM – current section

The TBM excavation configuration is studied in this section. The tunnel has an internal diameter of 8 m. Tunnel design is carried out for an anisotropic in-situ stress at a depth of 500 m embedded in argillite. The study has been carried out for two concepts: a standard concrete lining and the new double-layer concept. The latter consists of 45 cm of concrete on the inner side and 15 cm of compressible material on the outer side.

The main major stress field and the distribution of the normal lining force as well as the ground pressure acting on the lining after 100 years are presented in the following figure. It is shown that the use of compressible material drastically reduces the ground pressure applied on the lining, and consequently the internal lining force.

These calculations show that the use of compressible material on the outer side of the lining can be more efficient than the standard rigid concept in limiting the internal stresses of the concrete ring.

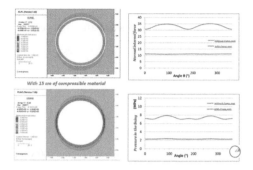

Figure 11. Comparison between the new concept and the standard segmental TBM liner: Major stress fields at 100 years on the left side, the normal internal force of the lining as well as the ground pressure at 100 years on the right side.

4.3 Results for tunnel intersections

For Cigeo project, the rigid lining solution does not ensure stability of the structure without a very significant concrete thickness (more than 1.5 m), which is not feasible. In this case, the double-layer concept could be a good alternative to significantly reduce the thickness of the cast-in-place concrete.

Figure 12 shows an overview of the modelling of the intersection in Flac3D, including a double layer for the lining. The compressible material is shown in red.

In this case study, the main gallery, in an anisotropic in-situ stress configuration, has an internal diameter of 8 m and the secondary gallery, in isotropic configuration, has an internal

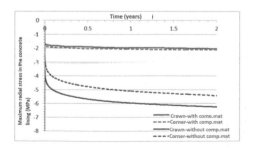

Figure 12. Flac3D modelling of an intersection with a double-layer lining.

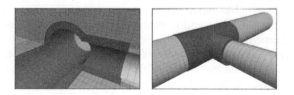

Figure 13. Time history main stress evolution in the concrete.

Figure 14. Comparison between the standard concrete lining (on the left) and the new concept (on the right) in terms of the main stress field produced in concrete after 100 years.

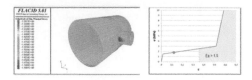

Figure 15. Strain field produced in the compressible material after 100 years.

diameter of 2 m. The concrete thickness is 115 cm and 50 cm respectively in these two galleries. The cast-in-place concrete is covered or not with 15 cm of compressible material on its outer side. This final lining is applied one year after excavation of the intersection. The following figures shows that the use of a compressible material drastically decreases the maximum stress of the concrete at the corner of the intersection. In addition, the maximum deformation of the compressible material after 100 years is approximately 12% at this point, with a value very much lower than the allowable maximum strain proposed for this material (Figure 15).

5 CONCLUSIONS

The "CONVERGENCES" engineering group, for the basic design of Cigeo galleries, applied advanced numerical modeling to justify a new concept for the deep tunnel in viscoplastic rock. The design involves a double-layer lining including an inner concrete ring covered on its outer side with a layer of compressible material. It is shown that this concept is very useful in optimizing the concrete lining thickness and in guaranteeing the functional dimensions of the gallery for long life requirement.

The feasibility of this solution was studied for both the TBM and conventional method, particularly for tunnel intersections. The double-layer lining can be considered a very good alternative in optimization of the cost and planning for tunneling in squeezing rock. Some experimental tests are however required to validate the feasibility of this solution in terms of manufacturing, handling, and implementation of the compressible material.

REFERENCES

Armand G. et al. 2013. Short-and long-term behaviors of drifts in the Callovo-Oxfordian claystone at the Meuse/Haute-Marne Underground Research Laboratory, *Journal of Rock Mechanics and Geotechnical Engineering* 5, pp 221–230.
Billig B. et al. 2007. DeCo Grout – Innovative Grout to cope with Rock Deformations in TBM Tunnelling – *ITA-AITES*.
Billig B. et al. 2008. Ausbausysteme für den machinellen Tunnelbau in druckhaftem Gebirge – *Beitrag für das Tunnelbautaschenbuch* 2008.
Bonnet-Eymard T., et al. 2011. Soutènement souple pour le creusement de galeries dans les argiles du CMHM – *Congres AFTES, Lyon 17-19 octobre 2011*.

Bonnet-Eymard T., Ceccaldi F., Richard L. 2011. Extension of the Andra underground laboratory: methods and equipment used for dry, dust-free works. In: *Proceedings of World Tunnel Congres 2011*. Helsinky, Finland.

Guayacan- Carrillo M-L., et al. 2015. Analysis of Long-Term Anisotropic Convergence in Drifts Excavated in Callovo-Oxfordian Claystone, *Rock Eng*.

Zghondi J. et al. 2015. Monitoring and behavior of an instrumented concrete lining segment of a TBM excavation experiment at the Meuse Haute-Marne Underground Research Laboratory, *10th International Conference on Mechanics and Physics of Creep, Shrinkage, and Durability of Concrete and Concrete Structures, CONCREEP-10*

Tunnels and Underground Cities: Engineering and Innovation meet Archaeology,
Architecture and Art, Volume 6: Innovation in underground engineering,
materials and equipment - Part 2 – Peila, Viggiani & Celestino (Eds)
© 2020 Taylor & Francis Group, London, ISBN 978-0-367-46871-2

Material properties and construction performance of ultra-high strength sprayed concrete

H. Takeda, T. Kawaguchi & H. Yoshimoto
Technology center, Taisei Corporation, Yokohama, Kanagawa, Japan

S. Fujiwara
Technical center, DC Co. Ltd., Kawasaki, Kanagawa, Japan

K. Sato
Business Development, Construction Chemicals, BASF Japan Ltd., Chigasaki, Kanagawa, Japan

ABSTRACT: This research focuses on the high performance wet-mix sprayed concrete (HPS) with compressive strength of over 95MPa at 28 days, developed as one of the solutions for supportive system in tunnel construction. To the best of our knowledge, this will be the highest achieved compressive strength in such an application once commercialized. To achieve ultra-high strength after application, it is essential to reduce the water to binder ratio (W/B) with enough margin of the final strength after the addition of accelerator. With such low W/B, the fresh concrete is often too viscous to be pumped and sprayed. To resolve this problem, we have developed a new binder material which will allow fresh concrete properties with good workability and ultra-high strength, suited for sprayed concrete. This paper reports the various material properties of HPS and verification results on manufacturing and workability conducted at a tunnel project.

1 INTRODUCTION

Recent tunnel construction in Japan often encounters situations involving overburden up to 1000m, weak and vulnerable ground conditions. Although the NATM approach in such condi-tions involves application of various auxiliary supporting systems, the new super high-performance sprayed concrete (HPS) was considered for one of the new addition to the supportive materials.

Ordinary ultra-high strength concrete is known to provide compressive strength equivalent or more than 100MPa and have been used in high rise RC buildings commercially (Jinnai et al. 2007, Tomioka et al. 2007). However, properties of these heavily rich mixes are known to be highly viscous and thus the handling and application is often conducted via concrete buckets. As for wet-mix sprayed concrete, the concrete is pumped through pipes and sprayed at the nozzle by compressed air, which requires appropriate fresh concrete properties for successful and high-quality application. Furthermore, it is crucial to attain thorough and homogenous mixing between the fresh concrete and accelerator to achieve stable and consistent strength development.

As mentioned, the viscosity of conventional ultra-high strength concrete is very high and pumping such material is a dangerous task not suited for spraying application. With such reasons, a new material with properties acceptable for spraying, yet achieves ultra-high strength was developed. This material was sprayed for trial at a full construction scale to verify its practicability for on-site application as a sprayed concrete.

2 CONCEPT AS A SUPPORTIVE MATERIAL

Implementation of this super-high performance sprayed concrete as a supportive system may potentially simplify the conventional support structure, shorten the construction process and im-prove safety of working environment during excavation. Figure 1 schematically presents the relationship between ground deformation and support reaction force, exhibited on the support structure composed of steel framings and sprayed concrete. As supportive solutions for con-struction under large deformation and high ground pressure, implementation of excess support multiplexing, deformation allocated abbreviated supports and high-strength shotcrete with com-pressive strength up to 50MPa have already been put to practical use.

The concept of this research is to prevent and or minimize the damage inflicted on conventional supportive structures by increasing the final strength of initial lining sprayed concrete and effectively function against larger ground pressure to suppress deformation and efficiently stabilize the ground. Larger deformation is often observed earlier after the excavation and simultaneously, so does the growth of strength and stiffness of applied sprayed concrete. As such, fine balance between strength and stiffness development under deformation pressure is a factor which must be strongly concerned. If the sprayed concrete is excessively rigid during a state without sufficient strength exhibited, stress exceeding the compressive strength may increase the risk and possibility of fracturing or even rupturing.

Though some mixes of high-strength sprayed concrete have been designed to provide good early strength development immediately after spray, excessive natural stress may increase the risk of damage to the concrete. On the contrary, by providing relatively slower stiffness development compared to strength development, the magnitude of deformation experienced may be slightly higher, however, the risk of destruction of the sprayed concrete at the early age can be reduced.

3 MATERIAL AND EXPERIMENTS

3.1 *Materials used*

Table 1 presents the constituent materials used for development of the super-high performance sprayed concrete.

The premix binder consists of cement, premix material comprising of blast furnace slag, gypsum-based binder and silica fume. All materials used were commercially available. Various cements were tested to determine the type which displayed the best flowability and workability properties. Generic aggregates for ordinary wet-mix sprayed concrete were selected. As for admixture, polycarboxylic acid-based superplasticizer which are known to be used for high-strength concrete was judged most appropriate. To prevent fracturing and rupturing, implementation of organic fiber was also considered.

The chemical composition of the used cement and mineral admixtures are shown in Table 2.

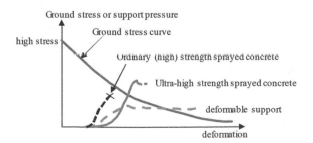

Figure 1.   Conceptual diagram of ground characteristic curve and support pressure characteristic curve

Table 1. Constituent materials

Material	Symbol	Properties of material
Cementitious materials	B1 (C+M1)	Premixed powder (C: cement, M1: blast furnace slag, gypsum and silica fume-based material), density: 3.08g/cm3
	B2 (C+M1 +M2)	Premixed powder (C: cement, M1: blast furnace slag, gypsum and silica-fume-based material and M2: silica fume), density: 3.02g/cm3
Fine aggregate	S	River sand, maximum size: 5mm, density: 2.63g/cm3, water absorption ratio:1.12%, F.M.:2.94
Coarse aggregate	G	River stone, maximum size: 15mm, density: 2.66g/cm3, water absorption ratio:0.29%, F.M.:6.18
Chemical admixture	SP	High range water reducing admixture, polycarboxylate type
Accelerator	AC1	Alkali-free, liquid type set accelerator, water-soluble aluminum salt, density:1.44–1.50g/cm3, pH:2.3
	AC2	Alkali-free, liquid type set accelerator, water-soluble aluminum salt, density:1.40–1.50g/cm3, pH:2.6
Fiber	F	Polypropylene fiber, cross section: 0.6mm*1.2mm, length: 65mm, density: 0.90g/cm3

Table 2. Chemical compositions of cement and mineral admixtures

Material	Composition (%)								
	SiO_2	Al_2O_3	Fe_2O_3	CaO	MgO	SO_3	Ig. loss	Alkali	Cl
C	-	-	-	-	0.78	2.05	0.56	0.62	0.010
M1	42.37	5.64	0.16	29.56	2.77	17.76	-	-	-
M2	93.70	-	-	0.35	0.40	0.18	-	-	-

3.2 Selection of binder

3.2.1 Considered binder formulas

Table 3 presents the mix formula of considered pre-mix binders. The primary factors of consideration were the types of binders and the water-binder ratio (W/B). During designing of the base mix, target mortar flow was set between 250 to 350mm, which was adjusted with the dosage of superplasticizer. As result, the values of superplasticizer dosage vary with each mix design.

Table 3. Mix formula of premix binders and proportions in base mortar mix

Mixture	Binder	W/B	S/B	Unit weight (kg/m³)			SP
	Type	(%)	(%)	W	B	S	(B×%)
M16-B1	B1	16	56	229	1429	807	1.36
M18-B1	B1	18	70	233	1297	908	1.12
M20-B1	B1	20	83	237	1190	988	0.96
M16-B2	B2	16	55	228	1425	788	1.44
M17-B2	B2	17	62	231	1356	841	1.44
M18-B2	B2	18	66	237	1313	863	1.20
M22-B2	B2	22	96	242	1086	1042	1.09

3.2.2 *Measurement variables and method of experiment*

3.2.2.1 Production of mortar

Prior to mixing the base mortar, the fine aggregates were adjusted to dry surface state. The powder-based materials and the fine aggregates were dry mixed for 15 seconds, followed by addition of water and superplasticizer. Wet mortar of 2L per batch was then mixed for 5 minutes using 5L ASTM mixer.

3.2.2.2 Measurement variables and methods

The temperature, mortar flow, flow time taken to reach 20cm, unit volume mass, funnel flow down time and compressive strength at 7days and 28days were measured. Measurement for mortar flow and compressive strength were conducted accordance to Japanese Industrial standard (JIS R 5201). The curing condition of the compressive strength test specimens were fixed at 20 degrees Celsius under sealed environment. The top surface of the specimens was polished, and the bottom was kept as the mold surface.

3.2.2.3 Evaluation method of rheological properties.

Measurement of rheological properties was conducted by evaluation of funnel flow test and by a rheometer. Using the funnel shown in Figure 2, the flowing time taken to reach 800ml was measured at each 100ml mark. Under the assumption the Buckingham-Reiner equation shown in equation (1) is applicable in the straight pipe section, using the values obtained, a flow curve graph is shown in Figure 2 can be attained. From the graph, plastic viscosity and yield value were determined by the method of least squares.

$$Q = \frac{\pi R^4 \Delta P}{8 l \eta_{pl}} \left\{ 1 - \frac{4}{3} \left(\frac{r_f}{R} \right) + \frac{1}{3} \left(\frac{r_f}{R} \right)^4 \right\} \tag{1}$$

Where Q = unit flow rate (m³/s), ΔP = pressure difference (N/m²) = $\rho g h$, ρ= density (kg/m³), g = gravitational acceleration (m/s²), h = sample height (m), r_f = plug flow radius (m) = $2l\tau_f/\Delta P$, τ_f = yield value (Pa), η_{pl} = plastic viscosity (Pa.s), R = radius (m), l = straight pipe section length (m).

3.2.3 *Mortar test results*

Table 4, Figure 3 and Figure 4 presents the examination results of the tested base materials. By observing the relationship between W/B, the apparent viscosity and the flow time to 20cm in Figure 3, when W/B is 18% or lower, the formulation using the binder B2 resulted in less viscosity than of the formulation using B1. Both binders showed almost equal compressive strength at 7days and 28days. Figure 4 presents the measurement results of M18-B2 obtained by rheometer. The connection between the shear rate and stress was proven to be linear, showing a behavior comparable to characteristics of Newtonian fluid, contrasting against ordinary mortar. As shown in the graph on the right side, although the speed dependency of viscosity is observed within the lower region, at speed above 5(1/s) the viscosity is almost constant at 11.7Pa.s. Overall, the viscosity value by the funnel was estimated at 7.6 Pa.s and the measurement by the rheometer was almost equal.

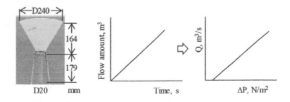

Figure 2.   Funnel used for rheology test and processing the measured data

Table 4. Test results of mortar

| Mixture | flow (mm) | 20cm flow time (s) | viscosity (Pa.s) | compressive strength, f'c (MPa) | |
				7days	28days
M16-B1	302.5	5.77	20.8	127.3	153.3
M18-B1	295.5	3.15	13.2	105.0	122.0
M20-B1	290.0	2.56	12.0	97.9	116.7
M16-B2	317.5	2.57	-	121.3	130.0
M17-B2	348.0	2.31	-	125.3	144.7
M18-B2	341.0	2.62	7.6	111.7	-
M22-B2	294.5	1.81	2.4	88.3	116.7

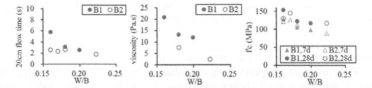

Figure 3.   Fresh and harden mortar properties according to binder type (B1, B2) with respective W/B

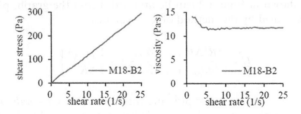

Figure 4.   Rheological properties of M18-B12 mortar

3.3   *Accelerator type, dosage and early strength development*

3.3.1 *Mix design and measured variables*
To determine the appropriate accelerator suited for concrete mix with very low W/B, the initial setting of mortar and the influence of strength development was investigated. Binder B1 was used for the base mortar and the W/B was set to 16%. The dosage of accelerator was set to 2.4%, 3.6% and 4.8% of the binder content. Items measured were base mortar flow, flow time taken to reach 20cm, unit volume mass, compressive strengths and initial and final setting time.

3.3.2 *Results*
There are no significant differences in initial and final setting time of the mortars due to the 2types of accelerator. They were 1m10s and 9m30s for AC1, 1m20s and 9m00s for AC2, respectively. Figure 5 shows the test results submitted to each accelerator type and dosage. As displayed in Figure 5, early strength development up to 3 hours and the dosage of the accelerator showed proportional properties, regardless of the type which was used. The sharpest strength development was observed between 3 and 24 hours after dosing of accelerator. However, on the contrary, lower accelerator dosage reflected higher strength development after 7 days. There are no significant differences between AC1 and AC2, however, the strength development after 7 days at higher dosage is slightly larger with AC1. With consideration of practicability it can be considered that the amount of accelerator added must be adjusted accordingly to best suit the application. Therefore, AC1 can be considered to allow minimum fluctuation of strength development and provide overall, higher quality sprayed concrete.

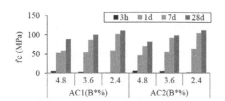

Mix proportion of base mortar
Binder: B1, W/B=16%, S/B=56%
SP: M16-AC1: 1.36% of binder
M16-AC2: 1.28% of binder

Fresh properties of base mortar

Mix	Flow (mm)	20cm flow time(s)	Air (%)
M16-AC1	316.5	4.81	2.1
M16-AC2	309.0	3.59	2.8

Figure 5.   Compressive strength (f'c) development submit to each accelerator type and dosage

3.4 *Properties of concrete*

Binder type B2 was selected after initial concrete test proving the acceptable effect of binder types to the fresh and hardened concrete properties. Laboratory concrete tests was conducted to confirm safe workability and sprayability properties with changes over time. Subsequently, with

W/B and accelerator dosage set as primary factors, practical spray trial was conducted in a mock up tunnel. Ultimately, the developed HPS was successfully mixed using on-site batching plant and sprayed with an conventional spray machine at an existing major tunnel project, proving its practicality and performance as a sprayable supportive material.

3.4.1 *Measuring variable and testing method*

3.4.1.1 Production of concrete in laboratory
Mix designs for laboratory test are presented in Table 5. Fine and coarse aggregates were adjusted to dry surface state. 30L Fresh concrete was mixed each batch with a 60L concrete mixer. Fresh concrete test was also conducted at a practical scale, using an on-site batching plant where the mixing time was adjusted accordingly.

3.4.1.2 Measuring variables and methods
Measured variables were; concrete temperature, slump flow, flow time taken reach 50cm, air volume, v-funnel flow down time, compressive strength and static modulus of elasticity. Test specimens for compressive strength tests were cured at 20 degrees Celsius under sealed environment. The static modulus of elasticity was measured with compressometer.

3.4.2 *Concrete properties over time*
Figure 6 displays the change in concrete characteristics over time. Fibers were also implemented in this test. As seen, the decrease in flow was stable regardless of presence of fibers and no drastic differences were observed. The v-funnel flow down time and flow time taken to reach 50cm showed more noticeable changes in comparison to the slump flow measurement. This can be considered due to the increase in viscosity of mortar component over time. With the reflection of mortar's viscosity change over time, the time margin acceptable for safe and successful application for HPS at 20 degrees Celsius can be approximated to be between 2.5~3.0 hours after batching. Furthermore, in the same 20 degrees Celsius environment as in the laboratory, at actual site conditions the concrete will be continuously in motion inside an agitator truck during delivery, thus it is estimated that actual changes in properties overtime is milder than this test result.

Table 5.   Mix design of concrete for laboratory test

Mixture	binder type	Gmax (mm)	W/B (%)	S/B (%)	s/a (%)	unit weight (kg/m³)				F (vol.%)	SP (B*%)
						W	B	S	G		
C18-B2-NF	B2	15	18	66	60	194	1077	708	478	-	1.12
C18-B2-F	B2	15	18	66	60	194	1077	708	478	0.4	1.12

3.4.3 Properties of sprayed concrete

To investigate the properties as a sprayed concrete, a practical trial was conducted using a spray machine and a mock-up tunnel. Figure 7 shows simple diagrams of the mock-up tunnel, portable concrete mixer and the spray machine which was used. $0.4m^3$ of fresh concrete was mixed per batch. Powder content, fine aggregates, water and the superplasticizer were mixed for 4 minutes to produce the mortar first. Coarse aggregates were then added and mixed for 1 minute before batching. The mix designs are presented in Table 6. For this trial, W/B was set to 17%, 18% and 19% with the use of binder B2. Fibers were not implemented at this stage. Measuring variables included basic fresh concrete properties, early strength development using penetration needle, core piece compressive strength, elastic modulus and small scale rebound ratio.

3.4.4 Outline of experimental results

Fresh concrete with same properties in the laboratory was able to be produced with the mobile batching plant and the concrete was able to be properly sprayed with a spraying machine. Figure 8 presents the strength development of base and sprayed concrete. The early age strength development was measured by a penetration needle. From the results, the early strength development pattern was determined equivalent to J2 (EN 14487). With W/B of 19% and AC3%, the 1 hour strength was insufficient, however, equivalent strength development was achieved at AC4%. Both base and sprayed compressive strengths for 7 and 28 days are shown in Figure 8b. As seen, 28 days compressive strength of the base concrete ranged between 120MPa and 140MPa. As for the sprayed concrete, regardless of the same specimen dimension and curing time, the compressive strength at 28 days ranged between 90 MPa to 100MPa, resulting in compressive strength ratio of approximately 0.7. Rebound ratio in ordinary concrete spray in Japan is around 20 to 30%. Though the test was conducted at a smaller scale, the rebound ratio for this spray trial measured 4.0%, significantly lower than the norm. This can be explained by the rather higher viscosity due to the rich powder content and the smaller portion of coarse aggregates compared to conventional sprayed concrete mix in Japan.

3.4.5 Stiffness development at young material age

3.4.5.1 Experiment methods of testing young material age

The initial compressive strength and elastic modulus of the sprayed concrete were investigated. The mix design selected was C17-B2, shown back in Table 6. The test specimens were produced by directly spraying concrete into an original cylindrical mold ($\varphi100\times200mm$) one end partially enclosed with wire mesh (Tani, T. et al. 2012). The measurement took place 3 to 24 hours, 7 days and 28days after addition of water. Test specimens for 7 days and 28days were extracted by core drilling from a sprayed box. Distortion in both longitudinal and transverse axis were measured (Figure 9a). Loading was measured following the specification JIS A 1149 with controlled loading speed of 0.6±0.4 MPa/s.

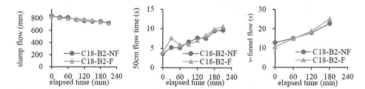

Figure 6.   Fresh concrete properties over time

Figure 7.   Diagram of mock up tunnel, mobile mixer and spray machine used for trial

3.4.5.2 Development of strength and elastic modulus

Table 7 presents the test results. Figure 9b shows the relationship between stress-strain affiliation at young material age. Seen from the graph, larger strain is observed at lower stress level particularly at 3 hours. Figure 9c shows the relationship between compressive strength and elastic modulus over time. HPS recorded compressive strength of 1.37MPa at 3 hours and showed drastic increase after 7 hour mark. Figure 9d shows the relationship between elastic modulus and compressive strength. For reference, Figure 9d also includes the past test results of RPC without addition of accelerator. Here the relationship in ordinary concrete (JSCE 2017) is also shown. From the same diagram, it can be seen that the elastic modulus with respect to compressive strength tend to be smaller at younger age compared to ordinary concrete, same for both RPC and HPS. This can be speculated to be due to the smaller amount of coarse aggregate and the unordinary constituent material of the matrix. However, it is necessary to investigate further in the future.

4 FIELD TRIAL ON A EXISTING TUNNELING PROJECT

4.1 *Mix design and concrete production trial using on-site batching plant*

Standard 1.35m³ mixer installed in conventional batching plant was used for this field trial. Each batch mixed 0.5m³ of HPS. Pre-mix trial was conducted using the mixer to adjust and set the mixing time. 210 seconds was determined to be the most appropriate mixing time for the flowability and strength development. Table 8 presents the mix design of the concrete. In order to prevent explosive destruction of ultra-high strength concrete at breakage, organic fiber with a volume ratio of 0.25% was used. The fiber was added into the drum of the agitator truck by the batching plant and mixed with medium speed for 60 seconds.

4.2 *On-site spray trial*

Concrete with and without fibers were sprayed inside of a progressing tunnel to confirm the sprayability and practicability of HPS. Experimental results of fresh concrete tests, sprayed concrete tests are displayed in Table 9, Table 10, Figure 10 and Figure 11, respectively.

Initially, the pumping rate was speculated to be too low and inefficient due to the high viscosity properties of conventional ultra-high strength concrete. However, as proven in Figure 11a, formulating a new low viscous binders allowed high workability and suppression of concrete pressure. Simultaneously, a build-up test was conducted to measure the highest possible spraying thickness. As seen in Figure 11b, the peak measured at 65cm from the surface before falling from the base root, proving the high and reliable adhesion and cohesion

Table 6. Mix proportion for Mock-up test

Mixture	Binder type	Gmax (mm)	W/B (%)	S/B (%)	s/a (%)	Unit weight (kg/m³)				SP (B*%)	AC (B*%)
						W	B	S	G		
C17-B2	B2	15	17	62	60	189	1116	690	474	1.12%	2.4
C18-B2	B2	15	18	63	60	196	1094	691	474	1.04%	2.8
C19-B2	B2	15	19	66	60	201	1057	702	482	1.04%	3.0, 4.0

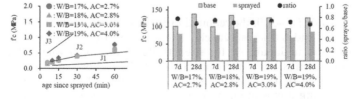

Figure 8. strength development of base and sprayed concrete. a) Early strength development of sprayed concrete measured by penetration needle, b) Compressive strength without fiber

Table 7. Test results of strength development and stiffness development of HPS(C17-B2)

Age	compressive strength f'c (MPa)	elastic modulus Ec (GPa)	Poisson's ratio
3h	1.37	1.07	0.01
5h	1.57	0.98	0.10
7h	2.58	2.28	0.15
24h	27.2	14.5	0.13
7d	79.3	27.9	-
28d	95.0	35.0	-

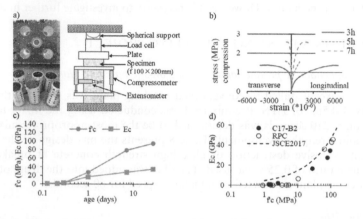

Figure 9. Test of stiffness development at young material age. a) Sampling mold and the test setup, b) Relationship between compressive stress and strain of very young concrete, c) Relationship between compressive strength, elastic modulus over time, d) Relationship between compressive strength and elastic modulus.

Table 8. Mix design of the concrete for on-site trial

Mixture	Binder type	Gmax (mm)	W/B (%)	s/a (%)	Unit weight (kg/m³)					SP (B*%)	AC (B*%)
					W	B	S	G	F		
C17-B2-F	B2	15	18	60	192	1064	722	485	2.25	1.2	1.8 ~ 2.4

Table 9. Test results of fresh concrete

mixture	sampling	slump flow mm	50cm flow time s	air content %	unit weight kg/l
C18-B2-F2	shipment	787.5	2.93	6.0	2.321
(with fiber)	arrival	849.0	3.41	-	-
C18-B2	shipment	622.0	8.99	6.5	2.361
(No fiber)	arrival	776.5	6.05	-	-

properties. As shown in Table 10, the rebound ratio for concrete with and without fibers were measured to be 5.3% and 8.0% respectively. These values are significantly smaller than the standard rebound ratio in Japan. As for strength development in Figure 10, base concrete experienced inferior results with implementation of fiber, however, the results from the sprayed concrete proved otherwise. This is due to the removal of air contained in the base concrete during spraying application. As result, the compressive strength ratio of base concrete and sprayed concrete surpassed 80%. Ultimately, the compressive strength of sprayed HPS at 28 days measured as 96.9MPa, surpassing our original goal.

Table 10. Test results of sprayed concrete

mixture	sampling	rebound ratio %	compressive strength (MPa)			elastic modulus (GPa)			split tensile strength (MPa)		
			7d	28d	91d	7d	28d	91d	7d	28d	91d
C18-B2-F	base	-	93.9	113.3	126.7	33.6	37.6	40.4	-	-	-
(with fiber)	sprayed	5.3	85.5	96.8	106.3	27.8	30.4	33.4	4.95	6.17	6.23
C18-B2	base	-	99.8	120.0	125.7	35.9	38.3	37.0	-	-	-
(No fiber)	sprayed	8.0	79.6	96.9	105.7	31.1	32.3	36.6	-	-	-

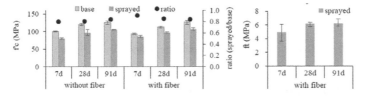

Figure 10. Compressive strength (f'c) and split tensile strength (ft) of base and sprayed concrete_

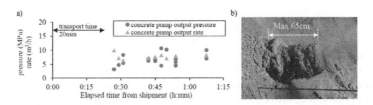

Figure 11. Verification of the construction performance of HPS by on site spray trial. a) Concrete output pressure and rate, b) Sprayed concrete build-up test

5 CONCLUSION

To successfully pump and spray concrete with super-high strength properties, the high viscosity due to the low W/B required to be significantly reduced. To resolve this viscosity problem, by formulating a synthetic binder with various composition ratio of cement, blast furnace slag, gypsum and silica fume, a concrete mix with acceptable workability, sprayability, strength and stiffness development was successfully developed. Verified by on-site trial, HPS was also proven to be manufacturable with a conventional batching plant. Furthermore, with successful application by a conventional spray machine, low rebound ratio of less than 10%, high adhesion and cohesion properties, HPS proved its practicability as a sprayed concrete. Ultimately, HPS provided compressive strength of 97MPa at 28 days, making it the highest strength developing sprayable concrete known today.

REFERENCES

Jinnai, H. et al. 2007. Adiabatic curing method for estimating 91-day strength in structure with 150MPa concrete. *Journal of advanced concrete technology* Vol.5 No.2: 161–170.
JSCE 2017. *Standard specifications for concrete structures, Design*, Japan Society of Civil Engineering.
Tani, T. et al. 2012. Study on Deformation Characteristics of Young-Age Tunnel Shotcrete, *Journal of MMIS Vol.128*: 113–120.
Tomioka, H. et al. 2007. Actual Production of High Strength Concrete Processed at Ready- Mixed Concrete Plant Using High Strength Admixture. *Summaries of technical papers of annual meeting, Architectural Institute of Japan* 1035: 69–70. (in Japanese)

Tunnels and Underground Cities: Engineering and Innovation meet Archaeology,
Architecture and Art, Volume 6: Innovation in underground engineering,
materials and equipment - Part 2 – Peila, Viggiani & Celestino (Eds)
© 2020 Taylor & Francis Group, London, ISBN 978-0-367-46871-2

Influence of consistency index on soil conditioning in EPB-TBM tunneling

S. Tarigh Azali
Executive Director at Line 6 Project, Tehran Metro and Suburban Railway Group of Companies, Tehran, Iran

E. Khorasani
PhD Candidate in Rock Mechanics, College of Engineering, University of Tehran, Tehran, Iran

J. Hassanpour
Assistant Professor, School of Geology, College of Science, University of Tehran, Tehran, Iran

ABSTRACT: In EPB-TBM tunneling, the excavated materials are conditioned for improving the workability of the soil for the purpose of applying pressure to the tunnel face. The workability is defined as the plasticity of the collected soil in the pressure chamber and is an index for the applicability of the EPB machine in different grounds. The effective agents in workability of soil are Consistency Index (Ic), water content, fine grains percentage and Foam Injection Ratio (FIR). The workability of soil can be estimated by the slump test, where the appropriate values for slump vary from 10 cm to 20 cm. In this research, the effect of consistency index on the soil conditioning and related workability are investigated by laboratory slump tests in three different water contents of 10% (I_C = 1 to 1.25), 15% (I_C = 0.75 to 1) and 20% (I_C = 0.5 to 0.75). The main variables of this study are water content and FIR. The results show with increasing I_C, the slump values of conditioned material decrease and by selecting a proper value of FIR in a specified I_C, it is possible to sufficiently condition the excavated soil.

1 INTRODUCTION

The use of mechanized methods in excavating tunnels has been widely developed in recent decades. The production of the Earth Pressure Balance-Tunnel Boring Machine (EPB-TBM) started for the first time by the Japanese in the middle of 1970 decade and the application of this machine has been increased for tunneling in the soil. The main property of these machines is a pressure chamber for repelling the ground and water pressure at the tunnel face. In this method, the bored materials are collected and compacted in a specific chamber, the boring chamber or pressure chamber, and then a cover is prepared that can protect the stability of the tunnel face. The important point is that some properties are necessary for the collected material in the pressure chamber. In other words, the collected soil in the chamber needs to show these two characteristics: applying pressure and the soil transition. It should be noticed that these properties (applying pressure and the soil transition) are considered as the workability of material in the mechanized boring literature. The appropriate workability of soil in boring by the EPB machine is provided when the collected soil in the boring chamber is converted to a plastic material. This type of material shows the ability of applying pressure in the rock face as well as the transition of soil by the screw conveyor. Thus, the workability defines the plasticity of the collected soil in the pressure chamber and is an index for the applicability of the EPB machine in different grounds (Thewes et al. 2010; Tarigh Azali & Moammeri 2012; Tarigh Azali et al. 2013). On the other hand, the workability of the soil varies in different water contents and grading of the soil. Accordingly, the effort is providing a well workability for the bored soil by injecting foams or polymers to the present materials in the pressure chamber. In the boring literature, adding the foams and polymers to the soil is considered as soil conditioning which is one of the main stages of boring by EPB machines.

Based on different studies, the workability of the soil can be estimated by the slump test (Quebaud et al. 1998; Leinala et al. 2000; Peila et al. 2009; Avunduk & Copur 2018). This test is very common in the concrete industry. The studies about the workability of soil in previous years indicated that the appropriate values for slump is from 10 cm to 20 cm (Quebaud et al. 1998; Jancsecz et al. 1999; Vinai et al. 2008; Thewes et al. 2010; Thewes and Budach 2010; Gharahbagh et al. 2014). It should be considered that, an appropriate condition of the plastic soil with well workability will not be provided in lower than 10 cm or higher than 20 cm slump. The study in the Oxford University about the foam and sand properties for the tunneling purposes by the Psomas (2001) was one of the initial studies in this field. Many studies have been carried out in different universities after that (Borghi 2006; Peila et al. 2008; Peila et al. 2013). One of the recent studies is going on form 2010 in the laboratory of Bochum University and there are several articles about the mentioned studies available in different journals and conferences (Thewes and Budach 2010). The main goal of the tests in this laboratory is the study of conditioning parameters on the soil workability.

The aim of the current study is investigating on the effects of consistency index on soil conditioning by slump tests. For this purpose, the followings are the sequences of the study in this paper:

- Designing the laboratory tests
- Providing the laboratory equipment and necessary materials
- Performing the laboratory tests
- Analysis of the results

2 REVIEWING OF THE FOAM INJECTION IN TUNNELING BY THE EPB MACHINE

These days there is not an EPB machine available that do not apply the soil conditioning system during the boring process. Figure 1 illustrates the foam injection system in an EPB machine. The foam injecting system includes the reservoir of water, compressed air and foam with the necessary pumps, the sensor for pressure measurement, injection nozzles, the pressure reducer valve, transition pipes to the systems of tunnel face, the pressure chamber and the screw conveyor.

In the past decade, containing at least 30% of fine grains in the soil volume has been a criterion for applying the EPB for tunneling in a ground. Nowadays, with the improvements of the soil conditioning technology by foam, applying the EPB machines in the grounds with 10% fine grains is also possible that is described in the instructions for selecting an appropriate boring machine by the German underground structures (DAUB 2010).

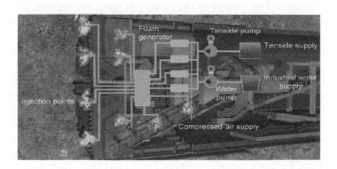

Figure 1. Schematic view of producing and injecting foam in EPB-TBM (Herrenknecht & Rehm 2002).

In general, foam is formed from three parts including foaming liquid, water and air. The main parameters used in the subject of foam injection are the concentration of the foaming liquid (C_f), foam expansion ratio (FER) and foam injection ratio (FIR) (Borghi 2006). The C_f describes the amount of surfactant in the solution of surfactant and water (Eq. 1).

$$C_f = \frac{Q_f}{Q_f + Q_w}.100 = \frac{Q_f}{Q_L}.100 \qquad [vol\%] \qquad (1)$$

where Q_f is the volume flow of surfactant [l/min], Q_w is the volume flow of water [l/min] and Q_L is the volume flow of liquid [l/min]. The FER describes the ratio of the total foam volume Q_F [l/min] and the liquid phase volume of the foam Q_L [l/min] (Eq. 2). So it means, the FER is an indicator for the wetness of a foam. The determining factor in this description is the volume of air Q_A [l/min].

$$FER = \frac{Q_L + Q_A}{Q_L} = \frac{Q_F}{Q_L} \qquad [-] \qquad (2)$$

And the FIR describes the volume of foam in relation to the volume of soil being excavated Q_S [l/min] (Eq. 3).

$$FIR = \frac{Q_F}{Q_S}.100 \qquad [vol\%] \qquad (3)$$

3 THE LABORATORY TESTS OF THE SOIL CONDITIONING

For laboratory studies, 15 samples of soil with specified grading were used to investigate on the effects of consistency index on soil conditioning (Tarigh Azali 2015). The essentials of the soil conditioning tests include three main frameworks:

• The required materials: the foaming liquid (surfactant) and the soil samples with specified grading properties;
• Equipment for producing foam bubble: foam generator, mixer and compressor;
• The soil conditioning tests: the soil grading, the Atterberg limits, the water content and the slump tests.

In this study, a foam producer agent is used which shows all properties in EFNARC standard (EFNARC 2005). These properties include the 390 second half life of foam and the 0.5 mm average size of the bubble. In addition, a foam generator machine is used for producing foam, which is capable of controlling the exact amount of air, water and the foam producing material by the taps, gages and the flow meters and conduct these materials into a cylinder that is full of glass particles. The output of the materials from this cylinder is foam bubbles.

For more clarification, all steps of the soil conditioning tests (providing sample, preparation, foam production, mixing and the slump test) are presented in the flow diagram (Figure 2). In addition, the photos of different stages of the test in the laboratory are presented in Figure 3. According to the flow diagram (Figure 2), the tests start by the soil grading step. Then foams with different FIRs are produced by the foam generator machine. The produced foam in the electrical mixer is blended by the soil with 10%, 15% and 20% water content for soil conditioning and eventually the slump test and the visual quality control were carried out for evaluation of the workability.

In this study, the foams with different FIRs including 20%, 30%, 40% and 50% are produced by the foam generator machine and are mixed with the soil samples and eventually the slump test is carried out for each sample. It should be noted that the values of FER and C_f are constant in all tests which are equal to 7.2% and 3.2%, respectively.

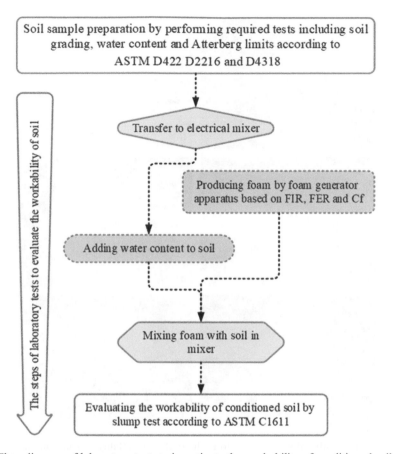

Figure 2. Flow diagram of laboratory tests to investigate the workability of conditioned soil.

3.1 The mechanical soil grading test

This test was carried out on the soil samples to specify the grading of them. The diagram of grading for the samples used in this study is presented in Figure 4. As can be seen, the contents of fine grains, sand and gravel are 40, 50 and 10 percent, respectively.

3.2 The consistency index test

In the presence of the clay minerals in the fine grained soils, the soil converts to a plastic material without crushing by a few percent of water. The resulted stickiness is related to the absorption of water by the clay particles. The soil converts to a hard material where the water content is lower than the liquid limit and higher than the plastic limit. Its behavior is like a plastic material and show lower viscosity. The soil acts as a liquid, if the water content increase to higher than the liquid limit. Therefore, the soil behavior is defined by the Consistency Index (I_c) in different water contents. The consistency index is determined based on the Liquid Limit (LL), Plastic Index (PI) and the percentage of the natural water content of the soil (w) by the following equation:

$$I_c = \frac{LL - w}{PI} = \frac{LL - w}{LL - PL} \qquad (4)$$

Figure 3. The steps of laboratory activates to evaluate the workability of soil using injecting foam: a) weighing the soil samples; b) mixing foam with water and air; c) foam generation; d) adding water content; e) mixing the produced foam with soil in electrical mixer; f) slump test.

In this equation, the LL and PL parameters are part of the inherent properties of the soil and are not changeable, but the water content can be changed easily and this parameter can affect the consolidation and consistency of the soil directly. In other words, by including the consistency index, in addition to the water content, the other inherent parameters such as the plastic index and liquid limit are also included. The soil properties can be classified according to Table 1 based on different criteria of consistency.

Based on the Atterberg definition tests, the liquid limit (LL) is about 30% and the plastic limit is about 12% in the soil samples. Since the tests were carried out on the samples with three different water contents: 10%, 15% and 20%, the consistency index for these values is in three ranges of 0.5 to 0.75, 0.75 to 1 and 1 to 1.25. These are classified in the categories of soft, stiff and very stiff, respectively based on the Table 1.

Table 1. Soil classification based on I_c.

Description of soil condition	Consistency index (Ic)
Hard	More than 1.25
Very stiff	1 to 1.25
Stiff	0.75 to 1
Soft	0.5 to 0.75
Very soft	0 to 0.5
Liquid	Less than 0

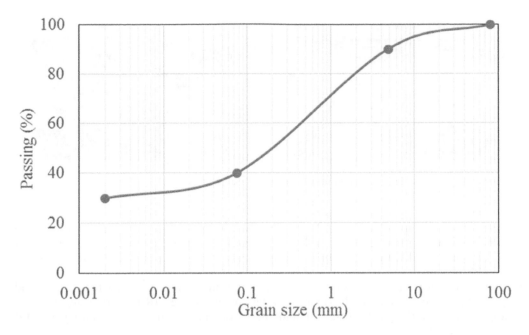

Figure 4. The grading of soil used in this study.

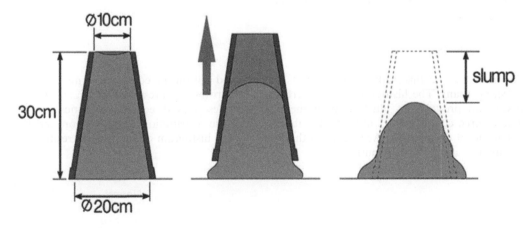

Figure 5. Schematic view of performing slump test on the conditioned materials.

3.3 *The slump test*

According to the result of all recent studies, the slump test can be used for evaluation of the workability of the conditioned materials. This test is approximately similar to the slump test that is used for the fresh concretes (ASTM C143 2003). The only difference is that the rod which is used for compacting the soil is not used in this test. The sequences of the test are illustrated in Figure 5. According to the figure, the conditioned soil that includes the material and foam is poured in the slump cylinder. After 1 minute the cylinder must rise without any compacting or mixing. Then the amount of the slump in the mixture is determined that can be between 0 and 30 cm.

Generally, according to this study and the previous studies (Peila et al. 2007), three different behaviors can be expected for the result of the slump test (Figure 6):

Figure 6.   Different behaviors of conditioned materials in slump test.

- Stiff behavior: This behavior can be observed where the slump is less than 10 cm, in the lack of enough water (usually) or foam (or both) and the high content of fine grains (Figure 6a). Besides, this behavior is not appropriate for the workability of soil in boring by EPB machine.
- Liquid behavior: This behavior happens where the slump value is higher than 20 cm and generally is due to the grain distribution, water content or the FIR ratio that might not be appropriate for the plastic behavior and the workability of the soil in boring by EPB machine (Figure 6b).
- Plastic behavior: This behavior includes 10 to 20 cm of slump with a regular shape of the mass in low or without water expulsion (Figure 6c). This behavior shows well workability of materials in boring by the EPB machine.

4   FINDINGS

As mentioned, in this study, 15 slump tests were carried out on the conditioned materials by injecting foam. The histogram of the performed slump tests is presented in Figure 7. The histogram provides an illustration of the data that describes three characteristics simply. These three characteristics are: the distribution type, the distribution centralization and the extension of distribution. According to the Figure 7, the histogram of slump tests results can be fitted by a normal distribution.

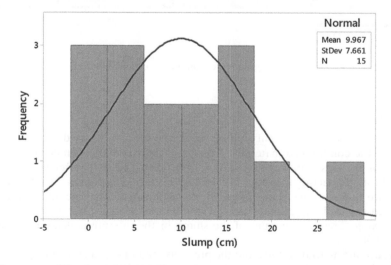

Figure 7.   Frequency of the slump tests results.

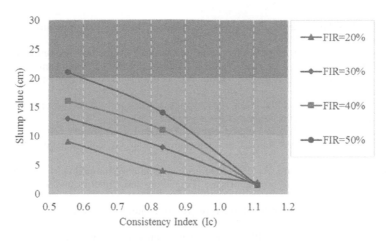

Figure 8. The relationship between slump values of conditioned materials and consistency index with various FIRs.

In Figure 8, the values of slump results versus consistency index (Ic) were drawn for different FIRs. As seen in this figure, the value of slump decreases with increasing the Ic. Moreover, for a constant Ic, with increasing the value of FIR, the slump values increase too. For consistency index more than 1, the FIR has not significant effect on the slump value of the conditioned materials. As mentioned, the appropriate range for slump value is 10 cm to 20 cm (green area on the Figure 8) and some tests have had suitable values of slump. Therefore, by changing the FIR in different values of I_c, it is possible to achieved the desired slump value.

5 CONCLUSIONS

In this research, the effect of consistency index (I_c) on the soil conditioning process was investigated. For this purpose, 15 slump tests were carried out on the conditioned soil which had a specified grading. The main variable in the slump tests were water content and Foam Injection Ratio (FIR). The water content assesses the consistency index, because the other parameters of Ic including liquid limit and plastic limit are inherent properties of the soil. The results of this study can be summarized as follows:

- With increasing the Ic, the slump value of materials would be decreased. This behavior can be ascribed to the consistency condition of soil, where the soil with higher consistency is stiffer and consequently has a lower value of slump.
- For the materials of this study, when the FIR is less than about 25%, the injected foam cannot sufficiently condition the materials, in any values of I_c.
- By selecting suitable FIR in a specified I_c, the conditioned materials would have appropriate slump value in the range of 10 cm to 20 cm.

REFERENCES

ASTM C143, Standard Test Method for Slump of Hydraulic-Cement Concrete 2003.
Borghi, F. 2006. Soil conditioning for pipe jacking and tunnelling (Doctoral dissertation, University of Cambridge).
Empfehlungen zur Auswahl von Tunnelvortriebsmaschinen, DAUB (Deutscher Ausschuss für unterirdisches Bauen 2010. Available online at: www.daubita.de/uploads/media/gtcrec14.pdf
Gharahbagh, E. A., Rostami, J., & Talebi, K. 2014. Experimental study of the effect of conditioning on abrasive wear and torque requirement of full face tunneling machines. Tunnelling and Underground Space Technology, 41, 127–136.

Herrenknecht, M., & Rehm, U. 2002. Newest Development in Mechanized Tunnelling. Journées d'études Internationales. Toulouse: AFTES

Peila, D., Oggeri, C., & Borio, L. 2008. Influence of granulometry, time and temperature on soil conditioning for EPBS applications. In Underground Facilities for Better Environment and Safety. In: Kanjlia, VK, Ramamurthy, T., Wahi, PP, Gupta, AC, (Editors), ITA-AITES World Tunnel Congress (WTC), Agra, India.

Peila, D., Oggeri, C., & Borio, L. 2009. Using the slump test to assess the behavior of conditioned soil for EPB tunneling. Environmental & Engineering Geoscience, 15(3),167–174.

Peila, D., Oggeri, C., & Vinai, R. 2007. Screw conveyor device for laboratory tests on conditioned soil for EPB tunneling operations. Journal of Geotechnical and Geoenvironmental Engineering, 133 (12),1622–1625.

Peila, D., Picchio, A., & Chieregato, A. 2013. Earth pressure balance tunnelling in rock masses: Laboratory feasibility study of the conditioning process. Tunnelling and Underground Space Technology, 35, 55–66.

Psomas, S. 2001. Properties of foam/sand mixtures for tunnelling applications (Doctoral dissertation, University of Oxford).

Quebaud, S., Sibai, M., & Henry, J. P. 1998. Use of chemical foam for improvements in drilling by earth-pressure balanced shields in granular soils. Tunnelling and underground space technology, 13 (2),173–180.

Specication and Guidelines for the Use of Specialist Products in Mechanized Tunneling (TBM) in Soft Ground and Hard Rock, EFNARC, Association House, U.K. 2005.

Tarigh Azali, S. 2015. Classification of alluvium and Pediment deposits for tunneling with EPB-TBM machine. Ph.D. Dissertation. Ferdowsi University of Mashhad. Iran.

Tarigh Azali, S. T., Ghafoori, M., Lashkaripour, G. R., & Hassanpour, J. 2013. Engineering geological investigations of mechanized tunneling in soft ground: A case study, East–West lot of line 7, Tehran Metro, Iran. Engineering Geology, 166, 170–185.

Tarigh Azali, S., & Moammeri, H. 2012. EPB-TBM tunneling in abrasive ground, Esfahan Metro Line 1. In ITA-AITES world tunnel congress (WTC), Bangkok, Thailand.

Thewes, M., & Budach, C. 2010. Soil conditioning with foam during EPB Tunnelling. Geomechanics and Tunnelling, 3(3),256–267.

Thewes, M., Budach, C., & Galli, M. 2010. Laboratory tests with various conditioned soils for tunnelling with earth pressure balance shield machines. Tunnel International Journal For Subsurface Use, (6), 21.

Vinai, R., Oggeri, C., & Peila, D. 2008. Soil conditioning of sand for EPB applications: A laboratory research. Tunnelling and underground space technology, 23(3),308–317.

Tunnels and Underground Cities: Engineering and Innovation meet Archaeology,
Architecture and Art, Volume 6: Innovation in underground engineering,
materials and equipment - Part 2 – Peila, Viggiani & Celestino (Eds)
© 2020 Taylor & Francis Group, London, ISBN 978-0-367-46871-2

Use of robots in road tunnel inspection

H. Terato & S. Yasui
Japan Construction Method and Machinery Research Institute, Fuji, Shizuoka, Japan

Y. Nitta
Public Works Research Institute, Tsukuba, Ibaraki, Japan

T. Masu
Ministry of Land, Infrastructure, Transport and Tourism, Noshiro, Akita, Japan

ABSTRACT: We proposes methods for using robots in the periodic inspection of road tunnels and associated evaluation methods. The main purpose of using a robot is to reduce working hours spent on the close visual inspection which is currently conducted by humans. For this purpose, the report proposes two methods: a robot conducting measurement after inspection work by humans (Use Case 1), and a robot conducting measurement before inspection work by humans (Use Case 2). To verify the effects of these methods, we conducted "on-site verification" at the actual inspection site and "elemental verification" that uses some specimens.

1 INTRODUCTION

With regard to road tunnels in Japan, it is prescribed in a standard specified by the Japanese government that inspection be performed by means of close visual inspection once every five years. A close visual inspection is an inspection whereby a person checks for deformation by getting close to the lining surface and checking it visually. This work is mostly carried out by humans, and is often accompanied by difficult work.

Amid these circumstances, the Ministry of Land, Infrastructure, Transport and Tourism and the Ministry of Economy, Trade and Industry are considering the introduction of robots for maintenance, management and inspection work for a range of structures including tunnels. In this article, we present the results of a verification that was conducted within the initiatives above, and evaluate and discuss them. The main content of this article is as follows:

1) Proposal of use cases of robots in current inspection work, and methods for verifying the effects;
2) Discussion of on-site verification of proposed use cases, and the results of element verification for checking the robots' performance; and
3) A summary of the results obtained in this article and an outline of issues that should be addressed in the future.

The purpose of this research is to contribute to the clarification of points to focus on in future development of tunnel inspection robots, and to the establishment of goals for development.

2 INSPECTION WORK AND INSPECTION ROBOTS

2.1 *Inspection work*

In a periodic inspection of road tunnels that is conducted once every five years, inspection by means of a close visual inspection, hammering test and examination by touch set forth in the

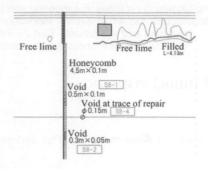

Free lime

Free lime   Filled
L=4.13m

Honeycomb
4.5m×0.1m

Void   S8-1
0.5m×0.1m

Void at trace of repair
φ0.15m   S8-4

Void
0.3m×0.05m
S8-2

Figure 1.   Example of a deformation expansion plan.

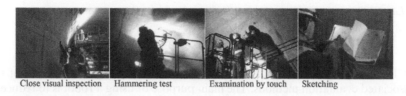

Close visual inspection   Hammering test   Examination by touch   Sketching

Figure 2.   Situation of inspection work.

"Periodic Road Tunnel Inspection Procedures (Ministry of Land, Infrastructure, Transport and Tourism, 2014)" (hereinafter referred to as the "procedures") is performed. Typically, the results of this inspection are recorded as a hand-drawn sketch, and on the basis of this sketch, a deformation expansion plan is created, which shows graphically the deformation situation of the tunnel. An example of a deformation expansion plan is shown in Figure 1. Also, each of the tasks are shown in Figure 2. Because these tasks, including creation of deformation expansion plans, are basically manual work performed using manpower, it is thought that improving the efficiency of inspection work and the quality of record through the use of robots will be important in the future. Hereinafter, inspection by means of these methods is referred to as the "current inspection."

2.2   Inspection robot

It is indicated in the procedures that the standard is for tunnel inspections to be performed by humans, such as by means of a close visual inspection and hammering test, as described above. Because of this, the issue that will arise when introducing a robot into tunnel inspections will be, "Can an inspection by a robot ensure the same or higher quality as an inspection by a human?" Even now, various types of robots are being developed, as well as a range of high-precision inspection technologies (Next-generation Social Infrastructure Robot Technologies and Robot Systems, 2018).

However, in an inspection by a human, complex tasks are carried out simultaneously. At present, substituting for all these tasks by using robots is probably difficult. Therefore, in this research, we performed verification for conducting inspections in a safe and efficient manner, by carrying out inspections that combine a robot and a human, and making use of the advantages of the two.

In this research, we propose use cases and evaluation methods for robots in inspection work. We targeted vehicle-type wall photographing technology (Figure 3) that is at a level of practical application, and conducted on-site verification. The features of each robot are shown below. In this paper, the main objective is to describe the method of performance evaluation of the robot, is not intended to show the performance of each robot, so avoid describing detailed specifications.

Figure 3.   Example of vehicle-type wall photographing technology.

- Robots A and B: Video camera
- Robot C: Laser
- Robot D: Line sensor camera
- Robot E: Single lens reflex camera

3   PROPOSAL OF USE CASES OF ROBOTS AND METHOD FOR VERIFYING THE EFFECTS

As described in the previous chapter, it is considered that it will be difficult to substitute for all human inspections by using robot inspections alone. Therefore, in this research, we propose use cases where robots can be used in actual inspection work. Through these proposals, it is anticipated that robot users and robot developers will be able to share the same images, and humans and robots will be able to complement each other with regard to the respective work performed by the two. Below are two use cases we established for this research and an outline of the methods for verifying the effects.

3.1   *Use case 1*

In use case 1, measurement by a robot is performed before the current inspection, and the current inspection is per-formed by referring to the results of the robot inspection in order to reduce the inspection time needed for the current inspection (close visual inspection, hammering test, chalk marking, sketching). Figure 4 is a schematic diagram of use case 1.

3.2   *Use case 2*

In use case 2, measurement by a robot is performed after the completion of a close visual inspection, hammering test and chalk marking in the current inspection, then deformation expansion plans, etc. are created. This use case aims to improve the efficiency of the inspection work by enabling the omission of sketching work. Figure 5 is a schematic diagram of use case 2.

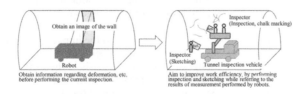

Figure 4.   Schematic diagram for use case 1.

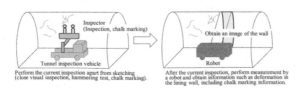

Figure 5.   Schematic diagram for use case 2.

3.3 *Outline of verification methods*

By performing inspections in accordance with the use cases above, we verified and evaluated how much the efficiency of the current inspection can be improved. Verification and evaluation were performed by means of on-site verification using an actual tunnel. The tunnel used for on-site verification was Niraone Tunnel (length: 615 m), located in Sagamihara City, Kanagawa Prefecture. The tunnel is a two-lane road tunnel that is in service, but is closed at night, from 7 p.m. until 7 a.m. the next day. For on-site verification, we simulated the actual inspection work using the hours during which the tunnel is closed.

The main purposes of this on-site verification are to evaluate the efficiency improvement effects resulting from introduction of robots into actual inspection sites, and also to check whether it is possible to apply robots in actual inspection work. In order to achieve these purposes, we conducted verification with the focus on the following points, and from the results, we evaluated the applicability of robots to actual inspection work.

1) Efficiency: Evaluate how the number of working people and work time for the current inspection can be reduced, and how the efficiency of the inspection can be improved, by using robots.
2) Accuracy: Evaluate how accurate inspection results with small errors can be obtained by using robots.
3) Safety: Because the work will be carried out under lane control, evaluate whether safety of not only inspection workers but also vehicles and pedestrians passing the tunnel is ensured.

Among the points above, quantitative evaluation was performed for efficiency and accuracy, and qualitative evaluation was performed for safety. Table 1 summarizes the thinking behind each evaluation.

With regard to accuracy, element verification using specimens that simulate a variety of crack widths (hereinafter referred to as "specimens") were performed as supplementary verification.

Table 1. Thinking regarding evaluation.

Use case	Verification item	Evaluation category	Evaluation method
Use case 1	Efficiency	Quantitative evaluation	- Evaluate whether referring to robot measurement results has improved the efficiency of inspection, by surveying the time and people needed for the work.
	Accuracy	Quantitative evaluation	- Evaluate the number of erroneous detections and detection omissions, by comparing deformation obtained from measurement by robots and deformation obtained from the usual inspection methods.
			- Evaluate the accuracy of obtaining the locations of cracks and points with abnormal sounds in a hammering test, using model specimens,
	Safety	Qualitative evaluation	- Evaluate with regard to the impact on passing vehicles, pedestrians and inspection operators (danger, impact on the human body, surprise), on the basis of the work situation resulting from robots, and by means of interviews.
Use case 2	Efficiency	Quantitative evaluation	- Measure the time required for sketching in the inspection, and evaluate on the basis of the work time reduction if the sketching time were saved.
			- Evaluate on the basis of the fact that sketching by robots reduces lane control time.
	Accuracy	Quantitative evaluation	- Evaluate the accuracy by comparing a deformation expansion plan created by the regular method and a deformation expansion plan created by a robot.
	Safety	Qualitative evaluation	- Evaluate with regard to the impact on passing vehicles, pedestrians and inspection operators (danger, impact on the human body, surprise), on the basis of the work situation resulting from robots, and by means of interviews.

Figure 6.   Appearance of crack specimen models.

This was conducted because in on-site verification, accuracy verification cannot be performed for a variety of different crack widths. For the specimens, multiple 15 cm x 15 cm mortar plates were made, and cracks with a width of 0.1 mm to 3.0 mm were created on the plates. A crack with one type of width was created on one mortar plate, and mortar plates with no cracks were also mixed in as specimens. The appearance of the specimens is shown in Figure 6.

4   VERIFICATION AND EVALUATION METHOD FOR USE CASE 1

Figure 7 shows the flow of verification targeting use case 1, and below, the procedure for on-site verification is described.

1. Perform measurement before the current inspection for a verification section that is common to all robots (Section 0, length: approx. 30 m) and verification sections unique to each robot (Robots A to E). Install specimens to a section other than those mentioned above.
2. For Section 0 and the specimens, output deformation expansion plans created on the basis of the results of measurement performed by all robots, and for Sections A to E, output a plan created on the basis of the results of measurement performed by the corresponding robot.
3. Perform the current inspection for Section 0 and Sections A to E. For Section 0, perform the usual current inspection with robot measurement results not referred to, and for Sections A to E, perform the current inspection by referring to the deformation expansion plan created by each robot for the corresponding section.
4. For Section 0 and Sections A to E, create deformation expansion plans on the basis of the results for the current inspection performed in 3).

4.1   Efficiency verification and evaluation method

With regard to work efficiency, we evaluated the work time reduction effect for the current inspection performed by using the results obtained by each robot. In this article, as shown in Figure 7, the current inspection time for Section 0 (the current inspection performed without referring to results obtained by robots) was compared with the current inspection time for each of Sections A to E (the current inspection performed by referring to results obtained by the robots). We set a different verification section for each robot because in the current inspection, chalk marking is performed on the lining surface, so in order to perform verification for multiple robots in one section, we would have to erase the chalk marks each time verification is performed, which is not realistic.

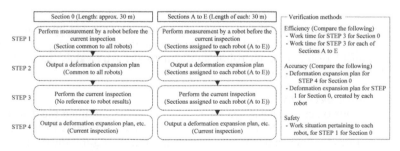

Figure 7.   Verification flow for use case 1.

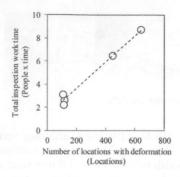

Figure 8.   Relation between the number of locations with deformation and the inspection time.

Under these conditions, what we have to be careful about is the fact that the amount and types of deformation vary between each of Sections A to E. If there is a large amount or many types of deformation, more time will be required for inspection. Consequently, if there are two different sections subject to verification, it will not be possible to make a direct comparison between the two. In order to correct the inspection time, we used the relation between the number of locations with deformation and the inspection time (people, time) that we obtained from the results of a separately conducted tunnel inspection work survey (Figure 8), and corrected the work time. As shown in the same figure, the relation between the number of locations with deformation and the inspection time is that they are roughly proportional. Using the relation above, we calculated the ratio of the number of locations with deformation in Sections A to E to the number of locations with deformation in Section 0, and obtained the total time required to inspect each section from Figure 8. As a result, we were able to obtain the inspection time for each of Sections A to E for when the inspection time for Section 0 was defined as 1. We attempted to normalize the work time for each section, by multiplying value obtained above by the time taken for the current inspection performed by referring to the results given by robots for Sections A to E.

4.2   *Accuracy verification and evaluation method*

We performed accuracy verification for use case 1 by comparing the deformation expansion plan for the current inspection for Section 0 obtained from STEP 4 in Figure 7 and the deformation expansion plan for Section 0 obtained from only the measurement results of each robot obtained from STEP 1. In addition, we conducted verification for specimens as well, using the same method.

The purpose of accuracy verification is to evaluate whether the use of robots leads to inspection results with high accuracy, and to check whether the number of errors in robot measurement results is small. We conducted the verification because we believed that if the accuracy of robot measurement results, and consequently the accuracy of de-formation expansion plans, is low, the results cannot be used as a reference for the current inspection, and the efficiency of the current inspection is unlikely to improve.

4.2.1   *Evaluation based on deformation expansion plans*

For the evaluation, we defined the deformation expansion plan created by current inspection as the true value, and evaluated to what extent the deformation expansion plans created from only the measurement results of each robot conform to the true value. Table 2 is a list of deformation detected by current inspection and by robots.

4.2.2   *Evaluation using specimens*

With regard to various types of cracks created on specimens, we obtained the detection rate of crack widths as shown in Figure 9. The figure shows the detection rates for all crack widths and the detection rates for each crack width. As indicated in the figure, the detection rates for cracks with a width less than 0.3 mm varied from 0 to about 90%, but they tend to be low in

Table 2. Types of deformation and the number of locations with deformation indicated on deformation expansion plans.

		Current inspection	Robot				
			A	B	C	D	E
Types of deformation and the number of locations with deformation	Crack	41	27	34	19	28	27
	Void	27	11	8	0	13	0
	Cold joint	1	0	0	0	0	0
	Honeycomb	3	0	1	0	3	0
	Level difference	1	0	0	0	0	0
	Exposed reinforcement steel	1	0	0	0	1	0
	Deteriorated water supply works	4	0	0	0	0	0
	Free lime	52	50	51	42	48	30
	Water leak	31	28	18	12	10	17
	All forms of deformation	161	116	112	73	103	74
Deformation detection rate (%)		-	72.0	69.6	45.3	64.0	46.0

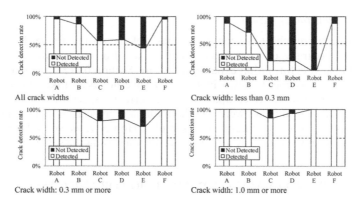

Figure 9. Accuracy verification results using specimens.

general. On the other hand, the detection rates for cracks with a width of 0.3 mm or more are mostly 80% or more.

4.3 Safety verification and evaluation method

With regard to safety, we checked the implementation status regarding on-site verification, focusing on the following points:

- Is there any danger of collision, etc. with passing vehicles, pedestrians and inspection operators when the robot conducts the work?
- Is anything generated that could have an impact (exposure, etc.) on the bodies of vehicle drivers, pedestrians and inspection operators when the robot conducts the work?
- Is there anything that could surprise vehicle drivers, pedestrians and inspection operators when the robot con-ducts the work?

Because it is difficult to evaluate the impact on the human body from on-site verification, we evaluated them by conducting interviews with the developers to hear their opinions. In the on-site verification, although it was judged that some technologies would require anti-glare measures for vehicle drivers and pedestrians owing to the use of flood-lights, measures could

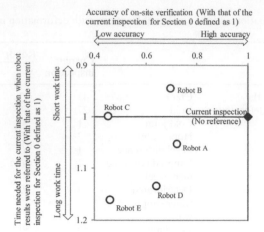

Figure 10.   Efficiency and accuracy verification results for use case 1.

be taken by, for example, modifying the equipment and improving the work methods, and there were no technologies for which safety was a fundamental problem.

4.4   *Evaluation results for use case 1*

Figure 10 is a distribution chart created on the basis of the results above. It shows a plot of the efficiency and accuracy of each robot on two axes. For the accuracy, the verification results obtained from a deformation expansion plot for Section 0 are shown, and the verification results for crack specimens shown in Figure 9 have not been taken into consideration. This is because the purpose was to compare with the current inspection in Section 0.

From the verification results, we were unable to find a clear relation between efficiency and accuracy, partly owing to the low number of samples. However, we should be able to obtain some kind of relation in the future by adding more samples using the same method.

5   VERIFICATION AND EVALUATION METHOD FOR USE CASE 2

Figure 11 shows the flow of verification targeting use case 2, and shown below are the procedures for on-site verification.

1) Perform the current inspection apart from sketching, for the verification section that is common to all robots (Section 0, length: approx. 30 m).
2) In Section 0, perform only the sketching in the current inspection.
3) In Section 0, perform measurement by each robot.
4) Create a deformation expansion plan on the basis of the results obtained from 2) and 3) above, for each case.

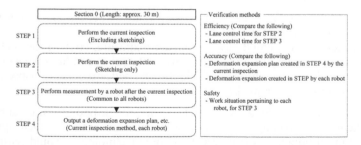

Figure 11.   Verification flow for use case 2.

Below we give details on efficiency, accuracy and safety evaluated on the basis of the above on-site evaluation results.

5.1 Efficiency verification and evaluation method

By using the results of the work time and the number of operators required for on-site verification performed in Section 0, we conducted a qualitative evaluation regarding to what extent fewer people and less time will be needed for sketching if sketch measurement by a robot is employed, compared to the number of people and amount of time needed for the sketching in conventional inspection work.

5.2 Accuracy verification and evaluation method

By comparing the deformation expansion plan created by robot measurement and the plan created by the current inspection (sketch drawn by a human), both targeting Section 0, we quantitatively evaluated the accuracy of sketches drawn by a robot.

With regard to comparison of inspection results, because measurement by robots detects deformation on the basis of chalk marks, it will be possible to obtain deformation expansion plans with high accuracy. Therefore, we did not perform detailed checking regarding inspection accuracy. Instead, we evaluated the accuracy for each section of lining (side wall, shoulder, crown) in the transverse direction of the tunnel on a six-point scale, giving two points for sections where the record is accurate, one point for sections with minor omissions in the record, and zero points for sections with many omissions in the record.

5.3 Safety verification and evaluation method

Discussion of safety is omitted because it was evaluated by using the same method as use case 1 (see 4.3).

5.4 Evaluation results for use case 2

Figure 12 is a distribution chart created on the basis of the results above. It shows a plot the inspection accuracy and work efficiency of each technology on two axes. The same as with use case 1, because the purpose of this article is to propose evaluation methods, we will not go into comparison of the performance of each technology.

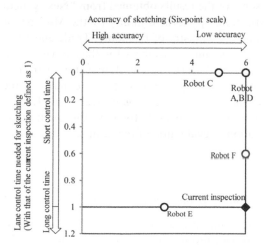

Figure 12.   Efficiency and accuracy verification results for use case 2.

From the same figure, we were able to ascertain the relation between the lane control time and accuracy, and were therefore able to obtain results that could be used, for example, when selecting inspection robots for inspection work.

6 CONCLUSION

In this article, we proposed some ways of using robots in road tunnel inspection, and indicated points of focus for evaluation. Furthermore, as a method of evaluation, we performed on-site verification, and conducted verification and evaluation on the basis of the points of focus on for evaluation. Below is a summary of the results obtained in this article.

– As ways to introduce robots into tunnel inspection, we established two kinds of use cases. One is a use case in which measurement by a robot is performed before the current inspection, and the other is a use case in which measurement by a robot is performed after the current inspection.
– In order to check the effects of the use cases above, we performed on-site verification using an actual tunnel.
– We set efficiency, accuracy and safety to be points of focus on for checking the effects, and indicated an evaluation method for each.
– On the basis of the on-site verification results, we evaluated the robots using the evaluation methods above.

In this article, we have only indicated the basic thinking regarding ways to use and evaluate robots. We have not discussed their validity or ways to use the evaluation results, because there was only a small number of samples. However, by indicating arrangement methods using the relation between accuracy and efficiency, it is anticipated that, in the future, it will lead to clarifying development goals for robots. We believe that besides the thinking regarding evaluation described in this article, it will also be important to create manuals and establish estimation standards in order to introduce robots into inspection work.

It would be desirable if data to enhance this research is collected in the future. However, on-site verification using actual sites requires costs and time, and is not easy to carry out. Therefore, we think that establishing simpler verification methods that use tunnel models, etc. will be also important.

ACKNOWLEDGMENT

This article summarizes some of the results obtained from "Next-generation Social Infrastructure Robot On-site Verification," performed jointly by the Ministry of Land, Infrastructure, Transport and Tourism and the Ministry of Economy, Trade and Industry. We would like to express our gratitude to all committee members for their advice regarding this verification, and all members of the Regional Development Bureau for their cooperation in the on-site verification. In particular, we are deeply grateful to the committee members of the "Tunnel Maintenance and Management Committee (Committee Chief: Kazuo Nishimura, Professor emeritus of Tokyo Metropolitan University)" for their thorough guidance, ranging from the method for on-site verification to evaluation of the results.

REFERENCES

Ministry of Land, Infrastructure, Transport and Tourism 2014. Periodic Road Tunnel Inspection Procedures.
Next-generation Social Infrastructure Robot Technologies and Robot Systems - On-site Verification Portal Site - 2018. <https://www.c-robotech.info/>, accessed on July 5, 2018.

*Tunnels and Underground Cities: Engineering and Innovation meet Archaeology,
Architecture and Art, Volume 6: Innovation in underground engineering,
materials and equipment - Part 2 – Peila, Viggiani & Celestino (Eds)*
© 2020 Taylor & Francis Group, London, ISBN 978-0-367-46871-2

Realisation of a single-shell drained segmental tunnel lining by means of a water-permeable annular gap fill

C. Thienert
STUVA e. V., Cologne, Germany

D. Edelhoff
IMM Maidl & Maidl Beratende Ingenieure GmbH & Co. KG, Bochum, Germany

E. Kleen
MC-Bauchemie Müller GmbH & Co. KG, Bottrop, Germany

N. Hörlein
PORR Bau GmbH, Vienna, Austria

ABSTRACT: In order to extend the scope of application of mechanised tunnelling with single-shell segment lining in solid rock, an annular gap fill material has been developed, which is capable of being pumped and acting as a drainage layer, with a k_F value $> 10^{-4}$ m/s. The particular relationships of construction material, process technology and construction were completely taken into account. The basic idea is the defined foaming of cement suspension with a foam generator, which produces an open-pored and permeable structure with sufficient strength to ensure the bedding of the segments. The general suitability of the method has been impressively demonstrated in the course of an extended test programme including site trials.

1 INTRODUCTION

1.1 *Initial situation*

The many advantages of construction process technology and ecology as well as constant further development of the machinery have led to the situation that mechanised tunnelling can now even be used at great depths under high groundwater pressures, for example for tunnelling through low mountain structures in Germany. Depending on the prevailing hydrogeological conditions, the tunnel lining on mechanised drives is mostly either constructed as a single-shell watertight or a double-shell drained construction (Figure 1).

Single-shell watertight linings are normally formed of a single tube of segments (see Figure 1, left), which has to be designed to resist the entire water pressure. The waterproofing consists of gaskets in the joints. The essential logistical and economic advantage is that no additional cast in-place concrete has to be placed on site. This variant can also be considered ecologically positive since there is practically no effect on the natural groundwater regime, neither in the competed state nor during the construction period. The economic limit of this construction method is often reached under groundwater pressures of about 5 bar.

Double-shell drained constructions are also often first lined with segments, followed by the installation of an additional inner lining of cast in-place concrete (see Figure 1, right). Ingress of water from fissures in the rock is prevented by an (umbrella) waterproofing membrane between the two linings in the upper part. The water from fissures flows through an adjacent fleece into longitudinal drainage pipes, where it is collected and drained away. Since practically no hydrostatic pressure builds up on a drained tunnel, the tunnel lining only has to be

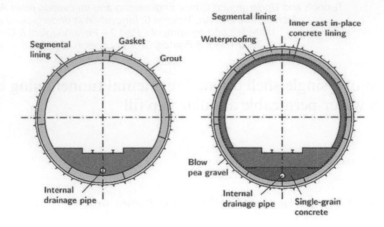

Figure 1. Single-shell and double-shell segment construction.

designed to resist earth pressure and a residual water pressure. This means that tunnels can also be constructed in ground formations with a high natural groundwater pressure head. The disadvantage of this construction method is the fact that the inner lining can only be completed after the actual driving of the tunnel, resulting in higher costs due to the additional material used and a longer construction time.

1.2 *Improvement approach*

Starting from the already described advantages and disadvantages of single-shell and double-shell construction, a construction would be desirable for mechanised tunnelling, with which the water pressure on the tunnel lining can be reduced to a technically manageable pressure head and which is nonetheless simple to construct. The result would be a single-shell drained construction method, representing the best possible balance between technical feasibility and economy.

As part of a publicly funded research project, the idea was pursued of achieving such a draining segment construction with a permeable material for the grouting of the annular gap. In this way, water from fissures can be discharged through an external drainage layer, which can be placed in one working step with the driving of the tunnel and the installation of the tunnel lining (Figure 2).

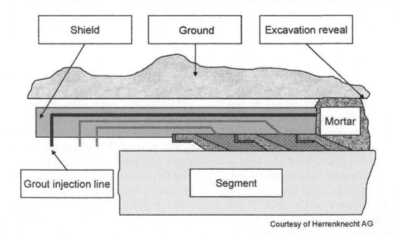

Courtesy of Herrenknecht AG

Figure 2. Annular gap grouting through the shield skin, Source: Herrenknecht AG.

The water permeability is to be achieved by an innovative approach, which intends the defined foaming of cement suspension. In this regard, various dependencies of the following areas have to be considered:
- Construction material,
- Process technology,
- Construction.

The results in all three areas are described below as well as the large-scale tests carried out to verify the practical applicability.

2 CONSTRUCTION MATERIAL

Annular gap grout is subject to various requirements [1], of which only a few are described here as representative. Essentially, the grout must first be suitable for pumping in order to travel to the installation location, the annular gap. Then the early strength has to develop as quickly as possible in order that excessive floating of the segment tube in the still fluid grout can be prevented. Otherwise the consequences can be concrete spalling, misaligned joints and leaks. Finally, the grout in the hardened state must show sufficient stiffness and strength in order to be able to transfer bedding stresses out of the surrounding rock mass to the segment lining. The permissible settlements at the surface also have to be considered.

The additional requirement of high permeability or porosity makes extensive further development necessary in the construction technology. After all, it is already known from experience in the past [2] that an open-pored structure is already disadvantageous for pumping and also that strength is negatively influenced by porosity. The produced improvement approach therefore omits coarse aggregate in order to ensure pumping suitability and is based on a cement suspension that only contains fine grains.

In order to be able to meet the contradictory requirements of high water permeability on the one hand and high (early) strength on the other, the use of a rapid hardening cement with a high strength class is unavoidable. As a result, a high-strength blast furnace cement CEM III/A 52,5 R is used for the present task, which (using the process technology described in more detail in Section 3) can achieve an early strength of about 0.4 MN/m^2 or about 2.0 MN/m^2 after 28 days and whose water permeability coefficient $k_F > 10^{-4}$ m/s (Figure 3). The use of blast furnace cement also hinders sintering, which is absolutely essential for the creation of a durably permeable pore structure.

Figure 3.   Water-permeable annular gap material.

In order to improve the properties of the cement suspension, bentonite is added as a stabiliser as well as plasticiser to reduce the water demand. With a water-cement ratio (W/C ratio) of W/C = 0.50, this means that the suspension is suitable for pumping, both regarding its consistency and its settling behaviour, and still develops the adequate strength stated above despite its high porosity.

3 PROCESS TECHNOLOGY

The basic precondition for the achievement of water-permeable annular gap material is the formation of a pore space with capillary pores in continuous contact. This aim can be implemented by defined foaming of cement suspension. For this purpose, a foam generator with a special surfactant-based air-entraining agent is used, similar to those used for soil conditioning with foams. A foam generator from MC-Bauchemie, which has already been available on the market for some time, was used for the research and development work carried out on this project. This has a gasification section of porous plastic and can produce foams with a monocellular structure. In addition, the danger of formation of the dreaded blockages can be considerably reduced thanks to the omission of so-called obstructing bodies.

The process technology concept for the newly developed permeable annular gap material, which was first set up on the process technology test rig of the STUVA (Figure 4), has the following steps, which are described in more detail below:
  • Mixing and pumping of the cement suspension,
  • Foaming of the cement suspension,
  • Addition of a setting accelerator.

3.1 *Mixing and pumping of the cement suspension*

Mixing is preferably undertaken in colloidal mixers (as are known from grouting technology) in a batching plant above ground. More simple machinery is however adequate at the cost of a longer mixing time. Then the material is transported in a mortar car by rail to the TBM and processed similarly to conventional grout. For the pumping of the cement suspension, progressive cavity pumps are suitable, as they cause little flow pulsing.

Figure 4.   Process technology test rig.

3.2 Foaming of the cement suspension

The cement suspension is foamed by the already mentioned foam generator. First a surfactant solution is injected into the pumped flow of the cement suspension. The addition of the surfactant is only possible after the actual mixing since otherwise uncontrolled foam formation would occur. In contrast to the production of conventional foam cements (for example in the production of insulation materials), only so much compressed air is injected as can be bound by the surfactant. Otherwise there is a danger of larger voids in the annular gap. In this connection, care should be taken that the compressed air is compressible. This makes adaptation of the mass-related compressed air flow necessary in order to achieve identical cement foam densities depending on the pressure level in the annular space – and thus identical properties of water permeability and strength.

3.3 Addition of a setting accelerator

Considering the positive experience with two-component systems for annular gap grouting (cement-bentonite suspension and water glass), a shotcrete accelerator is added immediately before the inflow of the cement foam. This setting accelerator prevents any decomposition of the cement foam and also contributes to the development of early strength. In order to avoid sintering, which could greatly reduce permeability over time or even seal the material, the use of an alkali-free accelerator is absolutely essential. The injection system can be either the well-known equipment (injection cover) from the two-component system mentioned above or else also ring jets similar to those used for dry-mix shotcrete spraying in conventional tunnelling.

4 LARGE-SCALE TRIALS

In order to demonstrate the overall functionality of the newly developed construction material and process technology under practical conditions, large-scale trials were considered to be essential. These entailed site trials on the site of the Emscher BA 40 (AKE BA40) sewer tunnel project and also a large-scale model test with a diameter of 2.00 m and an annular gap width of 11 cm.

4.1 Site trials

The AKE BA 40 project is currently underway between the cities of Bottrop and Oberhausen in Germany with two EPB machines as a segment tunnel with an inside diameter of 2.60 m. Shafts are provided at a spacing of about one kilometre along the route and represented the preferred area for site trials since the segment tubes are removed in any case after the excavation of the shaft. In order to minimise the effect on continued tunnel driving, an additional and nearly completely independent grouting system was designed and adapted to the existing systems on the TBM. This included pumps, supply pipes, grout containers and the electronic control system. The functional capability of all components was first tested above ground before being transported to the TBM along the tunnel railway and installed there during an interruption of tunnelling.

The first site trials took place in February 2017. The cement suspension was mixed in the site batching plant and delivered to the TBM in a grout container. For reasons of logistics, this had to be positioned behind the last backup, about 80 m behind the shield tail. The cement suspension was fed into the independent grout pumping system with a suction pump. It proved possible to fill all the hoses without problems despite the long lengths. All further components were installed in the area behind the control position of the TBM and next to the screw conveyor in the machine pipe, which was a great challenge considering the restricted space (Figure 5).

In the course actually injecting the material, however, a blockage occurred quite quickly so that the trial had to be broken off. In the analysis of the causes, two significant points were identified: firstly the cement was partially clumped due to damp from poor weather conditions, and a coupling in the hose was also leaking, leading to a grout filtration process at this

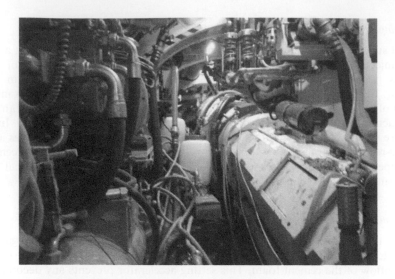

Figure 5. Site trial on the TBM.

Figure 6. Site trial in the target shaft.

location with solid material being deposited in the hose and blocking the flow section. Both these stated causes were thus primarily due to human error, or factors that are relatively easy to avoid. Nonetheless, a few detail points in the supply system were then improved where deposits of coarse cement particles were recorded.

Further site trials were carried out in June 2017 after the west TBM had reached the target shaft (Figure 6). An additional outer formwork was installed in the area in front of the diaphragm wall since the TBM had already driven about 8 m past. This time, most of the components for grouting and the control of the tested grouting system could be installed on the grillage inside the target shaft, so the elaborate installation inside the TBM could be omitted. Only a few hose and cable connections had to be run through the manhole of the pressure wall of the TBM. In order to offer reliable and rapid bedding for the segments in the invert, only the grout injection line in the crown were supplied with the new construction material, with the grout injection line in the invert (like all the rest of the tunnel drive) being supplied with conventional two-component grout. This trial was successful and it was possible to demonstrate that the control of the independent grouting equipment functioned reliably in combination with the machinery conditions, above all in the form of the machine thrust and through the grout injection line.

Figure 7.   Large scale model test.

4.2   *Large-scale model tests*

A large-scale model test was planned as an in-house construction and set up in the test hall of the STUVA (Figure 7). The objective of these tests was to demonstrate the complete and homogeneous filling of the annular gap. The basic idea is to simulate a tunnel drive with a sealed but moving face shutter representing the shield tail of a TBM. This has an integrated grout injection line of 50 mm diameter in the crown with an ovalised cross-section. The travelling face shutter is located in an annular gap of 11 cm between the two reinforced concrete jacked pipes of 2,000 mm and 1,400 mm diameters. The smaller pipe serves as the air-side formwork like a segment, and the larger pipe represents the formwork on the outside. The smaller pipe is only supported in an area of about 90° in the invert, so that the injected construction material (seen from the crown) can propagate tangentially by about 135° in both directions; this corresponds to a distance of about 2.5 m. The already completed tests have shown that the construction material propagates round the entire area despite the comparatively high cement content and the very rapid setting due to the addition of a setting accelerator as well as the small annular gap width of 11 cm.

5   CONSTRUCTIONAL ASPECTS

In the course of the research project, a concept was also produced to investigate under what conditions the newly developed construction material with its process technology can be used practically. One challenge was to balance the best possible technical feasibility, economy and protection of the groundwater regime. This includes in particular the drainage concept and the structural design. Both these aspects are now briefly described.

5.1   *Drainage concept*

In low mountain structures, the solid rock often contains fissures, which can create high water pressures due to impounding. The draining material injected into the annular gap collects the water and ideally discharges it into connected fissures in deeper strata. The arriving fissure water can be transported around and along the tunnel by the annular gap material. The discharge of the fissure water prevents its impounding in the fissures of the solid rock. In case of heavy water ingress and/or impounding in the annular gap, the pressure is reduced by draining the water inside the tunnel. This is necessary when the water pressure exceeds the application limits of the gaskets used in a single-shell tunnel lining. The water pressure can be reduced to a bearable level by the possibility of partial relief.

In order that the drainage water from the annular gap can be drained inside the tunnel, there has to be a connection through the tunnel lining. In principle, this can be through openings in

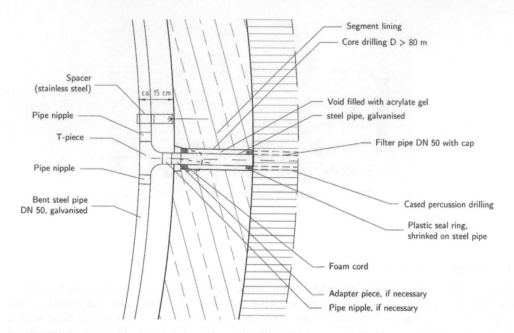

Figure 8.   Pressure limiting system in segment openings.

the segments, which are already provided when they are made, or a drilled through the segments after installation (Figure 8). The distribution and arrangement of the drilled holes can be derived from observations measured in the construction state, since fissure water quantities and fissure distributions can only be inadequately estimated in advance. The inlet openings follow the hydrostatic curve and, in order to ensure accessibility, placed in the sides and are fitted with a pressure valve similar to known pressure limiting systems. No radial drilling into the rock mass to create a drainage route is necessary due to the presence of the all-round permeable draining material. The water is transported through the structure in a longitudinal drainage pipe. The spacing of the pressure valves along the tunnel varies depending on the expected water ingress. Subsequent enlargement is possible.

5.2   *Structural design*

For the structural design of the segment tubes, an early strength of the annular gap material in the order of 0.5 N/mm^2 after 24 hours is necessary in order in particular to be able to safely resist the bedding stresses in the invert. The present mix design of the construction material reaches an early strength of about 0.4 MN/m^2. Therefore the current application concept intends that the draining annular gap material is only installed in the sides and the crown and that unfoamed material or another annular gap grout is installed in the invert. The fact that this is possible was demonstrated in the site trials described in Section 4.1.

6   CONCLUSION

Through the extensive research and development work of STUVA, IMM, MC-Bauchemie and PORR, a new approach has been developed for the extension of the scope of application of mechanised tunnelling with single-shell segment lining in solid rock. This approach is based on the principle of a material in the annular gap, which is capable of draining water with a k_F value $> 10^{-4}$ m/s, through which water from fissures can be drained inside the tunnel. The performance of many series of tests has been successfully demonstrated that the partially contradictory requirements of suitability for pumping, strength and water permeability could be successfully complied with overall. A patent has now been applied for based on the results of the research and development work.

ACKNOWLEDGEMENTS

The described research project was financially supported from 2015 to 2017 by the German Federal Ministry for Economic Affairs and Energy under a decision of the German Bundestag. Particular thanks are due to the Emscher BA 40 sewer tunnel project, which supported and furthered the research project in an impressive way despite some unavoidable interruptions of construction progress. Another essential contribution to the success of the project has been the specially adapted pumping concept from SEEPEX, which was made available for the extended series of tests.

REFERENCES

Thienert, C. (2011). Zementfreie Mörtel für die Ringspaltverpressung beim Schildvortrieb mit flüssigkeitsgestützter Ortsbrust [Cement-free mortar for annular gap grouting in shield tunnelling with slurry face support], Report No. 31, Institute for Geotechnics of Bergische Universität Wuppertal, Editor: Pulsfort, M., ISBN: 978-3-8440-0167-9, Shaker Aachen, Online-Ressource: http://elpub.bib.uni-wup pertal.de/servlets/DocumentServlet?id=1977

Könemann, F., Tauch, B. (2010). Entwicklung eines durchlässigen Ringspaltverpressmaterials für den Schildvortrieb – Stand der Entwicklung [Development of a water-permeable material for annular gap fill in shield tunnelling – State of development], Taschenbuch für den Tunnelbau 2011, pp. 337–363, Glückauf Essen

Thienert, C., Edelhoff, D., Kleen, E., Hörlein, N.: Realisation of a single-shell drained segmental tunnel lining by means of a water-permeable annular gap fill. Geomechanics and Tunnelling. 2018. Issue 5 of volume 11. Copyright Wilhelm Ernst & Sohn Verlag für Architektur und technische Wissenschaften GmbH & Co. KG. Reproduced with permission.

Tunnels and Underground Cities: Engineering and Innovation meet Archaeology,
Architecture and Art, Volume 6: Innovation in underground engineering,
materials and equipment - Part 2 – Peila, Viggiani & Celestino (Eds)
© 2020 Taylor & Francis Group, London, ISBN 978-0-367-46871-2

Glass Fibre Reinforced Plastic (GFRP) permanent rockbolts

A.H. Thomas
All2plan (formerly Minova), Copenhagen, Denmark

ABSTRACT: Glass Fibre Reinforced Plastic (GFRP) has been used for many years in tunnelling, primarily for temporary applications due to the ease of cutting. GFRP has many other advantages such as its light weight and its lower environmental impact. GFRP also has excellent durability characteristics. Permanent GFRP rock bolts have been used successfully in demanding situations. One obstacle to the more widespread adoption for permanent applications has been a concern over its performance in shear. Data from full scale shear tests on rockbolts is presented here. A new design method is described to account for the benefit of the bolts in resisting shearing on joints. An example calculation with a numerical model is presented, showing how this design method can be used to demonstrate that GFRP bolts function as well as steel bolts in a typical Nordic case in terms of resisting both tensile and shear loading from the rock mass.

1 INTRODUCTION

GFRP has been used in many applications in civil engineering and other fields for decades. However, it is used less often in tunnelling. At the same time this material offers many advantages. In terms of its structural performance, it is particularly suitable for linear reinforcing elements such as bars in concrete or rock bolts. In the following sections, this paper will elaborate on the properties of the material as well as its advantages. Design issues, notably shear, are discussed. The performance of GFRP rockbolts has been demonstrated on projects worldwide.

GFRP – sometimes known as fibreglass – consists of glass fibres, either as long strands, mats or chopped fibres, encased in a resin. It was first developed in 1936 in USA and the first known structural use was the hull of a boat in 1937. Having gained popularity in the 1950s, GFRP is often see in applications which draw on its high strength, light weight and flexibility. Common examples include storage tanks, boats, building cladding panels and pipes but it is also used in high performance applications such as wind turbine blades, aircraft fuselages and pole-vaulting poles. Many different forms of GFRP exist and, depending on its components and manufacture, GFRP can be considerably stronger than steel and cheaper and more flexible than carbon fibre.

As one might expect with a composite material, there can also be a considerable variation in quality, depending on the supplier. For the best quality, GFRP should be produced in a pultrusion process in a temperature controlled environment, using high quality glass fibres, having a high fibre content of about 75%. Hence the product is of high quality. The glass fibres are embedded in either Polyester- or Vinylester- or Epoxy resin matrix. This gives the rod a high tensile strength, especially in the longitudinal direction. The resin matrix fixes and protects the glass fibres. The most durable GFRP uses a Vinylester resin matrix. The fib report (fib 2005) provides an excellent overview of GFRP in general.

2 MECHANICAL BEHAVIOUR

Table 1 outlines the main properties of a GFRP rockbolt, alongside those of a standard 20 mm diameter steel bolt. The latter is a bolt size commonly used in tunnelling and it is the basis for rock support in the Q-system chart. As noted above, by the nature of the manufacturing

Table 1. Key mechanical properties of GFRP (FIREP K60-25) and steel bolts.

Parameter	GFRP	Steel	Notes
	K60-25	M20	
Internal diameter of solid bar/mm	21	20	
External diameter/mm	24		Including the thread
Cross sectional area of core/mm^2	350	299	For structural purposes
Tensile strength/MN/m^2	1000	600	Characteristic value
Material partial factor of safety, γm	1.9*	1.15	* GFRP as per fib (2005) section 3.5 for wet conditions & 100 year service life
Design tensile strength/MN/m^2	525	522	GFRP as per fib (200)
Elastic modulus/MN/m^2	50,000	200,000	Steel modulus assumed
Working load (Yield strength)/kN	183	137	Safe working load
Elongation/%	2	8	In a tensile test
Shear strength at 50°	347**	204	** Average value. The winding of glass fibres to form a thread on FIREP's K60-25 bolt enhances greatly its shear capacity
Maximum load/kN			
Density/kg/m^3	2200	7800	

process GFRP bolts or reinforcing bars are orthotropic and they have a higher strength in the longitudinal direction than in the transverse direction. In contrast steel is isotropic. This becomes relevant when the angle of loading is relevant – for example, in shear. In shear tests at 90°, GFRP typically has a strength which is 10% lower than steel. However, in rock applications, shearing occurs at lower angles where GFRP is stronger. Hence, the average values from tests at 50° have been stated in the table below – see also section 4.

Other advantages of GFRP include:

- Very durable (see section 5)
- Non-magnetic (no stray current corrosion)
- Easy to cut
- No electrical conductivity
- High thermal insulation
- Flexible (the elastic modulus is a quarter of steel)

Various coatings can be added to GFRP such as an anti-static coating or UV protection (for outdoor applications to protect the components from the harmful effects of ultraviolet radiation).

3 DESIGN

Several organizations and countries have published design standards for the use of GFRP, generally in the context of reinforcement for concrete – fib (2005), CNR (2006) and ACI 440.1R-03 (2003). In Table 1, the partial factor of safety for the material has been derived according to the fib code (fib 2005), based on Minova FIREP bolts and a 120 year design life.

3.1 Tensile behaviour

While GFRP is almost twice as strong as steel in the longitudinal direction, which is the primary load bearing direction for bolts and bars, when the relevant safety factors are applied,

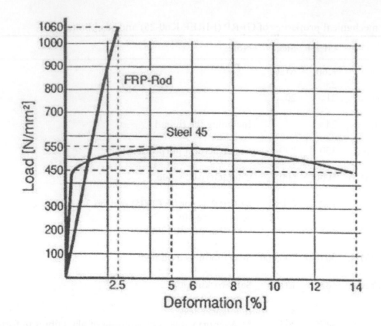

Figure 1. Typical pull-out test results for GFRP and steel bolts (Minova 2018).

the capacity as a working load for the bolts is basically the same for a GFRP and steel bolt of the same diameter. In other words, GFRP can replace steel in a rock support design without any significant changes to the design or construction.

In some locations, GFRP bolts are already being used as permanent rock support, e.g. Sydney, Australia. Salcher & Bertuzzi (2018) summarized the results of pull-out tests for permanent rockbolts on the Metro North West project in Sydney, concluding that GFRP bolts performed just as well as steel bolts in terms of pull-out stiffness and grout-bar bond strength.

4 SHEAR BEHAVIOUR

While rock bolts are designed to act primarily in tension, they can be loaded in shear too. This design case is rarely checked in typical rock tunnel design and in fact most numerical modelling software simulates rock bolts as tensile 1D elements – i.e. they cannot carry any shear load. Due to the high horizontal insitu stresses in the Hawkesbury sandstone, Sydney, Australia, is one of the very few places in the world where rockbolts are designed to carry shear. Consequently a number of authors have proposed design methods for this case. There is a consensus, supported by experimental evidence, that the mode of failure in this shear case is in fact local bending of the bolt, causing crushing of the grout and rock adjacent to the shearing plane and an extension of the bolt until it fails in tension – as shown in Figure 1.

Pells proposed a comprehensive analytical solution for this mode of failure in 2002. However, the original article is believed to have numerous typographical errors in the equations and Pells himself has questioned the theory's ability to match test results (Pells et al 2018). Furthermore, in the author's experience, the solution does not match the real test data well. The conceptual model above comprises a frictional enhancement and a resistance provided by the bolt in tension. The relative contributions vary with shear displacement and the angle of incidence.

Badelow et al (2005) set out a design method for determining the contribution of bolts to shear strength at joints, accounting for the bending and extension of the bolts. In the original paper, the radius of curvature of the bolt, R, is assumed to be constant. One might expect that in fact R is initially very large and then reduces as there is displacement along the joint and yielding of the

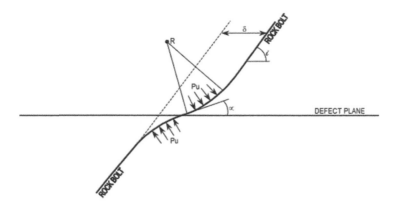

Figure 2.   Conceptual model of bolt bending during shearing (Badelow et al. 2005).

grout permits more localized bending. There is also an assumption – which is largely valid for steel – that the bolts have yielded, unless the deflections are very small. Hence the enhancement of shear force on the joint is expressed as a multiple of the yield force in the bolt. This is not necessarily true for GFRP. The enhancement can be expressed more generally as:

$$SF = (F + Fp).(\sin \alpha. \tan \phi j + \cos \alpha) \qquad (1)$$

where SF = total shear resistance, F = the force in the bolt due to extension, Fp = the pretension force in the bolt, α = angle of the bolt to the plane of shearing and Φj = joint friction angle

In an improvement to the method by Badelow et al. (2005), a new expression is proposed to determine the slope of the bolt, based on the theory of bending of piles under horizontal loads. This have been calibrated against full-scale shear tests on both steel and GFRP bolts at different angles to the plane of shearing. Considering one half of the bolt (i.e. P/2) – see

Figure 3, the slope at the joint provides the angle, α, a key input for the estimation of the shear enhancement, can be found from:

$$\tan(i - \alpha) = \frac{\left(\frac{P}{2}\right)}{2\,EI\,\beta^2} \qquad (2)$$

where P is the total shearing force acting on the block, E is the stiffness of the bolt, I is the second moment of inertia of the bolt and β defined by

$$\beta = \sqrt[4]{4EIk} \qquad (3)$$

where k is the lateral coefficient of reaction, which can be obtained from back-analyzing full scale shear tests on bolts.

Having determined the shear force that the bolt is capable to resisting, SF, this can be added as an increased joint cohesion in the analytical or numerical model of a tunnel, thereby enabling the shear case to be checked. So long as the model shows that the joint does not yield and the joint displacement is less than the value corresponding to the assumed shear capacity, it can be concluded that the bolts will carry the shear loads from the rock.

Figure 4 shows the predicted values of the normalized shear force which can be carried by a fully grouted Minova Firep K60-25 bolt with a joint friction angle of 20° and zero pretension. The bolt ultimate tensile capacity, FY, is 350 kN. These predictions from this new method agree well with data from 25 full scale shear tests for GFRP and steel.

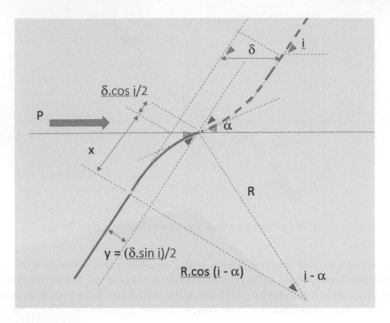

Figure 3. Conceptual model.

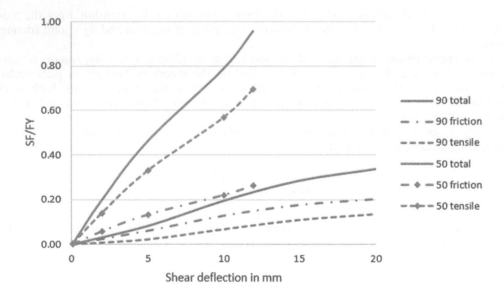

Figure 4. Normalized shear force enhancement vs displacement curves for various angles of incidence, i = 50° & 90° (FY =350 kN).

Figure 5 shows the results from some full scale shear tests and the prediction using this method.

This method offers a simple way to demonstrate that the GFRP bolts can carry the expected shear loads. A more thorough description of this new design method is planned in a forthcoming paper.

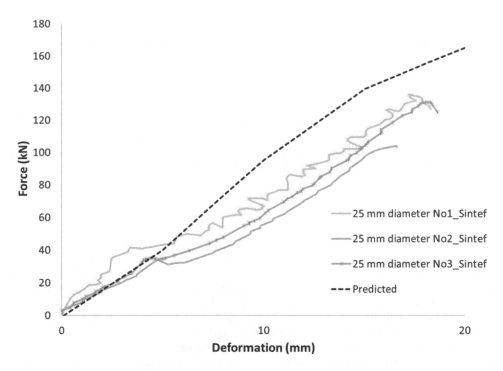

Figure 5.   Full scale shear test data from Sintef tests (at 90°) & prediction.

5   DURABILITY

GFRP is ideal for aggressive environments as it is very resistant to corrosion. The fib report gives an excellent overview of GFRP durability in general (fib 2005). Figure 6 shows that high quality, vinylester GFRP bolts can maintain their properties well over the lifetime of tunnels, which is typically 50 to 120 years (Jesel & Janicek 2017). These tests were performed by an independent laboratory using actual groundwater conditions for a specific project. The best results are obtained by fully grouting the GFRP bolts with resin rather than cement as the alkali in cement grout can reduce the GFRP capacity over time (fib 2005). In contrast, steel bolts are vulnerable to a variety of corrosion processes, even when coated. This suggests that GFRP is a better option than traditional galvanized steel in high risk environments such as subsea tunnels or where groundwater has a high chloride content. In one recent study,

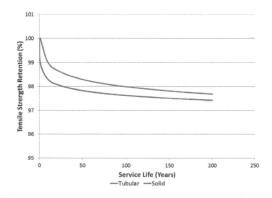

Figure 6 .   The reduction in tensile strength vs age (Jesel Janicek 2017).

3205

corrosion has been observed in galvanized rock bolts after only 2 years, despite being installed in ideal conditions (Sederholm & Reutersward 2013).

6 HEALTH & SAFETY

Safety and occupational health are rightly at the forefront of our work. Anyone who has lifted a steel rockbolt knows how heavy they are. There is still a lot of manual handling in bolting operations. GFRP has a much lower density (25% of steel) so using GFRP bolts reduces the potential harm from prolonged manual handling of rock bolts.

7 ENVIRONMENTAL IMPACT

As we become more aware of our negative impact on the environment, there is a growing need to reduce this impact. One of the ways to do this is to use products which have inherently less impact on the environment, such as GFRP. Kodymova et al. (2017) described a Life Cycle Assessment of GFRP and steel permanent rockbolts and concluded that GFRP bolts could have up to 40% less environmental impact than an equivalent twin coated steel bolt. These are the 2 main options for permanent rock bolts. In part this benefit arises from the lower weight of the GFRP bolts which weigh a quarter of the equivalent steel bolts. However, GFRP was better in all ten categories examined (see Figure 7) and not just in terms of embodied carbon (see Global Warming Potential category in the original paper).

As an aside, another good way to reduce the impact of construction is to source materials locally. The concept of "food miles" was first coined by Prof Tim Lang in the 1990s and it is now well-established in the public consciousness in terms of food shopping and our environmental impact. Of course, this only one element of the overall environmental impact but the attention on it as a major influence has helped to push positive trends towards more sustainable food production. Similarly, in the construction industry, we may need to start thinking about "materials miles" as an important aspect in enhancing the sustainability of our projects. The study above considered the use of the permanent bolts in a tunnel in Scandinavia with bolts sourced in Europe or from China. The negative impact of transporting products halfway around the world is clearly visible – see ratings for Global Warming Potential in Figure 7.

8 CASE STUDY – THE VEREINA TUNNEL

An excellent example of the use of permanent GFRP rockbolts in a demanding environment is the Vereina tunnel. Vereina is a narrow-gauge railway tunnel in eastern Switzerland. For the most part, it has a single track, while there are three double track areas where trains can

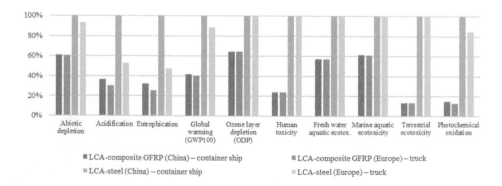

Figure 7. The characterization of a rock bolt - shipped from China and Europe (midpoint category, CML method) – Kodymova et al 2017.

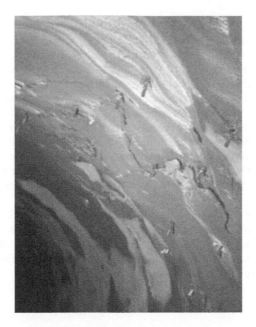

Figure 8.   Squeezing ground supported by GFRP bolts and sprayed concrete (Jesel & Janicek 2017).

Figure 9.   The Vereina tunnel in operation, showing the excellent condition of the rock support, which was installed more than 20 years ago (Jesel & Janicek 2017).

pass. Vereina is an essential part of the railway network in the Canton of Graubünden and, at 19.1 km long, it is the longest narrow gauge railway tunnel in Switzerland. With an overburden ranging up to 1500 m and a highly heterogeneous geology, many challenges for the tunnel excavation had to be overcome. Dramatic rock burst as well as extensive water inflow and squeezing were encountered – see Figure 8. The construction began in 1991 and tunnel was opened in November 1999.

The Vereina tunnel is well-known within the tunnelling fraternity as one of the largest uses of permanent sprayed concrete as a single shell tunnel lining in Switzerland. What is less well known is that it also featured GFRP bolts as the permanent rock support. The vinylester GFRP rock bolts from Minova via FiReP were chosen because of their excellent resistance to corrosion and high durability. The bolts were anchored using Minova's Lokset, fast-acting resin cartridges. In the very end of the bore hole, fast reactive resin cartridges were used, while further back a slower reactive resin was applied. After 10 minutes, the fast reacting resin had hardened and the bolt was manually prestressed to increase the effectiveness of the rock bolts at the face. Additional sprayed concrete and hollow GFRP rock bolts were added later to be able to cope with the requirements for the permanent support and guarantee long term stability. Altogether almost 100,000 GFRP rock bolts were installed. Over the course of the last 20 years, the rock support has been inspected regularly and the GFRP bolts are performing very well – see

Figure 9 – despite rock conditions which imposed significant tensile and shear loads on the bolts. Jesel & Janicek (2017) report that no significant repairs of the rock support have been required and that this has been independently checked. Similarly, the GFRP bolts in the nearby Furka tunnel have performed well. That project preceded the Vereina tunnel by 10 years, opening in 1982, and has a similar design.

9 CONCLUSIONS

GFRP rockbolts are a proven technology for permanent rock support, with excellent durability over a design life of more than 100 years. GFRP has also a significantly lower environmental impact than the normal steel options. A new analytical method has been developed to describe the behaviour rock bolts in shear, based on the concept by Badelow et al. (2005). This matches experimental data well. This method permits the shear enhancement provided by either steel or GFRP bolts to be incorporated in the joint properties of a standard numerical model for the design of a tunnel so that the capacity of the bolts in shear can be checked. As discussed, broadly speaking, steel bolts can be exchanged for GFRP bolts of the same diameter in rock support design. With the trend to use permanent rockbolts and an increasing emphasis on reducing our environmental impact, GFRP rock bolts offer an excellent solution.

ACKNOWLEDGEMENTS

The author would like to acknowledge his former colleagues in Minova and FiReP for their support in the background studies for this paper.

REFERENCES

ACI 440.1R-03 2003 Guide for the Design and Construction of Concrete Reinforced with FRP Bars. *American Concrete Institute*, Farmington Hills, MI, USA.

Badelow, F., Best, R., Bertuzzi, R. & Maconochie, D. 2005. Modelling of defect and rock bolt behaviour in geotechnical numerical analysis for Lane Cove Tunnel, *AGS AUCTA Mini-Symposium: Geotechnical Aspects of Tunnelling for Infrastructure Projects* – October 2005

CNR 2006 Guide for the design and construction of concrete structures reinforced with fiber-reinforced polymer bars. *CNR-DT 203/2006*, Italian National Research Council, Rome, Italy.

fib 2005, *FRP Reinforcement for reinforced concrete Structures*. Task Group 9.3 (Fiber Reinforced Polymer) Reinforcement for Concrete Structures, Lausanne, Switzerland.

Jesel, T. & Janicek, A. 2017 Furka and Vereina, narrow-gauge railway tunnels in Switzerland, 20/30 years after commissioning. Permanent rock support with non-metallic GFRP rock bolts. *Bergmekanikdag 2017*, Stockholm: Stiftelsen Bergteknisk Forskning

Kodymova, J., Thomas A.H. & Will, M. 2017. Life-cycle assessments of rock bolts. *Tunnelling Journal*, June/July, p 47-49.

Minova 2018. *Internal document on test data.*

Pells, P.J., Pells, S.E. & Pan, L. 2018. On the resistance provided by grouted rock reinforcement to shear along bedding planes and joints. *Australian Geomechanics* Vol 53 No. 2 June 2018

Salcher, M. & Bertuzzi, R. 2018. Results of pull tests of rock bolts and cable bolts in Sydney sandstone and shale. *Tunnelling and Underground Space Technology* 74 (2018): 60–70.

Sederholm, B. & Reutersward, P. 2013 Corrosion testing of different types of rock bolts. *BEFO Report 127*, Rock Engineering Research Foundation, Stockholm.

Tunnels and Underground Cities: Engineering and Innovation meet Archaeology,
Architecture and Art, Volume 6: Innovation in underground engineering,
materials and equipment - Part 2 – Peila, Viggiani & Celestino (Eds)
© 2020 Taylor & Francis Group, London, ISBN 978-0-367-46871-2

Two component backfilling in shield tunneling: Laboratory procedure and results of a test campaign

C. Todaro, L. Peila, A. Luciani, A. Carigi & D. Martinelli
Politecnico di Torino, Turin, Italy

A. Boscaro
U.T.T. Mapei SpA, Milan, Italy

ABSTRACT: The two-component backfilling system is becoming the most frequently used method to fill the annular gap created during shield machine advancement. It is based on the mixing of two fluids (mortar and accelerator) that flow through separated pipelines from the batching station to the machine tail and are mixed few centimeters before the output nozzles. The induced turbulence allows a good mixing and the obtained new material gels in few seconds. The mortar is easy to be pumped, stable and preserves its workability for long time. The mixing process allows an easy management of the gelling time and the hardening speed of the final product. The check of the material properties is important and should be carried out before starting tunneling to assess both which mechanical properties can be obtained and how to manage the final resistance and the gelling time. Since no standardized laboratory schemes are available, laboratory procedures for both mortar manufacturing and testing the obtained samples are presented. The results of tests on hardened two-component grout produced with different mix design are also presented and discussed.

1 INTRODUCTION

When tunnels are excavated with a full face shielded machine an annular void is created by the advancement of the shield between the soil and the segmental lining. The filling of the gap with a material of specified mechanical properties is an operational of paramount importance for the excavation process and for the final tunnel quality (ITATech, 2014). The size of this annular gap (herein called annulus) is function of the head overcut, the thickness of the shield and the size of the tail brushes (Thewes & Budach, 2009). The creation of this annulus is part of the tunnel construction process and the consequent annulus filling process must be conducted in a continuous way.

A perfect filling operation minimizes surface settlements (Peila et al., 2011; Pelizza et al., 2010), blocks segments in the designed position, bears the back-up load, ensures the uniform contact between ground and linings (avoiding punctual load on concrete) and increases the waterproofing of the tunnel segmental lining.

The backfilling grouting mixes consist in materials that should be easily transportable and should be stable from chemical and physical point of view. Different materials and technologies have been applied for this purpose, also in function of the geotechnical properties of the medium that embeds the tunnel (Thewes & Budach, 2009; Mähner & Hausmann, 2017).

In the last years, the use of two component grout is spreading in many job sites (Antunes, 2012) due to the technical advantages that this technology can provide: immediate hardening, easy transportation and pumpability, stability during time of the mixes when stand on the machine before injection. For this type of mix it is important to highlight that a suitable mix design must be chosen and tested before the start of the works in order to get a mortar and an hardened two component grout that can satisfy the following features:

- the gelling time must have an order of magnitude of seconds;
- the hardened material must become stiff and hard in a very short time;
- the mortar should be fluid and homogeneous in all the pumping phases;
- the hardened material must be homogenous in every point of the annulus;
- the fluid grout must not be washed out by water flow.

To get all the above-mentioned properties, tunnel engineers must pay attention to the mortar design because only with a careful additives metering it is possible to satisfy all those technical requirements and in the design technical documents a clear assessment of the fresh mortar and hardened grout properties must be highlighted. Furthermore the chosen material must be able to guarantee the quality of the mix and to completely fill the voids and the possible irregular shape. Moreover, the mix should not choke the injection lines.

Thanks to these advantages the use of two component mix is becoming more and more popular but, despite the large number of applications, a limited amount of researches have been carried out in these field are there are no fixed and standard requirements on the laboratory testing of mortar and hardened grout properties. Each designer prescribes different requirements in function of their personal technical document, ideas or calculations. Therefore, there is the clear need to define fixed and univocal test procedures that can allow the comparison of outcomes among different job sites.

In the following a laboratory test procedure for producing and testing both the mortar and the hardened grout is presented to provide a basis for discussion.

2 TWO COMPONENT MIXES GENERAL ASPECTS

The two-component mix is typically a super fluid grout, stabilized in order to guarantee its workability for a long time (from batching, to transport and injection), to which an accelerator admixture is added at the injection point into the void annulus around the segment lining (ITAtech, 2014). The two fluids (identified in the following as component A and component B), are made up of:

- component A: cement, bentonite and a retarding/fluidifying agent;
- component B: accelerator admixture, mainly composed from sodium silicate diluted in water.

The mix gels a few seconds after the addition of the accelerator, (normally 10-25 seconds, during which the TBM advances approximately 2-15 mm), exhibits a thixotropic consistency and starts developing mechanical strength almost instantaneously.

This system usually injects the components under pressure throughout the shield in the void annulus and the obtained mix is able to penetrate into any voids present in this area and into the surrounding ground also, depending on its permeability thus allowing a perfect backfilling of the whole system.

Furthermore, the retarding agent added to the grout has a plasticizing effect and is able to inhibit the mix from setting thereby guaranteeing its workability for very long time (up to 72 hours when necessary) after batching: this facilitates stockpiling grout in the mixer-containers that are bigger than the theoretical volume of material to be injected for every ring (Dal Negro et al., 2017). This is useful in avoiding one of the most common mistakes coming out on the job sites, that is, batching and stockpiling only the theoretical amount and not more. Therefor not being prepared for any possible uncertainty if, eventually, a bigger void is found that needs to be filled in, you would leave the crown unsupported for too long time leading to potentially consequences for the ground stability.

The bentonite content inside the mix (generally not less than 25 kg/m^3, depending on the bentonite quality) increases significantly the homogeneity and the impermeability of the hardened mix. Furthermore, it minimizes the bleeding and helps in achieving the thixotropic consistency when the flow stops because the void annulus is full and so helps in the gelling process, conferring higher impermeability to the system (less than 10-8 m/s) and allow to keep the internal humidity inside the mix itself after its hardening in a confined environment.

The formulation and the metering of single elements of component A is a very important operation in order to ensure the manufacturing of a suitable two-component hardened grout therefore also in laboratory for sample preparation a great care should be taken into account. Every single component that constitutes component A and the relative percentages of the various components are strongly linked with the behavior that the mortar will have in the fluid and hardened phase: for example the quantities of cement and of bentonite influence the final uniaxial compressive strength of the grout but, since, bentonite helps to keep also in suspension the cement particles, a surplus of bentonite in the mix design could lead to a not pasty mortar i.e. a not pumpable mix.

For what concerns the component B, once the percentage of sodium silicate inside the liquid is fixed, the main design parameter is his amount in the final two-component grout, i.e. the volume percentage of component B on the whole mix, that regulate the hardening mix.

In the scientific literature, there is a limited amount of informations about two component grout and the available ones concern case histories that describe practical experiences from different job sites. Thewes & Budach (2009) presented a good global overview of the most used mixes while Peila et al. (2011) and Boscaro et al. (2015) presented a global description of relevant case histories of applications and the tests of a two component mix. Peila et al. (2011) presented the test carried out in the mix used in the metro C line of the Rome metro. More tests can be found in Novin et al., (2015); Zarrin et al., (2015); Barnett, (2008); Pellegrini et al., (2009); Ivantchev & Del Rio, (2015); Thewes, (2013); Pelizza et al., (2010); Youn et al. (2016) and Dal Negro et al., (2017).

2.1 *Usually carried out laboratory tests*

The most frequently used laboratory tests on component A are (Peila et al., 2011; Zarrin et al., 2015; Barnett, 2008; Pellegrini et al., 2009; Antunes, 2012) density assessment, viscosity test and bleeding test while on the mix the gelling time assessment is very important. On hardened grout the uniaxial compressive test is usually performed. Also flexural test on the slabs as done for conventional mortar is frequently used.

The goal of the determination of the viscosity and the bleeding is to check if component A maintains his fluid state (i.e. it is pumpable) and if it remains homogeneous (i.e. water and cement do not separate). The assessment of the gel time has the goal to estimate the correct amount of component B be used to obtain a fast gelling while the compressive strength is the main mechanical parameter of the hardened grout that is usually assessed by the various researches.

2.2 *Laboratory preparation of component A*

Since the final quality of the grout is strongly affected by the quality of the component A, a standardized laboratory scale procedure for manufacturing has been developed. The order and the duration of each step of manufacturing has been defined in order to obtain, at laboratory scale, a final product similar to the one obtained on the construction site plant, in terms of specific weight, homogeneity, bleeding, consistency, fluidity and hardening time. On a job site, the component A is produced outside the tunnel (Peila et al., 2011; ITAtech, 2014; Dal Negro et al., 2017) in suitable areas called batching stations, where a high turbulence process allows the production of mortar. Normally job site mixers require few minutes to produce some cubic meter of component A. Although the proposed procedure to produce component A is supported by many previous tests, in order to get a better correlation and validate laboratory outcomes, a successive test campaign should be carried out using mortar coming from the job site batching plant. To ensure a good homogeneity of the final product, the following points should be respected:

- the ratio between the volume of the tank and the volume of the mortar should range between 2 and 3;
- the impeller should be featured with sloping blades able to create large turbulence and should be located in the center of the tank during the mortar preparation. In Figure 1 the chosen impeller is shown;
- the distance between the impeller and the bottom of the tank should be with a minimum that avoids impact during rotation;
- the ratio between the diameter of the tank and that of the impeller should be about 2 to obtain a good turbulence in the tank. The Tunnelling and Underground Space laboratory of Politecnico di Torino impeller has a diameter of 100 mm and the manufacturing procedure is summarized in Table 1.

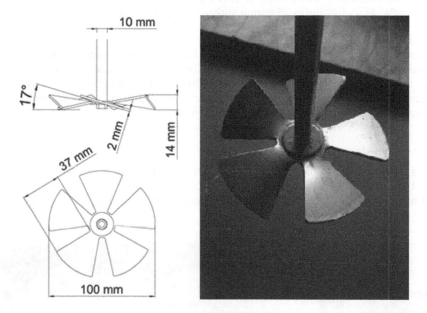

Figure 1.   Scheme and photo of the used impeller.

Table 1.   Proposed procedure for fresh component A manufacturing.

Operation	Impeller rotation speed (rpm)	duration (min)
fill the tank with water and start the mixer	800	/
add the bentonite increasing the propeller speed at a constant rate	from 800 to 2000	0.5
	2000	6.5
add the cement	2000	3
add the retarding/fluidifying agent	2000	2

3 CARRIED OUT TESTS

3.1 Tests on component A

3.1.1 Specific weight
The specific weight of the mix can be measured in accordance with the standard EN 1015-6. Three weight measurements are carried out after a time lap of 5 minutes in order to allow the air bubble ascent.

3.1.2 Flowability
The flowability is linked with the viscosity of the mix. Instead of performing standard viscosity tests using viscosimeter, the tests are usually carried out using the simple Marsh cone procedure that can be easily used directly on the job site since the test is quite simple. The flowability test consists in the measuring of the time needed to a fix volume of mortar to flow out from the standard nozzle of the Marsh cone (Figure 2). The tests can be carried out in accordance with the standard UNI 11152-13.

The flowability test should be performed at different times to understand how the mix behaves during time and therefore how the increase of viscosity can affect the pumpability of the mix. In this research it was chosen to check each mix design after the preparation (fresh mortar) after 24, 48 and 72 hours of curing without agitation of the mix.

3.1.3 Bleeding test
The bleeding value is a percentage index, calculated as the ratio between volumes of loss water and mortar when the mix stand still in a bucket. The loss water is the water expelled from the mortar in a specific time range (Figure 3). The bleeding determination can be performed in accordance with the standard UNI 11152–11. It can be useful to have measurements of the released water after 3 and 24 hours from manufacturing.

3.2 Tests on gelled grout

By mixing component A and component B the gelled grout is obtained in just a few seconds. The interval between the mixing of the components and the instant when the obtained material stops to be fluid is defined as gel time. The gel time is the key parameter

Figure 2. Flowability test device installed in laboratory. The filling of the Marsh cone and the measurements of the flow time phase are shown in the center and on the right respectively.

Figure 3. Bleeding test obtained by measuring water layer due to segregation of the mix after a fixed time.

that leads operators to define the volume ratio between component A and component B. Standard procedures are not available for the measurement of the gel time. In this study a new procedure has been adopted in order to make the operation repeatable. Two tank with volume of 0.4 l were adopted and a fixed amount of 200g of component A was used for all the tests. The proportional amount of component B was determined on the basis of the type of test to be carried out, based on the mix design. Bigger containers and bigger amount of components were used for trial tests of the gelling time but not acceptable results were obtained due to the not homogenous mixing and consequently to not homogeneous hardened material. The proposed procedure is the following: when the two tanks are ready they are quickly mixed by the operator by pouring them in the two tanks. The test time is stopped when the mix is flowing no more from one tank into the other one. It is fundamental to start the test pouring the component A inside the component B in order to have the required turbulence.

3.3 Tests on hardened grout: uniaxial compression tests

Because of a lack of standards for standardized compressive strength test on this type of product (more similar to a hard clay more than a weak concrete) the standard UNI EN 196-1 has been used since it has been proven to be simple to be applied. Although this regulation is not applicable to mortars with very short setting times, it can be adapted to the specific needs of two component grouts. According to this standard the samples to be produced need to have sized of 160x40x40mm.

The sample preparation is not easy with this type of mix since it requires to fill the mould in a very short time (before hardening has started). After filling the mould these are hermetically sealed (in order to avoid water losses due to evaporation) and are cured for 24 hours (with a fix air temperature of 25°). This maturation period is fundamental in order to be able to carry out the demoulding operation without damages the samples. The samples are than cured in clean drinkable water with a fixed temperature of 25°.

Uniaxial compressive tests are then performed on samples with different curing time: 3, 24 hours and 28 days. The compression tests carried out following the UNI EN 196-1 are easy to be applied for samples cured 28 days while for samples with curing time of only 24 hours, it is not easy to split them by flection on three points due to presence of surface weakness: a curing time of 24 hours, indeed, is not enough to provide enough strength to the hardened grout sample. Therefore, forces applied from the three-points flection test cause tools indentation inside the samples, without flexion. For this technical limitation, in order to obtain the two half-samples required from the UNI EN 196-1, a cutting blade was used. For samples with curing time of 3 hours, it is not possible directly to apply the UNI EN 196-1 and also it is not possible to obtain the two half-samples by cutting. Therefore, a special mould was designed with a waterproof plastic layer located on half-length of sample, in order to obtain the halfed samples.

Once obtained the half-samples, the uniaxial compression test campaign has been started by using a unconfined compression testing machine.

Two different types of laboratory devices and setup has been used to perform the test, first one for short term hardening is a press used for testing soils (Wykeham-Ferrance) with a speed of compression between 0.25 mm/min (for 3 hr sample) and 0.5 mm/min (for 24hr sample), 40x40mm plates were used to distribute load on the sample. For 28 days hardened sample the used press is Zwick/Roell with a speed of compression of 3 mm/min, compression device were used to distribute load on the sample.

4 PRELIMINARY LABORATORY TESTS CAMPAIGN

A preliminary laboratory test campaign has been developed on four different mix designs. Mix designs 2, 3 and 4 were manufactured using the same kind of water, high resistance cement (Buzzi CEM I 54.5 R) and sodium bentonite (type 1) changing for each mix design the accelerator/retarding agent supplier. Mix design 1 is the same of mix design 2 but using a type of bentonite provided from a different supplier (type 2). All the used mix designs are summarized in Table 2.

The tests have shown that the density of Component A is relatively constant for all the mix design and is mainly influenced by the amount of cement in the mix and by the water/cement ratio. In Table 3 a summary of the results are shown reporting the average values of a wide set of tests.

Regarding bleeding (Table 4), while the values on the short term (3 hours) are almost the same for all the mixes, even though with slightly higher values for mix designs 2 and 4, on the long term (24 hours) mix design 4 shows a much higher bleeding value than all the other.

This depends on the lower water-cement ratio. The worst behaviour of mix design 4 is confirmed by the viscosity value data (Table 5): mix design 4 has very higher flowing times than the other 3 mix that at 24 hours show a very similar behaviour. Furthermore, at 72 hours was not possible to carry out any more the test due to the density of the mix.

The uniaxial compressive tests show that all the mixes reach a compressive strength of about 1 MPa in 24 hours as summarized in Figure 5 while figures from 6 to 13 report in the detail obtained stress-deformation diagram for all mixes at different ages (3hr, 24hr and 28 days).

The gel time has been evaluated both for the mix design value of accelerator (Table 6) and for different values, in order to study the influence of the accelerator percentage on the variation of gel time. This information if very important for the job site management, where setting the correct gel time is of key importance for the correct balance between avoiding pipes stocking and to get a quick grout setting. It is possible to see that increasing the amount of accelerator there is an important increase of the gel time and that this behaviour is more or less similar for all the mix designs (Figure 8). It is interesting to observe that the effect of the accelerator is less effective on the mix design 4 that was also the one with more cement and less water.

Table 2. Used mix designs for the preliminary test campaign.

MIX DESIGN	1	2	3	4
Cement (kg)	230	230	230	260
Bentonite (kg)	30	30	28	28
	(type 2)	(type 1)	(type 1)	(type 1)
Water (kg)	853	853	819	819
Retarding agent (kg)	3.5	3.5	5	2.5
Accelerator (kg)	81	81	90	80

Table 3. Average values of the measured density of the 4 different mix designs.

mix design	density (kg/l)
1	1.15
2	1.17
3	1.19
4	1.22

Table 4. Average values of the measured bleeding of the 4 different mix designs.

mix design	bleeding (%) t = 3 h	bleeding (%) t = 24 h
1	0.52	2.80
2	1.04	2.09
3	0.70	2.40
4	1.21	8.50

Table 5. Average values of the measured flowing time of the 4 different mix designs.

mix design	flow time (s)			
	fresh mortar	24 h	48 h	72 h
1	32	36	39	43.8
2	34.5	41	43	51.2
3	38	40	40	42
4	36	60	82	Not possible to executed

Table 6. Average values of the measured gel time of the 4 different mix designs.

mix design	gel time (s)
1	11
2	8
3	10
4	7

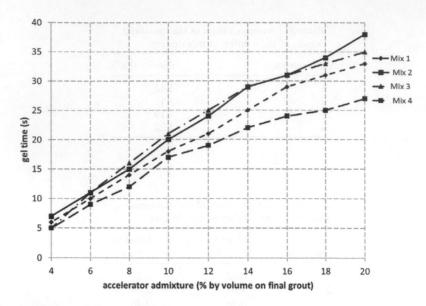

Figure 4.   Accelerator admixture vs gel time.

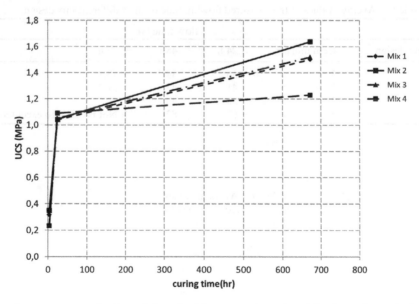

Figure 5.   Average measured values of the uniaxial compressive strength for the various mix designs.

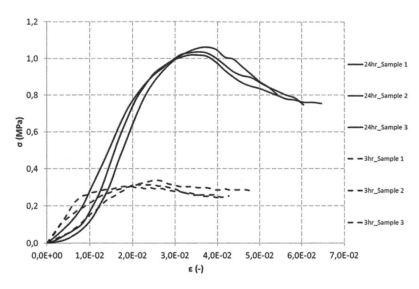

Figure 6.   Mix 1, 3hr and 24hr stress-deformation diagram.

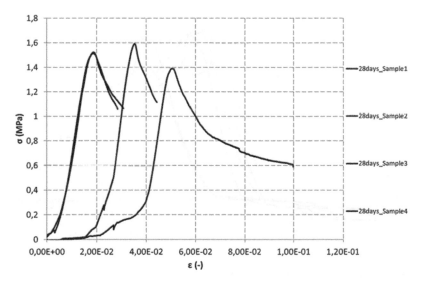

Figure 7.   Mix 1, 28days stress-deformation diagram.

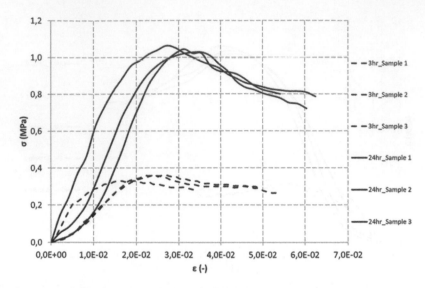

Figure 8. Mix 2, 3hr and 24hr stress-deformation diagram.

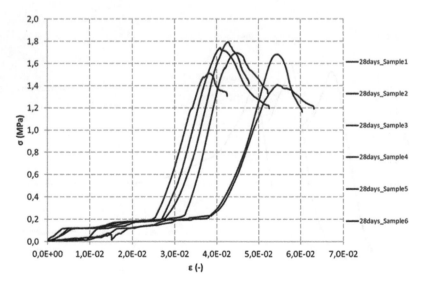

Figure 9. Mix 2, 28days stress-deformation diagram.

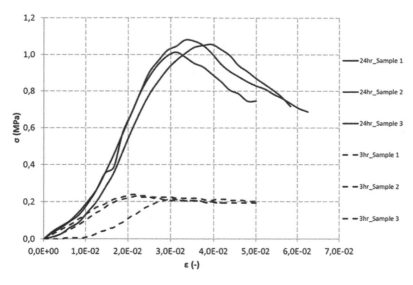

Figure 10. Mix 3, 3hr and 24hr stress-deformation diagram.

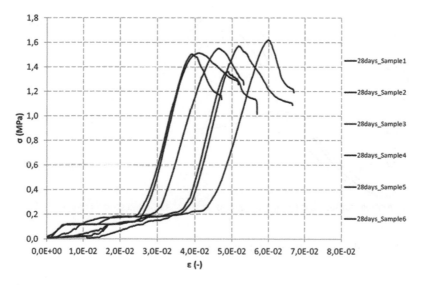

Figure 11. Mix 3, 28days stress-deformation diagram.

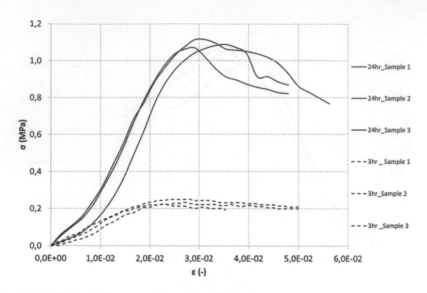

Figure 12.   Mix 4, 3hr and 24hr stress-deformation diagram.

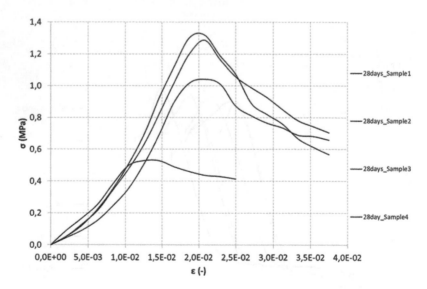

Figure 13.   Mix 4, 28days stress-deformation diagram.

5   CONCLUSIONS

The two-component backfilling system is becoming the most frequently used method to fill the annular gap created during shield machine advancement.

Since it is based on the mixing of two fluids (mortar and accelerator) that flow through separated pipelines from the batching station to the machine tail and are mixed few centimetres before the output nozzles, this technology permits to obtain in a simple way a material suitable for the backfilling operation.

In a tunnel project, a preliminary laboratory test phase is necessary in order to set-up the mix design for the two-component backfilling.

The procedure used in the Tunnelling and Underground Space Laboratory of Politecnico di Torino and the results of some standard tests on four different mixes are therefore presented to provide a standardized laboratory schemes and a contribution to the understanding of the behaviour of the mixes in function of their composition.

ACKNOWLEDGMENTS

The authors wish to thank the Giovanni Lombardi Foundation for funding this research.

REFERENCES

Antunes, P. 2012. Testing Procedures for Two-Component Annulus Grouts. *North American Tunneling Proceedings 2012*.

Boscaro, A., Barbanti, M., Dal Negro, E., Plescia, E. & Alexandrowicz, M. 2015. The first successful experience in Poland of tunnel excavation with EPB for the Metro Warsaw. In *ITA WTC World Tunnel Congress 2015, Dubrovnik (HR), 22-28 May 2015*.

Dal Negro, E., Boscaro, A., Barbero, E. & Darras, J. 2017. Comparison between different methods for backfilling grouting in mechanized tunneling with TBM: technical and operational advantages of the two-component grouting system. In *AFTES International Congress 2017, Paris (FR), 13-16 November 2017*.

ITATech Activity Group Excavation. 2014. ITAtech Guidelines on Best Practices For Segment Backfilling. *ITAtech report n°4 – May 2014*. N° ISBN: 978–2–9700858–5–0.

Ivantchev, A. & Del Rio, J. 2015. Two-component Backfill Grouting for Double Shield TBMs. In *ITA WTC World Tunnel Congress 2015, Dubrovnik (HR), 22-28 May 2015*.

Mäner, D. & Hausmann, M. 2017. New Development of an Annular Gap Mortar for Mechanized Tunnelling. In *AFTES International Congress 2017, Paris (FR), 13-16 November 2017*.

Novin, A., Tarighazali, S., Forughi, M., Fasihi, E. & Mirmehrabi, S. 2015. Comparison between simoltaneous backfilling methods with two components and single component grouts in EPB shield tunneling. In *ITA WTC World Tunnel Congress 2015, Dubrovnik (HR), 22-28 May 2015*.

Peila, D., Borio, L. & Pelizza, S. 2011. The behaviour of a two-component backfilling grout used in a Tunnel-Boring Machine. *ACTA GEOTECHNICA SLOVENIKA* 1: 5–15. ISSN 1854–0171. http://hdl.handle.net/11583/2435575.

Pelizza, S., Peila, D., Borio, L., Dal Negro, E., Schulkins, R. & Boscaro, A. 2010. Analysis of the Performance of Two Component Back-filling Grout in Tunnel Boring Machines Operating under Face Pressure. In *ITA-AITES World Tunnel Congress 2010, Vancouver (CA), 14–20 May 2010*.http://hdl.handle.net/11583/2370602.

Pellegrini, L. & Perruzza, P. 2009. Sao Paulo Metro Project – Control Of Settlements In Variable Soil Conditions Through EPB Pressure And Bicomponent Backfilling Grout. In *Rapid Excavation & Tunneling Conference 2009, Las Vegas, Nevada (US), 14-17 June, 2009*.

Thewes, M. & Budach, C. 2009. Grouting of the annular gap in shield tunneling – an important factor for minimisation of settlements and production performance. *Proceedings of the ITA-AITES World Tunnel Congress*.

Youn B., Shulte-Schrepping C. & Breitenbucher R. 2016. Properties and Requirements of Two-Component Grouts in Mechanized Tunneling. *In ITA WTC World Tunnel Congress 2016, San Francisco (US), 22–28 April*

Zarrin, A., Zare, S. & Jalali, S.M.E. 2015. Backfill grouting with two-component grout – Case study Tchran metro line 7 east-west lot. In *ITA WTC World Tunnel Congress 2015, Dubrovnik (HR), 22–28 May 2015*.

Tunnels and Underground Cities: Engineering and Innovation meet Archaeology,
Architecture and Art, Volume 6: Innovation in underground engineering,
materials and equipment - Part 2 – Peila, Viggiani & Celestino (Eds)
© 2020 Taylor & Francis Group, London, ISBN 978-0-367-46871-2

Digital twin of tunnel construction for safety and efficiency

R. Tomar, J. Piesk & H. Sprengel
DigitalTwin Technology GmbH, Cologne, Germany

E. Isleyen, S. Duzgun & J. Rostami
Colorado School of Mines, Golden, CO, USA

ABSTRACT: Virtual reality (VR) has been increasingly used in various industries for training and design purposes where the new staff gets a feel for the working conditions and designers can observe the impact of their design on the system and human behavior. They consist of IT components for status updates, connectivity, defined data structures and user interfaces that visualize relevant data. Tunnel industry has the potential to greatly benefit from digital twins & related business and product modelling. Digital twins simulate, visualise and optimise the tunnel construction process through real-time monitoring to increase the efficiency. This paper will focus on two aspects of tunnel construction process. One is the situation room where stakeholders can meet virtually and then monitor together the construction process in real-time. Second, virtually visualise the TBM performance in advance to have a better understanding of any future problems to come.

1 INTRODUCTION

Tunneling industry, much like any other field of engineering, has witnessed the trend towards the use of electronic and intelligent systems. This includes the tunneling equipment and machinery that has been increasingly automated and offer variety of sensory systems as well as CCTV and advanced communication and data collection systems through Programmable Logic Controls (PLC) as well as micro processors. The sophisticated monitoring systems used in tunneling during the construction and service life of the tunnel for different project has also grown significantly. During constructions machines such as Tunnel Boring Machines (TBM) offer a steady stream of data from various sensory systems. For example the 17.4 m diameter Hitachi TBM used for excavation of SR-99 in Seattle (Big Bertha) was collecting data from over 700 sensors during the construction phase. Similarly, tunnels are fitted with variety of sensory and monitoring systems for their operation. This includes the sensors measuring heat, flow (air or water), CO_2 or other chemicals as part of smoke detectors, all types of signaling and CCTVs for transportation tunnels. Sophisticated Supervisory Control and Data Acquisition (SCADA) systems that use computers, networked data communications and graphical user interfaces for high-level process supervisory management are common in many tunnels during their service life. As training of the operators and construction crew on increasingly complex and sophisticated machines become critical for success of the projects, use of modern training systems such as virtual reality has gained importance. In addition, the VR systems can be used to educate the public on a variety of issues such as construction of the tunnels as well as operation and maintenance issues. These tools can support the design engineers to offer visual communications to the public relative to the upcoming projects or existing infrastructures, which helps in appropriation of funds for construction of new projects or allocation of funds for upgrading and maintenance of the tunnels already in service. While many consulting companies have used animations to show the construction phase as well as the finished structure, the use of more comprehensive systems such as VR or AR could offer better results when dealing with variety of stakeholders in any projects.

VR/AR has been in use for several years in variety of industries for different purposes. Healthcare, military and aviation are examples of industries that are first to adopt VR/AR as a training and visualization environment. In the last decade, there have been a number of attempts to use VR/AR as a medium of safety training in mining and underground tunneling industry. It's an ideal training environment, since almost all of the real-world hazard scenarios can be experienced while eliminating possible negative consequences of training people at dangerous working conditions. Initial examples of VR-based safety training simulations in mining industry depended on desktop systems, with limited immersion. The recent technological improvements, affordable high speed computing, and increased ease of accessibility to VR equipment has allowed for more realistic and immersive training simulations to be developed. Stothard et al. (2008) provided a preliminary study on immersion experienced by mine workers during a VR simulation. The results showed that the VR-based training is more effective than conventional classroom methods as the virtual environments successfully immerse trainees for transfer of skills and knowledge. Another training simulation was developed to aid miners in improving required skills and knowledge for evacuation during mine fires (Orr et al., 2009). The preliminary tests indicated that the users improved their abilities, but more elaborate tests and supplementary materials are required to enhance the performance.

Qui and Wan (2010) introduced a tunnel safety monitoring method based on 3D virtual environments. The methodology provides an efficient data collection approach to visualize tunnel deformations. Zhang et al. (2010) compared the efficiency of 3D simulation-based and classroom-based training in operating an exhaust fan. Training using simulation resulted in more efficient transfer of skills. Physical modeling and animation of a fully mechanized mining face in virtual reality system were presented by Xiaoqiang et al. (2011). Tichon and Burgess-Limerick (2011) reviewed the usage of VR in safety training in the mining industry. It is evident that the potential of VR is large, however systematic evaluations are required to assess the real impact on the intended audience. Pedram et al. (2013) introduced a framework to evaluate the value of VR/AR in safety training. The framework assesses the trainee' performance and workplace' safety record by analyzing the impact of training sessions. The level of immersion experienced by the users in VR-based training in mining was assessed by Grabowski ad Jankowski (2015). Based on the coal miners' comments, wide field of view and motion tracking are offered as the best solution for training simulations. Pedram et al. (2016) also presented a framework to evaluate the role of VR as a training tool in the mining industry. The study showed that the VR outcomes many of the constraints associated with practical training and can be used as a complementary training approach in the mining industry. A classification of VR training systems was prepared by Hui (2017) and immersion experience between different systems were compared with a drilling training simulation. The result of this study showed that head-mounted display VR systems provide much better user experience.

Application of virtual/augmented/assisted reality within the tunnel construction is still in a concept phase. Researchers have discussed variety of approaches to use this system but there is no precedent for practical use of the full capabilities of the system that is implemented within the industry. However, there are many examples starting from the design to operation phase of certain tunneling projects. With VR/AR full tunnel design to operation phase can be modeled with Building Information Modelling (BIM) in 5D and more. Such model will be the complete digital twin of tunnel. "*Digital twin refers to a digital replica of physical assets (physical twin), processes and systems that can be used for various purposes. The digital representation provides both the elements and the dynamics of how an Internet of Things device operates and lives throughout its life cycle.*" (Wikipedia definition)

In design phase with right geological and survey data full tunnel construction process can be modeled and viewed via virtual or augmented reality. In this way the design of tunnel can be shared in real-time with other designers in different geological locations to get their expertise on board. With this type of model designers can develop an understanding of the uncertainties during the construction phase and implement design improvements to partially mitigate the related risks. This type of models can help the designer in making a decision on whether to go with mechanized or conventional tunneling, while allows for looking into safety measures that could enhance the operations during the construction phase.

Once modeled in design phase, the same model can be used as basis of tender document and offer contractors a better feel for the final product so that the can bid on the project with better feel for the scale of operation, timing of the various activities, and perhaps with bill of quantities. This will reduce the uncertainty during the tunnel construction phase. Tender documents can track the uncertainties and risk mitigation plans for the whole construction period. The same models can be updated during the construction period and used for the quality controls by the consultants and construction management team during the construction phase.

Another big application of VR/AR in tunnel industry is during the construction phase. For mechanized tunneling, not only the tunnel but also the TBM can be digitally visualized with all data in real-time while the sensory information can be displayed in pertinent sections. Also, the steering or navigation system can be viewed through the available data by the TBM manufacturer or surveying subcontractor. VR models allow for introduction of an avatar who can interact with other individuals and subsystems during the process of construction in real-time while being in different locations. This allows consultants and contractors can offer real time interaction with other parties in the virtual meeting room to inspect the ongoing process. This virtual meeting room will reduce the construction delays since in many cases the process can continue without disruptions in real life. As an example, inspection of the cutterhead for wear and damages or inspecting the excavation chamber can be done on the VR models that are updated by laser scanning of the areas. The data management of sensors from TBM can be done with VR in 3D view which is very intuitive for the engineers to understand the situation during every shift. Shift reports can be done via assisted reality using smart glass by voice enabled service, which makes it easier of every shift engineer to quickly assess the conditions at the site before traveling to the headings. Smart glass can also let engineer to take photographs for the reporting and let them to stream live video to all other concern person at the ground surface during any emergency situation. The use of VR/AR can lead to improved safety of the operation and quality of construction while increases the efficiency of the construction crew and their supervisory team.

For conventional tunneling, augmented/assistive reality will play a major role in improvement and streamlining of the construction process towards more efficient and safer working environment. One major application of augmented reality is in installation of tunnel support by providing the exact axis to operators for accurate installion of the support measures such as bolts or steel sets. This is similar to how a child can play with different objects geometry of when inserting them into their pertinent and properly shaped place. In this case, augmented reality will provide the operator the right axis view to place the support in the correct orientation. Another use of augmented reality in tunneling is to improve drilling of the blastholes for drill and blast operation by having the exact view of tunnel face with right points to drill without human error due to visual misconceptions. There is a very interesting use of assistive reality during the conventional tunneling, which involves monitoring the shotcrete process in real-time to provide the correct information for the operator on real time basis to avoid installing extra shotcrete or missing spots as compared to the designed ground support measures.

2 HARDWARE AND SOFTWARE REQUIREMENTS

For simulating the Tunnel Boring Machine AutoCAD 3D was used and then converted it into Unity 3D environment. Unity is a cros platform game engine. We chose this engine due to the fact that software can be used with VR glass, mobile devices and ob browser without working for all platforms independently.

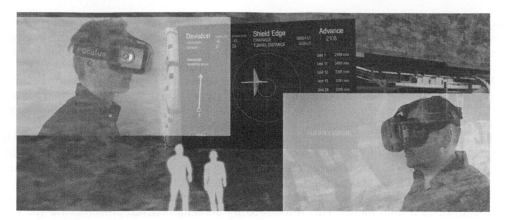

Figure 1.   Visualising TBM model with VR Glass.

System requirements for Unity 2018.2 – For development:

- OS: Windows 7 SP1+, 8, 10, 64-bit versions only; macOS 10.11+
- CPU: SSE2 instruction set support. SSE2 (Streaming SIMD Extensions 2) is one of the Intel SIMD (Single Instruction, Multiple Data) processor supplementary instruction sets.
- GPU: Graphics card with DX10 (shader model 4.0) capabilities.
- iOS: Mac computer running minimum macOS 10.12.6 and Xcode 9.0 or higher.
- Android: Android SDK and Java Development Kit (JDK); IL2CPP scripting backend requires Android NDK.
- Universal Windows Platform: Windows 10 (64-bit), Visual Studio 2015 with C++ Tools component or later and Windows 10 SDK.

For running Unity software:

- Desktop:
  - OS: Windows 7 SP1+, macOS 10.11+, Ubuntu 12.04+, SteamOS+
  - Graphics card with DX10 (shader model 4.0) capabilities.
  - CPU: SSE2 instruction set support.
- iOS player requires iOS 8.0 or higher.
- Android: OS 4.1 or later; ARMv7 CPU with NEON support or Atom CPU; OpenGL ES 2.0 or later.
- WebGL: Any recent desktop version of Firefox, Chrome, Edge or Safari.
- Universal Windows Platform: Windows 10 and a graphics card with DX10 (shader model 4.0) capabilities.

For visualizing with VR glass we have used are HTC Vive Room-Scale VR glass and Oculus Rift (Figure 1).

3   EXAMPLES OF VR IN TUNNELING APPLICATION

Following are examples of using VR in tunneling and generation of tunnel twins for variety of end applications.

a)   Underground mapping based on laser scanning

The necessity of increased efficiency in tunneling operations requires automated tools and applications. Mapping the underground demands efficient methods as it needs to be done regularly for operational monitoring. Laser scanning has become the main tool for acquiring geometrical information of the underground openings with high-precision in a short amount of time. The mapping also remarks the location and size of the underground utilities such as pipes, ventilation fans and equipment. In addition, the scanning technology reached to a point where it can be integrated with the day-to-day operations of tunneling without causing any delays.

Due to the technological improvements in AR/VR equipment, virtual environments exhibit opportunities for improving operational monitoring. Immersion in the tunnels' virtual equivalent with real scale, is a novel approach to monitor the status of underground excavation, tracking the location of the equipment and identifying the potential hazards.

The remainder of the section describes the construction of a virtual tunnel based on LiDAR scanning. Point cloud of Colorado School of Mines' Edgar Experimental Mine is collected with LiDAR (Figure 2). The raw data requires data cleaning and noise filtering. After pre-processing of the data, the point cloud is transformed into a polygon mesh (Figure 3).

At this stage, geometry of the tunnel is at real scale. Tunnel volume and other utilities can be identified as a solid model. This is followed by defining material properties for the surrounding rock around the tunnel to get realistic visual effects in real time rendering for VR (Figure 4). The scene is enlightened by only a headlight above the viewpoint to reflect tunnels' restricted vision conditions.

The developed virtual twin of Edgar Mine is available in the Mining Engineering Department of Colorado School of Mines. The VR infrastructure enables tracking the motion of the

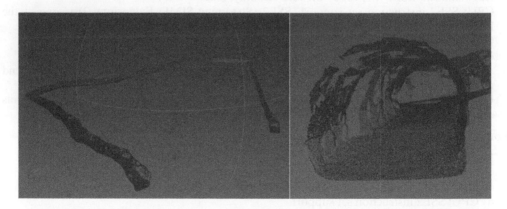

Figure 2.   Point Cloud of Edgar Experimental Mine (Left) and Inside of the Point Cloud (Right).

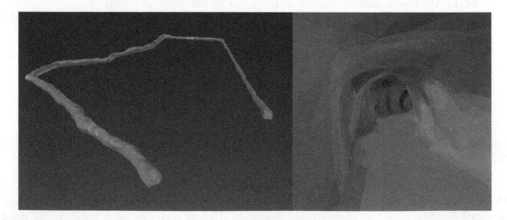

Figure 3.   Polygon Mesh; Entire Model (Left), Inside of the Model (Right).

Figure 4.   Rendered view of virtual tunnel.

user via infrared cameras and LED lights attached to the Oculus headset. This type of loco-motion is beneficial for the participants, since it reduces the effect motion sickness which is the most commonly experienced problem in VR by users.

b)   TBM example (s)

At DigitalTwin Technology we did develop the TBM and construction site virtual twin (Figure 5 to Figure 9) for the use cases like situation room, cutterhead inspection, tunnel segment inspection, and navigation visualization. Simulation was carried out using the AutoCAD 3D model of TBM.

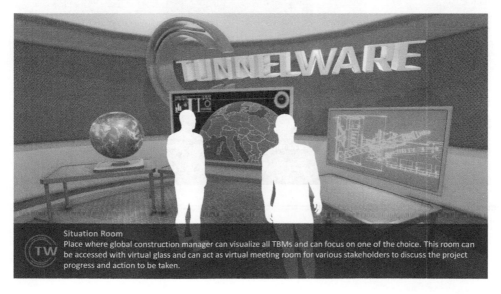

Figure 5.   Construction site situation room.

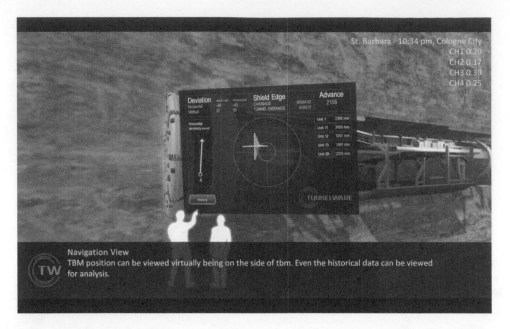

Figure 6. TBM Navigation View.

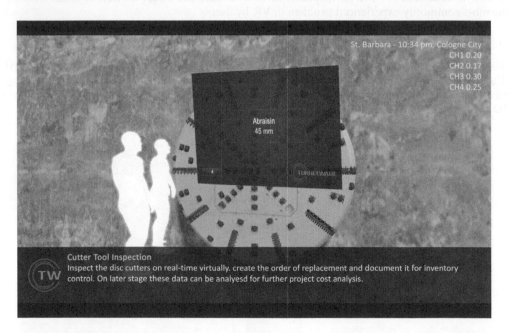

Figure 7. TBM Cutterhead Inspection.

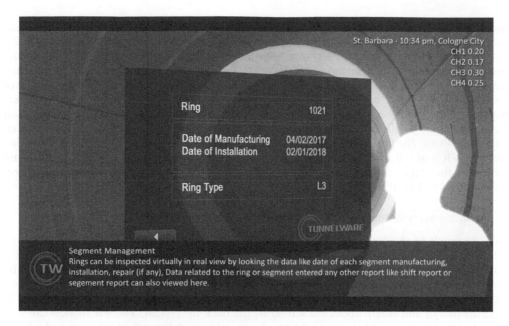

Figure 8. Tunnel segment inspection.

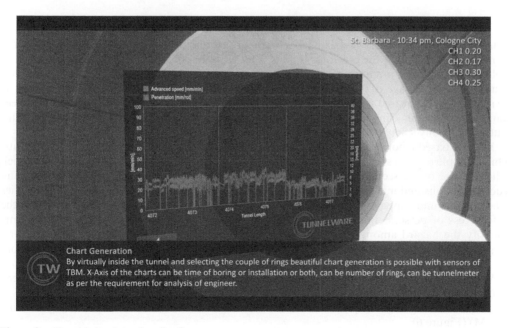

Figure 9. Tunnel site data visualization.

4 APPLICATIONS OF VR IN TUNNELING INDUSTRY

a) Dealing with public and owners

Majority of the tunneling projects involve public or various stakeholders that are not necessarily professionals with engineering and technical training and background in tunneling. In such cases, the capability to visualize the structure to them is very helpful to understand the magnitude and implication of the design parameters. Successful projects are those that involve the stakeholders, especially public, early and keep them updated on the progress of the activities and their continuous engagement often translates to continued support of the project, which are often cost in the range of tens or hundreds of million dollar range and require approval of constituencies for the budgeting purposes. Also, the ability to maintain project owners engaged in the design and progress of the construction phase is critical to have their buy in on the project and to inform them on the technical issues by using visual aid. This could even help in resolving disputes in the case of claims and change orders that is common in tunneling projects. VR and AR models are powerful visual tools that can be used in such scenarios and will allow the stakeholders to be updated and remain engaged, without overwhelming them with detailed technical information.

b) Machine manufacturers and operators

Virtual models of the machine can be easily built from the 3D models of the machine design either from CAD or Solidwork models. As such, it takes minimal amount of efforts for machine manufacturers to build the virtual models of the machine, as it has been practiced by various manufacturers in the past few years. These models can be critical for operators to become familiar with the design of the machine components and perhaps provide input on the machine design and surely about the setting of the back up system. The VR models can also help identifying the possible issues in the on site assembly and disassembly process and relative to the maintenance of the machine, as the transfer and installation of the parts into the desired location and maneuvering of the parts in available spaces in the tunnel can be practiced.

c) Labor training

Tunneling industry requires regular labor training sessions in order to maintain safe working conditions. As noted in the previous sections, VR/AR is an effective tool for training of personnel and it's superior to conventional classroom training methods, especially in high-risk work places. The advantage of VR/AR is that the participants perceive the virtual environment as real and show increased concentration levels. Also, the extreme conditions that the employees can face within tunneling industry, can become a part of training in virtual environments without bringing any safety hazard to participants.

VR/AR based labor training requires relevant scenarios developed for virtual environments. One example of such scenarios in tunneling is the inspection of failure risks around the tunnel. The purpose of the scenario would be to train the personnel who is responsible for assessing roof conditions and mitigating the risks in order to improve their decision-making capabilities since the ground support systems must be installed based on the stand up time to ensure the safe access of personnel and equipment to the work area. During the training, participants identify the hazard among several options and choose a support system to prevent failure. Displacement amounts are visible to the participants before and after the support installation, so that a self-evaluation of mitigation measures is possible by the participants.

VR/AR serves as a practice environment as the trainees have the opportunity to gain experience under the most realistic conditions possible in a virtual world. Another scenario is employee training to get acquainted with the working environment of a tunnel boring machine (TBM) (Figure 6).

In confined and hazardous working environments such as operating a TBMs, training in virtual environments is less costly compared to on-site trainings and more effective than traditional methods in terms of understanding potential on-site risks. In this scenario, participants receive information about the components of the TBM and site-related advice from their supervisors inside the virtual environment. It is very customary that machine manufacturers develop the machine design ahead of full assembly and the design CAD files can be made into the VR models and transferred over to the contractors so that they can start training of their crew with the actual model of the machine that is correct scale for inspection and repair. This

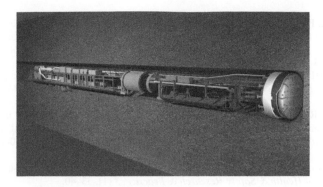

Figure 10.   TBM Simulation within simulated tunnel.

means that the operators, maintenance crew, and workers can be familiar with machine components and their location, assembly and disassembly at their own location ahead of machine delivery and operation.

5   CONCLUSIONS

Virtual and Augmented reality models are useful tools for variety of end uses in tunneling application as they have been utilized heavily in variety of other industries and end applications such as manufacturing, construction, and military uses. The use of VR/AR models can help the design team and let them immersed experience with the structure to allow for better planning of the work and for risk management. The twin models allows the project owner and management teams to track the progress of the project and be up to date relative to the original plans and as a powerful tool to inform public of the progress of the given project. The ability to project the critical data on in the model by combining the VR models with the GIS information allows for enhanced capabilities to observe the potential problems in the tunnel. This can be used as a tool for design modification relative to the actual subsurface geotechnical conditions and the condition of the structure during the construction, as well as training of the work force for the variety of tunneling methods, including conventional or mechanized tunneling and use of TBMs. It is clear that future tunneling projects will involve VR and AR in various stages of the projects starting from conceptual design, detailed engineering, construction, project management, and ultimately for operating and maintenance (O&M) of these critical infrastructures.

REFERENCES

C. Traxler, G. Hseina, A. Fuhrmann, K.Chemlina (2017). A Virtual Reality Monitoring Sys-tem For Tunnel Boring Machines. Proceedings of the World Tunnel Congress 2017

Grabowski, A., Jankowski, J. (2015). Virtual reality-based pilot training for underground coal miners. *Safety Science*, 72, 310–314.

Herrenknecht uses ESI's collaborative Virtual Reality solution IC.IDO to make engineering decisions (2013), ESI-Group.

Hobs 3D create virtual reality training system for Thames Tideway's Tunnel Boring Machine (2018), Hobs Studio.

Hui, Z. (2017). Head-mounted display-based intuitive virtual reality training system for the mining industry. *International Journal of Mining Science and Technology*, 27, 717–722.

Orr, T. J., Mallett, L. G., Margolis, K. A. (2009). Enhanced fire escape training for mine workers using virtual reality simulation. *Mining Engineering*, 61 (11), 41–44.

Pedram, S., Perez, P., Dowsett, B. (2013). Assessing the impact of virtual reality-based training on health and safety issues in the mining industry. *International Symposium for Next Generation Infrastructure*, 1–4 October 2013. Wollongong, Australia.

Pedram, S., Perez, P., Palmisano, S., Farrelly, M. A systematic approach to evaluate the role of virtual reality as a safety training tool in the context of the mining industry, in N. Aziz and B. Kininmonth (eds.), *Proceedings of the 16th Coal Operators' Conference*, Mining Engineering, University of Wollongong, 10–12 February 2016, 433–442.

Qui, D., Wan, S. (2010). Three-dimensional virtual scenery construction of tunnel and its application in safety monitoring. *IEEE International Conference on Progress in Informatics and Computing*, Shanghai, pp 809–812.

Stothard, P., Mitra, R., Kovalev, A. (2008). Assessing levels of immersive tendency and presence experienced by mine workers in interactive training simulators developed for the coal mining industry, in *Proceedings of SimTect 2008: Maximizing Organizational Benefits*, May 12–15 2008. Melbourne, Vic: SimTect.

Tichon, J., Burgess-Limerick, R. (2011). A review of virtual reality as a medium of safety related training in mining. *Journal of Health & Safety Research & Practice, 3* (1), 33–40.

Xiaoqiang, Z., An, W., Jianzhong, L. (2011). Design and application of virtual reality system in fully mechanized mining face. *Procedia Engineering, 26*, 2165–2172.

Zhang, S. Q., Stothard, P., Kehoe, J. E. (2010). Evaluation of underground virtual environment training: is a mining simulation or conventional power point more effective?, in Proceedings of SimTect 2010: Simulation-Improving Capability and Reducing the Cost of Ownership, May 31 – Jun 3 2010. Brisbane, QLD: SimTect.

Tunnels and Underground Cities: Engineering and Innovation meet Archaeology, Architecture and Art, Volume 6: Innovation in underground engineering, materials and equipment - Part 2 – Peila, Viggiani & Celestino (Eds)
© 2020 Taylor & Francis Group, London, ISBN 978-0-367-46871-2

Investigation on the geotechnical properties of a chemically conditioned spoil from EPB excavation, a case study

P. Tommasi
Institute of Environmental Geology and Geoengineering - Italian National Research Council, Rome, Italy

P. Lollino
Research Institute for Geo-Hydrological Protection - Italian National Research Council, Bari, Italy

A. Di Giulio & G. Belardi
Institute of Environmental Geology and Geoengineering - Italian National Research Council, Rome, Italy

ABSTRACT: The management of the spoil from0 a high-speed railway tunnel to be constructed with EPB technology required the geotechnical characterization of the conditioned soil after excavation. This paper describes the laboratory characterization before and after chemical conditioning of the two lithotypes that should be encountered more extensively by the EPB-TBM. The influence of chemicals on the geotechnical properties was investigated by simulating the conditions that muck will experience during its temporary stock on site ("ageing"), to meet environmental and geotechnical requirements for its re-use. The evolution of the index properties, the compactability and the shear strength was followed with measurements at 0, 7, 14 and 28 days (end of ageing) after the injection. In Author's knowledge this experimentation was the first attempt in Italy to characterize a chemically conditioned spoil at such extent.

1 INTRODUCTION

The use of EPB technology for TBM tunneling in soils has being increased in the last years, especially in urban areas, to mitigate problems related to the excavation process and to reduce the negative effects on the built environment. EPB technology requires soil conditioning with foams and polymers in order to reduce the risk of clogging in clay soils or to create a homogeneous consistent paste behind the tunnel face in coarse-grained soils. Several research works based on specific laboratory tests have been recently carried out to investigate the effects of soil conditioning with chemical additives. The influence of plastic fluidity, rheology and permeability of the conditioned soil on the reduction of volume loss (and hence settlements at ground surface) and on frictional overheating of the TBM were investigated through laboratory slump tests (e.g., Quebaud et al. 1998, Pena 2003, Vinai et al. 2008). Milligan (2000) pointed out that clay soils and gravels are less suitable for EPB technology than fine sands, thus requiring larger amounts of additives. Kuomoto and Houlsby (2001) first investigated the relationship between water content, additive percentage and undrained strength of the conditioned soil. More recent works investigated which ranges of foam injection parameters ensure the reduction of stickiness and risk of shield clogging in clayey soils as a function of the physical soil properties (Merritt and Mair 2006, Zumsteg and Puzrin 2012, Hollmann and Thewes 2013).

In the last decade the spoil of tunnels excavated with EPB technology has been increasingly re-used. Therefore characterization of the conditioned soil has become fundamental also for spoil management (Langmaack and Feng 2005). This is the object of the study presented in this paper, which describes the extensive geotechnical investigation conducted on the soils to be encountered along the layout of a high-speed railway tunnel in a metropolitan area in Italy.

The design group has proposed a beneficial re-use of some 5×10^5 m^3 of chemically treated spoil for constructing an embankment hosting re-creational activities. Before emplacement the spoil will be stocked in temporary heaps with bottom drainage, for an "ageing" period during which the spoil will be periodically harrowed. Drainage and harrowing during ageing aim to reduce excess humidity and to favour bio-chemical processes, thus allowing successive use in reclamation works. In this respect "ripening" of the muck for its definitive emplacement is considered to be attained when both environmental and geotechnical requisites are met. The former consists in the reduction of the concentration of chemical species down to acceptable values due to chemical and biological processes. The latter represents the suitability of the dried soil to be readily compacted and the development, after compaction, of the mechanical properties required by the design of the final embankment.

To simulate in laboratory the site conditions of the muck heaps, the conditioned soils was stored in bins having the same height of the heaps and periodically harrowed for the ripening time expected in the field. The two lithotypes that should be encountered more extensively by the EPB-TBM were characterized in the laboratory, both in natural conditions and after mixing with the selected foams. The influence of chemicals injected during excavation on the geotechnical soil properties was investigated both at different steps and at the end of ageing. Samples were retrieved from the bins at 0, 7, 14 and 28 days (end of ageing) after the injection. Measurements of the index properties and undrained strength, as well as Proctor compaction tests were performed at each sampling time; consolidated undrained triaxial tests with measurement of pore pressure were carried out at the end of the ageing. During the experimental activity, issues related to the scale effect from laboratory to field and the representativeness of the results were investigated. The experimental activity presented in this work is deemed to be one of the first attempts in Italy to approach the physical and mechanical characterization of a chemically conditioned spoil at such extent.

The experimental activity was conducted at the geotechnical laboratory of the Department of Structural and Geotechnical Engineering (DISG) of Sapienza University of Rome, where a foam generator similar to that used in the actual TBM has been used to prepare the conditioned soils, and at laboratories of the Institute for Environmental Geology and Geo-Engineering (IGAG) of National Research Council.

2 PRELIMINARY EXPERIMENTAL ACTIVITY

2.1 Geotechnical characterization of the soils

The designed tunnel crosses heterogeneous fluvio-lacustrine deposits, formed by sands, silty-sandy clays, clayey silts and gravels in clayey matrix. The fine-grained soil (Soil 1) and the gravelly soil (Soil 2) were retrieved on site from a shaft, while sands (Soil 3), were sampled from an excavation front. Most of the soil was taken with an excavator, but for Soil 1 it was possible to obtain also undisturbed samples that were used for measurement of the shear strength (section 4).

The physical and index properties of the soils, measured where possible, are presented in Table 1, while the sieve analyses are presented in Figure 1.

Considering that Soil 1 and Soil 2 were sampled at about 10 m distance from each other, the geotechnical characterization is clearly affected by the heterogeneity of the deposit. The grading curves are shown together with the ranges indicated by the EFNARC guidelines, which state the purpose of conditioning for different particle sizes: in particular, guidelines indicate that for Soil 1 conditioning is needed to reduce clogging risk, while for Soil 2 it is necessary to improve workability and reduce permeability.

A detailed analysis of a wide database of geotechnical measurements coming from the extensive investigation campaign carried out for the tunnel design was performed to check that the soils chosen were largely representative of the lithotypes expected during the excavation.

Table 1. Classification of the soil samples.

Soil	WL(%)	WP(%)	PI(%)	IC	Activity	γ(kN/m^3)	γ_S(kN/m^3)	w(%)
1	37.6	19.8	17.8	1.1	0.70		26.4	18.2
1	41.4	20.9	0.5	1.0	0.64	20.1	25.9	19.9
1	54.1	22.9	31.2	1.0	0.70		26.1	23.9
1	39.1	20.2	18.9	0.9	0.58		26.4	21.6
1	38.2	21.2	17.1	1.0	0.75	20.3	25.9	21.1
1	52.2	23.6	28.6	0.9	0.54	20.6	25.7	25.1
1	47.1	23.4	23.7	1.1	0.56		25.9	21.3
2	43.8	20.9	22.9	26.5			26.6	
2	32.8	20.5	12.3					

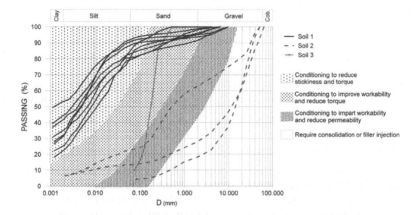

Figure 1. Grading curves of the soil samples and ranges of conditioning by EFNARC 2005.

2.2 *Mineralogical characterization*

The mineralogical characterization was carried out by XRPD while the quantities of clay minerals present in the size fraction -25 + 2 µm were measured by thermogravimetric analysis. Diffractometric spectra are shown in Figure 2.

The characterization of the clay minerals was carried out on the fraction less than 2 microns obtained by controlled sedimentation while the presence of mixed structure silt-clay was carried out on the size fraction -25+2 µm.

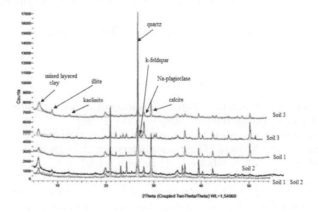

Figure 2. Diffractometric spectra of the different samples analyzed for the size fraction -63+25 micron with the main minerals present.

The minerals present in the samples of Soil 1, Soil 2 and Soil 3 are mainly: quartz, calcite, Na-plagioclase, K-feldspars, clay mineral such as kaolinite and illite and mixed layered clay mineral illite/smectite with prevalent illitic component. In Soil 3 there are more no-clay minerals and the illite/smectite phase is not present. In the size fraction -0.063+0.025 mm grains are present with a silty-clay mixed structure difficult to separate in their individual components. They are characterized by a small grain size, flaky mineral shapes and electrically charged surfaces. In fact, Hollmann and Thewes (2013) pointed out that the amount of clay in terms of grain size (<0.002 mm) in cohesive soils is not necessarily equivalent to the mass fraction of clay minerals.

These materials contain mainly illite and kaolinite. From a mineralogical perspective, illite and particularly kaolinite are considered non-swelling clays.

Illite is characterized by osmotic swelling according to Einstein (1996). In osmotic swelling, the cationic concentration of the mineral surfaces and the surrounding liquid is balanced. This results in water being absorbed at the mineral surfaces of the clay. Einstein (1996) concludes that also osmotic swelling is important for clay behavior. The smectite (montmorillonite) is present in subordinate quantities and this determines a low intercrystalline swelling component. The kaolinite behavior it's not clear. Experience gathered at construction sites shows that, besides illite, sometimes kaolinite-bearing clays can lead to massive clogging. Pimentel (2003) gives theoretical maximal values of strains and stresses for intracrystalline and osmotic swelling. Quartz, feldspar plagioclase and carbonate are uncritical for the development of clogging.

2.3 Selection of the lithotypes to be studied

Given the wide range of soils crossed by the designed tunnel, with significant variations observable also at the scale of the tunnel diameter, it was impossible to carry out the experimental activity on all the lithotypes, therefore it was decided to select two soils for the prosecution of the study. This choice was made considering the occurrence of each lithotype along the tunnel, on the basis of the geological profile available among the project documents. In detail, Soil 1, i.e. clayey silt and silty-sandy clay, accounts for the 52% of the total length of lithotypes crossed, while Soil 2 accounts for 47% of the total length, divided in a 36% of sandy gravel in clay matrix and a 11% of sandy gravel. Soil 3 accounts for a small length (about 300 m) located at the tunnel entrance.

One of the purposes of the research was the preparation of treated soil and leachate samples for biodegradation and toxicity analyses of the chemical products. The selection of Soil 1 and Soil 2 appeared to be safe in this respect, because the conditioning of fine soils for clogging prevention usually requires higher dosages of chemicals for unit of excavated soil, i.e. low foam expansion ratios and high foam injection ratio (Merritt 2004), than those used for the excavation of gravelly soils. Otherwise coarser soils, though treated with lighter foams (high foam expansion ratios), often require adoption of water retaining polymers injected with the foam or from a separate line.

A more extensive sampling was carried out on site after this selection, sending at the DISG laboratory about 2 m^3 of each soil.

2.4 Verification of the conditioning parameters

The conditioning parameters that can be regulated during injection of foams with EPB-TBM are the concentration factor (Cf), the foam expansion ratio (FER), the foam injection ratio (FIR) and the water added during excavation (w_{add}) as described in detail by Merritt (2004), EFNARC guidelines (2005) and other Authors. These parameters are strongly dependent on the type of soil to be excavated, and are often adjusted on site with a trial-and-error process, observing the performances of the EPB during each drive. The definition of products and dosages for the optimum treatment of Soil 1 and Soil 2 has been entrusted to the main European producers of foaming agents, with the aim of selecting among them two products to adopt in the final phase of the research. Each producer, provided with samples of Soil 1 and Soil 2, gave specific indication on products and dosages that were tested at DISG laboratory, thus evaluating the effectiveness of about 10 combinations for each soil by means of the slump test. According to the EFNARC guidelines (2005), the slump can be assumed to be an indicator of the proper consistency of the soil paste; acceptable values were considered in the range 12–18 cm, as suggested by several Authors (Quebaud et. al 1998, Boone et al. 2005, Vinai et al. 2008) for these types of soil.

At the end of this verification phase, five foaming agents and relevant parameters resulted the most effective, so that the final selection of two products (Foam A and Foam B in the following) was made according to the results of ecotoxicity evaluations. After this phase, two water retaining polymers (Polymer A and Polymer B in the following) were included in the study in accordance with design indications. The required concentrations of polymers were added to the water used to produce the foams after the introduction of the foaming agent along the same water line.

3 PREPARATION OF FULL-SCALE SAMPLES

Reproducing the spoil from a real excavation at laboratory scale is a difficult task, especially when production of lumps or cobbles by the TBM is expected, because the laboratory equipment allows to handle a limited quantity of soil and each standard geotechnical test has restrictions on particle size. Furthermore, in this case it was necessary to recreate a stocking area for the ageing of the conditioned soil in a relatively small space.

Limits for Soil 2 regarded the maximum particle size, which was chosen equal to 25 mm, namely that required by the modified Proctor test standard (UNI EN 13286-1/2). The elimination of the coarser fraction produced a fictitious increase in the content of fines, which presumably induced an underestimation of the shear strength measured in laboratory as compared with the real one. Similarly, Soil 1 is a stiff soil (see consistency indexes in Table 1) that would be cut by the excavation into large lumps that affect the mechanical behaviour of the spoil, which is worth investigating. Therefore, samples were especially prepared by cutting Soil 1 into lumps ranging in size from 1 to 25 mm. It should be noted that the same maximum size of clods was adopted by other researchers in laboratory tests carried out to support the mechanized excavation of clays in similar contexts (Merritt 2004, Merritt & Mair 2006). For sake of comparison, some samples of Soil 1 were grinded before mixing to destroy the lumps produced during in situ sampling and to appreciate possible difference in the evolution of the geotechnical properties with respect to the lumpy sample. Both soils were sieved after drying at room temperature to avoid modifications in mineralogical composition.

About the conditioning parameters, each soil was treated with two products and relevant dosages as reported in Table 2.

The two soils and the relevant foams were mixed in a cement mixer and immediately stored in bins of 0.3 m in diameter and 1 m in height, filled up to a height of 0.8 m, the same of the stock muck heaps. The bottom of the drums was equipped with a non-woven geosynthetics filter and drainage holes in order to collect samples of percolate water. The amount of soil required to fill each bin was about 80 kg.

The standard practice for in situ spoil management envisages a daily harrowing of the muck to facilitate aeration and drying; accordingly, in some of the bins soil was stirred with a

Table 2. Combinations of soils, products and conditioning parameters studied.

Soil	Combination	Products	Cf	FER	FIR	w_{add}
	(%)		(%)	(xx:1)	(%)	(% in weight)
1	P1	Foam A	2.0	8	90	10
1	P2	Foam B	1.6	6	80	10
		Polymer B	0.1			
2	P1	Foam A	3.0	12	70	2
		Polymer A	2.0			
2	P2	Foam B	1.8	8	70	2
		Polymer B	0.1			

Table 3. Conditions of preparation and/or ageing applied to each combination of Table 2.

ID	Description
H	Harrowed: the soil in the bin was stirred with a screwed tool every day during ageing
NH	Not harrowed: the soil in the bin was never stirred
D	Destructured: only Soil 1, the soil was grinded before mixing to destroy the lumps

screwed tool every day during ageing (H bins); for sake of comparison other bins were never touched for the whole ageing period (NH bins).

To summarize, 6 bins of Soil 1 were prepared: 3 with treatment P1 and 3 with P2, both of them with the conditions reported in Table 3; similarly, only 4 bins of Soil 2 were prepared, since destructuration was not needed for this soil type. Some duplicates were also needed, so that the whole experimental activity required a total of 18 bins.

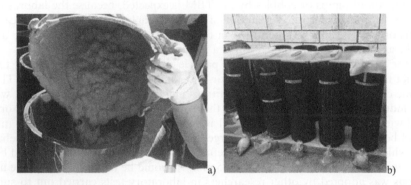

Figure 3. Soil 2 extracted from the cement mixer after conditioning and placed in the bin (a). Final array of the large-scale samples (b).

4 GEOTECHNICAL CHARACTERIZATION OF THE CONDITIONED SOIL

4.1 Tests on the conditioned soil

The geotechnical tests performed for each bin are summarized in Table 4. Liquid and plastic limits, w_L and w_P, as well as the undrained shear strength, c_u, were measured only on Soil 1.

4.2 Evolution of the physical and index properties

The variation of water content, w, measured during ageing for both soils, shown in Figure 4, indicates a general decrease of w with the progress of maturation.

Table 4. Summary of the measurements executed on each bin

Sample time	w	c_u	w_L (fall-cone)	w_L	w_P	Unit weight	Proctor
0	x	x	x				
1	x	x	x	x	x		
3	x	x	x	x	x		
7	x	x	x	x	x	x	x
12	x	x	x	x	x		
14						x	x
20	x	x	x	x	x		
28	x	x	x	x	x	x	x

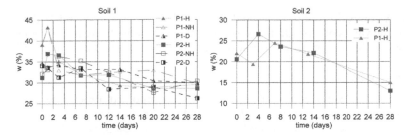

Figure 4.   Evolution of water content with the ageing time.

However, it can be noted that Soil 1 in a lumpy state and Soil 2 show a delayed increase in water content in the first days, presumably due to a delayed absorption of the conditioning water by the lumps in the first case and by the fine fraction in the second.

The standard ASTM procedure for measurement of the liquid limit cannot be applied without modifying the interstitial fluid (and therefore the foaming agent concentration in the conditioned soil); it follows that index properties were evaluated by means of fall cone tests, h, and standard plastic limits, w_P. From Figure 5 it can be observed that for those samples that have a "regular" initial water content (P1-D and P2-D) the corresponding value recorded with fall cone at day 0, i.e. 10 mm, corresponds to the liquid limit; w_P values instead do not show significant variations during the ripening process and oscillate around the value of the plastic limit of the natural soil ($w_P = 20\%$). The different values of index properties observed at different times for the differently treated soils are comparable with the variability due to the non-homogeneity of the soil.

Values of undrained strength calculated with fall cone (Figure 6a) show an increase of c_u with ageing, as expected due to the progressive drying of the soil and the corresponding increment of

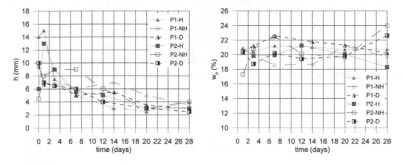

Figure 5.   Evolution of the fall cone sinking and plastic limit with the days of aging.

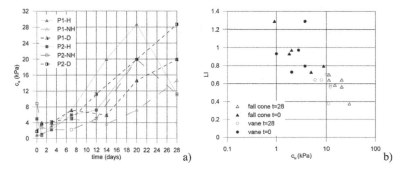

Figure 6.   Evolution of shear strength measurements (fall cone) with aging time (a) and comparison of fall cone and vane test measurements at t=0 and t=28 days (b).

consistency. A comparison of fall cone and vane test data executed at 0 and 28 days (Figure 6b) indicate that the values of strength measured with the two methods are in quite good agreement and consistent with those of natural soil with the same consistence (after Mitchell 1976).

4.3 Evolution of the compactability

Figure 7 shows the compaction curves obtained with modified Proctor tests on both soils (yellow dots) in untreated conditions and after 7, 14 and 28 days from conditioning with products P1 and P2. All the preparation conditions detailed in section 3, H, NH a D, were tested. In general, Soil 2 shows higher values of maximum dry density, γ_d, and lower water content for optimum compaction, w_{opt}, with respect to Soil 1, as expected given the relevant grain sizes. Maximum values of dry density equal to 1.9 and 2.05 Mg/m^3 for Soil 1 and Soil 2 respectively, are reached for water content of about 12–14 % and 8–10 %. Both the figures indicate that different treatments do not produce significant variations of compactability. Even the treatment itself does not affect soil compactability, since the compaction curve of the untreated soil lays within the conditioned soil curves, thus showing no significant differences in the response.

4.4 Shear strength after ageing

Specimens for consolidated undrained triaxial tests (CIU) were prepared from both Soil 1 and Soil 2 at the end of the ageing (namely, at day 28), after compaction at the optimum water content (estimated with the above standard Proctor tests). Tests on Soil 1 were executed on samples of 38.1 mm of diameter, while for Soil 2, given the presence of the gravel fraction, a diameter of 100 mm was adopted.

Figure 8 shows the observed curves of deviatoric stress as a function of axial strain as well as the curves of excess pore water pressure against axial strain for three specimens. The curves, as for all the tests performed on both soils, showed a strain-hardening and contractive behavior.

Figure 9 reports the q-p' values at peak (Figure 9a) and the corresponding shear strength envelopes resulting from the triaxial tests performed for Soil 1 (Figure 9b). Figure 10 shows the stress paths measured for Soil 2 (Figure 10a), along with the resulting peak strength envelope (Figure 10b). Both figures indicate that the peak states of the soils follow lines with very low dispersion, regardless of the type of conditioning, preparation and harrowing. In Figure 9b the maximum and minimum shear strength envelopes obtained for conditioned Soil

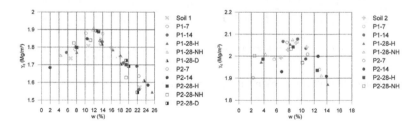

Figure 7.   Evolution of the compactability with the days of aging.

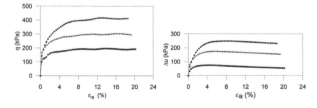

Figure 8.   Typical deviatoric stress-axial strain and excess pore pressure- axial strain curves obtained from triaxial test on conditioned Soil 1.

3242

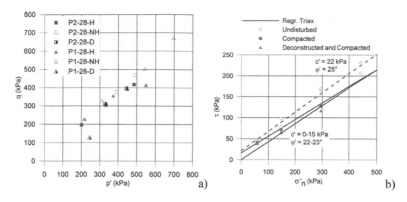

Figure 9. Peak strength states from CIU triaxial tests on Soil 1 at 28 days after conditioning (a); shear strength envelopes from CIU tests on conditioned soil, from direct shear tests on untreated compacted soil and from direct shear tests on undisturbed soils (b).

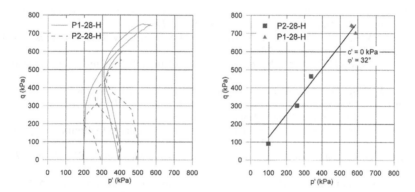

Figure 10. Stress-paths of consolidated CIU tests on Soil 2 at 28 days after conditioning (a) and shear envelope at peak (b).

1 (continuous lines) are also compared with the peak failure states resulting from direct shear tests carried out on untreated compacted soil (squares and triangles). The agreement between the strength envelopes obtained from conditioned and untreated soils indicates that, after 28 days from injection, the foaming agents do not produce appreciable modifications of the mechanical strength of the soil, while a moderate variation of the shear strength is induced by the compaction itself, as pointed out by the comparison with the results of shear tests carried out on undisturbed samples (circles, with dotted regression line).

5 CONCLUSIONS

A detailed geotechnical characterization has been carried out in the laboratory on samples of clay soils and gravels conditioned with chemical additives and later on exposed to ripening in order to simulate the in-situ ageing process of conditioned soils for reclamation purposes.

Based on the measurement of the physical properties of the conditioned soils during the ageing process, a clear influence of the variation of the water content of the conditioned soils with time can be observed on the variation of the soil consistence as well as of the undrained shear strength.

The testing results indicate for both the soils that, at the end of the ageing process, no relevant effect of the conditioning process is observed on the soil compactability.

In terms of shear behaviour, the failure envelopes of the unconditioned soils, after compaction process, are comparable with those of the same soils subjected to conditioning, at the end of the ageing process. It comes out that, after a suitable ageing process, no remarkable effect of the conditioning process on the mechanical behaviour of both the soils tested has been observed. For Soil 1, the shear strength envelope of the undisturbed soil is observed to be generally higher than those of the same soil subjected to conditioning at the end of ageing as well as of the same soil subjected only to compaction, this being a clear consequence of the destructuring process of the soil before compaction.

Open issues, as those regarding the representativeness of the soil samples within the bins subjected to ageing with respect to the actual field embankments of conditioned soil, as well as the simulation of the conditioning process in the laboratory with respect to the real TBM excavation process, still remain to be investigated with more detail in a future work.

REFERENCES

EFNARC, 2005. Specification and Guidelines for the use of specialist products for Mechanized Tunneling (TBM) in Soft Ground and Hard Rock. *EFNARC recommendations*.

Einstein, H.H., 1996. Tunnelling in difficult ground – swelling behaviour and identification of swelling rocks. *Rock Mechanics and Rock Engineering* 29 (3), 113–124.

Hollmann, F. S., Thewes, M., 2013. Assessment method for clay clogging and disintegration of fines in mechanized tunnelling. *Tunnelling and Underground Space Technology*, 37, 96–106.

Iannacchione, A., Vallejo, L., 2000. Shear Strength Evaluation of Clay-Rock Mixtures. ASCE Conference "Slope Stability 2000", pp. 209–223.

Koumoto, T., Houlsby, G. T., 2001. Theory and practice of the fall cone test. *Géotechnique*, 51, No. 8, 701–712

Langmaack, L. & Feng, Q., 2005. Soil conditioning for EPB machines: balance of functional and ecological properties. *Underground Space Use: Analysis of the Past and Lessons for the Future* – Erdem & Solak (eds), Taylor and Francis Group, London, 729–735.

Leschinsky, D., Richter, S., Fowler, J., 1992. Clay lumps under simulated hydraulic transport conditions. *Geotechnical Testing Journal* 15 (4), 393–398.

Maidl, B., Herrenknecht, M., Maidl, U., Wehrmeyer, G., 2012. *Mechanised Shield Tunnelling*. Ernst & Sohn, Berlin.

Merritt, A.S., 2004. Conditioning of clay soils for tunneling machine screw conveyors. A dissertation submitted for the degree of Doctor of Philosophy at the University of Cambridge. 292 pp.

Merritt, A.S., Mair, R.J. (2006). Mechanics of tunneling machine screw conveyors: model tests. *Geotechnique* 56, 9:605–615.

Milligan G., 2000. *Lubrication and soil conditioning in tunneling, pipe jacking and microtunnelling: a state-of-the-art review*. Geotechnical Consulting Group, London.

Mitchell, J.K., 1976. *Fundamentals of Soil Mechanics*. Wiley, New York, NY, 422 pp.

Peña, M.A., 2003. Soil conditioning for sands. *Tunnels and Tunneling International* (July), pp. 40–42.

Pimentel, E., 2003. Swelling behaviour of sedimentary rocks under consideration of micromechanical aspects and its consequences on structure design. In: Proc. Geotechnical Measurements and Modelling. Swets & Zeitlinger, Karlsruhe, Lisse, pp. 367–374.

Quebaud, S., Sibaï, M. and Henry, J.P.,1998. Use of Chemical Foam for Improvements in Drilling by Earth-Pressure Balanced Shields in Granular' Soils. *Tunnelling and Underground Technology*, Vol.13 No.2

Sass, I., Burbaum, U., 2008. A method for assessing adhesion of clays to tunneling machines. *Bulletin of Engineering Geology and Environment* 68 (1), 27–34.

Tabet, W.E., Cerato, A.B., Miller, G.A., 2012. Influence of Clod Size and Moisture Condition on the Shearing Behavior of Compacted Lean Clay. ASCE GeoCongress 2012, pp.1156–1164

Thewes, M., Burger, W., 2004. Clogging risks for TBM drives in clay. *Tunnels & Tunneling International*, 28–31.

Vinai, R., Oggeri, C., Peila, D., 2008. Soil conditioning of sand for EPB applications: A laboratory research. *Tunnelling and Underground Space Technology*.

Williamson, G.E., Traylor, M.T., Higuchi, M., 1999. Soil conditioning for EPB shield tunneling on the South Bay Ocean Outfall. Proceedings of RETC 1999, pp. 897–925.

Zumsteg, R., Plötze, M., Puzrin, A.M., 2012. Effect of soil conditioners on the pressure and rate dependent shear strength of conditioned clays. *Journal of Geotechnical and Geoenvironmental Engineering*, 138, 1138–1146

Zumsteg, R., Puzrin, A.M., 2012. Stickiness and adhesion of conditioned clay pastes. *Tunnelling and Underground Space Technology* 31, 86–96.

Tunnels and Underground Cities: Engineering and Innovation meet Archaeology,
Architecture and Art, Volume 6: Innovation in underground engineering,
materials and equipment - Part 2 – Peila, Viggiani & Celestino (Eds)
© 2020 Taylor & Francis Group, London, ISBN 978-0-367-46871-2

Flexural behavior of precast tunnel segments reinforced by macro-synthetic fibers

I. Trabucchi, A. Conforti, G. Tiberti, A. Mudadu & G.A. Plizzari
University of Brescia, Brescia, Italy

R. Winterberg
Elasto Plastic Concrete, Singapore, Singapore

ABSTRACT: The use of fiber reinforced conc0rete in tunnel linings, with or without conventional rebars, has increased in the two last decades, especially in segmental linings. The design process of segmental concrete linings in ground conditions generally refers to standard load cases of demolding, storage, transportation, embedded ground condition, grouting process and TBM thrust jack phase. Excluding the latter loading condition, a flexural bearing capacity of segments is generally required both at provisional and final stages. Within this framework, the present study investigates the possibility of using macro-synthetic fibers in precast tunnel segments by means of an experimental program on full-scale segments, which are part of a tunnel lining with an internal diameter of 3.5 m. To this aim, a preliminary material characterization on standards small samples was carried out in order to select a PFRC adequate for use in precast tunnel segments. The PFRC selected was used for casting two full-scale segments to be tested under three point bending tests.

1   INTRODUCTION

In the last two decades, Fiber Reinforced Concrete (FRC), with or without conventional rebars, was progressively adopted in several precast tunnel lining projects (ITA report n.16, 2016; ACI 544.7R-16, 2016; *fib* bulletin 83, 2017). Steel Fiber Reinforced Concrete (SFRC) has been generally used for tunnel linings even though there is a general growing interest in the scientific community on macro-synthetic fibers. Tiberti et al. (2015) have demonstrated the effectiveness of PFRC presenting a post-cracking class "2e" (according to *fib* Model Code 2010, 2013) in controlling local splitting crack phenomena under high concentrated forces applied by the Tunnel Boring Machine (TBM). Full-scale tests on precast tunnel segment carried out by di Prisco et al. (2015) and by Conforti et al. (2017), have evidenced the effectiveness of macro-synthetic fibers in combination with a minimum amount of conventional rebars (hybrid solution) for tunnel linings having small diameters (around 3 m, typical of hydraulic tunnels).

The design process of segmental concrete linings in ground conditions is principally governed by load cases (de-moulding, handling, grouting, positioning and the final stage) that can be described by rather simple analytical or numerical models principally based on beam theory (Blom, 2002; Di Carlo et al., 2016), as well as by TBM-thrust phase (during excavation process), which needs a proper numerical modelling (Trabucchi et al., 2015; Tiberti et al., 2017). In the former cases, the estimation of the flexural behavior of precast tunnel segment is of paramount importance, especially in segments reinforced by fibers only where there is a growing interest among practitioners and designers (ITA report n.16, 2016; ACI 544.7R-16, 2016; *fib* bulletin 83, 2017).

Within this framework, an experimental campaign on full-scale segments, which are part of a tunnel lining with an internal diameter of 3.5 m, was carried out for investigating the possibility of using macro-synthetic fiber reinforcement.

The research work was developed in two phases. In an initial phase, fracture properties of four PFRCs (Polypropylene Fiber Reinforced Concretes) were determined through 3 Point Bending Tests (3PBTs) carried out on notched beams according to EN14651 (2005). Based on the nominal residual post-cracking strengths exhibited on small samples, one PFRC was selected for casting two precast tunnel segments (second phase) in order to carry out full-scale flexural tests.

2 EXPERIMENTAL PROGRAM

2.1 *Mechanical characterization of PFRCs*

An experimental program was developed for the characterization of PFRCs adequate for tunnel elements. Continuous embossed polypropylene (PP) fibers 48 mm long with a diameter of 0.70 mm (aspect ratio of 68), tensile strength of 640 MPa, elastic modulus of 12 GPa and density of about 900 kg/m^3 were adopted. Fibers were added to a base concrete with a target mean cube compressive strength of about 60–65 MPa (typical of precast applications) in four different amounts: 4 kg/m^3, 6 kg/m^3, 8 kg/m^3 and 10 kg/m^3. All PFRCs were produced by a planetary concrete mixer.

Table 1 shows the mix proportions of PFRCs: it can be observed that they vary only for fiber content and dosage of polycarboxilated based superplasticizer. The latter was adjusted in each concrete type in order to reach a good concrete workability. For each matrix, twelve small beams (150x150x550 mm) according to EN14651 (2005) for determining post-cracking behavior, six cubes (150x150x150 mm) and six cylinders (150x300 mm) for evaluating concrete compressive strength were produced.

Figure 1a represents the typical nominal stress vs. CMOD (Crack Mouth Opening Displacement) curves of notched small beams of one investigated PFRC; the corresponding mean experimental curve is also reported. In order to compare the post-cracking performance of all PFRCs, the corresponding mean experimental curves are plotted in Figure 1b. It can be observed that all mixtures showed a stable post-cracking response characterized by a softening behavior with a load drop after the peak load, followed by an increment of the residual flexural tensile strength. For larger values of crack width, a soft and progressive decrease of the residual flexural tensile strength was observed. By using higher fiber contents, the load drop (after the peak) decreases and the increment of the residual flexural tensile strength becomes more pronounced. These enhanced post-cracking performances given by the adopted PP fibers are mainly due to their embossed shape, which significantly increases the bond between fiber and matrix and, in addition, to their higher elastic modulus and tensile strength.

Table 2 summarizes the mean mechanical properties of the four types of concrete (CV provided in brackets), in terms of cube ($f_{c,cube}$) and cylinder compressive strength (f_c). The limit of

Table 1. Mix-design of PFRCs.

Concrete	PFRC4	PFRC6	PFRC8	PFRC10
Portland cement CEM II/A-LL 42.5R [kg/m^3]	460			
Natural sand 0–4 mm [kg/m^3]	842			
Aggregate 4–12 mm [kg/m^3]	343			
Coarse aggregate 12–20 mm [kg/m^3]	599			
Water/cement ratio	0.37			
Superplasticizer (% of cement content)	0.70%	0.75%	0.90%	1.20%
PP fiber content [kg/m^3]	4	6	8	10
Fiber volume fraction V_f	0.44%	0.66%	0.88%	1.10%

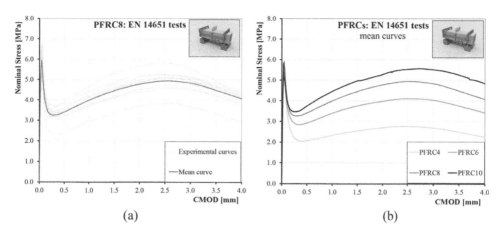

(a) (b)

Figure 1. Nominal stress vs. CMOD curves of PFRC8 notched samples with evidenced the corresponding mean experimental curve (a); mean curve comparisons of PFRCs (b).

Table 2. Mean mechanical properties of PFRCs measured on standard specimens (CV brackets).

Concrete	$f_{c,cube}$	f_c	f_L	$f_{R,1}$	$f_{R,2}$	$f_{R,3}$	$f_{R,4}$	Class.
	[MPa]	[MPa]	[MPa]	[MPa]	[MPa]	[MPa]	[MPa]	[-]
PFRC4	65.9	57.3	5.52	2.07	2.54	2.76	2.48	1.5d
	(0.03)	(0.04)	(0.11)	(0.16)	(0.19)	(0.19)	(0.19)	
PFRC6	65.8	54.6	5.61	2.92	3.75	4.10	3.75	2e
	(0.06)	(0.03)	(0.07)	(0.15)	(0.16)	(0.17)	(0.17)	
PFRC8	66.9	59.5	5.92	3.41	4.45	4.94	4.47	2.5e
	(0.06)	(0.02)	(0.11)	(0.11)	(0.11)	(0.10)	(0.09)	
PFRC10	62.6	52.0	5.82	3.74	4.97	5.53	5.26	3e
	(0.03)	(0.07)	(0.06)	(0.10)	(0.12)	(0.12)	(0.12)	

proportionality (f_L) and the mean values of residual flexure strengths ($f_{R,1}$, $f_{R,2}$, $f_{R,3}$, $f_{R,4}$, corresponding to CMOD values of 0.5, 1.5, 2.5 and 3.5 mm, respectively) are listed in Table 2 as well. A comparable result dispersion in terms of post-cracking mechanical properties can be observed, i.e. CV ranging from 0.10 to 0.19. Moreover, as expected, the CVs are lower in case of PFRC8 and PFRC10, characterized by a higher fiber content. Considering the characteristic values of flexural strengths equal to $f_{R,jk} = f_{R,j} (1-1.64 \cdot CV)$, the PFRCs can be classified according to *fib* Model Code 2010 (2013), as listed in Table 2.

2.2 *Precast tunnel lining geometry*

The precast tunnel lining considered herein corresponds to the Scilla tunnel, which is one of the main structures of a new power line between Sicilia and Calabria (Italy). This underground structure is a part of a large project who foresaw also the construction of a vertical shaft about 300 m deep. These two facilities allow the overcoming of a difference in altitude of about 625 m between the Favazzina electric station (sea level) to Scilla electric station (625 m a.s.l.) to achieve the crossing of electric cables.

The tunnel was excavated with a Tunnel Boring Machine (TBM). It is characterized by a total length of approximately 2800 m and it is made of precast elements. Each ring is composed by four different precast segments. This tunnel has an internal diameter (D_i) of 3500 mm and a thickness (t) of 200 mm (see Figure 2). Hence, the tunnel lining aspect ratio (the ratio of tunnel lining diameter to ring thickness) is equal to 17.5. The trapezoidal segment C1

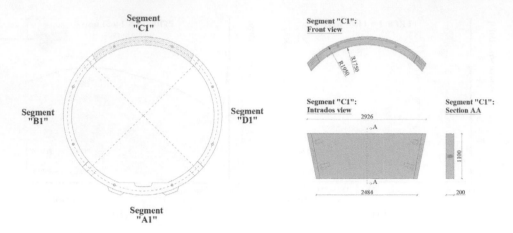

Figure 2. Tunnel lining section and segment C1 details (measurements in mm).

was considered (see Figure 2), having an average length (l) of approximately 2906 mm and a width (w) of 1100 mm.

According to results of 3PBTs reported in Section 2.1, PFRC8 presents a FRC-class 2.5e (according to *fib* Model Code 2010, 2013) and residual strengths ($f_{R,i}$) slightly less than PFRC10 (approximately 12%). Moreover, the residual strengths exhibited by PFRC8 satisfy the requirements currently suggested by MC2010 (2013) for substituting minimum shear reinforcement. Hence, a fiber dosage of PP fibers equal to 8 kg/m^3 was selected for casting two full-scale segments in light of a good balance between fibers amount and FRC post-cracking performance.

2.2.1 *Material properties*

Two full-scale tunnel segments (identified as "C1", Figure 2) were cast by using the same planetary-mixer and mix-design adopted in the initial material characterization (Table 1). The segments were cast in two batches. For each batch, one segment, six small beams (150x150x550 mm) for determining post-cracking behavior and 10 cubes (150x150x150 mm) were produced.

The segments were de-molded one day after casting. The full-scale segment flexural testes were carried out at 56 days in order to simulate typical concrete mechanical properties exhibited during transportation, handling and positioning of tunnel segments. Consequently, 3PBTs on small notched beams according to EN14651 (2005) were also developed at 56 days. The corresponding limit of proportionality (f_L)and the mean values of residual flexure strengths ($f_{R,1}$, $f_{R,2}$, $f_{R,3}$, $f_{R,4}$) are listed in Table 3 together with the cube compressive strength ($f_{c,cube}$). Moreover, the mean curves, as obtained from flexural tests on small notched samples tested in the first and second phase, are compared in Figure 3. It can be noticed that nominal strengths ($f_{R,i}$) measured on standard specimens cast with 1st and 2nd tunnel segment are

Table 3. Mean mechanical properties of PFRCs measured on standard specimens cast with 1st and 2nd tunnel segment, respectively.

Concrete	$f_{c,cube}$	f_L	$f_{R,1}$	$f_{R,2}$	$f_{R,3}$	$f_{R,4}$
	[MPa]	[MPa]	[MPa]	[MPa]	[MPa]	[MPa]
PFRC8 1st segment	68.9	5.19	3.21	4.08	4.58	4.67
	(0.03)	(0.07)	(0.11)	(0.12)	(0.11)	(0.12)
PFRC8 2nd segment	68.8	4.93	2.96	3.77	4.21	4.35
	(0.01)	(0.04)	(0.12)	(0.14)	(0.11)	(0.16)

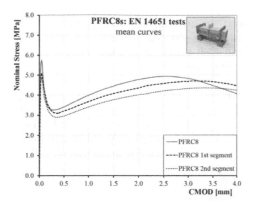

Figure 3.   Nominal stress vs. CMOD curves of PFRC8 notched samples with evidenced the corresponding mean experimental curve.

similar with differences of about 8%. These post-cracking performances are rather close to those obtained during the first phase of material characterization with a difference of about 10%; this confirms the consistency and repeatability of casting procedure adopted for PFRC8.

3   FLEXURAL TESTS

Two trapezoidal "C1" PFRC8 segments were tested in order to evaluate their flexural bearing capacity, which is required during temporary phases (de-molding, storage, transportation and positioning) and final stage (ITA report n.16, 2016; *fib* bulletin n.83, 2017; Di Carlo et al., 2016).

3.1   *Test set-up and instrumentation of flexural test*

Figure 4 summarizes the test set-up, providing the front and lateral views of the segment before testing. A three points bending test configuration was adopted, characterized by a net span of 1600 mm. The two supports were continuous on the entire segment width, while the load was applied at segment extrados by means of two steel plates (150x200 mm) placed on a

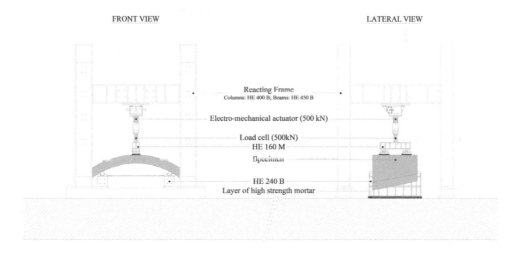

Figure 4.   Scheme of loading system adopted for three points flexural tests.

layer of high-strength mortar. In order to ensure a good distribution of the load along segment width, a stiff steel girder was adopted between the electro-mechanical actuator (loading capacity of 500 kN) and the steel plates. A displacement-controlled procedure was used by measuring the applied load by means of a load cell (Figure 4).

Figure 5 shows the instrumentation details and intrados view of segments. Three Linear Variable Differential Transducers (LVDTs) were placed on segment intrados to measure the mid-span deflection at segment center (D3) and sides (D1 and D2). Moreover, the local displacements of supports were measured by means of four LVDTs placed at the sides of each support (D4. D5, D6, D7, Figure 5); consequently, the actual mid-span deflection was evaluated by considering the vertical settlements at supports. The flexural crack opening at the maximum bending moment was measured by means of two Potentiometric Transducers (PTs), identified as W1 and W2 in Figure 5. The flexural tests were carried by using a quasi-static loading rate of 0.3 mm/min.

3.2 *Experimental results of flexural test*

Table 4 summarizes the main experimental results in terms:

- flexural cracking load (P_{cr});
- mid-span deflection at P_{cr} (δ_{cr});
- flexural initial peak load (P_{peak});
- mid-span deflection at P_{peak} (δ_{peak});
- flexural maximum load after initial peak load (P_{max});
- mid-span deflection at P_{max} (δ_{max}).

Figure 6a and b show the experimental curves of load vs. mid-span deflection response and load vs. relative displacement, respectively. The latter was calculated as mean value of the two PTs placed on the intrados surfaces of segments (W1 and W2, Figure 5). The former was obtained as mean value of the deflections measured by LVDTs D1, D2 and D3, by also considering the support settlements, as retrieved by LVDTs D4, D5, D6 and D7 (Figure 5).

It can be noticed that both segments exhibited a similar first cracking load (P_{cr}) as well as an initial flexural stiffness based on the mid-span deflections at P_{cr} (δ_{cr}, Table 4). After initial cracking, both segments present a further increase of the load up to P_{peak} (Table 4) due to typical

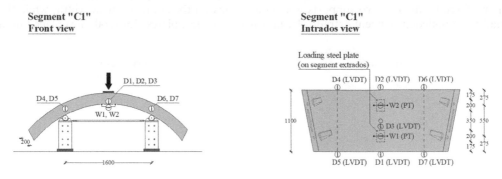

Figure 5.   Details of test set-up and instrumentation (measurements in mm).

Table 4.   Main experimental results of flexural tests on PFRC8 segments.

Specimen	P_{cr} [kN]	δ_{cr} [mm]	P_{peak} [kN]	δ_{peak} [mm]	P_{max} [kN]	δ_{max} [mm]	$\delta_{max}/\delta_{peak}$ [-]
PFRC8 1st segment	47.8	0.28	51.4	0.46	48.6	4.90	10.6
PFRC8 2nd segment	43.5	0.23	52.4	0.58	53.9	8.02	13.8

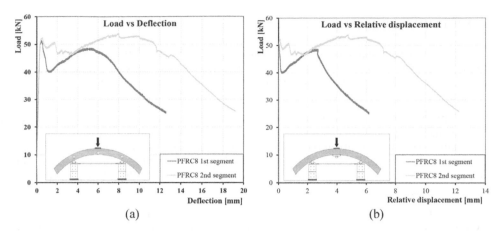

(a)   (b)

Figure 6.   PFRC8 tested segments: applied Load vs. Deflection (a); applied Load vs Relative displacement measured at intrados (b).

sectional re-distribution of stresses. Peak loads are approximately the same (51.4 vs 52.4 kN) for both specimens as well as the corresponding deflections (δ_{peak}, Table 4). It can be noticed that, even though the load after initial cracking increases up to P_{peak}, the flexural stiffness decreases, as confirmed by the doubling of δ_{peak} with respect to δ_{cr}. By further increasing the applied displacement by means of electro-mechanical actuator, the first segment presents a load drop after the peak load, followed by an increment up to P_{max}, while the second segment exhibit a smaller load decrease and a progressive increase of load up to P_{max} equal to 53.9 kN, which is higher than the corresponding P_{peak} (52.4 kN, Table 4). Both specimens present a remarkable ductility, since the corresponding ratios are higher $\delta_{max}/\delta_{peak}$ than 10. More in details, the second segment presents higher ductility as shown in Figure 6a. The same tendencies can be clearly pointed out in Figure 6b in terms of local behavior at mid-span with regard to tensile stresses occurring at the intrados of segments. In this latter case, the comparison of results is more difficult, since for the first segment, the bases of measurement used for PTs did not allow to detect all cracks.

The better flexural behavior (in terms of bearing capacity and ductility) of second segment can be qualitatively explained by analyzing the final crack pattern as shown in Figure 7. The second segment exhibits a more distributed crack pattern; in particular, the first segment has localized the deformation in almost a single crack, since the second crack is much smaller (Figure 7a). To the contrary, the second segment has three cracks having uniform and similar width (Figure 7b).

(a)   (b)

Figure 7.   Final crack pattern on front side exhibited by PFRC8 segments: 1st segment (a); 2nd segment (b).

4 CONCLUDING REMARKS

The main aim of this research concerns the evaluation of the structural applicability of macro-synthetic fibers in a precast tunnel lining having an internal diameter of 3.5 m and a thickness of 0.2 m. Based on the experimental study carried out in two phases, the following conclusions can be drawn:

- polypropylene fiber reinforcement can be successfully used as flexural and minimum shear reinforcement in precast tunnel segments having small diameter;
- the flexural bearing capacity and ductility of the investigated tunnel segments is affected by the performance of PFRC but also by the crack pattern development. In case of higher number of cracks presenting similar and well distributed crack width, the ductility in terms of deflection tends to be higher.

Further studies should be also carried out in order to compare the flexural behavior of PFRCs tunnel segments with that resulting from traditionally reinforced segments (RC, reinforced only by rebars) and in case of hybrid reinforcement solutions (a small amount of rebars in combination with fiber reinforcement).

ACKNOWLEDGEMENTS

The Authors are very grateful to TETRARENT S.r.L. for the grant of segment steel formworks, essential for the casting process. Moreover, the Authors would like to express their gratitude to Engineers Daniele Rivetta and Andrea Piardi, as well as the technicians Augusto Botturi, Domenico Caravaggi, Andrea Delbarba and Luca Martinelli, for the assistance in performing the experimental program.

REFERENCES

ACI Committee 544. 2016. *Report on Design and Construction of Fiber Reinforced Precast Concrete Tunnel Segments*; ACI 544.7R-16. American Concrete Institute, p. 36.

Blom C.B.M. 2002. *Design philosophy of concrete linings for tunnels in soft soils*; Ph.D Thesis, Delft University of Technology, The Netherlands, pp. 184.

Conforti, A., Tiberti, G., Plizzari, G.A., Caratelli, A. & Meda, A. 2017. Precast tunnel segments reinforced by macro-synthetic fibers; *Tunnelling and Underground Space Technology*, 63, pp.1–11, DOI: 10.1016/j.tust.2016.12.005.

Deutscher Ausschuss für Stahlbeton (DAfStb) Guideline. 2012. German Committee for Reinforced Concrete, *Steel fibre reinforced concrete; design and construction, specification, performance, production and conformity, execution of structures*; p. 48.

Di Carlo, F., Meda, A. & Rinaldi, Z. 2016. Design procedure of precast fiber reinforced concrete segments for tunnel lining construction; *Structural Concrete*, 17 (5), pp. 747–759, DOI: 10.1002/suco.201500194.

Di Prisco, M., Tomba, S., Bonalumi, P. & Meda, A. 2015. On the use of macro synthetic fibres in precast tunnel segments. *Proceedings of the International Conference Fibre Concrete 2015 (FC2015)*, Prague, Czech Republic, pp. 478–483.

EN 14651. 2005. *Test method for metallic fibre concrete – measuring the flexural tensile strength (limit of proportionally (LOP), residual)*; European Committee for Standardization, p 18.

EN 1992-1-1. 2004. *Eurocode 2: design of concrete structures – Part 1: general rules and rules for buildings*; European Committee for Standardization, p. 225.

fib bulletin No. 83. 2017. *Precast tunnel segments in fibre-reinforced concrete*; fib WP 1.4.1, p. 168, ISBN: 978-2-88394-123-6.

fib Model Code for Concrete Structures 2010, 2013. Ernst & Sohn, p. 434, ISBN 978-3-433-03061-5.

ITA report n. 16. 2016. *Twenty years of FRC tunnel segments practice: lessons learnt and proposed design principles*; p. 71, ISBN 978-2-970-1013-5-2.

Tiberti, G., Conforti, A. & Plizzari, G.A. 2015. Precast segments under TBM hydraulic jacks; *Tunnelling and Underground Space Technology*, 50, pp. 438-450, DOI: 10.1016/j.tust.2015.08.013.

Tiberti, G., Trabucchi, I. & Plizzari, G.A. 2017. Numerical study on the effects of TBM high-concentrated loads applied to precast tunnel segments; in: *Proceedings of the IV International Conference on Computational Methods in Tunneling and Subsurface Engineering (EUROTUN 2017)*, Innsbruck (Austria), 18–20 April, 2017, pp. 285–293, ISBN: 978-3-903030-35-0.

Trabucchi I. 2015. *Studio numerico di rivestimenti per gallerie realizzati in conci prefabbricati*; Master Thesis, University of Brescia, pp. 250.

Tunnels and Underground Cities: Engineering and Innovation meet Archaeology, Architecture and Art, Volume 6: Innovation in underground engineering, materials and equipment - Part 2 – Peila, Viggiani & Celestino (Eds)

Immersive tunnel monitoring by data driven navigation in 3D

C. Traxler & G. Hesina
VRVis Zentrum für Virtual Reality und Visualisierung Forschungs-GmbH, Vienna, Austria

K. Chmelina
GEODATA Ziviltechnikergesellschaft mbH, Vienna, Austria

ABSTRACT: We present a system for the analysis of tunnel monitoring data that provides data driven navigation in a 3D environment. Tunnel models are obtained from geo-referenced designs or from laser scans of ongoing constructions. Sensor values are visualized at the location where they were recorded, preserving their geospatial context. The monitoring data is analyzed, and critical values appear in a list sorted by their measurement date. When activated the viewpoint smoothly changes to the corresponding position in the 3D scene to allow a close-up inspection. Flying back and forth between such locations supports a better understanding of their spatial distribution and their overall effect on the tunnel. Moving back and forth in time enables interpretation of their development and their relations with other tunneling data such as the construction progress. First evaluations on real data show that this is a valuable method for the analysis of critical monitoring data.

1 INTRODUCTION

In this paper, we present a prototype system for the analysis of tunnel monitoring data that provides data driven navigation in an immersive 3D virtual environment. Our interactive viewer runs on the desktop or with a virtual reality (VR) system such as the HTC Vive or Oculus Rift, which provides a stereoscopic experience and a much better sense of scale. The central element of the scene, the 3D tunnel model, is either provided by a geo-referenced design or by actual measurements such as laser scans acquired during the excavation and lining process. The monitoring data originating from various sensors is also geo-referenced and visualized at the location where it was recorded, preserving its geospatial context. This is one advantage of a 3D visualization, another is the possibility to spatially arrange the display of different sensor types, which allows more comprehensive comparisons and a more accurate assessment of the overall situation. A time control allows the user to revisit the entire surveillance period and analyze various time series of sensors and hence their development.

It is especially important to early detect critical values and where they occur. For this our system provides data driven navigation. The monitoring data is analyzed, and critical values appear in a list sorted by the date of their occurrence. When double-clicking an entry, the user is smoothly guided to the corresponding location in the 3D scene by a camera animation and the date of the time control is updated accordingly. There he/she can closely inspect the critical situation in its full spatiotemporal context. It is also possible to switch on additional visualizations for a combined analysis such as other sensor values, geological sketches or the surrounding terrain.

Flying back and forth between locations where critical values occur supports a better understanding of their spatial distribution and their overall effect on the tunnel. Moving back and forth in time enables interpretation of their development and their relations with other temporal tunneling data such as the construction progress. In this paper, we focus on laser deformation measurements of the tunnel wall using prism targets. However, data driven navigation can be applied to various data types and properties.

In the following sections we give an overview of the related work before describing our visualization methods for deformation measurements and the benefits of data driven navigation.

2 RELATED WORK

A first 3D visualization system for underground construction was presented in Chmelina (2009), which demonstrated that complex processes such as deformations of the tunnel wall are more comprehensible when perceived in 3D. Deformation vectors were visualized by polylines emanating from the tunnel wall, which is one of the methods also used in this work. Later Chmelina et al. (2012) described a virtual environment for data interpretation supporting decision making in tunnel constructions. The focus here was on extensometer and micrometer measurements.

Traxler et al. (2017) presented a similar approach for tunnels built with tunnel boring machines. A variety of different sensor values can be combined by using methods of information visualization in 3D space. This includes the status of the cutting head, the progress of the excavation and ring lining, various deformation measurements of ring segments and more. Our approach presented in this paper is based on these research results and focused on data driven navigation to make the analysis more efficient. It also introduces new visualization methods for deformations of the tunnel wall and accumulative display techniques to overlook larger areas.

To our knowledge data driven navigation for tunnel monitoring in 3D environments is a new concept. Most approaches address abstract search spaces such as databases or the internet, which is not relevant for our work. Jankowski & Hachet (2015) gave a detailed overview of interaction techniques for 3D environments including navigation. Their paper summarizes techniques for smooth and non-irritating movements to points of interest, which we considered for our work. Gentle navigation is especially important in highly immersive VR experiences to avoid motion sickness of users.

Ferreira et al. (2015) worked on a framework for data driven decision making in urban scenarios. It uses a city map and simple 3D block models for buildings, which is sufficient for visibility analysis and to assess the impact of newly planned constructions on the city.

Ortner et al. (2016) introduced an interesting system that combines 3D visualization with 2D visual analytics. It focuses on a single monitoring task, – to detect and assess the criticality of cracks in tunnel walls lined with shotcrete. Cracks can be statistically analyzed with respect to their position, length, orientation and thickness. Clusters and outliers are easily detected. A linked view is established between the visual analytics and 3D viewer. In this way cracks with certain properties are highlighted in the 3D scene. It also supports data driven navigation by automatically navigating to an optimal viewing position for individual cracks or clusters of them.

3 IMMERSIVE TUNNEL MONITORING

3.1 *Visualization of deformation measurements*

The deformation of the tunnel wall is usually measured with a total station and prism reflectors. Typically, a couple of prisms is arranged around a tunnel profile and this setup is repeated along the tunnel axis every few meters. Readings are performed on a regular basis yielding a time series of spatial deformation vectors. The 3D space of our virtual environment gives us the opportunity to directly display these spatial vectors as polylines. Each line segment of such a polyline represents exactly one measurement and shows how far and in which direction the tunnel wall has displaced at the corresponding prism location and measurement date. The line segments are also color-coded to indicate the criticality of the value (ranging from green, over yellow and orange to red).

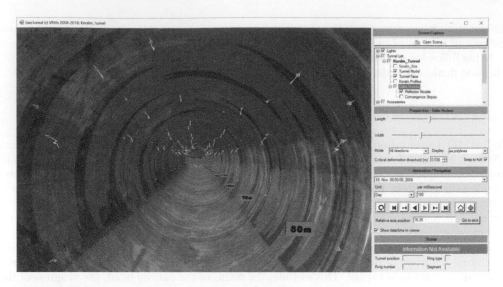

Figure 1.   Visualization of deformation vectors as polylines.

Usually the length of the deformation vectors is very small (mm-range), so that there is not much to see compared to the size of the tunnel wall. For that reason, the length of the line segments can be exaggerated whereby the user controls how much. Figure 1 shows visualizations of plenty of deformation measurements in our desktop viewer. The origins of the polylines are 3D models of prism reflectors, which are placed at the geo-referenced positions of the real ones. The data comes from the Austrian construction project "Koralm Tunnel". With the time control, the user can run an animation and adjust its playback speed to see how deformations develop with respect to the advancement of the tunnel face.

Another important geotechnical parameter for assessing the stability of the wall is the "tunnel convergence" and is derived from deformation vectors. Whenever a new vector is acquired the distance between pairs of prism reflectors is calculated and compared with their initial distance (reference measurement). Between subsequent measurements these distances can shrink, grow or stay constant. They reveal whether the tunnel wall contracts or expands between the pairs of prisms.

We visualize the convergence by an easily comprehensible information visualization method that brings animated 2D diagrams into 3D space. A line is drawn between the corresponding prisms and a small ruler is attached in the middle of it. The ruler is designed to show expanding and contracting forces by bars emanating from two zero points at ¼ and ¾ of its length. Thereby expansions are shown as blue bars extending to the outside from both zero points and contractions as red bars extending to the inside. The connecting lines are color-coded, whereby blue hues indicate expansions and red hues contractions. Additionally, numerical values are displayed below the rulers.

Figure 2 shows the visualization of several convergence measurements. To avoid visual clutter, rulers appear only when users view the lines between prisms from a close enough distance. Otherwise only the color-coded lines display the convergence values. Like with deformation vectors the time control can be used to see an animation of the measurement history and hence the temporal development of the tunnel convergence. This information visualization method is also used to display other quantities derived from deformation vectors such as the settlement and the lateral and longitudinal displacements.

We typically use small geometric objects to display sensor values in an accurate and intuitive way. This is required for close-up to mid-range inspections. But when the user views the tunnel from a larger distance he/she would hardly recognize these small details. For that reason, we also provide aggregated visualizations, which mediate values in a simpler and more abstract way. They are ideal when overlooking longer stretches of a tunnel or when looking far into the distance. For deformation vectors this is achieved by switching from polylines to color-coded spheres. Hence only the criticality of the values is indicated by the color, while the length and orientation of the deformation vectors are hidden in this mode. The spheres are snapped to the wall so that half of it is visible outside the tunnel (Figure 3 left), whereas polylines can only be viewed from inside.

Users can either explicitly change between the detailed and accumulated display of deformation vectors or choose an automatic mode, where the viewing distance decides what is shown. The accumulated view for the tunnel convergence just presents the color-coded lines connecting pairs of prisms. Here the rulers are always automatically hidden when too far away. The lines keep a certain thickness also when viewed from far away and are thus easily recognizable. Users can hide the tunnel model to view convergence lines from outside. To see the course of the tunnel, they can turn on polygons showing profiles at regular intervals and the curve of the axis (Figure 3 right).

3.2 *Monitoring by data driven navigation*

We shall not expect from users to easily detect critical values by exploring the scene for many different dates. A tunnel is usually very long as are the measurement time series. It is true that

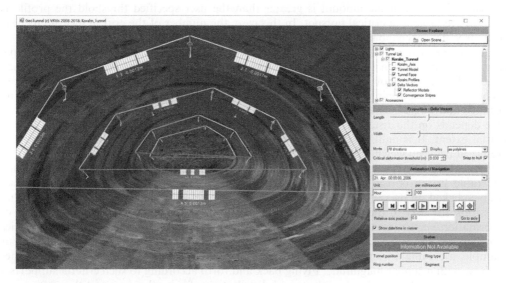

Figure 2.   Color-coded lines and rulers display tunnel convergence values.

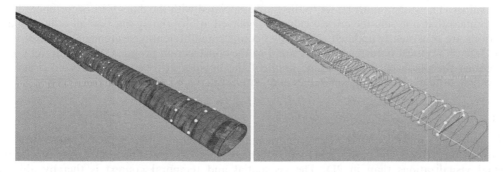

Figure 3.   Aggregated visualizations for deformation vectors (left) and tunnel convergence (right).

3D space improves detection by aggregated visualizations and fast forward animations. So, it is more efficient and reliable than sifting through endless lists of numerical values or studying graphs. Nevertheless, a mechanism that guides users to locations in space and time where critical values occur is an important asset for monitoring systems. This is called data driven navigation.

Critical values are detected in a pre-processing step. The user specifies thresholds for each data type in the scene file. When a scene is loaded into the viewer all the time series to be visualized are read in and analyzed. Afterwards critical values are inserted into a table in the graphical user interface (GUI) and sorted by the date of their occurrence. Beside the date other properties listed in the other columns of the table are the data type, its value and the tunnel position in meters. Users can sort the rows by all these properties. In our prototype we currently support three different data types, which are:

- Deformation vectors
- Tunnel convergence
- Tunnel profiles with clusters of critical values

The same principle can be applied to any other types of measurements but also to results of statistical analyses. In this way correlations, outliers, clusters or certain properties of distributions might be used for data driven navigation. Regarding this aspect we just realized a proof of concept by counting the number of critical values that occur on the same tunnel profile at the same date. When the amount is greater than the user specified threshold the profile is listed in the table as a critical hotspot. In this case, the number of the critical measurement is given as value in the table.

Each entry in the table is associated with a unique location in the 3D scene and a unique date. This information is used to guide users to the occurrences of critical values. When double-clicking on a row of the table a camera animation is started that takes the user from his/her current viewpoint to the corresponding hotspot. For each data type an ideal viewing distance and angle is calculated. The speed of the locomotion is constant so that the travel time depends on the distance. This avoids irritations of users and gives them a feeling for the spatial extent. It is unlikely that they lose their orientation and so the geospatial context is preserved between viewing positions.

Figure 4 shows a result of a data driven navigation in the 3D scene for the Koralm Tunnel. The table listing critical hotspots is shown in the GUI on the right-hand side. The user double-clicked on the critical value detected with prism reflector 4 at the tunnel position 89m. This took the user to the viewpoint shown in the 3D rendering for a close-up inspection of deformation vectors measured with the prism at this location. The different data properties provided by the table columns enable users to navigate based on multiple criteria. For example, users might want to see other critical values at close locations, or contemporary ones or those with a similar magnitude. Flying back and forth according to such criteria supports a better understanding of the spatiotemporal distribution of critical values and their effects on the tunnel.

Besides navigating to hotspots, it is also important to consider the date associated with critical values. The system must be set to the state associated with the date of its occurrence. This guarantees that not only the critical value is displayed appropriately but that all other visualizations are consistently updated. This allows comparing different measurements effectively at the date of a critical occurrence and relate it to the construction progress. Furthermore, users can start the playback from the time point of the hotspot in forward or reversed order to investigate the dynamic of sensor values and better understand why criticalities arise.

3.3 *Assessment by combined visualizations*

3D space has an advantage when simultaneously inspecting different data types such as deformation vectors and convergence values. The perspective gives a better perception of combined visualizations than in 2D. The geospatial and temporal context is thereby always

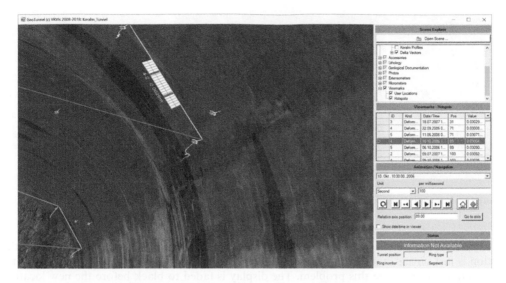

Figure 4.   Inspecting a critical deformation vector by data driven navigation.

preserved. Therefore, it is also more efficient to compare measurements which are spatiotemporally close together as explained in the previous section. This helps to recognize correlations and to better understand the dynamic of important parameters and quantities.

Beside sensor values, our viewer supports the display of additional information as shown in Figure 5. Sketches showing geological interpretations of the tunnel face can be viewed at the relevant geo-referenced position (Figure 5 left). In this way users can investigate how geologic features influence sensor measurements. The stratigraphy in which the tunnel is embedded can also be shown as terrain model (Figure 5 middle) to study the larger geological context. Another example are laser scans which are done during the construction. The scans allow comparing the planned tunnel wall with the constructed one (Figure 5 right).

3.4   Monitoring with a virtual reality system

Our viewer also supports virtual reality (VR) systems such as the HTC Vive or the Oculus Rift. So, we conducted a few experiments to investigate the pro and cons of VR monitoring. The head mounted displays (HMD) of these systems have a sufficient resolution and cover a large portion of the field of view. This allows highly immersive stereoscopic impression of the 3D scene with a good sense of depth. Users are aware of the real dimension of the tunnel. This is a clear advantage for a geospatial visualization like ours.

On the downside data driven navigation in VR frequently causes motion sickness. Current theories see the reason for that in the discrepancy between the sense of vision and the

Figure 5.   Additional information include geological sketches (left), terrain (middle) and laser scans (right).

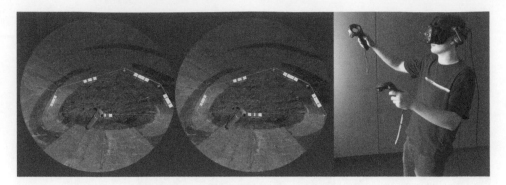

Figure 6. Immersive inspection of tunnel convergence measurements with a VR system.

vestibular system. The vision detects motion while the vestibular system does not, because the motion is just virtual, and the body remains on its spot. Many VR-applications use the beaming-metaphor to mitigate this problem. The display is faded to black before the new location is faded in again. However, the geospatial context gets lost with this method. Another solution just narrows down the field of view when moving. This is better than a blackout, but users can nevertheless easily lose their orientation.

Another problem for data driven navigation in VR arises from head tracking. A great advantage of an HMD is that users can look around in a natural way by turning their head. This, however makes it impossible to force them to look in the direction of a critical hotspot. So, in VR we just can guide them to a position. Additional visual clues might be useful to hint to the right viewing direction such as floating labels or arrows.

Figure 6 shows a monitoring session in VR. Users can move along the tunnel axis with the touchpad buttons of the VR controllers. Motion sickness is less an issue when users control the motion by themselves. Monitoring of tunnel constructions in VR demands further research, especially concerning data driven navigation but also interaction techniques to control the playback and adjust visualization properties.

4 CONCLUSIONS

4.1 *Summary*

In this paper we presented a prototype of a 3D virtual environment for immersive tunnel monitoring that provides data driven navigation. We focused on deformation measurements of the tunnel wall. A new information visualization method was introduced for tunnel convergence that allows to display animated 2D diagram in 3D space. We described aggregated visualization techniques to avoid visual clutter and overlook plenty of sensors from the distance.

The system detects critical values for different kinds of measurements and smoothly guides users to these hotspots. Users can simultaneously inspect different data types near a hotspot and recognize relations and accumulations. They can start animations and control the playback speeds to see how sensor values change forwards or backwards in time. The geospatial and temporal context is always preserved.

Preliminary user tests with real monitoring data from the Koralm Tunnel project showed that our viewer enables an efficient and reliable analysis of the data and a better understanding of the overall situation during the construction process.

4.2 *Limitations*

Currently our prototype system is limited to three types of data, deformation vectors, tunnel convergence and accumulation of critical values on a profile. But this is just a matter of analyzing additional time series and calculate optimal viewing positions for the corresponding sensor visualization in 3D space. Currently our data driven navigation approach does not consider occlusions. So, it can happen that the view towards a hotspot is obscured by another object such as the tunnel face. This would demand a visibility test to detect such occluders and subsequently hide them or make them transparent. Another limitation is that data driven navigation does not work very well when using a VR system. Motion sickness, the loss of the geospatial context and arbitrary viewing angles from head tracking are open problems that need to be tackled.

4.3 Future work

In the future we plan to extend data driven navigation to various other types of measurements such as extensometers, micrometers, and tunnel boring machine data. It should also work for results of basic statistical analyses such as detection of clusters and outliers. The occlusion problem should also be solved. We will work on additional aggregation visualizations for other data types to better overlook many sensor values from a distance. Our further research in this application domain will also include monitoring in VR. Here we will investigate new navigation and interaction techniques with a focus on those which are data driven.

ACKNOWLEDGEMENTS

VRVis is funded by BMVIT, BMDW, Styria, SFG and Vienna Business Agency in the scope of COMET – Competence Centers for Excellent Technologies (854174) which is managed by FFG.

REFERENCES

Chmelina, K. 2009. A virtual reality system for the visualisation of underground construction data. *Proceedings of the II Int. Conference on Computational Methods in Tunnelling, Euro: Tun*, (pp. 9–11).

Chmelina, K., Jansa, J., Hesina, G., & Traxler, C. 2012. A 3-d laser scanning system and scan data processing method for the monitoring of tunnel deformations. *Journal of Applied Geodesy, 6*, 177–185.

Ferreira, N., Lage, M., Doraiswamy, H., Vo, H., Wilson, L., Werner, H.,... Silva, C. 2015. Urbane: A 3D framework to support data driven decision making in urban development. *Proc. IEEE Conf. Visual Analytics Science and Technology (VAST)*, (pp. 97–104). doi:10.1109/VAST.2015.7347636.

Jankowski, J., & Hachet, M. 2015. Advances in interaction with 3D environments. *Computer Graphics Forum, 34*, pp. 152–190.

Ortner, T., Sorger, J., Piringer, H., Hesina, G., & Gröller, E. 2016. Visual analytics and rendering for tunnel crack analysis. *The Visual Computer, 32*, 859–869.

Traxler, C., Hesina, G., Fuhrmann, A., & Chmelina, K. 2017. A Virtual Reality Monitoring System for Tunnel Boring Machines. In M. Society for Mining, & E. Inc. (Ed.), *Proceedings of the World Tunnel Congress 2017 – Surface challenges – Underground solutions*. Bergen, Norway.

Tunnels and Underground Cities: Engineering and Innovation meet Archaeology,
Architecture and Art, Volume 6: Innovation in underground engineering,
materials and equipment - Part 2 – Peila, Viggiani & Celestino (Eds)
© 2020 Taylor & Francis Group, London, ISBN 978-0-367-46871-2

Investigation of deformation of shield tunnel based on large-scale model test

K. Tsuno, K. Kinoshita & T. Ushida
Railway technical research institute, Kokubunji, Japan

ABSTRACT: This research investigates the mechanical behavior of segmental linings of shield tunnels based on 1/3 scale model loading tests. The test equipment consists of frame, hydraulic jack, hydraulic cylinders, disc springs and others and takes into account the interaction between linings and surrounding ground by means of the disc springs. The test results show that the location of joints influences the mechanical behavior. The bolt axial force decrease with the occurrence of compressive failure and the deformation capacity is large as compared with that of cast-in-place RC linings. This research also carried out numerical calculation by means of 3D FEM, taking into account the joint parts such as segmental joint bolts and joint plates. It is observed that the calculation results correspond to those by model tests.

1 INTRODUCTION

As soft ground is frequently found in many Japanese cities on alluvial plain, there are some cases where shield tunnels are distorted because of the wide-area ground settlement caused by consolidation of surrounding ground, neighboring construction and others. Shield tunnels may be deformed because of large seismic motion. Shield tunnels are normally strong against the external force because they are circular shape and supported by surrounding ground. However, the knowledge of their mechanical behavior under large axial force has not been accumulated.

As segmental linings of shield tunnels consist of segments and joints, their mechanical behavior may be different from those of cast-in-place concrete linings. This research investigates the mechanical behavior of segmental linings based on 1/3 scale model loading tests. The test equipment can take into account the interaction between linings and surrounding ground and enable to carry out the loading test under the situation that axial force occurs in the model lining. The numerical analysis method is also investigated in this research.

2 MODEL LOADING TEST

2.1 Test equipment

Figure 1 and 2 shows the test equipment used in this research (Takahashi et al. 2006, Shimamoto et al., 2017). The test equipment consists of the reaction force frame, the hydraulic jack for applying load, the hydraulic cylinders for applying the reaction force and ground springs jigs. The maximum working pressure of the hydraulic jack was 500kN. This is equivalent to 5.6MPa, since the size of the loading plate was 300mm sq. There were 8 ground spring jigs simulating the interaction between tunnel lining and ground. Each spring jig had 20 conical springs, with a spring constant of 3kN/mm, which was assumed to be the stiffness of the diluvial formation. The corresponding subgrade reaction coefficient was 16MN/m^3.

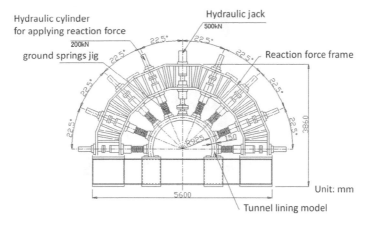

Figure 1.   Outline of test equipment.

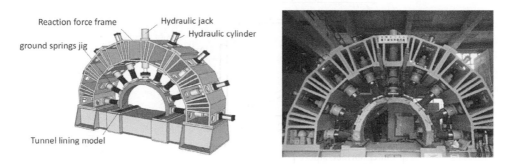

Figure 2.   Overview of model test.

2.2   *Test specimen*

Semicircular tunnel lining models were prepared for the loading tests. As shown in Figure 3, they consisted of segments connected by joint parts. Their inner diameter, thickness and depth were 1850mm, 150mm and 300mm, respectively. The segments were made of reinforced concrete with main reinforcing bars (D6@27mm), distributing bars (D3@60mm) and stirrups. The segments had been cured for 28 days after the concrete casting and their uniaxial compressive strength was 37.2N/mm^2.

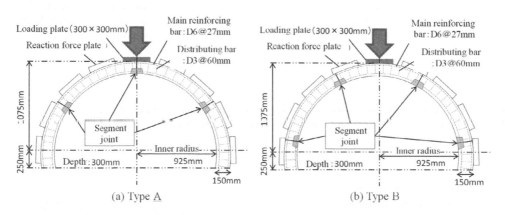

Figure 3.   Outline of tunnel lining model.

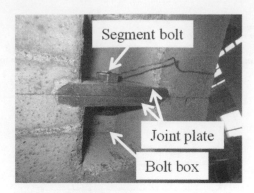

Figure 4.　Joint part.

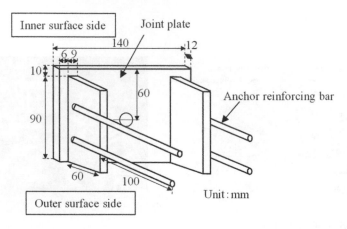

Figure 5.　Dimension of joint plate.

The joint parts consist of joint plates and segment bolts as shown in Figure 4. In a joint part, two joint plates were connected with one segment bolt (M20). The thickness of joint plates was 12mm. The joint plates were fixed with concrete by 4 anchor reinforcing bars as shown in Figure 5.

In this research, there were 2 types of segment rings, namely Type A and Type B, according to the locations of segment joints. One of the segment joints was located just at the crown part in Type A, while any segment joints were not located at the crown part in Type B. There were 3 tunnel lining models prepared for the loading tests, namely 1-ring model (Type A), 1-ring model (Type B) and 2-ring model. In the 2-ring model, two segment rings with Type A and B were connected with 6 ring joints, whose configuration and dimensions were same as those of segment joints.

2.3　Procedure of loading test

Vertical force acted on the crown part of tunnel lining model. Loading was applied controlling displacement to 0.2mm/step. The time for each loading step was 1miniute.

2.4　Test result

2.4.1　Load-displacement relationship
Figure 6 shows the relationship between loading displacement and load regarding 1-ring models. Initial stiffness of 1-ring model (Type B) was larger than that of 1-ring model (Type A).

3264

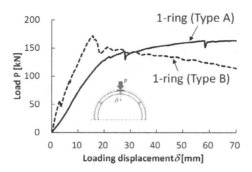

Figure 6. Load-displacement relationship (1-ring).

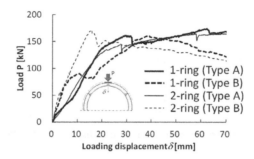

Figure 7. Load-displacement relationship (2-ring).

In the case of 1-ring model (Type B), load increased up to 15mm in loading displacement and then slightly decreased. On the other hand, in the case of 1-ring model (Type A), load continued to increase and the reduction of load was not observed. It was considered that the situation of deformation became 3 hinges arch in the range of large deformation. In this matter, the location of joints influenced the mechanical behavior of segmental lining.

Figure 7 shows that the relationship between loading displacement and load regarding the 2-ring model. This figure also shows the results of 1-ring models. The load-displacement curve of Type A ring corresponds to that of 1-ring model (Type A). Below 8mm in loading displacement, the load-displacement curve of Type B ring corresponds to that of 1-ring model (Type B). Then, load reduction was observed at 10mm. It was considered that the spliced effect of ring joints affected the load-displacement relationship.

2.4.2 Axial force of segment bolt
Figure 8 shows the axial force of segment bolt in the case of 1-ring model (Type A). The axial force of segment bolt at the crown part increased and reduced after the occurrence of diagonal cracks. The axial force of segment bolt started to reduce after the occurrence of compressive failure at arch shoulder. As described above, bolt axial force decreased with the occurrence of diagonal crack and compressive failure.

2.4.3 Distribution of bending moment
Figure 9 shows the distribution of bending moment when loading displacement was 25mm. The distribution of bending moment in the case of Type A differed from that of Type B. It was considered that the location of segment joints affected the bending moment in the tunnel lining. This figure also suggested that the results of 1-ring model corresponded to those of 2-ring model.

Figure 10 shows the configuration and timing of cracks in the tunnel lining models regarding the 2-ring model. It was observed that cracks and compressive failure occurred near segment joints. In addition, cracks were observed at the inner side of crown part and outer side

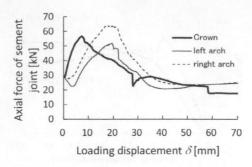

Figure 8.   Relationship between displacement and axial force of segment joint (1-ring (Type A)).

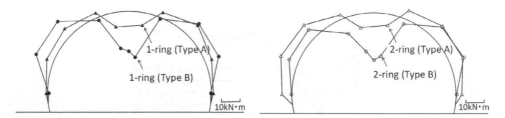

Figure 9.   Distribution of bending moment when loading displacement is 25mm.

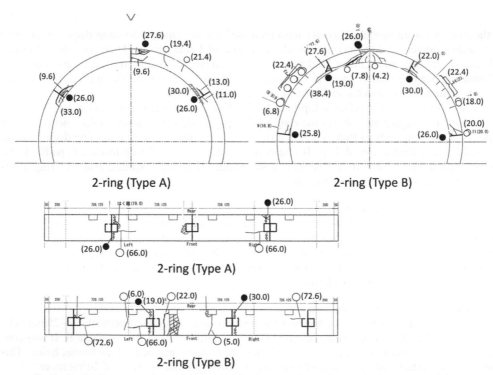

○ : Crack   ● : Compressive failure   ( ): Loading displacement (mm) when crack or compressive failure occur

Figure 10.   Progress of cracks (2-ring model).

of arch in the case of Type B. These results suggested that the tendency of crack progression was affected by the location of segment joints.

3 NUMERICAL ANALYSIS

3.1 Outline of numerical analysis

The model loading tests described in Chapter 2 were simulated by means of 3D Finite element analysis. Numerical analysis models are shown in Figure 11. The segments, joint plates and segment bolts were modeled with solid elements, shell elements and spring elements, respectively. The number of nodes and elements were 14,700 and 11,736, respectively.

Spring elements which do not work tensile side were arranged on the joint surfaces in order to take into account the opening of joint parts during loading. Spring elements were also arranged at the position of segment bolts and fixed with the shell elements for the joint plates. The shell elements of joint plates were fixed with the solid elements for the segments, on three sides, in which the joint plates contacted with the segment concrete.

To take into account the interaction between ground spring jigs and tunnel lining model, ground spring elements were arranged in the area of reaction force plates of the spring jigs.

Material properties were determined based on the element tests. Young's modulus, Poisson's ratio and unit weight of concrete were $22.5kN/mm^2$, 0.2 and $24.5kN/m^3$, respectively.

3.2 Calculation result

Figure 12 shows the relationship between displacement and load obtained by numerical analysis for the cases of 1-ring model (Type A and Type B). In the figure, the analytical results

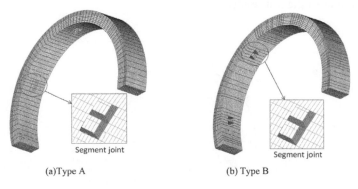

(a)Type A   (b) Type B

Figure 11.   Numerical analysis model.

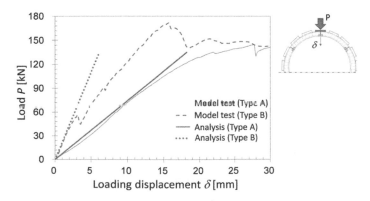

Figure 12.   Load – displacement relationship.

were compared with the results of model loading test. The analytical results corresponded to the test results up to 15.0mm of loading displacement in the case of 1-ring model (Type A) and up to 2.5mm in the case of 1-ring model (Type B). When loading displacement exceeds to the values described above, the stiffness in the model loading tests reduced as compared with the analytical results. In the case of 1-ring model (Type A), it was considered that the reduction of stiffness was caused by the occurrence of compressive failure at the arch part near the segment joint when loading displacement was 19.0mm. On the other hand, in the case of 1-ring model (Type B), the occurrence of cracks reduced the load when loading displacement was 2.4mm, and caused the reduction of stiffness.

The deformation of tunnel lining model obtained be numerical analysis were compared with those obtained by the model loading test in the case of 1-ring model (Type A) as shown in Figure 13. It was found that the analytical results roughly corresponded to the test results.

Figure 14 shows the incremental axial force of segment bolt at the crown part obtained by numerical analysis and model loading test in the case of 1-ring model (Type A). In the model test, the axial force at the crown part increased with the loading displacement up to 8.6mm in loading displacement, and then reduced. Below the load-level, in which axial force started to reduce, the axial force of segment bolt obtained by analysis roughly corresponded with the test results.

Figure 15 shows the distribution of von Mises (VM) strain obtained by numerical analysis. In the case of 1-ring model (Type A), large strain was observed near the segment joints, especially at arch shoulder part. In the case of 1-ring model (Type B), large strain was observed at the crown part as well as near the segment bolts. In the model loading test, cracks and compressive failure occurred near segment joints. In addition, cracks were observed at the crown part in the case of Type B. On these points, the distribution of strain obtained by numerical analysis corresponds to the test results.

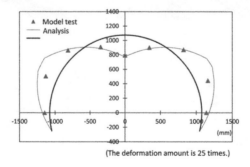

(The deformation amount is 25 times.)

Figure 13. Deformation of lining model (Type A).

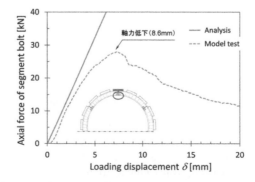

Figure 14. Axial force of segment bolt (Type A).

3268

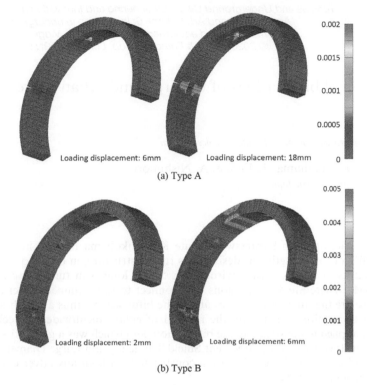

Figure 15. Distribution of strain.

4 CONCLUSION

The mechanical behavior of segmental linings were investigated based on 1/3 scale model loading tests and numerical analysis. The following results were obtained.

(1) The location of segment joints influenced the mechanical behavior of segmental lining, such as load-displacement relationship, distribution of bending moment in tunnel lining and tendency of crack progression. The spliced effect of ring joints affected the load-displacement relationship.
(2) The bolt axial force decreased with the occurrence of diagonal crack and compressive failure under the large-deformation situation.
(3) Numerical calculation by means of 3D FEM was carried out, taking into account the joint parts such as segmental joint bolts and joint plates. It was observed that the calculation results correspond to those by model tests.

REFERENCES

Takahashi, M., Tsuno, K. and Kojima, Y. 2006. Development of large-scale tunnel lining model test equipment, *Proceedings of annual conference of the Japan society of civil engineers*, 30(1): 516 517. (in Japanese)
Shimamoto, K., Yashiro, K. and Ito, N. 2017. Spalling prevention method for tunnel lining using polyuria resin, *Quarterly Report of RTRI*, 58(3): 217–222

Tunnels and Underground Cities: Engineering and Innovation meet Archaeology,
Architecture and Art, Volume 6: Innovation in underground engineering,
materials and equipment - Part 2 – Peila, Viggiani & Celestino (Eds)
© 2020 Taylor & Francis Group, London, ISBN 978-0-367-46871-2

Construction of bifurcations of underground urban expressway

K. Uchiumi
Metropolitan Expressway Company Limited., Tokyo, Japan

K. Matsubara, T. Takahama, T. Fujii & A. Nishimori
Obayashi Corporation, Tokyo, Japan

ABSTRACT: Metropolitan Expressway Route K7, Yokohama North Line, is 8.2km long expressway. 70% of total length was designed as tunnel structure, in order to minimize house transfer as well as to reserve local environment. 5.5km long twin tunnels were constructed with EPB TBMs. There are 4 bifurcations that connect to ramp tunnels under another contract, and there are the number of houses above the bifurcations, thus a trench less construction method was required. Therefore, the method of enlargement/widening technology was developed and applied in constructing the bifurcation, and which was a global first application in actual project. It consists of Enlargement Shield Tunneling and Large Diameter Pipe Roofing. This technology enabled the bifurcation construction without any adverse impact on the residential environment above.

1 INTRODUCTION

Metropolitan Expressway Route K7, Yokohama North Line, (hereinafter referred to as "Yokohama North Line"), which forms the main spine of the road network within Yokohama City. It is the north section of the Yokohama Circular Expressway, and an 8.2km long expressway between "Yokohama Kohoku JCT" on the Daisan Keihin Road and "Namamugi JCT" on the Yokohane Line (Otsuka, Mori, Saito, Matsubara, 23–28 May 2009).

Enhancing the connection between the Yokohane Line and the Daisan Keihin Road via the construction of the Yokohama North Line will improve transportation convenience to Haneda Airport and the Tokyo wan Aqua-line Expressway. And it will also improve access to various commercial facilities and event facilities in the central of Shin-Yokohama area.

2 PROJECT OVERVIEW

In order to minimize house transfer and preserve local environment, Yokohama North Line was designed as tunnel structure for 70% of total length.

Twin EPB TBMs (OD ϕ12.49m) were drove from Shin-Yokohama starter shaft to Koyasudai ventilation arrival shaft for 5.5km length simultaneously. After that, 4 nos. bifurcations were constructed at Namamugi and Kohoku side, which connect to ramp tunnels leading to entrances and exits, shown in Figures 1 & 2.

A trench less construction method was applied for bifurcation construction, because the surface was private land. The earth cover depth is 32 to 54m on the Kohoku side and 28 to 43m on the Namamugi side. The ground was hard and was mainly composed of mudstone (Km), sandy mudstone (Kms), and sand and sandstone (Ks).

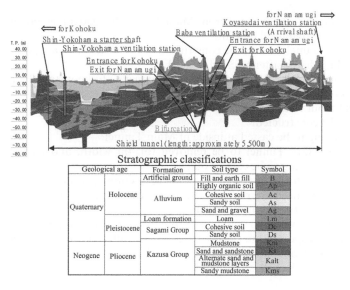

Stratographic classifications

Geological age		Formation	Soil type	Symbol
Quaternary	Holocene	Artificial ground	Fill and earth fill	B
		Alluvium	Highly organic soil	Ap
			Cohesive soil	Ac
			Sandy soil	As
			Sand and gravel	Ag
	Pleistocene	Loam formation	Loam	Lm
		Sagami Group	Cohesive soil	Dc
			Sandy soil	Ds
Neogene	Pliocene	Kazusa Group	Mudstone	Kni
			Sand and sandstone	Ks
			Alternate sand and mudstone layers	Kalt
			Sandy mudstone	Kms

Figure 1. Geological profile along tunnel route.

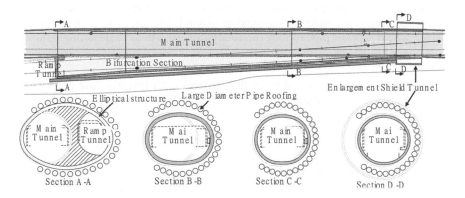

Figure 2. Bifurcation.

3 PLANNING FOR CONSTRUCTING BIFURCATIONS

3.1 *Points with bifurcation section construction method*

As mentioned above, the surface is private land, a trench less construction method was required for the tunnel widening work of the bifurcations from inside the main line shield tunnel (hereinafter referred to as "Main Tunnel"). One conceivable method of widening the tunnel would be NATM (New Austrian Tunneling Method).

However, the groundwater pressure of the Ks layer at Kohoku side was high value of maximum 0.5MPa. Also, Ks layer at Namamugi side was hydraulically connected to nearby alluvium, so that consolidation settlement of the alluvium could occur if the groundwater level was lowered. Therefore, preventing flooding were important issue. In addition, in order to minimize ground deformation, a proper construction method had to be selected, which would suppress loosening of ground.

3.2 *Planning for bifurcation section construction*

The bifurcations construction step is as follows and Refer to Figure 3;

Step1: Chemical grout to be injected from Main Tunnel for widening start point, extra water cut off wall, and widening end point.

Step2: Pipe roofing launch base to be constructed at the widening start point by Enlargement Shield Tunneling (hereinafter referred to as "EST")

Step3: Large Diameter Pipe Roofing (ϕ1,200mm diameter) to be installed along Main Tunnel axial direction from the pipe roofing launch base.

Step4: Chemical grout to be injected between the pipe roofing from inside the pipes, and water cutoff zone to be constructed.

Step5: Concrete lining to be casted at inside of Main Tunnel.

Step6: Segment of Main Tunnel to be removed for 4m width, and partial excavation surrounded by pipe roofing to be carried out.

Step7: Concrete lining to be casted at the widened section, and next block excavation to be consequently commenced after concrete strength reaches required value.

Step8: Completion of bifurcation construction.

Ground deformation would be minimized, because rigidity of the pipe roofing is high and the pipe roofing is supported constantly by concrete lining and unexcavated ground. Moreover, chemical grout from the pipe roofing would be more reliable than work from Main Tunnel, because drilling length is shortened.

4 EVALUATION OF EFFECT CALCULATED VIA NUMERICAL ANALYSIS

Numerical analyses were applied in order to estimate the effect caused by the trench less construction method. One is for effect on surrounding ground, and another is effect on unexcavated ground which would support pipe roofing (hereinafter referred to as "supporting ground").

4.1 The effect on the surrounding ground

In order to evaluate the effect on surrounding ground with investigating effect of the pipe roofing on unexcavated ground and concrete lining, 3D FEM step analysis was applied. The analysis process is shown in Figure 4.

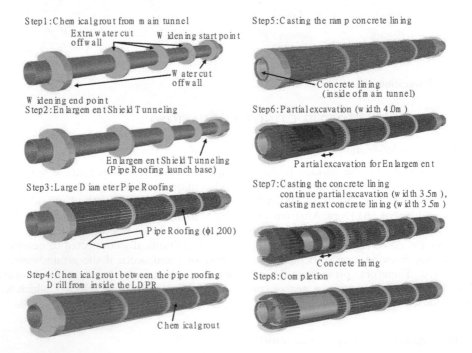

Figure 3.   Construction step.

 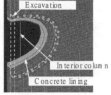 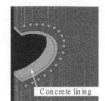

Step1:Excavation for Enlargement   Step2:Concrete lining   Step3:Next block concrete
                                   & Next block excavation   lining

Figure 4.   Numerical 3D FEM analysis step.

Two sections were selected for this analysis; one was the deepest and the other one was the shallowest of the ramp areas. The width of 3D FEM model was constructed as 5.5m, considering half of first excavation width of 4.0m and full of second excavation width of 3.5m.

Th ground, the pipe roofing, the concrete lining, and the interior columns were considered in 3D FEM model. Young's modulus for the ground was determined by using the Dynamic deformation characteristics ($G/G0 - \gamma$ curve) from the closest boring data. And the reduction in rigidity according to the strain generated in the ground was also taken into consideration.

The effect on surrounding ground was evaluated by comparing the maximum shear strain on the elasto-plastic boundary (hereinafter referred to as "critical shear strain") with the calculated shear strain. The results of this analysis are shown in Table 1. The calculated shear strain was smaller than critical shear strain, therefore the judgement was made that the effect on surrounding ground would be small.

4.2   The effect on the supporting ground

In order to estimate the effect on supporting ground, 2D FEM step analysis was applied. The 2D FEM model and the construction step are shown in Figure 5.

The ground was modelled as uniform plane strain elements, and the ground stability was evaluated for each of Kms and Ks layer respectively.

In addition, the spring element value of the concrete lining and interior columns which would support pipe roofing were calculated via additional frame model analysis and taken into consideration.

Table 1.   Result of Numerical 3D FEM analysis.

Cover depth	Shallowest point	Deepest point
Calculated Shear strain	Dc Ds layer: max shear strain 3.03×10^{-3} Ds Km Km layer: max shear strain 2.29×10^{-3} Kms	Ks Km Km layer: max shear strain 2.77×10^{-3} Ks Km layer: max shear strain 3.60×10^{-3} Km
Critical shear strain	Ds:18.00×10^{-3}, Km:8.00×10^{-3}, Kms:11×10^{-3}, Ks:19×10^{-3}	
Deformation of surface ground	8.2mm	7.8mm

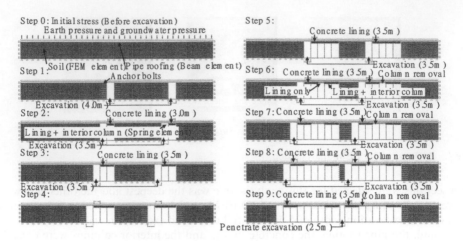

Figure 5. Numerical 2D FEM analysis step.

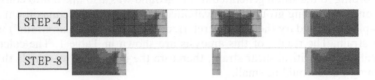

Figure 6. Example of strain distribution for numerical 2D FEM analysis.

Table 2. Result of Numerical 2D FEM analysis.

Analysis case	Layer	Value of spring element*	Distance between pipe roofing (m)	Shear strain		
				Max shear strain γ max	Critical shear strain γc	$\gamma max/\gamma c$
Case 1	Ks	Minimum	1.80	1.947E-03	4.706E-03	0.414
Case 2	Kms	Minimum	1.80	1.570E-03	6.107E-03	0.257

* For concrete lining and interior column.

The results of the effect on supporting ground are shown in Figure 6 and Table 2. The effect was evaluated by the critical shear strain as the 3D FEM analysis, although the value of critical shear strains differed between the 3D and 2D FEM analysis because of the difference in modeling.

In the 2D FEM analysis, the calculated shear strain was as small as about 40% of the critical shear strain, and the effect on supporting ground would be small. Therefore, the conclusion was that this construction method would be applicable.

5 PLANNING AND CONSTRACTION WITH ENLARGEMENT SHILED TUNNELING

5.1 *Issues with EST (Enlargement Shield Tunneling)*

An Enlargement Shield Tunneling Machine (hereinafter referred to as "ESTM") was used to enlarge the Main Tunnel (OD ϕ12.3m) to OD ϕ18.3m along length 11m, to construct the pipe roofing launch base. In this project, there are 4ramps, but only one number of the ESTM was manufactured and converted for 4 sections.

The issues noted with constructing the EST were as follow;

1) Inhibition of ground deformation and groundwater-level lowering
   When the ESTM advances, it is important to minimize the impact of any ground deformation and groundwater-level lowering, because the surface is private land.
2) Reliable construction of EST
   The ESTM advances in the circumferential direction along the Main Tunnel; therefore the machine orientation changes in every ring. In addition, the tunnel alignment is a circumferential direction with sharp curve, and upward and downward excavation take place. Although construction experience with the EST has been gained, this is the first time for such a large section to be applied for EST. Therefore, the careful examination was required for the workability of the segment erection and reliable alignment control.

5.2 *Planning with EST*

The planning for the EST set out to solve the above issues as follows;

1) Measures to inhibit ground deformation and groundwater-level lowering
   The ground is exposed during ESTM launch shaft construction and connection work of Main Tunnel and EST segment. Therefore, chemical grout was injected from Main Tunnel in advance to stop water inflow.

   Water inflow into ESTM can be cut off by chemical grout, hence open type ESTM could be selected. However, closed type EPB ESTM was adopted to inhibit of ground deformation and emergency flooding. 2 earth pressure gages were installed at the bulkhead of the ESTM, and the excavation was carried out with controlling the face pressure.
2) Segment erection method
   If segments were erected in ESTM, the workability would be decreased, because the machine orientation would change in every ring. Therefore, the segments were erected in the launch shaft, and thrusted by base jacks, as the pipe jacking method.
3) Alignment control method
   ESTM was designed as arc shape, in terms of circumferential direction thrusting (Figure 7). Alignment control might be difficult if only base jacks were equipped, therefore the ESTM was divided into two parts (front shield and rear shield), and equipped with articulation jacks, moving sledges, and grippers to ensure sensitive control of the alignment.

The excavation method was as follows;

First, only the front shield advanced to the certain stroke while controlling the alignment and using the articulation jacks.

Next, the rear shield and segments were thrusted by the base jacks while retracting the articulation jacks.

With the alignment control a 3-dimension survey was carried out by the total station. The ESTM was equipped with a clearance detection device to verify the distance between the ESTM and the Main Tunnel segment.

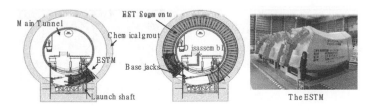

Figure 7.   Construction sequence for the EST.

5.3 Construction Outcome of EST

1) Excavation management of the ESTM

By the results of constructing 4 ESTs, the alignment was controlled within the tolerance (+/-100mm) for both up/down and left/right directions.

The face pressure was adjusted in terms of smooth discharge from the screw conveyor, as the ground was enforced with chemical grout. No much differences were observed in cutter torque and thrust force between upward and downward excavations, and those values were stable to the equipped capacity. The ESTM advance speed was about 5 to 10mm/min, and it took about 3 hours for the segment erection.

2) Effect of excavation due to the ESTM

Ground displacement and surface settlement was not observed during ESTM work. The stress and displacement of Main Tunnel segment were also small enough for those structures.

6 PLANNING AND CONSTRUCTION OF LARGE DIAMETER PIPE ROOFING

6.1 Issue with the LDPR (Large Diameter Pipe Roofing)

The issues noted with the planning and construction of the LDPR were as follows;

1) Reliable construction (Alignment control and narrow space construction)

Reliably alignment control of the LDPR was required, because it was located close to urban boundary line and also clearance had to be kept for next concrete lining work.

In addition, pipe roofing gantry and launching cradle had to be designed properly in terms of installation route of steel pipes. Because launching position of each pipes was different for height and direction, also the work space was too narrow for handling of such large and long steel pipes.

2) Ensuring reliable joint rigidity

Joint of pipe roofing is weak member of roofing structure which is important to inhibit any ground deformation during widening excavations. The pipe joints had to be fixed properly to keep required rigidity.

6.2 Planning for the LDPR

6.2.1 Alignment control

The tolerance of the LDPR alignment was set to be +/-50mm based on previous experience.

For accurate control of the pipe roofing, the deviation of micro tunneling machine were constantly monitored by laser transit during advance.

In addition, verification survey was carried out after each pipe installation (6 to 6.5m for each), including horizontal/vertical deviation of LDPR, thrusted length, and deviance of laser transit.

6.2.2 Pipe roofing gantry and launching cradle for micro tunneling machine

In order to shorten the construction period in each ramp, the work space was divided into 3 sections, upper, middle, and bottom, and 3 to 4 micro tunneling machines were used simultaneously as shown in Figure 8.

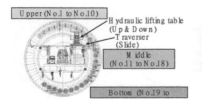

Figure 8.   Pipe roofing configuration.

In order to secure the logistics within the Main Tunnel, pipe roofing gantries were set up. For the upper construction, a traverser and a launching cradle with hydraulic lifting table were set on the gantries for smooth lateral movement and short time position adjustment.

Moreover, the pipe roofing gantries was utilized as "cargo handling facility" for efficient logistics within such a narrow space.

6.2.3 Measures against flooding with the arrival of the micro tunneling machine (micro tunneling machine pull back)

After completion of a unit of the LDPR work, main body of the micro tunneling machine were pulled back to the portal.

Chemical grout had been injected at LDPR arrival point previously from Main Tunnel. However, water inflow was concerned from front of the micro tunneling machine via tail void around steel pipes. Therefore, water inflow was checked through check valves in the micro tunneling machine before the main body of machine was pulled back. Injection ports of chemical grout were equipped to the machine in case for water inflow.

After pulling back the main body, steel plate was welded on the tip of the steel pipe and cement bentonite grout was injected to completely close off the tip.

6.3 Construction Outcome with the LDPR

6.3.1 Status of LDPR advance management

The face pressure was stabilized within +0.05MPa of natural groundwater pressure during the advance. The thrust force mostly equaled to the calculated value that was the combined resistance of the earth pressure and groundwater pressure on the cutting face, and friction between the skin plate and the ground (maximum 3,000kN). The advance speed was 40 to 50mm/min in the Km layer and 50 to 60mm/min in the Ks layer.

6.3.2 Status of LDPR alignment control management

The deviation was surveyed with every pipe and the direction was adjusted by the articulation jacks, so that the LDPR alignments were controlled within the tolerance (+/-50mm) over the entire length including curve sections.

6.3.3 Effect of advance from the LDPR

Face collapsing and excessive mucking out was not observed during LDPR work, and the ground displacement and surface settlement was not observed too.

7 PLANNING AND CONSTRACTION WITH WIDENING EXCAVATION

7.1 Issues with Widening Excavation

The external load (ground load + groundwater pressure) on the widening excavation section was supported by the rigid LDPR. However, the ground peeling off and falling was concerned because there were many cracks in the Km layer.

7.2 Measures to address the Widening Excavation issues

1) Shotcrete and anchor bolts

NATM Shotcrete was applied as ground support during widening excavation. Compressive strength was 18N/mm^2, concrete slump 10cm and lining thickness 100mm.It reduced work time under exposed ground and improved safety. Also, it restricted water inflow, suppressed loosening of ground, and support the ground in early time, so that the risk of surface settlement was reduced.

In order to prevent any ground collapse or the shotcrete peeling off, anchor bolts of D25 - 2.5m were drilled and installed at the side wall, and fixed with shotcrete.

2) Sewing bolts

Ground peeling off and falling were concerned between the widening excavation face and LDPR due to cracks in the Km layer, especially at the crown. Therefore, sewing bolts were drilled and installed from inside the LDPR (refer to Figure 9) before widening excavation.

Ground peeling off and falling was suppressed by prompt fixing of the sewing bolts with bearing plate after secondary shotcrete. ϕ25mm steel rods were installed up to targeting 100mm outside from the outer surface of the concrete lining. The bolt end was utilized as marker for the excavation end.

3) Planning of monitoring

In order to monitor ground deformation and tunnel stability continuously, measuring instruments were installed in the tunnels and at the surface.

4) Procedure of widening excavation

The excavation span was 1.2m, and weight of removal segment piece was less than 2ton according to handling machine capacity. The procedure of the widening excavation was as follows and shown in Figure 10.

Step1: Installation of vertical column for supporting Main Tunnel,

Step2: Removal of Main Tunnel segment, Widening excavation, Primary shotcrete, Installation of welded steel wire mesh, Secondary shotcrete, Sewing bolts fixed in place, and Drilling and installation of anchor bolts.

Step3: Concrete lining work for widened section, Installation of interior column for supporting concrete lining.

Step4: Installation of vertical columns for supporting Main Tunnel, Next widening excavation.

Step5: Excavation and Concrete lining work of Step2, 3 & 4 to be continued.

7.3 Feedback on construction of Widening Excavation

7.3.1 Expansion of excavation width

After the first span excavation with 4m width, the monitoring results were observed to be much lower than the analysis results at the design stage. Therefore, reverse analysis was carried, and the external load for the design was re-examined and reduced.

Consequently, the excavation width was changed to 8m, because the stress and displacement of the concrete lining and the LDPR would be small enough according to the additional analysis with the reduced external load.

7.3.2 Omission of interior columns for concrete lining

The stress and displacement were monitored in the concrete lining at each step for 8m width excavation. The stress and displacement were lower than the analysis. Therefore, the installation of the interior columns for concrete lining was omitted. In consequence, sufficient work space was secured, and construction steps were reduced, so that construction efficiency was improved.

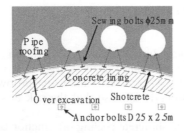 

Figure 9.   Sewing bolts detail.

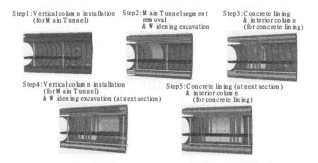

Step1 : Vertical column installation (for Main Tunnel)   Step2 : Main Tunnel segment removal & Widening excavation   Step3 : Concrete lining & interior column (for concrete lining)

Step4 : Vertical column installation (for Main Tunnel) & Widening excavation (at next section)   Step5 : Concrete lining (at next section) & interior column (for concrete lining)

Figure 10. Widening excavation steps (Longitudinal section).

Figure 11. Completion of bifurcation.

7.3.3 *Effect of Widening Excavation*

Even after changes of construction sequence with 8m width excavation and interior column omission, all of surface monitoring results (surface settlement, inclinometer, and groundwater level sensor) were within the allowable values, and the widening excavation has been completed safely. In addition, for monitoring in tunnel, the stress and displacement results were lower than the analysis.

Moreover, the ground support method by a combination of shotcrete, anchor bolts, and sewing bolts enabled the widening excavation to be successfully completed with good face stability and construction safely.

8   CONCLUSION

In March 2017, Metropolitan Expressway Route K7, Yokohama North Line opened, after various technical issues had been overcome and construction of the bifurcations completed without affecting the residential area (Figure 11).

Practical application of this widening technology can make a large contribution to constructions at great depths and with large section tunnels, and not only with road tunnels, but also as a key technology in the development of civil engineering businesses of the future such as the creation of new underground spaces.

REFERENCE

Otsuka, K & Mori, K & Saito, K & Matsubara, K 2009. Technical countermeasures against construction of large space road tunnel under private land. World Tunnel Congress in Budapest Hungary, ITA-AITES, 23–28 May 2009.

Tunnels and Underground Cities: Engineering and Innovation meet Archaeology, Architecture and Art, Volume 6: Innovation in underground engineering, materials and equipment - Part 2 – Peila, Viggiani & Celestino (Eds)

Riachuelo Lote 3 – Innovative method for the construction of the sea outfall

N. Valiante
Salini Impregilo S.p.A., Milan, Italy

M. Martini
JV Salini Impregilo S.p.A. – J.J. Chediack, Buenos Aires, Argentina

L. Gaggioli
Palmieri Group, Gaggio Montano, Italy

ABSTRACT: The "Desarrollo Sustentable de la Cuenca Matanza-Riachuelo - Lote 3" project includes a gravity outfall system to discharge treated wastewater into the River Plate, in Buenos Aires (Argentina). It consists of: 50 m deep upstream shafts; 10.5 km long sub-fluvial outfall tunnel; 1.5 km long diffuser tapered tunnel; 34 risers to discharge the effluent from the diffuser tunnel up to the river bed diffuser heads. An innovative method has been developed (the "Riser Concept") to install vertical risers by jacking pipe segments from inside the tunnel upward to the river bed and through soil formations. The method provides several benefits, of which: reduction of maritime activities; construction schedule not affected by marine conditions; minimization of navigation traffic disruptions; environmental benefits; improvement of efficiency during construction and operation. This is the first time that this method will be used and it will represent an important step forward in sea outfall technology.

1 INTRODUCTION

Marine outfalls are used to discharge treated wastewater to the environment. The released effluent can be positively buoyant if originated from a wastewater treatment plant or negatively buoyant if released from a desalinization plant. In both cases an outfall shaft is foreseen together with an inclined pipeline or tunnel to transport the effluent offshore, plus a diffuser section to uniformly release the effluent into the waterbody (Figure 1). The outfall tunnel is inclined upward to avoid air entrapment along the alignment. The diffuser section of the tunnel generally presents a variable cross section to ensure the velocity of the effluent remains high enough to avoid sedimentation and includes a series of vertical offtakes, the risers. The top of each riser is fitted with a rosette head presenting a number of nozzles (also known as ports) whereby the wastewater is released and diluted into the final receptor.

The construction of long sewerage interceptors and outfalls is becoming increasingly common for modern urban cities facing a large water front and under significant environmental burden imposed by decades of sustained growth.

It has been 10 years since Argentina's Supreme Court ordered the environmental restoration of the Riachuelo river, one of the world's most polluted. Since then, the government is implementing a comprehensive project to make the polluted Matanza-Riachuelo basin a safe and healthy place to live and work. The Riachuelo Sanitation Program, foresees the execution of in land tunnels, a treatment plant and an outfall system (capacity 27 m^3/s) that serves as an outlet for the water treated through the plant.

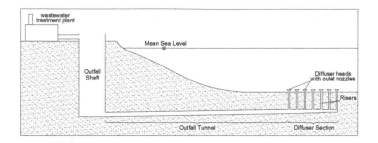

Figure 1. Schematic of marine outfall project.

The Project "Desarrollo Sustentable de la Cuenca Matanza-Riachuelo - Lote 3" (Riachuelo L3), is a critical part of this program and includes three components: deep shafts regulating the hydraulic head, a conveyance tunnel and a diffuser section with vertical risers (Figure 1). The outfall tunnel will be one of the longest of its kind in the world and for the first time, for the construction of the risers will be used an innovative method developed by Salini Impregilo.

2 THE PROJECT

The project "Riachuelo L3", is a sub-fluvial tunnel conveying treated wastewater from the treatment plant and to be diffused into the River Plate. The project comprises a series of out-fall shafts ("Cámara de Carga"), from which the submarine outfall tunnel ("Túnel Emisario") starts extending thereafter under the estuary of the river.

The original diffuser system consisted of: an approximately 10.5 km long outfall tunnel located 40 m below the riverbed and arriving into an off-shore transition shaft ("Cámara de Transición"), connected to a diffusion pipeline ("Tramo Difusor"), with a length of around 1.5 km, equipped with vertical risers. The diffusion pipeline had to be constructed on the previously dredged riverbed, being founded on pile foundations (Figure 2).

Salini Impregilo proposed an alternative solution, to minimize the critical off-shore activities and improve the hydraulic performance. The solution considered extending the length of the conveyance tunnel to 12 km and converting the last 1.5 km into a diffuser section with No. 34 regularly distributed 30 m long risers. On the other hand, the transition shaft, the pile foundations and the diffusion pipeline were eliminated (Figure 2).

The alternative solution included an innovative construction method (the "Riser Concept") for constructing the risers. Traditionally, in similar submarine outfall projects, the installation of risers is performed offshore, with the standard construction methodology implying various maritime operations. On the contrary, in the proposed alternative the risers (each formed from several segments) will be installed from inside the tunnel upwards to the riverbed of the River Plate, by means of a prototype vertical pipe-jacking equipment.

The alternative solution of the project results into several benefits:

1. Elimination of several works:
    - maritime offshore works, i.e. dredging of the bed sediment, construction of underwater deep foundation, underwater installation of pipes.
    - underground activities, i.e. connection between tunnel and transition shaft, connection between transition shaft and diffusion tunnel.

2. Elimination of various maritime operations:
    - Avoidance of potential construction delays, due to suspension of maritime operations, which can be inevitably induced by marine conditions.
    - Minimization navigation traffic disruptions.
    - Mitigation of negative environmental impacts caused by dredging, drilling activities, etc.

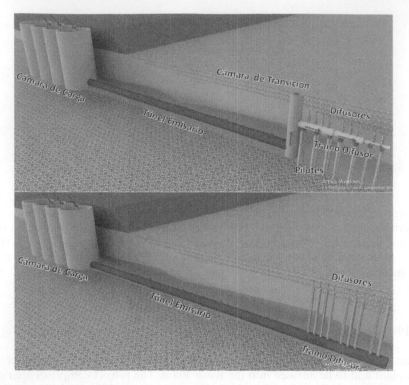

Figure 2. Schematic of Riachuelo Project – Initial solution (upper) and alternative solution (lower).

3. Allowing an improvement of efficiency in the construction and operational phases:
   - During construction: some complex activities are avoided and maritime operations are significantly reduced by eliminating several works.
   - During the operational life: the hydraulic performance inside the tunnel is much improved, as encountered hydraulic losses are less, with a reduction of energy consumption and/or a higher flow capacity.

3 THE RISER CONCEPT

The innovative construction method developed by Salini Impregilo, is a bottom-up system which allows installing the vertical riser from inside the tunnel, using a jacking equipment.

The method - which is patent pending - requires the initial installation of tunnel lining special segment rings: a positioning segmental ring and a launching segmental ring, to ensure the verticality of the riser. Considering the direction of the TBM advance, the first ring is the positioning segmental ring whilst the second one is the launching segmental ring, characterized by a keystone launching segment.

A displacement head (2) is fastened to keystone launching segment (1), which has previously been placed in the desired position for the installation of the riser (3). The correct position of keystone launching segment is adjustable by regulating the rotation of the special segment rings through special slotted holes for the longitudinal connection between the special rings. A sealing system (4), comprising a multi-lip gaskets and additional emergency seals, is provided to ensure the water-tightness between riser or displacement head and keystone launching segment (Figure 3). Once the jacking equipment (5) is positioned underneath keystone launching segment, the riser is connected to the displacement head and then driven through the soil by means of a thrust platform and the displacement head which also includes a hydro-demolition system injecting pressurized water to weaken the soil structure in the penetration area. The operations are repeated until the last riser segment is installed and connected to the tunnel to reach the desired elevation. Finally the displacement head is removed and a diffuser is installed at the top of the riser (Figure 4).

In the traditional approach, in submarine outfall tunnels the installation of the diffusion risers represents one of the most critical activities. In fact, this is performed through a multistage offshore work: initially, the bed sediment in correspondence of the riser head is dredged, then a jack up drilling vessel is floated into position with a drilling template used to ensure the correct positioning. For each drilling phase, if required, a permanent casing is placed and the annular void between the drilled hole and the casing is filled with grout. Once the required level is achieved, the riser is lowered into position, grouted, and capped. Connection of the offshore works to the underground works represents a challenging operation: "probe holes are drilled from the tunnel to ascertain the location of the pre-installed risers and to drain the risers of ballast water", then the off-take adits are excavated to expose the risers, which afterwards are cut and permanently linked to the tunnel through an elbow section (Eisenberg & Brooks, 1992). Additionally, the conventional approach for installing the risers produces environmental damage due to the dredging and drilling activities. Such damage may be destruction of habitats, suspension of sediments, resettling of fishes, and displacement of infauna and marine plants (Peter et al., 2017).

The most significant changes introduced by the "Riser Concept" solution is related to the avoidance of interferences with maritime weather conditions and operations, reducing the overall construction and environmental impact and improving the hydraulic efficiency of the outfall system.

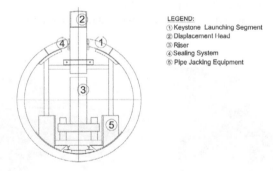

LEGEND:
① Keystone Launching Segment
② Displacement Head
③ Riser
④ Sealing System
⑤ Pipe Jacking Equipment

Figure 3.   "Riser concept" - Components.

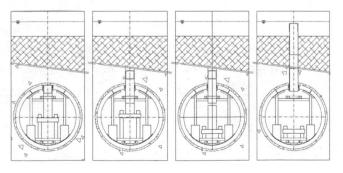

Figure 4.   Risers installation method.

4 PROJECT STATUS

4.1 *The construction of the outfall shafts and tunnel*

The shafts whereby the outfall tunnel starts consists of 4 circular shafts separated between each other by reinforced concrete diaphragm walls (Figure 5).

With a thickness of 1.20 m and an approximate depth of 58 m, the diaphragm walls presented an adequate overlap between primary and secondary panels which secured the water tightness of the vertical joints. The diaphragm walls key into an impervious layer of stiff Miocenic, clay enabling in a second phase the safe excavation of the shaft down to the tunnel level. By adopting a sequenced excavation, almost 37,000 m3 of soft ground were removed from inside the shaft, initially in dry and thereafter in underwater conditions followed by the construction of a 3.0 m thick concrete bottom plug also poured in underwater conditions to avoid the risk of the excavation bottom failure. Upon completion of this phase, the shaft was finally be emptied and the permanent 1.5 m thick bottom slab casted above the bottom plug. The construction of the shaft walls demanded almost 14,000 m3 of concrete and 1000 t of reinforcing steel.

Once completed, the outfall tunnel (Figure 5) will enter within the list of the longest TBM submarine/sub-fluvial tunnels constructed to date across the globe. In addition, the pressures at which the TBM is operating and will continue to operate until the end (> 4 bar) place it within the upper limit range of the EPB operability, considering in particular that such conditions remain unchanged for the entire tunnel length.

The operating conditions of outfalls (driven by large upstream heads) often result in internal pressures that are higher than the external pressures, resulting in tension loads acting on the tunnel lining. The tunnel lining of this project is no exception and Salini – Impregilo adopted a "simple" one-pass segmental lining composed of 5 segments 300 mm thick, forming an universal ring of 1400 mm length. An innovative solution was proposed by mainly relying on a load sharing mechanism between segment radial bolts (two T28 bolts per radial joint) and circumferential dowels (twenty per ring) and including several water tightness measures to prevent egress of effluent throughout the operating life.

Currently the tunnel is under construction with daily production peaks of up to 40 m reached to date.

Figure 5.   Construction of Outfall shafts (upper) and outfall tunnel & TBM (lower).

4.2 The design and construction of the riser system

4.2.1 Development of the prototypes

An extensive phase of design, construction and testing of prototypes was required, to develop the method proposed for the execution of the risers.

Initial studies were oriented to evaluate the requirements for the design of the different components, considering the geological conditions of the site. Several tests and procedures had to be identified adopting a multistage testing approach, to progressively validate each component of the system.

One of the most complex studied topic relates to forces originated by vertical jacking of the risers and the geotechnical behavior and mechanisms generated. The lack of geotechnical formulation available to represent the mechanism and stresses induced on the soil by the bottom-up jacking of the riser, produced the need to define various tests in order to reproduce the expected stress site conditions. Additionally, the anticipated soil materials had to be properly characterized and reproduced in the tests. The assessments of the riser-soil interaction and the forces and stresses expected were required in particular to design the jacking equipment and the risers segments.

Another important subject considered was the need for a robust and redundant sealing system between the riser and the tunnel and a number of tests to validate it. Certain constrains and requirements taken into account to develop the sealing system were related to the geometrical limitations (the tunnel segment is 300 mm thick) and the interface with the riser during installation (diameter, roundness, roughness, joints, etc.). The solution finally implemented included a series of 3 lip gaskets with grease chambers and emergency inflatable seals.

The design of the riser joints and segments had to consider many aspects: structural capacity (axial force 400 ton), construction requirements (various type of elements), geometrical limitations (tunnel diameter, riser OD diameter and overall riser length), interface with the sealing system (smooth and continuous surfaces), water-tightness of the joints (redundancy) and operational constrains (handling and assembling), durability (100 years design life). After various studies and investigations the solution adopted resulted in a pipe with a diameter of 0.71 m, a length of 1.80 m and 19 mm thickness (Figure 6). The adopted joint is a bell and spigot solution with o-rings and a connection achieved by means of locking spheres. The riser segments are made of stainless steel and the standard segment has a weight of approx. 650 kg.

To drive the riser through the soil material above the tunnel, a system capable to remove the very dense material and displace the weak soil, was considered. The solution includes a hydro-demolition system and a soil discharge line system. The hydro-demolition includes 3 high pressure water lines serving more than 20 frontal nozzles, to weaken the soil in front of the riser and a number of 5 washing nozzles in the excavation chamber, to facilitate the discharge of the excavated material. The soil discharge system includes a gate valve, a flexible pipeline and a washing water pipe to discharge the material. A prototype of the solution was constructed and tested.

Two riser jacking equipment and various services portals were designed considering in particular the complex logistic for supply of materials and services. Each riser jacking station includes a load distribution system (having 26 jacks), designed to control the distribution of the jacking loads on the tunnel segments. The equipment includes a riser handling devices to move and place the different riser segments. The first prototype of the jacking station has been constructed (Figure 7) and a real size test has been executed to verify the functionality of the entire method.

Figure 6.   Prototypes of riser standard segments.

Figure 7.   Riser equipment. Prototype of riser jacking station.

4.2.2 *Tests of the prototypes*

Various tests, investigations and trials were necessary to define the means and methods required for the "Riser Concept" solution. Some of these tests are still under execution to complete the study of this innovative method. Herewith a brief introduction to some of the most relevant tests carried out and ongoing, is provided.

4.2.2.1 SEALING SYSTEM TESTS

The sealing system is a prototype designed and constructed to ensure water tightness between the riser and the tunnel, during riser jacking. The system consists of: a guiding ring; a multi-lip sealing that comprises 3 lips with 3 grease chambers; an emergency sealing with 2 inflatable gaskets.

The purpose of testing the sealing system was to verify its performances under the most critical work conditions and to verify compatibility between the sealing and the riser segments.

The work conditions considered in the test included:

– Operative water pressure: 3.5 bars generated by water depth of 35 m.
– Construction tolerances for Riser Segments: ± 0.5 mm on the external diameter, which implies a maximum out-of-roundness of 1.0 mm.
– Riser joint as per Riser Design: external dimensions, mechanical machining, and maximum possible gap between two segments.
– Potential verticality misalignment of the Keystone Launching Segment due to operative tolerances during tunnel construction operations.
– Possible eccentricity of the Riser with respect to the vertical axis of the Sealing System.

A prototype test equipment was designed and constructed (Figures 8 and 9). The prototype comprised two connected riser elements reproducing the most critical geometry of the riser joint. The geometry of the pipes reproduced the possible riser construction tolerances: maximum external diameter, minimum external diameter, and maximum out-of-roundness.

The tests were carried out in static and dynamic conditions and considering a vertical and inclined configuration of the riser, to simulate the various possible misalignments. The test equipment was developed to allow the translational up/down movement of the riser element to implement dynamic tests that simulate the riser jacking process.

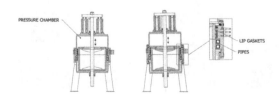

Figure 8.   Sealing system test layout.

During the test, working parameters were recorded and registered for further interpretations and analysis. By concluding the tests, all the sealing components showed positive outcomes in both testing states (static and dynamic) and on the extreme operative conditions simulated.

4.2.2.2 DISPLACEMENT HEAD TESTS

A very important study to complete was the design of a system capable to weaken and remove the material in front of the displacement head (DH). A displacement head test was defined for this purpose, which allowed to reproduce the expected geotechnical and geological conditions.

The scope of the DH test included:

– Verify the capacity of the DH and the hydro-demolition system to remove the stiff soil material and to enable the riser insertion.
– Control general operational aspects such as removal and discharge of the soil material.

For performing the DH Test, a specific test facility was designed and constructed (Figure 10). The facility comprised: a base frame that included a jacking station, a silo 3 m high and with a diameter of about 2 m, including a loading plate to reproduce the soil stresses expected, a DH mounted inside the Sealing System and kept in position by locking elements. During the test, working parameters were recorded and registered for further interpretations and analysis.

An extensive soil characterization campaign was required to reproduce the materials intended to be tested. Geological and geotechnical studies were carried out both to characterize the in-situ materials and reproduce the expected geotechnical conditions. Various materials testing embankments were prepared to define the soil compaction procedure and equipment.

The test facility was filled with the characterized soil materials and geotechnical instrumentations were used (load cells and piezometers) to monitor the soil behavior and stresses during the execution of the test. The test facility allowed to control the pre-loading of the soil material and to simulate various water pressures.

The DH was equipped with 3 loading cells to measure the forces acting on the top of the riser during the test. The hydro-demolition water pressure and the flow were monitored for each of the 3 waterlines and working conditions of the sealing system were also recorded during the test. Finally, the soil discharge system was also controlled (pressure and flow) to estimate the amount of soil removed at each stage.

During the tests various load conditions, hydraulic gradients, jacking forces and hydro-demolition conditions were simulated. The DH test proved the capability of the DH to demolish and remove the soil material, enabling the riser to advance and substitute the excavated materials.

Figure 9.   Equipment for sealing system tests.

Figure 10. Displacement head test facility and test execution.

4.2.2.3 REAL SIZE TEST

Certainly one of the most important tests considered to validate the components of the method was the "real size test" (RST), that included the construction of a portion of the tunnel in scale 1:1 and the simulation of the riser jacking using the prototypes constructed (riser jacking equipment, riser segments, sealing system, etc. see Figure 11).

The test was designed to simulate jacking of the riser, to confirm the expected forces on the riser's head and shaft, to monitor the behavior of the riser joints and to verify the overall functionality of the riser jacking equipment.

The test facility included a tunnel with an internal diameter of 4.3 m, a silo 10 m high filled with soil material, No. 2 keystone launching segments and various monitoring systems (soil stresses, jacking forces, riser jacking equipment, riser segments, biaxial clinometers, etc.).

The RST was not only a single test but encompasses a number of simulations and validations, which is divided in 3 main stages.

Figure 11. Real size test of the jacking equipment and components.

Figure 12.   Real size test. Riser completed from below (left) and from above (central and right).

– Preparative activities: to be performed prior riser jacking tests and for verifying adequate performance of the equipment.
– Riser jacking activities: relevant to the execution of risers, including installation of riser segments, connecting the riser to the tunnel structure, etc.
– Riser intervention activities: performed posterior riser jacking and including contingencies.

The test was successful. Two risers were executed: the first inclined and a second one vertical (Figure 12). The jacking forces measured during the tests were lower than expected, proving the suitability of the hydro-demolition to substantially reduce the forces acting on the riser's head and shaft. A very good control the verticality was achieved, proving a very good operational performances of the riser joints. Finally, various construction aspects (set-up of equipment, construction sequences, training of personnel, etc.) were also positively verified.

5   CONCLUSIONS

The "Riser Concept" solution developed for the Riachuelo L3 is a challenging and complex project involving many different disciplines and expertise and requiring a sensible constructability approach throughout the whole design, construction and testing process of the components.

The integration of construction knowledge and experience in planning, engineering, procurement and field operations, played a key role to achieve the satisfactory results accomplished to date. Different teams, working groups and experts had to be involved and coordinated, through a continuous design review and integration, to advance on the construction and testing of this innovative method and equipment. Besides, a multistage and extensive phase of testing was required to gradually validate each component of the method.

The solution developed includes many innovative topics and challenging design and construction solutions. Once the construction of Riachuelo L3 will be completed, the proposed method will represent an important step forward in sea outfall technology.

ACKNOWLEDGMENT

The authors wish to acknowledge the efforts put into the studies of the "Riser Concept" to all the Salini-Impregilo team involved and particularly H. Elgarhi & M. Lazzarino.

REFERENCES

Eisenberg, Y. & Brooks, P. 1992. Design and construction of the Boston outfall. Coastal Engineering Proceedings. 23, 3285–3304.
Peter, M., Salvatore, S., & Bruce, C. 2017. Marine Outfalls. In Dhanak, M.R. & Xiros, N.I. Springer Handbook of Ocean Engineering (pp. 711–740). Springer International Publishing.

Tunnels and Underground Cities: Engineering and Innovation meet Archaeology, Architecture and Art, Volume 6: Innovation in underground engineering, materials and equipment - Part 2 – Peila, Viggiani & Celestino (Eds)
© 2020 Taylor & Francis Group, London, ISBN 978-0-367-46871-2

Innovation through in-situ machining for repair and maintenance of silt flushing tunnel gates of underground Nathpa Jhakri Hydro Power Station (1500 MW), India

L.M. Verma, R. Kumar, A. Kumar & S.R. Sood
SJVN Ltd., Shimla, India

ABSTRACT: NJHPS has largest Desilting Complex, longest HRT & deepest Surge Shaft in the world. River Satluj carries heavy sediment load. To exclude sediments, Desilting Chambers have been provided. Each chamber is connected to SFT by 3 conduits. Free flow D-shaped SFT carries silted water back to river. 24 slide gates are provided for flushing out silt. Silted water strikes embedded parts of SFT Gates with high velocity and erodes them necessitating repair and maintenance. Same was earlier repaired by manual welding and grinding method. Due to space limitation, verticality & plainness of seal/guide faces were not achieved. This resulted in leakage from the gates and influenced generation. This paper describes techniques used earlier and reasons for their being unsuccessful. Further details of welding technique, in-situ machining of gate grooves, reduction in leakage, generation rise & financial gains is discussed. Specific emphasis is given to equipment and procedure of in-situ machining.

1 INTRODUCTION

Nathpa Jhakri Hydro Power Station (1500 MW) is run-of-river project located on the River Satluj in the foothills of the Himalayas is currently largest hydro power station in India. This project has many unique features like the largest underground Desilting Complex, the longest and largest Head Race Tunnel and Deepest Surge Shaft in the world. River Satluj carries heavy sediment load during high flow season. The annual suspended silt load carried by the river varies from 8 to 54 million m3. The concentration of suspended sediment in Satluj River during flood is high and goes beyond 5000 ppm. It contains about 17.48% of coarse, 24.99% medium and 57.53% fine sediment.

2 DESILTING ARRANGEMENT

To exclude sediment particles up to 0.2 mm size an Underground Desilting Arrangement comprising of four parallel Chambers each of size 15m wide, 27.5m deep and 525m long fed by four independent Intakes has been provided on the left bank of river Satluj. The overall efficiency of the arrangement as per the model studies is 37% and for the coarser sediment particles the efficiency of the arrangement is 90%. A discharge of 465 cumecs enters in chambers out of which 81 cumecs is meant to be utilized for flushing the settled sediments out of the chambers. The flushing arrangement consists of 3 No. silt flushing conduits running at the bottom of each desilting chamber, one each for carrying coarse, medium & fine particles. Flow area i.e. the width and the depth of the chamber has been fixed so as to arrive at a flow through velocity of about .30 m/s for removal of sediment coarser than .20mm. The length of the chamber 525m has been arrived at by the horizontal distance travelled by the particles within the time needed for the particles to settle down from the top layer of the flow to the bed of desilting basin.

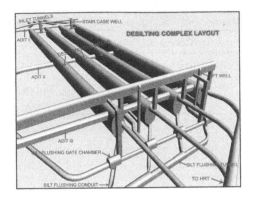

Figure 1. Desilting Chamber Layout.

3 SILT FLUSHING ARRANGEMENT

The silt settling at the bottom of hoppers is flushed by three flushing conduits running at the bottom of each chamber, one each for coarse, medium and fine particles. The flushing discharge can be controlled on the downstream of the chambers by installation of bonnet type silt flushing gate on individual conduits. The particle size in each of the fractions of coarse, medium & fine sediments varies between 60mm to 0.2mm, 0.2mm to 0.075mm and 0.075mm to 0.001mm respectively. The silt flushing arrangement comprises of the following components:-

3.1 *Silt Flushing Gates*

Twenty-four nos. bonneted slide gates 700mm X 1800mm (w X h) are provided in the outlet conduits of the desilting chambers for flushing the silt. These gates are with stainless steel skin plate and fabricated from structural steel and comprises of skin plate, horizontal girders, vertical stiffeners and end vertical girders etc. The gate bottom lip has been manufactured from corrosion and abrasion resistant stainless steel grade FV 520 B. The gates are sliding type with metal seals and seats. Each gate is operated by a double acting hydraulic hoist vertically mounted on the bonnet. The operating speed is 0.5 m/min. The gates are arranged in pairs (i.e. service and guard) and are operated under fully unbalanced head conditions. The regulation is done through service gates and can be partially opened whereas guard gates are kept either fully opened or closed so that their wear and tear, is minimum. The design head above sill is 58.28 m. The silt flushing gates embedded parts consists of a body and bonnet assembly, upstream, middle and downstream liner assemblies. These assemblies are fabricated from structural steel and stainless steel (FV520B grade sealing faces). In the gate groove, side seal face and guides are more prone to abrasion due to high velocity of silt laden water.

3.2 *Steel Liner*

Steel liners 1800mm u/s of guard gate, 1950mm d/s of service gate, 2750mm middle portion manufactured from abrasion/corrosion resistant steel (grade FV 520 B) has been provided. In addition to this, steel liner from abrasion resistant steel Hardox 400 of thickness 36 mm has been provided downstream upto additional gate.

3.3 *Material of Steel Liner*

3.3.1 *FV 520 B*
FV520B is an improved Stainless steel. The characteristics of FV520B are its good corrosion/ abrasion resistance, weldability and capacity of being hardened by low temperature treatment.

Table 1. (Chemical Composition).

Description	Value
C max %	0.07
Si Max %	0.7
Mn Max %	1.0
Cu %	1.2 – 2
Cr %	13.2 – 14.7
Ni %	5.0 – 6.0
Mo %	1.2 – 2.0
Nb %	0.2 – 0.7

Table 2. (Mechanical Properties).

Description	Value
Hardness	HB 375-425
Proof stress	1030 N/mm^2
Tensile strength (Rm)	1270 N/mm^2
Elongation	10%

3.3.2 *HARDOX 400*

HARDOX 400 is abrasion resistant steel with a hardness of about 400 HB, intended for applications where demands are imposed on abrasion resistance in combination with impact and/or good cold bonding properties. HARDOX 400 possesses very good weldability.

Table 3. (Chemical Composition).

Description	Value (Plate Thickness (mm))	Value (Plate Thickness (mm))
	(4-50)	(50-130)
C max %	0.25	0.32
Si Max %	0.7	0.70
Mn Max %	1.6	1.60
P Max %	0.025	0.025
S Max %	0.010	0.010
Cr Max %	1.00	1.40
Ni Max %	0.70	1.2
Mo Max %	0.80	0.80
B Max %	0.004	0.00

Table 4. (Mechanical Properties).

Description	Value
Hardness	HB 370-430
Yield strength (Re)	1000 N/mm^2
Tensile strength (Rm)	1250 N/mm^2
Elongation	10 – 16 %

3.4 *Silt Flushing Tunnel*

Each desilting chamber is connected to the silt flushing tunnel (SFT) by three silt flushing conduits. A free flow silt flushing tunnel with D- shaped cross-section 5.88m x 7.44m in a bed slope of 1 in 430 having a length of 1580m upto the portal and thereafter a cut & cover section of 1100m length & tunnel of 370m length has been provided to carry the silted water of desilting chambers to Satluj river.

4 EXISTING MAINTENANCE PROCEDURES/ACTIVITIES INVOLVED FOR REPAIR

In high flow season silt laden water mainly coarse particles strike the embedded parts of Silt Flushing Gates with a velocity of 24 m/s which erodes the side seal/bearing pad, guide faces and downstream steel liner, necessitating considerable repair and maintenance of SFT gates, embedded parts and steel liner every year.

Complete repair/refurbishment of all gates leaf eroded due to silt is being carried out successfully by welding and grinding as well as by hard coating using HVOF (High Velocity Oxyfuel) technique for Tungsten Carbide coating ranging from 300-500 microns of bottom most portion of the gate upto 700 mm height. Eroded/damaged seals (Al-Bronze and Neoprene) are dismantled from the gates, carrying out repair of seal base and fixing of new seals.

Figure 2.   Eroded and Damaged Gates.

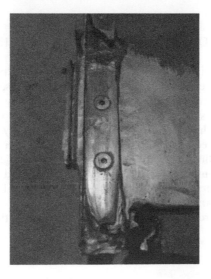

Figure-3.   Eroded and Damaged Al-Bronze seals.

Figure-4.   SFT gates after Hard coating.

Figure-5.   SFT gates after Hard coating.

5   DIFFICULTIES FACED DURING REPAIR/REFURBISHMENT WORK

- Due to space constraint, it is difficult to repair the eroded 2nd stage embedded parts of gate grooves. Every year erosion of gate groove S.S. plates (seal and guide faces) embedded in concrete takes place and the same is being repaired by manual welding and grinding method. But due to space limitation in the gate groove (maximum width is 75mm) and manual grinding method, required tolerances in verticality and plainness of seal/guide faces could not be achieved. This results into leakage of water from the SFT gates.
- During repair of eroded liners, it was observed that welding of Hardox/SS/MS plates is difficult due to humid and moist condition in gate conduits and in addition to this leakage from Guard gates hampers the welding of liner plates. Suitable liner plates, which do not require pre-heating and post-heating, can only be used.
- The welding of invert Hardox/SS/MS plates with the plates of the side liner, at the corner is a difficult job.

6 INITIAL MAINTENANCE PROPOSALS/METHODS TO REDUCE LEAKAGE

6.1 *Milling Machine*

There are 12 nos. of Service Gates provided in the Desilting Complex of NJHPS, Nathpa in order to flush out the silt. Each Gate groove has Sealing/Guide faces (SS GR 304) on the both sides and sill beam on the bottom (for details and drawings refer sketch below). The Aluminium Bronze metal seals/flats fitted on gates rest on these Sealing Faces and provide sealing. These Seal/Guide Faces and Sill beam gets damaged due to high velocity silted water and every year onsite repair is being carried out by welding & grinding method. However, by this method, true verticality i.e. Vertical plane cannot be achieved which results in water leakage. To make proper contact with the Aluminium Bronze metal seals with sealing face and for proper seating of gate on the sill beam to minimize the leakage of water machining is to be carried out on the Sealing faces and sill beam.

To achieve this milling machine is to be installed at the flange portion i.e. top of gate groove or at the bottom and milling head shall travel in the gate groove from top to bottom i.e. in a vertical height of 5176mm.

6.2 *Grout Holes*

Contact grouting should be done between the steel liner plates and the uneven concrete surface to fill in the gaps. For details of contact grouting Civil Design may be consulted.

6.3 *Material for the Liner*

Hardox 400 should be used as a standard material for steel liner throughout all the conduits due to its high hardness and good weldability. In case test material supplied by CPRI found more suitable, the same shall be used instead of Hardox 400.

6.4 *Repair of Guard Gates*

There is leakage from the guard gates, which creates lots of problems to carry out repair work downstream. Hence, it was suggested that guard gates may be repaired after depletion of desilting chambers one by one. This is necessary to create conditions to facilitate repair work d/s. All the above proposed methods of maintenance were tried but the results were not satisfactory. Therefore, leakage of water from the gates beyond permissible limits could not be stopped permanently and consequently influencing the power generation. To address the problem of not achieving the true verticality i.e. vertical plainness by the existing welding and manual grinding method, welding & in-situ machining of gate grooves was proposed.

7 WELDING & IN-SITU MACHINING OF GATE GROOVES – (CASE STUDY DESILTING CHAMBER NO.1)

The tender of welding & in-situ machining of gate grooves of 6 nos. SFT gates of Desilting Chamber No. 1 was been awarded to M/s Fleurdelis Technologies Pvt. Ltd., Bangalore, India. Desilting chamber No. 1 was depleted to carry out the following scope of work.

7.1 *Scope of Work*

Each gate has 6 nos. vertical and 2 nos. horizontal faces.

Total 6 nos. of SFT Gate Grooves, lintel seals and sill beams are required to be welded and machined as per specification given in Table 5 below (48 Nos. faces are required to be welded and machined).

Table 5. Scope of Work.

Sr. No.	Detailed work scope	Size (s) of the job	Total Qty.	Tolerance
Upstream guide face				
a)	Welding	140 mm x 600 mm + Eroded Patches	12 (2 per gate)	Surface finishing of Ra 3.2
b)	Machining	140 mm x 2000 mm		
Side sealing face				
a)	Welding	120 mm x 600 mm + Eroded Patches	12 (2 per gate)	Maintain 75±0.5mm distance between the Side Sealing face and
b)	Machining	120 mm x 2000 mm		Upstream Guide Face with Surface finishing of Ra 3.2
Side guide face				
a)	Welding	50 mm x 600 mm + Eroded Patches	12 (2 per gate)	Maintain 1030±1mm distance between the Side guide face with Sur-
b)	Machining	50 mm x 2000 mm		face finishing of Ra 3.2
Lintel Sealing face				
a)	Welding	200 mm x 830 mm	6	Maintain 326±1mm distance
b)	Machining	200 mm x 830 mm	(1 per gate)	between lintel sealing face and upstream liner with Surface finishing of Ra 3.2
Sill Beam				
a)	Welding	75 mm x 1030 mm	6	Maintain 5176mm distance between
b)	Machining	75 mm x 1030 mm	(1 per gate)	sill beam to top of the bonnet bush with Surface finishing of Ra 3.2

7.2 Working Area

The SFT is built in underground cavern (The total height of the Cavern is 8500mm from the ground Level.) in which the 5 Ton EOT crane is available at the height of the 6000 mm from the ground level.

For access to the SFT Gate Grooves there is 1no. of Main Hole (Dia 1500 mm approx.) in the downstream of the grooves.

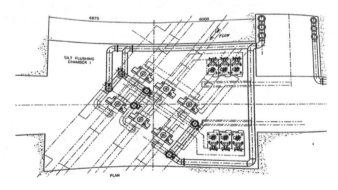

Figure-6. Layout of SFT Chamber No. 1.

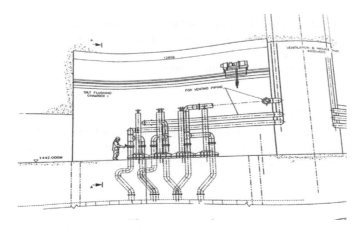

Figure-7.   Working Area in SFT Chamber No. 1.

7.3   *Requirements*

To carry out the above scope of work, following equipment were deployed at site:-

- On site slot mill machines
- On-site machining tools
- Measuring equipment
- Safety equipment
- Handling equipment

7.4   *Machining Procedure*

- Tight the draw bolts and mounts hydraulic motor.
- Inserts should be checked before giving cut.
- Start the machining on side guide face as per the required dimensions.
- Complete the pass and check the flatness and also measure the distance between 2 opposite faces.
- Check insert corners after each cut and change accordingly
- Follow the same procedure till the final cut
- Take the final reading and check with the required dimension.
- After side guide face machining start machining on upstream face by keeping shim plates according to the weld buildup.
- After each pass remove shim plates for giving feed, complete the machining till achieving 0.1mm variation between top face and machined face.
- After upstream face machining, start machining on side sealing face by keeping shim plates according to the weld buildup.
- After each pass add shim plate for giving feed, Complete the machining till we achieve 74.8 mm distance from two faces
- Set the Quill assembly vertically to do the bottom sill beam machining.
- Mount the 200mm Dia. Cutter, hydraulic motor and check the alignment.
- Complete the pass and check the dimensions between sill beam and top of the bonnet bush with distance meter.
- Complete the final cut till achieving distance 5174mm from the sill beam to top of the bonnet bush.
- After completing bottom face invert the Quill assembly vertically to mill top lintel sealing face.
- Mount the Shoulder and face mill Cutter, hydraulic motor and check the Quill alignment.

- By keeping shim plates according to the weld build up start the machining.
- After each pass add shim plate for giving feed, complete the machining till we achieve 325.8 mm distance from two faces.
- For blue matching reference take the final reading.

7.5 Benefits of In-Situ Machining

- Verticality of groove is maintained within tolerances so that leakages will be minimum.
- Sill beam is maintained within tolerances so that leakages will be minimum.
- The working of in-situ machining is fast and efficient as compared to manual.

7.6 Results

The performance of grooves of SFT gates after and before In-Situ machining of Chamber No. 1, is shown in Table 6 below:

Table 6. Performance before and after in-Situ machining.

Sr. No.	Chamber No. #1	Leakage before in-situ machining	Leakage after in-situ machining	*Allowable leakage as per IS:9349
		(Ltr/min)	(Ltr/min)	(Ltr/min/m/gate)
1	Conduit No#1 (Coarse)	102	48	57.50
2	Conduit No#2 (Fine)	130	54	57.50
3	Conduit No#3 (Medium)	104	24	57.50

* **Allowable Leakage= ((1925 X 2+950 X 2)/1000) x 10=57.50**(Ltr/min/m/gate)

8 CONCLUSION

As a result of in-situ machining, the leakage was achieved within the permissible limits as per the Indian Standard Codes. After this in-situ machining method was used in other chambers also which resulted in considerable reduction in leakage of water thereby contributing in increased power generation during the lean seasons.

Tunnels and Underground Cities: Engineering and Innovation meet Archaeology,
Architecture and Art, Volume 6: Innovation in underground engineering,
materials and equipment - Part 2 – Peila, Viggiani & Celestino (Eds)
© 2020 Taylor & Francis Group, London, ISBN 978-0-367-46871-2

Stability analysis for tunnel faces supported by means of polymer solutions

R. Verst & M. Pulsfort
University of Wuppertal, Wuppertal, Germany

ABSTRACT: Polymers in solution have been widely studied and applied by the oil and gas industry. An increasing number of projects where bentonite-based support fluids are replaced is also found in the construction of drilled shafts and in pipe jacking. Yet, so far, existing guidelines concerning stability calculations of slurry-supported earth walls cannot be applied directly. The usage of polymer solutions in geotechnical applications mainly relies on the expertise of polymer manufacturing companies. This paper discusses existing design approaches and presents adjustments within the framework of a calculative assessment of polymer support fluids for temporary tunnel face support formulated in terms of a critical stand-up time. The analytical assessment is supported by laboratory flow tests and rheometer tests with polyacrylamide copolymers. The experimental test results show good accordance with the analytical predictions. Possible improvements for existing formulations considering the influence of polymer properties on the time-dependent penetration are illustrated.

1 INTRODUCTION

Addressing possible collapse mechanisms using stability checks is crucial for understanding and assessing residual risks and the effectiveness of countermeasures in shield tunneling under fluid support where high groundwater pressure may be present (Anagnostou & Kovari 1994, ITA 2000). The functionality of bentonite-based slurries for face support, often complemented with air cushion regulation as well as lubrication, is well understood and implemented in preliminary design. The counterpressure necessary to support the excavation face is either transferred through formation of an external filter cake or by means of shear stresses along the infiltration length, depending on the yield strength of the slurry and the effective pore diameter in the grains. In addition to their capacity to build up a yield strength for face support, suitable bentonite slurries possess a flow viscosity low enough for pumping yet high enough to be capable to carry the excavated debris back to the starting shaft and reduce the risk of an uncontrolled loss of fluid.

However, the capacity of these slurries to load up and hold debris demands costly and space consuming separating plants to clean the slurry for discharge and reuse. Clogging and swelling effects make them less suitable for silty or clayey ground (Anagnostou 2007). Here, polymer solutions offer a promising alternative (Borghi 2006, Verst & Pulsfort 2017).

Within the field of tunneling a fivefold reduction of the jacking forces in pipe jacking is reported by Alexanderson (2001), when the lubricating fluid is changed from bentonite slurry to a polymer solution in clayey ground.

The oil and gas industry introduced highly viscous polymer solutions in the 1960s and have widely studied and applied the usage for polymer flooding in enhanced oil recovery and for more environmentally friendly water-based drilling, where problems with swelling soil and rock are often encountered (Sorbie 1991, Anderson et al. 2010, Asef & Farrokhrouz 2013).

The piling industry has registered an increasing number of projects with polymer fluid support in the past few years, especially for operational and economic advantages. Due to their lack of yield strength, there is typically no need for a separation plant and disposal costs are low as polymer solutions may be easily recycled and often allowed to be discharged into regular sewer systems (Lescmann et al. 2015, Lam & Jefferis 2018).

The negligible yield strength of polymer solutions however prevents a direct application of existing guidelines to analytically predict the temporary excavation stability for safety or parameter sensitivity calculations. Also, polymer solutions may behave quite differently in the porous structure of the penetrated ground. Polymer molecules, meaning large macromolecules, are much smaller in size compared to bentonite lamellas and much more diverse in structure, composition and thus flow behavior in solution compared to the specifically chosen clean clay mineral. Consequently, until now, experience from trial excavations and the expertise of polymer manufacturing companies form the basis of preliminary planning.

First design approaches exist for non-cohesive soils as given by Steinhoff (1993) and Lesemann et al. (2016) for drilled piles and diaphragm walls in soils. They predict the penetration behavior of polymer solutions in terms of the excavation time as the polymer fluid will not fully stagnate due to the negligible yield strength, yet high viscosity at low shear levels. Stability checks are therefore formulated time-dependently. Based on the failure mechanisms of the German design standard DIN 4126 proven for more than 30 years, simple capillary bundle models to represent the void system of the soil and ideal power-law behavior to reflect polymer rheology, these approaches give a first estimation of the penetration behavior of the supporting fluid.

The general penetration behavior at the tunnel face is essentially the same as for diaphragm walls if the successive horizontal soil excavation is neglected for a first estimation. Similarly as for bentonite-based fluid support (Anagnostou & Kovari 1994), stability checks for temporarily fluid-supported earth walls may therefore be adapted to assess face failure by adjusting the underlying earth pressure. The adaption of this concept for tunnels along with an implementation into a MATLAB-based graphical user interface for computational assessment will be presented hereafter.

However, practice proves this first estimation to overestimate fluid losses into the ground. It is assumed that the existing models may not adequately represent the so-called in-situ rheology depending on an interaction between polymer characteristics and the porous structure. Filtration or membrane effects and a reduction in permeability are assumed to favor the flow process. In addition, polymer chains of high molecular weight penetrated into the ground are assumed to increase cohesion and thus reduce earth pressure of granular soils (Goodhue/Holmes 1997, Lesemann & Schwab 2015).

Hence, 1D laboratory flow tests and rheometer tests will be added aiming and an empirical correction and possible simplification of the calculative framework.

2 CALCULATIVE ASSESSMENT

The analytical calculations are based on the three failure mechanisms of DIN 4126 (2013) taking under consideration the partial safety factors according to DIN EN 1997-1 (2014). They were established for bentonite suspensions, but the general failure concept remains the same for polymer solutions if formulated in its general form (Steinhoff 1993, Lesemann et al. 2016).

The first mechanism (Eq. 1) describes failure due to groundwater inflow. Stability is guaranteed if the stabilizing pressure $p_{F,stb}$ exceeds the destabilizing hydrostatic groundwater pressure $p_{w,dst}$ and can be formulated independently of time. An adaption for tunneling requires $p_{F,stb}$ to be defined in terms of the hydrostatic pressure of the support fluid $p_{fluid,stb}$ in addition to the pressure generated by air compression $p_{air,stb}$.

$$p_{w,dst} \leq p_{F,stb} = p_{fluid,stb} + p_{air,stb} \qquad (1)$$

The second failure mechanism (Eq. 2) considers the successive removal of single grains or small groups of grains from the earth wall as a local form of failure, the so-called internal stability as explained in Haugwitz & Pulsfort (2018).

$$f_{s0} \geq f_{s0,req} \leftrightarrow \eta_F \frac{\Delta p}{s(T)} \geq \frac{\gamma'' d}{\tan \phi'_d} \qquad (2)$$

This failure is assumed to occur when the gravitational force of the grains represented by their buoyant specific weight γ'' exceeds the stabilizing frictional force induced by the actual pressure gradient at the earth wall $f_{s0,DIN} = \Delta p/s$ multiplied by $\tan \varphi'$. Δp represents the pressure difference between p_F and p_w, φ' is the friction angle and η_F an optional correction factor. The approach therefore presumes pressure to drop linearly along the penetration length s for a given excavation depth z and time interval T by analogy with water.

Several researchers have discussed models to calculate the penetration length of polymer solutions. One approach derived from Steinhoff (1993) for the construction of diaphragm walls and drilled piles defines the penetration length $s(T) = l(z_i, T)$ as follows (Eq. 3).

$$s(T) = \left[4 \cdot \left(\frac{m+1}{3m+1} \cdot T \right)^m \cdot \frac{K \cdot \Delta p}{\kappa \cdot n} \cdot \left(\frac{8K}{n} \right)^{\frac{m-1}{2}} \right]^{\frac{1}{m+1}} \qquad (3)$$

Sorbie (1991) summarizes similar approaches with focus on the oil and gas industry. These models essentially represent laminar unidirectional flow within the porous structure by a bundle of horizontally ordered capillaries of uniform radius. Porosity n and permeability K represent the soil. κ and m (<1) describe the flow behavior of polymer solutions. They are material constants of the rheological power law model from Ostwald & de Waele (Eq. 4) with which the nonlinear relationship between shear stress τ and shear rate $\dot{\gamma}$ from rotational rheometer flow tests is accurately approximated. So equality of flow behavior between a capillary tube and the geometry of a rotational rheometer or viscosimeter is assumed. The power law model reflects the usefulness of polymer solutions as support fluids as it shows low viscosity η for pumping at higher velocities and a significant increase in viscosity when velocity (e.g. shear rates) decreases with progressing penetration of the fluid into the ground due to a decrease in pressure gradient. So penetration velocity will slow down significantly, as, according to power law, viscosity yields infinity, but penetration will never stop due to the lack of yield strength. Other models such as the Ellis model assume an upper limit to viscosity within their rheological formula. For reasons of security, Steinhoff (1993) therefore adds a limit to viscosity at $\eta_{max} = (\dot{\gamma} = 1.5)$, e.g. he assumes Newtonian flow behavior below $\dot{\gamma} = 1.5$ 1/s.

$$\eta(\dot{\gamma}) = \frac{\tau(\dot{\gamma})}{\dot{\gamma}} = \kappa \cdot \dot{\gamma}^{m-1} \qquad (4)$$

Considering the time-dependency of the penetration length for polymer solutions, it is evident that local stability is guaranteed for a longer period of time for a more viscous fluid with slower penetration. In addition, taking into account time-dependency may also include the successive horizontal soil excavation at the tunnel face resulting in a repeated reduction of the effective penetration length. This increase in safety is neglected so far. In addition, an increase in the local pressure gradient at the earth wall may be induced by a local reduction of the permeability near the earth wall due to filtration or membrane effects, as assumed by Goodhue et al (1997) or Lesemann et al. (2015). This effect could allow the local stability check to be formulated independently of the penetration length by introducing a fixed ratio $f_{s0,wall} = \Delta p_{wall}/s_{wall}$ as the relevant pressure gradient directly at the fluid-supported earth wall.

The third failure mechanism according to DIN 4126 (2013) (Eq. 5) describes the shear failure of a monolithic soil body, where the destabilizing horizontal earth pressure E_{ah} opposes the stabilizing pressure S_k.

$$\max\left(\frac{E_{ah,d}(\vartheta, T)}{S_{k,d}(\vartheta, T)}\right) \leq 1 \qquad (5)$$

This so-called external stability approach considers a prismatic failure body for the calculation of E_{ah} (Figure 1). Shear forces T integrated over the triangular edges account for the influence of friction and cohesion on the pseudo-spatial problem. The stabilizing force S_k must be equal to the flow forces transmitted to the granular structure and integrated along the effective penetration length within the respective failure body (Lesemann et al. 2016) (Eq. 6). As the penetration L of the support fluid depends upon time influencing S_k as well as the specific weight of the soil body above groundwater level, stability is a function of time, too. The maximal stable time T_{crit} is reached, when the incremental raise of L reaches $E_{ah}/S_k = 1$ for the respective critical slip surface angle ϑ_{crit}.

$$S_k(\vartheta, T) = \int f_{s0}(z, T)dA = \int \frac{\Delta p(z)}{l(z, T)}dA = \int_0^{z_l(T)} \int_0^{l(z,T)} \frac{\Delta p(z)}{l(z, T)}dxdz + \int_{z_l(T)}^{z_{max}} \int_0^{b(z,\vartheta)} \frac{\Delta p(z)}{l(z, T)}dxdz$$

$$= \int_0^{z_l(T)} \Delta p(z)dz + \int_{z_l(T)}^{z_{max}} \Delta p(z)\frac{b(z, \vartheta)}{l(z, T)}dz \qquad (6)$$

Anagnostou & Kovari (1994) and other researchers have adapted this failure approach from DIN 4126 to the tunnel face according to Horn (1961) by adding an overburden pressure p according to silo theory by Janssen (1895) to the prismatic failure body with a face area A equal to the square of the outside tunnel diameter D (Figure 2).

This ensures equality of hydrostatic pressure between square and circular failure face. The present approach however adjusts the equivalent width $l_s = d$ so that the tunnel face is equal in area to the now rectangular face (Eq. 7).

$$A = \frac{D^2}{4} \cdot \pi \rightarrow d = \frac{A}{D} = \frac{D}{4} \cdot \pi \qquad (7)$$

Global face stability can be formulated according to Equation 3, thereby assuming penetration of the support fluid into the ground from the rectangular failure face only. The successive advancement of the tunnel machine may again be seen as an additional safety factor

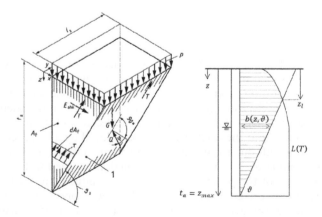

Figure 1. Failure body from DIN 4126 (2013) and visualization of corresponding effective penetration length.

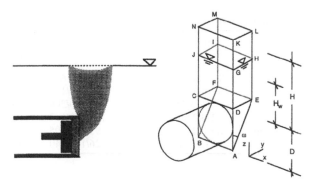

Figure 2. Visualization of face failure mechanism and corresponding failure body for tunnel face according to Horn (1961) from Anagnostou & Kovari (1994).

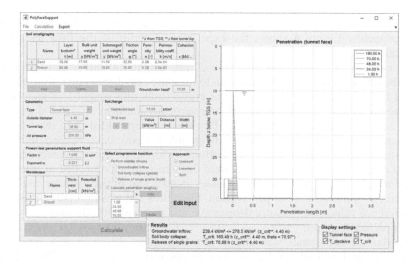

Figure 3. GUI for an estimation of the temporary tunnel face stability.

increasing f_{s0} and therefore S_k. Under the assumption that the slurry shield method is usually applied below groundwater level, fluid penetration does not influence the specific weight of the supported ground, as the small amounts of polymer of around 0.7 to 3 g/l used in solution are not able to significantly deviate the bulk specific weight.

Summarizing the above, it is possible to adapt stability checks for bentonite slurries to polymer fluid support to assess temporary tunnel face stability. However, due to the absence of a yield strength, local and global stability need to be reformulated to give a critical stand-up time T_{crit}.

On this basis, the aforementioned equations were implemented into a MATLAB graphical user interface (GUI) to perform exemplary stability checks (Figure 3) and sensitivity tests (Table 1) of selected variables. The GUI is based on a program for diaphragm walls and drilled piles developed at the University of Wuppertal and was adapted to perform stability checks for tunnel faces. The sensitivity tests are based on a tunnel with an outside diameter of 4.4 m, constructed in sand 30 m below ground surface with the groundwater head 20 m above the tunnel ridge. The fluid support was chosen as a polymer solution of high viscosity ($\kappa = 1.65$, $m = 0.22$) and complemented with air cushion regulation of 2.5 bar. Six parameters were varied to quantify their effect on the critical stand-up time according to both the external (global) and the internal (local) stability check.

The results given in Table 1 show high sensitivity for the permeability coefficient k_w and for the value of $\dot{\gamma}$ to account for the upper limit of viscosity. This outcome reflects the importance of experimental tests as the dependence of the properties of the long-chain polymer molecules on these parameters is unclear.

Table 1. Sensitivity tests.

Parameter		Variation of values	$T_{crit,global}$ h	$T_{crit,local}$ h
k_w	m/s	8e-3/8e-4/8e-5	14/160/1625	5/70/729
n	-	0.3/0.36/0.4	133/160/178	58/70/79
φ	°	30/32.5/35	122/160/206	57/70/86
m	-	0.1/0.22/0.7	151/160/197	66/70/88
κ	N s/m²	1/1.6/2	95/160/195	40/70/87
$\dot{\gamma}_{crit}$ *	1/s	15/1.5/0.15	26/160/940	12/70/401

* for η_{max}

3 EXPERIMENTAL SETUP AND DISCUSSION

Laboratory experiments were performed to validate the basis of the previously stated stability approaches. Figure 4 shows the experimental setup of the flow tests performed with a typical sand.

A vertical cylinder of acrylic glass (diameter 20 cm, height 100 cm) with 12 pressure transducers distributed along the height was chosen to measure one-dimensional flow as is assumed at the tunnel face. A perforated plate and filter gravel (bulk unit weight 1.78 g/cm³) were used to enclose the tested porous medium and ensure vertically uniform inflow and outflow at the bottom and top of the cylinder respectively. Dry sand was sprinkled onto the filter gravel to obtain high compaction (bulk unit weight 1.67 g/cm³, porosity 0.37, k_w 2e-3 m/s). The sand was slowly saturated with tap water from bottom to top until water connection was closed from inlet to outlet. Pressure level and pressure difference were controlled at inlet and outlet by means of air pressure regulators. Floor scales at inlet and outlet cylinder enabled control and measurement of the velocity of penetration. Each flow test was first performed with water to obtain reference values for pressure drop curve and permeability. Subsequently, water was replaced by polymer solution.

A polyacrylamide copolymer was chosen for testing, as this type is a typical polymer applied in drilling and as it allows the variation of polymer characteristics. Property variation comprised the chain length and the amount of polymer in solution. The respective properties are given in Table 2. The granular polymer material was slowly sprinkled into 5 l of colored deionized water at pH 7 according to a fixed size distribution of the polymer granules using a

Figure 4. Experimental setup of one-dimensional flow tests.

Table 2. Parameters of tested polymer solutions.

Test name	Concentration g/l	Polymer chain length -	Marsh time* s	κ^{**} N s/m	m^{**} -
S150	1.50	Short	58	0.36	0.57
L075	0.75	Long	107	0.72	0.28
L150	1.50	Long	144	1.38	0.30

* at 1000ml
** Power law parameter

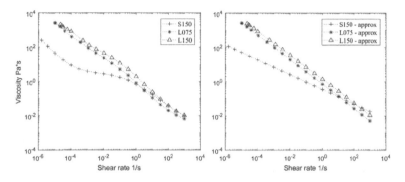

Figure 5. Rheometer test results and power law approximation.

magnetic and a mechanical stirrer at 500 rpm. The polymer fluid mixture was stirred for 60 min. Food coloring powder brilliant blue of 0.04 g/l was chosen to visualize penetration of the polymer fluid. Comparison of rheometer flow curves of the colored and uncolored fluid served as validation for equality of rheological performance.

The power-law parameters of the tested polymer solutions were approximated from flow curves obtained with rotational rheometer tests (Figure 5). As the penetration velocity in sand decreases gradually under a constant fluid pressure Δp, the measurement of shear stress was adapted to a stepwise decrease in shear rate from 1000 to 1e-6 1/s. Figure 5 (left) shows that the curves of the polymers with equal chain length are approximately parallel. The curves with equal concentration but different chain lengths display similar viscosity values at high shear rates. A Newtonian plateau at low shear rates is only visible for the shorter polymer at a shear rate of approximately 1e-1 1/s.

The results obtained from flow tests with sand are displayed in Figure 6 to Figure 11. Figure 6 shows the penetration curves over time calculated from scaling. Penetration velocity decreases with increasing molecular chain length (S150 to L150) and especially with increasing concentration of polymer in solution (L075 to L150). Approximations according to the approach from Steinhoff (1993) are given in dashed and dotted lines for two different critical shear rates from which point on Newtonian flow behavior is assumed. It can be seen that all curves may be approximated with acceptable accuracy, with highest conformity for the polymer solution with the lowest viscosity values (S150). The penetration length of the polymer solution with the highest viscosity values is greatly overestimated by the classical approach. The reason may be found in an overestimation of the critical shear rate as the approximation using a lower value for the critical shear rate as suggested by Steinhoff (1993) shows significantly better conformity.

Figure 7 to Figure 10 display normalized pressure gradient curves for different time steps (Figure 7 to Figure 9) and pressure drop levels (Figure 10). The pressure drop displayed as Δp^* thereby comprises the hydrostatic pressure Δp_{hyd} within the cylinder resulting from vertical flow displayed as a dotted line. $\Delta p = \Delta p^* - \Delta p_{hyd}$ therefore gives the decisive pressure difference causing the flow. The penetration lengths corresponding to the time steps given in hh: mm. are visualized as dashed lines. All flow tests show nonlinear pressure gradients along the

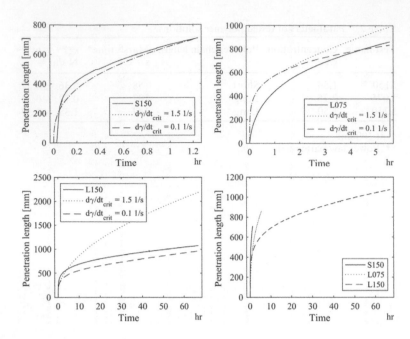

Figure 6. Penetration curves (-) and approximation.

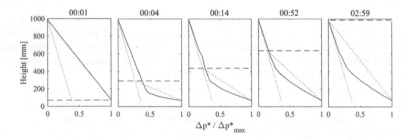

Figure 7. S150: normalized pressure gradient (-) for different penetration lengths (–) at Δp = 0.15 bar.

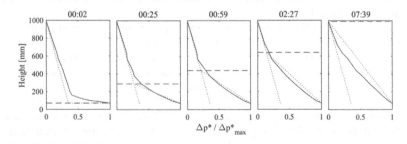

Figure 8. L075: normalized pressure gradient (-) for different penetration lengths (–) at Δp = 0.15 bar.

penetration. The respective curvatures remain the same for different pressure drop levels as can be seen in Figure 10. The nonlinearity may be a sign for filtration effects with a reduction in permeability, but without development of a membrane near the earth wall. This effect is thought to be a result of the type of chosen polymer. Typical copolymers of polyacrylamide possess thin and long chains. Bridged types of polymers with high viscosity in solution might perform differently.

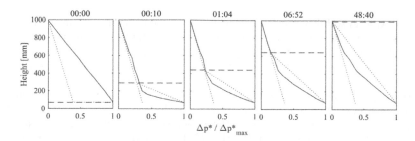

Figure 9.   L150: normalized pressure gradient (-) for different penetration lengths (–) at $\Delta p = 0.15$ bar.

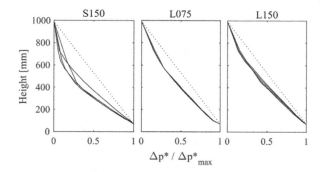

Figure 10.   Summary of normalized pressure gradients for pressure levels 0.15 bar, 0.3 bar, 0.5 bar and 0.7 bar at full penetration of the cylinder.

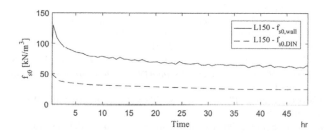

Figure 11.   L150: Comparison of pressure gradients $f_{s0,DIN}$ (–) and $f_{s0,wall}$ (-).

The increased pressure drop along the first 2 cm of penetration results in a significant increase of the respective pressure gradient $f_{s0,wall}$ compared to the value $f_{s0,DIN}$, where pressure is assumed to drop linearly along the penetration length. Figure 11 shows an exemplary comparison of both values for L150. An increase by a factor of $63.2/24.8 = 2.5$ can be observed as a minimum value even at full penetration of the cylinder (48h). Internal stability for a time slot of up to 48h may in this case be formulated on the safe side independently of the penetration length on the basis of $f_{s0,wall} = 63.2$ kN/m³.

4   CONCLUSION

Polymers in solution are used increasingly in geotechnical environments as they bring several advantages over bentonite slurries. Approaches have been stated which allow analytical stability checks essential to assess residual risks for tunnel faces under polymer fluid support to be formulated within the existing sets of rules and failure concepts similarly to bentonite slurries.

Internal and external stability checks, however, need to be phrased in terms of a critical stand-up time, as polymer solutions possess only a negligibly small yield strength.

Laboratory experiments show good agreement with the analytical approaches, especially for polymer solutions with lower viscosity values, due to a safe-side choice of the critical shear rate.

Moreover, experiments reveal a nonlinear course of the pressure gradient f_{s0} along the penetration length with an increased pressure drop $f_{s0,wall}$ along the first 2 cm of penetration compared to $f_{s0,DIN}$. These results show that the underestimation of the critical stand-up time according to the approach based on DIN 4126 (2013) may be reduced if the actual curvature of the pressure drop and the critical shear rate of the respective polymer solution with regard to the specific ground are taken into account.

Further research is therefore necessary to explain the interaction between different types of polymers and soils, with special focus on polymer structure and elasticity within the porous system given by the grain size distribution and density of the porous medium to assess filtration, permeability and cohesion effects and their impact on both $f_{s0,wall}$ and the time-dependent penetration.

REFERENCES

Anderson, R. L., Ratcliffe, I., Greenwell, H. C., Williams, P. A., Cliffe, S., & Coveney, P. V. 2010. Clay swelling—a challenge in the oilfield. *Earth-Science Reviews*, *98*(3–4), 201–216.

Alexanderson, M. 2001. Construction of a cable tunnel ID 2,6 Vibhavadi. *Proceedings of the 5th international symposium on microtunnelling*, 139–147. Bauma, Munich.

Anagnostou, G. 2007. Zur Problematik des Tunnelbaus in quellfähigem Gebirge. In Schweizerische Gesellschaft für Boden-und Felsmechanik (ed.), *Frühjahrstagung über "Quellprobleme in der Geotechnik"*, 1–10. Fribourg.

Anagnostou, G., Kovarí, K. 1994. The face stability of slurry-shield-driven tunnels. *Tunnelling and underground space technology*. 9(2),164–174.

Asef, M., Farrokhrouz, M. 2013. *Shale engineering: Mechanics and mechanisms*. CRC Press.

Borghi, F. 2006. Soil conditioning for pipe jacking and tunneling. University of Cambridge.

DIN 4126 2013. *Stability analysis of diaphragm walls*. Beuth Verlag.

DIN EN 1997-1 2014. *Eurocode 7: geotechnical design - part 1: general rules*. Beuth Verlag.

Goodhue, K., Holmes, M. 1997. *Polymeric earth support fluid compositions and method for their use*. U.S. Patent 5,663,123. Spring, Texas.

Haugwitz, H. G., Pulsfort, M. 2018. Pfahlwände, Schlitzwände, Dichtwände. Witt, K.-J. (ed), *Grundbau-Taschenbuch* 8 (3). Ernst & Sohn.

ITA WG Mechanized Tunnelling 2000. Recommendations and guidelines for tunnel boring machines (TBMs). 1–118. ITA - AITES, www.ita-aites.org.

Janssen, H. (1895). Versuche über Getreidedruck in Silozellen. *Zeitschrift des Vereins Deutscher Ingenieure*. 39(35),1045–1049.

Lesemann, H., Vogt, N., Pulsfort, M. 2016. Analytical stability checks for diaphragm wall trenches and boreholes supported by polymer solutions. *Proceedings of 13th Baltic Sea geotechnical conference*, 231–238. Lithuania.

Lesemann, H., Schwab, G. 2015. Vorträge zum 22. Darmstädter Geotechnik-Kolloquium am 12. März 2015. *Mitteilungen des Institutes und der Versuchsanstalt für Geotechnik der Technischen Hochschule Darmstadt* (94), 169–178.

Lam, C., Jefferis, S. 2018. *Polymer support fluids in civil engineering*. London: ICE Publishing.

Sorbie, K 1991. *Polymer-improved oil recovery*. 1991. Blackie and Son Ltd, Glasgow and London.

Steinhoff, J. 1993. *Standsicherheitsbetrachtungen für polymergestützte Erdwände*. Forschungs- und Arbeitsberichte aus den Bereichen Grundbau, Bodenmechanik und Unterirdisches Bauen an der Bergischen Universität - GH Wuppertal (13). Wuppertal.

Verst, R., Pulsfort, M. 2017. Reduction of soil swelling effects in pipe jacking and shield tunnelling using polymer solutions as support fluid. In EURO:TUN 2017, *proceedings of the IV international conference on computational methods in tunneling and subsurface engineering*. Innsbruck: Studia Universitätsverlag.

Tunnels and Underground Cities: Engineering and Innovation meet Archaeology,
Architecture and Art, Volume 6: Innovation in underground engineering,
materials and equipment - Part 2 – Peila, Viggiani & Celestino (Eds)
© 2020 Taylor & Francis Group, London, ISBN 978-0-367-46871-2

The development and in-situ use of fiber optic continuous strain monitoring for tunnel support elements

N. Vlachopoulos
Royal Military College of Canada, Kingston, Ontario, Canada

B. Forbes
Queen's University, Kingston, Ontario, Canada

ABSTRACT: An observational approach to tunnel design and construction is commonly employed in order to assess displacements as a result of the excavation process. Temporary tunnel support design decisions are based on such displacement data. In weak ground conditions, a correct and accurate evaluation of such measurements is critical to the safety, sustainability, optimization and to the economics of the construction process. Within this context, the development and application of ground support (i.e. forepoles, spiles, rock bolts etc.) outfitted with fiber optic technology is presented. This state-of-the-art optical technique monitors strain with a spatial resolution of 0.65 mm; allowing a continuous strain profile along the length of the ground support specimen to be determined. It has been shown that this novel technology allows the design engineer to respond and make relevant adjustments to the support regime and excavation process in response to current and potentially future ground conditions.

1 INTRODUCTION

1.1 *The use of fiber optic strain monitoring for Tunnelling Projects*

A rising demand for underground transportation and resource management has led to the development of many more subterranean projects (deep foundations, tunnels, utility corridors etc.) which are constructed at larger scales, over greater distances, increased depths, and within proximity to sensitive urban environments (i.e. reduced tolerances with respect to adjacent infrastructure). For such projects, engineering design of support is primarily based on the stress and strain that are developing within the support structures as a result of the surrounding ground conditions. These ground loads are distributed continuously and spatially and as such, an improved understanding of the continuous strain profile would provide better insight into the true behaviour of such support elements. Research currently being conducted at the Royal Military College of Canada and Queen's University focuses on such micro-scale geomechanical mechanisms and interactions with a view to determining the overall design implications for full-scale support design for tunnels. The use of fiber optics within the Geotechnical/Geological Engineering field is not a new concept. There are multiple projects that have utilized a particular type of fiber optic technology in the past, ranging from their use to monitor the construction and performance of embankments, tunnels, piles, mining operations and other geotechnical works. It is important to note that not all fiber optic technologies are similar as each type has their unique strengths and limitations. Historically, monitoring of such ground support members has been limited to electrical and mechanical techniques (e.g. foil-resistive strain gauges, inclinometers, linear variable displacement transducers). Such techniques provide discrete measurement points, implying that many sensors are required to obtain a full strain profile along the length of the support element (Vlachopoulos and Forbes, 2018).

These techniques provide a limited spatial resolution along the element, making such methods prone to misinterpretation, underestimation, and possibly omission of support response. For example, it not uncommon to observe a 'failed' rock bolt that has been subjected to both axial loads as well as bending (i.e. transverse loading(s)) (Figure 1a). As depicted in Figure 1b, a coarse monitoring arrangement could potentially miss such local mechanisms completely.

1.2   *The application of Continuous Strain Monitoring for Tunnel Support*

In the context of tunnelling and ground-support interaction, an abundance of investigation has been performed on such support elements that constitute and overall support regime of a tunnel (Figure 2) separately, in isolation; however, the combined effect of such assortments has not been investigated fully. As well, when support elements have been individually addressed in the past, the spatial resolution of monitoring has been quite coarse, limiting the understanding of related mechanisms involved. There is then an opportunity to optimize tunnel performance and support design within the framework of a state-of-the-art observational (i.e. well instrumented) approach whereby the behaviour of ground/soil, support elements, and geomaterial-support interactions can be explicitly determined. Through back-calculations from both laboratory and *in situ* measurements, fundamental behaviour and material properties can be derived for input into empirical and numerical simulations (Forbes et al., 2015).

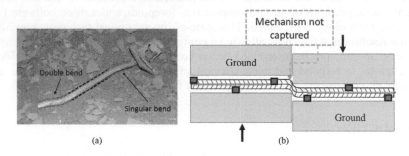

Figure 1.   (a) Permanently deformed rebar element after a rock burst event (Courtesy of Brad Simser). Lateral deformation of the rebar is clearly evident. (b) Sensors (red locations) are too sparsely arranged in order to capture local phenomenon.

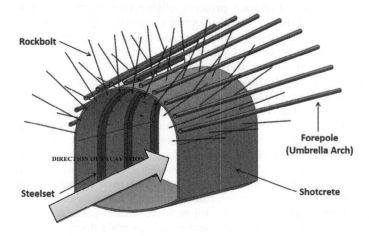

Figure 2.   An example of selected support elements shown are: forepoles, steel sets, rock bolts, and shotcrete.

2 OPTICAL MONITORING TECHNIQUE

An optical frequency domain reflectometry (OFDR) capable of monitoring strain with a spatial resolution of 0.65 mm along the length of a standard, low cost optical fiber has been the focus of the research herein. The operational accuracy of the strain monitoring technique (utilizing a Luna analyzer) that was developed is quite acceptable (better than +/- 10 microstrain). The optical technology monitors spectral shifts in the local Rayleigh back-scattering arising from alterations in the index of refraction of the optical fiber, which is inherently sensitive to strain. In this manner, an initial measurement can be recorded in order to reference all succeeding measurements and therefore determine changes in strain (Duncan, et al. 2007). In essence, thousands of individual, discrete transducers can there-fore be replaced by a single optical fiber (125μm in diameter), acting as both the transducer and lead. Furthermore, the optical fiber will not degrade (i.e. to a significant extent) or require recalibration over time as does its electrical counterpart. Such qualities make the optical technique ideal for monitoring longitudinal support elements, which would not only capture the performance of individual support members, but also provide unparalleled insight into the ground and ground-support behaviour ahead of the excavation face. In this manner the optical technique can be realized as a supplemental tool to current observa-tional monitoring techniques, which are primarily limited to measurements within the exca-vation (and monitoring at surface). As such, results from selected laboratory testing using the OFDR instrumentation are highlighted in the following sections of this paper (Forbes et al., 2015a).

3 LABORATORY TESTING USING FIBER OPTIC STRAIN MONITORING TECHNIQUE

To date, many configurations of testing that include axial, bending, and shear testing have been conducted utilizing multiple support elements. These support elements were tested as unique specimens as well as grouted within concrete (rock) samples. The support tested in the laboratory to date includes: Rebar (rock bolts), D-bolts, Cable bolts, Spiles and, Forepoles. Each sample preparation has its own unique challenges in terms of fitting the fiber optics in conjunction with a particular support element.

3.1 *Axial Pullout Tests and Two-way Shear Tests Involving Rebar (Rockbolts)*

Within selected tests, No. 6 Grade 60 rebars were prepared by instrumenting them with a fiber optic. Steel bars were modified with 2.5 mm by 2.5 mm diametrically opposing grooves as shown in Figure 3 (Vlachopoulos and Forbes, 2018).

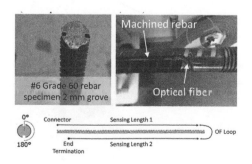

Figure 3. Photos and schematic depicting the grooves that were created during specimen preparation and the outfitting of the optical fiber. The optical fiber was looped around one end of the rebar specimen providing continuous strain monitoring along two sides (Sensing Length 1 & 2) of the sample.

In Figure 4 below, one can see selected results from the laboratory testing that has been conducted as part of this line of research. Figure 4a depicts results from an axial pullout test while Figure 4b depicts results from a 2-way shear test. More in-depth result from pullout tests involving the optimization of pullout length can be found in Vlachopoulos et al., 2018.

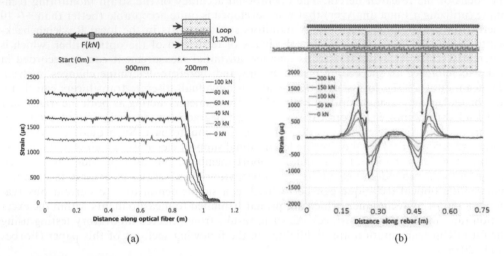

(a)　　　　　　　　　　　　　　(b)

Figure 4.　Selected results from a. pullout testing and b. 2-way shear configuration.

3.2 *Bending Testing of Forepole Elements*

Forepole support members (refer to Figure 2) act bi-directionally in both the radial and longitudinal direction. The radial support component will initiate in response to the state of the excavation sequence, which will transition from a cantilever reaction to a fixed, simple supported reaction. In the latter case, the radial load will be transferred to ground ahead of the excavation face and to the lining/steel-set support within the excavated region. As such, the ground properties and strength and stiffness of the lining/steel-set will play a major role in the supporting contribution of the forepole elements (Volkmann, & Schubert, 2007). This contribution to radial support was simplified to a symmetric-three-point bending apparatus to assess the merits of the OFDR technique (Figure 5). Two steel pipe sizes were tested as the forepole member: a. 114 mm outer diameter (OD) with 6 mm wall thickness and b. 21.3 mm OD with 2.8 mm wall thickness. The former is a commonly used forepole size *in situ*. The latter was chosen as it has a similar second moment area to rebar which was tested under three-point bending with the optical technique by (Hyett, A.J. et al., 2013).

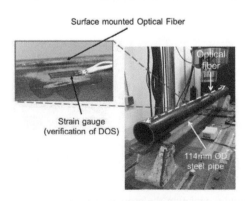

Figure 5.　Symmetric-three-point bending apparatus. Optical fiber runs along the top profile of the specimen.

The optical fiber was mounted to the steel members using a metal bonding adhesive in three separate methods: a) surface mounting, b) embedded and encapsulated in a 4 mm wide by 2 mm thick machined groove, and c) mounting to the inner profile of the pipe (only applicable to 114 mm OD). More details can be found in Forbes et al., 2015b. Incremental load was applied to the steel members using a platen piece as to ensure the load was not being directly transferred to the optical fiber, which was aligned along the top (i.e. compressive) profile of the specimens. Results depicting the strain profile along the supported length are displayed in Figure 6. The optical technique was capable of accurately and repeatedly capturing the strain at various levels of applied load.

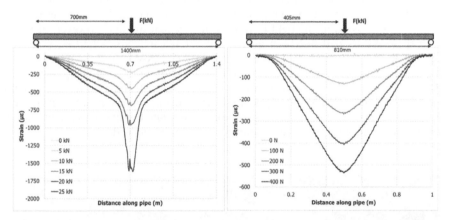

Figure 6.   Strain profiles captured at a spatial resolution of 1.25mm along the top (compressive) section of a steel pipe at various simple-supported loads. Left: 114mm OD, 6mm wall thickness, and support spacing of 1400mm, Right: 21.3mm OD, 2.8mm wall thickness, and support spacing of 810mm.

The majority of the laboratory testing that was conducted as part of this research was conducted within the Civil Engineering laboratory at the Royal Military College of Canada. A set-up of a rebar pull-out test utilizing the fiber optic technique is shown in Figure 7 below.

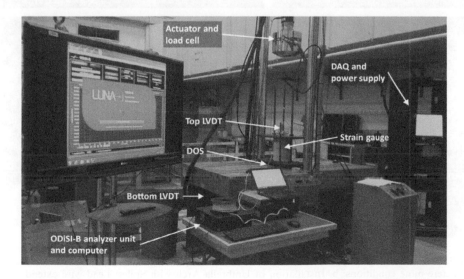

Figure 7.   Laboratory testing set-up utilizing fiber optic analyser, data acquisition (DAQ) and actuator.

4 FIELD TRIALS

4.1 *General – Proof of Concept in the Field*

As with any technology of this nature, it is encouraging to obtain excellent results within the controlled environment of the laboratory. The question now becomes how this technology can be employed in the harsh conditions associated with the field while limiting its impact on operations. To date, multiple successful field experiments have been conducted at 3 separate locations around the world. The authors are also in contact with other interested global parties who have shown an interest in employing such a technique within their operations (Vlachopoulos and Forbes, 2018). Below (Figure 8) are relevant photos from the in-situ installation of the fiber optic technology within support elements that were designed by the authors. The data amassed in the field to date is of excellent quality, however, at the time of publication these results had not been authorized for release. None-the-less, it is extremely encouraging that the technology developed and tested at RMC is functioning as expected within the austere field site conditions with no real interruption to tunnelling or mining operations. It should also be noted that a unique fiber optic instrumentation solution must be determined for each type of ground support element; this is a non-trivial undertaking due to the unique requirements and installation procedures associated with each type of support and site. To date, this technology has been employed at 4 sites world-wide.

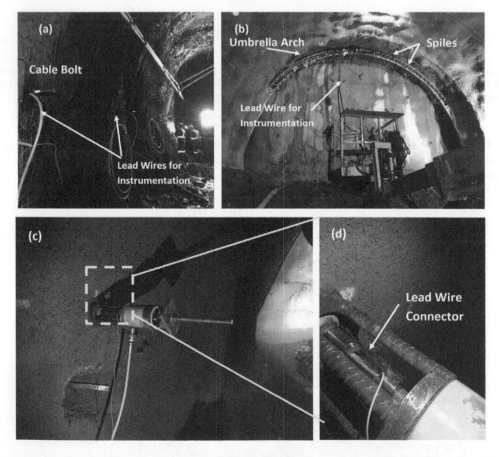

Figure 8.   Field Trials for Support Elements Outfitted with Optical Fiber. a. Optical cable bolts installed at in austere mine conditions; b. Installation of Umbrella Arch with Spiles; Lead wire extending from instrumented support member; c. and d. In-situ pullout test of rock bolt instrumented within fiber optics;

4.2 Case Study – Installation of Fiber Optic Spiles as part of Edmonton Light Rail Transit (LRT) (Forbes et al., 2018)

4.2.1 General

In the 2009 the city of Edmonton, Alberta, Canada agreed upon the development of a long-term LRT network of six lines: including the 13.1-kilometer Valley Line. Constructed by TransEd Partners, the Valley Line LRT Stage 1 construction project extends from the downtown Edmonton core. At an approximate cost of $1.8 Billion CAD, the low-floor LRT system includes: 11 stops, 1 transit station, 1 operations and maintenance facility, 1 interchange with Edmonton's existing high-floor LRT system (Churchill Interchange), and, 1 bridge spanning the Northern Saskatchewan river (Tawatinâ Bridge) (TransED LRT, 2017).

All stops and a majority of the of the Valley Line LRT rail are designated as surface operations. Only approximately 400 meters of the LRT system is designated as a mined excavation. The northern portal of the Valley Line LRT mined segment is located shortly after Quarters Stop and the southern portal is located prior to the Tawatinâ Bridge. The mined segment of the LRT system consists of a twin tunnel construction project to accommodate the north and south bound lines (with a 10.1-meter minimum center-to-center spacing between adjacent tunnels). Each tunnel is horseshoe shaped with an approximately 42.3-meter-squared face section being advanced using a top heading and bench sequential excavation method. The shallow tunnel has a maximum overburden of approximately 15.1-meters and transverses lacustrine clay and varying plasticity index glacial tills with intra-till sand pockets similar to those conditions discussed for Edmonton's North LRT expansion by Elwood and Martin (2016).

Various temporary support categories have been classified over the extent of the twin tunnel excavation. Within the region of interest of this monitoring project, ground support measures consist of: A sealing shotcrete layer, A steel-fiber reinforced shotcrete (SFRS) layer, Six-meter long self-drilling and grout injectable spiles, and, Lattice girders with reinforced continuity strips at the springline. Additional face support and probe/drainage measures (e.g., glass-fiber reinforced face bolts) are also employed when required.

4.2.2 Optical Fiber Instrumented Spiles

The spile members employed in the Valley Line LRT consist of a hollow steel bar with a 29.5-millimeter external dimeter and a 15-millimeter internal diameter. The external diameter of the spile is very accommodating to the lengthwise machined slots and looping grooves; however, transferring this optical fiber sensing technology to the spile members was found to be a non-trivial undertaking based on the spile installation procedure. Each six-meter long spile member consists of two three-meter long spile segments that are joined with a steel coupler only after the first segment (i.e., deepest within the borehole) has been drilled into the ground mass. In addition, the hollow core is required for water flow during self-drilling and cement grout flow once fully inserted. Accordingly, the centered position of the LC connector at the head of the support member, which is used to terminate the sensor (i.e., connect to the optical fiber sensor), makes it susceptible to damage and water/grout blockage during installation. However, there were no practical alternatives to the positioning of the LC connector that would not leave the connector prone to damage from the spile installation socket/drill. Therefore, it was decided to maintain the centered positioning of the LC connector, but to create water/grout ports (Figure 9). This necessitated a modification to the standard installation socket to allow the water/grout to circumvent the LC connector (and its installation protection piece) as shown in Figure 9. This installation/grouting technique was chosen over an alternative to insert the spile within a pre-drilled borehole and grout the spile using a grout tube as this could possibly impact the mechanistic behaviour of the spile: in addition to altering the spile installation procedure employed at the project. Furthermore, due to the uncertainty of the optical sensor surviving the self-drilling procedure, it was decided to only instrument the three-meter segment closest to the excavation.

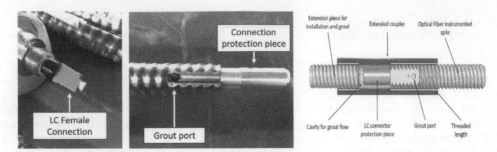

Figure 9. Head view of the instrumented spile member. Left: LC connector centered at the head of the spile. Centre: LC connector protection piece for installation and water/grout ports drilled into the side of the spile member. Right: Modified installation coupler to allow water/grout flow to circumvent the LC connector at the head of the instrumented spile. The coupler threads onto the head of the instrumented spile on one side while allowing the standard installation socket to be used on the opposing side.

4.2.3 *Installation and Monitoring*

Eight spiles were instrumented as part of the monitoring project at the Valley Line LRT tunnel. All eight spiles were installed in the southbound tunnel and were relatively uniformly spaced between 14-meters and 25-meters from the northern portal. The instrumented spiles were installed with the normal spile support pattern and were situated in proximity to the center position of the crown. The procedure developed to install the instrumented spiles and monitoring lead-wires included (This procedure is visualized in Figure 10):

i. Conduct the top heading advance (1-meter), spray the shotcrete sealing layer, install the lattice girder, and install the conventional spile round (spiles installed through girder) while leaving two open positions for an instrumented spile (two positions were left available in case difficulties were encountered during the first install attempt);
ii. Install and grout the instrumented spile member;
iii. Remove the installation coupler and the LC connector protection piece;
iv. Connect a ruggedized 10-meter optical fiber lead-wire to the LC connector at the spile head and record a measurement of the installation induced strain;
v. Run the ruggedized optical fiber lead-wire around the periphery of the lattice girder to the bench heading for ease of access to measurements, and;
vi. Spray SFRS layer, completely covering the instrumented spile head, lattice girder, and optical fiber lead-wire (except for the connection point at the bench used for subsequent measurements).

The selected sensor interrogation unit used to take strain measurements of the optical fiber sensors during the monitoring project was limited to one sensor. Accordingly, it was necessary to manually connect the sensing unit to the optical fiber lead-wires that had been run along the excavation periphery to the bench level. Measurements were taken daily, at minimum, for the instrumented spile members during an approximately 1.5-month duration and 36-meters of tunnel advance. Measurements were most often taken synchronous with surveying rounds after the top heading or bench advance.

4.2.4 *Spile Strain Measurements*

The strain distributions measured from one selected instrumented spile members is presented in Figure 11. The plots show: (a) The strain distribution measured along the entirety of the optical sensor (i.e., all three sensing lengths and the two connecting loops); and, (b) The resolved maximum (or principal) strain distribution along the length of the spile member. More details and results can be found in Forbes et al, 2018.

The strain distribution of Spile Sensor A behaves in a manner expected of an umbrella arch member, where the spile is observed to act similarly to a multi-span beam (i.e. John and Mattle, 2002). Referring to the lateral strain distribution plot, over the first three excavation

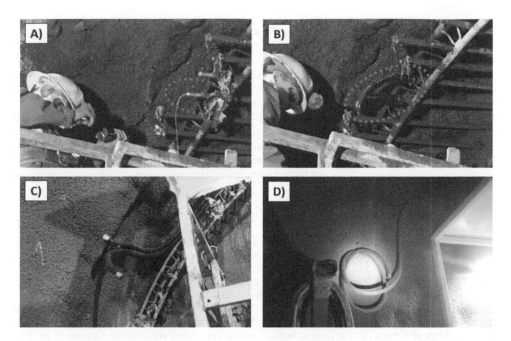

Figure 10. Installation procedure for instrumented spile members. A) Connection of the optical fiber lead-wire after removal of the installation coupler and LC protection piece. B) Attachment of the "ruggedized" conduit for the optical fiber lead-wire. C) Extension of the optical fiber lead-wire around the tunnel periphery to the bench heading level. D) Post SFRS view of the optical fiber lead-wire, accessible for strain measurements.

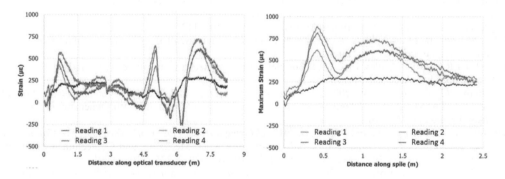

Figure 11. Strain distributions showing the strain measured along the entirety of the optical sensors (i.e., all three sensing lengths) and the resolved maximum strain, coaxial strain, and lateral strain along the length of the spile. Positive strain is extensile while negative strain is compressive.

advances the length of the spile which is subjected to bending-moment induced strain increases almost equally with the 1-meter long advancements of the tunnel.

5 CONCLUSION

The distributed optical strain sensing technique has been verified as a novel monitoring and geotechnical tool for capturing the performance of ground support members used in underground projects. The sensitive spatial resolution allows a continuous strain profile to be measured, overcoming the limitations of conventional, discrete strain measuring techniques, which in most cases will not fully capture the geomechanical behaviour of the support, especially when considering localized complexities. The results of using this instrumentation with ground support elements in the laboratory and the field as they relate to tunnelling projects have provided confidence for using and improving upon such a technique within the field. In addition, the optical technique can be realized as a novel tool with the capability to "see" and "sense" into the ground ahead of the working face, allowing the engineer to react and make adjustments to the support and excavation process in response to future ground conditions. As a monitoring solution, DOS provides unparalleled information concerning the behaviour and the interaction between the ground medium and the support elements which can be back-analyzed for predictive numerical model methods and ultimately support design optimization.

REFERENCES

Duncan, R.G. et al., 2007. *OFDR-Based Distributed Sensing and Fault Detection for Single and Multimode Avionics Fiber-Optics*. Proc. the 10th Joint DOD/NSA/FAA Conf. Aging Aircraft, Palm Springs, CA, pp. 10–14.

Elwood, D.E.Y., and C.D. Martin. 2016. *Ground response of closely spaced twin tunnels constructed in heavily overconsolidated soils*. Tunnelling and Underground Space Technology. 51: 226–237.

Forbes, B., Vlachopoulos, N., & Diederichs, M.S., 2015a. *Monitoring the Ground in Order to Optimize Support: Ground Support Elements Equipped with Optical Frequency Domain Reflectometry Technology*. In Proceedings of Eurock 2015 & Geomechanics Colloquium, Saltsburg, Austria, October 7–10, 2015.

Forbes, B., Vlachopoulos, N., Diederichs, M.S., 2015b. *Improving Ground Support Design with Distributed Strain Monitoring*. Proc. of the 68th Canadian Geotechnical Conference and 7th Canadian Permafrost Conference: GEOQuébec 2015, Québec City, QC.

Forbes, B., Vlachopoulos, N. and Diederichs, M.S., Hyett, A.J., 2018. *Spile support performance monitored in a shallow urban tunnel using distributed optical strain sensing*. 52nd US Rock Mechanics/Geomechanics Symposium: Seattle 2018,24–27 June 2018.

Hyett, A.J., Forbes, B.J., & Spearing A.J. 2013. *Enlightening Bolts: Determination of the Strain Profile along Fully Grouted Rock Bolts using Distributed Optical Sensing*. Proc. of the 32nd International Congress on Ground Control in Mining, Morgantown, WV, 107–112.

John, M., and B. Mattle. 2002. *Design of tube umbrella*. Magazine of the Czech Tunnelling Committee and Slovak Tunnelling Association, 11:3; 4–11.

TransEd LRT (2017). Valley Line LRT Booklet. http://transedlrt.ca/resources/september-2016-valley-line-lrt-booklet/ [Accessed 20 April 2017].

Volkmann, G., & Schubert, W. 2007. *Geotechnical Model for Pipe Roof Supports in Tunneling*. Proc. of the 33rd ITA-AITES World Tunneling Congress, Underground Space - the 4th Dimension of Metropolises. Prague, pp 755–760.

Vlachopoulos, N., Cruz, D., and Forbes, B. 2017. *Utilizing a Novel Fiber Optic Technology to Capture the Axial Response of Fully Grouted Rock Bolts*. Journal of Rock Mechanics and Geotechnical Engineering. Submitted 26 July 2017, Accepted 7 Nov 2017, JRMGE-2017–225, V10 N3 2018.

Vlachopoulos, N. and Forbes, B., 2018. *Geological Engineering - 'Smart' Ground Support: Continuous Strain Monitoring using Fiber Optics*. Geotechnical News, Newsletter of the Canadian Geotechnical Scociety. June 2018.

Tunnels and Underground Cities: Engineering and Innovation meet Archaeology,
Architecture and Art, Volume 6: Innovation in underground engineering,
materials and equipment - Part 2 – Peila, Viggiani & Celestino (Eds)
© 2020 Taylor & Francis Group, London, ISBN 978-0-367-46871-2

Microtunnelling in South Africa – a competitive solution

P. Vorster, M. Wainstein & H.J. Tluczek
GIBB – Engineering and Architecture, Johannesburg, Gauteng, South Africa

ABSTRACT: This paper presents an assessment of the options available for the new water abstraction works for Mthatha Dam, South Africa. The current abstraction works, which are in a critical state of repair, are the only source of water conveyance to the Water Treatment Works of Mthatha Town. Due to critical timeframes, a rapid and cost effective construction method is required to safeguard the water supply. The proposed abstraction works consist of a low lift pump station and two 150 m long tunnels which will be constructed under high water pressure and tunnelled into the water body. The options assessment compares a drill and blast lake tap methodology to the relatively new method, in a Sub-Saharan African context, of microbore tunnelling with an in-lake retrieval.

1 INTRODUCTION

Mthatha dam is a 38 m high dam situated in the Eastern Cape, South Africa. The town of Mthatha's only source of potable water is from Thornhill Water Treatment Works (WTW). Water is conveyed to the WTW from Mthatha dam via two 450 mm diameter steel pipes that run from the inlet tower and are embedded in crushed stone and concrete underneath the spillway slab. These pipes, which are the sole means of conveying raw water from the dam to the WTW are barely meeting current demand. Due to inaccessibility for maintenance, the outlet pipes are corroding and partial failure of the pipes and spillway slab has occurred.

As part of the regional bulk water supply scheme currently under construction, GIBB (Pty) Ltd. conducted a site inspection and observed that a 50 m long section of the spillway discharge channel floor slab had developed a longitudinal bulge 300 mm high along its centre (see Figure 1). This bulge is due to the build-up of pressure below the slab as a result of a rupture of the outlet pipes. The leak at this stage is small enough to allow enough water flow to the WTW, however there is an ever-increasing risk that a more significant leak could develop or that complete failure may occur thereby starving the WTW of its raw water supply. This would result in the whole of Mthatha being without potable water for a prolonged period. It has been observed that the leakage has increased over the last couple of months and that the supply pressure has decreased at the WTW.

The difficulty with the current situation is that repairs or replacement of the damaged pipes under the spillway slab will be onerous while contending with the dam overflow and without shutting down the supply to the WTW. It should also be noted that even if the pipes could be rehabilitated completely, the current design capacity of the pipes is insufficient to meet the future demands of Mthatha and the surrounding area. The need for remedial works to augment the supply of water is therefore considered critical.

GIBB engaged with the local municipality to present alternative options to abstract water from Mthatha dam. These options would not only provide raw water to the existing Thornhill WTW but would also supply raw water to the proposed Highbury WTW. The preferred option entailed a new abstraction point upstream of the dam wall with an in-lake tunnel abstraction point and a new Raw Water Pump Station (RWPS). Because of the critical time frames given to ensure water supply to the town of Mthatha, GIBB was given a mandate by the local municipality to consider state of the art design and construction methodologies for the proposed abstraction works.

Figure 1. Mthatha Dam spillway channel and outlet pipe failure.

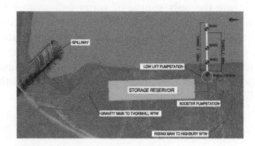

Figure 2. Proposed Mthatha dam extraction work layout.

2 PROPOSED ABSTRACTION WORKS

The proposed abstraction works will be situated on the left flank approximately 400 m upstream of the spillway. Water will be abstracted from Mthatha Dam via two 1.0 – 1.5 m diameter tunnels. The two tunnels will be situated 5 and 12 m below the Full Supply Level of the dam and will be 100 and 150 m long respectively. A Low Lift Raw Water Pump Station (RWPS) will supply 100 Ml/day to Highbury WTW and 115 Ml/day to Thornhill WTW. The water will initially be pumped into a storage reservoir from which a gravity main will supply Thornhill WTW while water for Highbury WTW will pass through a booster pump station to a rising main.

The RWPS will be located underground in a 20 m deep shaft with a 23 m diameter to be excavated into the rock. The proposed abstraction works layout can be seen in Figure 2.

Two construction options were considered for the tunnelling works of the abstraction system. These included a drill and blast excavation method, and a micro-bore option using a micro tunnel boring machine (MTBM).The drill and blast option would require a lake tap to daylight into the water body while the MTBM would utilise a rescue module for an in-lake retrieval. Both these options are discussed in more detail in the paper below. Several contractors were approached to advise on the construction cost and programme in order to draw a clear comparison between the two options.

3 DESIGN CONSIDERATIONS

There are a number of risks associated with each construction option which impact on the project in different ways. Based on the proposed development the following aspects have been considered for the two construction options:

- Topography
- Geological conditions
- Ground water conditions
- Health and Safety
- Construction Time
- Construction Complexity
- Resource Availability
- Construction Cost

3.1 Topography

The topography in the vicinity of the proposed abstraction works is characterised by gentle gradients. This results in shallow overburden as the tunnels enter the dam, especially for tunnel 2 (Figure 3). This would be problematic for the drill and blast option especially under submerged conditions. For this reason, the MTBM method would be the preferred option.

3.2 Geological conditions

Geological conditions have a significant influence on the excavation process and cost of a project. Besides bulk rock mass properties such as jointing, faults and fractures, several physical properties such as overburden and rock mass variability need to be assessed. A geotechnical investigation was carried out to assess the ground conditions on site. This included an electrical resistivity survey, five (5) rotary core boreholes, twelve (12) test pits and soil and rock laboratory testing. Based on the results of the investigation the proposed site is characterised by a thin soil mantle consisting of a medium dense clayey silty sand which is underlain by a moderately weathered to unweathered, widely jointed, thickly bedded, very fine grained, hard to very hard sandstone rock with localised siltstone lenses with interbeds of unweathered, widely jointed, very fine grained, very strong sandy mudstone. The geological profile is illustrated in Figure 3.

Rock laboratory tests were carried out on samples selected during borehole core logging. Selected results from the laboratory tests are given in the table below.

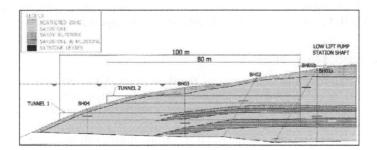

Figure 3.   Geological Profile.

Table 1.   Selected results from laboratory testing conducted.

Rock Type	UCS Average (Range)	Density Average (Range)
Sandstone	122 MPa (110 to 129)	2.52 g/cm^3 (2.49 to 2.55)
Sandy Mudstone	144 MPa (129 to 164)	2.68 g/cm^3 (2.68)

The rock mass was characterised using the classification system developed by Barton, Lien, and Lunde (1974) which expresses the quality of the rock mass as a Q-value. The unweathered sandstone, sandy mudstone and sandy siltstone all classified as Very (with Q values of 96, 95 and 89 respectively) while the weathered sandstone classified as Good (with a Q value of 33.40).

A suite of rock tests was carried out in order to determine the excavatability, abrasivity and drillability of the rock mass. The Drilling Rate Index (DRI) expresses the drillability or bore-ability of the intact rock mass – laboratory test were recorded as Very-High to Extremely-High. The Bit Wear Index (BWI) expresses the Bit wear rate - laboratory test results were recorded as Very-Low to Extremely-Low. The Cutter life Index expresses the life expectancy of the cutters – laboratory test results were recorded as Very-High to High. This all bodes well for drilling and boring through the rock mass.

In general, the geology of the site can be characterised as good, with no major fault zones and good quality rock. The rock strengths are within a small range which reduces the risk of mixed face tunnelling problems for the MTBM.

3.3 *Ground water conditions*

The current dam would have to remain in operation during construction as it is currently the only source of raw water supply to the town of Mthatha. This implies that the underground excavations would have to be excavated under full supply head with resulting high ground water pressure. For the drill and blast option, ground treatment ahead of the excavation face will be considered critical, which would entail the drilling of probe holes and pre-grouting of the rock mass where necessary. The probe and grouting holes can range in the order of 20–30 m in length. In-situ Lugeon testing was conducted in the boreholes drilled on site. The majority of the tests resulted in Lugeon values of zero with no flow. A maximum Lugeon value of 0.1 was recorded which is indicative of a non-permeable rock mass. However, the shallow slope and shallow overburden leading into the dam for the drill and blast tunnel option becomes problematic as the rock becomes more weathered and weaker near surface.

The MTBM will be equipped with an appropriate bulkhead to work under submerged conditions. The only time personnel will be exposed to the submerged conditions is when/if cutter heads need to be replaced on the MTBM as a result of cutter wear or change in geological conditions. This is done by skilled divers who are sent through the tunnel bulkhead, airlock and compressed air regulation unit. Divers will also be required to form the underground portal for the MTBM and to retrieve the MTBM once it daylights in the dam.

3.4 *Health & safety considerations*

The anticipated poor ground conditions at the dam entry with very thin overburden and high water pressures would make a conventional drill and blast tunnel, with a lake tap, a high risk option in terms of safety. Micro tunnelling on the other hand would be controlled remotely with fewer personnel thus reducing the safety risks on the project. If there is the need for cutterheads replacement in the MTBM, skilled divers will be required to replace the cutterheads.

3.5 *Construction time, complexity and resource availability*

It is generally accepted that the advance rate for tunnel construction is much higher for TBM boring compared to drill and blast tunnelling (Holen 1998). This is however dependent on the rock conditions and the accuracy of the geological investigation in order for an MTBM to be equipped with the appropriate equipment and cutter head. One of the most important aspects to consider when deciding between drill and blast tunnelling and tunnel boring is the start-up time. MTBM availability plays a large role in the construction programme because if a new machine has to be purchased the delivery time could be between 6–12 months (Holen 1998).

The duration of the construction programme is considered of paramount importance for this project as it has been initiated due to the requirement for the sustainable delivery of water. The mandate given to GIBB at the onset of the project was to consider the turnaround time for the commissioning of this project as critical.

Both construction options, drill and blast (D&B) with a lake-tap as well as micro bore using an in-lake retrieval involve specialist work which is not readily available in South Africa nor in many parts of the world.

The lake tap method is a traditional Norwegian method of underwater tunnel penetration. In Norway more than 600 lake taps have been successfully performed since the 1890's which indicates that it is a well proven technique. This method however not only requires specialised blasting and excavation techniques, but relies on a thorough evaluation of the engineering geological aspects and geometrical considerations to optimize the tunnel alignment in order to minimise the risks for the lake tap (Mathiesen 2013).

Micro boring under submerged conditions with an underwater retrieval has become an increasingly popular method of construction for sea outfalls with numerous successful projects around the world. The micro tunnelling machine is equipped to work under submerged conditions, is controlled remotely and bores until it daylights into the water body. The machine is then separated from the tunnel using the pistons in the rescue module and removed from the water body by crane.

A search was conducted for micro tunnelling machines available in South Africa and Africa as a whole. Two machines were located; a 1.1 m OD slurry machine which has been purchased by a South African contractor and a 2.0 m ID slurry machine which was operating in Ghana.

The programme provided by the contractors that were approached indicated that the Drill and Blast option would take 4.5 times longer to construct than the MTBM option. It was however noted that the rock strength on site would be too high for the machine currently stationed in South Africa and also prevents access to change cutting discs which increases the risks for this option. The contractor therefore opted that a new larger diameter machine (1.5 m OD) will have to be sourced, in which case the duration of the Drill and Blast option will still be 2 times longer than the combined construction and sourcing of the MTBM. Therefore, the MTBM option remains the preferred option due to its shorter programme.

3.6 *Construction cost*

A micro tunnelling solution is typically a costly exercise, and extremely so, where machines need to be imported from overseas. Erection and capital costs are usually higher for MTBM tunnelling whilst the costs for the excavation phase are less compared to D&B. It is because of this that there is usually a tipping point in tunnel length where a MTBM becomes more cost effective compared to drill and blast. For longer tunnels, the high TBM capital and installation costs can be distributed over more tunnel meters whereas for D&B tunnels additional length may cause prohibitive ventilation expenses (Holen 1998).

For this project, even though a MTBM still has to be sourced internationally, the option is still a competitive solution, even for such short tunnel lengths. The contractors that were approached to advise on construction cost for the two options indicated that with a MTBM, the mechanised tunnelling option was only 14% more expensive than the drill and blast option. As tunnel boring machines become more readily available the costs involved with using such machines for projects become more competitive, especially on the microbore front. With the increase in tunnel boring machine diameter it is expected that the capital cost would increase significantly and in turn reduce the feasibility of such an option.

4   CONSTRUCTION METHODOLOGY

Both tunnel construction methods will be conducted from the RWPS shaft which will have a diameter of 23 m and depth of 20 m. The construction methodology and the advantages and disadvantages of the respective options are discussed below.

4.1 Drill and Blast Tunnel

There are specific challenges that need to be addressed when conducting a drill and blast operation under high ground water pressures. These include a thorough planning and execution of the underground works. The following safety measures are required to ensure safe tunnel construction (Palmström 2002):

- Systematic 20 - 30m long exploratory probe drill holes ahead of the tunnel working face.
- Where poor rock masses are expected, longer exploratory core drill holes.
- High pressure pre-excavation grouting where water bearing zones or poor rock masses are expected or detected
- A high pumping capacity is required in case there are high water inflows
- Large enough capacity to apply shotcrete quickly after blasting in order to support poor stability rock masses.
- A drilling and charging pattern has to be designed which ensures the desired blasting result based on tunnel dimensions, tunnel geometry, final quality requirements, by taking into account geological and rock mechanical conditions, explosives availability and means of detonation, expected water leaks, vibration restrictions and drilling equipment.

The general drilling and blasting cycle is as follows (Satici 2006):

- Drilling probe holes and conducting pre-excavation grouting where necessary.
- Drilling blast holes and loading them with explosives.
- Detonating the blast, followed by ventilation to remove blast fumes.
- Scaling crown and walls to remove loose pieces of rock.
- Removal of the blasted rock (mucking).
- Installing initial ground support.
- Advancing rail, ventilation, and utilities.

As mentioned, the tunnel will daylight into the dam by using a lake tap. The principal of lake tapping involves advancing the tunnel under a lake, leaving a rock plug to the lake floor. A sump is then excavated at the end of the tunnel. The plug is then blasted so piercing the lake floor from below and the blasted material falls into the sump. The Norwegian lake tap principal is illustrated in Figure 4. When excavating the last part of the tunnel towards the intake with very low overburden the following principles need to be followed (Mathiesen 2013).

- Probing is required all the way through to the reservoir in order to verify the exact location of the tunnel in relation to the lakebed.
- Careful blasting in shorter rounds as the face approaches the final rock plug.
- A system of probe drilling through the final face to gather data for the final blast design.

A tunnel lining system will be installed as excavation progresses as well as tunnel support where deemed necessary. Two methods of lake tap can be used namely, the open or closed method. In the open method, the lake tap system is linked to the atmosphere which evens out the pressure after blasting. In the closed system the lake piercing is isolated from the atmosphere by a gate or valve. The two methods are illustrated in Figure 5. For this project, the closed system would be utilised with the tunnel closed off at the RWPS shaft excavation.

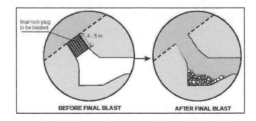

Figure 4.   Norwegian Lake tap principle (Palmström 2002).

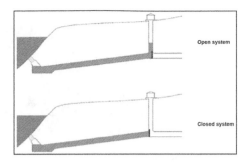

Figure 5. Open or Closed Lake tap systems (Mathiesen 2013).

4.2 *Micro Bore Tunnel*

In constructing the proposed abstraction works, the following key components should be considered for a full faced, slurry machine excavation using a pipe jacking advance system (Ulkan 2014). The process is described below and the setup shown in Figure 6:

- The RWPS will form the launch shaft for the micro tunnel machine and will be fitted with a thrust wall to provide a surface against which to jack. The shaft cross sectional area should be large enough to accommodate the MTBM assembly and jacking area.
- The jacking frame will be fitted into the launch shaft to thrust the machine forward.
- The initial alignment of the pipe jack will make use of guide rails within the thrust pit after which the tunnel excavation will be controlled remotely from a surface control room.
- The tunnel excavation will be advanced by the MTBM cutter head and the tunnel supported by a hydraulic support comprising a bentonite slurry suspension.
- The removal of the muck will be facilitated by hydraulic conveyance of the excavated material through a closed slurry circuit.
- The bentonite slurry suspension, injected via nozzles into the tunnels annulus gap, also aids in reducing friction between the pipeline and the surrounding ground which in turn reduces the jacking forces required during construction.
- The drive length of the tunnel for this project is relatively short and the performance of the cutting tools are unlikely to be compromised during excavation. However, in case the cutter heads do need to be replaced, this will be done by sending divers through the tunnel, bulkhead, airlock and compressed air regulation unit. An example of the airlock chamber is shown in Figure 7
- The MTBM will daylight into the reservoir at which stage divers will be sent down to remove any excess material around the machine. Using the rescue module bulkhead, the MTBM is separated from the trailing tunnel using hydraulic pistons. Shown in Figure 8
- The recovery of the MTBM is done using a barge fitted with a crane or by using air balloons to float the machine to the surface to be removed with a crane positioned on the shore.

Figure 6. General Pipe Jacking Jobsite layout with AVN machine (Schmaeh 2015).

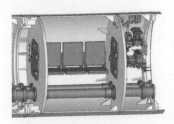

Figure 7.   Example of air lock chamber (Schmaeh 2015).

Figure 8.   Machine recovery and rescue module (Schmaeh 2015).

5   HEALTH, SAFETY AND RISK ASSESSMENT

Tunnelling and other underground works have always created difficult Health and Safety (H&S) issues that need to be addressed. Dust, gases (diesel fumes, methane, etc.), soot, oil mist, noise and vibration, chemicals, rockfalls, traffic accidents, visibility and working with

Table 2.   Drill and Blast Risk, Health and Safety Concerns.

POTENTIAL RISKS AND H&S CONCERNS	POSSIBLE REMEDIATION MEASURES
a) Settlement of the ground surface due to excavation	Control of excavation rate and timeous installation of primary rock support
b) Instability of tunnel crown excavation	
c) Excessive ground water inflow during excavation	Ensure adequate length and number of probe holes and pre-excavation grouting
d) Damage to dam and appurtenant structures due to blasting vibrations	Proper blast design to limit peak particle velocities to acceptable limits
e) Excessive damage to rock mass from blasting	Proper blast design and initiation sequence
f) Excessive convergence of tunnel	Ensure adequate design and timeous installation of primary rock support
g) Encountering zones of highly weathered soft to very soft rock	Control of excavation rate and timeous installation of primary rock support
h) Drilling into undetonated explosives	Follow proper blasting procedures with regard to initiation, making safe and post blast inspections
i) Rockfalls during excavation and after blasting	Follow proper excavation procedures with regard to making safe and timeous installation of primary support
j) Toxic gasses from blasting and mucking machinery	Choose explosive type to keep toxic blast fume level as low as possible and ensure adequate ventilation system is installed

Table 3. MTBM Risk, Health and Safety Concerns.

POTENTIAL RISKS AND H&S CONCERNS	POSSIBLE REMEDIATION MEASURES
Settlement of ground surface due to excavation Blowouts at ground surface during excavation	The tunnel excavation is supported by a hydraulic support comprising a bentonite slurry suspension which is governed by the earth pressure, no settlement or blowouts are expected
Instability at face of excavation	The proposed machine belongs to the category of closed, full-face excavation machines with a hydraulic slurry circuit which enables the control of face pressure during excavation.
Variable rock parameters encountered Mixed face tunneling	The proposed machine offered will have a full rock face, suitable for all anticipated rock types and will have access to change cutting tools for the appropriate geology
Encountering geological faults Highly weathered soft rock to very soft rock zones encountered Changing cutting gear	Low permeability rock mass is expected, however the machine is equipped with a shaft air lock bulkhead to prevent water inflow and allow for cutters to be inspected and changed. Divers sent through the tunnel, bulkhead, airlock and compressed air regulation unit is a standard low risk process that forms part of the tunneling process.
Uplift of MTBM due to shallow gradient before daylighting. MTBM Recovery from lake bed	A local depression will be dredged where the TBM is planned to daylight/enter the dam, the depression is to be filled with crushed rock into which the TBM can be driven for stability and ease of recovery. A specialist contractor is used for the underwater recovery of tunneling machines. Mthatha dam is seen as a stable environment compared to more unstable sea recovery conditions that have been performed.

explosives are but some of the challenges that need to be addressed (Myran 2014). In all cases, all the work procedures must be described in a quality plan, which details the sequence of work activities before the risk evaluation is prepared. The appropriate regulations for the work being conducted should be followed at all times. Contractors should establish standard procedures for the execution of the various work activities to be followed throughout the duration of the project. Table 2 below provides potential Risk and H&S concerns as well as possible remediation and preventative measures for the Drill and Blast option.

The use of tunnel boring machines has reduced or completely eliminated environmental hazards such as blasting fumes and diesel exhaust fumes, oil mist and rockfalls. However, there are still health and safety risks as well as risks with cost implications. Table 3 below provides potential Risk and H&S concerns as well as the possible remediation and preventative measures for the MTBM option.

6 CONCLUSION

There is an urgent need for a new water abstraction works at Mthatha dam due to the deteriorating state of the current outlet pipes situated under the spillway slab. There is an ever-increasing risk that the current outlet pipes could fail completely and leave the town of Mthatha without a water supply. Because of this risk the mandate was given to GIBB that the construction time was of paramount importance. An option assessment was conducted that took into account various factors to ensure not only the quickest construction period, but also took into account the construction complexity, cost, availability, H&S and risk. The conclusions drawn are summarised below:

- The geological conditions on site are favourable with good quality rock and no major fault zones which does not hinder either the drill and blast option or MTBM option.

- The topography of the site does however have a gentle gradient and therefore the rock overburden leading into the dam for the tunnels will be shallow and pass through weathered rock. This could cause problems for drill and blast tunnelling under submerged conditions and to conduct the final lake tap.
- Construction will occur under submerged conditions under high groundwater pressure where ground treatment and support ahead of the excavated face would be considered critical for the drill and blast option. This means that pre-excavation probe holes and grouting is a definite requirement throughout the excavation process. The MTBM will be equipped with the appropriate bulkhead to work under submerged conditions and will be operated remotely which will reduce the safety risk.
- The lake tap method for drill and blast option is a proven technique which has been conducted on numerous occasions in Norway. It is however still a high risk procedure and requires specialised skills to be outsourced. The MTBM sea outfall procedure has come to the fore in the past 20 years and has grown in popularity due to the greatly reduced risk and complexity of the construction process.
- The Health, Safety and Risk assessment indicates that there are fewer risks involved with the MTBM option for this project.
- Even with an internationally sourced MTBM the construction cost is only 14% more than the drill and blast option but the construction time is 2 times faster than the drill and blast option.

Due to the significantly shorter construction time proposed for the MTBM option, with reduced safety risk and construction complexity, and with only 14 % more construction cost, the MTBM option, even with the short tunnel drive and sourced internationally, is the preferred option to construct the two tunnels for the new abstraction works at Mthatha dam. It can be expected that with more MTBMs becoming available locally, that the MTBM option will become even cheaper in future.

REFERENCES

Barton, N.R. Lien, R. and Lunde, J 1974. Engineering Classification of Rock Masses for the Design of Tunnel Support, *Rock Mechanics and Rock Engineering* 6(4):189–236

Mathiesen, T.K. 2013. Lake Tap Design for Sisimiut Hydropower Plant, West Greenland, Publication no. 22, *Norwegian Hydropower Tunnelling II*, Norwegian tunnelling society

Holen, H. 1998. TBM vs. Drill & Blast Tunnelling, Publication no. 11, *Norwegian TBM Tunnelling*, Norwegian soil and rock engineering association

Handen, A.M. 2014. Norwegian TBM Tunnelling – Machines for hard rock and mixed face conditions, *Norwegian Tunnelling Technology*, Publication No. 23

Myran, T. 2014. Health and Work Environment, *Norwegian Tunnelling Technology*, Publication No. 23

Palmström, A. 2002. Sub-sea tunnels and lake taps in Norway - a short overview, www.rockmass.net

Satici, Ö. 2006. Drilling & Blasting as a Tunnel Excavation Method, Ankara

Schmaeh, P. 2015. Increasing requirements move boundaries in Pipe Jacking, Symposium and Exhibition, No Dig Berlin

Wainstein, M. Du Plessis, J. Tluczek, H.J. 2017. Considering first world alternatives in augmenting the supply of Mthatha Dam, South Africa, Hydro

Tunnels and Underground Cities: Engineering and Innovation meet Archaeology, Architecture and Art, Volume 6: Innovation in underground engineering, materials and equipment - Part 2 – Peila, Viggiani & Celestino (Eds)
© 2020 Taylor & Francis Group, London, ISBN 978-0-367-46871-2

Study on pull-out test of longitudinal joints of shield tunnel

Q. Wang, P. Geng, G. Zeng & X. Guo
Key Laboratory of Transportation Tunnel Engineering, Ministry of Education, Southwest Jiaotong University, Chengdu, PRC

X. Tu, X. Zhang & F. Li
State Grid Corporation of China, Beijing, PRC

M. Xiao
China Railway Siyuan Survey and Design Group Co., Ltd, Wuhan, PRC

ABSTRACT: The mechanical characteristics of the oblique longitudinal joints of shield tunnel under different load conditions was studied by 1:1 joint tensile performance test. Research indicates: the longitudinal joint of the inclined bolt will affect the stress distribution on the surface of the longitudinal joint and the outer surface of the segment during the drawing process. When the tension is perpendicular to the joint surface named vertical pulling condition the bearing capacity can be fully exerted because it can take proper cooperation with the concrete strength. The ultimate failure mode of the structure is the connection failure between the bolt, the sleeve and the concrete. Under the condition of drawing along the axis of bolt, cone destruction will occur around the orifice of the longitudinal seam face near the outer surface. While under the vertical pulling condition, the internal thread of the sleeve is destroyed because of being crushed.

1 INTRODUCTION

The shield tunnel is a fabricated structure in which the segments are bolted together. A large number of studies have shown that the joint is the weak part of the shield tunnel while subjected to force and deformation. Longitudinal deformation will occur in the shield tunnel when affected by earthquakes, longitudinal stratum deformation, construction loads and water level changes, etc. And the longitudinal deformation of tunnel is unfavorable to the safety of structure When the longitudinal deformation or deformation curvature reaches a certain value, the circumferential seam of the tunnel may open and the longitudinal joint may be damaged (Zhang Liancheng, Ye Fei, Zhang Tiezhu, 2009), which endangers the tunnel safety. For example, In the Wenchuan M=8.0 earthquake in 2008, longitudinal seam leakage and some joint bolts loose in the constructed shield tunnel of Chengdu Metro (Lin Gang, Luo Shipei, Ni Juan, 2009). Therefore, it is necessary to study the force characteristics of the longitudinal joint under the tensile load.

At present, experimental research on shield tunnels mainly includes the joint test and lining ring test. The joint test is mainly used to study the mechanical characteristics such as stiffness efficiency and joint opening angle. In addition to the joint test, many scholars have also carried out full-scale experimental research on the lining full ring of shield tunnels. The research contents include the overall bearing capacity, deformation characteristics, bending moment transmission, joint corner, waterproof performance, concrete crack distribution and structural failure modes, etc. (C.B.M. Blom, E.J. van der Horst, P.S. Jovanovic, 1999; Lu Liang, Lu Xilin, Fan Peifang, 2011). He Chuan, Feng Kun, et al. (2011) proposed a prototype test loading method and a loading test system for the segment structure, in view of the characteristics of large-section underwater shield tunnel. For the longitudinal joints of shield tunnels, some scholars have conducted experimental research. Ye Fei, Yang Pengbo, et al. (2015) carried out seams, staggered seams and the homogeneous cylinder model to research longitudinal model test in order to find out the effective value of longitudinal deformation performance and bending stiffness of shield tunnels.

The research contents of the above experiments are mostly about the stress and deformation characteristics of the lining ring and joints in shield tunnel. For longitudinal joints, similar model tests are often used for research. Due to the limitation of the similarity ratio of the test, the mechanical properties and destruction characteristics of the longitudinal joint detail structure cannot be characterized. Therefore, for the longitudinal oblique bolt joint structure of the shield tunnel, the 1:1 full-scale test of the longitudinal joint is carried out, to explore the mechanical characteristics and the failure mode of the longitudinal joint structure when the longitudinal joint is pulled.

2 TEST SUMMARY

2.1 Engineering backgrounds

The supporting project of the longitudinal joint tensile performance test - Sutong GIL integrated pipe gallery tunnel project is the key node project of "Huainan ~ Nanjing ~ Shanghai 1000 kV AC UHV transmission and transformation project" that under construction in China. The tunnel length is 5468m, the inner diameter of the segment is 10.5m, the outer diameter is 11.6m, the thickness is 0.55m, and the width is 2m. Each loop of the segment is divided into 8 blocks, including a capping block F, two adjacent blocks L1, L2 and 5 standard blocks B1 to B5, each of which is cast with C60 reinforced concrete. 22 M40 longitudinal oblique bolts in longitudinal direction, bolt length 70cm, thread length 22cm (37 turns), diameter 40mm, mechanical strength 10.9 (yield strength 900MPa). The relevant structure diagram is shown in Figure 1.

2.2 Test purpose

In the actual engineering, when the longitudinal deformation or deformation curvature reaches a certain value, the circumferential seam of the tunnel may open and the longitudinal joint may be damaged, which endangers the tunnel safety. Therefore, it is necessary to study the mechanical behavior, load-bearing characteristics and component bonding capability (including bolts and sleeves, sleeves and concrete) of the longitudinal joints when the longitudinal joint is pulled.

2.3 Test devices and measurement system

According to the prototype segment and the coupling bolt used in this test, the test loading system adopts the self-made bolt drawing system, and its structural design is shown in

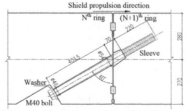

(a) Lining ring segment block diagram. (b) Longitudinal joint construction diagram (unit: mm).

Figure 1. Shield tunnel construction diagram.

Figure 2. Jack I is a through-heart hydraulic jack for providing the pulling force of the forward bolt. Jacks II and III are ordinary hydraulic jacks for providing the pulling force in the direction of the vertical joint surface. The measurement system includes TST3826E static strain test analysis system (measurement accuracy 0.1), differential variable pressure displacement measurement system (measurement accuracy 0.01 mm) and Sensor-Highway II type acoustic emission monitoring system produced by American Physical Acoustics (set preamplifier 35 dB, the noise threshold is 35 dB, the wave velocity is 410 m/s, the event definition value is 1310, the event blocking value is 2620, and the overdetermined position 131. The sensor is coupled to the surface of the specimen by Vaseline, and the broken lead test was carried out to eliminate the external interference) (Zhang Yamei, Wang Chao, et al, 2010). The measurement system is shown in Figure 3.

2.4 *Measuring point arrangement*

The measuring points in this test are mainly divided into three categories: The first type is the displacement measuring point, which is used to measure the displacement of the bolt during the pulling out process; The second type of measuring points is strain measuring points, including the surface strain measuring points of the segments and the bolt strain measuring points. The arrangement of the measuring points is shown in Figure 4: on the joint surface, centered on the sleeve, 5 pieces of strain gauges are evenly laid at 45°; 6 pieces of strain gauges are symmetrically arranged on each surface of the inner and outer surfaces of the segment along the sleeve; 2 pieces of strain gauges are symmetrically arranged in the middle of the screw and where near the thread separately, among whom L1-1 and L2-1 are adjacent to the inner surface of the segment, and L1-2 and L2-2 are adjacent to the outer surface of the segment. The third type of measuring points is acoustic emission measuring point, and three acoustic emission measuring points SF-1, SF-4 and SF-5 are arranged on the joint surface. A total of 9 acoustic emission measuring points are alternately arranged on the inner and outer surfaces along the sleeve. The arrangement of the measuring points is shown in Figure 5.

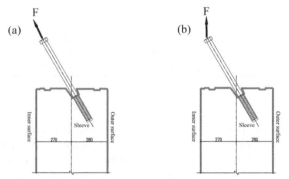

(a) Schematic diagram of the bolt loading system arranged along the axis.
(b) Schematic diagram of the vertical bolt loading system schematic.

Figure 2.   Loading system diagram. (unit: mm).

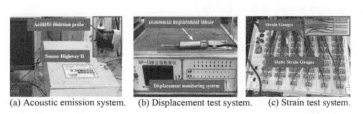

(a) Acoustic emission system.   (b) Displacement test system.   (c) Strain test system.

Figure 3.   Diagram of schematic measuring system.

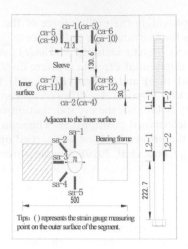
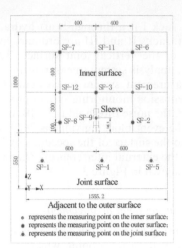

Figure 4. Strain gauge measuring point layout. (unit: mm).

Figure 5. Acoustic emission measuring point layout. (unit: mm).

Table 1. Loading steps.

Load step	Oil meter reading/ MPa	Load /kN	Load step	Oil meter reading/ MPa	Load /kN	Load step	Oil meter reading/ MPa	Load /kN	Load step	Oil meter reading/ MPa	Load /kN
1	1	18.61	10	10	176.18	19	19	333.74	28	28	491.30
2	2	36.12	11	11	193.68	20	20	351.25	29	29	508.81
3	3	53.63	12	12	211.19	21	21	368.75	30	30	526.32
4	4	71.13	13	13	228.70	22	22	386.26	31	31	543.82
5	5	88.64	14	14	246.20	23	23	403.77	32	32	561.33
6	6	106.15	15	15	263.71	24	24	421.27	33	33	578.84
7	7	123.66	16	16	281.22	25	25	438.78	34	34	596.34
8	8	141.16	17	17	298.73	26	26	456.29	35	35	613.85
9	9	158.67	18	18	316.23	27	27	473.80	36	36	631.36

2.5 Test cases

There are three groups of cases (Case-1, Case-2, Case-3) in the test. Case-1 is that the bolt is screwed into the sleeve for 37 turns (37 turns in total), i.e. screwed into 100%, and the test load is directly applied to the bolt, while the loading direction is the forward bolt; Case-2 is that the bolt is screwed into the sleeve 26 turns (37 turns in total), i.e. screwed in 70%, and the loading direction is the same with the Case-1; Case-3 is that the bolt is screwed into the sleeve for 37 turns (37 turns in total), i.e. screwed into 100%, and the test load is applied to the self-made bearing frame, while the loading direction is 30° with the bolt, perpendicular to the longitudinal joint surface. Each of the loading step is carried out by grading loading method. The test cases' loading steps are shown in Table 1.

3 TEST RESULTS AND ANALYSIS

3.1 *The stress distribution of segment*

Extract the stress distribution of the measuring points on the surface of the segment during the loading process. Due to space limitation, there only shows the stress distribution at the

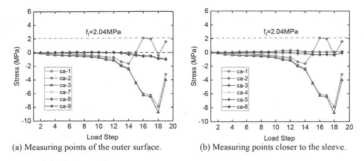

(a) Measuring points of the outer surface.　　(b) Measuring points closer to the sleeve.

Figure 6.　Case-2 segment stress distribution.

inner and outer surfaces of the Case-2 (i.e. 70% of the bolt is screwed into the sleeve and the load is arranged along the axis of the bolt) and the stress distribution of the measuring points ca-2 and ca-5 which near the sleeve in three cases (Figure 6).

It can be seen from Figure 6 that in the initial loading stage, as the load increases, the reaction force of the self-made bearing frame gradually increases, and each measuring point will withstand the gradually increasing compressive stress. The stress of the measuring points which closer to the joint surface are greater than that of the measuring points farther from the joint surface, and the outer surface is subjected to a compressive stress greater than the inner surface. As the loading process continues, the concrete near the sleeve is subjected to a large pulling force, and the compressive stresses of the measuring points ca-1, ca-2 and ca-3 which are closer to the joint surface on the outer surface are gradually reduced. In the measuring pointca-2, the stress even changes from compressive stress into tensile stress, and the tensile stress peak was 2.11 MPa, which exceeds the designed tensile strength of C60 concrete ($f_t = 2.04$ MPa). The development of cracks can be found on the surface of the joint and the outer surface of the segment gradually, and the joint gradually approaches the critical instability state. In the 19th loading step (corresponding to the loading force of 333.74 kN), the concrete in surface layer of the joint is taken out, and the outer surface of the segment has a through crack in the vicinity of the sleeve, which means the joint invalids.

It can be seen from Figure 7 that under the three cases, the change rule of the measuring point ca-2 during the loading process is similar. At the initial stage of loading, ca-2 is subjected to compressive stress, and then is subjected to tensile stress at the later stage of loading. In Case-3 (i.e. the bolt is 100% screwed in, and the vertical joint surface of the segment is loaded), the stress is the first to change, while Case-2 (i.e., the bolt is screwed 70%, the forward bolt is loaded) changes latter and the Case-1 (i.e., the bolt is 100% screwed in, the forward bolt is loaded) changes. Finally, the maximum tensile stress under each case is ranked as Case-1> Case-2> f_t>Case-3. It can be seen that the loading direction and the bolt screwed extent have a great influence on the anti-pull-out performance of the joint. The rule of the inner surface measuring point ca-5 is similar to that of ca-2. There are stress turning points at both

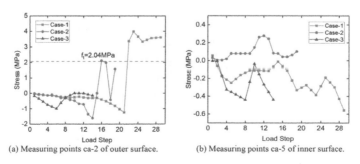

(a) Measuring points ca-2 of outer surface.　　(b) Measuring points ca-5 of inner surface.

Figure 7.　Stress distribution under three cases.

measuring points. However, the stresses of ca-5 in the whole process is smaller than that of ca-2, and the concrete is always in an elastic working state so that no cracks occur. The tensile bolt tensile test has a limited influence on the inner surface of the segment.

3.2 *The stress and displacement distribution of bolt*

Extract the stress and displacement distribution of the bolt during loading, as shown in Figure 8. It can be seen from Figure 8(a) and (b) that for Case-1 and Case-2, since the loading direction is along the axis of bolt, the stress of bolt gradually increases with the increase of the loading force. In the critical instability stage, due to the different screwed lengths (screwe 100% in Case-1 and 70% in Case-2), the maximum stress on the inclined bolts of the two cases is different. The maximum stress of the bolt in Case-1 is 583.10 MPa, while the maximum stress of the bolt in Case-2 is 486.22 MPa. But none of them exceeds the yield strength (900 MPa) of the bolt and the bolt is not damaged (Figure 9). So during the forward drawing process of the oblique bolt, the bolt does not fully exert its bearing capacity. At the same time, the stress value at the middle of the screw is greater than that at the thread. It can indicate that the bolt has a certain transfer efficiency during the forward drawing process (i.e. the ratio of the stress value at the thread to the stress value in the middle of the screw). The maximum transmission efficiency of Case-1 is 69.57%, and that of Case-2 is 66.45%. The stress curves at the same bolt section do not completely coincide.

It can be seen from Figure 8(c) that for the Case-3, since the loading direction is perpendicular to the joint surface of the segment and is at an angle of 30° with the bolt, the measuring points L1-1 and L2-1 on the side of the bolt adjacent to the inner surface are subjected to tensile stress and the measuring points L1-2 and L2-2 on the side adjacent to the outer surface are subjected to compressive stress during the loading process. In the second loading step, the bolt almost reached its yield strength. Then the stress of the bolt continued to reach the peak value of 1503.95 MPa in the sixth loading step, exceeding the bolt ultimate strength of 1000 MPa. The bolt is destroyed (Figure 9). Then the stress gradually lower down. It can be seen that in the vertical drawing process of the oblique bolt, the bolt has fully exerted its bearing capacity. The maximum transfer efficiency of the bolt is 44.77%.

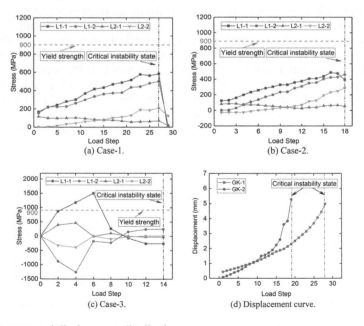

Figure 8.   Bolt stress and displacement distribution.

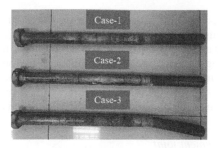

Figure 9.   Bolt status during the joint broken.

As shown in Figure 8(d), for the Case-1 and the Case-2, with the increase of forward load-ing force, the displacement of the bolt along its axis shows a nonlinear increase. However, the displacement growth rates of the two operating conditions are different. The growth rate of the Case-2 is higher than that of the Case-1. The measuring points in Case-3 fails during the loading process, so the displacement rule cannot be known.

3.3   *Acoustic emission information*

The acoustic emission characteristic parameters include change rate and accumulated value. The change value represents the change of the parameter within a certain time and the split-second damage state of the structure. The accumulated value reflects the accumulated value of the acoustic emission parameters of the structure during the loading process, which can gener-ally show the structural damage and damage station (Shen Xingzhu, 2018). According to the acoustic emission information collected by the acoustic emission sensor, the acoustic emission event and the acoustic emission energy changing with the loading time under three cases are plotted in Figure 10.

Overall, the damage process of Case-3 is much faster than Case-1 and Case-2. The reason is that the angle between the vertical loading force and the bolt cause a shear component between the bolt and the sleeve. So the thread on the side of the inner surface of the sleeve is gradually broken, and the surrounding concrete is also damaged by compression and shear. The coupling of bolt, sleeves and concrete is gradually lost, eventually leading to joint failure. The loading process of Case-2 is faster than Case-1. The reason is that both of them are loaded along the bolt, but the extent of screwing of the Case-2 is less than the Case-1. The coupling effect between the bolt and the sleeve is weaker than the Case-1. So the joint is dam-aged first.

In the initial stage of loading of three loading conditions, the acoustic emission events are less. The acoustic emission events of the joints in the subsequent loading process basically showed a steady growth trend. A small amplitude acoustic emission event rate, a cumulative number of acoustic emission events and an acoustic emission energy suddenly increase in the

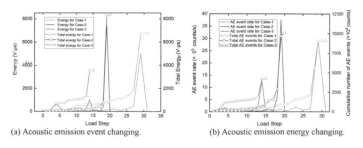

(a) Acoustic emission event changing.   (b) Acoustic emission energy changing.

Figure 10.   Acoustic emission parameters changing with loading curve.

fourth loading step of Case-3. At the 14th loading step, the acoustic emission event rate, the cumulative number of acoustic emission events and the acoustic emission energy suddenly increase. The peak energy was $1004.768 V \cdot \mu s$, and the joint structure was in a critical instability state. At the 19th loading step of Case-2, the acoustic emission event rate, the cumulative number of acoustic emission events and the acoustic emission energy suddenly increase. The peak energy was $7462.134 V \cdot \mu s$, and the joint structure was in a critical instability state. It can be seen that the failure of longitudinal oblique bolt joint under tension is sudden.

3.4 Final failure mode of segment joint

The ultimate failure of the longitudinal oblique bolt joint structure can be attributed to the failure of the coupling between the bolt and the sleeve, the sleeve and the concrete.

Positioning of the acoustic emission position through the acoustic emission probe array during the test (Li Haoran, Yang Chunhe, et al. 2014). The location of the sound source can be determined to show the damage state of joint and the development trend on each loading stage. Due to space limitations, there are the sound emission positioning points of Case-1at different loading steps, as shown in Figure 11.

It can be seen from Figure 11 that the longitudinal oblique bolt joint structure is a progressive failure process under tensile load. The acoustic emission signal is mainly concentrated in the periphery of the sleeve and the orifice. The acoustic emission signal which away from the position of the sleeve is less. As the load increases gradually, new cracks and through cracks are continuously generated. The coupling effects between the bolt and the sleeve, the sleeve and the concrete are gradually lost.

The acoustic emission information and the final failure state of the three cases are shown in Figure 12~13.

As shown in Figure 12 and Figure 13, the inclined sleeve cause the bottom of the sleeve is closer to the outer surface of the segment. The joint failure under three cases is mainly concentrated on the joint surface of the sleeve hole and the outer surface of the segment. Case- 1 and

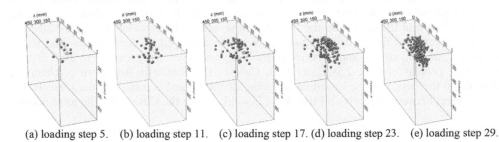

(a) loading step 5.   (b) loading step 11.   (c) loading step 17. (d) loading step 23.   (e) loading step 29.

Figure 11.   Acoustic emission positioning point map of Case-1 loading step.

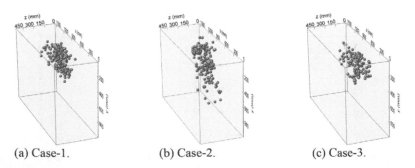

(a) Case-1.   (b) Case-2.   (c) Case-3.

Figure 12.   Acoustic emission positioning point map in the final destruction step of three cases.

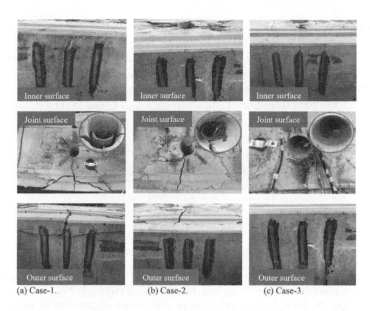

Figure 13.   Joint surface failure diagram of the final failure stage of three Cases.

Case-2 are the cases that pull the bolt along its axis. In the failure stage, the concrete of the joint surface near the outer surface is gradually taken out by the bolt. There occurs cone damage on the outer surface, and the bottom angle of the cone is about 5°~10°. The pre-embedded sleeve is not completely pulled out, but the internal thread has been broken and the joint has failed. For Case-3, due to the small loading force, there is no large through crack on the surface and the concrete is not pulled out. The existence of the shear component destroys one side of the sleeve, and the coupling ability is lost. Combined with the acoustic emission positioning point map, it can be seen that more cracks are generated inside the sleeve and the concrete, not on the surface concrete.

4   CONCLUSIONS

Aiming at the longitudinal oblique bolt joint structure of shield tunnel, the full-scale longitudinal oblique bolt joint pull-out test was carried out. Monitoring by displacement, stress and strain, acoustic emission, etc. the mechanical properties and failure modes of the longitudinal oblique bolt joints under different load conditions are obtained. The relevant conclusions are as follows:

(1) The drawing action of longitudinal inclined bolt joint has a limited influence area, which only affects the joint surface and the outer surface of the segment that near the sleeve. Excessive drawing force will cause cone deformation on the concrete surface of joint.
(2) The extent of bolting screwed into the sleeve affects the pull-out resistance of the longitudinal joint. The greater the extent of screwing, the stronger the coupling ability between the bolt and the sleeve will be, and the more the tensile strength of the longitudinal joint can be exerted.
(3) The sleeve connects the bolt and the segment during the loading process. The thread strength of the inner surface of the sleeve determines the coupling ability of the bolt and the sleeve. The roughness of the outer surface of the sleeve determines the coupling ability with the concrete.

ACKNOWLEDGEMENT

The study was approved by The National Key Research and Development Program of China (Grant No. 2016YFC0802201), National Natural Science Foundation of China (Grant No. 51578457, 51878566), Sichuan Science and Technology Key R&D Project (Grant No. 2018GZ0360), and The fund from State Grid Corporation of China (SHJJGC1700024).

REFERENCE

C.B.M. Blom, E.J. van der Horst, P.S. Jovanovic. Three-dimensional structural analyses of the shield-driven "Green Heart" tunnel of the high-speed line South[J]. *Tunnelling and Underground Space Technology incorporating Trenchless Technology Research*, 1999, 14(2).

He Chuan, Feng Kun, Su Zongxian. Development and application of prototype structure loading test system for large section underwater shield tunnel[J]. *Chinese Journal of Rock Mechanics and Engineering*, 2011,30(02):254–266.

Kawashima Kazuhiko. Earthquake-resistant design of underground structures [M]. *Tokyo: Kashima Publishing House*, 1994.

Lu Liang, Lu Xilin, Fan Peifang. Full-ring experimental study of the lining structure of Shanghai Changjiang tunnel[J]. *Journal of Civil Engineering and Architerture*, 2011(8):732–739.

Lin Gang, Luo Shipei, Ni Juan. Earthquake damage and treatment measures of subway structure[J]. *Modern Tunnel Technology*, 2009,46(04):36–41+47.

Li Haoran, Yang Chunhe, Liu Yugang, et al. Experimental study on acoustic and acoustic emission characteristics of salt rock under uniaxial loading[J]. *Chinese Journal of Rock Mechanics and Engineering*, 2014,33(10):2107–2116.

Shen Xingzhu. Study on mechanical characteristics of double-layer lining structure of shield tunnel[D]. *Southwest Jiaotong University*, 2018.

Ye Fei, Yang Pengbo, Mao Jiahua, et al. Analysis of Longitudinal Stiffness of Shield Tunnel Based on Model Test[J]. *Journal of Geotechnical Engineering*, 2015,37(01):83–90.

Zhao Wusheng, He Xianzhi, Chen Weizhong, et al. Seismic damage analysis of segments and joints between shield tunnel and shaft[J]. *Chinese Journal of Rock Mechanics and Engineering*, 2012, 31 (a02):3847–3854.

Tunnels and Underground Cities: Engineering and Innovation meet Archaeology, Architecture and Art, Volume 6: Innovation in underground engineering, materials and equipment - Part 2 – Peila, Viggiani & Celestino (Eds)
© *2020 Taylor & Francis Group, London, ISBN 978-0-367-46871-2*

A multiscale modelling method for crack propagation of shield tunnel lining

F.Y. Wang & H.W. Huang
Key Laboratory of Geotechnical and Underground Engineering of Ministry of Education; Department of Geotechnical Engineering, College of Civil Engineering, Tongji University, Shanghai, China

ABSTRACT: As a major defect, crack has a significant effect on the mechanical behavior of shield tunnel lining. However, there is an obvious multiscale phenomenon considering the size of crack and the tunnel lining structure. A novel multiscale modelling method has been proposed in which the damaged and the undamaged zones are firstly recognized according to the pre-analysis results and then simulated at different scales. After validated by the results of segmental loading test, the multiscale simulations under different loading paths have been carried out to investigate the effect of crack evolution on the performance of lining segment. The study indicates that the proposed multiscale model can reveal random propagation of multiple cracks and stiffness degradation of the lining segment. Moreover, the stiffness of the lining segment is not only dependent on the loading path but also the cracking features.

1 INTRODUCTION

Crack is a typical and common defect in the existing tunnel lining which can affect the mechanical behaviors in some cases (Ghannoum, 2018; Grégoire et al., 2013). As a major material of tunnel lining, concrete consisting of mortar and aggregate etc. exhibits heterogeneous behavior in mesoscale, especially for the concrete with crack. And the heterogeneous constituents have serious influence on the failure phenomenon such as crack (Hansen et al., 1991; Sun and Li, 2016). Besides, the components of concrete and the crack are of millimeter. Therefore, it is supposed that the tunnel lining in service with crack is characterized with a multiscale phenomenon. Herein, it is worth pointing out that mesoscale is defined as the scale, ranging from nanometer to millimeter, which affects the mechanical behavior of the material. Therefore, to accurately evaluate the performance of tunnel lining, the mesoscopic features including crack, aggregate and mortar etc. should be involved in the analysis.

For its simplification and computational efficiency, homogeneous model is usually utilized to obtain structure's response (Katebi et al., 2015). However, this kind of model is not available to consider the influence of constituents and random propagation of crack which is frequently found in the existing shield tunnel lining. Moreover, it is inefficient if the simulation of the entire tunnel lining involves mesoscopic features. Therefore suitable model should be proposed to take into account both the effects of mesoscopic features including the components of the concrete and the crack on the tunnel lining and the computational efficiency. Actually, most regions of the tunnel lining are under undamaged linear elastic stage. In this stage, the influence of mesostructures on the material behaviors can be neglected (Unger and Eckardt, 2011). Therefore, heterogeneous constituents and random propagation of crack are considered only in the damaged or cracked region, while the non-damaged region is still regarded as homogeneous material in the consideration of the computational efficiency.

The objective of this paper is to propose a multiscale modelling method for the tunnel lining which can consider the mesoscopic effects and analyze the mechanical behaviors based on the model. The paper is arranged as follows. In Section 2, the multiscale modelling method for tunnel lining is established and the reasonability of the model is illustrated by comparing the simulation with the test. In Section 3, loading path of the vulnerable section of the tunnel lining under surface loading is estimated with the typical stratum-structure model. And the multiscale modelling method is then used to analyze the mechanical behavior and evaluation features of multiple cracks of the segment under different loading paths. Finally, conclusions are drawn from the results in Section 4.

2   MULTISCALE MODELLING METHOD FOR SHIELD TUNNEL LINING

2.1   Implementation of multiscale model

Combining computational efficiency and accuracy, multiscale modelling method is proposed in this section. In this model, the regions where the crack randomly initiate and propagate are simulated by mesostructure cohesive zone modelling method. And the non-damaged regions are done by homogeneous macroscale method. The detail of multiscale modelling method includes the following procedures.

1. The potential damaged region in which crack may initiate and propagate is recognized by the pre-analysis of the whole region with macroscale homogeneous model.
2. As above mentioned, the damaged region is controlled by the mesoscopic features. There are two crucial steps to realize the simulation of the damaged region with incorporating the mesoscopic information. Firstly, the physical components of concrete are randomly generated in the damaged region by a bespoke in-house computer program. And the generated geometric model is meshed with triangular elements. After that cohesive elements are inserted into the damaged region to simulate the initiation and propagation of crack.
3. Fracture properties of cohesive element are firstly calibrated by uniaxial compression and tension tests. And then the fracture properties are assigned to the cohesive elements.
4. As for the whole model, the non-damaged region and the damaged region are connected together according to displacement compatibility equations at the interfaces. The model is applied with boundary conditions and solved. So far the process of the multiscale model is completed.

There are two crucial points in the realization of multiscale model, the recognition of the damaged region and the random propagation simulation of the multiple cracks. As for the first point, pre-analysis is performed with the homogeneous finite element (FE) model. In this model, the whole region is assumed to be trilinearly elastic-plastic material. And the potential damaged region is distinguished according to the equivalent plastic strain nephogram which is elaborated in the following section.

The mesoscopic features of the damaged region are needed to be reflected in the simulation. After the recognition of the damaged region, the random initiation and propagation of the crack in this region are simulated by incorporating the mesoscopic features. As we know, concrete is a kind of multiple media material, consisting of aggregate and mortar. Therefore, the simulation of concrete in mesoscale includes two aspects, the generation of random aggregates and the simulation method of crack. Based on this starting point, mesostructure cohesive zone modelling method (MCZM) is proposed. In this method, aggregate following Fuller's grading curve is assumed to be elliptic and have a uniform distribution in the potentially damaged zone. And the random propagation of crack is realized by inserting cohesive interface element (CIE) which has been used in the simulation of concrete block and beam (Rodrigues et al., 2016; Wang et al., 2015a; Wang et al., 2015b).

Before the process of MCZM is introduced, two primary assumptions are made, which aggregate and mortar are simulated as elastic materials and the propagation of crack merely propagate along the interface between the aggregate and the mortar or inside the mortar. Based on these two basic assumptions, random aggregate model including aggregate and mortar is primarily generated according to the gradation of concrete. Then the geometric model is meshed by commercial software ABAQUS. And cohesive interface elements (CIE)

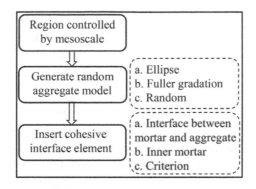

Figure 1.   Flow chart for mesostructure cohesive zone modelling method.

are inserted inside the mortar and also between the aggregate and the mortar. Material properties of aggregate, mortar and the interface are then given to the different domains. It is worth to point that the material properties of CIE between the aggregate and the mortar are different from those of CIE inside the mortar, to allow for the weakened interface between aggregate and mortar. The detailed procedure of mesostructure cohesive zone modelling method is shown in Figure 1.

Another main point for MCZM is to determine reasonable material parameters. For this purpose, mesostructure cohesive zone model of uniaxial tension and compression are performed to inverse the material properties based on the test results. When the simulation results are consistent with the test results, the material parameters are finally determined. Meanwhile, the results reveal that MCZM is accurate and reliable to simulate the random propagation of tensile and compressive crack. By this point, the damaged region simulated with MCZM is finished.

2.2   Validation of multiscale model by segmental loading test

In order to validate the proposed model, the results of the model are compared with the results of segmental loading test. The schematic diagram of segmental loading test is plotted in Figure 2. The equipment shown in the figure is segmental loading system TJGPJ-2000. The segment is hinged at the ends. And the segment is loaded by the vertical and the horizontal jacks. The loading path is controlled by setting the ratio of vertical load to the horizontal load.

Before the simulation, pre-analysis is conducted by homogeneous model to recognize the potential damaged region of the multiscale model. The segment in the homogeneous model is regarded as trilinearly elastic-plastic material. After the computation of the homogeneous model, the equivalent plastic strain nephogram is presented in Figure 3(a). Actually, the region with plastic strain is probably damaged. Therefore, this potential damaged region is simulated

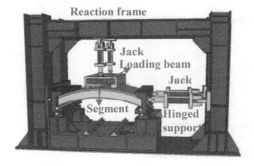

Figure 2.   Schematic diagram of segmental loading test.

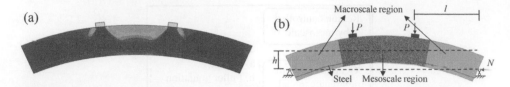

Figure 3. Realization of multiscale model: (a) Equivalent plastic strain; (b) Multiscale model.

with MCZM and the other regions are still assumed to be homogeneous. Finally, the multiscale model of segment shown in Figure 3(b) is generalized according to the aforementioned implementation process. The parameters in mesoscale and macroscale are presented in Table 1.

As for different segments, the propagation of crack is different because of the diversity of the concrete component in mesoscale. The final cracking features of the one experiment and the one simulation are shown in Figure 4. From the results of the test and the simulation, cracks initiate vertically and then propagate in the area between the loading beams. Actually, there are several cracks propagating with loading after the presence of the crack. Meanwhile, the propagation process of the multiple cracks in the test is in good agreement with that in the simulation. It's worth noting that the mesostructure of the segment, such as the aggregate distribution, directly influences the crack path. On the whole, the multiscale model adopted in this section is available to reflect random propagation of multiple cracks.

There are several distinct features in the propagation process of the crack. At the initial fracture stage, the crack expands vertically upwards, appearing as I-type crack. The cracks in this stage are mainly located in the tensile region of the segment, which is resulted from the tensile stress. At the end stage, Y-type crack emerges. The reason for the Y-type crack is that the crack move up to the compression region and the concrete is subjected to tension and compression. Furthermore, the Y-type crack is a mixed mode crack which is caused by the resultant of compression and tension.

Table 1. Material properties for multiscale model.

Materials	Elastic modulus /GPa	Tensile strength /MPa	Shear strength /MPa	Poisson's ratio	Fracture energy (I) kN/m	Fracture energy (II) kN/m
Aggregate	70	-	-	0.16	-	-
Mortar	25	-	-	0.22	-	-
Mortar-mortar interface	100000	4	25	-	0.030	8.00
Mortar-aggregate interface	100000	3	15	-	0.018	5.00
Concrete	32			0.18		

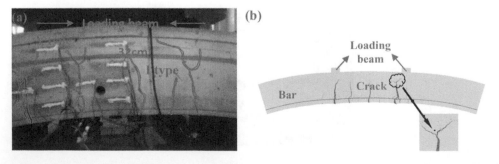

Figure 4. Cracking features at failure: (a) Loading test; (b) Numerical simulation (scaling factor of 20).

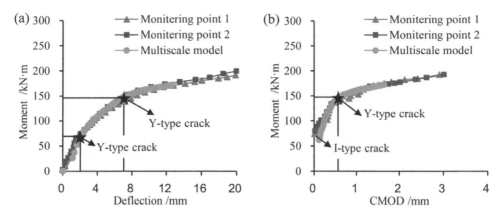

Figure 5. Comparison between numerical and experimental results: (a) Moment-deflection curves, (b) Moment-CMOD curves.

The crack growth laws of the loading experiment are basically consistent with those obtained by mutiscale model. To describe the mechanical behavior in macroscopic scale, the curves of internal force and deformation of the test and the simulation are plotted in Figure 5. The bending moment-deflection curves of two monitoring points during the test and one point of the multiscale model at the mid-span are shown in Figure 5(a). Besides, the maximum opening crack is selected as the analysis object. And the bending moment- the crack mouth opening displacement (CMOD) curves of are plotted in Figure 5(b). The star marks in the figure signify that Y-type crack appears when the moment reaches 145kN·m.

Figure 5 represents the bending moment in the mid-span section respect to the corresponding deflection and CMOD. And the star marks in the figure indicate the occurrence of I-type crack and Y-type crack, respectively. The curves obtained by the model are basically in line with those of the experiment. As shown, the relationship between internal force and deformation of non-crack segment is approximately linear in the first loading stage. And in the second loading stage, the slope of bending moment-deflection or CMOD curves nonlinearly decreases with the loading after the presence of the cracks. Finally, the slope of the curves decreases sharply with loading in the third loading stage. Furthermore, the bending moment-deflection curves illustrate that the stiffness slightly decreases with loading in the first two stages while the stiffness decreases rapidly after the emergence of Y-type crack.

3   MULTISCALE ANALYSIS OF SHIELD TUNNEL LINING

3.1   *Determination of the loading path*

In this section, the proposed multiscale model is utilized to analysis the mechanical behavior of the tunnel lining under different loading paths. To achieve this, the loading path of the tunnel lining under the surface loading is estimated by the typical stratum-structure model. Then, the proposed multiscale modelling method is conducted to evaluation the performance of the tunnel lining according to the estimating loading path.

Typical stratum-structure model has been extensively carried out by some relevant research (e.g. Katebi et al., 2015; Lambrughi et al., 2012; Mollon et al., 2012). The schematic of typical stratum-structure model is presented in Figure 6, in which the overburden of the tunnel is three times tunnel diameter (*D*). The dimension of the model is taken to minimize of the boundaries effects and improve the computational efficiency. The model extends to a depth of three times the tunnel diameter (*D*) below the tunnel invert and laterally to a distance of 5.5D from the tunnel centerline. Additionally, there are four steps in the model, initial geostatic stress, excavation of the tunnel, installation of the tunnel lining and

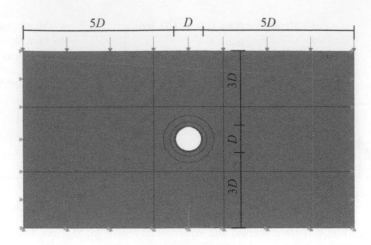

Figure 6.   Schematic diagram of the stratum-structure model.

Table 2.   Geotechnical properties for the soil layer in the model.

Soil type	Depth /m	Density γ (kN/m^3)	Elastic modulus E (kPa)	Cohesion c (kPa)	Friction angle φ (°)	Dilation angle (°)
Clay ⑤-1	42.0	18.3	20000	21.0	12.5	0

application of the surface loading. The stratum-structure model is used to estimate the loading path of the tunnel lining.

The soil is simulated with 4-node first order reduced integration elements (CPE4R). And the soil behavior is assumed to be governed by an elastic perfectly-plastic constitutive relation based on the Mohr-Coulomb criterion with a non-associative flow rule. To be simplified, only one soil layer is taken into account in the model and the parameters are listed in Table 2. The tunnel lining consists of six segments. Relative motion is allowed between adjacent segments and between segments and soil. Therefore, contact friction is adopted for this behavior. The node-surface contact is set up between segment and surrounding soil and between the adjacent segments. The normal and the tangential behaviors of the contact are governed by isotropic Coulomb friction model and hard contact, respectively.

The internal force along the circumferential section of the tunnel lining is obtained from the results of the model. Figure 7 represents the ratio of bending moment to axial force. Circumferential axis denotes the angle θ of section. From the figure, relative large ratio of the tunnel lining are located at 0°, 90°, 150°, 210° and 270°. And the ratio ranges from 0.1 to 0.3. Generally the bending capacity of the tunnel lining is much weaker than the compression. Therefore, it is supposed that the larger the ratio of bending moment to axial force, the more dangerous it is.

According to the calculation results, the loading path of the multiscale model is estimated as the following process. The ratio of bending moment to surface load and the ratio of axial force to surface load are denoted by $k_M=(M-M_{ini})/P_{load}$ and $k_N=(N-N_{ini})/P_{load}$, respectively. P_{load} is the surface load. M and N are internal force calculated by the stratum-structure model. After a series transformation, the Eq.(1) is obtained to determine the loading path of multiscale model with respect to the corresponding stratum-structure model.

$$P/N = (k_M/k_N + h)/l = (e + h)/l \tag{1}$$

where P, N, h and l are the physical parameters of the multiscale model which is shown in Figure 3, and $e=M/N$ is eccentricity distance.

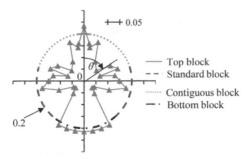

Figure 7.   Ratio of bending moment to axial force along circumferential section of the tunnel lining.

According to the results of the stratum-structure model and the Eq.(1), P/N of the vulnerable section is equal to 0.65. In order to investigate the effects on the performance of the tunnel lining, the values of P/N, 0.55, 0.65 and 0.75 are analyzed using the multiscale model.

3.2   *Results analysis*

The segment under different loading paths specified in the previous section is analyzed with the multiscale model. And crack and bearing capacity are significant indexes to the performance of the tunnel lining. Based on the analysis results, the cracking features and the effects of the crack on the bearing capacity of the tunnel lining under the specified loading paths are investigated

Figure 8 shows the cracking features under different loading paths with these values of P/N 0.55, 0.65 and 0.75, respectively. Actually, the value of P/N indicates eccentricity of the segment section. Multiple cracks are mainly distributed between the loading beams for different loading paths. In addition, there are more cracks when eccentricity is smaller. However, the crack mouth opening displacement of the segment with large eccentricity is larger for the same bending moment.

The curves of bending moment versus deflection and CMOD in the mid-span section are plotted in Figure 9. The curves shown in Figure 9 are divided into three stages according to the discussion in the previous section. In the first stage, loading path has little influence on the curves of the moment-deflection. In the last two stages, the effect of eccentricity is gradually significant with loading. Herein, the maximum bending moment is regarded as the bending bearing capacity. It's easy to find that the bearing capacity increases with the reduction of

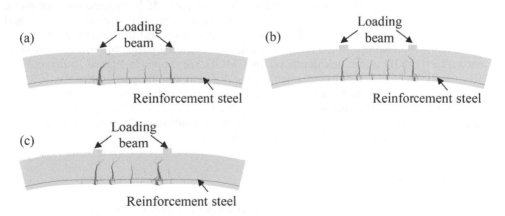

Figure 8.   Cracking features under different loading paths. (a) $P/N=0.55$ (scaling factor of 40), (b) P/N=0.65 (scaling factor of 10), (c) $P/N=0.75$ (scaling factor of 10).

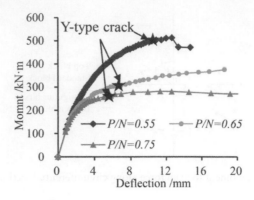

Figure 9.   Curves of moment versus deflection under different loading paths.

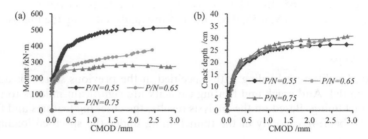

Figure 10.   Curves of bending moment versus CMOD and crack depth versus CMOD. (a) Moment-CMOD, (b) crack depth -CMOD.

eccentricity from the curves. And the curves also indicate that the stiffness of the segment degenerates to a large extent in the later loading stage. This situation is obvious for the large eccentricity.

From the moment-deflection curves, the stiffness of the segment is mainly dependent on the loading path and the development of the multiple cracks. Before the presence of the crack, the stiffness of the segment has little to do with the loading path. In the cracking stage, the stiffness declines with the increase of eccentricity. When Y-type cracks appear, the stiffness of the segment decreases sharply. Herein, the authors suggest that the bending moment at the presence of Y-type crack is regarded as the ultimate bearing capacity. And the stiffness of the segment depends on the cracking features and the loading path. If homogeneous monoscale model is utilized to simulate tunnel lining with crack, the stiffness of the segment is supposed to be reasonably reduced according to both the cracking features and the loading path.

To reveal the evolution law of the crack under different loading paths, the curves of bending moment versus CMOD and crack depth versus CMOD are shown in Figure 10 (a) and (b). There is little difference of cracking moment for the different loading paths. However, CMOD grows rapidly when the Y-type crack is present. And the gradient of moment to CMOD declines rapidly at the later loading stage for the large value of P/N. Although there are large differences of bending moment-CMOD under different loading paths, the curves of crack depth versus CMOD are coincident with each other.

4   CONCLUSIONS

The multiscale modelling method is proposed to reveal the random evolution of multiple cracks in the segment. The proposed model is available and effective to simulate the initiation and the random propagation of multiple cracks. Meanwhile material behavior and the bearing capacity obtained by the proposed model are analyzed in this study. According to the above analysis of segment under different loading paths, the bearing capacity and the characteristics of the crack evolution can be concluded as follows.

1. The bearing capacity of the segment increases with the reduction of eccentricity.
2. There are three stages for the cracking under the different loading paths, non-crack (linear state), I-type crack (non-linear stage) and Y-type crack (failure stage).
3. The stiffness of the segment is not only dependent on the loading paths but also the crack features. The influence of the loading path and the crack features on the stiffness of the segment gradually increases with loading.
4. The curves of the relation between crack opening mouth displacement and crack depth are closely consistent for different loading paths.

ACKNOWLEDGMENTS

The study has been supported by the Natural Science Foundation Committee Program of China (No. 51538009). This financial support is gratefully acknowledged.

REFERENCES

Ghannoum, M., 2018. Effects of heterogeneity of concrete on the mechanical behavior of structures at different scales. Grenoble: HAL.

Grégoire, D., Rojas-Solano, L.B., Pijaudier-Cabot, G., 2013. Failure and size effect for notched and unnotched concrete beams. *International Journal for Numerical & Analytical Methods in Geomechanics* 37: 1434–1452.

Hansen, A., Hinrichsen, E.L., Roux, S., 1991. Scale-invariant disorder in fracture and related breakdown phenomena. *Physical Review B Condensed Matter* 43: 665–678.

Katebi, H., Rezaei, A.H., Hajialilue-Bonab, M., Tarifard, A., 2015. Assessment the influence of ground stratification, tunnel and surface buildings specifications on shield tunnel lining loads (by FEM). *Tunnelling and Underground Space Technology incorporating Trenchless Technology Research* 49: 67–78.

Lambrughi, A., Rodríguez, L.M., Castellanza, R., 2012. Development and validation of a 3D numerical model for TBM–EPB mechanised excavations. *Computers & Geotechnics* 40: 97–113.

Mollon, G., Dias, D., Soubra, A.H., 2012. Probabilistic analyses of tunneling-induced ground movements. Acta Geotechnica 8: 181–199.

Rodrigues, E. A., Manzoli, O.L., Bitencourt, Jr., Luís, A. G., et al., 2016. 2D mesoscale model for concrete based on the use of interface element with a high aspect ratio. *International Journal of Solids & Structures* 94–95:112–124.

Sun, B., Li, Z., 2016. Multi-scale modeling and trans-level simulation from material meso-damage to structural failure of reinforced concrete frame structures under seismic loading. *Journal of Computational Science* 12: 38–50.

Wang, X., Yang, Z., Jivkov, A.P., 2015a. Monte Carlo simulations of mesoscale fracture of concrete with random aggregates and pores: a size effect study. *Construction & Building Materials* 80: 262–272.

Wang, X.F., Yang, Z.J., Yates, J.R., Jivkov, A.P., Zhang, C., 2015b. Monte Carlo simulations of mesoscale fracture modelling of concrete with random aggregates and pores. *Construction & Building Materials* 75: 35–45.

*Tunnels and Underground Cities: Engineering and Innovation meet Archaeology,
Architecture and Art, Volume 6: Innovation in underground engineering,
materials and equipment - Part 2 – Peila, Viggiani & Celestino (Eds)*
© *2020 Taylor & Francis Group, London, ISBN 978-0-367-46871-2*

Real-time sensing technology of excavated material in earth pressure balanced shield tunnelling construction

Q. Wang, X. Xie & B. Zhou
Department of Geotechnical Engineering, Tongji University, Shanghai, China

Z. Huang
Nanning Rail Transit Co., Ltd., Nanning, China

ABSTRACT: Settlement control has been always the main concern during shield tunnelling construction, especially in earth pressure balanced (EPB) shield tunnelling construction. However, owing to the lack of efficient real-time sensing methods of excavated material, settlement even sinkholes occasionally occurred in urban areas. Therefore, a near real-time sensing technology of excavated material in EPB shield was designed and developed with the weighing cells installed on the bottom of the muck cars, which solved the problem of data transmission due to the muck car moving by the means of wireless communication. The filed results in Nanning metro indicate that the measured error is less than 5% compared with the gantry crane weighing results and it is beneficial for the engineers to have a knowledge of the excavated material in real-time to make better decision for steering tunnelling parameters to control settlement.

1 INTRODUCTION

In urban areas, settlement control has been always the main concern in shield tunnelling construction, especially for earth pressure balanced (EPB) shield tunnelling construction. Usually, unexpected deformation (settlement and heave) or even sinkholes on the ground occurred due to the instability in tunnel face, which is related to the excavated material (Anagnostou and Kovari, 1996). Therefore the volume control of excavated material is crucial for EPB shield construction. Shield operators steer the rotation speed of the screw conveyer together with some other tunnel parameters to change the excavated material so as to keep balance between the actual volume and theoretical volume of excavated material (Liu *et al.*, 2011). However, there is a lack of an efficient real-time measuring actual excavated material method during shield tunnelling construction, which is suitable for different kinds of geological conditions.

Nowadays, the excavated material is mostly carried by the means of muck cars in EPB shield in China, and there are two kinds of measuring methods for the excavated material, which can be distinguished by the measuring locations. The front measurement refers to measure the excavated material near the tailskin of the EPB shield, which can be regarded as the near real-time sensing method. The end measurement occurs at the launching shaft, which is usually more reliable but can't reflect the real-time excavated situation of the shield machine. Thus, it's more significant to employ the front measurement methods for better construction management. The belt weigher and laser scanning method are typical approaches for the front measurement. In Dragon project (Galler, 2013), DB 320 tunnel in KCRC project in Hong Kong (Chen Jian, 2004) and Bund tunnel in Shanghai, the belt weigher method was adopted to measure the weight of excavated material (Figure 1). The filed results indicated that for the cohesion less soil, such as the soil in DB 320 tunnel, the belt weigher is appropriate while for the clay, such as the soft caly soil in Shanghai, the measurement error can't satisfy the

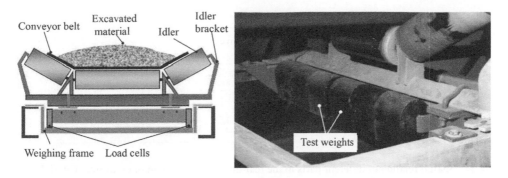

Figure 1. Conveyor belt used for measuring weight of the excavated material in EPB shield tunnelling.

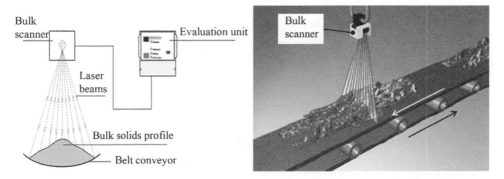

Figure 2. Laser scanning method applied for measuring volume of the excavated material in EPB shield.

engineering demand. The laser scanning method was implemented in the Dragon project and Bund tunnel, as illustrated in Figure 2. Although the laser scanning method can directly obtain the volume of excavated material, several drawbacks such as high cost, dust pollution to the lens, poor durability, limit its application. Meanwhile, it's difficult to obtain the profile at the bottom of the excavated material on the conveyor belt. Thus, it's vital to carry out a new method to measure the excavated material in real-time for EPB shield tunnelling construction.

2 REAL-TIME SENSING SYSTEM DESIGN OF EXCAVATED MATERIAL

2.1 System architecture

Our system aims to provide a solution for real-time sensing technology of excavated material in different geological condition and meeting the required precision for EPB shield tunnelling construction. The system architecture is demonstrated in Figure 3.

The soil is excavated by the cutter head and carried from the chamber to the muck cars via the screw conveyer and belt conveyer. The weighing modules installed at the bottom of the muck cars capture the change of the weight and then the signal will be transmitted from the slave wireless module to the master wireless module, and finally to the use's client (software used to illustrate the weight results) in the shield operation room, as depicted in Figure 4. In the upper computer in operation room, the measured results of excavated material are displayed with the cylinder stroke obtained from the shield machine PLC (Programmable Logic Controller) system. Operators can steer the tunnelling parameters based on the measured results of excavated material together with the backfill grouting so as to control the settlement.

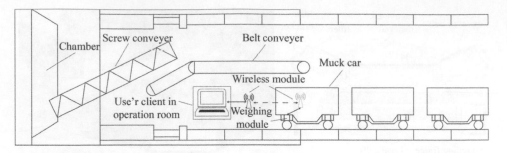

Figure 3. Spatial relation of different parts in the real-time sensing system of excavated material.

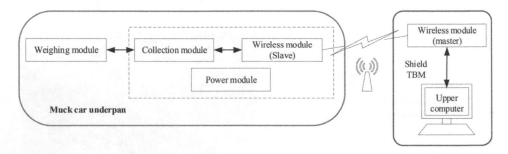

Figure 4. Data transmission of weight from weighing module to the shield TBM operation room.

2.2 Sensor selection

There are several requirements for the sensors used for real-time sensing of excavated material, as list in followings:

1) The sensors are capable of static load and dynamic load;
2) The measurement error should satisfy the engineering demand, which is less than 5%;
3) Wireless transmission is necessary for the measured data, which should have a good stability in the tunnel;
4) Re-development is essential for the data display software in the upper computer in order to coordinate with the PLC data;
5) High ingress protection (IP) to keep good performance in the wet and dusty environment.

Through investigations, most of the industrial weighing modules can meet our demands and the sensors from HBM are chosen. The RTNC3/22 sensor has a measuring range of 22 t and a IP level of IP68, which is capable of static and dynamic load very well. The measurement accuracy is 1/3000 when the sensors leave factory while the measurement accuracy can reach a level of 1/1000 when the sensors have been installed to the weighing module. The RTN/M2LBR22T module is used to protect the sensors.

2.3 Sensor installation

The weighing modules are installed on the top of the bogies of the muck cars, where all the load of the excavated material will be transferred to the wheels, as shown in Figure 5. There are four weighing modules in each side of the muck car. The detailed layout of the weighing module is demonstrated in Figure 6, which including four symmetrically arranged sensors, four horizontal positive stops and overturn-preventing devices. Meanwhile, the power module that uses rechargeable lithium battery, the junction box and slave wireless module are also included in the weighing module assembly. The power module is the power source of the collection module and slave wireless module, which need to be changed every certain interval (5 days in general). All the wiring connections are packaged in the junction box.

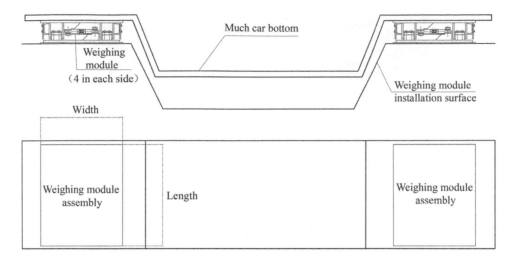

Figure 5.   Schematic diagram of weighing modules installation on the muck car.

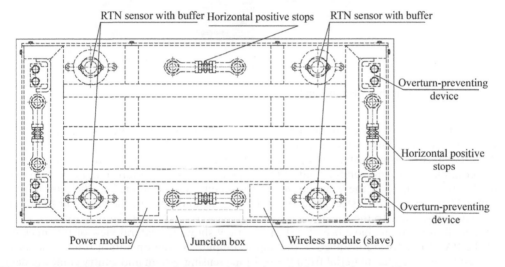

Figure 6.   Detailed layout of the weighing module in the weighing module assembly.

3   FIELD TEST

3.1   *General description of the project*

The filed test was conducted in a section of Nanning metro line 3, which used the CREG EPB shield TBM. The EPB shield would drive in the mudstone whose volume-weight is 2.2~2.4t/m³ and expansion coefficient is 1.35 (Wang *et al.*, 2017). The volume of a muck car is 13 m³, thus the theoretical weight of excavated material is 20~23t for a muck car. An empty muck car has a weight of 7t. Therefore a full load muck car will have a weight of 27~30t approximately.

3.2   *Weighing module assembly Installation and results*

In the field test, only one muck car has been used for the sake of economy. A total of eight sensors have been implemented in this test. The weighing module is shown in Figure 7. The measured data is transferred from the slave wireless module to the master wireless module

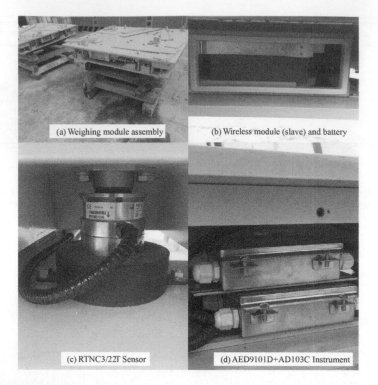

Figure 7.   Weighing module assembly and its contents.

that installed on the trolley of the shield machine, and then to the upper computer in the operation room by the means of optical cable.

Figure 8 illustrates the installation process of the weighing module. The first step is to weld an interim plate to strengthen the joint of the weighing module assembly and muck car unperpan (Figure 8 (a)). Then the weighing module assembly will be welded to the interim plate (Figure 8 (b)). Finally, four steel plates are welded to the underpan of the muck car to prevent muck car rollover due to the rise of the center of gravity of muck car (Figure 8 (c)).

When all installations have been finished, Figure 9 shows the lateral view of the muck car and the PANLE software installed in the upper computer in the operation room. The measured results of excavated material from the real-time sensing system and gantry crane are demonstrated in Figure 10. The maxmium value of measured difference between real-time sensing system and crane is 1.3t and the minmium value is 0.1t. The filed results indicate that the measured relative error is less than 5%, which satisfies the engineering demand. The wireless transmission performed quite well in terms of the stability and the rechargeable battery was updated every 3~5 days.

3.3   *Data integrate to PLC system*

In order to provide better service for the shield operators, the measured weight data will be integrated to the PLC system. When the installation process has been finished in section 3.2, the measured data can only be shown on the PANLE software, which is inconvenient for the operators to make decisions based on the measured weight and PLC data. Thus a conversion programme was developed to transfer the measured data from the PANLE software to the OPC (object linking and embedding (OLE) for process control) server.

Figure 11 shows the conversion programme and a new-developed module in the PLC data system. The measured excavated material weight data is demonstrated with the cylinder stroke, including the current ring and the history rings. Engineers can set up the target value of

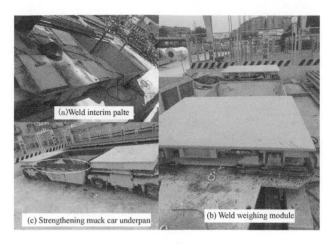

Figure 8.   Installation of weighing module.

Figure 9.   Muck car with weighing module and display software (PANLE) in operation room.

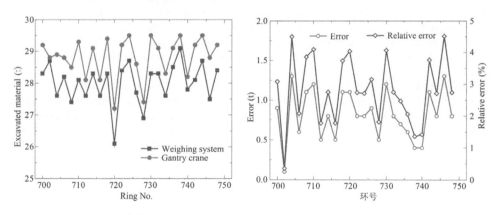

Figure 10.   Weight measured results comparison between our weighing system and gantry crane.

excavated material for each ring based on the theoretical calculation results and filed experience. When the measured value of excavated material weight has exceeded the target value, the system will give an alarm to the operators to take measures for steering tunnelling parameters.

As the muck cars has to be driven to the launch sharft when one ring has finished excavation, the master wireless module will lose connection with the slave one. Therefore, it's vital to create a kind of auto data reset logic. The master wireless module shall always try to communicate with the slave one. After 5 times, if the wireless communication is still in the failure

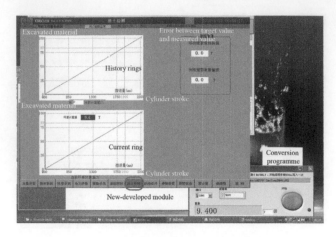

Figure 11. Conversion programme to transfer the measured excavated material weight data to PLC data system.

state, the system will regard it as the disconnection state and the measured data will be reset. In the disconnection state, once the wireless communication between the master module and slave module is successful, the system will start a new cycle to collect the measured data, which means a new ring has begun.

4 CONCLUSION

This paper propose a novel measured system of the excavated material for EPB shield construction, which realized the function of near real-time sensing technology of excavated material. The real-time measured system consists of weighing sensors, protection equipments, power module and wireless module. The installation and test of weighing module assembly have been conducted in Nanning metro and the filed results indicate that the measured error is less than 5% compared with the gantry crane weighing results. In order to help the operators make better decisions for steering tunnelling parameters to control settlement, the measured weight data has been integrated with the original PLC data.

The presented contents in this paper is just the first step for more efficient management of excavated material. We proved the feasibility of near real time measured data sensing and transmission in the tunnel. Measured error may come from the installation process of the weighing module assembly on the old muck car. New muck cars are designed and manufactured to fit the weighing system and are going to conduct the filed tests.

REFERENCES

G. Anagnostou, & K. Kovari. 1996. *Face stability in slurry and EPB sheid tunneling*, Tunnels & Tunnelling International 28(12): 453–458.

J. Chen. 2004. Excavated soil management technology in shield tunnelling construction, Tunnel construction, 25(6): 69–71.

R. Galler. 2013. *DRAGON-Development of resource efficient and advanced underground technologies*. Leoben, Austria. Available at: http://www.dragonproject.eu/en/.

X. Liu, S. Chen, H. Ma & R. Liu. 2011. *Optimal earth pressure balance control for shield tunneling based on LS-SVM and PSO*, Automation in Construction. 20(4): 321–327.

Q. Wang, X. Xie, Z. Huang & Y. Qi. 2017. *Protection of Adjacent Buildings Due to Mixshield Tunnelling in Mixed Ground with Round Gravel and Mudstone*, Proceedings of the World Tunnel Congress 2017 – Surface challenges – Underground solutions. Bergen, Norway, 14635-1–10.

Tunnels and Underground Cities: Engineering and Innovation meet Archaeology, Architecture and Art, Volume 6: Innovation in underground engineering, materials and equipment - Part 2 – Peila, Viggiani & Celestino (Eds)
© 2020 Taylor & Francis Group, London, ISBN 978-0-367-46871-2

3D simulations for the Brenner Base Tunnel considering interaction effects

T. Weifner & K. Bergmeister
Brenner Base Tunnel BBT SE, Innsbruck, Austria

ABSTRACT: The Brenner Base Tunnel will be the world's longest tunnel with a total length of 64 km. The tunnel system of the Brenner Base Tunnel consists of several tunnels; it is therefore of crucial importance to consider how the single tunnels act together during and after their construction. In order to evaluate these interaction effects, several 3D finite element calculations with advanced non-linear material laws were carried out, which are illustrated in the contribution. The determination of the material parameters for the used material laws for the rock mass and concrete is described and the finite element model including the concrete shells is explained. The calculations show that advanced material models are appropriate and loading/unloading effects can be simulated realistically. The calculations show further that the influence of the excavation to the concrete shells of the existing tunnel is quite relevant and appropriate measures must be taken to avoid risks of damage.

1 INTRODUCTION

The Brenner Brenner Base Tunnel (BBT) includes two single track tunnels (called also main tunnels) and an exploratory tunnel placed 12 m below in between the main tunnels, c.f. Bergmeister (2012), Insam et al., in prep). For safety reasons, the main tunnels are connected each 333 m with by-passes.

The system includes also 3 underground emergency stations, two connection tunnels to the existing Innsbruck railway bypass tunnel and also rescue and access tunnels.

While passenger trains arriving from Italy will pass the Innsbruck railway station (tunnel length 55 km), freight trains will drive directly from Italy (Fortezza) to the lower Inn valley (Tulfes) and reach the main tunnel via the connection tunnel to the already existing Innsbruck railway bypass tunnel and cross the Alps in an underground tunnel of 64 km length, c.f. Bergmeister (2011).

The existing Innsbruck railway bypass tunnel –a single two-track tunnel – is connected with the two main tunnels by two so-called connecting tunnels. Due to the transition from left hand traffic in Italy to right hand traffic in Austria an overpassing of the western connecting tunnel over the eastern connecting tunnel is necessary (cf. Figure 1, red circle). The vertical distance between both rail levels of the two connection tunnels is quite small at the projected intersection point of the two tunnel axes and the difference of the rail levels in the two tunnels at this point is only 15.061 m. The connection tunnels are excavated by drill and blast and using the New Austria Tunnel Method (NATM), cf. Rabcewicz (1965).

This situation was investigated with 2D and 3D finite element calculations in order to know the best excavation sequence to minimize the load effects on the sprayed concrete shells.

In the last years beside 2D finite element (FE) modelling also 3D numerical modelling of tunnel excavations becomes more and more of common. Most of the case studies in the literature regarding finite element modelling of NATM tunnels refer to swallow tunnels, e.g. Svoboda & Masin (2011), Umar et al. (2018).

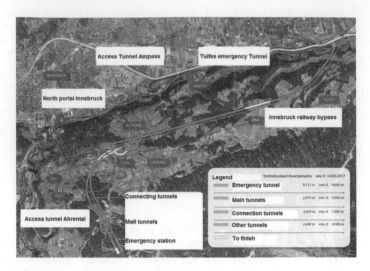

Figure 1.   Ground plan of Brenner Base Tunnel in the zone of Innsbruck with the overpass of the western over the eastern connection tunnel (red circle).

In Svoboda & Masin (2011) 2D and 3D finite elements of the Heathrow express trial tunnel in London (UK) and the Dobrovekeho tunnel in Brno (CZ) are presented and this contribution shows that it is recommendable to start with 2D calculation first and to continue with 3D calculations after some knowledge from the 2D calculations. The same approach to carry out first 2D calculations and then 3D analyses is also followed in this contribution. The calculations were made with the software TOCHNOG using various material laws (Hypoplasticity, Hardening Soil model, Mohr-Coulomb model).

The contribution of Umar et al. (2018) threads tunnel interaction with a 3-dimensional numerical model of the overpassing of the transport tunnel Siel Wallring (D) over two already existing subway tunnels. The simulations were carried out with the software SOFISTIK using the Hardening Soil model.

In Volderauer et al. (2012) are shown 3D finite element calculations for a deep staying tunnel, the Gotthard Base Tunnel (GBT). The full face excavation of the tunnel zone with an overburden of 960 m was carried out using tunnel excavators with lengths of rounds of 1.34 m to 1.0 m and the model was generated with a size of 100 m in depth, 200m in height and 118 m in length. In the 3-dimensional simulations with the software ABAQUS the material models of Mohr-Coulomb and Drucker-Prager were used; both models were integrated with an approach for stress-dependent Young's moduli.

The first tube of the Chamoise Tunnel (F) with an overburden of 400 m was built in the early 1980 with sequential excavation. Ca 10 years later the second tube was built with full-face excavation. Numerical back analyses using 3D finite elements calculations of both tubes comparing full-face with sequential excavation were carried out with the software CESAR and with the material model of Mohr-Coulomb, cf. Putz-Perrier et al. (2014).

2   FINITE ELEMENT CALCULATIONS

2.1   General remarks

The finite element (FE) calculations were carried out with the software PLAXIS 2D and PLAXIS 3D. PLAXIS is geotechnical finite element software, specifically developed for the 2D and 3D analysis of deformation, stability and groundwater flow. The basic soil elements of PLAXIS 2D are triangular elements with 12 nodes, while the basic soil elements of PLAXIS 3D are tetrahedral elements with 10 nodes. In addition to the soil elements, in both software versions, special types of elements are used to model structural behavior. For the calculation of concrete shells 6-node plate elements are available in PLAXIS 3D, c.f. PLAXIS BV (2018).

2.2 Material properties

Both tunnels are situated entirely in the Innsbruck Quartzphyllite. The rock mass properties of the Innsbruck Quartzphyllite for the connection tunnel were described in the technical report "Rock mass classes, rock mass behavior types - connection tunnel" BBT SE (2013). This report was elaborated by the geotechnical planning team of the BBT SE under the advice of the tunnel consultant Dr. M. John. In this report the Innsbruck Quartzphyllite is characterized by the rock mass behavior type BT 2 after the Guideline for the Geotechnical Design of Underground Structures with Conventional Excavation, c.f. Austrian Society for Geomechanics (2010).

The Innsbruck Quartzphyllite consists mainly of quartz-rich phyllites (quartzphyllites), mica and quartzite schists with greenschist, marble, orthogneiss and graphite phyllite inclusions. From a volumetric point of view, the most important minerals are quartzite, mica and chlorite in varying proportions with the addition of feldspar, resulting in phyllite, quartz phyllite, micaschists and quartzite schists, c.f. Bergmeister & Töchterle (2013).

In BBT SE (2013) were also given the rock mass parameters for the material model of Mohr-Coulomb, cf. Mohr (1914), which are summarized in Table 1.

The Hardening Soil model, Schanz et al. (1999), is an advanced model for the simulation of soil behavior. As for the Mohr-Coulomb model, with the Hardening Soil model limiting states of stress are described by means of the friction angle φ', cohesion c and the dilatancy angle ψ. However, soil stiffness is described much more accurately by using three different input stiffnesses: the triaxial stiffness E_{50}, the triaxial unloading stiffness E_{ur} and the oedometer loading stiffness E_{oed}. In contrast to the Mohr-Coulomb model, the Hardening Soil model also accounts for stress-dependency of stiffness moduli. This means that all stiffnesses increase with pressure.

For the Hardening Soil model the elastic parameters E_{ur}, E_{50}, E_{oed}, ν and the parameters φ', c, ψ, m, K_o^{NC} were chosen according to the technical report BBT SE (2013). The parameter $R_F=0.9$ was set to a standard value.

2.3 Initial stress state

The overburden pressure was simulated in the model using an upper loading plate with high density.

The density of the plate was set in order to obtain an initial stress of 5.2 MPa on the upper border of the model. This corresponds to an overburden of ca 200 m on the upper border of the model and ca 220 m above the crown of the connection tunnel.

Compared to the assumption of vertical fixities in the upper boundary as used e.g. in Volderauer et al. (2012), an upper loading plate has the advantage that the vertical displacements of the upper boundary of the model are not vanishing, which is a more realistic assumption. With vertical fixities in the upper boundary the model size must be larger in height. In fact, a FE calculation with a similar model, but with the total height of the soil body (200 m) instead

Table 1. Rock mass parameters, Mohr-Coulomb model.

Parameter	E	ν	φ'	c	ψ
Unit	MPa	-	°	MPa	°
Value	6000	0.2	37	0.8	0

Table 2. Rock mass parameters, Hardening Soil model.

Parameter	Value	Unit
E_{50}	3000	MPa
E_{oed}	2600	MPa
E_{ur}	6000	MPa
ν	0.2	-
m	0.5	-
φ'	37	°
c	0.8	MPa
ψ	0	°
R_F	0.9	-
K_0^{NC}	0.4	-

of the upper loading plate carried out by the first author found that the displacements were the same in both models. Of course, the bending stiffness of the upper plate must be set to an appropriate value in consideration of the bending stiffness of the upper soil body (in this case with a thickness of 200 m).

Next to the connection tunnels an exploration drilling with 10 hydraulic fracturing tests was made in order to determine the magnitude and orientation of the horizontal stress components.

The hydraulic fracturing tests showed $k_0=1.5$ in the direction of the maximum horizontal stress and $k_0=0.8$ for the minimum horizontal stress. The reason of these different values can be explained with the topography of the mountain plateau next to the exploration drilling position, cf. BBT SE (2013). Considering the angle between the orientation of tunnel axis and the maximum horizontal stress the coefficient of lateral earth pressure in the direction of the tunnel axis could be determined with $k_0=0.9$, while in the transversal direction to the tunnel axis k_0 could be determined to 1.1.

Due to the k_0 values close to 1 in the primary calculation stage a k_0 consolidation step with $k_0=1$ was applied. Further calculations with $k_0=0.8$ and $k_0=1.2$ –which were carried out to capture the influence of a non-hydrostatic state– did not show significantly different results in regard to the investigated interaction effects of the two tunnels and are not followed up in this contribution.

2.4 Concrete

The outer sprayed concrete liners (SPC 20/25) were modelled with a linear elastic material law. A hardened concrete of the class C 20/25 has a Young's modulus of 30,000 MPa. Due to the very pronounced creep and shrinking behavior of the sprayed concrete, only a reduced Young's modulus of 15,000 MPa was considered, cf. Bergmeister et al. (2012). The parameters of the sprayed concrete are summarized in Table 3.

Table 3. Parameters of the sprayed concrete SPC 20/25.

Parameter	Value	Unit
E_{cm}	15,000	MPa
ν	0.2	-
f_{ck}	25	MPa
f_{ctm}	2.2	MPa

3 2D CALCULATIONS

The overpassing zone of the two connection tunnels was investigated first with 2D calculations in order to find the most suitable excavation sequence. In the zone of the intersection of the two tunnels several sections were evaluated to find the determinant section with the highest interaction. The determinant section was found to be at a distance of about 23 m from the projected intersection with a horizontal distance of the two tunnels of ca 4.5 m.

For the simulation of the time dependent installation of the support a stress release of 90 % of the initial stress was carried out. The stress release factor α of 90% was determined considering longitudinal displacement profiles, cf. Vlachopoulos & Diederichs (2009). This contribution includes an equation to calculate the convergence at a certain distance from the tunnel front considering the tunnel radius and the radius of the plastic zone. Considering that the tunnel section of the connection tunnel is not circular, the equivalent tunnel radius was obtained approximately by assuming a circular section with the same surface as the non-circular connection tunnel.

By varying the release factor α in FE calculations, the appropriate release factor α could be obtained in order to get the convergence c determined before. The calculations led to a release factor α of ca 90 % considering the length of round of 1.70 m.

The software PLAXIS allows to model construction phases (staged construction). In PLAXIS the input of a stress release of 90% is done using the factor $M_{Stage}=0.9$ in the staged construction procedure.

The construction sequence of the NATM tunnels is the following: First a partial section of the tunnel is excavated. In the next stage the outer lining with a thickness of 30 cm and a Young's modulus of 15,000 MPa is installed.

The calculations were carried out with the Hardening Soil model and the parameters of Table 2.

In order to find a construction sequence which leads to less stress on the tunnel shells as possible, four excavation variants (variant 1 to 4) were investigated: In the case of variant 1 after the excavation of the top heading, bench and invert of the upper connection tunnel the ring closure of the tunnel liner is carried out immediately, while with variant 2 after the excavation of the top heading, bench and invert of the upper connection tunnel no ring closure is done, the ring closure is carried out just in a later step after the excavation of the top heading of the lower connection tunnel (cf. Figure 2).

With variant 3 first the excavation of the top heading of the upper tunnel is done and thereafter the excavation of the top heading of the lower connection tunnel. In further excavation stages are excavated the bench and the invert of the upper tunnel and at last the excavation of the bench and the invert of the lower connection tunnel (cf. Figure 2).

In the case of variant 4 first the excavation of the top heading of the lower connection tunnel is done and thereafter the excavation of the top heading of the upper tunnel. In further excavation stages are excavated the bench and the invert of the lower connection tunnel and at last the excavation of the bench and the invert of the upper tunnel (cf. Figure 2).

The construction stages and the resulting maximum normal forces in the spared concrete shell are summarized in Figure 2 for the variants 1, 2, 3 and 4.

Comparing the normal force after the installation of the sprayed concrete shell in the top heading of the lower connection tunnel (Figure 2) it can be observed that the maximum normal force is much lower for variant 2 (3341 kN/m in stage 4) than for variant 1 (6971 kN/m in stage 3).

The maximum normal force is also lower in the final stage for variant 2 with 4147 kN/m compared to variant 1 (7306 kN/m, cf. stage 5 in Figure 2)

For the variants 3 and 4 the maximum normal force are also higher than for variant 2 (variant 3: 7818 kN/m, variant 4: 7558 kN/m, cf. Figure 2, stage 5).

It can be therefore summarized that variant 2 can be recommended for execution, because with this variant the normal forces are much lower than with all other variants.

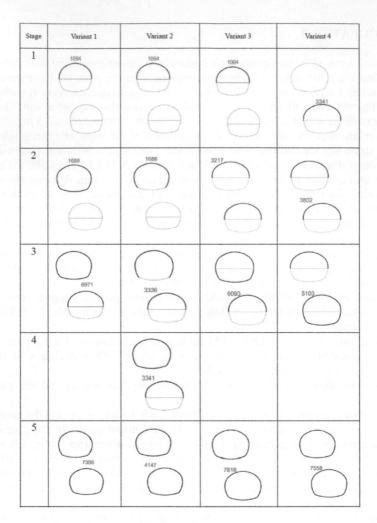

Figure 2.   Maximum normal forces in the outer shell in kN/m: variants 1, 2, 3, 4.

4   3D CALCULATIONS

4.1   *Excavation stages*

The 2D calculations showed that the construction sequence of variant 2 leads to the lowest normal forces in the tunnel shell due to interaction effects of both tunnel advances; therefore variant 2 was further investigated with a 3D analysis.

The finite element model has a length of 200 m, a width of 100 m and a height of 60m. The two connection tunnels are modelled with elastic linings in the longitudinal direction of the model space and are represented in Figure 3.

In the first stage of the calculations the primary stress state was calculated. In the next stages the excavation of the tunnels and their support was simulated.

The excavation stages were simulated as in the following: First the excavation of the top heading of the upper connection tunnel with a round length of 1.7 m for the whole tunnel length was simulated.

For each stage of the excavation the round closed to the face was simulated without support, while for the next rounds behind the first round a support by the outer sprayed concrete shell with a thickness of 30 cm was assumed.

Figure 3.   Finite element model of the connection tunnels with upper loading plate.

In the following the excavation of the bench and invert of the upper connection tunnel for the whole tunnel length took place stepwise.

The ring closure of the upper connection tunnel (the upper tunnel) was carried out for the whole tunnel length just in a later step after the excavation of the top heading of the lower connection tunnel.

In the last step the excavation of the bench and invert of the lower connection tunnel took place.

For all stages of excavation a round length of 1.70 m was assumed and the first round after the tunnel face was modelled unsupported, the next round was assumed to be supported by a liner with a thickness of 30 cm and a Young's modulus of 5,000 MPa. During the advancement the stiffness of the liners of the single rounds was increased with increasing distance to the tunnel face stepwise first to 10,000 MPa and then to the final value of 15,000 MPa.

Due to the fact that all single excavation stages were modelled, no determination of a stress release factor like for the 2D calculations was necessary.

4.2   Calculation results

4.2.1 Displacements

In Figure 4 are shown the maximum displacements for the last calculation stage for the Mohr-Coulomb model and the Hardening Soil model.

The maximum horizontal displacements reach 16 mm on the tunnel abutments of the upper connection tunnel with the model of Mohr-Coulomb. On the top heading of the upper connection tunnel the maximum displacement reaches 12 mm. Considering the top heading of the

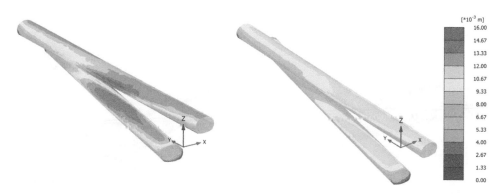

Figure 4.   Total displacements in mm. Left: material model of Mohr-Coulomb. Right: Hardening Soil model.

lower connection tunnel there are locally reached displacements of 15 mm on two small zones directly below the abutments of the upper connection tunnel (cf. Figure 4).

With the Hardening Soil model the maximum displacements reach 11 mm on the tunnel abutments of the upper connection tunnel and 12 mm on the top heading of the upper connection and the lower connection tunnel (cf. Figure 4). The displacement plot shows less pronounced peaks than with the Mohr-Coulomb model.

With both models a zone with lower displacements on the top heading of the lower tunnel directly in the projection of the axis of the upper tunnel can be observed (cf. Figure 4).

4.2.2 *Comparison with measured displacements*

During the excavation of the connection tunnels measurements in 9 points per section were carried out. The measured displacements reached 11.4 mm on the western abutment of the upper connection tunnel and 17 mm on the top heading.

In the lower connection tunnel was measured a maximum settlement of 18.2 mm on the top heading and a maximum horizontal displacement of 15 mm on the western abutment.

Considering the calculation results with the model of Mohr Coulomb, the horizontal displacements on the tunnel abutments were overestimated, while the settlements on the top heading were underestimated. With the Hardening Soil model the horizontal displacements on the tunnel abutments are similar to the measured values; the calculated settlements on the top heading, however, are too low.

The ratio of horizontal displacements (abutment) to vertical displacements (top heading) is 1.33 for the upper tunnel with the Mohr-Coulomb model and 1.09 with the Hardening Soil model, while the measurements show a ratio of 0.67, which means that –different to the calculation results– the measured vertical displacements are higher than the horizontal displacements; for the lower tunnel can be observed qualitatively similar values.

Possible reasons of these disaccordances could be a lower ratio of horizontal to vertical stress level ($k_0 < 1$) compared to $k_0 = 1$ assumed in the calculations (which is, however rather unlikely considering the results of the fracturing tests described in Section 2.3) or the anisotropy of the Quartzphyllite, cf. e.g. Ramamurthy et al. (1994). A zone of lower displacements on the top heading of the lower tunnel directly in the projection of the axis of the upper tunnel as observed in Figure 4 could not be observed in the measurements.

4.2.3 *Normal forces*

The normal forces in transversal direction for the connection tunnels for the last calculation stage are shown for the material model of Mohr Coulomb in Figure 5 and for the Hardening Soil model in Figure 6.

It can be observed that with the Mohr-Coulomb model the normal force reaches a local maximum of 4740 kN/m on the bottom lining (cf. Figure 5). Considering the top headings the maximum values of the normal force are reached in the middle part of the upper connection

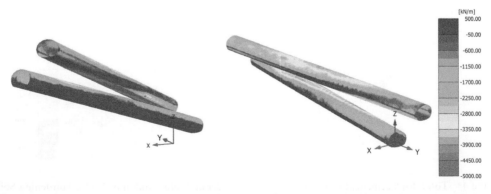

Figure 5. Normal forces in transversal direction, material model of Mohr-Coulomb.

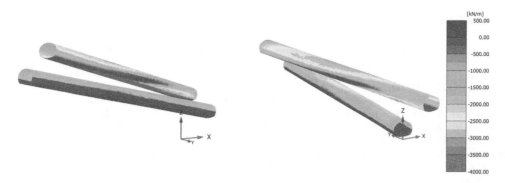

Figure 6. Normal forces in transversal direction, Hardening Soil model.

tunnel with 3942 kN/m and in two zones before and after the intersection of the two tunnels with 4123 kN/m (cf. Figure 5). These last two values are in good agreement with the 2D calculation (cf. N_{max}=4147 kN/m in Figure 2), but the latter does not show the local maximum of 4740 kN/m on the bottom lining like in Figure 5.

For the Hardening Soil model can be observed a similar behavior, but with lower normal forces: The normal force reaches a local maximum of 3844 kN/m on the bottom lining and a maximum on the tunnel crown of 3263 kN/m (cf. Figure 6).

Summarizing, it can be stated, that the maximum normal forces are below 5000 kN/m (Mohr-Colomb model) and 4000 kN/m (Hardening Soil model) and confirm therefore the results of the 2D calculations, with a maximum force of 4142 kN/m for variant 2, while for the other variants 1, 3 and 4 much higher normal forces of 7306 kN/m to 8818 kN/m were achieved.

However, considering the normal forces of 4000 kN/m, using a sprayed concrete shell with 30 cm thickness a concrete class SPC 20/25 was not sufficient, therefore a concrete class SPC 25/30 was necessary.

5 CONCLUSIONS

The 2D calculations of the overpass of the western connecting tunnel over the eastern connecting tunnel showed that the appropriate excavation sequence can reduce the stress on the tunnel shell significantly.

The 2D calculations showed further, that by using an excavation sequence without immediate ring closure, both connection tunnels could be excavated without damage of the sprayed concrete outer shells.

The 3D calculations of the overpass showed that advanced material models are appropriate in the case of the interaction of tunnels with loading/unloading effects due to tunnel excavation can be simulated more realistically than with simpler material laws like Mohr-Coulomb. The 3D calculations showed in agreement with the 2D calculations that the appropriate excavation sequence without immediate ring closure leads to normal forces which do not exceed the bearing capacity of the used outer tunnel shells with 30 cm thickness.

The excavation of the two connection tunnels of the Brenner Base Tunnel took place in winter 2016/2017. By using an excavation sequence without immediate ring closure as suggested by the authors, both connection tunnels could be excavated without damage of the sprayed concrete outer shells.

REFERENCES

Austrian Society for Geomechanics 2010. Guideline for the Geotechnical Design of Underground Structures with Conventional Excavation.

BBT SE 2013. Technischer Bericht Gebirgsarten, Gebirgsverhaltenstypen Verbindungstunnel. 01 GH3 GP001-GTB-0002 06.

Bergmeister, K. 2011. *Brenner Basistunnel – der Tunnel kommt*. Lana: Tappeiner Verlag.

Bergmeister, K. 2012. Brenner Base Tunnel under Construction. *Tunnel* 2012 (01): 18–30.

Bergmeister, K. & Töchterle, A. 2013. Brenner Base Tunnel: Importance of Preliminary Prospection. *Tunnel* 2013 (01): 12–23.

Bergmeister, K., Weifner, T. & Collizzolli, M. 2012. Effects of the geometric position of the Brenner Base Tunnel on rock mass plastification and the tunnel's shotcrete shell. *Beton- und Stahlbetonbau* 107(11): 735–748.

Insam, R., Wahlen, R. & Wieland, G., in prep. Brenner Base Tunnel – Interaction between underground structures, complex challenges and strategies. *ITA-AITES World Tunnel Congress 2019*, Naples, Italy.

Mohr, O. 1914. Abhandlungen aus dem Gebiete der Technischen Mechanik. Ernst und Sohn.

PLAXIS BV 2018. PLAXIS Reference Manual.

Putz-Perrier, M., Gilleron, N., Bourgeois, E. & Saitta A. 2014. Full-face versus sequential excavation – A case study of the Chamoise Tunnel (France) *Geomechanics and Tunneling* 7(5): 469–480.

Ramamurthy, T., Rhao, G.V. & Singh, J. 1993. Engineering behavior of phyllites. *Engineering Geology* 33 (3): 209–225.

Rabcewicz, L. v. 1965: The New Austrian Tunneling Method, part three. *Waterpower* 17(1):19–24.

Schanz, T., Vermeer, P.A. & Bonnier, P.G. 1999. The Hardening Soil model: Formulation and verification. *Beyond 2000 in Computational Geotechnics – 10 Years of PLAXIS: 281–296*. Rotterdam: Balkema.

Svoboda, T. & Masin, D. 2011. Comparison of displacement field predicted by 2D and 3D finite element modelling of shallow NATM tunnels in clays. *Geotechnik* 34(2): 115–126.

Umar, A., Blaschko, M. & Grubert, T. 2018. Numerical modelling for crossing over two existing subway tunnels on the example of transport Siel Wallring. *Geotechnik* 41(3): 197–211.

Volderauer, C., Marcher, T. & Galler, R. 2012. 3-dimensional numerical calculations considering geotechnical measurement data. *Berg- und Hüttenmännische Monatshefte* 157(12): 451–458.

Vlachopoulos, N. & Diederichs, M.S. 2009. Improved longitudinal displacement profiles for convergence confinement analysis of deep tunnels. *Rock Mechanics and Rock Engineering* 42: 131–146.

Tunnels and Underground Cities: Engineering and Innovation meet Archaeology,
Architecture and Art, Volume 6: Innovation in underground engineering,
materials and equipment - Part 2 – Peila, Viggiani & Celestino (Eds)
© 2020 Taylor & Francis Group, London, ISBN 978-0-367-46871-2

Artificial neural networks in streamlining serviceability assessment of single shell TBM tunneling

R. Wenighofer & R. Galler
Chair of Subsurface Engineering, Montanuniversität Leoben, Austria

ABSTRACT: Artificial neural networks (=ANN) is the upcoming technology of the recent years. It improves the performance of search in internet, enables autonomous cars and enhances interaction with electronic devices. However, the reliability of ANN's heavily depends on the data base neural networks are trained on. In tunneling there are rare applications of neural networks although they can mitigate labor intensive, tedious and error-prone information handling for e.g. as-built documentation and geotechnical measurement. Tunneling constitutes a harsh environment for gathering the indispensable representative data. The contribution shows the necessity of constituting a data set with sufficient quantity and representative quality for the training of networks. It is a decisive prerequisite for successful application of machine learning which can be applied for the assessment of serviceability of single-layer lining. Additionally, its role in crack detection in geotechnical tests is emphasized. The contribution points out different ANNs ready to be readapted for tunneling purposes. The data for setting up the data base needed for network training is gathered in Austrian TBM projects using 3D laser scanners and image acquisition systems during geotechnical testing.

1 INTRODUCTION

Design, construction, operation and maintenance of underground structure are increasingly permeated by the current digitalization process synonymously known as Building Information Modelling (BIM). This means that a growing number of construction processes turns away from the maze of drawing (=dwgs) versions, gain online accessibility and procure lifelong data storage for the structure. From industrialized tunnelling using TBM's even a bigger amount of digital data results and necessitates that steadily more complex algorithms are employed which save tedious and labor-intensive hours of work to process it.

The nomenclature of these algorithms comprises frequently the designations of machine learning and computational or artificial intelligence. Being an approach relinquished after successful implementation for recognition of handwritten characters their development nowadays parallels the growing processing power of new CPU- and GPU-architectures.

The contribution firstly provides insight into several machine learning approaches designed for different tasks already employed but not maintained in tunnelling. Then it points out its utilization in tunnel inspection and serviceability ascertainment. It spotlights the importance of data sets to be assembled, preserved and maintained to refine them and render them more applicable, sustainable and flexible for future projects.

2 CURRENT STATE OF THE ART IN DATA PROCESSING

Current digitalization spawns a tremendous quantity of data to be stored and handled. Essentially, machine learning caters to the needs of data abundance to make sense of it. Web search engines, robust email spam filters and voice recognition software are examples, which take advantage of enormous data aggregation steadily adjusting itself for better performance. Toolkits like Matlab, python-based open source libraries like Tensor Flow or Keras are today available for a broad community of people and not only for data processing experts. The toolkits allow the development of customized solutions to tackle this age of data inundating also underground construction. They benefit this area by removing the barrier of reinventing and reimplementing crucial software components like optimization computation as well as they allow for centring efforts around data preparation and finding the proper algorithms for analyzation.

The three pillars of supervised, unsupervised and reinforcement learning constitute machine learning which is put to use in various applications in underground construction as exemplified hereinafter.

3 APPLICATIONS OF MACHINE LEARNING IN TUNNELLING

3.1 *Supervised machine learning technique – Artificial Neural Networks*

Supervised learning mainly tries to achieve the goal of a correct classification in case of existing labelled data. Consequently, it presupposes the existence of a desired output that can be used to train the model and to predict the output from new data collected.

Very early Leu et al. (2001) presented a supervised learning approach using artificial neural networks (ANN). ANNs embody numerous single artificial neurons which allow for interrelating heterogenous input factors with an output e.g. a continuous function or a classification. Figure 1 left illustrates a single artificial neuron combining the input factors X_i, W_i and the activation function f to form the output signal y. Leu et al. derived a classification model of three warning levels regarding tunnel support stability which is based on data acquisition of 470 rounds of a NATM (New Austrian Tunnelling Method) excavation. The data involved consists of several factors like rock type, discontinuity properties, support timing and support types and the ANN finally delivers an output of three categories. Besides the focus on ascertaining the most appropriate machine learning the study of Leu et al. (2001) reveals that the warning levels determined by the model partly depend on other factors than the involved experts considered.

3.2 *Unsupervised machine learning technique – clustering*

Clustering belongs to the unsupervised machine learning methods, which are, inter alia, used in Galende-Hernandez et al. (2018). Therein the employment of machine learning purposes the prediction of design parameters in a RMR-system (Rock Mass Rating) dealing with noisy data and a large number of variables. The data is registered by monitor while drilling (MWD) of jumbos during face drilling in drill&blast tunnelling. The processing described by Galende-Hernandez et al. (2018) conquer the challenging task of complexity reduction and data representation influencing the entire chain of downstream algorithms to obtain a rock classification.

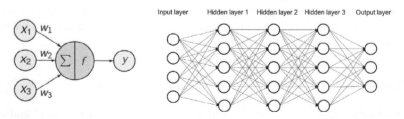

Figure 1. Left: single artificial neuron, input signal X_i, weighting factors W_i, activation function f, output signal y; Right: sketch of a deep learning neural network with several hidden layers.

4 APPLICATION OF MACHINE LEARNING SYSTEMS IN TUNNEL MONITORING AND MAINTENANCE

4.1 Tunnel inspection using Artificial Neural Networks

Machine learning does not only cover tunnel excavation aspects but also enriches tunnel inspection for maintenance e.g. in shielded mechanized tunnelling (Attard et al. 2018 and Huang et al. 2018). Reasonably, tunnel inspection exploits image acquisition systems, which aim to facilitate and accelerate checking the surface of tunnel lining by naked eyes. The tunnel inspection system developed in Huang et al. (2018) meets the challenge that seepage is one of the severe ingredients of lining deterioration in mechanized tunnelling and needs corrective actions prior to other defects (Yuan et al. 2013). The camera system takes advantage of an extensive ANN – a so called deep neural network (Figure 1 right) – to solve the complex task of identifying cracks and areas of leakage and discriminating them from interfering cables, segmental joints, hand-written marks etc. Since the camera system covers only visible aspects of lining degradation and defects it does not allow for monitoring lining deformation and can be solely a technological constituent of tunnel inspection.

4.2 Analysis of tunnel deformation using 3D laser scanning

However, deformations like ovalization of the cross-section ranks among the most informative indicators of the condition of the tubbing lining. Figure 2 shows a cross section measured by a 3D laser scanning system.

The ovalization directly affects the serviceability of the tunnel support by causing cracks, joint offsets and impermissible longitudinal joint rotations which worsen the waterproof function of the lining (Huang, et al., 2016). For determining tunnel deformation 3D laser scanners have turned out to be appropriate surveying instruments enabling fast acquisition of a point cloud with high resolution and accuracy. So, they permit that crack development due to bending moments can be inferred from overall geometry alteration (Figure 3).

Moreover, the 3D point clouds render possible to measure segmental joint offsets and rotations (Wenighofer et al, 2016). The dependence of serviceability as well as waterproofness of tunnels on segmental joint rotation and offset is well underscored recently by Huang et al. (2017) and earlier by the STUVAtec Teams of Authors (2005), who submitted the recommendations on proving water tightness. They issued the procedure of proving water tightness of the lining by diminishing the groove basic gap while increasing the water pressure until leakage occurs. This test is conducted using different joint offsets and is similarly carried out if, in contrast, opening the groove basic gap is the touchstone by which to assess the water tightness of the specific lining.

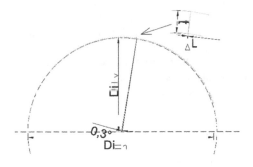

Figure 2. Horizontal convergence of a cross section indicating two segments (red, green) and joint rotation.

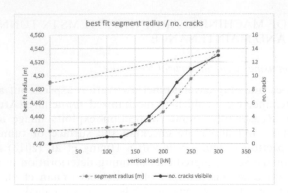

Figure 3.   Correlation of the overall pre-cast concrete segment and increasing number of cracks.

4.2.1 *Application of 3D-Laser Scanners in mechanized tunnelling*

Mechanized single-layer linings in shielded tunnelling are generally sealed using preformed elastomer gaskets. They are fixed along the annular and longitudinal segmental joints of the lining for prevention of water seepage. As a huge number of segments constitutes a single-layer lining the length of the segmental joints to be sealed rapidly amounts to several kilometres given a tunnel length of one kilometre. Consequently, safeguarding serviceability of single-layered tubbing linings involve arduous manual inspections for cracks and monitoring the proper installation of the segments. The latter requires measuring the permissible offsets and joint gap widths by hand which are often hardly accessible from the platforms of the backup system of the TBM. In contrast, contactless measurement systems provide detailed information on segment rotation and deformation as can be seen as Δ_L in Figure 2 as proposed by Huang et al. (2017). For this reason, one can go so far as to introduce a 3D laser scanner system mounted on the backup system closely behind the shield of a TBM for the documentation of correct installation of the segments. The system scans a 360° view after each stroke of the TBM and accumulates a tremendous amount of data in a short period of time. However, for full exploitation of the scans automated working sequences are indispensable in order to cope with the data amount and the speed of the mechanized excavation process (Wenighofer et al., 2016).

4.2.1.1 CHALLENGES IN USING 3D LASER SCANNING FOR TUNNEL SERVICEABILITY MONITORING

Using 3D laser scanning for monitoring the overall serviceability of the single-layered lining faces the challenges of discriminating single segments in the cylindric tubbing lining since segmental joints represent weakness zones of this support. Therefore, offset, opening and closure of the gaps occur due to overload or eccentric load onto the lining along the joints. Monitoring this entails onerous and complex processing of the 3D laser scans whose goal is to represent the cylindric shape of the segment by a point cloud. That includes the removal of invalid points around the recesses for the screws, the erector, interfering installations etc. This renders handling the point cloud arduous and stimulates the development of advanced point cloud processing methods. In Wenighofer et al. (2016) point cloud processing has turned out to be powerful knowing the exact geometry of the segments. Though, it is prospectively less applicable in situations of interfering cables and pipes.

5   TRANSFER LEARNING STRATEGY

Tunnelling often falls back on unrolling the tunnel surface due to the incapability of representing three-dimensional structures as it is generally the case e.g. when checking the geometry of the cross section. Thereby, the complex geometry converts to a 2D representation like images thus, allowing for easier handling due to the cylindric shape of segmental linings.

5.1 *Geometry recognition using point clouds and neural networks*

Object recognition in image processing can make use of deep learning networks such as AlexNet or GoogLeNet. These elaborate networks are designed to solve the tough task of categorizing thousands of object classes based on a data set of millions of labelled images the networks are trained on. Repurposing them for different classes signifies that parts of the network layers have to be exchanged and retrained.

This necessarily means to provide a large data set of these other classes that need to be recognized. This approach also is known as transfer learning. As an example of the data set Figure 4 shows unrolled images which are part of the images designed to readapt the pretrained deep learning networks mentioned. The data set comprises 23k images generated from several unrolled 3D laser scans from different tunnel projects. The images are interpolated from scattered point cloud data. Essentially, unrolling the cylindrically shaped single-layered lining and thereby substituting perpendicular distances from the cylindric surface by grey scale values turns image recognition into geometry recognition. The classes illustrated by Figure 4 cater to several requirements of monitoring construction quality. E.g. joint localisation contributes to continuous offset and rotation angle determination. The first one is usually measured by hand whereas the latter is not monitored at all. By identification of the intrados the estimation of the radius of segments is endeavoured which is indicative of cracks due to exceeded concrete tensile strength. They are triggered by strong bending moments.

5.2 *Readaptation of existing deep learning networks for 3D laser scans*

Readapting a deep learning network generally is more recommendable than generating it from scratch. The approach used from generation of the training data set to its application is shown by Figure 5 and tremendously profits from public provision of already trained networks. As deep learning networks require specific input like certain width and height of images the strategy of transfer learning is confined to generate data sets of the particular size

Figure 4.   Dataset for readapting pretrained deep neural networks: Top left: intrados, top right: annular joint, bottom left: recess for erector, bottom right: recess for bolts.

Figure 5.   Learning strategy applied on AlexNet and GoogLeNet in the described approach.

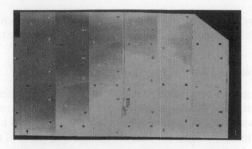

Figure 6.   Unrolled 3D laser scan colored accordingly to its deviation from an ideal cylindric surface.

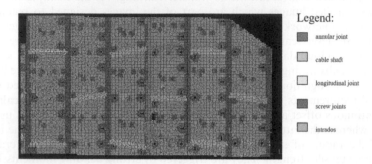

Legend:

■ annular joint

□ cable shaft

□ longitudinal joint

■ screw joints

■ intrados

Figure 7.   Result of sliding window with classifications using readapted GoogLeNet according to the legend.

needed by the respective network. Though, using interpolation from scattered point cloud data this restriction is not noteworthy. Consequently, the readaptation of the networks GoogLeNet from Google Research (Szegedy et al., 2015) and AlexNet from Krizhevsky, et al., (2012) is based on representations (Figure 4) of such as recesses in the segments, their joints and intrados.

For attaining a high success rate the reutilization of the networks is steadily optimized using different sizes of the patches finally resulting in a square size of 0.30 m. A separate test data set is introduced in order to estimate the success rate applying a sliding window all over the unrolled 3D laser scan (Figure 6). At each position of the sliding window a patch is generated and classified by the readapted network GoogLeNet leading to the result shown by Figure 7. It shows good accordance of recognition with the classes learnt as indicated by the legend. Since there are also cables installed a set of patches containing cables has to be introduced in the training data set. The cables are recognized indicated by orange markers. Sporadic false detection is discovered especially near the recesses.

5.3   Readaptation of existing deep learning networks for crack detection

Another implementation of deep learning networks is detection of cracks during load tests applied on tubbing segments where increasing loads are put onto the segments. During the tests development of crack widths is monitored. Thus, a camera system employed in the pauses between the growing load stages generates images showing a growing number of cracks to appear. The camera takes images of the front of the annular joint of the segments. Crack widths are crucial to the serviceability of pre-cast concrete segments since impermissible crack widths are likely to deteriorate the reinforcement. During the tests they result from bending moments which are caused by increasing load stages. The data set for readapting a deep neural network is based on the images of the monitoring camera system. It includes two classes of cropped images, one showing cracks and one showing plain concrete. The retrained

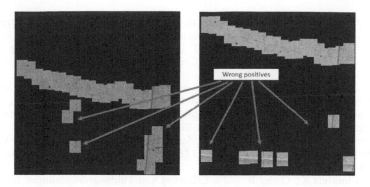

Figure 8. Binarization of images into cropped patches with cracks or plain concrete.

network is the AlexNet being repurposed from 1000 classes to be distinguished to binarize the image containing a crack or not. Hence, the layout of the results varies fairly from the network reutilized for the 3D laser scans. Applying a sliding window patches of the images are cropped and classified by the readapted network which is illustrated by Figure 8. Equally to the application on the unrolled 3D laser scans the selected strategy enables testing the network on patches which do not constitute the training data set. Figure 8 highlights that readapted deep learning networks can be powerful aids of crack detection since the patches follow right the route of the crack. However, Figure 8 also discloses that features not being part of the training data set like the rope can cause false positives. Consequently, readapting deep learning networks necessitate a large data set being a bedrock of their successful reutilization. They also require intense study to figure out the most appropriate deep learning network structure.

6 CONCLUSION

The contribution reveals a broad applicability of artificial neural networks on a large range of purposes in tunnelling. It emphasizes the successful utilization in various questions facilitating and substituting manual activities. However, ANNs entail a detailed investigation of training and testing data sets which are not easily accessible in tunnelling. Failing storage of the data sets, their maintenance and availability for following projects cause confinement of the found solutions to a specific task or project which is not sustainable. Other sectors already organize their data sets in large data basis like ImageNet which enables networks even exceeding recognition capabilities of humans. A recent approach of improved data handling in tunnelling is called BIM which means thinking one step further in sustainable data management. It should not be dissipated to solely impress by colourful 3D models.

REFERENCES

Attard L., Debono C.J., Valentino G., Castro M. 2018. Tunnel inspection using photogrammetric techniques and image processing: A review. *ISPRS Journal of Photogrammetry and Remote Sensing* 144: 180–188.

Galende-Hernández M., Menéndez M., Fuente M.J., Sainz-Palmero G.I. 2018. Monitor-While-Drilling-based estimation of rock mass rating with computational intelligence: The case of tunnel excavation front. *Automation in Construction* 93: 325–338.

Huang H., Li Q., Zhang D. 2018. Deep learning based image recognition for crack and leakage defects of metro shield tunnel. *Tunnelling and Underground Space Technology* 77: 166–176.

Huang H., Zhang Y., Zhang D., Ayyub B. M. 2017. Field data-based probabilistic assessment on degradation of deformational performance for shield tunnel in soft clay. *Tunnelling and Underground Space Technology* 67: 107–119.

Huang, H. W., Zhang, D. M. 2016. Resilience analysis of shield tunnel lining under extreme surcharge: Characterization and field application. *Tunnelling and Underground Space Technology* 51: 301–312.

Krizhevsky A., Sutskever I., Hinton G. E. 2012. ImageNet Classification with Deep Convolutional Neural Networks. *NIPS 12 Proc. 25th Int. Conf. on Neural Information Processing Systems* (1): 1097–1105.

Leu, S.-S., Chen, C.-N., Chang, S.-L. 2001. Data mining for tunnel support stability: neural network approach. *Automation in Construction* 10: 429–441.

STUVAtec Team of Authors. 2005. STUVA Recommendations for Testing and Application of sealing Gaskets in segmental Linings. *Tunnel* 8(2005): 8–21.

Szegedy Ch., Liu W., Jia Y., Sermanet P., Reed S, Anguelov, D., Erhan, D., Vanhoucke V., Rabinovich, A. 2015. Going Deeper with Convolutions, *2015 IEEE Conf. on Comp. Vision and Pattern Recognition (CVPR)*: 1–9.

Wenighofer R., Chmelina K., Galler R. 2016. Erfassung von Tübbingverformungen bei TVM-Vortrieben, *Terrestrisches Laserscanning 2016 (TLS 2016)* 85/2016: 59–73.

Yuan, Y., Jiang, X., Liu, X. 2013. Predictive maintenance of shield tunnels. *Tunnelling and Underground Space Technology* 38: 69–86.

Tunnels and Underground Cities: Engineering and Innovation meet Archaeology,
Architecture and Art, Volume 6: Innovation in underground engineering,
materials and equipment - Part 2 – Peila, Viggiani & Celestino (Eds)
© 2020 Taylor & Francis Group, London, ISBN 978-0-367-46871-2

Digitalization of data management and quality assurance in ground and tunneling works

L. Winkler
Technische Universität Wien, Vienna, Austria

P. Maroschek
eguana GmbH, Vienna, Austria

F. Weber
Züblin Spezialtiefbau GmbH, Vienna, Austria

M. Ouschan
eguana GmbH, Vienna, Austria

ABSTRACT: Nowadays engineers do their individual evaluations and analysis on site, con-solidating several types of data from different sources. In order to transform these inconsistent processes, the researchers of eguana and the TU Wien developed a solution for automated documentation and evaluation processes. This paper illustrates how digital data management for construction processes is established, considering the life cycle of the data. Based on the requirements of the stakeholders, the research group developed an ideal model of a digital data management system. These new hard- and software solutions build the foundation for a consistent, interface-independent data flow. Site engineers evaluated the model and tested the systems successfully on tunnel sites. In conclusion, the article shows the direct and indirect potential of digitalization on sites, not only saving time in the documentation and analysis of data, but also helping to evaluate non value-added processes in the construction.

1 STATE OF THE ART

Digitalization facilitates a variety of opportunities for a better understanding of construction processes, quality assurance and optimization of construction works. Especially in tuneling and ground engineering works, the accuracy of data is of utmost importance. Modern con-struction machines enable engineers–through data logging–to take a deeper look into con-struction processes, but only give a small glimpse in terms of the overall picture. This article will focus on the construction process of drilling and grouting. The described requirements, set-up processes and experiences can be adapted to other construction methods used in ground engineering. The authors applied the same procedure for digital data management and quality assurance for jet grouting, drilling and water draining.

The purpose of the intensive and detailed recordkeeping process is to track pay items and quantities and the verification that goals are being met. Direct testing of the grouting success with verification holes are generally not practicable due to cost. Therefore, the resulting data from the construction process and the system- and material tests of the grouting program need to be assessed in order to find anomalies, define additional grouting instructions and check where the parameters have been met. The tracking of material and time makes it pos-sible to allocate the measured quantities to the unit prices and to process the invoice. Contract modifications or supplements are only justified with the evidence of the financial failure of the contract. Apart from these reasons, data monitoring and records justify equitable adjustments of the grouting program and procedures and establish communication of interim results to others (Owen 2017).

Bruce (Bruce 2012) published, that superior levels of technology in the grouting industry combine at least four major components. Powerful monitoring systems should comprise a real-time data display of information from the ongoing operations on site, a central database to store, link and analyze the information, a customized function to display the stored information up to the minute in visualizations and individual queries for records and analytical capabilities.

Dreese (Dreese et al. 2003) analyzed the state of the art in monitoring systems of grouting and determined three levels. These levels were adapted to the common technology levels and use (Owen 2017). On sites, these levels can vary from hand-written protocols to web portals with interactive and automatic interfaces, depending on the specifications and the grouting method.

Level 1: Dipstick and Gauge Technology is the most rudimentary level of recordkeeping. It consists of manually measured volumes with a graduated dipstick in the agitator and the pressure is read from a dial pressure gauge with intervals of 5 to 15 minutes. As an example of best practice, site engineers calculate the average grout take and enter them manually onto a large master wall chart that shows the hole-location and a spreadsheet program for further analyses. In 2003, the experts considered this technology outdated and no longer viable for use in guide specifications of public contracts due to the time consuming, laborious work and frequently occurring human errors (Owen 2017). At minimum, experts require an electronic flow meter and a pressure transducer for instantaneous readouts. However, automation is absent and all field records, pay items and as-built grouting profiles are generated manually at this level.

Level 2: Data Collector and Display Technology automates the data collection and has a display unit for recording flow rates and pressure. To fulfill this level, a data recorder logs injection flow rates and pressure and displays it at the instrument location. Advanced control units automatically correct data, calculate parameters like cumulative grout take or effective injection pressure and track pay items such as grouting time and volume of material injected. Proprietary products, which requires manipulation from an engineer collect, store and plot data. Other records and analysis must be manually entered and generated. At grand projects, onsite personnel cannot oversee the information gathered during the process. Finding patterns or anomalies without time delay is difficult to achieve (Dreese et al 2003). The data monitoring of level two on-site is further described in the section "Previous Procedure".

Level 3: Integrated Systems Technology incorporates all field instruments in real time through a direct computer interface and provides the highest quality project record with minimal operator effort and interference. Web applications provide on-site and off-site analyses. Fully integrated systems include the features:

- Automatic tracking of data with geological unit
- Scalable visual presentation of data
- Link to the used materials and mix
- Quality assurance for produced material on site and link to production data
- Process management for the construction machinery, personnel and materials
- Semi-automatic verification and approval of measurements
- Customizable displays to meet the different needs of the stakeholders
- Query functions in the database and graph drawing to analyze correlations and patterns
- User-defined plots for periodic summaries
- Automatic billing of quantity for transparent invoicing
- Links to other data and systems within the site
- Full on-site and off-site capability

Economic and technical advantages of Advanced Integrated Analytical Systems (AIA Systems), as described at level three, can be fully realized with grouting projects that have high-risk profiles or costs equal to or exceeding €850.000. The state of the art at most grouting sites is level two. This level should be the minimum requirement for projects with costs up to €300.000, including the acquisition and setup costs.

The current three stages of data management on-site from the point of view of digitalized construction processes has already been defined (Winkler et al. 2017). These are analog data management, partial digital data management and complete digital data management, which are linked respectively to the three levels of data monitoring described above. Direct interfaces between the logging, storage and processing of data are fundamental components of digital data management on- and off-site. Isolated solutions already exist (Zhong et al. 2015) without interfaces between steps. Indirect interfaces are too time and human-resource intensive to guarantee sufficient and accurate quality control.

2 REQUIREMENTS OF THE STAKEHOLDERS

A survey was conducted at the conference "Future questions of construction processes" (Goger et al. 2018) with 200 decision-makers within the German-speaking construction industry, regarding the topic of digital cooperative management systems for construction processes relating to level three technology. The stakeholder groups related to the process of data monitoring are the owner, planner/inspector and contractor. Figure 1 shows that the planner and inspectors are in support of the development of complete digital data monitoring systems, with an overall approval of 55%. However, tried and true processes are still preferred over new digital methods. Only 11% of the stakeholders trust digital outputs more than handwritten protocols. Another big challenge for the implementation and development of digital monitoring systems at the construction site are the limitations regarding contracts. The systems specified in current contracts are not state of the art and digital development is not included in the project costs.

The state-of-the art technology was analyzed at a Level 3 system in 2010 for an implemented grout curtain project, the Center Hill Dam, in Tennessee and certain advantages were identified (Bruce 2012):

- *Fast and reliable information*
- *Rapid response to encountered foundation conditions and grouting results*
- *Informed decision making by project team*
- *Reduction in project schedule*
- *Reduction in project cost*
- *Increase in confidence of grouting results*
- *Greater success in meeting or exceeding project goals*

This list does not include efficiency control, simplification of administration and fulfillment of quality, which were also identified as expected benefits of digital cooperative management systems by the survey respondents mentioned above. The two benefits identified most by the stakeholders were processes that are more efficient and simplification of their

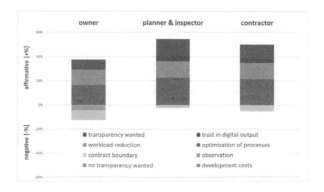

Figure 1. Survey Results to digital cooperative management system for construction processes.

own work. The functionality of the features, implemented in new digital systems, should therefore focus on a clear overview and time-saving components. The experiences of a digital monitoring system at the construction site "Tunnel Kriegsstraße", shows the additional advantages of time-saving components and the implementation of more consistent procedures due to a central system.

For requirements engineering, the system context and the definition of use cases is the important key to get from the vision to the technical approach of a system. Digital data management and quality assurance of a system via web-application has to fulfill different functional and non-functional requirements from different stakeholders. Expert interviews, conducted with several stakeholders in underground and tunnel works showed that location independency, quality and comprehensibility of the system documentation and the links to other systems are the most important non-functional attributes. Defining the system boundaries and the context are therefore the first step.

2.1 *System Context*

The context diagram in Figure 2 depicts the link to stakeholders, already existing software, technical processes and infrastructure to a digital data-management system. For the digitalization process, the types of interfaces to the application therefore have to be user-interfaces for people and soft- and wireless hardware interfaces to other systems. Stakeholders, material testing results and machinery must provide the data for the application. This could be logged data as production-, additional process- and material-testing data or data from the stakeholders for organization such as the input of the implementation and work planning.

2.2 *Use Cases*

A digital management system has to fulfill the features listed at level three with a focus on wireless and direct interfaces between the different systems used on-site. The optimal use for such a system is not only in the technical evaluation, the documentation of parameters and verification of goals, but also in the organizational processes. Features include forecasting of material and labor disposition, as well as time needed. Digital instead of analog approval for as-built data speeds up inspections and user interfaces for material- and direct testing leads to a paperless and fully transparent construction site.

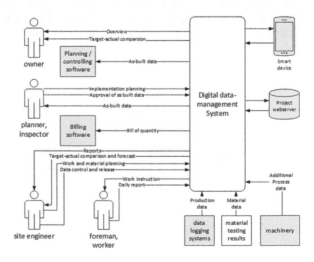

Figure 2.   Context diagram for digital data-management system for ground- and tunnel engineering.

3   SET-UP OF DIGITALIZATION PROCESS

The digitalization of the construction industry at tunneling and ground works is advancing at an ever-increasing pace. Digitalization is more than just the conversion of analog processes. It rather questions the construction process as a whole and how it can be transformed through digital systems. Digitalization changes the interactions between all involved parties and further the business models of construction companies (Krause et al. 2018).

3.1   *Digitalization of construction processes*

Digitalization on the construction site comes with very specific challenges in terms of robustness of hardware, data upload connectivity and speed, as well as a user-friendly approach. Employees at the construction site only accept technologies if they are easy-to-use, simple, and designed to meet their requirements (Schreyer 2016).

Today it is important to include software and hardware developers on a construction site team. The focus on User Interfaces (UI) and User Experience (UX) Design is crucial to digitalize construction processes successfully. This system requires agile development as an approach to ensure continuous improvements and feature development as unforeseen needs arise. This method establishes the digitalization process on-site and is able to deal with changing requirements, even in later stages. These needs are defined through collaborative efforts of the development team and the participating parties on site. Agile development leads to an iterative, vital approach for the successful implementation of a complex system, such as those described in chapter 4.

The understanding of construction methods and the documenting of the state of smooth constructions processes is necessary to examine their potential for digitalization. Civil engineers, as part of the development team use different tools, such as workshops, expert-interviews, surveys and project specifications to draw and model the current on-site status (Bergsmann 2012, Goger et al. 2018). Business Process Model Notation (BPMN 2.0) is a semantic basis to communicate between developers and site personnel.

Nowadays state of the art sensors and data-logging systems are improving the quality of the construction processes on-site, but site engineers cannot fully understand and analyze this raw data. To perform sophisticated analysis and evaluations they have to use spreadsheet programs, which require hours of analyses and are prone to error, leaving engineers with a superficial understanding of the data produced.

To address this issue, the project team of eguana GmbH developed a generic, modular digital data management system for construction processes to meet the requirements discussed in Chapter 2. It enables site engineers and other stakeholders to analyze and evaluate the data of several construction processes simultaneously and in real-time.

3.2   *System Architecture*

Data are gathered through the integration of existing infrastructure, e.g. machines, data loggers, sensors etc., as well as plan data. Therefore, the developed system allows secure upload of data automatically from construction machines, data loggers or from workers via web interface. Specifically developed data loggers allow the connection or integration of all relevant sensors and further enable the integration of specific logic, which is important to the respective construction processes. The uplink could be performed through traditional mobile connections (2G/3G/4G) or wireless network technologies (Wifi, LoRa, XBee, etc.).

The system simultaneously stores the data in object-relational databases on several servers to ensure data security even in break down situations and data storage over years as required by many clients.

3.3 Evaluation and Analysis

The collected and uploaded data is being processed and analyzed automatically in terms of integrity and process specific criteria. If there are any deviations from predefined construction parameters, faults or errors, managers are informed through automated notifications and the respective data sets are highlighted.

Specific views of the production process data sets enable deeper analysis and evaluation. These visualizations have been developed under consideration of modern UI/UX principles (EN 2006) and through feedback from analytical process optimizations with planners, project managers, site engineers and foremen. User-defined plots of periodic summaries give the stakeholders access to more sophisticated analysis and reporting. Database query functions have been developed to show correlations and patterns and to create abstractions of construction processes. Through these query functions, complex databases are simplified and can be performed by the site manager and further understood though different tables and graph drawings. The type of material is linked directly to the production data. Material delivery, stock, as well as material consumption are tracked throughout the production period to ensure quality, detect material losses and faulty sensors as early as possible.

The collected data is further processed and analyzed in terms of main construction and sub- or side processes. Since most of the sub-processes, such as repairs or interruptions cannot be digitally recorded, and a semi-automated process association has been developed to generate accurate data of sub-processes and the proportion of time used for each. Due to this approach, managers only need to edit data in case of special events, like downtimes. Considering this, allows the system to document the construction process in an accurate way.

The focus of the visualistion is on the fast navigation and understanding of information. This enables stakeholders to oversee the construction site and to focus in on a specific production process. Throughout all levels of visualisation, the status of grouting work is illustrated through specific colour codes. Thereby the status of all subsections are analysed and the results forwarded to the higher level of visualisation. Figure 5 shows an example of such status visualisations. Inspections make it easy to identify specific or query-dependent areas. The system's identification of areas where specific criteria have not been met or the process has been aborted allows for quick responses by inspectors. The system highlights these areas by colour-coding within interactive diagrams indicating specific information such as material consumption or completion.

Detailed analysis of main construction, sub- and side processes can be used to identify further potential for optimization, as well as measure the effect of optimization approaches on an ongoing

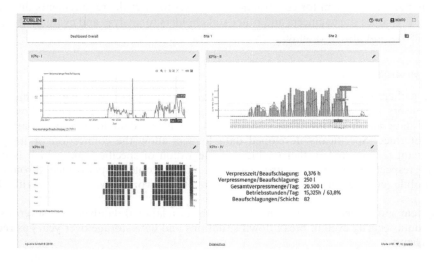

Figure 3. Dashboard.

basis. Figure 3 shows the dashboard with personalized data analysis, such as target comparisons of process times, trends in quantities used of time and daily key performance indicators.

Once the construction is done, the database functions as a feedback loop for all participants through the entire life cycle of a groundwork. This also provides valuable information for planning and calculating future projects.

Furthermore, the collected data can automatically generate bill of quantity. Specific protocols provide a detailed, transparent and comprehensible documentation. Through the implementation of specific interfaces connecting planning and controlling tools, the system organizes the data, and provides the appropriate format for software like RibITWO or BIM-software.

The access through web browsers does not require specific software licenses or local software installation, which is an advantage, as most common data monitoring systems at level two are proprietary. It allows remote access for multiple users on various devices, at any location, 24 hours per day. Furthermore, it enables the access to construction data of several construction sites at once, which enables managers and stakeholders to keep track of various projects.

Different stakeholders, e.g. planners, project managers, site engineers, foremen etc., have different areas of interest when looking at the data. The developer configures what aspects each stakeholder sees and what features they can use, based on their individual needs.

Thus the developed system not only focuses on technical aspects, but also on operational aspects. It frees up the engineer's time to concentrate on the evaluation, as the documentation, organization of- and data handling is automated. The self-explanatory and easy-to-use approach makes it easy for stakeholders to discuss the information with each other without redundancies.

4 EXPERIENCE ON THE CONSTUCTION SITE

On injection sites, timing is of the essence when collecting and processing all necessary data in order to quickly respond to errors to prevent damages and extra costs. The big challenge is the large amount of data generated during the production of the holes and the injection process itself. Quickly sorting out the essential parts for evaluation, is difficult with standard technology of today.

4.1 Introduction

The construction site "Tunnel Kriegsstrasse" in Karlsruhe in Germany was as an example of the implementation of the new digitalization process described in chapter 3. The project is under construction and will be finished by 2021.

The consortium of the companies Züblin and Schleith are building a 1.6 km road tunnel in open construction in the area of the Kriegsstrasse. After completion of the civil engineering structure, a tram-line and bicycle lanes will be created on the surface above the tunnel.

The construction project is divided into nine construction pits with a size up to 6,300 m² per pit. The tight time schedule has the completion of the last excavation planned in late 2020. A horizontal soft gel injection at a depth of up to 25 metres below the top ground surface acts as the sealing bottom for each pit. Figure 4 shows a schematic section of the injection works. As a vertical construction pit enclosure, intersected bored pile walls and diaphragm walls with sheet pile inserts are produced. Two of the nine excavation pits have already been completed, the other excavation pits are currently under construction.

4.2 Why new data management?

It is necessary to inject about 33,500 m² of sole foundation with soft gel for the entire construction project. Following the drilling works and the installation of approximately 12,500 pieces injection pipes, different injection horizons are executed with cement-bentonite suspension (the cover) and soft gel for the horizontal construction pit sealing. In total, there are about 37,500 injection points in the course of the entire construction project.

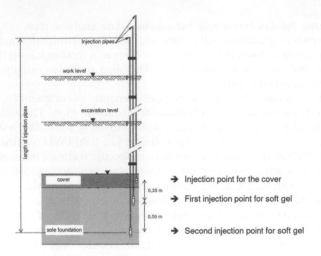

Figure 4.   Schematic section of the soft gel sole foundation.

All drilling works and injection operations of the cement suspension for the cover and the soft gel must be documented in detail. This is to ensure the required manufacturing quality of the injection, to detect mistakes early, and to get clear records for the client. Therefore, the contractor will produce approximately 50,000 digital records during the manufacture of the soft gel injection for the construction site "Tunnel Kriegsstrasse". Due to this amount of information, a good data management system is essential. With the new digital system, the monitoring and archiving of injection site data will be handled without printing any paper.

In the periods of the grouting, 14 pumps are operating at the same times, twenty-four hours a day, seven days a week. For an optimal result, it is necessary that site engineers monitor the injection work in continuous operation. Therefore operators get several hundred protocols per working day. The joint venture "Tunnel Kriegsstrasse", where the company Züblin Spezialtiefbau executes the injection work, has decided, together with the owner, to use eguana SCALES for the data management. Before using this data management software, the standard report of soft gel injection was a manually entered into proprietary software to convert the data to comma-separated values, storing and analyzing them in Microsoft Excel.

4.3   From the past to the future - a comparison

This chapter will compare the method of manual entry on previous construction sites and the digital method. It will focus mainly on the measurable time savings and the consistent documentation and analyses procedures by the inspector and the contractor Züblin Sezialtiefbau. Other benefits of digital data management are already described in chapter two and three.

4.3.1 Previous procedure

A pump operator managed up to 14 injection pumps, controlled the injection system directly on-site and monitored all data such as flow and pressure during the injection. After completion of an injection point, the data were stored locally on the control computer.

Engineers downloaded the individual files, one per injection point, intermittently once or twice a day via a external storage medium from the computer. In addition, the data logging system generated a file with a daily summary of all operations.

On a network drive, all data were sorted by date. Afterwards, site engineers had to open and check every file individually and evaluate the process. The result of the data was a diagram with the pressure, flow and the injection quantity depending on the time of each injection. The technicians saved this diagram as a pdf printout and stored a hardcopy version in a folder.

These data were submitted digitally and as a hardcopy to the client on a daily basis. An additional problem of this method is that the data are not available immediately after the end of injection and inconsistencies during shift changes and downtimes can occur. Therefore, there is a high time delay, such as in the evening or the next day when the construction site manager gets the reports. An ongoing control of the progress and the parameters was not possible.

4.3.2 *The new method with eguana SCALES*

An important fact of the new method is that the operating process is not affected by the digital system. In addition to the control technology, the construction site must have internet access, so the new data monitoring has access to the data logging system. After completion of an injection-point, the datasets are immediately entered into the database. In order to be able to display the data in a clear manner via browser, these are displayed in tabular form and overlaid on planning documents. A graphic representation with different status of grouting points is depicted in Figure 5. A click on an individual point makes the information about quality and quantity of a certain grouting process visible.

Authorized people have access to eguana SCALES. There are different access rights for the stakeholders.

The data can be evaluated chronologically or other filter options such as quality, completion status or material quantity. Quick navigation from one report to the next is possible with the tabular based program. A hardcopy printout is possible, but not necessary. Notes can be added directly (for example: problem with pump, recording error). The submission to the owner and the construction supervision takes place online. As the supervision and the contractor are using the same application, the data control, release and the approval are done in a consistent way, without redundant datasets. The site manager has to approve the report, and only then the report is visible for the client.

In addition to the injection reports as a diagram, the progress of the scheduled work is shown in a plan view with eguana MAPS. By color coding the injection points, the current progress can be determined at a glance.

4.4 *Conclusion*

Within the framework of this highly innovative project, Züblin Spezialtiefbau as a part of the Züblin/Schleith joint venture has therefore opted for using the latest technologies. The

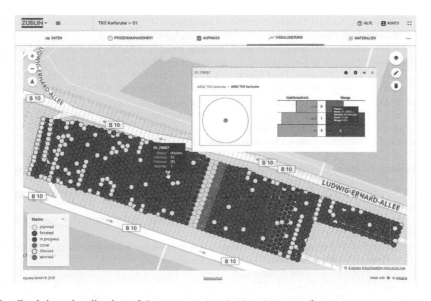

Figure 5. Real time visualization of the construction field and process data.

monitoring and construction site communication is largely digitized by means of eguana SCALES and comprises the four major components of a powerful data monitoring system discussed in Chapter 1. The cooperation between construction supervision and the consortium takes place via this innovative monitoring system. All works are tracked digitally - without unnecessary paper waste and manual, analog protocols. In order to have an overview of the numerous works, eguana MAPS allows a perfect presentation of all incoming reports. The current status of the execution is quickly understandable via color codes in a digital map. Construction site areas, which are still in progress or are problematic can be quickly and easily identified.

By august 2018, about 6,000 m² of soft gel injection for the Tunnel Kriegsstrasse in Karlsruhe has been done and checked with high quality by the help of eguana SCALES in a short time span. Züblin Spezialtiefbau has gained a competitive edge by using this innovative system.

5 FUTURE OF CONSTRUCTION PROCESSES

Literature research, expert interviews and a survey turned out, that high-developed data monitoring systems have to fulfil many different features to check that the goals are being met. New systems have the power not only to provide these features, but also to establish a standardized documentation and evaluation procedure and support organization processes.

The system described has been designed in terms of user interfaces (UI) and user experience (UX) to meet the specific needs of geotechnical engineers and construction managers at best.

Thereby reducing the efforts spent on evaluation and analysis up to 72%, compared to common grouting software. Due to the development of algorithms for the calculation of grouting process distributions and resource consumptions, bill of quantities can be generated automatically. This leads to an improvement of accuracy of 2.1% in average. Through the development of specific user roles and corresponding access permissions, all stakeholders are able to access data through a web browser, remotely and in real-time. This leads to an ease of communication and reduced reaction or response time.

Agile development is capable to adapt new requirements arising during the construction. Digital work planning, approval of as-built data are an approach to digitalize monitoring systems fully. Further improvements, like the whole testing procedure of materials, which is substantial for the quality assurance are being adapted in the following years.

REFERENCES

Bergsmann, S. 2012. End-to-End-Geschäftsprozessmanagement, Organisationselement – Integrationsinstrument – Managementansatz, ISBN 978-3-7091-0839-0, Springer, Vienna

Bruce, D.A. 2012. Computer Monitoring in the Grouting Industry. At the Geo-Institute GeoCongress, *State of the Art and Practice in Geotechnical Engineering Conference*, March 25-29, San Francisco, CA

Dreese, T.L., Wilson, D.B., Heenan, D.M.Cockburn J. 2003. State of the Art in Computer Monitoring and Analysis of Grouting. At the Third International Conference on Grouting and Ground Treatment, February 10-12, New Orleans, Louisiana, United States

EN ISO 9241-110. 2006. Ergonomics of human-system interaction – Part 110: Dialogue principles, European committee for standardization, Brussels

Goger, G.Winkler, L. 2018. Conference Book from the First Colloquium *Zukunftsfragen des Baubetriebs, Mai*, TU Wien, Vienna

Krause, St.Pellens, B. 2018. Betriebswirtschaftliche Implikationen der digitalen Transformation, ISBN 978-3658187507, Springer Fachmedien Wiesbaden

Owen, P.E. 2017. Engineering and Design GROUTING TECHNOLOGY, EM 1110-2-3506, Department of the army U.S. Army Corps of Engineers, Washington DC

Schreyer, M. 2016. BIM – Einstieg kompakt für Bauunternehmer. BIM-Methoden für die Bauausführung, ISBN 978-3-410-25702-8, Beuth Verlag GmbH

Winkler, L.Maroschek P.Piskernik M. 2017. Cooperative knowledge management systems for grouting works at tunnel sites, At the 4th Arabian Tunnelling Conference 2017, UAE

Zhong, D.H.Yan, F.G.Li M. C. and Huang C.X., Fan, K.Tang, J.F. 2015. A Real-Time Analysis and Feedback System for Quality Control of Dam Foundation Grouting Engineering. At Rock Mechanics and Rock Engineering, DOI 10.1007/s00603-014-0686-6, Springer.Verlag Wien

Tunnels and Underground Cities: Engineering and Innovation meet Archaeology,
Architecture and Art, Volume 6: Innovation in underground engineering,
materials and equipment - Part 2 – Peila, Viggiani & Celestino (Eds)
© 2020 Taylor & Francis Group, London, ISBN 978-0-367-46871-2

New advancement in durable segments for combined sewer overflow tunnels, USA

R. Winterberg
BarChip Inc., Tokyo, Japan

M.R. Garbeth
Super Excavators Inc., Menomonee Falls, WI, USA

ABSTRACT: The Blacksnake Creek and stormwater runoff in St. Joseph, MO, is currently piped along with sewage in a 100-year old pipe not large enough to carry all the stormwater and sewage to the wastewater treatment plant and it overflows to the Missouri River after most rainstorms. The Blacksnake Creek Stormwater Separation Project will convey stormwater directly to the Missouri River. This will reduce water quantity in the existing sewer during storms and the quantity of combined stormwater and wastewater overflowing to the river. Currently, a 2.75 m ID and 2 km long segmental tunnel is under construction as part of the Separation Project. This project is an America's First, using segments solely reinforced with macro synthetic fibre (MSF). This paper addresses the solutions to the technical challenges of the project, the design of the segments and the benefits associated with the use of MSF.

1 INTRODUCTION

1.1 *Fibre reinforced concrete tunnel segments*

Fibre reinforced concrete (FRC) is becoming widely utilized in segmental linings due to the improved mechanical performance, robustness and durability of the segments. Further, significant cost savings can be achieved in segment production and by reduced repair or reject rates during temporary loading conditions. The replacement of traditional rebar cages with fibres further allows changing a crack control governed design to a purely structural design with more freedom in detailing.

Macro synthetic fibres (MSF) are non-corrosive and thus ideal for segmental linings in critical environments. Although fibre reinforcement for segments is relatively new, recent publications such as the ITAtech guideline (ITAtech 2016), the British PAS 8810 (2016) and the FIB state-of-the-art report (FIB 2017) have now given more credibility to this reinforcement type and the basis for design.

It seems that the industry globally has picked up MSF for segments, mainly for the durability reasons the synthetic fibres provide, as compared to steel fibres. A major reference for the use of MSF as sole structural reinforcement for precast tunnel segments is the Santoña Laredo General Interceptor Collector in Northern Spain (Winterberg et al. 2018a). The experience gained in the Santoña-Laredo project showed that macro synthetic fibre reinforced segments perform very robust and satisfactory even under difficult conditions (Winterberg et al. 2018b).

Also the US started to embrace this technology and there are a couple of innovative people among the design consultants, contractors and owners supporting this drive. After successful completion of the final lining of the starter and end tunnel, and successful trials for the Euclid

Creek tunnel in the NEORSD sanitation project (Wotring et al., 2016), now the first MSF segmental tunnel is on its way in St. Joseph, Missouri.

This America's First is even more significant as the new ACI guideline on FRC tunnel segments basically exclusively deals with steel fibre (ACI 544.7R 2016). That shows that innovative drivers can overrule conservative practice for beneficial and sustainable solutions. In this particular case, all parties involved had the common target to make it a world's first in North America.

1.2 *Project description*

The Blacksnake Creek Stormwater Separation Improvement Project is required as part of the City of St. Joseph, Missouri's Combined Sewer Overflow (CSO) Long Term Control Plan in order to improve water quality as mandated by the Federal Clean Water Act. The CSO control mandate is described in the Combined Sewer Overflow Control Policy as administered by the Environmental Protection Agency (EPA).

Blacksnake Creek is currently directed into the City's combined sewer system through a double box culvert, and the creek flow is conveyed through the sewer system to the Water Protection Facility and unnecessary treated 365 days of the year. During wet weather events, storm water runoff exceeds the capacity of the combined sewer system and causes combined sewer overflows to the Missouri River. The overflows are a mix of stormwater and sanitary sewage and result in adverse water quality problems. To combat the discharge, the project will intercept and redirect Blacksnake Creek stream flows away from the City's combined sewer system to a new and dedicated stormwater conveyance system that flows directly to the Missouri River, thus reducing the frequency, volume, and impacts of combined sewer overflows to the river.

The Tunnel Bid Package (TBP) was a critical portion of the project, as the topography of the area did not allow for construction of a new and dedicated stormwater canal via traditional trench excavation without significant impacts to the public. The TBP primarily consisted of the following components and key facts:

- Contractor: Super Excavators, Inc. Menomonee Falls, WI
- Bid Award Amount: $27,991,000 USD
- Duration: 2018 to 2019
- 2.74 m ID by 2 km long segmentally lined tunnel with macro synthetic fibre reinforcement.
- 11 m ID Baffle Drop Structure
- 15 m Reinforced Concrete Box Culvert
- Energy Dissipation Structure

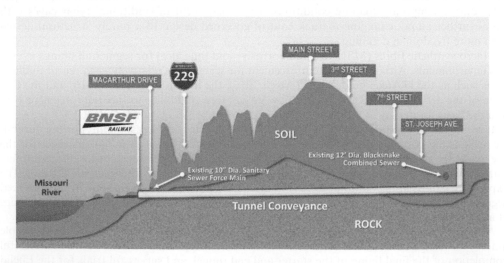

Figure 1. Tunnel alignment.

2   GROUND CONDITIONS

The tunnel is to be constructed in soft ground and mixed ground conditions, including soils and shale and combinations of both at interfaces. The soils to be encountered vary from silty clay, silty sand, clayey sand, and sandy clay. The sections with higher sand content may lead to great instability in the excavation and could result in greater groundwater inflows and flowing ground.

The shale to be encountered is low in strength, has low to medium-low durability, is easily excavatable, and has low hydraulic conductivity. Groundwater along the tunnel varies along with the topography from 12 m to 45 m of groundwater head.

The general geography and topography of the project site is located in north-western St. Joseph, Missouri between St. Joseph Avenue and the Missouri River. The overburden along the tunnel alignment consists primarily of fill material and alluvial deposits with alternating beds of silt, clay, silty-clay and clayey silt. The minimum overburden on the project occurs at the launch shaft with approximately 9 meters of cover. As you progress away from the launch shaft at the peak of the topography on Highland Avenue between Main Street and 2nd Street there is 36 meters of cover and by the time you reach the receiving shaft this reduces down to 18 meters. The primary rock conditions predominantly consists of shale and claystone along the tunnel horizon. A limestone bed is present above the tunnel crown for approximately 1,280 meters.

The percentage of the alignment for the various different ground conditions consists of: soft-ground 23% of alignment, rock 63% of alignment and mixed-ground 14%. The tunnel will transition between soft ground and rock along each end of the tunnel alignment and mixed-faced conditions shall be anticipated for reaches along these subsurface transitions. The soil to rock transitions on each end of the alignment may be gradual. Meaning, that rock or soil may be present for a length, and then not present in the excavation face over these transitional mixed-face areas. It is anticipated that the tunnel groundwater inflow will not exceed a steady state of flow greater than 50 GPM.

3   TBM DESIGN AND LAUNCH

3.1   *TBM design*

Considering the ground conditions along the tunnel alignment and the contract specifications, the tunnel must be excavated with an EPB tunnel boring machine in order to migrate risks. The TBM was newly commissioned, engineered by LOVSUNS in Canada, and manufactured and fully tested at their parent company Liaoning Censcience Industry Co. Ltd. Facilities in China. Details of the TBM are presented in Table 1.

Figure 2.   TBM factory acceptance in China.

Table 1. Technical data of the EPB TBM.

Parameter	Value
Excavation Diameter	3,314 mm
Total Length	78 m
Min. Turning Radius	250 m
Max Speed - Cutting Head	9.21 rpm
Maximum Torque	981 kNm
Main Drive Motors	4 No.
Thrust Capacity	900 Tonne
Thrust Cylinders	12 No.

3.2 TBM launch

Prior to the TBM arriving onsite in late June of 2018, a cradle was placed inside the secant pile launch shaft. This cradle consisted of two I-Beams cast in a concrete floor to fit the curvature of the machine. All the shaft and tunnel utilities were established inside the shaft prior to lowering any machine components (Figure 3 right). A special electrical and hydraulic umbilical assembly was designed and used to help launch the TBM. Note the secant launch shaft diameter is 15 m and the entire TBM assembly is 78 m of length.

The first stage was receiving the 18 truckloads of the disassembled machine from Port of Seattle in Washington, and transporting to St. Joseph, MO. The shipment took roughly three weeks at completion. It took one month to ship the TBM from China to Seattle by boat.

The entire machine was methodically aligned in two rows consisting of Row # 1, Main TBM Parts: cutterhead, telescopic shield, gripper shield, and tail skin; and Row # 2, Backup System: all the gantries and backup equipment including the electrical substation (Figure 3 left). The TBM cutterhead, gripper shield and telescopic shields were nearly completely assembled at the surface due to the ease of assembly at the surface versus the shaft. A 500 Tonne mobile crane was mobilized to the project site and used to lower the heavy components into the shaft.

Once the major components were lowered down into the shaft, one crew began hooking up the electrical and hydraulic lines while another crew continued to work on assembly of the gantries. The gantries included such items as: the operator's cabin, electrical substation, air compressor, main drive motors, grease pumps, ground conditioning system, ventilation cassette and cable festoons.

Figure 3. TBM Carrie's main parts (left) and launch cradle in the secant shaft (right).

Figure 4.   Launching the TBM (left) and half-ring installation for TBM propulsion (right).

The special hydraulic and electrical umbilicals were used to link the gantries on the surface to the main TBM body in the launching cradle on the bottom of the shaft. At this point of time, the TBM could begin the functional testing and troubleshooting process.

After finishing assembly of the initial configuration for the machine, the excavation has started. During this process, the primary propulsion cylinders pushed off of beams welded to the I-Beam in the cradle. This allowed for the advancement of the cutterhead, telescopic shield and gripper shield to advance forward and penetrate the secant shaft. A short screw conveyor was utilized to control the material in front of the TBM (Figure 3 right).

After the TBM was launched, and most of the components installed and in the ground as mentioned above, the next step was to disconnect the complete electrical and hydraulic cables/ hoses, which allowed the following tasks to be completed:

- Remove the short launch screw conveyor
- Install the longer screw conveyor
- Install the tail shield
- Install the segment transporter belly pan
- Re-hook up the electrical and hydraulic cables/hoses.

During this stage, a precast concrete segmental tunnel liner "half-ring" structure has been used to push and advance the TBM further into the excavation (Figure 4 right).

Once the tailskin reaches the secant shaft wall the full ring will begin to be installed. A support bracket will be braced off the secants to hold the first segment ring in place while the TBM advances forward. The gantries will be lowered down into the shaft one by one until all nine are buried directly behind the TBM. Then, full production tunnelling can begin.

4   SEGMENTAL LINING

4.1   Introduction

The lining for the Blacksnake Creek Stormwater Separation Tunnel consists of a precast concrete segmental tunnel lining. The segmental lining is composed of six fibre reinforced concrete segments with rubber gasket frames inserted between the segments and adjacent rings to create a watertight lining. When assembled, the six trapezoidal segments form a ring with an internal diameter of 2743 mm (Figure 5).

The universal rings of 1219.2 mm length have a taper of 12.7 mm in order to accommodate a turning radius of 300 meters. Kinematic control during installation is provided by means of drifting bolts and alignment indicators on the segments. The radial joints are connected with galvanised steel bolts.

The precast concrete segments are produced by CSI Tunnel Systems Inc.'s plant in Macedonia, Ohio, and were shipped by truck to the project site in St. Joseph, Missouri. Segments are installed within the tunnel during the excavation process by completing a segment ring with each advancement. Each individual segment is placed using a segment erector located in the shield trailing the TBM and segments are manually bolted together. Annular grouting is performed through grout ports built into each segment after the rings are erected.

BarChip 54 Macro Synthetic Fibres were used to reinforce the precast concrete segments as an alternative to steel fibre or traditional deformed steel bar reinforcement. This alternative was chosen to reduce material costs while ensuring compliance with American Iron and Steel provisions of the Contract Documents. The use of synthetic fibre reinforcement further eliminates the risk of corrosion and increases the service lift of the tunnel lining. This, in turn, reduces future maintenance cost of the tunnel. The Blacksnake Tunnel will be the first tunnel in North America to be lined with precast concrete segments that are solely reinforced with synthetic fibre reinforcing.

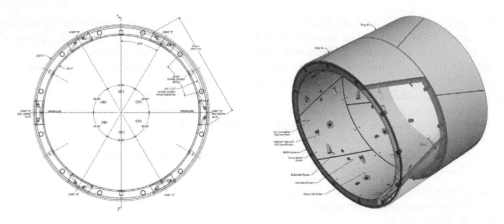

Figure 5.   Ring segmentation (left) and tunnel ring assembly (right).

Figure 6.   MSFRC segments in the Ohio factory (left) and on site (right).

4.2 Structural design

To better accommodate the fibre reinforced concrete solution, smaller segments were adopted in order to limit the segments' aspect ratio. The tunnel has an internal diameter of 2.743 m with a segment thickness of 190.5 mm. The segmentation was selected to be six trapezoidal segments (Figure 5). This yields a segment aspect ratio of 8.1, which is well below the acknowledged limit of 10 to ensure segment robustness for temporary load cases (ITAtech 2016).

The FRC segments are made using a concrete class C40/50 (f'_c 48 MPa) at 28 days age, with a stripping strength of 14 MPa. The residual strength was specified to 3.2 MPa at l/150 according to ASTM C1609. BarChip 54 is the chosen fibre, based on the good experience made in the Euclid Creek tunnel project (Wotring et al. 2016). A dose rate of 7 kg/m³ of this fibre was sufficient to exceed the residual strength specification.

The design approach adopted for the FRC segmental lining in ultimate limit state (ULS) is the use of Normal Force-Bending Moment interaction diagrams or Moment-Thrust Capacity Limit Curves. The factored design load couples acting on the section must remain within the N–M envelope (Nitschke and Winterberg, 2016). The FRC material properties are herein derived from the beam tests, which are eventually used as the basis to determine the stress-strain relationship of the concrete on the tension side. The idealized stress-strain diagram enables setting up the capacity limit curves, which are obtained by equilibrium iterations on the given cross-section.

Cooperating with the contractor's design team, L-7 Services from Denver, CO, the Technical Department of BarChip Inc. (formerly known as Elastoplastic Concrete or EPC) was asked to conduct a feasibility study for the use of macro synthetic fibres as the sole reinforcement for the lining.

The structural design for ULS also comprised a simulation of 1% ground loss. The initial check for 1% ground loss simulation showed a slight outlier (Figure 7 right), however, the factored load couple was later revised to a lower value of bending moment to fit the capacity envelope.

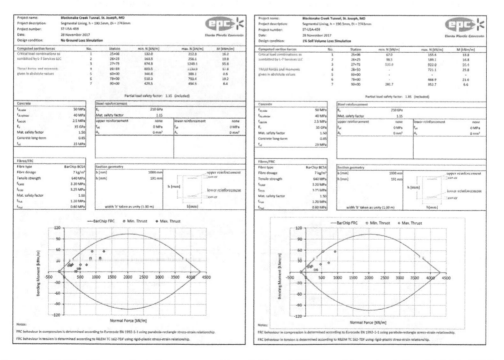

Figure 7. N-M section analysis for no ground loss (left) and 1% ground loss simulation (right) in ULS.

All checks for temporary load cases as well as the checks for serviceability limit states (SLS) have been performed with Finite Element Analysis using Atena (JKP 2017).

5 TUNNEL DRIVE

Full tunnel production has just started at the time of submittal of this paper. Experiences during the tunnel drive, i.e. segment/ring installation, evolution of TBM thrust forces, handling of eccentric thrust forces and behaviour of the segments, etc. will be presented in detail in the presentation of this paper. However, the experiences of the TBM launch and installation of the first rings are described here.

After receiving the TBM from Liaoyang, China to St. Joseph, MO (United Stated), performing the complex assembly and functionality testing on the surface, and receiving the fibre reinforced concrete segmental liner sections, now it's time to discuss the approach to the initial tunnel drive.

As previously discussed, space in the launch shaft is very limited compared to the entire length of the TBM, thus requiring the launch process to occur in phases. This process was explained in greater detail under the launching section of this paper.

The initial tunnel drive launched in soft ground, and then transitions into mixed ground conditions fairly quickly. To control alignment and grade during the launching process, and throughout the tunnel drive, a TAC's system is implemented. The TAC's system provides continuous information about how the machine axis is aligned with respect to its designed alignment, and in what direction it is moving. The system also determines how the six piece segmental liner will be installed, dependent on the position of the rings with the taper, which determines the straightness or curve of the tunnel drive.

To date, there are roughly 30 segment sections installed (approx. 36.5 meters). The initial tunnel drive installation has proven that the fibre reinforced segments are robust enough to withstand the temporary load cases, i.e. transportation, hoisting, handling and installation, without experiencing damage to the segments. In addition, the half-ring segment sections used in the shaft for the launching process were inspected after the initial push and no damage

Figure 8. Segment erector inside the tail can (left); erected segments in the tunnel (right).

occurred to the segments. We haven't experienced high forces during the initial tunnelling drive due to the fact we haven't encountered the rock section thus far. Also, the forces while launching through the concrete secant shaft were insignificant. We don't expect any issues to arise when the rock section is encountered due to the fact the segments and TBM are designed to handle the anticipated ground conditions.

This particular TBM uses a bulkhead type mechanical segment erector inside the tail can to erect the six piece segmental ring. It utilizes a ball type pick up system, where the ball head bolt is screwed into a threaded socket in the centre of the segments (Figure 8 left). The erector has a safe working load of 130% of the segment weight. The rotational speed is fully variable even when loaded between 0–2 RPM. The erector is capable to reach segments over the two inner most rows of the tail seal brushes.

It is anticipated based on the performance of the TBM, robustness of the segments, and ground conditions that are expected to be encountered, that the advancement should be between 14.6 to 18.2 meters per 10 hour shift once full production mining commences.

6 CONCLUSIONS

Being the first contractor in North America to utilize a Lovsuns TBM, and macro synthetic fibre as the sole reinforcement of the precast concrete segments, is a very exciting aspect to this project. Launching the TBM with an extensive amount of umbilical cords, in a relative small size shaft compared to the overall length of the TBM, and a unique launching process has been challenging but a lot of knowledge has been gained for future launching of this TBM on other projects.

The durability and performance of the macro synthetic fibre reinforced segmental lining has outperformed expectations to date. The indicating factors of the performance of the machine, durability and robustness of the segments, and performance of the crew, will meet the project goals once production mining commences.

Ongoing research and continuous developments on macro synthetic fibre and macro synthetic fibre reinforced concrete have made it today being a modern and cost-efficient construction material. Eliminating durability issues with regard to corrosion of the primary reinforcement yields significant advantages for the design, since it is no longer governed by serviceability limits.

The experience gained so far in the drive of the Blacksnake tunnel project shows that macro synthetic fibre reinforced segments perform very robust and satisfactory even under difficult conditions during the TBM launch phase.

The successful completion of this "America's First" project will build further confidence in macro synthetic fibre reinforced segmental linings. The success and gained experience of this project will lead to the implementation of this technology in other segmental tunnel projects.

These types of utility tunnelling projects (e.g. sewage, power, irrigation or gas transfer tunnels) are widely existing in the world market and they present a huge opportunity for MSF reinforced concrete linings, benefiting from the given advantages.

REFERENCES

ACI 544.7R-16 Report on Design and Construction of Fiber-Reinforced Precast Concrete Tunnel Segments, ACI Committee 544, *American Concrete Institute*, Farmington Hills, MI, 2016
ASTM C1609/C1609M-12, Standard Test Method for Flexural Toughness of Fiber-Reinforced Concrete (Using Beam with Third-Point Loading), *ASTM International*, West Conshohocken, PA, 2012
FIB 2017. Precast Tunnel Segments in Fibre-Reinforced Concrete, State-of-the-art report fib WP 1.4.1. fib Bulletin 83, *Fédération internationale du béton (fib)*, October 2017
ITAtech 2016. Guidance for Precast Fibre Reinforced Concrete Segments – Vol. 1 Design Aspects. ITAtech Report No. 7, *ITAtech Activity Group Support*, April 2016, www.ita-aites.org
JKP 2017. Design report on MSF reinforced concrete segments for the Blacksnake Creek tunnel. JKP Static, Budapest, Hungary, November 2017

Nitschke, A. & Winterberg, R. (2016). Performance of Macro Synthetic Fiber Reinforced Tunnel Linings. *Proc. World Tunnel Congress 2016*, San Francisco, USA, 22–28 April 2016

PAS 8810 2016. Tunnel design – Design of concrete segmental tunnel linings – Code of practice. *The British Standards Institution*, BSI Standards Ltd

Winterberg, R., Mey Rodríguez, L., Justa Cámara, R. & Sualdea Abad, D. (2018a). Segmental Lining Design using Macro Synthetic Fibre Reinforcement. *Proc. FRC 2018, Fibre Reinforced Concrete: from Design to Structural Applications*, Desenzano, Italy, 28–30 June 2018

Winterberg, R., Justa Cámara, R. & Sualdea Abad, D. (2018b). Santoña–Laredo General Interceptor Collector – Challenges and Solutions. *Proc. FRC 2018, Fibre Reinforced Concrete: from Design to Structural Applications*, Desenzano, Italy, 28–30 June 2018.

Wotring, D.A., Vitale, M.G. & Gabriel, D.A., 2016. Synthetic-Fiber-Reinforced Concrete Segmental Lining - Laboratory and Field Testing Program and Results. *Proc. World Tunnel Congress 2016*, San Francisco, USA, 22–28 April 2016

Tunnels and Underground Cities: Engineering and Innovation meet Archaeology,
Architecture and Art, Volume 6: Innovation in underground engineering,
materials and equipment - Part 2 – Peila, Viggiani & Celestino (Eds)
© 2020 Taylor & Francis Group, London, ISBN 978-0-367-46871-2

Intelligent attitude control during shield tunneling in soft soils based on big data

B. Wu, M. Hu, W. Xu, P. Dong & B. Xue
SHU-UTS SILC Business School, Shanghai, China
SHU-SUCG Research Centre for Building Industrialization, Shanghai, China

ABSTRACT: To assure high quality of tunnel, controlling shield attitude is crucial issue while tunneling. Current control models mainly rely on the expert experience rules or geometrical axis fitting and neglect the important fact that shield attitude control strategies not only impact the direction of tunnel axis but also impact the disturbances of surrounding soil, so it is not reasonable to consider only a single control target. This paper proposes a new big data-driven Shield Attitude Control method (DSAC) to optimize shield posture Adjustment strategy, which consider tunnel axis control and settlement control comprehensively. DSAC mine historical engineering data from different projects by machine learning to establish two sub-models, one for setting shield attitude control target, the other for embodying the relationship between control variables (shield oil pressures) and control objectives (shield attitude). We used 550 thousand data records in Shanghai Metro line as training sample to create DSAC model and applied it to control other shield attitude of Shanghai Metro Line 18, The engineering application show the performance of tunnel axis and settlement are improved compared with the traditional control strategy method.

1 INTRODUCTION

The control of thrust axis of shield is the key step in tunneling. In the process of shield propulsion, the control strategy of the shield directly affects the quality of tunneling. In terms of this issue, many scholars have studied the tunneling strategy based on rules obtained from tunneling experience or using geometric fitting method. At the same time, the accuracy of setting the shield oil pressures will directly affect the effect of the strategy execution. In recent years, many papers have conducted in-depth theoretical analysis on the load calculation of oil pressures by means of force analysis and mathematical modeling.

According to the tunneling experience, Rui Li considers the influence of the grouting and propulsion speed on the propulsion control of the shield. And, according to the amount of axial deviation in the advancing process of shield, the strategy of tunneling is developed. Juan Hu proposed that the incomplete anastomosis of the segment joint would have an important impact on the shield attitude, by establishing the finite element model, the relationship between the attitude of the shield and the force of the segment is analyzed. Finally, the control strategy of shield propulsion is summarized under the standard requirement of segment axis deviation. Gang Chen proposed that, under the condition that the shield tail gap meets the attitude adjustment, the shortest path is taken as the optimization objective to fit the cubic curve, which is used as the control strategy of the shield. Rohola Hasanpour proposed a numerical simulation method to establish the relationship between geological conditions and shield tunneling parameters. Through nonlinear regression analysis, the sensitivity of geological parameters was analyzed to quantify the influence of geological factors on the shield propulsion strategy. However, the data universality of the objects analyzed by the above methods is relatively low. And, the methods are mostly based on the construction experience, and the factors affecting the advancement are relatively single, which leads to the poor practical application of the analysis results. To implement the propulsion strategy of the shield and refine the given pressure of the shield, it is also necessary to establish a relationship model between the attitude change and the magnitude of the thrust action. Bing Gao et analyzed the force of the shield and established the dynamic equation and kinematics equation of the shield attitude. However, since the force analysis is directed to a certain state in the shield propulsion, this method can only be used to statistically describe the propulsion hydraulic pressure of the shield, so it cannot be applied to the tunneling site. Li et al. based on the shield driving attitude type and the experience of setting the subdivision pressure, the fuzzy control rule is established to realize the shield automatic rectifying. Based on the data-driven method, Wei Sun used the random forest model to predict the tunneling speed and jack thrust pressure by taking the soil distribution on the excavation surface and the operation data of shield construction as input. However, the above method only considers the motion characteristics of the shield at a certain moment, does not consider the characteristics of the tunnel design axis, and whether the shield control will affect the surrounding soil environment. Therefore, such methods may cause large settlements in the soil surrounding the tunnel during the propulsion process.

Based on the tunneling monitoring data and shield operation data of Shanghai Metro Line 18, this paper based on big data proposes a big data-driven Shield Attitude Control method (DSAC). The method establishes the shield tunneling attitude optimization control model by determining the relationship between the shield oil pressure setting and the attitude adjustment, and learning shield attitude control more excellent section of the strategy. This model realizes automatic sets the oil pressures which decides the shield attitude during the process of the shield propelling along the axis.

2 DSAC METHOD

This paper based on historical big data proposes a big data-driven Shield Attitude Control method (DSAC), which establishes a tunneling attitude optimization control model through data-driven technology. The model is divided into two parts. The first one is an oil pressure-attitude model based on historical operational data. The second part trains control strategy model based on the shield tail axis deviation control excellent section and the design axis data.

According to the real-time attitude of the shield, combined with the excellent deviation control strategy that has been learned, the control strategy model gives the target position that the shield tail should reach in the next operating unit. According to the target position provided by the control strategy model, the oil pressure of the shield tail reaches the target position will be given by oil pressures-attitude model. Finally, aiming at the center of the shield excavation face to eliminate the horizontal deviation y and the vertical deviation z, then the M point is close to the axis position P, as shown in Figure 1. The framework of the method is shown in Figure 2.

In this paper, the fully-connected neural network is used to train the oil pressure-attitude relationship. In the control strategy model with time series characteristics, the LSTM model with long-term and short-term memory effects is used for training.

2.1 Oil pressure-attitude model

The complex relationship between control quantity and attitude cannot be described by using simple linear equations. Therefore, this paper uses a fully-connected neural network to describe the relationship between the shield oil pressure and the shield attitude change. The fully connected neural network structure is shown in Figure 3. A given training set has d input attributes and l output attributes. $D = \{(x_1, y_1), (x_2, y_2), \cdots \cdots (x_m, y_m)\}, xi \in R^d, yi \in R^l$. Using the BP algorithm to train the data set.

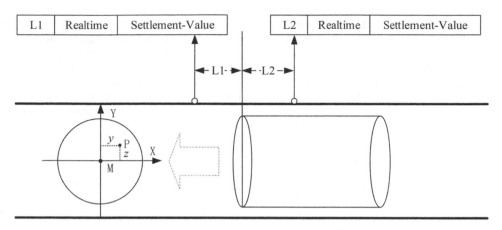

Figure 1. Schematic diagram of deviation control and settlement monitoring points.

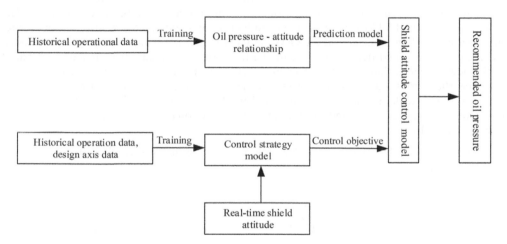

Figure 2. DSAC method framework.

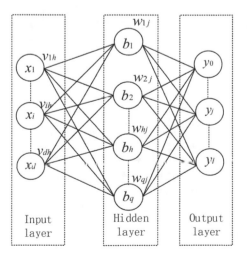

Figure 3. Fully connected neural network structure.

The oil pressure-attitude model aims to construct the relationship between the shield oil pressure and shield attitude. The input characteristics of the model are set as follows: t time each partition oil pressure, actual oil pressure of each partition, positive earth pressure, shield incision deviation, shield tail deviation, shield slope value, shield rolling angle value. Output characteristics: $t + 1$ time shield incision deviation, shield tail deviation, shield slope value, shield rolling angle value. Model time lag is 1 minute. It can be seen from the input that the model considers the actual oil pressure of the jack, and also considers the position characteristics of the shield and the pressure characteristics of the excavation surface.

2.2 *Control strategy model*

For the problem of shield control strategy, the selected model needs to be able to learn the characteristics of design axis and excellent segment operation parameter sequence in historical advancement. When a similar design axis data sequence is encountered again, the model can output a correspondingly excellent shield trajectory sequence.

LSTM is a variant of RNN, presented by German scholars Hochreiter and Schmidhuber in 1997. Unlike traditional neural networks, LSTM has a memory effect on historical data and has proven its superiority in processing sequence data in a large number of experiments. The main feature of the LSTM network is the introduction of a state unit C whose network structure is shown in Figure 4. Since the LSTM model contains state units throughout the model, and the gate control strategy allows the model to constantly update its own state unit based on the new sequence characteristics. Therefore, when the model predicts the $t + 1$ time output sequence, the outputs based on the data characteristics at time t and combined with the trained historical sequence data. The core of the LSTM model is the door mechanism, which includes input gates, forgetting gates, and output gates. The functional design of each type of door is described as follows:

(1) Forgotten door. The role is to discard unnecessary data in historical data. The mathematical model is shown in (1):

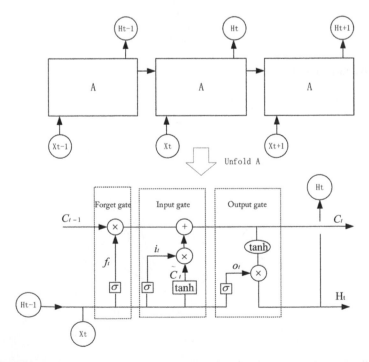

Figure 4.   LSTM Model Structure.

$$f_t = \sigma(W_f \bullet [h_{t-1}, x_t] + b_f) \tag{1}$$

Where f_t is the threshold of the forgotten gate, and the information retained after the forgotten gate is $f_t \otimes C_{t-1}$.

(2) Enter the door. The role is to filter out new information and update status information. The mathematical model is shown in (2)(3):

$$\begin{cases} i_t = \sigma(Wi \bullet [h_{t-1}, x_t] + b_i) \\ \widetilde{C_t} = \tanh(W_c \bullet [h_{t-1}, x_t] + b_c) \end{cases} \tag{2}$$

$$C_t = f_t \bullet C_{t-1} + i_t \bullet \widetilde{C_t} \tag{3}$$

Where i_t is the input of the input gate at time t, and the information retained after the input gate is $i_t \otimes C_t$, and C_t represents the state information after the input gate and the forgetting gate.

(3) Output gate. The role is to determine the output value in conjunction with the updated state information, as shown in (4):

$$\begin{cases} o_t = \sigma(W_o[h_{t-1}, x_t] + b_o) \\ h_t = o_t \bullet \tanh(C_t) \end{cases} \tag{4}$$

Where O_t is the component of the hidden layer at time t, and ht is the state of the output layer at time t.

This paper uses multivariate LSTM time series prediction models, its model structure is shown in Figure 5. The model includes an input layer, a hidden layer, and an output layer. Wherein, the input layer is the t time design axis data, and the time sequence of the shield operation data; The hidden layer is a structure including a plurality of LSTM units, and the output layer outputs the $t + 1$ time predicted data in a format of input data. Through the BPTT algorithm iteration, adjust the weight between each layer, reduce the error until the output error converges to a certain range.

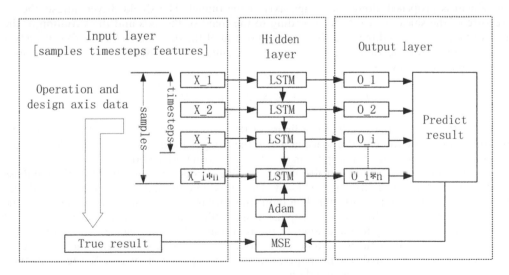

Figure 5. Multivariate LSTM model structure.

In order to use the LSTM model to learn the trajectory data of the better segment from the original data set, the original data is filtered based on the constraint that the tunnel segment assembly deviation conforms to the specification. Since the ultimate goal of the control strategy is to enable the assembled segments of the tunnel to be laid along the design axis, the deviation data of the shield tail in the horizontal and vertical directions is extracted as the trajectory information. While studying the law of the deviation trajectory of the horizontal and vertical directions of the shield tail, the stroke change X along the design axis direction is introduced to determine the spatial position of the shield excavation surface. The model takes the t time trajectory feature as input and the $t + 1$ time feature as output. The input characteristics of the model are set as follows: t time design axial direction stroke change X, shield tail horizontal direction deviation, shield tail vertical direction deviation; The output characteristics are: $t + 1$ time design direction stroke change X, shield tail horizontal direction deviation, shield tail vertical direction deviation. To make the data multivariate, consolidate the input dataset into 3D data [sample value, time step, feature], According to the operation data acquisition frequency of each ring, every 25 samples as a time step, and data status information is passed between each step so that the training results for each group are passed on at all times.

3 ENGINEERING VERIFICATION

3.1 *Project introduction*

This paper studies based on the tunneling data of Shanghai Metro Line 18. The project has 1025 rings. The tunneling has generated more than 1 million operational data.

The verification process is mainly divided into two parts. The first part uses the data of 1–1000 rings to train the two sub-models separately to determine the model with better prediction effect. In the process of training the oil pressure-attitude and control strategy model, the data set is divided into two parts, 70% for the training model and 30% was used to verify the training effect of the model. In the second part, the trained model is used to verify the attitude control of shield on the 1000–1025 rings.

3.2 *Data preprocessing*

3.2.1 *Design axis data processing*
The shield is propelled along the design axis of the tunnel. The shield driver adjusts the oil pressure of the shield in real time according to the trend of the axis and the variation of the horizontal and vertical directions. In the original data of the design axis (X, Y, Z) is the absolute coordinate of the tunnel axis, the relative change of the axis coordinate is obtained according to the following formula:

$$P_r(\mathrm{x,y,z}) = P_r(\mathrm{X,Y,Z}) - P_{(\mathrm{r}-1)}(\mathrm{X,Y,Z}) \tag{5}$$

Where $P_\mathrm{r}(\mathrm{x,y,z})$ is the relative coordinate of the r ring to the $r - 1$ ring, $P_\mathrm{r}(\mathrm{X,Y,Z})$ is the absolute coordinate value of the space in which the r ring is located, and $P_{(\mathrm{r}-1)}(\mathrm{X,Y,Z})$ is the absolute coordinate value of the space in which the $r - 1$ ring is located.

The real-time operating data of the shield machine is collected and stored in the database by the sensor device installed on the shield. The initial operational data contains shield segment assembly data and shield tunneling data. In order to better analyze the variation of the regional oil pressure following the propulsion axis, filter the data generated by the shield assembly piece according to the change in the partition jack stroke. Since the sensors of the shield machine collect data in a harsh underground tunneling environment, ineffective data will inevitably be generated. In this paper, the in-group squared error and K-Means method are used to identify invalid detection data.

$$SSE = \sum_{i=1}^{k} (p_i - m_i)^2 \qquad (6)$$

Where pi is the sample in the ith cluster and mi is the centroid value of the i th cluster. Using SSE as an indicator, as the number of clusters increases, the slope of the SSE curve gradually decreases. Determine the critical k value of the slope become slowly as the optimal cluster number. At the same time, SSE is used as the objective function for optimizing K-Means clustering. According to the characteristics that the invalid data is accidental, the cluster containing the smaller number of samples is regarded as the abnormal cluster, and is eliminated in the data set.

3.3 Verification results

In order to illustrate the effectiveness of the verification results, this paper uses RMSE (Root Mean Square Error) and MAPE (Mean Absolute Percent Error) as the accuracy evaluation criteria for predictive model results. R^2 indicates the degree of the model predicts the actual value.

$$RMSE = \sqrt{\frac{\sum_{i=0}^{n} (y_i - \widehat{y_i})^2}{n}} \qquad (7)$$

$$MAPE = \sum_{i=1}^{n} \left| \frac{y_{i+1} - \widehat{y_i}}{y_{i+1}} \right| \cdot \frac{100}{n} \qquad (8)$$

$$R^2 = 1 - \sqrt{\left(\frac{\sum_{i=0}^{n} (y_i - \widehat{y_i})^2}{\sum_{i=0}^{n} y_i^2} \right)} \qquad (9)$$

Where yi is the true value of the i th sample, $\widehat{yi}$ is the predicted value of the i th sample, and n is the number of samples. When the values of RMSE and MAPE are smaller, the larger the R^2 value, the more accurate the prediction effect of the model is. Test the pre-processed data set and record the parameters of each sub-model with better results. The oil pressure-attitude model has an epoch of 30000, a batch size of 500, a hidden layer of 3 layers, and a hidden layer node number of 80; The input step size of the control strategy model is 25, the epoch of the training model is 12000, the batch size is 110, the number of hidden layers is 1 layer, and the number of hidden layer nodes is 25. The predicted results of the test are shown in Table 1. Through the model evaluation index, it can be found that the fully connected BP neural network has higher accuracy in the oil pressure-attitude model. In the control strategy model, the multivariate LSTM model has a memory effect on the sequence data and has good results on this problem.

Table 1. Sub model experiment results.

oil pressure-attitude model

Model	RMSE	MAPE	R^2
BP Neural Network	1.752	3.4%	0.993

control strategy model

Model	RMSE	MAPE	R^2
Multivariate LSTM	3.653	6%	0.961

Figure 6 shows the performance of the fully connected neural network in the oil pressure-attitude problem on the test set. The solid line is the actual deviation value of the incision (Figures 6a,6b) and the shield tail (Figures 6c,6d) in the horizontal (Y) direction and the vertical (Z) direction, and the broken line is the incision (Figures 6a,6b) and the shield tail (Figures 6c,6d) predicted deviation values in the horizontal (Y) and vertical (Z) directions. By comparing the predicted value with the true value. We have found that in the case of considering the front earth pressure, the shield slope and the shield rotation angle, the attitude of the shield can be accurately quantified according to the given oil pressure and actual output oil pressure of the shield machine.

Figure 7. The overall effect of using the multivariate LSTM model to learn about excellent shield propulsion strategies. Where (Figures 7a,7b) is the comparison of the predicted value of the trajectory control strategy (broken line) and the true value (solid line) in the horizontal (Y) and vertical (Z) directions of the shield tail. The results show that the proposed values of the shield propulsion trajectory in the subsequent tunneling can be given better in the horizontal and vertical directions according to the law of historical trajectory control. And the predicted trajectory has less error with the actual data.

Through verification, the model learns the historical excellent propulsion trajectory data, and gives the target position that the deviation of the shield tail in the horizontal Y and vertical direction Z should be reached in the next operation unit.

According to the trained oil pressure-attitude model and control strategy model, the project site verification was carried out on the 1000–1025 rings of the 18th line. The shield driver obtains the target position of the next operating unit shield according to the control strategy model, the target position is combined with real-time surface pressure information and oil pressure-attitude model, the shield driver obtains the setting of the oil pressure in each zone in the next operating unit. Through the test, the deviation of the shield tail in the horizontal and

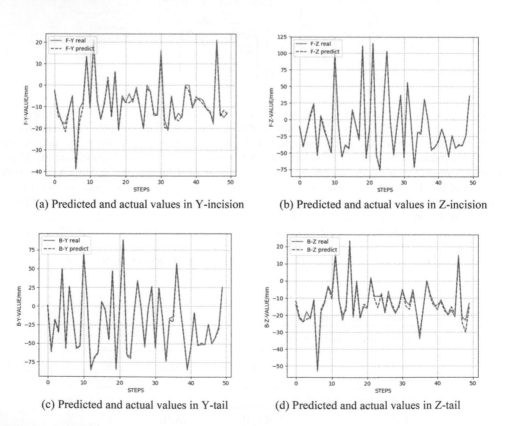

(a) Predicted and actual values in Y-incision

(b) Predicted and actual values in Z-incision

(c) Predicted and actual values in Y-tail

(d) Predicted and actual values in Z-tail

Figure 6.   Comparison of oil pressure-attitude model predictions with real values.

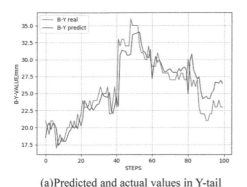

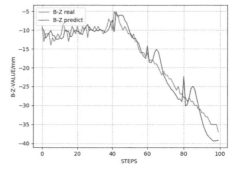

| (a)Predicted and actual values in Y-tail | (b) Predicted and actual values in Z-tail |

Figure 7.   Comparison of predicted and actual values of the control strategy model.

Table 2.   Monitoring of soil settlement around the excavation face.

Ring Number	1025	1020	1015	1010	1005	1000
Single maximum settlement/mm	-1.49	-1.49	-1.91	-0.89	-0.89	-0.89
L1 settlement/mm	-1.49	-1.49	-1.21	-0.5	-0.5	-0.5
L2 settlement/mm	0.31	0.31	-0.95	0.33	0.33	0.33

vertical directions is controlled within 50mm. It shows that the method can better control the shield machine to advance along the tunnel design axis. At the same time, the settlement of the surrounding soil during the propulsion process was monitored. The monitoring points and monitoring contents are shown in Figure 1. The monitoring results show that the control method will not disturb the settlement of the surrounding soil. Due to the limited space, the 6 rings soil settlement monitoring data was selected for display. The results are shown in Table 2.

4   CONCLUSION

This paper proposes a big data-driven Shield Attitude Control method (DSAC) based on big data. This method makes up for the defect that the oil pressure of jack cannot be quantified based on construction experience. At the same time, the method describes the oil pressure-attitude relationship and control strategy relatively accurately. At the same time, it avoids the defect that only the force analysis and the fuzzy control method are used, and the surrounding soil settlement is not paid attention to. In the future, we will consider the design of automatic control system based on two sub-models combined with hardware devices to realize automatic control of shield propulsion.

REFERENCES

David, E. 1986. Learning representations by back-propagating errors. *NATURE 323*

Fan, Z. 2016. Time series forecasting for building energy consumption using weighted Support Vector Regression with differential evolution optimization technique[J]*Energy and Buildings* 126: 94–103

Gang, C. 2017. Design of deviation correction curve for shield construction. *Shanghai construction technology*. 1:1–28

Hinton, G. 2014. Dropout: A simple way to prevent neural networks from overfitting[J]. *The Journal of Machine Learning Research*15(1):1929–1958.

Juan, H. 2016. Study on the influence of shield tunneling posture on the stress of composite construction pipes. *Wuhan university of engineering*16:17–21

Lin, W. 2015. Back propagation neural network with adaptive differential evolution algorithm for time series forecasting [J]. *Expert Systems with Applications*42 (2):855–863.

Lintao, W. 2014. Research on the key technology of shield tunneling attitude control [D]. *Zhejiang university*. 7:41–44

Rui, L. 2018. Control technology of forming segment axis in subway tunnel construction. *Engineering construction technology* 04:233–235.

Xiaoli, S. 2014. Measurement and rectifying curve design of shield tunnel axis. *Urban survey* 2:138–140

Yingcong, D. 2011. Zoning modeling and performance evaluation of shield propulsion system [D]. *Shanghai jiao tong university*.21:332–337

Yueqiang, L. 2014. Design of automatic rectifying control system for shield machine. *Journal of Beijing university of information technology*, 4:10–15

Tunnels and Underground Cities: Engineering and Innovation meet Archaeology,
Architecture and Art, Volume 6: Innovation in underground engineering,
materials and equipment - Part 2 – Peila, Viggiani & Celestino (Eds)
© 2020 Taylor & Francis Group, London, ISBN 978-0-367-46871-2

The composite lining total safety factor design method and cases study

Mingqing Xiao & Chen Xu
China Railway Siyuan Survey and Design Group Co., Ltd., Wuhan, Hubei province, China

ABSTRACT: In this paper, the composite lining total safety factor design method based on load structure model was established, and the influence of different section shapes on safety was studied. The conclusions are as follows. (1) The representative value of surrounding rock pressure based on the elastoplastic theory is used as the design load, which is suitable for safety and economy. (2) The load structure models established in this paper can calculate the safety factors of shotcrete layer, bolt-surrounding rock bearing arch, anchor bolt and secondary lining, and can obtain the total safety factors of tunnel construction and operation period. (3) The height-span ratio can significantly affect the safety factor. (4) The influence of tunnel depth on surrounding rock pressure and safety factor is relatively large. It is suggested that the corresponding supporting parameters should be adopted according to different burial depths, so as to improve the economy.

1 INSTRUCTIONS

Since the birth of NATM in 1950s, the composite lining has gradually become a common form of support for the railway and highway traffic tunnels, but the design methods of the composite lining and the sampling method of load are different in the world so far. Many scholars have put forward various design methods, such as engineering analogy method, ground structure method, characteristic curve method and so on. Engineering analogy method is a common method in design, but it cannot give clear safety factor, and the reliability is greatly reduced when the shape or the span are different. The ground structure method is theoretically applicable to the analysis of various tunnel shapes and various geological conditions, but it is not convenient for designers, and the calculation parameters of the stratum and the stress release rate are difficult to be selected, besides, the rock bolt is difficult to simulate. The characteristic curve method is also difficult to apply because of the uncertainty of calculation parameters and support time, in addition, the influence of section shape cannot be considered.

Many scholars have done useful research on the calculation and evaluation method of the safety factor of tunnel composite lining. The strength reduction finite element method is used to calculate the safety factor of surrounding rock as the criterion of stability analysis, which is mainly applied to the shotcrete and rock bolt support (Zheng, Y.R. et al. 2008); The reliability calculation model of secondary lining has been established, but there is a problem that the pressure value and reduction factor of surrounding rock pressure are difficult to determine (Jing, S.T et al. 2002). The joint action model of applying the load on the boundary of the stratum combined with the load structure method and the ground structure method. This model avoids the problem of the stress release rate, but there is a certain difficulty in the value of the boundary load, and the total safety factor of the composite lining cannot be given (Zhao, S.Y. 2007, Song, Y.X. et al. 2015).

Because of the complexity and randomness of rock and soil properties, ground stress, boundary conditions, construction process, and so on, it makes it very difficult to determine or calculate the pressure of the surrounding rock in tunnel (Qu, H.F. et al. 2007). The main calculation methods of tunnel surrounding rock load include: Empirical methods based on the surrounding rock classification system, such as Q system and RMR classification (Palmström, A. et al. 2000 & 2006); Empirical methods based on the lab tests and field tests such as the Terzaghi's theory and Protojiakonov's Theory (Wong, H. et al. 2006, Guan, Z.C. et al. 2007); Theoretical analysis methods based on interaction between surrounding rock and support system, such as the Kastner formula and convergence-constraint method (Osgoui, R. 2007, Fraldi, M. et al. 2010); Based on 1046 tunnel collapse data, Chinese railway industry (2016) considered the collapse soil as a loose body and established a statistical formula for the load of deep buried tunnels, which are related to the surrounding rock grade, the excavation span, but has nothing to do with the depth of the buried depth.

Due to the different calculation methods of surrounding rock pressure, the design of support parameters will also have a great impact. Selecting a representative value as the design load, which can synthetically consider various factors, envelop all kinds of unfavorable loads, and also has appropriate economic nature, is particularly important for the support parameters rational selection of the tunnel.

Xiao, M.Q. and others put forward the load structure model and the safety factor calculation method for the primary support, and established the total safety factor design method of the composite lining (2018). This method is used to calculate the safety factor of two typical tunnel forms in China, and to study the influence of the shape change of the section on the safety factor. This study can provide ideas for the quantitative design and safety check of composite lining.

2 THE REPRESENTATIVE VALUES OF SURROUNDING ROCK PRESSURE

2.1 *Classification of surrounding rock in China*

In the Code for Design of Railway Tunnel of China, surrounding rocks are divided into six grades. The physical and mechanical indexes of surrounding rocks at all levels are shown in the table 1. At the same time, the general comparison with Q system is given.

2.2 *The calculation method of the representative values of surrounding rock pressure*

In order to solve the uncertainty problem of supporting pressure in actual construction, the representative values of surrounding rock pressure is adopted, which can envelop a variety of unfavorable conditions and have proper economy. It is suitable for the cases with buried depth greater than 15 times the tunnel diameter, without taking into account the rock burst and expansion pressure. The calculation formula are as follows:

Top pressure:

$$q = \alpha\gamma(R_{pd} - a) \tag{1}$$

Lateral pressure:

$$e = \beta\lambda q \tag{2}$$

Table 1. Physical and mechanical indicators of surrounding rocks at all levels.

Grade of surrounding rock	Bulk density γ(kN/m³)	Elastic reaction coefficient K_e (MPa/m)	Elastic modulus E(GPa)	Poisson ratio υ	Internal friction angle $\varphi(°)$	Cohesion c(MPa)	Estimated Q value
I	26 ~ 28	1800 ~ 2800	>33	< 0.2	>60	>2.1	1000~400
II	25 ~ 27	1200 ~ 1800	20 ~ 33	0.2 ~ 0.25	50 ~ 60	1.5 ~ 2.1	400~100
III	23 ~ 25	500 ~ 1200	6 ~ 20	0.25 ~ 0.3	39 ~ 50	0.7 ~ 1.5	100~10
IV	20 ~ 23	200 ~ 500	1.3 ~ 6	0.3 ~ 0.35	27 ~ 39	0.2 ~ 0.7	10~1
V	17 ~ 20	100 ~ 200	1 ~ 2	0.35 ~ 0.45	20 ~ 27	0.05 ~ 0.2	1~0.01
VI	15 ~ 17	< 100	< 1	0.4 ~ 0.5	< 22	< 0.1	0.01~0.001

$$R_{pd} = R_0 \left\{ \frac{[p_0(1+\lambda) + 2c\cot\varphi](1-\sin\varphi)}{2p_i + 2c\cot\varphi} \right\}^{\frac{1-\sin\varphi}{2\sin\varphi}} \times \left\{ 1 + \frac{p_0(1-\lambda)(1-\sin\varphi)\cos 2\theta}{[p_0(1+\lambda) + 2c\cot\varphi]\sin\varphi} \right\} \quad (3)$$

Where γ = surrounding rock bulk density; λ = lateral pressure coefficient of surrounding rock; α & β are respectively the adjustment coefficient of the top and lateral surrounding rock pressure, which are generally not less than 1.2, and are adjusted according to the surrounding rock occurrence, such as the horizontal rock, the α is more than 1, and the β is less than 1. R_{pd} is the radius of the plastic zone at the 45° position when the supporting force p_i is equal to 0; p_0 = initial stress; c = cohesive force; φ = internal friction angle; θ = the angle between the tunnel horizontal axis; R_0 = the tunnel excavation radius (not round, circumradius replaced); a = the distance from the tunnel excavation boundary at the position of 45°to the circular center;

For grade V surrounding rock, the reduction of the space effect of stratum and the bearing capacity of the grouting ring should be considered. The calculation in this paper is approximately 1.8 times the pressure of grade IV with the same burial depth.

2.3 Safety and economic evaluation of representative values of surrounding rock pressure

2.3.1 Safety
Because the representative value of surrounding rock pressure is the whole weight of the surrounding rock between the boundary of the plastic zone and the excavation boundary when the supporting force is 0, it is an extreme case, so it has enough safety.

2.3.2 Economic
1) The relationship between the representative value of surrounding rock pressure and the minimum supporting force

When $\lambda = 1$, for the circular tunnel, the slip surface of the surrounding rock loose area is a pair of logarithmic spiral (Zheng, Y.R. et al. 2012). Assuming that the strength in the loosened area has been greatly reduced, it can be considered that the sliding rock mass has no self-supporting ability so that the total weight of the sliding rock mass is borne by the supporting resistance P_{imim}. Thus:

$$p_{i\min} = \frac{\gamma(R_{\max} - R_0)}{2} \quad (4)$$

Where R_{max} is the radius of loosening zone corresponding to P_{imim}
The loosening radius R_{max} is determined by the tangential stress $\sigma_\theta = p_0$:

$$R_{\max} = R_0 \left\{ \frac{(p_0 + c\cot\varphi)(1-\sin\varphi)}{(P_{i\min} + c\cot\varphi)(1+\sin\varphi)} \right\}^{\frac{1-\sin\varphi}{2\sin\varphi}} \quad (5)$$

The minimum supporting force can be solved by the formula (4) (5). The relationship between P_r/P_{imin} (P_r = representative value of the surrounding rock pressure) and different equivalent radius when 400m buried is shown in Figure 1. We know: (1) The equivalent circle radius has little effect on the ratio of the surrounding rock grade III and IV, respectively in 3.30 and 3.00; (2) The ratio of grade V is positively correlated with the equivalent circle radius, and the ratio is always greater than 1.0; (3) The representative value of the surrounding rock pressure is always greater than the minimum supporting force, and the most is about 3 times of the minimum supporting force which has right economy.

2) Expansion ratio of plastic zone before and after support by the P_r

The representative value is used as supporting force, while the bearing capacity of 3m range reinforcement ring in grade V is considered. The range of plastic zone is less than that of no

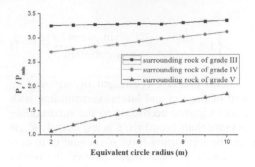 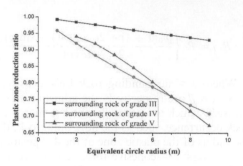

Figure 1. Relationship between (P_r/P_{imin}) and radius.

Figure 2. Relationship between (R_s/R_p) and radius.

support, and the relationship between the ratio R_s/R_p and the equivalent radius of the tunnel is shown in Figure 2. Thus, (1) the greater the equivalent radius of the tunnel, the smaller the R_s/R_p, the more obvious the control effect of the supporting force on the expansion of the plastic zone; (2) When the equivalent radius is less than 9m, the range of the actual plastic zone can be controlled to 67~95% of the maximum plastic zone, which is close to the range of the maximum plastic zone. It can be also explained that it has the appropriate economy.

3   THE COMPOSITE LINING TOTAL SAFETY FACTOR DESIGN METHOD

3.1   *Load structure model of composite lining and calculation method of safety factor*

1) Model 1: Calculation of the shotcrete layer (Shotcrete and steel arch, abbreviation: the shotcrete layer, the same below)
   The load structure model of the shotcrete layer is shown in Figure 3. The shotcrete layer is simulated by the beam element. The interaction between the structure and the stratum is simulated by a tension free spring and a tangential spring. After calculating the internal force of the shotcrete layer, the structural safety factor (K_1) is checked by using the damage stage method according to the 'Code for Design on Railway Tunnel'.
2) Model 2: Calculation of bolt-surrounding rock bearing arch
   In this model (Figure 4), the external end of the bolt diffuses pressure to the inside of the tunnel at a certain angle (such as 45°). The connecting line formed by the intersection point of the adjacent bolts pressure diffusion is the outer line of the bearing arch, and the inner line of the bearing arch is the outer surface of the shotcrete layer. Radial spring is used to simulate the interaction between surrounding rock and bearing arch, and elastic support is used at arch foot.
   After calculating the internal force, its safety factor K_2 is also calculated by the damage stage method. However, the ultimate strength of surrounding rock within the bearing arch only considers the strength increased after support, that is, the supporting force provided by bolt, shotcrete and secondary lining is taken as σ_3 (detailed calculations are given below), and then σ_1 of surrounding rock is the ultimate strength which is obtained according to Mohr-Coulomb strength criterion.
3) Model 3: Calculation of bolt
   The parameters of bolts (length, spacing and diameter) need to be calculated in combination with model 2, and different methods are adopted according to support types.
   (1) Calculation model of bolt in shotcrete-bolt combined support
      When shotcrete-anchor combined support is adopted, the bolt mainly provides part of the lateral limiting force for the bearing arch of model 2 (i.e. part of σ_3, the other part of σ_3 is provided by shotcrete layer and secondary lining). The length of the bolt is determined according to the stress requirement of the bearing arch, and the spacing and diameter (strength) of the bolt are calculated according to the lateral limiting force

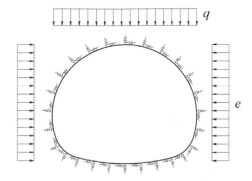

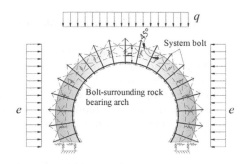

Figure 4. Model 2 (bolt-surrounding rock bearing arch).

Figure 3. Model 1 (shotcrete layer).

provided by bolts (e.g. As shown in Figure 5), the safety factor of yield strength and pull-out strength of the bar itself should not be less than 2.0 and 2.5 respectively.

In the figure, k_s is the safety factor of yield bearing capacity; k_g is the safety factor of pull-out capacity; σ_{31} is the lateral limiting force provided by bolts; f_y is the yield strength of steel bar; d is the diameter of bolt bar; f_{rb} is the ultimate cohesional strength between mortar anchorage and stratum; d_g is the external diameter of mortar anchorage; l_g is the anchorage length between anchor bar and mortar; b and s are the circumferential and longitudinal spacing of bolts respectively.

It should be noted that σ_3 is composed of σ_{31} (provided by bolt), σ_{32} (provided by shotcrete layer) and σ_{33} (provided by secondary lining). During construction stage, σ_{33} can not be included. σ_{32} and σ_{33} can be calculated approximately according to the following formula.

$$\sigma_{32} = 0.5K_1 \cdot q \tag{6}$$

$$\sigma_{33} = 0.5K_3 \cdot q \tag{7}$$

In formula, K_1 and K_3 are safety factors of shotcrete layer and secondary lining respectively.

(2) Calculating model of bolt in anchor bolt-based support

When only bolts are used for support, the strength of bolt should not be less than the requirement of "minimum support force P_{imin}" (Formula 4) besides meeting the σ_3 requirement of the bearing arch.

4) Model 4: Calculation of the secondary lining

The calculation model of the secondary lining is basically the same as the model 1, but because the waterproof layer between the shotcrete layer and the secondary lining does not transfer the shear stress, the interaction is only simulated by the tension free spring, and the safety factor is calculated by the damage stage method.

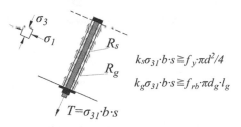

Figure 5. Model 3 (bolt).

3.2 Calculation of the total safety factor of composite lining

3.2.1 Calculation method of total safety factor

According to the above model 1, model 2 and model 4, the safety factors of the shotcrete layer, the bearing arch and the secondary lining are calculated respectively. The total safety factor of the composite lining can be calculated as follows:

Construction stage (no secondary lining): $K_c=K_1+K_2$;

Operation stage: when using durability bolts, $K_{op}=K_1+K_2+K_3$; when using non durability bolts, $K_{op}=K_1+K_3$.

3.2.2 The reasonableness analysis of the calculation method of total safety factor

1) Analytical model for ultimate bearing capacity of composite structure

The bolt-surrounding rock bearing arch is located in the outermost layer of the whole structure, even if all of it enters the plastic state, as long as the shotcrete layer and the secondary lining do not appear overall instability, it will not lose stability alone. Therefore, it is necessary to analyze the bearing capacity of the shotcrete layer-secondary lining composite structure.

The schematic diagram of the composite structure consisting of the shotcrete layer and the secondary lining is shown in Figure 6 (a). When one section of the shotcrete layer or secondary lining reaches the damage stage, it is assumed that it can maintain the bearing capacity of the damage stage. The internal force in the damaged area is applied to the damaged position as boundary condition, and then the load is increased until the whole structure is damaged, as shown in Figures 6 (b) and 6 (c), which correspond to two failure modes of large eccentric compression failure and small eccentric compression failure.

2) Influence of damaged sequence of composite structure on bearing capacity

By increasing the load, the load proportionality coefficient K_d (i.e. the ratio of damaged load to design load) of composite structure in the whole failure stage can be obtained, which connotation is different from that of safety factor.

According to the failure order of the shotcrete layer and secondary lining, comparisons with the results of the total safety factor method can be divided into three situations:

The first: when the shotcrete layer and the secondary lining reach the most unfavorable strength at the same time, then $K_d=K_1+K_3$;

The second: when the shotcrete layer has reached the design strength before the secondary lining, but the secondary lining is located on the inside of it, the shotcrete will not be completely unstable and can continue to carry the load until the most unfavorable bearing section of the secondary lining reaches the damage stage, at this time, $K_d>K_1+K_3$;

The third: when the secondary lining reached the most unfavorable strength before the shotcrete layer, the plastic zone of the secondary lining needs to keep developing or form a new plastic zone, then the shotcrete layer can reach the damaged stage, at this time, $K_d>K_1+K_3$.

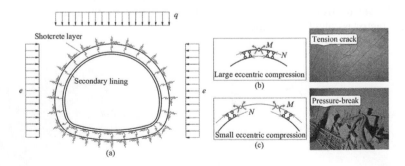

Figure 6. Schematic diagram of ultimate bearing capacity of shotcrete layer and secondary lining composite structure.

For the third case, the secondary lining has multiple plastic zones, which may exceed the normal use requirements, for example, for a high–speed railway tunnel, the plastic zone of the arch wall may cause fall-block due to the train vibration. Therefore, the position of the first plastic zone of the secondary lining should be controlled by a reasonable section shape or strength matching, so that it does not appear in the arch wall; For the second case, the reasonable strength matching between the shotcrete layer and the secondary lining should be reasonably controlled to prevent the sudden damage of the secondary lining caused by the excessive damage zone of the shotcrete layer.

In summary, the most reasonable design should be that the shotcrete layer and the secondary lining reach the most unfavorable cross section strength at the same time, and the load proportionality coefficient K_d is basically equal to or slightly higher than K_1+K_3. When the K_d is greatly higher than K_1+K_3, the design parameters or section shape of tunnel should be adjusted. Therefore, the method of safety factor addition provides an objective function for the optimization design of the whole structure, and can be verified and optimized by the bearing capacity model of shotcrete layer-secondary lining composite structure.

Similarly, the overall bearing capacity model of bolt-surrounding rock bearing arch, shotcrete and secondary lining can also be established, but because K_2 is based on the damage stage of shotcrete layer, bolt and secondary lining, and reaches the maximum with the failure state of model 1, 3 and 4, the three-layer structure can be simplified to the two-layer structure shown in Figure 6.

3) Analysis of bearing capacity of bolt

Whether the bolt will exert its full bearing capacity is related to its strain. It is necessary to analyze the relative deformation of the two ends of the bolt when the composite structure reaches the overall failure.

For the bolt with steel bar, the strain should attain 2‰ until the bolt reaches the design strength, that is, the yield strength of the steel bar.

Assuming that the axial force of the bolt is distributed in a triangle, as long as the relative stratum displacement at the two ends of the bolt is not less than 1‰L (L is the bolt length), the bolt bearing capacity can be fully played, otherwise, only part of the bearing capacity can be counted. Therefore, the reasonable length of bolt should match the deformation capacity of shotcrete layer and secondary lining composite structure.

When the whole anchor bolt is in the plastic zone, the average strain ε can be calculated as follows:

$$\varepsilon = \Delta/(R_0 + L) \tag{8}$$

In the form: Δ is the maximum displacement of the tunnel wall.

When the outer end of the bolt is located in the elastic zone and the inner end is located in the plastic zone, the average strain in the plastic zone is higher than that in the elastic zone. The average strain can be calculated by the following formula:

$$\varepsilon = \Delta/R_p \tag{9}$$

3.3 *Recommended value for safety factor of composite lining*

The total safety factor of the composite lining should be larger than the single structure because the primary support and the secondary lining are not closely fitting with each other, and have different materials and different deformability.

It is suggested that the overall safety factor $K_{op} \geq 3.0 \sim 3.6$, and the safety factor of the primary support in the construction period $K_c \geq 1.8 \sim 2.1$. The safety factor can be adjusted according to the specific conditions of surrounding rock and the quality control of construction.

4 CASES STUDY

In order to compare and analyze the influence of different section shapes on safety, the safety factors of two types of tunnel structures in China is checked by the design method established in this paper, including a three lane highway tunnel with a flat section shape and a single track railway tunnel with a relatively lanky cross-section shape.

The physical and mechanical indexes of grade III, IV and V surrounding rock in the calculation cases are the lower 1/3 sub values of the current 'Code for Design of Railway Tunnel' in China, as shown in Table 2.

4.1 Case of tunnel with relatively small height-span ratio, taking a three lane highway tunnel as an example

4.1.1 Section shape and support parameters of the three lane highway tunnel

A three-lane highway tunnel in China has been drawn up with scheme 1 and scheme 2 as shown in Figures 7 (a) and 7 (b) respectively. For the scheme 1, the excavation span is 17.27m, height 11.54m, high-span ratio 0.67, and the excavation span of scheme 2 is the same as the scheme 1, but the excavation height increases 1m, and the height span ratio changes to 0.73. The change of the safety factor of the two kinds of sections is compared.

The same supporting parameters are adopted in the two sections, as shown in Table 3.

4.1.2 Calculation results of three lane highway tunnel

In terms of structural stress characteristics: (1) The shotcrete layer is mainly damaged by small eccentricity in the vault position. (2) The control points of the two sections in the bolt-surrounding rock bearing arch model are all at the arch foot position. (3) The secondary lining is mainly controlled by the arch foot, followed by the vault and the foot of wall. When the section is heightened 1m, the failure feature of the arch is changed from large eccentric compression to small eccentric compression failure.

The calculation results of the composite structure model show that the shotcrete layer reaches the failure state before the secondary lining in all cases.

When the composite structure is destroyed, for the scheme 1, the average strain of the rock bolt of grade III, IV and V is 2.8‰, 4.6‰, 4.1‰ respectively. The average strain of the rock bolt of scheme 2 is 3.1‰, 5.2‰, 4.5‰ respectively. The average strain of the two schemes is more than 1‰, so the anchor bolt can give full play to bearing capacity.

The calculated safety factors are shown in Table 4 and Table 5.

Table 2. Mechanics parameters of surrounding rock of class III, IV and V.

Grade	γ(kN/m³)	K_e (MPa/m)	E(GPa)	υ	φ(°)	c(MPa)
III	23.7	733	10.67	0.28	42.7	0.97
IV	21.0	300	2.87	0.33	31.0	0.37
V	18.0	133	1.33	0.41	22.3	0.10

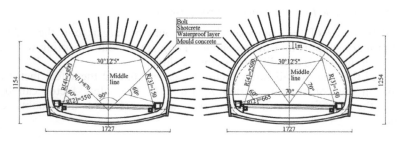

Figure 7. Two schemes of three lane highway tunnel.

Table 3. Support parameters for three lane highway tunnel.

Grade of surrounding rock	Shotcrete Arch wall (cm)	Invert arch (cm)	System anchor bolt Length (m) @ Spacing (m)	Steel frame Specifications/ Position/ Spacing (m)	Arch wall (cm)	Secondary lining Invert arch (cm)	Reinforcement (mm)
III	19	-	3.0@1.2×1.2	Grid steel/Arch wall/1.2	45	-	-
IV	25	10	3.0@1.0×1.0	I18/Full-ring/1.0	50	50	Φ22@200
V	27	27	3.5@1.0×0.7	I20a/Full-ring/ 0.7	60	60	Φ25@200

(1) The greater the tunnel burial depth, the smaller the safety factor of the structure. (2) The total safety factors of the two sections are all less than 3.0 under the condition of grade V surrounding rock with 800m depth. While the total safety factors of the grade IV with depth of less than 400m and grade III are high. Even if the permanent support of the bolt is not considered, the support parameters still have a certain optimum space. (3) The cross section of scheme 2 is more circular and continuous, therefore, the structural stress is improved and the safety factor is slightly higher than that of scheme 1. (4) The safety factor in construction period (K_c) of the two schemes are both less than 1.80 under the condition of the grade V surrounding rock, thus the safety of the construction period should be ensured by the timely application of the secondary lining or the reinforcement of the surrounding rock. (5) The results calculated by the composite structure model show that the strength matching relationship between the shotcrete layer and the secondary lining is reasonable, and the load proportionality coefficient K_d is basically equal to or slightly higher than $K_1 + K_3$.

Table 4. Safety factor of three lane highway tunnel (scheme 1).

G	H(m)	K_1	K_2 K_{2c}	K_{2op}	K_3	K_c	K_{op}	K_{ig}	K_d
III	400	5.05	12.53	17.46	11.10	17.58	33.61	16.15	16.70
	800	3.19	7.84	11.09	7.30	11.03	21.58	10.49	10.51
IV	400	1.90	1.40	2.14	2.86	3.3	6.9	4.76	5.27
	800	1.19	0.87	1.35	1.83	2.06	4.37	3.02	3.28
V	400	1.17	0.52	0.98	1.92	1.69	4.07	3.09	3.71
	800	0.71	0.32	0.61	1.22	1.03	2.54	1.93	2.31

Note: In the table, K_{2c} and K_{2op} indicate the safety factors of the bolt-rock bearing arch during construction and operation respectively, while K_{ig} indicates the safety factors when the anchor action is ignored (the same below).

Table 5. Safety factor of three lane highway tunnel (scheme 2).

G	H(m)	K_1	K_2 K_{2c}	K_{2op}	K_3	K_c	K_{op}	K_{ig}	K_d
III	400	6.37	17.77	24.00	13.74	24.14	44.11	20.11	25.45
	800	3.32	9.55	13.33	8.34	12.87	24.99	11.66	13.76
IV	400	1.99	1.52	2.36	3.16	3.51	7.51	5.15	5.85
	800	1.21	0.92	1.45	1.98	2.13	4.64	3.19	3.55
V	400	1.19	0.55	1.05	1.99	1.74	4.23	3.18	3.99
	800	0.72	0.33	0.64	1.23	1.05	2.59	1.95	2.42

4.2 Case of tunnel with relatively large height-span ratio, taking a single track railway tunnel as an example

4.2.1 Section shape and support parameters of the single track railway tunnel

A single track railway tunnel in China has been drawn up with scheme 1 and scheme 2 as shown in Figures 8 (a) and 8 (b) respectively. For the scheme 1, the excavation span is 8.34m, height 10.15m, high span ratio 1.22, and the excavation height of scheme 2 is the same as the scheme 1, but the excavation span increases 1m, and the height span ratio changes to 1.09 The change of the safety factor of the two kinds of sections is compared.

The same supporting parameters are adopted in the two sections, as shown in Table 6.

4.2.2 Calculation results of single track railway tunnel

In terms of structural stress characteristics: (1) The shotcrete layer is damaged by large eccentricity in the side-wall position. (2) In the bolt-surrounding rock bearing arch model, the control sections of scheme 1 and scheme 2 are all located in the side wall area. (3) The control sections of the secondary lining are all located in the vault area, and subjected to large eccentric compression failure.

The shotcrete layer reaches the failure state before the secondary lining in all cases.

When the composite structure is destroyed, for the scheme 1, the average strain of the rock bolt of grade III, IV and V is 1.1‰, 1.9‰, 2.6‰ respectively. The average strain of the rock bolt of scheme 2 is 1.3‰, 2.1‰, 2.5‰ respectively. The average strain of the two schemes is more than 1‰, so the anchor bolt can give full play to its bearing capacity.

The calculated safety factors are shown in Table 7 and Table 8.

(1) The total safety factor of this type single track railway tunnel is high before and after the section span being widened. Even if the bolt is ignored, the safety factor is still greater than 3.0, and the support parameters can be weakened properly. (2) When the height-span ratio is reduced from 1.22 to 1.09, the total safety factor of grade III and IV surrounding rocks

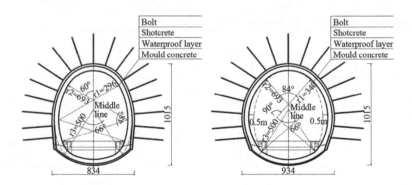

Figure 8. Two schemes of single track railway tunnel.

Table 6. Support parameters for single track railway tunnel.

Grade of surrounding rock	Shotcrete Arch wall (cm)	Invert arch (cm)	System anchor bolt Length (m) @ Spacing (m)	Steel frame Specifications/ Position/ Spacing (m)	Secondary lining Arch wall (cm)	Invert arch (cm)	Reinforcement (mm)
III	8	-	2.5@1.2×1.5	-	35	40	-
IV	12	-	3.0@1.2×1.2	-	40	40	-
V	23	10	3.0@1.2×1.0	Grid steel/Arch wall/1.0	45*	45*	Φ20@250

Table 7. Safety factor of single track railway tunnel (scheme 1).

| G | $H(m)$ | K_1 | K_2 | | K_3 | K_c | K_{op} | K_{ig} | K_d |
			K_{2c}	K_{2op}					
III	400	0.55	36.69	54.74	18.17	37.24	73.46	18.72	34.69
	800	0.37	23.74	37.18	13.50	24.11	51.05	13.87	22.36
IV	400	0.17	3.09	9.05	11.11	3.26	20.33	11.28	13.77
	800	0.11	1.95	5.86	7.30	2.06	13.27	7.41	8.64
V	400	0.43	0.44	1.74	5.09	0.87	7.26	5.52	5.65
	800	0.27	0.28	1.04	2.98	0.55	4.29	3.25	3.55

Table 8. Safety factor of single track railway tunnel (scheme 2).

| G | $H(m)$ | K_1 | K_2 | | K_3 | K_c | K_{op} | K_{ig} | K_d |
			K_{2c}	K_{2op}					
III	400	0.98	23.93	40.74	22.27	24.91	63.99	23.25	42.86
	800	0.52	16.39	26.03	15.52	16.91	42.07	16.04	26.09
IV	400	0.39	3.03	7.73	9.60	3.42	17.72	9.99	14.28
	800	0.19	1.95	5.03	6.29	2.14	11.51	6.48	8.85
V	400	0.80	0.63	2.67	7.02	1.43	10.49	7.82	10.3
	800	0.49	0.39	1.69	4.47	0.88	6.65	4.96	6.41

decrease slightly, while that of grade V surrounding rocks increases greatly. (3) For grade V, the safety factor (K_c) of the two schemes during construction period is small, and the primary support parameters need to be strengthened. (4) For the tunnels of scheme 1 with grade IV and V surrounding rock, the K_d is equal to or slightly higher than K_1+K_3, which indicates that the strength matching relation between the shotcrete layer and the secondary lining is relatively reasonable, while the K_d of other cases is higher than K_1+K_3, which indicates that the matching relation should be adjusted properly.

5 CONCLUSION

1. Using the representative value of surrounding rock pressure as the design load, it provides a way to solve the problem of uncertainty of surrounding rock pressure in the design, and a unified evaluation method for tunnel safety can be established accordingly. It is suggested that the deadweight of the plastic zone of the surrounding rock based on the Kastner's equation is used as the representative value of the pressure of the top surrounding rock. This value can reasonably reflect the comprehensive influence of the physical and mechanical indexes of surrounding rock, buried depth, section size and other factors, and has appropriate safety and economy.
2. The calculation models of shotcrete layer (containing steel frame), bolt-surrounding rock bearing arch, anchor bolt, secondary lining and their respective safety factor calculation methods are established. The calculation methods and standard of total safety factor of tunnel construction and operation period are put forward. It is suggested that the total safety factor of operation period is not less than 3.0 ~ 3.6 and the primary support in the construction stage is not less than 1.8 ~ 2.1.
3. The conclusion that the height-span ratio can significantly affect the safety factor is concluded. Therefore, under the condition of certain support parameters, reasonable section shape should be selected according to the pressure distribution pattern of surrounding rock.
4. The depth of the tunnel has a great influence on the surrounding rock pressure and safety factor. It is suggested that the corresponding support parameters should be adopted

according to the depth to ensure that the tunnels with different buried depth have basic close safety factor, so as to improve the economy.

5. In order to make each supporting component reach the failure state at the same time and give full play to their bearing capacity, the strength matching relationship between the supporting structures should be satisfied.

REFERENCES

Zheng, Y.R. & Qiu, C.Y. 2008. Exploration of Stability Analysis Methods for Surrounding Rocks of Soil Tunnel. *Chinese Journal of Rock Mechanics and Engineering*, (10):1968–1980.

Jing, S.T. & Zhu Y.Q. 2002. Reliability of tunnel structure. Beijing: China Railway Press, 175–183.

Song, Y.X. & Liu, Y. 2015. Research on the calculation method of composite lining internal force for railway tunnel. *Journal of Railway Engineering Society* 32(08):73–75+110.

Zhao, S.Y. & Zheng, Y.R. 2007. An analysis of the design method of underground tunnel lining. *Journal of Logistical Engineering University* (04):29–33.

Qu, H.F. & Yang, C.C. et al. 2007. Research and development on surrounding rock pressure of road tunnel. *Chinese Journal of Underground Space and Engineering* (03):536–543.

Palmström, A. 2000. Recent developments in rock support estimates by the RMI. *Journal of Rock Mechanics and Tunneling Technology* 6(1):1–19.

Palmström, A. & Broch, E. 2006. Use and misuse of rock mass classification systems with particular reference to the Q-system. *Tunneling and Underground Space Technology* 21 (6): 575–593.

Wong, H. & Subrin, D. et al. 2006. Convergence-confinement analysis of a bolt-supported tunnel using the homogenization method. *Canadian Geotechnical Journal* 43 (5): 462–483.

Guan, Z.C. & Jiang, Y.J. et al. Reinforcement mechanics of passive bolts in conventional tunneling. *International Journal of Rock Mechanics and Mining Sciences* 44 (4): 625–636.

Osgoui, R. 2007. Development of an elasto-plastic analytical model for design of grouted rock bolts in tunnels with particular reference to poor rock masses. Ankara, Turkey: Middle East Technical University.

Fraldi, M. & Guarracino, F. 2010. Analytical solutions for collapse mechanisms in tunnels with arbitrary cross sections. *International Journal of Solids and Structures* 47(2):216–223.

National Railway Administration of People's Republic of China. 2016. Code for Design on Railway Tunnel (TB10003-2016). Beijing: *China Railway Press.*

Xiao, M.Q. 2018. Discussion on Design Method of General Safety Factor of Composite Lining Tunnel. *Journal of Railway Engineering Society* (01):84–88.

Xiao, M.Q. & Wang, S.F. et al. 2018. Study on Design Method of Primary Support of Tunnel Based on the Load-Structure Method. *Journal of Railway Engineering Society* (04). 60–64.

Zheng, Y.R. Zhu, H.H. et al. 2012. The Stability Analysis and Design Theory of Surrounding Rock of Underground Engineering. Beijing: *China Communications Press* 367–373.

REVISION NOTES

(1) In accordance with the requirements, the introduction of surrounding rock classification of railway tunnels in China is added, and the comparison is made with the commonly used Q system, as detailed in Section 2.1 of the paper.

(2) The total safety factor method of composite lining has been improved and some contents have been revised, including:

① The original load structure model of shotcrete&bolt support-surrounding rock combined arch is modified to the calculation model of bolt-surrounding rock bearing arch, and the original load structure model of bolt is modified to make the calculation method more reasonable.

② According to the revised method, the cases in this paper are recalculated and analyzed, while the main arguments and conclusions remain unchanged.

We remain at your disposal should you need any question or support.

Xu Chen
China Railway Siyuan Survey and Design Group Co., Ltd., Wuhan
Email: xc1505@126.com

*Tunnels and Underground Cities: Engineering and Innovation meet Archaeology,
Architecture and Art, Volume 6: Innovation in underground engineering,
materials and equipment - Part 2 – Peila, Viggiani & Celestino (Eds)
© 2020 Taylor & Francis Group, London, ISBN 978-0-367-46871-2*

Pressure infiltration of sandy foam during EPB shield tunnelling in saturated sand

T. Xu
Ghent University, Ghent Belgium

A. Bezuijen
Ghent University, Ghent Belgium
Deltares, Delft The Netherlands

Z. Lu
Macau University, Taipa, Macau, China

ABSTRACT: Experiments on infiltration of pressurised sandy foam into saturated sand have been carried out in a laboratory setup that provides a comparable hydraulic gradient as in real tunnels. The sandy foam used was comparable to the excavated soil (foam-water-sand mixture) that can be expected in the mixing chamber of an EPB (Earth Pressure Balance) shield. It appears that a higher FER_m (effective foam expansion ratio) is more effective to form an impermeable layer in the sand. Furthermore, the permeability of sand for water flow through the foam decreases with the FIR (foam injection ratio). In practice, a FIR of 35 - 40 is recommended for a sandy ground and the situation with high sand fraction of the sandy foam should be avoided in the field because there will be no impermeable or low permeable layer formed at the tunnel face.

1 INTRODUCTION

With additional soil conditioning including chemical additives (e.g. foam), EPB shields have been used in tunnelling in sandy grounds, e.g. Botlek Rail Tunnel in the Netherlands, Port of Miami Tunnel in America, Metro Rio Line 4 in Brazil and Tehran Metro, Line 7 in Iran (Bezuijen, 2002; Merritt et al., 2013; Maidl and Pierri, 2014; Amoun et al., 2017). Experimental investigations on the properties of foam-soil mixtures were carried out in different laboratories (Bezuijen et al., 1999; Mair et al., 2003; Borio et al., 2011; Bezuijen, 2011; Mori et al., 2018). However, few studies have been carried out on the infiltration of pressurised foam into the ground, which has significance for the tunnel face stability, because due to infiltration the support pressure at the tunnel face and thus the tunnel face stability will be reduced. Just recently, Galli (2016) conducted experiments of the infiltration of clear foam into saturated sand. It was found that the infiltration process can be evaluated qualitatively by a regression analysis. Though a flow model for porous media to the foam infiltration process would probably provide a sound description of the dynamics, a power-law model ($z_p(t) = a \cdot t^{0.166}$, $z_p(t)$ (cm) is the infiltration distance, a the fitting factor and t (s) the time) with intercept zero showed a good agreement with the experimental results. The authors of this article have also performed the experiment of infiltration of clear foam into the sand to simulate the infiltration of foam in front of the tunnel face, and the findings are presented in Xu et al. (2018).

In a field situation, the cutter head will cut away the soil and there is a gap between cutter head and the ground, see Figure 1. Consequently, a mixture of foam and cut soil (containing

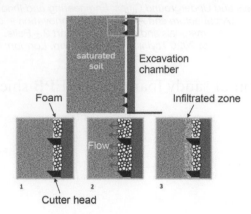

Figure 1. Supposed principle of foam infiltration and pore water replacement (modified after Galli, 2016).

pore water) will be present in the gap (Maidl, 1995; Galli, 2016). In this case, the mechanism of foam infiltration into the sand can be different from clean foam infiltration (Xu and Bezuijen, 2018). In this study, therefore, experiments on infiltration of foam from pressurised foam-water-sand mixture (sandy foam) into saturated sand were performed. The experimental results will be discussed with the results of clear foam infiltration into saturated sand. A micro scale sketch of the infiltration process and a micro scale theory of sandy foam will be used to explain the mechanism of the infiltration. This experimental study is also a work after Budach (2012) and Galli (2016) in order to establish a standardised and comparable test method to estimate performance of the foam conditioning for EPB tunnelling in saturated sandy ground.

2 DEFINITIONS

In this paper three basic definitions are used:

FER (foam expansion ratio): the volume of foam (Q_F) divided by the volume of surfactant solution (water and surfactant) (Q_L). This parameter indicates how much air is used for each volume of water and foaming agent.

With respect to the FER, it is noted that this is related to the absolute pressure and it depends if the foam is made under atmospheric condition or kept under a different pressure. In the experiment, the foam is made under atmospheric pressure and then pressurised under the desired pressure of 150 kPa (atmospheric pressure + air pressure of 50 kPa), the value of FER thus is calculated by:

$$FER_p = \frac{p_a}{p}(FER_a - 1) + 1 \tag{1}$$

where FER_p is the FER at the desired pressure, FER_a the FER at atmospheric pressure, p the desired air pressure, p_a the atmospheric pressure.

FER_m (effective foam expansion ratio): the total volume of foam volume in the mixture of soil and foam (air + water and surfactant) divided by the total volume of water + surfactant volume. Normally this is less than the FER due to the pore water in the soil (Bezuijen, 2012).

FIR (foam injection ratio): Is in EPB tunnelling defined as: the volume of foam injected into the soil (Q_F) divided by the volume of the soil (containing pore water) removed (Q_S). In our experiments the soils is not excavated and we calculate the FIR of the sandy foam mixture as the volume of the foam at 150 kPa divided by the volume that the sand would have at the same porosity as the sand column in the test.

3 EXPERIMENTS

3.1 Setup

The setup used for these experiments is shown in Figure 2, which has been described in detail in Xu and Bezuijen (2018). Four pore pressure transduces (PPTs) k1 to k4 were used to measure the pore water pressures in the foam and sand during the test. The amount of discharged water was measured continuously with an electronic balance. During a test, all data were measured with a frequency of 1 Hz. The setup is comparable to the setup used by Talmon et al. (2013) and Galli (2016), but an extra flow resistance is installed (the small cylinder in Figure 2) to reduce the flow in the beginning of the experiment, so that the flow velocity is comparable to what can be expected in front of a EPB shield when drilling through fine saturated sand. According to Xu and Bezuijen (2018), the hydraulic gradient over the sand in the setup is $i = \Delta\phi/L_s = 5/3 \approx 1.7$ ($\Delta\phi$ is the difference in piezometric head over the sand, L_s the equivalent thickness of the sand column). For a real tunnel with a diameter of, for example, 10 m, the hydraulic gradient at the centre of the tunnel face is approximately, $i = \Delta\phi/R = 5/5 = 1$ according to Bezuijen (2006). The hydraulic gradient in the setup is very comparable to the one in the field. A test started by opening the bottom valve and lasted one hour.

3.2 Test materials

A foam generator FG3 (Budach and Thewes, 2015) at Ruhr-University Bochum, which is a reproduction of an original foam generator of an EPB TBM but suitable for low foam rates, was

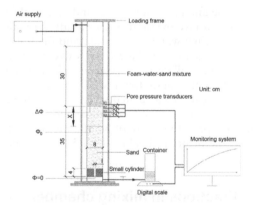

Figure 2. Scheme of the setup and principle of pressure infiltration of sandy foam.

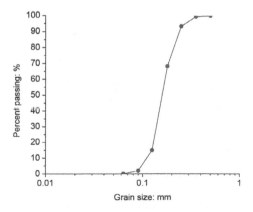

Figure 3. Grain size distribution of S80 sand.

utilised to generate foam for this experimental study. Since the influence of the foaming agent is not a topic in this study, one agent, Condat CLB F5/TM in one concentration of 3%, which is recommended to use predominantly in highly permeable sands (CONDAT LUBRIFIANTS, 2015), was used through all the tests. Properties of the foam itself used in this experimental study have been reported by Budach (2012) and Galli (2016).

The Euroquarz S80 sand was used to make the sand column. The grain size distribution of S80 is shown as in Figure 3. d_{10}, d_{50} (average grain diameter), d_{60}, d_{90} are 0.11, 0.16, 0.17 and 0.24 mm, respectively. The permeability of the sand at a saturated density of 2060 kg/m^3 is 1×10^{-4} m/s. The saturated density of the sand varied between 2050 and 2080 kg/m^3 in the various tests.

Measurements at the Botlek Rail Tunnel during construction have shown that the porosity of the muck in the mixing chamber is just a bit higher than the maximum porosity of the sand. Figure 4 shows a graph of fractions of gas, water and sand in foam conditioned soil excavated in Botlek Rail Tunnel (Bezuijen, 2006). In this study, the fractions of air, water and sand, and the FIR of the sandy foams refer to these measurements, see Table 1.

The foam was first made at atmospheric pressure and was then added to the sand sample with a desired water content in a beaker. Afterwards, the foam and sand sample were mixed very carefully by hand until the mixture was uniform. During mixing it was assumed that no bubbles collapsed. In this study, this mixture of foam-water-sand is called 'sandy foam'. In the sandy foam, the sand particles were suspended by the foam.

3.3 Procedure

To get a good impression of the amount of air in the sandy foam, the sandy foam was brought up to a certain height of 30 cm, above the sand surface at atmospheric pressure. The FER$_m$ of the sandy foam that has the original water of the foam and the pore water can be calculated. Then when the pressure is increased to 50 kPa but the valve below is still closed, the volume and thus the height of the mixture will decrease. The air volume will decrease to 1/1.5 the original volume. Sand and water in the sandy foam will keep the same volume. The amount of air in the sandy foam therefore can be calculated when the volume change was measured. It was found that the ratios of the air fraction at atmospheric pressure to the air fraction at atmospheric pressure plus the applied pressure of 50 kPa are between 0.96/1.5 to 1/1.5, which are more or less the same as the theoretical value of 1/1.5. It can therefore be said that hardly

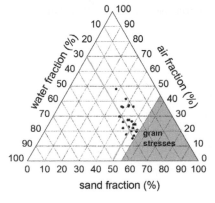

Fractions in mixing chamber

wet density = 1625 kg/m³

Figure 4. Volume fractions in mixing chamber at the Botlek Rail Tunnel in the Netherlands (Bezuijen, 2012).

Table 1. Properties of foam and sandy foam used in the tests.

Test date	FER	FER$_m$	FIR (%)	Air fraction (V%)	Water fraction (V%)	Sand fraction (V%)	Density (kg/m^3)
5-jan	5	1.5	21.9	15	30	55	1757.5
5-feb		1.7	32.1	20	30	50	1625
5-mrt		2	40.2	25	25	50	1575
5-apr		2	53.5	30	25	45	1442.5
5-mei		2.4	70.3	35	25	40	1310
5-jun		3	80.3	40	20	40	1260
10-jan	10	1.5	19.5	15	30	55	1757.5
10-feb		1.7	28.6	20	30	50	1625
10-mrt		2	35.7	25	25	50	1575
10-apr		2	47.6	30	25	45	1442.5
10-mei		2.4	62.5	35	25	40	1310
10-jun		3	71.4	40	20	40	1260
15-jan	15	1.5	18.8	15	30	55	1757.5
15-feb		1.7	27.5	20	30	50	1625
15-mrt		2	34.4	25	25	50	1575
15-apr		2	45.9	30	25	45	1442.5
15-mei		2.4	60.2	35	25	40	1310
15-jun		3	68.8	40	20	40	1260
20-jan	20	1.5	18.4	15	30	55	1757.5
20-feb		1.7	27.1	20	30	50	1625
20-mrt		2	33.8	25	25	50	1575
20-apr		2	45.1	30	25	45	1442.5
20-mei		2.4	59.2	35	25	40	1310
20-jun		3	67.6	40	20	40	1260

any bubbles collapsed during the mixing with sand and pressurising mixture. The test started by opening the bottom valve and lasted one hour or was stopped when all the foam in the sandy foam had infiltrated into the sand.

4 RESULTS AND DISCUSSIONS

With the measured discharges and pore pressures, the permeability of sand for water through the foam bubbles, k_f, can be determined according to Darcy's law:

$$k_f = \frac{\Delta L_s Q}{\Delta \phi_p A} \qquad (2)$$

where ΔL_s is the thickness of sand between two adjacent PPTs, A the cross-sectional area of the sand in the large diameter cylinder, Q the discharge and $\Delta \phi_p$ the corresponding difference in piezometric head over the sand between two adjacent PPTs (k1-k2, k2-k3 and k3-k4). Eq. (6.2) is only valid if the water is incompressible. For the water with foam, deviations are possible because the pore water (with the foam) is compressible. However, in a steady state situation where the pressures are more or less constant (as at the end of the test), the permeability can be calculated.

Figure 5 shows the result of permeability change with FIR. It is shown that the fit curves are likely 'S' lines. The drop of permeability is between FIR of ~ 35 - 70. When FIR < 35, the permeability of sand for foam flow is more or less equal to the permeability of sand for water except for the FER = 5. This indicates that the permeability of sand for foam flow depends on properties of the original foam. It is likely that a minimum FIR of 35 is required before a less

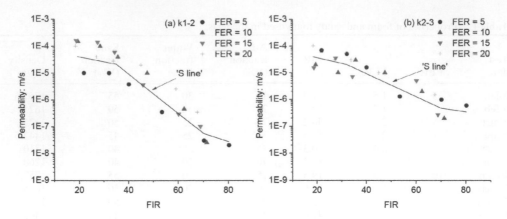

Figure 5. Permeability change with the ratio of air fraction/sand fraction.

permeable face is formed. This is comparable to the recommendation of a FIR = 40 by EFNARC (2005) for drilling in a sandy ground.

Consequently, in the case of tunnelling a low permeable ground, since a large amount of pore water remains in the soil, the FIR in the excavated soil (mixture of foam-water-sand) will be small. To reduce the permeability of the face, a higher FIR is necessary than to just prevent grain-grain contact. From the figure it can be seen that FIR = 70 is enough for a nearly impermeable face to be formed. Refer to the recommendation by EFNARC (2005), therefore FIR of 35 - 40 is recommended for a sandy ground. Further tests on other soils have to be carried out in order to gain a usable range of FIR for various soils.

A possible mechanism for the influence of the FIR for the sandy foam experiments is presented in Figure 6. Again the lower halve indicates the original sand sample. The upper halve is the sandy foam for a low FIR. With a low FIR, only a limited water flow will cause that there is a skeleton of loosely packed grains above the original sand bed. This is difficult to draw in a 2D picture, but in 3D the grains of the upper halve are more or less connected or there is effective stress because bubbles are trapped in between the sand grains. The trapped

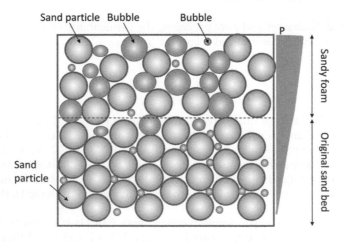

Figure 6. Mechanism for the influence of the FIR for the sandy foam experiments.

larger bubbles can now not migrate to the sand surface to form a low permeability layer. There can be some reduction of the sand permeability due to small grains invading into the sand. But a clean pressure drop as for 'clean foam' is not observed. With a higher FIR there will not be enough sand to prevent all migration of the bubbles.

Analogous to the depth filtration of slurry containing sand (or sandy slurry), there will be a sand layer formed on the boundary of the sand and the sandy foam. With a high sand fraction, there will be a dense sand layer formed, the foam bubbles thus will be blocked by the grains and the foam flow will be reduced. On the other hand, with a lower sand fraction, there will be loose sand layer formed, and thus more foam bubbles will flow out from the sandy foam into the sand, resulting in a low permeable layer formed close the boundary. This means the sand fraction is of importance for the plastering at the tunnel face.

In the extreme case there is 55% of sand in the sandy foam. This means that the porosity of the sandy foam is only 45% thus close to the porosity of loosely packed sand (0.45). For these situations (55 or 50 % sand fractions), there will be a grain skeleton very quickly and there is no foam available to create a low permeable area between k1 and k2. The tests end very quickly, because there is no foam above the newly formed sand skeleton. At lower sand fractions, more foam can flow from the sandy foam into the sand and this can form a low permeable layer between k1 and k2, the discharge reduces and the pressure drop over k1-2 increases. In all cases the pressures measured at k2 and k3 are the same, thus the foam does not reach this layer.

The situation with the high sand fraction should be avoided in the field, because there is the risk that the mixing chamber is blocked by the sand.

5 CONCLUSIONS

The experiments of infiltration of foam from sandy foam into saturated sand have been presented, and discussed together with the experiment of infiltration of clean foam. Conclusions may be drawn:

(a) A higher FER_m (effective foam expansion ratio) is more effective to form an impermeable layer in the sand. Furthermore, the permeability of sand for water flow through the foam decreases with the FIR (foam injection ratio).
(b) In practice, a FIR of 35 - 40 is recommended for a sandy ground and the situation with high sand fraction of the sandy foam should be avoided in the field because there will be no impermeable or low permeable layer formed at the tunnel face.
(c) Analogous to the depth filtration of slurry containing sand (or sandy slurry), high sand content of the sandy foam will impede the impermeable or low permeable layer formation on the boundary between the sandy foam and the sand. Moreover, in the situation with high sand content there will be the risk that the mixing chamber is blocked by the sand. Therefore, this situation should be avoided in the field.

ACKNOWLEDGEMENT

The first author would like to acknowledge the scholarship funded by China Scholarship Council and the CWO Mobility Fund of Ghent University.

REFERENCES

Amoun, S., Sharifzadeh, M., Shahriar, K., Rostami, J., Azali, S. T., 2017. Evaluation of tool wear in EPB tunneling of Tehran Metro, Line 7 Expansion. *Tunnelling and Underground Space Technology* 61, 233–246.

Bezuijen, A., Schaminée, P. E. L., and Kleinjan, J. A., 1999. Additive testing for earth pressure balance shields. *Proc. 12th Eur. Conf. on Soil Mech. and Geotech. Engrg.*, Amsterdam, Balkema, Rotterdam, pp. 1991–1996.

Bezuijen, A., 2002. The influence of permeability on the properties of a foam mixture in a TBM. 4th Int. Symp. on Geotechnical Aspects of Underground Construction in Soft Ground - IS Toulouse 2002.

Bezuijen, A., Pruiksma, J. P., et al., 2006. Pore pressures in front of tunnel, measurements, calculations and consequences for stability of tunnel face. *Tunnelling. A Decade of Progress. GeoDelft 1995-2005*, 35–41.

Bezuijen, A., 2011. Foam used during EPB tunnelling in saturated sand, description of mechanisms. *Proceedings WTC 2011*, Helsinki.

Bezuijen, A., 2012. Foam used during EPB tunnelling in saturated sand, parameters determining foam consumption, *Proceedings WTC 2012*, Bangkok.

Budach, C., 2012. Extended application ranges of EPB-Shields in coarse grained soils. Dissertation, Ruhr-Universität Bochum.

Borio, L., Picchio, A., Peila, D., Pelizza, S., 2011. Study of the permeability of foam conditioned soils with laboratory tests. *Proceedings of the ITA-AITES World Tunnel Congress in Helsinki*.

Budach, C., Thewes, M., 2015. Application ranges of EPB shields in coarse ground based on laboratory research. *Tunnelling and Underground Space Technology*, 50, 296–304.

CONDAT LUBRIFIANTS, 2015. New high-tech foaming agent for TBM. https://www.condat-lubricants.com/product/sealant-foam-lubricant-tunnel-boring/foaming-agents-epb/.

EFNARC (ed.) (2005), Specification and guidelines for the use of specialist products for mechanised tunnelling (TBM) in soft ground and hard rock.

Maidl, U., 1995. Erweiterung der Einsatzbereiche der Erddruckschilde durch Bodenkonditionierung mit Schaum. Doctoral thesis, Ruhr-Universität Bochum, AG Leitungsbau und Leitungsinstandhaltung, Germany.

Maidl, U., Pierri, J., 2014. Innovative hybrid EPB tunnelling in Rio de Janeiro. Geomech. Tunn. 7, 55–63.

Merrit, A., Jefferis, S., Storry, R., Brais, L., 2013. Soil conditioning laboratory trials for the Port of Miami Tunnel, Miami, Florida, USA. In: *Proceedings of the ITA-AITES World Tunnel Congress in Geneva*.

Xu, T., Bezuijen, A., 2018. Pressure infiltration characteristics of bentonite slurry. *Géotechnique*. https://doi.org/10.1680/jgeot.17.T.026.

Xu, T., Bezuijen, A., Thewes, M., 2018. Pressure infiltration characteristics of foam for EPB shield tunnelling in saturated sand – Part 1: clean foam. *Géotechnique*. In preparation.

Tunnels and Underground Cities: Engineering and Innovation meet Archaeology,
Architecture and Art, Volume 6: Innovation in underground engineering,
materials and equipment - Part 2 – Peila, Viggiani & Celestino (Eds)
© 2020 Taylor & Francis Group, London, ISBN 978-0-367-46871-2

Research on new prefabricated technology of metro station

X.R. Yang, M.Q. Huang & F. Lin
Beijing Urban Construction Design & Development Group Co. Ltd., Beijing, China

National Engineering Laboratory for Green & Safe Construction Technology in Urban Rail Transit, Beijing, China

ABSTRACT: New fully prefabricated technology of metro station is first researched in china to reduce resources consumption, shorten the construction period and decrease environmental impacts for large-scale metro construction. For now, five metro stations have been built in Changchun, northeast of China. This paper introduces the application status of prefabricated technology in underground engineering in the whole world, and tells the different characteristics of prefabricated technology applied to above-ground buildings and underground engineering. In addition, combined with the first successful application of new prefabricated technology in metro station, this paper further conducts a thorough analysis and discussion on the key points of new prefabricated technology for large underground structure. And economic and social benefits achieved are discussed at last.

1 INTRODUCTION

Prefabricated above-ground structure has the advantages of high efficiency, high quality, high lifespan and environmentally friendly, and it has mature and complete technology system. So it has been widely used in Occident and Japan (Gu, 2014). In China, prefabricated above-ground buildings also have made gratifying achievements after a long period of research and application by key enterprises. The related technology and management systems have been established (Chinese Technical specification, JGJ1-2014) and extensive applications have been applied in the above-ground buildings and bridge engineering.

For prefabricated technology in underground engineering, prefabricated linings were firstly applied in shield tunnel in the late19th and early 20th century (Liu, 1991). After more than 100 years of development, shield tunnel, usually with circular section, has been applied in many fields like metro, highway, municipal engineering, etc. The other applications of prefabricated linings originated in the 1980s. In order to overcome problems of cast-in-place concrete construction in cold weather, prefabricated technology were researched and applied in open-cut metro running tunnels, principal parts of stations and auxiliary aisles in the former Soviet Union (Frarov, 1994; Li, 1995).

Later, prefabricated technology were used in underground engineering in Netherlands, France and Japan (Beilasov, et al. translated by Qian, et al., 2012). But limited to technological conditions at that time, connection joints used cast-in-place concrete (see Figure 1). These joints have difficulty in waterproofing, which affects construction progress and entirety of structural waterproofing and limits the advantages of construction prefabricated structure in low temperature. In addition, interaction between strata and prefabricated underground structure expected to be further researched. So prefabricated technology is more for use in simple structure with small cross-section like metro running tunnel and municipal pipeline. At present, research on prefabricated technology of large underground structure, especially the full prefabricated technology, is still blank.

Figure 1.   Caption of a typical figure.

Prefabricated underground structures have different performance characteristics from pre-fabricated above-ground structures (Wang, et al., 2012; Chen, et al., 2015; Li, 2017; Bao, 2017; Ma, 2017):

a. The above-ground structure, surrounded by air without any constraints, bears not only vertical load, but also lateral wind load and seismic force. And theses lateral loads will cause whiplash effect. At this point, self-vibration characteristics decided by shape, quality and stiffness of structures play a decisive role. For underground structure, surrounded by strata, apart from self-weight, pedestrian load, equipment load and other vertical loads, the more important loads come from all around water-soil pressure. Meanwhile, underground structure also suffers all-around constrains from surrounding strata. In other words, strata is both load and carrier. Even under earthquake, underground structure carried by the sur-rounding strata has synchronous vibration and co-deformation with strata. Motion charac-teristics of soil play a principle role, not the self-vibration characteristics of underground structure. And the shape, quality and stiffness of the structure can't change vibration char-acteristics of structure and strata.

b. Waterproofing targets and measures are totally different between above-ground structure and underground structure because of different using environment. The waterproofing principle of above-ground structures likes "umbrellas and raincoats". It changes path of water flow for retaining and draining water and avoids continuous hydrostatic pressure. While the waterproofing principle of underground structure is similar to "submarine", and both structure itself and the connection joint need carrying high groundwater load. Which has essential difference from "umbrellas and raincoats".

c. Above-ground structure uses cast-in-place joints with rebar splicing to connect prefabri-cated components. This connection of integral rigidity conforms to using environment and mechanical characteristics of above-ground structure. For above-ground prefabricated structure, the reinforcement welding or embedded coupling sleeve for connection, struc-tural calculation model and principle of prefabrication are same with cast-in-place struc-ture. While for prefabricated underground structure, its nonrigid joints and constraints from strata make the structural bearing system extremely complex and its mechanical behaviour and properties have great difference with integral cast-in-place underground structure.

d. The prefabricated components of above-ground structure are small with less reinforcement and have free space for hoisting and assembly. In addition, rigid joints of above-ground prefabricated structure are easy to implement. While for prefabricated underground structure with large components and high reinforcement ratio, construction environmental conditions should be fully considered in every stages from structural selection, joint type, waterproofing measures to hoisting, positioning and assembly of components. Because prefabrication-assembly for underground structure is similar to building blocks in a box. So, prefabricated technologies for above and under ground structure are two different technology systems. Prefabricated technology for above-ground structure has been matured, while prefabricated technology for underground structure is still at a preliminary stage and needs to be improved and systemic studied.

With the rapid development of rail transit construction in China, social environmental awareness increases continuously. Moreover, long and tight construction period, large resource consumption, decrease of young labourers of civil engineering causing shortage of skilled labour and not guaranteed structure quality bring great challenges to traditional construction technology of metro station. Above situations are particularly prominent in the northeast of China. Like Changchun city, located in the northeast region, where is so cold that 4~5 months' winter break is needed for metro construction. Which causes huge deadline pressure and is hard to guarantee construction quality under low temperature. To solve these problems in cold regions, Yuanjiadian station on Changchun Metro Line 2 was selected as test section to carry out research and application of prefabricated technology in underground engineering in 2012. For now, 5 prefabricated metro stations have been completed and remarkable results have been achieved.

2 PROJECT OVERVIEW OF FULL PREFABRICATED UNDERGROUND STATION

The five cut-and-cover stations, located in Changchun Metro Line 2 (see Figure 2), are all supported by anchor-pile system. And all those horseshoe-shaped two-storey stations are 20.5m-wide and 17.45m-high. The full prefabricated station structure is assembled by seven 2m width prefabricated components (see Figure 3), and the station entrance-exit are also assembled by prefabricated components (see Figure 4) without any concrete wet spraying. Except two components in the arch crown are staggered assembling in the longitudinal direction, the others are assembling with straight joints. Mortise and tenon joint with grout is used between prefabricated components. And multi-waterproof measures are set up in joints. Except arch crown, no more outer waterproof layers are equipped. By now, there is no any water leakage in the construction-completed station with covering soil during one year and a half including two rainy seasons. And stabilities of structural deformation perform well for long-term monitoring.

Figure 2.   Five prefabricated stations on Changchun Metro Line 2.

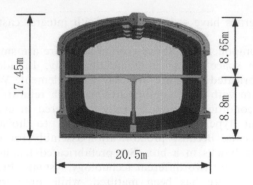

Figure 3. Sectional diagram of prefabricated station on Changchun Metro Line 2.

Figure 4. Station entrance-exit.

3 KEY TECHNOLOGIES OF PREFABRICATED STATIONS ON CHANGCHUN METRO LINE 2

A series of core supporting researches are carried out in this prefabricated metro station project, which contains mechanical behaviour and connection technology of joints, mechanical behaviour of structural system, seismic performance of prefabricated structure, waterproofing technology, closed cavity thin-walled components of lightweight, integrated technology of prefabricated component manufacturing, new-style production lines in tunnel kiln, laser point cloud monitoring equipment for prefabricated components, assembly technology of prefabricated structure, equipment for assembling and controlling system, grout technology and equipment for joints and multi-specialty integration design of prefabricated station. And then complete technology system of prefabricated metro station, covering design, construction and component manufacturing, is formed. Next, key technologies will be introduced.

3.1 Connection joints between prefabricated components

The joints connecting prefabricated components, affecting comprehensive stress state and load-bearing characteristics of structure, manufacturing technique of components, construction assembly process, and waterproofing performance, are the core issue of prefabricated structure. Review study on joints technology, such as rebar splicing by grout-filled coupling sleeve and composite assembly technology of above-ground structure, prefabricated joint technology of shield tunnel, cast-in-place and opening inserting joint technology in the former Soviet Union, has been investigated. Based on the advantages-disadvantages and adaptability of these previous joint technologies and characteristics of prefabricated underground structure, mortise and tenon joint with grout is developed. This kind of joint takes advantages of prefabricated technology as simple process, accurate positioning and rapid assembly (see Figure 5).

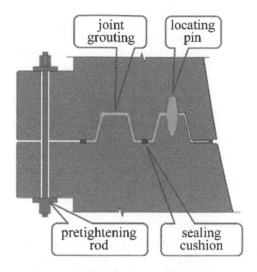

Figure 5. Schematic diagram of mortise and tenon joint with grout.

Figure 6. Specimen loading layout image.

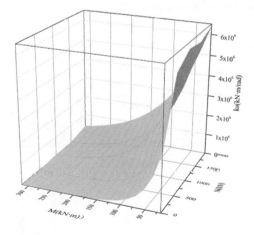

Figure 7. Variable stiffness of the joint.

Mortise and tenon joint with grout has a typical characteristic of variable stiffness and presents various stiffness under different loading conditions. Which leads to increased complexity of prefabricated underground structure, but makes main contributions to effectively regulate and optimize structural inner force amplitude (especially bending moment). A series of studies with theoretical analysis and 1:1 prototype test (see Figure 6) have been carried out to discuss mechanical behaviour, stiffness, bearing capacity, reasonable geometric type, grouting mechanism and waterproofing property of mortise and tenon joint with grout. Thus obtaining constitutive relations and design method of bearing capacity of joint, which provides the theoretical basis of prefabricated technology of metro station.

Based on experimental and theoretical analysis, joint's characteristic of variable stiffness (see Figure 7) and bearing capacity are indicated, and calculation methods of stiffness and bearing capacity for the joints are also obtained.

3.2 Structural system and mechanical behaviour

"Prefabrication" and "Nonrigid joint" of full prefabricated underground structure are the key difference of structural system and mechanical behaviour from cast-in-place underground structure. Therefore, "structural stability" and "mechanical behaviour of structural system with variable stiffness" are the most focused.

Stability of structural system involves many aspects. Firstly, stable mechanical equilibrium should be formed between structures and surrounding soils, and more necessary capacity of structure needs to be reserved. Secondly, structure surrounded with soils should keep statically determinate property and carrying components as baseboard, side wall, arch crown, floor slab needs to form statically determinate or statically indeterminate system. Moreover, local instability problems, such as bearing capacity of single component, load and deformation capacity of joint, must be avoided.

Mechanical behaviour of structural system with variable stiffness is relatively complex. Firstly, Nonrigid joint effectively increases deformability of structural system. Secondly, stiffness and bearing capacity of joint are not only related to geometric type, but also concerned with acting axial force. And the diversities of geometric type and acting force lead to the difference of stiffness of every joint, which has a nonlinear constitutive relationship with acting bending moment and axial force. Therefore, mechanical behaviour of structural system should consider all the stages from assembly construction to normal service stages under normal working condition and seismic cases. Behavioural characteristics of the joint and mechanical behaviour of the whole structure in the loading evolution process have been investigated with multiple iterations of joint's property.

Figure 8.   Different assembly stages of prefabricated structure.

Based on experiment, field test and theoretical analysis, we finally understood key mechanical laws and proposed design methods of prefabricated structure. Figure 8 displays different assembly stages of prefabricated metro station structure.

3.3 Seismic performance

Seismic behaviour of underground structure is better than above-ground structure because of stratigraphic constraints. And governing factor of structural reaction under earthquake is movement characters of soil. Vibration characteristics of soil is the main concerns by previous studies, while research on seismic behaviour of underground structure especially prefabricated underground structure is still in its starting stage.

Increasing structural ductility is an important way to improve seismic behaviour of underground structure. Under earthquake, nonrigid joint of prefabricated structure increases deformability of the whole structure, which can decrease bending moment of joint and adapt to stratum deformation, and the structural ductility of prefabricated structure is better than traditional cast-in-place structure. In addition, rational structural system can avoid forming plastic hinges.

Based on the above concepts, there are three key points in the study of seismic performance for prefabricated underground structure.

a. Analysis for seismic behaviour of structural system under multi conditions.

Extensive numerical simulations of station structure needs conducting under different geological conditions, joint properties and earthquake magnitude. And performance, internal force and deformation law of structural system need to be investigated. Parallel simulations of different computing software will be validated when necessary.

b. Analysis for seismic behaviour and structure of joint.

Seismic behaviour of joint reflects in two aspects, bearing capacity and deformability (angle of rotation). Prototype test is carried out to study the above two aspects. Moreover, characteristic of joint depends on structure of joint, and the geometric type and additional structural measures directly affect deformability and failure mode of joint.

c. Comparative analysis of prefabricated and cast-in-place structure of the same type under static and dynamic conditions.

Difference characteristics of prefabricated and cast-in-place structure are discussed to validate technical advantages and characteristics of prefabricated structure.

Research on main behavioural characteristics and relative technical index of prefabricated metro station structure in Changchun has fully verified that seismic behaviour meets performance requirements under earthquake and prefabricated structure has an advantage of reducing internal force amplitude. Moreover, the completed stations have already experimented the 4.5 magnitude Songyuan earthquake once and 3.0 magnitude earthquakes several times, and the performance of prefabricated structure indicates well under field real-time monitoring.

3.4 Waterproofing technology

Based on the characteristics of prefabricated technology, waterproofing target of prefabricated metro station structure in Changchun is established: removing traditional outer waterproofing layer and reaching or exceeding standard specifications of waterproofing for metro station. Prefabricated component itself can solve the problem of structural self-waterproof. So joint waterproofing is the key.

There are many crisscross joints distributed in the whole structure. So reliable waterproofing system should be built after reasonable planning and designing the integral sealing and waterproofing system firstly. Then the path setting of such system requires rational planning, especially parts like sides, corners, mortise and groove, cross and reserved holes, in order to ensure no leakage of joints and no disordered water string in longitudinal direction.

Joint waterproofing with high-standard configuration is essential. Our design philosophy is multi-channel waterproofing line and active sealing. Whose specific measures are "two cushions-one grout-one caulking" (see Figure 5). Specifically, "two cushions" means that two hydro-expansive rubber sealing cushions are set up in the joint face; one grout means that the gap between two joint faces is tightly filled with grout; one caulking means that caulking, with the function of drainage, should be performed in the grooves of joint surface (see Figure 9).

A series of waterproofing tests for joint are carried out based on prefabricated metro station in Changchun. The results show that only one hydro-expansive rubber sealing cushion can bear 80m head pressure without any leakage under the worst location of the joint (10mm joint opening and 5mm dislocation of the other sealing cushion). And grouting section filling with the developed modified epoxy resin has possibility of leakage in theory. In addition, caulking in the inner face of joint has the function of both drainage and surface sealing.

3.5 Lightweight design of prefabricated components

Generally, underground structure bears huge water-soil pressure and so components of underground structure are large and have high reinforcement ratio, which is different from above-ground structure. Even the prefabricated underground structure is divided into multi components, each one is still large and heavy. Which increases difficulty of assembly and hoisting device. Therefore, lightweight design of components are the important measure to improve constructability of prefabricated structure.

Reducing the value of concrete is an effective way for lightweight of concrete components. Closed cavity filled with light materials setting in the internal is proposed according to mechanical characteristics of concrete cross-section. Figure 10 displays axonometric drawing of one closed cavity thin-walled component.

Closed cavity with light materials has remarkable effect on lightweight of component. For example, before lightweight, every 2m segment ring weights 360t, of which the heaviest and lightest component weight 65t and 35.35t respectively. After lightweight, one 2m segment ring weights 300t, achieving average weight reduction of 16.67%. Lightweight of component is not only convenient for construction, but also saves concrete and reinforcement consumption. Another unexpected benefit has been found in component manufacturing stage, which is that hydration heat and cooling down time of closed cavity thin-walled component reduce a lot. Which avoids cracking when cooling and improves productivity.

Lightweight design of component increases difficulty in research, design and manufacturing without doubt. The internal force and stress transfer mechanism and process in the endings, cross-beam, longitudinal ribs and cavity of component are very complex because of the existence of the internal closed cavity. Therefore, mechanical behaviour of closed cavity thin-walled components and interactions among internal force, stress – structural scale – component performance need to

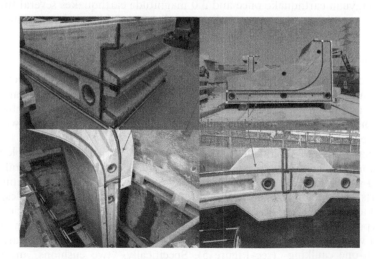

Figure 9. Waterproofing measure on each component.

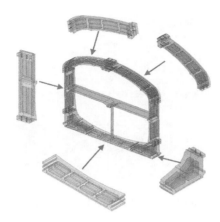

Figure 10. Axonometric drawing of closed cavity thin-walled components.

Figure 11. Prefabricated closed cavity thin-walled components in the factory.

be discussed, thus obtaining component design method. Various studies on light materials, molding method, prevention of water absorbency, economical efficiency of core-mold in closed cavity have been conducted and it turns out the expected effect has been achieved after use.

In the end, lightweight design technology of prefabricated components is formed, which reduces the weight of components and optimizes the structural system as well. Figure 11 shows the prefabricated closed cavity thin-walled components in the factory.

3.6 *Manufacturing technology for large prefabricated components*

We have put forward one complete set technique of high accuracy manufacturing for prefabricated components with flexible mold, built new-style production lines in tunnel kiln and invented laser point cloud monitoring equipment for large prefabricated components (see Figure 12). Above series of technologies realize efficient and high-quality production of the prefabricated components.

3.7 *Construction technology and equipment development*

As a new construction method for prefabricated metro station, all the processes and steps on assembly technology cannot be ignored. Key construction stages must include scientific and reasonable process arrangement of assembly construction, hoisting of large prefabricated component, locating approach, assembly line of joint, tensioning and joint width control.

Development and adoption of special construction equipment (see Figure 13) and aided assembly accessories are also important. It covers special assembly stair vehicle, support screw jack, locating rod for component, auxiliary tensioning equipment, special grouting equipment for joint, measuring and positioning system for component, automatic assembly control system, etc.

Figure 12.   Demostration on manufacturing techonoly of components.

Figure 13.   Special construction equipment.

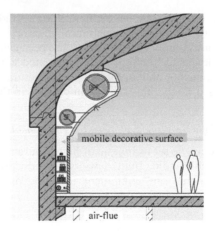

Figure 14.   Compatibility of decoration and pipelines.

3.8   *Multi-specialty integration design of prefabricated station*

There are numerous equipment pipelines with complex system need to be placed in limited space of the underground station. And crowned section of prefabricated station is quite different from rectangle section of standard station. So, we conducted system design for pipelines and made them reasonably arranged in limited space – optimizing and utilizing vertical space, classifying equipment space, redistribution and combination each module, in order to make full use of space and optimize station scale (see Figure 14). In addition, traditional hangers

Figure 15.   Embedded channel.

Figure 16.   Station hall of Jianshe St on Changchun Metro Line 2.

and bolted connection are replaced by embedded channels to make post-installation of inner pipelines and equipment convenient (see Figure 15).

Research of this part integrates multi-specialty systems and creates new design concept of pipelines and new decoration concept of exposure of structure itself. Which forms satisfactory space vision and using effects. Figure 16 exhibits the decoration rendering of station hall in Jianshe St station.

Above research achievements have been successfully applied in practical project (Changchun Metro Line 2). And with the development of research work on open-cut prefabricated underground structure, construction system of open-cut prefabricated underground station, covering three technical supports – design, component manufacturing, is gradually formed.

4   TECHNICAL, ECONOMIC AND SOCIAL BENEFITS

The prefabricated underground structure shows the following six main advantages:

a. significantly improving engineering quality.
b. substantially increasing construction efficiency.
c. enhancing security of engineering construction.
   As we all know construction risk of underground engineering is high. While prefabricated technology will achieve greater security by greatly reducing construction operation steps.

d. small influence on environment.

Site construction steps are prominently reduced or even cancelled after using prefabricated technology. And construction noise and dust are decreased remarkably.

e. labour saving.

At present, shortage of site construction workers becomes more prominent with China's demographic dividend is diminishing gradually and cost of labour raises continuously. Prefabricated technology used in underground engineering could reduce demands for labour.

f. overcome winter impossible construction problems in cold region.

Preliminary analysis of technical and economic benefits has been carried out based on prefabricated metro station in Changchun. Compared with traditional open-cut cast-in-place metro station, one prefabricated metro station has the following benefits:

a. saving 4~6 months construction period (20%~30%);
b. reducing site construction workers from 130~150 to 30 at peak times;
c. saving steel consumption 800t, wood consumption 800m^3, reducing construction waste quantity 50%;
d. reducing construction space 1000m^2;
e. having a strong advantage of construction safety, noise and air pollution.

5 CONCLUDING REMARKS

The successful application of prefabricated technology for underground structure in Changchun Metro opens one window for a new technology. The advantages of prefabricated technology in underground structure will generate a strong boost to the development of urban rail transit in China, and become an important force in the field of construction industrialization in the future.

REFERENCES

Bao, Z.Q. 2017. Discussion on the common quality problems and preventive measures of prefabricated building construction. *Building Materials and Decoration* 27(53): 54.

Qian, Q.H. & Qi, C.Z. 2012. The essence of the Construction of Russian Underground Railway. *China Railway Press*. Beijing.

Chen, L., Jin, X., Wang, R.B. & Zhang, M.J. 2015. Problems and Countermeasures of Composite Slab Concrete Shear Wall structure Construction. *Construction techniques* 16: 57–59.

Frarov. 1994. *New Conception of Underground Railways*. Moscow.

Gu, T.C. 2014. Development status of prefabricated architecture at home and abroad. *Standardization of Engineering Construction* 8: 48–51.

Li, H.M. 2017. Analysis of common quality problems and preventive measures in assembly construction. *Standardization in China* 8: 68/85.

Li, T.H. 1995. Design and Construction experience of single Arch Subway Station in Minsk. *Metro and Light Rail* 2: 44–48.

Liu, J.H. & Hou, X.Y. 1991. *Shield tunneling*. Beijing: China Railway Press.

Ma, S.D. 2017. Analysis of key techniques of assembled concrete structure in Building. *Jiangxi Building Materials* 23: 63.

Wang, G., Shen, X.P., Zhang, W.L. & Ma, W. 2012. Experimental study on the force transfer performance of horizontal and vertical connections of superimposed wall panels. *Industrial buildings* 4: 51–55.

National industry standard. Technical specification for fabricated concrete structures: JGJ1-2014. *China Construction Industry Press*. Beijing.

Tunnels and Underground Cities: Engineering and Innovation meet Archaeology, Architecture and Art, Volume 6: Innovation in underground engineering, materials and equipment - Part 2 – Peila, Viggiani & Celestino (Eds)
© 2020 Taylor & Francis Group, London, ISBN 978-0-367-46871-2

Seismic deformation method for longitudinal bending deformation of underground structure

N. Yasuda
Kyoto university, Kyoto, Japan

T. Asakura
Research group for tunnel engineering, Tokyo, Japan

ABSTRACT: It is recognized that underground structures deforms with surrounding ground when an earthquake occurs. Therefore, seismic deformation method (SDM), where inertia of the ground is ignored and forced displacements are applied to the ground as the seismic force, is often used for seismic design. The most simplified SDM model to evaluate the longitudinal deformation of the underground structure is the elastic foundation model, where the ground is modeled as a Winkler foundation and the structure is modeled as a beam. This model is often used because of the simplicity. However, theoretical support is not sufficient. The seismic deformation method to evaluate the longitudinal bending deformation of cylindrical underground structure is reconsidered through the exact solutions based on the elastodynamics. Conventional seismic deformation method that assumes the structure as a beam overestimates the maximum axial strain of the structure because beam theory ignores the equilibrium of force in the longitudinal direction.

1 INTRODUCTION

It is recognized that underground structures deform with surrounding ground when an earthquake occurs. Therefore, seismic deformation method (SDM), where inertia of the ground is ignored and forced displacements are applied to the ground as the seismic force, is often used for the seismic design of the underground structure. The most simplified SDM model to evaluate the longitudinal bending deformation is the elastic foundation model, where the ground is modelled as a Winkler foundation and the structure is modelled as a beam (Jiang and Kuribayashi, 2000; Hashash et al. 2001). This model is often used because of the simplicity. However, the validity of the model is not sufficient.

In this paper, the elastic foundation model to evaluate the longitudinal bending deformation of cylindrical underground structure is reconsidered through the exact solution based on the elastodynamics (Mow and Pao, 1971) and shell theory (Flügge, 1973).

2 PROBLEM DEFINITION

Consider a plane harmonic shear wave propagates at an angle ϕ with respect to an axis of an infinite cylindrical underground structure, as shown in Figure 1. The surrounding ground is postulated to be an infinite elastic, homogeneous, isotropic medium, and the structure is treated as an elastic cylindrical shell (Flügge, 1973). The mean radius of the structure is R, and the thickness is h. Because the circular cylinder has azimuthal symmetry, it can be assumed that the incident shear wave propagates in the direction of the wave number vector k^{inc} on the x-z plane. The displacement vector of the incident shear wave u^i_{s1} is on the plane perpendicular to

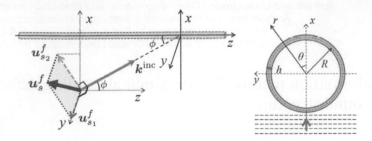

Figure 1. A cylindrical underground structure under oblique incidence of shear wave.

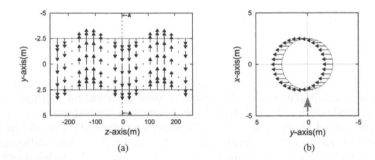

(a)   (b)

Figure 2. Longitudinal bending deformation under S_1 wave incidence: (a) longitudinal section and (b) cross section AA.

the wave number vector $\boldsymbol{k}^{\text{inc}}$. u_s^f can be decomposed into two independent vectors. One is a S_1 wave vector u_{s1}^f which is parallel to y axis, and the other is a S_2 wave vector u_{s2}^f.

The deformation of the structure can be expressed by the sum of the deformation modes. When the wavelength of the incident wave is enough long compared to the radius of the structure, the 0th mode (compression-extension deformation), the 1st mode (longitudinal bending deformation) and the 2nd mode (ovaling deformation of cross section) are predominant. In the following, Longitudinal bending deformation under S_1 wave incidence as shown in Figure 2 is only considered. The solution under S_2 wave incidence can be derived in the same way.

3 THEORY

3.1 Equilibrium of forces in the cylindrical coordinate system

Displacement of the ground and the structure at the ground-structure interface can be expressed as follows:

$$\left. \begin{array}{l} u_r^1 = U_r^1 \sin \theta e^{i(\gamma z - \omega t)} \\ u_\theta^1 = U_\theta^1 \cos \theta e^{i(\gamma z - \omega t)} \\ u_z^1 = -i\, U_z^1 \cos \theta e^{i(\gamma z - \omega t)} \end{array} \right\}, \tag{1}$$

$$\gamma = \frac{2\pi}{L_z}. \tag{2}$$

Where U_r^1, U_θ^1 and U_z^1 are complex constants. Superscript 1 is used to denote the 1st mode. ω is angular frequency of the incident wave. L_z is wavelength in longitudinal direction. In Eq. (1), U_z^1 is multiplied by $(-i)$ because there is the phase difference between the radial displacement and the axial displacement. This multiplication makes the following ground and structure stiffness matrixes symmetric.

3436

Surface loading of the ground and the structure at the ground-structure interface can be expressed similarly:

$$\left.\begin{array}{l} f_r^1 = F_r^1 \sin\theta e^{i(\gamma z-\omega t)} \\ f_\theta^1 = F_\theta^1 \cos\theta e^{i(\gamma z-\omega t)} \\ f_z^1 = -iF_z^1 \sin\theta e^{i(\gamma z-\omega t)} \end{array}\right\}. \tag{3}$$

Where F_r^1, F_θ^1 and F_Z^1 are complex constants.

Stiffness of the ground and the structure, which are defined as the relationship between displacement and surface loading at the ground-structure interface, can be expressed as follow:

$$\begin{pmatrix} F_r^1 \\ F_\theta^1 \\ F_z^1 \end{pmatrix} = \begin{pmatrix} K_{rr}^1 & K_{r\theta}^1 & K_{rz}^1 \\ K_{r\theta}^1 & K_{\theta\theta}^1 & K_{\theta z}^1 \\ K_{rz}^1 & K_{\theta z}^1 & K_{zz}^1 \end{pmatrix} \begin{pmatrix} U_r^1 \\ U_\theta^1 \\ U_z^1 \end{pmatrix}. \tag{4}$$

Considering Eq. (1) to Eq. (4), rigorous equilibrium of forces at the ground-structure interface can be derived as follow:

$$\begin{pmatrix} K_{rr(2)}^1 & K_{r\theta(2)}^1 & K_{rz(2)}^1 \\ K_{r\theta(2)}^1 & K_{\theta\theta(2)}^1 & K_{\theta z(2)}^1 \\ K_{rz(2)}^1 & K_{\theta z(2)}^1 & K_{zz(2)}^1 \end{pmatrix} \begin{pmatrix} U_{r(2)}^1 \\ U_{\theta(2)}^1 \\ U_{z(2)}^1 \end{pmatrix} = \begin{pmatrix} K_{rr(1)}^1 & K_{r\theta(1)}^1 & K_{rz(1)}^1 \\ K_{r\theta(1)}^1 & K_{\theta\theta(1)}^1 & K_{\theta z(1)}^1 \\ K_{rz(1)}^1 & K_{\theta z(1)}^1 & K_{zz(1)}^1 \end{pmatrix} \begin{pmatrix} U_{r(2)}^{1,e} & U_{r(2)}^1 \\ U_{\theta(2)}^{1,e} & -U_{\theta(2)}^1 \\ U_{z(2)}^{1,e} & U_{z(2)}^1 \end{pmatrix}. \tag{5}$$

Where Superscript e is used to denote the excavated ground with no structure, and subscript (1) and (2) are used to denote the ground and the structure, respectively.

3.2 Equilibrium of forces in the Cartesian coordinate system.

Considering the transformation from cylindrical coordinate system to Cartesian coordinate system, Displacements in Eq. (1) can be transformed as followings:

$$\begin{pmatrix} u_x^1 \\ u_y^1 \\ iu_z^1 \end{pmatrix} = \begin{pmatrix} \frac{1}{2}\sin2\theta & -\frac{1}{2}\sin2\theta & 0 \\ \sin^2\theta & \cos^2\theta & 0 \\ 0 & 0 & \sin\theta \end{pmatrix} \begin{pmatrix} U_r^1 \\ U_\theta^1 \\ U_z^1 \end{pmatrix} e^{i(\gamma z-\omega t)}, \tag{6}$$

$$\begin{pmatrix} u_x^1 \\ u_y^1 \\ iu_z^1 \end{pmatrix} = \begin{pmatrix} \sin2\theta & 0 & 0 \\ -\cos2\theta & 1 & 0 \\ 0 & 0 & \sin\theta \end{pmatrix} \begin{pmatrix} \dfrac{(U_r^1 - U_\theta^1)}{2} \\ \dfrac{(U_r^1 + U_\theta^1)}{2} \\ U_z^1 \end{pmatrix} e^{i(\gamma z-\omega t)}. \tag{7}$$

Surface loading in Eq. (3) can be transformed similarly. Moreover, stiffness in Eq. (4) can be transformed as follow:

$$\begin{pmatrix} \dfrac{F_r^1 - F_\theta^1}{2} \\ \dfrac{F_r^1 + F_\theta^1}{2} \\ F_z^1 \end{pmatrix} = \begin{pmatrix} \dfrac{K_{rr}^1 - 2K_{r\theta}^1 + K_{\theta\theta}^1}{2} & \dfrac{K_{rr}^1 - K_{\theta\theta}^1}{2} & \dfrac{K_{rz}^1 - K_{\theta z}^1}{2} \\ \dfrac{K_{rr}^1 - K_{\theta\theta}^1}{2} & \dfrac{K_{rr}^1 + 2K_{r\theta}^1 + K_{\theta\theta}^1}{2} & \dfrac{K_{rz}^1 + K_{\theta z}^1}{2} \\ \dfrac{K_{rz}^1 - K_{\theta z}^1}{2} & K_{rz}^1 + K_{\theta z}^1 & K_{zz}^1 \end{pmatrix} \begin{pmatrix} \dfrac{U_r^1 - U_\theta^1}{2} \\ \dfrac{U_r^1 + U_\theta^1}{2} \\ U_z^1 \end{pmatrix} \tag{8}$$

Suppose the following equations:

3437

$$U_r^1 \cong U_\theta^1, \tag{9}$$

$$F_r^1 \cong F_\theta^1, \tag{10}$$

displacements, surface tractions, and stiffness can be approximated as follows:

$$\begin{pmatrix} u_y^1 \\ iu_z^1 \end{pmatrix} = \begin{pmatrix} U_y^1 \\ U_z^1 \sin\theta \end{pmatrix} e^{i(\gamma z - \omega t)}, \tag{11}$$

$$\begin{pmatrix} f_y^1 \\ if_z^1 \end{pmatrix} = \begin{pmatrix} F_y^1 \\ F_z^1 \sin\theta \end{pmatrix} e^{i(\gamma z - \omega t)}, \tag{12}$$

$$\begin{pmatrix} F_y^1 \\ F_z^1 \end{pmatrix} = \begin{pmatrix} K_{yy}^1 & K_{yz}^1 \\ K_{zy}^1 & K_{zz}^1 \end{pmatrix} \begin{pmatrix} U_y^1 \\ U_z^1 \end{pmatrix}, \tag{13}$$

$$\left. \begin{aligned} K_{yy}^1 &= \tfrac{1}{2}\left(K_{rr}^1 + 2K_{r\theta}^1 + K_{\theta\theta}^1\right) \\ K_{zy}^1 &= 2K_{yz}^1 = K_{rz}^1 + K_{\theta z}^1 \end{aligned} \right\}. \tag{14}$$

Based on the above considerations, under the assumptions Eq. (9) and Eq. (10), rigorous equilibrium of forces in the cylindrical coordinate system in Eq. (5) can be approximated as follow:

$$\begin{pmatrix} K_{yy(2)}^1 & K_{yz(2)}^1 \\ K_{zy(2)}^1 & K_{zz(2)}^1 \end{pmatrix} \begin{pmatrix} U_{x(2)}^1 \\ U_{z(2)}^1 \end{pmatrix} = \begin{pmatrix} K_{yy(1)}^1 & K_{yz(1)}^1 \\ K_{zy(1)}^1 & K_{zz(1)}^1 \end{pmatrix} \begin{pmatrix} U_{x(1)}^{1,e} - U_{x(2)}^1 \\ U_{z(1)}^{1,e} - U_{z(2)}^1 \end{pmatrix}. \tag{15}$$

Incidentally, the elastic foundation model for longitudinal bending deformation is expressed as follow:

$$E_{(2)} I \frac{\partial}{\partial z^4} u_{x(2)} = K_B \left(u_{x(1)}^e - u_{x(2)} \right). \tag{16}$$

Where $E_{(2)}$ is Young's modulus of the structure and I is the moment of inertia of area. K_B is the Winkler foundation constant. From comparison of Eq. (15) and Eq. (16), the following equation have to be satisfied in order for the elastic foundation model to be validated:

$$K_{rz}^1 + K_{\theta z}^1 \cong 0 \left(K_{yz}^1 = K_{zy}^1 = 0 \right). \tag{17}$$

4   NUMERICAL RESULT AND DISCUSSION

Input data for numerical calculations are given in Table 1 and Table 2. The buried pipeline in soft ground is assumed because the damaged structures are often constructed in such a ground. The ratio of the Young's modulus of the ground and the structure is 0.01:1. Frequency of the incident shear wave is 1.00Hz and 5.00Hz.

Figure 3 shows comparison of absolute value of ground stiffness components. Vertical axis K^* is the normalized stiffness that is divided by $\mu_{(1)}/R$. Except when the frequency is 5.00Hz and L_z is short, non-diagonal component $K_{rz(1)}^1 + K_{\theta z(1)}^1 \left(K_{yz(1)}^1 \text{ or } 2K_{zy(1)}^1 \right)$ is smaller than diagonal components $K_{rr(1)}^1 + 2K_{r\theta(1)}^1 + K_{\theta\theta(1)}^1 \left(2K_{yy(1)}^1 \right)$ and $K_{zz(1)}^1$. Therefore, non-diagonal

Table 1. Ground properties.	
The ground parameter	Value
Young's modulus	0.2GPa
Poisson's ratio	0.400
Density	2000kg/m³

Table 2. Structure properties.	
Structure parameter	Value
Young's modulus	20.0GPa
Poisson's ratio	0.200
Density	2000kg/m³
Mean radius	2.63m
thickness	0.250m

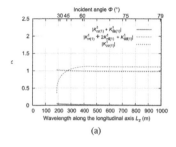 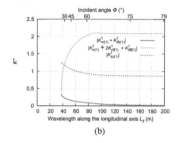

(a)   (b)

Figure 3.   Comparison of absolute value of ground stiffness components: (a) frequency is 1.00Hz and (b) frequency is 5.00Hz.

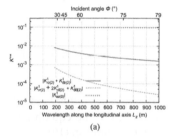 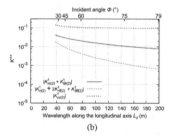

(a)   (b)

Figure 4.   Comparison of absolute value of structure stiffness components: (a) frequency is 1.00Hz and (b) frequency is 5.00Hz.

components of the ground stiffness can be ignored, and the surrounding ground can be treated as elastic foundation springs in axial and axis orthogonal directions

Figure 4 shows comparison of absolute value of structure stiffness components. Vertical axis K^{**} is the normalized stiffness that is divided by $\mu_{(2)}/R$. Non-diagonal component $K^1_{rz(2)} + K^1_{\theta z(2)} \left(K^1_{yz(2)} \text{ or } 2K^1_{zy(2)} \right)$ is larger than diagonal components $K^1_{rr(2)} + 2K^1_{r\theta(2)} + K^1_{\theta\theta(2)} + \left(K^1_{yy(2)} \right)$. Therefore, unlike the ground stiffness, non-diagonal components of the structure stiffness can not be ignored.

Figure 5 shows comparison of axial orthogonal displacement on the middle surface of the structure. Amplitude of axial orthogonal displacement by incident S_1 wave is assumed to be 1.00. The incident angle ϕ for each frequency is the value when the axial strain becomes maximum. $u^e_{y(1)}$ is the excavated ground displacement at the ground-structure interface. $u_{y(s_1)}$ is the rigorous solution calculated by Eq. (5). $u_{y(s'_1)}$ is the approximated solution calculated by Eq. (15). $u_{y(b_1)}$ is the solution calculated by Eq. (16). $u_{y(s_1)}, u_{y(s'_1)}$ and $u_{y(b_1)}$ show relatively good agreement for both (a) and (b).

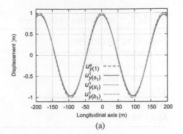
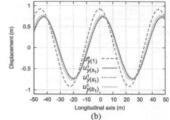

(a)   (b)

Figure 5. Comparison of axis orthogonal displacement on the middle surface of the structure: (a) frequency is 1.00Hz ($\phi = 4°$) and (b) frequency is 5.00Hz ($\phi = 33°$).

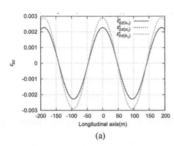
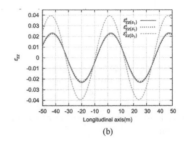

(a)   (b)

Figure 6. Comparison of axis strain on the middle surface of the structure: (a) frequency is 1.00Hz ($\phi = 4°$) and (b) frequency is 5.00Hz ($\phi = 33°$).

Figure 6 shows comparison of axial strain on the middle surface of the structure. Amplitude of axial orthogonal displacement by incident shear wave is assumed to be 1.00 $\varepsilon_{zz(s_1)}$ and $\varepsilon_{zz(s_1')}$ are calculated by the $\partial u_{z(2)}/\partial z$, and $\varepsilon_{zz(b_1)}$ is calculated by $-R\partial^2 u_{y(2)}/\partial z^2$. $\varepsilon_{zz(s_1)}$ and $\varepsilon_{zz(s_1')}$ show good agreement. On the other hand, $\varepsilon_{zz(s_1)}$ and $\varepsilon_{zz(b_1)}$ does not show good agreement. $\varepsilon_{zz(b_1)}$ overestimate the axial strain. This is because beam theory ignores the equilibrium of force in the axial direction as can be seen from a comparison of Eq. (15) and Eq. (16).

5 CONCLUSION

In this paper, the elastic foundation model to evaluate the longitudinal bending deformation of cylindrical underground structure is reconsidered. The solutions derived by elastic foundation model is compared to that derived by the solutions considering rigorous equilibrium of forces at the ground-structure interface. The results show that the seismic deformation method can evaluate the longitudinal bending deformation of cylindrical underground structure when the structure is treated as an elastic shell, with surrounding ground treated as elastic foundation springs in axial and axis orthogonal directions. On the other hand, conventional seismic deformation method that assumes the structure as a beam overestimates the maximum axial strain of the structure because beam theory ignores the equilibrium of force in the axial direction.

REFERENCES

Flügge, W. 1973. *Stress in Shells*. Berlin and New York: Springer-Verlag.
Hashash, Y., Hook, J., Schmidt, B., et al. 2001. Seismic design and analysis of underground structures. *Tunnelling and Underground Space Technology* 16, 247–293.
Jiang, T., Kuribayashi, E. 2000. Improvement and suggestion on seismic design method of buried pipeline with flexible joints. *Structural Eng./Earthquake Eng., JSCE* 17, 33–38.
Mow, C., Pao, Y. 1971. *The Diffraction of Elastic Waves and Dynamic Stress Concentrations*. Rand Corporation.

*Tunnels and Underground Cities: Engineering and Innovation meet Archaeology,
Architecture and Art, Volume 6: Innovation in underground engineering,
materials and equipment - Part 2 – Peila, Viggiani & Celestino (Eds)
© 2020 Taylor & Francis Group, London, ISBN 978-0-367-46871-2*

Site assembly technologies of super-large slurry TBM

L. Ye, D.J. Wang, K.W. He, G.W. Gu, C. Ye & J. Wang
China Railway Engineering Equipment Group Co., Ltd., Zhengzhou, Henan, China

ABSTRACT: Su'ai tunnel, as an important subsea tunnel in Shantou City, is a two-bore and two-way highway tunnel with six lanes which employed a super-large slurry TBM with the diameter of 15.03m. Due to the fact that the number of shield blocks and the weight of each block is larger than regular diameter TBMs, there are many technical difficulties during the site assembly. This paper presents the assembly process of super-large slurry TBM in launching shaft and discusses key technologies and matters worthy of attention in the whole process.

1 INTRODUCTION

Su'ai tunnel is the first two-bore and two-way subsea tunnel with the diameter of 15m in China, among which the shield section is 3km. The shield machine launches from the backfill cofferdam area in South Bank and passes through the Su'ai Bay and will be received at the east of Huaqiao Park in North Bank. It is designed a two-way tunnel with six lanes and adopts two 15m-class super-large slurry TBM to realize simultaneous advance (Feng 2013). The distance between East and West lines is 23.3m-29.7m. The machine will go through the complex ground conditions such as soft ground with high water pressure and upper-soft lower-hard ground which all will increase construction difficulty. The quality and efficiency of assembly before tunneling is directly related to the safety tunneling and the construction period of the whole project.

The CREG shield machine is applied to the West line, whose cutterhead is equipped with atmospheric cutter change device and possesses the telescopic and swinging function. The machine is 168m long including the main machine of 15.6m, 1 connection bridge, 5 gantries and 1 trailer platform with the total weight of 4100t. There are many overweight and super-large parts, which bring serious challenges to the site assembly. How to complete the assembly work safely and efficiently with high quality is a difficult problem which is worth discussing and studying.

Li Shishi et al. studied the assembly technology from the aspects of lifting, welding and disassembly for the example of a 15.88m EPB TBM. In addition, the Robbins EPB TBM applied to the Mexico metro tunnel project reduces the 70–80% construction period by the method of Onsite First Time Assembly(OFTA). Sun Liguang proposed that the assembly of two-bore small-diameter EPB TBM could adopt the single-shaft underground translation construction to solve the site limitation problem in two-bore tunnel project. The above researches focus on the launching scheme of EPB TBM, while the site assembly of super-large diameter slurry TBM has not been studied in detail.

This paper takes the site assembly scheme of the largest diameter shield machine in China as an example, describes in detail the process of underground assembly at the construction site, conducts in-depth research on the assembly technology of key components, and provides technical reference for the later construction site assembly of super-large shield machine.

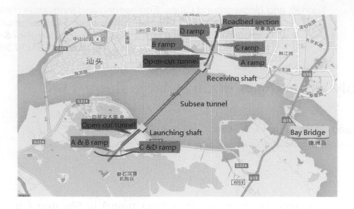

Figure 1. Layout of Shantou Su'ai tunnel.

2 ASSEMBLY CONDITION

2.1 *Site condition*

2.1.1 *Ground layout*
This initial assembly adopts a main shaft (top clear space is 18m x 18m, bottom clear space is 20m x 22m) and an auxiliary shaft (top clear space is 12.5m x 14m), among which the gantry 1, shield body, main drive, cutterhead and segment erector will be lifted to the main shaft and the connection bridge and other gantries will be lifted to the auxiliary shaft. Only need to move the gantry to complete the assembly of the whole machine. The two working shafts work simultaneously, which greatly shortens the assembling time.

This launching scheme saves space and reduces cost compared with Nanjing Yangtze tunnel which has one launching shaft and three shafts for lifting the gantries(Liu 2009); Compared with the single-shaft launching of Wuhan Yangtze tunnel(Hu & Li 2008), this launching scheme increases the assembly shaft which could reduce the assembly time; Compared with the single-shaft launching scheme of the long main tunnel and tailrace tunnel in Chongqing drainage tunnel project(Tang 2006), this scheme can reduce the moving distance of the main machine and reduce the construction risk.

It should be noted that due to the site limitation of the main shaft reinforcement area, the main machine parts must be arranged into the site in proper order to avoid the problem of insufficient space for large parts.

2.1.2 *Underground launching platform*
As shown in Figure 3, due to that the center line of the launching section has a slope of 3% and is on the horizontal curve, the circular arc launching platform is poured under the main shaft at a slope of 1.5%. There exists a gap about 1 meter between the launching platform in longitudinal and horizontal direction, leaving room for welding between the middle shield and the tail shield. On both sides of the middle launching platform, the placed grooves for lifting hydraulic cylinder should be reserved, which is convenient for the cylinder to adjust the attitude of the bottom shield body.

More than 100 meters tunnel for connecting gantries is reserved at the back of the main shaft. Sinking design is adopted at the front of the tunnel and circular arc launching platform is poured at a slope of 3% between the center axis and the middle line of the portal. The rear part is 4m higher than front part, and it pre-embedded steel plates every 8.4m and then laid steel rails.

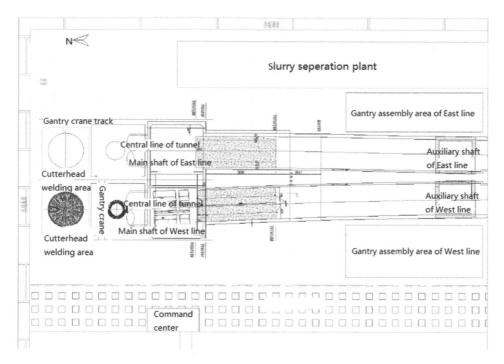

Figure 2. Launching construction site plan.

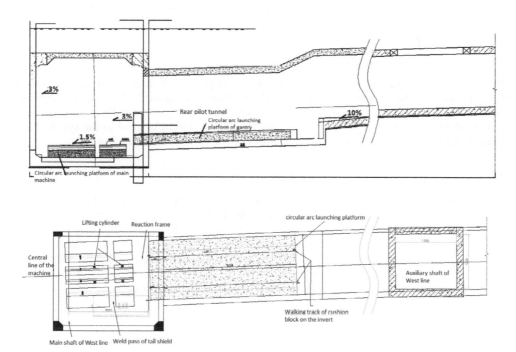

Figure 3. Underground launching platform.

Table 1. Actual and measured wear data of cutting tools.

Equipment	Parameters
Lifting tools	325T+325Tgantry crane, 400Tcrawler crane, 260Tcrawler crane, 25Ttruck crane
Wire rope	43mm ring structure wire rope, 90mm ring structure wire rope, 60mm ring structure wire rope, 120mm ring structure wire rope
Shackle	110T shackle, 150T shackle, 200T shackle, 400T shackle, 50T/20T shackle

2.2 Assembly tool

2.2.1 Lifting tools
The on-site lifting equipment is shown in Table 1, and the selection is mainly based on the single heaviest part (the weight of cutterhead 540t) and its size.

3 ASSEMBLY PROCESS

The site assembly adopts multi-working faces synchronous operation. Due to the limitations of the shaft, site assembly is divided into gantry assembly, main machine assembly and commissioning of whole machine.

The overall assembly process of shield machine is shown in Figure 4. The cutterhead assembly welding and the gantry 1 assembly are carried out simultaneously on the ground. Subsequently, the main machine and gantry 1 went down from the front main shaft, while the connection bridge and rear gantries went down from the auxiliary shaft. After the whole machine is assembled, it will start the system commissioning.

3.1 Lifting of gantry 1 in main shaft
After the front and rear assembly is completed on the ground, it adopts separated lifting for gantry 1, and lays the temporary walking track, as shown in Figure 5. After the underground assembly is completed, two winches are used to drag the gantry 1 to the fixed position in the shaft, as shown in Figure 6. When pulling back, attention should be paid to the synchronicity of two winches to prevent the left and right wheel of the gantry from jumping rail.

3.2 Lifting of rear gantries in auxiliary shaft
The rear gantries are lifted by crawler crane after the assembly is completed on the ground. The connection bridge is divided into three sections for assembly. Before going down, it should weld the launching pedestal at the bottom of each section of connection bridge. The process of going down and moving is shown in Figure 7 and it also use the winch with the pulley block to increase traction. Gantry 2–5 adopt the same scheme to assemble successively. All you need to do is to weld the sliding block at the bottom of the gantries, and apply the lubrication oil on the rail when moving forward to reduce sliding resistance.

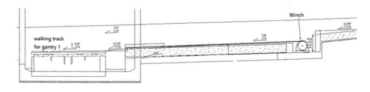

Figure 4. Temporary walking track of gantry 1.

Figure 5. The lifting and connection of gantry 1.

Figure 6. Lifting and moving of connection bridge and gantries.

When the main machine is well located, the gantry 1, the connection bridge and the gantry 2 will be pulled forward by the winch to connect with the main machine through towing cylinder. And then, gantry 3–5 are connected in turn.

3.3 *Lifting of shield body and main drive*

3.3.1 *Lifting of bottom shield body*
The shield body is divided into 10 parts in the circumferential direction. When the rear gantries are lowered to the auxiliary shaft, the main machine shield body also starts to go down. The temporary walking track of gantry 1 needs to be removed in advance. The gantry crane is used to place the Part 6 on the track of the launching platform. The Center Line of the Part 6 are positioned by the total station instrument based on the center line of the tunnel. At the same time, the adjustment is made with the pre-installed jack at the bottom. Then assemble the bottom shield body of Part 5, 7, 4 and 8 successively.

Figure 7. The assemble of bottom shield body.

Figure 8. Lifting of main drive.

3.3.2 *Lifting of main drive*

After the overall installation of the main drive is completed on the ground, it should be turned over and lowered to the shaft. The main drive is equipped with a specialized auxiliary structure for turning over and lifting. Pay attention to the use of hawser cable during the installation to avoid collision and damage of the seal, as shown in Figure 9.

The telescopic cylinders should be installed after the primary connection between the main drive and shield and then start to assemble the top shield body. It should be noted that the platform, pipe and other auxiliary equipment inside the main machine must be installed and positioned before the installation of top part, so as to avoid difficult lifting of accessories after the top sealing.

3.4 *Cutterhead assembly*

3.4.1 *Cutterhead welding*

Before the cutterhead goes down, it must ensure that the cutterhead welding is completed and the welding quality is completely qualified to reach the designed flatness and roundness. During welding, it shall follow the principle of first welding the longitudinal weld and then the transverse weld, first welding the inside weld and then the outside weld, and first welding the main cutterhead arm and then the auxiliary cutterhead arm. In addition, when welding, make sure that the flaw detection are conducted one by one, and the cutterhead can only be lifted after the final inspection is qualified.

3.4.2 *Cutterhead welding*

Before the lifting of cutterhead, it must make sure that the bolts of the main drive are completely tightened, and the zero position of the main drive is consistent with the lifting angle of the cutterhead, so that the bolt and locating pin can be connected smoothly.

When lifting the cutterhead, use the gantry crane to horizontally lift the main and auxiliary lifting lugs of the cutterhead and lower the cutterhead from the welding auxiliary structure. Then, separately lift the main lifting lug on the top of cutterhead to turn over the cutterhead until it is completely upright and connected with the main drive, as shown in Figure 11. And when the cutterhead positioning is completed, tighten the bolts.

Figure 9. Measurement before cutterhead welding and positioning of anti-deformation auxiliary structure.

Figure 10.   Turning over and lifting of cutterhead.

3.5   *Moving forward the main machine*

As the main shaft is small, the main machine should be moved forward to reserve space for segment erector and tail shield. Before moving forward, it should be installed the temporary reaction support structure and temporary pump station. Then, two sets of thrust cylinders in the lower part of the shield body are used to push the main machine forward along the track for about 3 meters. Subsequently, remove the temporary reaction support structure and install the bottom half of the launching reaction frame.

3.6   *Lifting of tail shield and segment erector*

Due to the space limitation of shaft, the assembly of segment erector and tail shield must be carried out step by step, which not only could shorten the assembly time, but also minimize the diameter of shaft and reduce the engineering cost.

3.6.1   *Lifting of tail shield bottom part*
After the tail shield bottom part is lowered into the shaft, the gantry crane is used to adjust so that its center line is parallel to the main machine axis, the front of the tail shield is level with the rear part of the middle shield. Meanwhile, the total station is used to retest the circumference circle of the tail shield bottom part and fine-tune it by using multiple jacks prepaid at the bottom, as shown in Figure 12. After the positioning of the bottom part is qualified, weld the shape steel to support at the bottom of the bottom part.

3.6.2   *Lifting of segment erector*
The segment erector is lowered to the shaft after the positioning of tail shield bottom part is completed. Before that, the segment erector including main beam and rotating ring has been assembled by gantry crane on the ground. Then the segment erector is lifted and connected with the main machine by using of gantry crane. During lifting, pay attention to the using of hawser cable for traction. After the segment erector is bolted with the shield body, the lifting hook can be released.

Figure 11.   Positioning of tail shield bottom part.

Figure 12.   Lifting of segment erector.

Figure 13.   Assembly of tail shield top part.

3.6.3 *Lifting of tail shield other parts*

The other parts of tail shield are lifted according to the order of left part, right part and top part after the installation of segment erector. After the overall installation of tail shield, the whole station is used to measure the tail shield in full circle. For points that do not meet the design requirements, the jacks are used on the inside or outside of the tail shield for fine-tuning until the design requirements are met.

It is required to weld the segmented welds symmetrically. In the welding process, the flaw detection is performed and the unqualified position is repaired. Finally, the internal support of the tail shield will be cut off and then the welding work of 5 tail brushes can be carried out.

4   COMMISSIONING AND LAUNCHING

4.1   *Commissioning*

The cables and pipes should be immediately connected and fixed in place according to the label, when completing the assembly of whole machine. Hydraulic pipes should be cleaned and sealed in advance. Calibrate the point to point control and feedback after the system is energized and start the commissioning of each subsystem in the order of no-load first and then load.

4.2   *Partial segment erection*

After the commissioning of the whole machine, the erection of partial segment and the remove of tunnel portal are carried out. The first ring of partial segment is directly welded with the reaction frame, and the second ring is erected with staggered joints. Transition plate is used to weld between segments of each ring, and fix between the rings to ensure the stability of thrust for launching.

4.3 Portal seal installation

In order to prevent the slurry losing from the shield body and the portal gap, two steel rings of portal are set to install the seal. The first steel ring is the steel brush sealing ring. When launching, the steel wire brush is applied with the grease which will be continuously injected through the grease port.

4.4 Launching

Before launching, it is necessary to remove the reinforced portal at the front of the cutterhead. It is used pneumatic pick to remove the portal from top to bottom and from the middle to both sides by manual. Once the portal is removed and the commissioning is completed, slurry circulation system is carried out to form stable pressure in excavation chamber. According to the characteristics of ground conditions, tunneling parameters of shield machine are set for steady tunneling.

5 CONCLUSIONS

The process of site assembly and commissioning of super-large shield machine is complex, which must be fully prepared, fully deployed and well planned. This paper takes the site assembly process of super-large slurry TBM in Su'ai Project as an example, focusing on the technical summary of the launching scheme, and elaborating on the key technologies such as lifting scheme, cutterhead welding, commissioning and launching during the assembly process.

REFERENCES

Feng, H.H. et al. 2013. Simple program analysis of shield tunnel in Shantou Sue Lumiere passage, *Shield Equipment & Project* 2013(09): 70–72.
Hu, B. & Li, Y.J. 2008. Shield machine launching technology of Wuhan Yangtze River Tunnel, *Soil Eng. and Foundation* 22(2):21–25.
Li, S.S. et al. 2014. Site assembly technologies of large-diameter shield machine for urban rail transport, *Journal of South China University of Technology* 2014(11): 14–22.
Liu, Z.X. 2009. Assembly technology of large-diameter TBM, *Railway Construction Technology* 2009(4): 60–64.
Shao, D.G. 2010. Mexico's urban challenge, *Tunnels & Tunnelling International* 2010(3): 28–30.
Sun, L.G. 2016. Shield machine launching technology under high-voltage cable of metro station, *Henan Building Materials* 2016(3): 277–280.
Tang, Z. 2006. Erection, launching and arrival technology of slurry shield machine in deep shafts, *Tunnel Construction* 26(4): 37–39.

A BIM-based emergency analysis method and application for utility tunnel

G. Yu, Z. Mao, M. Hu & L. Shi
SHU-UTS SILC Business School, Shanghai University, Shanghai, China

ABSTRACT: In recent years, the decision-making of emergency response tends to be complicated due to the expansion of the construction scale and the increase of internal complexity of the utility tunnel. However, the support provided by existing research to emergency managers remains on the assignment of on-duty personnel and the matching of emergency plans. The design or the study of utility tunnel, as a closed space underground, needs more information to guide the emergency response, such as the location of related emergency control equipment, and the planning of emergency repair path. Therefore, this paper proposes an emergency analysis method based on BIM (BIM_EAM). A modified geometry network model (MGNM) is constructed based on BIM, and an emergency analysis algorithm is proposed to guide the emergency response in a dynamic manner. In this paper, the BIM_EAM method is integrated into the AI-steered operation and maintenance platform of the utility tunnel, to verify its effectiveness.

1 INTRODUCTION

The utility tunnel, as an important infrastructure and lifeline of urban normal function, is a comprehensive corridor for underground urban pipelines, that is, to construct a tunnel space under the city, which integrates all kinds of engineering pipelines such as electric power, communication, gas, heating, water supply and drainage, and has special access ports and monitoring system (Chang and Lin 2014). With the expansion of the construction scale of utility tunnel, the hidden danger of underground space is increasing, which leads to the difficulty of emergency management such as rescue and relief work, accident disposal and so on. Meanwhile, due to the complexity of the concentrated laying of public pipelines and equipment, the uncertainties of emergencies are also increasing. Previous researches mainly focused on the planning, design and construction stages of the utility tunnel (Canto-Perello and Curiel-Esparza 2003), such as the risk and compatibility of laying different pipelines inside, but very few researches on emergency.

With more and more operation and maintenance management systems being developed and utilized in large-scale infrastructure in recent years, how to provide an effective emergency analysis method on the system has also attracted more and more attention (Tang and Ren 2012). Therefore, some emergency analysis methods have been proposed to assist emergency team to make decisions. However, most of the functions provided by these methods remain on the assignment of on-duty personnel and the matching of emergency plans (Lee et al. 2018). The emergency information provided by these methods is still very limited.

Research shows that in the first few hours after a disaster, disaster management plays the most important role in reducing post-disaster losses (Diehl and van der Heide 2005). As an underground enclosed space, the internal environment of the utility tunnel is uncertain after the disaster, which will lead to the emergency response personnel may be exposed to the hazardous environment (Kwan and Lee 2005). Meanwhile, complex pipes and equipment are installed inside the utility tunnel. However, the emergency response personnel often encounter unavoidable events such as visual interference, blocked area, which makes it difficult for them to quickly locate related emergency control equipment under the circumstance of disaster spreading (Jeon et al. 2011). Therefore, strengthening situational awareness is essential to minimize the damage caused by disasters. It is necessary to improve the safety of emergency response and reduce post-disaster losses by letting emergency response personnel know the location of relevant emergency control equipment and planned path before entering the utility tunnel. However, the existing emergency response methods are failed to provide such information, which makes the implementation of emergency response difficult and complicated.

To provide emergency management personnel with the above information, part of the research based on CAD puts forward some emergency analysis methods and achieved the emergency path planning in the emergency situation (Inoue et al. 2008). However, CAD lacks detailed geometric, spatial and semantic information within the infrastructure (Canto-Perello, Curiel-Esparza, and Calvo 2013). This makes the calculation result lack accuracy. In addition, the relevant control equipment and other information are unable to be calculated in such methods.

BIM, as a unified information exchange platform of AEC industry, contains the detailed information of facilities and equipment in the infrastructure, which effectively makes up for the deficiency of CAD in the expression of indoor information (Volk, Stengel, and Schultmann 2014). Although some researchers have established the operation and maintenance platform of utility tunnel by using BIM, few researchers have developed the emergency analysis method of utility tunnel by using BIM. Therefore, we propose an emergency analysis method based on BIM (BIM_EAM). BIM_EAM analyzes and calculates the related emergency control equipment and makes the path planning, by using BIM as the information source. The rest of this paper is organized as follows. Section 2 elaborates on the motivation for this study and Section 3 introduces the proposed method. In Section 4, the proposed method is implemented on the intelligent operation and maintenance platform of utility tunnel to further verify its effectiveness. The 5 section draws a conclusion.

2 MOTIVATION

Up to now, most of the emergency analysis methods of utility tunnel remains on the assignment of on-duty personnel and the matching of emergency plans. These methods ignore the information needed by the first responder and the emergency team under emergency situation. This results in the first responder having to spend a lot of time assessing the related control equipment and making corresponding plans upon arrival at the site.

Some researches attempt to develop emergency analysis methods in combination with CAD, to help emergency team and first responders make quick decisions (Inoue et al. 2008). These researches parse emergency information from CAD and abstract it as a geometric network model (GNM) in advance, and then provide emergency analysis function for emergency team in conjunction with the monitoring data of operation and maintenance system (Liu and Zlatanova 2011). For example, the research developed a GNM-based navigation system that can be accessed by portable information terminals, analyzing the user's path to the nearest exit by combining existing GNM with the user's current location (Inoue et al. 2008). However, CAD is unable to provide detailed three-dimensional information and partial semantics of facilities and equipment (Canto-Perello et al. 2013). There are still challenges in applying the above emergency analysis method to the utility tunnel, considering the characteristics of three dimensional distribution and complexity of the equipment and facilities in the utility tunnel.

In recent years, as a new technology in the field of AEC, building information model (BIM) has been widely implemented in the whole life cycle of buildings (Volk et al. 2014). Unlike CAD, BIM, which contains detailed geometric and semantic information of facilities in buildings, is considered to be the most ideal information model for representing and managing building information throughout the life cycle (Wang 2012). As noted in the research, BIM follows the generative modeling approach and focuses on the architectural environment, which makes it usually composed of volumetric and parametric primitives (Nagel, Stadler, and Kolbe 2009).

Therefore, using BIM to replace the original CAD as the data source of the emergency analysis method can effectively make up for the lack of three-dimensional information expression in the existing methods. However, emergency managers need to know more information under emergency situation of utility tunnel, such as related emergency control equipment, which was not mentioned in previous researches. To complete the analysis of this information, we need to make further improvements to the previous emergency analysis method. In this paper, the BIM is used as the data source and a model is constructed to express the emergency information of the utility tunnel. Based on the accurate expression, this method further provides an emergency analysis algorithm to calculate the necessary information we proposed.

3 PROPOSED BIM_EAM FOR EMERGENCY ANALYSIS OF UTILITY TUNNEL

First, based on the traditional geometry network model (GNM), BIM_EAM proposes a modified GNM model (MGNM) to store, calculate, and express the emergency control information of the utility tunnel. The model is defined as follows:

$$MGNM = (V, E, I) \tag{1}$$

$V = \{v_1, v_2, \cdots, v_i\}$ are MGNM node sets. The nodes in MGNM are divided into three types: escape node, sensor node and control node. The escape node is the door and space inside the utility tunnel. The control node is the node that the first responder needs to operate in the emergency response process, such as the upstream and downstream valves of the pipeline, the firefighting equipment inside the facility and so on. The sensor node is the sensor inside the utility tunnel. $E = \{e_1, e_2, \cdots, e_i\}$ are MGNM edge sets and the specific composition of it are shown in Table 1. The edges in MGNM are divided into two categories: topological relations and indoor paths. The "indoor path" represents the accessible paths within the utility tunnel, such as corridor, door to space, corridor to door. The "topological relation" represents the topological relationship between different control nodes and between sensors and control nodes. "I" is a set of MGNM semantic information, which is further divided into I_v and I_e, representing the semantics of nodes and edges respectively. In addition to defining the emergency topology of the utility tunnel, MGNM also constructs the corresponding semantic model, as shown in Figure 1A). Topological relations and semantic information enable MGNM to have an emergency analysis capability of utility tunnel. The BIM_EAM method uses the rich built-in functions in the Revit API as shown in Figure1B), to parse the

Table 1. The nodes and edges of MGNM.

MGNM elements	TYPE	NAME
Node	Escape node	Space
		Door
	Control node	Pipe
		Valve
		Fan control box
		Booster pump control
	Sensor	Sensor devices
Edge	Horizontal path	Corridor
		Corridor-control
		Corridor-door
		Space-door
	Vertical path	stair
	Topological relations	Sensor-control node
		Control node -control node

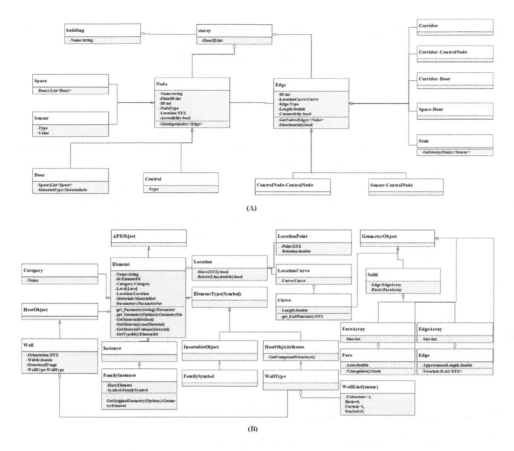

Figure 1. A) MGNM semantic model. B) UML diagram of Revit API.

BIM model and dynamically generates the node set, edge set, and semantic set needed in the description of emergency topological relations. For example, the Revit API is used to call the geometric information of the walls and floors to calculate the path of the corridor.

Using the constructed MGNM model, this method provides a utility tunnel emergency analysis algorithm (UEAA) for quickly diagnosing related emergency control equipment and calculating the recommended emergency repair path in the emergency disposal process.

The algorithm flow is shown below. The core idea of the algorithm is to access all control nodes on MGNM via the "topological relation" edge by taking the "sensor" node which receives the outlier as the starting point. Then, starting from this control node, the path from the control node to the nearest entrance is obtained by "horizontal path" and "vertical path" on MGNM. This allows the repair personnel to access all relevant control nodes in the shortest possible time from the most suitable entrance.

4 THE APPLICATION OF THE PROPOSED EMERGENCY ANALYSIS METHOD

In this paper, Shanghai Lingang New Town North Island West Road utility tunnel (hereinafter referred to as LN) as an application object to verify the proposed emergency analysis method. LN is in the main urban area of Lingang New Town, which is 973 meters in length. Double cabin arrangement is adopted inside the utility tunnel. The main pipelines inside the LN include power lines, information pipelines, gas pipelines and drainage pipelines. At the same time, the utility tunnel is equipped with a variety of emergency response related equipment, such as pipeline valves, fan control box, pump control box and so on. This paper integrates the proposed method into the intelligent operation and maintenance platform of LN and verifies the effectiveness of this method through an emergency scenario.

4.1 Structure of the system

Intelligent operation and maintenance platform of utility tunnel makes use of advanced sensor technology, monitoring technology, network technology, intelligent method, computing technology to comprehensively perceive the pipeline, equipment, environment and operation, and to comprehensively monitor and control the utility tunnel. We use data fusion and big data mining technology to integrate and excavate the relevant operation information, to guarantee the safety and efficient utilization of the utility tunnel services and build a new comprehensive maintenance mechanism relying on Internet + technology. The platform architecture, such as figure 2, is composed of six layers: support layer, data layer, intermediate service layer, Business Services Layer, access layer and user layer.

The developed emergency analysis function interface is shown in Figure 3. In the next section, we further propose a specific emergency scenario to illustrate the effectiveness of BIM_EAM.

4.2 Scenario

In this scenario, there was a fire inside the LN fifth partition. The corresponding sensor will receive the outlier at the moment. Based on the BIM_EAM, we get the MGNM model of

---

Algorithm1.   utility tunnel emergency analysis algorithm.

Input: sensor outliers
Output: related control equipment, optimal rush repair path
1.   run the breadth search priority algorithm form the sensor node that receives outlier to find all the control nodes
2.   Get v_n,$L(v)$, $L(e)$,node-edge connectivity, entrance-door list $L1$ from MGNM
3.   **For each** $v_j \in$ control nodes, $s_i \in$list $L1$
4.   Computing $D_H = (s_i, v_j)$
5.   Record $D_H(s_1, v_1)$, $D_H(s_1, v_2)$,..., $D_H(s_i, v_j)$
6.   **End For**
7.   Computing $\min\{D_H(s_i, v_j)|\ v_j \in$Allcontrols, $s_i \in$list $L1\}=D_H(s, v_{hj})$
8.   Set s=v_j, P_0=$D_H(s, v_{hj})$
9.   Computing $T_H = (s, v_i)$ via optimal path algorithm,i=1,2,...,k
10.   Record $T_H(s, v_1)$, $T_H(s, v_2)$,..., $T_H(s, v_i)$
11.   **End If**
12.   Loop1:
13.   Computing $\min\{T_H(s, v_i)|\ v_i \in$list $L1\}=T_H(s, v_{hj})$
14.   Set $u_0 = s, u_{k+1} = t$,record $u_1 = v_{h_1}, u_2 = v_{h_2}\ldots u_j = v_{h_j}$
15.   **If** j<k **then**
16.   Set listV=listV\$\{v_{h_j}\}$ j=j+1,goto loop1
17.   **End If**
18.   Loop2:
19.   Computing $P_w = P_{u_w u_{w+1}}$ via optimal path algorithm
20.   Record $P_{v_1 u_1}$, $P_{u_1 u_2}\ldots P_{u_w u_{w+1}}$, get $P = P_0, P_1 \cdots P_w$
21.   **If** w<k **then**
22.   Set w=w+1 goto loop2
23.   **End If**
24.   **For each** $v \in L1$
25. $P(v_i) \leftarrow \min(P(v))$
26.   **End for**
27.   **Return** $P = P_0, P_1 \cdots P_w$, control nodes

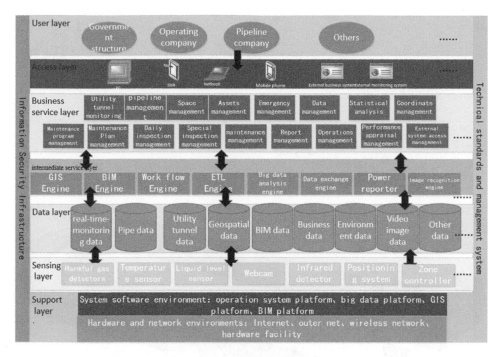

Figure 2. platform's architecture.

Figure 3. Emergency analysis function interface.

utility tunnel fifth partition by BIM, as shown in figure 4A). The connectivity of the MGNM nodes and edges will be updated according to the outlier of the sensor, as shown in Figure 4B.

Then, we use UEAA algorithm to traverse MGNM to find all the relevant control nodes with the sensor as the starting node. As shown in Figure 5, it can be calculated that the equipment to be operated in this case are fire door, firefighting equipment and so on, according to the outlier of this sensor node. Next, update the connectivity of the inner edge of the model combined with the current indoor environment. We access all edge sets and point sets and find

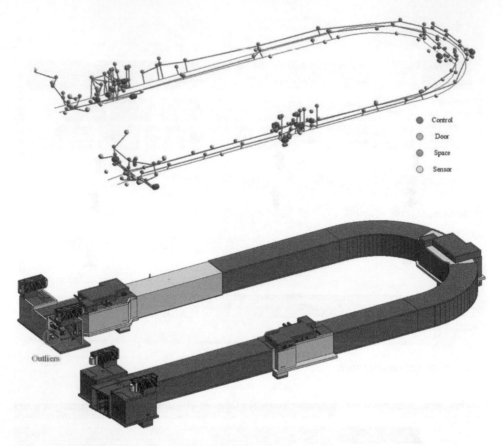

Figure 4.   A) MGNM of LN fifth partition B) Location of sensor node that receives outlier.

Figure 5.   Illustrative example of BIM_EAM.

the best repair path through the proposed UEAA algorithm, on the updated MGNM. In this case, the path and equipment in the fire area are not allowed to be accessed. Therefore, the system guides the repair personnel to operate the exhaust device and the fire extinguishing equipment first. BIM_EAM will automatically plan the optimal path from the entrance to the exhaust and fire extinguishing equipment for the repair personnel, wait until the data returned by the sensor to return to the normal range, and then further plan the path after entering. This information will be provided to the first responder visually, so that the first responder can simulate the operation before entering the utility tunnel, effectively reducing the emergency risk, and speeding up the process of emergency decision-making.

5 CONCLUSION

In this paper, a BIM-based emergency analysis method (BIM_EAM) for utility tunnel is proposed. A modified geometry network model (MGNM) is established by using BIM, through which important emergency information such as the location of relevant control equipment and planned emergency repair path are calculated. BIM_EAM has been applied to the intelligent operation and maintenance platform of the utility tunnel project of Beidao West Road in Lingang New City, Shanghai.

The main contributions of this paper are as follows: (1) A new emergency analysis method is proposed, which can be used to discover the internal control equipment of the utility tunnel and make plans for emergency team. (2) The emergency analysis method is implemented on the intelligent operation and maintenance platform of the utility tunnel, and the effectiveness of the method is verified by the actual emergency scenario.

Finally, to further improve the emergency analysis function, future research will be committed to the combination of MGNM and outdoor environment for a more comprehensive emergency analysis ability of utility tunnel.

REFERENCES

Canto-Perello, J. and J. Curiel-Esparza. 2003. "Risks and Potential Hazards in Utility Tunnels for Urban Areas." *Proceedings of the Institution of Civil Engineers: Municipal Engineer*.

Canto Perello, Julian, Jorge Curiel-Esparza, and Vicente Calvo. 2013. "Criticality and Threat Analysis on Utility Tunnels for Planning Security Policies of Utilities in Urban Underground Space." *Expert Systems with Applications*.

Chang, Jia-Ruey and Ho-Szu Lin. 2014. "Preliminary Study on Application of Building Information Modeling to Underground Pipeline Management." in *Recent Developments in Evaluation of Pavements and Paving Materials*.

Diehl, Stefan and Jene van der Heide. 2005. "Geo Information Breaks through Sector Think." *Geo-Information for Disaster Management SE - 7*.

Inoue, Yutaka, Akio Sashima, Takeshi Ikeda, and Koichi Kurumatani. 2008. "Indoor Emergency Evacuation Service on Autonomous Navigation System Using Mobile Phone." in *Proceedings of the 2nd International Symposium on Universal Communication, ISUC 2008*.

Jeon, Gyu Yeob, Ju Young Kim, Won Hwa Hong, and Godfried Augenbroe. 2011. "Evacuation Performance of Individuals in Different Visibility Conditions." *Building and Environment*.

Kwan, Mei Po and Jiyeong Lee. 2005. "Emergency Response after 9/11: The Potential of Real-Time 3D GIS for Quick Emergency Response in Micro-Spatial Environments." *Computers Environment & Urban Systems* 29(2):93–113.

Lee, Pin-Chan, Yiheng Wang, Tzu-Ping Lo, and Danbing Long. 2018. "An Integrated System Framework of Building Information Modelling and Geographical Information System for Utility Tunnel Maintenance Management." *Tunnelling and Underground Space Technology*.

Liu, L. and S. Zlatanova. 2011. "Towards a 3D Network Model for Indoor Navigation." *Urban and Regional Data Management*.

Nagel, Claus, Alexandra Stadler, and Thomas H. Kolbe. 2009. "Conceptual Requirements for the Automatic Reconstruction of Building Information Models from Uninterpreted 3D Models." *Proceedings of the Academic Track of the Geoweb 2009 - 3D Cityscapes Conference in Vancouver, Canada, 27-31 July 2009*.

Tang, Fangqin and Aizhu Ren. 2012. "GIS-Based 3D Evacuation Simulation for Indoor Fire." *Building and Environment*.

Volk, Rebekka, Julian Stengel, and Frank Schultmann. 2014. "Building Information Modeling (BIM) for Existing Buildings - Literature Review and Future Needs." *Automation in Construction*.

Wang, Xiangyu. 2012. "BIM Handbook: A Guide to Building Information Modeling for Owners, Managers, Designers, Engineers and Contractors." *Australasian Journal of Construction Economics and Building*.

Tunnels and Underground Cities: Engineering and Innovation meet Archaeology,
Architecture and Art, Volume 6: Innovation in underground engineering,
materials and equipment - Part 2 – Peila, Viggiani & Celestino (Eds)
© 2020 Taylor & Francis Group, London, ISBN 978-0-367-46871-2

Numerical simulation and case analysis of new construction method for multi-arch tunnel

J. Zhai & X. Xie
Tongji University, Shanghai, China

ABSTRACT: Through numerical simulation, the vertical displacement and horizontal displacement of surrounding rock during excavation of multi-arch tunnel are revealed. The stress mechanism of surrounding rock, lining structure and middle partition wall of multi arch tunnel is analyzed. Based on the project example of Bailongpo tunnel of Wuyi expressway in Yunnan, the construction technology without middle drift and monitoring method of multi-arch tunnel are described. The results show that: (1) The construction technology without middle drift is feasible. (2) The deformation characteristics of tunnel have obvious space-time effect, and the better the quality of surrounding rock, the smaller the deformation value and the shorter the stability time. (3) The bias effect of the following tunnel is very obvious. (4) The middle wall must have enough strength and stiffness to ensure the stability. It is the key and foundation for the construction of a double-arch tunnel.

1 INSTRUCTIONS

As a new type of tunnel structure, multi-arch tunnel can better meet the design requirements in the complex topography and geological conditions area。 But a large number of engineering practices have proved that the construction method such as three headings and middle pilot heading would bring many problems. (ZHANG, 2006) There are too many construction procedures and process interference between them, and the temporary support often lead to a large amount of demolition, and also the poor integrity of the middle wall leading to poor waterproof effect, crack development. (LIU,2006)

The construction technology without middle drift has the characteristics of simple working procedure and little disturbance to surrounding rock. (XIE,2017) It is helpful to shorten the construction period and reduce the cost, and overcomes some shortcomings of traditional multi-arch tunnel. And the waterproofing and drainage system of the left and right sides of the tunnel is independent of each other, so that the leakage disease of the middle wall can be eliminated. More and more engineering practices show that the construction method without central guide tunnel is more suitable for the structural characteristics of multi-arch tunnel.

2 GENERAL STATUS

The Wuyi Expressway in Yunnan Province is the Western Ring Road of the Economic Circle Expressway in central Yunnan. The route is 104.3 kilometers in length. It is constructed according to the standard of two-way six-lane expressway. The subgrade is 33.5 meters wide and the design speed is 100 kilometers per hour. Bailongpo tunnel is located in the section K72+440~K72+740 of Wuyi Expressway. The length of the tunnel is 300 meters. The surrounding rock in the K72+500-K72+600 section is grade IV and the rest is grade V。 Due to the restrictions of local topography and traffic requirements, a multi-arch tunnel is adopted.

The site of the tunnel belongs to the mountain mausoleum landform. The terrain is undulating, the valley is developed, and the topography is steep and complex. The foundation of tunnel entrance is strongly weathered and moderately weathered sandstone. Because of the insufficient bearing capacity of the foundation of the strongly weathered section, the corresponding section of the tunnel has been strengthened. The surrounding rock of the tunnel is sandstone of Chengjiang Formation of Lower Sinian with medium-fine grain structure. The massive structure is haord and brittle and the joints are developed.

3 NUMERICAL SIMULATION AND ANALYSIS

3.1 construction process

As shown in Figure 1, the left tunnel of the tunnel is the leading hole, and the excavation method takes the three-step construction technique as the basic mode. While the CRD construction technique is used for excavation of the right tunnel. The process of construction is as follows:

I→1→II→2→III→3→4→5→IV→6,7a,7b→V→8,7c→VI→9,7d→VII→10→11→12

3.2 numerical model

We established the numerical model for Bailongpo tunnel according the geological survey. The size of this model is 200 m×90 m. There are a total of 3652 nodes and 3941 units. In this paper the finite-element method is used to calculate. The calculation parameters are shown in Table 1.

3.3 results and analysis

The calculation result are shown in the Figure 2. After all excavation steps, the displacement of the tunnel is mainly concentrated at the left arch waist and the two arch shoulders. The displacement of the left arch waist is 5.5 cm, and the displacement of the arch shoulder is 3 cm~3.5 cm. The displacement of the right hole is located at the right arch waist and the left arch shoulder, which are 2 cm and 3.5 cm respectively. The maximum displacement of the

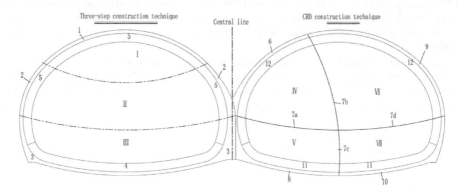

Figure 1. Construction process.

Table 1. Calculation parameters.

category	unit weigth (kN/m³)	E (Mpa)	Poision's ratio	Phi (°)	Cohesion (Mpa)
surrounding rock	23	50	0.35	20	0.05
initial lining	25	600	0.2	-	-
second lining	25	600	0.2	-	-

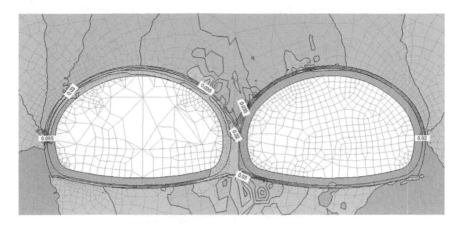

Figure 2. Cloud picture of displacement.

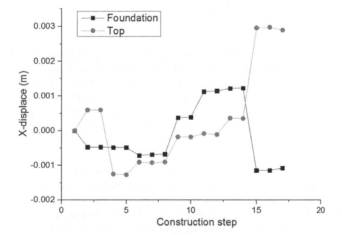

Figure 3. Horizontal displacement of the middle wall.

middle wall and the top and foundation of the wall are 2 cm and 3 cm respectively. The variations of the lateral displacement of the top and foundation of the middle wall with excavation steps are shown in Figure 3. Due to the eccentric pressure effect, the horizontal displacement of the top of the middle wall gradually turns to the positive direction of the x-axis and increases sharply after the excavation of the right tunnel. At the same time, the displacement of the foundation turns to the negative direction of x-axis sharply. This situation is quite unfavorable to the stability of the middle wall.

As shown is Figure 4. There is a large tensile stress concentration between the second lining and the initial lining on the middle wall and the left arch, witch are 0.8 Mpa and 1.4 Mpa respectively. It will cause the crack between the two linings and the lining of the first hole and the initial lining.

4 DIFFICULTIES AND MEASURES IN CONSTRUCTION

4.1 *Stability assurance of partition wall*

For the multi-arch tunnel, the load passes through the arch structure to the partition wall, so that the stress will concentrate on the partition wall. As the weakest position, the construction

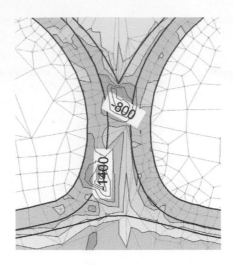

Figure 4.   Stress distribution of the middle wall.

technology and quality of the middle partition wall are very important in the whole tunnel construction. Whether the design of the middle partition wall is reasonable or not is related to the smooth completion of the tunnel. The design of partition wall is shown in Figure 5. The numerical simulation results show that the partition wall has overturning risk after excavation, and there is a large tensile stress concentration between the second lining and the initial lining at the upper part of the middle partition wall and the arch foot of the preceding tunnel.

As shown in Figure 5–7. In order to prevent the overturning of the partition wall during the construction of the following tunnel, steel diagonal bracing is applied in time during the construction of the following tunnel, and small conduit grouting reinforcement is applied to the base of the partition wall in the front tunnel. The small conduit should be embedded in 50 cm of shotcrete bottom of tunnel initial support. Cement slurry is used as grouting material. The grouting water cement ratio is 0.5:1.1 and the grouting pressure is 0.6 Mpa-1.0 Mpa.

As shown in Figure 5, 8. In order to prevent the partition wall from cracking, two layers of bar-mat reinforcement are added to the initial lining of the preceding tunnel, and the horizontal anchor bolt is applied at the same time.

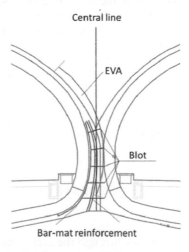

Figure 5.   Design of the partition wall.

Figure 6.   Steel diagonal bracing.

Figure 7.   Small conduit injection.

Figure 8.   Bar-mat reinforcement.

Figure 9.　Construction process of EVA.

Table 2.　Technical indicators of EVA.

category	unit	index
dissolved flow rate	g/10min	2.0±0.7
content of VA	% (m/m)	15.5±1.5
tensile strength	MPa≥	≥3.0
hue	≤	-16
cleanliness		40
elongation at break	%≥	600

4.2　*Structural protection of existing tunnel*

Because of the combination of mechanical excavation and controlled blasting, the excavation of the following tunnel will inevitably disturb the structure of the existing tunnel. As shown is Figure 4, 9. In order to reduce the influence on the existing structure during the excavation process, EVA foam is added between the initial support and the second lining. EVA material has the characteristics of light weight, elasticity, sound insulation and good seismic performance. Adding EVA material can greatly play a role in shock absorption, and is conducive to protecting the existing rigid tunnel lining structure. The technical indicators of EVA are shown in Table 2.

5　STUDY ON FIELD MEASUREMENT

5.1　*Analysis of vault sink monitoring data*

The layout of measuring points is shown in the Figure 10. The features of vault subsidence along the tunnel are shown in Figure 11. The average arch crown subsidence of the left tunnel is 15 mm, and the average right tunnel subsidence is 21.6 mm, which indicates that the excavation of the right tunnel has little effect on the first left tunnel under the precondition of adopting appropriate protective measures and construction technology. It is also indicated that the right hole is subject to severe bias.

　　The deformation of surrounding rock is also related to the geological conditions. The worse the geological condition is, the larger the settlement of the vault will be. For the entrance section, although the buried depth is shallow and the upper load is small, because of the high weathering degree of rock mass, the lithology is extremely poor, the vault settlement is large, the maximum value is 35 mm.

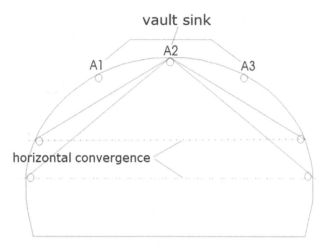

Figure 10. layout of measuring points.

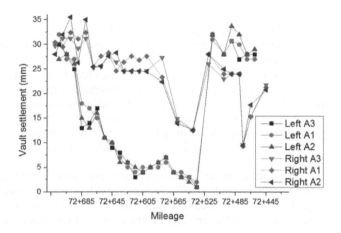

Figure 11. Vault sink monitoring data.

5.2 *Analysis of horizontal convergence monitoring data*

The features of horizontal convergence along the tunnel are shown in Figure 12. The average convergence deformation of the left tunnel is 21 mm, and that of the right tunnel is 4 mm. It shows that the influence of the later driving right tunnel on the first left tunnel is not significant. The convergence deformation of the right tunnel is obviously larger than that of the left tunnel about 15 mm, indicating that the tunnel has received a larger bias during the construction of the right tunnel. According to the relation curve between the convergence deformation of horizontal surveying line and practice, it can be seen that the horizontal convergence of the left cave body is basically less than 1 mm. The existing tunnel structure has been well protected.

6 DISCUSSION & CONCLUSION

(1) Multi-arch tunnel without middle guide tunnel is a new type of tunnel construction method. The multi-arch tunnel is constructed as two independent single tunnels. The first tunnel is excavated, and the second tunnel is excavated after the first tunnel lining is completed. Attention should be paid to the protection of the existing structure and the stability

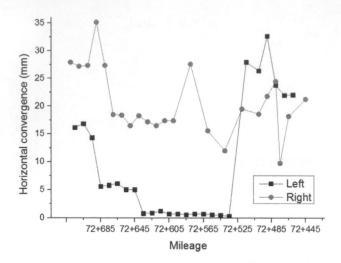

Figure 12. Horizontal convergence monitoring data data.

of the partition wall in the back tunnel excavation, which requires that the middle partition wall has sufficient strength and stiffness and adopt strict construction methods for the excavation of the back tunnel.

(2) The construction method of non-central guide tunnel reduces many construction procedures in traditional multi-arch tunnel construction, eliminates a large number of temporary construction support due to excavation of the pilot tunnel, reduces the labor consumption, the use of construction consumables, reduces the construction cost, and accelerates the construction of multi-arch tunnel. According to statistics, the total cost of labor saving in the application of Bailongpo tunnel is 2.6 million yuan, and the construction period is shortened by 181 d.

(3) According to the monitoring results of vault settlement and horizontal convergence deformation, the construction method of non-center guide tunnel is feasible. Compared with the middle guide tunnel method, although the deformation control ability of surrounding rock is weak at the entrance section, the construction procedure is simple and the construction period is short.

(4) Comparing the deformation characteristics of the left and right tunnel, it can be seen that the condition of the left tunnel is obviously better than that of the right tunnel, and the eccentric pressure effect of the back tunnel is very obvious, which should be fully considered in design and construction.

(5) The partition wall is the connection core of two tunnels, so the middle partition wall must have enough strength and stiffness to ensure the stability of the middle wall, which is the key and foundation for the construction of double-arch tunnel.

REFERENCES

Zhang Z. Q., He C. 2006. Study on mechanical behaviours of designing and construction for center pillar of double-arched tunnel. *Chinese Journal of Rock Mechanics and Engineering*. 25 (08): 1632–1638.

Liu T et al. 2006. Model test and 3D numerical simulation study on excavation of double-arch tunnel. *Chinese Journal of Rock Mechanics and Engineering*. 25(09):1802–1808.

Li E. B. et al. 2007. Monitoring and control of construction deformation of urban shallow-buried large-span double-arch tunnel under complex condition. *Chinese Journal of Rock Mechanics and Engineering*. 26(04):833–839.

Xie C. H. 2017. Study on construction method of urban multi arch tunnel without center heading. *Journal of China & Foreign Highway*. 37(04):214–218.

Tunnels and Underground Cities: Engineering and Innovation meet Archaeology,
Architecture and Art, Volume 6: Innovation in underground engineering,
materials and equipment - Part 2 – Peila, Viggiani & Celestino (Eds)
© 2020 Taylor & Francis Group, London, ISBN 978-0-367-46871-2

Experimental study on a novel fire-protection cement-based composite material for tunnel linings

T. Zhang, Z. Yan, Z. Xiao & H. Zhu
Department of Ceotechnical Engineering, Tongji University, Shanghai, China

ABSTRACT: Concrete lining structures are prone to spall induced by high temperature, which can be extremely dangerous for semi-closed tunnels, considering the relatively high humidity. This paper presents a novel active fire protection (AFP) system achieving by cement-based system consisted of ammonium polyphosphate-pentaerythritol-melamine (APP-PER-EN) composite with both fiber phase and particle phase. One-side heating experiments under open flame indicated that the thermally-triggered system functioned though melting, connecting and overflowing, which demonstrated a great retardant effect. Thermo-gravimetric analysis (TGA) and differential scanning calorimeter (DSC) were conducted to characterize the thermal behavior of the APP-PER-EN composite. Therefore, we believe that the novel system will exhibit a promising prospect for prolonging the life of lining structures after experiencing high-temperature environment.

1 INTRODUCTION

Reinforced concrete materials have been widely used in both high-rise buildings and deep-buried tunnels due to their high strength, good mechanical performance and favorable durability under normal conditions. However, as more and more skyscrapers and super tunnels occurred all around the world, the durability of structures in an extreme environment, especially under fire disaster, has been highlighted.

Many studies have been carried out about the mechanical properties of ordinary Portland cement (OPC) concrete structures at high temperature. At macro level, nearly all Portland-cement-based concretes lose their load-bearing capacity at temperatures above 600 °C. Similarly, flexural strength, splitting tensile strength and modulus of elasticity of concrete decreases with the increase of temperature [Schrefler et al., 2002; Boström et al., 2006; Ma et al., 2015].

The common physical and chemical changes in OPC concrete during heating are multiple processes and reactions of water evaporation, decomposition of $Ca(OH)_2$, α-β phase transition of quartz (siliceous) aggregates and pore structure changes as shown in Figure 1 [Khoury et al., 2010; Xing et al., 2011]. Besides, it should be noticed that unsealed cement paste behaves differently from moist sealed cement pasted above 100 °C, the former is dominated by dehydration processes, while the latter is dominated by hydrothermal chemical transformations and reactions [Alonso et al., 2004].

When it comes to the micro-structure of concrete, transient thermal creep (TTc) accommodates the thermal strain incompatibilities between the paste and the aggregates, which ensures the Portland cement does not break down at relatively low temperatures (<100 °C) as it theoretically should [Rickard et al., 2016]. However, when the temperature rises to above 200 °C, the different thermal strains for cement matrix and aggregates will produce a stress between cement matrix and aggregates, causing the development of micro-cracks in the interfacial transition zone (ITZ) [Razafinjato et al., 2016].

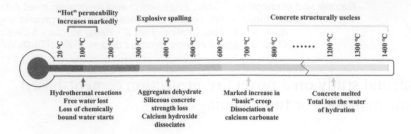

Figure 1. Physical and chemical changes in OPC concrete during heating.

In order to reduce losses caused by fire incidents, a number of different kinds of passive fire protection (PFP) have been developed to insulate the concrete structure from fire in the previous decades. Current PFPs may be broadly divided into the following categories:

(a) Build a fire barrier near the surface of structure components, such as installing non-parasitic protection system board or ceramic fiber mats. Recently, fiber-metal laminates (FMLs), which are hybrid composite materials built up from interlacing layers of thin metals and fiber reinforced adhesives, have been highlighted due to the optimum mechanical properties [Sinmazçelik et al., 2011; Christke et al., 2016]. However, the bonding property between the composite laminae and the metallic layer is a problem for the overall laminate performance.

(b) Incorporate fibers systems are adopted in concrete for releasing internal evaporative pressure caused by high thermal loading, such as polypropylene (PP) fiber, basalt textile or steel fiber [Yan et al., 2013; Rambo et al., 2015; Yurddaskal et al., 2017]. Fibers are effective in providing channels for the release of internal pressure via melting at high temperatures and connecting internal pores [Kalifa, et al., 2001; Zeiml et al., 2006].

(c) Various intumescent organic coating systems based on aluminosilicate source and fire-protecting mortars formulated with lightweight aggregates have been a common solution. Suitable aluminosilicate source materials include metakaolin and fly ash, with the latter being more commonly used in high volume applications [Temuujin et al., 2010; Abdollahnejad et al., 2015; Najimi et al., 2018; Qiu et al., 2018]. In react to the fire, these by-products increase several times in volume to create a highly expanded refractory carbonaceous char with low thermal conductivity [Formosa, 2011; Rickard et al., 2016].

However, all the aforementioned methods are all passive modes of protection and have several disadvantages. In terms of structural behavior, the coating systems are parasitic, which means they contribute to the weight of a structure without adding to its strength. Besides, the surface of components is covered by coating, which can hinder crack detection and structural maintenance. Moreover, regularly repainting of fire-protection coating every year to make sure that the coating is in validity period can be a burden for structure operation maintenance.

Therefore, this paper presents a novel active-mode fire-resistant system achieving by cement-based composite system consisted of ammonium polyphosphate-pentaerythritol-melamine (APP-PER-EN) composite with both fiber phase and particle phase. One-side heating experiments under open flame were conducted to demonstrate the thermal and flame-retardant behavior of the novel system. Besides, thermo-gravimetric analysis (TGA) and differential scanning calorimeter (DSC) were also conducted to characterize the thermal behavior of the composite.

2 MECHANISM OF FIRE-PROTECTION COMPOSITE SYSTEM

Sweating is the most general and effective form of heat dissipation, in particular when the ambient temperature increases or activity intensifies. Based on the heat self-regulation mechanism of the sweat gland, a novel fire-protection cement-based composite is designed to

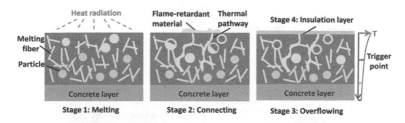

Figure 2.   The reaction of novel fire-protection system under fire condition.

improve the fire resistance of engineering structures. Under normal condition, the concrete layer and retardant layer can sustain the work loading together. Under fire condition, the fire-protection function is achieved through the following stages (as illustrated in Figure 2).

(a) Stage 1: Melting. Between 100–160 °C, the flame-retardant fibers melt and absorb a substantial amount of heat;
(b) Stage 2: Connecting. At this stage, micro channels caused by the melting fibers form. Meanwhile, micro-cracks can also be induced by the thermal load. The connection between micro-cracks and micro channels can facilitate the outflow of the retardant agent.
(c) Stage 3: Overflowing. Above 300 °C, the thermally triggered composite particles start to expand and foam. As a result, the retardant material produced by foaming fills the internal cracks and functions as a thermal resistance. Meanwhile, the retardant material can overflow to the heating surface through cracks and gradually forms a fire insulation layer.
(d) Stage 4: Formation of thermal insulation layer. The fire insulation layer outside of the fire-protection layer can block the heat transfer into the internal concrete. Besides, at a certain thickness, the fire-protection composite layer forms a "honeycomb" structure, which can further reduce the thermal conductivity.

3   EXPERIMENTS

3.1   Preparation of retardant materials

The retardant particles and fibers are made of ammonium polyphosphate (APP), pentaery-throtol (PER) and melamine (EN) with polyethylene as matrix. The retardant polymerization process was conducted by melt compounding in a twin-screw extruder followed by injection molding technique. Considering the confidentiality of materials/compositions (filed as technology disclosure [Yan, 2017]), detailed information on specific ratio will not discuss here. With the cooperation of winch and pelletizer, the size of particles can be about 0.8–1.0 mm diameter sphere, while the length of fibers to be around 10 mm with 0.8–1.0 mm diameter.

3.2   Sample preparation

Double-layer samples were 300-mm-length, 200-mm-wide, and 100-mm-thick with 30-mm-thick retardant layer and 70-mm-thick concrete layer. These samples were prepared by firstly pouring concrete with special mixture ratio (according to Table 1) into molds. And pouring the mixture of retardant materials (particles and fibers) 12 h after the casting of concrete layer. In order to capture the temperature data on the interface of double-layer, a 30-mm-length steel bar was inserted in the center of retardant layer and was removed after 24 h of curing. All the test samples were cured for 28 days in a room with controlled temperature and humidity.

3.3   One-side heating test

High-temperature test was applied by covering the furnace with samples, and the gaps between concrete samples and furnace mouth were filled with flame-retardant cotton. The furnace was assembled with a methane burner. By controlling the methane gas flow rate to 4.0 L/

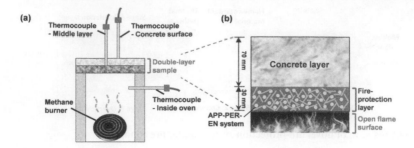

Figure 3. Distribution of (a) K-type thermocouples and (b) the sample during one-side heating test.

h, the temperature inside furnace maintained at approximately 800 °C. In order to simulate real fire condition of one-side heating, the fire-protection layer was exposed to temperature inside furnace, while the concrete layer at room temperature. One-side heating test contained six samples totally, and the basic parameters of experimental samples were listed in Table 1. Moreover, temperature data on the surface of concrete layer and inside furnace can be easily acquired by K-type thermocouples, while the statistics on the interface of double-layer was captured by insert thermocouples in the reserved hole located in the center of fire-protection layer as shown in Figure 3 (a). Copper powder was used to guarantee the close contact between sensors and samples.

3.4 *Characterization techniques*

Morphologies of samples were recorded with scanning electron microscope (SEM; KYKY, China) operated at 20 kV. Thermogravimetry (TG) and differential scanning calorimeter analysis (DSC) data were collected in the same time with a synchronous thermal analyzer (STA; NETZSCH, Germany) from 25 to 800 °C at a heating rate of 20 °C/min in the flow of air.

4 RESULTS

4.1 *The effect of APP-PER-EN system with maximum admixture*

To demonstrate the retardant effect of fire-protection system, maximum admixture of fiber-phase and particle-phase APP-PER-EN composite were firstly adopted in sample F0 and P0, respectively. As shown in Figure 4 (a), after 100-min one-side heating, the temperature of middle interface reached nearly 350 °C in reference concrete sample (C0), coupled with slightly spalling. In comparison, temperature of middle interface was 95.8 °C and 155.9 °C lower in sample F0 and P0 respectively at the end of the high-temperature test. Meanwhile, the temperature rising rate of fire-protection samples were significantly declined before and after the "temperature residence period" at approximately 100 °C.

For the sample with maximum admixture of fiber-phase APP-PER-EN composite (F0), although there was no apparent "temperature residence period", the rising rate of temperature on the interface was slow enough to extend the time to reach 100 °C to as much as 50 min. The prolongation of time when the temperature was lower than 100 °C (nearly triple longer than sample C0) left enough time for heat dissipation during the initial stage of the fire. The whole temperature curve of the interface presented a "double-fold line" style, with the increase in temperature rising rate after reached 100 °C, which means 100 °C was still a key point. The reason why no residence period was found can be explained by (i) the increased permeability in consequence of melting fibers and (ii) by the introduction of additional interfacial transition zones between the fibers and the cement paste.

When it comes to the sample with maximum admixture of particle-phase APP-PER-EN composite (P0), a super long "temperature residence period" occurred on the interface at approximately 100 °C lasted about 45 min. Besides, although it cost less time to reach 100 °C in sample

P0, it spent longer time to rise again over 100 °C. The temperature rising rate after "breaking" 100 °C is nearly the same for sample P0 and F0. Moreover, the temperature of middle interface in sample P0 was 60.1 °C lower than that of sample F0 at the end of the heating test. Similarly, we can also witness a significant decline (36.3 °C) in temperature on the surface of concrete, which can be partly explained by larger maximum admixture of particle-phase composite.

4.2 The influence of admixture with both fiber-phase and particle-phase composite

As aforementioned in 4.1, fiber-phase composite slowed down temperature rising rate in the same time shrunk the "temperature residence period" at approximately 100 °C, while fiber-phase composite prolonged the residence period but with higher temperature rising rate. Moreover, the maximum admixture of particle-phase composite was over double than that of fiber-phase composite according to test-in-place. Therefore, samples with admixture of both fiber-phase and particle-phase composite were also adopted (as listed in Table 1) to acquire better flame-retardant effect.

Sample PF-1 was designed to attempt to achieve maximum admixture of particle-phase composite under the circumstances that fiber-phase composite mixture was 200 g. Due to the small diameter of particle-phase composite, the maximum admixture of particle can reach to as high as 600 g, which was proved by test-in-place. In order to exclude the influence of total amount of APP-PER-EN composite, one-side heating temperature curves of sample PF-1 was compared with sample P0 under the same admixture level (APP-PER-EN composite of 800 g) as presented in Figure 5(a).

For sample PF-1, the rising rate (when the temperature was lower than 100 °C) was slowed down to some degree compared with sample P0. Meanwhile, an apparent "temperature resi-

Table 1. Basic parameters of experimental samples.

| Sample number | Specimen size ($l \times w \times h$; mm) | | Quality ratio of concrete layer | Additive ingredients in Retardant layer (g) | |
	Concrete layer	Retardant layer		Fiber	Particle
Reference-C0	300×200×100	None	Cement/Water/	None	None
Fiber-F0	300×200×70	300×200×30	Sand/Aggregate	300	None
Particle-P0	300×200×70	300×200×30	=1: 0.5:2.5:2.3	None	800
PF-1	300×200×70	300×200×30		200	600
PF-2	300×200×70	300×200×30		200	400
PF-3	300×200×70	300×200×30		200	200

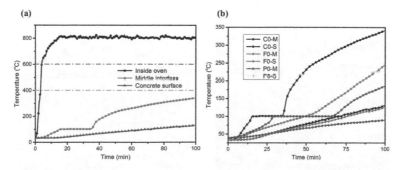

Figure 4. One-side heating temperature curves of (a) reference samples and (b) the comparison among samples C0, F0 and P0 (M-Middle interface of double layer; S-Surface of concrete layer).

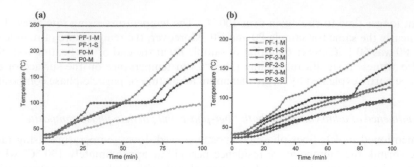

Figure 5. One-side heating temperature curves of (a) the comparison among samples PF-1, P0, and F0 and (b) the comparison among samples PF-1, PF-2 and PF-3 (M-Middle interface of double layer; S-Surface of concrete layer).

dence period" (lasted for 26 min) occurred at approximately 100 °C. Under the combined action of the above-mentioned two factors, the temperature of samples can maintain a low-level temperature (<100 °C) for a long time (even up to 75 min), which provide sufficient time for the heat dissipation, leading to a decline of 88.2 °C and 28.1 °C at the end of heating test compared with sample F0 and P0 respectively.

Figure 6 presented SEM micrographs of sample PF-1 after one-side heating test. And the holes left by formation of particle-phase composite were highlighted with yellow circles. Due to the interaction and char formation of APP-PER-EN composite, a number of new thermal pathway were formed, leading to the increase of permeability.

4.3 The impact of ratio between fiber-phase and particle-phase composite

To figure out the influence of ratio between two phases, sample PF-1, PF-2 and PF-3 were adopted. As shown in Figure 5 (b), generally speaking, with the increase of ratio (fiber/particle), the "temperature residence period" turned to be shorter and the rising rate of temperature turned to be slower. Furthermore, from the comparison between sample PF-1 and PF-3, it can be concluded that, under certain ratio, a better retardant effect may be acquired even with fewer amount of total admixture. However, due to the discreteness of the experiment results, more effort should be paid to find a better ratio between two phases in the research afterwards.

4.4 Thermal analysis of flame retardants

The char formation and interaction of ammonium polyphosphate-pentaerythritol-melamine (APP-PER-EN) composite plays an important role in flame-retardant structure. Therefore, thermo-gravimetric analysis (TGA) and differential scanning calorimeter (DSC) were conducted to characterize the thermal behavior of the APP-PER-EN composite. Figure 7 presented TG and DSC curves of these basic components.

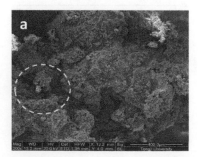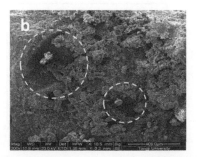

Figure 6. SEM micrographs of sample PF-1 after one-side heating test.

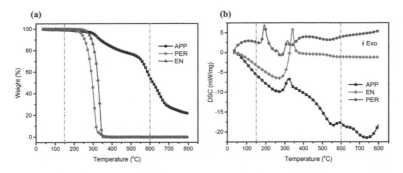

Figure 7. (a) TG and (b) DSC curves of flame retardants for ammonium polyphosphate (APP), pentaerythritol (PER) and melamine (EN).

APP as the carbonization catalyst and acid source, began to decompose at 287.1 °C, liberating phosphoric acid and amine gas at the same time. The resulting acid triggered a series of reactions starting with the dehydration of the carbonific compound and its subsequent charring.

PER as the carbonization agent, experiencing a double-stage thermal degradation. An endothermic peak during 240.7–306.3 °C and a small endothermic peak during 319.5–381.4 °C were attributed to the transformation of the crystal structure and the melting of PER, respectively.

EN as the spumific agent sublimed partly and decomposed partly during 278.4–343.9 °C, giving out large quantities of nonflammable gases, such as NH_3 and CO_2 which were helpful in forming "honeycomb" char structure layer over the substrate [Wang et al., 2006].

Furthermore, it should be noticed that the overlap decomposition temperature range for both PER and EN ensured good synergism in the flame-retardant system [Wang et al., 2005]. Overall, these thermodynamic properties confirmed that the retardant system worked efficiently.

5 CONCLUSIONS AND DISCUSSIONS

A novel active fire protection (AFP) system achieving by cementitious composite system consisted of APP-PER-EN composite with both fiber phase and particle phase was introduced. The novel composite system functioned through melting, connecting, overflowing, and finally an insulation layer will form which can further reduce the thermal conductivity.

One-side heating tests using open flame were conducted to demonstrate the flame-retardant effect of the composite under structural level. The flame-retardant characteristics of both fiber-phase and particle-phase composite with maximum admixture were analyzed. Besides, the influence of ratio between fiber and particle was preliminary discussed. Moreover, the thermal behavior of composite system was revealed by TGA and DSC methods.

Furthermore, in the future study, more attention still needs to be paid on: (i) bonding properties of the interface between concrete layer and fire-protection layer, (ii) mechanical properties (especially the strength characteristics) of the whole concrete structure with a certain depth of fire-protection layer and (iii) suitable ratio between fiber-phase and particle-phase composite to achieve better flame-retardant effect.

ACKNOWLEDGEMENTS

The authors acknowledge financial supports from the National Natural Science Foundation of China (51578410, 51478345), the Research Program of State Key Laboratory for Disaster Reduction in Civil Engineering, and the Fundamental Research Funds for the Central Universities (300102218511).

REFERENCES

Abdollahnejad, Z., Pacheco-Torgal, F., Félix, T., Tahri, W., & Aguiar, J. B. 2015. Mix design, properties and cost analysis of fly ash-based geopolymer foam. *Construction & Building Materials* 80: 18–30.

Alonso, C., & Fernandez, L. 2004. Dehydration and rehydration processes of cement paste exposed to high temperature environments. *Journal of Materials Science* 39(9): 3015–3024.

Boström, L., & Larsen, C. K. 2006. Concrete for tunnel linings exposed to severe fire exposure. *Fire Technology* 42(4): 351–362.

Christke, S., Gibson, A. G., Grigoriou, K., & Mouritz, A. P. 2016. Multi-layer polymer metal laminates for the fire protection of lightweight structures. *Materials & Design* 97: 349–356.

Formosa, J., Chimenos, J. M., Lacasta, A. M., Haurie, L., & Rosell, J. R. 2011. Novel fire-protecting mortars formulated with magnesium by-products. *Cement & Concrete Research* 41(2): 191–196.

Kalifa, P., Chéné, G., & Gallé, C. 2001. High-temperature behavior of HPC with polypropylene fibers: from spalling to microstructure. *Cement & Concrete Research* 31(10): 1487–1499.

Khoury, G. A. 2010. Effect of fire on concrete and concrete structures. *Progress in Structural Engineering & Materials* 2(4) 429–447.

Ma, Q., Guo, R., Zhao, Z., Lin, Z., & He, K. 2015. Mechanical properties of concrete at high temperature—a review. *Construction & Building Materials* 93: 371–383.

Najimi, M., Ghafoori, N., & Sharbaf, M. 2018. Alkali-activated natural pozzolan/slag mortars: a parametric study. *Construction & Building Materials* 164: 625–643.

Qiu, X., Li, Z., Li, X., Zhang, Z., Qiu, X., & Li, Z., et al. 2018. Flame retardant coatings prepared by layer by layer assembly: a review. *Chemical Engineering Journal* 334: 108–122.

Rambo, D. A. S., Silva, F. D. A., Filho, R. D. T., & Gomes, O. D. F. M. 2015. Effect of elevated temperatures on the mechanical behavior of basalt textile reinforced refractory concrete. *Materials & Design* 65: 24–33.

Razafinjato, R. N., Beaucour, A. L., Hebert, R. L., Ledesert, B., Bodet, R., & Noumowe, A. 2016. High temperature behaviour of a wide petrographic range of siliceous and calcareous aggregates for concretes. *Construction & Building Materials* 123: 261–273.

Rickard, W. D. A., Gluth, G. J. G., & Pistol, K. 2016. In-situ thermo-mechanical testing of fly ash geopolymer concretes made with quartz and expanded clay aggregates. *Cement & Concrete Research* 80 (4): 33–43.

Schrefler, B. A., Brunello, P., Gawin, D., Majorana, C. E., & Pesavento, F. 2002. Concrete at high temperature with application to tunnel fire. *Computational Mechanics* 29(1): 43–51.

Sinmazçelik, T., Avcu, E., Bora, M. Ö., & Çoban, O. 2011. A review: fiber metal laminates, background, bonding types and applied test methods. *Materials & Design* 32(7): 3671–3685.

Temuujin, J., Minjigmaa, A., Rickard, W., Lee, M., Williams, I., & Van, R. A. 2010. Fly ash based geopolymer thin coatings on metal substrates and its thermal evaluation. *Journal of Hazardous Materials* 180(1): 748–752.

Wang, Z., Han, E., & Ke, W. 2005. Influence of nano-LDHs on char formation and fire-resistant properties of flame-retardant coating. *Progress in Organic Coatings* 53(1) 29–37.

Wang, Z., Han, E., & Ke, W. 2006. Effect of nanoparticles on the improvement in fire-resistant and anti-ageing properties of flame-retardant coating. *Surface & Coatings Technology* 200(20) 5706–5716.

Xing, Z., Beaucour, A. L., Hebert, R., Noumowe, A., & Ledesert, B. 2011. Influence of the nature of aggregates on the behaviour of concrete subjected to elevated temperature. *Cement & Concrete Research* 41(4): 392–402.

Yan, Z. G., Zhu, H. H., & Ju, J. W. 2013. Behavior of reinforced concrete and steel fiber reinforced concrete shield TBM tunnel linings exposed to high temperatures. *Construction & Building Materials* 38 (2): 610–618.

Yan, Z. G., Zhu, H. H., Ding, W. Q., Shen, Y. Dec. 8 2017. A new kind of microcapsule self-fireproof concrete tunnel lining. *National invention patent*, Ref.: CN201510657437.7, China.

Yurddaskal, M., & Celik, E. 2017. Effect of halogen-free nanoparticles on the mechanical, structural, thermal and flame retardant properties of polymer matrix composite. *Composite Structures* 183: 381–388.

Zeiml, M., Leithner, D., Lackner, R., & Mang, H. A. 2006. How do polypropylene fibers improve the spalling behavior of in-situ concrete. *Cement & Concrete Research* 36(5): 929–942.

Tunnels and Underground Cities: Engineering and Innovation meet Archaeology,
Architecture and Art, Volume 6: Innovation in underground engineering,
materials and equipment - Part 2 – Peila, Viggiani & Celestino (Eds)
© 2020 Taylor & Francis Group, London, ISBN 978-0-367-46871-2

Deep learning-based instance segmentation for water leakage defects of Metro Shield Tunnel

S. Zhao & H. Huang

Department of Geotechnical Engineering, Tongji University, Shanghai, China

ABSTRACT: Recently, computer vision (CV)-based techniques have been widely applied for inspecting shield tunnel lining defects. However, this inspection method will produce a lot of image data. How to efficiently detect water leakage from the large image database therefore becomes a new challenge. To effectively improve recognition of water leakage of metro shield tunnel lining, a Mask Region-based Convolutional Neural Network (Mask R-CNN)-based method is employed in this paper. This approach efficiently detects water leakage in an image while simultaneously generating a high-quality segmentation mask for each water leakage. To realize the proposed method, a database including 4525 images is developed. Then, the Mask R-CNN is modified and trained using this database to detect and segment water leakage in images. The segmentation results show that the trained model shows quite better performances and can rapidly and accurately recognize water leakage of metro shield tunnels.

1 INTRODUCTION

Metro shield tunnel is an important civil infrastructure and is wildly used in urban underground railway transit system. Along with the long-term service, the tunnel affected by the environmental disturbance will deteriorate and generate cracks, water leakage and other defects with service time. To ensure the safety, human-based onsite inspection is regularly conducted to evaluate tunnels' conditions. However, it significantly depend on the trained inspectors and is very time-consuming and tedious. Moreover, inspecting metro tunnels is usually carried out at midnight with limited time (Ai et al., 2015; Huang et al., 2017a), which inevitably result erroneous evaluation concerning the tunnel's conditions.

Because automation of visual inspections can address the limitations of the ordinary human-oriented visual approach, researchers have been attracted to the development of CV-based methods for structural defects inspection. Motivated by the accidents that were caused by falling parts of the inner wall of concrete tunnels in recent years, Yao et al. (2003) developed an autonomous mobile robot mounted with a URS/VC semi-ring to inspect the concrete tunnel. Lee et al. (2007) proposed an inspection system for the rapid measurement of cracks in tunnel linings. Yu et al. (2007) designed an auto inspection system for detecting concrete cracks in a tunnel. The system consists of a mobile robot system, which acquire image data with a Charged Couple Device (CCD) camera, and a crack detection system, which extract crack information from the acquired image using image processing. To meet the requirements of subway tunnel inspection in China, Huang et al. (2017b) developed a Moving Tunnel Inspection (MTI-100) equipment. A real project test certified that the accuracy, stability, repeatability, labor intensity and efficiency of the MTI-100 is quite suitable for practical tunnel inspection. Although CV-based method is automated for tunnel inspection, this method brings a lot of image data. How to process the large amounts of image data and recognize defects in them is becoming a challenging problem. Despite a number of image processing techniques (IPTs) (Abdel-Qader et al., 2003; Nishikawa et al., 2012; German et al., 2012;

Sinha et al., 2003) have been applied for processing image data, these IPTs can't adapt the extensively varying real-world situations (e.g., lighting and shadow changes).

Deep learning was first put forward in 2006. It discovers intricate structure in large data sets by using the backpropagation algorithm to indicate how a machine should change its internal parameters that are used to compute the representation in each layer from the representation in the previous layer (LeCun et al., 2015). Convolutional neural network (CNN), a deep learning-based architecture, is designed to process data that come in the form of multiple arrays, such as images. There have been numerous applications of CNNs in civil engineering to detect structural defects infrastructure (Soukup and Huber-Mörk, 2014; Cha et al., 2017a; Huang et al., 2018; Xue and Li, 2018; Cha et al., 2017b) and proved to be a very effective method.

Recently, deep CNNs, especially Fast/Faster R-CNN (Girshick, 2015; Ren et al., 2015) and Fully Convolutional Network (FCN) (Long et al., 2015), have significantly improved accuracy of image classification, where the goal is to decide what object the image contain, object detection, where the goal is to classify individual objects and localize each using a bounding box, and semantic segmentation, where the goal is to classify each pixel into a fixed set of categories without differentiating object instances. Instance segmentation combines the tasks of object detection and semantic segmentation. Its goal is to detect objects in an image while simultaneously generating a high-quality segmentation mask for each instance.

Mask R-CNN, proposed by He et al. (2017), is a framework for instance segmentation and have achieved top results on the Microsoft COCO (Lin et al., 2015) instance segmentation task. Our goal in this work is to modify and train Mask R-CNN to detect water leakage in an image while simultaneously generating a high-quality segmentation masks for different water leakage in images.

2 DATA SET

CNNs are used to detection and semantic segmentation for objects in images. However, these tasks rely on a large number of labelled data and the computing power of the computer. The previous issue of CNNs was the requirement for a vast amount of labeled data, which came with a high-computational cost, but this issue was overcome through the development of labelling techniques and parallel computations using graphic processing units.

However, there are few well-annotated open source datasets that contain information about tunnel defects at present. Therefore, it was crucial to collect a sufficient amount of images containing water leakage for this research.

2.1 Image acquisition of metro shield tunnel lining

Subway tunnel inspection are often conducted at midnight and must be completed within 2–3 hours in order to guarantee the daily operation of the metro. Therefore, it is not easy to collect large amounts of image of tunnel lining surface in a short period of time. Moreover, China also cannot provide the train or power source for inspection, which is different from Japan and Europe. As a result of these characteristics of subway tunnel inspection, efficiency and high precision are must be considered for acquiring images. However, the existing tunnel inspection equipment cannot meet the requirement.

To meet the requirements of subway tunnel inspection and image acquisition, Huang et al. (2017b) developed a Moving Tunnel Inspection (MTI-100) equipment. The novel equipment MTI-200a is updated from the version of MTI-100. The MTI system mainly consists of 6 high-resolution linear CCD cameras, 12 light emitting diodes (LEDs), computer, monitor, encoder and battery, as shown in Figure 1. The weight of the MTI is relatively light and easy to move. The MTI equipment are able to scan over 13 meters of a tunnel's surface at a time. Despite the influence of lighting and vibration, the captured image's quality is relatively good. Therefore, there is no need for image optimization, such as contrast enhancement image denoising. The captured image can be cropped and labeled directly.

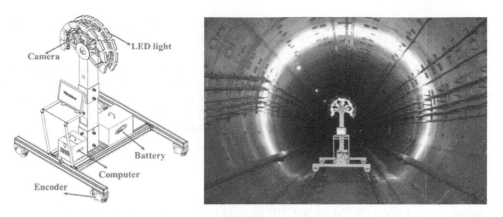

Figure 1.   The sketch map of MTI-200a and MTI-200a in tunnel.

2.2 *Establishment of image datasets*

The inspection tasks was carried out in Shanghai metro line 1, 2, 4, 7, 8, 10, and 12. Images are stored in 1,000×7,448 pixels within each camera. Images with 3000×7448 pixels are obtained by image stitching through the special tool. Then, the images are cropped into 3000×3724 pixels and the image containing water leakage is selected. To ensure that the size

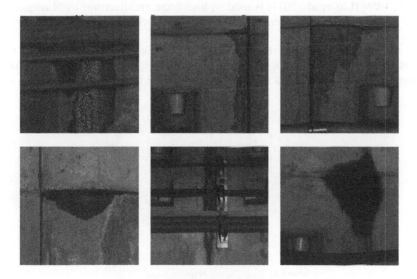

Figure 2.   The cropped images containing water leakage.

Figure 3.   Water leakage annotated by Labelme.

3477

of image is corresponding to the size of water leakage, the selected image is further cropped to 5 scales, which is 800×800 pixels, 1200×1200 pixels, 1600×1600 pixels, 2000×2000 pixels and 2400×2400 pixels. The cropped defect images are shown in Figure 2.

Then, the Water Leakage in COntext (WLCO) dataset is created. The format of annotation file of WLCO is similar to Microsoft COCO. All the images in WLCO are annotated by Labelme (Figure 3) and the annotation file is converted to a format similar to the COCO dataset through Python. The WLCO dataset can be used not only for the detection and semantic segmentation of water leakage but also for instance segmentation of water leakage.

3 METHODOLOGY

To detect and localize water leakage of image while simultaneously generating a high-quality segmentation mask for it, the Mask R-CNN method is used. The original Mask R-CNN consists of Faster R-CNN for classification and bounding box regression and FCN or predicting segmentation masks. This Mask R-CNN is modified to achieve water leakage's instance segmentation. The details of the Mask R-CNN and its modification are explained in this section. For clarity, we differentiate between: (i) the convolutional backbone architecture used for feature extraction over an entire image, and (ii) the network head for bounding-box recognition (classification and regression) and mask prediction.

3.1 *Backbone architecture*

In this paper, FPN (Lin et al., 2017) is used as backbone architecture. FPN uses a top-down architecture with lateral connections to build an in-network feature pyramid from a single-scale input.

Figure 4 shows the structure of FPN. With a coarser-resolution feature map, the spatial resolution is upsampled by a factor of 2 (using nearest neighbor upsampling for simplicity). The upsampled map is then merged with the corresponding bottom-up map (which undergoes

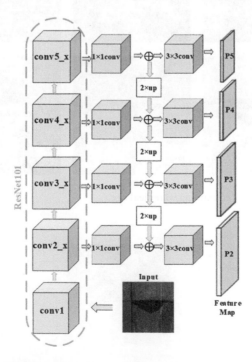

Figure 4.   The structure of FPN.

a 1×1 convolutional layer to reduce channel dimensions) by element-wise addition. This process is iterated until the finest resolution map is generated. To start the iteration, a 1×1 convolutional layer is attached on conv5_x (C5) to produce the coarsest resolution water leakage feature map. Finally, a 3×3 convolution is appended on each merged leakage feature map to generate the final leakage feature map in order to reduce the aliasing effect of upsampling. This final set of leakage feature maps is called {P2, P3, P4, P5}, corresponding to {C2, C3, C4, C5} that are respectively of the same spatial sizes.

3.2 Head architecture

Figure 5 shows the structure of Mask R-CNN including backbone architecture and head architecture. The head network is used to generate proposals by Region Proposal Networks (RPN), Region of Interests (RoIs) by RoI Align, and high-quality segmentation mask on each RoI by FCN. There are 2-time classification and bounding box regression, one in the RPN, and the other in the full connected layer.

3.2.1 RPN

RPN is a sliding-window class-agnostic object detector. In the original RPN design, a small subnetwork is evaluated on dense 3×3 sliding windows, on top of a single-scale convolutional feature map, performing object/non-object binary classification and bounding box regression. This is realized by a 3×3 convolutional layer followed by two sibling 1×1 convolutions for classification and regression.

We adapt RPN by replacing the single-scale feature map with our FPN. We attach a head of the same design (3×3conv and two sibling 1×1 convs) to each level on our feature pyramid. Because the head slides densely over all locations in all pyramid levels, it is not necessary to have multi-scale.

3.2.2 RoI Align

RoI pooling has the problem that misalignments between the RoI and the extracted features. This is caused by quantizations in the process that the coordinate on image converts to the

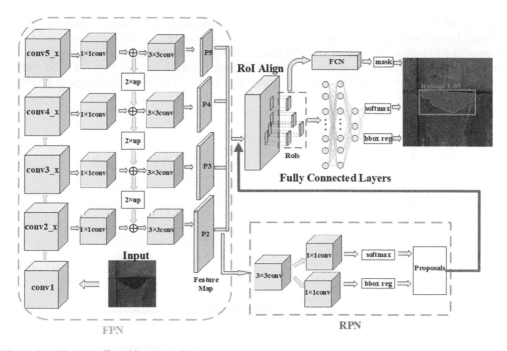

Figure 5. The overall architecture of the Mask R-CNN.

coordinate on feature map and RoI is subdivided into spatial bins, e.g., on a continuous coordinate x by computing [x/16], where 16 is a feature map stride and [·] is rounding; likewise, quantization is performed when dividing into bins (e.g., 7×7). While this misalignment problem may not impact classification, which is robust to small translations, it has a large negative effect on predicting pixel-accurate masks.

To address this misalignment problem, the RoIAlign layer is used instead of RoIPool. From RoIPool to RoIAlign, the change is simple, i.e., the quantization of the RoI boundaries or bins (i.e., [x/16] is instead by x/16) is avoided.

4 EXPERIMENTS

The experiments were conducted on a computer with one Intel Core i7-5820K CPU, 64GB Random Access Memory and two GeForce GTX 1080 GPU (24GB graphics memory). The proposed method was implemented based on the Detectron. Detectron is Facebook AI Research's software system that implements state-of-the-art object detection algorithms. The calculation software environment was set with python 2.7.14, CUDA 8.0 and cuDNN6.0.

The 4525 images is divided into two parts, 3620 images in training set, and 905 images in validation set. The Mask R-CNN is trained on training set. The input image is resized such that its shorter side has 800 pixels. Synchronized SGD (LeCun et al., 1998) is used to train the model on 2 GPUs. The weight decay and momentum are set 0.0001 and 0.9, respectively. The learning rate is 0.005 for the first 30k iteration and is decreased by 10 at the 40k iteration. The initial loss is 6.909. This value drops sharply to 1.968 after 20 iteration and then drops slowly after 20000 iteration. After about 45,000 iteration per GPU, the loss function converges to 0.024, as shown in Figure 6.

Next step, 6 new images containing water leakage neither in training set nor in validation set are fed into trained model. The detection and segmentation results is shown as Figure 7.

5 CONCLUSIONS

The goal of this paper is to carry out instance segmentation for images containing water leakage. Image acquisition of metro shield tunnel is performed by metro tunnel inspection equipment MTI-200a. A total of 4525 images containing water leakage are chosen by naked human eyes and the images are annotated by Labelme. Then the annotation file is converted to a format similar to the COCO dataset through Python. The Water Leakage in COntext

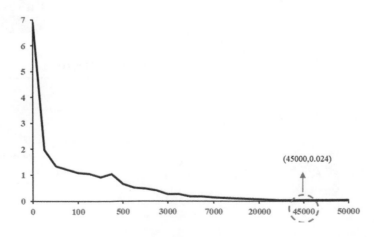

Figure 6.   The loss of training.

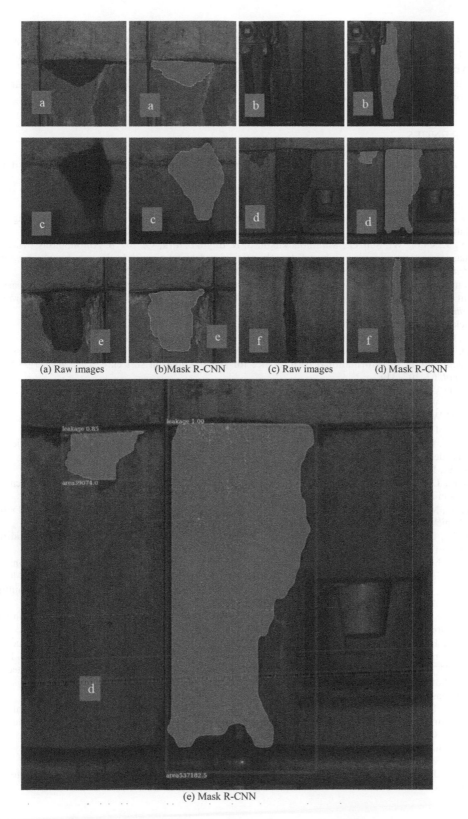

(a) Raw images   (b)Mask R-CNN   (c) Raw images   (d) Mask R-CNN

(e) Mask R-CNN

Figure 7. Part of original image and images' detection and segmentation results.

(WLCO) dataset is created and the Mask R-CNN is trained on the training set. After about 45,000 iterations, the Mask R-CNN model is converged. The trained Mask R-CNN model achieved good results on the WLCO instance segmentation task.

In the future, the Mask R-CNN will be modified and trained to detect and segment various types of defects, such as cracks and spalling.

REFERENCES

Ai, Q. Yuan, Y., Bi, X. 2015. Acquiring sectional profile of metro tunnels using charge-coupled device cameras. *Struct. Infrastruct. Eng.* 12 (9), 1–11.

Abdel-Qader, I., Abudayyeh, O. & Kelly, M. E. 2003. Analysis of edge-detection techniques for crack identification in bridges, *Journal of Computing in Civil Engineering*, 17(4),255–263.

Cha, Y.-J., Choi, W. & Büyüköztürk, O. 2017a. Deep learning-based crack damage detection using convolutional neural networks, *Computer-Aided Civil and Infrastructure Engineering*, 32, 361–378.

Cha, Y. J., Choi, W., Suh G., Mahmoudkhani, S. & Büyüköztürk, O. 2017b. Autonomous structural visual inspection using region-based deep learning for detecting multiple damage types, *Computer-Aided Civil and Infrastructure Engineering*, https://doi.org/10.1111/mice.12334

German, S., Brilakis, I. & DesRoches, R. 2012. Rapid entropy-based detection and properties measurement of concrete spalling with machine vision for post-earthquake safety assessments, *Advanced Engineering Informatics*, 26(4),846–858.

Girshick, R. 2015. Fast R-CNN. *In ICCV*.

He, K., Gkioxari, G., Dollar, P. & Girshick, R. 2017. Mask R-CNN, *In CVPR*.

Huang, H.W., Zhang, Y.J., Zhang, D.M., Ayyub, B.M., 2017a. Field data-based probabilistic assessment on degradation of deformational performance for shield tunnel in soft clay. *Tunn. Undergr. Space Technol.* 67, 107–119.

Huang, H.W., Sun, Y., Xue, Y.D., Wang, F. 2017b. Inspection equipment study for subway tunnel defects by grey-scale image processing. *Adv. Eng. Inf.* 32, 188–201.

Huang, H.W., Li, Q.T., Zhang D.M. 2018. Deep learning based image recognition for crack and leakage defects of metro shield tunnel [J]. *Tunnelling and Underground Space Technology*, 77: 166–176.

LeCun, Y., Bottou, L., Orr, G.B., Müller, K.-R., 1998. Efficient backprop. Neural Networks: Tricks of the Trade, LNCS 1524, 9–50.

LeCun, Y., Bengio, Y. & Hinton, G. 2015. Deep learning, *Nature*, 521, 436–444.

Lee, S.Y., Lee, S.H., Shin, D.I., Son, Y.K., Han, C.S., 2007, Development of an inspectionsystem for cracks in a concrete tunnel lining, *Can. J. Civil Eng.* 34, 966–975.

Lin, T.-Y., Maire, M., Belongie, S., Bourdev, L., Girshick, R., Hays, J., Perona, P., Ramanan, D., Dollár, P., Zitnick, C. L., 2015. Microsoft COCO: Common Objects in Context. arXiv: 1405.0312, 1–12.

Lin, T.-Y., Dollár, P., Girshick, R., He, K.M., Hariharan, B., Belongie, S., 2017. Feature pyramid networks for object detection. arXiv: 1612.03144, 1–10.

Long, J., Shelhamer, E., Darrell, T. 2015. Fully convolutional networks for semantic segmentation. *The 28th IEEE Conference on Computer Vision and Pattern Recognition*. Boston, MA, USA, 3431–3440.

Nishikawa, T., Yoshida, J., Sugiyama, T. & Fujino, Y. 2012. Concrete crack detection by multiple sequential image filtering, *Computer-Aided Civil and Infrastructure Engineering*, 27(1),29–47.

Ren, S., He, K., Girshick, R. and Sun J. 2015. Faster R-CNN: Towards real-time object detection with region proposal networks. *In NIPS*.

Sinha, K., Fieguth, P.W. & Polak, M. A. 2003. Computer vision techniques for automatic structural assessment of underground pipes, *Computer-Aided Civil and Infrastructure Engineering*, 18(2),95–112.

Soukup, D. & Huber-Mörk, R. 2014. Convolutional neural networks for steel surface defect detection from photometric stereo images, in *Proceedings of 10th International Symposium on Visual Computing*, Las Vegas, NV, 668–677.

Xue, Y.D., Li, Y.C. 2018. A Fast Detection Method via Region-Based Fully Convolutional Neural Networks for Shield Tunnel Lining Defects, *Computer-Aided Civil and Infrastructure Engineering*, https://doi.org/10.1111/mice.12367

Yao, F., Shao, G., Takaue, R., Tamaki, A. 2003. Automatic concrete tunnel inspection robot system, *Adv. Robotics*, 17, 319–337.

Yu, S. N., Jang, J. H. & Han, C. S. 2007. Auto inspection system using a mobile robot for detecting concrete cracks in a tunnel, *Automation in Construction*, 16(3),255–261.

Tunnels and Underground Cities: Engineering and Innovation meet Archaeology, Architecture and Art, Volume 6: Innovation in underground engineering, materials and equipment - Part 2 – Peila, Viggiani & Celestino (Eds)
© 2020 Taylor & Francis Group, London, ISBN 978-0-367-46871-2

Rectangular TBM application in utility tunnel

K.T. Zheng, X.J. Zhuo, Y.L. Feng, J. Zhang & S.L. Cheng
China Railway Engineering Equipment Group Co., Ltd., Zhengzhou, Henan, China

ABSTRACT: With the rising demand for infrastructure construction and the development of underground space exploration, there has been a great upsurge in utility tunnel construction which causes a great attention about the construction equipment – rectangular TBM in recent years. This paper presents the development of rectangular TBM application in utility tunnel and analyzes the existing structures to specify its structural features and working principle. An optimized jacking force formula is introduced in the process of TBM design. In addition, it also summarizes the advantages and disadvantages of rectangular TBM from a series of application cases. Furthermore, the development trend and application prospect are also proposed in this paper.

1 INTRODUCTION

Utility tunnel is a passage built in underground to carry public utility pipelines such as electricity cables, telecommunication cables, radio and television, water supply pipes, sewer pipes, heat distribution pipeline and gas pipeline etc. Different from conventional direct burial laying method, utility tunnel is an advanced laying method. General Office of the State Council issued "Guidance on Promoting the Construction of Urban Underground Pipe Gallery" in August 2015. On the condition that the underground infrastructure construction doesn't match with the rapid urbanization development, the Guidance clearly puts forward to promote the construction of urban underground utility tunnel and to coordinate the design, construction and maintenance of municipal pipelines to decrease disruption to traffics caused by recurrent excavation, dense overhead network and frequent pipeline accidents.

Underground utility tunnels offer the advantage of hosting multiple services' infrastructures inside an accessible and safe space, and of allowing regular inspections, maintenance and easy replacement. This solution has been implemented since many decades in Europe (see Figure 1). For example, the first utility tunnel was built in France in 1833. Besides, there are over 600km utility tunnels which have been constructed in Japan since 1926 (Shan 2016). However, China is making a rapid progress on underground utility tunnel despite the late start.

The rectangular cross section is the most ideal and economic cross section for utility tunnel from functional standpoint. As for the utility tunnel using shield method, the rectangular cross section can save more than 20% underground space under the condition of the same effective space and greatly reduce the overburden of tunnel (see Figure 2).

Rectangular TBM technology has been flourished in recent years. Zhengzhou Hongzhuan Road undercrossing Zhongzhou Avenue tunnel opened to traffic in December 2014, and Singaporean Thomson MRT Line T221 made a breakthrough in November 2016. The successful applications of the rectangular TBM show that it has strong adaptability, economy and security, and has been widely accepted in the market. Therefore, it is anticipated that rectangular TBM will play a very significant role in utility tunnel in the near future through its environmental protection, safety, high efficiency, smaller disruption to traffics, and little environmental impact etc. Further, it is destined to become the main equipment in the construction of utility tunnel (see Figure 3).

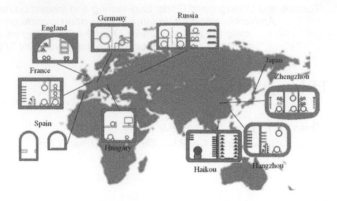

Figure 1. Utility tunnels at home and abroad.

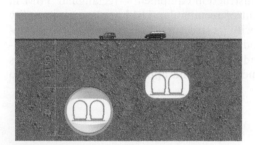

Figure 2. Comparison between rectangular and circular sections.

Figure 3. Rectangular TBM used in utility tunnel.

2 OVERVIEWS OF RECTANGULAR TBM

Rectangular TBM emerged at an appropriate moment in order to tackle engineering constraints (including the width and overburden etc.), reduce the volume of earthwork excavation, and improve the utilization rate of the tunnel section.

Japan lies in a leading position in the development and application of rectangular TBM and at the same time, our country also formed our own characteristics through research and development. The following are the different types of rectangular TBM and their features.

2.1 Japanese rectangular TBM

Open type rectangular TBM was firstly used to construct the utility tunnel (Sakaemachi utility tunnel by Chubu Electric Power) in Japan in 1965. Since there are various problems appeared in practical applications of open type shield machine, it has not been used any more. Subsequently, due to the further development of closed type shield machine technologies, closed type rectangular TBM has been paid abundant attention again in the construction of utility tunnel.

2.1.1 Wagging cutterhead shield machine

Since a wagging cutterhead shield machine (Kazunari & Takashi 2002) adopts a cutterhead which wags by a certain angle, the length of each spoke can be set for its wagging range. Accordingly, the uncut parts are reduced to a degree that they are not problems practically, compared to the parts left uncut by the circular cutterhead. The parts which are not cut with the rectangular shield are cut with the overcutter shown in Figure 5 to realize the full-face excavation. The machine is suitable for soft ground such as sandy soil.

The wagging cutterhead does not need to be rotated a full circle and then make it possible to be driven through the telescopic cylinders. Accordingly, the structure of the wagging

Figure 4.   10.24m wide and 6.87m high Wagging cutterhead shield machine.

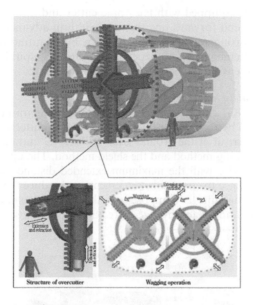

Figure 5.   Wagging cutting operation.

cutterhead consisting of the jacks and pins is simpler than the precise structure of the conventional rotary cutterhead consisting of a motor and gears. Besides, the wagging cutter shield machine also has the relative advantage of low cost.

2.1.2 *Paddle shield machine*

Unlike conventional shield machine, the cutter of paddle shield machine (Ooki 2016) is composed of a circle of small bits which are arranged horizontally on axis. The machine is equipped with several such cutters in vertical direction to excavate soil in the rectangular cross section by rotating them (see Figure 7).

Figure 6.   Paddle shield machine.

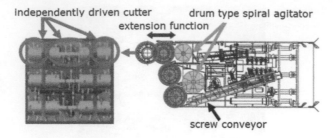

Figure 7.   Structure schematic drawing.

Paddle shield machine is equipped with telescopic cutter and independent agitation device. The upper cutter could extend forward one segment width in advancing direction. Therefore, it can complete pre-supporting in the low overburden by continuous extension forward to effectively control the settlements. The independent drum type spiral agitator is arranged in the excavation chamber to ensure the better agitating performance.

2.1.3 *R-SWING cutting shield machine*

R-SWING (Roof & SWING) cutting construction method (Niihara et al. 2011) developed in Japan in 2011, is conducted by rectangular TBM with swinging cutterhead driven by cylinder to realize the launching and arrival at the ground level. R-SWING cutting construction method is also suitable for the pipe jacking method and the shield method. The upper part of the machine is equipped with telescopic device with the maximum extended distance of 1.5m. And the device could extend to bore in advance to control the settlements and reduce the environmental impacts.

The basic unit of the machine is composed of the main machine and upper telescopic device. This machine could adapt to any size of rectangular cross section by combination of the several basic units.

Figure 8.   R-SWING cutting shield machine.

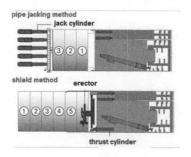

Figure 9.   R-SWING cutting construction method appropriate for pipe jacking method and shield method.

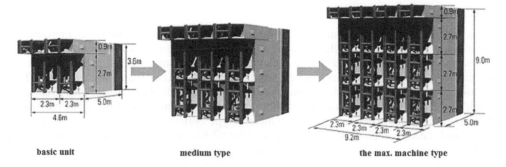

basic unit   medium type   the max. machine type

Figure 10.   Example of combinations of basic units.

2.1.4 *All potential rotary cutter shield machine*

All potential rotary cutter method (Sakane et al. 2012) could complete excavation of arbitrary cross section tunnels by the cooperation of the rotation of cylinder and oscillating axle and the revolution of cutterhead. Unlike conventional single axis rotation structure of circular cutterhead, a smaller cutterhead with high speed revolution assembled in all potential rotary cutter shield machine can effectively tackle the hard ground and cut the underground obstacles such as wooden stake. This construction method could achieve the full-face excavation by controlling the rotation of three axes (see Figure 12). At the same time, the oscillating axle could change the cutterhead's radius of gyration to realize the full-face excavation of rectangular cross section, Horseshoe-shape cross section or circular cross section (see Figure 13).

Figure 11. All potential rotary cutter shield machine.

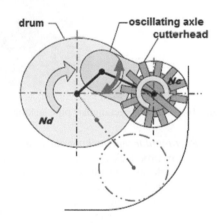

Figure 12.   Schematic of three axes control.

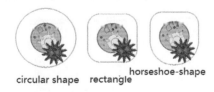

circular shape   rectangle   horseshoe-shape

Figure 13.   Circular and non-circular cross sections.

2.2 Chinese rectangular TBM

Due to rapidly advancing urbanization and rising demands for mobility in metropolitan regions, there is a constant development in tunnelling equipment especially rectangular TBM in China in recent years. The grid type rectangular pipe jacking machine prototype test was made in Hangtou Town, Nanhui District, Shanghai in 1995. After that, the rectangular TBM goes through the evolution and development of four types including cutterhead plus overcutters, multiple cutterheads, eccentric multi-shaft type and combined cutterhead (used for rectangular and quasi-rectangular cross sections) or planetary type cutterhead (only used for rectangular section), and the section dimension ranges from 2.5m×2.5m to 10.42m×7.57m. According to the excavation form, rectangular TBMs can generally be divided into two categories: One is wagging cutting by eccentric copy cutterhead to realize full face excavation of rectangular cross section. And the other is parallel shafts type cutterhead which could achieve excavation by the layout of front and rear cutterheads and blind area processing unit.

2.2.1 DPLEX (or eccentric multi-shaft type) rectangular pipe jacking machine
The DPLEX rectangular pipe jacking machine is used to construct tunnels of traffic lanes in Weft 4th Road and Shenzhuang North Road undercrossing Zhongzhou Avenue Project in Zhengzhou. The machine combines one main cutterhead with four eccentric swinging type cutterheads to complete the excavation of rectangular cross section (see Figure 14).

Figure 14.   DPLEX rectangular pipe jacking machine.

2.2.2 Rectangular pipe jacking machine with cutterheads on different planes
Rectangular pipe jacking machine with cutterheads on different planes (Li et al. 2016; Wang et al. 2013) adopts the three front and three back cutterheads arranged with parallel shafts, combined with scrapers on the shield and high pressure water flushing interface with universal joint in blind section to achieve full face excavation of rectangular section. After the successful application in Zhengzhou Hongzhuan Road undercrossing Zhongzhou Avenue Project, the machine has been popularized energetically and successfully applied in many projects including the underground passageway of Heiniucheng Road in Tianjin Metro Line 11, tunnel of IV entrance at Wuchang station in Wuhan Metro Line 7, and Singaporean Thomson MRT Line lot T221.

Figure 15.   Rectangular pipe jacking machine with three front and three back cutter heads arranged with parallel shafts (Jia 2014).

2.3 Design concepts analysis

Through the comparisons among above several rectangular TBMs, it can be found that the Japanese design concept of rectangular TBM is focused on universality, which is beneficial for remanufacturing. For example, R-SWING shield machine could enlarge and shrink the size of machine by increasing or decreasing the basic unit; Paddle shield machine adopts the universal motor and gear box, which reduce the cost effectively; all potential rotary cutter shield machine could be reused in tunnels of any other sections.

However, Chinese design concept of rectangular TBM places an emphasis on the practical application of the large section tunnel, as well as the further upgrading and improvement, such as continuous innovation on tunnel lining theory to transform rectangular pipe jacking machine into rectangular shield machine.

3 CASES STUDY OF RECTANGULAR TBM IN CHINESE UTILITY TUNNEL

3.1 Rectangular shield machine application in utility tunnel

Zhengzhou International Eco-aquapolis 19th Avenue Project ranges from the 12th South Road to 15th South Road. The north-south path has a higher relief in the south than the north with the road slop of 1.4‰-4.5‰. The 1065m long utility tunnel mainly composed of silty clay is arranged in the greenbelt in east side.

Due to the relatively long of the utility tunnel, the project intends to use shield method and up to now, the equipment is still in the stage of design and manufacture. The utility tunnel is distributed into three cabins to satisfy storage functions of gas, heat distribution pipeline and water & electricity simultaneously (see Figure 16). In order to meet the layout of the backup and the requirement of muck transportation, the partitions of utility tunnel will be made after the completion, and segmnt lining adopts the prefabricated segments (shown in the Figure 17).

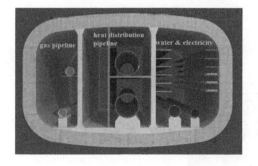

Figure 16.   Diagram of the utility tunnel.

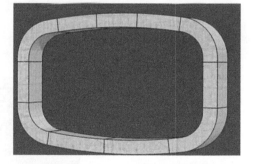

Figure 17.   Ring distributions of utility tunnel.

As for the construction of utility tunnel, rectangular shield machine has unique advantages such as higher construction efficiency, trenchless technologies, and long-distance excavation. At the same time, the shortcomings are also existed, for example ventilation shaft and partitions have to be done after the completion and the cabin needs thicker segments to meet stress requirement under the conditions of large span and no partition.

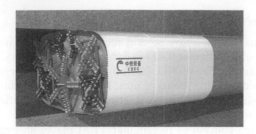

Figure 18.   Rectangular shield machine for utility tunnel.

3.2   *Rectangular pipe jacking machine application in utility tunnel*

Haikou Yehai Avenue East extension section of utility tunnel (see Figure 19), pass underneath high-speed railway bridge, highway bridge and river, which are six sections in total and each section is less than 150m. The geology is mainly composed of silty clay.

   Due to that the utility tunnel consists of multiple short sections, it's decided to use rectangular pipe jacking method. The utility tunnel is composed of both high voltage cabin and water & electricity cabin (see Figure 20). The cabin and its partitions are constructed simultaneously by prefabricated pipes which are pushed by thrust cylinder from launch pit (see Figure 21).

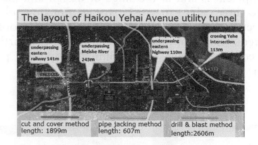

Figure 19.   Haikou Yehai Avenue utility tunnel.

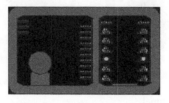

Figure 20.   Schematic of the utility tunnel.

Figure 21.   Rectangular pipe jacking machine used in utility tunnel.

Different from the shield method, except for frontal resistance in advancing direction and the friction between the shield and tunnel wall, rectangular pipe jacking machine still has to overcome the frictional resistance between prefabricated pipe and excavation soil which will enlarge gradually with the increase of advancing distance. Therefore, the jacking force of this pipe jacking machine (8.02×4.62m) can be calculated by the max. advancing distance of 141m in this project.

According to Technical Specification for Pipe Jacking of Water Supply and Sewerage Engineering (CECS246: 2008), the jacking force of the conventional circular pipe jacking machine can be calculated by:

$$F_0 = \pi D_1 L f_k + N_F \tag{1}$$

Where F_0 = the total jacking force (kN); D_1= the outer diameter of pipe (m); L = the length of prefabricated pipe (m); f_k = the average friction between the outer side of pipe and the excavation soil (kN/m^2); and N_F = frontal resistance of pipe jacking machine (kN).

On the basis of above formula, the computational formula of jacking force of rectangular pipe jacking machine can be optimized into:

$$F_0 = L_{circumference} \cdot L \cdot fk + N_F \tag{2}$$

Where $L_{circumference}$ = the circumference of shield of rectangular pipe jacking machine can be set to 24.34m; f_k = according to the Specification (CECS246:2008-Table 12.6.14) and geological conditions of this project, the value can be defined as 10 kN/m^2.

Due to the adopted equipment is EPB pipe jacking machine, the frontal resistance can be calculated by:

$$N_F - \gamma HS \tag{3}$$

Where γ = the density of soil (kN/m^3) and setting the value of 20 kN/m^3; H = the overburden (m) and setting the value of 6m (the max. overburden of the project); S = the area of tunnel face (m^2) and setting the value of 36.83 m^2.

Consequently, the jacking force can be obtained by:

$$F_0 = [24.34 \times 141 \times 10] + [20 \times 6 \times 36.83] = 38739 kN \tag{4}$$

The rated jacking force of the machine (8.02×4.62m) is designed as (24×300T) 72000kN and the safety coefficient is 1.86 which could satisfy the requirements.

Figure 21 shows the structure of rectangular pipe jacking machine. As for utility tunnel, the adoption of rectangular pipe jacking machine has the following merits including high construction efficiency, trenchless technologies and synchronized construction of cabin and partitions. However, the machine is only appropriate for the tunnel within 200m. With regard to long utility tunnel, there are several working shafts need to be constructed, which can also be ventilation shafts in the later period.

4 FUTURE DEVELOPMENT OF RECTANGULAR TBM

Rectangular section, as an ideal section for utility tunnel is promoted in the worldwide. On the basis of limited space of city area and narrow construction jobsite, the adoption of rectangular section has unique merits and meaning.

As the main construction equipment of rectangular section tunnel, the technologies of rectangular TBM are also in constant innovation and development according to the characteristics of working conditions and market requirements. The following are the development directions of rectangular TBM.

1. Design and research cutterhead types suitable for different grounds and furtherly study the various types of erectors.
2. Innovate the key technologies further such as shield attitude control, synchronized back grouting, segment lining, based on the deformation law of rectangular tunnel.

5 CONCLUSIONS

In early 2017, construction plans of underground utility tunnel are released in succession throughout the country. Sanya city, as a scenic city on South China's island province of Hainan, its construction plan of underground utility tunnel has formed initially, and 22.4km long underground utility tunnels will be constructed in 2017. There are eight new projects of underground utility tunnels with a total length of 96.8km in Guangzhou city, Guangdong Province. Hunan Province plans to construct 157.31km underground utility tunnels in 2017. 10 underground utility tunnels with the total length of 57km are intended to be built in Urumqi, Xinjiang in this year, which involves elevated road and main road in the north of the city (Zhongya Avenue-Suzhou Road) etc. In addition, Xiongan New Area, the key component of a massive "mega-region" developing around Beijing, Tianjin and Hebei, is approved in April 2017, and its construction will promote the further development of underground utility tunnel.

There is an old saying that good tools are prerequisites to the successful execution of a job. Rectangular TBM is the key equipment of underground excavation method for utility tunnel. Benefited from the strong support of national policy, the rise of utility tunnel provides the strong market demands, which will further promote the development of rectangular tunneling machine technology. Therefore, we should believe that the rectangular TBM has a very broad application prospect in the construction of utility tunnel.

REFERENCES

Jia, L.H. 2014. Key technologies for design of super-large rectangular pipe jacking machine. *Tunnel Construction* 34(11): 1098–1106.

Kazunari, K. & Takashi, M. 2002. Development of rectangular shield. *KOMATSU Technical Report* 47 (148): 46–54.

Li, J.B. et al. 2016. Rectangular TBM with multiple cutterheads in the same plane used for pebble stratum with rich water. *Chinese Patent CN205714206U*: 11–23.

Niihara et al. 2011. The application of all potential rotary cutter shield machine with rectangular cross section. *Proceedings of the Sixth China-Japan Conference on Shield Tunnelling*: 115–122.

Ooki, C. 2016. Closed type rectangular shield machine (paddle shield) method. *Tunnels and Underground*: 65–72.

Shan, J.G. 2016. The present status and features of utility tunnel construction in Japan. *Shield Tunnelling Technology* (6): 40–43.

Sakane, et al. 2012. Rectangular cross section tunnel constructed by R-SWING cutting method. *Civil Construction* 53(3): 84–88.

Wang, X.J. 2015. Application of rectangular pipe jacking machine in undercrossing Zhongzhou Avenue project in Zhengzhou. *Urban Construction Theory Research* (35).

Wang, X.T. et al. 2013. Study on rectangular pipe jacking machine cutterhead design. *Urban Construction Theory Research* (3): 183–185.

Tunnels and Underground Cities: Engineering and Innovation meet Archaeology,
Architecture and Art, Volume 6: Innovation in underground engineering,
materials and equipment - Part 2 – Peila, Viggiani & Celestino (Eds)
© 2020 Taylor & Francis Group, London, ISBN 978-0-367-46871-2

Shield machine cutterhead tools real-time wear monitoring system

X.J. Zhuo, Y.T. Lu, L. Ye & H. Sun
China Railway Engineering Equipment Group Co., Ltd., Zhengzhou, Henan, China

ABSTRACT: In order to reduce the risk of manual overhaul, improve construction efficiency, and assist the TBM driver in deciding whether the cutting tools need to be changed or not, it is of great significance to realize the visual management of the wear conditions for cutting tools. This paper introduces resistor array type wear sensor, studies how the sensor transmits signal to the Human Machine Interface (HMI) for real-time display, and presents a scraper and ripper real-time wear monitoring system in detail which is designed specifically for CREG(China Railway Engineering Equipment Group Co., Ltd.) TBMs. Furthermore, the system was applied to a metro project of Beijing New Airport Line Project. The results show that the wear sensor is of high accuracy and versatility, and abrasion magnitude could be displayed on the HMI in real time, so as to significantly improve the tunneling efficiency.

1 INTRODUCTION

Cutting tools maintenance is an important link in the application of tunnel boring machine. At present, TBM driver can't observe the wear conditions of cutting tools in tunneling, and thus the machine must be stopped to inspect the cutterhead and cutting tools. In addition, the crews need much time to prepare for inspections, which would greatly influence the construction period (Song 2013). What's more, it will bring certain safety risks to the crews due to that the cutterhead and cutting tools are located in the unsupported area (Shen 2007). Therefore, it is a key issue to realize visual management of the cutting tools abrasion for research and development of shield machine industry.

Existing wear monitoring systems for cutting tools mainly include hydraulic oil type wear monitoring system (Wang 2011) and ultrasonic positioning type wear monitoring system. Hydraulic oil type wear monitoring system has low fault rate and is easy for maintenance operations, but it can't monitor the wear conditions in real time and only when the cutting tools failures, it has signal output. Furthermore, it has poor measurement accuracy. By contrast, ultrasonic positioning type monitoring system can realize real-time monitoring for cutting tools (Shang 2018). However, the system has high communication requirements and thus how to transmit the signal well from the transmitting terminal to the receiving terminal is its technical difficulty. In addition, this system has high requirements for protection system and limited power supply duration.

This paper introduces the cutting tools real-time wear monitoring system which could help the driver properly estimate whether the cutting tools need to be changed by overall consideration of tunnelling rings and chainage. The system provides reliable basic data and technical support for the application of shield machine, contributes to avoid major accidents in construction, and possesses important economic and social benefits.

This real-time wear monitoring system is designed for cutterhead structure of CREG shield machines. The system could reduce labor consumption and construction risks, realize visual management of cutting tools abrasion to help driver estimate whether cutting tools need to be changed, improve the construction efficiency of shield machine, and possess important significance for research and development of new generation shield machine with the features of intelligence, high efficiency, and safety (Salgado et al. 2013).

Based on the Principle of Resistance Change, the real-time wear monitoring system for cutting tools collects signals, converts into serial interface signals, transmits to slip ring and then receiving terminal. And finally, the signals will be directly collected by Programmable Logic Controller (PLC) and displayed on the HMI.

2 OVERALL DESIGN OF SYSTEM

The whole system consists of sensor subsystem, control and data transmission subsystem and standard signal output subsystem. Among them, sensor subsystem including sensor and amplifier can be protected by protective device in cutterhead arm to guarantee that the sensor won't be damaged and has good communication signal. The function of data transmission

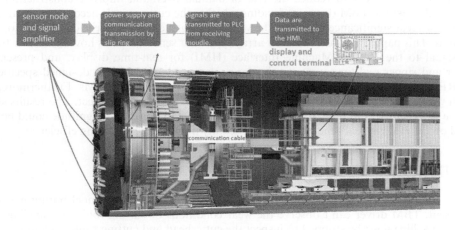

Figure 1.   Total block diagram of scraper wear monitoring system.

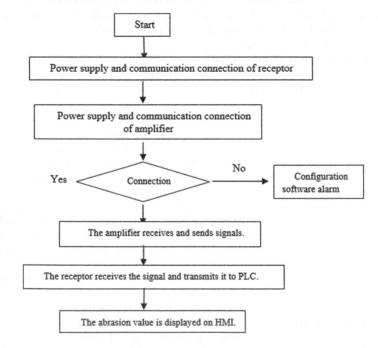

Figure 2.   Flow chart of system operation.

subsystem will be realized by slip ring. As for signal output subsystem, 6 standard current signals output from receiving modules will be transmited to the HMI to display real-time abrasion value of cutting tools. The total block diagram of scraper wear monitoring system is shown in Figure 1.

The sensor of the system uses the Principle of Resistance Change to monitor the abrasion value of cutting tools, and the outside of the circuit board is filled with epoxy resin adhesive for insulation. Taking the electric slip ring as the middle bridge, the receptor adopts 220V power supply. On the one hand, the receptor converts the voltage of 220V into 24V power supply and provides power to the amplifier and sensor. On the other hand, the receptor receives the sensor signal send by the amplifier through the electric slip ring. The amplifier sends the sensor signal at a frequency of 300ms to the receptor which will convert the serial interface signal of 485 into the signal of 4–20mA. The converted signal can be directly received by AI module of PLC and processed by the corresponding program. And then, abrasion value can be transmitted to HMI to provide references for the driver. The Figure 2 shows the working flow chart of the scraper and ripper real-time wear monitoring system.

3 SYSTEM COMPOSITION

3.1 *Communication system in complex working conditions*

Electric slip ring, as a precision electric transmission device for two data signals of relative rotating mechanism, could achieve the power supply of cutterhead sensor and the transmission of communication data. It is especially suitable for the condition in which unlimited continuous rotation is required and data needs to be transmitted from the rotating position to the fixed position.

Based on the structure of CREG EPB TBM, the electric slip ring is designed in the backside of rotary fluid coupling and the rotor of electric slip ring is synchronous with the cutterhead. It needs to open a M20 through-hole on the rotary fluid coupling directly to excavation chamber. As shown in Figure 3, the outgoing line of rotor inside slip ring is connected to the amplifier through rotary fluid coupling and excavation chamber. In addition, the middle part of the rotor outgoing line is connected by aviation plug, which is convenient for the assembly and disassembly of the cutterhead. In addition, the outgoing line of stator outside the slip ring will be connected to the receptor in the shield. The parameters of electric slip ring are as follows:

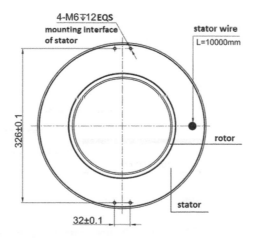

Figure 3. Electric slip ring structure diagram.

Table 1. Parameters of electric slip ring.

Items	Descriptions
Ring content	6
Current	3 circuits (power supply), 10A/circuit; 3circuits (signal)
Voltage	0~380VAC/240VDC
Operating speed	0~10rpm
Temperature	-20~+80°C
Protection class	IP67
Insulation resistance	1000VDC
Electric contact material	precious metal

3.2 Protection system of sensor in complex working conditions

In complex working conditions, in order not to affect the strength of the cutting tools, it is necessary to ensure that the sensor embedded in the cutter body is firm and reliable with high waterproof grade and could satisfy the requirements of accuracy measurement. The structural design and installation method of sensor are the key conditions to meet the above requirements.

As shown in the Figure 4, the front part of sensor, namely wear sensing head is 4mm in diameter and 48mm in length and the back part is circuit area with the diameter of 8mm. When operating, the sensing head will be worn out along with the cutter ring.

As shown in left part of Figure 5 (bounded by the middle line a-b), the outgoing line (four-core shielded wire) at the end of each sensor is 200–250mm in length and connected to a waterproof connector to form the overall structure of the sensor. The right part including the half of waterproof connector will be placed in the cutterhead arm and the cable on the right part will be connected to the amplifier in cutterhead arm from scrapper/ripper's structure by way of hydraulic protection pipe.

The opening size of scraper and ripper will influence the measurement accuracy of sensor. Openings need to be completed first according to the locations of holes shown in the Figures 6–7 to install the sensor. The through-hole has an inclined angle, and the principle of determining the inclination angle of the through-hole is that the axis of the through-hole is parallel to the side surface of the cutter ring, and the distance between the above two parts needs to be as small as possible. The diameter of the through-hole is 5mm in the sensing head

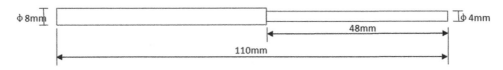

Figure 4. Sensor size.

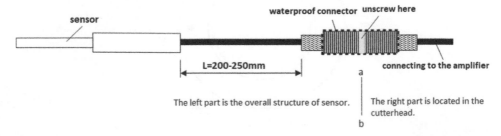

Figure 5. Sensor and connecting line's size and shape.

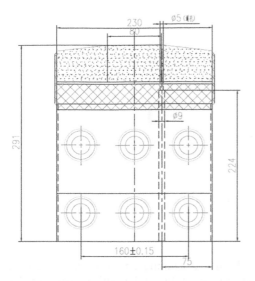

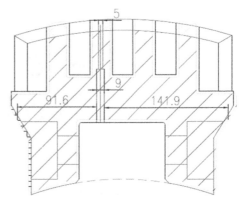

Figure 7. Opening location and size of ripper.

Figure 6. Opening location and size of scraper.

area, 1mm larger than the sensing head, and the diameter of corresponding sensor tailstock is 9–10mm, and the length of each section is consistent with that of the sensor. It is particularly important to note that the diameter of the front area of the sensor must not exceed more than 1mm compared with sensing head, otherwise the measurement error will be affected. Figures 6–7 show the openings locations of scraper and ripper and the middle part of two cutter rings is selected to install the sensor. If the diameter of the through-hole in the front part of the cutting tool is greater than 5mm, the measurement error of the sensor will be increased. Therefore, the through-hole should be inserted with the steel column first to plug the hole, and then opening the hole again according to the size.

The installation and protection of sensor for scrapers and rippers: (1) As shown in the Figure 7, ripper is fixed on the cutterhead structure by bolting and the sensor bottom is designed as plug (see Figure 8), which will be put in the mounting hole of cutterhead structure and then tighten the bolts to complete the installation of the ripper (including cutting tools wear monitoring system). (2) Figure 9 shows the installation of sensor in scraper by potting. In addition, Figure 10 shows the installation structure of scraper (bolting) and the protection design of sensor's bare wires (see the right part of Figure 10).

Figure 8. Installation of sensor in rippers.

Figure 9. Installation of sensor in scrapers.

Figure 10. Installation method and protection design of scrapers.

3.3 *Installation process of sensor*

The sensor is fixed with WL-036 epoxy resin adhesive (including epoxy resin and curing agent). The epoxy resin and curing agent are mixed in the ratio of 1:1 with stirring for 2–4 minutes, and then let stand for 3 to 5 minutes until large bubbles overflow. Slowly pour the mixed adhesive along the wall of hole (to prevent from forming blowhole) into the hole, and also note that the stirring and standing time should not exceed 10 minutes, otherwise, the adhesive will start curing, which will make the flowability poor. When the adhesive is about 15mm away from the port, the sensor is inserted slowly and then the adhesive will cover the top of the sensor and flush with the upper surface of the hole.

Before pouring, the oil stains should be removed from the mounting hole with alcohol wipes to ensure the bonding strength of epoxy resin adhesive and cutting tool materials. When pouring, the hole below shall be sealed with adhesive tapes and a strong magnet plate (see Figure 11) to prevent the adhesive flowing out. When two-thirds of the pouring is completed, the sensor is inserted into the bottom of the hole and the installation is completed after 24 hours of standing.

3.4 *Selection of CPU and IO module*

The selection of CPU is the key to data resource control and transmission, and the communication interface, program block data limitation, whether there is PROFINET master-slave interface, etc. should be taken into account. With the consideration of above parts and relevant experience, this project finally adopts BK9103 of BECKHOFF as the main controller. The CPU occupied with less space could connect multiple communication modules and save time to improve efficiency and flexibility and fully meet the requirements of the project.

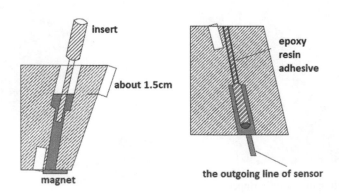

Figure 11. The schematic diagram of sensor's pouring process and the state after pouring.

The IO module is determined by the required number of input points (digital quantity and analog quantity) and output points. DI is a digital input module and when selecting, comprehensive consideration is made according to the number of input points and the distance of buttons (switches). DO is a digital output module that mainly controls external equipment. AI is an analog input module, which is determined by the number of points and the signal of sensor (current or voltage). AO is an analog output module, which is generally divided into voltage and current and the common examples include 4–20mA and 0–10V. In this project, IO module adopts KL3458 of BECKHOFF, which has 8 input points (4–20mA).

4 THE OPERATING PRINCIPLE OF SYSTEM

4.1 The measurement principle of wear sensor

The operating principle of resistance array type wear sensor is to use resistance array as the main body of internal circuit board by parallel connection method and then calculate each resistance value (Segreto et al. 2013) according to the designed measurement accuracy in advance and a linear relationship between the total resistance and abrasion value. Also due to the liner relationship between the change of total resistance and abrasion value, abrasion value of cutting tools can be calculated by using the change of total resistance. The design of above installation method based on this principle can realize the simultaneous wear of cutting tool and sensor and the abrasion value can be calculated according to the change of total resistance of the sensor. The principle formula is as follows:

The wear range of the sensor is 0–L, and the total resistance is R. When the abrasion value is Xmm, the total resistance R of the sensor changes into:

$$\Delta R = R * X/L = K_L * X \tag{1}$$

Where K_L=resistance per unit length; when the conductor material is uniformly distributed, it is constant. The output (resistance) of the sensor is linearly dependent on the input (displacement).

The sensitivity of the sensor is:

$$S = dR/dX \tag{2}$$

4.2 Power supply and communication transmission principle of slip ring

Due to that the cutterhead and cutting tools are in relative motion with the shield, the communication connection of the electric slip ring is designed so that the output end of the sensor directly wired to the receiving end in the shield (Kuldip et al. 2015).

Electrical slip ring can realize the power supply and communication of 360°unrestrictedly rotated cutterhead and is better at transmitting power and signal. This is because the power supply and signal transmission are concentrated in the slip ring, which can completely avoid the interference problem of power and signal, and signals, and can prevent electromagnetic interference to internal slip ring from outside. An M20 waterproof joint is designed at the rotor outlet of the electric slip ring to prevent water inflow flowing into the shield, and the electric slip ring adopts the protection grade of IP67. The inner brush of the electric slip ring is made of precious metal alloy material. Owing to the brush consisting of multiple steel wires, it can ensure multi-point contact with the metal ring and improve the conductivity efficiency. At the same time, the contact pressure is reduced due to the elasticity of multiple steel wires. These characteristics guarantee the stability and durability of the slip ring. If the signal transmission of the slide ring is abnormal, an aviation plug can be connected at the rotor outlet of the slide ring to quickly disassemble and repair the electric slip ring.

The electric slip ring consists of stator, rotor, brush assembly and fixed support. The connection part of the stator line and rotor line of the electric slip ring is treated by pouring adhesive with the protection grade of IP67. In order to ensure sufficient contact between the brush and the slip ring, the conductive ring is designed as a "V" type ring. In addition, the brush transmits signal and conducts electric current by sliding contact between elastic pressure and "V" type ring groove.

5 TEST RESULTS OF SYSTEM

The scraper wear monitoring system was applied to New Airport Line Subway Project in Beijing in 2017 (Cigezhuang station-1# ventilation shaft). The tunnel is 43km long in total including the elevated line and the ground line18.2km, U-shaped slot and the underground line 24.8km. There are three underground stations in total, namely the New Airport North Terminal Station, Cigezhuang Station and Caqoiao Station. The ground conditions from Cigezguang Station to 1# ventilation shaft are quaternary alluvial-diluvial beds, mainly composed of coarse grained soils such as pebbles, gravels and sands, followed by fine grained soils such as cohesive soils and silty soils.

The monitoring period of the system is two months, and the sampling will be conducted per 300ms. Abnormal conditions occurred at wear monitoring point No. 4 and No. 6 when excavated to 115 rings. The abrasion value of 40mm were measured at monitoring point No. 4, and the current and total resistance obtained from the monitoring point after breakthrough respectively are up to 20mA and infinite. It is preliminarily speculated that the monitoring point No. 4 is damaged, and the inspection by the opening in cutterhead arm (see Figure 12) shows that the sensor line in the cutterhead is broken, which is consistent with the prediction. The current of monitoring point No. 6 is 0mA measured by the multimeter, and the resistance between the second orange line and the third blue line of the four-core line is only 1.7Ω, which

Table 2.   Actual and measured wear data of cutting tools.

Time of collection	Wear monitoring point No.1 (mm)	Wear monitoring point No.2 (mm)	Wear monitoring point No.3 (mm)	Wear monitoring point No.4 (mm)	Wear monitoring point No.5 (mm)	Wear monitoring point No.6 (mm)
2017.12.07.09:10	4.87	7.24	2.48	40	0.18	0
2017.12.08.09:11	4.97	7.35	2.52	40	0.2	0
2017.12.09.09:12	5.01	7.42	2.59	40	0.22	0
2017.12.10.09:13	5.12	7.55	2.61	40	0.27	0
2017.12.11.09:14	5.22	7.61	2.65	40	0.31	0
2017.12.12.09:15	5.34	7.68	2.67	40	0.35	0
2017.12.13.09:16	5.39	7.92	2.71	40	0.38	0
2017.12.14.09:17	5.43	8.13	2.76	40	0.41	0
2017.12.15.09:18	5.49	8.22	2.8	40	0.43	0
2017.12.16.09:19	5.72	8.35	2.86	40	0.47	0
2017.12.17.09:20	6.23	8.39	2.91	40	0.49	0
2017.12.18.09:21	6.34	8.42	2.96	40	0.51	0
2017.12.19.09:22	6.42	8.49	3.11	40	0.53	0
2017.12.20.09:23	6.51	8.52	3.19	40	0.58	0
2017.12.21.09:24	6.65	8.61	3.21	40	0.61	0
2017.12.22.09:25	6.69	8.63	3.28	40	0.63	0
2017.12.23.09:26	6.71	8.69	3.32	40	0.66	0
Abrasion value after breakthrough*	7.50	8.25	5.25	40	6.25	0
Actual abrasion value**	9.6	9.80	7.2	10.0	9.1	9.2

* Abrasion value after breakthrough refers to the measuring results by amplifier and receptor after the connection of power supply communication;
** The actual abrasion value refers to the abrasion value obtained by manual measurement.

Figure 13. Proof procedure.

Figure 12. The location of opening in cutterhead arm.

were determined the short circuit, so the abrasion value of monitoring point No. 6 shown on the HMI is 0mm.

5.1 Measurement data

Table 2 shows the data collected by HMI and the following data are obtained by manual measurement and verification after breakthrough:

(1) The measurement data of current: the 1st circuit of 7.0mA, the 2nd circuit of 7.3mA, the 3rd circuit of 6.1mA, the 4th of 20mA, the fifth circuit of 6.5mA, the 6th circuit of no reading (infinite). (2) Resistance measurement data are shown in Table 3.

Measurement results: The 4th and 6th circuits are damaged, and the rest of sensor circuits operates properly.

5.2 The determination and analysis of sensor

Determine the cause of sensor damage:

Measurement process: The resistance between the second orange line and the third blue line is measured using the multimeter (both ends represent the total resistance of the sensor circuit board).

Measurement standard: The range of resistance between the four-core line of sensor output end respectively are 2–3Ω between 1st line and 2nd line, 2–3Ω between 3rd line and 4th line, and 10–50 Ω between 2nd line and 3rd line.

According to the readings of the other four intact sensors and the linear relationship between the current value and the abrasion value, the abrasion values can be obtained: 1st circuit 7.50mm, 2nd circuit 8.25mm, 3rd circuit 5.25mm and 5th circuit 6.25mm.

Table 3. Testing results of resistances after breakthrough.

	Resistance between conductors 1,2	Resistance between conductors 3,4	Resistance between conductors 2,3
1# Sensor	1.8Ω	2.0Ω	17.6Ω
2# Sensor	1.8Ω	2.0Ω	18.1Ω
3# Sensor	2.5Ω	2.5Ω	12Ω
4# Sensor	Infinite	Infinite	Infinite
5# Sensor	2.3Ω	2.1Ω	16.1Ω
6# Sensor	1.7Ω	2.1Ω	1.7Ω

6 CONCLUSIONS

The system is proved to be of high sensor precision and high versatility by industrial test, suitable for EPB TBM and Slurry TBM. When using in the slurry TBM, the acquisition device should be placed in the rigid joint between the backside of submerged wall and cutterhead to effectively reduce the fault rate.

However, the protection of the system also needs to be strengthened to prevent the aviation plug or the wire from loosening or breaking due to vibration (Wang et al. 2014), and to avoid the short circuit caused by water. In addition, the epoxy resin adhesive is suitable for the environment of above 5°C, and the temperature of cutting tools can be up to 60–80°C in the process of tunneling, so sensor and signal transmission would not be affected by using epoxy resin adhesive to fix sensors. However, the field test period of this system is only 2 months, and the abrasion value of scraper and ripper is relatively small, so it is difficult to prove the long-term stability and durability of the system. Therefore, it needs to strengthen the protection in the late industrial test and optimize system continuously and extend the test cycle to verify the durability of electrical slip ring, and the stability of epoxy resin adhesive under the harsh conditions such as high temperature, high pressure and severe vibration.

This system is designed to monitor the abrasion value of scraper and ripper according to the characteristics of CREG shield machines. The system can measure the wear conditions of scrapers and rippers to avoid entering the excavation chamber for checking the cutting tools so as to reduce labor consumption and construction risks, realize the visual management of the wear conditions of cutting tools to help the driver judge whether to replace the cutting tools. The system also provides reliable basic data and technical support for optimization design of shield construction, and is of great significance for research and development of a new generation shield machine with the features of intelligence, high efficiency and safety.

REFERENCES

Gao, L. 2010. Optimization design of belt conveyor centralized control system based on PLC technology, *Management & Technology of SME* 2010(02): 240–241.
Kuldip, S.S. et al. 2015. Optimization of machining parameters to minimize surface roughness using integrated ANN-GA Approach, *Procedia CIRP*.
Salgado, I.C. et al. 2013. Tool wear estimation for different workpiece materials using the same monitoring system, *Procedia Engineering*.
Segreto, T. et al. 2013. Multiple sensor monitoring in nickel alloy turning for tool wear assessment via sensor susion, *Procedia CIRP*.
Shang, W. & Wang, B.Q. 2018. Ultrasonic testing technology of wear for cutting tools, *Construction Mechanization*: 2018(1).
Shen, B. 2007. Measurement and calculation of synchronous grouting clearance at the end of shield tunneling, Underground engineering construction and risk prevention technology -2007 third Shanghai international tunnel engineering symposium.
Song, W.Q. et al. 2013. Tool Wear detection using Lipschitz Exponent and HarmonicWavelet, *Mathematical Problems in Engineering*.
Wang, D.S. & Mi, N. 2010. Research on remote controller of belt conveyor control system based on CAN bus, *China Mine Engineering* 39(6).
Wang, G.F. et al. 2014. Vibration sensor based tool condition monitoring usingvsupport vector machine and locality preserving projection. *Sensors & Actuators: A. Physical*.
Wang, M. 2011. Hob cutter with wear indication, 201110165313: 2011–11–23.
Xiao, Z. & Lin, T. 2010. Frequency control system design and implementation based on fuzzy control of the belt, *Electric Drive* 39(6).
Zhou, X.J. et al. 2014. Design of automatic control system of jigging machine based on PLC, *Coal Mine Machinery* 2014(10): 54–56.

Tunnels and Underground Cities: Engineering and Innovation meet Archaeology,
Architecture and Art, Volume 6: Innovation in underground engineering,
materials and equipment - Part 2 – Peila, Viggiani & Celestino (Eds)
© 2020 Taylor & Francis Group, London, ISBN 978-0-367-46871-2

Author Index

Printed and bound by CPI Group (UK) Ltd, Croydon, CR0 4YY

18/10/2024

01776250-0011